Group Sequential and Adaptive Methods for Clinical Trials

In clinical trials, monitoring accumulating data at regular intervals is essential for balancing ethical and financial considerations against scientific rigor. This comprehensive second edition reflects the remarkable evolution in adaptive clinical trial methodology over the past two decades. Since publication of the first edition, these approaches have transformed from theoretical concepts to widely accepted practices, now endorsed by regulatory authorities, including formal FDA guidance.

Features

- Eleven entirely new chapters, including optimal design construction, delayed "pipeline" data accommodation, and multiple comparison procedures
- Eight dedicated chapters on adaptive methods including sample size re-assessment, seamless Phase II/III trials, multi-arm multi-stage trials, and enrichment designs
- Exact methods for binary and Poisson data with applications to vaccine trials
- Expanded discussion of binding versus non-binding futility boundaries for enhanced decision-making
- A chapter with a thoroughly updated review of Bayesian approaches to sequential trial design and analysis
- Incorporates response-adaptive treatment assignment in group sequential designs
- Practical computational methods and software guidance for implementing the techniques

This definitive resource serves statisticians, clinical researchers, pharmaceutical scientists, and regulatory professionals involved in clinical trial design and analysis. The text balances theoretical foundations with practical implementation guidance, making it invaluable for both experienced practitioners and those new to adaptive trial methodology. With twice the content of the first edition, *Group Sequential and Adaptive Methods for Clinical Trials, Second Edition* provides the comprehensive understanding needed to design more efficient, ethical, and scientifically rigorous clinical trials, ultimately accelerating the delivery of effective treatments to patients.

Chapman & Hall/CRC Biostatistics Series
Series Editors: Mark Chang, *Boston University, USA*

ROC Analysis for Classification and Prediction in Practice
Christos Nakas, Leonidas Bantis and Constantine Gatsonis

Controlled Epidemiological Studies
Marie Reilly

Statistical Methods in Health Disparity Research
J. Sunil Rao

Case Studies in Innovative Clinical Trials
Edited by Binbing Yu and Kristine Broglio

Value of Information for Healthcare Decision Making
Edited by Anna Heath, Natalia Kunst, and Christopher Jackson

Probability Modeling and Statistical Inference in Cancer Screening
Dongfeng Wu

Development of Gene Therapies
Strategic, Scientific, Regulatory, and Access Considerations
Edited by Avery McIntosh and Oleksandr Sverdlov

Bayesian Precision Medicine
Peter F. Thall

Statistical Methods for Dynamic Disease Screening and Spatio-Temporal Disease Surveillance
Peihua Qiu

Causal Inference in Pharmaceutical Statistics
Yixin Fang

Applied Microbiome Statistics
Correlation, Association, Interaction and Composition
Yinglin Xia and Jun Sun

Association Models in Epidemiology
Study Design, Modeling Strategies, and Analytic Methods
Hongjie Liu

Likelihood Methods in Survival Analysis
with R Examples
Jun Ma, Malcolm Hudson, and Annabel Webb

Biostatistics for Bioassay
Ann Yellowlees and Matthew Stephenson

Power and Sample Size in R
Catherine M. Crespi

R for Health Technology Assessment
Gianluca Baio, Howard Thom, and Petros Pechlivanoglou

Advanced Statistical Analytics for Health Data Science with SAS and R
Ding-Geng Chen and Jeffrey Wilson

Group Sequential and Adaptive Methods for Clinical Trials
Second Edition
Christopher Jennison and Bruce W. Turnbull

For more information about this series, please visit: www.routledge.com/Chapman--Hall-CRC-Biostatistics-Series/book-series/CHBIOSTATIS

Group Sequential and Adaptive Methods for Clinical Trials

Second Edition

Christopher Jennison and Bruce W. Turnbull

CRC Press
Taylor & Francis Group
Boca Raton London New York

CRC Press is an imprint of the
Taylor & Francis Group, an **informa** business
A CHAPMAN & HALL BOOK

Second edition published 2026
by CRC Press
2385 NW Executive Center Drive, Suite 320, Boca Raton FL 33431

and by CRC Press
4 Park Square, Milton Park, Abingdon, Oxon, OX14 4RN

CRC Press is an imprint of Taylor & Francis Group, LLC

First edition published by CRC Press 2000

Library of Congress Cataloging-in-Publication Data
Names: Jennison, Christopher, 1956- author | Turnbull, Bruce W. author
Title: Group sequential and adaptive methods for clinical trials /
Christopher Jennison and Bruce W. Turnbull.
Other titles: Group sequential methods with applications to clinical trials
Description: Second edition. | Boca Raton, FL ; Abingdon, Oxon : CRC Press,
2026. | Series: Chapman & Hall/CRC biostatistics series | Includes
index. | Summary: "This definitive resource serves statisticians,
clinical researchers, pharmaceutical scientists, and regulatory
professionals involved in clinical trial design and analysis. The text
balances theoretical foundations with practical implementation guidance,
making it invaluable for both experienced practitioners and those new to
adaptive trial methodology. With twice the content of the first edition,
this volume provides the comprehensive understanding needed to design
more efficient, ethical, and scientifically rigorous clinical trials,
ultimately accelerating the delivery of effective treatments to
patients"-- Provided by publisher.
Identifiers: LCCN 2025029314 (print) | LCCN 2025029315 (ebook) | ISBN
9781584888475 hardback | ISBN 9781041161288 paperback | ISBN
9781584888482 ebook
Subjects: LCSH: Clinical trials--Statistical methods
Classification: LCC R853.C55 J46 2026 (print) | LCC R853.C55 (ebook) |
DDC 610.72/7--dc23/eng/20250919
LC record available at https://lccn.loc.gov/2025029314
LC ebook record available at https://lccn.loc.gov/2025029315

ISBN: 978-1-584-88847-5 (hbk)
ISBN: 978-1-041-16128-8 (pbk)
ISBN: 978-1-584-88848-2 (ebk)

DOI: 10.1201/9781584888482

Typeset in Nimbus font
by KnowledgeWorks Global Ltd.

Publisher's note: This book has been prepared from camera-ready copy provided by the authors.

To

Debbie and Marty

Contents

Preface to the First Edition

In one of our favorite books *Notes from a Small Island*[1], the American writer Bill Bryson reports that the term *statistics* was coined in 1834 by Sir John Sinclair, a "local worthy" of Thurso in the northern tip of Scotland, although Bryson did hasten to add that "things have calmed down there pretty considerably since". A check of an etymological dictionary might suggest a more classical-than-Gaelic origin of the word. Be that as it may, like the word *statistics*, the writing of this book has had a lengthy and much-travelled history. For one of us (BWT), it started in 1971, when as a fledgling assistant professor at Stanford University Medical School, he was responsible for the data management of an important clinical trial. The sequential nature of the study became rapidly apparent, when the Principal Investigator, a leading surgeon, would stop by the office every few days to report on a patient who had just relapsed or died that day and to ask whether he could report the results yet. While much of this book was written in Ithaca in upstate New York or on Claverton Down, Bath, U.K., various parts have origins from far and wide—from Bradford on Avon in Wiltshire to Tucson, Arizona, from North Ryde in New South Wales to Oak Ridge, Tennessee.

We wish to thank Alan Peacock for the pastoral cover illustration; Stuart Barber for assistance with the calculations for Table 10.3; Catalin Starica for assistance with the example in Section 11.6; and Rebecca Betensky, John Bryant, Jonathan Denne and Patricia Grambsch for helpful comments on an early draft of the manuscript. We are also grateful to support from grants from the U.S. National Cancer Institute (R01 CA-66218) and from the U.K. Science and Engineering Research Council/Engineering and Physical Sciences Research Council.

We are indebted to many people who have strongly influenced us; in particular at Cornell, to Bob Bechhofer who introduced us to the fascination of statistics as an academic subject and to Jack Kiefer for our first introduction to sequential analysis which he taught in a graduate seminar. BWT is also indebted to Rupert Miller and Toby Mitchell who taught him, by example, how to bring together theory and practice of statistics in real-world biomedical applications. Sadly, Bob Bechhofer, Jack Kiefer, Rupert Miller and Toby Mitchell have all passed on—taken before their time. BWT must also credit Dr. Larry Clark of the University of Arizona Cancer Center for supplying a long and taxing stream of important, difficult and highly motivating statistical questions that he encountered in the day to day conduct of the clinical trials and epidemiological studies in which he was engaged. Finally and above all, we are indebted to the patience of our families, Debbie, Marty, Kenneth and Douglas, who were amazed when we announced that the book was finally complete! Things have calmed down here pretty considerably since.

Bath and Ithaca
July 1999

Christopher Jennison
Bruce Turnbull

[1] Bill Bryson, *Notes from a Small Island*, Avon Books, New York, 1995, p. 292.

Preface to the Second Edition

A lot has happened since the publication of the first edition of our book *Group Sequential Methods with Applications to Clinical Trials* in 1999. There has been an explosion in the literature on the topic of *adaptive design* of clinical trials. The group sequential methods in our first edition could be described as adaptive in the sense that a trial could be stopped early if the accumulated evidence for efficacy, harm or futility was compelling. They could also adapt to changes in estimates of nuisance parameters such as response variance or baseline hazard rate. However adaptive designs now cover much more in the way of mid-course changes in the original design—changes in the maximum sample size, patient population, treatments, endpoints, etc., triggered by viewing current unblinded data. These new methods had to be statistically rigorous and clinically convincing.

At first, such proposals met with some skepticism. However, the flexibility afforded by adaptive trial designs has led gradually to their use being widely accepted over the past twenty years. For example, there are now two Guidances from the U.S. Food and Drug Administration (FDA, 2019a, 2020) on this subject, and in June 2025 the International Council for Harmonisation released a draft guideline ICH E20 (ICH, 2025) on Adaptive Designs for Clinical Trials.

We have made a slight change to the title of our book. When we started, we envisioned there would be relevance to a variety of fields where sequential experimentation is employed, such as in engineering design and quality control. However, it has become evident from the citation indices that, apart from some interests in the fields of epidemiology and animal trials, our audience has been overwhelmingly concerned with the design and analysis of clinical trials.

In this new edition, we have made many changes and updates throughout with much new material. As examples, we now include a full discussion of the role of binding and non-binding futility boundaries for decision making, and the chapter on Bayesian approaches to the design and analysis of clinical trials has been updated and expanded. In addition, there are eleven completely new chapters. We have added a chapter on constructing optimal group sequential designs and their use as benchmarks for other types of design. Another new chapter shows how to accommodate delayed or "pipeline" data. These are outcomes that are not yet available at the time of an interim analysis, but it is recognized that they will become available shortly thereafter. In most trials there will be more than one outcome of interest—co-primary, secondary, etc. In the past few years, there have been numerous advances in the field of multiple comparisons. A new chapter details how these methods can be incorporated into the group sequential structure. There follow eight new chapters on the principles and applications of adaptive methods in group sequential trials. These chapters comprise the general ideas behind the framework followed by various applications, including sample size re-assessment, seamless Phase II/III trials, use of intermediate or surrogate endpoints, multi-arm multi-stage (MAMS) trials, and enrichment trials. There follows a much updated chapter on the subject of response adaptive treatment assignment, and we conclude with a chapter on numerical computations.

With so much added material, some from the first edition has been omitted. In particular, for the procedures described in the various chapters, entries in tables of needed constants now concentrate only on a false positive rate of $\alpha = 0.05$ for two-sided tests and $\alpha = 0.025$

for one-sided tests. These values are the ones typically recognized by regulatory authorities. However, if other values of α are needed, there are now a number software packages, both commercial and non-commercial, available for the computations.

Writing such a long volume has been rather like painting the Forth Bridge. By the time we reached the end of the manuscript, inevitably interesting new material had appeared. However, we had to stop sometime and so we apologize for what the reader may find has been omitted!

Bath and Ithaca
September 2025

Christopher Jennison
Bruce Turnbull

Glossary

AE	adaptive enrichment
AIDS	acquired immunodeficiency syndrome
ALR	approximate likelihood ratio
AR	adaptive randomization
ASD	adaptive seamless design
ASN	average sample number
AUC	area under the curve
BH	Benjamini-Hochberg
BHAT	Beta-blocker Heart Attack Trial
cdf	cumulative distribution function
CHMP	Committee for Medicinal Products for Human Use
CHW	Cui, Hung & Wang
CME	conditional moment estimator
COPD	chronic obstructive pulmonary disease
CRP	conditional rejection probability
CTP	closed testing procedure
CV	cardiovascular
DMC	data monitoring committee
DRGST	delayed response group sequential test
DSMB	data and safety monitoring board
ECMO	ExtraCorporeal Membrane Oxygenation
EMA	European Medicines Agency
esf	error spending function
FDA	Food and Drug Administration
FDR	false discovery rate
FEV$_1$	initial second forced expiration volume
FWER	family-wise error rate
GLS	generalized least squares
GSD	group sequential design
GST	group sequential test
HIV	human immunodeficiency virus
ICH	International Council for Harmonisation
IDMC	independent data monitoring committee
ITN	inferior treatment number
JT	Jennison & Turnbull
LD	Lan & DeMets
LR	likelihood ratio
LSD	least significant difference
LSE	least squares estimate
MACE	major adverse cardiovascular event
MAMS	multi-arm multi-stage
MCP	multiple comparisons procedure
MI	myocardial infarction

MLE	maximum likelihood estimate
MRC	Medical Research Council
MSE	mean squared error
NDA	new drug application
NIMB	net incremental monetary benefit
NPC	nutritional prevention of cancer
OBF	O'Brien & Fleming
ORR	observed response rate
ORS	oral rehydration solution
OS	one-sided (in connection with error spending test)
OS	overall survival (in connection with time-to-event data)
PCI	percutaneous coronary intervention
PCR	polymerase chain reaction
PFS	progression-free survival
PTSD	post traumatic stress disorder
PW	play the winner
RAR	response adaptive randomization
RBO	relative betting odds
RCI	repeated confidence interval
RPSFT	rank preserving structural failure time
RPW	randomized play the winner
SCP	stochastically curtailed procedure
SDI	spinal deformity index
SPRT	sequential probability ratio test
SSR	sample size re-assessment (or re-estimation)
TB	tuberculosis
TSE	Thall, Simon & Ellenberg
UMVCUE	uniformly minimum variance conditionally unbiased estimator
UMVUE	uniformly minimum variance unbiased estimator
WT	Wang & Tsiatis
$\mathcal{I}$	information level
J	number of treatment arms in a multi-armed trial
K	number of groups (analyses, looks, stages) in a group sequential procedure
R	ratio of maximum information level for a group sequential test to required by a fixed sample test with the same α and β
$\boldsymbol{X}^T$	transpose of vector or matrix $\boldsymbol{X}$
$t_{\nu,\alpha}$	upper α tail point of a Student's t-distribution with ν degrees of freedom
z_α	upper α tail point of a standard normal distribution, i.e., $z_\alpha = \Phi^{-1}(1-\alpha)$
α	Type I error probability
β	Type II error probability
$\boldsymbol{\beta}$	vector of regression coefficients
Φ	standard normal cumulative distribution function
ϕ	standard normal density function

Vectors and matrices are indicated by boldface type.

Authors

Christopher Jennison is a professor of statistics in the Department of Mathematical Sciences, University of Bath, UK. He has worked on the analysis of clinical trials at the Dana-Farber Cancer Institute, Boston, and collaborated with medical researchers and served on Independent Data Monitoring Committees in areas such as cancer research, cardiology, gynaecology, and complementary medicine.

Bruce Turnbull is an emeritus professor of statistics in the School of Operations Research & Industrial Engineering and in the Department of Statistical Science, Cornell University, USA. He has a long experience of designing, monitoring, and analyzing clinical trials and has served on the Data Safety Monitoring Boards for international, multi-center trials for the treatment or prevention of cancer, heart disease, and AIDS. He has also served as a consultant to pharmaceutical companies and health-related government review panels.

Chapter 1

Introduction

1.1 About This Book

This book describes statistical methodology for the group sequential design and analysis of studies in which interim analyses will be performed. The principal context in which we illustrate the methods will be that of clinical trials. However, applications can be found in almost any area where an experiment or survey can be carried out in phases or where its duration is long enough to permit periodic interim analyses. In the medical field, this includes animal studies and also epidemiological studies; in industrial applications, there are established sequential and group sequential designs for acceptance sampling in quality control and for life testing in reliability. Clearly, interim analyses may be called for in any long-term follow-up study, be it a field test of product reliability, a parolee recidivism evaluation of correctional programs, or a surveillance program for monitoring the occurrence of chronic disease in an occupational cohort. The examples described in Section 1.4 will illustrate the breadth of possible applications.

Since most of our examples will be stated in terms of clinical trials, we need to explain a little of the terminology for those unfamiliar with this field. A new therapy may go through various stages of testing in human subjects after showing promise in laboratory and animal studies before it is accepted by the medical community. Phase I trials are exploratory and are concerned with aspects of clinical pharmacology and toxicity. An objective is to find a suitable dose level that avoids unacceptable adverse side effects. Usually the sample consists of between 10 and 50 healthy volunteers, often pharmaceutical company employees or medical students. Phase II studies are of moderate size, involving 100 to 300 patients with the disease in question, and are concerned with evaluating both efficacy and safety aspects. Phase II can be subdivided into Phase IIa and Phase IIb. Phase IIa studies, sometimes termed "proof of concept" studies, are for the purpose of demonstrating clinical efficacy or biological activity. Phase IIb studies are conducted to determine the optimal dose of a new drug that is effective subject to satisfying any safety concerns. A Phase III trial is the definitive evaluation for the proposed new therapy in which effectiveness is verified and the occurrence of any long-term adverse side effects is monitored. Patients suffering from the disease in question are randomly assigned either to the new treatment or to the current standard (or placebo) treatment. Phase III trials often involve more than 1000 patient volunteers and last three to five years, or maybe longer depending on recruitment rates and the necessary follow-up time. Sometimes, more than two treatments will be compared in the same study. It is in the comparative Phase III study that the statistical design and analysis come under the most attention and scrutiny (although, of course, careful statistical planning and evaluation are called for in any research endeavor). Phase IV refers to the additional testing and monitoring of experience with the new treatment after it has been accepted and approved for general use, sometimes referred to as post-marketing surveillance. This categorization is not a strict one, and a trial's purposes can overlap the boundary definitions, especially in Phases II and III. Sometimes these stages are partitioned into finer categories (see Burdette & Gehan, 1970). This nomenclature was introduced for *therapeutic* trials but

DOI: 10.1201/9781584888482-1

is now also used in disease *prevention* trials. There, the subjects in Phases II and III might be patients at high risk; for example, in a trial for the prevention of skin cancer by dietary supplementation, subjects might have a prior history of skin lesions or high sun exposure. Another example would be a trial for the prevention of COVID-19 infection by vaccination.

We think of clinical trials as being conducted to find improved or superior therapies. However, many trials are *equivalence* or non-inferiority studies, designed to show that a new treatment's efficacy or safety is "the same as" or "at least no worse than" that of a standard treatment; see Treadwell et al. (2012). Such trials are common in pharmaceutical applications where, for example, it may be desired to compare two formulations of the same drug. For a text on bioequivalence studies in pharmaceutical trials, we refer the reader to Chow & Liu (2008). Dunnett & Gent (1977) give an example of a randomized trial in the area of patient management, originally reported by Spitzer et al. (1974), with the objective of determining whether certain aspects of the health care provided by nurse-practitioners could be considered equivalent to conventional care methods requiring a greater commitment of the physician's time. It should be noted that if the goal is to demonstrate equivalence, then the trial should be designed as such. A clear distinction must be made between failure to find evidence of a treatment difference, perhaps due to the low power of a small study, and the positive demonstration of "equivalence" by confirming that the treatment difference lies within some acceptable region; see Freiman et al. (1978), Simon (1986) and Durrleman & Simon (1990). We look specifically at designs for equivalence studies in Chapter 7.

In medical trials, the term "endpoint" is often used to denote a response variable. These variables can be of various types. A binary response of success or failure is common, especially in Phase II trials. A quantitative response variable is one measured on a continuous scale and is often modeled by a normal distribution, perhaps on some transformed scale, e.g., logarithmic. Examples include the plasma level of a particular biochemical, or the forced expiratory volume in one second (FEV_1) as a measure of lung function in a trial of respiratory disorder. A survival endpoint refers to the time it takes for some event of interest to occur, such as death, failure or relapse. It is an elapsed time, usually measured from the date of each patient's entry to the study. Of course, for some patients the event of interest has not occurred by the time the analysis is performed, possibly because they have withdrawn from the trial or died from causes unrelated to the trial. In this case, we say the patient's survival time is *right censored* and note the partial information that the time to event exceeds the observed follow-up time. Some endpoints are more complex: a multivariate response may contain several endpoints; there may be repeated measurements over time for a longitudinal endpoint or growth curve data; or data may concern repeated occurrences of a recurrent event. We illustrate a variety of response types with examples in Section 1.4.

This book is not intended to be a comprehensive treatment of all statistical aspects of medical trials. There are many admirable books on this subject, including Burdette & Gehan (1970), Pocock (1983), Chow & Liu (2014), Friedman et al. (2015), Shih & Aisner (2016), Piantadosi (2017) and Kim, Bretz, Cheung & Hampson (2021). In addition, there are several collections of papers by well-known authors on various aspects of clinical trials: these include volumes edited by Tygstrup, Lachin & Juhl (1982), Miké & Stanley (1982), Shapiro & Louis (1983) and Buyse, Staquet & Sylvester (1984). We shall not be covering topics such as organization, protocols, eligibility requirements, randomization techniques, ethics, blinding, placebos and data management. These and many other related issues are well covered in the above references. Similarly, there are many textbooks concerning design and analysis of epidemiological studies: just a few are Breslow & Day (1980, 1987), Kleinbaum, Kupper & Morgenstern (1982), Checkoway, Pearce & Kriebel (2004) and Rothman, Greenland & Lash (2008). For a textbook on the design and analysis of long-term animal experiments, the reader is referred to Gart et al. (1986); see also Kilkenny et al. (2010). Finally, there are many excellent books on general medical statistics, including Altman (1991), Fleiss (1999) and Armitage, Berry & Matthews (2013).

We shall assume the reader has some familiarity with the classical design of multi-factor experiments, an important topic in the repertoire of every consulting statistician not only in agricultural and industrial applications, for which the subject was first developed, but also in biomedical experiments. Classic books on design include Cochran & Cox (1992) and a more modern treatment is provided by Box, Hunter & Hunter (2005). Although the practicalities of long-term Phase III trials usually preclude complicated designs and organizers typically aim to make the protocol as straightforward as possible, it is certainly possible to use some simple factorial designs. Beck et al. (1991) describe the use of a 2^2 design to investigate the effects of diet and blood pressure control on glomerular filtration function in a renal disease trial. Blot et al. (1993) report a one-half replicate of a 2^4 factorial design used to test effects of combinations of four nutrient groups in prevention of cancer incidence and cancer mortality among populations in Linxian County, China.

1.2 Why Sequential Methods?

In any experiment or survey in which data accumulate steadily over a period of time, it is natural to monitor results as they occur with a view to taking action such as early termination or making some modification to the study design. The duration of an experiment might be measured in terms of hours in the context of industrial acceptance sampling. In contrast, a clinical trial concerning disease prevention or an epidemiological cohort study of occupational exposure may run on a time scale of tens of years.

The many reasons for conducting interim analyses of accumulating data can be loosely categorized into three classes: *ethical*, *administrative* and *economic*. In trials involving human subjects, there is an ethical need to monitor results to ensure that individuals are not exposed to unsafe, inferior or ineffective treatment regimens. Even in negative trials where there appears to be no difference in the performance of two therapies, there is an ethical imperative to terminate a trial as soon as possible so that resources can be allocated to study the next most promising treatment waiting to be tested. Ethical considerations prescribe that accumulating data be evaluated frequently, not only with regard to internal comparisons of safety and efficacy, but also in the light of new information from outside the trial. In animal trials, ethics are also an important consideration, and interim monitoring is indicated there too.

Included in the administrative reasons for interim analyses is the need to ensure that the study is being executed as planned, that the subjects or experimental units are from the correct population and satisfy eligibility criteria, and that the test procedures or treatments are as prescribed in the protocol. An early examination of interim results will sometimes reveal the presence of problems which can be remedied before too much expense is incurred. As an example, in an accelerated life test of electronic components in helicopter parts, a rash of early failures indicated that the stress conditions had been incorrectly calibrated. In another application area, the Nutritional Prevention of Cancer (NPC) trial was a randomized clinical trial that investigated the effect of a nutritional supplement of selenium on the prevention of skin cancer, and a discovery that plasma selenium levels were not rising as expected in some patients in the supplement group indicated a possible non-compliance problem. Failure of some subjects to receive the prescribed amount of selenium supplement would have led to a loss of power to detect a significant benefit, if present, and action was therefore taken to initiate a run-in period whereby potential non-compliers could be identified and eliminated from the study prior to randomization. Another administrative reason for early examination of study results is to check on assumptions made in designing the trial. In a trial where the primary response variable is quantitative, the sample size is often set assuming this variable to be normally distributed with a certain variance. For binary response data, sample size calculations rely on an assumed value for the background incidence rate; for time-to-event data when individuals enter the trial at staggered intervals, an estimate of the subject accrual rate is important in determining the appropriate accrual period. An early interim analysis can reveal inaccurate assumptions in time for adjustments to be made to the design. We shall describe in Chapter 18 how such adjustments can be made in an objective fashion, based on statistical considerations.

Sequential statistical methods were originally developed in order to obtain economic benefits. For a trial with a positive result, early stopping means that a new product can be exploited sooner. If a negative result is indicated, early stopping ensures that resources are not wasted, referred to as "stopping for futility" or "abandoning a lost cause" (Gould, 1983). Sequential methods typically lead to savings in sample size, time and cost when compared with standard fixed sample procedures. The early sequential analysis literature stressed this aspect, as we shall do in this book. Interim analyses also enable informed management decisions to be made concerning the continuing allocation of limited research and development funds. Thus, a manager of a large diverse research program might require interim progress reports for this purpose.

In pharmaceutical industry clinical trials, there have been moves toward the formal requirement of interim analyses and also regulations governing the reporting of their results. In the U.S., the Food and Drug Administration (FDA, 1985) published regulations for new drug applications (NDAs) which included (Section 314.146.b.7) the requirement that the analysis of a Phase III trial "assess ... the effects of any interim analyses performed". This statement was further elaborated in a Guideline (FDA, 1988, p. 64):

> "The process of examining and analyzing data accumulating in a clinical trial, either formally or informally, can introduce bias. Therefore all interim analyses, formal or informal, by any study participant, sponsor staff member, or data monitoring group should be described in full even if treatment groups were not identified. The need for statistical adjustment because of such analyses should be addressed. Minutes of meetings of the data monitoring group may be useful (and may be requested by the review division)."

The Guideline also proposed that the plan for interim analyses appear on the protocol cover sheet for the NDA. Responses to the Guideline by representatives of the pharmaceutical industry were published (Enas et al., 1989; PMA Biostatistics and Medical Ad Hoc Committee on Interim Analysis, 1993). The Food and Drug Administration (1985) also required periodic monitoring and reporting of adverse drug experiences in Phase IV trials.

The FDA guidelines were updated by publication of "E9 Statistical Principles for Clinical Trials" in the *Federal Register*, Vol. 63, No. 179, p. 49583–49598 (FDA, 1998). This document was prepared under the auspices of the International Conference on Harmonization of Technical Requirements for Registration of Pharmaceuticals for Human Use. It gives a comprehensive list of recommendations for all aspects of statistical principles and methodology applied to clinical trials in the pharmaceutical industry. Of direct relevance for this book, Section III.4 of the document advocates the use of group sequential designs and Section IV gives detailed recommendations for trial conduct, including trial monitoring, interim analysis, early stopping, sample size adjustment and the role of an independent data and safety monitoring board (DSMB). Commonly used alternate terms are "data monitoring committee" (DMC) or "independent data monitoring committee" (IDMC).

Whereas continuous monitoring is desirable, it is often impractical. In fact, many of the possible benefits of sequential monitoring can be obtained by examining the data at periodic intervals, 2 or 3 times during the course of the study and almost all of the possible benefits can be achieved with 5 or 10 interim analyses. Such schemes are called "multi-stage" or "group sequential" and many studies can be naturally phased in this way. DSMBs of large multi-center clinical trials will only meet to examine the collected data at periodic intervals. In a rodent carcinogenicity study, the sample size in each stage might be the maximum number of animals that can be accommodated in the laboratory's cages and started on treatment at one time. In industrial acceptance sampling procedures, such as those prescribed in MIL-STD-105E (1989) (now replaced by ASQ/ANSI (2018a,b)), successive equally sized batches of items are inspected and a lot is accepted or rejected as soon as there is sufficient evidence to make a decision. This book is about these *group sequential* procedures.

When initially planned, the group sequential design for a clinical trial often depends on assumptions about some nuisance variables. For example, as mentioned above, the overall sample size may depend on the response variance or the response rate in the control group, which can only be estimated at the start of the trial. Generalizations of the group sequential design paradigm were proposed whereby such impediments could be accommodated, especially in the situation where treatment assignments at interim analyses remain blinded. However, it was realized that there was a practical need to allow more substantial mid-course changes in the design, driven by accruing outcome data both internal and external to the trial, while maintaining statistical validity. This has led over the last 25 years or so to a vast new literature in the theory and application of such *adaptive group sequential* procedures. This is the subject of the second part of our book.

1.3 A Short History of Sequential, Group Sequential and Adaptive Methods

Introductory texts on general statistics mention little or nothing about sequential design and analysis, even though interim analyses of experimental data are common in many branches of scientific research. As stated by Armitage (1993, p. 392):

> "The classical theory of experimental design deals predominantly with experiments of predetermined size, presumably because the pioneers of the subject, particularly R. A. Fisher, worked in agricultural research, where the outcome of a field trial is not available until long after the experiment has been designed and started. It is interesting to speculate how differently statistical theory might have evolved if Fisher had been employed in medical or industrial research."

The development of the subject of our book can be roughly divided into three periods: approximately (i) "sequential" (1945–1975), (ii) "group sequential" (1975–1999) and (iii) "adaptive" (1999 to present).

1.3.1 Sequential Methods (1945–1975)

The sequential approach has been a natural way to proceed throughout the history of experimentation. Perhaps the earliest proponent was Noah who, at weekly intervals, released a dove from the Ark in order to test for the presence of dry land during the subsidence of the Flood. The 17th and 18th century work of Huyghens, James Bernoulli, DeMoivre, Laplace and others who studied gambling systems might also be considered precursors; see, for example, Barnard (1946, p. 7). The formal application of sequential procedures started in the late 1920s in the area of statistical quality control in manufacturing production. Shewhart (1931) introduced control charts for process control. Dodge & Romig (1929) defined a two-stage acceptance sampling plan for components which could be tested and classified as effective or defective. Their plan has six parameters: the sample sizes for each stage (n_1 and n_2), acceptance numbers (c_1 and c_2), and rejection numbers (d_1 and d_2, where $d_1 > c_1 + 1$ and $d_2 = c_2 + 1$). In implementing the plan, one first takes an initial sample of n_1 items and if this contains c_1 defectives or fewer the lot is accepted, but if d_1 defectives or more are found the lot is rejected. Otherwise, the decision is deferred until a second sample of size n_2 is inspected, and the lot is then accepted if the total cumulative number of defectives is less than or equal to c_2 and rejected if this number is greater than or equal to d_2. This idea is easily generalized to that of a multi-stage or multiple sampling plan in which up to K stages are permitted (Bartky, 1943). The multi-stage plans subsequently developed by the Columbia University Research Group (Freeman et al., 1948) went on to form the basis of the U.S. military standard for acceptance sampling, MIL-STD-105E (1989). A similar problem arises in the early stages of drug screening and in Phase II clinical trials, where the aim is to identify drugs with at least a certain level of therapeutic efficacy.

The modern theory of sequential analysis has stemmed from the work of Abram Wald (1947) in the U.S. and George Barnard (1946) in the United Kingdom, who were participating in industrial advisory groups for war production and development from about 1943. In particular, the major influence on the subject has been the work of Wald (1947) related to his sequential probability ratio test (SPRT). Wallis (1980) gives a fascinating account of how the idea of sequential testing was discussed by members of the Statistical Research Group, an Office of Scientific Research and Development activity at Columbia University, during the Second World War. Wald was initially skeptical as to whether such an approach was feasible, but

> "On the second day, however, he phoned that he had found that such tests do exist and are more powerful, and furthermore he could tell us how to make them. He came over to the office and outlined his sequential probability ratio to us."

Suppose observations are from a distribution whose probability density or mass function is known apart from a parameter θ. In the basic form of the SPRT for testing a simple null hypothesis H_0: $\theta = \theta_0$ against a simple alternative H_1: $\theta = \theta_1$, successive observations are taken as long as the likelihood ratio remains in a certain interval (a, b), otherwise the experiment is stopped and the appropriate hypothesis selected. The constants a and b can be chosen so that the probabilities of a Type I error (selecting H_1 when H_0 is true) and of a Type II error (selecting H_0 when H_1 is true) are approximately equal to pre-specified values α and β, respectively. This procedure typically leads to lower sample sizes than fixed sample tests. In fact, Wald & Wolfowitz (1948) proved the SPRT has the theoretical optimal property that, among all tests with error probabilities not exceeding α and β, it attains the smallest possible expected sample size or "average sample number" (ASN) when either H_0 or H_1 is true. However, the SPRT is an "open" procedure, i.e., the sample size is not bounded, and a consequence of this is that the distribution of sample size can be skewed and have a large variance. Also the ASN can be large when θ is not equal to θ_0 or θ_1. These disadvantages led to consideration of non-parallel, curved boundaries for which the critical boundary values a and b are no longer constant but depend on the cumulative sample size, n, and which are "closed" in the sense that $a_{n^*} = b_{n^*}$ for some n^*, ensuring an upper limit on the sample size. One simple modification discussed by Wald is to truncate the SPRT at a certain sample size; Aroian (1968) and Aroian & Robison (1969) showed how to compute the operating characteristic (the probability of rejecting H_0 as a function of θ) and ASN curves for such plans using numerical integration.

Another class of sequential tests is based on triangular continuation regions that approximately minimize the maximum ASN; see Kiefer & Weiss (1957), Anderson (1960), Lai (1973) and Lorden (1976). Good discussions of generalizations of the SPRT and corresponding theoretical optimality properties can be found in Siegmund (1985), Wetherill & Glazebrook (1986) and Lai (2002). The mathematical statistics journals now contain a vast literature on the subject of sequential analysis; a detailed chronology up to 1990 can be found in Ghosh (1991).

The SPRT and related procedures are concerned primarily with the problem of selecting one of two competing hypotheses. The problem of selecting from among more than two hypotheses is more difficult and has attracted less attention. A classic book in the area is that by Bechhofer, Kiefer & Sobel (1968). An important application is in the field of communication theory, where problems include target detection in multiple-resolution radar (Marcus & Swerling, 1962) and infrared systems (Ernisse et al., 1997), signal acquisition (Veeravalli & Baum, 1996) and pattern recognition (Fu, 1968). For the multi-hypothesis problem, Baum & Veeravalli (1994) developed a procedure that generalizes the SPRT, which they term the MSPRT.

Another important area is that of the sequential *design* of experiments in which the experimenter is allowed to choose the type of observation to take at each stage. It is desirable to choose that type of observation which yields the most information for the ultimate decision of which hypothesis to accept, but the best type of observation to choose usually depends on the unknown true hypothesis. Accruing observations can help guide the selection of observations in an optimal manner; see Chernoff (1972, Sections 13 to 15) and Blot & Meeter (1973). The so-called two-armed or multi-armed bandit problems (see Berry & Fristedt, 1985; Gittins, Glazebrook & Weber, 2011) are special cases of this problem. Zelen (1969) introduced the "play-the-winner" rule for allocating treatments to subjects entering a medical study on the basis of previously observed responses, and various modifications to this rule have since been proposed. A highly controversial application of the play-the-winner design was the ECMO (ExtraCorporeal Membrane Oxygenation) study; see the article by Ware (1989) and accompanying discussion, where the issues are explored in detail. We shall discuss some aspects of data-dependent treatment allocation in Chapter 30, especially as it might be applied in medical trials. Note that these "response adaptive designs" should be distinguished from those designs that are now termed simply as "adaptive", which are the subject of Chapters 22 to 29.

1.3.2 Group Sequential Methods (1975–1999)

As noted above, early multi-stage or "group sequential" designs were proposed for industrial acceptance sampling with a binary response, effective or defective. Two-stage and three-stage procedures for normal responses were developed by Armitage & Schneiderman (1958), Schneiderman (1961), Dunnett (1961) and Roseberry & Gehan (1964). A review of specifically two-stage procedures is provided by Hewett & Spurrier (1983). It is noteworthy that, in this early work, repeated numerical integration is used to calculate operating characteristics and ASN curves of multi-stage procedures for normal data. This device now provides the key tool in the construction of many group sequential tests and forms the basis of methods designed to handle unpredictable group sizes and for the calculation of significance levels and confidence intervals following a group sequential test. Because of the discreteness of the sample space, calculations for binary responses are more straightforward, and there is a large literature on group sequential designs for drug screening experiments and Phase II clinical trials; see, for example, Schultz et al. (1973), Elashoff & Beal (1976), Herson (1979), Lee et al. (1979), Fleming (1982), Jennison & Turnbull (1983) and Duffy & Santner (1987). Group sequential methods are by no means restricted to binary or normal responses, however, and in later chapters we shall describe methods for monitoring t-statistics, hazard ratio estimates for survival data, multivariate test statistics and more.

Armitage (1954, 1958b, 1975) and Bross (1952, 1958) pioneered the use of sequential methods in the medical field, particularly for comparative clinical trials. Initially these plans were fully sequential and did not receive widespread acceptance, perhaps because continuous assessment of study results was often impractical. However, the very influential 1967 Greenberg Report (see Heart Special Project Committee, 1988) stressed the importance of periodic interim monitoring of trial data by an advisory board (DSMB): the trial should not continue beyond a time when either treatment benefit can be established or there appears to be a harmful trend which is unlikely to be reversed.

In the discussion reported in Cutler et al. (1966), Shaw proposed the use of group sequential methods for a clinical trial, using the term "block sequential analysis". The shift to formal group sequential methods for clinical trials did not occur until the 1970s. Elfring & Schultz (1973) specifically used the term "group sequential design" to describe their procedure for comparing two treatments with binary response; McPherson (1974) suggested that the repeated significance tests of Armitage, McPherson & Rowe (1969) might be used to analyze clinical trial data at a small number of interim analyses; Canner (1977) used Monte Carlo simulation to find critical values of a test statistic in a survival study with periodic analyses. However, a major impetus for group sequential methods came from Pocock (1977), who gave clear guidelines for the implementation of group sequential experimental designs attaining Type I error and power requirements. Pocock also demonstrated the versatility of the approach, showing that the nominal significance levels of repeated significance tests for normal responses can be used reliably for a variety of other responses and situations.

In an article that appeared shortly after Pocock's paper, O'Brien & Fleming (1979) proposed a different class of group sequential tests based on an adaptation of a truncated SPRT. These tests have conservative stopping boundaries at very early analyses and give a decision rule similar to the fixed sample test if the last stage is reached, features which have turned out to be appealing to practitioners. Later papers by Slud & Wei (1982) and Lan & DeMets (1983) were very important, as they showed that group sequential methods can be employed when group sizes are unequal and even unpredictable, a common situation in clinical trials but not a problem usually encountered in industrial sampling applications. The papers by Pocock (1977), O'Brien & Fleming (1979) and Lan & DeMets (1983) have been particularly influential and together form the starting point for subsequent methodological research and the basis of current practice in clinical trial design. Of course, these authors built on foundations laid by others, particularly the work of Armitage (1975).

It became clear in the 1980s, with the growing number of large-scale clinical trials and increasing use and influence of policy advisory boards or DSMBs to police the progress and conduct of a trial, that formal statistical procedures for interim monitoring of efficacy and safety data were required. Protocols of nearly all major long-term clinical trials now contain statements of formal group sequential procedures for statistical monitoring of accruing results. The practical demands have led to rapid progress in statistical methodology since 1980. The development and acceptance of group sequential methods also benefited from concurrent advances in biostatistical methods, especially in survival analysis (see Kalbfleisch & Prentice, 2002; Cox & Oakes, 1984; Miller, 1998), and from the availability of high-powered computers, facilitating both the data management and data analyses necessary for interim analyses.

These developments can be illustrated in the simplest setting of a comparison between two treatments. A group sequential test monitors a statistic summarizing the difference in primary responses between the two treatment groups at a series of times during the trial. If the absolute value of this statistic exceeds some specified critical value, the trial is stopped and the null hypothesis of no difference between the groups is rejected, usually regarded as a "positive" result. The critical values form a boundary for the sequence of test statistics, the null hypothesis being rejected if this boundary is crossed. If the statistic stays within the testing boundary until the trial's planned termination, then the null hypothesis is accepted (a "negative" result). Armitage, McPherson & Rowe (1969) pointed out that if the usual fixed sample critical values are used at each analysis, the false positive, or Type I, error probability is greatly increased over the nominal level of, say, $\alpha = 0.05$. Early developments by Pocock (1977), O'Brien & Fleming (1979), Fleming, Harrington & O'Brien (1984) and Wang & Tsiatis (1987) concerned methods of adjusting the critical values to maintain the overall false positive rate at an acceptable level; further calculations determine the sample size needed for the group sequential test to attain a desired power.

Subsequent researchers modified the two-sided testing boundary by inserting an "inner wedge" to permit early stopping to accept the null hypothesis; see, for example, Gould & Pecore (1982), Whitehead & Stratton (1983, Sec. 5) and Emerson & Fleming (1989, Sec. 4).

It should be noted that the idea of an "inner wedge" originated in the fully sequential methods for binary data proposed by Bross (1952) and Armitage (1957).

As different testing boundaries were being investigated, theoretical research and simulation studies extended group sequential methodology to endpoints other than a simple normal response. Notably, increasingly more general results were obtained justifying normal approximations to the joint distribution of test statistics arising in the analysis of survival data. The work of Slud & Wei (1982), Lan & DeMets (1983) and Kim & DeMets (1987a) permitted methods to be extended to the often more realistic setting in which group sizes, or more generally increments in information, are unequal and unpredictable, the typical situation in clinical trials with survival endpoints and staggered entry.

The group sequential tests referred to above are for the "two-sided" problem in which the null hypothesis of no difference between treatments A and B is tested against the alternative of a difference in either direction. Sometimes it is more appropriate to test the hypothesis of treatment equality against the one-sided alternative that one particular treatment, A say, is superior, and this leads to consideration of "one-sided" group sequential tests. Early stopping may arise with a decision in favor of treatment A or to accept the null hypothesis, which now represents the case that treatment A is equal or inferior to treatment B. Some problems have the further asymmetry that early stopping is only appropriate under one conclusion. Suppose, for example, treatment A is a new therapy for a non-life-threatening condition and treatment B is a standard therapy. Gould (1983) remarks that in such a trial, if interim findings indicate that A is better than B, the study should continue to completion in order to provide adequate information on safety, tolerability and secondary aspects of the efficacy of treatment A; see also Berry & Ho (1988). A similar asymmetry occurs in a Phase II trial when a safety endpoint is being monitored.

Although group sequential tests are formulated in terms of tests between particular hypotheses, it is possible to draw more general inferences on termination of a trial. These include calculation of interval estimates (e.g., Siegmund, 1978, 1985; Tsiatis, Rosner & Mehta 1984; Kim & DeMets, 1987b), of P-values (e.g., Fairbanks & Madsen, 1982; Madsen & Fairbanks, 1983) and of point estimates (e.g., Whitehead, 1986a). Since different outcomes to a sequential test have varying numbers of observations, the monotone likelihood ratio property does not apply and there is some freedom in defining the sample space orderings which underlie the construction of confidence intervals and P-values. This has led to a substantial literature defining and assessing different proposals.

A recurrent problem in sequential analysis is the dominant role of the stopping rule. All the methods mentioned in the preceding paragraphs rely for their justification on adherence to a statistical stopping rule. However, it cannot be guaranteed that a monitoring committee will abide by such a rule in practice. We have seen that there are many reasons to monitor a trial, and there are just as many to decide to stop or continue a trial. In medical applications, these might include side effects, financial cost, quality of patients' lives or the availability of new promising treatments unknown at the start of the trial. The decision to stop the study is *not* primarily statistical; rather, it is highly complex, involving many subjective factors. Meier (1975, p. 26) claims the problem is political rather than medical, legal or statistical. Because of its complex and subjective nature, the decision to terminate or continue a trial can be a cause of considerable controversy. One contentious example was the UGDP study (Kolata, 1979); others include the Physicians' Health Study, the NSABP trial on tamoxifen for breast cancer prevention and the CDC study in Thailand on the use of short-course zidovudine (AZT) to prevent mother-to-child transmission of HIV.

Standard sequential statistical analysis is not well equipped to handle a non-rigid stopping rule and more flexible approaches have been proposed. Methods developed within the frequentist paradigm include stochastic curtailment and repeated confidence intervals. The former approach monitors the probability that a specific fixed sample test will achieve a significant result; see Lan, Simon & Halperin (1982), Choi, Smith & Becker (1985) and

Spiegelhalter, Freedman & Blackburn (1986). Repeated confidence intervals (Jennison & Turnbull, 1984, 1989) allow interval estimates to be constructed for the parameter of interest, such as the treatment difference, at any stage independently of any stopping rule. The Bayesian approach (see, for example, Spiegelhalter, Freedman & Parmar, 1994) also affords the possibility of flexible monitoring of trial results, and we shall contrast and compare the Bayesian and frequentist approaches in Chapter 13. There has been related research into the decision-theoretic basis for group sequential designs by, for example, Bather (1985) and Eales & Jennison (1992, 1995).

1.3.3 Adaptive Sequential Methods (1999–2025)

An extensive and detailed treatment of the history of adaptive group sequential methods for clinical trials is provided by Bauer et al. (2016) and the accompanying published discussion. Thus we shall only give a much abbreviated version here. The methodology tries to answer the question: How can we redesign a clinical trial in mid-course using the interim, unblinded outcome data, without compromising the overall false positive (Type I error) rate? Of particular interest is the option to increase the planned sample size, i.e., sample size re-assessment (SSR). The initial papers in this area included Bauer (1989) and Bauer & Köhne (1994), who first proposed the use of *combination tests*, an idea borrowed from the literature of meta-analysis (see Chapter 23). Separately, Proschan & Hunsberger (1995) introduced the *conditional error function* approach to tackle this question; we shall see in Section 23.5 that this method turns out to be related to that of Bauer.

However, we have dated this "era" as starting in 1999, because of the appearance of three important papers that added to the earlier work of Bauer and his co-workers. In the U.S., proposals by Fisher (1998) and, more prominently, Cui, Hung & Wang (1999) instigated discussion at meetings and conferences held that year. In Europe, the paper of Lehmacher & Wassmer (1999) played a similar role. This gave rise to broader publicity and awareness of the new methodology. Controversy revolved around the fact that the methods did not follow established statistical principles of sufficiency and invariance. Lack of sufficiency leads to inefficiency; see, for example, Jennison & Turnbull (2003). Lack of invariance can lead to a paradox whereby a small change in the order in which the observations arrive can produce a large change in conclusions; see Burman & Sonesson (2006). The conversation then concerns the question of what compromises should be made in order to obtain the very real benefits of flexibility that are afforded by adaptive designs. As stated by Maurer (2016, p. 363):

> "The early enthusiasm, skepticism, or outright rejection by clinical researchers and statisticians, with regard to adaptive designs, has largely given way to a more balanced view, with a clearer picture of advantages and disadvantages depending on their particular use."

As exemplified by Bretz et al. (2006) and Schmidli et al. (2006), the benefits afforded by adaptive design methodology extend beyond application to sample size re-assessment. Other examples include: permitting mid-course changes in target sub-populations (enrichment); adding or dropping treatments at interim analyses in multi-arm trials; enabling a "seamless" transition from Phase II to Phase III trials, perhaps with different endpoints in the two phases. In settings where multiple hypotheses are involved, the benefits can also include efficiency gains over non-adaptive procedures as, for example, in enrichment designs; see Rosenblum (2015) and Burnett & Jennison (2021).

In 2019, the U.S. FDA published its Guidance for Industry on "Adaptive Designs for Clinical Trials of Drugs and Biologics" (FDA, 2019a). This guidance outlines principles for the use of adaptive designs along with examples of situations where they might be employed. A distinction is made between *pre-planned* (or *prospective*) adaptations, whereby the rules for design modifications under all eventualities are specified in advance, and flexible adaptations in which more arbitrary mid-course changes can be made. The former type is clearly preferred under the FDA Guidance (FDA, 2019a, Sec. II B):

> "In nearly all situations, potential adaptive design modifications should be planned and described in the clinical trial protocol (and in a separate statistical analysis plan) prior to initiation of the trial."

An advantage of pre-planned adaptations is that the statistical properties of any proposed design, such as false positive rates, power, selection probabilities, expected sample sizes, etc., under a variety of scenarios, can be computed when designing the study, if necessary by extensive simulations. Pre-planned adaptations are also favored in the adaptive designs extension (Dimairo et al., 2020) to the Consolidated Standards Of Reporting Trials (CONSORT) 2010 statement (Moher et al., 2010).

The draft guideline ICH E20 on Adaptive Designs for Clinical Trials released by the International Council for Harmonisation in June 2025 sets out five key principles, the second of which is "Adequacy of Trial Planning" (ICH, 2025. Sec. 3.2). This principle states:

> "Adequate planning is important for all clinical trials to ensure the design is pre-specified, conduct and analysis are appropriate, and results are reliable and interpretable. ... Prior to initiation of a trial with an adaptive design, further aspects should be specified and justified in addition to the typical components of trial planning. These include the number and timing of interim analyses, type of adaptation, statistical methods for producing interim results, anticipated rule governing the adaptation decision, statistical methods for the primary analysis aligned to each targeted estimand, and approaches to maintain trial integrity. ... All relevant details pertinent to the planning of an adaptive trial design should be appropriately documented."

Since implementation of an adaptive design will be the responsibilty of the Independent Data Monitoring Committee (IDMC), the following guidance is of prime importance:

> "Sponsors should discuss the type of adaptations and anticipated adaptation rules in detail with the IDMC to confirm its understanding and support. This ensures the IDMC is prepared to review interim results and make adaptation recommendations during the trial while also protecting individual trial participants' safety."

Although the requirements laid down by regulators may be demanding, the combination test methodology can be employed to make both planned and unplanned adaptations and the flexibility to handle unplanned design modifications and respond to unexpected events is undoubtedly appealing.

An innate problem with the analysis and interpretation of clinical trial data revolves around the issues of *multiplicity* from various sources. As well as taking multiple looks at the data, there may be multiple endpoints, multiple treatments, multiple centers, multiple readers or raters, multiple patient sub-populations, etc. Such issues will arise and be considered in this book. Needless to say, there is a long history on the statistical treatment of multiple comparisons; the reader is referred to books by Hochberg & Tamhane (1987), Hsu (1996), and Bretz, Hothorn & Westfall (2010), amongst others.

1.3.4 Comment

This section is by no means intended to be a complete survey and chronology of statistical methodology for sequential design and analysis. For the earlier history, the reader is referred to the book by Wetherill & Glazebrook (1986) and the volume edited by Ghosh & Sen (1991), in particular the first chapter (Ghosh, 1991). Good references for adaptive design methods include the books by Chow & Chang (2012) and Wassmer & Brannath (2016). However, much of the development of *group sequential* and *adaptive* methods will be reviewed in detail in the chapters that follow.

1.4 Types of Response Variables

In a clinical trial, data may be collected on many variables (perhaps hundreds!). These are typically classified into two categories: (i) covariates, either observed at baseline or collected during the study and possibly time-varying, and (ii) response (outcome) variables or "endpoints". In the terminology of regression theory these are termed independent (or explanatory) variables and dependent variables, respectively. To follow the development of the methods in this volume it is useful to have in mind some typical examples of response variables.

1.4.1 Acute Binary Response: Comparison of Proportions

Here, two treatments are compared with respect to a binary outcome measure taking one of two possible values, "success" or "failure". Let p_A and p_B denote the probabilities of success on the two treatments. The parameter of interest can be the difference $p_A - p_B$, the ratio p_A/p_B or the log odds ratio

$$\log\left\{\frac{p_A(1-p_B)}{p_B(1-p_A)}\right\}.$$

Here, as in later chapters, we use $\log(x)$ to denote the natural logarithm of x.

Kim & DeMets (1992) cite the example of the Thrombolytic Intervention in Myocardial Infarction Trial (The TIMI Study Group, 1985) which compared the effectiveness of two agents in breaking down blood clots in coronary vessels. Here, the dichotomous outcome is determined by whether or not the agent succeeds in re-opening these vessels.

The randomized trial described by Martinez et al. (1996) is an equivalence trial comparing two methods of treating acute diarrhea in young children in rural Mexico. One treatment is an easily available, home based rice-powder beverage, the other the standard oral rehydration solution (ORS) recommended by the World Health Organization. The response, successful rehydration or not, is known within a few hours. Because of the practical advantages of the rice-powder beverage, the trial organizers hoped to demonstrate that it has a success probability equal to, or at least not too much lower than, the standard ORS.

In vaccine trials, an extremely large number of subjects are randomized to receive the vaccine or a placebo injection. Usually the infection probability for any particular individual is very low. Then the problem can be approximated by one of comparing two Poisson distributions.

1.4.2 Comparison of Proportions with Strata

Consider the monitoring of incidence data in a prospective stratified intervention study. A pilot study under the direction of Dr L. Clark conducted in Qidong county in the People's Republic of China investigated the effect of a dietary supplement of selenium on the prevention of liver cancer. The full design called for a study population of 40 townships of approximately 30,000 people each. On the basis of demographic and other characteristics, the 40 townships were to be grouped in 20 matched pairs. One township in each pair was to be chosen at random to receive a supplement of selenium in the salt supply. The response variable was the number of deaths from liver cancer during the study period. Here the summary parameter of interest might again be the log odds ratio, assumed approximately constant across the strata (township pairs), i.e.,

$$\theta = \log\left\{\frac{p_{Aj}(1-p_{Bj})}{p_{Bj}(1-p_{Aj})}\right\}, \quad j = 1, \ldots, 20,$$

where p_{Aj} and p_{Bj} are the probabilities of dying of liver cancer for individuals on treatments A and B, respectively, in the jth township pair. The study was planned to last for 10 years but with interim reports to be made annually.

Similar data arose in the study U.S. Air Force Project Ranch Hand II, which assessed the effects of the herbicide Agent Orange. Here, each case, a pilot exposed to Agent Orange, was matched to several controls consisting of unexposed pilots of otherwise similar characteristics. It was mandated that reports be made by the Air Force to the U.S. Congress annually for 20 years.

1.4.3 Normally Distributed Responses

The normal distribution can often be used to approximate the distribution of a continuous outcome, sometimes after a logarithmic or power transformation. Examples of such a response might be change of tumor size from a baseline measurement, blood pressure, viral load, lung function, or concentration of some chemical in the blood or urine. In order to gain the full benefits of sequential analysis, the response should be available shortly after treatment is administered, at least relative to the study duration.

In a two-sample problem, the parameter of interest is the difference in the mean responses of the two treatments. Facey (1992) describes a placebo-controlled efficacy trial investigating a new treatment for hypercholesterolemia where the primary response, reduction in total serum cholesterol level over a four week period, is assumed to be normally distributed.

A one-sample problem in which the parameter of interest is simply the mean response can occur in a Phase I or II trial when there is no control or comparison group. Comparisons within pair-matched patients or between two measurements on the same patient in a crossover study (see, for example, Jones & Kenward, 2014) lead to essentially the same analysis as the simple one-sample problem. Armitage (1975, p. 119) gives an example where responses represent differences in recovery times for pairs of patients treated with two competing hypotensive agents. Recovery times were measured in minutes and thus responses can be considered to be immediate relative to the length of the trial.

1.4.4 Time-to-event Responses

A "time-to-event" or "survival" endpoint such as time to death or time to disease progression is very common in Phase III clinical trials for chronic diseases such as cancer or AIDS. Patients can enter the trial at different calendar dates and the outcome measure is the elapsed time from the patient's entry, the date of randomization say, to the occurrence of the event of interest. Response times are often of the same order of magnitude as the duration of the study, leading to right censoring when a patient has not experienced the event by the time of an interim or final analysis. Other patients' response times may be censored because of early withdrawal from the study for reasons unrelated to treatment.

Response time data also arise in industrial life testing and reliability studies. Cox & Oakes (1984, p. 7) give an example of an experiment where 60 springs were tested under cycles of repeated loading and failure time was measured in units of 10^3 cycles. The springs were subjected to different stress levels and at the lower stress levels, where failure times are longer, testing was abandoned before many units had failed, again giving rise to censored observations. The reliability and life-testing plans in MIL-STD-781C (1977) involve sequential monitoring of such failure time data.

1.4.5 Recurrent Event Data

In some studies the event of interest can occur repeatedly over time in the same patient. Studies of asthma attacks or epileptic seizures are obvious examples. The outcome data for each patient consist of a series of failure times which can be viewed as the realization of a stochastic point process. In the Nutritional Prevention of Cancer Trial (see Clark et al., 1996; Jiang, Turnbull & Clark, 1999), over 1000 high-risk patients were randomly allocated to selenium supplementation or placebo and followed for up to ten years. Each patient was monitored at a dermatology clinic where the time from entry to each new occurrence of a squamous or basal cell carcinoma was recorded. In another application, Prentice, Williams & Peterson (1981) discuss a study of the frequencies of infections in patients following bone marrow transplantation.

In engineering, such repeated event data also arise in experiments and field studies concerning the reliability of repairable systems; see, for example, Ascher & Feingold (1984).

1.4.6 Longitudinal Data

Sometimes the outcome for an individual subject consists of a series of repeated measurements. In a trial of treatments for prevention of osteoporosis in post-menopausal women, we may wish to compare a control group with various intervention groups taking estrogen, calcium or other supplements, measuring patient status by the Spinal Deformity Index (SDI). If the patients enter the study at staggered times during the accrual period and then undergo spine examinations at irregular intervals, the data for each patient will consist of a series of SDI values, taken at varying intervals and with unequal numbers for each patient.

Wu & Lan (1992) describe a sequential design for the Lung Health Study (Anthonisen, 1989), a three-armed study of the effect of an intensive anti-smoking program, with or without bronchodilators, on chronic obstructive pulmonary disease. Almost 6000 smokers with early airway obstruction were recruited over two years by 10 clinical centers. The primary outcome variable is the series of measurements of initial second forced expiration volume (FEV_1) taken at baseline and at each of five annual follow-up visits.

Armitage, Stratton & Worthington (1985) give an example of sequential testing in a clinical trial of toothpastes, in which the series of annual increments in the counts of decayed, missing or filled teeth forms the response variable.

1.4.7 Multivariate Data

In almost any trial, many outcome variables are examined. As an example, in the Nutritional Prevention of Cancer Trial mentioned in Section 1.4.5, the DSMB was provided with information on some 200 endpoints at each interim analysis. Usually a single most important variable, mortality for example, is designated of primary interest. However, there will sometimes be several endpoints of equal interest *a priori*. Variables of a similar type can be combined to give a univariate score or summary statistic, as O'Brien (1984) suggests for a randomized trial of treatments for diabetes in which nerve function was measured by 34 electromyographic variables.

In other instances, variables may be combined to reduce the outcome measure to a small number of components, each of a different type, and for which further reduction to a univariate statistic is inappropriate. Suppose, for example, we reduce a set of variables to a bivariate outcome whose two components measure overall "efficacy" and "safety". We may wish to treat these variables differently in a sequential design, stopping early to recommend the new treatment if it appears to be both safe and superior to the standard in terms of efficacy, but stopping early to reject the new treatment if it has low efficacy or is substantially less safe than the standard; see Jennison & Turnbull (1993c). Group sequential methods for multiple endpoints are treated in Chapters 19 and 20.

1.5 Chapter Organization: A Roadmap

There are four main threads of development through this book.

- Essentials of group sequential tests. These include: timing and number of interim analyses; issues of testing (Type I error, power and sample size) and estimation upon termination; handling unknown nuisance parameters such as response variance or baseline incidence rate; futility boundaries (binding or non-binding); accommodating unpredictable group sizes; allowing for delayed ("pipeline") observations.
- Application to a wide range of types of response distribution: normal, binary, time-to-event, Poisson, multivariate, etc.
- Extensions beyond the basic two parallel group comparison paradigm. Multiple endpoints (primary, co-primary, secondary, etc.); multi-arm multi-stage (MAMS) trials; seamless Phase II/III designs.
- Methods to enable mid-trial changes (adaptations) in the design based on unblinded interim results without compromising statistical validity. These changes could be prospectively planned or unplanned.

These threads are intertwined and we give here a brief roadmap of our chapter organization.

In Chapter 2, we introduce the subject of our book by describing the basic group sequential procedures for comparing results from two parallel treatment arms. Specifically, outcomes are assumed to be immediately available and normally distributed with known variance. It is desired to test the null hypothesis, H_0, of equality of the two mean responses against a two-sided alternative. Observations are to be taken in up to $K \geq 2$ equally sized groups, but the trial can stop early to reject H_0 if the observed difference is sufficiently large. We start with this simple situation, unrealistic for many applications, since historically this was the first to be considered and because it serves to illustrate the ideas of Type I and Type II error probabilities and choice of sample size and stopping criteria that are fundamental to group sequential procedures.

In Chapter 3, we describe a *unified formulation* of group sequential problems based on sequences of standardized test statistics and information levels. These statistics are said to have the *canonical joint distribution.* The information levels (which depend on the group sizes) do *not* need to be equally spaced. This formulation will be used throughout this book. It has the benefit of encompassing many common situations with normally distributed data, including two-treatment parallel designs, balanced or unbalanced, the one-treatment-arm designs common in Phase I and II trials, paired sample designs, crossover designs, general linear models used to adjust for covariates or factor variables, and longitudinal studies involving repeated measures or growth curve data. The same formulation also applies approximately to parametric models amenable to standard maximum likelihood analysis and to some special data types, in particular survival data.

We return to two-sided tests in Chapter 4 and show how the unified formulation is applied in a wide variety of situations to derive group sequential procedures with Type I and Type II error probabilities equal to specified values, α and β. The general method of implementation is explained in a simple "five-step summary". Details are given for the designs and response types described in Chapter 3. We continue Chapter 4 by explaining the "significance level approach", a key device in broadening the applicability of the group sequential tests defined in the idealized setting of Chapter 2 to include other situations. With this approach, we shall be able to treat such issues as unequal group sizes, linear model data, the log-rank test for survival data, and group sequential t-tests for normal data of unknown variance. All of these applications to different response distributions are presented in the context of a two-sided test. However, the same basic principles allow us to extend the use of other types of group sequential test from the simple case of normal data with known variance to the full range of response variables.

One-sided tests are introduced in Chapter 5. Here, there is an upper "efficacy" boundary and a lower "futility" boundary which coincide at the final analysis. The lower boundary can be treated in two ways, binding or non-binding. With a binding futility boundary the trial stops immediately with a negative result if this lower boundary is crossed. When a futility boundary is designated as non-binding, its purpose is to provide guidance and the trial may be allowed to continue even if this boundary is crossed. Since this continuation of the trial increases the chance of a false positive finding (Type I error), the upper boundary must be adjusted upwards accordingly. Methods for designing one-sided tests are extended to the case of unknown variance based on successive t-statistics. Finally in this chapter we present an alternate one-sided test, namely the triangular test proposed by Whitehead & Stratton (1983).

In Chapter 6 we introduce two-sided tests with an inner wedge allowing early stopping to accept the null hypothesis. Early acceptance of H_0 was not permitted in the two-sided tests of Chapters 2 and 4, but even though the ethical motivation may not be not so strong, there are sound financial reasons to curtail a study heading toward a negative conclusion; such curtailment can also benefit future patients by releasing experimental resources for the earlier exploration of further promising treatments.

The goal of some clinical trials, particularly in pharmaceutical applications, is to establish "equivalence", or at least "non-inferiority", of a new treatment relative to an existing treatment, rather than to prove the superiority of the new treatment. This goal leads to a reversal of the usual roles of null and alternative hypotheses, and so yields another class of group sequential procedures. We describe group sequential equivalence tests in Chapter 7, illustrating these with examples of tests for "average equivalence" and "individual equivalence".

Chapters 8 to 11 continue with generic methods, usually described for the standard case of normal observations but readily applicable to other responses. A modification to the basic group sequential design is needed to construct testing boundaries when group sizes, or more generally increments in information between analyses, are unequal and unpredictable. Such unpredictability is not uncommon and is certainly to be expected in survival studies, where observed information depends on subject accrual, dropout and baseline event rates. In Chapter 8 we describe "error spending" methods for dealing with these problems in both one-sided and two-sided tests. These error spending methods require numerical computation for their implementation, a topic discussed briefly in Chapter 8 and at greater length in Chapter 31. In Chapter 9, we discuss the full statistical analysis to be performed upon conclusion of a group sequential study, including the construction of P-values and confidence intervals that retain their standard frequentist properties despite the unusual sample space.

Chapter 10 concerns sequences of "repeated confidence intervals" and "repeated P-values" which enable results to be reported and assessed at any interim stage, irrespective of whether or not the study design calls for termination. These sequences possess the important feature that their validity does not depend on the stopping rule used. As well as providing data summaries with well-defined properties when a formal stopping rule is not adhered to, repeated confidence intervals and P-values are suitable for presentation at professional meetings during the course of a trial as they automatically protect against "over-interpretation" of interim results. Chapter 11 explores the idea of stochastic curtailment, an informal approach to interim monitoring based on predictions of the outcome of a single, fixed-sample analysis to be conducted at the end of the study when a specified time or sample size is reached. At any interim stage, it is natural to consider the probability that a particular decision will eventually be made when the trial reaches its planned conclusion, given trends seen in the data thus far. If a certain decision is inevitable, it is reasonable to stop or "curtail" the trial. Stopping a trial because a particular decision, while not inevitable, is highly likely is referred to as "stochastic curtailment".

Chapter 12 is concerned with finding the optimal choice of stopping boundaries so as to minimize expected sample size subject to specified Type I and II error constraints. Although the aim is to find an optimal design with specific frequentist properties, the techniques used involve sequential Bayes decision theory, as well as dynamic programming and numerical computation. While the resulting procedures could be used directly, we see their key role as providing benchmarks with which to assess the efficiency of designs that may be attractive for other reasons, for example, the error spending designs presented in Chapter 8. When discussing optimality properties, it is natural to consider the Bayesian approach to sequential monitoring: this is discussed in Chapter 13.

So far, the group sequential designs we have formulated have assumed that the response is immediate, or at least rapidly observed relative to the intervals between patients being recruited to the trial. However, in many clinical trials, the primary response is measured some time after the start of treatment. Thus, at each interim analysis there are subjects who have been treated but whose response is yet to be observed: we describe these patients as being "in the pipeline". In Chapter 14, we discuss the approach of Hampson & Jennison (2013), who proposed a new form of group sequential test to treat the issue of a delayed response.

The next three chapters take up the thread of applying group sequential procedures defined for an idealized form of normal data to other response types. In Chapter 15, we derive the general theory for normal linear models which underlies the "unified formulation" of Chapter 3 and the "canonical joint distribution" for test statistics presented there. This theory extends to correlated normal observations and the mixed effects models which often arise in longitudinal and repeated measures studies. We also present theory for the case of unknown variance, and this provides the exact joint distribution of sequences of t-statistics, justifying more general use of the group sequential t-tests presented earlier. We go on to address general parametric models satisfying the conditions for standard maximum likelihood theory, and these include generalized linear models as an important special case. The theory shows that standardized statistics have, asymptotically, the same multivariate normal joint distribution as statistics from a normal linear model, supporting the extension of the unified formulation of Chapter 3. We shall see in Chapter 17 that the same canonical joint distribution holds for sequentially observed log-rank statistics and parameter estimates in Cox's (1972) proportional hazards regression model for survival data, even though these are not obtained by standard maximum likelihood methods.

Chapter 16 concerns binary data. We describe methods of exact calculation for group sequential tests with small or moderate sample sizes for which the normal approximations of Chapter 15 may be inaccurate. This chapter also discusses more complex applications involving binary data, such as inference for logistic regression models and stratified 2×2 tables. The latter, commonly found in epidemiological cohort studies and matched pair case-control studies, lead to group sequential versions of the Mantel-Haenszel (Mantel & Haenszel, 1959) and McNemar (1947) tests. When the probability of an event of interest is very small and sample sizes are large, procedures based on a Poisson approximation are appropriate and these are described here. Such a situation commonly occurs in vaccine trials.

In many clinical trials, the primary response is a survival time or time to occurrence of an event such as failure, relapse or remission of symptoms. In Chapter 17 the group sequential log-rank test, already introduced in Section 4.5, is generalized to allow the effects of covariates through use of Cox's (1972) proportional hazards regression model; this is illustrated with an extensive example. Also in this chapter, methods are described for drawing inferences from Kaplan-Meier estimates (Kaplan & Meier, 1958) of survival probabilities and quantiles. As an example, a repeated confidence interval for a two year survival probability or median survival time might provide a useful data summary for consideration by a DSMB at an interim analysis.

In setting the sample size for a study, the power calculation may require knowledge of an unknown parameter, such as a response variance or event rate. In Chapter 18, we discuss studies with an initial "internal pilot" phase, the results of which are used to estimate parameters affecting sample size and, hence, to modify the planned overall sample size.

In Chapter 19, we consider the situation where the response variable is multivariate and the stopping boundaries define regions in a multi-dimensional space. Chapter 20 treats the multiplicity issues when making inferences about individual components of the response vector. These components (or endpoints) may be variously categorized as primary, co-primary, secondary, tertiary, etc. We first consider a single-stage design, introducing the concepts of family-wise error rate, closed testing procedures, intersection tests, graphical approaches to multiple testing, and adjusted P-values. We then add the extra multiplicity layer, namely that of multiple looks at the data, in order to accommodate these features into group sequential designs.

Because of the considerable expense in setting up a large clinical trial, there are economies of scale and savings in time if several competing treatments can be compared in the same trial. Accordingly, it is increasingly common for trials to be conducted with three or more treatment arms. This introduces another source of multiplicity. Methodology for such multi-arm multi-stage (MAMS) trials is the subject of Chapter 21. In such trials, it can become apparent at an interim stage that some treatments are performing poorly and should be dropped. Decisions to drop treatments can be regarded as "adaptive" in nature and the design of such multi-arm trials is best handled using adaptive methods: we return to this topic in discussing *adaptive* MAMS designs in Chapter 27.

Chapters 22 to 30 all concern the theory, methods and application of adaptive designs in clinical trials. These methods allow for mid-course changes in the design, based on unblinded interim data, including outcomes for the primary endpoint, without compromising statistical validity, in particular the Type I error or false positive rates. In Chapter 22, we start with a motivating example taken from Cui, Hung & Wang (1999) and then describe how the methods presented by Lehmacher & Wassmer (1999) produce a natural bridge from group sequential design (GSD) to adaptive design (AD). In Chapter 23, we introduce two of the basic building blocks of adaptive designs. The first is the combination test (Bauer & Köhne, 1994), which facilitates the combination of data from before and after an adaptation. The second is the conditional Type I error function; see Proschan & Hunsberger (1995), and Müller & Schäfer (2001, 2004). The most common application of adaptive designs is in mid-course re-assessment of the maximum sample size (or, more generally, maximum information) in the light of interim unblinded efficacy data. Indeed, this was the original motivation for the work of Cui, Hung & Wang (1999). We discuss this application in detail in Chapter 24, including such sample size re-assessment rules as the "promising zone" approach of Mehta & Pocock (2011) and generalizations of this.

An important and highly productive feature of adaptive designs is that they facilitate the design and analysis of trials in which multiple hypotheses are under investigation. We describe the general framework for adaptive designs with multiple hypothesis tests in Chapter 25. We present several applications which exploit this framework in Chapters 26 to 28. The first of these, described in Chapter 26, is the adaptive seamless design (ASD). As outlined in Section 1.1, a Phase II trial compares several treatments or dose levels to a control, and the subsequent Phase III trial is a confirmatory study intended to demonstrate the superiority against control of the treatment selected based on the safety and efficacy results from Phase II. In an ASD, there is a single overarching protocol for the combined study. An advantage is that the time (or "white space") between the two phases can be significantly reduced. There is also greater continuity: momentum and enthusiasm for subject recruitment at the trial centers is more likely to be maintained. There can also be a gain in statistical efficiency from the use of Phase II data alongside Phase III data in the final analysis.

A second application of adaptive design is to the selection of treatment arms in MAMS trials. Following the discussion of *sequential* MAMS trials in Chapter 21, we describe *adaptive* MAMS trials in Chapter 27. At each stage, treatment arms can be dropped (or even added) and sample sizes re-assessed based on unblinded interim efficacy data and on any other available information. In fact, the ASDs of Chapter 26 can be considered as a special case of a MAMS trial where the number of experimental treatments is reduced to one at a single interim analysis.

A third application of adaptive designs is to the selection of sub-populations that may experience particular benefit from an experimental treatment. Such "enrichment" designs are described in Chapter 28. Here the target population is heterogeneous, comprising several subgroups. The object is to discover which subgroups (some, none or all) receive benefit from the new treatment. If it is discovered at an early stage that the new treatment is ineffective in a certain sub-population, then that subgroup can be dropped from further consideration. Then, in future stages, resources are spent only on recruiting from the remaining subgroups, "enriching" the sample from these groups.

In Chapter 29 we describe a case study with survival endpoints, based on the GATSBY trial (Thuss-Patience et al., 2017) that illustrates elements of the methods presented in Chapters 23 to 28. This example also illustrates the process of designing an adaptive trial and the simulation studies that may be conducted to assess design options.

Response-adaptive designs are the subject of Chapter 30. These should be considered as separate and distinct in nature from the adaptive designs previously considered. There are complete books devoted to response-adaptive designs; see, for example, Hu & Rosenberger (2006) and Atkinson & Biswas (2014). These methods have their origins in the bandit problems of Robbins (1952). We include this chapter to highlight how such designs can be incorporated into the group sequential designs of the first half of this book. In a parallel design comparing two treatments, it is not unusual for unequal randomization or dropout of some subjects to produce unbalanced numbers of observations on the two treatment arms. In Chapter 30, we consider designs with intentional imbalance between treatments, achieved by modifying treatment allocation at each stage in response to previously observed outcomes. Such designs are said to use stage-wise data-dependent or response-adaptive sampling rules and their usual purpose is an ethical one, namely to reduce the number of subjects assigned to the inferior treatment. Since the identity of this treatment is initially unknown an adaptive approach is required.

Finally, in Chapter 31, we present details of the numerical computations required to design and evaluate group sequential tests. These computational methods were used to obtain the many tables and figures appearing throughout the book.

1.6 Bibliography and Notes

Since the classic book of Armitage (1975), there have been a number of books on the specific topic of sequential statistical procedures for clinical trials. One early book is that by Whitehead (1997) which concentrated mostly on the theory and application of the triangular tests we shall describe in Sections 5.7 and 6.3. More recent books are those by Proschan, Lan & Wittes (2006) on group sequential tests and by Wassmer & Brannath (2016) on group sequential and adaptive designs.

A number of case studies can be found in the book edited by Peace (1992) and in the proceedings of two workshops, one on practical issues in data monitoring sponsored by the U.S. National Institutes of Health in 1992, the other on early stopping rules held at Cambridge University in 1993. These proceedings were published as special issues of the journal *Statistics in Medicine* (Issues 5 and 6 of Volume 12 in March 1993 and Issues 13 and 14 of Volume 13 in July 1994). The review paper by Fleming (1992) and accompanying discussion contain descriptions of several case studies of sequentially monitored trials in cancer and AIDS. In particular, Fleming describes an experience which illustrates the perils and undesirable consequences of an unstructured approach to interim monitoring, and another in which premature release of interim results had a harmful effect. Another lesson to be learned from the experiences reported in Fleming (1992) is that group sequential methods are important in indicating when a trial should *not* be stopped.

Although the procedures in Chapters 2 to 7 can be applied with a minimum of calculation, this is not the case for those in later chapters including error spending tests and inference on termination of a group sequential test. The description of numerical integration methods given in Chapter 31 can serve as a basis for writing special purpose software. To assist the reader, we have made available some of the R programs used to produce tables and figures in this book at the website `https://people.bath.ac.uk/mascj/`. A number of software packages are available to aid in applying group sequential and adaptive methods. However, the landscape changes rapidly and new packages appear regularly. An extensive survey of software available in 2015 appeared in Section 6 of Bauer et al. (2016). Currently, non-commercial R packages include:

`seqmon (https://CRAN.R-project.org/package=seqmon)` by Schoenfeld & Zheng (2024),

`gsDesign (https://CRAN.R-project.org/package=gsDesign)` by Anderson (2025),

`rpact (https://CRAN.R-project.org/package=rpact)` by Wassmer & Pahlke (2025).

Commercially available computer software packages include `East` from Cytel Inc. (www.cytel.com) and `Addplan` from ICON plc (www.iconplc.com). The statistical software SAS facilitates sequential design and testing in its `SEQDESIGN` and `SEQTEST` procedures; readers are also referred to the book *Adaptive Design Theory and Implementation Using SAS and R* by Chang (2016).

Chapter 2

Two-sided Tests: Introduction

2.1 Two-sided Tests for Comparing Two Treatments with Normal Response of Known Variance

The name *two-sided test* is given to a test of a null hypothesis against a two-sided alternative. It is alternately termed a "non-directional superiority" hypothesis test. Such a test is appropriate when comparing two effective treatments to assess if one is superior to the other. An example is provided by the VA Cooperative Study No. 591 (Schnurr et al., 2015), which was a randomized controlled trial of treatments for post-traumatic stress disorder (PTSD) comparing two of the most effective treatments, Prolonged Exposure (PE) and Cognitive Processing Therapy (CPT). The primary outcome was improvement in PTSD symptoms, measured after a treatment period of 12 weeks. The paper reported: "Despite solid evidence that both treatments are effective, there is limited evidence about their effectiveness relative to one another. The primary objective is to compare the effectiveness of PE and CPT for reducing PTSD symptom severity in a healthcare system that offers both treatments."

In this chapter we restrict attention to testing for a difference in the mean response of two treatments when observations are normally distributed with common, known variance. Denoting the difference in means by θ, the null hypothesis H_0: $\theta = 0$ states that responses follow the same distribution under both treatments. The alternative hypothesis, H_A: $\theta \neq 0$, contains two cases, $\theta < 0$ and $\theta > 0$, which correspond to one treatment being superior to the other and vice versa.

In this comparison, the standardized test statistic Z is distributed symmetrically about 0 under H_0, and a fixed sample test rejects H_0 if $|Z| > c$ for some constant c. The sign of Z determines which treatment is to be preferred when H_0 is rejected. A test's *Type I error probability* is defined to be the probability of wrongly rejecting the null hypothesis,

$$Pr_{\theta=0}\{|Z| > c\}.$$

The *power* of a test is the probability of rejecting the null hypothesis when it does not hold, $Pr_\theta\{|Z| > c\}$, for values of $\theta \neq 0$. The power depends on θ, increasing as θ moves away from 0, but it is convenient to state a power requirement at a specific value or values of θ, for example,

$$Pr_{\theta=\delta}\{|Z| > c\} = Pr_{\theta=-\delta}\{|Z| > c\} = 1 - \beta, \tag{2.1}$$

where δ represents a treatment difference the investigators would hope to detect with high probability. The small probability β is referred to as the *Type II error probability* at $\theta = \pm\delta$. In practical terms, it is not desirable to reject H_0 in favor of $\theta < 0$ when actually $\theta > 0$ as this would imply a recommendation of the inferior of the two treatments. With this in mind, the above power requirement can be replaced by

$$Pr_{\theta=\delta}\{Z > c\} = Pr_{\theta=-\delta}\{Z < -c\} = 1 - \beta. \tag{2.2}$$

DOI: 10.1201/9781584888482-2

Fortunately, for the values of c and δ that we shall consider, the probabilities $Pr_{\theta=\delta}\{Z < -c\}$ and $Pr_{\theta=-\delta}\{Z > c\}$ are extremely small and the two requirements (2.1) and (2.2) are equal for all practical purposes.

We present the standard non-sequential test for this problem in Section 2.2, and in the remainder of the chapter we describe and compare group sequential tests that achieve specified Type I and Type II errors. There is no loss of generality in introducing two-sided tests for the particular problem of a two-treatment comparison. We shall explain in Chapter 4 how the same forms of group sequential tests can be used in other problems including paired comparisons, crossover trials and a two-treatment comparison with allowance for the effects of covariates.

The assumption of normal observations with known variance can also be relaxed. We discuss application of the same tests to general parametric models in Section 3.4, and consider tests for binary data and survival data in subsequent sections. Binary responses and survival data are also discussed in greater detail in Chapters 16 and 17, respectively. The assumption of a known variance for normal observations is often unrealistic but we start by deriving tests under this assumption for simplicity and because the resulting tests can, in fact, be adapted to more realistic problems. We discuss two-sided group sequential t-tests for normal data with unknown variance in Section 4.6 and consider the problem of re-computing sample size during the course of a study to meet a stated power requirement in Sections 18.4 and 23.3.

2.2 A Fixed Sample Test

Let X_{Ai} and X_{Bi}, $i = 1, 2, \ldots$, denote the responses of subjects allocated to two treatments, A and B. Suppose responses of subjects receiving treatment A are normally distributed with variance σ^2 and mean μ_A, which we write as $X_{Ai} \sim N(\mu_A, \sigma^2)$, $i = 1, 2, \ldots$. Likewise, suppose $X_{Bi} \sim N(\mu_B, \sigma^2)$, $i = 1, 2, \ldots$, and all observations are independent. Consider the problem of testing the null hypothesis of no treatment difference H_0: $\mu_A = \mu_B$ against the two-sided alternative $\mu_A \neq \mu_B$ with Type I error probability α and power $1 - \beta$ at $\mu_A - \mu_B = \pm\delta$.

If n subjects are allocated to each treatment, the standardized statistic

$$\begin{aligned} Z &= \frac{1}{\sqrt{(2n\sigma^2)}}\left(\sum_{i=1}^{n} X_{Ai} - \sum_{i=1}^{n} X_{Bi}\right) \\ &\sim N((\mu_A - \mu_B)\sqrt{\{n/(2\sigma^2)\}}, 1). \end{aligned}$$

Thus $Z \sim N(0, 1)$ under H_0 and the symmetric two-sided test with Type I error probability α rejects H_0 if $|Z| > \Phi^{-1}(1 - \alpha/2)$, where Φ denotes the standard normal cumulative distribution function (cdf). To satisfy the power requirement, we also need

$$Pr\{|Z| > \Phi^{-1}(1 - \alpha/2)\} = 1 - \beta$$

when $Z \sim N(\pm\delta\sqrt{\{n/(2\sigma^2)\}}, 1)$. As explained in Section 2.1, we can ignore the very small probabilities that $Z < -\Phi^{-1}(1 - \alpha/2)$ when $\mu_A - \mu_B = \delta$ or $Z > \Phi^{-1}(1 - \alpha/2)$ when $\mu_A - \mu_B = -\delta$. Thus we need

$$E(Z) = \Phi^{-1}(1 - \alpha/2) + \Phi^{-1}(1 - \beta)$$

at the positive alternative, $\mu_A - \mu_B = \delta$, and equating this expected value to $\delta\sqrt{\{n/(2\sigma^2)\}}$ we find the necessary sample size to be

$$n_f(\alpha, \beta, \delta, \sigma^2) = \{\Phi^{-1}(1 - \alpha/2) + \Phi^{-1}(1 - \beta)\}^2 2\sigma^2/\delta^2 \qquad (2.3)$$

subjects per treatment arm. In practice, n_f must, of course, be rounded to an integer sample size.

As an example, suppose $\sigma^2 = 4$, so $X_{Ai} \sim N(\mu_A, 4)$ and $X_{Bi} \sim N(\mu_B, 4)$, $i = 1, 2, \ldots$, and it is required to test H_0: $\mu_A = \mu_B$ with Type I error $\alpha = 0.05$ and power $1 - \beta = 0.9$ at $\mu_A - \mu_B = \pm 1$. From (2.3), the fixed sample test for this problem needs $\{\Phi^{-1}(0.975) + \Phi^{-1}(0.9)\}^2 \times 2 \times 4 = 84.1$ subjects on each treatment, which we round up to 85. The null hypothesis is rejected if

$$\left| \sum_{i=1}^{85} X_{Ai} - \sum_{i=1}^{85} X_{Bi} \right| \geq 1.96 \sqrt{(85 \times 2 \times 4)} = 51.1$$

and accepted otherwise. We shall return to this example to illustrate and compare different types of group sequential test.

2.3 Group Sequential Tests

The key feature of a group sequential test, as opposed to a fully sequential test, is that the accumulating data are analyzed at intervals rather than after every new observation. Continuous data monitoring can be a serious practical burden, and the introduction of group sequential tests has led to much wider use of sequential methods. Their impact has been particularly evident in clinical trials, where it is standard practice for a monitoring committee to meet at regular intervals to assess various aspects of a study's progress and it is relatively easy to add formal interim analyses of the primary patient response. Not only are group sequential tests convenient to conduct, they also provide ample opportunity for early stopping and can achieve most of the benefits of fully sequential tests in terms of lower expected sample sizes and shorter average study lengths.

Although earlier suggestions for group sequential medical studies had been made by Elfring & Schultz (1973), McPherson (1974) and Canner (1977), the major impetus to group sequential testing came with the papers of Pocock (1977) and O'Brien & Fleming (1979). These two papers describe group sequential two-sided tests which are easy to implement and can be readily adapted to a variety of response distributions. In the basic two-treatment comparison, a maximum number of groups, K, and a group size, m, are chosen, subjects are allocated to treatments according to a constrained randomization scheme which ensures m subjects receive each treatment in every group and the accumulating data are analyzed after each group of $2m$ responses. For each $k = 1, \ldots, K$, a standardized statistic Z_k is computed from the first k groups of observations, and the test terminates with rejection of H_0 if $|Z_k|$ exceeds a critical value c_k. If the test continues to analysis K and $|Z_K| < c_K$, it stops at that point and H_0 is accepted. The critical values $\{c_1, \ldots, c_K\}$ are chosen to achieve a specified Type I error probability, and different types of group sequential test give rise to different sequences. The group size, m, is determined separately by a power condition.

We introduce the Pocock and O'Brien & Fleming tests in Sections 2.4 and 2.5, and we compare properties of these two types of test in Section 2.6. We present two other types of test in Section 2.7, the Wang & Tsiatis (1987) family of tests, which include Pocock and O'Brien & Fleming tests as special cases as well as other tests with boundaries of intermediate shapes, and the "Haybittle-Peto" test, which uses a very simple criterion for early stopping. We present tabulated properties of a range of tests to assist in choosing the most suitable type of test and number of groups of observations for a particular application.

Although the assumption of equal numbers of subjects in each group can be relaxed a little, the tests described in this chapter do require approximate equality of group sizes. More flexible methods for handling unequal and unpredictable group sizes will be described in Chapter 8. A possibility which we do not consider at this stage is early stopping to *accept* the null hypothesis; tests with this property will be discussed in Chapter 6.

All the tests introduced in this chapter can be viewed as *repeated significance tests* since at each analysis the null hypothesis is rejected if a non-sequential test of H_0 is significant at or below a certain level. In the case of the Pocock test, the same significance level is applied at each analysis, whereas for other tests, the constants c_k decrease and the nominal significance levels increase over the course of the study. This representation provides a convenient mechanism for generalizing the tests. In Chapter 4 we shall extend the basic methods to other two-sided testing problems, to slightly unequal group sizes and to data from other response distributions, by applying tests of the null hypothesis at the same sequences of significance levels. Our method of specifying the group size required to attain a given power also generalizes easily. A group sequential test's maximum sample size is calculated by multiplying the group size needed in a fixed sample test by a factor, which we provide, appropriate to the type of test, the number of groups and the specified error probabilities. The same factors apply to other two-sided testing problems and are approximately correct when group sizes are slightly unequal or when non-normal response distributions give rise to approximately normal test statistics.

2.4 Pocock's Test

2.4.1 Definition of Pocock's Test

Pocock (1977) adapted the idea of a "repeated significance test" at a constant nominal significance level to analyze accumulating data at a relatively small number of times over the course of a study. Patient entry is divided into K equally sized groups containing m subjects on each treatment, and the data are analyzed after each new group of observations has been observed.

For simplicity, we shall first consider the case where treatment allocation within each group is random, subject to the constraint that m subjects receive each treatment. More sophisticated randomization methods, such as stratification with respect to prognostic factors, can be used instead, in which case the analysis falls under the general treatment for linear models described later on in Section 3.3. For the testing problem of Section 2.2, observations $X_{Ai} \sim N(\mu_A, \sigma^2)$ and $X_{Bi} \sim N(\mu_B, \sigma^2)$, $i = 1, 2, \ldots$, are available from treatment groups A and B, respectively. Pocock's test uses the standardized statistic after each group of observations,

$$Z_k = \frac{1}{\sqrt{(2mk\sigma^2)}} \left(\sum_{i=1}^{mk} X_{Ai} - \sum_{i=1}^{mk} X_{Bi} \right), \quad k = 1 \ldots, K. \tag{2.4}$$

If, at the kth analysis, $k = 1, \ldots, K$, the absolute value of Z_k is sufficiently large, the study stops with rejection of H_0. If H_0 has not been rejected by the final analysis, K, it is accepted. It is not appropriate simply to apply a level-α two-sided test at each analysis since the multiple looks at the data would lead to a Type I error well in excess of α; for example, Armitage, McPherson & Rowe (1969, Table 2) report that the probability under H_0 that $|Z_k|$ exceeds $\Phi^{-1}(0.975) = 1.96$ in *at least* one analysis $k \in \{1, \ldots, 5\}$ is 0.142, nearly three times the 0.05 significance level applied at each individual analysis.

Pocock's test retains the general form of a repeatedly applied significance test, rejecting H_0 at stage k if $|Z_k| \geq C_P(K, \alpha)$, $k = 1, \ldots, K$, but the critical value $C_P(K, \alpha)$ is chosen to give overall Type I error α, i.e.,

$$Pr_{\mu_A - \mu_B = 0}\{\text{Reject } H_0 \text{ at analysis } k = 1,\ k = 2, \ldots, \text{ or } k = K\} = \alpha.$$

Table 2.1 *Pocock tests: constants $C_P(K, \alpha)$ for two-sided tests with K groups of observations and Type I error probability α*

	$C_P(K,\alpha)$		
K	$\alpha = 0.01$	$\alpha = 0.05$	$\alpha = 0.10$
1	2.576	1.960	1.645
2	2.772	2.178	1.875
3	2.873	2.289	1.992
4	2.939	2.361	2.067
5	2.986	2.413	2.122
6	3.023	2.453	2.164
7	3.053	2.485	2.197
8	3.078	2.512	2.225
9	3.099	2.535	2.249
10	3.117	2.555	2.270
11	3.133	2.572	2.288
12	3.147	2.588	2.304
15	3.182	2.626	2.344
20	3.225	2.672	2.392

Formally, the test is

$$\begin{array}{ll} \text{After group } k = 1, \ldots, K-1 & \\ \quad \text{if } |Z_k| \geq C_P(K,\alpha) & \text{stop, reject } H_0 \\ \quad \text{otherwise} & \text{continue to group } k+1, \\ \text{after group } K & \\ \quad \text{if } |Z_K| \geq C_P(K,\alpha) & \text{stop, reject } H_0 \\ \quad \text{otherwise} & \text{stop, accept } H_0. \end{array} \qquad (2.5)$$

This is an example of the general group sequential test in which H_0 is rejected after group k if $|Z_k| \geq c_k$, and in this case the critical values c_k, $k = 1, \ldots, K$, are constant and equal to $C_P(K, \alpha)$. The constants $C_P(K, \alpha)$ displayed in Table 2.1 for $\alpha = 0.01$, 0.05 and 0.1 were computed numerically using the joint distribution of the sequence of statistics $Z_1, \ldots, Z_K$. Details of this computation, which need not concern us at this point, are described briefly in Section 3.5 and more fully in Chapter 31. Values for $K = 1$, the non-sequential case, are included in the table for comparison.

Note that at each analysis, the test rejects H_0 if the two-sided significance level of a non-sequential test of H_0 using the data available at that time is below $\alpha' = 2[1 - \Phi\{C_P(K, \alpha)\}]$. Hence, this is a repeated significance test with constant "nominal significance level" α'. As an example, if $K = 5$ and $\alpha = 0.05$, the nominal significance level applied at each analysis is $\alpha' = 2\{1 - \Phi(2.413)\} = 0.0158$. A nominal significance level below the overall Type I error probability is essential in order to avoid the "multiple-looks" problem mentioned earlier.

Table 2.2 *Pocock tests: constants $R_P(K, \alpha, \beta)$ to determine group sizes for two-sided tests with K groups of observations, Type I error probability α and power $1-\beta$*

	$R_P(K,\alpha,\beta)$					
	$1-\beta=0.8$			$1-\beta=0.9$		
K	$\alpha=0.01$	$\alpha=0.05$	$\alpha=0.10$	$\alpha=0.01$	$\alpha=0.05$	$\alpha=0.10$
1	1.000	1.000	1.000	1.000	1.000	1.000
2	1.092	1.110	1.121	1.084	1.100	1.110
3	1.137	1.166	1.184	1.125	1.151	1.166
4	1.166	1.202	1.224	1.152	1.183	1.202
5	1.187	1.229	1.254	1.170	1.207	1.228
6	1.203	1.249	1.277	1.185	1.225	1.249
7	1.216	1.265	1.296	1.197	1.239	1.266
8	1.226	1.279	1.311	1.206	1.252	1.280
9	1.236	1.291	1.325	1.215	1.262	1.292
10	1.243	1.301	1.337	1.222	1.271	1.302
11	1.250	1.310	1.348	1.228	1.279	1.312
12	1.257	1.318	1.357	1.234	1.287	1.320
15	1.272	1.338	1.381	1.248	1.305	1.341
20	1.291	1.363	1.411	1.264	1.327	1.367

2.4.2 Satisfying the Power Requirement

The test defined by (2.5) has Type I error probability α for any group size. It is the power requirement

$$Pr_{\mu_A-\mu_B=\pm\delta}\{\text{Reject } H_0\} = 1-\beta$$

that determines the appropriate group size. The maximum sample size that the group sequential test may need depends on K, α and β and is proportional to σ^2/δ^2. Since the sample size of a fixed sample test with the same Type I error α and power $1-\beta$ at $\mu_A-\mu_B=\pm\delta$ is also proportional to σ^2/δ^2, it is sufficient to specify the ratio $R_P(K,\alpha,\beta)$ of the group sequential test's maximum sample size to the fixed sample size, and this ratio will then apply for all δ and σ^2. We refer to this ratio of sample sizes as the "inflation factor" and we shall see that the same ratios can be used in other two-sided testing problems and with other data types. The values of $R_P(K,\alpha,\beta)$ in Table 2.2 were found by the same type of numerical calculation used to determine the $C_P(K,\alpha)$. Again, details of their derivation can be found in Chapter 31. For brevity, we have tabulated $C_P(K,\alpha)$ and $R_P(K,\alpha,\beta)$ only for selected values of K; accurate estimates of constants for $K=13, 14$ and $16,\ldots,19$ can be obtained by linear interpolation both here and for other tests to be presented later.

The sample size per treatment arm, $n_f(\alpha,\beta,\delta,\sigma^2)$, required by a fixed sample test with Type I error α and power $1-\beta$ at $\mu_A-\mu_B=\pm\delta$ is given by (2.3). A Pocock group sequential test with K groups needs a maximum of $R_P(K,\alpha,\beta)\,n_f(\alpha,\beta,\delta,\sigma^2)$ observations per treatment arm and groups of $m=R_P(K,\alpha,\beta)\,n_f(\alpha,\beta,\delta,\sigma^2)/K$ observations on each treatment. Rounding m to an integer value will have a slight effect on the test's actual power, and it is usual to round m upward to obtain a conservative test, i.e., with power a little larger than $1-\beta$.

2.4.3 An Example

To illustrate Pocock's test, we return to the example of Section 2.2 where it was required to test H_0: $\mu_A = \mu_B$ with Type I error $\alpha = 0.05$ and power $1-\beta = 0.9$ at $\mu_A - \mu_B = \pm 1$ when observations have variance $\sigma^2 = 4$. From (2.3), the fixed sample size is $n_f(0.05, 0.1, 1, 4) =$ 84.1 observations per treatment arm. From Table 2.2, $R_P(5, 0.05, 0.1) = 1.207$. Hence, a Pocock test with five groups of observations has maximum sample size per treatment arm

$$R_P(5, 0.05, 0.1)\, n_f(0.05, 0.1, 1, 4) \;=\; 1.207 \times 84.1 \;=\; 101.5$$

and the necessary group size is $m = 101.5/5 = 20.3$ per treatment, which we round up to 21. The test stops with rejection of the null hypothesis at analysis k, $k = 1, \ldots, 5$, if

$$\left| \sum_{i=1}^{21k} X_{Ai} - \sum_{i=1}^{21k} X_{Bi} \right| \geq 2.413 \sqrt{(21k \times 2 \times 4)} \;=\; 31.28 \sqrt{k}.$$

If the null hypothesis has not been rejected by analysis 5, it is accepted at that point.

The boundary of the Pocock test for this example is illustrated in Figure 2.1. The first diagram shows the critical values applied to the standardized statistics Z_k, which are constant over the five analyses for the Pocock test. In the second diagram, the same test is expressed in terms of the unstandardized differences of sample sums on treatments A and B,

$$D_k \;=\; \sum_{i=1}^{21k} X_{Ai} - \sum_{i=1}^{21k} X_{Bi}, \quad k = 1, \ldots, 5,$$

and in this representation the Pocock boundary points lie on a square-root-shaped curve, $C_P(K, \alpha)\sqrt{\{42\, k\sigma^2\}}$.

Note that the maximum total number of subjects over both treatment arms that may be required using Pocock's test is 210. This is somewhat greater than the 170 needed for a fixed sample test, but the benefit is a smaller expected number of subjects for the group sequential test when $|\mu_A - \mu_B|$ is sufficiently large. Numerical comparisons of expected sample size will be seen in Section 2.6.

2.5 O'Brien & Fleming's Test

2.5.1 Definition of O'Brien & Fleming's Test

As an alternative to the repeated significance test with constant nominal significance levels, O'Brien & Fleming (1979) proposed a test in which the nominal significance levels needed to reject H_0 at each analysis increase as the study progresses. Thus, in their test it is more difficult to reject H_0 at the earliest analyses but easier later on. As in Pocock's test, subjects are allocated to treatments in groups, and the accumulating data are analyzed after complete groups of data become available. The O'Brien & Fleming test is also defined in terms of the standardized statistics Z_k of (2.4), and H_0 is rejected after group k if $|Z_k| \geq c_k$ for a sequence of critical values $c_1, \ldots, c_K$. Formally, the test is

$$\begin{array}{lll}
\text{After group } k = 1, \ldots, K-1 & & \\
\quad \text{if } |Z_k| \geq C_B(K,\alpha)\sqrt{(K/k)} & \text{stop, reject } H_0 & \\
\quad \text{otherwise} & \text{continue to group } k+1, & \\
& & (2.6) \\
\text{after group } K & & \\
\quad \text{if } |Z_K| \geq C_B(K,\alpha) & \text{stop, reject } H_0 & \\
\quad \text{otherwise} & \text{stop, accept } H_0. &
\end{array}$$

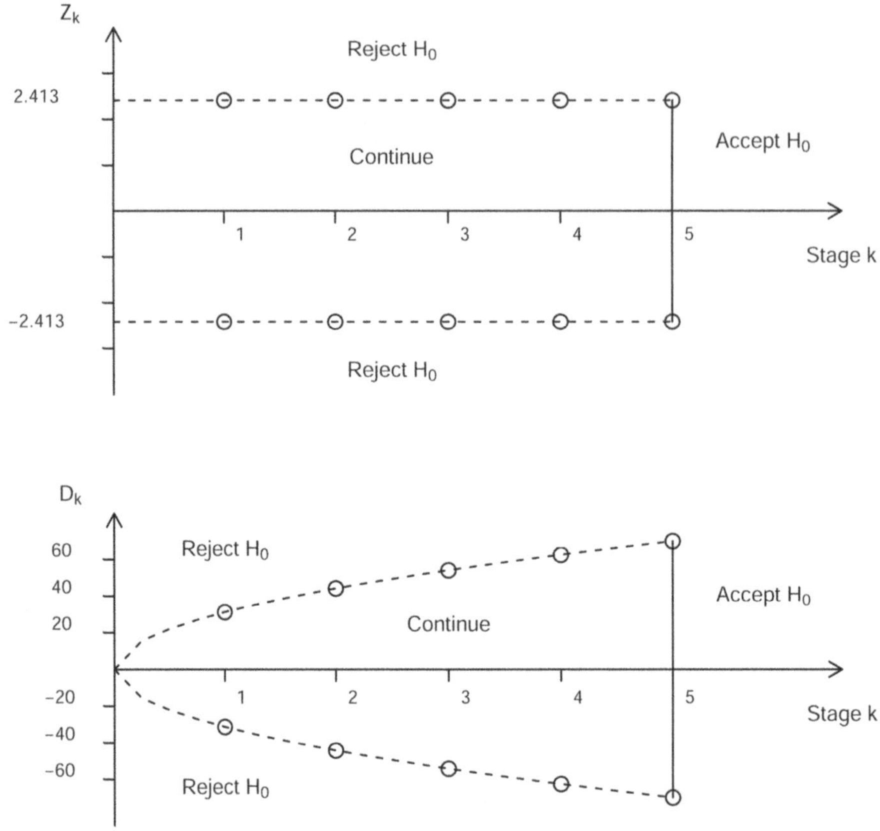

Figure 2.1 *A Pocock test for five groups of observations*

Thus, $c_k = C_B(K, \alpha)\sqrt{(K/k)}$, $k = 1, \ldots, K$. Values of $C_B(K, \alpha)$ which ensure an overall Type I error probability α are provided in Table 2.3. In terms of nominal significance levels, H_0 is rejected at analysis k, $k = 1, \ldots, K$, if the two-sided significance level of H_0 is below $\alpha'_k = 2[1 - \Phi\{C_B(K, \alpha)\sqrt{(K/k)}\}]$. Since $C_B(K, \alpha)\sqrt{(K/k)}$ decreases with increasing k, α'_k increases. For $K = 5$ and $\alpha = 0.05$ the nominal significance levels applied at analyses 1 to 5 are $\alpha'_1 = 0.000005$, $\alpha'_2 = 0.0013$, $\alpha'_3 = 0.0084$, $\alpha'_4 = 0.0225$ and $\alpha'_5 = 0.0413$.

Since $c_K = C_B(K, \alpha)$, comparing $C_B(K, \alpha)$ with $C_B(1, \alpha)$ shows how much larger the critical value for $|Z_K|$ is than the critical value that would be applied in a fixed sample analysis of the final data set. It is evident that this increase is quite small for O'Brien & Fleming tests.

2.5.2 Satisfying the Power Requirement

Table 2.4 contains constants $R_B(K, \alpha, \beta)$ for use in sample size calculations and these play the same role as the $R_P(K, \alpha, \beta)$ for the Pocock test. The maximum sample size on each treatment arm is $R_B(K, \alpha, \beta)\, n_f(\alpha, \beta, \delta, \sigma^2)$ where $n_f(\alpha, \beta, \delta, \sigma^2)$, given by (2.3), is the number of observations per treatment needed by a fixed sample size test with the same Type I error and power. Thus, the necessary group size is $m = R_B(K, \alpha, \beta)\, n_f(\alpha, \beta, \delta, \sigma^2)/K$ per treatment arm, suitably rounded.

In general, O'Brien & Fleming's test requires a smaller group size than Pocock's test to satisfy the same power requirement. To understand why this should be the case, note that from (2.4), the effect of $(\mu_A - \mu_B)$ on $E(Z_k)$ increases with k and so power is gained primarily

Table 2.3 *O'Brien & Fleming tests: constants $C_B(K, \alpha)$ for two-sided tests with K groups of observations and Type I error probability α*

	$C_B(K, \alpha)$		
K	$\alpha = 0.01$	$\alpha = 0.05$	$\alpha = 0.10$
1	2.576	1.960	1.645
2	2.580	1.977	1.678
3	2.595	2.004	1.710
4	2.609	2.024	1.733
5	2.621	2.040	1.751
6	2.631	2.053	1.765
7	2.640	2.063	1.776
8	2.648	2.072	1.786
9	2.654	2.080	1.794
10	2.660	2.087	1.801
11	2.665	2.092	1.807
12	2.670	2.098	1.813
15	2.681	2.110	1.826
20	2.695	2.126	1.842

Table 2.4 *O'Brien & Fleming tests: constants $R_B(K, \alpha, \beta)$ to determine group sizes for two-sided tests with K groups of observations, Type I error probability α and power $1 - \beta$*

	$R_B(K, \alpha, \beta)$					
	$1 - \beta = 0.8$			$1 - \beta = 0.9$		
K	$\alpha = 0.01$	$\alpha = 0.05$	$\alpha = 0.10$	$\alpha = 0.01$	$\alpha = 0.05$	$\alpha = 0.10$
1	1.000	1.000	1.000	1.000	1.000	1.000
2	1.001	1.008	1.016	1.001	1.007	1.014
3	1.007	1.017	1.027	1.006	1.016	1.025
4	1.011	1.024	1.035	1.010	1.022	1.032
5	1.015	1.028	1.040	1.014	1.026	1.037
6	1.017	1.032	1.044	1.016	1.030	1.041
7	1.019	1.035	1.047	1.018	1.032	1.044
8	1.021	1.037	1.049	1.020	1.034	1.046
9	1.022	1.038	1.051	1.021	1.036	1.048
10	1.024	1.040	1.053	1.022	1.037	1.049
11	1.025	1.041	1.054	1.023	1.039	1.051
12	1.026	1.042	1.055	1.024	1.040	1.052
15	1.028	1.045	1.058	1.026	1.042	1.054
20	1.030	1.047	1.061	1.029	1.045	1.057

from the later analyses. O'Brien & Fleming boundaries are wide early on but narrow at the last few analyses, where they are able to gain more power than Pocock boundaries with the same group sizes. Thus, for a fixed power requirement, the O'Brien & Fleming test needs a smaller group size.

2.5.3 An Example

We return to our previous example of testing H_0: $\mu_A = \mu_B$ with Type I error $\alpha = 0.05$ and power $1-\beta = 0.9$ at $\mu_A - \mu_B = \pm 1$ when observations have variance $\sigma^2 = 4$. From (2.3), the fixed sample size required is $n_f(0.05, 0.1, 1, 4) = 84.1$ observations per treatment arm and, from Table 2.4, $R_B(5, 0.05, 0.1) = 1.026$. Hence, an O'Brien & Fleming test with five groups of observations has maximum sample size per treatment arm

$$R_B(5, 0.05, 0.1)\, n_f(0.05, 0.1, 1, 4) \;=\; 1.026 \times 84.1 \;=\; 86.3$$

and the necessary group size is $m = 86.3/5 = 17.3$ per treatment, which rounds up to 18. The test stops to reject the null hypothesis at analysis k, $k = 1, \ldots, 5$, if

$$\left| \sum_{i=1}^{18k} X_{Ai} - \sum_{i=1}^{18k} X_{Bi} \right| \geq 2.040\, \sqrt{(5/k)}\, \sqrt{(18k \times 2 \times 4)} \;=\; 54.74$$

and accepts H_0 if it has not been rejected by analysis 5.

The maximum total number of subjects over both treatment arms that may be required by O'Brien & Fleming's test is 180, a small increase over the 170 needed for a fixed sample test. Just as for Pocock's test, the expected number of subjects is reduced substantially when $|\mu_A - \mu_B|$ is sufficiently large.

The boundary of the O'Brien & Fleming test for this example is illustrated in Figure 2.2. In the first diagram, where the test is described in terms of critical values for the standardized statistics, Z_k, the boundary values decrease markedly at successive analyses. The second diagram shows the same test expressed in terms of the unstandardized differences of sample sums,

$$D_k \;=\; \sum_{i=1}^{18k} X_{Ai} - \sum_{i=1}^{18k} X_{Bi}, \quad k = 1, \ldots, 5,$$

and on this scale the O'Brien & Fleming test has horizontal boundaries.

2.6 Properties of Pocock and O'Brien & Fleming Tests

2.6.1 Comparing Group Sequential Tests

A number of properties can be considered when comparing group sequential tests. Qualitative differences between the two types of test are apparent in Figures 2.1 and 2.2: the Pocock (1977) test has narrower boundaries initially, affording a greater opportunity for very early stopping, whereas the O'Brien & Fleming (1979) test has the narrower boundaries at later analyses and a smaller maximum sample size. For quantitative comparisons, a number of criteria are available. Error probabilities are a crucial property of any test, and it is important to compare properly "matched" tests. However, once the Type I and Type II error probabilities are specified, a test's power is largely determined over the range of possible parameter values, and there is little to be gained by investigating any remaining differences between tests. A principal reason for using a sequential test is to reduce sample size. In the example we have been considering, the sample size distribution depends on the value of $\mu_A - \mu_B$, and one should consider the sample size over a range of values of $\mu_A - \mu_B$ when

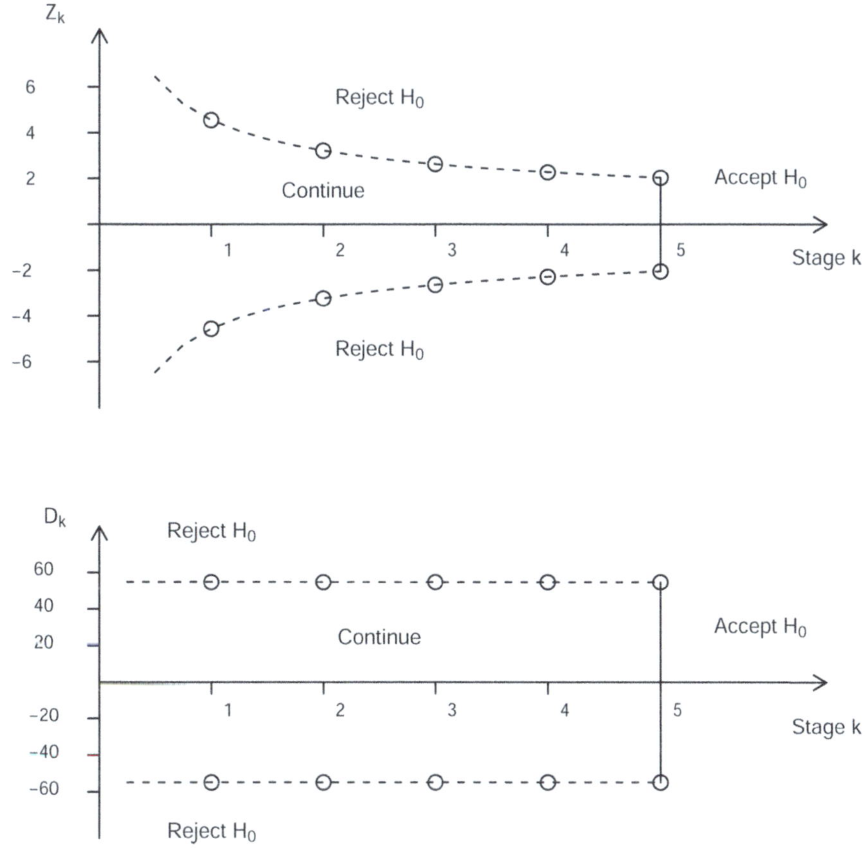

Figure 2.2 *An O'Brien & Fleming test for five groups of observations*

drawing comparisons. While it is convenient to summarize the sample size distribution under a particular value of $\mu_A - \mu_B$ by its expectation, other features of the distribution should not be ignored. In particular, the maximum possible sample size is a useful indication of what may happen in the worst case, and this sample size will occur quite often since it must arise whenever the null hypothesis is accepted.

We start by comparing the Pocock and O'Brien & Fleming tests in the previously studied example with five groups of observations. In Section 2.6.4, we explain how these comparisons generalize to other examples with five groups of observations. In Section 2.6.5 we present properties of tests with different numbers of groups which can be referred to when choosing a suitable number of groups of observations for a specific study.

2.6.2 *An Example with Five Groups*

Table 2.5 shows properties of the fixed sample test and the Pocock and O'Brien & Fleming tests with five groups of observations for the example described in Sections 2.2, 2.4.3 and 2.5.3. All values shown here and in subsequent examples were obtained by numerical computation and are accurate to the number of decimal places shown. Differences between the tests in power at $|\mu_A - \mu_B| = 1$ are due to the rounding of group sizes to integer values, and this rounding is also partly responsible for differences in power at other values of $|\mu_A - \mu_B|$. These differences in power are of minor interest, and we shall concentrate on comparisons of the sample size distributions of different tests.

Table 2.5 *Fixed sample, Pocock and O'Brien & Fleming tests: power functions, expected sample sizes and standard deviations of the sample size. Tests have five groups of observations and are designed to have Type I error rate* $\alpha = 0.05$ *and power* $1 - \beta = 0.9$ *at* $|\mu_A - \mu_B| = 1$. *Observations on each treatment are normal with variance* $\sigma^2 = 4$.

	Fixed sample test	*Pocock test*	*O'Brien & Fleming test*
		$Pr\{\text{Reject } H_0\}$	
$\|\mu_A - \mu_B\| = 0.0$	0.050	0.050	0.050
0.5	0.371	0.351	0.378
1.0	0.903	0.910	0.912
1.5	0.998	0.999	0.999
		$E\{\text{Number of observations}\}$	
$\|\mu_A - \mu_B\| = 0.0$	170.0	204.8	178.7
0.5	170.0	182.3	167.9
1.0	170.0	116.9	129.8
1.5	170.0	70.1	94.4
		St. dev. {Number of observations}	
$\|\mu_A - \mu_B\| = 0.0$	0.0	26.1	8.6
0.5	0.0	50.8	24.7
1.0	0.0	57.9	35.5
1.5	0.0	34.1	25.7

The fixed sample size of 170 subjects provides a benchmark against which we can compare sequential tests. Both group sequential tests have higher expected sample sizes at $\mu_A - \mu_B = 0$ than this baseline value, but this is to be expected since, in order to accept H_0, the group sequential tests must continue to their maximum sample sizes, 210 for Pocock's test and 180 for O'Brien & Fleming's test. However, the ethical imperative for stopping a clinical trial early is greatest when the treatment difference is large, since an early decision avoids the randomization of further subjects to an inferior treatment. The group sequential tests offer reductions in expected sample size over the fixed sample test when $|\mu_A - \mu_B|$ is sufficiently far from zero, and the Pocock test has the advantage of lower expected sample sizes under extreme values of $|\mu_A - \mu_B|$.

Table 2.5 also shows standard deviations of the sample size, and we see that values for the Pocock test are uniformly higher than those for the O'Brien & Fleming test. This difference between the two tests is easily explained, as the Pocock test offers greater opportunity to stop early with a low sample size, and when it continues to the final analysis, it has a higher maximum sample size.

2.6.3 Significance of the Maximum Sample Size

Continuing with the preceding example, Table 2.6 shows the complete distribution of the sample size upon termination of the Pocock and O'Brien & Fleming tests. Both tests give rise to considerable random variation in sample size and a substantial probability of continuing to observe all five groups even if $\mu_A - \mu_B \neq 0$.

Table 2.6 *Sample size distributions for Pocock and O'Brien & Fleming tests for five groups of observations. Tests are designed to have Type I error rate* $\alpha = 0.05$ *and power* $1-\beta = 0.9$ *at* $|\mu_A - \mu_B| = 1$. *Observations on each treatment are normal with variance* $\sigma^2 = 4$.

Pocock test

		Probability of sample size =				
		42	84	126	168	210
$\lvert\mu_A - \mu_B\rvert =$	0.0	0.016	0.012	0.009	0.007	0.956
	0.5	0.055	0.071	0.075	0.076	0.723
	1.0	0.214	0.268	0.210	0.138	0.171
	1.5	0.507	0.352	0.110	0.025	0.006

O'Brien & Fleming test

		Probability of sample size =				
		36	72	108	144	180
$\lvert\mu_A - \mu_B\rvert =$	0.0	0.000	0.001	0.008	0.017	0.974
	0.5	0.000	0.015	0.078	0.134	0.773
	1.0	0.001	0.134	0.354	0.282	0.229
	1.5	0.010	0.472	0.415	0.089	0.013

In some applications, the greater variance in the Pocock test's sample size may be unappealing and its higher maximum sample size may be regarded as a serious disadvantage. Suppose our example concerns the treatment of a rare disease and subjects are expected to enter the study at a rate of 40 per year. The maximum study duration is then 4.25 years for the fixed sample test, 4.5 years for the O'Brien & Fleming test and 5.25 years for the Pocock test. Despite the savings in *average* sample size, a cautious investigator might be deterred from using the Pocock test by the possibility of waiting 5.25 years to reach a conclusion when a fixed sample study can guarantee a conclusion one year earlier.

We have approached the testing problem by specifying Type I and Type II error probabilities, supposing it will be feasible to work with the group sizes determined by the calculations of Sections 2.4.2 and 2.5.2. This approach will not be realistic in some studies where time constraints or limitations on resources impose a maximum duration and a maximum sample size for any test. If sequential tests are to be used in such situations, they must sacrifice some of the available power in order to reduce expected sample size. If in our example the maximum sample size were fixed at 170 subjects altogether, i.e., 85 on each treatment arm, the fixed sample test with Type I error probability 0.05 would achieve power 0.903 at $\mu_A - \mu_B = \pm 1$. Group sequential tests with five groups of observations would have to take group sizes of $m = 17$ per treatment arm: the five-group O'Brien & Fleming test with $m = 17$ and Type I error probability 0.05 would achieve power 0.896 at $\mu_A - \mu_B = \pm 1$, and the five-group Pocock test with $m = 17$ and Type I error probability 0.05 would have a power of only 0.84 at $\mu_A - \mu_B = \pm 1$. Since, in many clinical trials, it is difficult to achieve a sample size large enough to guarantee minimal power requirements, this point should not be taken lightly.

2.6.4 Generalizing from a Single Example

There are important relationships between tests of a given type with the same number of groups and the same Type I error rate. In the two-treatment comparison with observations $X_{Ai} \sim N(\mu_A, \sigma^2)$ and $X_{Bi} \sim N(\mu_B, \sigma^2)$, $i = 1, 2, \ldots$, consider Pocock tests of H_0: $\mu_A = \mu_B$ with K groups of observations and Type I error probability α. Suppose also that the group size m is chosen to give power $1 - \beta$ at $\mu_A - \mu_B = \pm\delta$. The procedure described in Section 2.4.2 for satisfying a power requirement at $\mu_A - \mu_B = \pm\delta$ defines the group size to be proportional to σ^2/δ^2. We now compare two tests defined for different values of δ and σ^2. Test 1 is for observations X_{Ai} and X_{Bi} with variance σ_1^2 and achieves power $1 - \beta$ at $\mu_A - \mu_B = \pm\delta_1$ with groups of m_1 observations per treatment arm. Test 2 is for observations X'_{Ai} and X'_{Bi} with variance σ_2^2 and attains power $1 - \beta$ at $\mu_A - \mu_B = \pm\delta_2$ with groups of m_2 observations per treatment arm. We denote the sequence of standardized statistics for Test 1 by $\{Z_1^1, \ldots, Z_K^1\}$, where

$$Z_k^1 = \frac{1}{\sqrt{(2m_1k\sigma_1^2)}} \left(\sum_{i=1}^{m_1k} X_{Ai} - \sum_{i=1}^{m_1k} X_{Bi} \right), \quad k = 1 \ldots, K,$$

and denote those for Test 2 by $\{Z_1^2, \ldots, Z_K^2\}$, where

$$Z_k^2 = \frac{1}{\sqrt{(2m_2k\sigma_2^2)}} \left(\sum_{i=1}^{m_2k} X'_{Ai} - \sum_{i=1}^{m_2k} X'_{Bi} \right), \quad k = 1 \ldots, K.$$

It follows from results we shall present in Section 3.1 that for any value of ξ, the joint distribution of the sequence $\{Z_1^1, \ldots, Z_K^1\}$ when $\mu_A - \mu_B = \xi\delta_1$ and observations have variance σ_1^2 is the same as that of $\{Z_1^2, \ldots, Z_K^2\}$ when $\mu_A - \mu_B = \xi\delta_2$ and the variance is σ_2^2. Since Z_k^1 and Z_k^2 are compared to the same critical value, $c_k = C_P(K, \alpha)$, we see that for each $k = 1, \ldots, K$ the probability that Test 1 rejects H_0 at stage k when $\mu_A - \mu_B = \xi\delta_1$ and $\sigma^2 = \sigma_1^2$ is equal to the probability that Test 2 rejects H_0 at stage k when $\mu_A - \mu_B = \xi\delta_2$ and $\sigma^2 = \sigma_2^2$. Application of this result in the case $\xi = 0$ confirms that both tests have the same Type I error probability. Consideration of the case $\xi = 1$ shows that Test 1 has the same power at $\mu_A - \mu_B = \pm\delta_1$ when $\sigma^2 = \sigma_1^2$ as Test 2 has at $\mu_A - \mu_B = \pm\delta_2$ when $\sigma^2 = \sigma_2^2$, and this justifies our method for determining group sizes to meet a power requirement for general δ and σ^2. The power functions of the two tests also match up at other points, the power of Test 1 at $\mu_A - \mu_B = \xi\delta_1$ when $\sigma^2 = \sigma_1^2$ equaling that of Test 2 at $\mu_A - \mu_B = \xi\delta_2$ when $\sigma^2 = \sigma_2^2$.

In practice, the need for integer group sizes may mean that m_2 cannot be exactly equal to $m_1\sigma_2^2\delta_1^2/(\sigma_1^2\delta_2^2)$ and the sequences $\{Z_1^1, \ldots, Z_K^1\}$ and $\{Z_1^2, \ldots, Z_K^2\}$ have slightly different distributions. In order to obtain more general comparisons of different types of test, it is helpful to suppress such problems and evaluate tests assuming that groups of m observations are permissible for all values of m. Results for one pair of values for δ and σ^2 can then be generalized to other pairs, and only slight departures from this idealized case will arise from rounding to integer group sizes.

Another consequence of the above relation between $\{Z_1^1, \ldots, Z_K^1\}$ and $\{Z_1^2, \ldots, Z_K^2\}$ is that the distribution of the number of groups before termination is the same for both Test 1 and Test 2 under the given conditions, $\mu_A - \mu_B = \xi\delta_1$ and $\sigma^2 = \sigma_1^2$ for Test 1 and $\mu_A - \mu_B = \xi\delta_2$ and $\sigma^2 = \sigma_2^2$ for Test 2. Hence, the expected sample sizes of the two tests differ by a factor $m_2/m_1 = \sigma_2^2\delta_1^2/(\sigma_1^2\delta_2^2)$ in such situations. This relationship allows us to quote expected sample sizes at values of $\mu_A - \mu_B$ equal to stated multiples of the value δ at which the power requirement is set. Expected sample sizes can then be deduced for tests of the same type and with the same number of groups at other values of δ and σ^2 although, again, rounding to integer group sizes will produce small perturbations.

Table 2.7 *Pocock tests: maximum and expected sample sizes of tests with K groups of observations, Type I error probability $\alpha = 0.05$ and power $1-\beta$ at $\mu_A - \mu_B = \pm\delta$. Values are expressed as percentages of the fixed sample size, $n_f(\alpha, \beta, \delta, \sigma^2)$.*

	$\alpha = 0.05$				
	Sample size as percentage of fixed sample size				
K	*Maximum sample size*	*Expected sample size at $\mu_A - \mu_B =$* 0	$\pm 0.5\delta$	$\pm\delta$	$\pm 1.5\delta$
$1-\beta = 0.8$					
1	100.0	100.0	100.0	100.0	100.0
2	111.0	109.4	103.9	85.3	65.0
3	116.6	114.3	106.7	81.9	56.0
4	120.2	117.5	108.8	80.5	52.2
5	122.9	119.8	110.4	79.9	50.1
10	130.1	126.3	115.3	79.5	46.2
15	133.8	129.7	118.1	80.0	45.1
20	136.3	131.9	120.0	80.5	44.7
$1-\beta = 0.9$					
1	100.0	100.0	100.0	100.0	100.0
2	110.0	108.4	100.9	77.6	59.2
3	115.1	112.8	102.6	72.1	48.2
4	118.3	115.6	104.1	69.7	43.7
5	120.7	117.7	105.2	68.5	41.2
10	127.1	123.4	109.0	66.6	36.7
15	130.5	126.4	111.2	66.4	35.4
20	132.7	128.4	112.8	66.5	34.8

There is nothing specific to Pocock's test in the above arguments. Properties of other two-sided tests can be presented for general δ and σ^2 in the same way, and we shall follow a similar approach in describing properties of one-sided tests in later chapters.

2.6.5 *Choosing the Number of Groups*

As well as choosing the type of test to use, an experimenter must decide on the number of groups in which observations are to be taken. A larger number of groups usually implies smaller expected sample sizes at the values of $\mu_A - \mu_B$ of most interest, but this advantage has to be balanced against the effort required to carry out more frequent interim analyses. Tables 2.7 and 2.8 show how maximum and expected sample sizes vary with the number of groups in Pocock and O'Brien & Fleming tests. Results are for tests with $\alpha = 0.05$ and power $1-\beta = 0.8$ and 0.9 at $\mu_A - \mu_B = \pm\delta$. Entries in the table are 100 × the ratios of maximum and expected sample sizes to $n_f(\alpha, \beta, \delta, \sigma^2)$, the number of observations required for a fixed sample size test with the same Type I error and power. It should be remembered that, since n_f depends on β, the entries for $1-\beta = 0.8$ and 0.9 have been divided by different values of n_f. The tabulated values do not take account of the small effect of rounding group sizes to integer values, but apart from this, they apply to all pairs of δ and σ^2.

Table 2.8 *O'Brien & Fleming tests: maximum and expected sample sizes of tests with K groups of observations, Type I error probability $\alpha = 0.05$ and power $1-\beta$ at $\mu_A - \mu_B = \pm\delta$. Values are expressed as percentages of the fixed sample size, $n_f(\alpha, \beta, \delta, \sigma^2)$.*

		$\alpha = 0.05$			
	Sample size as percentage of fixed sample size				
K	*Maximum sample size*	*Expected sample size at* $\mu_A - \mu_B =$ 0	$\pm 0.5\delta$	$\pm\delta$	$\pm 1.5\delta$
$1-\beta = 0.8$					
1	100.0	100.0	100.0	100.0	100.0
2	100.8	100.5	99.0	90.2	71.9
3	101.7	101.2	98.3	85.6	68.0
4	102.4	101.7	98.0	83.1	64.1
5	102.8	102.1	97.9	81.8	61.9
10	104.0	103.1	97.9	79.1	58.1
15	104.5	103.5	97.9	78.3	56.9
20	104.7	103.7	97.9	77.9	56.3
$1-\beta = 0.9$					
1	100.0	100.0	100.0	100.0	100.0
2	100.7	100.5	98.2	85.1	63.3
3	101.6	101.1	96.9	79.9	61.0
4	102.2	101.6	96.4	76.7	57.3
5	102.6	101.9	96.1	75.0	54.8
10	103.7	102.8	95.6	71.8	50.8
15	104.2	103.2	95.5	70.8	49.5
20	104.5	103.4	95.5	70.3	48.9

It is clear from these tables that the introduction of one or two interim analyses is quite effective in reducing expected sample sizes at $\mu_A - \mu_B = \pm\,\delta$ and $\pm\,1.5\delta$, and even if interim analyses require little effort, designs with only four or five analyses will often be sufficient. Whereas expected sample sizes of O'Brien & Fleming tests at the larger values of $|\mu_A - \mu_B|$ continue to decrease as K increases, those for Pocock tests start to increase again. (The tabulated ratios vary smoothly with K, and expected sample sizes for values of $K < 20$ not given in the table are well approximated by interpolation.) Also, for Pocock tests, the maximum sample size and expected sample sizes when $\mu_A - \mu_B$ is near zero increase significantly with the number of groups. Thus, while it may be reasonable to choose as many as 10 groups for an O'Brien & Fleming test, there is little reason to take more than about 5 groups in a Pocock test. The same patterns are observed in results for $\alpha = 0.01$ and 0.1, with similar implications for choice of the number of groups.

2.6.6 Discussion

Both types of group sequential test considered so far have their advantages. Pocock tests have lower expected sample sizes than O'Brien & Fleming tests when $|\mu_A - \mu_B|$ is large and the ethical imperative for early stopping is greatest. Against this, Pocock tests have rather high maximum sample sizes and their expected sample sizes are higher than the fixed sample test when $|\mu_A - \mu_B|$ is small. O'Brien & Fleming tests sacrifice little of the available power for a given maximum sample size, and since the nominal significance level applied at the final analysis is close to the Type I error probability, the confusing situation in which the null hypothesis is accepted by the group sequential test when a fixed sample analysis of the final set of data would have rejected H_0 is unlikely to arise.

The O'Brien & Fleming tests apply extremely low nominal significance levels at the first few analyses. There are practical reasons to prefer group sequential tests with wide early boundaries that make it unlikely that a study will terminate at a very early stage. Problems in data quality most often arise at the start of a study as data recording procedures are introduced and participants gain experience in administering new treatments. It is difficult to check distributional assumptions on a small set of data. Most importantly perhaps, investigators are aware that the broader scientific community is likely to be skeptical of conclusions based on only a few observations. On the other hand, these issues will be of lesser concern if the sample sizes, even at the earlier stages, are quite substantial. In such circumstances, for ethical reasons a DSMB may be reluctant to continue a trial upon observing a high Z-value at an early stage, even though an O'Brien & Fleming boundary has not been crossed. The tests that will be described in the following section can be considered as compromises—enjoying many of the benefits of both the Pocock and the O'Brien & Fleming tests.

2.7 Other Tests

2.7.1 Wang & Tsiatis Tests

Wang & Tsiatis (1987) proposed a family of two-sided tests indexed by a parameter Δ which offers boundaries of different shapes, including Pocock and O'Brien & Fleming tests as special cases. The test with parameter Δ rejects H_0 after group k if $|Z_k| \geq c_k = C_{WT}(K, \alpha, \Delta)(k/K)^{\Delta - 1/2}$, $k = 1, \ldots, K$. Formally, the test is

$$
\begin{array}{lll}
\text{After group } k = 1, \ldots, K-1 & & \\
\quad \text{if } |Z_k| \geq C_{WT}(K, \alpha, \Delta)(k/K)^{\Delta - 1/2} & \text{stop, reject } H_0 & \\
\quad \text{otherwise} & \text{continue to group } k+1, & \\
\text{after group } K & & (2.7)\\
\quad \text{if } |Z_K| \geq C_{WT}(K, \alpha, \Delta) & \text{stop, reject } H_0 & \\
\quad \text{otherwise} & \text{stop, accept } H_0. &
\end{array}
$$

For consistency with our descriptions of other tests, we have adopted a different parameterization from that used by Wang & Tsiatis. However, the role of Δ is unchanged, so the test with parameter Δ defined by (2.7) is still the Wang & Tsiatis test with parameter Δ.

Taking $\Delta = 0.5$ gives Pocock's test and $\Delta = 0$ gives O'Brien & Fleming's test; values of Δ between 0 and 0.5 give tests with boundaries of intermediate shapes. The constants $C_{WT}(K, \alpha, \Delta)$ are chosen to ensure a specific Type I error α. Since $C_{WT}(K, \alpha, 0.5) = C_P(K, \alpha)$ and $C_{WT}(K, \alpha, 0) = C_B(K, \alpha)$, values for these cases can be obtained from Tables 2.1 and 2.3. Constants for $\alpha = 0.05$ and $\Delta = 0.1$, 0.25 and 0.4 are given in Table 2.9; constants for $\alpha = 0.01$ and $K \leq 5$, in Wang & Tsiatis' parameterization, can be found in Wang & Tsiatis (1987).

The constants $R_{WT}(K, \alpha, \beta, \Delta)$ listed in Table 2.10 determine the group size needed to satisfy a power requirement in the same way as the constants $R_P(K, \alpha, \beta)$ and

Table 2.9 *Wang & Tsiatis tests: design constants $C_{WT}(K, \alpha, \Delta)$ for two-sided tests with K groups of observations and Type I error probability $\alpha = 0.05$*

	$C_{WT}(K, \alpha, \Delta)$		
		$\alpha = 0.05$	
K	$\Delta = 0.1$	$\Delta = 0.25$	$\Delta = 0.4$
1	1.960	1.960	1.960
2	1.994	2.038	2.111
3	2.026	2.083	2.186
4	2.050	2.113	2.233
5	2.068	2.136	2.267
6	2.083	2.154	2.292
7	2.094	2.168	2.313
8	2.104	2.180	2.329
9	2.113	2.190	2.343
10	2.120	2.199	2.355
11	2.126	2.206	2.366
12	2.132	2.213	2.375
15	2.146	2.229	2.397
20	2.162	2.248	2.423

Table 2.10 *Wang & Tsiatis tests: constants $R_{WT}(K, \alpha, \beta, \Delta)$ to determine group sizes for two-sided tests with K groups of observations, Type I error probability $\alpha = 0.05$ and power $1 - \beta$*

	$R_{WT}(K, \alpha, \beta, \Delta)$					
			$\alpha = 0.05$			
	$1-\beta = 0.8$			$1-\beta = 0.9$		
K	$\Delta = 0.1$	$\Delta = 0.25$	$\Delta = 0.4$	$\Delta = 0.1$	$\Delta = 0.25$	$\Delta = 0.4$
1	1.000	1.000	1.000	1.000	1.000	1.000
2	1.016	1.038	1.075	1.014	1.034	1.068
3	1.027	1.054	1.108	1.025	1.050	1.099
4	1.035	1.065	1.128	1.032	1.059	1.117
5	1.040	1.072	1.142	1.037	1.066	1.129
6	1.044	1.077	1.152	1.041	1.071	1.138
7	1.047	1.081	1.159	1.044	1.075	1.145
8	1.050	1.084	1.165	1.046	1.078	1.151
9	1.052	1.087	1.170	1.048	1.081	1.155
10	1.054	1.089	1.175	1.050	1.083	1.159
11	1.055	1.091	1.178	1.051	1.085	1.163
12	1.056	1.093	1.181	1.053	1.086	1.166
15	1.059	1.097	1.189	1.055	1.090	1.172
20	1.062	1.101	1.197	1.058	1.094	1.180

$R_B(K, \alpha, \beta)$ for Pocock and O'Brien & Fleming tests. A test with Type I error rate α and power $1 - \beta$ at $\mu_A - \mu_B = \pm\delta$ requires a maximum sample size per treatment of $R_{WT}(K, \alpha, \beta, \Delta)\, n_f(\alpha, \beta, \delta, \sigma^2)$, where $n_f(\alpha, \beta, \delta, \sigma^2)$, given by (2.3), is the sample size per treatment required by a fixed sample size test with the same Type I error rate α and power $1 - \beta$ at $\mu_A - \mu_B = \pm\delta$. Thus, the necessary group size is $m = R_{WT}(K, \alpha, \beta, \Delta)\, n_f(\alpha, \beta, \delta, \sigma^2)/K$ per treatment arm, suitably rounded.

As an illustration, consider the five-group test with $\Delta = 0.25$ applied to the example of Section 2.2, testing H_0: $\mu_A = \mu_B$ with Type I error $\alpha = 0.05$ and power $1 - \beta = 0.9$ at $\mu_A - \mu_B = \pm 1$ when observations have variance $\sigma^2 = 4$. The fixed sample test requires $n_f(0.05, 0.1, 1, 4) = 84.1$ observations per treatment arm; hence the Wang & Tsiatis test with $\Delta = 0.25$ has a maximum sample size per treatment of

$$R_{WT}(5, 0.05, 0.1, 0.25)\, n_f(0.05, 0.1, 1, 4) = 1.066 \times 84.1 = 89.7.$$

Dividing by 5 and rounding to an integer gives a group size of 18 per treatment arm, the same as that required by an O'Brien & Fleming test but smaller than the value 21 of a Pocock test; without the effects of rounding, the Wang & Tsiatis group size would lie between those of the O'Brien & Fleming and Pocock tests. The boundary also has an intermediate shape. Critical values for the Pocock test are constant at 2.413; for the O'Brien & Fleming test, $c_k = C_B(5, 0.05)(5/k)^{0.5} = 2.040\,(5/k)^{0.5}$, $k = 1, \ldots, 5$, giving the sequence of critical values 4.562, 3.226, 2.634, 2.281, 2.040; for the Wang & Tsiatis test with $\Delta = 0.25$, $c_k = C_{WT}(5, 0.05, 0.25)(k/5)^{-0.25} = 2.136\,(k/5)^{-0.25}$, $k = 1, \ldots, 5$, and the sequence of critical values is 3.194, 2.686, 2.427, 2.259, 2.136. Note that the Wang & Tsiatis test reduces the first two very high critical values, $c_1 = 4.562$ and $c_2 = 3.226$, of the O'Brien & Fleming test, with only a small increase in the final critical value, c_5.

Table 2.11 shows maximum and expected sample sizes of Wang & Tsiatis tests with $\Delta = 0.1$, 0.25 and 0.4, Type I error $\alpha = 0.05$ and power $1 - \beta = 0.8$ when $\mu_A - \mu_B = \pm\delta$. Table 2.12 gives the same quantities for $1 - \beta = 0.9$. As before, sample sizes are expressed as percentages of the corresponding fixed sample sizes $n_f(\alpha, \beta, \delta, \sigma^2)$. Comparison of these tables with Tables 2.7 and 2.8 shows that Wang & Tsiatis tests with $\Delta = 0.1$ have lower expected sample sizes at $|\mu_A - \mu_B| = \delta$ and 1.5δ than O'Brien & Fleming tests, with only a small increase in maximum sample size. Tests with $\Delta = 0.4$ offer similar low expected sample sizes at $|\mu_A - \mu_B| = \delta$ and 1.5δ to Pocock tests but with significantly lower maximum sample size and expected sample sizes at $|\mu_A - \mu_B| = 0$ and 0.5δ. In general, as Δ increases, expected sample sizes at higher values of $|\mu_A - \mu_B|$ decrease but at the cost of larger maximum sample size and expected sample sizes at values of $|\mu_A - \mu_B|$ closer to zero. Thus, the Wang & Tsiatis family introduces a useful extra degree of freedom in choosing a group sequential test to balance the conflicting aims of low maximum sample size and low expected sample size over a range of values of $\mu_A - \mu_B$.

2.7.2 Haybittle-Peto Tests

Haybittle (1971) and Peto et al. (1976) suggested a simple form of sequential monitoring in which H_0 is rejected at an analysis $k < K$ only if $|Z_k| \geq 3$. If the number of interim analyses is small, early stopping is unlikely under H_0, and it is approximately correct to carry out a fixed sample analysis at the final stage with no allowance for the interim analyses. One can, however, adjust the final critical value for $|Z_K|$ to give an overall Type I error exactly equal to α, and this is the approach we shall adopt. Formally, the Haybittle-Peto test with

Table 2.11 *Wang & Tsiatis tests: maximum and expected sample sizes of tests with K analyses, Type I error probability* $\alpha = 0.05$ *and power* $1 - \beta = 0.8$ *at* $\mu_A - \mu_B = \pm\delta$. *Values are expressed as percentages of the fixed sample size,* $n_f(\alpha, \beta, \delta, \sigma^2)$.

$\alpha = 0.05,\ 1 - \beta = 0.8$

		Sample size as percentage of fixed sample size				
	K	*Maximum sample size*	*Expected sample size at* $\mu_A - \mu_B =$ 0	$\pm 0.5\delta$	$\pm\delta$	$\pm 1.5\delta$
$\Delta = 0.1$						
	1	100.0	100.0	100.0	100.0	100.0
	2	101.6	101.1	99.0	88.2	69.0
	3	102.7	102.0	98.5	84.1	64.7
	4	103.5	102.7	98.4	81.8	61.5
	5	104.0	103.1	98.3	80.4	59.4
	10	105.4	104.2	98.4	77.9	55.5
	15	105.9	104.7	98.5	77.1	54.2
	20	106.2	105.0	98.5	76.7	53.7
$\Delta = 0.25$						
	1	100.0	100.0	100.0	100.0	100.0
	2	103.8	103.0	99.7	86.0	66.1
	3	105.4	104.4	99.6	82.0	60.0
	4	106.5	105.2	99.7	79.9	57.1
	5	107.2	105.8	99.8	78.7	55.2
	10	108.9	107.3	100.2	76.2	51.3
	15	109.7	108.0	100.4	75.4	50.0
	20	110.1	108.3	100.5	75.0	49.4
$\Delta = 0.4$						
	1	100.0	100.0	100.0	100.0	100.0
	2	107.5	106.2	101.6	85.1	64.8
	3	110.8	109.1	102.8	81.2	56.9
	4	112.8	110.8	103.6	79.3	53.4
	5	114.2	112.0	104.2	78.3	51.4
	10	117.5	114.9	105.8	76.3	47.5
	15	118.9	116.2	106.5	75.7	46.2
	20	119.7	116.9	107.0	75.4	45.5

Table 2.12 *Wang & Tsiatis tests: maximum and expected sample sizes of tests with K analyses, Type I error probability $\alpha = 0.05$ and power $1 - \beta = 0.9$ at $\mu_A - \mu_B = \pm\delta$. Values are expressed as percentages of the fixed sample size, $n_f(\alpha, \beta, \delta, \sigma^2)$.*

$\alpha = 0.05,\ 1 - \beta = 0.9$					
	Sample size as percentage of fixed sample size				
K	*Maximum sample size*	*Expected sample size at* $\mu_A - \mu_B =$ 0	$\pm 0.5\delta$	$\pm\delta$	$\pm 1.5\delta$
$\Delta = 0.1$					
1	100.0	100.0	100.0	100.0	100.0
2	101.4	101.0	97.9	82.5	61.0
3	102.5	101.8	96.9	77.7	57.0
4	103.2	102.4	96.4	74.9	54.2
5	103.7	102.8	96.1	73.2	52.0
10	105.0	103.9	95.8	70.0	47.9
15	105.5	104.3	95.7	69.0	46.6
20	105.8	104.6	95.7	68.5	46.0
$\Delta = 0.25$					
1	100.0	100.0	100.0	100.0	100.0
2	103.4	102.6	98.0	79.5	59.0
3	105.0	103.9	97.4	74.5	51.9
4	105.9	104.7	97.1	71.9	49.1
5	106.6	105.3	97.0	70.4	47.3
10	108.3	106.7	96.9	67.2	43.3
15	109.0	107.3	97.0	66.2	41.9
20	109.4	107.6	97.0	65.7	41.3
$\Delta = 0.4$					
1	100.0	100.0	100.0	100.0	100.0
2	106.8	105.5	99.3	77.8	58.7
3	109.9	108.1	99.6	72.4	48.9
4	111.7	109.7	100.0	69.9	45.1
5	112.9	110.8	100.2	68.4	42.9
10	115.9	113.4	101.2	65.5	38.8
15	117.2	114.6	101.7	64.5	37.4
20	118.0	115.2	102.0	64.1	36.6

Table 2.13 *Haybittle-Peto tests: constants* $C_{HP}(K,\alpha)$ *and* $R_{HP}(K,\alpha,\beta)$ *for two-sided tests with* K *groups of observations and Type I error probability* $\alpha = 0.05$

		$\alpha = 0.05$	
K	$C_{HP}(K,\alpha)$	$R_{HP}(K,\alpha,\beta)$	
		$1-\beta = 0.8$	$1-\beta = 0.9$
1	1.960	1.000	1.000
2	1.967	1.003	1.003
3	1.975	1.007	1.007
4	1.983	1.011	1.010
5	1.990	1.015	1.014
6	1.997	1.019	1.017
7	2.003	1.023	1.021
8	2.010	1.027	1.024
9	2.016	1.030	1.027
10	2.021	1.033	1.030
11	2.027	1.037	1.033
12	2.032	1.040	1.035
15	2.046	1.048	1.043
20	2.068	1.061	1.055

a maximum of K groups of observations is

$$\begin{array}{ll} \text{After group } k = 1,\ldots,K-1 & \\ \quad \text{if } |Z_k| \geq 3 & \text{stop, reject } H_0 \\ \quad \text{otherwise} & \text{continue to group } k+1, \\ \text{after group } K & \\ \quad \text{if } |Z_K| \geq C_{HP}(K,\alpha) & \text{stop, reject } H_0 \\ \quad \text{otherwise} & \text{stop, accept } H_0. \end{array} \tag{2.8}$$

Thus, the test applies critical values $c_1 = \ldots = c_{K-1} = 3$ and $c_K = C_{HP}(K,\alpha)$.

We shall restrict attention to the case $\alpha = 0.05$ where the choice of $c_k = 3$ for $k = 1,\ldots,K-1$ is most appropriate. (Calculations show that for $K \geq 7$, in which case $c_1 = \ldots = c_6 = 3$, the probability under H_0 of rejecting H_0 at or before the sixth analysis is 0.0107, so it is not possible to achieve an overall Type I error as low as 0.01.) Table 2.13 gives values of $C_{HP}(K,\alpha)$ for $\alpha = 0.05$ and constants $R_{HP}(K,\alpha,\beta)$ which determine the group size needed to achieve power $1-\beta$ at $|\mu_A - \mu_B| = \delta$ in the usual way. Table 2.14 shows maximum and expected sample sizes of Haybittle-Peto tests expressed as percentages of the corresponding fixed sample sizes, $n_f(\alpha,\beta,\delta,\sigma^2)$. Expected sample sizes at $|\mu_A - \mu_B| = \delta$ are a little higher than those of O'Brien & Fleming tests, but at $|\mu_A - \mu_B| = 1.5\delta$ the Haybittle-Peto test has the smaller expected sample size for most values of K. Although the Haybittle-Peto test uses high critical values, $c_k = 3$, up to the penultimate analysis, these are less extreme than the first few O'Brien & Fleming critical values (for example, the five-group O'Brien & Fleming test has $c_1 = 4.562$ and $c_2 = 3.226$). Thus the low expected sample sizes at larger values of $|\mu_A - \mu_B|$ are not surprising.

Table 2.14 *Haybittle-Peto tests: maximum and expected sample sizes of tests with K analyses, Type I error probability $\alpha = 0.05$ and power $1-\beta$ at $\mu_A - \mu_B = \pm\delta$. Values are expressed as percentages of the fixed sample size, $n_f(\alpha, \beta, \delta, \sigma^2)$.*

		$\alpha = 0.05$			
	Sample size as percentage of fixed sample size				
K	*Maximum sample size*	*Expected sample size at $\mu_A - \mu_B =$* 0	$\pm 0.5\delta$	$\pm\delta$	$\pm 1.5\delta$
$1-\beta = 0.8$					
1	100.0	100.0	100.0	100.0	100.0
2	100.3	100.2	99.2	92.5	75.7
3	100.7	100.5	98.9	89.2	68.1
4	101.1	100.8	98.8	87.2	64.0
5	101.5	101.1	98.8	85.9	61.5
10	103.3	102.5	99.3	83.0	55.6
15	104.8	103.7	100.0	81.8	53.3
20	106.1	104.8	100.7	81.3	51.9
$1-\beta = 0.9$					
1	100.0	100.0	100.0	100.0	100.0
2	100.3	100.2	98.7	88.2	66.6
3	100.7	100.4	98.1	83.4	58.0
4	101.0	100.7	97.8	80.6	53.5
5	101.4	100.9	97.6	78.8	50.8
10	103.0	102.2	97.6	74.5	44.7
15	104.3	103.2	97.9	72.7	42.3
20	105.5	104.1	98.3	71.8	41.0

Figure 2.3 illustrates the differences in boundary shape of three tests in the Wang & Tsiatis family and the Haybittle-Peto test. The figure shows critical values for standardized statistics Z_k in tests with 10 groups of observations and Type I error rate $\alpha = 0.05$. At one extreme, the Haybittle-Peto test and the Wang & Tsiatis test with $\Delta = 0$, which is also the O'Brien & Fleming test, have very wide early boundaries and a final critical value close to that of 1.96 applied in a fixed sample analysis. The Wang & Tsiatis test with $\Delta = 0.5$, which is also the Pocock test, has constant critical values $c_k = 2.555$, $k = 1, \ldots, 10$, while the test with $\Delta = 0.25$ is intermediate to these extremes, its critical values decreasing steadily to $c_{10} = 2.199$ at the final analysis.

Although Haybittle-Peto tests do not attain the maximum possible reductions in expected sample size, this is not always the key issue. In some studies, investigators wish to continue sampling in order to learn more about aspects of a new treatment other than its primary efficacy. A new treatment will not necessarily be approved as soon as it has been shown to be effective in a single study: further data may be required to demonstrate the treatment's safety relative to a standard treatment or placebo, or it may be necessary to reproduce evidence of efficacy under other experimental conditions. It is also important to consider the contribution a study will make to meta-analyses in which data from a number

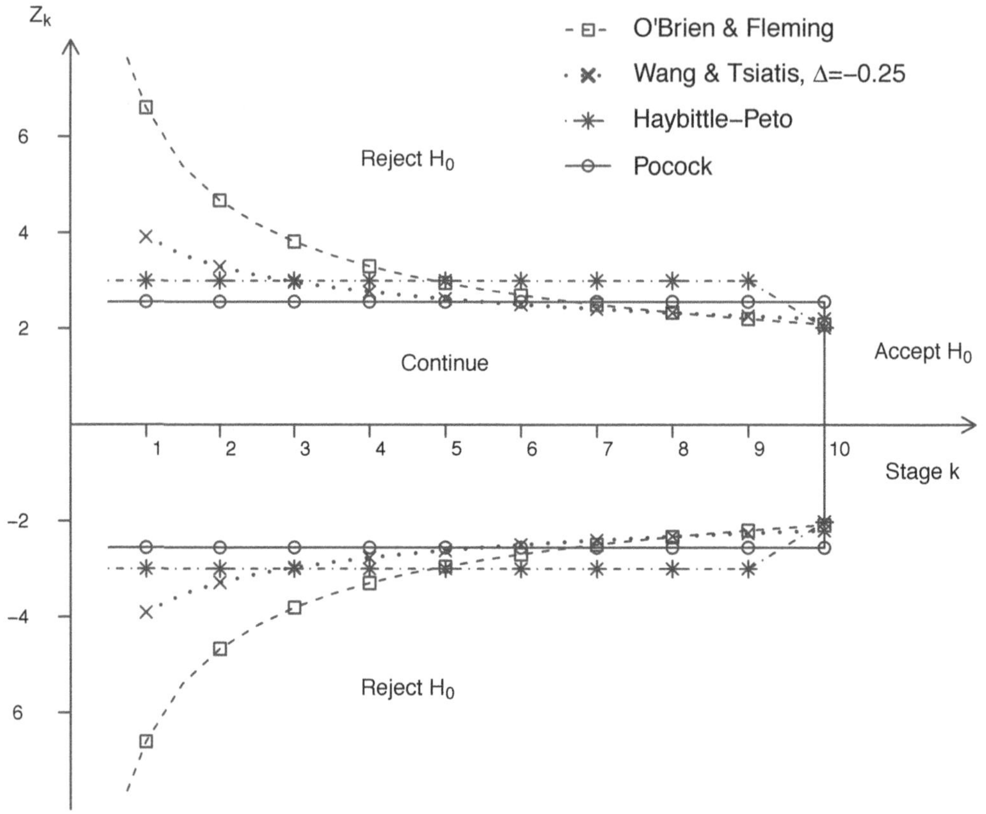

Figure 2.3 *Boundaries of four two-sided tests for ten groups of observations: Pocock, O'Brien & Fleming, Haybittle-Peto, and Wang & Tsiatis with* $\Delta = 0.25$

of studies are analyzed together. This is particularly relevant to treatments for diseases, such as many types of cancer, for which effective therapies are available and extremely large sample sizes are needed to detect the modest improvements that a new treatment is likely to offer. In such cases, the investigators' objective is really to gather as much information as possible on all aspects of a treatment, and there is little incentive for early stopping apart from the ethical need to cease randomizing patients to a clearly inferior treatment.

There has been some discussion of "administrative analyses" in clinical trials; see, for example, Enas et al. (1989) and Fleming & DeMets (1993). These are analyses at which the monitoring committee reviews various aspects of the trial such as patient recruitment, compliance to the specified treatment regimens, data recording procedures and data quality. Although there may be no intention to terminate the study on such an occasion, a regulatory body can require that a formal stopping rule for these analyses be stated in the protocol. The Haybittle-Peto criterion of early stopping at analysis k only if $|Z_k| \geq 3$ has the advantage of fulfilling this requirement with little effect on subsequent analyses. A similar approach may be taken in monitoring a secondary endpoint when the main objective is to gather data on a primary response. For example, a Haybittle-Peto rule could be applied to the safety variable in a study where treatment efficacy is the primary endpoint but where there is an ethical need to cease allocating subjects to a treatment which has proved to cause excessive toxicity.

2.8 Conclusions

In this chapter we have presented easily implemented two-sided tests and provided constants which define the tests and determine the group sizes required to satisfy a power condition. We have done this for the most common cases of Type I error $\alpha = 0.05$ and power $1-\beta = 0.8$ and 0.9 for all the tests considered. Constants for other cases can be computed using the methods described in Chapter 31. The tables of maximum and expected sample sizes aid in choosing the test best suited to a particular application and in selecting an appropriate number of groups of observations. Other forms of group sequential tests have been proposed; see, for example, Fleming, Harrington & O'Brien (1984).

In Section 2.6.6, we discussed the practical benefits of tests having wide early boundaries, such as the O'Brien & Fleming tests. However it was also stated there that a DSMB may be unwilling for ethical reasons to continue a trial upon observing a high Z-value at an early stage even though the stopping boundary has not been crossed. For example, at the first analysis of a four-stage O'Brien & Fleming test with Type I error $\alpha = 0.05$, the critical value is 4.048 corresponding to a nominal two-sided significance of less than 0.0001. As long as the sample size is large enough that validity of model assumptions and the phenomenon of "bad news traveling faster than good news" are not of concern, the DSMB may feel compelled to stop the trial in the face of a Z_1-value as great as $Z_1 = 4$, say. If this is the case, such a group sequential design is not the appropriate one to use. It is common practice to hold a "kick-off" meeting between the DSMB and the study sponsors, before the DSMB has seen any trial data and before a "firewall" is imposed between the DSMB and the study sponsors. We recommend that at this meeting the participants discuss the choice of criteria for stopping the trial. In particular, members of the DSMB should indicate circumstances under which they would be uncomfortable in agreeing to continue the trial. Of course, not all eventualities can be foreseen—see also the comments in Section 10.1. However, it is invaluable to have such a discussion among the members of the DSMB, with written records kept.

We believe the tests of Wang & Tsiatis with parameter Δ in the region of 0.25, have much to be commended. They have wide boundaries at early stages but they are not overly extreme. These tests provide much of the benefit of the Pocock test with respect to low expected sample sizes at alternatives that are important. They also enjoy many of the benefits of the O'Brien & Fleming test, which has a low maximum sample size and critical values close to those of the fixed test if the final analysis is reached.

The Wang & Tsiatis tests form a rich class, including the Pocock and O'Brien & Fleming tests as special cases. The experimenter is free to select a suitable member of the family, balancing reductions in expected sample size when the difference between treatments, θ, is away from zero against a higher maximum sample size, which will usually be needed if $\theta = 0$. Optimality results of Pocock (1982) and Eales & Jennison (1995) show there is not much opportunity for further reductions in expected sample size at values of θ away from zero for tests of the form discussed in this chapter. However, we shall see in Chapter 6 that allowing early termination to *accept* H_0 can reduce expected sample size when θ is close to zero.

Although we have introduced group sequential tests in the particular context of a two-sample comparison, these tests are readily generalized to other two-sided testing problems and other data types; we shall describe such generalizations in Chapter 4. The one limitation of the tests considered so far is their need for at least approximately equal group sizes. We shall show in Section 4.2 that small differences between groups, suitably handled, have negligible effects on error probabilities, but if larger differences are unavoidable the specialized methods for unequal and unpredictable group sizes presented in Chapter 8 should be used.

Chapter 3

A Unified Formulation and the General Calculation

3.1 Canonical Joint Distribution of Z-Statistics, Parameter Estimates and Score Statistics

In Chapter 2 we presented two-sided tests for a two-treatment comparison when normally distributed observations with a common, known variance are collected in equally sized groups. The tests were defined in terms of significance levels for testing the null hypothesis after each new group of observations, and power was guaranteed by setting the maximum sample size as a multiple of that required by a fixed sample test. We shall see that group sequential tests described in this manner can be applied to other testing problems with little or no modification. The reason for this lies in the common form of the joint distribution of the sequence of standardized test statistics, $\{Z_1, \ldots, Z_K\}$.

Suppose a group sequential study with up to K analyses yields the sequence of standardized statistics $\{Z_1, \ldots, Z_K\}$. We say that these statistics have the *canonical joint distribution* with information levels $\{\mathcal{I}_1, \ldots, \mathcal{I}_K\}$ for the parameter θ if:

$$\begin{aligned} &\text{(i)} \quad (Z_1, \ldots, Z_K) \text{ is multivariate normal,} \\ &\text{(ii)} \quad E(Z_k) = \theta\sqrt{\mathcal{I}_k}, \quad k = 1, \ldots, K, \quad \text{and} \\ &\text{(iii)} \quad \mathrm{Cov}(Z_{k_1}, Z_{k_2}) = \sqrt{(\mathcal{I}_{k_1}/\mathcal{I}_{k_2})}, \quad 1 \leq k_1 \leq k_2 \leq K. \end{aligned} \tag{3.1}$$

We shall explain as this chapter develops how this canonical joint distribution arises in a great many situations.

Note that (3.1) is the joint distribution of $\{Z_1, \ldots, Z_K\}$ when no stopping rule is applied. We shall refer to this as the marginal distribution of $\{Z_1, \ldots, Z_K\}$. Properties of group sequential tests in which early stopping does occur follow from this distribution; for details of the relationship between this marginal distribution and the distribution that arises under a stopping rule, see Sections 3.5 and 9.2 and Chapter 31.

Defining group sequential tests in terms of the standardized statistics $\{Z_1, \ldots, Z_K\}$ is not the only possibility. We could, instead, work with the sequence of parameter estimates $\hat{\theta}^{(k)}$. The information $\mathcal{I}_k$ at analysis k is defined as the inverse of the variance of the maximum likelihood estimate of θ (again, this variance is defined on the assumption of no early stopping before analysis k). Thus, the estimates and standardized statistics, also known as the "Wald statistics", satisfy the relation

$$Z_k = \hat{\theta}^{(k)}/\sqrt{\mathrm{Var}\{\hat{\theta}^{(k)}\}} = \hat{\theta}^{(k)}\sqrt{\mathcal{I}_k}, \quad k = 1, \ldots, K.$$

Another option is to define a test in terms of the sequence of score statistics for θ. For a general parametric model, the score statistic is the derivative with respect to θ of the marginal log-density of the data (so based on the density in the absence of early stopping)

 DOI: 10.1201/9781584888482-3

evaluated at $\theta = 0$. In the case of normally distributed data, the sequence of score statistics is simply

$$S_k = Z_k \sqrt{\mathcal{I}_k} = \hat{\theta}^{(k)} \mathcal{I}_k, \quad k = 1, \ldots, K,$$

and the same relation holds asymptotically for general parametric models. The above relationships between $\hat{\theta}^{(k)}$, Z_k and S_k mean that the joint distribution (3.1) of the sequence Z_1 to Z_K implies similar joint distributions for the two other forms of statistic. These properties can be summarized together as:

> Each sequence $\{\hat{\theta}^{(1)}, \ldots, \hat{\theta}^{(K)}\}$, $\{Z_1, \ldots, Z_K\}$ and $\{S_1, \ldots, S_K\}$ is multivariate normal, with
>
> $$\begin{aligned} \hat{\theta}^{(k)} \sim N(\theta, \mathcal{I}_k^{-1}) \quad &\text{and} \quad \mathrm{Cov}(\hat{\theta}^{(k_1)}, \hat{\theta}^{(k_2)}) = \mathcal{I}_{k_2}^{-1}, \\ Z_k \sim N(\theta \sqrt{\mathcal{I}_k}, 1) \quad &\text{and} \quad \mathrm{Cov}(Z_{k_1}, Z_{k_2}) = \sqrt{(\mathcal{I}_{k_1}/\mathcal{I}_{k_2})}, \\ S_k \sim N(\theta \mathcal{I}_k, \mathcal{I}_k) \quad &\text{and} \quad \mathrm{Cov}(S_{k_1}, S_{k_2}) = \mathcal{I}_{k_1}, \end{aligned} \tag{3.2}$$
>
> for $k = 1, \ldots, K$ and $1 \leq k_1 \leq k_2 \leq K$.

To illustrate the connections between these three sequences of statistics, consider a study yielding data $X_i \sim N(\theta, 1)$, $i = 1, 2, \ldots$, with n_k observations available at analysis k. Then, the maximum likelihood estimate of θ at analysis k is

$$\hat{\theta}^{(k)} = \frac{X_1 + \ldots + X_{n_k}}{n_k}.$$

Since $\mathrm{Var}(\hat{\theta}^{(k)}) = 1/n_k$, the information at analysis k is $\mathcal{I}_k = \{\mathrm{Var}(\hat{\theta}^{(k)})\}^{-1} = n_k$ and the standardized or Wald statistic for testing H_0: $\theta = 0$ is

$$Z_k = \hat{\theta}^{(k)} \sqrt{\mathcal{I}_k} = \frac{X_1 + \ldots + X_{n_k}}{\sqrt{n_k}}.$$

The score statistic is

$$S_k = Z_k \sqrt{\mathcal{I}_k} = X_1 + \ldots + X_{n_k},$$

the sample sum at analysis k, and it is straightforward to check this is the derivative of the log-density of the data with respect to θ evaluated at $\theta = 0$. The joint distributions of the three sequences of statistics are simple to derive directly in this case and we leave it as an exercise for the reader to confirm that these agree with (3.2) above.

Many authors have defined group sequential boundaries in terms of score statistics $\{S_1, \ldots, S_K\}$. In Chapter 2, where we introduced two-sided tests, the O'Brien & Fleming test for equally sized groups was defined by comparing each $|S_k|$, $k = 1, \ldots, K$, against a constant critical value. Wang & Tsiatis defined their tests by comparing $|S_k|$ against boundary values proportional to k^{Δ}, which explains why critical values for Z_k are proportional to $k^{\Delta - 1/2}$ in our description of these tests. Figure 3.1 shows Pocock and O'Brien & Fleming tests for four groups of observations designed to achieve two-sided Type I error rate $\alpha = 0.05$ with boundaries expressed in terms of both Z_k and S_k, $k = 1, \ldots, 4$. In each case, critical values are plotted against information levels $\mathcal{I}_k$ chosen to give power $1 - \beta = 0.9$ at $\delta = 0.2$. It should be noted that these information levels are higher for the Pocock test since its maximum information level is a larger multiple of the information needed for a fixed sample test.

Whitehead (1997) defines tests in terms of the sequences of pairs $(S_k, \mathcal{I}_k)$ using the symbols Z for S_k and V for $\mathcal{I}_k$ in his notation. The sequence $\{S_1, \ldots, S_K\}$ has a certain mathematical appeal as its increments S_1, $S_2 - S_1$, $S_3 - S_2$, etc., are independent. This sequence can also be viewed as a Brownian motion with drift θ observed at times

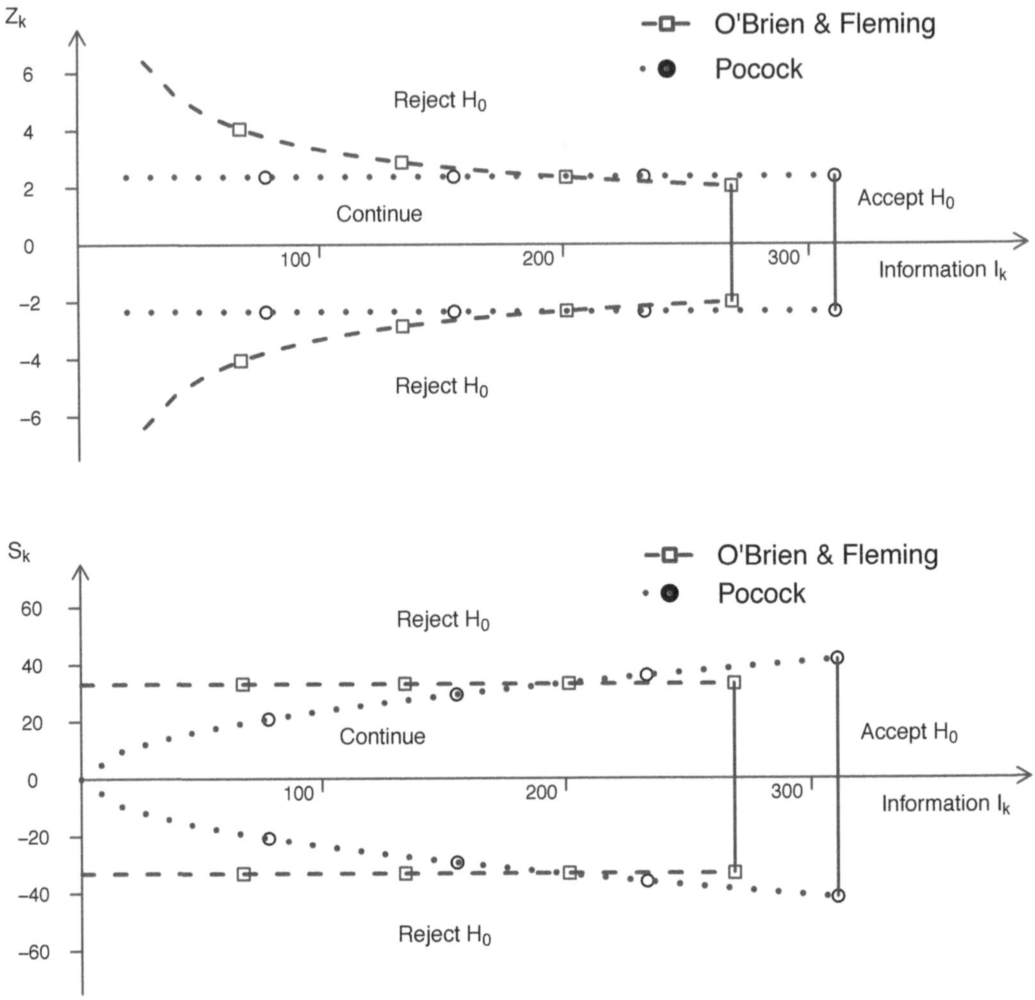

Figure 3.1 *Pocock and O'Brien & Fleming tests for four groups of observations expressed in terms of* Z_k *and* S_k

$\{\mathcal{I}_1, \ldots, \mathcal{I}_K\}$. Lan & Zucker (1993) and Proschan, Lan & Wittes (2006) note the relationship between the three processes $\{\hat{\theta}^{(k)}\}$, $\{Z_k\}$ and $\{S_k\}$ and their link with Brownian motion, pointing out the consequent freedom to express a group sequential test derived for one sequence of statistics in terms of another sequence.

The independent increments of the $\{S_k\}$ process indicate that it is Markov. It follows that the other two processes are also Markov and this property has important implications in simplifying calculations for group sequential tests. The properties of a group sequential test remain the same whether it is expressed through a boundary for $\hat{\theta}^{(k)}$, Z_k or S_k, and numerical computations of error probabilities and expected sample sizes are just as easy to implement using one statistic as another. We have chosen to express tests in terms of the standardized statistics Z_k as these are used commonly in applied work. Also, the method that we shall describe in Section 4.2 for adapting tests designed for equally spaced information levels to other information sequences is most simply defined in terms of standardized statistics. But, whichever sequence of statistics is used to define a test, it is a simple exercise to use the relations

$$S_k = Z_k\sqrt{\mathcal{I}_k} = \hat{\theta}^{(k)}\mathcal{I}_k$$

to re-express the test in terms of another set of statistics.

It should be noted that (3.1) and (3.2) specify the distribution of $\{Z_1, \ldots, Z_K\}$ given a specific sequence $\{\mathcal{I}_1, \ldots, \mathcal{I}_K\}$. When, for some reason, there is an element of randomness in the sample sizes observed at analyses 1 to K, we consider properties of tests conditional on the sequence of $\mathcal{I}_k$ values actually observed. Thus, as a group sequential test progresses, it is important to ensure that the value of $\mathcal{I}_k$ is not influenced by the values of statistics $Z_1, \ldots, Z_{k-1}$ seen previously, as this could destroy the property (3.1) on which error rate calculations are based. We shall consider this issue for group sequential tests in greater depth in Section 8.5 and Chapter 30. Certain adaptive designs make a feature of modifying group sizes, and therefore information levels, in response to observed responses; in Chapters 22 and 23 we shall describe some of the special methods proposed to control error probabilities in such circumstances.

In the following sections we shall illustrate how the canonical joint distributions for sequences of statistics arise in a selection of common examples. We start in Section 3.2 with simple examples of a normal response. In Section 3.3 we proceed to general normal linear models, in which the parameter of interest is one of several model parameters; this includes the case of correlated observations arising, say, in the group sequential analysis of longitudinal data. In Section 3.4 we consider other parametric models and also discuss the important case of right-censored survival data, analyzed using a group sequential log-rank test or by repeatedly fitting Cox's proportional hazards regression model. Our aim in this Chapter is to explain the application of general theory for group sequential data: the underlying theory will be presented in detail later on in Chapter 15.

3.2 Some Simple Examples with Normal Responses

3.2.1 Parallel Two-treatment Comparison

We generalize the example of a two-treatment comparison presented in Chapter 2 and allow different variances and unequal numbers of subjects on the two treatment arms. Thus, we have responses $X_{Ai} \sim N(\mu_A, \sigma_A^2)$, $i = 1, 2, \ldots$, for subjects allocated to treatment A and $X_{Bi} \sim N(\mu_B, \sigma_B^2)$, $i = 1, 2, \ldots$, for those on treatment B. For $k = 1, \ldots, K$, let n_{Ak} and n_{Bk} denote the cumulative number of observations on treatments A and B, respectively, at the time of the kth analysis. The natural estimate of $\mu_A - \mu_B$ is

$$\begin{aligned}\bar{X}_A^{(k)} - \bar{X}_B^{(k)} &= \frac{1}{n_{Ak}}\sum_{i=1}^{n_{Ak}} X_{Ai} - \frac{1}{n_{Bk}}\sum_{i=1}^{n_{Bk}} X_{Bi} \\ &\sim N(\mu_A - \mu_B, \frac{\sigma_A^2}{n_{Ak}} + \frac{\sigma_B^2}{n_{Bk}}).\end{aligned}$$

Note that the distribution of $\bar{X}_A^{(k)} - \bar{X}_B^{(k)}$ stated here is the marginal distribution for given n_{Ak} and n_{Bk}, i.e., the distribution arising in the absence of any early stopping. It is not, for example, the conditional distribution of $\bar{X}_A^{(k)} - \bar{X}_B^{(k)}$ given that the test continues up to analysis k, which arises in calculations of a group sequential test's error probabilities. We shall follow the same practice elsewhere, so, unless explicitly stated otherwise, the "distribution" of an estimate or test statistic should be taken to mean its marginal distribution.

We say that the parallel two-treatment design is *balanced* in the special case that $n_{Ak} = n_{Bk}$ for all $k = 1, \ldots, K$. In the general case of a possibly unbalanced design, we define the information for $\mu_A - \mu_B$ to be

$$\mathcal{I}_k = (\sigma_A^2/n_{Ak} + \sigma_B^2/n_{Bk})^{-1},$$

the reciprocal of our estimate's variance, and use this to create the standardized statistic at analysis k for testing H_0: $\mu_A = \mu_B$,

$$Z_k = (\bar{X}_A^{(k)} - \bar{X}_B^{(k)})\sqrt{\mathcal{I}_k}, \quad k = 1, \ldots, K.$$

The vector $(Z_1, \ldots, Z_K)$ is multivariate normal since each Z_k is a linear combination of the independent normal variates X_{Ai} and X_{Bi}, $i = 1, 2, \ldots$, and marginally,

$$Z_k \sim N(\theta\sqrt{\mathcal{I}_k}, 1), \quad k = 1, \ldots, K,$$

where $\theta = \mu_A - \mu_B$. It remains to establish the covariance of the Z_ks. For $k_1 \leq k_2$,

$$\begin{aligned}
\text{Cov}(Z_{k_1}, Z_{k_2}) &= \text{Cov}(\{\bar{X}_A^{(k_1)} - \bar{X}_B^{(k_1)}\}\sqrt{\mathcal{I}_{k_1}}, \{\bar{X}_A^{(k_2)} - \bar{X}_B^{(k_2)}\}\sqrt{\mathcal{I}_{k_2}}) \\
&= (\frac{1}{n_{Ak_1}}\frac{1}{n_{Ak_2}} n_{Ak_1}\sigma_A^2 + \frac{1}{n_{Bk_1}}\frac{1}{n_{Bk_2}} n_{Bk_1}\sigma_B^2)\sqrt{\mathcal{I}_{k_1}}\sqrt{\mathcal{I}_{k_2}} \\
&= (\mathcal{I}_{k_2})^{-1}\sqrt{\mathcal{I}_{k_1}}\sqrt{\mathcal{I}_{k_2}} = \sqrt{(\mathcal{I}_{k_1}/\mathcal{I}_{k_2})},
\end{aligned}$$

as required. Thus, $\{Z_1, \ldots, Z_K\}$ have the canonical joint distribution with information levels $\{\mathcal{I}_1, \ldots, \mathcal{I}_K\}$ for $\theta = \mu_A - \mu_B$.

3.2.2 Testing the Mean of a Single Population

Suppose observations $X_i \sim N(\mu, \sigma^2)$, $i = 1, 2, \ldots$, are independent, σ^2 is known, and we wish to test the hypothesis H_0: $\mu = \mu_0$. If n_k observations are available at analysis k, we estimate μ by

$$\bar{X}^{(k)} = \frac{1}{n_k}\sum_{i=1}^{n_k} X_i \sim N(\mu, \frac{\sigma^2}{n_k})$$

and define $\mathcal{I}_k = \{\text{Var}(\bar{X}^{(k)})\}^{-1} = n_k/\sigma^2$, the information for μ at analysis k. The standardized test statistics are then

$$Z_k = (\bar{X}^{(k)} - \mu_0)\sqrt{\mathcal{I}_k}, \quad k = 1, \ldots, K.$$

Since each Z_k is a linear combination of the independent normal variates X_i, the vector $(Z_1, \ldots, Z_K)$ is multivariate normal. Marginally,

$$Z_k \sim N(\theta\sqrt{\mathcal{I}_k}, 1), \quad k = 1, \ldots, K,$$

where $\theta = \mu - \mu_0$. Finally, for $k_1 \leq k_2$,

$$\begin{aligned}
\text{Cov}(Z_{k_1}, Z_{k_2}) &= \text{Cov}(\{\bar{X}^{(k_1)} - \mu_0\}\sqrt{\mathcal{I}_{k_1}}, \{\bar{X}^{(k_2)} - \mu_0\}\sqrt{\mathcal{I}_{k_2}}) \\
&= \frac{1}{n_{k_1}}\frac{1}{n_{k_2}} n_{k_1}\sigma^2\sqrt{\mathcal{I}_{k_1}}\sqrt{\mathcal{I}_{k_2}} = \sqrt{(\mathcal{I}_{k_1}/\mathcal{I}_{k_2})}
\end{aligned}$$

and we see that $\{Z_1, \ldots, Z_K\}$ have the canonical joint distribution with information levels $\{\mathcal{I}_1, \ldots, \mathcal{I}_K\}$ for $\theta = \mu - \mu_0$.

3.2.3 Paired Two-treatment Comparison

In a two-treatment comparison, it can be advantageous to control the variance in response attributable to known prognostic factors by using a "matched pairs" design. Subjects are paired so that both subjects in the same pair have similar values of the prognostic factors; one subject in each pair is randomly selected to receive treatment A and the other receives treatment B. Let X_{Ai} and X_{Bi} denote the responses of the subjects in pair i receiving treatments A and B, respectively. We suppose the differences within pairs are normally distributed,

$$X_{Ai} - X_{Bi} \sim N(\mu_A - \mu_B, \tilde{\sigma}^2), \quad i = 1, 2, \dots, \tag{3.3}$$

and the variance, $\tilde{\sigma}^2$, is known. If the variance of an individual observation is σ^2 and the correlation between the responses of subjects in the same pair is ρ, then $\tilde{\sigma}^2 = 2(1-\rho)\sigma^2$. Thus, if the matching of subjects within pairs achieves a moderately large positive correlation, $\tilde{\sigma}^2$ can be significantly less than $2\sigma^2$, the variance of the difference in response between two randomly selected subjects allocated to treatments A and B. We consider the problem of testing the null hypothesis H_0: $\mu_A = \mu_B$ in a group sequential test where observations are taken in up to K groups. If n_k pairs of observations are available at the kth analysis, we estimate $\theta = \mu_A - \mu_B$ by

$$\frac{1}{n_k}\sum_{i=1}^{n_k}(X_{Ai} - X_{Bi}) \sim N(\mu_A - \mu_B, \frac{\tilde{\sigma}^2}{n_k}),$$

the information for θ is $\mathcal{I}_k = n_k/\tilde{\sigma}^2$, the reciprocal of this estimate's variance, and the standardized test statistic is

$$Z_k = \frac{1}{\sqrt{(n_k\tilde{\sigma}^2)}}\sum_{i=1}^{n_k}(X_{Ai} - X_{Bi}), \quad k = 1, \dots, K. \tag{3.4}$$

The differences $X_{Ai} - X_{Bi}$ play the same role as the observations X_i in the single-population situation of Section 3.2.2, and essentially the same argument shows that $\{Z_1, \dots, Z_K\}$ have the canonical joint distribution with information levels $\{\mathcal{I}_1, \dots, \mathcal{I}_K\}$ for $\theta = \mu_A - \mu_B$.

3.2.4 Two-period Crossover Trial

If between-patient variation is high but it is possible to observe a single subject's response to two or more treatments, the necessary sample size may be reduced by using a crossover design in which inferences are based on within-patient comparisons. Here, we describe group sequential analysis of a two-treatment comparison in a two-period crossover trial.

In a two-treatment, two-period crossover trial each subject is allocated treatments A and B in a randomly chosen order. After observing a subject's response to each treatment, the difference (Response on treatment A) – (Response on treatment B) is calculated. Let X_i, $i = 1, 2, \dots$, denote the values of this difference for subjects receiving treatment A first and Y_i, $i = 1, 2, \dots$, the values for subjects who receive treatment B first. The commonly used normal model is then $X_i \sim N(\theta + \phi, \sigma^2)$ and $Y_i \sim N(\theta - \phi, \sigma^2)$, $i = 1, 2, \dots$, where θ represents the treatment difference and ϕ a period effect. We consider a group sequential test of H_0: $\theta = 0$ when the variance σ^2 is assumed known.

For each $k = 1, \dots, K$, suppose observed values $X_1, \dots, X_{n_{Xk}}$ and $Y_1, \dots, Y_{n_{Yk}}$ are available at the kth analysis. We can estimate θ by

$$\frac{1}{2}\left(\bar{X}^{(k)} + \bar{Y}^{(k)}\right) = \frac{1}{2}\left(\frac{1}{n_{Xk}}\sum_{i=1}^{n_{Xk}} X_i + \frac{1}{n_{Yk}}\sum_{i=1}^{n_{Yk}} Y_i\right) \sim N(\theta, \frac{\sigma^2}{4n_{Xk}} + \frac{\sigma^2}{4n_{Yk}}),$$

the information for θ is the reciprocal of the above variance,

$$\mathcal{I}_k = 4\left(\frac{\sigma^2}{n_{Xk}} + \frac{\sigma^2}{n_{Yk}}\right)^{-1}, \tag{3.5}$$

and the standardized statistic for testing H_0: $\theta = 0$ is

$$Z_k = \frac{1}{2}(\bar{X}^{(k)} + \bar{Y}^{(k)})\sqrt{\mathcal{I}_k}. \tag{3.6}$$

It is straightforward to check that $(Z_1, \ldots, Z_K)$ is multivariate normal,

$$Z_k \sim N(\theta\sqrt{\mathcal{I}_k}, 1), \quad k = 1, \ldots, K,$$

and

$$\mathrm{Cov}(Z_{k_1}, Z_{k_2}) = \sqrt{(\mathcal{I}_{k_1}/\mathcal{I}_{k_2})}, \quad 1 \leq k_1 \leq k_2 \leq K.$$

So we see, once again, that $\{Z_1, \ldots, Z_K\}$ have the canonical joint distribution with information levels $\{\mathcal{I}_1, \ldots, \mathcal{I}_K\}$ for θ.

In this example, the distribution of responses depends on two parameters, the treatment effect θ and a period effect ϕ. Since it is possible to write down an estimate of θ directly, we did not go through the process of fitting a full model for both parameters. In more complex crossover designs, it may be necessary to fit the full model in order to find an estimate of the parameter of interest. The same is true of other situations where the treatment effect is one element of the parameter vector in a general normal linear model, which is fitted repeatedly to accumulating data. We consider the group sequential analysis of such normal linear models in the next section.

3.3 Normal Linear Models

3.3.1 Theory for Independent Observations

In the general normal linear model, we suppose observations X_i, $i = 1, 2, \ldots,$ are independent and normally distributed, with expectations depending linearly on a parameter vector $\boldsymbol{\beta} = (\beta_1, \ldots, \beta_p)^T$. Consider a study in which data are collected group sequentially in a maximum of K groups and the first k groups contain a total of n_k observations, $k = 1, \ldots, K$. Let $\boldsymbol{X}^{(k)} = (X_1, \ldots, X_{n_k})^T$ denote the vector of observations available at the kth analysis. With independent observations, at each interim analysis, $k = 1, \ldots, K$, we have

$$\boldsymbol{X}^{(k)} \sim N(\boldsymbol{D}^{(k)}\boldsymbol{\beta}, \boldsymbol{I}_{n_k}\sigma^2),$$

where $\boldsymbol{D}^{(k)}$ denotes the design matrix for the first n_k observations and $\boldsymbol{I}_{n_k}$ the $n_k \times n_k$ identity matrix. Assuming $\boldsymbol{\beta}$ to be estimable at each analysis, its least squares estimate based on the first k groups of observations is

$$\widehat{\boldsymbol{\beta}}^{(k)} = (\boldsymbol{D}^{(k)T}\boldsymbol{D}^{(k)})^{-1}\boldsymbol{D}^{(k)T}\boldsymbol{X}^{(k)}$$

and it is well known that this estimate is normally distributed with expectation $\boldsymbol{\beta}$ and variance-covariance matrix

$$\mathbf{Var}(\widehat{\boldsymbol{\beta}}^{(k)}) = (\boldsymbol{D}^{(k)T}\boldsymbol{D}^{(k)})^{-1}\sigma^2.$$

In Section 15.3 we shall establish the joint distribution of the sequence of estimates of $\boldsymbol{\beta}$, proving that $\{\widehat{\boldsymbol{\beta}}^{(1)}, \ldots, \widehat{\boldsymbol{\beta}}^{(K)}\}$ is multivariate normal with

$$E(\widehat{\boldsymbol{\beta}}^{(k)}) = \boldsymbol{\beta}, \quad k = 1, \ldots, K, \tag{3.7}$$

and

$$\mathbf{Cov}(\widehat{\boldsymbol{\beta}}^{(k_1)}, \widehat{\boldsymbol{\beta}}^{(k_2)}) = \mathbf{Var}(\widehat{\boldsymbol{\beta}}^{(k_2)}), \quad 1 \le k_1 \le k_2 \le K. \tag{3.8}$$

Suppose an experimenter is mainly interested in β_1, the first element of the parameter vector $\boldsymbol{\beta}$, and wishes to test the null hypothesis H_0: $\beta_1 = 0$ against a two-sided alternative, with Type I error α and power $1 - \beta$ at $\beta_1 = \pm\delta$. Such a test might be conducted when β_1 represents a treatment effect and the other elements of $\boldsymbol{\beta}$ are covariates which affect response but are not themselves of primary interest. The group sequential test of H_0: $\beta_1 = 0$ is based on the sequence of estimates, $\{\hat{\beta}_1^{(1)}, \ldots, \hat{\beta}_1^{(K)}\}$. At the kth analysis, the variance of $\hat{\beta}_1^{(k)}$ is the $(1,1)$ element of the variance matrix

$$\mathbf{Var}(\widehat{\boldsymbol{\beta}}^{(k)}) = (\boldsymbol{D}^{(k)T}\boldsymbol{D}^{(k)})^{-1}\sigma^2$$

and $\mathcal{I}_k$, the "Fisher information" for β_1 after k groups of observations, is the reciprocal of this variance. Thus,

$$\mathcal{I}_k = \{\mathrm{Var}(\hat{\beta}_1^{(k)})\}^{-1} = \frac{1}{\{(\boldsymbol{D}^{(k)T}\boldsymbol{D}^{(k)})^{-1}\}_{11}\sigma^2}, \quad k = 1, \ldots, K.$$

The standardized statistic for testing H_0: $\beta_1 = 0$ after the kth group of observations is

$$Z_k = \frac{\hat{\beta}_1^{(k)}}{\sqrt{\mathrm{Var}(\hat{\beta}_1^{(k)})}} = \hat{\beta}_1^{(k)}\sqrt{\mathcal{I}_k}, \quad k = 1, \ldots, K.$$

It follows from the multivariate normality of $\{\widehat{\boldsymbol{\beta}}^{(1)}, \ldots, \widehat{\boldsymbol{\beta}}^{(K)}\}$ that $(\hat{\beta}_1^{(1)}, \ldots, \hat{\beta}_1^{(K)})$ is multivariate normal and properties (3.7) and (3.8) imply that

$$E(\hat{\beta}_1^{(k)}) = \beta_1, \quad k = 1, \ldots, K,$$

and

$$\mathrm{Cov}(\hat{\beta}_1^{(k_1)}, \hat{\beta}_1^{(k_2)}) = \mathrm{Var}(\hat{\beta}_1^{(k_2)}), \quad 1 \le k_1 \le k_2 \le K.$$

Hence, $(Z_1, \ldots, Z_K)$ is multivariate normal with

$$E(Z_k) = \beta_1\sqrt{\mathcal{I}_k}, \quad k = 1, \ldots, K,$$

and

$$\mathrm{Cov}(Z_{k_1}, Z_{k_2}) = \sqrt{(\mathcal{I}_{k_1}/\mathcal{I}_{k_2})}, \quad 1 \le k_1 \le k_2 \le K,$$

i.e., the sequence of standardized statistics has the canonical joint distribution (3.1) with information levels $\{\mathcal{I}_1, \ldots, \mathcal{I}_K\}$ for $\theta = \beta_1$.

Test statistics with the same joint distribution also arise when testing a hypothesis of the form H_0: $\boldsymbol{c}^T\boldsymbol{\beta} = \gamma$, where $\boldsymbol{c}$ is a $p \times 1$ vector and γ a scalar. Suppose a two-sided test is required with Type I error α and power $1 - \beta$ at $\boldsymbol{c}^T\boldsymbol{\beta} = \gamma \pm \delta$. Define $\theta = \boldsymbol{c}^T\boldsymbol{\beta} - \gamma$ and let

$$\hat{\theta}^{(k)} = \boldsymbol{c}^T\widehat{\boldsymbol{\beta}}^{(k)} - \gamma, \quad k = 1, \ldots, K.$$

It follows from the joint distribution of $\{\widehat{\boldsymbol{\beta}}^{(1)}, \ldots, \widehat{\boldsymbol{\beta}}^{(K)}\}$ that $(\hat{\theta}^{(1)}, \ldots, \hat{\theta}^{(K)})$ is multivariate normal,

$$E(\hat{\theta}^{(k)}) = \theta, \quad k = 1, \ldots, K,$$

$$\mathrm{Var}(\hat{\theta}^{(k)}) = \boldsymbol{c}^T(\boldsymbol{D}^{(k)T}\boldsymbol{D}^{(k)})^{-1}\boldsymbol{c}\sigma^2, \quad k = 1, \ldots, K,$$

and

$$\mathrm{Cov}(\hat{\theta}^{(k_1)}, \hat{\theta}_1^{(k_2)}) = \mathrm{Var}(\hat{\theta}_1^{(k_2)}) \quad \text{for } k_1 \le k_2.$$

The standardized test statistic at analysis k is

$$Z_k = \frac{\boldsymbol{c}^T \widehat{\boldsymbol{\beta}}^{(k)} - \gamma}{\sqrt{\mathrm{Var}(\boldsymbol{c}^T \widehat{\boldsymbol{\beta}}^{(k)})}} = \hat{\theta}^{(k)} \sqrt{\mathcal{I}_k}, \quad k = 1, \ldots, K,$$

where now

$$\mathcal{I}_k = \{\mathrm{Var}(\hat{\theta}^{(k)})\}^{-1} = \{\boldsymbol{c}^T (\boldsymbol{D}^{(k)T} \boldsymbol{D}^{(k)})^{-1} \boldsymbol{c} \sigma^2\}^{-1}$$

is the Fisher information for $\boldsymbol{c}^T \boldsymbol{\beta}$ at analysis k. Thus, the vector of standardized statistics $(Z_1, \ldots, Z_K)$ is multivariate normal,

$$E(Z_k) = (\boldsymbol{c}^T \boldsymbol{\beta} - \gamma)\sqrt{\mathcal{I}_k}, \quad k = 1, \ldots, K,$$

and

$$\mathrm{Cov}(Z_{k_1}, Z_{k_2}) = \sqrt{(\mathcal{I}_{k_1}/\mathcal{I}_{k_2})}, \quad 1 \le k_1 \le k_2 \le K.$$

With $\theta = \boldsymbol{c}^T \boldsymbol{\beta} - \gamma$, we once again have the canonical joint distribution (3.1).

This general theory explains the common features seen in the examples of Sections 3.2.1 to 3.2.4. Since the theory applies in generality, it justifies application of the group sequential tests derived for one simple situation to a great many other statistical models. We shall rely on this fact as we develop different types of group sequential test.

3.3.2 Examples

3.3.2.1 Crossover Trial

The two-treatment, two-period crossover trial of Section 3.2.4 provides an interesting example to represent in general linear model form. With observations X_i, $i = 1, 2, \ldots$, and Y_i, $i = 1, 2, \ldots$, as previously defined and parameter vector $\boldsymbol{\beta} = (\theta, \phi)^T$, where T denotes transpose, the observation vector and design matrix at analysis k are

$$\boldsymbol{X}^{(k)} = \begin{pmatrix} X_1 \\ \vdots \\ X_{n_{X1}} \\ Y_1 \\ \vdots \\ Y_{n_{Y1}} \\ \vdots \\ X_{n_{X,k-1}+1} \\ \vdots \\ X_{n_{Xk}} \\ Y_{n_{Y,k-1}+1} \\ \vdots \\ Y_{n_{Yk}} \end{pmatrix} \quad \text{and} \quad \boldsymbol{D}^{(k)} = \begin{pmatrix} 1 & 1 \\ \vdots & \vdots \\ 1 & 1 \\ 1 & -1 \\ \vdots & \vdots \\ 1 & -1 \\ \vdots & \vdots \\ 1 & 1 \\ \vdots & \vdots \\ 1 & 1 \\ 1 & -1 \\ \vdots & \vdots \\ 1 & -1 \end{pmatrix}.$$

A little algebra shows that

$$\widehat{\boldsymbol{\beta}}^{(k)} = (\boldsymbol{D}^{(k)T} \boldsymbol{D}^{(k)})^{-1} \boldsymbol{D}^{(k)T} \boldsymbol{X}^{(k)} = \begin{pmatrix} (\bar{X}^{(k)} + \bar{Y}^{(k)})/2 \\ (\bar{X}^{(k)} - \bar{Y}^{(k)})/2 \end{pmatrix}$$

where

$$\bar{X}^{(k)} = \frac{1}{n_{Xk}} \sum_{i=1}^{n_{Xk}} X_i \quad \text{and} \quad \bar{Y}^{(k)} = \frac{1}{n_{Yk}} \sum_{i=1}^{n_{Yk}} Y_i, \quad k = 1, \ldots, K.$$

Thus $\hat{\theta}^{(k)} = (\bar{X}^{(k)} + \bar{Y}^{(k)})/2$, $\mathcal{I}_k = 4(\sigma^2/n_{Xk} + \sigma^2/n_{Yk})^{-1}$ and $Z_k = \hat{\theta}^{(k)}\sqrt{\mathcal{I}_k}$, in agreement with the results of Section 3.2.4.

3.3.2.2 *Two-treatment Comparison Adjusting for Covariates*

We now have the necessary theoretical tools to consider group sequential monitoring of a clinical trial with randomized treatment allocation and allowance for baseline variables when making treatment comparisons.

One commonly used method for allocating treatments in clinical trials is the "randomly permuted block design". For a block size of $2m$, the sequence of treatment labels is created by combining strings of length $2m$, each of which is a random permutation of m As and m Bs. Patients are assigned the treatments denoted by these labels in the order of their admission to the study. Treatment allocation may also be balanced with respect to a small number of patient characteristics, each with two or three levels. If the total number of such categories is fairly small, a randomly permuted block design can be used within each category defined by a set of such "stratification" variables. When there are several stratification variables, the strategy of "minimization" is sometimes applied, choosing each new subject's treatment to equalize, as far as possible, the numbers on treatments A and B within each level of each stratification variable. More detailed descriptions of these and other treatment allocation schemes can be found in textbooks on the general practice of clinical trials; see, for example, Chapter 5 of Pocock (1983) or Chapter 2 of Whitehead (1997).

When testing for a treatment effect, it is good practice to model the effects of stratification variables and possibly other baseline variables not involved in the treatment allocation rule. Suppose responses are assumed to be independent and normally distributed with the ith patient's response

$$X_i \sim N(\sum_{j=1}^{p} d_{ij}\beta_j, \sigma^2), \quad i = 1, 2, \ldots . \tag{3.9}$$

Setting $d_{i1} = 0$ if patient i receives treatment A and $d_{i1} = 1$ if patient i receives treatment B defines β_1 as the treatment effect. Other terms in $\sum d_{ij}\beta_j$ may include an overall constant or intercept term, additive effects associated with particular levels of categorical variables, and linear terms in suitably transformed baseline covariates.

If n_k responses have been observed at the time of the kth analysis, the data vector is

$$\boldsymbol{X}^{(k)} = (X_1, \ldots, X_{n_k})^T \sim N(\boldsymbol{D}^{(k)}\boldsymbol{\beta}, \boldsymbol{I}_{n_k}\sigma^2),$$

where $\boldsymbol{\beta} = (\beta_1, \ldots, \beta_p)^T$ and $\boldsymbol{D}^{(k)}$ is the $n_k \times p$ matrix with (i,j)th element d_{ij}. The maximum likelihood estimate of $\boldsymbol{\beta}$ at this point is

$$\widehat{\boldsymbol{\beta}}^{(k)} = (\boldsymbol{D}^{(k)T}\boldsymbol{D}^{(k)})^{-1}\boldsymbol{D}^{(k)T}\boldsymbol{X}^{(k)} \sim N(\boldsymbol{\beta}, (\boldsymbol{D}^{(k)T}\boldsymbol{D}^{(k)})^{-1}\sigma^2)$$

and the estimated treatment effect is

$$\hat{\beta}_1^{(k)} \sim N(\beta_1, \mathcal{I}_k^{-1}),$$

where $\mathcal{I}_k = [\{(\boldsymbol{D}^{(k)T}\boldsymbol{D}^{(k)})^{-1}\}_{11}\sigma^2]^{-1}$. Thus, the standardized statistic for testing H_0: $\beta_1 = 0$ is $Z_k = \hat{\beta}_1^{(k)}\sqrt{\mathcal{I}_k}$.

3.3.3 Theory for Correlated Observations

The unified formulation (3.1) also applies to normal linear models for correlated observations. Suppose $\boldsymbol{X}^{(k)}$ and $\boldsymbol{D}^{(k)}$, $k = 1, \ldots, K$, and $\boldsymbol{\beta}$ are as previously defined, but now

$$\boldsymbol{X}^{(k)} \sim N(\boldsymbol{D}^{(k)}\boldsymbol{\beta}, \boldsymbol{\Sigma}^{(k)}\sigma^2),$$

where $\boldsymbol{\Sigma}^{(k)}$ and σ^2 are known. The maximum likelihood estimate of $\boldsymbol{\beta}$ at analysis k is

$$\widehat{\boldsymbol{\beta}}^{(k)} = (\boldsymbol{D}^{(k)^T}\boldsymbol{\Sigma}^{(k)^{-1}}\boldsymbol{D}^{(k)})^{-1}\boldsymbol{D}^{(k)^T}\boldsymbol{\Sigma}^{(k)^{-1}}\boldsymbol{X}^{(k)},$$

which has a multivariate normal distribution with mean equal to $\boldsymbol{\beta}$ and variance $(\boldsymbol{D}^{(k)^T}\boldsymbol{\Sigma}^{(k)^{-1}}\boldsymbol{D}^{(k)})^{-1}\sigma^2$.

The proof in Section 15.3 that the sequence of estimates $\{\widehat{\boldsymbol{\beta}}^{(1)}, \ldots, \widehat{\boldsymbol{\beta}}^{(K)}\}$ is multivariate normal with the expectations and covariances given in (3.7) and (3.8) holds in the correlated as well as the uncorrelated case. It therefore follows, as in Section 3.3.1, that if a group sequential test of H_0: $\boldsymbol{c}^T\boldsymbol{\beta} = \gamma$ is based on the sequence of standardized statistics

$$Z_k = \frac{\boldsymbol{c}^T\widehat{\boldsymbol{\beta}}^{(k)} - \gamma}{\sqrt{\mathrm{Var}(\boldsymbol{c}^T\widehat{\boldsymbol{\beta}}^{(k)})}}, \quad k = 1, \ldots, K,$$

the information for $\theta = \boldsymbol{c}^T\boldsymbol{\beta} - \gamma$ at analysis k is

$$\mathcal{I}_k = \{\mathrm{Var}(\boldsymbol{c}^T\widehat{\boldsymbol{\beta}}^{(k)})\}^{-1} = \{\boldsymbol{c}^T(\boldsymbol{D}^{(k)T}\boldsymbol{\Sigma}^{(k)^{-1}}\boldsymbol{D}^{(k)})^{-1}\boldsymbol{c}\,\sigma^2\}^{-1}$$

and $\{Z_1, \ldots, Z_K\}$ have the canonical joint distribution with information levels $\{\mathcal{I}_1, \ldots, \mathcal{I}_K\}$ for θ.

3.3.4 Sequential Comparison of Slopes in a Longitudinal Study

As an example with correlated responses, we consider a longitudinal study in which repeated measurements are made on each patient over a period of time. Suppose a patient's mean response is expected to change linearly over time, and interest focuses on the effect of treatment on the rate of change. A simple linear regression model for observations $X_{i1}, \ldots, X_{ip}$ on patient i recorded at times $t_{i1}, \ldots, t_{ip}$ is

$$X_{ij} = a_i + b_i t_{ij} + \epsilon_{ij}, \quad j = 1, \ldots, p,$$

where the ϵ_{ij} are independent and each is distributed as $N(0, \sigma_e^2)$. If slopes are expected to vary between patients, we may suppose $b_i \sim N(\mu_A, \sigma_b^2)$ for patients on treatment A and $b_i \sim N(\mu_B, \sigma_b^2)$ for those on treatment B. We can then write

$$X_{ij} = a_i + (\mu_{T(i)} + \delta_i)t_{ij} + \epsilon_{ij}, \quad j = 1, \ldots, p,$$

where $T(i) \in \{A, B\}$ denotes the treatment received by patient i and the δ_i are independent and each is distributed as $N(0, \sigma_b^2)$. If n patients have been treated and p_i measurements recorded on patient i for $i = 1, \ldots, n$, we have a *random effects* model in which the data $\{X_{ij};\, i = 1, \ldots, n,\, j = 1, \ldots, p_i\}$ are multivariate normal with

$$E(X_{ij}) = a_i + \mu_{T(i)}t_{ij}, \quad i = 1, \ldots, n,\ j = 1, \ldots, p_i,$$

$$\mathrm{Var}(X_{ij}) = t_{ij}^2\sigma_b^2 + \sigma_e^2, \quad i = 1, \ldots, n,\ j = 1, \ldots, p_i,$$

and

$$\text{Cov}(X_{ij}, X_{i'j'}) = \begin{cases} 0 & \text{if } i \neq i', \\ t_{ij}t_{ij'}\sigma_b^2 & \text{if } i = i' \text{ and } j \neq j'. \end{cases}$$

Re-arranging the X_{ij}s in a single-column vector, this model can be expressed as a normal linear model in which repeated observations on the same subject are positively correlated.

When a longitudinal study of this type is analyzed group sequentially, both the total number of subjects and the numbers of measurements on each subject may increase between analyses. If $\boldsymbol{X}^{(k)}$ denotes the observation vector at analysis k, containing responses from n_k different subjects, we can write

$$\boldsymbol{X}^{(k)} \sim N(\boldsymbol{D}^{(k)}\boldsymbol{\beta}^{(k)}, \boldsymbol{\Sigma}^{(k)}\sigma_e^2),$$

where $\boldsymbol{\beta}^{(k)} = (\mu_A, \mu_B, a_1, \ldots, a_{n_k})^T$ is the parameter vector for "fixed effects" at this analysis, $\boldsymbol{D}^{(k)}$ is the appropriate design matrix, and the matrix $\boldsymbol{\Sigma}^{(k)}$ involves the ratio σ_b^2/σ_e^2. The maximum likelihood estimate of $\boldsymbol{\beta}$ is then

$$\widehat{\boldsymbol{\beta}}^{(k)} = (\boldsymbol{D}^{(k)T}\boldsymbol{\Sigma}^{(k)-1}\boldsymbol{D}^{(k)})^{-1}\boldsymbol{D}^{(k)T}\boldsymbol{\Sigma}^{(k)-1}\boldsymbol{X}^{(k)},$$

and in order to investigate the treatment effect $\theta = \mu_A - \mu_B$, we extract $\hat{\theta}^{(k)} = \boldsymbol{d}_k^T\widehat{\boldsymbol{\beta}}^{(k)}$, where the contrast vector $\boldsymbol{d}_k = (1, -1, 0, \ldots, 0)^T$. The Fisher information for θ at this analysis is $\mathcal{I}_k = \{\text{Var}(\boldsymbol{d}_k^T\widehat{\boldsymbol{\beta}}^{(k)})\}^{-1}$ and the standardized statistic for testing H_0: $\theta = 0$ is $Z_k = \hat{\theta}^{(k)}\sqrt{\mathcal{I}_k}$.

Although the parameter vector $\boldsymbol{\beta}^{(k)}$ grows as intercepts a_i are added for additional subjects, successive estimates of μ_A and μ_B, and hence of θ, still follow the standard joint distribution (see the discussion of this point in Section 15.3). Consequently, $\{Z_1, \ldots, Z_K\}$ has the canonical joint distribution with information levels $\{\mathcal{I}_1, \ldots, \mathcal{I}_K\}$ for $\theta = \mu_A - \mu_B$.

We shall see in Chapter 4 that, if a group sequential test is to satisfy given Type I error and power requirements, it must be designed so that the information levels $\mathcal{I}_k$ reach specified values. The formula

$$\mathcal{I}_k = \{\text{Var}(\boldsymbol{d}_k^T\widehat{\boldsymbol{\beta}}^{(k)})\}^{-1} = \{\boldsymbol{d}_k^T(\boldsymbol{D}^{(k)T}\boldsymbol{\Sigma}^{(k)-1}\boldsymbol{D}^{(k)})^{-1}\boldsymbol{d}_k\,\sigma_e^2\}^{-1}$$

allows numerical computation of $\mathcal{I}_k$ for particular data sets but provides little insight into just how $\mathcal{I}_k$ depends on the numbers and times of measurements on each subject available at analysis k.

Suppose that, for $i = 1, \ldots, n_k$, we have $p_i(k)$ measurements on patient i taken at times $t_{i1}, \ldots, t_{ip_i(k)}$ available at analysis k. Lan, Reboussin & DeMets (1994) discuss this problem and note that $\hat{b}_i^{(k)}$, the usual least squares estimate of b_i from the $p_i(k)$ measurements on subject i, has expectation $\mu_{T(i)}$ and variance

$$V_i(k) = \sigma_b^2 + \frac{\sigma_e^2}{\sum_{j=1}^{p_i(k)}(t_{ij} - \bar{t}_{i.}(k))^2},$$

where $\bar{t}_{i.}(k) = (t_{i1} + \ldots + t_{ip_i(k)})/p_i(k)$. They then propose estimating μ_A and μ_B by the weighted averages of estimates $\hat{b}_i^{(k)}$ for subjects receiving treatments A and B, respectively, weights being inversely proportional to the variance of each $\hat{b}_i^{(k)}$. It can be shown that this method does, in fact, produce the maximum likelihood estimates of μ_A and μ_B, and so we can use it to find a more helpful expression for $\mathcal{I}_k$. We have

$$\begin{aligned}\text{Var}(\hat{\mu}_A^{(k)}) &= \text{Var}\left(\frac{\sum_{i=1}^{n_k} I(T(i) = A)\,\hat{b}_i^{(k)}/V_i(k)}{\sum_{i=1}^{n_k} I(T(i) = A)/V_i(k)}\right) \\ &= \left\{\sum_{i=1}^{n_k} I(T(i) = A)\{V_i(k)\}^{-1}\right\}^{-1},\end{aligned}$$

where I is the indicator variable, and similarly

$$\text{Var}(\hat{\mu}_B^{(k)}) = \left\{ \sum_{i=1}^{n_k} I(T(i) = B)\{V_i(k)\}^{-1} \right\}^{-1}.$$

Since $\hat{\mu}_A^{(k)}$ and $\hat{\mu}_B^{(k)}$ are based on separate, uncorrelated sets of variables, we conclude that

$$\mathcal{I}_k = \{\text{Var}(\hat{\theta}^{(k)})\}^{-1} = \{\text{Var}(\hat{\mu}_A^{(k)}) + \text{Var}(\hat{\mu}_B^{(k)})\}^{-1}.$$

We have tacitly assumed in describing this example that σ_b^2 and σ_e^2 are known. In practice, this is unlikely to be the case and these variance components will have to be estimated during the study. An estimate of the ratio σ_b^2/σ_e^2 can be substituted in $\boldsymbol{\Sigma}^{(k)}$ for use in computing $\hat{b}_i^{(k)}$ at each analysis. However, σ_b^2 and σ_e^2 also govern the variance of $\hat{b}_i^{(k)}$ and the value of $\mathcal{I}_k$ for each $k = 1, \ldots, K$. If the ratio σ_b^2/σ_e^2 were known, we would have a normal linear model with variance known up to a single scale factor, a situation we shall address in Section 4.6; applying methods for this case with estimates of σ_b^2/σ_e^2 substituted at each analysis provides an approximate solution to this problem.

3.4 General Response Distributions

3.4.1 Parametric Models

In Section 15.6 we consider the group sequential analysis of observations whose distribution depends on a parameter vector $\boldsymbol{\beta}$ for which maximum likelihood estimates are obtained at analyses $k = 1, \ldots, K$. There we state an asymptotic result proved by Jennison & Turnbull (1997a) that, under certain regularity conditions concerning the statistical model and the covariate values for observed data, the sequence of estimates $\{\widehat{\boldsymbol{\beta}}^{(1)}, \ldots, \widehat{\boldsymbol{\beta}}^{(K)}\}$ is approximately multivariate normal with

$$E(\widehat{\boldsymbol{\beta}}^{(k)}) = \boldsymbol{\beta}, \quad k = 1, \ldots, K,$$

and

$$\mathbf{Cov}(\widehat{\boldsymbol{\beta}}^{(k_1)}, \widehat{\boldsymbol{\beta}}^{(k_2)}) = \mathbf{Var}(\widehat{\boldsymbol{\beta}}^{(k_2)}), \quad 1 \leq k_1 \leq k_2 \leq K,$$

when sample sizes are sufficiently large. Thus, the joint distribution of successive estimates of the parameter vector matches that seen for normal linear models in Section 3.3. Furthermore, the inverse of the Fisher information matrix for $\boldsymbol{\beta}$ at analysis k provides a consistent estimate of $\mathbf{Var}(\widehat{\boldsymbol{\beta}}^{(k)})$.

Precise conditions under which this theory applies depend on the particular statistical model in question, but in all cases they reduce to the conditions required to derive standard maximum likelihood theory at each individual analysis. Thus, the scope of the group sequential theory for parametric models is as broad as that of the underlying fixed sample maximum likelihood theory. The sample size necessary for these results to hold accurately is problem specific, and in a group sequential analysis it is advisable to ensure the first group of observations is sufficiently large that the asymptotic approximation holds adequately at the first analysis.

We shall illustrate the general theory with two applications to binary response data in Section 4.4. Many further applications arise in the group sequential analysis of generalized linear models, including the important examples of logistic regression for a binary response (see Section 16.6 for an example), probit regression, multiplicative models for Poisson count data, and normal linear models with non-linear link functions.

3.4.2 *Further Special Results*

Even when standard maximum likelihood methods are not applicable, it is sometimes possible to show that sequences of parameter estimates $\{\hat{\theta}^{(1)}, \ldots, \hat{\theta}^{(K)}\}$, standardized statistics $\{Z_1, \ldots, Z_K\}$, or score statistics $\{S_1, \ldots, S_K\}$ follow the canonical joint distributions (3.2) approximately for large sample sizes. The major endpoint in many clinical trials is survival time, right censored for those patients alive at the time of an analysis. The proportional hazards regression model proposed by Cox (1972) is widely used for modeling the dependence of survival on treatment and other baseline covariates but, since this model is only semi-parametric, maximum likelihood estimation is not a viable option. Nevertheless, sequences of parameter estimates obtained by maximum partial likelihood estimation (Cox, 1975) do approximately follow the same canonical distribution given in (3.2) as maximum likelihood estimates for parametric models. Thus, these estimates can be used to construct group sequential tests in exactly the same way. We shall describe the Cox model in detail and illustrate its group sequential application in Chapter 17.

The log-rank test is commonly used for comparing two survival distributions without adjustment for covariates. Under a proportional hazards assumption, the sequence of log-rank statistics obtained from accumulating data has, approximately, the canonical joint distribution (3.2) of a sequence of score statistics. We describe the group sequential log-rank test in detail in Section 4.5.

If data arise as multiple 2×2 tables with a common odds ratio and the number of tables increases with the number of observations, standard maximum likelihood theory is not applicable due to the increasing number of nuisance parameters. Here, it is convenient to base tests on the Mantel-Haenszel estimate of the odds ratio, and we shall see in Section 16.4 that the sequence of such estimates is approximately multivariate normal with the standard covariance structure (3.2).

For normal data, it is usually necessary to estimate both the mean and variance and, hence, a group sequential t-test is required. We explain in Section 4.6 how the two-sided tests previously defined for normal data with known variance can be adapted to give group sequential t-tests with approximately the same Type I error probability, and power dependent on the true mean and variance. The general theory which underlies these group sequential t-tests is presented later in Section 15.4.

3.5 Calculations for Group Sequential Tests: Armitage's Iterated Integrals

Since we shall continue to present properties of group sequential tests such as their Type I error rate, power and expected sample size, it is instructive to explain how the results of this chapter enable these calculations to be performed. We shall re-visit this material in Chapter 9 and discuss numerical computation in greater depth in Chapter 31. The methods we describe are applicable to any of the models for accumulating data for which the canonical joint distribution seen in (3.1) and (3.2) applies and, as we have seen previously in this chapter, this covers a great many types of data and analyses. Furthermore, we use the same techniques for other hypothesis testing problems with one-sided, non-inferiority and equivalence hypotheses.

For a general group sequential procedure, let $\mathcal{C}_1, \ldots, \mathcal{C}_K$ be subsets of the real line $\Re = (-\infty, \infty)$ representing the continuation regions at analyses 1 to K. That is, if stage k has been reached, the trial stops if $Z_k \notin \mathcal{C}_k$ and otherwise it continues to stage $k+1$. The sets $\mathcal{C}_k$ need not be intervals; at the final analysis, $\mathcal{C}_K = \emptyset$, the empty set, to ensure termination by this stage. The complement of the set $\mathcal{C}_k$ is the stopping region at analysis k, and this may be partitioned into several subsets indicating the action to be taken upon stopping. If the only actions are to accept or reject the null hypothesis H_0, we have two sets, $\mathcal{A}_k$ and $\mathcal{B}_k$ say, where we stop and accept H_0 at stage k if $Z_k \in \mathcal{A}_k$ and we stop to reject H_0 if $Z_k \in \mathcal{B}_k$. Thus, $\mathcal{A}_k$, $\mathcal{B}_k$ and $\mathcal{C}_k$ form a partition of the real line.

As an example, all the two-sided tests of Chapter 2 have

$$\mathcal{A}_k = \emptyset, \quad \mathcal{B}_k = (-\infty, -a_k) \cup (a_k, \infty), \quad \mathcal{C}_k = (-a_k, a_k) \quad \text{for } k = 1, \ldots, K-1,$$

and

$$\mathcal{A}_K = (-a_K, a_K), \quad \mathcal{B}_K = (-\infty, -a_K) \cup (a_K, \infty) \quad \text{and} \quad \mathcal{C}_K = \emptyset.$$

In later chapters we shall describe group sequential two-sided tests with early stopping to accept H_0, one-sided tests, and equivalence tests. These procedures can also be defined in terms of sets $\mathcal{A}_k$, $\mathcal{B}_k$ and $\mathcal{C}_k$, $k = 1, \ldots, K$, but the sets have different forms.

Suppose $Z_1, \ldots, Z_K$ have the canonical joint distribution (3.1) and define the stopping time T

$$T = \min\{k\colon Z_k \notin \mathcal{C}_k\}.$$

For $k = 1, \ldots, K$ and $-\infty < z < \infty$, we define

$$G_k(z;\, \theta) = Pr_\theta\{T \geq k \text{ and } Z_k \leq z\}$$

and

$$g_k(z;\, \theta) = \frac{\partial}{\partial z}\, G_k(z;\, \theta).$$

Thus, $g_k(z;\, \theta)$ represents the sub-density of Z_k, allowing for the possibility that stopping occurs before analysis k and no value of Z_k is observed.

The recursive formulae of Armitage, McPherson & Rowe (1969) can be used to calculate $g_1(z;\, \theta)$ to $g_K(z;\, \theta)$ in turn. These formulae are

$$g_1(z;\, \theta) = \phi(z - \theta\sqrt{\mathcal{I}_1}) \tag{3.10}$$

and, for $k = 2, \ldots, K$,

$$g_k(z;\, \theta) = \int_{\mathcal{C}_{k-1}} g_{k-1}(u;\, \theta)\, \frac{\sqrt{\mathcal{I}_k}}{\sqrt{\Delta_k}}\, \phi\left(\frac{z\sqrt{\mathcal{I}_k} - u\sqrt{\mathcal{I}_{k-1}} - \Delta_k\theta}{\sqrt{\Delta_k}}\right) du \tag{3.11}$$

where $\phi(z) = (2\pi)^{-1/2}\exp(-z^2/2)$ denotes the standard normal density.

These equations follow directly from the joint distribution of $\{Z_1, \ldots, Z_K\}$ given by (3.1). Equation (3.10) is immediate and (3.11) follows by noting that (3.2) implies the sequence of score statistics $\{S_1, S_2, \ldots\}$ has independent normal increments with

$$S_k - S_{k-1} = Z_k\sqrt{\mathcal{I}_k} - Z_{k-1}\sqrt{\mathcal{I}_{k-1}} \sim N(\theta\Delta_k,\, \Delta_k),$$

where $\Delta_k = \mathcal{I}_k - \mathcal{I}_{k-1}$ is the increment in information between analyses $k-1$ and k. Hence

$$Z_k \,|\, Z_{k-1} = z_{k-1} \sim N\left(\frac{z_{k-1}\sqrt{\mathcal{I}_{k-1}} + \theta\Delta_k}{\sqrt{\mathcal{I}_k}},\, \frac{\Delta_k}{\mathcal{I}_k}\right).$$

We can obtain all the properties of a group sequential test from the subdensities $g_k(z;\, \theta)$, $k = 1, \ldots, K$. For example, the distribution of the stopping time is given by

$$Pr_\theta\{T = k\} = \int_{\mathcal{A}_k \cup \mathcal{B}_k} g_k(z;\, \theta)\, dz, \quad k = 1, \ldots, K.$$

Similarly, in a two-sided test, the probability of accepting H_0 is

$$Pr_\theta\{\cup_{k=1}^K \mathcal{A}_k\} = \sum_{k=1}^K \int_{\mathcal{A}_k} g_k(z;\, \theta)\, dz.$$

Thus, we have the tools we need to calculate the Type I error rate, power function and expected sample size function for all forms of group sequential test.

At first sight, it may appear that calculating the properties of a group sequential test will require evaluation of complicated K-fold integrals. However, the iterative definition of the sub-densities $g_k(z;\,\theta)$, $k = 1, \ldots, K$, implies these can be computed by a succession of $K-1$ univariate integrations and computations can be carried out rapidly. We describe the methods we use for the numerical evaluation of these integrals in Chapter 31.

Emerson & Fleming (1990) note that the formula

$$g_k(z;\,\theta) \;=\; g_k(z;\,0)\exp(\theta z\sqrt{\mathcal{I}_k} - \theta^2\mathcal{I}_k/2) \quad \text{for } k = 1, \ldots, K$$

can be used to convert a sub-density g_k evaluated under one value of θ for computations at another θ. This result clearly holds for $k = 1$. If we assume the result holds for $k-1$, then use of (3.11) and some algebraic manipulation shows the result holds for k, so the general result follows by induction. This is an example of a likelihood ratio identity as defined by Siegmund (1985, Propositions 2.24 and 3.2), and we shall use this result again in Chapter 9. The relation can also be used in simulation studies to obtain properties of a sequential test under several values of θ from simulations performed at a single θ.

Chapter 4

Two-sided Tests: General Applications

4.1 Applying Two-sided Tests with Equal Group Sizes

4.1.1 General Form of the Tests

Suppose it is required to test a null hypothesis H_0: $\theta = 0$ with two-sided Type I error probability α and power $1-\beta$ at $\theta = \pm\delta$. We consider group sequential tests in which up to K analyses are permitted and standardized statistics Z_k, $k = 1, \ldots, K$, are available at these analyses. For a general testing problem, we assume $\{Z_1, \ldots, Z_K\}$ have the canonical joint distribution (3.1) with information levels $\{\mathcal{I}_1, \ldots, \mathcal{I}_K\}$ for θ. We start by considering the case of equal group sizes or more generally equally spaced information levels, so $\mathcal{I}_k = k\mathcal{I}_1$. In this case the joint distribution (3.1) becomes

$$\begin{array}{rl} \text{(i)} & (Z_1, \ldots, Z_K) \text{ is multivariate normal,} \\ \text{(ii)} & E(Z_k) = \theta\sqrt{\mathcal{I}_k}, \quad k = 1, \ldots, K, \quad \text{and} \\ \text{(iii)} & \mathrm{Cov}(Z_{k_1}, Z_{k_2}) = \sqrt{(k_1/k_2)}, \quad 1 \le k_1 \le k_2 \le K. \end{array} \tag{4.1}$$

Each of the tests introduced in Chapter 2 has the form

$$\begin{array}{ll} \text{After group } k = 1, \ldots, K-1 & \\ \quad \text{if } |Z_k| \ge c_k & \text{stop, reject } H_0 \\ \quad \text{otherwise} & \text{continue to group } k+1, \\ \text{after group } K & \\ \quad \text{if } |Z_K| \ge c_K & \text{stop, reject } H_0 \\ \quad \text{otherwise} & \text{stop, accept } H_0. \end{array} \tag{4.2}$$

Since the null hypothesis is rejected at stage k if $|Z_k|$, a standard normal variate under H_0, exceeds the critical value c_k, such tests can be interpreted as "repeated significance tests", applying the two-sided significance level $2\{1 - \Phi(c_k)\}$ to the data at analysis k.

A test's Type I error probability is

$$Pr_{\theta=0}\{|Z_k| \ge c_k \text{ for some } k = 1, \ldots, K\}, \tag{4.3}$$

evaluated when $\{Z_1, \ldots, Z_K\}$ follow the null distribution given by (4.1) with $\theta = 0$. Note that this distribution depends on K, but not on the final information level $\mathcal{I}_K$ which determines the sequence $\mathcal{I}_k = (k/K)\mathcal{I}_K$, $k = 1, \ldots, K$. Thus (4.3) can be computed under any convenient set of K equally spaced information levels, for example, $\mathcal{I}_k = k$, $k = 1, \ldots, K$. This explains why the boundary values $\{c_k\}$ in the tests of Chapter 2 can be chosen to ensure the Type I error probability is equal to the specified value, α, independently of the sample size and Type II error rate. Each type of test, Pocock (1977), O'Brien & Fleming (1979), etc., uses a different sequence of critical values $\{c_1, \ldots, c_K\}$ but all are chosen to ensure the Type I error probability equals the specified value α when (4.1) holds with $\theta = 0$.

DOI: 10.1201/9781584888482-4

Once the number of analyses K and the critical values, $\{c_1, \ldots, c_K\}$ have been set, the remaining design parameter, $\mathcal{I}_{max} = \mathcal{I}_K = K\mathcal{I}_1$, can be chosen to satisfy the Type II error requirement that

$$Pr_{\theta=\pm\delta}\left\{\bigcup_{k=1}^{K} (|Z_j| < c_j \text{ for } j = 1, \ldots, k-1 \text{ and } |Z_k| \geq c_k)\right\} = 1 - \beta \tag{4.4}$$

where $\{Z_1, \ldots, Z_K\}$ now follow the distribution (4.1) with $\theta = \pm\delta$. Note that this distribution depends on δ and $\mathcal{I}_K$ only through $\delta\sqrt{\mathcal{I}_K}$. Hence the Type II error requirement fixes the value of $\delta\sqrt{\mathcal{I}_{max}} = \delta\sqrt{\mathcal{I}_K}$.

A fixed sample test of H_0: $\theta = 0$ may be based on a standardized statistic with distribution $Z \sim N(\theta\sqrt{\mathcal{I}})$ and the test with Type I probability α rejects H_0 if $|Z| > \Phi^{-1}(1-\alpha/2)$. To achieve power $1-\beta$ when $\theta = \pm\delta$, the information $\mathcal{I}$ must satisfy

$$\Phi(\Phi^{-1}(1-\alpha/2) + \delta\sqrt{\mathcal{I}}) + \Phi(\Phi^{-1}(1-\alpha/2) - \delta\sqrt{\mathcal{I}}) = 1 - \beta, \tag{4.5}$$

where Φ denotes the standard normal cdf. As in Section 2.1, we ignore the very small probabilities that $Z < -\Phi^{-1}(1-\alpha/2)$ when $\theta = \delta$ or $Z > \Phi^{-1}(1-\alpha/2)$ when $\theta = -\delta$ and, hence, obtain the solution

$$\mathcal{I}_{f,2} = \{\Phi^{-1}(1-\alpha/2) + \Phi^{-1}(1-\beta)\}^2/\delta^2. \tag{4.6}$$

Here, we use the subscript 2 in $\mathcal{I}_{f,2}$ to distinguish the two-sided case from the one-sided case, which we shall consider in Chapter 5.

Note that for a fixed sample test with information $\mathcal{I}$ for θ, the power at $\theta = \pm\delta$ is a function of $\delta\sqrt{\mathcal{I}}$. Thus if a group sequential test with K analyses requires maximum information $\mathcal{I}_K = \mathcal{I}_{max}$ to satisfy a pair of Type I and Type II error constraints while a fixed sample test needs information $\mathcal{I}_{f,2}$, we can define the "inflation factor"

$$R = \mathcal{I}_{max}/\mathcal{I}_{f,2},$$

which does not depend on δ (although it does depend on K and the Type I and Type II error probabilities, α and β). This ratio plays an important role in the design of group sequential tests. Indeed, in Tables 2.2, 2.4, 2.10 and 2.13, we have already seen the factors $R_P(K, \alpha, \beta)$, $R_B(K, \alpha, \beta)$, $R_{WT}(K, \alpha, \beta, \Delta)$ and $R_{HP}(K, \alpha, \beta)$, appropriate to Pocock, O'Brien & Fleming, Wang & Tsiatis, and Haybittle-Peto tests, respectively. Although these factors were introduced as multipliers of the sample size of each treatment group in a parallel arm comparison, the general theory implies the same multiples of $\mathcal{I}_{f,2}$ provide the maximum information levels for group sequential tests to achieve power $1-\beta$ in general testing problems.

4.1.2 Five-step Summary

The general method for conducting a group sequential two-sided test can be summarized in the following five steps:

(1) Identify the statistical model for the data to be observed and the parameter θ which is to have value zero under the null hypothesis. Specify δ, α and β. The test of H_0: $\theta = 0$ will have Type I error probability α and power $1-\beta$ at $\theta = \pm\delta$.

(2) Choose the type of test, Pocock, O'Brien & Fleming, etc., and the maximum number of analyses, K.

(3) Define the standardized test statistics Z_k, and find an expression for the information levels $\mathcal{I}_k$ in terms of the numbers and types of observations available at each analysis, $k = 1, \ldots, K$.

(4) Find the inflation factor R for the chosen test and number of analyses, K. Calculate the information levels required at each analysis, $\mathcal{I}_k = (k/K)R\mathcal{I}_{f,2}$, $k = 1, \ldots, K$, where $\mathcal{I}_{f,2}$ is given by (4.6). Design the study so that these levels will be attained.

(5) Run the study applying the stopping rule (4.2).

When the trial is conducted, it is likely that the observed information levels will differ from the sequence $\mathcal{I}_k = (k/K)R\mathcal{I}_{f,2}$, $k = 1, \ldots, K$, assumed in Step 4. However, we recommend making this simplifying assumption about the $\mathcal{I}_k$ at the design stage and we shall see that this approach is robust to variations between the planned and observed information levels.

4.1.3 Examples

Paired Comparison

As a first example, consider the paired comparison of two treatments described in Section 3.2.3. Pairs of observations X_{Ai} and X_{Bi} yield differences with mean $\mu_A - \mu_B$ and we define $\theta = \mu_A - \mu_B$. Suppose $|\mu_A - \mu_B| = 1$ represents a clinically important treatment difference and we decide to test H_0: $\theta = 0$ with Type I error probability $\alpha = 0.05$ and power $1 - \beta = 0.9$ at $\theta = \pm 1$ using an O'Brien & Fleming test with five groups of observations.

The standardized test statistics are given by (3.4) and information levels at analyses $k = 1, \ldots, 5$, are $\mathcal{I}_k = n_k/\tilde{\sigma}^2$ where n_k is the number of pairs of observations in the first k groups and $\tilde{\sigma}^2 = \mathrm{Var}(X_{Ai} - X_{Bi})$. Suppose it is known that $\tilde{\sigma}^2 = 6$. Equation (4.6) gives

$$\mathcal{I}_{f,2} = (1.960 + 1.282)^2/1^2 = 10.51$$

and, taking $R = R_B(5, 0.05, 0.1) = 1.026$ from Table 2.4, we find the desired information levels to be

$$\mathcal{I}_k = n_k/6 = (k/5)1.026 \times 10.51, \quad k = 1, \ldots, 5.$$

This requires $n_k = 12.94\,k$, i.e., 12.94 pairs of observations per group, which we round up to 13.

The critical values for an O'Brien & Fleming test are $c_k = C_B(5, 0.05)(5/k)^{1/2}$, and we find $C_B(5, 0.05) = 2.040$ from Table 2.3. Thus, at analyses $k = 1, \ldots, 5$, the test stops to reject H_0: $\theta = 0$ if

$$\left| \frac{1}{\surd(13k \times 6)} \sum_{i=1}^{13k} (X_{Ai} - X_{Bi}) \right| \geq 2.040\sqrt{\frac{5}{k}},$$

i.e., if $\left|\sum_{i=1}^{13k}(X_{Ai} - X_{Bi})\right| \geq 40.29$. If H_0 has not been rejected by analysis 5 it is accepted and the study terminates at this point.

Crossover Trial

For our second example, we take a two-treatment, two-period crossover trial, as described in Section 3.2.4. With the notation of that section, suppose it is required to test H_0: $\theta = 0$ with Type I error probability $\alpha = 0.05$ and power $1 - \beta = 0.8$ at $\theta = \pm 0.6$. We shall use a Wang & Tsiatis test, as defined in Section 2.7.1, with $\Delta = 0.25$ and four groups of observations. The standardized test statistics $\{Z_1, \ldots, Z_K\}$ are given by (3.6).

Assuming we know $\sigma^2 = 9$ and equal numbers will be allocated to each sequence of treatments so $n_{Xk} = n_{Yk} = n_k$, say, for $k = 1, \ldots, 4$, the information levels given in (3.5) are $\mathcal{I}_k = 2n_k/9$. Equation (4.6) gives $\mathcal{I}_{f,2} = (1.960 + 0.842)^2/0.6^2 = 21.81$, and we find $R = R_{WT}(4, 0.05, 0.2, 0.25) = 1.065$ from Table 2.10. Hence, we require

$$\mathcal{I}_k = 2n_k/9 = (k/4)1.065 \times 21.81, \quad k = 1, \ldots, 4,$$

i.e., $n_k = 26.1\,k$ for $k = 1, \ldots, 4$. Rounding up to integer values, we see that each group should consist of 27 subjects on each treatment sequence.

A Wang & Tsiatis test with $\Delta = 0.25$ has $c_k = C_{WT}(4, 0.05, 0.25)(k/4)^{0.25-1/2}$, $k = 1, \ldots, 4$, where, from Table 2.9, $C_{WT}(4, 0.05, 0.25) = 2.113$. Applying the rule (4.2), the test stops to reject H_0 at analysis k if

$$\left| \frac{1}{2}\left(\bar{X}^{(k)} + \bar{Y}^{(k)}\right)\sqrt{\frac{2n_k}{9}} \right| \geq 2.113\,(k/4)^{-0.25} = 2.988\,k^{-0.25}.$$

Substituting $n_k = 27\,k$ and simplifying, this condition becomes

$$\left| \sum_{i=1}^{27k} X_i + \sum_{i=1}^{27k} Y_i \right| \geq \sqrt{2} \times 3 \times \sqrt{27k} \times 2.988\,k^{-0.25} = 65.87\,k^{0.25}.$$

If H_0 has not been rejected by analysis 4, it is accepted at this point.

In both these examples, information is proportional to sample size and standard sample size formulae provide an alternative route to calculating the group sizes. In the second example, the crossover trial yields observations $X_i \sim N(\theta + \phi, \sigma^2)$ and $Y_i \sim N(\theta - \phi, \sigma^2)$. The number of subjects following each treatment sequence needed for a fixed sample test of H_0: $\theta = 0$ with Type I error α to have power $1 - \beta$ at $\theta = \pm\delta$ is

$$n_f = \{\Phi^{-1}(1 - \alpha/2) + \Phi^{-1}(1 - \beta)\}^2\sigma^2/(2\delta^2) = 98.11.$$

It follows that the maximum number of observations of each type required for the group sequential test is $R\,n_f$ where $R = R_{WT}(4, 0.05, 0.2, 0.25) = 1.065$. Thus, the number of observations of each type per group is $R\,n_f/K = 1.065 \times 98.11/4 = 26.1$, which rounds to 27.

This example of a crossover trial also serves to demonstrate another feature of sequential testing. In deriving a group sequential test, we have assumed a statistical model before observing any data. This particular model does not include a treatment-by-period interaction term, which could be present due to a "carry-over" effect of the first treatment into the second period. It may be desirable to check for the presence of such an interaction, using the separate responses of each subject on treatments A and B, before interpreting evidence of a treatment difference θ. Such a test has low power if the between-subject variation is high, as is usually the case when a crossover design is adopted; thus group sequential analysis is most appropriate when an interaction is known to be unlikely *a priori*. Even so, it may be wise to use a test with wide early boundaries, so that extreme evidence of a treatment main effect is required to terminate the study before there is at least a modest chance of detecting a treatment by period interaction.

4.2 Applying the Tests with Unequal Increments in Information

In the examples of Section 4.1.3, equal group sizes are required to produce the equally spaced information levels assumed in Section 4.1. In other cases, information will depend on the values of baseline variables that appear in the model or, in the case of time-to-event data, information depends on the number of observed events. In practice, it is often not possible to guarantee such equally spaced information levels. In a large multi-center trial, the dates of interim analyses are usually set in advance for administrative reasons and it is only to be expected that observed information levels will deviate somewhat from their target values. We now present an approximate but simple way of dealing with this problem. A more sophisticated treatment which can handle highly unpredictable information levels will be described in Chapter 8.

Our objective remains as before, to test a null hypothesis H_0: $\theta = 0$ with two-sided Type I error probability α and power $1 - \beta$ at $\theta = \pm\delta$ in a group sequential test with up to K analyses. We assume the standardized statistics at these analyses, $\{Z_1, \ldots, Z_K\}$, have the canonical joint distribution (3.1) with information levels $\{\mathcal{I}_1, \ldots, \mathcal{I}_K\}$ for θ but now these information levels may be unequally spaced.

At the design stage, some assumptions must be made about the sequence of information levels that will be observed, and it is often sensible to suppose these will be equally spaced. One can then follow the five steps of the procedure laid out in Section 4.1.2, defining statistics Z_k, information levels $\mathcal{I}_k$, and target values for the $\mathcal{I}_k$. Even though one expects observed information levels to depart from their intended values, analyses should still be planned so that target levels will be met as closely as possible. In particular, the final information level, $\mathcal{I}_K$, should be close to its target, $R\mathcal{I}_{f,2}$, since the value of $\mathcal{I}_K$ has a major influence on a test's power.

As analyses are conducted during the study, the observed information levels are found using the numbers of observations actually recorded at each stage. Standardized statistics Z_k are calculated from the data available at each analysis and the test is conducted using the rule (4.2) just as before. Note that, since each Z_k, $k = 1, \ldots, K$, has a marginal $N(0,1)$ distribution under H_0, the criterion for terminating with rejection of H_0 at analysis k is still that a single hypothesis test based on Z_k should reject H_0 at a two-sided significance level less than or equal to $2\{1 - \Phi(c_k)\}$. This "significance level" interpretation can be helpful in implementing a group sequential test, and we shall see later that it provides a useful starting point for generalizing a group sequential test for normal data to other response distributions.

For each k, we assume the value of $\mathcal{I}_k$ may be influenced by external factors, such as the accrual rate of suitable subjects, but not by the previously observed test statistics $Z_1, \ldots, Z_{k-1}$. In this case, the sequence $\{Z_1, \ldots, Z_K\}$ has the standard distribution (3.1) conditional on the observed information levels $\{\mathcal{I}_1, \ldots, \mathcal{I}_K\}$, and we can compute a test's properties conditional on $\{\mathcal{I}_1, \ldots, \mathcal{I}_K\}$.

Differences between the observed information levels and their target values lead to deviations from the planned Type I error probability and power. As an example, consider the parallel two-treatment comparison of Chapter 2 where responses $X_{Ai} \sim N(\mu_A, \sigma^2)$ and $X_{Bi} \sim N(\mu_B, \sigma^2)$, $i = 1, 2, \ldots$, are observed. Suppose it is known that $\sigma^2 = 4$, and we wish to test H_0: $\mu_A = \mu_B$ against $\mu_A \neq \mu_B$ with Type I error probability $\alpha = 0.05$ and power $1 - \beta = 0.9$ at $\mu_A - \mu_B = \pm 1$ using a group sequential test with five groups of observations. We saw in Chapter 2 that, if groups contain m observations on each treatment, the Pocock test requires $m = 21$, and the O'Brien & Fleming test and Wang & Tsiatis test with $\Delta = 0.25$ both require $m = 18$. Suppose one such test is planned and, when the data are analyzed, cumulative sample sizes on the two treatments are equal at each analysis, taking the values n_k, $k = 1, \ldots, K$, but these differ from the planned sample sizes $m \times k$. At the kth analysis we simply calculate the standardized test statistic from the data that are available

$$Z_k = \left(\frac{1}{n_k} \sum_{i=1}^{n_k} X_{Ai} - \frac{1}{n_k} \sum_{i=1}^{n_k} X_{Bi} \right) \sqrt{\frac{n_k}{2\sigma^2}}, \quad k = 1, \ldots, K,$$

and stop to reject H_0 if $|Z_k| > c_k$ using the original sequence of critical values $\{c_1, \ldots, c_K\}$.

Table 4.1 shows the Type I error probability and power actually achieved by these tests for a variety of sequences of cumulative sample sizes on each treatment, $\{n_1, \ldots, n_5\}$. Results were obtained by numerical computation. The first example for each test follows the original design, the Type I error rate is exactly 0.05 and power at $\mu_A - \mu_B = \pm 1$ is a little greater than 0.9 because of upward rounding of the group sizes to integer values. Examples 2 and 3 for each test have equal group sizes and since, under the null hypothesis, $\{Z_1, \ldots, Z_5\}$ still have the standard joint distribution (4.1) for equal group sizes with $\theta = 0$,

Table 4.1 *Type I error probability and power achieved by group sequential tests when group sizes are not equal to their design values. Tests are two-treatment comparisons with* $\sigma^2 = 4$ *and the total numbers of observations in the first k groups on each treatment arm are* $n_{Ak} = n_{Bk} = n_k$, $k = 1, \ldots, K$. *The first example for each test has group sizes equal to their design values.*

	$\{n_k;\ k = 1, \ldots, K\}$	*Type I error*	*Power at* $\mu_A - \mu_B = \pm 1$
Pocock test			
1	21, 42, 63, 84, 105	0.050	0.910
2	18, 36, 54, 72, 90	0.050	0.860
3	23, 46, 69, 92, 115	0.050	0.934
4	30, 50, 55, 86, 105	0.046	0.909
5	12, 31, 57, 81, 105	0.054	0.909
6	13, 42, 56, 78, 99	0.051	0.892
7	26, 40, 63, 96, 110	0.049	0.923
O'Brien & Fleming test			
	$\{n_k;\ k = 1, \ldots, K\}$	*Type I error*	*Power at* $\mu_A - \mu_B = \pm 1$
1	18, 36, 54, 72, 90	0.050	0.912
2	16, 32, 48, 64, 80	0.050	0.877
3	20, 40, 60, 80, 100	0.050	0.937
4	26, 39, 50, 76, 90	0.049	0.911
5	10, 27, 55, 66, 00	0.051	0.012
6	11, 38, 59, 65, 83	0.049	0.888
7	27, 40, 57, 73, 96	0.051	0.928
Wang & Tsiatis test, $\Delta = 0.25$			
	$\{n_k;\ k = 1, \ldots, K\}$	*Type I error*	*Power at* $\mu_A - \mu_B = \pm 1$
1	18, 36, 54, 72, 90	0.050	0.901
2	16, 32, 48, 64, 80	0.050	0.864
3	20, 40, 60, 80, 100	0.050	0.929
4	26, 39, 50, 76, 90	0.049	0.901
5	10, 27, 55, 66, 90	0.052	0.901
6	11, 38, 59, 65, 83	0.048	0.875
7	27, 40, 57, 73, 96	0.050	0.919

the Type I error rate is equal to 0.05; power is affected by the group sizes, increasing as n_K increases. In Examples 4 and 5, the final sample size is exactly as planned and power for each test is close to the value seen for Example 1. The unequal group sizes have a greater effect on the Type I error rate: if the cumulative sample sizes are bunched together, as in each Example 4, test statistics at the five analyses are more highly correlated and the Type I error rate decreases, whereas the patterns of cumulative sample sizes in each Example 5 reduce correlations and increase the Type I error rate. The effects on Type I error are smallest for the O'Brien & Fleming test in which values c_k are high for small k and there is little probability of rejecting H_0 at the early analyses. Finally, Examples 6 and 7 for each test show haphazard variations in group size; in each case the Type I error probability is very close to 0.05 and, since the final sample size is close to its design value, power is close to 0.9.

It is evident from (3.1) that the observed information levels $\{\mathcal{I}_1, \ldots, \mathcal{I}_K\}$ determine the joint distribution of $\{Z_1, \ldots, Z_K\}$ and, hence, a test's attained Type I error probability and power. Thus, the results of Table 4.1 also apply to other applications of the same tests when the same sequences $\{\mathcal{I}_1, \ldots, \mathcal{I}_5\}$ arise.

Table 4.2 shows further results from a systematic study of the effects of variations in observed information on attained error probabilities. Studies were designed for Pocock or O'Brien & Fleming tests with Type I error probability 0.05, power 0.9 at $\theta = \pm\delta$, and 2, 5 or 10 groups of observations. The target information levels are $\mathcal{I}_k = (k/K)\mathcal{I}_{max}$, $k = 1, \ldots, K$, where $\mathcal{I}_{max}$ is the multiple of $\mathcal{I}_{f,2}$ needed to meet the power requirement. Information levels actually observed are $\mathcal{I}_k = \pi(k/K)^r \mathcal{I}_{max}$, $k = 1, \ldots, K$. Since $\mathcal{I}_K = \pi\,\mathcal{I}_{max}$, π controls the final information level, and we see that this factor plays the major role in determining power. The attained power lies within 0.001 of its nominal level in all cases where $\pi = 1$ but power increases as π, and hence $\mathcal{I}_K$, increases: in all cases with $\pi = 0.9$, power is close to the value 0.868 attained by a fixed sample test with 0.9 times the required information, $\mathcal{I}_{f,2}$, and when $\pi = 1.1$, power is close to the value 0.925 achieved by a fixed sample test with information $1.1 \times \mathcal{I}_{f,2}$.

The exponent r governs the spacing of the $\mathcal{I}_k$: information levels are equally spaced if $r = 1$, bunched together toward $\mathcal{I}_{max}$ if $r < 1$, and more widely separated at the higher values of k if $r > 1$. As an example, values $\mathcal{I}_k = 10k$, $k = 1, \ldots, 5$, when $r = 1$ become $\mathcal{I}_k = \{14, 24, 33, 42, 50\}$ if $r = 0.8$, and $\mathcal{I}_k = \{7, 16, 26, 38, 50\}$ if $r = 1.25$. The Type I error probability is reduced below its nominal value for $r < 1$ and increased for $r > 1$, but these effects are slight, with the attained Type I error rate lying between 0.046 and 0.055 in all cases.

We conclude from these results that this approximate treatment of unequal group sizes will be satisfactory in many applications. The O'Brien & Fleming test and Wang & Tsiatis tests with low values of Δ are particularly insensitive to variations in early group sizes, as they allocate very little Type I error probability to the first few analyses. Proschan, Follmann & Waclawiw (1992) have investigated the maximum possible increase in Type I error probability under certain restrictions on the observed group sizes, finding this error rate to be robust to many types of departure from equal group sizes but capable of rising to unacceptably high levels in some extreme cases. Thus, if very unequal group sizes are anticipated, the alternative methods which we shall describe in Chapter 8 should be considered.

The tests we have described are in no way limited to the response types considered thus far. We illustrate some of the wide variety of experimental settings and data types to which they can be applied in the remainder of this chapter. In Section 4.3 we state the theoretical results which provide a basis for the group sequential analysis of general parametric models and we explain how these may be used to implement group sequential tests. We present tests for binary data in Section 4.4, introduce the group sequential log-rank test for comparing two survival distributions in Section 4.5, and describe group sequential t-tests in Section 4.6.

Table 4.2 *Properties of two-sided tests designed assuming information levels* $\mathcal{I}_k = (k/K)\mathcal{I}_{max}$, $k = 1, \ldots, K$, *but applied with* $\mathcal{I}_k = \pi(k/K)^r \mathcal{I}_{max}$, $k = 1, \ldots, K$. *Attained Type I error probabilities and power are shown for Pocock and O'Brien & Fleming tests with* K *analyses, designed to achieve Type I error rate* $\alpha = 0.05$ *and power* $1 - \beta = 0.9$ *at* $\theta = \pm\delta$.

			Pocock tests		*O'Brien & Fleming*	
	r	π	*Type I error*	*Power*	*Type I error*	*Power*
$K = 2$						
	0.80	0.90	0.048	0.868	0.050	0.867
		1.00	0.048	0.901	0.050	0.900
		1.10	0.048	0.926	0.050	0.925
	1.00	0.90	0.050	0.867	0.050	0.868
		1.00	0.050	0.900	0.050	0.900
		1.10	0.050	0.926	0.050	0.925
	1.25	0.90	0.052	0.865	0.051	0.868
		1.00	0.052	0.899	0.051	0.900
		1.10	0.052	0.925	0.051	0.925
$K = 5$						
	0.80	0.90	0.047	0.867	0.048	0.866
		1.00	0.047	0.901	0.048	0.899
		1.10	0.047	0.927	0.048	0.925
	1.00	0.90	0.050	0.866	0.050	0.867
		1.00	0.050	0.900	0.050	0.900
		1.10	0.050	0.927	0.050	0.925
	1.25	0.90	0.053	0.864	0.052	0.868
		1.00	0.053	0.899	0.052	0.900
		1.10	0.053	0.926	0.052	0.926
$K = 10$						
	0.80	0.90	0.046	0.866	0.047	0.865
		1.00	0.046	0.901	0.047	0.899
		1.10	0.046	0.928	0.047	0.924
	1.00	0.90	0.050	0.865	0.050	0.867
		1.00	0.050	0.900	0.050	0.900
		1.10	0.050	0.927	0.050	0.925
	1.25	0.90	0.055	0.863	0.053	0.868
		1.00	0.055	0.899	0.053	0.901
		1.10	0.055	0.926	0.053	0.926

4.3 Other Parametric Models

4.3.1 General Theory

In Section 15.6 we consider the group sequential analysis of observations whose distribution depends on a parameter vector $\boldsymbol{\beta}$ for which maximum likelihood estimates are obtained at analyses $k = 1, \ldots, K$. There we state an asymptotic result proved by Jennison & Turnbull (1997a) that, under certain regularity conditions concerning the statistical model and the covariate values for observed data, the sequence of estimates $\{\widehat{\boldsymbol{\beta}}^{(1)}, \ldots, \widehat{\boldsymbol{\beta}}^{(K)}\}$ is approximately multivariate normal with

$$E(\widehat{\boldsymbol{\beta}}^{(k)}) = \boldsymbol{\beta}, \quad k = 1, \ldots, K,$$

and

$$\mathbf{Cov}(\widehat{\boldsymbol{\beta}}^{(k_1)}, \widehat{\boldsymbol{\beta}}^{(k_2)}) = \mathbf{Var}(\widehat{\boldsymbol{\beta}}^{(k_2)}), \quad 1 \le k_1 \le k_2 \le K,$$

when sample sizes are sufficiently large. Thus, the joint distribution of successive estimates of the parameter vector matches that seen for normal linear models in Section 3.3. Furthermore, the inverse of the Fisher information matrix for $\boldsymbol{\beta}$ at analysis k provides a consistent estimate of $\mathbf{Var}(\widehat{\boldsymbol{\beta}}^{(k)})$.

Precise conditions under which this theory applies depend on the particular statistical model in question, but in all cases they reduce to the conditions required to derive standard maximum likelihood theory at each individual analysis. Thus, the scope of the group sequential theory for parametric models is as broad as that of the underlying fixed sample maximum likelihood theory. The sample size necessary for these results to hold accurately is problem specific, and in a group sequential analysis it is advisable to ensure the first group of observations is sufficiently large that the asymptotic approximation holds adequately at the first analysis.

We illustrate the general theory with two simple applications to binary response data in Section 4.4. Many further applications arise in the group sequential analysis of generalized linear models, including the important examples of logistic regression for a binary response (see Section 16.6 for an example), probit regression, multiplicative models for Poisson count data, and normal linear models with non-linear link functions.

4.3.2 Implementing Two-sided Group Sequential Tests

Suppose it is required to test the hypothesis H_0: $\boldsymbol{c}^T\boldsymbol{\beta} = \gamma$ with two-sided Type I error probability α and power $1-\beta$ at $\boldsymbol{c}^T\boldsymbol{\beta} = \gamma \pm \delta$. We assume that at each analysis $k = 1, \ldots, K$, we have an estimate $\widehat{\boldsymbol{\beta}}^{(k)}$ and an estimated variance matrix $\widehat{\mathbf{Var}}(\widehat{\boldsymbol{\beta}}^{(k)})$. Many statistical computer packages routinely produce such estimates after fitting a model to data, and it is straightforward to use output from such packages for group sequential testing. Our estimate of $\theta = \boldsymbol{c}^T\boldsymbol{\beta} - \gamma$ at analysis k is $\hat{\theta}^{(k)} = \boldsymbol{c}^T\widehat{\boldsymbol{\beta}}^{(k)} - \gamma$, and we approximate its distribution as

$$\hat{\theta}^{(k)} \sim N(\theta, \boldsymbol{c}^T\widehat{\mathbf{Var}}(\widehat{\boldsymbol{\beta}}^{(k)})\boldsymbol{c}).$$

Thus, the estimated information for θ is $\mathcal{I}_k = \{\boldsymbol{c}^T\widehat{\mathbf{Var}}(\widehat{\boldsymbol{\beta}}^{(k)})\boldsymbol{c}\}^{-1}$ and the standardized test statistic $Z_k = \hat{\theta}^{(k)}\sqrt{\mathcal{I}_k}$ is distributed, approximately, as $N(\theta\sqrt{\mathcal{I}_k}, 1)$.

It follows from the joint distribution of $\{\widehat{\boldsymbol{\beta}}^{(1)}, \ldots, \widehat{\boldsymbol{\beta}}^{(K)}\}$ that the statistics $\{Z_1, \ldots, Z_K\}$ have the canonical joint distribution (3.1) with information levels $\{\mathcal{I}_1, \ldots, \mathcal{I}_K\}$ for θ. We can therefore obtain a group sequential test with the required properties by following the five steps listed in Section 4.1.2 in the usual way.

In order to attain its desired power, a study must be designed to yield the appropriate sequence of information levels $\mathcal{I}_k$, where, as specified in (3.2),

$$\mathcal{I}_k = \{\mathrm{Var}(\hat{\theta}^{(k)})\}^{-1}, \quad k = 1, \ldots, K.$$

Achieving this requires an understanding of the relationship between information and sample size for the assumed statistical model. The following sections address this issue for tests based on binary data, survival data, and normal observations with an unknown variance. In general, formulae or tables that provide sample sizes needed to attain specified Type I error probability and power in a fixed sample trial can be used in planning a group sequential study. Assuming information is proportional to sample size, one sets the maximum sample size to be the corresponding fixed sample size multiplied by the factor R, i.e., one of the constants $R_P(K, \alpha, \beta)$, $R_B(K, \alpha, \beta)$, etc., as appropriate to the form of group sequential test being used. If analyses are arranged so that $\mathcal{I}_K = R\mathcal{I}_{f,2}$, where $\mathcal{I}_{f,2}$ is given by (4.6), and $\mathcal{I}_1, \ldots, \mathcal{I}_K$ are equally spaced, then the target Type I error and power will be attained accurately. Deviations from the desired pattern of information levels will have roughly the same effect as we saw for normal data in Section 4.2. If it is particularly difficult to control information levels, the more flexible methods of Chapter 8 may be preferred. These methods can also be used to adapt the maximum sample size or the timing of interim analyses as one learns about the variance of a normal response or the event rate in study with a time-to-event endpoint.

4.4 Binary Data: Group Sequential Tests for Proportions

4.4.1 The One-sample Problem

Binary data arise when each observation can take one of two possible values. An example of a binary outcome is whether or not a subject responds to a given treatment; another is whether a patient experiences severely toxic side-effects. Coding responses as $X = 0$ or $X = 1$, we obtain a Bernoulli distribution, denoted $B(p)$, in which the value one occurs with probability p and the value zero with probability $1 - p$. We first consider the case of a single Bernoulli distribution where p is to be compared with a standard value by testing the null hypothesis H_0: $p = p_0$.

Let X_1, X_2, ... denote a sequence of $B(p)$ observations. If data are collected group sequentially with total numbers of observations $n_1, \ldots, n_K$ at analyses 1 to K, the usual estimate of p at analysis k,

$$\hat{p}^{(k)} = \frac{1}{n_k} \sum_{i=1}^{n_k} X_i,$$

has expectation p and variance $p(1-p)/n_k$. The general theory discussed in Section 4.3.1 implies that the sequence of estimates $\{\hat{p}^{(1)}, \ldots, \hat{p}^{(K)}\}$ is approximately multivariate normal with

$$\mathrm{Cov}(\hat{p}^{(k_1)}, \hat{p}^{(k_2)}) = \mathrm{Var}(\hat{p}^{(k_2)}), \quad 1 \le k_1 \le k_2 \le K.$$

If, also, p is close to p_0,

$$\mathrm{Var}(\hat{p}^{(k)}) \simeq p_0(1-p_0)/n_k, \quad k = 1, \ldots, K.$$

Let $\theta = p - p_0$. We define the information for θ at analysis k to be the reciprocal of the null variance of $\hat{p}^{(k)}$, $\mathcal{I}_k = n_k/\{p_0(1-p_0)\}$, and the standardized statistics are $Z_k = (\hat{p}^{(k)} - p_0)\sqrt{\mathcal{I}_k}$, $k = 1, \ldots, K$. Then, for p close to p_0, the sequence $\{Z_1, \ldots, Z_K\}$ has, approximately, the canonical joint distribution (3.1) with information levels $\{\mathcal{I}_1, \ldots, \mathcal{I}_K\}$ for θ. Hence, the group sequential tests of Chapter 2 can be applied directly to create a group sequential two-sided test of H_0.

As an example, suppose it is desired to test H_0: $p = 0.6$ against the two-sided alternative $p \neq 0.6$ with Type I error rate $\alpha = 0.05$ and power $1 - \beta = 0.9$ at $p = 0.4$ and $p = 0.8$. We shall do this using a Pocock test with four groups of observations. Using (4.6) with $\delta = 0.2$, we find the information for p required by a fixed sample test with these error probabilities to be

$$\mathcal{I}_{f,2} = \{\Phi^{-1}(0.975) + \Phi^{-1}(0.9)\}^2/0.2^2 = 262.7,$$

and hence the group sequential test's final information level should be

$$R_P(4, 0.05, 0.1) \times 262.7 = 1.183 \times 262.7 = 310.8.$$

Since $\mathcal{I}_k = n_k/\{p_0(1-p_0)\} = n_k/(0.6 \times 0.4)$, we solve $n_4/0.24 = 310.8$ to obtain $n_4 = 74.6$, and rounding this to a multiple of four gives the sample size requirement $n_4 = 76$, i.e., four groups of 19 observations. The standardized test statistics are

$$Z_k = (\hat{p}^{(k)} - 0.6)\sqrt{\mathcal{I}_k} = (\hat{p}^{(k)} - 0.6)\sqrt{(19\,k/0.24)}, \quad k = 1, \ldots, 4,$$

and we reject H_0 at analysis k if $|Z_k| \geq C_P(4, 0.05) = 2.361$, i.e., if

$$|\hat{p}^{(k)} - 0.6| \geq 0.265/\sqrt{k}, \quad k = 1, \ldots, 4.$$

If H_0 has not been rejected by the fourth analysis, it is accepted.

We shall return to this example in Section 16.1. One point that merits further discussion is the dependence of $\mathrm{Var}(\hat{p}^{(k)})$, and hence $\mathcal{I}_k$, on the value of p. In the above treatment, we used the approximation

$$\mathrm{Var}(\hat{p}^{(k)}) = p_0(1 - p_0)/n_k = 0.24/n_k$$

in both Type I error and power calculations, whereas the lower variance of $0.16/n_k$ when $p = 0.8$ actually leads to greater power at this alternative than the stated value of 0.9. Another feature of binary data is their discreteness, and this is not well modeled by a normal approximation when sample sizes are small. We shall show in Chapter 16 that exact calculation of properties of group sequential tests for binary responses is quite straightforward and this can be used to fine-tune tests obtained under normal approximations.

4.4.2 Comparing Two Proportions

We now suppose observations X_{Ai} and X_{Bi} are Bernoulli responses of subjects receiving two different treatments with $X_{Ai} \sim B(p_A)$ and $X_{Bi} \sim B(p_B)$, $i = 1, 2, \ldots$. We shall consider tests of the null hypothesis of no treatment difference, H_0: $p_A = p_B$, when data are collected group sequentially.

Denote the total numbers of observations on treatments A and B at analyses $k = 1, \ldots, K$ by n_{Ak} and n_{Bk}, respectively. Let $\bar{p} = (p_A + p_B)/2$ and define the estimates of p_A and p_B,

$$\hat{p}_A^{(k)} = \frac{1}{n_{Ak}} \sum_{i=1}^{n_{Ak}} X_{Ai} \quad \text{and} \quad \hat{p}_B^{(k)} = \frac{1}{n_{Bk}} \sum_{i=1}^{n_{Bk}} X_{Bi}, \quad k = 1, \ldots, K.$$

Under H_0, $p_A = p_B = \bar{p}$ and the information for $p_B - p_A$ at analysis k is the reciprocal of $\mathrm{Var}(\hat{p}_B^{(k)} - \hat{p}_A^{(k)})$,

$$\mathcal{I}_k = \{\bar{p}(1 - \bar{p})(n_{Ak}^{-1} + n_{Bk}^{-1})\}^{-1}.$$

Estimating the common response probability $\bar{p}$ under H_0 by

$$\tilde{p}_k = \frac{\sum_{i=1}^{n_{Ak}} X_{Ai} + \sum_{i=1}^{n_{Bk}} X_{Bi}}{n_{Ak} + n_{Bk}},$$

we obtain the estimated information levels

$$\widehat{\mathcal{I}}_k = \{\tilde{p}_k(1-\tilde{p}_k)(n_{Ak}^{-1}+n_{Bk}^{-1})\}^{-1} \qquad (4.7)$$

at analyses $k = 1, \ldots, K$.

We base our test of H_0 on the standardized statistics

$$Z_k = (\hat{p}_B^{(k)} - \hat{p}_A^{(k)})\sqrt{\widehat{\mathcal{I}}_k}, \quad k = 1, \ldots, K.$$

The general theory of Section 4.3.1 implies that, if $p_B - p_A$ is small, the sequence of statistics $\{Z_1, \ldots, Z_K\}$ follows, approximately, the canonical joint distribution (3.1) with information levels $\{\widehat{\mathcal{I}}_1, \ldots, \widehat{\mathcal{I}}_K\}$ for $\theta = p_B - p_A$. Hence, we can apply any of the group sequential tests of Chapter 2.

As an example, suppose we wish to test H_0: $p_A = p_B$ against a two-sided alternative with Type I error probability $\alpha = 0.05$ and power $1 - \beta = 0.8$ at $|p_B - p_A| = 0.2$ using an O'Brien & Fleming test with eight groups of observations. From (4.6), the information required by a fixed sample test with these error probabilities is

$$\mathcal{I}_{f,2} = \{\Phi^{-1}(0.975) + \Phi^{-1}(0.8)\}^2/0.2^2 = 196.2,$$

and the final information level needed by the group sequential test is, therefore,

$$R_B(8, 0.05, 0.2) \times 196.2 = 1.037 \times 196.2 = 203.5.$$

It is evident from (4.7) that information depends on the value of $\bar{p}$, which is unknown at the design stage. However, since $\mathcal{I}_k$ varies slowly as a function of $\bar{p}$ for values away from zero and 1, a highly accurate estimate of $\bar{p}$ is not usually necessary. We shall continue this example assuming the worst case value, $\bar{p} = 0.5$, so any discrepancy will be in the direction of too large a group size and attained power greater than stipulated. Under this assumption, and supposing $n_{Ak} = n_{Bk}$ for each $k = 1, \ldots, 8$, we find

$$\mathcal{I}_8 = \{0.25 \times 2/n_8\}^{-1} = 2n_8,$$

where n_8 denotes the common value of n_{A8} and n_{B8}. Solving this equation with $\mathcal{I}_8 = 203.5$ gives $n_8 = 101.7$, which we round to 104 to obtain a multiple of 8, and we see that eight groups of 13 observations per treatment should be planned.

The standardized test statistic at analysis k is then

$$Z_k = (\hat{p}_B^{(k)} - \hat{p}_A^{(k)})\sqrt{\widehat{\mathcal{I}}_k} = (\hat{p}_B^{(k)} - \hat{p}_A^{(k)})\sqrt{\{6.5\,k/(\tilde{p}_k(1-\tilde{p}_k))\}},$$

and H_0 is rejected at analysis k if

$$|Z_k| \geq C_B(8, 0.05)\sqrt{(8/k)} = 2.072\sqrt{(8/k)},$$

i.e., if

$$|\hat{p}_B^{(k)} - \hat{p}_A^{(k)}| \geq \frac{2.30\sqrt{\{\tilde{p}_k(1-\tilde{p}_k)\}}}{k}.$$

If H_0 has not been rejected by the eighth analysis, it is accepted.

As for the one-sample problem, we defer discussion of the variance-mean relationship and the possibility of exact calculation of a test's properties until Chapter 16. Another issue, which we shall return to in Section 18.4, is the adaptive re-computing of sample sizes in response to increasingly accurate estimates of $\bar{p}$. The important conclusion for the moment is that the general theory for parametric models is easily applied to specific non-normal data types.

4.5 The Group Sequential Log-rank Test for Survival Data

The most important outcome in many clinical trials is the length of time from treatment to an event such as death, relapse or remission. Suppose the primary outcome is time until death, then survival time is right censored if a subject is alive at the time of an analysis since the subject's ultimate survival time is still unknown. In group sequential studies, this effect is seen repeatedly at successive analyses, but with the length of follow-up on each patient increasing as the study progresses. Follow-up also varies between patients according to their time of entry to the study. A separate form of "competing risk" censoring can occur when patients are lost to follow-up, for example, if a subject moves to another part of the country and cannot be traced by the investigators.

A key element of survival analysis is the hazard rate of a distribution. Let T be a continuous, positive random variable with probability density function $f(t)$, $t \geq 0$. The corresponding survival function is defined as

$$S(t) = Pr\{T > t\} = \int_t^{\infty} f(u)du, \quad t \geq 0,$$

and the hazard rate is

$$h(t) = \frac{f(t)}{S(t)}, \quad t \geq 0.$$

Note that $h(t)$ can be interpreted as the limit, as δt tends to zero, of

$$\frac{1}{\delta t} Pr\{T \in (t, t + \delta t) \,|\, T > t\}.$$

A convenient assumption in modeling survival data is that the hazard rates for different observations are affected multiplicatively by explanatory variables. In the proportional hazards model for a two-treatment comparison, we assume the hazard rate is $h(t)$ for subjects receiving treatment A and $\lambda h(t)$ for those receiving treatment B. Interest then focuses on the hazard ratio, λ, while the unknown function $h(t)$ is usually of secondary interest. The "log-rank" test of Mantel (1966) and Peto & Peto (1972) is commonly used to test the null hypothesis of no difference in survival between two treatments, which occurs when $\lambda = 1$.

In a group sequential study, let d_k, $k = 1, \ldots, K$, denote the total number of uncensored failures observed when analysis k is conducted. Assuming no ties, we denote the survival times of these subjects by $\tau_{1,k} < \tau_{2,k} < \ldots < \tau_{d_k,k}$, where the $\tau_{i,k}$ represent elapsed times between entry to the study and failure. At analysis k, let $r_{iA,k}$ and $r_{iB,k}$ be the numbers of patients on treatments A and B, respectively, who are known to have survived up to time $\tau_{i,k}$ since entering the study. Then, the log-rank score statistic at analysis k is

$$S_k = \sum_{i=1}^{d_k} \left\{ \delta_{iB,k} - \frac{r_{iB,k}}{r_{iA,k} + r_{iB,k}} \right\}, \tag{4.8}$$

where $\delta_{iB,k} = 1$ if the failure at time $\tau_{i,k}$ was on treatment B and $\delta_{iB,k} = 0$ otherwise.

The variance of S_k when $\lambda = 1$ can be approximated by the sum of the conditional variances of the $\delta_{iB,k}$ given the numbers $r_{iA,k}$ and $r_{iB,k}$ at risk on each treatment just before time $\tau_{i,k}$. This variance is also the information for $\theta = \log(\lambda)$ when $\lambda = 1$, so we have

$$\mathcal{I}_k = \widehat{Var}(S_k) = \sum_{i=1}^{d_k} \frac{r_{iA,k}\, r_{iB,k}}{(r_{iA,k} + r_{iB,k})^2}. \tag{4.9}$$

For λ close to unity and, thus, $\theta = \log(\lambda)$ close to zero, we can use the approximation $S_k \sim N(\theta\mathcal{I}_k, \mathcal{I}_k)$, given a sufficiently large $\mathcal{I}_k$. Furthermore, conditional on the observed information sequence $\{\mathcal{I}_1, \ldots, \mathcal{I}_K\}$, the joint distribution of $\{S_1, \ldots, S_K\}$ approximates the standard form for a sequence of score statistics given in (3.2). Alternatively, defining estimates $\hat{\theta}^{(k)} = S_k/\mathcal{I}_k$ or standardized statistics $Z_k = S_k/\sqrt{\mathcal{I}_k}$, $k = 1, \ldots, K$, yields sequences $\{\hat{\theta}^{(1)}, \ldots, \hat{\theta}^{(K)}\}$ or $\{Z_1, \ldots, Z_K\}$ with the standard joint distributions, also stated in (3.2). Whichever sequence of statistics one chooses to work with, a group sequential test can be developed in the usual way. These joint distributions of sequences of log-rank and associated statistics have been established asymptotically by Harrington, Fleming & Green (1982) and Tsiatis (1982a), and their small sample accuracy is demonstrated in simulation studies by Gail, DeMets & Slud (1982), Jennison & Turnbull (1984), DeMets & Gail (1985) and Jennison (1992).

As an illustrative example, consider the design of an O'Brien & Fleming test of $H_0\colon \lambda = 1$ with five analyses, to achieve two-sided Type I error probability 0.05 and power 0.8 at $\lambda = 1.5$ or 0.667, i.e., at $\theta = \pm 0.405$. Following the general approach laid out in Sections 4.1 and 4.2, we first find the information for θ required by the corresponding fixed sample test,

$$\mathcal{I}_{f,2} = \{\Phi^{-1}(0.975) + \Phi^{-1}(0.8)\}^2/0.405^2 = 47.85.$$

Hence, the target for the final information level is

$$47.85 \times R_B(5, 0.05, 0.2) = 47.85 \times 1.028 = 49.19,$$

where the value 1.028 of $R_B(5, 0.05, 0.2)$ is taken from Table 2.4. The study should therefore be designed to produce observed information levels as close as possible to the values $\mathcal{I}_k = 49.19(k/5)$, $k = 1, \ldots, 5$. In practice, it will be difficult to attain these levels accurately since observed information depends on the realized pattern of deaths and censoring times. However, if the numbers at risk in each treatment group remain nearly equal over time, as is to be expected if λ is close to unity, then $r_{iA,k} \approx r_{iB,k}$ for each i and we can use the approximation $\mathcal{I}_k \approx d_k/4$. Thus, a total of about $4 \times 49.19 \approx 197$ deaths is needed by the end of the study, and the sample size for each treatment arm and study duration should be chosen to achieve this, bearing in mind the likely rates of competing risk and end-of-study censoring.

Implementation of the O'Brien & Fleming test is straightforward. Taking the constant $C_B(5, 0.05) = 2.040$ from Table 2.3, we see that the test should stop with rejection of H_0 at analysis k if

$$|Z_k| = |S_k/\sqrt{\mathcal{I}_k}| \geq 2.040\sqrt{(5/k)}, \quad k = 1, \ldots, 5,$$

and H_0 is accepted if it has not been rejected by the fifth analysis. As explained in Section 4.2, the Type I error probability will be close to 0.05 as long as the information levels $\mathcal{I}_1, \ldots, \mathcal{I}_5$ are approximately equally spaced, but the attained power depends primarily on the final information level $\mathcal{I}_5$.

When a study's monitoring committee is scheduled to meet at fixed calendar times, it is quite possible that increases in information levels between analyses will be highly uneven. In this case the "error spending" approach which we present in Chapter 8 offers an attractive alternative to the significance level method we have described here since it yields tests that satisfy the Type I error requirement precisely, conditional on the observed sequence of information levels. We shall illustrate the application of an error spending test to survival data in Section 8.6.4 and discuss methods for survival endpoints more generally in Chapter 17.

4.6 Group Sequential *t*-Tests

4.6.1 Implementation

When observations follow a normal linear model, it is usually necessary to estimate the variance, σ^2, as well as the parameter vector $\boldsymbol{\beta}$. In the setting described in Section 3.3, estimates $\widehat{\boldsymbol{\beta}}^{(1)}, \ldots, \widehat{\boldsymbol{\beta}}^{(K)}$ are obtained at analyses 1 to K. Let $\boldsymbol{V}_1, \ldots, \boldsymbol{V}_K$ be such that $\mathbf{Var}(\widehat{\boldsymbol{\beta}}^{(k)}) = \boldsymbol{V}_k \sigma^2$ and denote by s_k^2 the usual unbiased estimates of σ^2 at analyses $k = 1, \ldots, K$. Then, the s_k^2 follow $\sigma^2 \chi^2_{n_k - p}/(n_k - p)$ marginal distributions. At each analysis $k = 1, \ldots, K$, the estimate of $\boldsymbol{c}^T \boldsymbol{\beta}$ is

$$\boldsymbol{c}^T \widehat{\boldsymbol{\beta}}^{(k)} \sim N(\boldsymbol{c}^T \boldsymbol{\beta}, \boldsymbol{c}^T \boldsymbol{V}_k \boldsymbol{c} \sigma^2),$$

the Z-statistic for testing H_0: $\boldsymbol{c}^T \boldsymbol{\beta} = \gamma$ when σ^2 is known is

$$Z_k = \frac{\boldsymbol{c}^T \widehat{\boldsymbol{\beta}}^{(k)} - \gamma}{\sqrt{\{\boldsymbol{c}^T \boldsymbol{V}_k \boldsymbol{c} \sigma^2\}}},$$

and the t-statistic for testing H_0 is

$$T_k = \frac{\boldsymbol{c}^T \widehat{\boldsymbol{\beta}}^{(k)} - \gamma}{\sqrt{\{\boldsymbol{c}^T \boldsymbol{V}_k \boldsymbol{c}\, s_k^2\}}}. \tag{4.10}$$

Suppose an investigator wishes to test the hypothesis H_0: $\boldsymbol{c}^T \boldsymbol{\beta} = \gamma$ in a two-sided group sequential test with Type I error probability α. In adapting one of the known variance tests of Chapter 2 to the unknown variance problem, we follow the approach suggested by Pocock (1977). We take the two-sided significance levels, $2\{1 - \Phi(c_k)\}$, defined for the Z-statistics (here Φ denotes the standard normal cumulative distribution function) but apply these to the t-statistics $T_1, \ldots, T_K$. Since T_k has a marginal $t_{n_k - p}$ distribution, we reject H_0 at analysis k if

$$|T_k| \geq t_{n_k - p, 1 - \Phi(c_k)}, \quad k = 1, \ldots, K,$$

where $t_{\nu,q}$ denotes the upper q tail-point of a t-distribution on ν degrees of freedom, that is $Pr\{T > t_{\nu,q}\} = q$ when $T \sim t_\nu$. If H_0 has not been rejected by analysis K, it is accepted.

Here $\{c_1, \ldots, c_K\}$ can be any of the sequences of critical values from Chapter 2. Since these were derived for a sequence of Z-statistics, this form of test satisfies the Type I error requirement only approximately. Nevertheless, investigations have shown that this approximation is remarkably accurate.

Guaranteeing power at a specified value of $\boldsymbol{c}^T \boldsymbol{\beta} - \gamma$ is more difficult since power depends on the unknown value of σ^2, and so one is obliged to design a study choosing group sizes on the basis of an estimated value of σ^2. However, we can find a good approximation to this power for any given value of σ^2 and, hence, check that adequate power will be achieved at a stated alternative $\boldsymbol{c}^T \boldsymbol{\beta} - \gamma = \pm\delta$ if σ^2 lies in some plausible range.

The information for $\boldsymbol{c}^T \boldsymbol{\beta}$ at the final analysis is

$$\{\mathrm{Var}(\boldsymbol{c}^T \widehat{\boldsymbol{\beta}}^{(K)})\}^{-1} = (\boldsymbol{c}^T \boldsymbol{V}_K \boldsymbol{c} \sigma^2)^{-1}.$$

By analogy with the case of known variance, we expect the group sequential t-test's power at a given alternative to be approximately that of a fixed sample t-test with $1/R(K, \alpha, \beta)$ times this information for $\boldsymbol{c}^T \boldsymbol{\beta}$, where $R(K, \alpha, \beta)$ is the appropriate sample size factor for the test as tabulated in Chapter 2. Thus, we need to find the power of a fixed sample test with estimator $\widehat{\boldsymbol{\beta}}$ for which

$$\mathrm{Var}(\boldsymbol{c}^T \widehat{\boldsymbol{\beta}}) = R(K, \alpha, \beta)\, \mathrm{Var}(\boldsymbol{c}^T \widehat{\boldsymbol{\beta}}^{(K)}).$$

If there are $n_K - p$ degrees of freedom to estimate σ^2 at analysis K, the fixed sample test statistic with these degrees of freedom has a t_{n_K-p} distribution under H_0 and when $\boldsymbol{c}^T\boldsymbol{\beta} - \gamma = \delta$ it has a

$$T(n_K - p; \delta/\sqrt{\{R(K,\alpha,\beta)\,\boldsymbol{c}^T\boldsymbol{V}_K\boldsymbol{c}\,\sigma^2\}})$$

distribution, where $T(\nu;\xi)$ denotes a noncentral t-distribution with ν degrees of freedom and noncentrality parameter ξ. The probability of rejecting H_0 with a large, negative t-statistic is negligible when $\boldsymbol{c}^T\boldsymbol{\beta} - \gamma = \delta > 0$. So, this fixed sample test has power

$$Pr\{T(n_K - p; \delta/\sqrt{\{R(K,\alpha,\beta)\,\boldsymbol{c}^T\boldsymbol{V}_K\boldsymbol{c}\,\sigma^2\}}) \geq t_{n_K-p,\alpha/2}\}, \tag{4.11}$$

and this is our approximation to the power of the group sequential t-test when $\boldsymbol{c}^T\boldsymbol{\beta} - \gamma = \delta$ and the variance is σ^2. By symmetry, the power at $\boldsymbol{c}^T\boldsymbol{\beta} - \gamma = -\delta$ is the same. If $n_K - p$ is large, we may choose to disregard the error in estimating σ^2, leading to the cruder but simpler estimate of power

$$\Phi\left(\frac{\delta}{\sqrt{\{R(K,\alpha,\beta)\,\boldsymbol{c}^T\boldsymbol{V}_K\boldsymbol{c}\,\sigma^2\}}} - \Phi^{-1}(1-\alpha/2)\right). \tag{4.12}$$

Since the factor $R(K,\alpha,\beta)$ depends on the power $1-\beta$, which we are in the process of calculating, these definitions appear to be somewhat circular. However, it is evident from Tables 2.2, 2.4 and 2.10 that the factors $R_P(K,\alpha,\beta)$, $R_B(K,\alpha,\beta)$ and $R_{WT}(K,\alpha,\beta)$ vary only slightly with β, and it should suffice to work with a factor appropriate to a plausible value of the power, $1-\beta$. Alternatively, the calculation can be repeated a second time, using the factor R appropriate to the power computed in a first iteration.

Table 4.3 shows the Type I error rate and power attained by group sequential t-tests defined in the above manner in situations where formula (4.11) gives the value 0.8. Thus, in each example the value of δ is implicitly determined through the noncentrality parameter in (4.11). Of course, the test cannot be designed to attain power precisely without knowledge of σ^2, but the results do demonstrate the accuracy of (4.11) to assess the power that will be attained under plausible values of σ^2. The two values of ν_K for each combination of type of test and number of groups, K, arise when groups are of 3 and 5 observations, respectively, and two degrees of freedom are lost in fitting model parameters at each analysis. It follows from the general results we shall present in Section 15.4 that the null joint distribution of the sequence of t-statistics, $T_1,\ldots,T_K$, is the same for all tests of H_0: $\boldsymbol{c}^T\boldsymbol{\beta} = \gamma$ with the same error degrees of freedom, $\nu_1,\ldots,\nu_K$, and $(\boldsymbol{c}^T\boldsymbol{V}_k\boldsymbol{c})^{-1} \propto k$, $k = 1,\ldots,K$; hence, the tabulated Type I errors apply with the same generality. Further, the joint distribution of $T_1,\ldots,T_K$ at an alternative $\boldsymbol{c}^T\boldsymbol{\beta} - \gamma = \delta$ depends only on the value of $\delta/\sqrt{\{\boldsymbol{c}^T\boldsymbol{V}_K\boldsymbol{c}\,\sigma^2\}}$, so the attained powers reported in Table 4.3 also apply to a general class of problems.

Departures from the desired Type I error rates are quite minor and most substantial when the degrees of freedom for σ^2 are low. It is to be expected that the greatest discrepancies should occur for the Pocock tests since these afford the most opportunity for stopping at very early analyses when degrees of freedom may be very low. Note that if a precise Type I error probability is required, a test with a particular boundary shape can be constructed using numerically exact computations; see Sections 15.4 and 31.6.2 and Jennison & Turnbull (1991b). Formula (4.11) predicts power of 0.8 for all these examples, so we can see this formula provides an accurate estimate of power as long as the degrees of freedom are reasonably high but it does over-estimate power when degrees of freedom are low. This discrepancy is mainly attributable to the use of ν_K in setting the target information levels, whereas t-tests at early analyses have fewer degrees of freedom than this.

The final column of Table 4.3 shows the approximation to the power obtained using formula (4.12). Not surprisingly, this simpler formula gives good results when degrees of freedom are high but is unreliable when the final degrees of freedom are as low as 10 or 20.

Table 4.3 *Properties of group sequential t-tests designed to test H_0: $c^T\beta = \gamma$ with two-sided Type I error probability $\alpha = 0.05$. Tests are based on Pocock, O'Brien & Fleming, and Wang & Tsiatis tests for known variance. Degrees of freedom for σ^2 at analysis k are $\nu_k = (\nu_K + 2)(k/K) - 2$ and $Var(c^T\hat{\beta}^{(k)}) = (K/k)Var(c^T\hat{\beta}^{(K)})$, $k = 1, \ldots, K$. The penultimate column shows actual power at the alternative for which the approximation (4.11) gives power 0.8 and the final column shows the power predicted by the formula (4.12).*
Type I error probability and power were estimated by simulation, using 1 million replicates; standard errors are 0.0002 for Type I error and 0.0004 for power.

Pocock test

K	ν_K	*Type I error*	*Attained power*	*Known σ^2 power*
3	7	0.058	0.744	0.905
3	13	0.054	0.772	0.858
5	13	0.058	0.762	0.858
5	23	0.055	0.779	0.833
8	22	0.057	0.772	0.834
8	38	0.055	0.785	0.820

O'Brien & Fleming test

K	ν_K	*Type I error*	*Attained power*	*Known σ^2 power*
3	7	0.054	0.795	0.905
3	13	0.052	0.798	0.858
5	13	0.054	0.796	0.858
5	23	0.052	0.798	0.833
8	22	0.053	0.798	0.834
8	38	0.052	0.800	0.820

Wang & Tsiatis test, $\Delta = 0.25$

K	ν_K	*Type I error*	*Attained power*	*Known σ^2 power*
3	7	0.056	0.783	0.905
3	13	0.054	0.791	0.858
5	13	0.056	0.789	0.858
5	23	0.053	0.794	0.833
8	22	0.055	0.793	0.834
8	38	0.053	0.795	0.820

4.6.2 *Examples*

4.6.2.1 *Two-treatment Comparison*

To illustrate the above methods, consider the two-treatment comparison of Chapter 2 but now with σ^2 unknown. Suppose observations $X_{Ai} \sim N(\mu_A, \sigma^2)$ and $X_{Bi} \sim N(\mu_B, \sigma^2)$, $i = 1, 2, \ldots$, and it is required to test H_0: $\mu_A = \mu_B$ against the two-sided alternative $\mu_A \neq \mu_B$ with Type I error probability $\alpha = 0.01$ using an O'Brien & Fleming test with four groups of observations. Supposing each group contains m observations per treatment, the t-statistics at successive analyses are

$$T_k = \frac{\sum_{i=1}^{mk} X_{Ai} - \sum_{i=1}^{mk} X_{Bi}}{\sqrt{2mks_k^2}}, \quad k = 1, \ldots, 4,$$

where

$$s_k^2 = \frac{\sum_{i=1}^{mk}(X_{Ai} - \bar{X}_A^{(k)})^2 + \sum_{i=1}^{mk}(X_{Bi} - \bar{X}_B^{(k)})^2}{2(mk-1)}$$

and $\bar{X}_A^{(k)}$ and $\bar{X}_B^{(k)}$ denote the means of $X_{A1}, \ldots, X_{A,mk}$ and $X_{B1}, \ldots, X_{B,mk}$, respectively. From Table 2.3, we have $C_B(4, 0.01) = 2.609$, so $c_k = 2.609\sqrt{(4/k)} = 5.218/\sqrt{k}$ for $k = 1, \ldots, 4$. Thus, the O'Brien & Fleming test applies two-sided significance levels $2\{1 - \Phi(5.218/\sqrt{k})\}$ at analyses $k = 1, \ldots, 4$, and the group sequential t-test rejects H_0 at analysis k if

$$|T_k| \geq t_{2mk-2, 1-\Phi(5.218\, k^{-1/2})}, \quad k = 1, \ldots, 4,$$

and H_0 is accepted if it has not been rejected by the fourth analysis.

By design, this test's Type I error probability will be close to 0.01, but its power depends on m and the unknown σ^2. Suppose our experimental resources make a total of 64 observations a convenient sample size so we could take $m = 8$ observations per treatment in each group, giving a total of 32 observations on each treatment at the final analysis and thus the variance of $\hat{\mu}_A - \hat{\mu}_B$ would be $\sigma^2/16$. In using the approximation (4.11), $\delta = \mu_A - \mu_B$ but σ^2 is unknown, so we can only evaluate power at given values of the ratio $(\mu_A - \mu_B)/\sigma$. For instance, power is approximately

$$Pr\{T(62; \sqrt{\{16/1.01\}}) \geq t_{62,0.005}\} = 0.903$$

when $\mu_A - \mu_B = \pm\sigma$. Here, we have used $R_B(4, 0.01, 0.1) = 1.010$ since an initial calculation with $R = 1$ showed power to be approximately 0.9. Simulations of the group sequential t-test show these approximate error probabilities to be highly accurate: based on one million simulations, estimates of Type I error and power at $\mu_A - \mu_B = \sigma$ are 0.0101 and 0.9021, respectively, with standard errors 0.0001 and 0.0003. In judging whether power close to 0.9 when $\mu_A - \mu_B = \pm\sigma$ is acceptable we have to consider likely values of σ^2 or, if acting cautiously, an upper bound for σ^2. If we do regard this power as acceptable, the study can be conducted with $m = 8$. If not, power should be re-evaluated under other values of m until a suitable group size is found.

A more direct approach to sample size calculation is possible if a reliable initial estimate of σ^2, σ_0^2 say, is available. Formula (4.12) provides a good approximation to the sample size required to meet the stipulated power condition when $\sigma^2 = \sigma_0^2$ and the adequacy of this sample size can then be checked using the more accurate formula (4.11) and adjustments made if necessary. The calculation based on (4.12) is really just the sample size calculation for a Z-test when σ^2 is known to be equal to σ_0^2 and reduces to the condition

$$\begin{aligned}(\boldsymbol{c}^T \boldsymbol{V}_k \boldsymbol{c} \sigma_0^2)^{-1} &= R(K, \alpha, \beta)\{\Phi^{-1}(1-\alpha/2) + \Phi^{-1}(1-\beta)\}^2/\delta^2 \\ &= R(K, \alpha, \beta)\mathcal{I}_{f,2}.\end{aligned}$$

4.6.2.2 Two-treatment Comparison Adjusted for Covariates

We are now able to re-visit the example of Section 3.3.2.2, where previously we assumed σ^2 to be known. Continuing in our earlier notation, suppose the goal in this example is to test the null hypothesis of no treatment effect, H_0: $\beta_1 = 0$, with Type I error probability $\alpha = 0.05$ using an O'Brien & Fleming test with six analyses. Suppose it is also desired that the test should have power 0.8 to reject H_0 if $\beta_1 = \pm 0.5$. Estimates

$$\hat{\beta}_1^{(k)} \sim N(\beta_1, \{(\boldsymbol{D}^{(k)T}\boldsymbol{D}^{(k)})^{-1}\}_{11}\sigma^2)$$

are available and the t-statistics at successive analyses are

$$T_k = \frac{\hat{\beta}_1^{(k)}}{\sqrt{[\{(\boldsymbol{D}^{(k)T}\boldsymbol{D}^{(k)})^{-1}\}_{11} s_k^2]}}, \quad k = 1, \ldots, 6,$$

where

$$s_k^2 = (\boldsymbol{X}^{(k)} - \boldsymbol{D}^{(k)}\widehat{\boldsymbol{\beta}}^{(k)})^T(\boldsymbol{X}^{(k)} - \boldsymbol{D}^{(k)}\widehat{\boldsymbol{\beta}}^{(k)})/(n_k - p)$$
$$\sim \sigma^2\chi^2_{n_k-p}/(n_k - p), \quad k = 1, \ldots, 6.$$

Table 2.3 gives $C_B(6, 0.05) = 2.503$. Hence,

$$c_k = 2.503\sqrt{(6/k)} = 6.131/\sqrt{k}, \quad k = 1, \ldots, 6,$$

and the O'Brien & Fleming test applies the two-sided significance levels $2\{1 - \Phi(6.131/\sqrt{k})\}$ at analyses $k = 1, \ldots, 6$. Comparing the sequence of t-statistics against these significance levels gives the rule for the group sequential t-test, which is to reject H_0 at analysis k if

$$|T_k| \geq t_{n_k-p, 1-\Phi(6.131\, k^{-1/2})}, \quad k = 1, \ldots, 6,$$

and to accept H_0 if it has not been rejected by the sixth analysis.

The power of this test depends primarily on the sequence of values of $\text{Var}(\hat{\beta}_1^{(k)})$, which in turn depend on the unknown variance σ^2. If, as before, we believe 1.2 to be a plausible value for σ^2, we can use the sample size that satisfies the power condition in a known variance test for $\sigma^2 = 1.2$ as a starting point for designing a group sequential t-test. If treatment allocation is well balanced with respect to covariates, we can expect

$$\text{Var}(\hat{\beta}_1^{(k)}) \approx 4\sigma^2/n_k, \quad k = 1, \ldots, 6,$$

where n_k is the number of patients, divided equally between treatments A and B, at analysis k. Under this assumption, the known variance test for $\sigma^2 = 1.2$ requires

$$\left(\frac{4 \times 1.2}{n_6}\right)^{-1} = R_B(6, 0.05, 0.2)\,\{\Phi^{-1}(0.975) + \Phi^{-1}(0.8)\}^2/0.5^2$$
$$= 1.032 \times (1.960 + 0.842)^2/0.5^2 = 32.41,$$

which is achieved by setting $n_k = 26\,k$, $k = 1, \ldots, 6$, to give a final sample size $n_6 = 156$. In this case,

$$\text{Var}(\hat{\beta}_1^{(6)}) \approx 4\sigma^2/156 = \sigma^2/39.$$

Suppose also that $p = 6$, representing one parameter for the treatment effect and five for covariate terms. Then, using $R_B(6, 0.05, 0.2) = 1.032$ for the sample size factor in formula (4.11), we obtain the approximate power at $\beta_1 = \pm 0.5$ as

$$Pr\{T(156 - 6; 0.5/\sqrt{\{1.032\sigma^2/39\}}) \geq t_{156-6, 0.025}\}.$$

With $\sigma^2 = 1.2$, this expression is 0.796, admirably close to the target of 0.8. However, the power varies steadily with σ^2, the same formula giving power 0.86 for $\sigma^2 = 1.0$ but only 0.70 if $\sigma^2 = 1.5$. Thus, if σ^2 is unknown, one should consider power under a range of plausible values of σ^2 when choosing sample size, and bear in mind that only the power under the largest plausible value of σ^2 is guaranteed with any certainty.

One way to try and achieve specified power at a given alternative when σ^2 is unknown is to re-compute sample size as the study progresses, using current variance estimates. In Chapter 18 we shall discuss methods of this type for producing a single test on termination as well as group sequential designs which test the null hypothesis and update the target sample size concurrently.

Chapter 5

One-sided Tests

5.1 Introduction

The name *one-sided test* is given to a test of a hypothesis against a one-sided alternative, for example, a test of the null hypothesis H_0: $\theta = 0$ against the alternative H_A: $\theta > 0$. If a fixed sample test based on a statistic Z rejects H_0 when $Z > c$, for some constant c, the test's Type I error probability is

$$Pr_{\theta=0}\{Z > c\},$$

and its power is the probability of rejecting H_0 under values of θ in the alternative hypothesis, i.e., $Pr_\theta\{Z > c\}$ for values $\theta > 0$.

In some problems, a one-sided alternative is appropriate because departures from H_0 in the other direction are either implausible or impossible. In other cases, θ may lie on either side of $\theta = 0$ but the alternative is chosen in the direction in which departures from $\theta = 0$ are of interest. Suppose θ measures the improvement in response achieved by a new treatment over a standard, and the new treatment will only be deemed acceptable if it produces superior responses; then the main aim of a study will be to distinguish between the cases $\theta \leq 0$ and $\theta > 0$. However, the probability of deciding in favor of the new treatment when $\theta \leq 0$ is greatest at $\theta = 0$, so it suffices to specify the Type I error rate at $\theta = 0$. It matters little whether one regards the null hypothesis in such a study as $\theta \leq 0$ or $\theta = 0$. We shall adopt the former formulation and so discuss tests of H_0: $\theta \leq 0$ against $\theta > 0$ designed to achieve Type I error probability α when $\theta = 0$ and power $1 - \beta$ at a specified alternative value $\theta = \delta$ where $\delta > 0$. When a two-sided test with the conventional Type I error rate of 0.05 is used to compare a new treatment against a standard, the probability of declaring the new treatment to be superior when $\theta = 0$ is 0.025. In order to maintain comparability between one-sided and two-sided tests it is customary to set the Type I error probability to be $\alpha = 0.025$ and it is standard practice for regulators to require this significance level as evidence of the superiority of a new treatment in a Phase III trial.

In testing H_0: $\theta \leq 0$, a fixed sample test based on a statistic Z distributed as $N(\theta\sqrt{\mathcal{I}}, 1)$ attains Type I error probability α by rejecting H_0 if $Z > \Phi^{-1}(1-\alpha)$. Hence, in order to attain power $1 - \beta$ when $\theta = \delta$, the sample size must be chosen to give an information level

$$\mathcal{I}_{f,1} = \{\Phi^{-1}(1-\alpha) + \Phi^{-1}(1-\beta)\}^2/\delta^2, \tag{5.1}$$

the subscript 1 in $\mathcal{I}_{f,1}$ denoting a one-sided test. As before, Φ denotes the standard normal cdf. As for two-sided tests, we shall express the maximum and expected information levels of group sequential one-sided tests as multiples of the fixed sample information level, $\mathcal{I}_{f,1}$.

We shall present group sequential one-sided tests for a parameter θ when standardized statistics $Z_1, \ldots, Z_K$ available at analyses $k = 1, \ldots, K$ follow the canonical joint distribution (3.1). In particular $Z_k \sim N(\theta\sqrt{\mathcal{I}_k}, 1)$, where $\mathcal{I}_k$ is the information for θ at analysis k. As explained in Chapter 4, this joint distribution arises, either exactly or approximately, in a great many applications. The group sequential one-sided tests we shall define are directly applicable to all these situations.

DOI: 10.1201/9781584888482-5

A general one-sided group sequential test of H_0: $\theta \le 0$ versus $\theta > 0$ is defined by pairs of constants (a_k, b_k) with $a_k < b_k$ for $k = 1, \dots, K-1$ and $a_K = b_K$. It takes the form:

$$\begin{array}{lll} \text{After group } k = 1, \dots, K-1 & & \\ \quad \text{if } Z_k \ge b_k & \text{stop, reject } H_0 & \\ \quad \text{if } Z_k \le a_k & \text{stop, accept } H_0 & \\ \quad \text{otherwise} & \text{continue to group } k+1, & \quad (5.2) \\ \text{after group } K & & \\ \quad \text{if } Z_K \ge b_K & \text{stop, reject } H_0 & \\ \quad \text{if } Z_K < a_K & \text{stop, accept } H_0. & \end{array}$$

The condition $a_K = b_K$ ensures that the test terminates at analysis K. We note that the stopping and continuation regions can be written in the form discussed in Section 3.5 with

$$\mathcal{A}_k = (-\infty, a_k), \quad \mathcal{B}_k = (b_k, \infty), \quad \mathcal{C}_k = (a_k, b_k), \quad \text{for } k = 1, \dots, K,$$

where $\mathcal{C}_K = \emptyset$. Hence we can use Armitage's iterated integral technique, as described in Section 3.5, to compute operating characteristics of the test, such as its Type I error rate and power function, and the distribution of the stopping time under specified values of θ.

The upper boundary $\{b_1, \dots, b_K\}$ is often termed the *efficacy* boundary as the new treatment will be deemed efficacious (a "positive" result) if it is crossed. The lower boundary $\{a_1, \dots, a_K\}$ may be referred to as the *futility* boundary since, when this boundary is crossed, it is unlikely that a positive result will occur, and continuing the trial is deemed to be futile. Gould (1983) refers to this decision as "abandoning a lost cause". The futility boundary may be regarded as either *binding* or *non-binding*. If it can be assumed that the trial will stop as soon as this boundary is crossed, the lower boundary is considered "binding". However, if the monitoring committee is allowed to consider additional information when deciding to recommend termination of the trial at an interim analysis, the lower stopping boundary may be applied flexibly, so there is no guarantee that this aspect of the stopping rule will be followed rigidly. In this case, the lower boundary is said to be "non-binding". When the lower boundary is non-binding, the upper (efficacy) boundary must be set so that the Type I error probability is no greater than the specified value α, even if the trial may continue after the lower boundary is crossed.

Most of the early proposals for one-sided group sequential tests assumed a binding futility boundary. We consider this case in the remainder of this section and in Sections 5.2, 5.5, 5.6 and 5.7, but elsewhere we shall consider both possibilities, with an emphasis on the non-binding case.

After the impact of the two-sided group sequential tests of Pocock (1977) and O'Brien & Fleming (1979), proposals for one-sided group sequential tests soon followed. DeMets & Ware (1980, 1982) presented tests adapted from Pocock and O'Brien & Fleming two-sided tests and Wald's (1947) sequential probability ratio test. Subsequently, Jennison (1987), Eales & Jennison (1992) and Barber & Jennison (2002) explored the extent of possible reductions in expected sample size by deriving optimal group sequential one-sided tests.

We shall concentrate in this chapter on a family of tests with boundaries of a certain parametric form. In Section 5.2 we describe the family of tests proposed by Emerson & Fleming (1989), and extended by Pampallona & Tsiatis (1994) to include asymmetric tests with unequal Type I and Type II error probabilities. Tests in this family are indexed by a parameter Δ that appears as a power in the formulae for their stopping boundaries, and we shall refer to this as the *power family* of group sequential one-sided tests. As Δ is increased, tests provide greater opportunity for early stopping, resulting in lower expected sample size, but the maximum sample size also increases. We provide tables of properties of a selection of tests in the power family to guide choice of an appropriate test, bearing in mind the effort of more frequent monitoring or the increase in maximum sample size that accompanies reductions in average study length. We present versions of the power family tests with a non-binding futility boundary in Section 5.3 and we illustrate three different ways of applying a lower boundary in an example in Section 5.4.

Although the power family tests are designed for a specific sequence of information levels, they can still be used when observed information levels differ from their planned values. In Section 5.5 we present a method for doing this that parallels the significance level approach seen in Section 4.2 for two-sided tests, and which maintains Type I error rates quite accurately. Later, in Section 8.3, we shall present a more precise strategy for handling unpredictable information sequences based on "error spending" functions, and the "Rho family" of error spending tests defined in Section 8.3.2 offers a similar range of options to the power family.

The significance level approach can also be used to create group sequential, one-sided t-tests from the power family tests, and we describe this construction in Section 5.6. These group sequential t-tests maintain a specified Type I error probability to a high degree of accuracy. Just as for the two-sided tests of Section 4.6, power at a given value of θ depends on the actual value of σ^2 and thus, for this form of test, power should be assessed under a range of plausible values of σ^2 when determining an appropriate sample size.

Returning to the case of known variance, in Section 5.7 we describe the group sequential test proposed by Whitehead & Stratton (1983) and developed further by Whitehead (1997, Chapter 4). When used with Whitehead's "Christmas tree adjustment", this test maintains specified Type I error and power to quite a high accuracy for general sequences of observed information levels. Coupled with the first edition of Whitehead's book, published in 1983, and the software package PEST, this work was highly influential in promoting the use of group sequential designs.

5.2 The Power Family of One-sided Group Sequential Tests

5.2.1 *Definition of the Tests*

Tests in the power family follow the general form of group sequential one-sided tests stated in (5.2). The tests are designed for equally spaced increments in information, i.e., $\mathcal{I}_k = (k/K)\mathcal{I}_K$, $k = 1, \ldots, K$, and the test with parameter Δ has critical values for $\{Z_1, \ldots, Z_K\}$ equal to

$$\begin{aligned} b_k &= \tilde{C}_1(K, \alpha, \beta, \Delta)(k/K)^{\Delta - 1/2} \quad \text{and} \\ a_k &= \delta\sqrt{\mathcal{I}_k} - \tilde{C}_2(K, \alpha, \beta, \Delta)(k/K)^{\Delta - 1/2}, \quad k = 1, \ldots, K. \end{aligned} \tag{5.3}$$

In order that $a_K = b_K$, the final information level must be

$$\mathcal{I}_K = \frac{\{\tilde{C}_1(K, \alpha, \beta, \Delta) + \tilde{C}_2(K, \alpha, \beta, \Delta)\}^2}{\delta^2}. \tag{5.4}$$

The constants $\tilde{C}_1(K, \alpha, \beta, \Delta)$ and $\tilde{C}_2(K, \alpha, \beta, \Delta)$ are chosen to ensure the Type I error and power conditions. Thus, we require

$$Pr_{\theta=0}\left\{ \bigcup_{k=1}^{K} (a_j < Z_j < b_j \text{ for } j = 1, \ldots, k-1 \text{ and } Z_k \geq b_k) \right\} = \alpha \tag{5.5}$$

and

$$Pr_{\theta=\delta}\left\{ \bigcup_{k=1}^{K} (a_j < Z_j < b_j \text{ for } j = 1, \ldots, k-1 \text{ and } Z_k \leq a_k) \right\} = \beta \tag{5.6}$$

when the boundary values are given by (5.3) and $\{Z_1, \ldots, Z_K\}$ follow the canonical joint distribution (3.1) with information levels $\mathcal{I}_k = (k/K)\mathcal{I}_K$, $k = 1, \ldots, K$, for $\mathcal{I}_K$ as specified in (5.4). These probabilities can be computed using the techniques of Section 3.5 and the constants $\tilde{C}_1(K, \alpha, \beta, \Delta)$ and $\tilde{C}_2(K, \alpha, \beta, \Delta)$ may be found by a 2-dimensional search.

It is important to note that the constants $\tilde{C}_1(K, \alpha, \beta, \Delta)$ and $\tilde{C}_2(K, \alpha, \beta, \Delta)$ do not depend on δ. To see this, note that since $\mathcal{I}_k = (k/K)\mathcal{I}_K$, $k = 1, \ldots, K$, the boundary values can be written without explicit mention of δ as

$$\begin{aligned} b_k &= \tilde{C}_1(K, \alpha, \beta, \Delta)(k/K)^{\Delta - 1/2} \quad \text{and} \\ a_k &= \{\tilde{C}_1(K, \alpha, \beta, \Delta) + \tilde{C}_2(K, \alpha, \beta, \Delta)\}\,(k/K)^{1/2} \\ &\quad - \tilde{C}_2(K, \alpha, \beta, \Delta)\,(k/K)^{\Delta - 1/2}, \quad k = 1, \ldots, K. \end{aligned} \tag{5.7}$$

For any value of θ, the joint distribution of $\{Z_1, \ldots, Z_K\}$ is multivariate normal with correlations depending on the ratios between information levels. Under $\theta = 0$, $E(Z_k) = 0$, $k = 1, \ldots, K$. Furthermore, when $\theta = \delta$,

$$E(Z_k) = \delta\sqrt{\mathcal{I}_k} = \{\tilde{C}_1(K, \alpha, \beta, \Delta) + \tilde{C}_2(K, \alpha, \beta, \Delta)\}\,(k/K)^{1/2}, \quad k = 1, \ldots, K,$$

and thus the joint distribution of $\{Z_1, \ldots, Z_K\}$ when calculating the Type II error probability in (5.6) is determined by the values of $\tilde{C}_1(K, \alpha, \beta, \Delta)$ and $\tilde{C}_2(K, \alpha, \beta, \Delta)$.

Since the joint distribution of $\{Z_1, \ldots, Z_K\}$ under $\theta = 0$ depends only on the ratios $\mathcal{I}_k/\mathcal{I}_K$, $k = 1, \ldots, K$, applying the boundaries defined in Equation (5.7) to a sequence of Z-statistics with equally spaced information levels $\mathcal{I}_k = (k/K)\mathcal{I}_K^*$, $k = 1, \ldots, K$, produces a test with Type I error probability exactly α, regardless of the value of $\mathcal{I}_K^*$. Of course, the power when $\theta > 0$ does depend on $\mathcal{I}_K^*$ and it follows from Equation (5.4) that, when information levels are $\mathcal{I}_k = (k/K)\mathcal{I}_K^*$, $k = 1, \ldots, K$, power $1 - \beta$ will be achieved at $\theta = \tilde{\delta}$ where,

$$\mathcal{I}_K^*\,\tilde{\delta}^2 = \{\tilde{C}_1(K, \alpha, \beta, \Delta) + \tilde{C}_2(K, \alpha, \beta, \Delta)\}^2,$$

i.e., at $\theta = \delta\sqrt{\{\mathcal{I}_K/\mathcal{I}_K^*\}}$.

There is a similarity between these tests and the Wang & Tsiatis tests presented in Section 2.7.1. The upper boundary can be viewed as a repeated significance test of $\theta = 0$ with critical values for the Z_k proportional to $k^{\Delta - 1/2}$, $k = 1, \ldots, K$, while the lower boundary arises as a repeated significance test of $\theta = \delta$, this hypothesis being rejected if $Z_k - \delta\sqrt{\mathcal{I}_k}$, the standardized statistic with mean zero if $\theta = \delta$, is negative and less than a critical value proportional to $k^{\Delta - 1/2}$.

From Equations (5.1) and (5.4), we see that the ratio of the test's maximum information level to that required for a fixed sample test is

$$\tilde{R}(K, \alpha, \beta, \Delta) = \frac{\mathcal{I}_K}{\mathcal{I}_{f,1}} = \frac{\{\tilde{C}_1(K, \alpha, \beta, \Delta) + \tilde{C}_2(K, \alpha, \beta, \Delta)\}^2}{\{\Phi^{-1}(1-\alpha) + \Phi^{-1}(1-\beta)\}^2}.$$

In Tables 5.1, 5.2 and 5.3, we present values of $\tilde{C}_1(K, \alpha, \beta, \Delta)$, $\tilde{C}_2(K, \alpha, \beta, \Delta)$ and $\tilde{R}(K, \alpha, \beta, \Delta)$, along with properties of the resulting tests for $\Delta = -0.5$, -0.25, 0 and 0.25, $\alpha = 0.025$ and $1 - \beta = 0.8$, 0.9 and 0.95. Entries in these tables were computed numerically using the methods described in Chapter 31. Interpolation may be used to obtain constants for values of $K < 20$ not present in these tables. Constants for other cases can be found in Emerson & Fleming (1989) for $\Delta = 0$, 0.1, 0.2, 0.3, 0.4 and 0.5 and $\alpha = \beta = 0.05$ and 0.01, and in Pampallona & Tsiatis (1994) for the same Δ and α values and $1 - \beta = 0.8$, 0.9 and 0.95.

We have defined tests in terms of the standardized statistics $Z_1, \ldots, Z_K$. Emerson & Fleming and Pampallona & Tsiatis define the same tests in terms of the score statistics

Table 5.1 *Constants* $\tilde{C}_1(K, \alpha, \beta, \Delta)$, $\tilde{C}_2(K, \alpha, \beta, \Delta)$ *and* $\tilde{R}(K, \alpha, \beta, \Delta)$ *for power family one-sided tests with shape parameter* Δ. *Also shown are expected sample sizes at* $\theta = 0$, $\delta/2$ *and* δ *expressed as percentages of the corresponding fixed sample size. Tests are for* K *groups of observations, Type I error probability* $\alpha = 0.025$ *at* $\theta = 0$ *and power* $1 - \beta = 0.8$ *at* $\theta = \delta$.

	K	$\tilde{C}_1$	$\tilde{C}_2$	$\tilde{R}$	*Expected sample size, as percentage of fixed sample size, at* $\theta = 0$	$\theta = \delta/2$	$\theta = \delta$
$\Delta = -0.5$							
	1	1.960	0.842	1.000	100.0	100.0	100.0
	2	1.947	0.870	1.011	70.8	89.5	97.6
	3	1.933	0.902	1.024	69.1	85.5	89.7
	4	1.928	0.921	1.034	65.1	82.5	86.8
	5	1.926	0.933	1.042	62.9	80.9	84.9
	10	1.927	0.966	1.067	59.7	77.8	81.3
	15	1.930	0.981	1.080	58.7	76.9	80.2
	20	1.932	0.990	1.088	58.2	76.4	79.7
$\Delta = -0.25$							
	1	1.960	0.842	1.000	100.0	100.0	100.0
	2	1.935	0.903	1.027	67.4	86.5	93.9
	3	1.930	0.934	1.045	63.3	82.2	86.5
	4	1.930	0.954	1.060	61.1	79.8	83.6
	5	1.931	0.969	1.071	59.2	78.1	81.8
	10	1.941	1.005	1.106	55.6	75.0	78.3
	15	1.947	1.022	1.123	54.6	74.1	77.2
	20	1.950	1.032	1.133	54.1	73.6	76.7
$\Delta = 0$							
	1	1.960	0.842	1.000	100.0	100.0	100.0
	2	1.934	0.948	1.059	65.6	83.8	88.2
	3	1.941	0.986	1.092	58.6	78.9	82.8
	4	1.949	1.010	1.116	56.0	76.5	79.7
	5	1.956	1.026	1.133	54.4	75.0	78.0
	10	1.977	1.070	1.183	51.0	71.8	74.5
	15	1.988	1.091	1.207	49.8	70.8	73.4
	20	1.994	1.103	1.222	49.2	70.2	72.8
$\Delta = 0.25$							
	1	1.960	0.842	1.000	100.0	100.0	100.0
	2	1.982	1.000	1.133	66.3	82.6	83.9
	3	2.007	1.067	1.204	56.9	76.9	78.6
	4	2.026	1.106	1.250	52.8	74.1	75.7
	5	2.041	1.133	1.283	50.6	72.4	73.9
	10	2.082	1.201	1.373	46.3	68.9	70.2
	15	2.102	1.231	1.415	44.9	67.7	68.8
	20	2.115	1.249	1.442	44.2	67.1	68.2

Table 5.2 *Constants $\tilde{C}_1(K, \alpha, \beta, \Delta)$, $\tilde{C}_2(K, \alpha, \beta, \Delta)$ and $\tilde{R}(K, \alpha, \beta, \Delta)$ for power family one-sided tests with shape parameter Δ. Also shown are expected sample sizes at $\theta = 0$, $\delta/2$ and δ expressed as percentages of the corresponding fixed sample size. Tests are for K groups of observations, Type I error probability $\alpha = 0.025$ at $\theta = 0$ and power $1 - \beta = 0.9$ at $\theta = \delta$.*

	K	$\tilde{C}_1$	$\tilde{C}_2$	$\tilde{R}$	*Expected sample size, as percentage of fixed sample size, at* $\theta = 0$	$\theta = \delta/2$	$\theta = \delta$
$\Delta = -0.5$							
	1	1.960	1.282	1.000	100.0	100.0	100.0
	2	1.958	1.286	1.002	80.6	96.2	97.3
	3	1.952	1.305	1.010	74.7	90.1	86.8
	4	1.951	1.317	1.017	70.2	87.5	84.1
	5	1.952	1.327	1.023	68.6	85.8	81.9
	10	1.958	1.352	1.043	64.9	82.5	78.1
	15	1.963	1.364	1.053	63.8	81.5	77.0
	20	1.966	1.372	1.060	63.3	81.0	76.5
$\Delta = -0.25$							
	1	1.960	1.282	1.000	100.0	100.0	100.0
	2	1.954	1.302	1.009	73.4	92.5	91.9
	3	1.953	1.326	1.024	70.1	87.4	83.4
	4	1.956	1.342	1.035	66.2	84.6	80.2
	5	1.960	1.353	1.044	64.1	83.0	78.4
	10	1.974	1.383	1.072	60.6	79.8	74.7
	15	1.982	1.397	1.087	59.5	78.8	73.6
	20	1.987	1.406	1.096	59.0	78.3	73.0
$\Delta = 0$							
	1	1.960	1.282	1.000	100.0	100.0	100.0
	2	1.957	1.338	1.033	68.6	88.4	84.7
	3	1.968	1.366	1.058	63.7	83.9	79.3
	4	1.978	1.385	1.077	61.1	81.4	75.9
	5	1.987	1.400	1.091	59.3	79.8	74.0
	10	2.011	1.438	1.132	55.6	76.6	70.4
	15	2.024	1.456	1.153	54.4	75.6	69.3
	20	2.032	1.467	1.165	53.8	75.1	68.7
$\Delta = 0.25$							
	1	1.960	1.282	1.000	100.0	100.0	100.0
	2	2.002	1.398	1.100	67.2	85.7	79.4
	3	2.032	1.449	1.153	59.3	80.9	73.9
	4	2.053	1.481	1.189	55.9	78.4	71.0
	5	2.070	1.503	1.214	54.0	76.8	69.2
	10	2.115	1.560	1.285	50.1	73.5	65.4
	15	2.136	1.587	1.319	48.8	72.4	64.1
	20	2.150	1.604	1.341	48.1	71.8	63.4

Table 5.3 *Constants $\tilde{C}_1(K, \alpha, \beta, \Delta)$, $\tilde{C}_2(K, \alpha, \beta, \Delta)$ and $\tilde{R}(K, \alpha, \beta, \Delta)$ for power family one-sided tests with shape parameter Δ. Also shown are expected sample sizes at $\theta = 0$, $\delta/2$ and δ expressed as percentages of the corresponding fixed sample size. Tests are for K groups of observations, Type I error probability $\alpha = 0.025$ at $\theta = 0$ and power $1 - \beta = 0.95$ at $\theta = \delta$.*

	K	$\tilde{C}_1$	$\tilde{C}_2$	$\tilde{R}$	*Expected sample size, as percentage of fixed sample size, at*		
					$\theta = 0$	$\theta = \delta/2$	$\theta = \delta$
$\Delta = -0.5$							
	1	1.960	1.645	1.000	100.0	100.0	100.0
	2	1.960	1.645	1.000	88.6	98.7	95.7
	3	1.959	1.653	1.004	77.5	92.8	83.3
	4	1.961	1.661	1.010	74.5	90.1	80.8
	5	1.964	1.667	1.014	72.4	88.3	78.4
	10	1.974	1.686	1.031	68.5	85.0	74.5
	15	1.981	1.696	1.040	67.3	83.9	73.3
	20	1.985	1.702	1.046	66.8	83.4	72.7
$\Delta = -0.25$							
	1	1.960	1.645	1.000	100.0	100.0	100.0
	2	1.960	1.651	1.003	79.6	95.8	88.7
	3	1.964	1.666	1.014	74.1	89.9	79.8
	4	1.969	1.677	1.023	69.7	87.3	76.3
	5	1.974	1.686	1.031	67.9	85.6	74.4
	10	1.992	1.710	1.055	64.2	82.4	70.6
	15	2.001	1.723	1.067	63.0	81.3	69.4
	20	2.007	1.731	1.075	62.5	80.8	68.9
$\Delta = 0$							
	1	1.960	1.645	1.000	100.0	100.0	100.0
	2	1.967	1.673	1.020	72.1	91.4	80.2
	3	1.982	1.695	1.041	68.0	86.7	75.1
	4	1.994	1.712	1.057	64.8	84.0	71.5
	5	2.003	1.724	1.069	62.7	82.4	69.5
	10	2.031	1.758	1.105	59.0	79.2	65.8
	15	2.045	1.774	1.122	57.8	78.2	64.6
	20	2.054	1.784	1.133	57.2	77.6	64.1
$\Delta = 0.25$							
	1	1.960	1.645	1.000	100.0	100.0	100.0
	2	2.015	1.731	1.079	68.5	87.8	74.5
	3	2.047	1.772	1.122	61.8	83.3	68.8
	4	2.070	1.798	1.151	58.8	80.8	65.9
	5	2.087	1.818	1.173	56.9	79.3	64.0
	10	2.134	1.870	1.233	53.1	76.0	60.2
	15	2.156	1.895	1.263	51.7	74.8	58.9
	20	2.171	1.910	1.281	51.1	74.3	58.2

$S_k = Z_k\sqrt{\mathcal{I}_k}$, $k = 1,\ldots,K$, (see Section 3.1) and their $C_j(K,\alpha,\beta,\Delta)$ are, therefore, $K^{1/2-\Delta}$ times our $\tilde{C}_j(K,\alpha,\beta,\Delta)$ for $j = 1$ and 2. We have expressed tests in terms of the Z_k for several reasons. First, the $\tilde{C}_j(K,\alpha,\beta,\Delta)$ vary slowly with K, facilitating the use of interpolation for values of K omitted from our tables. Secondly, the $\tilde{C}_j(K,\alpha,\beta,\Delta)$ play a specific role in the final analysis if the test reaches that stage: H_0 is rejected if Z_K, a standard normal variable under $\theta = 0$, exceeds $\tilde{C}_1(K,\alpha,\beta,\Delta)$ and H_0 is accepted if $Z_K - \delta\sqrt{\mathcal{I}_K}$, a standard normal variable under $\theta = \delta$, is less than $-\tilde{C}_2(K,\alpha,\beta,\Delta)$. Finally, our definition of the test is easily adapted to give the methods we recommend for dealing with departures from planned sequences of information levels and for creating group sequential t-tests for a normal response with unknown variance.

In Tables 5.1, 5.2 and 5.3, we see a number of instances where $C_1(K,\alpha,\beta,\Delta) < z_\alpha = 1.960$ and thus $a_K = b_K = C_1(K,\alpha,\beta,\Delta) < z_\alpha$. This occurs when there is a sufficiently high probability that sample paths which would have led to $Z_K > z_\alpha$ are terminated due to crossing the futility boundary. However, even though the Type I error rate is controlled at level α, this form of decision rule is typically discouraged because rejection of H_0 is unconvincing when the final Z_k is less than z_α. A simple way to avoid this problem is to increase a_K and b_K to z_α in cases where $C_1(K,\alpha,\beta,\Delta) < z_\alpha$; a more elaborate solution would be to re-compute $C_1(K,\alpha,\beta,\Delta)$and $C_2(K,\alpha,\beta,\Delta)$ with this constraint in place.

As an illustration of designs in the power family, Figure 5.1 shows the boundaries of two power family tests with four groups of observations, expressed on both the Z_k and S_k scales. Each test has Type I error probability $\alpha = 0.025$ when $\theta = 0$, and power $1-\beta = 0.9$ at $\theta = \delta = 0.2$. The tests have parameter values $\Delta = -0.5$ and $\Delta = 0.0$ and, since $\tilde{R}(K,\alpha,\beta,\Delta)$ increases with Δ, the required information levels are greater for the test with $\Delta = 0.0$. For comparison, the tests with parameter values $\Delta = -0.5$ and $\Delta = 0.0$ have maximum information $\mathcal{I}_5 = 273.7$ and 293.2, respectively, while the corresponding fixed sample test requires information $\mathcal{I}_{f,1} = (1.282 + 1.960)^2/0.2^2 = 262.7$.

The power family designs can be extended to allow different shape parameters for the upper and lower boundaries. A test with parameters Δ_1 and Δ_2 is defined by constants $\tilde{C}_1(K,\alpha,\beta,\Delta_1,\Delta_2)$ and $\tilde{C}_2(K,\alpha,\beta,\Delta_1,\Delta_2)$, and has critical values

$$\begin{aligned} b_k &= \tilde{C}_1(K,\alpha,\beta,\Delta_1,\Delta_2)(k/K)^{\Delta_1-1/2} \quad \text{and} \\ a_k &= \delta\sqrt{\mathcal{I}_k} - \tilde{C}_2(K,\alpha,\beta,\Delta_1,\Delta_2)(k/K)^{\Delta_2-1/2}, \quad k = 1,\ldots,K. \end{aligned} \tag{5.8}$$

In order that $a_K = b_K$, the final information level should be

$$\mathcal{I}_K = \frac{\{\tilde{C}_1(K,\alpha,\beta,\Delta_1,\Delta_2) + \tilde{C}_2(K,\alpha,\beta,\Delta_1,\Delta_2)\}^2}{\delta^2}. \tag{5.9}$$

As before, the constants $C_1(K,\alpha,\beta,\Delta_1,\Delta_2)$ and $\tilde{C}_2(K,\alpha,\beta,\Delta_1,\Delta_2)$ are chosen to ensure the Type I error and power conditions

$$Pr_{\theta=0}\left\{\bigcup_{k=1}^{K}(a_j < Z_j < b_j \text{ for } j = 1,\ldots,k-1 \text{ and } Z_k \geq b_k)\right\} = \alpha$$

and

$$Pr_{\theta=\delta}\left\{\bigcup_{k=1}^{K}(a_j < Z_j < b_j \text{ for } j = 1,\ldots,k-1 \text{ and } Z_k \leq a_k)\right\} = \beta.$$

The values of Δ_1 and Δ_2 determine the degree of early stopping for efficacy and futility, respectively. One might, for example, choose a larger value for Δ_2 to facilitate early termination for futility but assign a smaller value to Δ_1 so that more extreme evidence would be required to stop early for efficacy.

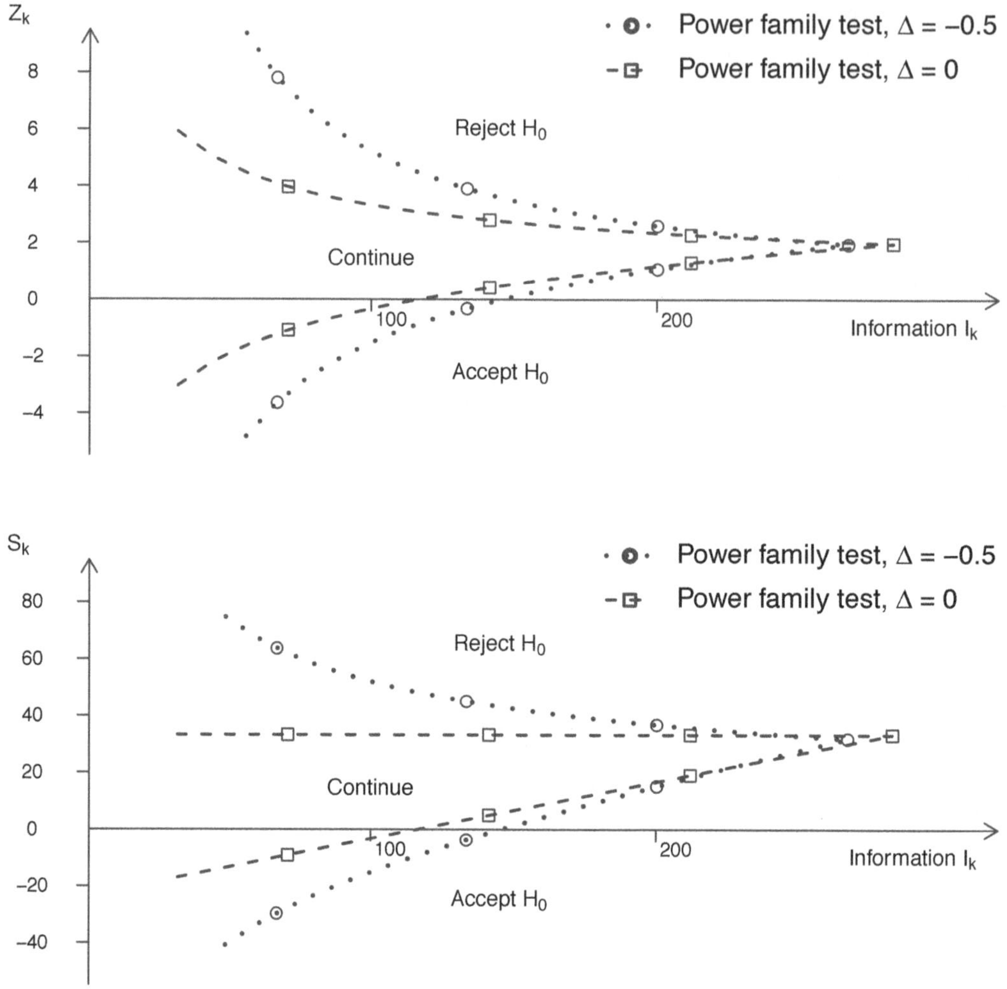

Figure 5.1 *Two power family one-sided tests for four groups of observations*

5.2.2 *Properties of Power Family Tests*

We shall follow the practice of other authors and refer to sample size when discussing properties of group sequential tests. We note, however, that the results in Tables 5.1, 5.2 and 5.3 really concern information rather than sample size when the two are not directly proportional. As we saw for two-sided tests in Chapters 2 and 4, the ratios of the maximum and expected sample sizes to the sample size required for a fixed sample test with the same error probabilities are invariant to the value of δ.

It is evident from Tables 5.1, 5.2 and 5.3 that, for a fixed shape parameter Δ, the maximum sample size of a group sequential test, determined by $\tilde{R}(K, \alpha, \beta, \Delta)$, increases with the number of analyses K, while the expected sample sizes at $\theta = 0$, $\delta/2$ and δ decrease. Only minor reductions in expected sample size are gained by increasing the number of analyses, K, beyond 5 or 10, and we would expect five analyses or fewer to be sufficient in most instances.

Larger values of Δ also increase the maximum sample size and, up to a point, decrease expected sample sizes. We commented in Section 2.6.3 on the significance of maximum sample size, noting that investigators might be deterred by a group sequential design that

Table 5.4 *Minimum possible expected sample sizes for group sequential one-sided tests with error probabilities* $\alpha = 0.025$ *at* $\theta = 0$ *and* $\beta = 0.1$ *at* $\theta = \delta$*,* K *equally sized groups, and maximum sample size equal to* R *times the corresponding fixed sample size. Tabulated values are the minimum expected sample sizes under the stated values of* θ *expressed as a percentage of the fixed sample size. Note that different tests are used to minimize* $\{E_{\theta=0}(N)+E_{\theta=\delta}(N)\}/2$ *and* $E_{\theta=\delta/2}(N)$.

Minima of $\{E_{\theta=0}(N) + E_{\theta=\delta}E(N)\}/2$

	R						
K	1.01	1.05	1.1	1.15	1.2	1.3	*Minimum over R*
2	80.8	74.7	73.2	73.1	73.7	75.8	73.0 at $t = 1.13$
5	72.2	65.2	62.2	60.7	59.8	59.0	58.8 at $t = 1.40$
10	69.2	62.2	59.0	57.4	56.3	55.1	54.2 at $t = 1.61$
20	67.8	60.6	57.5	55.7	54.6	53.3	51.7 at $t = 1.82$

Minima of $E_{\theta=\delta/2}(N)$

	R						
K	1.01	1.05	1.1	1.15	1.2	1.3	*Minimum over R*
2	92.0	87.2	85.7	85.4	85.7	87.0	85.4 at $t = 1.14$
5	86.5	80.6	78.2	77.0	76.5	76.2	76.2 at $t = 1.30$
10	84.2	78.1	75.5	74.2	73.4	72.8	72.7 at $t = 1.39$
20	83.0	76.8	74.1	72.7	71.9	71.1	70.8 at $t = 1.47$

could require substantially more observations than a fixed sample size study. The four values of Δ in our tables offer a range of options for balancing savings in expected sample size against higher maximum sample sizes.

Although Emerson & Fleming (1989) and Pampallona & Tsiatis (1994) did not consider tests with negative values of Δ, our results show that such tests offer substantial reductions in expected sample size with maximum sample size just a little greater than that of the fixed sample test. On the other hand, maximum sample size increases rapidly if Δ is increased beyond 0.25, with at best only minor reductions in expected sample sizes and, therefore, we have not included these tests in our tables. As an example, the test with $\alpha = 0.025$, $1 - \beta = 0.8$, $K = 5$ and $\Delta = 0.5$ would have a maximum sample size 1.694 times the fixed sample size compared to a ratio of 1.283 for the test with $\Delta = 0.25$, an expected sample size at $\theta = \delta$ equal to 73.1% of the fixed sample size, only marginally lower than the 73.9% for the $\Delta = 0.25$ test, and expected sample sizes at $\theta = 0$ and $\theta = \delta/2$ equal to 51.5% and 74.3% of the fixed sample size, both higher than the values 50.6% and 72.4% for the test with $\Delta = 0.25$.

It is useful to know just how small the expected sample size of a group sequential test can be. Table 5.4 shows properties of optimal group sequential tests computed using the methods of Barber & Jennison (2002); see Chapter 12 for more about these computations. These results are for group sequential one-sided tests with Type I and Type II error probabilities $\alpha = 0.025$ and $\beta = 0.1$ at $\theta = 0$ and $\theta = \delta$, respectively. The first section of the table shows the minimum possible average of the expected sample sizes at $\theta = 0$ and $\theta = \delta$ subject to a fixed number, K, of equally spaced analyses and a fixed maximum sample size of R times the corresponding fixed sample size. The second section of the table shows the minimum possible average expected sample size at $\theta = \delta/2$. A different testing boundary is used in the two minimizations, so one should not necessarily expect a single test to be close to optimal in both cases. In each row of the table, minimum values decrease initially as R increases and then, eventually, increase again. The final column shows the smallest minimum expected sample size for each row as the inflation factor R is varied.

It is clear from Table 5.4 that substantial reductions in expected sample size can be achieved by tests with maximum sample size only a little greater than the fixed sample size. It is for this reason that we investigated power family tests with the negative values of Δ that lead to lower maximum sample sizes. Direct comparisons can be made between Table 5.4 and the results in Table 5.2 for power family tests with $\alpha = 0.025$ and $\beta = 0.1$. The power family tests are seen to be fairly efficient, for their maximum sample sizes, in reducing $\{E_{\theta=0}(N)+E_{\theta=\delta}E(N)\}/2$ and $E_{\theta=\delta/2}(N)$, although they do fall several percentage points short of optimality. As an example, the test with $K = 5$ and $\Delta = 0.0$ has maximum sample size $R = 1.091$ times the fixed sample size, and its expected sample sizes $\{E_{\theta=0}(N) + E_{\theta=\delta}E(N)\}/2$ and $E_{\theta=\delta/2}(N)$ are, respectively, 4.0 and 1.3 greater (in the units of the table) than the lowest possible values for $R = 1.091$. We shall see in Section 8.3 and Tables 8.7 to 8.9 that some useful improvements on power family tests are obtained by another family of group sequential one-sided tests defined through an error spending function.

At first sight, these remarks appear to be at odds with the comment of Emerson & Fleming (1989, p. 913) that the power family tests are 99% efficient at $\theta = 0$ or δ relative to the set of all possible group sequential tests. The explanation is that Emerson & Fleming's remark refers to the minimum expected sample size at $\theta = 0$ or δ over *all* values of the parameter Δ with no constraint on the maximum sample size, whereas we have drawn comparisons between tests with specific maximum sample sizes, setting these to be low multiples of the fixed sample size.

Notwithstanding these comparisons with "optimal" benchmarks and error spending tests, the power family provides a selection of easily implemented, group sequential designs within which one should have no trouble in finding a suitable test for a particular problem.

5.3 Non-binding Futility Boundaries

In many studies, it is desirable that the decision to abandon the trial for futility be somewhat flexible. For example, even though the lower boundary has been crossed, the trial's monitoring committee may prefer to continue if they retain confidence in the new treatment despite the negative results so far, or if they see other potential benefits of the new therapy. Continuing the trial after the futility boundary has been crossed increases the chance that the upper efficacy boundary may be crossed. Thus if a trial is designed assuming the futility boundary is binding but, in practice, this boundary is treated as guidance rather than a rigid rule, the Type I error probability will be inflated above the specified level α. Indeed the FDA's Guidance on Adaptive Designs (FDA, 2019a, Sec. V A) warns that

> "if a trial continues despite meeting prespecified binding futility rules, the Agency will likely consider that trial to have failed to provide evidence of efficacy, regardless of the outcome at the final analysis."

In the defining Equations (5.5) and (5.6), it is clear that the two boundaries $\{a_k\}$ and $\{b_k\}$ are interdependent. In order to have a flexible futility boundary, we need a new construction whereby the upper boundary is crossed under $\theta = 0$ with probability at most α, regardless of how any lower boundary is applied. Regulators reviewing the report of a positive trial will then have the reassurance that a somewhat arbitrary, possibly *post hoc*, choice of futility boundary has not influenced the statistical significance of the result.

In creating a group sequential test with a non-binding futility boundary, we no longer use Equations (5.5) and (5.6). Instead, we require the upper boundary $\{b_1, \ldots, b_K\}$ to satisfy

$$Pr_{\theta=0}\{Z_k > b_k \text{ for some } k = 1, \ldots, K\} = \alpha. \tag{5.10}$$

This places just one constraint on the K constants $\{b_1, \ldots, b_K\}$. We may, for example, consider the power family with parameter Δ and, assuming equally spaced information levels, set

$$b_k = \tilde{C}_1'(K, \alpha, \Delta)(k/K)^{\Delta - 1/2}, \quad k = 1, \ldots, K. \tag{5.11}$$

For given K, α and Δ, Equation (5.10) may be solved for $\tilde{C}_1'(K, \alpha, \Delta)$ using the iterated integral formulae of Section 3.5. In contrast to a design with a binding futility boundary, the constant $\tilde{C}_1'$ does not depend on the value of β, nor on any lower boundary values $\{a_k\}$. Moreover, as long as the observed information levels are equally spaced, $\tilde{C}_1'$ and the boundary values given by (5.11) do not depend on the final information level $\mathcal{I}_K$. Thus we can choose the value of $\mathcal{I}_K$ independently of β, as in the construction of a two-sided test; see the comment following Equation (4.3).

In choosing the sample size, or more precisely the final information level $\mathcal{I}_K$, we can still aim to achieve power $1 - \beta$ under $\theta = \delta$. However, there are different ways to do this depending on how we expect a futility boundary may be used.

No early stopping to accept H_0

If the choice of futility boundary is to be completely arbitrary, we proceed as if no early acceptance of H_0 is possible, so there is no futility boundary and H_0 can only be accepted at the final stage. We choose $\mathcal{I}_K$ to satisfy

$$Pr_{\theta=\delta}\{Z_k > b_k \text{ for some } k = 1, \ldots, K\} = 1 - \beta, \tag{5.12}$$

corresponding to Equation (5.5) when $a_1 = \ldots = a_{K-1} = -\infty$ and $a_K = b_K$. Since the upper boundary values $\{b_k\}$ have already been specified, if we assume equally spaced information levels, $\mathcal{I}_k = (k/K)\mathcal{I}_K$, the left hand side of (5.12) depends only on $\mathcal{I}_K$ and the value of $\mathcal{I}_K$ required to give power $1 - \beta$ when $\theta = \delta$ can be found by solving this equation. As usual we may utilize the methods of Section 3.5.

Taking a power family efficacy boundary with parameter Δ given by (5.11), we define the inflation factor to be

$$\tilde{R}'(K, \alpha, \beta, \Delta) = \mathcal{I}_K / \mathcal{I}_{f,1},$$

where $\mathcal{I}_K$ is the final information level needed to satisfy (5.12). In Tables 5.5, 5.6 and 5.7, we present values of $\tilde{C}_1'(K, \alpha, \Delta)$ and $\tilde{R}'(K, \alpha, \beta, \Delta)$, along with properties of the resulting tests for $\Delta = -0.5$, -0.25, 0 and 0.25, $\alpha = 0.025$ and $1 - \beta = 0.8$, 0.9 and 0.95. Entries in these tables were computed numerically using the methods described in Chapter 31.

Note that all the expected sample sizes in Tables 5.5, 5.6 and 5.7 are computed assuming there is no possibility to stop early for futility. Any arbitrary or *post hoc* rule used to declare futility and accept H_0 will only decrease the false positive rate below α. Such a rule will, however, reduce the power at $\theta = \delta$ below $1 - \beta$ as well as reducing expected sample size.

Non-binding futility boundaries from the power family

In practice, applying no futility boundary may lead to larger sample sizes than desired when H_0 is true. Also, the monitors of a trial might prefer to have some futility boundary $\{a_k, k = 1, \ldots K-1\}$ in place as a guideline. We use the term "guideline" because the trial can be stopped with acceptance for H_0 even if the futility boundary is not crossed and conversely the trial may be continued even if it is crossed. For this reason the futility boundary is termed *non-binding*. We emphasize that, by Equation (5.10), the nominal Type I error rate α is preserved whether or not this guideline is followed, with any early stopping for futility reducing the actual Type I error probability. The example that follows in Section 5.4 illustrates the degree of conservatism that may occur when a non-binding futility boundary is actually adhered to: in this example the Type I error probability rate falls from 0.025 to

Table 5.5 *Constants $C_1'(K,\alpha,\Delta)$ and $R'(K,\alpha,\beta,\Delta)$ for power family one-sided tests with no early stopping to accept H_0 and shape parameter Δ for the upper boundary. Also shown are expected sample sizes at $\theta = 0$, $\delta/2$ and δ expressed as percentages of the corresponding fixed sample size. Tests are for K groups of observations, Type I error probability $\alpha = 0.025$ at $\theta = 0$ and power $1-\beta = 0.8$ at $\theta = \delta$.*

	K	C_1'	R'	*Expected sample size, as percentage of fixed sample size, at* $\theta = 0$	$\theta = \delta/2$	$\theta = \delta$
$\Delta = -0.5$						
	1	1.960	1.000	100.0	100.0	100.0
	2	1.960	1.000	100.0	99.9	98.7
	3	1.964	1.001	100.1	98.9	91.6
	4	1.971	1.004	100.2	98.3	89.1
	5	1.978	1.006	100.4	98.0	87.5
	10	2.006	1.012	101.0	97.5	84.6
	15	2.022	1.015	101.2	97.4	83.7
	20	2.034	1.017	101.4	97.4	83.2
$\Delta = -0.25$						
	1	1.960	1.000	100.0	100.0	100.0
	2	1.962	1.001	100.1	99.5	95.4
	3	1.975	1.005	100.4	98.4	88.7
	4	1.988	1.009	100.8	98.0	86.4
	5	1.999	1.013	101.0	97.8	84.9
	10	2.034	1.021	101.8	97.5	82.1
	15	2.054	1.025	102.1	97.4	81.2
	20	2.067	1.027	102.3	97.4	80.8
$\Delta = 0$						
	1	1.960	1.000	100.0	100.0	100.0
	2	1.977	1.008	100.6	99.0	90.2
	3	2.004	1.017	101.5	98.3	85.6
	4	2.024	1.024	102.1	98.0	83.1
	5	2.040	1.028	102.5	97.9	81.8
	10	2.087	1.040	103.5	97.9	79.1
	15	2.110	1.045	104.0	97.9	78.3
	20	2.126	1.047	104.2	98.0	77.9
$\Delta = 0.25$						
	1	1.960	1.000	100.0	100.0	100.0
	2	2.038	1.038	103.4	99.7	86.0
	3	2.083	1.054	104.9	99.7	82.0
	4	2.113	1.065	105.8	99.7	79.9
	5	2.136	1.072	106.5	99.8	78.7
	10	2.199	1.089	108.1	100.2	76.2
	15	2.229	1.097	108.8	100.4	75.4
	20	2.248	1.101	109.2	100.6	75.0

Table 5.6 *Constants* $C_1'(K, \alpha, \Delta)$ *and* $R'(K, \alpha, \beta, \Delta)$ *for power family one-sided tests with no early stopping to accept* H_0 *and shape parameter* Δ *for the upper boundary. Also shown are expected sample sizes at* $\theta = 0$, $\delta/2$ *and* δ *expressed as percentages of the corresponding fixed sample size. Tests are for* K *groups of observations, Type I error probability* $\alpha = 0.025$ *at* $\theta = 0$ *and power* $1 - \beta = 0.9$ *at* $\theta = \delta$.

	K	C_1'	R'	*Expected sample size, as percentage of fixed sample size, at* $\theta = 0$	$\theta = \delta/2$	$\theta = \delta$
$\Delta = -0.5$						
	1	1.960	1.000	100.0	100.0	100.0
	2	1.960	1.000	100.0	99.9	97.4
	3	1.964	1.001	100.1	98.4	87.3
	4	1.971	1.003	100.2	97.5	84.7
	5	1.978	1.005	100.4	97.0	82.7
	10	2.006	1.011	100.9	96.1	79.1
	15	2.022	1.014	101.2	95.9	78.1
	20	2.034	1.016	101.3	95.8	77.6
$\Delta = -0.25$						
	1	1.960	1.000	100.0	100.0	100.0
	2	1.962	1.001	100.0	99.3	92.2
	3	1.975	1.005	100.4	97.5	83.9
	4	1.988	1.009	100.7	96.8	81.0
	5	1.999	1.012	101.0	96.4	79.3
	10	2.034	1.020	101.7	95.7	75.9
	15	2.054	1.024	102.0	95.6	74.8
	20	2.067	1.026	102.2	95.5	74.3
$\Delta = 0$						
	1	1.960	1.000	100.0	100.0	100.0
	2	1.977	1.007	100.6	98.2	85.1
	3	2.004	1.016	101.4	96.9	79.9
	4	2.024	1.022	101.9	96.4	76.7
	5	2.040	1.026	102.3	96.1	75.0
	10	2.087	1.037	103.3	95.6	71.8
	15	2.110	1.042	103.7	95.5	70.8
	20	2.126	1.045	103.9	95.5	70.3
$\Delta = 0.25$						
	1	1.960	1.000	100.0	100.0	100.0
	2	2.038	1.034	103.0	98.0	79.5
	3	2.083	1.050	104.4	97.4	74.5
	4	2.113	1.059	105.3	97.1	71.9
	5	2.136	1.066	105.9	97.0	70.4
	10	2.199	1.083	107.5	96.9	67.2
	15	2.229	1.090	108.1	97.0	66.2
	20	2.248	1.094	108.5	97.0	65.7

Table 5.7 *Constants $C'_1(K,\alpha,\Delta)$ and $R'(K,\alpha,\beta,\Delta)$ for power family one-sided tests with no early stopping to accept H_0 and shape parameter Δ for the upper boundary. Also shown are expected sample sizes at $\theta = 0$, $\delta/2$ and δ expressed as percentages of the corresponding fixed sample size. Tests are for K groups of observations, Type I error probability $\alpha = 0.025$ at $\theta = 0$ and power $1-\beta = 0.95$ at $\theta = \delta$.*

	K	C'_1	R'	*Expected sample size, as percentage of fixed sample size, at* $\theta = 0$	$\theta = \delta/2$	$\theta = \delta$
$\Delta = -0.5$						
	1	1.960	1.000	100.0	100.0	100.0
	2	1.960	1.000	100.0	99.8	95.7
	3	1.964	1.001	100.1	97.8	83.4
	4	1.971	1.003	100.2	96.6	80.9
	5	1.978	1.005	100.4	96.0	78.6
	10	2.006	1.011	100.9	94.8	74.8
	15	2.022	1.014	101.1	94.4	73.6
	20	2.034	1.015	101.3	94.3	73.0
$\Delta = -0.25$						
	1	1.960	1.000	100.0	100.0	100.0
	2	1.962	1.001	100.0	99.0	88.7
	3	1.975	1.005	100.4	96.6	79.9
	4	1.988	1.008	100.7	95.7	76.5
	5	1.999	1.011	100.9	95.1	74.6
	10	2.034	1.019	101.6	94.1	70.9
	15	2.054	1.023	101.9	93.8	69.8
	20	2.067	1.025	102.1	93.7	69.2
$\Delta = 0$						
	1	1.960	1.000	100.0	100.0	100.0
	2	1.977	1.007	100.5	97.4	80.3
	3	2.004	1.015	101.3	95.6	75.1
	4	2.024	1.021	101.8	94.8	71.6
	5	2.040	1.025	102.1	94.3	69.6
	10	2.087	1.036	103.1	93.5	66.1
	15	2.110	1.040	103.5	93.2	65.0
	20	2.126	1.043	103.7	93.2	64.4
$\Delta = 0.25$						
	1	1.960	1.000	100.0	100.0	100.0
	2	2.038	1.032	102.8	96.5	74.0
	3	2.083	1.047	104.1	95.3	68.5
	4	2.113	1.056	105.0	94.8	65.7
	5	2.136	1.062	105.5	94.5	63.9
	10	2.199	1.078	107.0	94.0	60.4
	15	2.229	1.085	107.6	93.9	59.2
	20	2.248	1.088	108.0	93.9	58.7

0.022 and maximum and expected sample sizes are 2 or 3 percent higher than for a design with a binding futility boundary.

The proposed non-binding futility boundary can be used to establish the target maximum information level $\mathcal{I}_K$ so that power $1-\beta$ at $\theta=\delta$ will be achieved if this lower boundary is obeyed. It is convenient to take a non-binding lower boundary in the power family form although, given the different roles of the efficacy and futility boundaries, it is natural to allow these to have different shape parameters.

Suppose we assume equally spaced information levels $\mathcal{I}_k = (k/K)\mathcal{I}_K$, $k=1,\ldots,K$, and define the efficacy boundary by (5.11) with shape parameter $\Delta=\Delta_1$ where $\tilde{C}_1'(K,\alpha,\Delta_1)$ is set to give Type I error probability α under $\theta=0$. We note that the resulting upper boundary $(b_1,\ldots,b_K)$ does not depend on the value of $\mathcal{I}_K$. The non-binding futility boundary with shape parameter Δ_2 has the form

$$a_k = \delta\sqrt{(k/K)\mathcal{I}_K} - \tilde{C}_2^*(K,\alpha,\beta,\Delta_1,\Delta_2)(k/K)^{\Delta_2-1/2}, \quad k=1,\ldots,K. \qquad (5.13)$$

The constant $\tilde{C}_2^*(K,\alpha,\beta,\Delta_1,\Delta_2)$ and the maximum information level $\mathcal{I}_K$ are determined by the constraint that $a_K=b_K$, which can be written as

$$\mathcal{I}_K = \{\tilde{C}_1'(K,\alpha,\Delta_1)+\tilde{C}_2^*(K,\alpha,\beta,\Delta_1,\Delta_2)\}^2/\delta^2,$$

and the Type II error probability requirement (5.6). We may then define the inflation factor

$$\tilde{R}^*(K,\alpha,\beta,\Delta_1,\Delta_2) = \mathcal{I}_K/\mathcal{I}_{f,1},$$

where $\mathcal{I}_{f,1}$ is given by (5.1). In Tables 5.8, 5.9 and 5.10, we present values of $\tilde{C}_1'(K,\alpha,\Delta_1)$, $\tilde{C}_2^*(K,\alpha,\beta,\Delta_1,\Delta_2)$ and $\tilde{R}^*(K,\alpha,\beta,\Delta_1,\Delta_2)$, along with properties of the resulting tests when the efficacy and futility boundaries have a common shape parameter $\Delta_1=\Delta_2=\Delta$. Results are presented for $\Delta=-0.5$, -0.25, 0 and 0.25, $\alpha=0.025$, and $1-\beta=0.8$, 0.9 and 0.95, and expected sample sizes are shown for the case where the non-binding futility stopping rule is adhered to. Entries in these tables were computed numerically using the methods described in Chapter 31

A feature of the procedures in this section is that the critical value at any analysis, either interim or final, cannot be less than z_α, the critical value of a fixed sample test with Type I error rate α, so the issue discussed in Section 5.2.1 of H_0 being rejected at the final analysis when $z_K > z_\alpha$ does not arise. This property follows from (5.10) since, for any $k^* \in \{1,\ldots,K\}$,

$$Pr_{\theta=0}\{Z_{k^*} > b_{k^*}\} < Pr_{\theta=0}\{Z_k > b_k \text{ for some } k=1,\ldots,K\} = \alpha.$$

By (3.1), each Z_{k^*} has a standard normal distribution in the absence of an early stopping rule and, hence, $b_{k^*} > z_\alpha$. In particular, $b_K > z_\alpha$.

Finally we note that stopping early for futility can be based on the ideas of stochastic curtailment, whereby a trial is terminated when, based on current data, it appears unlikely that H_0 will ultimately be rejected. This approach can be employed in a binding or, more usually, a non-binding fashion. We shall discuss stochastic curtailment rules in Chapter 11, in particular in Section 11.5.

5.4 An Example Comparing Three Futility Stopping Strategies

We illustrate the design and analysis of a group sequential trial using a one-sided test from the power family in a balanced, parallel arm, two-treatment comparison with normal responses with a common, known variance. In the notation of Section 3.2.1, suppose responses $X_{Ai} \sim N(\mu_A, \sigma^2)$ and $X_{Bi} \sim N(\mu_B, \sigma^2)$ are observed on treatments A and B,

Table 5.8 *Constants* $\tilde{C}_1'(K, \alpha, \Delta)$, $\tilde{C}_2^*(K, \alpha, \beta, \Delta, \Delta)$ *and* $\tilde{R}^*(K, \alpha, \beta, \Delta, \Delta)$ *for power family one-sided tests with non-binding futility boundaries and common shape parameter* $\Delta_1 = \Delta_2 = \Delta$. *Also shown are expected sample sizes at* $\theta = 0$, $\delta/2$ *and* δ *expressed as percentages of the corresponding fixed sample size. Tests are for* K *groups of observations, Type I error probability* $\alpha = 0.025$ *at* $\theta = 0$ *and power* $1 - \beta = 0.8$ *at* $\theta = \delta$.

	K	$\tilde{C}_1'$	$\tilde{C}_2^*$	$\tilde{R}^*$	*Expected sample size, as percentage of fixed sample size, at* $\theta = 0$	$\theta = \delta/2$	$\theta = \delta$
$\Delta = -0.5$							
	1	1.960	0.842	1.000	100.0	100.0	100.0
	2	1.960	0.870	1.021	71.3	90.3	98.6
	3	1.964	0.902	1.047	70.2	87.3	91.9
	4	1.971	0.921	1.066	66.6	85.0	89.9
	5	1.978	0.934	1.081	64.7	83.9	88.5
	10	2.006	0.968	1.127	62.2	82.1	86.6
	15	2.022	0.984	1.151	61.5	81.8	86.4
	20	2.034	0.993	1.167	61.3	81.8	86.5
$\Delta = -0.25$							
	1	1.960	0.842	1.000	100.0	100.0	100.0
	2	1.962	0.903	1.046	68.3	88.0	95.9
	3	1.975	0.934	1.078	64.8	84.8	89.6
	4	1.988	0.955	1.104	62.9	83.0	87.6
	5	1.999	0.970	1.123	61.2	81.8	86.3
	10	2.034	1.008	1.179	58.2	79.9	84.5
	15	2.054	1.026	1.209	57.4	79.6	84.3
	20	2.067	1.037	1.228	57.1	79.6	84.3
$\Delta = 0$							
	1	1.960	0.842	1.000	100.0	100.0	100.0
	2	1.977	0.949	1.091	67.1	86.3	91.5
	3	2.004	0.988	1.140	60.4	82.3	87.1
	4	2.024	1.012	1.174	58.0	80.4	84.7
	5	2.040	1.029	1.200	56.6	79.3	83.5
	10	2.087	1.074	1.273	53.5	77.3	81.5
	15	2.110	1.096	1.310	52.5	76.8	81.2
	20	2.126	1.109	1.333	52.1	76.7	81.1
$\Delta = 0.25$							
	1	1.960	0.842	1.000	100.0	100.0	100.0
	2	2.038	1.002	1.177	68.4	85.8	87.8
	3	2.083	1.069	1.266	59.0	81.0	83.6
	4	2.113	1.109	1.323	54.9	78.6	81.4
	5	2.136	1.137	1.365	52.6	77.2	80.0
	10	2.199	1.206	1.477	48.5	74.5	77.4
	15	2.229	1.237	1.531	47.1	73.7	76.6
	20	2.248	1.256	1.564	46.4	73.3	76.3

Table 5.9 *Constants* $\tilde{C}_1'(K, \alpha, \Delta)$, $\tilde{C}_2^*(K, \alpha, \beta, \Delta, \Delta)$ *and* $\tilde{R}^*(K, \alpha, \beta, \Delta, \Delta)$ *for power family one-sided tests with non-binding futility boundaries and common shape parameter* $\Delta_1 = \Delta_2 = \Delta$. *Also shown are expected sample sizes at* $\theta = 0$, $\delta/2$ *and* δ *expressed as percentages of the corresponding fixed sample size. Tests are for* K *groups of observations, Type I error probability* $\alpha = 0.025$ *at* $\theta = 0$ *and power* $1 - \beta = 0.9$ *at* $\theta = \delta$.

	K	$\tilde{C}_1'$	$\tilde{C}_2^*$	$\tilde{R}^*$	*Expected sample size, as percentage of fixed sample size, at* $\theta = 0$	$\theta = \delta/2$	$\theta = \delta$
$\Delta = -0.5$							
	1	1.960	1.282	1.000	100.0	100.0	100.0
	2	1.960	1.286	1.003	80.7	96.3	97.4
	3	1.964	1.305	1.017	75.1	90.8	87.6
	4	1.971	1.318	1.029	70.9	88.5	85.3
	5	1.978	1.327	1.040	69.5	87.3	83.5
	10	2.006	1.353	1.074	66.3	85.0	80.9
	15	2.022	1.366	1.093	65.5	84.6	80.4
	20	2.034	1.374	1.105	65.2	84.5	80.3
$\Delta = -0.25$							
	1	1.960	1.282	1.000	100.0	100.0	100.0
	2	1.962	1.302	1.014	73.6	92.9	92.5
	3	1.975	1.326	1.037	70.8	88.6	84.7
	4	1.988	1.342	1.055	67.2	86.3	82.1
	5	1.999	1.353	1.069	65.2	85.0	80.7
	10	2.034	1.384	1.112	62.2	82.9	78.1
	15	2.054	1.400	1.135	61.3	82.5	77.7
	20	2.067	1.409	1.150	61.0	82.4	77.5
$\Delta = 0$							
	1	1.960	1.282	1.000	100.0	100.0	100.0
	2	1.977	1.338	1.046	69.2	89.6	86.1
	3	2.004	1.367	1.081	64.6	85.8	81.4
	4	2.024	1.386	1.107	62.3	83.8	78.5
	5	2.040	1.401	1.127	60.6	82.5	77.0
	10	2.087	1.441	1.184	57.2	80.3	74.5
	15	2.110	1.460	1.213	56.2	79.8	74.0
	20	2.126	1.471	1.231	55.8	79.6	73.8
$\Delta = 0.25$							
	1	1.960	1.282	1.000	100.0	100.0	100.0
	2	2.038	1.399	1.124	68.2	87.7	81.5
	3	2.083	1.451	1.188	60.5	83.5	76.8
	4	2.113	1.483	1.231	57.2	81.4	74.3
	5	2.136	1.505	1.262	55.3	80.1	72.8
	10	2.199	1.564	1.347	51.6	77.5	69.8
	15	2.229	1.591	1.389	50.3	76.7	68.9
	20	2.248	1.608	1.415	49.6	76.4	68.5

Table 5.10 *Constants* $\tilde{C}_1'(K,\alpha,\Delta)$, $\tilde{C}_2^*(K,\alpha,\beta,\Delta,\Delta)$ *and* $\tilde{R}^*(K,\alpha,\beta,\Delta,\Delta)$ *for power family one-sided tests with non-binding futility boundaries and common shape parameter* $\Delta_1 = \Delta_2 = \Delta$. *Also shown are expected sample sizes at* $\theta = 0$, $\delta/2$ *and* δ *expressed as percentages of the corresponding fixed sample size. Tests are for* K *groups of observations, Type I error probability* $\alpha = 0.025$ *at* $\theta = 0$ *and power* $1-\beta = 0.95$ *at* $\theta = \delta$.

	K	$\tilde{C}_1'$	$\tilde{C}_2^*$	$\tilde{R}^*$	*Expected sample size, as percentage of fixed sample size, at* $\theta = 0$	$\theta = \delta/2$	$\theta = \delta$
$\Delta = -0.5$							
	1	1.960	1.645	1.000	100.0	100.0	100.0
	2	1.960	1.645	1.000	88.6	98.7	95.7
	3	1.964	1.653	1.007	77.7	93.0	83.6
	4	1.971	1.661	1.015	74.8	90.6	81.3
	5	1.978	1.667	1.023	72.9	89.1	79.2
	10	2.006	1.687	1.049	69.4	86.6	76.1
	15	2.022	1.697	1.065	68.4	86.0	75.4
	20	2.034	1.704	1.075	68.1	85.8	75.2
$\Delta = -0.25$							
	1	1.960	1.645	1.000	100.0	100.0	100.0
	2	1.962	1.651	1.004	79.7	96.0	88.8
	3	1.975	1.667	1.020	74.4	90.5	80.4
	4	1.988	1.678	1.034	70.2	88.2	77.3
	5	1.999	1.686	1.045	68.6	86.9	75.7
	10	2.034	1.712	1.080	65.2	84.5	72.8
	15	2.054	1.725	1.099	64.3	83.9	72.1
	20	2.067	1.733	1.111	63.9	83.8	71.9
$\Delta = 0$							
	1	1.960	1.645	1.000	100.0	100.0	100.0
	2	1.977	1.673	1.026	72.4	91.9	80.8
	3	2.004	1.696	1.053	68.6	87.8	76.3
	4	2.024	1.712	1.075	65.6	85.5	73.0
	5	2.040	1.725	1.091	63.6	84.2	71.3
	10	2.087	1.760	1.138	60.1	81.9	68.5
	15	2.110	1.777	1.163	59.1	81.3	67.8
	20	2.126	1.787	1.178	58.7	81.1	67.5
$\Delta = 0.25$							
	1	1.960	1.645	1.000	100.0	100.0	100.0
	2	2.038	1.731	1.093	69.1	89.0	75.7
	3	2.083	1.773	1.144	62.5	85.1	70.6
	4	2.113	1.800	1.178	59.6	83.0	68.0
	5	2.136	1.819	1.204	57.8	81.7	66.4
	10	2.199	1.872	1.275	54.1	79.1	63.1
	15	2.229	1.898	1.311	52.8	78.3	62.2
	20	2.248	1.913	1.333	52.2	77.9	61.7

respectively, where $\sigma = 10$. The parameter of interest is the effect size $\theta = \mu_A - \mu_B$ and we wish to test the hypothesis H_0: $\theta \le 0$, or equivalently, $\mu_A \le \mu_B$.

We shall design a power family test with $\Delta = 0$ and four equally sized groups of observations to test H_0: $\mu_A \le \mu_B$ against the one-sided alternative $\mu_A > \mu_B$ with Type I error probability $\alpha = 0.025$ and power $1 - \beta = 0.9$ at $\mu_A - \mu_B = \delta = 5.0$. A fixed sample test requires information

$$\mathcal{I}_{f,1} = \{\Phi^{-1}(0.975) + \Phi^{-1}(0.9)\}^2/5.0^2 = 0.420$$

for $\theta = \mu_A - \mu_B$. Since the estimate of $\mu_A - \mu_B$ from n subjects on each arm has variance $2\sigma^2/n = 200/n$, information is equal to $n/200$ and the fixed sample design needs $n = 200 \times 0.420 = 84.03$, which rounds up to 85 subjects per arm. As alternative designs, we consider three versions of a one-sided, power family, group sequential design, each with four stages, Type I error rate $\alpha = 0.025$ and power $1 - \beta = 0.9$ when $\theta = \delta = 5$. In case (a) the futility boundary is assumed to be binding, in case (b) there is no early stopping for futility, and in case (c) a non-binding futility boundary is provided as guidance to the trial's monitoring committee. As previously discussed, in cases (b) and (c) the Type I error rate will be no greater than $\alpha = 0.025$ whatever rule is actually applied to decide whether to stop for futility.

(a) The futility boundary is binding

In the case of a binding futility boundary, we have $\tilde{R}(4, 0.025, 0.1, 0) = 1.077$ from Table 5.2 and the group sequential trial should be designed to achieve a final information level $\mathcal{I}_4 = \tilde{R}(4, 0.025, 0.1, 0.0)\,\mathcal{I}_{f,1} = 1.077 \times 0.420 = 0.453$. Thus we require a maximum sample size of $n = 200 \times 0.453 = 90.6$ observations per arm, which can be achieved with four equally sized groups of 23 subjects per arm.

Denoting the total number of subjects observed on each arm at analysis k by n_k, the standardized test statistics at the four analyses are

$$Z_k = \frac{1}{\sqrt{(2 n_k \sigma^2)}} \left(\sum_{i=1}^{n_k} X_{Ai} - \sum_{i=1}^{n_k} X_{Bi} \right), \quad k = 1, \ldots, 4, \tag{5.14}$$

and the observed information levels are

$$\mathcal{I}_k = n_k/(2\sigma^2) = n_k/200, \quad k = 1, \ldots, 4.$$

Substituting constants $\tilde{C}_1(4, 0.025, 0.1, 0) = 1.978$ and $\tilde{C}_2(4, 0.025, 0.1, 0) = 1.385$ from Table 5.2. into (5.7), we see that at analyses $k = 1$, 2 and 3, the test stops to reject H_0 if

$$Z_k \ge \tilde{C}_1(4, 0.025, 0.1, 0)\,(k/4)^{-1/2} = 1.978\,(k/4)^{-1/2}, \tag{5.15}$$

it stops to accept H_0 if

$$\begin{aligned} Z_k \le\; & \{\tilde{C}_1(4, 0.025, 0.1, 0) + \tilde{C}_2(4, 0.025, 0.1, 0)\}\,(k/4)^{1/2} \\ & - \tilde{C}_2(4, 0.025, 0.1, 0)\,(k/4)^{-1/2} \\ =\; & 3.363\,(k/4)^{1/2} - 1.385\,(k/4)^{-1/2}, \end{aligned} \tag{5.16}$$

and otherwise the study continues to the next analysis. If the trial reaches the fourth analysis it terminates at this stage, rejecting H_0 if $Z_4 \ge 1.978$ and accepting H_0 otherwise. The boundary values $(a_1, \ldots, a_4)$ and $(b_1, \ldots, b_4)$ are shown in the third column of Table 5.11 and displayed in Figure 5.2.

Table 5.11 *Examples of power family, one-sided tests using three futility stopping strategies. Constants and boundary values are for tests with $K = 4$ stages and shape parameter $\Delta = 0$ in designs (a) and (b), and common shape parameter $\Delta_1 = \Delta_2 = 0$ in design (c). Tests are designed to have Type I error rate $\alpha = 0.025$ and power $1 - \beta = 0.9$ at $\theta = 5$. In maximum and expected sample sizes, N denotes the number of subjects per treatment arm. In design (c), values of $E(N)$ are presented for the case where the suggest futility boundary is strictly obeyed.*

		Type of futility boundary		
		(a) Binding	*(b) None*	*(c) Non-binding*
Constants	C_1	1.978	2.024	2.024
	C_2	1.385	—	1.386
	R	1.077	1.022	1.107
from		Table 5.2	Table 5.6	Table 5.9
Upper boundary	b_1	3.957	4.049	4.049
	b_2	2.798	2.863	2.863
	b_3	2.284	2.337	2.337
	b_4	1.978	2.024	2.024
Lower boundary	a_1	−1.089	$-\infty$	−1.068
	a_2	0.419	$-\infty$	0.451
	a_3	1.313	$-\infty$	1.353
	a_4	1.978	2.024	2.024
Type I error rate		0.025	0.025	0.022
Power when $\theta = \delta$		0.905	0.907	0.909
Maximum N		92	88	96
$E_{\theta=0}(N)$		52.2	87.7	54.0
$E_{\theta=\delta/2}(N)$		69.6	82.8	72.8
$E_{\theta=\delta}(N)$		64.6	65.6	67.6

(b) No early stopping for futility

Continuing with the same example, suppose we wish to define a test with no lower boundary at all. This may be because we would like to allow an arbitrary stopping rule for futility while maintaining a Type I error probability of at most α. We define the upper, efficacy boundary from the power family according to (5.11). With shape parameter $\Delta = 0$, $K = 4$ equally sized groups, $\alpha = 0.025$ and $\beta = 0.1$, Table 5.6 gives the constant $\tilde{C}'_1(4, 0.025, 0) = 2.024$ and inflation factor $\tilde{R}'(4, 0.025, 0.1, 0) = 1.022$. Thus, the group sequential test should be designed to attain information level $\mathcal{I}_4 = \tilde{R}'(4, 0.025, 0.1, 0)\,\mathcal{I}_{f,1} = 1.022 \times 0.420 = 0.430$. This implies a maximum sample size of $n = 85.9$ observations per arm, which can be achieved with four equally sized groups of 22 subjects per arm.

At analyses $k = 1, \ldots, 4$, the test statistic Z_k and information level $\mathcal{I}_k$ are computed as before. We stop to reject H_0 at analysis k if $Z_k \geq b_k = 2.024(k/4)^{-1/2}$ and, if analysis 4 is reached and $Z_4 < 2.024$, H_0 is accepted. The boundary values $(b_1, \ldots, b_4)$ and a_4 are shown in the fourth column of Table 5.11 and displayed in Figure 5.2. By design, if the trial is terminated early with acceptance of H_0 for any reason, then the Type I error probability will be reduced, as will the power at $\theta = \delta$.

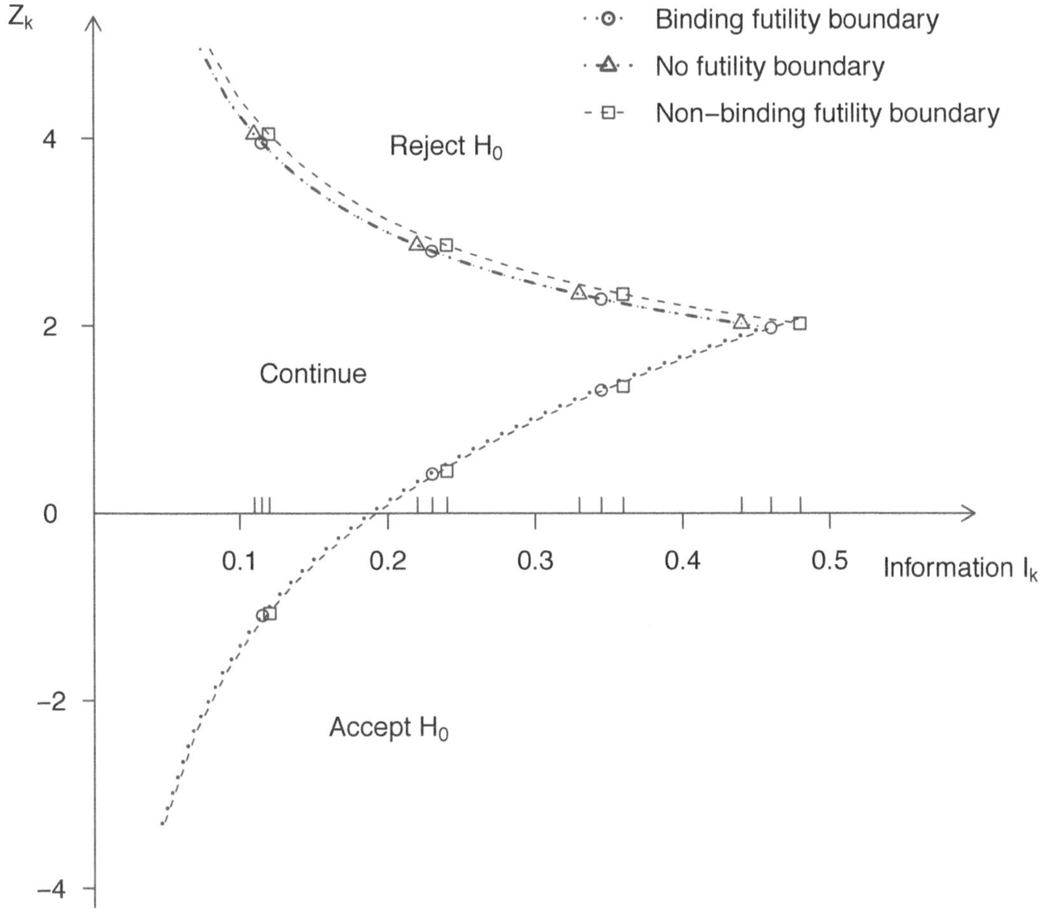

Figure 5.2 *One-sided tests with four groups of observations for the example of Section 5.4*

(c) Non-binding futility boundary provided as guidance

Finally, suppose we wish to design a power family test with $K = 4$ analyses that controls the Type I error rate at $\alpha = 0.025$ whatever futility boundary is employed, but we specify a lower boundary as guidance for the study monitors and, if this boundary is applied strictly, power of $1 - \beta = 0.9$ will be achieved if $\delta = 5.0$. Thus we are to have a non-binding lower boundary.

With shape parameters $\Delta_1 = 0$ for the efficacy boundary and $\Delta_2 = 0$ for the non binding futility boundary, Table 5.9 gives the constants $\tilde{C}_1'(4, 0.025, 0) = 2.024$ and $\tilde{C}_2^*(4, 0.025, 0.1, 0, 0) = 1.386$, and inflation factor $R^*(4, 0.025, 0.1, 0, 0) = 1.107$. Thus, the target for the final information level is $\mathcal{I}_4 = R^*(4, 0.025, 0.1, 0, 0)\, \mathcal{I}_{f,1} = 1.107 \times 0.420 = 0.465$. This implies a maximum of $n = 93.1$ observations per arm, which can be achieved with four equally sized groups of 24 subjects per arm. Note that since the futility boundary is non-binding, the upper boundary is the same as for case (b) when there was no lower boundary at all, but analyses occur after different numbers of observations.

Again, the test statistic Z_k and information level $\mathcal{I}_k$ are computed at analyses $k = 1, \ldots, 4$. The test stops to reject H_0 at analysis k if $Z_k \geq b_k = 2.024(k/4)^{-1/2}$ and if analysis 4 is reached and $Z_4 < 2.024$, H_0 is accepted. The lower boundary, which is for guidance only, suggests stopping to accept H_0 at stage k if

$$
\begin{aligned}
Z_k &\leq \{(\tilde{C}_1'(4, 0.025, 0) + \tilde{C}_2^*(4, 0.025, 0.1, 0)\} (k/4)^{1/2} \\
&\quad - \tilde{C}_2^*(4, 0.025, 0.1, 0, 0)\, (k/4)^{-1/2} \\
&= 3.410\, (k/4)^{1/2} - 1.386\, (k/4)^{-1/2}, \quad k = 1, \ldots, 3.
\end{aligned}
$$

However, if the lower boundary is crossed and it is decided that the study should continue, the Type I error rate is still controlled at level α because of property (5.10). The boundary values $(a_1, \ldots, a_4)$ and $(b_1, \ldots, b_4)$ are shown in the fifth column of Table 5.11 and displayed in Figure 5.2.

Differences between the three types of design can be seen in Figure 5.2 and in the properties listed in Table 5.11. In the table, designs are implemented with equally sized groups with integer sample sizes. In case (c), Type I error rate, power and expected sample sizes are calculated assuming the futility boundary is strictly obeyed. For all three types of futility boundary, rounding up the group sizes to integer values increases power at $\theta = 5$ beyond 0.9. A comparison between cases (a) and (c) shows that treating the futility boundary as non-binding rather than binding has a small cost in terms of maximum and expected sample sizes, which can be attributed to the fact that the Type I error condition is satisfied conservatively. In case (b), expected sample sizes are calculated assuming no early stopping for futility and the high values of $E(N)$ when $\theta = 0$ or $\delta/2$ show why, in practice, it is important to provide an opportunity for early stopping. While it is an option to provide trial monitors with a design of type (b) and suggest they apply their own criteria to decide when to stop for futility, we would recommend the more structured approach of case (c) where a guideline is provided but monitors are given freedom in the way they apply this suggested boundary.

5.5 Adapting Power Family Tests to Unequal Increments in Information

5.5.1 A Method that Protects the Type I Error Rate

Although the power family tests are designed for the specific sequence of information levels, they may still be applied for other observed information levels, but error probabilities may then differ from their intended values. Consider a design with Type I error rate α, power $1 - \beta$ at $\theta = \delta$, K analyses, shape parameter Δ and a binding futility boundary. Suppose constants $\tilde{C}_1(K, \alpha, \beta, \Delta)$ and $\tilde{C}_2(K, \alpha, \beta, \Delta)$ and inflation factor $\tilde{R}(K, \alpha, \beta, \Delta)$ have been derived to define a test appropriate to the information levels $\mathcal{I}_k = (k/K)\, \tilde{R}(K, \alpha, \beta, \Delta)\, \mathcal{I}_{f,1}$, $k = 1, \ldots, K$. When observed information levels differ from this sequence of values, one might simply apply the stopping rule (5.2) with critical values a_k and b_k obtained by substituting the observed information levels $\mathcal{I}_1, \ldots, \mathcal{I}_K$ into (5.3). However, we recommend the following approach which is more successful in maintaining the specified Type I error rate α. Note that, although we describe this method for the case of a binding futility boundary, the same prescription can be applied to designs with no futility boundary or a non-binding futility boundary.

Let $(a_1, \ldots, a_K)$ and $(b_1, \ldots, b_K)$ be as defined in (5.7), so they give a boundary on the Z-scale derived for the case of equally spaced information levels but with no explicit mention of the values $\mathcal{I}_1, \ldots, \mathcal{I}_K$. In adapting the power family test to the observed information levels we simply apply the stopping rule (5.2) with these values of a_k and b_k. Thus, the sequence of standardized statistics, $Z_1, \ldots, Z_K$, is compared with a fixed sequence of critical values which do not depend on the observed $\mathcal{I}_1, \ldots, \mathcal{I}_K$.

Since the joint distribution of $\{Z_1, \ldots, Z_K\}$ under $\theta = 0$ depends only on the ratios between observed information levels, the Type I error probability will be exactly α if the actual information levels, $\mathcal{I}_1, \ldots, \mathcal{I}_K$, are equally spaced, just as for two-sided tests in Section 4.2, while unequal spacing will cause minor perturbations in the Type I error rate. Attained power is subject to greater variability, depending primarily on the final information level, $\mathcal{I}_K$, with a subsidiary dependence on the spacing of information levels.

Table 5.12 shows the results of applying this rule with information sequences following the patterns considered for two-sided tests in Table 4.2. Target information levels are $\mathcal{I}_k = (k/K)\mathcal{I}_{max}$, $k = 1, \ldots, K$, where $\mathcal{I}_{max} = \tilde{R}(K, \alpha, \beta, \Delta)\,\mathcal{I}_{f,1}$, and observed information levels are $\mathcal{I}_k = \pi(k/K)^r \mathcal{I}_{max}$, $k = 1, \ldots, K$, the factor π controlling the final information level and the power r the spacing between levels. We see that the Type I error probability remains within 0.001 of $\alpha = 0.025$ in all cases. The attained power is affected primarily by the overall scale of information and is low for $\pi = 0.9$ and high for $\pi = 1.1$. For comparison, a fixed sample test with 0.9 times the required information, $\mathcal{I}_{f,1}$, has power 0.871 and a fixed sample test with information $1.1 \times \mathcal{I}_{f,1}$ has power 0.923. In all cases where $\pi = 1.0$, so $\mathcal{I}_K$ exactly equals its target $\mathcal{I}_{max}$, power is within 0.024 of its nominal value, and it is often much closer. If, instead, we had obtained values a_k and b_k by substituting observed values of $\mathcal{I}_k$ into (5.3), we would have had to resolve the problem of $a_K \neq b_K$ when $\mathcal{I}_K \neq \mathcal{I}_{max}$; using b_K as the critical value to protect the Type I error still leads to variations in attained Type I error from 0.022 to 0.028 in the examples of Table 5.12.

The preceding proposal and its extension to group sequential t-tests is described by Jennison & Turnbull (2001a). Emerson & Fleming (1989) propose an alternative treatment of general information sequences using boundary values on the score statistic scale, but this approach maintains the Type I error rate less accurately. Emerson & Fleming also suggest using their approach to apply a test when the number of analyses differs from that planned: this can lead to quite large discrepancies in Type I error (their Table 7 shows Type I error probabilities ranging from 0.029 to 0.074, when the intended value is 0.05) and we would recommend use of tests defined through error spending functions (see Section 8.3), which preserve the Type I error rate exactly in such situations.

5.5.2 An Example

As an illustrative example, consider design (c) of Section 5.4. In this case, the study is designed to have four groups each comprising 24 patients per treatment arm. The study stops with rejection of H_0 if

$$Z_k \geq b_k = 2.024\,(k/4)^{-1/2}$$

and the futility boundary is crossed if

$$Z_k \leq a_k = 3.410(k/4)^{1/2} - 1.386\,(k/4)^{-1/2}.$$

Since the futility boundary is non-binding, the trial monitors can exercise discretion in deciding whether to recommend terminating the trial when the futility boundary is crossed.

Suppose that when the study is conducted, the cumulative numbers of responses per treatment arm at each analysis are $n_1 = 18$, $n_2 = 44$, $n_3 = 73$ and $n_4 = 95$ instead of the planned values 24, 48, 72 and 96. The stopping boundaries remain as described but with the observed values $n_1, \ldots, n_4$ used in (5.14) to define $Z_1, \ldots, Z_4$.

Calculations show that, if the futility boundary is followed strictly, the attained Type I error rate remains at 0.023 but power at $\mu_A - \mu_B = 5.0$ is 0.903 and expected sample sizes of 51.1, 70.9 and 67.0 under $\theta = 0$, $\delta/2$ and δ are a little lower than those shown in Table 5.11. Although the group sizes are unequal, the final sample size is close to the target for analysis 4 and, in view of earlier evidence of robustness to variations in group sizes, it is no surprise that both error probabilities are close to their nominal values.

Table 5.12 *Properties of power family one-sided tests designed for information levels* $\mathcal{I}_k = (k/K)\mathcal{I}_{max}$, $k = 1, \ldots, K$, *but applied with* $\mathcal{I}_k = \pi(k/K)^r \mathcal{I}_{max}$, $k = 1, \ldots, K$. *Attained Type I error probabilities and power are shown for tests with a binding futility boundary,* K *analyses, and shape parameter* $\Delta = -0.5$ *and* 0, *designed to achieve Type I error rate* $\alpha = 0.025$ *and power* $1 - \beta = 0.9$ *at* $\theta = \delta$.

			$\Delta = -0.5$		$\Delta = 0$	
	r	π	*Type I error*	*Power*	*Type I error*	*Power*
$K = 2$						
	0.80	0.90	0.025	0.868	0.025	0.874
		1.00	0.025	0.901	0.025	0.906
		1.10	0.025	0.925	0.025	0.930
	1.00	0.90	0.025	0.868	0.025	0.868
		1.00	0.025	0.900	0.025	0.900
		1.10	0.025	0.925	0.025	0.925
	1.25	0.90	0.025	0.866	0.024	0.856
		1.00	0.025	0.898	0.024	0.889
		1.10	0.025	0.924	0.024	0.915
$K = 5$						
	0.80	0.90	0.025	0.872	0.025	0.880
		1.00	0.025	0.904	0.025	0.911
		1.10	0.025	0.928	0.025	0.934
	1.00	0.90	0.025	0.868	0.025	0.868
		1.00	0.025	0.900	0.025	0.900
		1.10	0.025	0.925	0.025	0.925
	1.25	0.90	0.024	0.859	0.024	0.847
		1.00	0.024	0.893	0.024	0.881
		1.10	0.024	0.918	0.024	0.908
$K = 10$						
	0.80	0.90	0.025	0.875	0.025	0.883
		1.00	0.025	0.906	0.025	0.913
		1.10	0.025	0.930	0.025	0.936
	1.00	0.90	0.025	0.868	0.025	0.868
		1.00	0.025	0.900	0.025	0.900
		1.10	0.025	0.925	0.025	0.925
	1.25	0.90	0.024	0.856	0.025	0.842
		1.00	0.024	0.890	0.025	0.876
		1.10	0.024	0.916	0.025	0.903

5.6 Group Sequential One-sided t-Tests

5.6.1 Methodology

We now consider the situation described in Section 3.3 where observations follow a normal linear model with parameter vector $\boldsymbol{\beta}$ and variances depend on a scale factor σ^2. We suppose estimates $\widehat{\boldsymbol{\beta}}^{(1)}, \ldots, \widehat{\boldsymbol{\beta}}^{(K)}$ are obtained at analyses 1 to K, $Var(\widehat{\boldsymbol{\beta}}^{(k)}) = \boldsymbol{V}_k \sigma^2$ and s_k^2 is the usual unbiased estimate of σ^2 at analysis k based on $n_k - p$ degrees of freedom. In this section we shall define group sequential t-tests of a hypothesis of the form H_0: $\boldsymbol{c}^T\boldsymbol{\beta} \leq \gamma$ against the one-sided alternative $\boldsymbol{c}^T\boldsymbol{\beta} > \gamma$. Our aim is to create tests with Type I error probability α and to assess power at an alternative where $\boldsymbol{c}^T\boldsymbol{\beta}$ exceeds γ by a specified amount δ.

As for the group sequential two-sided t-tests described in Section 4.6,

$$\boldsymbol{c}^T\widehat{\boldsymbol{\beta}}^{(k)} \sim N(\boldsymbol{c}^T\boldsymbol{\beta}, \boldsymbol{c}^T\boldsymbol{V}_k\boldsymbol{c}\sigma^2),$$

the Z-statistic for testing H_0: $\boldsymbol{c}^T\boldsymbol{\beta} \leq \gamma$ if σ^2 is known is

$$Z_k = \frac{\boldsymbol{c}^T\widehat{\boldsymbol{\beta}}^{(k)} - \gamma}{\sqrt{\{\boldsymbol{c}^T\boldsymbol{V}_k\boldsymbol{c}\sigma^2\}}},$$

and the t-statistic for testing H_0 is

$$T_k = \frac{\boldsymbol{c}^T\widehat{\boldsymbol{\beta}}^{(k)} - \gamma}{\sqrt{\{\boldsymbol{c}^T\boldsymbol{V}_k\boldsymbol{c}\, s_k^2\}}}.$$

We shall adapt tests constructed for known σ^2 for use when σ^2 is unknown. However, since observed information is proportional to $1/\sigma^2$, it will not be possible to guarantee both Type I and II error probabilities using pre-specified group sizes, and we shall give priority to maintaining the Type I error rate close to its target value. We follow a similar strategy to that applied in Section 5.5 in dealing with unequally spaced information levels when σ^2 is known. Our starting point is the stopping rule (5.2) with critical values a_k and b_k, $k = 1, \ldots, K$, which may be for a design with a binding or non-binding futility boundary, defined without explicit reference to δ or $\mathcal{I}_k$ by Equation (5.7).

In adapting to the case of unknown variance, we replace Z_k in (5.2) by T_k and set critical values for T_k to maintain the same marginal probability under $\boldsymbol{c}^T\boldsymbol{\beta} = \gamma$ (ignoring the effect of possible earlier stopping) of being above or below the boundary at each analysis. This gives the rule:

After group $k = 1, \ldots, K-1$

if $T_k \geq t_{n_k-p,1-\Phi(b_k)}$	stop, reject H_0	
if $T_k \leq t_{n_k-p,1-\Phi(a_k)}$	stop, accept H_0	
otherwise	continue to group $k+1$,	(5.17)

after group K

if $T_K \geq t_{n_K-p,1-\Phi(b_K)}$	stop, reject H_0
if $T_K < t_{n_K-p,1-\Phi(a_K)}$	stop, accept H_0.

Here, $t_{\nu,q}$ is the upper q tail-point of a t-distribution on ν degrees of freedom, i.e., $Pr\{T > t_{\nu,q}\} = q$ when $T \sim t_\nu$.

The same "significance levels" are applied to the T_k in rule (5.17) as to the Z_k in rule (5.2). Thus, our proposed group sequential t-test can be regarded as the "repeated significance test" analogue of the original Z-test. The comparison is a little forced here since the lower boundary does not necessarily lie at extreme values of the null distribution of Z_k. Nevertheless, there are good reasons why this construction should produce a test with Type I error probability close to the target value. First, if the degrees of freedom of the T_k are high, there is little difference between our group sequential t-test and the Z-test proposed in Section 5.5 for handling unpredictable information levels; hence, the Type I error rate will be close to its intended value as long as information levels are close to equally spaced. This is a key property as it overcomes the problem that information levels depend on the unknown σ^2. Second, if the degrees of freedom are low, there is strong empirical evidence that matching the marginal probabilities under H_0 of being above and below each boundary point to those of the original Z-test will maintain the overall Type I error probability to considerable accuracy.

We note in passing that there is another way in which the "significance level approach" can be used to create a group sequential one-sided t-test from a test for known σ^2. When introducing power family tests in Section 5.2, we observed that the upper and lower boundaries could be viewed as repeated significance tests of $\theta = 0$ and $\theta = \delta$, respectively, tests being applied to Z_k to create the upper boundary and to $Z_k - \delta\sqrt{\mathcal{I}_k}$, the standardized statistic for testing $\theta = \delta$, to create the lower boundary. (In our current problem, $\boldsymbol{c}^T\boldsymbol{\beta} - \gamma$ takes the place of θ and $\mathcal{I}_k$ is $(\boldsymbol{c}^T\boldsymbol{V}_k\boldsymbol{c}\sigma^2)^{-1}$.) A group sequential t-test can be created by applying significance tests at the same sequences of levels to the T_k to obtain the upper boundary, and to the t-statistics for testing $\boldsymbol{c}^T\boldsymbol{\beta} - \gamma = \delta$,

$$T_k' = \frac{\boldsymbol{c}^T\widehat{\boldsymbol{\beta}}^{(k)} - \gamma - \delta}{\sqrt{\{\boldsymbol{c}^T\boldsymbol{V}_k\boldsymbol{c}\, s_k^2\}}}, \quad k = 1, \ldots, K,$$

to produce the lower boundary. In some ways, this may appear a more natural analogue of the original Z-test. However, there are difficulties with this approach: there is no guarantee that the boundaries will meet up at the last analysis, and both Type I and Type II error rates vary substantially with the observed sequence of information levels. For these reasons, we shall not pursue this approach further.

Returning to our proposed test, we can repeat the arguments of Section 4.6 to find an approximation to the test's power at a specified alternative and value of σ^2. By analogy with the case of known σ^2, for given $\boldsymbol{V}_1, \ldots, \boldsymbol{V}_K$ we expect the group sequential t-test's power at a certain alternative to be approximately that of a fixed sample t-test with $1/\tilde{R}(K, \alpha, \beta, \Delta)$ times the maximal information. Thus, we need to find the power of a fixed sample test with estimator $\widehat{\boldsymbol{\beta}}$ for which $Var(\boldsymbol{c}^T\widehat{\boldsymbol{\beta}}) = \tilde{R}(K, \alpha, \beta, \Delta)\boldsymbol{c}^T\boldsymbol{V}_K\boldsymbol{c}\sigma^2$. The approximation to the power when $\boldsymbol{c}^T\boldsymbol{\beta} - \gamma = \delta$, corresponding to formula (4.11) for a two-sided test, is

$$Pr\{T(n_K - p;\ \delta/\sqrt{\{\tilde{R}(K, \alpha, \beta, \Delta)\boldsymbol{c}^T\boldsymbol{V}_K\boldsymbol{c}\sigma^2\}}) \geq t_{n_K-p,1-\alpha}\}, \tag{5.18}$$

where $T(\nu; \xi)$ denotes a noncentral t random variable with ν degrees of freedom and noncentrality parameter ξ. The simpler but cruder approximation that can be used when $n_K - p$ is large is

$$\Phi\left(\frac{\delta}{\sqrt{\{\tilde{R}(K, \alpha, \beta, \Delta)\,\boldsymbol{c}^T\boldsymbol{V}_K\boldsymbol{c}\sigma^2\}}} - \Phi^{-1}(1-\alpha)\right). \tag{5.19}$$

Table 5.13 shows the Type I error probabilities of our group sequential one-sided t-tests and the power they attain in situations where formula (5.18) gives the value 0.8. As in Table 4.3, the degrees of freedom are those which arise when groups consist of three or five observations and two degrees of freedom are used in fitting the linear model at each analysis. Departures from the intended Type I error rates are minor and decrease as the degrees of freedom increase. Formula (5.18) is seen to provide a good estimate of power, especially

Table 5.13 *Properties of group sequential t-tests designed to test H_0: $c^T\beta \leq \gamma$ with one-sided Type I error rate $\alpha = 0.025$. Tests are based on power family tests for known variance with $\Delta = -0.5$ and $\Delta = 0$. Degrees of freedom for σ^2 at analysis k are $\nu_k = (\nu_K + 2)(k/K) - 2$ and $Var(c^T\hat{\beta}^{(k)}) = (K/k)Var(c^T\hat{\beta}^{(K)})$, $k = 1, \ldots, K$. The penultimate column shows actual power at the alternative for which the approximation (5.18) gives power 0.8, and the final column shows the power predicted for each test by formula (5.19).*
Type I error and power were estimated by simulation, using 1 million replicates; standard errors are 0.0002 for Type I error and 0.0004 for power.

$\Delta = -0.5$

K	ν_K	*Type I error*	*Attained power*	*Known σ^2 power*
3	7	0.028	0.819	0.905
3	13	0.026	0.811	0.858
5	13	0.028	0.817	0.858
5	23	0.026	0.809	0.833
8	22	0.027	0.813	0.834
8	38	0.027	0.807	0.820

$\Delta = 0$

K	ν_K	*Type I error*	*Attained power*	*Known σ^2 power*
3	7	0.029	0.828	0.905
3	13	0.027	0.817	0.858
5	13	0.029	0.818	0.858
5	23	0.027	0.811	0.833
8	22	0.028	0.813	0.834
8	38	0.027	0.808	0.820

when degrees of freedom are high. Because tests with $\Delta = -0.5$ have less opportunity for early stopping than those with $\Delta = 0.0$, the slightly higher discrepancies in error rates in the second case are as one might expect.

Since our definition of group sequential t-tests does not rely on equally spaced information levels, these tests may also be used with unequal group sizes or information increments. We have simulated tests with sequences of information levels of the form

$$\mathcal{I}_k = (k/K)^r \mathcal{I}_{max}, \quad k = 1, \ldots, K, \tag{5.20}$$

to investigate the effects of such unequal spacing. Our findings are similar to those reported in Section 5.5 for the known variance problem. With $r = 0.8$ and 1.25 in (5.20), the Type I error and power of a test differ only slightly from those of the same test with $r = 1$, i.e., with equally spaced information levels and $\mathcal{I}_K = \mathcal{I}_{max}$. Thus, there is no difficulty and no serious loss of accuracy in applying these tests with both an unknown variance and unequal information increments.

5.6.2 Example: A Balanced Two-treatment Comparison

To illustrate the above methods, we revisit the example of a balanced parallel two treatment comparison design described in Section 5.4. The responses, denoted X_{Ai} and X_{Bi} for $i = 1, 2, \ldots$ are assumed to be normally distributed with means μ_A and μ_B, respectively, and have a common, unknown variance σ^2. Our aim is to design a power family test with $\Delta = 0$ and four groups of observations to test H_0: $\mu_A \leq \mu_B$ against the one-sided alternative $\mu_A > \mu_B$ with Type I error probability $\alpha = 0.025$. We further suppose there is evidence that σ is likely to be in the region of 10 and, in so far as it is possible, we would like to achieve power of about 0.9 at $\mu_A - \mu_B = 5.0$.

If each group contains m observations per treatment arm, the t-statistics for testing H_0 at successive analyses are

$$T_k = \frac{\sum_{i=1}^{mk} X_{Ai} - \sum_{i=1}^{mk} X_{Bi}}{\sqrt{\{2mks_k^2\}}}, \quad k = 1, \ldots, 4,$$

where

$$s_k^2 = \frac{1}{2mk-2}\left\{\sum_{i=1}^{mk}(X_{Ai} - \bar{X}_A^{(k)})^2 + \sum_{i=1}^{mk}(X_{Bi} - \bar{X}_B^{(k)})^2\right\},$$

with $\bar{X}_A^{(k)}$ denoting the mean of $X_{A1}, \ldots, X_{A,mk}$ and $\bar{X}_B^{(k)}$ the mean of $X_{B1}, \ldots, X_{B,mk}$.

Suppose the study is designed with a binding futility boundary, then Table 5.2 gives $\tilde{C}_1(4, 0.025, 0.1, 0) = 1.978$, $\tilde{C}_2(4, 0.025, 0.1, 0.0) = 1.385$ and $\tilde{R}(4, 0.025, 0.1, 0.0) = 1.077$. Thus, for $k = 1, \ldots, 4$, the test stops to reject H_0 at analysis k if

$$T_k \geq t_{2mk-2, 1-\Phi(b_k)}$$

and to accept H_0 at analysis k if

$$T_k \leq t_{2mk-2, 1-\Phi(a_k)},$$

where, from (5.7),

$$b_k = 1.978 \times (k/4)^{-1/2}$$

and

$$a_k = (1.978 + 1.385) \times (k/4)^{1/2} - 1.385 \times (k/4)^{-1/2}.$$

Evaluating these boundary values, we find $(b_1, \ldots, b_4) = (3.956, 2.797, 2.284, 1.978)$ and $(a_1, \ldots, a_4) = (-1.088, 0.419, 1.313, 1.978)$. The degrees of freedom for $T_1, \ldots, T_4$ are 44, 90, 136 and 182. Hence, the group sequential t-test has upper boundary values $(4.364, 2.867, 2.310, 1.991)$ and lower boundary $(-1.102, 0.421, 1.320, 1.991)$. With only 44 degrees of freedom for error at the first analysis, the upper boundary for T_1 is noticeably higher than b_1 but at later analyses the boundaries for T_2, T_3 and T_4 are only slightly wider than those for Z_2, Z_3 and Z_4 in the case of known variance.

This test will have Type I error probability close to 0.025, but its power depends on the value chosen for m and the unknown σ. In the known variance case, a group size of $m = 23$ produced power 0.9 at $\mu_A - \mu_B = 5.0$ when $\sigma = 10$. The expression (5.18) gives approximate power

$$Pr\{T(184-2\,;\, 5.0/\sqrt{\{1.077 \times (2\sigma^2/92)\}}) \geq t_{184-2,\,0.975}\}$$

for the group sequential t-test with the same group size, and this is equal to 0.902 when $\sigma = 10$. Simulations show the test's actual Type I error rate is 0.0252 and its power at $\mu_A - \mu_B = 5.0$ when $\sigma = 10$ is 0.9022 (standard errors of these estimates are 0.00005 and

0.0001, respectively). If these properties are regarded as satisfactory, the group size $m = 23$ can be used.

If, however, investigators suspect the estimated value of 10 for σ may be inaccurate, a larger group size can be chosen to safeguard power. If, for example, it is thought that σ could be as high as 12, a group size of $m = 33$ would be appropriate since (5.18) then gives

$$Pr\{T(264-2;\, 5.0/\sqrt{\{1.077 \times (2 \times 12^2/132)\}}) \geq t_{264-2,0.975}\} = 0.901$$

as an approximation to the power at $\mu_A - \mu_B = 5.0$ if $\sigma = 12$ and simulations show the test's actual Type I error rate is 0.0251 and its power at $\mu_A - \mu_B = 5.0$ when $\sigma = 10$ is 0.9018.

5.7 Whitehead's Triangular Test

5.7.1 Definition

This test is an adaptation to the group sequential setting of the fully sequential test referred to by Lorden (1976) as a "2-SPRT". It is more naturally explained in terms of the score statistics $S_1, \ldots, S_K$. Recall from Section 3.1 that $S_k = Z_k\sqrt{\mathcal{I}_k}$, and $(S_1, \ldots, S_K)$ are multivariate normal with $S_k \sim N(\theta\mathcal{I}_k, \mathcal{I}_k)$, $k = 1, \ldots, K$, and independent increments S_1, $S_2 - S_1, \ldots, S_K - S_{K-1}$. This joint distribution is that of a Brownian motion with drift θ, $\{S(\mathcal{I}), \mathcal{I} > 0\}$, observed at times $\mathcal{I} = \mathcal{I}_1, \ldots, \mathcal{I}_K$; here $S(\mathcal{I}) \sim N(\theta\mathcal{I}, \mathcal{I})$, $\mathcal{I} > 0$, and the continuous S process has independent increments.

The origins of Whitehead's test lie in a sequential test of H_0: $\theta \leq 0$ against $\theta > 0$ in which the S process is monitored continuously. This continuous test has upper and lower boundaries

$$d(\mathcal{I}) = \frac{2}{\delta}\log\left(\frac{1}{2\alpha}\right) + \frac{\delta}{4}\mathcal{I} \quad \text{and} \quad c(\mathcal{I}) = -\frac{2}{\delta}\log\left(\frac{1}{2\alpha}\right) + \frac{3\delta}{4}\mathcal{I}$$

for $\mathcal{I}$ in the range

$$0 < \mathcal{I} \leq \frac{8}{\delta^2}\log\left(\frac{1}{2\alpha}\right).$$

Termination occurs the first time $S(\mathcal{I}) \geq d(\mathcal{I})$ or $S(\mathcal{I}) \leq c(\mathcal{I})$, although, since the process is continuous, the relevant condition is always met with equality. The test rejects H_0 if $S(\mathcal{I}) = d(\mathcal{I})$ at termination and accepts H_0 if $S(\mathcal{I}) = c(\mathcal{I})$. Lorden (1976) gave a likelihood ratio argument to show that the Type I error probability of this test is exactly equal to α and power at $\theta = \delta$ is exactly $1 - \alpha$ under continuous monitoring.

Whitehead & Stratton (1983) adapted the continuous test to discrete monitoring using a result of Siegmund (1979) concerning the expected overshoot of a straight line boundary by a discretely monitored Brownian motion. They subtract the expected overshoot from the continuous boundary to obtain a discrete boundary with approximately the same exit probabilities. If information levels are equally spaced and $\mathcal{I}_K = \mathcal{I}_{max}$, so $\mathcal{I}_k = (k/K)\mathcal{I}_{max}$ at analyses $k = 1, \ldots, K$, the upper and lower boundaries for S_k are, respectively,

$$d_k = \frac{2}{\delta}\log\left(\frac{1}{2\alpha}\right) - 0.583\sqrt{\frac{\mathcal{I}_{max}}{K}} + \frac{\delta}{4}\frac{k}{K}\mathcal{I}_{max} \tag{5.21}$$

and

$$c_k = -\frac{2}{\delta}\log\left(\frac{1}{2\alpha}\right) + 0.583\sqrt{\frac{\mathcal{I}_{max}}{K}} + \frac{3\delta}{4}\frac{k}{K}\mathcal{I}_{max}. \tag{5.22}$$

If the boundaries are to meet with $d_K = c_K$ at the final analysis, we find, by solving a quadratic equation, that $\mathcal{I}_K$ must take the value

$$\mathcal{I}_{max} = \left[\sqrt{\frac{4 \times 0.583^2}{K} + 8\log\left(\frac{1}{2\alpha}\right)} - \frac{2 \times 0.583}{\sqrt{K}}\right]^2 \frac{1}{\delta^2}. \tag{5.23}$$

For given K and α, taking $\mathcal{I}_{max}$ from (5.23) and boundary values from (5.21) and (5.22) gives a group sequential one-sided test of H_0: $\theta \leq 0$ with K analyses, achieving a Type I error rate approximately equal to α and power approximately $1 - \alpha$ at $\theta = \delta$. The formulae for c_k and d_k indicate that the boundary values $(c_1, d_1), \ldots, (c_K, d_K)$ define a triangular continuation region in the $(S, \mathcal{I})$ plane.

Whitehead & Stratton (1983) also explain how to create a test with, approximately, Type I error probability α at $\theta = 0$ and power $1 - \beta$ at $\theta = \delta$ when $\alpha \neq \beta$. They first note that a fixed sample test with Type I error α and power $1 - \beta$ at $\theta = \delta$ has power $1 - \alpha$ at $\theta = \xi\delta$ where

$$\xi = \frac{2\Phi^{-1}(1-\alpha)}{\Phi^{-1}(1-\alpha) + \Phi^{-1}(1-\beta)}.$$

The group sequential test is then conducted as defined above, but with $\tilde{\delta} = \xi\delta$ in place of δ in the formulae for $\mathcal{I}_{max}$ and the boundary values. Since the power curves of the group sequential test and fixed sample test follow each other closely, the group sequential test also has power close to $1 - \beta$ at $\theta = \delta$.

If the power family tests of Section 5.2 are expressed in terms of the score statistics $\{S_1, \ldots, S_K\}$, boundary values a_k and b_k for Z_k become $c_k = a_k\sqrt{\mathcal{I}_k}$ and $d_k = b_k\sqrt{\mathcal{I}_k}$ for S_k. The power family test with $\Delta = 0$ also has a triangular continuation region in the $(S, \mathcal{I})$ plane, but this is not the same as Whitehead's triangular test: the boundary of the power family test is wider early on and slopes to a point more rapidly, ending at a lower maximum information level. This can be seen by considering again the example at the end of Section 5.2.1.

Suppose we wish to construct a one-sided test with $K = 4$ stages, Type I error rate $\alpha = 0.025$ and power $1 - \beta = 0.9$ at $\theta = \delta = 0.2$. Figure 5.1 showed the boundary for two power family tests with these specifications. Because the triangular tests here require $\alpha = \beta$, we replace δ by $\xi\delta = 2 \times 1.960\,(1.960 + 1.282)^{-1}0.2 = 0.2419$, as the value for θ at which the power is to be $1 - \alpha = 0.975$. For fully sequential or continuous monitoring, the 2-SPRT with maximum information $-8\log(2\alpha)/0.2419^2 = 409.70$ guarantees the Type I and II error probability requirements. Using 0.2419 for δ in Equation (5.23), the maximum information for the test of Whitehead & Stratton (1983) with four stages is 323.06. The boundaries (using the S scale) for both tests are shown in Figure 5.3. For comparison, the triangular test in the power family ($\Delta = 0$), which has $\mathcal{I}_{max} = 282.84$, is also displayed; it is the same as that shown in Figure 5.1.

5.7.2 *Properties of Whitehead's Triangular Test for Equally Spaced Information Levels*

Table 5.14 shows properties of Whitehead's test when information levels are equally spaced and boundaries meet at the final analysis. The first two columns show the actual Type I error rate and power, which differ slightly from the specified values due to the approximations used in deriving the tests. However, we see that all entries are close to their intended values. The results for small numbers of groups are surprisingly good since there is little reason to expect Siegmund's boundary crossing result to apply accurately in these cases. Results for $K = 1$ are for the extreme case where, after a rather large initial group, the boundaries defined by (5.21) and (5.22) meet up and, thus, the test terminates. The information level at which this occurs is not quite equal to that required by a fixed sample test, and hence

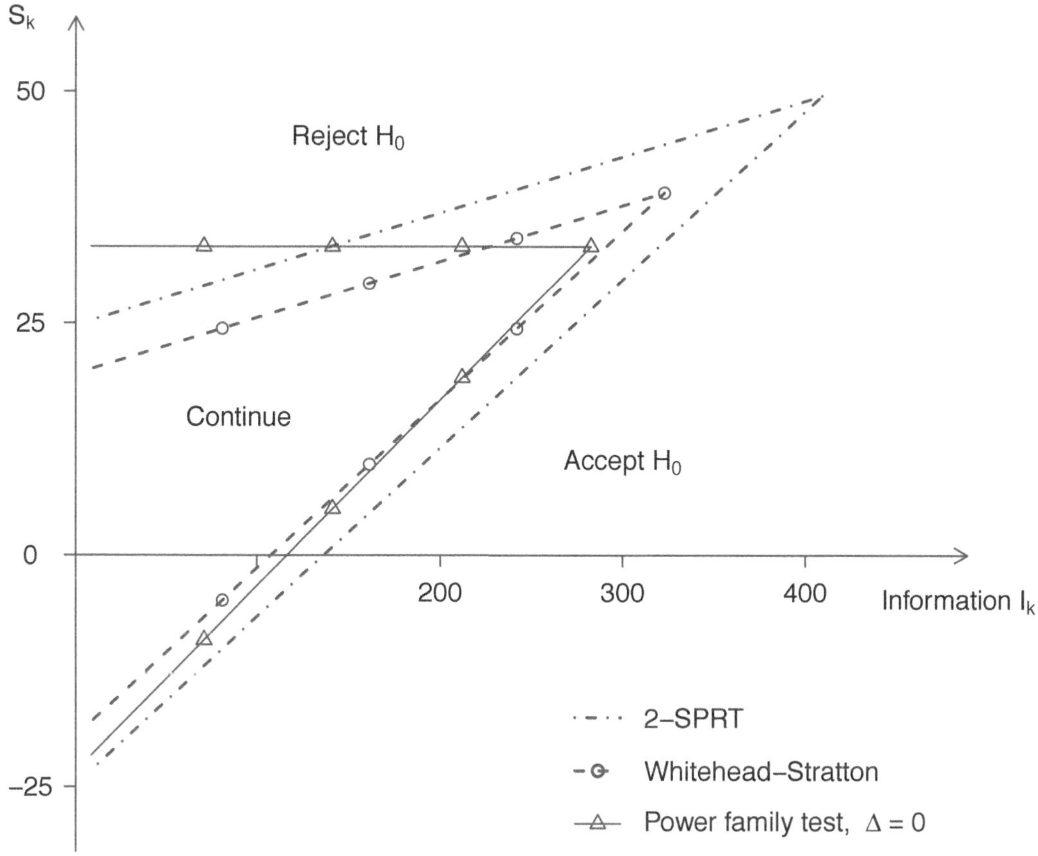

Figure 5.3 *One-sided triangular tests for four groups of observations*

the attained error probabilities are not exactly equal to their target values. However, the differences are surprisingly small and the results show that (5.23) gives a value of $\mathcal{I}_{max}$ within a few percent of the fixed sample size and Type I error and power are in fact very close to their nominal values. These observations are in keeping with remarks made by Whitehead (1997, p. 128) on the accuracy of his proposed approximations.

The third column of Table 5.14 contains values of $R_{Wh}(K, \alpha, \beta)$, the ratio between each group sequential test's maximum information level and the information, $\mathcal{I}_{f,1}$, required by a fixed sample test with the same error probabilities. (Although $R_{Wh}(K, \alpha, \beta)$ depends on K and α but not on β, we have retained the more general notation.) Expected sample sizes on termination under $\theta = 0$, $\delta/2$ and δ are shown in the remaining columns.

The results follow the standard pattern, expected sample sizes decreasing and maximum sample size increasing as the number of analyses grows. In general, the maximum sample sizes tend to be high for the expected sample sizes achieved. For example, comparing Table 5.14 with Tables 5.1, 5.2 and 5.3, we see that the corresponding power family one-sided tests either have a similar maximum sample size and lower expected sample sizes or comparable expected sample sizes and a lower maximum sample size.

5.7.3 Applying Whitehead's Triangular Test with Unequal Increments in Information

Whitehead (1997, Sec. 4.13) makes further use of Siegmund's expected overshoot result to adapt the triangular test to general information sequences. At analysis k, the boundaries

Table 5.14 *Properties of Whitehead's triangular tests with equal group sizes. Tests are for K groups of observations, and designed to have Type I error probability $\alpha = 0.025$ at $\theta = 0$ and power $1-\beta = 0.8$, 0.9 and 0.95 at $\theta = \delta$. Constants $R_{Wh}(K, \alpha, \beta)$ are ratios of the maximum sample size to that of a fixed sample test with error probabilities α and β. Expected sample sizes at $\theta = 0$, $\delta/2$ and δ are expressed as percentages of the corresponding fixed sample size.*

K	*Type I error*	*Power at* $\theta = \delta$	R_{Wh}	*Expected sample size, as percentage of fixed sample size, at* $\theta = 0$	$\theta = \delta/2$	$\theta = \delta$
$1-\beta = 0.8$						
1	0.027	0.797	0.973	97.3	97.3	97.3
2	0.027	0.800	1.115	67.9	83.6	81.1
3	0.026	0.802	1.186	61.9	80.1	77.3
4	0.025	0.803	1.230	59.7	78.2	75.3
5	0.025	0.803	1.261	58.2	76.9	74.0
10	0.025	0.804	1.342	54.5	73.8	70.7
15	0.025	0.804	1.379	53.2	72.7	69.5
20	0.025	0.804	1.402	52.6	72.1	68.9
$1-\beta = 0.9$						
1	0.027	0.897	0.973	97.3	97.3	97.3
2	0.027	0.899	1.115	67.9	84.9	76.1
3	0.026	0.901	1.186	61.9	81.6	71.6
4	0.025	0.902	1.230	59.7	79.8	69.5
5	0.025	0.902	1.261	58.2	78.6	68.0
10	0.025	0.902	1.342	54.5	75.5	64.6
15	0.025	0.902	1.379	53.2	74.4	63.4
20	0.025	0.902	1.402	52.6	73.8	62.7
$1-\beta = 0.95$						
1	0.027	0.948	0.973	97.3	97.3	97.3
2	0.027	0.949	1.115	67.9	85.5	71.7
3	0.026	0.950	1.186	61.9	82.3	66.4
4	0.025	0.950	1.230	59.7	80.6	64.2
5	0.025	0.951	1.261	58.2	79.3	62.7
10	0.025	0.951	1.342	54.5	76.3	59.1
15	0.025	0.951	1.379	53.2	75.2	57.8
20	0.025	0.951	1.402	52.6	74.6	57.2

of the original continuous test are brought in by an amount $0.583\surd(\mathcal{I}_k - \mathcal{I}_{k-1})$. Thus, for a general sequence $\{\mathcal{I}_1, \ldots, \mathcal{I}_K\}$, at analysis k the test stops to reject H_0 if

$$S_k \geq d_k = \frac{2}{\delta} \log\left(\frac{1}{2\alpha}\right) - 0.583\surd(\mathcal{I}_k - \mathcal{I}_{k-1}) + \frac{\delta}{4}\mathcal{I}_k,$$

stops to accept H_0 if

$$S_k \leq c_k = -\frac{2}{\delta} \log\left(\frac{1}{2\alpha}\right) + 0.583\surd(\mathcal{I}_k - \mathcal{I}_{k-1}) + \frac{3\delta}{4}\mathcal{I}_k$$

and continues to stage $k+1$ otherwise. At stage 1, $\mathcal{I}_0$ is taken to be zero. If S_k continues to fall in the continuation region, a stage will eventually be reached at which $c_K \geq d_K$ and the

Table 5.15 *Properties of Whitehead's triangular test applied with information levels $\mathcal{I}_k = \pi(k/K)^r \mathcal{I}_{max}$, $k = 1, \ldots, K$, where $\mathcal{I}_{max}$ is such that boundaries meet up exactly at analysis K for information levels $\mathcal{I}_k = (k/K)\mathcal{I}_{max}$, $k = 1, \ldots, K$. Attained Type I error probabilities and power are shown for tests designed to achieve Type I error rate $\alpha = 0.025$ and power $1 - \beta = 0.9$ at $\theta = \delta$.*

		$K = 2$		$K = 5$		$K = 10$	
r	π	*Type I error*	*Power*	*Type I error*	*Power*	*Type I error*	*Power*
0.80	0.90	0.025	0.868	0.025	0.887	0.025	0.895
	1.00	0.027	0.894	0.026	0.900	0.025	0.902
	1.10	0.028	0.906	0.026	0.902	0.025	0.902
1.00	0.90	0.025	0.873	0.024	0.889	0.024	0.896
	1.00	0.027	0.899	0.025	0.902	0.025	0.902
	1.10	0.028	0.913	0.026	0.904	0.025	0.902
1.25	0.90	0.024	0.878	0.024	0.891	0.024	0.896
	1.00	0.026	0.907	0.025	0.904	0.025	0.903
	1.10	0.028	0.923	0.025	0.907	0.025	0.903

test must then terminate. Clarification of the decision rule at this final stage is needed. We suggest retaining the condition $S_K \geq d_K$ for rejecting H_0 and accepting H_0 if $S_K < d_K$, thereby gaining as much power as possible while still protecting the Type I error probability. Because of the appearance of a plot of the resulting boundary, Whitehead refers to this as the "Christmas tree" adjustment. Table 5.15 shows the Type I error probability and power attained by Whitehead's triangular test under a variety of information sequences.

This test differs from previous methods in that it can be applied without an initial estimate of the maximum number of analyses, K, or of the information levels which will arise at each analysis. We have investigated attained error rates under observed information sequences of the form used in Sections 4.2 and 5.5, i.e., $\mathcal{I}_k = \pi(k/K)^r \mathcal{I}_{max}$, $k = 1, \ldots, K$, where now $\mathcal{I}_{max}$ is given by (5.23). If $\pi = r = 1$, the boundaries meet up exactly at analysis K and the test is sure to terminate at this point. For some combinations of π and r, the test has $c_K < d_K$ and we enforce termination at analysis K by rejecting H_0 if $S_K \geq d_K$ and accepting H_0 otherwise. This might be necessary if the number of analyses is limited due to budgetary restrictions; in other circumstances, it may be possible to continue sampling until the boundaries meet and then terminate the test. For certain other values of π and r, the test terminates at analysis $K - 1$ since $c_{K-1} > d_{K-1}$ and, following our stated rule, we reject H_0 in such situations if and only if $S_{K-1} \geq d_{K-1}$.

It is clear from Table 5.15 that attained Type I error and power are close to their nominal values in all the situations considered. Results for tests with power 0.8 and 0.95, not shown here, were equally impressive. Interestingly, power is not so adversely affected by the lower information levels when $\pi = 0.9$, possibly because there is little probability of continuing to the final stage and truncation of the final sample size has only a limited effect.

An appealing feature of Whitehead's triangular test is that only elementary calculations are needed to determine boundary values, and yet the test satisfies error probability requirements accurately under quite general information sequences. The alternative error spending approach, which we shall present in Chapter 8, involves the numerical computation of successive boundary points. Against this, the triangular test does not manage to achieve the best possible reductions in expected sample size given its rather high maximum sample size. (The sample size at which boundaries meet up depends on the observed pattern of group sizes, but the maximum sample sizes for equally sized groups are already quite high.) Although the probability that the test proceeds to its maximum sample size may be small, this eventuality does have to be allowed for when designing a study and it is hard not to favor a test which has similar properties but a smaller maximum sample size.

Chapter 6

Two-sided Tests with Early Stopping under the Null Hypothesis

6.1 Introduction

We return to the two-sided testing problem of Chapters 2 and 4 but now consider tests which allow early stopping to accept, as well as to reject, the null hypothesis. Following the general treatment of normal data described in Chapter 4, we suppose standardized statistics $Z_1, \ldots, Z_K$ are available at analyses $k = 1, \ldots, K$ and follow the canonical joint distribution (3.1), in particular, $Z_k \sim N(\theta\sqrt{\mathcal{I}_k}, 1)$ where $\mathcal{I}_k$ is the information for θ available at analysis k. Our objective is to test, group sequentially, the null hypothesis H_0: $\theta = 0$ with Type I error probability α and power $1 - \beta$ at some specified alternative $\theta = \pm\delta$.

A general group sequential test with possible early stopping to accept H_0 is defined by pairs of constants (a_k, b_k) with $0 \le a_k < b_k$ for $k = 1, \ldots, K-1$ and $a_K = b_K$. It has the form

$$\begin{array}{lll} \text{After group } k = 1, \ldots, K-1 & & \\ \quad \text{if } |Z_k| \ge b_k & \text{stop, reject } H_0 & \\ \quad \text{if } |Z_k| < a_k & \text{stop, accept } H_0 & \\ \quad \text{otherwise} & \text{continue to group } k+1, & (6.1) \\ \text{after group } K & & \\ \quad \text{if } |Z_K| \ge b_K & \text{stop, reject } H_0 & \\ \quad \text{if } |Z_k| < a_K & \text{stop, accept } H_0. & \end{array}$$

Since $a_K = b_K$, termination at analysis K is ensured. At analysis k, the set of values $\{Z_k\colon |Z_k| < a_k\}$ forms an "inner wedge" within the continuation region $\{Z_k\colon |Z_k| < b_k\}$ where early stopping to accept H_0 occurs. It may be that such early stopping is not deemed appropriate at the first few analyses, in which case one can set $a_1 = 0$, $a_2 = 0$, etc., to preclude this possibility.

The tests introduced in Chapter 2 have no inner wedges at all and, if $\theta = 0$, sampling must continue up to analysis K with probability at least $1 - \alpha$ in order to satisfy the Type I error requirement. Consequently, the expected sample sizes of these tests at $\theta = 0$ show no reduction below the corresponding fixed sample size. In fact, we see from Tables 2.7 and 2.8 that expected sample sizes under $\theta = \delta/2$ are also above the fixed sample size for the Pocock test and only a little below the fixed sample size in the case of the O'Brien & Fleming test. We shall see that early stopping to accept H_0 can lead to reductions in expected sample size at all values of θ, including $\theta = 0$. Although the ethical motivation for stopping a clinical trial early is greatest when a treatment difference is found and randomization of patients to an inferior treatment can be avoided, there are sound financial reasons to curtail a study when it is evident that there is no treatment difference to detect. Moreover, such curtailment can benefit future patients by releasing resources for the earlier exploration of further promising treatments.

DOI: 10.1201/9781584888482-6

Gould & Pecore (1982) noted the importance of early stopping to accept the null hypothesis, and Gould (1983) proposed a method of "abandoning lost causes" in which early termination occurs *only* to accept H_0. In Gould & Pecore's inner wedge design, the stopping rule has the form (6.1) with $a_k = a$ and $b_k = b$, $k = 1, \ldots, K$, where a and b are selected constants. Thus, early stopping to accept H_0 occurs at any analysis where the two-sided significance level of a test of H_0 is greater than $2\Phi(-a)$, irrespective of how much data this test is based on. Calculations show the Gould & Pecore tests have rather high maximum sample sizes, especially when there are five or more analyses, and they are not as effective in reducing expected sample size under H_0 as the family of tests we shall present in Section 6.2.

We shall focus on the tests described by Pampallona & Tsiatis (1994) which are indexed by a parameter Δ, occurring as a power in formulae for the boundary points a_k and b_k, $k = 1, \ldots, K$. (Emerson & Fleming (1989) proposed similar tests which achieve the power condition conservatively.) We shall refer to this family of tests as the *power family* of two-sided inner wedge tests. In Sections 6.2.1 and 6.2.2, we define the tests and present their properties. Jennison & Turnbull (2001a) show how the "significance level" approach can be used to apply these tests when observed information levels depart from their planned values and to create group sequential t-tests for normal data with unknown variance; we describe these extensions in Sections 6.2.3 and 6.2.4, respectively.

An alternative way to construct two-sided tests with early stopping to accept H_0 is to combine two one-sided tests. We present the group sequential test of this form described by Whitehead & Stratton (1983) and Whitehead (1997, Sec. 4.7) in Section 6.3.

6.2 The Power Family of Two-sided Inner Wedge Tests

6.2.1 *Definition of the Tests*

Consider a group sequential study in which statistics Z_k observed at analyses $k = 1, \ldots, K$ follow the canonical joint distribution (3.1) and suppose we wish to test H_0: $\theta = 0$ against H_A: $\theta \neq 0$ with Type I error probability α and power $1 - \beta$ at $\theta = \pm\delta$. A fixed sample test based on a statistic Z distributed as $N(\theta\sqrt{\mathcal{I}}, 1)$ can attain these requirements if $\mathcal{I}$ is equal to the value $\mathcal{I}_{f,2}$ given by (4.6). The power family inner wedge tests are group sequential and designed for equally spaced information levels, $\mathcal{I}_k = (k/K)\mathcal{I}_K$, $k = 1, \ldots, K$. For the test with parameter Δ, the critical values a_k and b_k to be used in (6.1) are

$$\begin{aligned} b_k &= \tilde{C}_{W1}(K, \alpha, \beta, \Delta)(k/K)^{\Delta - 1/2} \text{ and} \\ a_k &= \delta\sqrt{\mathcal{I}_k} - \tilde{C}_{W2}(K, \alpha, \beta, \Delta)(k/K)^{\Delta - 1/2}, \quad k = 1, \ldots, K. \end{aligned} \tag{6.2}$$

A final information level

$$\mathcal{I}_K = \frac{\{\tilde{C}_{W1}(K, \alpha, \beta, \Delta) + \tilde{C}_{W2}(K, \alpha, \beta, \Delta)\}^2}{\delta^2}$$

ensures that $a_K = b_K$ and the test terminates properly.

The constants $\tilde{C}_{W1}(K, \alpha, \beta, \Delta)$ and $\tilde{C}_{W2}(K, \alpha, \beta, \Delta)$ do not depend on δ and are chosen to ensure the Type I error and power conditions. These constants determine $\mathcal{I}_K$ and, hence, the ratio of the test's maximum information level to that required for a fixed sample test,

$$\tilde{R}_W(K, \alpha, \beta, \Delta) = \frac{\mathcal{I}_K}{\mathcal{I}_{f,2}} = \frac{\{\tilde{C}_{W1}(K, \alpha, \beta, \Delta) + \tilde{C}_{W2}(K, \alpha, \beta, \Delta)\}^2}{\{\Phi^{-1}(1 - \alpha/2) + \Phi^{-1}(1 - \beta)\}^2}.$$

Values of $\tilde{C}_{W1}(K, \alpha, \beta, \Delta)$, $\tilde{C}_{W2}(K, \alpha, \beta, \Delta)$ and $\tilde{R}_W(K, \alpha, \beta, \Delta)$ and some properties of these tests are provided in Tables 6.1, 6.2 and 6.3 for $\Delta = -0.5, -0.25, 0$ and 0.25, $\alpha = 0.05$

Table 6.1 *Constants $\tilde{C}_{W1}(K, \alpha, \beta, \Delta)$, $\tilde{C}_{W2}(K, \alpha, \beta, \Delta)$ and $\tilde{R}_W(K, \alpha, \beta, \Delta)$ for power family two-sided inner wedge tests with shape parameter Δ. The first opportunity for early stopping to accept H_0 occurs at analysis k^*. Also shown are expected sample sizes at $\theta = 0$, $\pm\delta/2$ and $\pm\delta$ expressed as percentages of the corresponding fixed sample size. Tests are for K groups of observations, Type I error probability $\alpha = 0.05$ at $\theta = 0$ and power $1 - \beta = 0.8$ at $\theta = \pm\delta$.*

K	$\tilde{C}_{W1}$	$\tilde{C}_{W2}$	$\tilde{R}_W$	k^*	*Expected sample size, as percentage of fixed sample size, at* $\theta = 0$	$\theta = \delta/2$	$\theta = \delta$
$\Delta = -0.5$							
1	1.960	0.842	1.000	1	100.0	100.0	100.0
2	1.949	0.867	1.010	1	90.7	94.6	98.1
3	1.933	0.901	1.023	2	79.6	87.1	89.8
4	1.929	0.919	1.033	2	79.0	85.7	87.2
5	1.927	0.932	1.041	3	75.9	83.4	85.1
10	1.928	0.964	1.066	5	73.2	80.6	81.6
15	1.931	0.979	1.078	8	72.2	79.6	80.5
20	1.932	0.988	1.087	10	71.8	79.1	80.0
$\Delta = -0.25$							
1	1.960	0.842	1.000	1	100.0	100.0	100.0
2	1.936	0.902	1.026	1	83.3	89.9	94.2
3	1.932	0.925	1.040	2	78.8	85.8	86.9
4	1.930	0.953	1.059	2	74.7	82.4	83.8
5	1.934	0.958	1.066	3	74.8	82.0	82.2
10	1.942	0.999	1.102	5	70.8	78.4	78.7
15	1.948	1.017	1.120	7	69.7	77.4	77.6
20	1.952	1.027	1.131	9	69.2	76.9	77.1
$\Delta = 0$							
1	1.960	0.842	1.000	1	100.0	100.0	100.0
2	1.935	0.948	1.058	1	78.3	86.0	88.4
3	1.950	0.955	1.075	1	78.5	84.7	83.6
4	1.953	0.995	1.107	2	72.2	80.2	80.2
5	1.958	1.017	1.128	2	71.0	78.8	78.4
10	1.980	1.057	1.175	4	67.9	75.9	75.0
15	1.991	1.075	1.198	6	67.0	74.9	73.9
20	1.998	1.087	1.212	8	66.6	74.5	73.4
$\Delta = 0.25$							
1	1.960	0.842	1.000	1	100.0	100.0	100.0
2	1.982	1.000	1.133	1	75.9	83.9	84.0
3	2.009	1.059	1.199	1	73.2	80.9	79.0
4	2.034	1.059	1.219	1	71.8	79.3	76.5
5	2.048	1.088	1.253	2	68.8	76.9	74.6
10	2.088	1.156	1.341	3	65.3	73.7	71.0
15	2.109	1.180	1.379	4	64.3	72.6	69.7
20	2.122	1.195	1.402	6	63.7	72.1	69.1

Table 6.2 *Constants* $\tilde{C}_{W1}(K,\alpha,\beta,\Delta)$, $\tilde{C}_{W2}(K,\alpha,\beta,\Delta)$ *and* $\tilde{R}_W(K,\alpha,\beta,\Delta)$ *for power family two-sided inner wedge tests with shape parameter* Δ. *The first opportunity for early stopping to accept* H_0 *occurs at analysis* k^*. *Also shown are expected sample sizes at* $\theta = 0$, $\pm\delta/2$ *and* $\pm\delta$ *expressed as percentages of the corresponding fixed sample size. Tests are for* K *groups of observations, Type I error probability* $\alpha = 0.05$ *at* $\theta = 0$ *and power* $1-\beta = 0.9$ *at* $\theta = \pm\delta$.

K	$\tilde{C}_{W1}$	$\tilde{C}_{W2}$	$\tilde{R}_W$	k^*	*Expected sample size, as percentage of fixed sample size, at* $\theta = 0$	$\theta = \delta/2$	$\theta = \delta$
$\Delta = -0.5$							
1	1.960	1.282	1.000	1	100.0	100.0	100.0
2	1.960	1.282	1.000	2	100.0	99.9	97.4
3	1.952	1.305	1.010	2	83.4	90.9	86.9
4	1.952	1.316	1.016	3	83.1	89.4	84.2
5	1.952	1.326	1.023	3	80.3	87.2	82.0
10	1.958	1.351	1.042	6	76.9	84.1	78.2
15	1.963	1.363	1.053	9	76.0	83.1	77.1
20	1.967	1.370	1.060	12	75.7	82.7	76.5
$\Delta = -0.25$							
1	1.960	1.282	1.000	1	100.0	100.0	100.0
2	1.957	1.294	1.006	1	95.6	97.2	92.2
3	1.954	1.325	1.023	2	80.8	88.5	83.5
4	1.958	1.337	1.033	2	81.3	87.4	80.4
5	1.960	1.351	1.043	3	77.2	84.8	78.5
10	1.975	1.379	1.071	5	74.7	82.0	74.8
15	1.982	1.394	1.085	8	73.4	80.9	73.7
20	1.988	1.403	1.094	10	73.0	80.4	73.1
$\Delta = 0$							
1	1.960	1.282	1.000	1	100.0	100.0	100.0
2	1.958	1.336	1.032	1	85.4	91.2	84.8
3	1.971	1.353	1.051	2	79.7	86.7	79.4
4	1.979	1.381	1.075	2	75.8	83.6	76.0
5	1.990	1.385	1.084	3	75.8	82.9	74.1
10	2.013	1.428	1.127	5	71.6	79.4	70.6
15	2.026	1.447	1.148	7	70.5	78.3	69.4
20	2.034	1.458	1.160	9	69.9	77.8	68.8
$\Delta = 0.25$							
1	1.960	1.282	1.000	1	100.0	100.0	100.0
2	2.003	1.398	1.100	1	79.3	87.1	79.4
3	2.037	1.422	1.139	1	78.5	85.0	74.1
4	2.058	1.443	1.167	2	73.5	81.4	71.1
5	2.073	1.477	1.199	2	71.4	79.7	69.3
10	2.119	1.521	1.261	4	68.6	76.8	65.6
15	2.140	1.551	1.297	5	67.3	75.7	64.3
20	2.154	1.565	1.316	7	66.7	75.1	63.7

Table 6.3 *Constants* $\tilde{C}_{W1}(K,\alpha,\beta,\Delta)$, $\tilde{C}_{W2}(K,\alpha,\beta,\Delta)$ *and* $\tilde{R}_W(K,\alpha,\beta,\Delta)$ *for power family two-sided inner wedge tests with shape parameter* Δ. *The first opportunity for early stopping to accept* H_0 *occurs at analysis* k^*. *Also shown are expected sample sizes at* $\theta = 0, \pm\delta/2$ *and* $\pm\delta$ *expressed as percentages of the corresponding fixed sample size. Tests are for* K *groups of observations, Type I error probability* $\alpha = 0.05$ *at* $\theta = 0$ *and power* $1-\beta = 0.95$ *at* $\theta = \pm\delta$.

K	$\tilde{C}_{W1}$	$\tilde{C}_{W2}$	$\tilde{R}_W$	k^*	*Expected sample size, as percentage of fixed sample size, at* $\theta = 0$	$\theta = \delta/2$	$\theta = \delta$
$\Delta = -0.5$							
1	1.960	1.645	1.000	1	100.0	100.0	100.0
2	1.960	1.645	1.000	2	100.0	99.8	95.7
3	1.959	1.653	1.004	2	88.2	93.6	83.3
4	1.961	1.661	1.010	3	84.5	90.7	80.8
5	1.964	1.666	1.014	3	85.2	89.9	78.4
10	1.974	1.685	1.030	6	80.2	86.1	74.5
15	1.981	1.695	1.040	9	78.8	85.0	73.3
20	1.985	1.702	1.046	12	78.2	84.4	72.8
$\Delta = -0.25$							
1	1.960	1.645	1.000	1	100.0	100.0	100.0
2	1.962	1.644	1.001	2	100.0	99.0	88.7
3	1.964	1.666	1.014	2	83.6	90.5	79.8
4	1.970	1.674	1.022	3	83.4	88.9	76.4
5	1.975	1.685	1.031	3	80.2	86.7	74.5
10	1.992	1.709	1.054	6	76.9	83.6	70.6
15	2.002	1.721	1.066	9	76.1	82.7	69.5
20	2.008	1.729	1.075	11	75.5	82.2	68.9
$\Delta = 0$							
1	1.960	1.645	1.000	1	100.0	100.0	100.0
2	1.970	1.667	1.018	1	92.9	94.6	80.3
3	1.983	1.691	1.039	2	80.9	87.9	75.1
4	1.995	1.708	1.055	2	79.8	85.9	71.5
5	2.004	1.718	1.066	3	77.1	84.0	69.5
10	2.032	1.753	1.102	5	73.8	80.9	65.8
15	2.046	1.767	1.119	7	73.1	80.0	64.7
20	2.055	1.778	1.131	10	72.4	79.5	64.1
$\Delta = 0.25$							
1	1.960	1.645	1.000	1	100.0	100.0	100.0
2	2.015	1.730	1.079	1	83.0	89.2	74.5
3	2.052	1.729	1.100	2	80.9	86.2	68.6
4	2.072	1.776	1.140	2	75.1	82.6	65.8
5	2.088	1.799	1.163	2	74.3	81.5	64.0
10	2.136	1.845	1.220	4	70.5	78.2	60.2
15	2.159	1.867	1.247	6	69.5	77.1	58.9
20	2.173	1.880	1.264	8	69.0	76.6	58.2

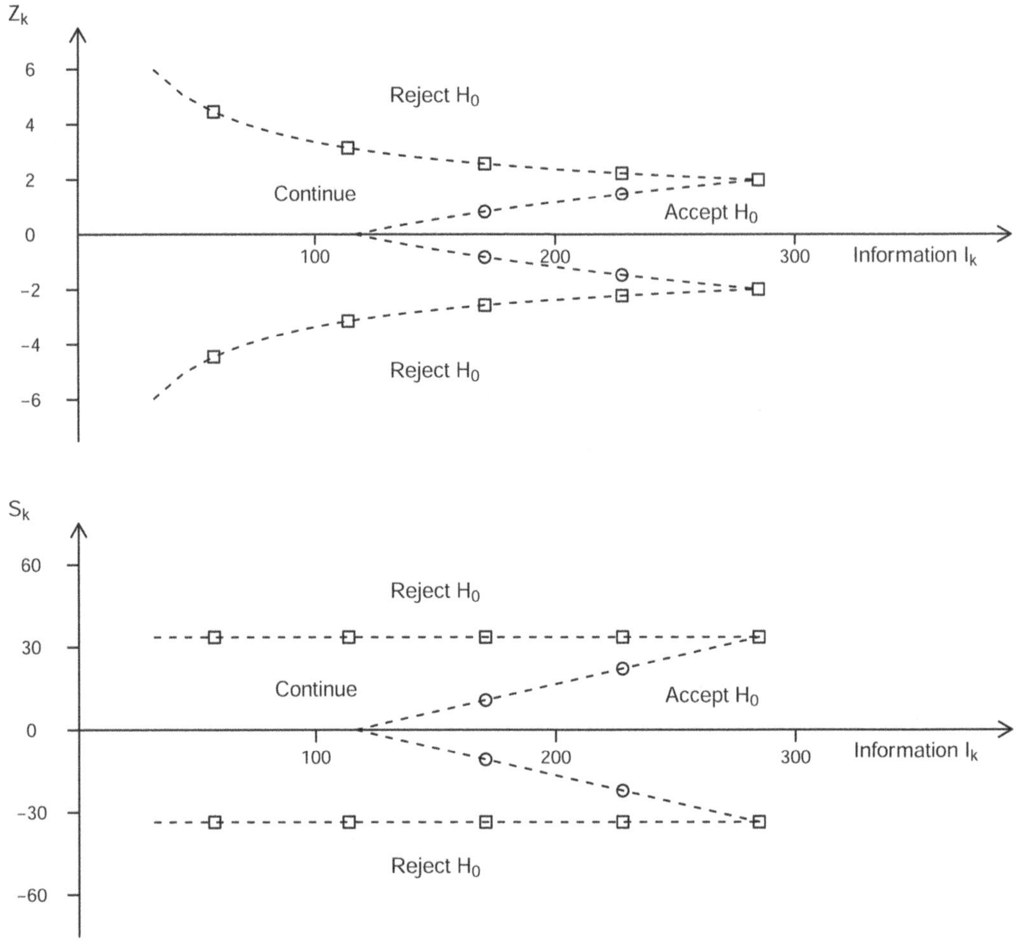

Figure 6.1 *A power family inner wedge test with* $\Delta = 0$ *for five groups of observations*

and $1 - \beta = 0.8$, 0.9 and 0.95; interpolation may be used to obtain constants for values of $K < 20$ not present in the tables. For many of the tabulated tests, the first few values of a_k are negative, which we interpret as indicating there is no opportunity to stop and accept H_0 at these analyses, a reasonable feature since acceptance of H_0 requires evidence that θ is both less than δ and greater than $-\delta$ and insufficient information may be available at early analyses to limit θ to such a narrow range. In the tables, the value k^* indicates the first analysis at which the test can stop to accept H_0.

As in Chapter 5, we have defined tests in terms of the standardized statistics $Z_1, \ldots, Z_K$, whereas Pampallona & Tsiatis (1994) defined the same tests in terms of the score statistics $S_k = Z_k \sqrt{\mathcal{I}_k}$, $k = 1, \ldots, K$, and their $C_j(K, \alpha, \beta, \Delta)$ are $K^{1/2-\Delta}$ times our $\tilde{C}_{Wj}(K, \alpha, \beta, \Delta)$ for $j = 1$ and 2. Constants for other cases with $\alpha = 0.05$ and 0.01, $1 - \beta = 0.8$, 0.9 and 0.95, and $\Delta = 0$, 0.1, 0.2, 0.3, 0.4 and 0.5 can be found in Pampallona & Tsiatis (1994). Also, as in Chapter 5, we follow the standard practice of referring to expected and maximum sample size when discussing properties of group sequential tests, although it would be more correct to talk of information when the two are not directly proportional.

Figure 6.1 shows the boundary of the power family inner wedge test with $\Delta = 0$ and five groups of observations, expressed on both the Z_k and S_k scales. The test has Type I error probability $\alpha = 0.05$ and power $1 - \beta = 0.9$ at $\theta = \pm 0.2$.

6.2.2 Properties of Power Family Inner Wedge Tests

It is clear from Tables 6.1 to 6.3 that the inner wedge is successful in allowing early stopping to accept H_0, and expected sample sizes under $\theta = 0$ of around 70% to 75% of the corresponding fixed sample size can be obtained in tests with 5 or 10 analyses. For designs with power 0.9 at $\theta = \pm\delta$, these reductions in expected sample size are broadly comparable with those at the alternative $\theta = \pm\delta$. In view of the savings that can be achieved in study costs and in time, we would recommend that early stopping to accept the null hypothesis be considered in any group sequential two-sided test.

The ratio $\tilde{R}_W(K, \alpha, \beta, \Delta)$ of the maximum sample size to the fixed sample size increases with Δ. We have added negative values of Δ, not considered by Pampallona & Tsiatis(1994), to allow a choice of tests with lower maximum sample sizes, but we have stopped at an upper limit of $\Delta = 0.25$ since higher values require substantial increases in maximum sample size and yield little reduction, or sometimes even an increase, in expected sample size.

As the number of analyses K increases, the maximum sample size increases and expected sample sizes decrease. As for other types of test, there is little additional benefit to be gained in planning more than 5 or 10 analyses.

6.2.3 Adapting Power Family Inner Wedge Tests to Unequal Increments in Information

The power family inner wedge tests are easily adapted for use with unpredictable sequences of information levels. Our adaptation follows similar lines to that of the power family, one-sided tests of Chapter 5. We first note that the a_k and b_k defined by (6.2) for the case of equally spaced information levels can be rewritten as

$$\begin{aligned} b_k &= \tilde{C}_{W1}(K, \alpha, \beta, \Delta)(k/K)^{\Delta - 1/2} \text{ and} \\ a_k &= \{\tilde{C}_{W1}(K, \alpha, \beta, \Delta) + \tilde{C}_{W2}(K, \alpha, \beta, \Delta)\}(k/K)^{1/2} \\ &\quad - \tilde{C}_{W2}(K, \alpha, \beta, \Delta)(k/K)^{\Delta - 1/2}, \quad k = 1, \ldots, K. \end{aligned} \tag{6.3}$$

This alternative definition has the advantage that the information levels $\mathcal{I}_k$ do not appear explicitly, and we shall use this in adapting the tests to general information sequences. Note that this definition automatically implies $a_K = b_K$.

Suppose a study gives rise to the sequence of standardized statistics $\{Z_1, \ldots, Z_K\}$ and observed information levels $\{\mathcal{I}_1, \ldots, \mathcal{I}_K\}$ for θ. In our adaptation of the inner wedge test, we apply the original stopping rule (6.1) but with a_k and b_k, $k = 1, \ldots, K$, defined by (6.3). Since $Z_1, \ldots, Z_K$ follow the canonical joint distribution (3.1), the joint distribution of the Z_ks under H_0: $\theta = 0$ depends only on the ratios between the observed $\mathcal{I}_k$s, and hence the Type I error probability remains at precisely α if observed information levels are equally spaced. Unequal spacing causes minor variations in Type I error. We have chosen this adaptation to preserve the Type I error probability as closely as possible, and it is only to be expected that variations in power will be greater. The attained power depends primarily on the final information level, $\mathcal{I}_K$, with additional dependence on the spacing of earlier information levels.

Table 6.4 shows the Type I error probability and power obtained when our rule is applied with information levels departing from their planned values. The target information levels are $\mathcal{I}_k = (k/K)\mathcal{I}_{max}$, $k = 1, \ldots, K$, where $\mathcal{I}_{max} = \tilde{R}_W(K, \alpha, \beta, \Delta)\mathcal{I}_{f,2}$ and observed information levels are $\mathcal{I}_k = \pi(k/K)^r\mathcal{I}_{max}$, $k = 1, \ldots, K$. Here, π controls the final information level and r the spacing between successive information levels. The Type I error probability remains within 0.001 of $\alpha = 0.05$ in all cases. Power is affected most by the overall scale of information. For the low information levels arising when $\pi = 0.9$, power is low and varies around the value 0.868 achieved in a fixed sample test with 0.9 times the required information, $\mathcal{I}_{f,2}$. Conversely, $\pi = 1.1$ gives rise to higher information levels

Table 6.4 *Properties of power family inner wedge tests designed for information levels* $\mathcal{I}_k = (k/K)\mathcal{I}_{max}$, $k = 1, \ldots, K$, *but applied with* $\mathcal{I}_k = \pi(k/K)^r \mathcal{I}_{max}$, $k = 1, \ldots, K$. *Attained Type I error probabilities and power are shown for tests with* K *analyses and shape parameter* $\Delta = -0.5$ *and* $\Delta = 0$, *designed to achieve two-sided Type I error rate* $\alpha = 0.05$ *and power* $1 - \beta = 0.9$ *at* $\theta = \pm\delta$.

			$\Delta = -0.5$		$\Delta = 0$	
	r	π	*Type I error*	*Power*	*Type I error*	*Power*
$K = 2$						
	0.80	0.90	0.050	0.868	0.051	0.874
		1.00	0.050	0.900	0.051	0.905
		1.10	0.050	0.925	0.051	0.930
	1.00	0.90	0.050	0.868	0.050	0.868
		1.00	0.050	0.900	0.050	0.900
		1.10	0.050	0.925	0.050	0.925
	1.25	0.90	0.050	0.868	0.049	0.857
		1.00	0.050	0.900	0.049	0.890
		1.10	0.050	0.925	0.049	0.915
$K = 5$						
	0.80	0.90	0.051	0.872	0.050	0.878
		1.00	0.051	0.904	0.050	0.909
		1.10	0.051	0.928	0.050	0.932
	1.00	0.90	0.050	0.868	0.050	0.868
		1.00	0.050	0.900	0.050	0.900
		1.10	0.050	0.925	0.050	0.925
	1.25	0.90	0.049	0.860	0.049	0.851
		1.00	0.049	0.893	0.049	0.885
		1.10	0.049	0.919	0.049	0.911
$K = 10$						
	0.80	0.90	0.051	0.874	0.050	0.882
		1.00	0.051	0.906	0.050	0.912
		1.10	0.051	0.930	0.050	0.935
	1.00	0.90	0.050	0.868	0.050	0.868
		1.00	0.050	0.900	0.050	0.900
		1.10	0.050	0.925	0.050	0.925
	1.25	0.90	0.049	0.857	0.050	0.845
		1.00	0.049	0.890	0.050	0.879
		1.10	0.049	0.916	0.050	0.907

and higher attained powers, varying around the power 0.925 of a fixed sample test with information $1.1 \times \mathcal{I}_{f,2}$.

Unequal increments in information can also be handled using an "error spending" approach. We shall present such methods for two-sided tests with an inner wedge in Section 8.2.5.

In "equivalence testing" problems where one hopes to show the treatment difference θ is close to zero, it is important to limit the probability of concluding $\theta = 0$ when, really, θ differs from zero by a clinically significant amount δ. Thus, in adapting a test of H_0: $\theta = 0$ to unpredictable information levels, the priority is to preserve a stated power at $\theta = \pm\delta$. This can be achieved by applying the "significance level" approach to Z-statistics for testing $\theta = \delta$ and $\theta = -\delta$, namely $Z_k - \delta\sqrt{\mathcal{I}_k}$ and $Z_k + \delta\sqrt{\mathcal{I}_k}$ where $Z_k \sim N(\theta\sqrt{\mathcal{I}_k}, 1)$. We shall discuss such methods in detail, and also related group sequential t-tests, in Chapter 7.

6.2.4 Applying Power Family Inner Wedge Tests When Variance is Unknown

Consider the setting of Section 3.3 in which observations follow a normal linear model with parameter vector $\boldsymbol{\beta}$ and variances depend on a scale factor σ^2. Group sequential estimates $\widehat{\boldsymbol{\beta}}^{(1)}, \ldots, \widehat{\boldsymbol{\beta}}^{(K)}$ are obtained at analyses 1 to K, $Var(\widehat{\boldsymbol{\beta}}^{(K)}) = \boldsymbol{V}_k\sigma^2$ and s_k^2 is the usual unbiased estimate of σ^2 at analysis k on $n_k - p$ degrees of freedom. We shall define group sequential two-sided t-tests of a null hypothesis H_0: $\boldsymbol{c}^T\boldsymbol{\beta} = \gamma$. Our main aim is for tests to have Type I error probability α, but we also wish to assess power at alternatives of the form $\boldsymbol{c}^T\boldsymbol{\beta} - \gamma = \pm\delta$.

As before, we have

$$\boldsymbol{c}^T\widehat{\boldsymbol{\beta}}^{(k)} \sim N(\boldsymbol{c}^T\boldsymbol{\beta}, \boldsymbol{c}^T\boldsymbol{V}_k\boldsymbol{c}\sigma^2)$$

and the t-statistic for testing H_0 is

$$T_k = \frac{\boldsymbol{c}^T\widehat{\boldsymbol{\beta}}^{(k)} - \gamma}{\sqrt{\{\boldsymbol{c}^T\boldsymbol{V}_k\boldsymbol{c}\,s_k^2\}}}.$$

The group sequential inner wedge tests already introduced in this chapter are for the case of known σ^2. In adapting these tests to an unknown variance we begin with the stopping rule (6.1) and constants a_k and b_k, $k = 1, \ldots, K$, defined by (6.3). We replace Z_k in (6.1) by T_k and choose critical values for T_k to preserve the marginal probability under H_0 of being above or below each boundary point at analysis k. The new stopping rule is therefore:

$$\begin{array}{lll}
\text{After group } k = 1, \ldots, K-1 & & \\
\quad \text{if } |T_k| \geq t_{n_k-p,1-\Phi(b_k)} & \text{stop, reject } H_0 \\
\quad \text{if } |T_k| < t_{n_k-p,1-\Phi(a_k)} & \text{stop, accept } H_0 \\
\quad \text{otherwise} & \text{continue to group } k+1, \\
\text{after group } K & \\
\quad \text{if } |T_K| \geq t_{n_K-p,1-\Phi(b_K)} & \text{stop, reject } H_0 \\
\quad \text{if } |T_K| < t_{n_K-p,1-\Phi(a_K)} & \text{stop, accept } H_0,
\end{array}$$

where $a_K = b_K$. Here, $t_{\nu,q}$ is such that $Pr\{T > t_{\nu,q}\} = q$ when $T \sim t_\nu$.

This construction is closely related to that used in Section 5.6 to create group sequential one-sided t-tests. Essentially the same arguments imply that the resulting tests have Type I error probability close to the target value α, especially if information levels are approximately equally spaced and the degrees of freedom for each s_k^2 are not too low. Also, a similar approach can be taken to estimate a test's power. The approximation analogous

Table 6.5 *Properties of power family, inner wedge t-tests designed to test* H_0: $c^T\beta = \gamma$ *with two-sided Type I error probability* $\alpha = 0.05$. *Tests are based on power family, inner wedge tests for known variance with* $\Delta = -0.5$ *and* $\Delta = 0$. *Degrees of freedom for* σ^2 *at analysis k are* $\nu_k = (\nu_K + 2)(k/K) - 2$ *and* $Var(c^T\hat{\beta}^{(k)}) = (K/k)Var(c^T\hat{\beta}^{(K)})$, $k = 1, \ldots, K$. *The penultimate column shows actual power at the alternative for which the approximation* (6.4) *gives power* 0.8, *and the final column shows the power predicted for each of the tests by formula* (6.5). *Type I error and power were estimated by simulation, using 1 million replicates; standard errors are* 0.0002 *for Type I error and* 0.0004 *for power.*

$\Delta = -0.5$

K	ν_K	*Type I error*	*Attained power*	*Known* σ^2 *power*
3	7	0.055	0.819	0.905
3	13	0.053	0.812	0.858
5	13	0.055	0.818	0.858
5	23	0.053	0.809	0.833
8	22	0.054	0.813	0.834
8	38	0.053	0.807	0.820

$\Delta = 0$

K	ν_K	*Type I error*	*Attained power*	*Known* σ^2 *power*
3	7	0.058	0.820	0.905
3	13	0.054	0.813	0.858
5	13	0.058	0.821	0.858
5	23	0.055	0.812	0.833
8	22	0.056	0.816	0.834
8	38	0.054	0.809	0.820

to (5.18) gives

$$Pr\{T(n_K - p;\, \delta/\sqrt{[\tilde{R}_W(K,\alpha,\beta,\Delta)\, \boldsymbol{c}^T\boldsymbol{V}_K\boldsymbol{c}\sigma^2]}) \geq t_{n_K-p,\alpha/2}\} \tag{6.4}$$

as an estimate of the power at $\boldsymbol{c}^T\boldsymbol{\beta} - \gamma = \pm\delta$ for a given value of σ^2. The cruder approximation for large $n_K - p$ is

$$\Phi\left(\frac{\delta}{\sqrt{\{\tilde{R}_W(K,\alpha,\beta,\Delta)\, \boldsymbol{c}^T\boldsymbol{V}_K\boldsymbol{c}\sigma^2\}}} - \Phi^{-1}(1-\alpha/2)\right). \tag{6.5}$$

Table 6.5 shows the Type I error probability of our group sequential inner wedge t-tests and the power they attain in situations where (6.4) gives the value 0.8. The degrees of freedom are those which arise when each group contains three or five observations and two degrees of freedom are used in fitting the linear model at each analysis. Results follow the pattern previously seen for group sequential t-tests. There are only minor departures from the target Type I error rates and formula (6.4) gives a reliable approximation to attained power, particularly when the degrees of freedom are fairly high. Once again, the definition of group sequential t-tests does not rely on equally spaced information levels and these tests may also be used with unequal group sizes or information increments.

6.2.5 *Example: Two-treatment Comparison*

We illustrate the above methods with an example of a parallel two-treatment comparison conducted group sequentially with five groups of observations. Suppose that observations $X_{Ai} \sim N(\mu_A, \sigma^2)$ and $X_{Bi} \sim N(\mu_B, \sigma^2)$, $i = 1, 2, \ldots$, and we wish to test H_0: $\mu_A = \mu_B$ against the two-sided alternative $\mu_A \neq \mu_B$ with Type I error probability $\alpha = 0.05$ using a power family inner wedge test with $\Delta = 0.25$. We further suppose there is reason to believe σ^2 is in the region of 1.0, and we would like the test to have power of about 0.95 to detect a difference of $\mu_A - \mu_B = \pm 0.6$, bearing in mind our uncertainty about σ^2. If groups contain m observations per treatment, the sequence of t-statistics is

$$T_k = \frac{\sum_{i=1}^{mk} X_{Ai} - \sum_{i=1}^{mk} X_{Bi}}{\sqrt{\{2mks_k^2\}}}, \quad k = 1, \ldots, 5,$$

where

$$s_k^2 = \frac{\sum_{i=1}^{mk}(X_{Ai} - \bar{X}_A^{(k)})^2 + \sum_{i=1}^{mk}(X_{Bi} - \bar{X}_B^{(k)})^2}{2(mk-1)}.$$

From Table 6.3, the constants defining this inner wedge test are $\tilde{C}_{W1}(5, 0.05, 0.05, 0.25) = 2.088$, $\tilde{C}_{W2}(5, 0.05, 0.05, 0.25) = 1.799$ and $\tilde{R}_W(5, 0.05, 0.05, 0.25) = 1.163$. For $k = 1, \ldots, 5$, the test stops at analysis k to reject H_0 if

$$|T_k| \geq t_{2mk-2, 1-\Phi(b_k)}$$

and to accept H_0 if

$$|T_k| < t_{2mk-2, 1-\Phi(a_k)},$$

where, from (6.3), $b_k = 2.088 \times (k/5)^{-0.25}$ and

$$a_k = (2.088 + 1.799) \times (k/5)^{0.5} - 1.799 \times (k/5)^{-0.25}.$$

We expect the Type I error probability of this test to be close to 0.05, but its power will depend on m and the unknown σ^2. If σ^2 were known to be equal to 1.0, the corresponding five-group test for known variance would need a final information level $\tilde{R}_W(5, 0.05, 0.05, 0.25)\,\mathcal{I}_{f,2}$, i.e.,

$$1.163 \times \{\Phi^{-1}(0.975) + \Phi^{-1}(0.95)\}^2/0.6^2 = 41.98.$$

Five groups of m observations on each treatment give an estimate of $\mu_A - \mu_B$ with variance $2\sigma^2/(5m)$ and, therefore, information $5m/(2\sigma^2)$ for $\mu_A - \mu_B$. Equating this to 41.98 when $\sigma^2 = 1.0$ gives the value $m = 16.8$. We know the corresponding group sequential t-test will require a slightly higher sample size, but rounding up to the next integer, $m = 17$, is evidently sufficient since (6.4) then gives 0.950 as the approximate power at $|\mu_A - \mu_B| = 0.6$ if $\sigma^2 = 1.0$. Simulations with one million replications show this test's actual Type I error rate to be 0.051 and its power at $|\mu_A - \mu_B| = 0.6$ when $\sigma^2 = 1.0$ is 0.951 (standard errors of these estimates are less than 0.0005).

The above power calculation relies on σ^2 being equal to its predicted value of 1.0. If the experimenter desires greater confidence that the power condition will actually be met, the sample size should be calculated using an upper bound on plausible values of σ^2. Suppose, for example, it is thought that σ^2 could be as high as 1.4; the analogous calculation to that shown above gives the value $m = 23.5$ and the rounded value $m = 24$ appears adequate since formula (6.4) then gives 0.952 as the approximate power at $|\mu_A - \mu_B| = 0.6$ if $\sigma^2 = 1.4$ (the estimated power from one million simulations is 0.953, again with a standard error of less than 0.0005).

6.3 Whitehead's Double Triangular Test

Sobel & Wald (1949) describe a way of combining two one-sided sequential probability ratio tests to create a two-sided test with early stopping to accept the null hypothesis. Whitehead & Stratton (1983) and Whitehead (1997, Sec. 4.7) have used this technique to combine two triangular tests, producing a "double triangular" test.

Suppose standardized statistics $Z_1, \ldots, Z_K$ at analyses $k = 1, \ldots, K$ follow the canonical joint distribution (3.1) and we require a test of H_0: $\theta = 0$ with Type I error probability α and power $1 - \beta$ at $\theta = \pm\delta$. The double triangular test is formed from two of Whitehead's one-sided triangular boundaries for testing the null hypothesis $\theta = 0$ with Type I error rate $\alpha/2$ and power $1 - \beta$ at $\theta = \delta$ and $\theta = -\delta$, respectively. Sampling continues until both tests have reached a conclusion, then H_0 is rejected if either of the tests has rejected $\theta = 0$ and otherwise H_0 is accepted.

The test is most naturally described in terms of the score statistics $S_k = Z_k\sqrt{\mathcal{I}_k}$, $k = 1, \ldots, K$. For equally spaced information levels, the one-sided triangular test of $\theta = 0$ versus $\theta > 0$ has upper and lower boundaries for S_k at analysis k given by

$$d_k = \frac{2}{\tilde{\delta}} \log\left(\frac{1}{\alpha}\right) - 0.583\sqrt{\frac{\mathcal{I}_{max}}{K}} + \frac{\tilde{\delta}}{4}\frac{k}{K}\mathcal{I}_{max} \tag{6.6}$$

and

$$c_k = -\frac{2}{\tilde{\delta}} \log\left(\frac{1}{\alpha}\right) + 0.583\sqrt{\frac{\mathcal{I}_{max}}{K}} + \frac{3\tilde{\delta}}{4}\frac{k}{K}\mathcal{I}_{max}\,, \tag{6.7}$$

where

$$\tilde{\delta} = \frac{2\Phi^{-1}(1-\alpha/2)\,\delta}{\Phi^{-1}(1-\alpha/2) + \Phi^{-1}(1-\beta)}$$

and

$$\mathcal{I}_{max} = \left[\sqrt{\frac{4 \times 0.583^2}{K} + 8\log\left(\frac{1}{\alpha}\right)} - \frac{2 \times 0.583}{\sqrt{K}}\right]^2 \frac{1}{\tilde{\delta}^2}.$$

The upper and lower boundaries for S_k of the test of $\theta = 0$ versus $\theta < 0$ are $c'_k = -c_k$ and $d'_k = -d_k$, respectively. Combining these tests in the specified manner gives the formal stopping rule:

$$\begin{array}{lll}
\text{After group } k = 1, \ldots, K-1 & & \\
\quad \text{if } |S_k| \geq d_k & \text{stop, reject } H_0 & \\
\quad \text{if } S_j \leq c_j \text{ for some } j \leq k \text{ and} & & \\
\quad\quad S_j \geq -c_j \text{ for some } j \leq k & \text{stop, accept } H_0 & \\
\quad \text{otherwise} & \text{continue to group } k+1, & \quad (6.8)\\
\text{after group } K & & \\
\quad \text{if } |S_K| \geq d_K & \text{stop, reject } H_0 & \\
\quad \text{otherwise} & \text{stop, accept } H_0. &
\end{array}$$

For more general information sequences, $\mathcal{I}_{max}/K$ is replaced by $\mathcal{I}_k - \mathcal{I}_{k-1}$ in the second terms of Equations (6.6) and (6.7) and $(k/K)\mathcal{I}_{max}$ is replaced by $\mathcal{I}_k$ in the third terms. The first part of rule (6.8) defines the test until a stage K is reached where $c_K \geq d_K$; in keeping with our suggestion in Section 5.7.3 for the one-sided test, we recommend that H_0 be rejected at this final stage if $|S_K| \geq d_K$ and accepted otherwise.

Table 6.6 shows properties of the double triangular test when information levels are equally spaced and boundaries meet at the final analysis. The Sobel-Wald construction guarantees Type I error probability less than or equal to the sum of the Type I error

Table 6.6 *Properties of Whitehead's double triangular tests with equal group sizes. Tests are for K groups of observations, and designed to have Type I error probability $\alpha = 0.05$ at $\theta = 0$ and power $1 - \beta = 0.8$, 0.9 and 0.95 at $\theta = \pm\delta$. Constants $R_{DT}(K, \alpha, \beta)$ are ratios of the maximum sample size to that of a fixed sample test with the same error probabilities. Expected sample sizes at $\theta = 0$, $\delta/2$ and δ are expressed as percentages of the corresponding fixed sample size.*

K	*Type I error*	*Power at* $\theta = \delta$	R_{DT}	*Expected sample size, as percentage of fixed sample size, at* $\theta = 0$	$\theta = \delta/2$	$\theta = \delta$
$1 - \beta = 0.8$						
1	0.053	0.797	0.973	97.3	97.3	97.3
2	0.053	0.800	1.115	80.1	85.6	81.2
3	0.052	0.802	1.186	84.3	87.8	78.8
4	0.051	0.803	1.230	75.9	82.1	75.9
5	0.051	0.803	1.261	74.2	80.5	74.4
10	0.050	0.804	1.342	71.1	77.7	71.2
15	0.050	0.804	1.379	70.1	76.6	70.1
20	0.050	0.804	1.402	69.2	75.9	69.5
$1 - \beta = 0.9$						
1	0.053	0.897	0.973	97.3	97.3	97.3
2	0.053	0.899	1.115	80.1	86.3	76.2
3	0.052	0.901	1.186	84.3	87.8	72.4
4	0.051	0.902	1.230	75.9	82.7	69.8
5	0.051	0.902	1.261	74.2	81.2	68.2
10	0.050	0.902	1.342	71.1	78.4	64.8
15	0.050	0.902	1.379	70.1	77.4	63.6
20	0.050	0.902	1.402	69.2	76.7	63.0
$1 - \beta = 0.95$						
1	0.053	0.948	0.973	97.3	97.3	97.3
2	0.053	0.949	1.115	80.1	86.5	71.7
3	0.052	0.950	1.186	84.3	87.5	66.8
4	0.051	0.950	1.230	75.9	82.9	64.3
5	0.051	0.951	1.261	74.2	81.4	62.8
10	0.050	0.951	1.342	71.1	78.6	59.2
15	0.050	0.951	1.379	70.1	77.5	58.0
20	0.050	0.951	1.402	69.2	76.9	57.3

probabilities of the two one-sided tests, and power at $\theta = \pm\delta$ is that of the relevant one-sided test in each case. Thus, the slight inaccuracies in Type I error rate and power in the first two columns of Table 6.6 are due only to approximations made in the original one-sided tests. The third column of the table contains values of $R_{DT}(K, \alpha, \beta)$, the ratio between each group sequential test's maximum information level and the information, $\mathcal{I}_{f,2}$, required by a fixed sample test with error probabilities α and β. The final three columns show expected information levels on termination under $\theta = 0$, $\pm\delta/2$ and $\pm\delta$. As for the single triangular test of Section 5.7, results for $K = 1$ represent the case where a large first information level $\mathcal{I}_1$ leads to $d_1 = c_1$ and termination of the test. This time, the value of $\mathcal{I}_1$ is not quite so close to the information level needed for a fixed sample test with error probabilities exactly α and β,

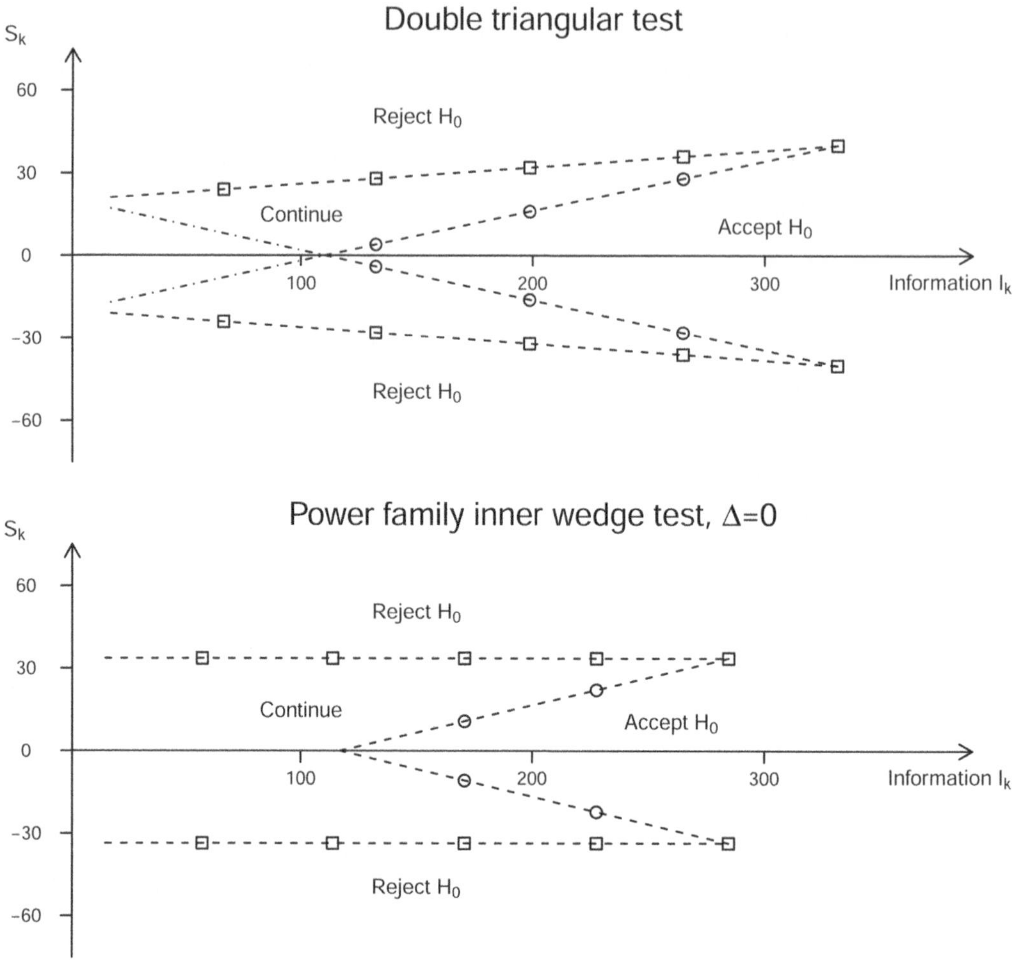

Figure 6.2 *Comparison of two inner wedge tests with five groups of observations*

but it is still fairly close, and the results for this case testify to the robustness of Whitehead & Stratton's scheme for converting a continuous boundary to discrete monitoring.

A comparison with Tables 6.1 to 6.3 shows the double triangular tests' maximum sample sizes to be similar to those of power family tests with $\Delta = 0.25$ in the case $1 - \beta = 0.8$ and higher for $1 - \beta = 0.9$ and 0.95. Apart from some instances when only two analyses are performed, the power family tests with $\Delta = 0.25$ yield lower or roughly equal expected sample sizes across the range of θ values for $1-\beta = 0.8$ and 0.9 and similar expected sample sizes for $1 - \beta = 0.95$, when their maximum sample sizes are lower.

Figure 6.1 in Section 6.2.1 showed the boundary of a power family inner wedge test with $\Delta = 0$, five groups of observations, Type I error rate $\alpha = 0.05$ and power $1 - \beta = 0.9$ at $\theta = \pm 0.2$. Figure 6.2 compares this boundary, on the score statistic scale, with that of Whitehead's double triangular test meeting the same specifications. In constructing the double triangular test we set power $1 - \alpha = 0.95$ at

$$\theta = \tilde{\delta} = 2 \times 1.960 \times 0.2/(1.960 + 1.282) = 0.2418.$$

The double triangular test has

$$\mathcal{I}_{max} = R_{DT}(5, 0.05, 0.1)\, \mathcal{I}_{f,2} = 1.261 \times 262.76 = 331.3,$$

somewhat higher than the value $\tilde{R}_W(5, 0.05, 0.1, 0)\, \mathcal{I}_{f,2} = 1.084 \times 262.76 = 284.8$ for the power family inner wedge test with $\Delta = 0$. These values of $\tilde{R}_W(5, 0.05, 0.1, 0)$ and $R_{DT}(5, 0.05, 0.1)$ are taken from Tables 6.2 and 6.6, respectively.

The Sobel-Wald construction is unusual in that the stopping rule produced is not a function of a sufficient statistic for θ. Suppose, for example, Z_k lies below the lower boundary of the one-sided test of $\theta = 0$ versus $\theta > 0$ but in the continuation region of the test of $\theta = 0$ versus $\theta < 0$. Then, early stopping to accept H_0 occurs if the upper boundary of the test of $\theta = 0$ versus $\theta < 0$ was crossed previously, i.e., if $Z_j \geq -c_j$ for some $j < k$. However, given the value of Z_k, the conditional distribution of $Z_1, \ldots, Z_{k-1}$ is independent of θ, so it is difficult to justify why a conclusion should depend on the values of $Z_1, \ldots, Z_{k-1}$ once Z_k is known. The double triangular test can be simplified by ignoring the inner boundary until c_k becomes positive, and then stopping to accept H_0 at an analysis k if $-c_k \leq S_k \leq c_k$, in which case the test has the same form as that defined on the Z scale in (6.1). Calculations show this modification has almost no effect on the Type I error probability, the only change to values in Table 6.6 being an increase from 0.052 to 0.053 for $K = 3$. There are small increases in power, the largest being 0.009, 0.007 and 0.005 for $K = 3$ and $1 - \beta = 0.8$, 0.9 and 0.95, respectively. In view of the sound theoretical reasons for defining statistical procedures in terms of sufficient statistics, we recommend making this modification in applying Whitehead's double triangular test.

Chapter 7

Tests for Equivalence and Non-inferiority

7.1 Introduction

Sometimes the goal of a clinical trial is to establish *equivalence* between two treatments rather than the superiority in efficacy of one over the other. For example, if a new therapy is less toxic or less expensive than the standard, it may not be necessary to prove it is also more effective; instead, it can suffice to demonstrate that it is equally effective or that it is less effective than the standard by at most a small amount. Dunnett & Gent (1977) give an example of a trial in which an objective was to determine whether certain aspects of health care could be delivered equally well by a system involving a triage process by nurse-practitioners as by conventional primary physician care methods. Other examples include: lumpectomy versus radical mastectomy for breast cancer, a local rice-based beverage versus WHO oral rehydration solution for diarrhea in infants in Mexico, and low-dose AZT treatment versus standard dose during pregnancy and delivery to prevent mother-to-child HIV transmission in a developing country. Fleming (2008) gives further examples. Studies with such an objective are referred to as "one-sided equivalence" trials or "non-inferiority" trials.

A "two-sided equivalence" problem can arise in the area of bioequivalence testing. Here, a pharmaceutical manufacturer hopes to demonstrate that a new preparation of a drug has the same bioavailability properties as a standard, within a small tolerance limit, as a step toward proving that the new and standard preparations have equal therapeutic effects. Demonstrating "bioequivalence" in this way can greatly reduce the amount of experimentation required for approval of a new drug.

Chinchilli (1996) classifies equivalence problems into the following categories

Population equivalence: The responses for both treatments have the same probability distribution;

Average equivalence: The responses have the same mean;

Individual equivalence: The responses are "approximately" the same for a large proportion of subjects when they receive either treatment.

In Sections 7.2, 7.3 and 7.4 we shall be concerned principally with average bioequivalence, while we consider tests of individual equivalence in Section 7.5.

Of course, to conclude equivalence, it is not sufficient that the data fail to reject a null hypothesis of equality. This might simply be due to a lack of power of the study, and a *practically significant* difference could still exist despite the lack of *statistically significant* evidence. Thus, if a hypothesis test of equality is to be employed as a means of establishing equivalence, this test's Type II error probability, representing the probability of wrongly declaring equivalence when a practically significant difference exists, must be restricted to an acceptable level. An alternative approach is to test a specific hypothesis of non-equality or "inequivalence" and conclude equivalence if this hypothesis is rejected. In this case, the error requirements have a more familiar form since priority is given to the test's Type I error

DOI: 10.1201/9781584888482-7

rate, i.e., the probability of rejecting the hypothesis of inequivalence when this hypothesis is true. We shall illustrate both these approaches: in Sections 7.2 and 7.3 we adopt the second method, creating a hypothesis of inequivalence in order to test for one-sided equivalence, and in Section 7.4 we take the first approach, basing a test for two-sided equivalence on a test of a null hypothesis of no treatment difference.

7.2 Non-inferiority Trials: One-sided Tests of Equivalence

Let μ_A and μ_B denote the true means of the primary response variable for a new and standard treatment, respectively, where higher responses are more favorable. Suppose previous studies have shown that the new treatment has fewer toxic side effects than the standard. In view of this known advantage of the new treatment, we can define a hypothesis of inequivalence as H_I: $\mu_A - \mu_B \leq -\delta_N$, where δ_N is a positive constant and a decrease in mean response up to δ_N is deemed acceptable for the new treatment in view of its reduced toxicity. This difference δ_N is termed the *non-inferiority margin*. The hypothesis of equivalence is then H_E: $\mu_A - \mu_B > -\delta_N$, and it is appropriate to conduct a one-sided equivalence test to choose between H_I and H_E. Note that H_E includes instances where μ_A is considerably greater than μ_B since in this one-sided case we intend "equivalence" of the two treatments to mean that the new treatment is at least as good, overall, as the standard.

In essence, this problem is the same as the one-sided hypothesis testing problem of Chapter 5, and if we define $\theta = \mu_A - \mu_B + \delta_N$, the methods of Chapter 5 can be applied directly. Then, $\theta = 0$ occurs at the upper limit of H_I, $\mu_A - \mu_B = -\delta_N$ and so the Type I error rate of the test of H_0: $\theta \leq 0$ becomes the maximum probability of wrongly declaring equivalence. The power at $\theta = \delta_N$ is of particular interest as this represents the probability of declaring equivalence when μ_A and μ_B are exactly equal. Since the new treatment has other advantages over the standard, a high power to decide in favor of the new treatment is desirable when both treatments are equal with respect to the primary response .

The choice of non-inferiority margin is a crucial element of a non-inferiority trial. Usually, the margin δ_N is smaller than the effect size δ that would be specified in a superiority trial of the new treatment against the standard. This is especially true if the standard treatment has been demonstrated to be superior to a placebo in a previous trial. Unless δ_N is small, there will be doubt that the new treatment is indeed superior to placebo. This problem can be exacerbated if there are to be repeated non-inferiority trials with a succession of new treatments, each being compared with the treatment most recently judged non-inferior. Fleming (2008, p. 325) terms this phenomenon "bio-creep".

7.3 Simultaneous Testing of Superiority and Non-inferiority

Consider a trial in which investigators hope to demonstrate that a new treatment is superior to a control but, if that is not achievable, it is still of value to show the new treatment is non-inferior with non-inferiority margin δ_N. They therefore wish to perform two hypothesis tests:

Test for superiority	$H_{0S}: \mu_A - \mu_B \leq 0$	vs	$\mu_A - \mu_B > 0,$
Test for non-inferiority	$H_{0N}: \mu_A - \mu_B \leq -\delta_N$	vs	$\mu_A - \mu_B > -\delta_N.$

Suppose the two hypothesis are tested separately, each with Type I error probability α. Since the hypotheses are nested with $H_{0N} \subset H_{0S}$, considering separately the cases $\mu_A - \mu_B \leq -\delta_N$, $-\delta_N < \mu_A - \mu_B \leq 0$ and $\mu_A - \mu_B > 0$, one can check that the *family-wise Type I error rate*, defined as the probability of rejecting at least one true null hypothesis, is at most α whatever the value of $\mu_A - \mu_B$, and thus there is no need to adjust the tests to account for the simultaneous testing of two hypotheses. This is an example of an *intersection-union* test and thus there is no need to make any adjustment to account for the simultaneous testing of two hypotheses; see, for example, Morikawa & Yoshida (1995), Dunnett & Gent (1996) and CPMP (2000). An important proviso in conducting such a study is that the non-inferiority margin δ_N must be pre-specified and not chosen in a *post hoc* manner after observing the study data; see, for example, Ng (2003).

The trial could be conducted using a group sequential test of H_{0S}: $\mu_A - \mu_B \leq 0$ with an additional analysis at the end of the trial to test H_{0N}. It is convenient to describe this approach in terms of a $(1 - 2\alpha)$-level confidence interval for the parameter $\psi = \mu_A - \mu_B$, calculated on conclusion of the trial. We shall describe methods for calculating such an interval in Section 9.5. Denoting the confidence interval by (ψ_L, ψ_U), key properties are that

$$P_\psi\{\psi_L > \psi\} = P_\psi\{\psi_U < \psi\} = \alpha$$

and $\psi_L \geq 0$ if and only if H_{0S} is rejected by the group sequential test. The outcomes of the two hypothesis tests are then:

If $\psi_L \geq 0$	Reject both H_{0S} and H_{0N},
If $-\delta_N \leq \psi_L < 0$	Reject H_{0N} only,
If $\psi_L < -\delta_N$	Reject neither hypothesis.

It is clear that rejection of H_{0S} implies rejection of H_{0N} and thus the family-wise Type I error rate is controlled at level α.

If this approach is applied using a one-sided group sequential test of H_{0S}, as defined in Chapter 5, the resulting procedure may be inefficient for testing H_{0N}. A more appropriate approach to designing a group sequential stopping rule is to consider this as a three-decision problem with outcomes: superiority, non-inferiority (only), and inferiority. This leads naturally to consideration of an inner wedge design broadly similar to those described in Chapter 6. We refer the reader to Öhrn & Jennison (2010) for a detailed description of such designs. Since the non-inferiority margin δ_N is typically smaller than the effect size δ at which power is set in the test for superiority, the stopping boundaries are typically not symmetric about the horizontal axis and the upper continuation region for testing between superiority and non-inferiority comes to an end first, while the lower region continues to higher sample sizes to allow differentiation between non-inferiority and inferiority.

Öhrn & Jennison (2010, Sec. 2) show how one can optimize an inner wedge design for a specified objective and they compare the resulting procedures with a design suggested by Lai, Shih & Zhu (2006). The need for different sample sizes to test for non-inferiority and superiority has prompted proposals in which the group sizes are allowed to depend on previously observed data, adapting the trial to the more appropriate goal. Two-stage adaptive designs are described by Brannath et al. (2003), Shih, Quan & Li (2004) and Koyama, Sampson & Gleser (2005), while multi-stage procedures are presented by Wang et al. (2001). Our general discussion of adaptive methods for sample size modification is postponed until later chapters, in particular Chapter 24. However, in comparing adaptive methods with inner wedge designs with fixed group sizes, Öhrn & Jennison (2010, Sec. 3) conclude there is little to be gained by the additional complexity of adaptive designs.

7.4 Two-sided Tests of Equivalence: Application to Comparative Bioavailability Studies

7.4.1 The Testing Problem

As mentioned above, two-sided equivalence tests arise most frequently in bioequivalence studies, so we shall frame our discussion here in that context. Westlake (1979) reviews comparative bioavailability trials for testing the bioequivalence of two competing formulations of a drug. In a typical trial, a single dose is administered to a subject and blood samples are drawn at various times over the next 24, 36 or 48 hours. This leads to a sequence of drug concentration levels in the blood, or "drug profile", in which the drug level usually rises to a peak during the initial absorption phase and then decays. A univariate statistic summarizing this response might be the "area under the curve" (AUC) or the peak concentration level (C_{max}). It is often reasonable to assume these quantities are log-normally distributed. The experimental design may utilize two parallel patient samples, as in Section 3.2.1 but it is more usual to employ a two-period crossover design as described in Section 3.2.4. The crossover design has the advantage of removing the between-subject component of variability, resulting in a much more sensitive test, and this in turn permits use of smaller sample sizes. A "washout" interval between the administrations of the two drugs in a given patient should be of sufficient length to eliminate any carry-over effect of the drug given first. In some cases it is possible to ascertain the presence of carry-over effects by measuring concentration levels during the washout period. Another possible complication is the existence of treatment-by-period interactions. More sophisticated designs have been proposed which need to be analyzed by more complex linear models incorporating carry-over and interaction effects; see, for example, Jones & Kenward (2014, Sec. 2.5). In general, group sequential versions of these designs can be adapted from those proposed in this section using techniques for normal linear models described in Section 3.3 and Chapter 15. Here we shall confine ourselves to the standard 2×2 crossover design of Section 3.2.4.

Consider a group sequential two-treatment, two-period crossover trial. For $i = 1, 2, \dots$, let X_i denote the natural logarithm of the ratio of response on treatment A to response on treatment B for the ith subject receiving treatment A first. Similarly, define Y_i to be the logarithm of the same ratio for the ith subject receiving treatment B first. Response here may be defined as the AUC, C_{max}, or other summary feature of the drug profile. The usual model treats the responses as log-normally distributed with

$$X_i \sim N(\theta + \phi, \sigma^2) \quad \text{and} \quad Y_i \sim N(\theta - \phi, \sigma^2), \quad i = 1, 2, \dots, \tag{7.1}$$

where θ is the treatment difference and ϕ a period effect.

In a two-sided test of equivalence, treatments A and B are to be considered equivalent if $|\theta| < \delta$. Here, δ should be "determined from the practical aspects of the problem in such a way that the treatments can be considered for all practical purposes to be equivalent if their true difference is unlikely to exceed the specified δ" (Dunnett & Gent, 1977, p. 594). We impose the error probability requirements

$$Pr_{\theta=\pm\delta}\{\text{Declare equivalence}\} \le \beta \tag{7.2}$$

and

$$Pr_{\theta=0}\{\text{Do not declare equivalence}\} \le \alpha. \tag{7.3}$$

Some authors would interchange the symbols α and β here, but we use the above choice for consistency with the notation for two-sided tests in Chapter 6. The important point is that β in (7.2) represents the "consumer's risk" since wrongly declaring equivalence may lead to an unsuitable preparation being allowed onto the market. Values of β and δ must be chosen to satisfy the appropriate regulatory agency. Recommended choices are often $\beta = 0.025$ or 0.05 and $\delta = \log(1.25)$ so that $|\theta| < \delta$ implies that, on the original scale (before taking logarithms), the treatment effect ratio lies in the range 0.8 to 1.25. The probability α in (7.3) is the "manufacturer's risk" and, since the benefits of proving equivalence are so great, one would expect a small value of α to be chosen to ensure a high probability of a positive conclusion when the two treatments really are equivalent.

7.4.2 Using the Power Family of Inner Wedge Tests

If σ^2 is known, we may directly utilize the techniques of Chapter 6 to design a group sequential crossover trial to test for two-sided equivalence. The number of analyses, K, and the shape parameter Δ of the power family test must be chosen. Suppose for now that the study can be organized so that equal numbers of observations following each treatment sequence accrue between successive analyses. Tests are defined by the inflation factor $\tilde{R}_W(K, \alpha, \beta, \Delta)$ and constants $\tilde{C}_{W1}(K, \alpha, \beta, \Delta)$ and $\tilde{C}_{W2}(K, \alpha, \beta, \Delta)$. Boundary values are defined by (6.2) and the stopping rule (6.1) is applied with the modification that the decisions "accept H_0" and "reject H_0" are replaced by "declare equivalence" and "declare non-equivalence", respectively. If $\alpha = 0.05$ and $\beta = 0.05$, constants in Table 6.3 can be used. Table 7.1 provides constants for the case $\alpha = 0.05$ and $\beta = 0.025$ and shows properties of the resulting designs.

As an example, suppose we specify $\alpha = 0.05$, $\beta = 0.025$, $\delta = \log(1.25) = 0.223$, $\Delta = 0$ and a balanced design with $K = 4$ equally sized groups. Suppose also that the within-subject coefficient of variation (CV) of the responses on the original antilog scale is known to be 24%, which Hauck, Preston & Bois (1997, p. 90) suggest is as a "moderate range" value for a CV for AUC. It follows that $\sigma^2 \approx 2 \times 0.24^2 = 0.115$.

After n_k observations on each treatment sequence, the estimate

$$\hat{\theta}^{(k)} = (\bar{X}^{(k)} + \bar{Y}^{(k)})/2$$

has variance $\sigma^2/(2n_k)$, so the information for θ is $\mathcal{I}_k = 2n_k/\sigma^2$. From (4.6), the information needed for a fixed sample size test is

$$\mathcal{I}_{f,2} = (1.960 + 1.960)^2/0.223^2 = 309.0,$$

which requires $309.0\,\sigma^2/2 = 309.0 \times 0.115/2 = 17.8$ subjects in each treatment sequence. For our sequential test we multiply $\mathcal{I}_{f,2}$ by $\tilde{R}_W(4, 0.05, 0.025, 0) = 1.041$ to obtain the maximum information level 321.7, and solving $2n_4/0.115 = 321.7$ gives the final sample size per treatment sequence $n_4 = 18.5$. Setting $n_4 = 20$ to give a multiple of 4, we see the study should be designed with five subjects on each treatment sequence in each of the four groups. The test is implemented by applying rule (6.1) to the standardized statistics

$$Z_k = \hat{\theta}^{(k)}\sqrt{\mathcal{I}_k} = (\bar{X}^{(k)} + \bar{Y}^{(k)})\sqrt{\{n_k/(2\sigma^2)\}}, \quad k = 1, \ldots, 4,$$

taking acceptance of H_0 to indicate declaration of equivalence. The critical values a_k and b_k are obtained from (6.2) using $\tilde{C}_{W1}(4, 0.05, 0.05, 0) = 2.005$ and $\tilde{C}_{W2}(4, 0.05, 0.05, 0) = 1.994$ from Table 7.1. With $\delta = 0.223$ this gives

$$(a_1, a_2, a_3, a_4) = (-1.91,\ 0.12,\ 1.30,\ 2.16)$$

and

$$(b_1, b_2, b_3, b_4) = (4.01,\ 2.84,\ 2.32,\ 2.005).$$

Since a_1 is negative, it is not possible to stop and declare equivalence at the first analysis. Due to the rounding of n_4 to give an integer group size, the values of a_4 and b_4 are not exactly equal; although the difference is slight, one could choose to use the lower value, b_4, as the critical value for $|Z_4|$ to protect against the key error of wrongly declaring equivalence. (This minor difficulty can be avoided altogether by following the method for dealing with unequal group sizes explained in the next section.) The result of applying this test is a noticeable saving in expected total sample size across the range of θ values: expected samples sizes are 30.7, 32.6 and 25.2 under $\theta = 0$, $\pm\delta/2$ and $\pm\delta$, respectively, compared to the 36 subjects required by a fixed sample procedure.

Table 7.1 *Constants $\tilde{C}_{W1}(K,\alpha,\beta,\Delta)$, $\tilde{C}_{W2}(K,\alpha,\beta,\Delta)$ and $\tilde{R}_W(K,\alpha,\beta,\Delta)$ for power family two-sided inner wedge tests with shape parameter Δ. The first opportunity for early stopping to accept H_0 occurs at analysis k^*. Also shown are expected sample sizes at $\theta = 0$, $\pm\delta/2$ and $\pm\delta$ expressed as percentages of the corresponding fixed sample size. Tests are for K groups of observations, with Type I error probability $\alpha = 0.05$ at $\theta = 0$ and power $1-\beta = 0.975$ at $\theta = \pm\delta$ appropriate to two-sided equivalence testing.*

K	$\tilde{C}_{W1}$	$\tilde{C}_{W2}$	$\tilde{R}_W$	k^*	Expected sample size, as percentage of fixed sample size, at $\theta = 0$	$\theta = \delta/2$	$\theta = \delta$
$\Delta = -0.5$							
1	1.960	1.960	1.000	1	100.0	100.0	100.0
2	1.960	1.960	1.000	2	100.0	99.7	93.7
3	1.962	1.962	1.002	2	93.2	95.2	80.0
4	1.966	1.966	1.006	3	86.1	91.6	77.7
5	1.970	1.969	1.010	4	86.3	90.5	75.2
10	1.983	1.983	1.024	7	81.6	86.8	71.1
15	1.991	1.991	1.032	10	80.4	85.7	69.9
20	1.997	1.996	1.038	13	79.9	85.2	69.3
$\Delta = -0.25$							
1	1.960	1.960	1.000	1	100.0	100.0	100.0
2	1.962	1.959	1.001	2	100.0	98.7	85.1
3	1.969	1.969	1.009	2	86.7	91.7	76.6
4	1.977	1.976	1.017	3	84.3	89.4	72.7
5	1.983	1.982	1.023	3	83.4	88.0	70.8
10	2.003	2.002	1.044	6	79.0	84.5	66.9
15	2.014	2.013	1.055	9	77.8	83.4	65.7
20	2.021	2.020	1.062	12	77.3	82.9	65.2
$\Delta = 0$							
1	1.960	1.960	1.000	1	100.0	100.0	100.0
2	1.977	1.957	1.007	1	99.9	96.5	75.8
3	1.990	1.989	1.030	2	82.5	88.6	71.3
4	2.005	1.994	1.041	2	83.6	87.2	67.5
5	2.014	2.011	1.054	3	78.7	84.7	65.4
10	2.044	2.038	1.084	5	76.3	81.9	61.6
15	2.059	2.055	1.101	8	74.7	80.6	60.5
20	2.068	2.063	1.111	10	74.4	80.2	59.9
$\Delta = 0.25$							
1	1.960	1.960	1.000	1	100.0	100.0	100.0
2	2.023	2.021	1.064	1	86.8	90.4	70.1
3	2.059	2.028	1.087	2	81.2	86.2	64.0
4	2.081	2.067	1.120	2	76.9	83.2	61.2
5	2.100	2.072	1.133	2	76.9	82.3	59.2
10	2.148	2.123	1.187	4	72.6	78.9	55.4
15	2.172	2.147	1.214	6	71.3	77.7	54.1
20	2.186	2.161	1.230	8	70.7	77.2	53.4

7.4.3 Adapting to Unequal Group Sizes and Unknown Variance

To satisfy practical needs, methods must be able to handle unequal numbers on each treatment at any stage in a parallel design or unequal numbers on each sequence (AB or BA) during a crossover design. Later, in Section 8.2.5, we shall discuss how unequal increments in information can be handled using an "error spending" approach to create two-sided tests with an inner wedge. Here, we describe an approximate approach using power family inner wedge tests.

It is also important to deal with normal responses of unknown variance and power family inner wedge tests can be adapted to this problem too. In dealing with either unequal increments in information or unknown variance, special modifications are needed to ensure the probability of wrongly declaring equivalence remains within a specified limit. In Chapters 5 and 6 we adapted power family tests in a way which preserved the Type I error rate by expressing the stopping boundary in terms of significance levels under H_0: $\theta = 0$. We shall follow a similar approach, working with significance levels against the hypotheses $\theta = \delta$ and $\theta = -\delta$ in order to preserve the *power* of the original test. Our development here parallels that of Jennison & Turnbull (2001a).

Suppose we wish to construct a two-sided equivalence test satisfying

$$Pr_{\theta=\pm\delta}\{\text{Declare equivalence}\} \le \beta$$

for specified δ and β, using a group sequential design with K analyses based on a power family inner wedge test with shape parameter Δ. Supposing, for now, that σ^2 is known, we can also plan to satisfy a second error condition

$$Pr_{\theta=0}\{\text{Do not declare equivalence}\} \le \alpha,$$

noting that this condition will not be met precisely if actual information levels differ from their design values.

We take as our starting point the two-sided test of H_0: $\theta = 0$ with Type I error probability α and power $1-\beta$ at $\theta = \pm\delta$. Thus, we plan to achieve information levels

$$\mathcal{I}_k = (k/K)\, \tilde{R}_W(K, \alpha, \beta, \Delta)\, \mathcal{I}_{f,2}$$

at analyses 1 to K. If these information levels arise exactly, the test defined by the stopping rule (6.1) with a_k and b_k, $k = 1, \ldots, K$, given by (6.2) will have Type I error probability α and power $1-\beta$ at $\theta = \pm\delta$. Thus, the associated equivalence test will have exactly the desired error probabilities.

When observed information levels differ from their planned values, we use the significance level approach to maintain properties under $\theta = \pm\delta$. Consider first the case $\theta = \delta$: since $Z_k - \delta\sqrt{\mathcal{I}_k}$ has a standard normal distribution under $\theta = \delta$, we can proceed by expressing the boundary as a sequence of critical values for $Z_k - \delta\sqrt{\mathcal{I}_k}$ in a form which does not explicitly involve the information levels $\mathcal{I}_1, \ldots, \mathcal{I}_K$. However, we must simultaneously treat the case $\theta = -\delta$. Fortunately, a typical test generates little probability of crossing the lower boundary arms if $\theta = \delta$ or the upper arms if $\theta = -\delta$, and there is no serious loss of accuracy in defining the upper section by considering $\theta = \delta$ and the lower section by considering $\theta = -\delta$.

To simplify notation, we omit the arguments of $\tilde{C}_{W1}$ and $\tilde{C}_{W2}$ in the following derivation. For $\theta = \delta$, the upper part of the stopping boundary defined by (6.1) with a_k and b_k as specified in (6.2) is based on comparing

$$Z_k \quad \text{versus} \quad \begin{cases} \tilde{C}_{W1}(k/K)^{\Delta-1/2} & \text{and} \\ \delta\sqrt{\mathcal{I}_k} - \tilde{C}_{W2}(k/K)^{\Delta-1/2}, & \end{cases}$$

i.e., comparing

$$Z_k - \delta\sqrt{\mathcal{I}_k} \quad \text{versus} \quad \begin{cases} -\delta\sqrt{\mathcal{I}_k} + \tilde{C}_{W1}(k/K)^{\Delta-1/2} & \text{and} \\ -\tilde{C}_{W2}(k/K)^{\Delta-1/2}, \end{cases}$$

at each analysis $k = 1, \ldots, K$. Under the planned sequence of information levels, we shall have $\mathcal{I}_k = (k/K)\{\tilde{C}_{W1} + \tilde{C}_{W2}\}^2/\delta^2$ and we can write this comparison as

$$Z_k - \delta\sqrt{\mathcal{I}_k} \quad \text{versus} \quad h_k \text{ and } g_k,$$

where

$$h_k = -\{\tilde{C}_{W1} + \tilde{C}_{W2}\}(k/K)^{1/2} + \tilde{C}_{W1}(k/K)^{\Delta-1/2}$$

and

$$g_k = -\tilde{C}_{W2}(k/K)^{\Delta-1/2}, \quad k = 1, \ldots, K.$$

These formulae for g_k and h_k avoid explicit mention of $\mathcal{I}_k$. Also, g_K is automatically equal to h_K. By symmetry, the lower section of the stopping boundary arises from comparing $Z_k + \delta\sqrt{\mathcal{I}_k}$ with $-g_k$ and $-h_k$ at analyses $k = 1, \ldots, K$. Incorporating the additional feature that stopping to accept H_0 should not occur before analysis k^*, the first stage at which $g_k + \delta\sqrt{\mathcal{I}_k}$ is positive, we obtain the full stopping rule:

$$\begin{array}{lll}
\text{After group } k = 1, \ldots, k^* - 1 & & \\
\quad \text{if } Z_k - \delta\sqrt{\mathcal{I}_k} \geq h_k \text{ or} & & \\
\quad\quad Z_k + \delta\sqrt{\mathcal{I}_k} \leq -h_k & \text{stop, reject equivalence} & \\
\quad \text{otherwise} & \text{continue to group } k+1, & \\
\text{after group } k = k^*, \ldots, K-1 & & \\
\quad \text{if } Z_k - \delta\sqrt{\mathcal{I}_k} \geq h_k \text{ or} & & \\
\quad\quad Z_k + \delta\sqrt{\mathcal{I}_k} \leq -h_k & \text{stop, reject equivalence} & \\
\quad \text{if } Z_k - \delta\sqrt{\mathcal{I}_k} < g_k \text{ and} & & (7.4) \\
\quad\quad Z_k + \delta\sqrt{\mathcal{I}_k} > -g_k & \text{stop, declare equivalence} & \\
\quad \text{otherwise} & \text{continue to group } k+1, & \\
\text{after group } K & & \\
\quad \text{if } Z_K - \delta\sqrt{\mathcal{I}_K} \geq h_K \text{ or} & & \\
\quad\quad Z_K + \delta\sqrt{\mathcal{I}_K} \leq -h_K & \text{stop, reject equivalence} & \\
\quad \text{otherwise} & \text{stop, declare equivalence.} &
\end{array}$$

The condition to stop and reject equivalence at analysis k can be re-expressed as

$$|Z_k| > h_k + \delta\sqrt{\mathcal{I}_k} = \tilde{C}_{W1}(k/K)^{\Delta-1/2} + \delta\sqrt{\mathcal{I}_k} - \{\tilde{C}_{W1} + \tilde{C}_{W2}\}(k/K)^{1/2}$$

and that to declare equivalence as

$$|Z_k| < g_k + \delta\sqrt{\mathcal{I}_k} = \delta\sqrt{\mathcal{I}_k} - \tilde{C}_{W2}(k/K)^{\Delta-1/2}.$$

The above test can now be applied using the information levels $\mathcal{I}_1, \ldots, \mathcal{I}_K$ actually observed. Table 7.2 shows the Type I error probability and power achieved by such tests with $K = 2$, 5 and 10 analyses and shape parameter $\Delta = -0.5$ and 0, under sequences of information levels of the form $\mathcal{I}_k = \pi(k/K)^r\mathcal{I}_{max}$, where $\mathcal{I}_{max} = \tilde{R}_W(K, \alpha, \beta, \Delta)\,\mathcal{I}_{f,2}$. It

Table 7.2 *Properties of power family two-sided equivalence tests designed for information levels* $\mathcal{I}_k = (k/K)\mathcal{I}_{max}$, $k = 1, \ldots, K$, *but applied with* $\mathcal{I}_k = \pi(k/K)^r \mathcal{I}_{max}$, $k = 1, \ldots, K$, *in a manner which preserves power, i.e., the probability of correctly rejecting equivalence when* $\theta = \pm\delta$. *Attained Type I error probabilities and power are shown for tests with* K *analyses and shape parameter* $\Delta = -0.5$ *and* $\Delta = 0$, *designed to achieve Type I error probability* $\alpha = 0.05$ *and power* $1 - \beta = 0.975$ *at* $\theta = \pm\delta$.

			$\Delta = -0.5$		$\Delta = 0$	
	r	π	*Type I error*	*Power*	*Type I error*	*Power*
$K = 2$						
	0.80	0.90	0.079	0.975	0.076	0.975
		1.00	0.050	0.975	0.048	0.974
		1.10	0.031	0.975	0.030	0.974
	1.00	0.90	0.079	0.975	0.079	0.975
		1.00	0.050	0.975	0.050	0.975
		1.10	0.031	0.975	0.031	0.974
	1.25	0.90	0.079	0.975	0.083	0.975
		1.00	0.050	0.975	0.054	0.975
		1.10	0.031	0.975	0.034	0.975
$K = 5$						
	0.80	0.90	0.077	0.975	0.072	0.975
		1.00	0.049	0.975	0.046	0.975
		1.10	0.031	0.975	0.028	0.975
	1.00	0.90	0.079	0.975	0.078	0.975
		1.00	0.050	0.975	0.050	0.975
		1.10	0.031	0.975	0.032	0.975
	1.25	0.90	0.082	0.975	0.091	0.975
		1.00	0.052	0.975	0.059	0.975
		1.10	0.033	0.975	0.039	0.975
$K = 10$						
	0.80	0.90	0.076	0.975	0.070	0.975
		1.00	0.048	0.975	0.044	0.975
		1.10	0.030	0.975	0.027	0.975
	1.00	0.90	0.079	0.975	0.078	0.975
		1.00	0.050	0.975	0.050	0.975
		1.10	0.031	0.975	0.032	0.975
	1.25	0.90	0.084	0.975	0.095	0.975
		1.00	0.054	0.975	0.063	0.975
		1.10	0.034	0.975	0.041	0.975

is clear from the table that, as intended, power remains close to its nominal value, 0.975, in all cases. Although similar two-sided tests were originally planned in the examples of Table 6.4 in Section 6.2.3 and similarly perturbed information sequences were observed, we have worked toward a different objective in adapting these tests to unequal group sizes and the resulting tests have quite different properties. Whereas the tests in Table 6.4 maintain a fixed Type I error probability and have variable power at $\theta = \pm\delta$, our new equivalence tests closely control power, the probability of correctly rejecting equivalence when $\theta = \pm\delta$, and channel variability into the Type I error rate, the probability of wrongly rejecting equivalence when $\theta = 0$.

It remains to consider adaptation of power family equivalence tests to the case of unknown variance in a way which maintains power at $\theta = \pm\delta$ as closely as possible. We shall illustrate this in the context of a crossover trial where we suppose data follow the normal linear model (7.1). If observations $X_1, \ldots, X_{n_{Xk}}$ and $Y_1, \ldots, Y_{n_{Yk}}$ are available at analysis k, the estimate of θ is

$$\hat{\theta}^{(k)} = (\bar{X}^{(k)} + \bar{Y}^{(k)})/2$$

where

$$\bar{X}^{(k)} = \sum_{i=1}^{n_{Xk}} X_i/n_{Xk} \quad \text{and} \quad \bar{Y}^{(k)} = \sum_{i=1}^{n_{Yk}} Y_i/n_{Yk}, \quad k = 1, \ldots, K.$$

The sequence of estimates $\{\hat{\theta}^{(1)}, \ldots, \hat{\theta}^{(K)}\}$ is multivariate normal with

$$\mathrm{Var}(\hat{\theta}^{(k)}) = \sigma^2(n_{Xk}^{-1} + n_{Yk}^{-1})/4, \quad k = 1, \ldots, K,$$

and

$$\mathrm{Cov}(\hat{\theta}^{(k_1)}, \hat{\theta}^{(k_2)}) = \mathrm{Var}(\hat{\theta}^{(k_2)}), \quad 1 \le k_1 \le k_2 \le K.$$

The usual estimate of σ^2 at analysis k is

$$s_k^2 = \{\sum_{i=1}^{n_{Xk}} (X_i - \bar{X}_k)^2 + \sum_{i=1}^{n_{Yk}} (Y_i - \bar{Y}_k)^2\}/(n_{Xk} + n_{Yk} - 2)$$

on $\nu_k = n_{Xk} + n_{Yk} - 2$ degrees of freedom.

In the known variance problem, the Z-statistics $Z_k - \delta\sqrt{\mathcal{I}_k}$ and $Z_k + \delta\sqrt{\mathcal{I}_k}$ had standard normal distributions under the hypotheses $\theta = \delta$ and $\theta = -\delta$, respectively. With σ^2 unknown, we now use the t-statistics

$$T_k^+ = \frac{\hat{\theta}^{(k)} - \delta}{\sqrt{\{s_k^2\,(n_{Xk}^{-1} + n_{Yk}^{-1})/4\}}}$$

and

$$T_k^- = \frac{\hat{\theta}^{(k)} + \delta}{\sqrt{\{s_k^2\,(n_{Xk}^{-1} + n_{Yk}^{-1})/4\}}}$$

which have t_{ν_k} distributions under the same two hypotheses. To create a group sequential t-test, we replace $Z_k - \delta\sqrt{\mathcal{I}_k}$ by T_k^+ and $Z_k + \delta\sqrt{\mathcal{I}_k}$ by T_k^- in rule (7.4) and convert the g_k and h_k into percentiles of the t_{ν_k} distribution rather than the standard normal distribution.

The test is then:

$$
\begin{array}{lll}
\text{After group } k = 1, \dots, k^* - 1 & & \\
\quad \text{if } T_k^+ \geq t_{\nu_k, 1-\Phi(h_k)} \text{ or} & & \\
\quad\quad T_k^- \leq -t_{\nu_k, 1-\Phi(h_k)} & \text{stop, reject equivalence} & \\
\quad \text{otherwise} & \text{continue to group } k+1, & \\
\text{after group } k = k^*, \dots, K - 1 & & \\
\quad \text{if } T_k^+ \geq t_{\nu_k, 1-\Phi(h_k)} \text{ or} & & \\
\quad\quad T_k^- \leq -t_{\nu_k, 1-\Phi(h_k)} & \text{stop, reject equivalence} & \\
\quad \text{if } T_k^+ < t_{\nu_k, 1-\Phi(g_k)} \text{ and} & & (7.5) \\
\quad\quad T_k^- > -t_{\nu_k, 1-\Phi(g_k)} & \text{stop, declare equivalence} & \\
\quad \text{otherwise} & \text{continue to group } k+1, & \\
\text{after group } K & & \\
\quad \text{if } T_K^+ \geq t_{\nu_K, 1-\Phi(h_K)} \text{ or} & & \\
\quad\quad T_K^- \leq -t_{\nu_K, 1-\Phi(h_K)} & \text{stop, reject equivalence} & \\
\quad \text{otherwise} & \text{stop, declare equivalence,} &
\end{array}
$$

where, as before, $t_{\nu,q}$ is the value such that $Pr\{T > t_{\nu,q}\} = q$ when $T \sim t_\nu$ and Φ denotes the standard normal cdf.

The prime objective of this construction is to achieve power close to $1 - \beta$ at $\theta = \pm\delta$. A target for the Type I error rate under $\theta = 0$ is implicit in the design since a value of α is needed to determine the constants $\tilde{C}_{W1}(K, \alpha, \beta, \Delta)$ and $\tilde{C}_{W2}(K, \alpha, \beta, \Delta)$ on which g_k and h_k depend. However, the test's actual Type I error probability will depend strongly on the observed sample sizes $n_{X1}, \dots, n_{XK}$ and $n_{Y1}, \dots, n_{YK}$ and the unknown value of σ^2.

The following approximation to the Type I error probability under a given σ^2 is useful as it allows one to check that this error rate is acceptably low over a plausible range of σ^2 values. We note that a fixed sample test based on estimates $\hat{\theta} \sim N(\theta, V\sigma^2)$ and $s^2 \sim \sigma^2 \chi^2_\nu / \nu$ achieves power $1 - \beta$ at $\theta = \pm\delta$ by rejecting equivalence if

$$T^+ = \frac{\hat{\theta} - \delta}{\sqrt{\{Vs^2\}}} \geq -t_{\nu,\beta} \quad \text{or} \quad T^- = \frac{\hat{\theta} + \delta}{\sqrt{\{Vs^2\}}} \leq t_{\nu,\beta},$$

and has Type I error probability equal to

$$Pr_{\theta=0}\{T^+ \geq -t_{\nu,\beta} \text{ or } T^- \leq t_{\nu,\beta}\} = 2\,Pr\{T(\nu; \delta/\sqrt{\{V\sigma^2\}}) \leq t_{\nu,\beta}\}.$$

Here, as before, $T(\nu; \xi)$ denotes a noncentral t random variable with ν degrees of freedom and noncentrality parameter ξ. Setting $\nu = \nu_K$ in this expression and, by analogy with the known variance case, replacing $V\sigma^2$ by $\tilde{R}_W(K, \alpha, \beta, \Delta)$ times the variance of $\hat{\theta}^{(K)}$, we obtain the approximation

$$2\,Pr\{T(\nu_K; \delta/\sqrt{\{\tilde{R}_W(K, \alpha, \beta, \Delta)\,\sigma^2(n_{XK}^{-1} + n_{YK}^{-1})/4\}}) \leq t_{\nu_K,\beta}\} \qquad (7.6)$$

to the group sequential t-test's Type I error probability under $\theta = 0$.

Table 7.3 shows the power at $\theta = \pm\delta$ and attained Type I error probability of group sequential t-tests for equivalence when groups are equally sized and consist of four or six subjects on each treatment sequence. The degrees of freedom ν_k are the numbers of observations at each analysis minus two for fitting the linear model. In each example, the value of σ^2 is chosen such that the approximation (7.6) gives a value of 0.05 for the Type I error rate. It is clear from the table that power and Type I error both agree well with their predicted values.

Table 7.3 *Properties of power family two-sided t-tests designed to test for equivalence, $\theta \approx 0$, with power $1-\beta = 0.975$ to reject equivalence when $\theta = \pm\delta$. Tests are based on power family tests for known variance with $\Delta = -0.5$ and $\Delta = 0$. The degrees of freedom for σ^2 at analysis k are $\nu_k = (\nu_K + 2)(k/K) - 2$ and $Var(\hat{\theta}^{(k)}) = (K/k)Var(\hat{\theta}^{(K)})$, $k = 1, \dots, K$. Values of δ and σ^2 are chosen such that the approximation (7.6) gives the value 0.05 for the Type I error probability, i.e., the probability of failing to declare equivalence when $\theta = 0$. Power and Type I error probability were estimated by simulation using 1 million replicates; standard errors for both are 0.0002.*

$\Delta = -0.5$

K	ν_K	*Power at* $\theta = \pm\delta$	*Type I error*
3	10	0.974	0.049
3	16	0.975	0.049
5	18	0.974	0.048
5	28	0.974	0.049
8	30	0.974	0.048
8	46	0.974	0.048

$\Delta = 0$

K	ν_K	*Power at* $\theta = \pm\delta$	*Type I error*
3	10	0.973	0.047
3	16	0.974	0.048
5	18	0.973	0.046
5	28	0.974	0.047
8	30	0.973	0.047
8	46	0.974	0.047

In further simulations with the same group sizes but σ^2 increased or decreased by a factor of two, we found power at $\theta = \pm\delta$ to remain within 0.002 of the intended value $1-\beta = 0.975$. Additional perturbations in information levels of the form studied in Table 7.2 had little or no impact on power at $\theta = \pm\delta$. As one would expect, changes in σ^2 had considerable effects on the Type I error rate. We conclude that these designs are robust in satisfying constraints on the probability of wrongly declaring equivalence. Furthermore, they provide a framework within which the producers of a new drug formulation can design and conduct a study with suitably high probability of finding equivalence when it is present.

7.4.4 An Example

Chow & Liu (2008, Sec. 3.6) present an example of a comparison of bioavailability between two formulations of a drug using a 2×2 crossover design with 24 subjects, 12 on each treatment sequence. Responses are AUC values calculated from blood samples taken at a sequence of times in a 32-hour period after dosing. We label the test formulation "treatment A" and the reference formulation "treatment B". The original study was not

Table 7.4 *AUC equivalence test data, adapted from Chow & Liu* (2008). *The table shows AUC values from* 2 × 2 *a crossover design with* 12 *subjects on each treatment sequence obtained in groups* 1 *and* 2 *of a two-stage group sequential study.*

	Sequence AB			*Sequence BA*		
		Period			*Period*	
Group	*Subject*	I	II	*Subject*	I	II
1	2	74.825	37.350	1	74.675	73.675
	3	86.875	51.925	4	96.400	93.250
	7	81.865	72.175	5	101.950	102.125
	8	92.700	77.500	6	79.050	69.450
	9	50.450	71.875	11	79.050	69.025
	10	66.125	94.025			
	13	122.450	124.975			
	14	99.075	85.225			
2	17	86.350	95.925	12	85.950	68.700
	18	49.925	67.100	15	69.725	59.425
	21	42.700	59.425	16	86.275	76.125
	22	91.725	114.050	19	112.675	114.875
				20	99.525	116.250
				23	89.425	64.175
				24	55.175	74.575

designed sequentially, but for the purpose of illustration we shall analyze the data as if they had been obtained in a two-stage group sequential study. Table 7.4 shows the AUC values for the 24 subjects where we have supposed that results from 13 subjects were available after the first stage, 8 receiving the sequence AB and 5 the sequence BA.

In our analysis, we define X_i, $i = 1, \ldots, 12$, to be the logarithm of the ratio of the response on treatment A to the response on treatment B for the ith subject receiving treatment A first. Similarly we define Y_i, $i = 1, \ldots, 12$, to be the logarithm of the same ratio for the ith subject receiving treatment B first. We suppose these variables follow the model (7.1) and that a test is required to satisfy the power condition specified in (7.2) with $\alpha = 0.05$, $\delta = \log(1.25) = 0.223$ and $\beta = 0.05$. We shall construct a group sequential test of equivalence from the power family two-sided inner wedge test with $K = 2$ and $\Delta = 0$. Using the constant $\tilde{R}_W(2, 0.05, 0.05, 0) = 1.018$ from Table 6.3, we see that formula (7.6) gives an approximate Type I error probability of 0.05 if observations X_i and Y_i have variance $\sigma^2 = 0.084$. Of course, this variance is unknown initially but this value could have been used as a prior estimate in constructing the design.

Taking $\tilde{C}_{W1}(2, 0.05, 0.05, 0) = 1.970$ and $\tilde{C}_{W2}(2, 0.05, 0.05, 0) = 1.667$ from Table 6.3, we find

$$h_1 = -(1.970 + 1.667)\, 0.5^{0.5} + 1.970 \times 0.5^{-0.5} = 0.214,$$

$$g_1 = -1.667 \times 0.5^{-0.5} = -2.357$$

and

$$h_2 = g_2 = -1.667.$$

Hence, following rule (7.5), the group sequential version of the trial should stop to reject equivalence after stage 1 if $T_1^+ \geq t_{11,1-\Phi(0.214)} = 0.219$ or $T_1^- \leq -0.219$, and stop to accept equivalence after stage 1 if both $T_1^+ < t_{11,1-\Phi(-2.357)} = -2.765$ and $T_1^- > 2.765$.

From the data, we find

$$\hat{\theta}^{(1)} = 0.0277, \;\; s_1^2 = 0.0886, \;\; T_1^+ = -2.304 \;\; \text{and} \;\; T_1^- = 2.956.$$

The large value of T_1^- indicates that $\theta \leq -\delta$ is implausible but T_1^+ is not sufficiently negative to preclude the possibility $\theta \geq +\delta$ and the test continues to the second stage. At stage 2, the rule is simply to reject equivalence if either $T_2^+ \geq t_{22,1-\Phi(-1.667)} = -1.742$ or $T_2^- \leq 1.742$ and to accept equivalence otherwise. The final data set gives $\hat{\theta}^{(2)} = -0.0286$ and $s_2^2 = 0.0745$. Hence, we have $T_2^+ = -4.519$ and $T_2^- = 3.493$ which provide strong evidence of $\theta < \delta$ and $\theta > -\delta$, respectively, and the decision is to accept equivalence at this stage. It is of interest to note that the final estimate of σ^2 is close to the value of 0.084 for which we supposed the study to be designed.

The finding of bioequivalence agrees with that found by Chow & Liu (2008, Sec. 3.6) using various other methods. Since this test compares the overall distributions of responses under the two treatments pooled over patients, the conclusion is that the two treatments have "average bioequivalence", not as strong a statement of "individual bioequivalence" which we shall examine in the next section.

7.5 Individual Bioequivalence: A One-sided Test for Proportions

Suppose we wish to compare a test formulation of a drug to a reference formulation, which we label A and B, respectively, with respect to a continuous response variable Y. In the procedure for testing individual equivalence ratios or "TIER" (Anderson & Hauck, 1990), the two treatments should be declared equivalent if the probability of the within-subject similarity of response given by

$$p = Pr\{C_L < (Y_A/Y_B) < C_U\} \tag{7.7}$$

exceeds a given critical fraction, p_0 say. Here Y_A and Y_B represent the responses to test and reference formulations *in the same patient*, and C_L and C_U are pre-specified limits. For example, the "75/75" rule proposed by Purich (1980) would use $C_L = 0.75, C_U = 1.25$ and $p_0 = 0.75$. Another common choice is $C_L = 0.8$ and $C_U = 1.25$. Suppose we have data from a two-period crossover design in which patient i provides responses Y_{Ai} and Y_{Bi} under the two treatments. Assuming there are no period effects, we define the binary outcome variable for the ith patient as

$$X_i = \begin{cases} 1 & \text{if } C_L < (Y_{Ai}/Y_{Bi}) < C_U, \\ 0 & \text{otherwise.} \end{cases}$$

We may consider the patients' binary responses $X_1, X_2, \ldots$ as a sequence of independent Bernoulli observations with success probability p. Two-sided group sequential tests for a proportion p based on the normal approximation were described in Section 4.4.1. Here, we require a one-sided test between the hypotheses

$$H_0\colon p \leq p_0 \quad \text{and} \quad H_1\colon p > p_0,$$

where, as in Section 7.2, we set up the null hypothesis to denote inequivalence. Proceeding as in Section 4.4.1, we define

$$\hat{p}^{(k)} = \frac{1}{n_k}\sum_{i=1}^{n_k} X_i, \quad k = 1, \ldots, K,$$

to be the cumulative proportion of patients with responses in the equivalence range among those available up to stage k. Defining $\theta = p - p_0$, the information for θ at analysis k is

$\mathcal{I}_k = n_k/\{p_0(1-p_0)\}$, and the sequence of standardized statistics $Z_k = (\hat{p}^{(k)} - p_0)\sqrt{\mathcal{I}_k}$, $k = 1, \ldots, K$, has approximately the canonical joint distribution (3.1) when p is close to p_0. Hence, the methods of Chapter 5 can be applied directly to construct a group sequential one-sided test of H_0 versus H_1.

As an example, suppose we take $p_0 = 0.75$ and wish to test H_0: $p = 0.75$ against a one-sided alternative $p > 0.75$ with Type I error probability $\alpha = 0.025$ and power $1 - \beta = 0.8$ at $p = 0.98$. We shall do this using a power family test with $\Delta = 0$ and $K = 2$ stages, as defined in Section 5.2. We obtain the values $\tilde{C}_1(2, 0.025, 0.2, 0) = 1.934$ and $\tilde{C}_2(2, 0.025, 0.2, 0) = 0.948$ from Table 5.1 and, using these in (5.4), we see that the test's final information level should be

$$\mathcal{I}_2 = (1.934 + 0.948)^2/(0.98 - 0.75)^2 = 157.0.$$

Setting $n_2/\{p_0(1-p_0)\} = 157.0$ with $p_0 = 0.75$, we find that a maximum sample size of 29.4 is required and round this up to $n_2 = 30$, so 15 subjects are required in each of the two stages. From (5.7) with $K = 2$, at the first analysis the critical values for

$$Z_1 = \frac{\hat{p}^{(1)} - 0.75}{\sqrt{\{(0.75 \times 0.25)/15\}}}$$

are $a_1 = 0.697$ and $b_1 = 2.735$. Thus, with $n_1 = 15$, the test continues to stage 2 if

$$0.697 < \left(\frac{1}{15}\sum_{i=1}^{15} X_i - 0.75\right)\left(\frac{15}{0.75 \times 0.25}\right)^{1/2} < 2.735,$$

i.e., if

$$12.42 < \sum_{i=1}^{15} X_i < 15.84,$$

which occurs if 13, 14 or 15 patient responses fall in the equivalence range. With 12 such responses or fewer, the procedure stops and accepts the hypothesis of inequivalence. Since $\sum_{i=1}^{15} X_i$ cannot exceed 15.84, it is not possible to stop at stage 1 to declare equivalence. At stage 2, the critical value for Z_2 is $a_2 = b_2 = 1.934$, and equivalence is accepted if

$$\left(\frac{1}{30}\sum_{i=1}^{30} X_i - 0.75\right)\left(\frac{30}{0.75 \times 0.25}\right)^{1/2} > 1.934$$

i.e., if

$$\sum_{i=1}^{30} X_i > 27.09,$$

which requires at least 28 out of the 30 patients to have responses in the equivalence range; otherwise equivalence is rejected.

In this example, it is easy to see that the exact probability of accepting equivalence is simply $Pr\{B(30, p) \geq 28\}$, where $B(30, p)$ denotes a binomial random variable with parameter $n = 30$ and success probability p. Thus, when $p = 0.75$ the test's exact Type I error probability is 0.011 and its exact power when $p = 0.98$ is 0.978. In fact, using exact calculation, we see that the specified power of 80% is actually attained at $p = 0.95$, rather than at $p = 0.98$. Clearly, the initial approximation to the power is poor here, which is not surprising since the variances of $\hat{p}^{(1)}$ and $\hat{p}^{(2)}$ change substantially as p moves from 0.75 to 0.98 but the approximation assumes the variance under $p = 0.75$ to apply in all these cases. We shall discuss exact procedures for binary data in greater detail in Chapter 16.

Returning to our pedagogic example of the AUC equivalence data given in Table 7.4, suppose we use the 75/75 rule so that $C_L = 0.75$ and $C_U = 1.25$ with $p_0 = 0.75$. Suppose also that the two-stage plan described above is specified, but it turns out that results from $n_1 = 13$ rather than 15 patients are available at analysis 1. We must adapt the test to account for the unequal group sizes using the method described in Section 5.5. We see from Table 7.4 that of the 13 patients in stage 1, the response ratios of four, namely patients 2, 3, 9 and 10, fall outside the critical range. Thus $\hat{p}^{(1)} = 9/13$ and

$$Z_1 = \frac{\hat{p}^{(1)} - 0.75}{\sqrt{\{0.75 \times 0.25/13\}}} = -0.480.$$

Since $Z_1 < a_1 = 0.697$, the test terminates at stage 1 with H_0 accepted and equivalence rejected. (Had the test continued to stage 2, the proportion $\hat{p}^{(2)}$ would have been 16/24 corresponding to $Z_2 = -0.943$, again below the critical value $a_2 = b_2 = 1.934$.) The property of individual equivalence is more restrictive than that of average equivalence, so we should not be surprised that a hypothesis of individual equivalence fails to be accepted here even though the weaker hypothesis of average bioequivalence was accepted in the analysis of Section 7.4.4.

Methods and concepts relating to individual bioequivalence have been developed further by several authors including Schall (1995) and Ju (1997). Although the 75/75 rule gained some popularity, it is no longer formally used by the FDA; for an assessment of this rule, see Dobbins & Thiyagarajan (1992).

7.6 Bibliography and Notes

The reversal of the roles of null and alternative hypotheses in equivalence testing has been the cause of some confusion. Given the unusual uses of familiar concepts, there is much to be said for making a clean start, defining requirements of a test operationally, i.e., in terms of the probabilities of particular decisions under particular θ values, and constructing a suitable test from first principles.

In the fixed sample case, a number of authors have proposed basing the decision to accept or reject equivalence on whether or not a confidence interval for the treatment difference lies in an "equivalence region"; see, for example, Westlake (1972, 1976 and 1981), Metzler (1974), Kirkwood (1981) and Rocke (1984). The paper by Rocke also contains a compelling example, using data from Selwyn, Dempster & Hall (1981), in which equivalence is declared between two drugs and yet a simple hypothesis test would clearly reject the (inappropriate) null hypothesis of no difference between the two drugs: the large sample sizes show that there is a small difference between the drugs, but it is also clear that this difference is not large enough to be of practical importance.

Confidence intervals can also be used in a sequential setting. Durrleman & Simon (1990) and Jennison & Turnbull (1993a) define group sequential one-sided equivalence tests in terms of repeated confidence intervals (RCIs), which we shall present in Chapter 10. We discuss two-sided equivalence tests based on RCIs proposed by Jennison & Turnbull (1993b) in Section 10.3.4. As a consequence of the flexibility of RCIs to cope with non-adherence to a formal stopping rule, the related tests attain their nominal error rates conservatively.

If problems in implementing the stopping rule are not envisaged, conventional group sequential tests will usually be preferred to tests based on RCIs. In addition to those described in this chapter, Hauck, Preston & Bois (1997) have proposed tests with early stopping just to accept equivalence so there are no outer boundaries at interim analyses and equivalence can only be rejected at the final analysis. These tests might be considered the complement of the two-sided tests described in Chapters 2 and 4 in which there is no early stopping to accept equality. Other forms of testing boundary, primarily with two stages and with opportunity to accept or reject equivalence at each stage, are presented by Gould (1995a).

Chapter 8

Information Monitoring: The Error Spending Approach

8.1 Unpredictable Information Sequences

The tests described in Chapters 2 to 7 are all originally designed for a fixed maximum number, K, of equally sized groups of observations which give rise to equally spaced information levels $\{\mathcal{I}_1, \ldots, \mathcal{I}_K\}$. We have explained how the "significance level approach" can be used to adapt these tests to unequal information sequences in a way which maintains the Type I error rate close to its nominal value. However, this method breaks down as information increments become more and more uneven. Also, the significance level approach requires the maximum possible number of analyses, K, to be fixed at the design stage. In this chapter, we shall describe a more flexible way of dealing with unpredictable information sequences which guarantees a Type I error rate exactly equal to α for any observed information sequence. This method has the further advantage of not requiring the maximum number of analyses to be fixed in advance.

There are sound practical reasons why observed information levels may be unpredictable. Administratively, it is convenient to schedule interim analyses at fixed calendar times, but if patients are recruited at an uneven rate, the number of new observations between analyses will vary. When testing a parameter in a linear model, as discussed in Section 3.3, increments in information depend on covariate values of the new subjects. In the group sequential log-rank test of Section 4.5, the information $\mathcal{I}_k$ is approximately one quarter of the total number of deaths recorded by analysis k. Clearly, methods for handling unpredictable information sequences are an important part of the statistical methodology for such trials.

In the following, we shall suppose that a group sequential test concerns a parameter θ, and estimates $\hat{\theta}^{(k)}$ and information levels $\mathcal{I}_k = \{Var(\hat{\theta}^{(k)})\}^{-1}$, $k = 1, 2, \ldots$, arise at successive analyses. The maximum possible number of analyses may not be completely determined at the start of the study; for example, sampling may continue until $\mathcal{I}_k$ reaches some pre-specified value if the stopping rule does not imply earlier termination. Nevertheless, it is convenient to refer to the maximum number of analyses that would arise in a particular realization if there were no early stopping, and we denote this number by K.

We assume the sequence of estimates of θ and associated test statistics follow the joint distributions presented in Chapter 3 conditionally on the observed information sequence $\{\mathcal{I}_1, \ldots, \mathcal{I}_K\}$. In particular, we suppose the sequence of standardized statistics $\{Z_1, \ldots, Z_K\}$ for testing $\theta = 0$ has the joint distribution (3.1), either exactly for data following a normal linear model with known variance or approximately for other data types, the basis of this assumption lying in the general theory we shall present in Chapter 15. However, there is a new subtlety that must be noted as we broaden the range of experimental designs. The desired conditional distribution of $\{Z_1, \ldots, Z_K\}$ given $\{\mathcal{I}_1, \ldots, \mathcal{I}_K\}$ will fail to hold if, in the course of a study, future information levels are allowed to depend on current or previous estimates of θ. Therefore, for the methods of this chapter, we shall require

DOI: 10.1201/9781584888482-8

$\mathcal{I}_{k+1}$ to be conditionally independent of $\{\hat{\theta}^{(1)}, \ldots, \hat{\theta}^{(k)}\}$ given $\{\mathcal{I}_1, \ldots, \mathcal{I}_k\}$. This has the practical implication that a trial management committee should be blinded to estimates of the parameter under investigation, or related test statistics, when planning further patient accrual or the times of future analyses. It is however permissible, and indeed desirable, to respond to observed information levels and to be aware of factors which are liable to affect future information levels, such as the number of patients recruited and the overall failure rate in a survival study.

Our aim in this chapter is to describe group sequential "error spending" tests which attain a stated Type I error rate exactly, despite unpredictable information levels. We present methods for two-sided tests in Section 8.2 and for one-sided tests in Section 8.3. In each case, we describe "maximum information" study designs in which a target for the maximum information level is chosen to satisfy a power condition. This power will be attained to a high accuracy as long as the specified maximum information level is reached eventually. In practice, if information levels are lower than anticipated, it may be necessary to increase patient accrual or follow-up to achieve this target and so maintain power. We shall also describe "maximum duration" trials which do not impose such strong constraints on information levels but, in consequence, do not guarantee a pre-specified power. Both maximum information and maximum duration trials attain their stated Type I error probabilities exactly, provided observed information levels are not influenced by previous estimates of θ: in Section 8.5 we present examples to illustrate how the Type I error rate can be inflated if future analysis times are allowed to depend on a current estimate of θ.

Another situation where successive information levels will be unpredictable and unequal is when they depend on some unknown nuisance parameter. Examples include comparing normal means when variances are unknown, or comparing two proportions, p_1 and p_2 say, when their average $\overline{p} = (p_1 + p_2)/2$ is unknown. Estimates of these nuisance parameters based on data accumulated so far may be used to estimate the current information level. This leads naturally to an approximate procedure based on the error spending approach which we describe in Section 8.6: Mehta & Tsiatis (2001) and Tsiatis (2006) refer to such tests as information-based monitoring procedures. Other approaches to this problem will be discussed in Chapter 18.

We conclude the chapter in Section 8.7 with details of the numerical computations needed to construct error spending group sequential tests.

8.2 Two-sided Tests

8.2.1 Spending Type I Error

We introduce the idea of error spending in the context of a two-sided test of the hypothesis H_0: $\theta = 0$, using the exact method for guaranteeing Type I error suggested by Slud & Wei (1982). In this case, the maximum number of analyses, K, is fixed before the study commences and the Type I error is partitioned into probabilities $\pi_1, \ldots, \pi_K$, which sum to α. As information levels $\mathcal{I}_1, \ldots, \mathcal{I}_K$ are observed, critical values c_k for the standardized statistics Z_k, $k = 1, \ldots, K$, are calculated such that conditionally on $\mathcal{I}_1, \ldots, \mathcal{I}_k$,

$$Pr_{\theta=0}\{|Z_1| < c_1, \ldots, |Z_{k-1}| < c_{k-1}, |Z_k| \geq c_k\} = \pi_k. \tag{8.1}$$

The test proceeds according to the familiar stopping rule (4.2), rejecting H_0 at analysis k if $|Z_k| \geq c_k$, $k = 1, \ldots, K$, or stopping to accept H_0 if it has not been rejected by analysis K. Thus, π_k represents the probability of stopping at analysis k to reject H_0 when this hypothesis is true, also termed the "error spent at stage k". This should not be confused with the nominal significance level $2\{1 - \Phi(c_k)\}$ applied at analysis k in the significance level approach to handling unequal group sizes, as described in Section 4.2. Since $\pi_1 + \ldots + \pi_K = \alpha$, the overall Type I error rate is exactly α, as required.

The condition $Pr\{|Z_1| \geq c_1\} = \pi_1$ when $\theta = 0$ implies that the first critical value is simply

$$c_1 = \Phi^{-1}(1 - \frac{\pi_1}{2}).$$

Computation of subsequent critical values necessitates solving (8.1) for c_k with $k = 2, \ldots, K$ in turn. Note that at the kth stage, c_k depends on $\mathcal{I}_1, \ldots, \mathcal{I}_k$ but not on the as yet unobserved $\mathcal{I}_{k+1}, \ldots, \mathcal{I}_K$. We describe the calculation of c_k, which involves the joint distribution of $\{Z_1, \ldots, Z_k\}$, in Section 8.7 and in the more general discussion of numerical methods in Chapter 31. Proschan (1999) discusses several mathematical properties of the critical values $c_1, \ldots, c_K$ and presents reassuring results on their behavior when successive information levels are very close.

Although Slud & Wei's method contains the key idea of error spending, it has certain limitations. Since the maximum number of analyses is fixed, it is difficult to adapt to unexpectedly high or low rates of accrual of information and, hence, to meet a desired power requirement. Also, one may prefer to vary the amount of error spent at analyses in response to the observed information levels; for example, if little information is available at the first analysis due to slow recruitment or delayed patient response, very little power will be gained from Type I error used up at this stage and it would seem reasonable to reduce π_1, saving Type I error for later, more informative analyses. The method introduced by Lan & DeMets (1983) has the potential to overcome these problems. In their original paper, Lan & DeMets propose spending Type I error as a function of the observed information levels and, combined with a sampling rule which aims to continue to a pre-specified maximum information level, this approach can satisfy a power condition quite accurately. We describe this form of maximum information trial in Section 8.2.2. If sampling to a maximum information level is not feasible, it may still be desirable to adapt to irregularities in the calendar times of analyses. Lan & DeMets (1989a) present a group sequential test in which Type I error is spent as a function of calendar time, and we describe tests of this form in Section 8.2.3.

8.2.2 Maximum Information Trials

8.2.2.1 Spending Type I Error Probability in a Maximum Information Trial

In a maximum information trial, the Type I error probability is partitioned according to an error spending function, $f(t)$, which is non-decreasing and satisfies $f(0) = 0$ and $f(t) = \alpha$ for $t \geq 1$. The value $f(t)$ indicates the cumulative Type I error that is to be spent when a fraction t of the maximum anticipated information has been obtained. Thus, in implementing this approach, it is necessary to specify a target, $\mathcal{I}_{max}$, for the maximum information. Once this target has been reached, the procedure will terminate, either rejecting or accepting H_0. It is implicitly assumed that the target $\mathcal{I}_{max}$ will eventually be reached if the trial does not terminate with an early decision. Thus if, for example, patients enter a study at a lower rate than anticipated, organizers must take steps to extend the study's duration. In a survival study, where information depends primarily on the number of observed deaths, it may be necessary to recruit more subjects, extend the follow-up period, or both.

The error spending function $f(t)$ and the target information level $\mathcal{I}_{max}$ must be selected before the study commences. Then, the "information fraction" at analysis k is $t_k = \mathcal{I}_k/\mathcal{I}_{max}$ for $k = 1, \ldots, K$, and the Type I error probabilities allocated to each analysis are

$$\begin{aligned} \pi_1 &= f(t_1) = f(\mathcal{I}_1/\mathcal{I}_{max}), \\ \pi_k &= f(t_k) - f(t_{k-1}) = f(\mathcal{I}_k/\mathcal{I}_{max}) - f(\mathcal{I}_{k-1}/\mathcal{I}_{max}), \quad k = 2, 3, \ldots . \end{aligned} \tag{8.2}$$

Critical values c_k are computed successively to satisfy (8.1), exactly as in Slud & Wei's method. Note that c_k does not depend on the as yet unobserved $\mathcal{I}_{k+1}, \mathcal{I}_{k+2}, \ldots$. The stopping rule has the form (4.2), only now the maximum possible number of analyses, K, depends on the observed values $\mathcal{I}_1, \mathcal{I}_2, \ldots$. Specifically, K is the smallest value k for which $\mathcal{I}_k \geq \mathcal{I}_{max}$. Since by definition $f(\mathcal{I}_K/\mathcal{I}_{max}) = \alpha$, we have $\pi_1 + \ldots + \pi_K = \alpha$ and the overall Type I error rate is exactly α, as required.

There have been a number of suggestions for the form of the error spending function $f(t)$. Lan & DeMets (1983) propose a function that yields critical values close to those of the O'Brien & Fleming (1979) test when group sizes are equal and adapting this proposal to a two-sided test gives the error spending function

$$f(t) = \min\{4 - 4\,\Phi(z_{\alpha/4}/\sqrt{t}),\, \alpha\}, \tag{8.3}$$

where, $z_{\alpha/4}$ denotes $\Phi^{-1}(1-\alpha/4)$. For a close analogue of the Pocock (1977) boundary, Lan & DeMets (1983) suggest

$$f(t) = \min\{\alpha \log\left[\,1 + (e-1)\,t\,\right],\, \alpha\}. \tag{8.4}$$

Hwang, Shih & De Cani (1990) introduce a family of error spending functions indexed by a parameter γ,

$$f(t) = \begin{cases} \alpha\,(1-e^{-\gamma t})/(1-e^{-\gamma}) & \text{for } \gamma \neq 0 \\ \alpha\, t & \text{for } \gamma = 0, \end{cases} \tag{8.5}$$

where γ may be positive, negative or equal to zero. Note that, since the functions (8.4) and (8.5) are linear in α, they can be applied in spending a total error probability α in both one-sided and two-sided tests.

A useful family of error spending functions, indexed by the parameter $\rho > 0$, is given by

$$f(t) = \min\{\alpha t^{\rho},\, \alpha\}. \tag{8.6}$$

We term this the "Rho family". The cases $\rho = 1$, 1.5 and 2 are studied by Kim & DeMets (1987a). Jennison & Turnbull (1989, 1990) note that tests using certain values of ρ produce boundaries similar to those of Pocock and O'Brien & Fleming tests. We shall concentrate on tests in this family for the remainder of this section.

8.2.2.2 Achieving a Specified Power

The joint distribution under $\theta = 0$ of $Z_1, \ldots, Z_K$, conditional on $\mathcal{I}_1, \ldots, \mathcal{I}_K$, depends only on the ratios $\mathcal{I}_1/\mathcal{I}_{max}, \ldots, \mathcal{I}_K/\mathcal{I}_{max}$. Hence, the critical values $c_1, \ldots, c_K$ are functions of these ratios. In contrast, the distribution of $Z_1, \ldots, Z_K$ under $\theta \neq 0$ depends on the absolute information levels $\mathcal{I}_1, \ldots, \mathcal{I}_K$. The value of $\mathcal{I}_{max}$ is the major factor in determining power, but there is also some dependence on the actual sequence of observed information levels. For design purposes, it is convenient to anticipate a maximum number of analyses, K, and to suppose information levels will be equally spaced up to the final value $\mathcal{I}_{max}$, i.e.,

$$\mathcal{I}_k = \frac{k}{K}\,\mathcal{I}_{max}, \quad k = 1, \ldots, K. \tag{8.7}$$

Although the test will be implemented using the sequence of information levels actually observed, we shall see that variations in these levels affect the achieved power only slightly. Under assumption (8.7), the value of $\mathcal{I}_{max}$ can be chosen to meet a given power requirement. Alternatively, if the maximum possible information level is fixed by feasibility and cost considerations, the methods we describe can be used to evaluate the study's approximate power.

Table 8.1 *Constants $R_{LD}(K, \alpha, \beta, \rho)$ for Lan & DeMets "maximum information" two-sided tests using Type I error spending functions $f(t) = \min\{\alpha t^\rho, \alpha\}$. Implementation with K equally spaced information levels leading to $\mathcal{I}_{max} = R_{LD}(K, \alpha, \beta, \rho)\, \mathcal{I}_{f,2}$ attains Type I error probability $\alpha = 0.05$ and power $1 - \beta = 0.8$ or 0.9 at $\theta = \pm\delta$.*

	$R_{LD}(K,\alpha,\beta,\rho)$					
		$1-\beta=0.8$			$1-\beta=0.9$	
K	$\rho=1$	$\rho=2$	$\rho=3$	$\rho=1$	$\rho=2$	$\rho=3$
1	1.000	1.000	1.000	1.000	1.000	1.000
2	1.082	1.028	1.010	1.075	1.025	1.009
3	1.117	1.045	1.020	1.107	1.041	1.018
4	1.137	1.056	1.027	1.124	1.051	1.025
5	1.150	1.063	1.032	1.136	1.058	1.030
6	1.159	1.069	1.036	1.144	1.063	1.033
7	1.165	1.073	1.039	1.150	1.067	1.036
8	1.170	1.076	1.041	1.155	1.070	1.039
9	1.174	1.079	1.043	1.159	1.073	1.040
10	1.178	1.081	1.045	1.162	1.075	1.042
11	1.180	1.083	1.046	1.164	1.077	1.043
12	1.183	1.085	1.048	1.166	1.078	1.044
15	1.188	1.088	1.050	1.171	1.082	1.047
20	1.193	1.092	1.054	1.176	1.085	1.050

Under (8.7) the value of $\mathcal{I}_{max}$ needed to achieve power $1 - \beta$ at $\theta = \pm\delta$ is proportional to $1/\delta^2$ and can be expressed as a constant times $\mathcal{I}_{f,2}$, given by (4.6), the information required by a fixed sample test with two-sided Type I error probability α and power $1 - \beta$ at $\theta = \pm\delta$. We denote this constant by $R_{LD}(K, \alpha, \beta, \rho)$ for a Lan & DeMets test defined by the error spending function (8.6) with parameter ρ. Values of $R_{LD}(K, \alpha, \beta, \rho)$ are listed in Table 8.1 for selected values of K, $\alpha = 0.05$, $\beta = 0.8$ and 0.9, and $\rho = 1$, 2 and 3. Values for other combinations of parameters can be computed using the numerical methods described in Section 8.7 and Chapter 31. As the maximum possible number of analyses may not be known a priori, an estimated value should be used in determining $\mathcal{I}_{max}$. Since $R_{LD}(K, \alpha, \beta, \rho)$ increases with K, one may wish to use an upper bound for the maximum number of analyses as a conservative choice.

8.2.2.3 Properties of Tests with $f(t) = \min\{\alpha t^\rho, \alpha\}$

Tables 8.2 and 8.3 show properties of selected maximum information tests based on error spending functions of the form (8.6). As in previous chapters, we follow other authors in referring to the tests' maximum and expected sample sizes but note that our results really concern information levels and this is the interpretation that should be made in situations where information and sample size are not directly proportional. The tabulated tests are designed for equally spaced information levels, and results are for the case where these information levels are actually observed. In practice, one would expect variations about the planned information sequence to occur, leading to a larger final information level since, by definition, this must be greater than or equal to $\mathcal{I}_{max}$, and this will perturb the actual power of the test (sometimes termed the "attained" or "post-hoc" power) and expected

Table 8.2 *Properties of maximum information, two-sided tests with Type I error spending functions* $f(t) = \min\{\alpha t^{\rho}, \alpha\}$ *applied with* K *equally spaced information levels, culminating in* $\mathcal{I}_K = \mathcal{I}_{max} = R_{LD}(K, \alpha, \beta, \rho)\mathcal{I}_{f,2}$. *Maximum and expected sample sizes are expressed as percentages of the corresponding fixed sample size. All tests attain Type I error probability* $\alpha = 0.05$ *and power* $1 - \beta = 0.8$ *at* $\theta = \pm\delta$.

$\alpha = 0.05,\ 1 - \beta = 0.8$

		Sample size as percentage of fixed sample size				
	K	*Maximum sample size*	*Expected sample size at* $\theta =$ 0	$\pm 0.5\delta$	$\pm\delta$	$\pm 1.5\delta$
$\rho = 1$						
	1	100.0	100.0	100.0	100.0	100.0
	2	108.2	106.9	102.1	85.0	64.8
	3	111.7	109.9	103.4	81.2	56.5
	4	113.7	111.6	104.2	79.5	53.0
	5	115.0	112.7	104.8	78.5	51.0
	10	117.8	115.1	106.1	76.6	47.4
	15	118.8	116.0	106.6	76.1	46.3
	20	119.3	116.5	106.9	75.8	45.8
$\rho = 2$						
	1	100.0	100.0	100.0	100.0	100.0
	2	102.8	102.1	99.3	86.7	67.0
	3	104.5	103.5	99.2	82.3	60.4
	4	105.6	104.4	99.2	80.1	57.1
	5	106.3	105.1	99.3	78.8	55.0
	10	108.1	106.6	99.7	76.2	51.0
	15	108.8	107.2	99.9	75.4	49.7
	20	109.2	107.5	100.0	75.1	49.1
$\rho = 3$						
	1	100.0	100.0	100.0	100.0	100.0
	2	101.0	100.7	98.9	89.5	70.7
	3	102.0	101.4	98.3	84.6	64.7
	4	102.7	102.0	98.0	82.1	61.0
	5	103.2	102.4	97.9	80.6	58.8
	10	104.5	103.4	97.9	77.8	54.6
	15	105.0	103.9	98.0	76.9	53.3
	20	105.4	104.2	98.0	76.5	52.7

sample sizes. Nevertheless, results for this idealized situation give a good indication of the properties of the error spending tests and provide guidance for the choice of error spending function and frequency of analyses.

Comparisons with Tables 2.7 and 2.8 show that the parameter values $\rho = 1$ and 3 in the error spending function (8.6) yield tests with similar properties to Pocock and O'Brien & Fleming tests, respectively. In fact, these error spending tests have some advantages: setting $\rho = 1$ gives tests with smaller maximum sample sizes than Pocock's tests and roughly the same or sometimes lower expected sample sizes, and $\rho = 3$ achieves useful reductions in

Table 8.3 *Properties of maximum information, two-sided tests with Type I error spending functions* $f(t) = \min\{\alpha t^\rho, \alpha\}$ *applied with* K *equally spaced information levels, culminating in* $\mathcal{I}_K = \mathcal{I}_{max} = R_{LD}(K, \alpha, \beta, \rho)\, \mathcal{I}_{f,2}$. *Maximum and expected sample sizes are expressed as percentages of the corresponding fixed sample size. All tests attain Type I error probability* $\alpha = 0.05$ *and power* $1 - \beta = 0.9$ *at* $\theta = \pm\delta$.

	$\alpha = 0.05,\ 1 - \beta = 0.9$					
		Sample size as percentage of fixed sample size				
	K	*Maximum sample size*	*Expected sample size at* $\theta =$ 0	$\pm 0.5\delta$	$\pm\delta$	$\pm 1.5\delta$
$\rho = 1$						
	1	100.0	100.0	100.0	100.0	100.0
	2	107.5	106.1	99.6	77.7	58.7
	3	110.7	108.8	100.1	72.2	48.5
	4	112.4	110.3	100.5	69.8	44.5
	5	113.6	111.3	100.7	68.4	42.3
	10	116.2	113.6	101.5	65.7	38.4
	15	117.1	114.4	101.8	64.9	37.3
	20	117.6	114.8	102.0	64.5	36.7
$\rho = 2$						
	1	100.0	100.0	100.0	100.0	100.0
	2	102.5	101.9	97.9	80.5	59.6
	3	104.1	103.2	97.1	75.0	52.3
	4	105.1	104.0	96.8	72.2	48.9
	5	105.8	104.6	96.7	70.5	46.8
	10	107.5	106.0	96.6	67.2	42.6
	15	108.2	106.5	96.6	66.2	41.3
	20	108.5	106.8	96.6	65.7	40.7
$\rho = 3$						
	1	100.0	100.0	100.0	100.0	100.0
	2	100.9	100.6	98.1	84.1	62.4
	3	101.8	101.3	96.8	78.2	56.7
	4	102.5	101.8	96.2	75.1	53.2
	5	103.0	102.1	95.9	73.3	50.9
	10	104.2	103.1	95.4	69.8	46.5
	15	104.7	103.6	95.4	68.7	45.1
	20	105.0	103.8	95.3	68.2	44.5

expected sample size over the O'Brien & Fleming tests, especially at $\theta = \pm 1.5\delta$. Thus, there is little reason to opt for the more complicated error spending functions (8.3) and (8.4) designed to mimic O'Brien & Fleming and Pocock tests more closely.

The family of tests defined by (8.6) parallels the Wang & Tsiatis (1987) family of two-sided tests when observed information levels are equally spaced. We have noted the relation between the error spending tests with $\rho = 1$ and 3 and Pocock and O'Brien & Fleming tests, both of which belong to the Wang & Tsiatis family. As an intermediate example, the error spending test with $\rho = 2$ behaves very similarly to the Wang & Tsiatis tests with $\Delta = 0.25$, whose properties are shown in Tables 2.11 and 2.12.

8.2.2.4 Example 1: Comparison of Two Normal Means When Variances are Known

Consider the parallel two-treatment comparison of Section 3.2.1 in which the responses $X_{Ai} \sim N(\mu_A, \sigma_A^2)$ and $X_{Bi} \sim N(\mu_B, \sigma_B^2)$, $i = 1, 2, \ldots$, are recorded for subjects receiving treatments A and B, respectively. Suppose we wish to test H_0: $\mu_A = \mu_B$ with two-sided Type I error probability $\alpha = 0.05$ and power $1 - \beta = 0.9$ at $\mu_A - \mu_B = \pm 1$ when it is known that $\sigma_A^2 = \sigma_B^2 = 4$. If we have n_A and n_B observations from subjects on treatments A and B respectively, the estimate

$$\widehat{\mu}_A - \widehat{\mu}_B = \bar{X}_A - \bar{X}_B$$

has variance $4\,(n_A^{-1} + n_B^{-1})$ and the information for $\mu_A - \mu_B$ is the reciprocal of this variance, $(n_A^{-1} + n_B^{-1})^{-1}/4$. From (4.6), a fixed sample test with the desired error probabilities needs information

$$\{\Phi^{-1}(0.975) + \Phi^{-1}(0.9)\}^2 = 10.51,$$

which can be achieved with 85 subjects on each treatment.

Suppose we decide to conduct this test group sequentially using the error spending function $f(t) = \min\{\alpha t^2, \alpha\}$, i.e., the member of family (8.6) with $\rho = 2$, and we wish to have a maximum of 10 analyses. From Table 8.1, $R_{LD}(10, 0.05, 0.1, 2) = 1.075$, so we set

$$\mathcal{I}_{max} = 1.075 \times 10.51 = 11.30$$

as our target for the final information level. This value can be achieved with 91 observations per treatment arm and the study should be planned accordingly. For example, we could aim to collect responses from 10 subjects on each treatment per stage and then $\mathcal{I}_{max}$ would be reached after 10 groups, even if slightly fewer responses were obtained in practice. If, for some reason, we obtain more observations per group, a maximum of eight or nine groups could suffice. Although the power achieved at $\mu_A - \mu_B = \pm 1$ will depend on the precise sequence of information levels obtained when the study is conducted, it is a defining property of the error spending test that the Type I error probability will be exactly $\alpha = 0.05$, whatever the observed information sequence.

We can also evaluate the likely power in situations where the maximum possible sample size is determined by external constraints. Suppose, in the above example, the same error spending function is used to apportion a Type I error of $\alpha = 0.05$ but financial costs limit the total number of observations to 160. If it is anticipated that these observations will arise in 10 groups of 8 per treatment arm, the final information level will be

$$\mathcal{I}_{10} = \frac{80}{2 \times 4} = 10.0,$$

so the test should be implemented with $\mathcal{I}_{max} = 10.0$. To find where power 0.9 is achieved by this test we simply calculate the value of δ for which Equation (4.6) with $\beta = 0.1$ gives $\mathcal{I}_{f,2} = 10.0/1.075 = 9.30$. This value is $\delta = 1.06$, so the error spending test has power 0.9 at $\mu_A - \mu_B = \pm 1.06$. A similar calculation using $R_{LD}(10, 0.05, 0.2, 2) = 1.081$ in place of 1.075 and solving Equation (4.6) with $\beta = 0.2$ shows that this error spending test with $\mathcal{I}_{max} = 10.0$ has power 0.8 at $\mu_A - \mu_B = \pm 0.92$. Once again, any departure from the assumed sequence of information levels will affect the power actually achieved, but the Type I error probability of $\alpha = 0.05$ will be maintained exactly in all cases.

8.2.2.5 Variations on the Maximum Information Design

A change in the definition of the maximum information design is required to handle one important eventuality. Suppose in the preceding example external constraints impose a limit of 10 groups of observations and we expect 9 patients on each treatment arm in each group, leading to a final information level

$$\mathcal{I}_{10} = \frac{90}{2 \times 4} = 11.25.$$

We set $\mathcal{I}_{max} = 11.25$ and anticipate power close to 0.9 at $\mu_A - \mu_B = \pm 1$, as our previous calculations showed this is achieved by 10 equal groups with $\mathcal{I}_{max} = 11.30$. Now suppose fewer subjects than expected are actually recruited and the study ends at the 10th analysis with $\mathcal{I}_{10} < 11.25$. The error spending procedure was originally defined on the assumption that $\mathcal{I}_{max}$ would eventually be reached and, since that is not the case here, a modification is needed if the full Type I error probability is to be spent.

We refer to the problem of reaching the last possible analysis with information less than $\mathcal{I}_{max}$ as "under-running". In order for the overall Type I error probability to equal α, the final boundary value, c_K, is now obtained by solving (8.1) for $k = K$ with $\pi_K = \alpha - f(\mathcal{I}_{K-1}/\mathcal{I}_{max})$. Of course, the low final information level will result in reduced power: if, for example, patient numbers alternate between 8 and 9 per treatment arm in the 10 groups with a total of 85 per treatment and $\mathcal{I}_{10} = 10.62$, power at $\mu_A - \mu_B = \pm 1$ will be reduced to 0.89.

By definition, $f(t) = \alpha$ for $t \geq 1$, and so the analogous problem of "over-running" is handled automatically, the procedure necessarily terminating at any stage when $\mathcal{I}_k > \mathcal{I}_{max}$. In this case, the power will usually be greater than the design value. In our example, if there are 12 patients per treatment arm in each group, the test will terminate by analysis 8, where $\mathcal{I}_8 = 12.0$, attaining power of 0.91 at $\mu_A - \mu_B = \pm 1$.

Another practical difficulty arises when the information levels $\mathcal{I}_k$ depend on an unknown nuisance parameter which can only be estimated as the trial proceeds. For instance, in the comparison of two Bernoulli distributions discussed in Section 4.4.2, under H_0: $p_A = p_B$ the information levels depend on the common success probability. The error spending approach can be applied with information levels calculated using estimates of the nuisance parameter. If these estimates change greatly during the study, it is even possible (although usually unlikely) for the resulting information levels to decrease between successive analyses; if this does happen, we recommend setting the boundary at infinity and continuing to sample the next group of observations.

In some instances, it may be necessary to modify a study's recruitment or follow-up period if information is to approach its target, $\mathcal{I}_{max}$. If it is feasible to make such adjustments, the study's course can be re-considered at appropriate points in the light of current predictions of future information levels. Scharfstein, Tsiatis & Robins (1997) and Scharfstein & Tsiatis (1998) describe a general approach to "information-based design and monitoring" in which simulation is used to assess the range of plausible information levels at future analyses and they illustrate its application to a Phase III clinical trial studying the effects of drug treatments on HIV patients' CD4+ cell counts. However, as noted in Section 8.1, it is important to remember that, while the timing of future analyses may depend on observed information levels, adjustments to the study design should not be influenced by estimates of the treatment effect.

8.2.2.6 A Systematic Study of Attained Power

Although differences between actual information levels and those used in design calculations affect the power of error spending group sequential tests, the effects are often small. The results in Table 8.4 are for studies planned assuming $\tilde{K}$ analyses at equally spaced information levels but conducted with $\mathcal{I}_k = \pi(k/\tilde{K})^r \mathcal{I}_{max}$, $k = 1, \ldots, \tilde{K}$. These information sequences are of the form considered in previous chapters, the factor π controlling the final information level and r the spacing between successive information levels. In most cases the

Table 8.4 *Power attained by maximum information, two-sided tests using Type I error spending functions* $f(t) = \min\{\alpha t^{\rho}, \alpha\}$, *with* $\rho = 1$, 2 *and* 3. *Tests have Type I error probability exactly equal to* $\alpha = 0.05$. *The tests are designed to achieve power* 0.9 *under information levels* $\mathcal{I}_k = (k/\tilde{K})\mathcal{I}_{max}$, $k = 1, \ldots, \tilde{K}$, *but levels* $\mathcal{I}_k = \pi(k/\tilde{K})^r \mathcal{I}_{max}$, $k = 1, \ldots, \tilde{K}$, *are actually observed.*

			$\tilde{K}$			
	r	π	2	5	10	15
$\rho = 1$						
	0.80	0.9	0.870	0.875	0.878	0.879
		1.0	0.899	0.899	0.899	0.900
		1.1	0.921	0.914	0.902	0.904
	1.00	0.9	0.871	0.875	0.878	0.879
		1.0	0.900	0.900	0.900	0.900
		1.1	0.923	0.917	0.907	0.904
	1.25	0.9	0.872	0.876	0.878	0.879
		1.0	0.902	0.901	0.901	0.901
		1.1	0.925	0.920	0.912	0.902
$\rho = 2$						
	0.80	0.9	0.869	0.873	0.876	0.877
		1.0	0.899	0.899	0.899	0.899
		1.1	0.922	0.915	0.902	0.904
	1.00	0.9	0.870	0.874	0.876	0.877
		1.0	0.900	0.900	0.900	0.900
		1.1	0.923	0.918	0.907	0.904
	1.25	0.9	0.871	0.874	0.877	0.878
		1.0	0.902	0.901	0.901	0.901
		1.1	0.925	0.921	0.913	0.903
$\rho = 3$						
	0.80	0.9	0.868	0.871	0.874	0.875
		1.0	0.899	0.899	0.899	0.899
		1.1	0.923	0.916	0.902	0.904
	1.00	0.9	0.869	0.872	0.874	0.875
		1.0	0.900	0.900	0.900	0.900
		1.1	0.924	0.919	0.908	0.904
	1.25	0.9	0.869	0.872	0.875	0.876
		1.0	0.901	0.901	0.901	0.901
		1.1	0.925	0.921	0.914	0.903

Table 8.5 *Power attained by maximum information, two-sided tests using Type I error spending functions $f(t) = \min\{\alpha t^\rho, \alpha\}$, with $\rho = 1$, 2 and 3. Tests have Type I error probability exactly equal to $\alpha = 0.05$. The tests are designed to achieve power 0.9 under $\tilde{K}$ equally spaced information levels, $\mathcal{I}_k = (k/\tilde{K})\mathcal{I}_{max}$, $k = 1, \ldots, \tilde{K}$, but K analyses with equally spaced information levels $\mathcal{I}_k = (k/K)\mathcal{I}_{max}$, $k = 1, \ldots, K$, actually occur.*

			$\tilde{K}$			
			2	5	10	15
$\rho = 1$						
	K	2	0.900	0.915	0.921	0.923
		5	0.883	0.900	0.906	0.909
		10	0.875	0.893	0.900	0.902
		15	0.873	0.891	0.898	0.900
$\rho = 2$						
	K	2	0.900	0.909	0.913	0.915
		5	0.891	0.900	0.904	0.906
		10	0.886	0.895	0.900	0.902
		15	0.884	0.894	0.898	0.900
$\rho = 3$						
	K	2	0.900	0.906	0.909	0.910
		5	0.894	0.900	0.903	0.905
		10	0.891	0.897	0.900	0.901
		15	0.889	0.895	0.899	0.900

maximum possible number of analyses, K, is equal to the planned number $\tilde{K}$, but there are a few exceptions: for $\tilde{K} = 10$, $\pi = 1.1$ and $r = 0.8$, we find $\mathcal{I}_9 > \mathcal{I}_{max}$ and so the test terminates after only $K = 9$ analyses; similarly, for $\tilde{K} = 15$, $\pi = 1.1$ and all three values of r, $\mathcal{I}_{14} > \mathcal{I}_{max}$ and at most $K = 14$ analyses are possible. The error spending tests maintain their nominal Type I error rates exactly in all cases, in contrast to tests defined by the significance level approach which attained Type I error rates only approximately (see Table 4.2). The attained power remains within 0.002 of its target value, 0.9, in all cases where $\pi = 1$, i.e., when the test concludes with $\mathcal{I}_K = \mathcal{I}_{max}$. If $\pi = 0.9$ under-running occurs, the test concludes with $\mathcal{I}_K < \mathcal{I}_{max}$ and, not surprisingly, power falls below 0.9. However, since analyses occur at lower information levels, less cumulative Type I error is spent at all analyses other than the final one and the loss of power is smaller than that seen in Table 4.2 for significance level tests, especially for the larger values of K. Conversely, $\pi = 1.1$ leads to over-running with $\mathcal{I}_K > \mathcal{I}_{max}$ and attained power in excess of 0.9 but, again, the deviations from 0.9 are smaller than those for significance level tests.

An important feature of the error spending method is its ability to vary the maximum possible number of analyses in response to observed information levels. The effects of applying a test with observed information levels considerably greater or smaller than anticipated are shown in Table 8.5. Here, tests are designed for $\tilde{K}$ equally space information levels but the target $\mathcal{I}_{max}$ is reached at analysis K. Again, the Type I error rate remains exactly equal to its specified value. Variations do occur in the attained power, and this lies above its target value when there are fewer analyses than anticipated and below its target when increments in information are small and more analyses are required. The effects on attained power are smaller for the error spending functions with $\rho = 2$ and 3, which spend relatively small amounts of power at early analyses. In a typical study, one would also expect to see unequal increments in information. However, comparisons between cases in Table 8.4 with the same π and $\tilde{K}$ and different values of r show that once the number of analyses and final information level are fixed, the additional effects on attained power due to the spacing of observed information levels are liable to be quite small.

Overall, the variations in attained power seen in Tables 8.4 and 8.5 are quite modest. It is this insensitivity of the attained power to the number of analyses and observed information levels that allows us to design a trial as if information levels will be equally spaced, even though we realize increments in information will be unequal and the number of analyses may not be as planned.

8.2.3 *Maximum Duration Trials*

It is common practice to organize a clinical trial by setting dates for the beginning and end of patient recruitment, for interim analyses, and for the ultimate termination of the trial. Thus, the maximum duration of the trial, T_{max} say, is specified, but the information that will have accrued in this time is not known at the outset. If this final information level is highly unpredictable, it may be unrealistic to define a group sequential design in terms of a target information level $\mathcal{I}_{max}$ since that level may not be approached at all, or it may be obtained at quite an early stage of the study. It is still appealing to spend error as a function of the ratio of current information to that which will be available at the final planned analysis. Since this final information level, $\mathcal{I}_K$ say, is unknown, at each analysis k an estimated value $\widehat{\mathcal{I}}_K^{(k)}$ must be used in calculating the cumulative error to be spent up to this point,

$$f(\mathcal{I}_k/\widehat{\mathcal{I}}_K^{(k)}), \quad k = 1, \ldots, K.$$

Lan & Lachin (1990) and Kim, Boucher & Tsiatis (1995) describe the application of this method to survival studies and demonstrate how to calculate boundaries satisfying (8.1) with increments in Type I error

$$\begin{aligned} \pi_1 &= f(\mathcal{I}_1/\widehat{\mathcal{I}}_K^{(1)}), \\ \pi_k &= f(\mathcal{I}_k/\widehat{\mathcal{I}}_K^{(k)}) - f(\mathcal{I}_{k-1}/\widehat{\mathcal{I}}_K^{(k-1)}), \quad k = 2, \ldots, K. \end{aligned} \tag{8.8}$$

A simple, practical alternative is to spend Type I error as a function of elapsed *calendar* time. In place of (8.2) or (8.8), we partition the Type I error α into

$$\begin{aligned} \pi_1 &= f(T_1/T_{max}), \\ \pi_k &= f(T_k/T_{max}) - f(T_{k-1}/T_{max}), \quad k = 2, \ldots, K, \end{aligned} \tag{8.9}$$

where T_k is the elapsed calendar time from the start of the trial to the kth analysis and f is a non-decreasing function with $f(0) = 0$ and $f(t) = \alpha$ for $t \geq 1$. The use of the quantities T_k/T_{max} in place of information fractions allows Type I error to be spent at a rate which is responsive to the increments in information between analyses, as long as information accrues fairly evenly over time. Even though $\mathcal{I}_K$ is unknown at any interim stage k, the covariances $\text{Cov}(Z_{k_1}, Z_{k_2}) = \sqrt{(\mathcal{I}_{k_1}/\mathcal{I}_{k_2})}$ for $k_1 < k_2 \leq k$ are available at that time, and these are all that is needed to find c_k by solving (8.1) with π_k given by (8.9). If the dates of analyses are fixed at the design stage, this determines the elapsed times $T_1, \ldots, T_K$ and T_{max}, and hence the partition of α into $\pi_1, \ldots, \pi_K$ is actually fixed in advance, just as in Slud & Wei's (1982) method.

The power attained in a maximum duration trial with $\pi_1, \ldots, \pi_K$ defined by (8.8) or (8.9) is determined by the observed information levels. Thus, if these are unpredictable, only a rough initial estimate of power will be available when the study commences. In a survival study with no interim analyses, this problem is often addressed by conducting the single analysis after a certain number of events have been observed: for a group sequential trial, a similar solution is offered by a maximum information design in which the final analysis is conducted after a pre-specified number of observed events.

As an illustration of a maximum duration trial with error spent as a function of elapsed calendar time, we consider the Beta-blocker Heart Attack Trial (BHAT) example given by Lan & DeMets (1989a) and also discussed by DeMets et al. (1984), Lan, Rosenberger & Lachin (1993) and Lan, Reboussin & DeMets (1994). BHAT was planned as a trial of 48 months duration, starting in June 1978. In total, 3837 patients were randomized to either placebo or propranolol treatment arms. The data were monitored at approximate six-month intervals, analyses taking place in May 1979, October 1979, March 1980, October 1980, April 1981 and October 1981. No formal stopping rule was established at the start of the study, but the O'Brien & Fleming test with an overall two-sided Type I error probability of 0.05 was selected early in the trial to serve as a guide to the Policy and Data Monitoring Board. The study was terminated in October 1981 due to an observed benefit of propranolol treatment.

In BHAT, the cumulative numbers of deaths at the six interim analyses were 56, 77, 126, 177, 247 and 318. With overall mortality as the primary endpoint, outcomes can be monitored using the log-rank statistic (see Section 4.5). The sequence of standardized log-rank statistics at the interim analyses reported by Lan & DeMets (1989a) was (1.68, 2.24, 2.37, 2.30, 2.34, 2.82). As explained in Section 4.5, we can estimate the information level $\mathcal{I}_k$ at analysis k by one quarter of the total number of deaths, d_k, observed by that time.

We shall reconstruct how this trial would have progressed had it been run as a group sequential test, spending Type I error probability as a function of calendar time. Suppose the error spending function was $f(t) = \alpha \min(t, 1)$ with $\alpha = 0.05$. Then, at analysis k, we evaluate $f(T_k/T_{max})$ where T_k is the time in months since the start of the trial and $T_{max} = 48$, the trial's maximum planned duration. Hence, $T_1 = 11$, $T_2 = 16$, $T_3 = 21$, $T_4 = 28$, $T_5 = 34$ and $T_6 = 40$, and $\pi_1 = (11/48) \times 0.05$, $\pi_2 = ((16/48) - (11/48)) \times 0.05$, etc. The boundary used at the first analysis in May 1979 is found by solving

$$Pr\{|Z_1| \geq c_1\} = \pi_1 = \frac{11}{48} \times 0.05 = 0.0115$$

when $Z_1 \sim N(0, 1)$, and we obtain $c_1 = 2.53$ from standard normal tables. Since the observed log-rank statistic was 1.68, the trial continues to the next analysis. At the second analysis in October 1979, we find the boundary value c_2 by solving

$$Pr\{|Z_1| < 2.53,\ |Z_2| \geq c_2\} = \pi_2 = (\frac{16}{48} - \frac{11}{48}) \times 0.05 = 0.0052$$

when Z_1 and Z_2 are standard normal variables with correlation $\sqrt{(\mathcal{I}_1/\mathcal{I}_2)}$, which we approximate by

$$\sqrt{\{(d_1/4)/(d_2/4)\}} = \sqrt{(56/77)} = 0.853.$$

Numerical computations using the methods described in Section 8.7 give the solution $c_2 = 2.59$. Again, the observed log-rank statistic is between $-c_2$ and c_2 and the study continues. At the third analysis in March 1980, we find c_3 by solving

$$Pr\{|Z_1| < 2.53,\ |Z_2| < 2.59, |Z_3| \geq c_3\} = \pi_3 = (\frac{21}{48} - \frac{16}{48}) \times 0.05 = 0.0052$$

when Z_1, Z_2 and Z_3 are standard normal variables with $Corr(Z_1, Z_2) = \sqrt{(56/77)} = 0.853$, $Corr(Z_1, Z_3) = \sqrt{(56/126)} = 0.667$ and $Corr(Z_2, Z_3) = \sqrt{(77/126)} = 0.782$, since the numbers of deaths observed at analyses 1, 2 and 3 were 56, 77 and 126, respectively. This calculation gives $c_3 = 2.63$ and, as the observed log-rank statistic is 2.37, the study continues.

Proceeding in this way, we obtain boundary values $c_4 = 2.50$, $c_5 = 2.51$ and $c_6 = 2.47$, and find the boundary is crossed at the sixth analysis when $Z_6 = 2.82$ is observed. Thus, under this scheme, the study would have terminated in October 1981 with rejection of the null hypothesis of no treatment difference.

The numerical values $c_1, \ldots, c_6$ in the above example are fairly close, fluctuating in the range 2.47 to 2.63. This is not surprising as we have already noted that the error spending function (8.6) with $\rho = 1$ produces a boundary similar to that of Pocock's repeated significance test for which the c_k are all equal.

8.2.4 A Note on "Information Time"

We have defined error spending tests in terms of observed information levels $\mathcal{I}_1, \mathcal{I}_2, \ldots$. Some authors describe the error spending approach in terms of "information time", defined as the ratio of current information to a value for the information level at the final possible analysis. Since the final information level can have more than one interpretation, it is important to state what is meant by this: is it the current estimate of the final information $\mathcal{I}_K$ or a pre-specified target for this information level? For completeness, we shall explain the relation between our exposition and the "information time" treatment.

In the case of maximum information trials, we have taken the cumulative Type I error spent up to analysis k to be

$$f(\mathcal{I}_k/\mathcal{I}_{max}).$$

This may be viewed as $f(t_k)$, where $t_k = \mathcal{I}_k/\mathcal{I}_{max}$ is the information time at analysis k, with denominator equal to the target information level that was specified when designing the trial.

In the first form of maximum duration trial, we set the Type I error spent by analysis k to be

$$f(\mathcal{I}_k/\widehat{\mathcal{I}}_K^{(k)}).$$

This can be viewed as $f(\hat{t}_k)$, where the information time $\hat{t}_k = \mathcal{I}_k/\widehat{\mathcal{I}}_K^{(k)}$ has the current estimate of the final information level in its denominator. In the second form of maximum duration trial, the ratios of calendar times T_k/T_{max}, $k = 1, \ldots, K$, take the place of information time.

Once the cumulative Type I error probability at analysis k is determined, calculation of the error spending boundary at analysis k involves the joint distribution of $Z_1, \ldots, Z_k$, in particular their covariances,

$$\text{Cov}(Z_{k_1}, Z_{k_2}) = \sqrt{(\mathcal{I}_{k_1}/\mathcal{I}_{k_2})} \quad \text{for } k_1 < k_2.$$

Since

$$\frac{\mathcal{I}_{k_1}}{\mathcal{I}_{k_2}} = \frac{\mathcal{I}_{k_1}/\mathcal{I}_{max}}{\mathcal{I}_{k_2}/\mathcal{I}_{max}},$$

these covariances can be expressed in terms of information times, $t_k = \mathcal{I}_k/\mathcal{I}_{max}$. However, when estimated information times are used, the equation

$$\frac{\mathcal{I}_{k_1}}{\mathcal{I}_{k_2}} = \frac{\mathcal{I}_{k_1}/\widehat{\mathcal{I}}_K^{(k)}}{\mathcal{I}_{k_2}/\widehat{\mathcal{I}}_K^{(k)}}$$

implies that the values of $\hat{t}_{k'}$ for each $k' < k$ should be updated to include the current estimate of $\mathcal{I}_K$, i.e., one should set $\hat{t}_{k_1} = \mathcal{I}_{k_1}/\widehat{\mathcal{I}}_K^{(k)}$ and $\hat{t}_{k_2} = \mathcal{I}_{k_2}/\widehat{\mathcal{I}}_K^{(k)}$ when calculating the value of $\text{Cov}(Z_{k_1}, Z_{k_2}) = \hat{t}_{k_1}/\hat{t}_{k_2}$ for $k_1 < k_2 \leq k$ to be used in computing the boundary c_k at analysis k. This is in contrast to the formula (8.8) for the Type I error probability π_k allocated to analysis k, in which the term $f(\mathcal{I}_{k-1}/\widehat{\mathcal{I}}_K^{(k-1)})$ does not contain the updated estimate of $\mathcal{I}_K$ since this represents the cumulative Type I error spent up to analysis $k-1$ and that cannot now be changed.

Our conclusion from the above discussion is that if the phrase "information time" is to be used, the denominator of this fraction should be clearly defined. However, we find it less confusing to define the parameter t in the error spending function $f(t)$ directly in terms of raw information levels and simply avoid using the terminology "information time".

8.2.5 *Two-sided Tests with Early Stopping to Accept H_0*

The error spending approach can also be used to create two-sided tests with an "inner wedge" to allow early stopping in favor of H_0. Both inner wedge error spending tests and one-sided error spending tests, which we discuss in the next section, are defined through a *pair* of error spending functions and, for this reason, similar issues arise in their implementation. We shall not investigate inner wedge tests in detail but the brief description we present here is complemented by the fuller treatment of one-sided tests which follows. Our description of inner wedge tests assumes a target information level $\mathcal{I}_{max}$ will ultimately be reached and so is appropriate to maximum information, rather than maximum duration, trials.

If it is desired to test H_0: $\theta = 0$ with Type I error probability α and power $1-\beta$ at $\theta = \pm\delta$, an inner wedge test is defined by a Type I error spending function $f(t)$ and a Type II error spending function $g(t)$, where $f(0) = g(0) = 0$ and f and g increase monotonically to $f(t) = \alpha$ and $g(t) = \beta$ for $t \geq 1$. The Type I error probability is partitioned as

$$\begin{aligned}\pi_{1,1} &= f(\mathcal{I}_1/\mathcal{I}_{max}),\\ \pi_{1,k} &= f(\mathcal{I}_k/\mathcal{I}_{max}) - f(\mathcal{I}_{k-1}/\mathcal{I}_{max}), \quad k = 2, 3, \ldots,\end{aligned}$$

and the Type II error probability as

$$\begin{aligned}\pi_{2,1} &= g(\mathcal{I}_1/\mathcal{I}_{max}),\\ \pi_{2,k} &= g(\mathcal{I}_k/\mathcal{I}_{max}) - g(\mathcal{I}_{k-1}/\mathcal{I}_{max}), \quad k = 2, 3, \ldots.\end{aligned}$$

Suppose standardized statistics $Z_1, \ldots, Z_K$ follow the canonical joint distribution (3.1), given $\mathcal{I}_1, \ldots, \mathcal{I}_K$. Then, the stopping rule is of the form (6.1) and the critical values (a_k, b_k) can be calculated to satisfy

$$Pr_{\theta=0}\{a_1 < |Z_1| < b_1, \ldots, a_{k-1} < |Z_{k-1}| < b_{k-1}, |Z_k| \geq b_k\} = \pi_{1,k}$$

and

$$Pr_{\theta=\delta}\{a_1 < |Z_1| < b_1, \ldots, a_{k-1} < |Z_{k-1}| < b_{k-1}, |Z_k| < a_k\} = \pi_{2,k},$$

for $k = 1, 2, \ldots$. The target information level $\mathcal{I}_{max}$ should be chosen so that the final boundary points coincide under a plausible information sequence assumed for design purposes. If, for example, one anticipates K equally spaced information levels, then a_K should be equal to b_K when $\mathcal{I}_k = (k/K)\mathcal{I}_{max}$, $k = 1, \ldots, K$.

In the inner wedge tests of Chapter 6, it is not usually possible to stop in favor of H_0 at the first few analyses and we commented on the desirability of this feature. This property can be incorporated in an error spending test by choosing a value t^*, around 0.3 or 0.5 say, and setting $g(t)$ equal to 0 for $t \leq t^*$ before it rises toward β once t passes t^*.

An extreme case of an inner wedge boundary is when the outer boundary is infinite at all interim looks and hence early stopping is only possible to accept H_0. Gould (1983) considers this situation in the context of "abandoning a lost cause". He argues that in many trials for non-life-threatening conditions, when interim findings suggest a positive outcome, the trial will be carried on to completion to provide adequate information on safety endpoints and to permit analysis of patient subgroups. Only if the new treatment seems ineffective will early termination of the trial be justified. This can be viewed as a special case of the above test in which the error spending function for Type I error is given by $f(t) = 0$ for $0 < t < 1$ and $f(t) = \alpha$ for $t \geq 1$, and hence $b_1 = \ldots = b_{K-1} = +\infty$.

8.3 One-sided Tests with a Binding Futility Boundary

8.3.1 *The Error Spending Formulation*

The error spending method we have presented in the context of two-sided tests can be adapted to one-sided problems of the form considered in Chapter 5. Suppose, as usual, that we have estimates $\hat{\theta}^{(k)}$ of a scalar parameter θ and information levels $\mathcal{I}_k = \{Var(\hat{\theta}^{(k)})\}^{-1}$ at analyses $k = 1, \ldots, K$, and the standardized statistics

$$Z_k = \hat{\theta}^{(k)} \sqrt{\mathcal{I}_k}, \quad k = 1, \ldots, K,$$

have the canonical joint distribution (3.1) conditionally on the sequence $\{\mathcal{I}_1, \ldots, \mathcal{I}_K\}$. As for other error spending tests, we do not suppose the maximum number of analyses to be completely determined initially and we use K to denote the maximum number of analyses that would occur in a particular realization of the study in the absence of early stopping to accept or reject the null hypothesis. We consider the one-sided problem of Chapter 5 in which it is desired to test H_0: $\theta \leq 0$ against the alternative H_A: $\theta > 0$ with Type I error probability α and power $1 - \beta$ at $\theta = \delta$ where $\delta > 0$.

Just as for two-sided tests with an inner wedge, we define the upper and lower group sequential boundaries through error spending functions for both Type I and Type II error probabilities. In doing this, we anticipate meeting the power requirement quite closely, so a maximum duration design in which attained power is allowed to vary widely, depending on observed information levels, is not ideal and we restrict attention in this section to maximum information trials.

In designing a maximum information trial, we specify a target information level $\mathcal{I}_{max}$ which we expect to reach eventually if the study does not terminate with an early decision. We define two error spending functions $f(t)$ and $g(t)$, for Type I and Type II errors, respectively, which are non-decreasing and satisfy $f(0) = 0$, $g(0) = 0$, $f(t) = \alpha$ for $t \geq 1$ and $g(t) = \beta$ for $t \geq 1$. We shall explain in Section 8.3.2 how to choose the target value $\mathcal{I}_{max}$ so that the power requirement will be satisfied under a design assumption of equally spaced information levels terminating at $\mathcal{I}_{max}$. The Type I error probability α is partitioned between analyses as

$$\begin{aligned} \pi_{1,1} &= f(\mathcal{I}_1/\mathcal{I}_{max}), \\ \pi_{1,k} &= f(\mathcal{I}_k/\mathcal{I}_{max}) - f(\mathcal{I}_{k-1}/\mathcal{I}_{max}), \quad k = 2, 3, \ldots, \end{aligned} \tag{8.10}$$

and the Type II error β is partitioned as

$$\begin{aligned} \pi_{2,1} &= g(\mathcal{I}_1/\mathcal{I}_{max}), \\ \pi_{2,k} &= g(\mathcal{I}_k/\mathcal{I}_{max}) - g(\mathcal{I}_{k-1}/\mathcal{I}_{max}), \quad k = 2, 3, \ldots. \end{aligned} \tag{8.11}$$

The test is defined through critical values $(a_1, b_1), \ldots, (a_K, b_K)$ for the statistics $Z_1, \ldots, Z_K$, and we use these in the standard form of stopping rule (5.2) for a one-sided group sequential test. In place of (8.1) we now have two equations,

$$Pr_{\theta=0}\{a_1 < Z_1 < b_1, \ldots, a_{k-1} < Z_{k-1} < b_{k-1}, Z_k \geq b_k\} = \pi_{1,k} \tag{8.12}$$

and

$$Pr_{\theta=\delta}\{a_1 < Z_1 < b_1, \ldots, a_{k-1} < Z_{k-1} < b_{k-1}, Z_k < a_k\} = \pi_{2,k} \tag{8.13}$$

to define the stopping boundary at analysis k. As before, the critical values are calculated successively, starting with a_1 and b_1 at the first analysis.

Our objective when drawing the test to a conclusion at the end of the error spending process is to achieve a Type I error of exactly α, letting the attained power absorb the effects of variations in observed information levels. If neither boundary is crossed first, the test continues until an analysis with information level $\mathcal{I}_k \geq \mathcal{I}_{max}$; this becomes the final possible analysis, K, and we solve (8.12) to obtain b_K and set a_K equal to b_K.

Under-running, as discussed in Section 8.2.2.5, can occur if external constraints impose a limit on the study duration which forces termination at an analysis K with $\mathcal{I}_K < \mathcal{I}_{max}$. If this occurs, the remaining Type I error should be spent by taking b_K to be the solution to (8.12) with

$$\pi_{1,K} = \alpha - f(\mathcal{I}_{K-1}/\mathcal{I}_{max}). \tag{8.14}$$

Then, setting $a_K = b_K$ ensures termination at this analysis.

It is also possible, under some information sequences, that the test can be terminated at an analysis k with $\mathcal{I}_k < \mathcal{I}_{max}$ with overall Type I and II error probabilities within their permitted bounds. To check if this is the case for a particular analysis k, Equations (8.12) and (8.13) should be solved with $\pi_{1,k} = \alpha - f(\mathcal{I}_{k-1}/\mathcal{I}_{max})$ and $\pi_{2,k} = \beta - g(\mathcal{I}_{k-1}/\mathcal{I}_{max})$; then, if $a_k > b_k$, we define this to be the final analysis, K, and set b_K equal to the value b_k just computed and $a_K = b_K$. If this check fails, the values a_k and b_k should be re-calculated using $\pi_{1,k}$ and $\pi_{2,k}$ defined in the usual way by (8.10) and (8.11).

8.3.2 The Rho Family of Error Spending Tests

There have been just a few proposals for error spending functions in one-sided tests. Jennison (1987) used a four-parameter family of error spending functions to approximate optimal group sequential boundaries. Eales & Jennison (1992) created error spending functions from group sequential boundaries defined for fixed information sequences by calculating the cumulative error at each analysis and specifying the error to be spent at intermediate information levels by interpolation. Chang, Hwang & Shin (1998) used the Gamma family of error spending functions proposed by Hwang, Shih & De Cani (1990) (see Equation 8.5) to define one-sided tests with Type I and Type II error spending functions and they report the use of such an error spending design in a clinical trial comparing treatments for coronary heart disease. Pampallona, Tsiatis & Kim (2001) created error spending functions which interpolate the cumulative Type I and Type II error probabilities at each analysis for tests in the families proposed by Pampallona & Tsiatis (1994).

Here, we shall define tests using the simple error spending functions

$$f(t) = \min\{\alpha t^{\rho_1}, \alpha\} \quad \text{and} \quad g(t) = \min\{\beta t^{\rho_2}, \beta\}, \tag{8.15}$$

where $\rho_1 > 0$ and $\rho_2 > 0$. We shall refer to these as the "Rho family" of error spending tests. At the design stage, we assume there will be K equally spaced information levels ending with $\mathcal{I}_K = \mathcal{I}_{max}$, as in (8.7), and we determine the value of $\mathcal{I}_{max}$ for which (8.12) and (8.13) give $a_K = b_K$ so that the test terminates at analysis K with error rates exactly α and β. Of course, when applying this test with a general sequence of information levels, the upper and lower boundaries will not necessarily meet at analysis K and the rules given in Section 8.3.1 must be followed to conclude the test with its specified Type I error probability For simplicity, in the remainder of this chapter, we shall consider the case of a common index $\rho_1 = \rho_2 = \rho$, although spending Type I and Type II error probabilities at different rates is certainly possible and may be preferred in some trials.

For now, we assume the futility boundary will be treated as binding. With equally spaced information levels as in (8.7), the value of $\mathcal{I}_{max}$ needed to meet Type I error and power requirements is a constant times $\mathcal{I}_{f,1}$, the information required by a fixed sample one-sided

Table 8.6 *Constants $R_{OS}(K, \alpha, \beta, \rho)$ for maximum information, one-sided tests with binding futility boundaries, using error spending functions $f(t) = \min\{\alpha t^\rho, \alpha\}$ and $g(t) = \min\{\beta t^\rho, \beta\}$. Implementation with K equally spaced information levels leading to $\mathcal{I}_K = R_{OS}(K, \alpha, \beta, \rho)\, \mathcal{I}_{f,1}$ attains Type I error probability $\alpha = 0.025$ and power $1-\beta = 0.8$, 0.9 or 0.95 at $\theta = \delta$.*

	$R_{OS}(K,\alpha,\beta,\rho)$					
	$1-\beta=0.8$		$1-\beta=0.9$		$1-\beta=0.95$	
K	$\rho=2$	$\rho=3$	$\rho=2$	$\rho=3$	$\rho=2$	$\rho=3$
1	1.000	1.000	1.000	1.000	1.000	1.000
2	1.043	1.015	1.044	1.016	1.044	1.016
3	1.070	1.029	1.072	1.031	1.072	1.032
4	1.087	1.039	1.089	1.042	1.089	1.043
5	1.098	1.047	1.100	1.049	1.101	1.051
6	1.106	1.052	1.108	1.055	1.109	1.057
7	1.112	1.056	1.115	1.059	1.115	1.061
8	1.117	1.060	1.119	1.063	1.120	1.065
9	1.120	1.063	1.123	1.066	1.124	1.068
10	1.124	1.065	1.126	1.068	1.127	1.070
11	1.126	1.067	1.129	1.070	1.129	1.072
12	1.128	1.068	1.131	1.072	1.132	1.074
15	1.133	1.072	1.136	1.075	1.136	1.078
20	1.138	1.076	1.141	1.079	1.141	1.082

test given by (5.1). We denote this constant by $R_{OS}(K, \alpha, \beta, \rho)$ for a test defined by the error spending functions (8.15) with $\rho_1 = \rho_2 = \rho$. Values of $R_{OS}(K, \alpha, \beta, \rho)$ are listed in Table 8.6 for selected values of K, $\alpha = 0.025$, $1-\beta = 0.8$, 0.9 and 0.95, and $\rho = 2$ and 3. As for two-sided tests, an estimated value of K, the maximum possible number of analyses, should be used in choosing $\mathcal{I}_{max}$ when designing a study. We have not shown constants for tests using $\rho = 1$ as these have rather high maximum information levels, around 14% higher than a fixed sample test when $K = 2$, rising to 32% higher for $K = 10$, and we saw in Chapter 5 that most of the benefits of group sequential testing can be achieved with considerably lower maximum information levels.

8.3.3 Properties of the Tests

We see from the values of $R_{OS}(K, \alpha, \beta, \rho)$ in Table 8.6 that error spending functions of the form (8.15) with $\rho = 2$ and 3 require only modest increases in maximum sample size over the fixed sample test; for example, if $K = 5$, increases of around 5% and 10% are necessary for $\rho = 3$ and 2, respectively. Tables 8.7, 8.8 and 8.9 present expected sample sizes of these tests (more precisely, the expected information on termination) when information levels coincide with their equally spaced design values. We can contrast these results with properties of the parametric "power family" tests shown in Tables 5.1, 5.2 and 5.3. When comparing designs with the same number of analyses and approximately equal maximum sample sizes, we see that the Rho family error spending tests offer significantly lower expected sample sizes at $\theta = \delta$ than the parametric tests while having similar or lower expected sample sizes at $\theta = 0$ and $\theta = \delta/2$. Furthermore, reference to Table 5.4 and results we shall present in Chapter 12 when discussing optimal group sequential tests show that Rho family error spending tests

Table 8.7 *Properties of maximum information, one-sided error spending tests with binding futility boundaries, using spending functions* $f(t) = \min\{\alpha t^\rho, \alpha\}$ *and* $g(t) = \min\{\beta t^\rho, \beta\}$ *applied with* K *equally spaced information levels, leading to* $\mathcal{I}_K = \mathcal{I}_{max} = R_{OS}(K, \alpha, \beta, \rho)\, \mathcal{I}_{f,1}$*. Maximum and expected sample sizes are expressed as percentages of the corresponding fixed sample size. All tests attain Type I error probability* $\alpha = 0.025$ *and power* $1 - \beta = 0.8$ *at* $\theta = \delta$*.*

	$\alpha = 0.025,\ 1-\beta = 0.8$					
		Sample size as percentage of fixed sample size				
	K	*Maximum sample size*	*Expected sample size at* $\theta =$ 0	0.5δ	δ	1.5δ
$\rho = 2$						
	1	100.0	100.0	100.0	100.0	100.0
	2	104.3	70.2	87.0	85.2	67.4
	3	107.0	63.5	81.9	79.8	60.9
	4	108.7	60.1	79.2	77.0	57.5
	5	109.8	58.0	77.6	75.2	55.5
	10	112.4	53.9	74.2	71.7	51.3
	15	113.3	52.6	73.1	70.6	50.0
	20	113.8	51.9	72.5	70.0	49.4
$\rho = 3$						
	1	100.0	100.0	100.0	100.0	100.0
	2	101.5	75.2	90.9	88.6	70.9
	3	102.9	68.8	85.6	82.9	64.9
	4	103.9	65.3	82.8	79.9	61.3
	5	104.7	63.1	81.1	78.1	59.1
	10	106.5	59.0	77.6	74.5	54.8
	15	107.2	57.6	76.4	73.3	53.5
	20	107.6	57.0	75.8	72.7	52.9

have close to the minimum possible expected sample sizes given their maximum sample sizes.

When the error spending tests are implemented with a general information sequence, departures from the planned information levels do not change the Type I error probability, but the attained power is subject to perturbations similar to those seen for two-sided error spending tests. Suppose tests are designed using error spending functions (8.15) with $\mathcal{I}_{max}$ chosen so that power 0.9 will be achieved if $\mathcal{I}_k = (k/\tilde{K})\mathcal{I}_{max}$, $k = 1, \ldots, \tilde{K}$. We have calculated the power attained for actual information sequences $\mathcal{I}_k = \pi(k/\tilde{K})^r \mathcal{I}_{max}$, $k = 1, \ldots, \tilde{K}$, for the values of ρ, $\tilde{K}$, r and π appearing in Table 8.4 and found this to remain within 0.002 of 0.9 whenever $\pi = 1$, and thus the planned information level $\mathcal{I}_{max}$ is met at the final analysis. Attained power falls below or above its target when there is under-running ($\pi = 0.9$) or over-running ($\pi = 1.1$), respectively, with similar values to those seen for the two-sided case in Table 8.4. If, instead, information levels are equally spaced but do not fall at their planned values, the attained power follows the pattern seen in Table 8.5 for two-sided tests with the same K and $\tilde{K}$. Comparisons with Table 5.12 show that, overall, error spending tests control attained power more closely than tests defined by the significance level approach, and, of course, the error spending tests have additional flexibility to handle variable numbers of analyses.

Table 8.8 *Properties of maximum information, one-sided error spending tests with binding futility boundaries, using spending functions* $f(t) = \min\{\alpha t^{\rho},\ \alpha\}$ *and* $g(t) = \min\{\beta t^{\rho},\ \beta\}$ *applied with* K *equally spaced information levels, leading to* $\mathcal{I}_K = \mathcal{I}_{max} = R_{OS}(K, \alpha, \beta, \rho)\,\mathcal{I}_{f,1}$. *Maximum and expected sample sizes are expressed as percentages of the corresponding fixed sample size. All tests attain Type I error probability* $\alpha = 0.025$ *and power* $1 - \beta = 0.9$ *at* $\theta = \delta$.

	$\alpha = 0.025,\ 1-\beta = 0.9$					
		Sample size as percentage of fixed sample size				
	K	*Maximum sample size*	*Expected sample size at* $\theta =$ 0	0.5δ	δ	1.5δ
$\rho = 2$						
	1	100.0	100.0	100.0	100.0	100.0
	2	104.4	70.2	88.4	80.2	60.3
	3	107.2	63.6	83.5	74.4	53.0
	4	108.9	60.3	80.9	71.3	49.7
	5	110.0	58.2	79.3	69.5	47.6
	10	112.6	54.2	76.0	65.8	43.3
	15	113.6	52.8	74.9	64.6	41.9
	20	114.1	52.2	74.4	64.0	41.3
$\rho = 3$						
	1	100.0	100.0	100.0	100.0	100.0
	2	101.6	74.6	91.6	83.9	62.6
	3	103.1	68.3	86.5	77.7	57.1
	4	104.2	64.8	83.8	74.4	53.6
	5	104.9	62.7	82.1	72.5	51.4
	10	106.8	58.5	78.7	68.7	47.0
	15	107.5	57.2	77.5	67.4	45.6
	20	107.9	56.5	77.0	66.8	44.9

Although one can easily define other forms of error spending function, our investigations have shown (8.15) to be a good choice, producing one-sided group sequential tests which are efficient in reducing expected sample size across a range of θ values. As for two-sided error spending tests, robustness of the attained power to the number of analyses and the spacing of information levels enables us to design a study assuming a certain number of equally spaced information levels, even when the observed information sequence is liable to be quite different.

8.3.4 *Early Stopping Only to Accept* H_0

At the end of Section 8.2.5, we discussed two sided tests in which there was only early stopping to accept H_0. The same idea can be applied to a one-sided test when it is desirable to continue a trial to completion to provide adequate information on safety endpoints and patient subgroups, even if interim findings are positive. Then, only if results appear negative, will early termination be considered in order to "abandon a lost cause". In the formulation of Section 8.3.1, the Type II error spending function $g(t)$ must be specified but, since no Type I error is to be spent at interim analyses, the Type I error spending function is simply $f(t) = 0$ for $0 < t < 1$ and $f(t) = \alpha$ for $t \geq 1$. Equation (8.12) then implies

Table 8.9 *Properties of maximum information, one-sided error spending tests with binding futility boundaries, using spending functions* $f(t) = \min\{\alpha t^{\rho}, \alpha\}$ *and* $g(t) = \min\{\beta t^{\rho}, \beta\}$ *applied with* K *equally spaced information levels, leading to* $\mathcal{I}_K = \mathcal{I}_{max} = R_{OS}(K, \alpha, \beta, \rho)\, \mathcal{I}_{f,1}$. *Maximum and expected sample sizes are expressed as percentages of the corresponding fixed sample size. All tests attain Type I error probability* $\alpha = 0.025$ *and power* $1 - \beta = 0.95$ *at* $\theta = \delta$.

	$\alpha = 0.025,\ 1-\beta = 0.95$					
		Sample size as percentage of fixed sample size				
	K	*Maximum sample size*	*Expected sample size at* $\theta =$ 0	0.5δ	δ	1.5δ
$\rho = 2$						
	1	100.0	100.0	100.0	100.0	100.0
	2	104.4	70.6	89.3	75.4	56.4
	3	107.2	64.2	84.6	69.3	47.7
	4	108.9	60.8	82.0	66.1	44.4
	5	110.1	58.8	80.5	64.2	42.4
	10	112.7	54.8	77.2	60.3	38.0
	15	113.6	53.5	76.1	59.0	36.6
	20	114.1	52.8	75.6	58.4	35.9
$\rho = 3$						
	1	100.0	100.0	100.0	100.0	100.0
	2	101.6	74.6	92.2	79.2	57.5
	3	103.2	68.4	87.1	72.9	51.5
	4	104.3	64.9	84.4	69.5	48.4
	5	105.1	62.7	82.8	67.4	46.2
	10	107.0	58.6	79.4	63.4	41.7
	15	107.8	57.3	78.3	62.1	40.3
	20	108.2	56.6	77.7	61.5	39.6

that $b_1 = \ldots = b_{K-1} = +\infty$. Since some sample paths that could have led to a value $Z_K > \Phi^{-1}(1-\alpha)$ terminate on crossing the futility boundary, the calculation will give a final boundary point $b_K < \Phi^{-1}(1-\alpha)$. However, it is difficult to make a convincing argument for a positive treatment effect θ when a simple fixed sample analysis of the data does not reject the null hypothesis $\theta \le 0$ and we would recommend setting $b_K = \Phi^{-1}(1-\alpha)$ in such a design.

8.4 One-sided Tests with No Futility Boundary or a Non-binding Futility Boundary

In Section 5.3 we discussed designs in which the lower, futility boundary is considered "non-binding" on the basis that the trial will not necessarily be stopped if this boundary is crossed. Then, in order to preserve the Type I error probability α, the upper boundary $\{b_k, \ldots, b_K\}$ was constructed with the property

$$Pr_{\theta=0}\{Z_k > b_k \text{ for some } k = 1, \ldots, K\} = \alpha,$$

ensuring that the Type I error probability is no greater than α, even if the lower boundary is ignored completely. In defining an error spending test with a non-binding futility boundary, we specify a Type I error spending function $f(t)$ which, as usual, is non-decreasing with $f(0) = 0$ and $f(t) = \alpha$ for $t \geq 1$. We can, for example, choose a function from the Rho family described in Section 8.3.2. As before, the Type I error probability α is partitioned between analyses as

$$\pi_{1,1} = f(\mathcal{I}_1/\mathcal{I}_{max}) \text{ and } \pi_{1,k} = f(\mathcal{I}_k/\mathcal{I}_{max}) - f(\mathcal{I}_{k-1}/\mathcal{I}_{max}), \quad k = 2, 3, \ldots, \tag{8.16}$$

and now upper boundary values $\{b_1, \ldots, b_K\}$ are determined by the single set of equations

$$Pr_{\theta=0}\{Z_1 < b_1, \ldots, Z_{k-1} < b_{k-1} \text{ and } Z_k \geq b_k\} = \pi_{1,k}, \quad k = 1, 2, \ldots, \tag{8.17}$$

which can be solved successively at analyses $k = 1, \ldots, K$.

Designing for the case of no futility boundary at all

The maximum information level $\mathcal{I}_{max}$ should be set to achieve desired power. Suppose power $1 - \beta$ is specified at $\theta = \delta$ and a design is required which will achieve this power if there is no early stopping at all for futility. At the design stage, we assume there will be K equally spaced information levels ending with $\mathcal{I}_K = \mathcal{I}_{max}$, that is $\mathcal{I}_k = (k/K)\mathcal{I}_{\max}$, $k = 1, \ldots, K$. When the ratios $\mathcal{I}_k/\mathcal{I}_K$, $k = 1, \ldots, K - 1$ are fixed, the constants $b_1, \ldots, b_K$ computed in (8.17) do not depend on $\mathcal{I}_{max}$ and we can search for the value $\mathcal{I}_{max}$ such that

$$Pr_{\theta=\delta}\{Z_k > b_k \text{ for some } k = 1, \ldots, K\} = \beta \tag{8.18}$$

to meet the power requirement. As before, when the test is implemented using the sequence of observed information levels, the power achieved at $\theta = \delta$ may differ slightly from $1 - \beta$ but the Type I error rate under $\theta = 0$ will still be exactly α.

With equally spaced information levels, the value of $\mathcal{I}_{max}$ needed to meet Type I error and power requirements is a constant times $\mathcal{I}_{f,1}$, the information required by a fixed sample one-sided test given by (5.1). For the case of no futility boundary, using a Rho family Type I error spending function

$$f(t) = \min\{\alpha t^\rho, \alpha\},$$

we denote this constant by $R_{NoFB}(K, \alpha, \beta, \rho)$. Values of $R_{NoFB}(K, \alpha, \beta, \rho)$ are listed in Table 8.10 for $\alpha = 0.025$, $1 - \beta = 0.8$, 0.9 and 0.95, $\rho = 2$ and 3, and selected values of K. As for two-sided tests, an estimated value of K, the maximum possible number of analyses, should be used in choosing $\mathcal{I}_{max}$ when designing a test.

Properties of error spending tests with no early stopping for futility are shown in Table 8.11 for designs with Type I error rate $\alpha = 0.025$ and power $1 - \beta = 0.9$ at $\theta = \delta$. Expected sample sizes under $\theta = \delta$ are close to those seen in Table 8.8 for tests that do have a futility boundary. However, in the absence of a futility boundary expected sample sizes for $\theta = 0$ or $\theta = \delta/2$ are close to the fixed sample size, demonstrating that failure to allow early stopping for futility may lead to unnecessary consumption of resources in studies that are destined to conclude with a negative result. While it is possible to overcome this problem by giving a monitoring committee responsibility to decide if and when to stop a trial for futility, making such a decision can be difficult in the absence of a futility stopping rule. Furthermore, such interventions will reduce power below that of the original study design. We therefore recommend specifying a non-binding futility boundary to be used as a guideline when considering early stopping for futility, and setting $\mathcal{I}_{\max}$ so that the desired power will be achieved if this guideline is followed. We shall return to this topic in Section 11.5.

Designing for the case of a non-binding futility boundary

Suppose a study is designed with a Type I error spending function $f(t)$ and upper boundary values $\{b_1, \ldots, b_K\}$ are determined by the Equations (8.17). A non-binding futility

Table 8.10 *Constants $R_{NoFB}(K, \alpha, \beta, \rho)$ for maximum information, one-sided tests using Type I error spending function $f(t) = \min\{\alpha t^{\rho}, \alpha\}$ and no early stopping to accept H_0. Implementation with K equally spaced information levels culminating in $\mathcal{I}_K = \mathcal{I}_{max} = R_{NoFB}(K, \alpha, \beta, \rho)\,\mathcal{I}_{f,1}$ will attain Type I error probability $\alpha = 0.025$ and power $1-\beta = 0.8$, 0.9 or 0.95 at $\theta = \delta$. If early stopping to accept H_0 occurs, the Type I error rate will be below $\alpha = 0.025$ and power will decrease.*

	$R_{NoFB}(K, \alpha, \beta, \rho)$					
	$1-\beta = 0.8$		$1-\beta = 0.9$		$1-\beta = 0.95$	
K	$\rho = 2$	$\rho = 3$	$\rho = 2$	$\rho = 3$	$\rho = 2$	$\rho = 3$
1	1.000	1.000	1.000	1.000	1.000	1.000
2	1.028	1.010	1.025	1.009	1.023	1.009
3	1.045	1.020	1.041	1.018	1.038	1.017
4	1.056	1.027	1.051	1.025	1.048	1.023
5	1.063	1.032	1.058	1.030	1.055	1.028
6	1.069	1.036	1.063	1.033	1.060	1.031
7	1.073	1.039	1.067	1.036	1.063	1.034
8	1.076	1.041	1.070	1.039	1.066	1.036
9	1.079	1.043	1.073	1.040	1.069	1.038
10	1.081	1.045	1.075	1.042	1.071	1.040
11	1.083	1.046	1.077	1.043	1.072	1.041
12	1.085	1.048	1.078	1.044	1.074	1.042
15	1.088	1.050	1.082	1.047	1.077	1.045
20	1.092	1.053	1.085	1.050	1.080	1.048

boundary $\{a_1, \ldots, a_K\}$ can be added by specifying a Type II error spending function $g(t)$, a non-decreasing function with $g(0) = 0$ and $g(t) = \beta$ for $t \geq 1$, where t represents the ratio of the current information to a target information level $\mathcal{I}_{max}$. At each analysis k, the Type II error probability to be spent is defined as

$$\pi_{2,1} = g(t_1),$$

$$\pi_{2,k} = g(t_k) - g(t_{k-1}), \quad k = 2, 3, \ldots,$$

where $t_k = \mathcal{I}_k/\mathcal{I}_{max}$, and the boundary value a_k is computed as the solution to

$$Pr_{\theta=\delta}\{a_1 < Z_1 < b_1, \ldots, a_{k-1} < Z_{k-1} < b_{k-1} \text{ and } Z_k \leq a_k\} = \pi_{2,k}. \tag{8.19}$$

The target information level $\mathcal{I}_{max}$ can be chosen so that the test has power $1-\beta$ under $\theta = \delta$ in the case that the futility boundary is strictly enforced. Then, if the futility boundary is applied flexibly, the Type I error rate will remain equal to α but the power actually achieved will vary around the original value.

For planning purposes, it is convenient to assume there will be K equally spaced information levels, so $\mathcal{I}_k = (k/K)\mathcal{I}_{max}$ and $t_k = k/K$, $k = 1, \ldots, K$. Suppose we use Rho family error spending functions $f(t) = \min\{\alpha t^{\rho}, \alpha\}$ and $g(t) = \min\{\beta t^{\rho}, \beta\}$. Then, with equally spaced information levels, the value of $\mathcal{I}_{max}$ needed to meet Type I error and power requirements is a constant times $\mathcal{I}_{f,1}$, the information required by a fixed sample one-sided test given by (5.1). We denote this constant for tests with a non-binding futility

Table 8.11 *Properties of maximum information, one-sided error spending tests with Type I error spending function* $f(t) = \min\{\alpha t^{\rho}, \alpha\}$ *and no early stopping to accept* H_0*, applied with* K *equally spaced information levels culminating in* $\mathcal{I}_K = \mathcal{I}_{max} = R_{NoFB}(K, \alpha, \beta, \rho)\, \mathcal{I}_{f,1}$*. Maximum and expected sample sizes are expressed as percentages of the corresponding fixed sample size. All tests attain Type I error probability* $\alpha = 0.025$ *and power* $1 - \beta = 0.9$ *at* $\theta = \delta$*.*

$\alpha = 0.025,\ 1 - \beta = 0.9$

		Sample size as percentage of fixed sample size				
	K	*Maximum sample size*	*Expected sample size at* $\theta =$ 0	0.5δ	δ	1.5δ
$\rho = 2$						
	1	100.0	100.0	100.0	100.0	100.0
	2	102.5	102.2	97.9	80.5	59.6
	3	104.1	103.6	97.1	75.0	52.3
	4	105.1	104.6	96.8	72.2	48.9
	5	105.8	105.2	96.7	70.5	46.8
	10	107.5	106.7	96.6	67.2	42.6
	15	108.2	107.4	96.6	66.2	41.3
	20	108.5	107.7	96.7	65.7	40.7
$\rho = 3$						
	1	100.0	100.0	100.0	100.0	100.0
	2	100.9	100.8	98.1	84.1	62.4
	3	101.8	101.6	96.8	78.2	56.7
	4	102.5	102.1	96.2	75.1	53.2
	5	103.0	102.6	95.9	73.3	50.9
	10	104.2	103.7	95.5	69.8	46.5
	15	104.7	104.1	95.4	68.7	45.1
	20	105.0	104.4	95.3	68.2	44.5

boundary by $R_{NBF}(K, \alpha, \beta, \rho)$ and list values in Table 8.12 for $\alpha = 0.025$, $1 - \beta = 0.8$, 0.9 and 0.95, $\rho = 2$ and 3, and selected values of K. Once again, an estimated value of K should be used in choosing $\mathcal{I}_{max}$ when designing a test.

Properties of error spending tests with power $1 - \beta = 0.9$ at $\theta = \delta$ and non-binding futility boundaries are shown in Table 8.13. These tests have Type I error rate probability $\alpha = 0.025$ if the futility boundary is ignored and there is no early stopping at all for futility; if, instead, the futility boundary is applied and the study stops to accept H_0: $\theta \le 0$ whenever this boundary is crossed, the designs are conservative with Type I error rates ranging between 0.023 and 0.025. A consequence of this is that the inflation factors R_{NBF} are higher than the corresponding factors R_{OS} for designs with binding futility boundaries. Also, the expected sample sizes in Table 8.13, which are calculated supposing the futility boundary will be adhered to, are up to 1 or 2% higher than those shown in Table 8.8 for designs with binding futility boundaries. We have found similar small increases in expected sample sizes for tests with power 0.8 and 0.95 at $\theta = \delta$. Since the differences in performance between designs with binding and non-binding futility boundaries are so small, we see no serious disadvantage in running a study with a non-binding futility boundary.

Table 8.12 *Constants* $R_{NBF}(K, \alpha, \beta, \rho)$ *for maximum information, one-sided tests using Type I error spending function* $f(t) = \min\{\alpha t^\rho,\ \alpha\}$ *and Type II error spending function* $g(t) = \min\{\beta t^\rho,\ \beta\}$ *when the futility boundary is non-binding. Implementation with* K *equally spaced information levels culminating in* $\mathcal{I}_K = \mathcal{I}_{max} = R_{NBF}(K, \alpha, \beta, \rho)\, \mathcal{I}_{f,1}$ *will attain Type I error probability* $\alpha = 0.025$ *and, when the futility boundary is applied rigidly, power* $1 - \beta = 0.8$, 0.9 *or* 0.95 *at* $\theta = \delta$.

	$R_{NoFB}(K, \alpha, \beta, \rho)$					
	$1-\beta=0.8$		$1-\beta=0.9$		$1-\beta=0.95$	
K	$\rho=2$	$\rho=3$	$\rho=2$	$\rho=3$	$\rho=2$	$\rho=3$
1	1.000	1.000	1.000	1.000	1.000	1.000
2	1.056	1.019	1.055	1.020	1.054	1.020
3	1.093	1.039	1.093	1.041	1.091	1.041
4	1.117	1.054	1.117	1.056	1.114	1.057
5	1.133	1.065	1.133	1.068	1.130	1.068
6	1.145	1.073	1.144	1.076	1.142	1.077
7	1.153	1.080	1.153	1.083	1.150	1.084
8	1.160	1.085	1.160	1.088	1.157	1.089
9	1.165	1.089	1.165	1.093	1.162	1.094
10	1.170	1.093	1.169	1.096	1.166	1.097
11	1.173	1.096	1.173	1.099	1.170	1.100
12	1.176	1.098	1.176	1.102	1.173	1.103
15	1.183	1.104	1.183	1.107	1.180	1.109
20	1.189	1.109	1.189	1.113	1.186	1.115

8.5 Data-Dependent Timing of Analyses

The flexible nature of the error spending approach lends itself well to accommodating irregular, unpredictable and unplanned interim analysis schedules. However, there are still some restrictions on how a trial may be conducted if the conditional distribution of $\{Z_1, \ldots, Z_K\}$ given the observed information sequence $\{\mathcal{I}_1, \ldots, \mathcal{I}_K\}$ is to have the standard form (3.1), as we have assumed in calculating properties of error spending tests. Allowing the timing or number of future analyses to depend on the current and past responses can introduce additional dependencies between the Z_ks and $\mathcal{I}_k$s, in which case error probabilities may no longer be guaranteed.

The following example, given by Jennison & Turnbull (1991a), illustrates how the Type I error probability can be affected if $\mathcal{I}_k$ is allowed to depend on $Z_1, \ldots, Z_{k-1}$. Suppose observations are taken in three groups, each contributing one unit of information so the statistics $Z_1 \sim N(\theta, 1)$, $Z_2 \sim N(\theta\sqrt{2}, 1)$ and $Z_3 \sim N(\theta\sqrt{3}, 1)$ will become available. A two-sided test with Type I error probability $\alpha = 0.05$ conducted as a maximum information trial with $\mathcal{I}_{max} = 3$ and error spending function $f(t) = \min(\alpha t, \alpha)$ rejects H_0 if $|Z_1| \geq 2.394$, $|Z_2| \geq 2.294$ or $|Z_3| \geq 2.200$. Suppose now the investigators decide that if $|Z_1| < 1.2$, the second analysis will be omitted and the second and third groups of observations pooled, in which case the critical value for $|Z_3|$ will be 2.076, the value arising for the given error

Table 8.13 *Properties of maximum information, one-sided error spending tests with Type I error spending function* $f(t) = \min\{\alpha t^\rho, \alpha\}$, *Type II error spending function* $g(t) = \min\{\beta t^\rho, \beta\}$, *and non-binding futility boundaries, applied with* K *equally spaced information levels culminating in* $\mathcal{I}_K = \mathcal{I}_{max} = R_{NBF}(K, \alpha, \beta, \rho)\, \mathcal{I}_{f,1}$. *Maximum and expected sample sizes are expressed as percentages of the corresponding fixed sample size. All tests attain Type I error probability* $\alpha = 0.025$ *and, when the futility boundary is applied rigidly, power* $1 - \beta = 0.9$ *at* $\theta = \delta$.

		$\alpha = 0.025,\ 1 - \beta = 0.9$				
		Sample size as percentage of fixed sample size				
	K	*Maximum sample size*	*Expected sample size at* $\theta =$ 0	0.5δ	δ	1.5δ
$\rho = 2$						
	1	100.0	100.0	100.0	100.0	100.0
	2	105.5	70.7	89.2	80.8	60.7
	3	109.3	64.4	84.8	75.3	53.5
	4	111.7	61.1	82.4	72.4	50.3
	5	113.3	59.2	80.9	70.6	48.2
	10	116.9	55.2	77.9	67.1	43.9
	15	118.3	53.9	76.9	65.9	42.6
	20	118.9	53.3	76.3	65.3	41.9
$\rho = 3$						
	1	100.0	100.0	100.0	100.0	100.0
	2	102.0	74.8	92.0	84.1	62.7
	3	104.1	68.8	87.2	78.2	57.3
	4	105.6	65.4	84.7	75.1	54.0
	5	106.8	63.3	83.2	73.3	51.8
	10	109.6	59.4	80.2	69.7	47.5
	15	110.7	58.1	79.2	68.5	46.1
	20	111.3	57.5	78.6	68.0	45.5

spending function when only Z_1 and Z_3 are observed. Thus, the overall test is as follows:

If $|Z_1| \geq 2.394$, reject H_0.

If $1.2 \leq |Z_1| < 2.394$, continue and observe Z_2 then

- if $|Z_2| \geq 2.294$, reject H_0
- if $|Z_2| < 2.294$, observe Z_3 then
  - if $|Z_3| \geq 2.200$, reject H_0
  - if $|Z_3| < 2.200$ accept H_0.

If $|Z_1| < 1.2$, take both the second and third groups of observations then

- if $|Z_3| \geq 2.076$ reject H_0
- if $|Z_3| < 2.076$ accept H_0.

Jennison & Turnbull (1991a) use numerical integration to show that this test has Type I error probability 0.0525, a small increase over the intended value of 0.05. A similar four-stage procedure in which groups 2 and 3 are pooled if $|Z_1| < 1.2$ and groups 3 and 4 are pooled if $|Z_2| < 1.2$ has Type I error probability 0.0522. Although these effects are small, they do demonstrate the possibility of increases in Type I error probability and suggest a need for caution in the application of error spending tests.

Lan & DeMets (1989b) use simulation to investigate the effects of increasing the frequency of future analyses when Z_k starts to approach an error spending boundary. They report no serious distortion of Type I error rate or power, although perturbations similar to those in Jennison & Turnbull's (1991a) examples cannot be ruled out as these are of the same order as the standard errors of the simulation results.

Lan & DeMets (1989b) go on to suggest that a monitoring committee may decide to switch from group sequential to continuous monitoring if it appears at an interim analysis that a testing boundary is likely to be crossed soon. Betensky (1998) has addressed the computational problems involved in implementing this suggestion but the potential benefits of such a strategy are unclear. Comparisons with the examples of Jennison & Turnbull (1991a) and Lan & DeMets (1989b) suggest this practice will produce a slightly elevated Type I error and little change in power (one would expect an increase in power to accompany the higher Type I error rate, but this will be offset by the decrease in power due to spending some Type I error probability earlier). The switch to continuous monitoring will not automatically yield useful gains in terms of earlier stopping: such gains diminish as additional analyses are added, and the results of Tables 8.2, 8.3, 8.7, 8.8, 8.9, 8.11 and 8.13 show only slight reductions in expected sample size as the number of planned interim analyses is increased beyond 5 or 10.

Proschan, Follman & Waclawiw (1992) present a thorough study of a variety of group sequential designs when analysis times are allowed to depend on observed responses. They conclude that the Type I errors of error spending tests are robust to departures from the specified protocol, as long as error is spent as a smooth function of information or calendar time. In other situations, they observe that Type I error can be inflated by a factor of two or more when rather devious strategies are followed in implementing tests based on the Slud & Wei (1982) method or tests defined through discontinuous error spending functions; see also Proschan & Hunsberger (1995), Graf & Bauer (2011) and Wang et al. (2013).

Fleming & DeMets (1993) note the importance of providing a full and clear description of any form of error spending test in the study protocol in order to ensure credibility of a study's conclusions. They cite the extreme form of misuse in which the error spending function is changed during a study so that all the two-sided Type I error is spent the first time $|Z_k|$ exceeds $z_{\alpha/2}$, leading to a gross inflation of the Type I error rate comparable to that of repeated fixed sample tests at significance level α discussed in Chapters 1 and 2.

Fleming, Harrington & O'Brien (1984) observe that it may be desirable to extend a study if information accrues too slowly and suggest spending fractions of the Type I error probability at the extra interim analyses. They reiterate the need to keep any decision to change the study duration independent of the observed responses: cautious experimenters might even avoid legitimate changes in design if there is a perceived danger of compromising their study's credibility. Interestingly, the maximum information designs of Section 8.2.2 share the same motivation as Fleming, Harrington & O'Brien's proposal but, since the rules for continuing to a target information level are specified at the outset, the suggestion that the design is being changed as the study progresses does not arise. If one of the usual forms of error spending function is used, such a maximum information design falls into the category of procedures recommended by Proschan, Follman & Waclawiw (1992) for their robustness to any misuse.

8.6 Information Monitoring to Accommodate Nuisance Parameters

8.6.1 The General Method

A maximum information, error spending trial design, as described in Section 8.2.2, is well suited to the situation where information levels $\mathcal{I}_1$, $\mathcal{I}_2$, ... are not known precisely due to their dependence on one or more nuisance parameters. These nuisance parameters can be estimated based on the accrued data at each analysis k, and then an estimate $\hat{\mathcal{I}}_k$ of the information is available. The target information level $\mathcal{I}_{max}$ is fixed in advance by the Type I and Type II error probabilities, the form of the error spending function and the anticipated pattern of information levels, and so does not depend on the values of the nuisance parameters.

As the trial proceeds, the error to be spent at each stage is calculated as previously described but with each $\mathcal{I}_k$ replaced by $\hat{\mathcal{I}}_k$ in the relevant equation, namely Equation (8.2) for a two-sided test, Equations (8.10) and (8.11) for a one-sided test, and Equation (8.16) for a one-sided test with a non-binding futility boundary. Just as in Section 8.2.2, the trial stops at any stage k where the estimated information $\hat{\mathcal{I}}_k$ exceeds $\mathcal{I}_{max}$. In this regard, the method is in the same spirit as the fully sequential procedure of Chow & Robbins (1965) for finding a fixed width confidence interval for a normal mean when the variance is unknown, where stopping occurs when the estimated standard error of $\hat{\theta}$ falls below a given threshold.

Nuisance parameters occur naturally in a number of applications discussed elsewhere in this book. These include longitudinal data with unknown correlation parameters (Section 3.3.4), responses that follow a generalized linear model (Section 15.6), and survival data modeled by a Cox regression model (Section 17.4). Here, we give detailed examples of two simpler situations: the comparison of two normal means when the variance is unknown and the analysis of two dichotomous variables with unknown baseline response rate.

8.6.2 Information Monitoring for Normal Responses with Unknown Variance

8.6.2.1 Applying the Significance Level Approach to t-Statistics

Consider the parallel two-treatment comparison of Section 3.2.1 in which responses $X_{Ai} \sim N(\mu_A, \sigma_A^2)$ and $X_{Bi} \sim N(\mu_B, \sigma_B^2)$, $i = 1, 2, \ldots$, are recorded on subjects receiving treatments A and B, respectively. Suppose we assume a common variance, $\sigma_A^2 = \sigma_B^2 = \sigma^2$. Then with n_A observations on treatment A and n_B on treatment B, the information for $\theta = \mu_A - \mu_B$ is

$$\mathcal{I} = \frac{1}{Var(\hat{\theta})} = \frac{1}{Var(\bar{X}_A - \bar{X}_B)} = \left\{ \frac{\sigma^2}{n_A} + \frac{\sigma^2}{n_B} \right\}^{-1}. \qquad (8.20)$$

As seen in Equation (4.6), a fixed sample, two-sided test of H_0: $\theta = 0$ against $\theta \neq 0$ with Type I error rate α and power $1 - \beta$ at $\theta = \pm\delta$ needs information $\mathcal{I}_{f,2} = (z_{\alpha/2} + z_\beta)^2/\delta^2$. A group sequential test requires maximum information $\mathcal{I}_{max} = R\mathcal{I}_{f,2}$, where the inflation factor R is determined by the error spending function and the number of anticipated analyses; see, for example, Table 8.1.

Suppose there are n_{Ak} and n_{Bk} observations on treatments A and B, respectively, at analysis k. Let

$$\bar{X}_{Ak} = \frac{1}{n_{Ak}} \sum_{i=1}^{n_{Ak}} X_{Ai}, \quad \bar{X}_{Bk} = \frac{1}{n_{Bk}} \sum_{i=1}^{n_{Bk}} X_{Bi}$$

and

$$s_k^2 = \frac{\sum_{i=1}^{n_{Ak}} (X_{Ai} - \bar{X}_{Ak})^2 + \sum_{i=1}^{n_{Bk}} (X_{Bi} - \bar{X}_{Bk})^2}{n_{Ak} + n_{Bk} - 2}. \qquad (8.21)$$

Then the estimated information for θ at analysis k is

$$\widehat{\mathcal{I}}_k = \left\{ \frac{s_k^2}{n_{Ak}} + \frac{s_k^2}{n_{Bk}} \right\}^{-1}.$$

The investigators can monitor the estimated information at interim analyses and modify recruitment to increase the chance that the target $\mathcal{I}_{max} = R\,\mathcal{I}_{f,2}$ will be reached if there is no early stopping.

We have presented calculations for error spending tests based on a sequence of normally distributed statistics $\{Z_1, \ldots, Z_K\}$. We use these same calculations here but we interpret the upper and lower boundaries at analysis k as *significance levels* for Z_k and apply these significance levels to the t-statistic

$$T_k = \frac{\bar{X}_{Ak} - \bar{X}_{Bk}}{s_k \sqrt{n_{Ak}^{-1} + n_{Bk}^{-1}}}. \tag{8.22}$$

This is analogous to the implementation of group sequential t-tests described in Section 4.6. We illustrate this process in the following example.

8.6.2.2 Example 2: Comparison of Two Normal Means when Variance is Unknown

Consider again Example 1 of Section 8.2.2. Responses X_{Ai} and X_{Bi} $i = 1, 2, \ldots,$ are recorded on subjects receiving treatments A and B, respectively. We assume these responses are normally distributed with means μ_A and μ_B, respectively and common variance σ^2. However, we now treat σ^2 as an unknown nuisance parameter. We wish to test H_0: $\mu_A = \mu_B$ against $\mu_A \neq \mu_B$ with two-sided Type I error probability $\alpha = 0.05$ and power $1 - \beta = 0.9$ at $\theta = \mu_A - \mu_B = \pm 1$.

Suppose it is decided to conduct this test group sequentially using the error spending function $f(t) = \min\{\alpha t^2, \alpha\}$, from the Rho family defined in (8.6) with $\rho = 2$, and we intend to have a maximum of $K = 10$ analyses. The target information level is calculated as $\mathcal{I}_{max} = R_{LD}(10, 0.05, 0.1, 2) \times \mathcal{I}_{f,2} = 1.075 \times 10.51 = 11.30$.

We shall describe the calculations for a hypothetical trial as it evolves stage by stage. Table 8.14 presents a worksheet summarizing these calculations

Start of study: Stage $k = 0$

Suppose an initial estimate of the variance σ^2 is $s_0^2 = 4.0$. We assume there will be equal sample sizes on the two treatment arms so $n_{Ak} = n_{Bk} = n_k$, say, at each stage k. Setting $\mathcal{I}_{max} = 11.30 = n/(2s_0^2)$ leads to an initial target maximum sample size $\hat{n}_{max}^{(0)} = 90.4$, which we round up to 91 per arm. (This is the same calculation as in Example 1.) Since we are planning a study with $K = 10$ analyses, we decide to conduct the first analysis after $m_1 = 10$ observations have been taken on each arm.

Interim analysis 1

At the first interim analysis with $n_{A1} = n_{B1} = n_1 = m_1 = 10$, we observe $\hat{\theta}_1 = \bar{X}_{A1} - \bar{X}_{B1} = 1.16$ and the pooled variance estimate is $s_1^2 = 4.5$, somewhat higher than the initial guess of $s_0^2 = 4.0$. We estimate the information for θ at this analysis as $\hat{\mathcal{I}}_1 = n_1/(2\,s_1^2) = 1.11$. Substituting this value into Equation (8.2), we calculate the Type I error probability to be spent at this look as $\pi_1 = f(\hat{\mathcal{I}}_1/\mathcal{I}_{max}) = 0.05(1.11/11.30)^2 = 0.00048$. Thus the boundary at stage 1 is obtained by solving Equation (8.1) with $k = 1$. This yields boundary values $\pm c_1$ on the Z-scale where $c_1 = \Phi^{-1}(1 - 0.00024) = 3.49$. The t-statistic given by (8.22) is $T_1 = 1.16/\sqrt{\{2 \times 4.5/10\}} = 1.22$ with $\nu_1 = 10 + 10 - 2 = 18$ degrees of freedom. We compare the value of $|T_1|$ with the critical value $t_{18,1-\Phi(3.49)} = 4.25$ and, since this threshold is not exceeded, the study continues to the next stage.

Table 8.14 *Example 2 Worksheet: Information monitoring of normal responses with unknown variance*

Analysis k	$k=0$	$k=1$	$k=2$	$k=3$	$k=4$
Interim data					
Group size per arm, m_k	—	10	11	10	9
Cumulative sample size, n_k	0	10	21	31	40
$\bar{X}_{Ak} - \bar{X}_{Bk}$	—	1.16	1.39	1.48	1.52
Pooled variance estimate, s_k^2	4.0	4.5	4.4	4.1	4.6
Interim analyses					
$\nu_k = 2n_k - 2$	—	18	40	60	78
Estimated information, $\hat{\mathcal{I}}_k$	0	1.11	2.39	3.78	4.35
$T_k = (\bar{X}_{Ak} - \bar{X}_{Bk})\sqrt{\hat{\mathcal{I}}_k}$	—	1.22	2.15	2.88	3.17
Error spent at analysis k, π_k	—	0.00048	0.00175	0.00337	0.00181
Boundary for $\|Z_k\|$, c_k	—	3.49	3.11	2.87	2.86
Boundary for $\|T_k\|$, $t_{\nu_k,1-\Phi(c_k)}$	—	4.25	3.33	2.97	3.01
Decision	—	Continue	Continue	Continue	Stop, Reject H_0
Next group size:					
$\hat{n}_{max}^{(k)} = 2s_k^2\,\mathcal{I}_{max}$	90.4	101.7	99.4	92.6	—
$m_{k+1} = (\hat{n}_{max}^{(k)} - n_k)/(K-k)$	10	11	10	9	—

With our new estimate $s_1^2 = 4.5$ we re-estimate the sample size needed to achieve information $\mathcal{I}_{max} = 11.30$. This gives a new target of $\hat{n}_{max}^{(1)} = 2\,s_1^2\,\mathcal{I}_{max} = 101.7$ per arm. Since we already have 10 observations per arm and 9 further analyses are planned, we take the next group size to be $m_2 = (101.7 - 10)/9 = 10.2$, which we round up to $m_2 = 11$.

Interim analysis 2

At the second analysis with $n_{A2} = n_{B2} = n_2 = m_1 + m_2 = 21$, we observe $\hat{\theta}_2 = \bar{X}_{A2} - \bar{X}_{B2} = 1.39$ and the pooled estimate of the variance is $s_2^2 = 4.4$. The estimated information for θ is now $\hat{\mathcal{I}}_2 = n_2/(2s_2^2) = 2.39$. From Equation (8.2), we calculate the Type I error probability to be spent at this analysis as $\pi_2 = 0.05(2.39/11.30)^2 - \pi_1 = 0.00175$. We obtain c_2 by solving (8.1) with $k = 2$ and the value $c_1 = 3.49$ computed at analysis 1. Using numerical integration (see Section 8.7) the boundary value $c_2 = 3.11$ is obtained. The t-statistic given by (8.22) is $T_2 = 1.39/\sqrt{\{2 \times 4.4/21\}} = 2.15$ with degrees of freedom $\nu_2 = 21 + 21 - 2 = 40$. We compare the value of $|T_2|$ with the critical value $t_{40,1-\Phi(3.11)} = 3.33$ and, since this threshold is not exceeded, the study continues to the third stage.

With our new estimate $s_2^2 = 4.4$ we re-estimate the sample size needed to achieve information $\mathcal{I}_{max} = 11.30$ and obtain the new target $\hat{n}_{max}^{(2)} = 2\,s_2^2\,\mathcal{I}_{max} = 99.4$ per arm. Since we already have 21 observations per arm and 8 further analyses are planned, we take the next group size to be $m_3 = (99.4 - 21)/8 = 9.8$, which we round up to $m_3 = 10$.

Interim analysis 3

At the third analysis with $n_{A3} = n_{B3} = n_3 = 31$, we observe $\hat{\theta} = \bar{X}_{A3} - \bar{X}_{B3} = 1.48$ and the pooled estimate of the variance is $s_3^2 = 4.1$. The estimated information for θ is now $\hat{\mathcal{I}}_3 = n_3/(2s_3^2) = 3.78$. From Equation (8.2), we calculate the Type I error probability to be spent at this analysis as $\pi_3 = 0.05(3.78/11.30)^2 - (\pi_1 + \pi_2) = 0.00337$. Using the values $c_1 = 3.49$ and $c_2 = 3.11$, we solve (8.1) with $k = 3$ to obtain the boundary value $c_3 = 2.87$. The t-statistic given by (8.22) is $T_3 = 1.48/\surd\{2 \times 4.1/31\} = 2.88$ with degrees of freedom $\nu_3 = 60$. We compare the value of $|T_3|$ with the critical value $t_{60,1-\Phi(2.87)} = 2.97$ and, since this threshold is not exceeded, the study continues to the fourth stage.

With our new estimate $s_3^2 = 4.1$ we re-estimate the sample size needed to achieve information $\mathcal{I}_{max} = 11.30$ and obtain the new target of $\hat{n}_{max}^{(3)} = 2\, s_3^2\, \mathcal{I}_{max} = 92.6$ per arm. Since we already have 31 observations per arm and 7 further analyses are planned, we take the next group size to be $m_4 = (92.6 - 31)/7 = 8.8$, which we round up to $m_4 = 9$.

Interim analysis 4

At the fourth analysis with $n_{A4} = n_{B4} = n_4 = 40$, we observe $\hat{\theta} = \bar{X}_{A4} - \bar{X}_{B4} = 1.52$ and the pooled estimate of the variance is $s_4^2 = 4.6$. The estimated information for θ is now $\hat{\mathcal{I}}_4 = n_4/(2s_4^2) = 4.35$. From Equation (8.2), we calculate the Type I error probability to be spent at this analysis as $\pi_4 = 0.05(4.35/11.30)^2 - (\pi_1 + \pi_2 + \pi_3) = 0.00181$. Using the values $c_1 = 3.49$, $c_2 = 3.11$ and $c_3 = 2.87$, we solve (8.1) with $k = 4$ to obtain the boundary value $c_4 = 2.86$. The t-statistic given by (8.22) is $T_4 = 1.52/\surd\{2 \times 4.6/40\} = 3.17$ with degrees of freedom $\nu_4 = 78$. We compare the value of $|T_4|$ with the critical value $t_{78,1-\Phi(2.86)} = 3.01$ and, since this threshold is exceeded, the study terminates with rejection of H_0.

Remarks on this example

This form of "information monitoring" design was introduced by Mehta & Tsiatis (2001). These authors compared the t-statistics T_k with boundary values c_k based on Z-statistics rather than converting them to the significance levels $t_{\nu_k,1-\Phi(c_k)}$ of the t distribution. While such an approach may be justified when sample sizes are large and t-statistics are approximately normal, in simulation studies with moderately large sample sizes we have found it is important to follow the significance level approach to avoid inflating the Type I error rate.

Our simulations also showed that re-estimating sample size based on estimates of σ^2 leads to a downward bias in the final variance estimate. This is because over-estimates of σ^2 lead to larger future sample sizes which are likely to reduce the final estimate of σ^2, while there is less chance to correct under-estimates of σ^2 that generate smaller future sample sizes. Since the estimated variance appears in the denominator of the t-statistic, downward bias in this estimate can produce some inflation of the Type I error rate. We shall return to this issue in Chapter 18 and in Section 23.3 we shall describe how to avoid this source of Type I error rate inflation altogether by using a "combination test" to combine data across stages after updating sample size.

The amounts of Type I error probability to be spent at each analysis depend on the sequence of estimated information levels $\hat{\mathcal{I}}_1, \hat{\mathcal{I}}_2, \ldots$, which, in turn, depend on $s_1^2, s_2^2, \ldots$. Questions may therefore arise when solving Equation (8.1) since at analysis k, it would be possible to update estimates $\hat{\mathcal{I}}_1, \ldots, \hat{\mathcal{I}}_{k-1}$ using the current variance estimate s_k^2. However, since analyses 1 to $k-1$ have already occurred, it is too late to change the Type I error probability assigned to these analyses. Furthermore, since the statistics $Z_1, \ldots, Z_k$ in Equation (8.1) are standardized, their joint distribution under $\theta = 0$ does not depend on σ^2, and Equation (8.1) can be solved using the covariances

$$\mathrm{Cov}(Z_{k_1}, Z_{k_2}) = \surd(\mathcal{I}_{k_1}/\mathcal{I}_{k_2}) = \surd(n_{k_1}/n_{k_2}), \quad k_1 < k_2,$$

which are known precisely.

We discovered in our simulations that the estimated information level may decrease as more responses are observed, so $\hat{\mathcal{I}}_k < \hat{\mathcal{I}}_{k-1}$ for some k. This is a consequence of the high variance of the χ-squared distribution. As an example, if analyses 1 and 2 have cumulative sample sizes of $n_1 = 10$ and $n_2 = 20$ per treatment arm, there is a probability of 0.013 that $\hat{\mathcal{I}}_2 < \hat{\mathcal{I}}_1$. A pragmatic solution is to stipulate that the error spending rule allocates no Type I error probability to any analysis k at which $\hat{\mathcal{I}}_k < \hat{\mathcal{I}}_{k-1}$ and to continue automatically to the next analysis. If a trial is designed with a fixed total number of analyses K and $\hat{\mathcal{I}}_K < \hat{\mathcal{I}}_{K-1}$, the method described in Section 8.2.2.5 should be followed so all the remaining error probability $\alpha - f(\hat{\mathcal{I}}_{K-1})$ is spent at analysis K using the fact that $\mathrm{Cov}(Z_{K-1}, Z_K) = \sqrt{(\mathcal{I}_{K-1}/\mathcal{I}_K)} = \sqrt{(n_{K-1}/n_K)}$.

The rule for choosing the next group size after re-estimation of σ^2 and $\mathcal{I}_{max}$ does not have to be exactly as described in our example. One might, for example, postpone any adjustment of sample size until later analyses when the estimate s_k^2 is less variable. However, for the reasons discussed in Section 8.5, it is essential that the choice of group size does not depend on the estimated treatment effect $\hat{\theta}_k$.

It is not necessary to assume a common variance for responses on treatments A and B. If the variances σ_A^2 and σ_B^2 are assumed to be unequal, they can be estimated separately by the sample variances s_{Ak}^2 and s_{Bk}^2 at analysis k. Information is then estimated as $\hat{\mathcal{I}}_k = (s_{Ak}^2/n_{Ak} + s_{Bk}^2/n_{Bk})^{-1}$, and T_k can be defined as the usual non-pooled t-statistic with degrees of freedom computed using the Welch-Satterthwaite approximation. Also, the sample sizes on the two arms do not need to be equal: the numbers at each stage on each arm can be chosen arbitrarily, again subject to the provisos of Section 8.5.

8.6.3 Information Monitoring for Comparing Two Treatments with Binary Responses

8.6.3.1 Example 3: One-sided Test of Superiority for a Two sample Binary Problem

Consider a trial comparing a novel intervention, treatment B, to the standard practice, treatment A, in which the primary endpoint is re-admission to hospital or death within 30 days. Let p_A and p_B denote the 30-day re-admission probabilities for treatments A and B, respectively, and define the treatment effect as $\theta = p_A - p_B$. Suppose the historical re-admission rate for the standard treatment is known to be 0.25 and, in order to establish superiority, we wish to test H_0: $\theta \leq 0$ against the alternative $\theta > 0$. We specify the Type I error probability $\alpha = 0.025$ and aim to achieve power $1 - \beta = 0.9$ when $\theta = \delta = 0.05$, so the new treatment reduces the re-admission rate by 5%. From (5.1), a fixed sample test requires information

$$\mathcal{I}_{f,1} = \{\Phi^{-1}(1-\alpha) + \Phi^{-1}(1-\beta)\}^2/\delta^2 = 4203.0. \tag{8.23}$$

With equal allocation and sample sizes $n_A = n_B = n$ on the two treatment arms we have

$$\mathcal{I}_{f,1} = \{Var(\hat{\theta})\})^{-1} = \left(\frac{p_A(1-p_A)}{n} + \frac{p_B(1-p_B)}{n}\right)^{-1}. \tag{8.24}$$

Substituting $p_A = 0.25$ and $p_B = 0.2$ into (8.24), we see that a sample size of $n = 1461$ per treatment arm is required to achieve $\mathcal{I}_{f,1} = 4203.0$ in a fixed sample design.

Suppose we decide to conduct a one-sided test using Rho family error spending functions as defined in (8.15) with $\rho_1 = \rho_2 = 3$ and we plan to observe subjects in $K = 5$ equally sized groups. Taking the inflation factor $R_{OS}(5, 0.025, 0.1, 3) = 1.049$ from Table 8.6, we find the target maximum information level is

$$\mathcal{I}_{max} = R_{OS}(5, 0.025, 0.1, 3) \times \mathcal{I}_{f,1} = 1.049 \times 4203.0 = 4409.9, \tag{8.25}$$

which is achieved by a sample size of $n = 1533$ per treatment arm when $p_A = 0.25$ and $p_B = 0.2$.

If we were to proceed under the assumption that $p_A = 0.25$, we could run the trial with five groups of 307 patients on each treatment. Then, solving Equations (8.12) and (8.13)

Table 8.15 *Example 3 Worksheet: Information monitoring with binary responses*

Analysis k	$k=1$		$k=2$		$k=3$	
Interim data						
Treatment arm	A	B	A	B	A	B
Sample sizes, n_{Ak} and n_{Bk}	310	306	612	602	955	950
Re-admissions, y_{Ak} and y_{Bk}	73	70	152	147	240	232
Estimated rates, $\hat{p}_{Ak}$ and $\hat{p}_{Bk}$	0.2355	0.2288	0.2484	0.2442	0.2513	0.2442
Interim analyses						
$\hat{\theta}_k = \hat{p}_{Ak} - \hat{p}_{Bk}$	0.0067		0.0042		0.0071	
Standard error of $\hat{\theta}_k$	0.0340		0.0247		0.0198	
$\hat{\mathcal{I}}_k = 1/\text{Var}(\hat{\theta}_k)$	864.1		1635.0		2555.6	
Test statistic, $Z_k = \hat{\theta}_k \sqrt{\hat{\mathcal{I}}_k}$	0.198		0.169		0.359	
$t_k = \hat{\mathcal{I}}_k/\mathcal{I}_{max}$	0.196		0.371		0.580	
Cumulative Type I error, $\alpha\, t_k^3$	0.00019		0.00127		0.00487	
Cumulative Type II error, $\beta\, t_k^3$	0.00075		0.00510		0.01946	
Lower boundary value, a_k	−1.704		−0.577		0.417	
Upper boundary value, b_k	3.556		3.048		2.636	
Decision	Continue		Continue		Stop for futility	

with $\delta = 0.05$, values of $\pi_{1,k}$ from (8.10) with $f(t) = \alpha\, t^3$, and values of $\pi_{2,k}$ from (8.11) with $g(t) = \beta\, t^3$ yields boundaries

$$(a_1, a_2, a_3, a_4, a_5) = (-1.671, -0.415, 0.501, 1.275, 2.012),$$

$$(b_1, b_2, b_3, b_4, b_5) = (3.540, 2.974, 2.605, 2.306, 2.012). \tag{8.26}$$

Note that the choice of $\mathcal{I}_{max}$ results in convergence of the two boundaries with $a_5 = b_5$.

In reality, the re-admission rate on treatment A may differ from the historical value of 0.25. Furthermore, the numbers of patients in each group may differ from the planned values. We therefore estimate the information level $\mathcal{I}_k$ at analysis k as

$$\hat{\mathcal{I}}_k = \left(\frac{\hat{p}_{Ak}(1-\hat{p}_{Ak})}{n_{Ak}} + \frac{\hat{p}_{Bk}(1-\hat{p}_{Bk})}{n_{Bk}} \right)^{-1},$$

where the estimates $\hat{p}_{Ak}$ and $\hat{p}_{Bk}$ are based on data from cumulative sample sizes n_{Ak} and n_{Bk} on treatments A and B. The actual boundary values depend on these observed information levels. We shall describe the calculations for a hypothetical study as it evolves stage by stage. A worksheet summarizing these calculations is given in Table 8.15.

Start of study: Stage k=0.

As described above, we start by assuming the historical re-admission rate $p_A = 0.25$ for the standard treatment and assume the desired improvement of $\theta = 0.05$ so the re-admission rate for the new treatment is $p_B = 0.2$. Hence, our initial estimate of the necessary maximum sample size is

$$4409.9 \times (0.2 \times 0.8 + 0.25 \times 0.75) = 1532.4$$

subjects per arm. Since we are planning for five equally sized groups, we aim to perform the first analysis after approximately 307 subjects have been observed on each arm.

Interim analysis 1

At the first analysis, suppose we observe $y_{A1} = 73$ re-admissions out of $n_{A1} = 310$ subjects on treatment A, while on treatment B we observe $y_{B1} = 70$ re-admissions out of $n_{B1} = 306$ subjects. Thus $\hat{p}_{A1} = 0.2355, \hat{p}_{B1} = 0.2288$ and $\hat{\theta}_1 = \hat{p}_{A1} - \hat{p}_{B1} = 0.0067$. The estimated variance of $\hat{\theta}_1$ is

$$\hat{p}_{A1}(1 - \hat{p}_{A1})/n_{A1} + \hat{p}_{B1}(1 - \hat{p}_{B1})/n_{B1} = 0.001157.$$

We use the reciprocal of this quantity as our estimated information, $\hat{\mathcal{I}}_1 = 864.1$, and the information fraction at analysis 1 is $t_1 = 864.1/4409.9 = 0.196$. Applying the stipulated error spending functions $f(t) = 0.025 \min\{t^3, 1\}$ and $g(t) = 0.1 \min\{t^3, 1\}$, the Type I and Type II error probabilities to be spent at this analysis are $\pi_{1,1} = 0.025 \times 0.196^3 = 0.00019$ and $\pi_{2,1} = 0.1 \times 0.196^3 = 0.00075$, respectively.

It is straightforward to solve (8.12) and (8.13) to obtain the pair of boundary values $a_1 = 0.05 \times \sqrt{864.1} - \Phi^{-1}(1 - 0.00075) = -1.704$ and $b_1 = 1 - \Phi^{-1}(1 - 0.00019) = 3.556$. Finally we compute the test statistic

$$Z_1 = \hat{\theta}_1/\sqrt{\mathrm{Var}(\hat{\theta}_1)} = 0.0067/0.0340 = 0.198.$$

Since Z_1 lies between the upper and lower boundaries, the trial continues. We might at this stage re-estimate the sample size based on the updated estimates of p_A and p_B. However $\hat{p}_{A1}$ and $\hat{p}_{B1}$ are close to the values assumed initially and have quite large standard errors, so we choose to continue with the goal of 307 subjects per arm in the second stage.

Interim analysis 2

At the second analysis, suppose we observe cumulative sample sizes $n_{A2} = 612$ and $n_{B2} = 602$ with cumulative re-admissions $y_{A2} = 152$ and $y_{B2} = 147$ on treatment arms A and B. Following the same sequence of calculations as at analysis 1, we obtain $\hat{\theta}_2 = (152/612) - (147/602) = 0.0042$ and $\hat{\mathcal{I}}_2 = 1/\mathrm{Var}(\hat{\theta}_2) = 1635.0$, so the information fraction is $t_2 = 1635.0/4409.9 = 0.371$.

The Type I error probability to be spent at this analysis is

$$\pi_{1,2} = 0.025 \times (0.371^3 - 0.196^3) = 0.00109$$

and the Type II error probability to be spent is

$$\pi_{2,2} = 0.1 \times (0.371^3 - 0.196^3) = 0.00434.$$

Solving (8.12) and (8.13) for $k = 2$ with $a_1 = -1.704$, $b_1 = 3.556$ and these values of $\pi_{1,2}$ and $\pi_{2,2}$ gives the boundary values $(a_2, b_2) = (-0.577, 3.048)$. Since the test statistic $Z_2 = 0.0042/0.0247 = 0.169$ lies between these two values, the trial continues.

At this point, we note that the information fraction $t_2 = 0.371$ is below the value of 0.4 that was intended after 2 out of 5 analyses. Based on the current estimates of p_A and p_B, a revised estimate of the necessary maximum sample size is

$$4409.9 \times (0.2482 \times 0.7518 + 0.2442 \times 0.7558) = 1637.1$$

subjects per arm. This will be achieved if the remaining three groups contain 344 subjects on each treatment arm, so recruitment is increased accordingly.

Interim analysis 3

At the third analysis, suppose we observe cumulative sample sizes $n_{A3} = 955$ and $n_{B3} = 950$ with cumulative re-admissions $y_{A3} = 240$ and $y_{B3} = 232$ on arms A and B. Following the same sequence of calculations as before, we obtain $\hat{\theta}_3 = (240/955) - (232/950) = 0.0071$ and $\hat{\mathcal{I}}_3 = 1/\mathrm{Var}(\hat{\theta}_3) = 2555.6$, so the information fraction is $t_3 = 2555.6/4409.9 = 0.580$.

The Type I error probability to be spent at this analysis is

$$\pi_{1,3} = 0.025 \times (0.580^3 - 0.371^3) = 0.00359$$

and the Type II error probability to be spent is

$$\pi_{2,3} = 0.1 \times (0.580^3 - 0.371^3) = 0.01436.$$

Solving (8.12) and (8.13) for $k = 3$ with existing values of (a_1, b_1) and (a_2, b_2) and these values of $\pi_{1,3}$ and $\pi_{2,3}$ gives the boundary values $(a_3, b_3) = (0.417, 2.636)$. We see that the test statistic $Z_3 = 0.0071/0.0198 = 0.359$ has now fallen below the lower boundary and the study stops for futility, having failed to reject H_0.

Remarks on this example

It may be that the new treatment is observed to have certain beneficial features compared to the standard treatment, such as fewer adverse events, shorter hospital stays, improved quality of life, or lower cost. As a consequence, the trial's monitoring committee would like to see the trial continue past analysis 3. Since the observed re-admission rate for the new treatment is slightly lower than for the standard (although not to a practically or statistically significant degree), there is no ethical impediment to continuing the study. However, if the trial continues and the upper boundary is crossed at a later analysis, it will not be possible to reject H_0: $\theta \leq 0$ and claim a positive result. This is because the Type I error rate was calculated assuming a binding lower boundary and allowing rejection of H_0 after crossing this futility boundary would inflate the Type I error probability above $\alpha = 0.025$.

With hindsight, the sponsor should perhaps have considered a group sequential design with a non-binding futility boundary since then it would have been possible to continue the study beyond analysis 3 and subsequently reject H_0 if later results supported such a conclusion. We describe a one-sided test with a non-binding lower boundary in Example 4 below. Another option would be to design the trial with the objective of demonstrating non-inferiority: we shall describe such a design in Example 5.

8.6.3.2 *Example 4: Two-sample Binary Problem with a Non-binding Futility Boundary*

As in Example 3, we consider a clinical trial with a group sequential design which will test H_0: $\theta \leq 0$ against $\theta > 0$ with Type I error rate $\alpha = 0.025$ and power $1-\beta = 0.9$ at $\theta = 0.05$. However, we now assume the lower futility boundary is to be treated as non-binding. This may be because the sponsor wishes the futility boundary to be considered as a guideline and to have the capability to make a valid positive decision if the upper efficacy boundary is exceeded after first crossing the lower boundary. Alternatively, a regulatory body may insist that the futility boundary be treated as non-binding to avoid any possibility of Type I error inflation.

No futility boundary at all

We follow the approach described in Section 8.4 to define an appropriate error spending design. Suppose first of all that no futility boundary is specified and we spend Type I error probability according to the Rho family function with $\rho_1 = 3$. Then, the upper boundary values, $b_1, b_2, \ldots$, are computed by solving recursively the set of Equations (8.17) with error probabilities $\pi_{1,k}$ as defined in (8.16). If a maximum of $K = 5$ analyses are conducted with equal group sizes, we find these boundary values to be

$$(b_1, b_2, b_3, b_4, b_5) = (3.540, 2.974, 2.605, 2.306, 2.045). \qquad (8.27)$$

Although the values of b_2 to b_5 are, by definition, higher than those in Example 3, we see that values of b_2, b_3 and b_4 agree to 3 decimal places and the only difference is in b_5, where we now have $b_5 = 2.045$ compared to $b_5 = 2.012$ in Example 3. If power 0.9 is to be achieved when $\theta = 0.05$, the target for the maximum information is

$$R_{NoFB}(5, 0.025, 0.1, 3) \times \mathcal{I}_{f,1} = 1.030 \times 4203.0 = 4327.9,$$

where the value of $R_{NoFB}(5, 0.025, 0.1, 3)$ is taken from Table 8.10.

Non-binding futility boundary

Since power will decrease if early stopping for futility may possibly occur, we recommend specifying a non-binding futility boundary and increasing the sample size accordingly. In designs with an additional non-binding futility boundary, the upper boundary is calculated as before and critical values on the Z-scale remain as in (8.27) in the case of equal group sizes. With an error spending function specified for the Type II error, the lower boundaries are found by solving the set of Equations (8.19) recursively for $k = 1, 2, \dots$. Suppose we choose a Rho family error spending function for Type II error with $\rho_2 = 3$, then in order to achieve power $1 - \beta = 0.9$ at $\theta = 0.05$ if the futility boundary is actually adhered to, the target for the maximum information is

$$R_{NBF}(5, 0.025, 0.1, 3) \times \mathcal{I}_{f,1} = 1.068 \times 4203.0 = 4487.3,$$

where the value of $R_{NBF}(5, 0.025, 0.1, 3)$ is taken from Table 8.12.

In practice, the boundary or boundaries for both the above cases will be created by applying error spending functions to the observed information levels. The Type I error rate will be protected, conservatively in the case of a non-binding futility boundary, and power will vary around the target value, depending on the observed information levels.

8.6.3.3 *Example 5: Non-inferiority Design for the Two-sample Binary Problem*

Consider the situation described in Examples 3 and 4 but suppose now that the goal is to demonstrate non-inferiority of the new treatment B relative to the standard treatment A. The motivation for this may be that the new treatment is less expensive or easier to administer, or has fewer side-effects than the current treatment. While the new treatment is not expected to reduce the re-admission rate, it is hoped to match the rate for the standard treatment.

Following the methods of Section 7.2, we formulate the problem as a one-sided hypothesis test. The null hypothesis of inequivalence is H_I: $p_A - p_B \le -\delta_N$ and the hypothesis of equivalence is H_E: $p_A - p_B > -\delta_N$. Here δ_N is the specified non-inferiority margin, which we shall take to be 0.05, the same as the value of δ in the previous two examples, as this will facilitate comparisons between the testing problems.

As usual, we first consider the fixed sample design. With $\theta = p_A - p_B$, we wish to test the null hypothesis $H_I : \theta \le -0.05$ against $H_E : \theta > -0.05$ with Type I error rate $\alpha = 0.025$ if $\theta = -0.05$ and power $1 - \beta = 0.9$ when $\theta = 0$. The fixed sample test requires information

$$\mathcal{I}_{f,1} = \{\Phi^{-1}(1-\alpha) + \Phi^{-1}(1-\beta)\}^2/\delta^2 = 4203.0,$$

just as for the superiority tests in Examples 3 and 4. With equal allocation to the two treatment arms and sample sizes $n_A = n_B = n$, say, we have

$$\mathcal{I}_{f,1} = \{Var(\hat{\theta})\}^{-1} = \left(\frac{p_A(1-p_A)}{n} + \frac{p_B(1-p_B)}{n}\right)^{-1}.$$

We solve this with $\mathcal{I}_{f,1} = 4203.0$ and re-admission rates $p_A = p_B = 0.25$, the values at which power 0.9 is required. This results in a sample size of $n = 1577$ per treatment arm for this fixed sample test, somewhat higher than the 1461 required for the superiority test.

Suppose we again choose a one-sided error spending design using spending functions of the form (8.15) with $\rho_1 = \rho_2 = 3$ and we plan for $K = 5$ equally sized groups. In a design with a binding futility boundary, the inflation factor is $R_{OS}(5, 0.025, 0.1, 3) = 1.049$ and, hence, the target maximum information level is $\mathcal{I}_{max} = 4409.9$, as in (8.25). If $p_A = p_B = 0.25$, 1654 subjects per treatment will be needed to achieve this information level so five groups of 331 subjects per treatment arm should be planned. If the observed information levels are exactly equal to $\mathcal{I}_{max}/5,\ 2\mathcal{I}_{max}/5,\ \ldots,\ \mathcal{I}_{max}$, the boundary values for the sequence of Z-statistics will be those given in (8.26). In practice, the actual boundary values will depend on the sequence of estimated information levels observed as the trial progresses.

The key difference in this non-inferiority trial is that the Z-statistic for testing the null hypothesis H_I: $\theta = p_A - p_B \le -0.05$ at analysis k is

$$Z_k = \frac{\hat{p}_{Ak} - \hat{p}_{Bk} + 0.05}{\sqrt{\{\hat{p}_{Ak}(1-\hat{p}_{Ak})/n_{Ak} + \hat{p}_{Bk}(1-\hat{p}_{Bk})/n_{Bk}\}}} = (\hat{\theta}_k + 0.05)\sqrt{\hat{\mathcal{I}}_k}.$$

Suppose group sizes are a little smaller than planned and we observe exactly the same values of n_{Ak}, n_{Bk}, y_{Ak} and y_{Bk} in the first three stages as were shown in Table 8.15. Then the only difference for the subsequent entries in the table are in the test statistics where we now have $Z_k = (\hat{\theta}_k + 0.05)\sqrt{\hat{\mathcal{I}}_k}$. The new values are

$$(Z_1,\ Z_2,\ Z_3) = (1.667,\ 2.191,\ 2.886),$$

so the study continues at analyses 1 and 2 but terminates at analysis 3 with acceptance of the equivalence hypothesis H_E as $Z_3 > b_3 = 2.636$.

Remarks on this example

It may be difficult for the sponsors of a new treatment to choose between running a trial to test for superiority and a trial designed to test for non-inferiority since the benefits of proving superiority are greater but the test for non-inferiority may be more likely to succeed. Another option is to test both hypotheses in the same study. The fact that the null hypotheses are nested, with H_I: $\theta \le -\delta_N$ implying H_0: $\theta \le 0$, is helpful here. Another consideration is that an appropriate value for δ_N may be quite small so that a non-inferiority trial with adequate power at $\theta = 0$ requires a larger sample size than a suitably powered superiority trial. In Chapter 25 we shall discuss adaptive methods that allow sample size to be adjusted depending on which hypothesis test is to be pursued in light of interim data.

8.6.4 Information Monitoring for Comparing Two Treatments with Survival Responses

8.6.4.1 Group Sequential Analysis of a Time-to-event Endpoint

In our final example of information monitoring, we consider a study with a time-to-event endpoint. We shall assume a proportional hazards model and apply a group sequential log-rank test as described in Section 4.5.

In comparing a new treatment against a standard, suppose the hazard rate for subjects on the standard treatment is λ times that on the new treatment and let $\theta = \log(\lambda)$. We wish to test the null hypothesis H_0: $\lambda \le 1$ against $\lambda > 1$ or, equivalently, H_0: $\theta \le 0$ against $\theta > 0$. The log-rank statistic S_k at analysis k is defined in Equation (4.8) and we shall use the large sample approximation to the distribution of S_k described in Section 4.5, which is applicable when λ is close to unity and, thus, θ is close to zero. According to the large sample theory, the variance of S_k given by Equation (4.9) is also the information $\mathcal{I}_k$ for θ and in a study with K analyses, we have, approximately,

$$\hat{\theta}_k = S_k/\mathcal{I}_k \sim N(\theta,\ \mathcal{I}_k^{-1})$$

and

$$Z_k = S_k/\sqrt{\mathcal{I}_k} \sim N(\theta\sqrt{\mathcal{I}_k}, 1)$$

for $k = 1, \ldots, K$. Furthermore, the sequences $\{\theta_1, \ldots, \theta_K\}$ and $\{Z_1, \ldots, Z_K\}$ follow the standard joint distributions stated in (3.2). In planning the trial, it is convenient to use the approximation $\mathcal{I}_k \approx d_k/4$ where d_k denotes the number of events observed at analysis k. This approximation is accurate if the numbers at risk in each treatment group remain nearly equal over time, as is to be expected if sample sizes per arm are equal and λ is close to unity.

An error spending test is particularly useful here since information depends on the number of observed events. Careful planning may be required in order to attain information levels close to their desired values at interim and final analyses. In a large, multi-national trial, meetings of the data monitoring committee must be scheduled well in advance to ensure availability of the participants, so patterns of recruitment and event rates over time need to be modeled to predict when required information levels will be met and to set dates for interim analyses accordingly. We noted in Section 8.5 that information levels at interim analyses should not be influenced by previous treatment effect estimates $\hat{\theta}_k$. Thus, it is important that members of the data monitoring committee, who have seen these estimates, should not be involved in setting analysis times. Rather, this should be handled by a trial steering committee whose members have knowledge of recruitment and the total number of events but are blinded to treatment effect estimates.

8.6.4.2 Example 6: One-sided test of Superiority for a Two-sample Survival Study

Consider a trial investigating a new treatment intended to reduce the risk of cardiac events for patients in a certain high risk category. The primary endpoint is the time to occurrence of a major adverse cardiovascular event (MACE), defined as the occurrence of myocardial infarction, stroke or cardiovascular death. A proportional hazards model is assumed for the time to the first MACE and, in planning the trial, it is anticipated that the hazard rate for a patient on the control arm will be 0.0015 per month.

Let $h(t)$ denote the hazard rate for an event when a patient receives the new treatment and suppose the hazard rate is $\lambda\, h(t)$ for a patient receiving the current standard treatment on the control arm. We wish to test H_0: $\lambda \le 1$ against $\lambda > 1$ with Type I error probability $\alpha = 0.025$ and power$1 - \beta = 0.9$ if $\lambda = 1.25$. Equivalently, with $\theta = \log(\lambda)$, we shall test H_0: $\theta \le 0$ against $\theta > 0$ with power 0.9 if $\theta = \log(1.25) = 0.223$. (We have chosen these definitions so that θ is positive if the new treatment is effective: had we defined λ to be the ratio of the hazard rate on the new treatment to that on the control, power would be attained at a hazard ratio of $1/1.25 = 0.8$.)

The study is to have an error spending design with four analyses using one-sided error spending functions from the Rho family, as defined in (8.15), with $\rho_1 = \rho_2 = 2$. From (5.1), the information for θ required in a fixed sample test is

$$\mathcal{I}_{f,1} = \{\Phi^{-1}(1-\alpha) + \Phi^{-1}(1-\beta)\}^2/0.223^2 = 211.0.$$

Assuming the futility boundary is binding and taking the inflation factor $R_{OS}(4, 0.025, 0.1, 2) = 1.089$ from Table 8.6, the target maximum information level is

$$\mathcal{I}_{max} = R_{OS}(4, 0.025, 0.1, 2) \times \mathcal{I}_{f,1} = 1.089 \times 211.0 = 229.8$$

so, using the approximation that information is equal to the number of observed events divided by 4, we require $4 \times 229.8 = 919$ events. It is estimated that under current treatment the hazard rate per month is 0.0015. If this rate is reduced by a factor 0.8 for subjects on the new treatment, the average event rate is 0.00135 per month and the study needs to observe patients for a total of $919/0.00135 = 681,000$ person months to reach this target.

Table 8.16 *Example 6 Worksheet: Information monitoring of time-to-event responses*

Analysis k	$k=1$		$k=2$		$k=3$		$k=4$	
Interim data								
Month of study	22		43		54		64	
Treatment arm	A	B	A	B	A	B	A	B
Numbers recruited	4403	4408	8424	8421	8424	8421	8424	8421
Numbers of events	104	130	211	255	311	366	422	503
Interim analyses								
Information, $\mathcal{I}_k$	56.4		110.9		161.8		222.9	
Log-rank statistic, Z_k	1.766		2.131		2.187		2.733	
$t_k = \mathcal{I}_k/\mathcal{I}_{max}$	0.245		0.483		0.704		0.970	
Cum. Type I error	0.00037		0.00281		0.00873		0.025	
Cum. Type II error	0.00148		0.01124		0.03491		0.1	
Lower boundary, a_k	−1.297		−0.042		0.975		1.970	
Upper boundary, b_k	3.375		2.796		2.435		1.970	
Decision	Continue		Continue		Continue		Stop, Reject H_0	

Suppose it is anticipated that 465 subjects will be recruited per month. If recruitment continues for 36 months, after which patients are followed for a further 24 months, 16,740 subjects will be enrolled in the trial and calculations show that, with $\lambda = 1.25$, around 684,000 person months will be observed. Furthermore, if analyses are conducted with data collected at 21, 39, 50 and 60 months after the start of the trial, and if $\lambda = 1.25$, the expected numbers of events at these analyses will be approximately 230, 460, 705 and 925.

In practice, recruitment and event rates will often differ from predictions. Suppose in this case recruitment is slower at around 400 per month and the event rate for a patient on the control arm is a little lower at 0.0014 per month. In response to the actual recruitment and the number of observed events, recruitment is extended to 42 months and analyses take place after 22, 43, 54 and 64 months. Hypothetical data from such a study and their analysis, as described in Section 4.5, are summarized in Table 8.16. Since the standardized log-rank statistic, Z_k, lies within the continuation region at the first three analyses, the trial continues up to the fourth and final analysis and at this point H_0 is rejected and the new treatment is declared to be superior to the standard.

In this example, it is notable that the interim statistics Z_2 and Z_3 were somewhat greater than $\Phi^{-1}(0.975) = 1.96$ but not sufficiently high to imply early stopping for efficacy in the group sequential design. Had the results at analysis 3 been a little better with $Z_3 > 2.435$, early stopping at this point would still not have reduced the sample size as recruitment was complete by analysis 2, but a positive outcome after 50 months rather than 60 months would have led to earlier dissemination of the findings with patients outside the trial benefiting from the new treatment sooner.

8.7 Computations for Error Spending Tests

For the procedures of the preceding chapters the boundaries are computed at the design stage before any trial data are observed. In contrast, the boundary values for an error spending test are computed at each analysis, once the observed information level is known. These computations are based on the methods of Section 3.5 and the key quantities are the sub-distribution functions

$$G_k(z;\,\theta) \;=\; Pr_\theta\{T \geq k \text{ and } Z_k \leq z\}$$

and their derivatives

$$g_k(z;\theta) \;=\; \frac{\partial}{\partial z}\,G_k(z;\,\theta)$$

for $k = 1, \ldots, K$.

We first consider two-sided tests and the computations needed to obtain critical values c_k by solving Equation (8.1). Here, the sub-distribution function is

$$G_k(z;\theta) \;=\; Pr_\theta\{|Z_1| < c_1, \ldots, |Z_{k-1}| < c_{k-1},\, Z_k \leq z\}.$$

In calculating the boundary values, we solve successively the equations

$$G_k(-c_k;\,0) \;=\; \pi_k/2, \quad k = 1, \ldots, K, \tag{8.28}$$

where the π_k are pre-specified in the case of a Slud & Wei (1982) test or obtained from Equation (8.2) for an error spending test with spending function f. For $k = 1$ we obtain $c_1 = \Phi^{-1}(1 - \pi_1/2)$ which can be computed at analysis 1 once $\mathcal{I}_1$ is observed. At each analysis $k = 2, \ldots, K$, when information levels $\mathcal{I}_1, \ldots, \mathcal{I}_k$ have been observed and critical values $c_1, \ldots, c_{k-1}$ are already computed, Equation (8.28) can be solved to obtain the critical value c_k. Here we use the fact that the sequence $\{Z_1, \ldots, Z_K\}$ specified by (3.1) is Markov and, as noted in Section 3.5, this allows use of the recursive formulae of Armitage, McPherson & Rowe (1969). Specifically, for $k = 2, \ldots, K$,

$$Z_k\sqrt{\mathcal{I}_k} - Z_{k-1}\sqrt{\mathcal{I}_{k-1}} \;\sim\; N(\theta\Delta_k, \Delta_k),$$

where $\Delta_k = \mathcal{I}_k - \mathcal{I}_{k-1}$ is the increment in information between analyses $k-1$ and k, and this distribution is independent of $Z_1, \ldots, Z_{k-1}$. Once c_k has been computed, the decision to stop or continue at analysis k can be made.

The attained power at a given value of θ can also be computed from the sub-densities $g_1, \ldots, g_K$ since it is equal to

$$\pi_1(\theta) + \ldots + \pi_K(\theta)$$

where

$$\pi_k(\theta) \;=\; \int_{-\infty}^{-c_k} g_k(z;\theta)\,dz + \int_{c_k}^{\infty} g_k(z;\theta)\,dz, \quad k = 1, \ldots, K.$$

We describe numerical methods for computing these integrals in Chapter 31.

Calculations for one-sided error spending tests follow a similar pattern. In the case of a design with a binding futility boundary, the continuation regions are $\mathcal{C}_k = (a_k, b_k)$, $k = 1, \ldots, K$, and the sub-distribution function at analysis k is

$$G_k(z;\,\theta) \;=\; Pr_\theta\{a_1 < Z_1 < b_1, \ldots, a_{k-1} < Z_{k-1} < b_{k-1},\, Z_k \leq z\}.$$

The derivative $g_k(z;\,\theta)$ of $G_k(z;\,\theta)$ is computed as in (3.10) and (3.11). Once the information level $\mathcal{I}_k$ has been observed at analysis k, the required increments $\pi_{1,k}$ and $\pi_{2,k}$ in the Type I and Type II error probabilities are given by Equations (8.10) and (8.11) and the boundary values a_k and b_k are obtained by solving

$$\int_{b_k}^{\infty} g_k(z;0) \;=\; \pi_{1,k} \quad \text{and} \quad \int_{-\infty}^{a_k} g_k(z;\delta) \;=\; \pi_{2,k}, \quad k = 1, \ldots, K.$$

Calculations for one-sided error spending tests with no futility boundary are as above but with $\pi_1 = \ldots = \pi_{K-1} = 0$, so $b_1 = \ldots = b_{K-1} = -\infty$, and $\pi_K = \beta$.

For a one-sided error spending test with a non-binding futility boundary, we define two versions of the sub-distribution function,

$$G_k(z;\,0) \;=\; Pr_{\theta=0}\{Z_1 < b_1,\,\dots\,,\,Z_{k-1} < b_{k-1},\,Z_k \le z\}$$

and

$$G_k(z;\,\delta) \;=\; Pr_{\theta=\delta}\{a_1 < Z_1 < b_1,\,\dots\,,\,a_{k-1} < Z_{k-1} < b_{k-1},\,Z_k \le z\}$$

with derivatives $g_k(z;\,0)$ and $g_k(z;\,\delta)$, respectively. We then compute boundary values a_k and b_k by solving

$$\int_{b_k}^{\infty} g_k(z;0) \;=\; \pi_{1,k} \quad \text{and} \quad \int_{-\infty}^{a_k} g_k(z;\delta) \;=\; \pi_{2,k}, \quad k=1,\dots,K.$$

8.8 Bibliography and Notes

The methods in this chapter solve the problem of constructing a group sequential test with specified Type I error rate α when information levels at each analysis are unpredictable. In earlier chapters we saw that the "significance level" approach proposed by Pocock (1977) and Whitehead's (1997) "Christmas tree" adjustment offer approximate solutions to this problem. The proposals by Slud & Wei (1982) and Lan & DeMets (1983) provide an exact solution. However, the methods of Lan & DeMets are more responsive to observed information levels, varying the cumulative Type I error probability at each analysis according to the observed information.

In their 1983 paper, Lan & DeMets used an error spending function to create a design with a single efficacy boundary, in which the null hypothesis H_0: $\theta \le 0$ is rejected if this boundary is crossed; their focus was on Type I error and they did not consider power. Our descriptions of procedures with "inflation factors" and "information monitoring" to achieve desired power and to deal with nuisance parameters draw on a number of papers in this area. The extension of the methodology to one-sided tests with two error spending functions, for Type I probability α under $\theta = 0$ and Type II error probability β under $\theta = \delta$, was a crucial development for the design of Phase III clinical trials. Jennison (1987) and Eales & Jennison (1992) proposed error spending functions for tests with $\alpha = \beta$ and symmetric stopping rules about $\delta/2$ on the $\hat{\theta}$ scale. Chang, Hwang & Shin (1998) proposed the use of Type I and Type II error spending functions for general α and β and we have followed their approach of deriving the inflation factor for a trial under the assumption of equally spaced information levels. Pampallona, Tsiatis & Kim (2001) present similar proposals in a paper arising from a 1995 Harvard University Technical Report.

Stallard & Facey (1996) offer an alternative construction for the lower boundary in an error spending design. They define a rule for spending the probability under $\theta = 0$ of accepting H_0 during the course of the study. The upper and lower boundaries automatically meet at the final analysis when, under $\theta = 0$, the cumulative probability of crossing the upper boundary is α and the probability of crossing the lower boundary is $1 - \alpha$.

We noted in Section 8.2.2 that Lan & DeMets (1983) proposed error spending functions that lead to tests with boundaries similar to those of Pocock (1977) and O'Brien & Fleming (1979) designs. While mimicking such influential designs has its appeal, these boundaries are somewhat arbitrary and the wide early boundaries of O'Brien & Fleming tests are not always desirable. It is just as reasonable to start by defining a mathematically appealing form of error spending function and checking that the resulting group sequential tests have good properties. Barber & Jennison (2002) evaluated the performance of one-sided tests based on several error spending functions and concluded that error spending functions in the Rho family and Hwang, Shih and De Cani's (1990) Gamma family error spending functions are highly efficient in reducing the expected information level on termination for a given maximum information, $\mathcal{I}_{max}$, and number of analyses, K. We shall return to this topic of efficiency in Chapter 12.

Chapter 9

Analysis Following a Sequential Test

9.1 Introduction

So far we have concentrated on hypothesis tests, establishing a framework for *designing* a group sequential study by performing "pre-data" calculations. However, once data have been collected, the hypothesis testing paradigm may no longer be the most useful one for statistical analysis; see, for example, the discussion in Cutler et al. (1966). Whereas in the context of industrial acceptance sampling it may well be that decisions are made and actions taken as soon as a study is terminated, in medical studies a more complete analysis is usually required than the simple "accept" or "reject" decision of a hypothesis test. In this chapter we discuss the construction of point estimates, P-values and confidence intervals. In particular, an interval estimate for the treatment effect θ will allow consideration of the *magnitude* of the treatment difference and indicate its practical importance, irrespective of whether a hypothesis test has declared statistical significance; see Simon (1993) and Braitman (1993).

There are two situations in which interval estimates might be required in a group sequential study. The first is upon conclusion of the trial after the test statistic has crossed a stopping boundary, as described in previous chapters. This is the case we shall consider in this chapter. Alternatively, one may wish to give an interval estimate of the treatment effect at an interim stage of the study, regardless of whether the design calls for termination or not. This situation calls for a "repeated confidence interval" sequence, the subject of Chapter 10.

In neither of the above situations is it appropriate to compute a "naive" confidence interval, treating the data as if they had been obtained in a study with a fixed sample size. Tsiatis, Rosner & Mehta (1984) investigate naive 90% confidence intervals calculated on termination of a five-stage Pocock test with Type I error probability 0.05 and find their coverage probabilities to vary between 0.846 and 0.929, depending on the true parameter value.

Note that in this chapter we shall be concerned with frequentist properties of interval estimates. The topic of credible intervals in the Bayesian paradigm will be discussed in Chapter 13.

9.2 Distribution Theory

We have seen in Chapter 3 that many common problems produce a sequence of test statistics $Z_1, \ldots, Z_K$ with the canonical joint distribution (3.1), conditional on observed information levels $\mathcal{I}_1, \ldots, \mathcal{I}_K$. We consider problems of this type in this chapter, although it should be evident that the key ideas extend to other situations; in Chapter 16 we shall develop similar methods for the discrete distributions arising in the analysis of binary data.

Assuming that (3.1) applies, define

$$Y_1 = Z_1 \sqrt{\mathcal{I}_1}, \;\; \Delta_1 = \mathcal{I}_1$$

DOI: 10.1201/9781584888482-9

and

$$Y_k = Z_k\sqrt{\mathcal{I}_k} - Z_{k-1}\sqrt{\mathcal{I}_{k-1}}, \quad \Delta_k = \mathcal{I}_k - \mathcal{I}_{k-1}, \quad k = 2, \ldots, K.$$

As an example, in the one sample problem described in Section 3.2.2, when $\sigma^2 = 1$ the information increment Δ_k is the number of observations in group k and Y_k is the sum of these Δ_k observations. It follows from (3.1) that the Y_k are independent and $Y_k \sim N(\Delta_k\,\theta,\,\Delta_k)$, $k = 1, \ldots, K$. We shall use this fact and the relation

$$\begin{pmatrix} Z_1 \\ Z_2 \\ \vdots \\ Z_K \end{pmatrix} = \begin{pmatrix} 1/\sqrt{\mathcal{I}_1} & 0 & \ldots & 0 \\ 1/\sqrt{\mathcal{I}_2} & 1/\sqrt{\mathcal{I}_2} & \ldots & 0 \\ \vdots & \vdots & & \vdots \\ 1/\sqrt{\mathcal{I}_K} & 1/\sqrt{\mathcal{I}_K} & \ldots & 1/\sqrt{\mathcal{I}_K} \end{pmatrix} \begin{pmatrix} Y_1 \\ Y_2 \\ \vdots \\ Y_K \end{pmatrix}$$

in the following calculations.

Let $\boldsymbol{Z}_k = (Z_1, \ldots, Z_k)$ be the vector of the first k standardized statistics. In the group sequential tests of Chapters 2 to 8, stopping occurs at analysis $T = \min\{k\colon Z_k \notin \mathcal{C}_k\}$, where $\mathcal{C}_k$ is the continuation region at analysis k and $\mathcal{C}_K = \emptyset$, the empty set, to ensure termination by analysis K. For $k = 1, \ldots, K$, define

$$\mathcal{A}_k = \{\boldsymbol{z}_k\colon z_i \in \mathcal{C}_i,\ i = 1, \ldots, k-1, \text{ and } z_k \notin \mathcal{C}_k\},$$

the set of sample paths for Z_1, Z_2, ... that terminate precisely at analysis k. Thus, $\boldsymbol{Z}_k \in \mathcal{A}_k$ implies that $T = k$. In view of the above representation of $Z_1, \ldots, Z_K$ in terms of independent normal variables $Y_1, \ldots, Y_K$, we can write the joint density of $(Z_1, \ldots, Z_k)$ at a point $\boldsymbol{z}_k$ in $\mathcal{A}_k$ as

$$p_{T,\boldsymbol{Z_T}}(k, \boldsymbol{z}_k;\, \theta) = \prod_{i=1}^{k} \frac{\sqrt{\mathcal{I}_i}}{\sqrt{\Delta_i}}\, \phi\left(\frac{y_i - \Delta_i\theta}{\sqrt{\Delta_i}}\right),$$

where $\phi(x) = \exp(-x^2/2)/\sqrt{(2\pi)}$ is the standard normal density function. Using the facts $z_k\sqrt{\mathcal{I}_k} = y_1 + \ldots + y_k$ and $\mathcal{I}_k = \Delta_1 + \ldots + \Delta_k$, we find

$$\begin{aligned} p_{T,\boldsymbol{Z_T}}(k, \boldsymbol{z}_k;\, \theta) &= \prod_{i=1}^{k} \frac{\sqrt{\mathcal{I}_i}}{\sqrt{(2\pi\Delta_i)}} \exp\left\{-\sum_{i=1}^{k} \frac{y_i^2 - 2\Delta_i\theta y_i + \Delta_i^2\theta^2}{2\Delta_i}\right\} \\ &= h(k, \boldsymbol{z}_k, \mathcal{I}_1, \ldots, \mathcal{I}_k) \exp(\theta z_k\sqrt{\mathcal{I}_k} - \theta^2\mathcal{I}_k/2), \end{aligned} \tag{9.1}$$

where

$$h(k, \boldsymbol{z}_k, \mathcal{I}_1, \ldots, \mathcal{I}_k) = \prod_{i=1}^{k} \frac{\sqrt{\mathcal{I}_i}\, \exp\{-y_i^2/(2\Delta_i)\}}{\sqrt{(2\pi\Delta_i)}}.$$

The form of this joint density function shows that the pair (T, Z_T) is a sufficient statistic for θ (since T determines $\mathcal{I}_T$), and hence the maximum likelihood estimator (MLE) of θ, found by maximizing (9.1), is $\hat{\theta} = Z_T/\sqrt{\mathcal{I}_T}$; see Chang (1989, p. 249). Thus, in the one-sample problem of Section 3.2.2, the MLE of $\theta = \mu - \mu_0$ is $\bar{X}_T - \mu_0$, implying that $\bar{X}_T$ is the MLE of μ. In the two-sample problem considered in Chapter 2 and Section 3.2.1, the MLE of $\theta = \mu_A - \mu_B$ is $Z_T/\sqrt{\mathcal{I}_T} = \bar{X}_{AT} - \bar{X}_{BT}$.

For $k = 1, \ldots, K$, we denote the sub-density of the terminal statistic Z_k when the test stops at analysis k by $p(k, z;\, \theta)$. Thus, $p(k, z;\, \theta)$ is non-zero only for values of $z \notin \mathcal{C}_k$ and its integral over a set $E \in \mathcal{C}_k^c$ gives the probability of stopping at analysis k with $Z_k \in E$. We obtain the kth of these sub-densities by integrating $p_{T,\boldsymbol{Z_T}}(k, \boldsymbol{z}_k; \theta)$ over sequences

$z_1, \ldots, z_{k-1}$ that give sample paths $\boldsymbol{z}_k = (z_1, \ldots, z_{k-1}, z) \in \mathcal{A}_k$. Thus, for $z \notin \mathcal{C}_k$, we have

$$p(k, z; \theta) = \int \cdots \int_{\mathcal{B}_k(z)} h(k, (z_1, \ldots, z_{k-1}, z), \mathcal{I}_1, \ldots, \mathcal{I}_k) \times$$
$$\exp(\theta z \sqrt{\mathcal{I}_k} - \theta^2 \mathcal{I}_k / 2) \, dz_1 \ldots dz_{k-1}, \tag{9.2}$$

where $\mathcal{B}_k(z)$ is the set of vectors $\boldsymbol{z}_k = (z_1, \ldots, z_k) \in \mathcal{A}_k$ for which $z_k = z$. It is evident from Equation (9.2) that

$$p(k, z; \theta) = p(k, z; 0) \exp(\theta z \sqrt{\mathcal{I}_k} - \theta^2 \mathcal{I}_k / 2). \tag{9.3}$$

This is an example of a likelihood ratio identity; see Siegmund (1985, Propositions 2.24 and 3.2). Emerson & Fleming (1990) note the usefulness of this relation in using a sub-density evaluated under one value of θ for computations at another θ. Such a relation can also be used in simulation studies to obtain properties of a sequential test under several values of θ from simulations performed at a single θ.

Although it might appear from (9.2) that the computation of $p(k, z; \theta)$ involves a difficult multivariate normal integral, the recursive formulae of Armitage, McPherson & Rowe (1969) can be used to calculate $p(1, z; \theta)$ to $p(K, z; \theta)$ in turn. These formulae are

$$p(k, z; \theta) = \begin{cases} g_k(z; \theta) & \text{if } z \notin \mathcal{C}_k \\ 0 & \text{if } z \in \mathcal{C}_k, \end{cases} \tag{9.4}$$

where

$$g_1(z; \theta) = \phi(z - \theta \sqrt{\mathcal{I}_1}) \tag{9.5}$$

and, for $k = 2, \ldots, K$,

$$g_k(z; \theta) = \int_{\mathcal{C}_{k-1}} g_{k-1}(u; \theta) \frac{\sqrt{\mathcal{I}_k}}{\sqrt{\Delta_k}} \phi\left(\frac{z\sqrt{\mathcal{I}_k} - u\sqrt{\mathcal{I}_{k-1}} - \Delta_k \theta}{\sqrt{\Delta_k}} \right) du. \tag{9.6}$$

Thus, the computation is very much simplified, requiring only a succession of $K - 1$ univariate integrations. In Chapter 31 we shall describe in detail methods for evaluating these integrals numerically.

As an example, consider a two-sided Wang & Tsiatis (1987) test, as defined in Section 2.7.1, with parameter $\Delta = 0.25$ and four equally sized groups of observations. For a test with Type I error rate $\alpha = 0.05$, we take $C_{WT}(4, 0.05, 0.25) = 2.113$ from Table 2.9 and find the continuation regions at analyses 1 to 3 to be $\mathcal{C}_1 = (-2.988, 2.988)$, $\mathcal{C}_2 = (-2.513, 2.513)$ and $\mathcal{C}_3 = (-2.271, 2.271)$. The sub-densities $p(k, z; 0)$ for $k = 1$, 2, 3 and 4 under the null hypothesis $\theta = 0$ are plotted in Figure 9.1. Since the four sub-densities overlap, those for $k = 1$, 2 and 3 have been multiplied by 6, 3 and 2 respectively to aid visibility. Continuing this example, suppose the test had been designed to achieve power $1 - \beta = 0.8$ at $\theta = \pm\delta$ for some specified δ. Taking the value $R_{WT}(4, 0.05, 0.2, 0.25) = 1.065$ from Table 2.10, the maximum information would have been set as

$$\mathcal{I}_4 = R_{WT}(4, 0.05, 0.2, 0.25) \{\Phi^{-1}(0.975) + \Phi^{-1}(0.8)\}^2 / \delta^2 = 8.36/\delta^2.$$

Hence, $\mathcal{I}_k = 2.09\, k/\delta^2$ for $k = 1, \ldots, 4$. Substituting these values in (9.5) and (9.6) we obtain the four sub-densities under $\theta = \delta$, the positive value of θ at which power is 0.8, that are shown in Figure 9.2 (with no adjustments needed to aid visibility this time).

The area under sub-density k is the probability of stopping at analysis k,

$$Pr_\theta\{T = k\} = \int_{z \notin \mathcal{C}_k} p(k, z; \theta) \, dz, \quad k = 1, \ldots, K. \tag{9.7}$$

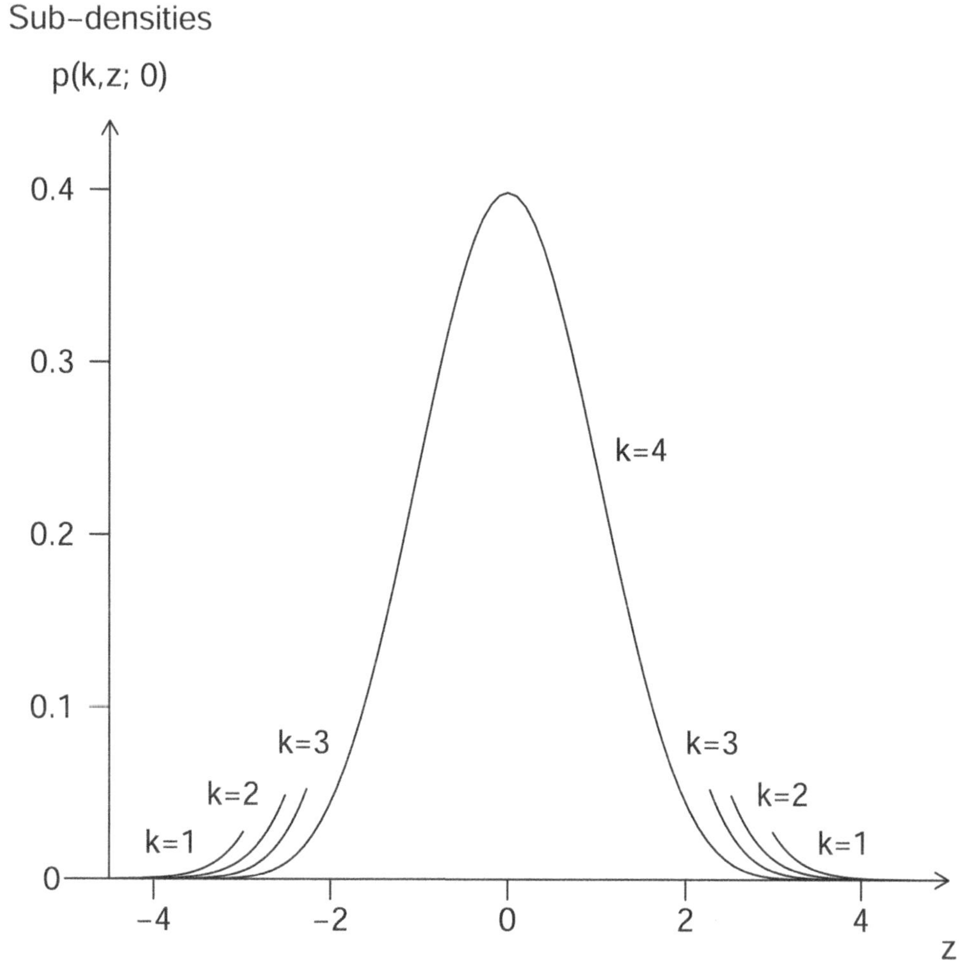

Figure 9.1 Sub-densities $p(k, z; \theta)$ when $\theta = 0$ for a four stage Wang & Tsiatis test with $\Delta = 0.25$, two-sided Type I error probability 0.05 and power 0.8 at $\theta = 1$. For $k = 1$, 2 and 3 the sub-densities are very small and the plot shows these sub-densities multiplied by 6, 3 and 2 respectively.

In our example, when $\theta = 0$ these probabilities are 0.0028, 0.0109, 0.0167 and 0.9696 at analyses 1 to 4, respectively, and these are the areas under the curves in Figure 9.1. For $\theta = \pm\delta$, the corresponding probabilities are 0.0934, 0.3520, 0.2995 and 0.2551, distributed over z values according to the sub-densities shown in Figure 9.2.

The expected number of analyses and the expected final information level are, respectively,

$$E(T) = \sum_{k=1}^{K} k \, Pr_\theta\{T = k\} \quad \text{and} \quad E(\mathcal{I}_T) = \sum_{k=1}^{K} \mathcal{I}_k \, Pr_\theta\{T = k\}.$$

In our four-stage example, calculations show $E(T) = 3.953$ and $E(\mathcal{I}_T) = 8.261/\delta^2$ when $\theta = 0$, whereas $E(T) = 3.003$ and $E(\mathcal{I}_T) = 6.27/5\delta^2$ when $\theta = \pm\delta$.

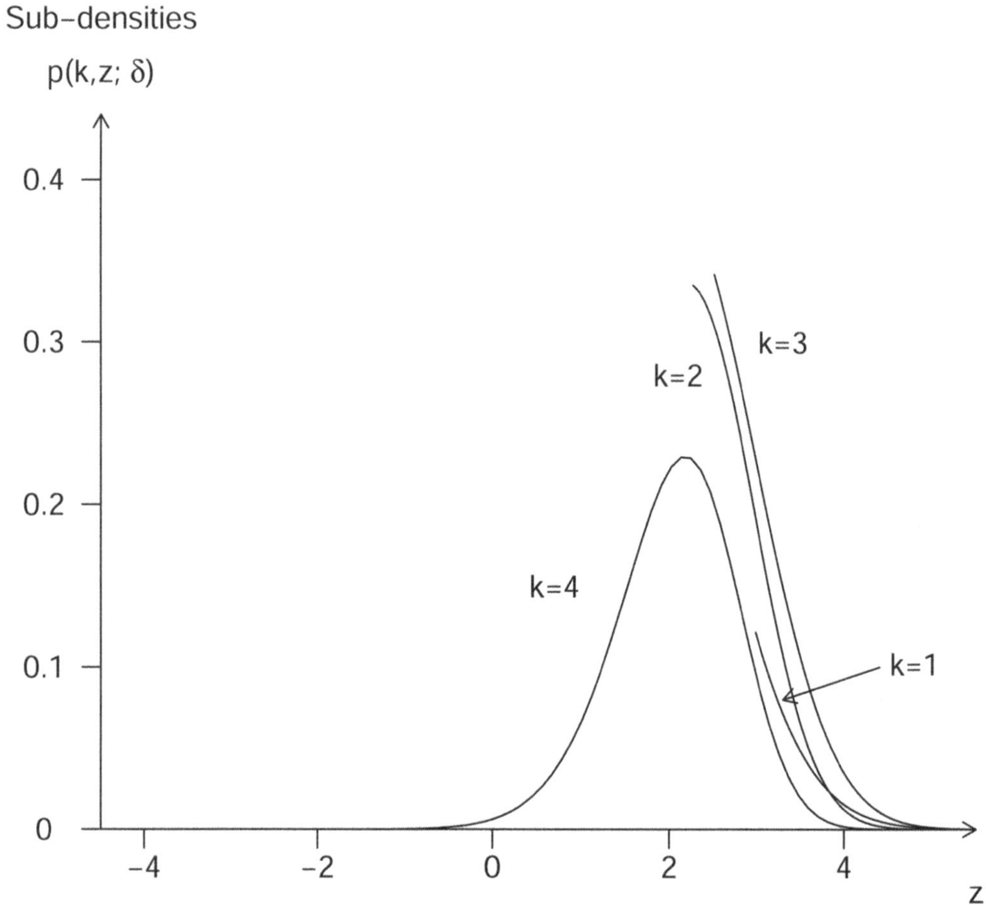

Figure 9.2 *Sub-densities $p(k, z;\, \theta)$ when $\theta = 1$ for a four-stage Wang & Tsiatis test with $\Delta = 0.25$, two-sided Type I error probability 0.05 and power 0.8 at $\theta = 1$*

9.3 Point Estimation

9.3.1 Bias of the Maximum Likelihood Estimator

We now consider point estimation of the parameter θ. In the previous section we derived the MLE, $\hat{\theta} = Z_T/\sqrt{\mathcal{I}_T}$. Since in the one-sample normal problem $\hat{\theta} = \bar{X}_T$ is the MLE of $\theta = \mu - \mu_0$ and in the two-sample problem $\hat{\theta} = \bar{X}_{AT} - \bar{X}_{BT}$ is the MLE of $\theta = \mu_A - \mu_B$, we can also call $\hat{\theta} = Z_T/\sqrt{\mathcal{I}_T}$ the "sample mean" estimator.

The sampling distribution of $\hat{\theta}$ has density

$$f(\hat{\theta};\, \theta) \;=\; \sum_{k=1}^{K} p(k, \hat{\theta}\sqrt{\mathcal{I}_k};\, \theta)\sqrt{\mathcal{I}_k},$$

using the notation of the previous section. We illustrate this in the example of the four-stage, two-sided Wang & Tsiatis test with $\Delta = 0.25$ discussed in Section 9.2. The density $f(\hat{\theta};\, \theta)$ is shown in Figure 9.3 when $\theta = 0$ and in Figure 9.4 when $\theta = \delta = 1$. The components of $f(\hat{\theta};\, \theta)$ arising from termination at each analysis k are indicated in the two figures and these correspond to contributions from the four sub-densities $p(k, z;\, \theta)$ shown in Figure 9.1 for

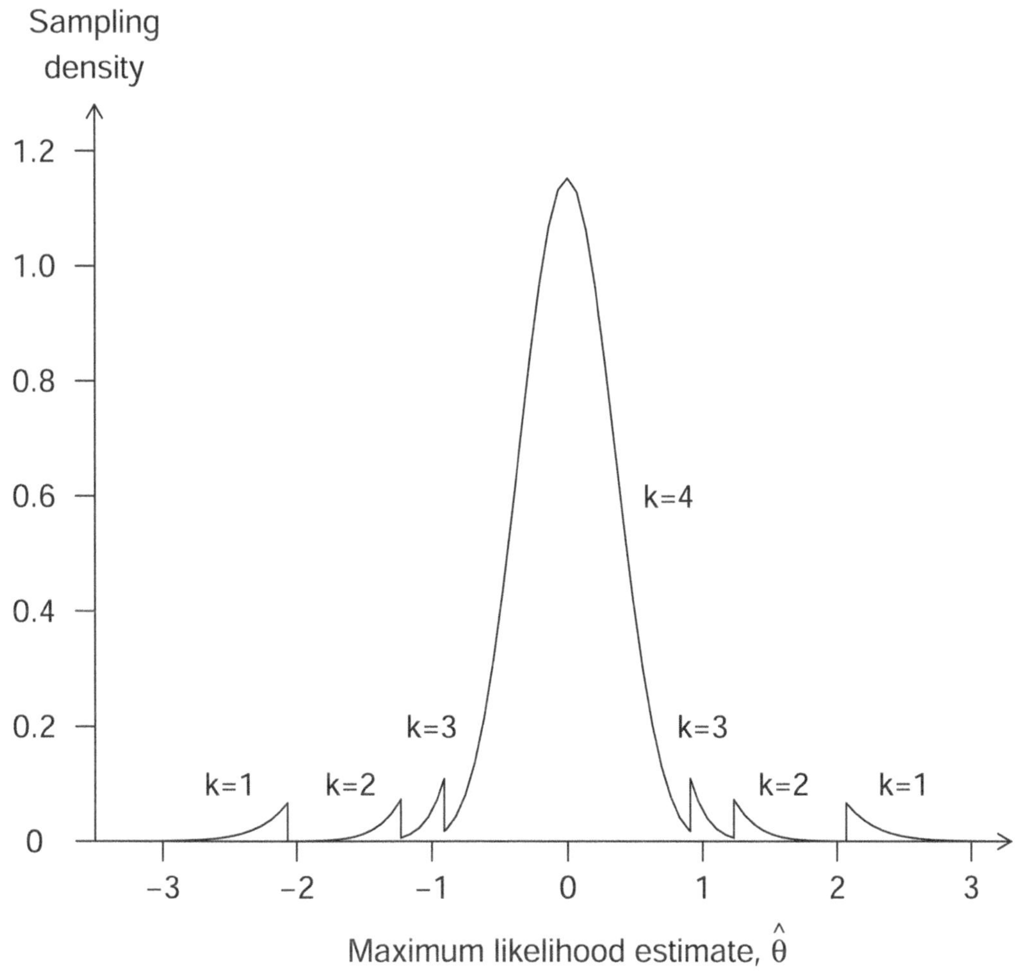

Figure 9.3 *Sampling density of $\hat{\theta}$, $f(\hat{\theta};\,\theta)$, when $\theta = 0$ for a four-stage Wang & Tsiatis test with $\Delta = 0.25$, two-sided Type I error probability 0.05 and power 0.8 at $\theta = 1$*

the case $\theta = 0$ and in Figure 9.2 for $\theta = 1$. Note that for $\theta = 1$, the probability of stopping with Z_k below the lower boundary at analysis $k = 1$, 2 or 3 is extremely small and the left hand part of the sub-density $p(k, z;\, \theta)$ is not visible in Figure 9.4.

We can see that the density of $\hat{\theta} = Z_T/\sqrt{\mathcal{I}_T}$ is multi-modal with a peak for each value of $T \in \{1, 2, 3, 4\}$. This density is *not at all* like that of a normal distribution with mean θ, as it would be in the analysis of a fixed sample design. In view of the nature of this sampling density, it is not surprising that $\hat{\theta}$ is a biased estimator of θ. The expectation of $\hat{\theta}$ can be calculated as

$$E_{\theta}(\hat{\theta}) \;=\; \sum_{k=1}^{K} \int_{z \notin \mathcal{C}_k} \frac{z}{\sqrt{\mathcal{I}_k}}\, p(k, z;\, \theta)\, dz. \tag{9.8}$$

Note that this expectation relates to the unconditional or marginal distribution of $\hat{\theta}$. We do not consider estimation conditional on the analysis at which a trial stops or on whether H_0 is rejected as these events are highly informative about θ and far from "ancillary";

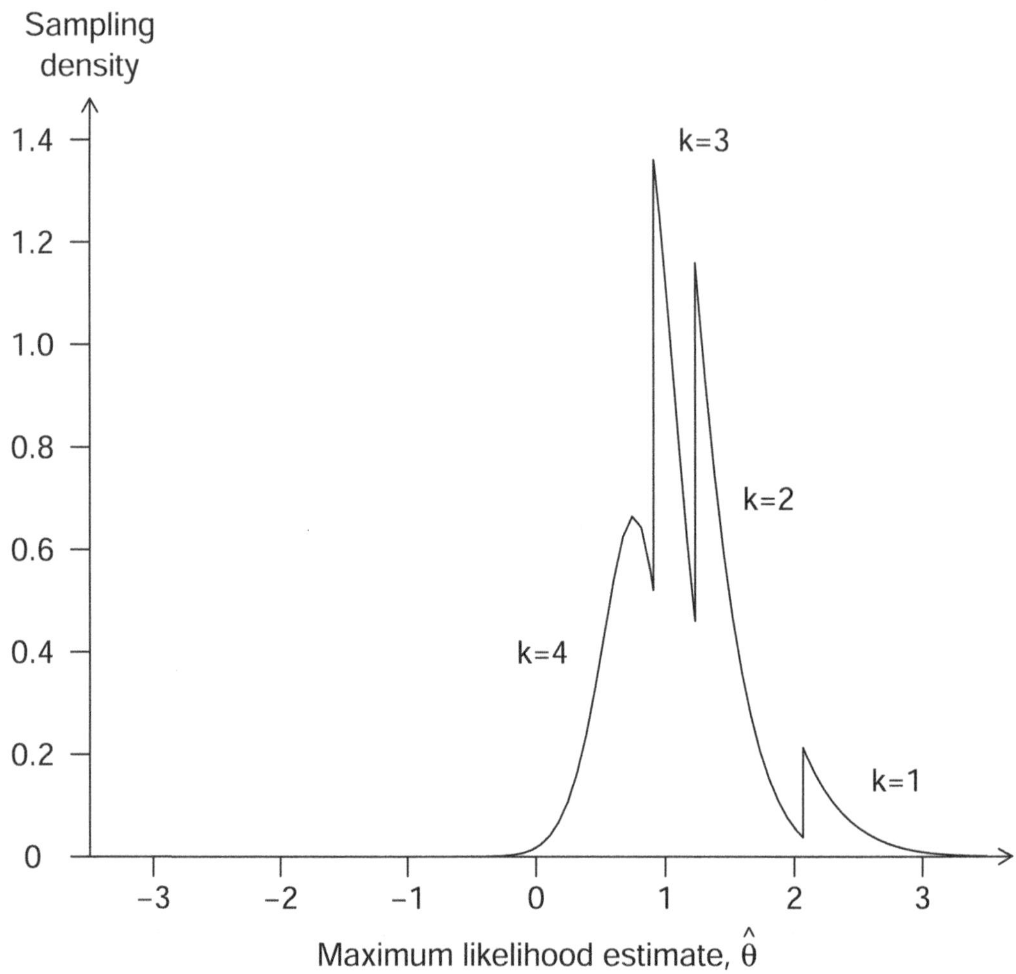

Figure 9.4 *Sampling density of* $\hat{\theta}$, $f(\hat{\theta}; \theta)$, *when* $\theta = 1$ *for a four-stage Wang & Tsiatis test with* $\Delta = 0.25$, *two-sided Type I error probability* 0.05 *and power* 0.8 *at* $\theta = 1$

see Sections 2.2 and 2.3 of Cox & Hinkley (1974). Other authors have considered such conditional inference; see, for example, Ohman Strickland & Cassella (2003).

For the two-sided testing boundaries with no inner wedge of Chapters 2 and 4, the MLE typically overestimates the magnitude of θ, giving a positive bias for $\theta > 0$ and a negative bias for $\theta < 0$. To see why this should occur, consider the case $\theta > 0$. If $\hat{\theta}_k = Z_k/\sqrt{\mathcal{I}_k}$ is sufficiently greater than θ at an early analysis, the test stops and $\hat{\theta} = \hat{\theta}_k$. However, if $\hat{\theta}_k$ is less than θ by a similar amount, the test can continue, allowing later data to dilute the effect of early results on the final estimate of θ. Pinheiro & DeMets (1997, Sec. 2.2) give an explanation of this phenomenon based on an analytic expression for the bias of the MLE and Whitehead (1986a, Sec. 9) tabulates the bias of the MLE of θ for the triangular tests of Section 5.7.

For the simple case of a two-stage test with continuation interval $\mathcal{C}_1 = (a, b)$ at analysis 1, Emerson (1988) obtains an analytic expression for this bias. It follows from Equations (9.6) and (9.8) that

$$E_\theta(\hat{\theta}) = \int_{-\infty}^{a} \frac{z_1}{\sqrt{\mathcal{I}_1}} \phi(z_1 - \theta\sqrt{\mathcal{I}_1})\, dz_1 + \int_{b}^{\infty} \frac{z_1}{\sqrt{\mathcal{I}_1}} \phi(z_1 - \theta\sqrt{\mathcal{I}_1})\, dz_1$$

$$+ \int_a^b \int_{-\infty}^{\infty} \frac{z_2}{\sqrt{\Delta_2}} \phi\left(\frac{z_2\sqrt{\mathcal{I}_2} - z_1\sqrt{\mathcal{I}_1} - \Delta_2\theta}{\sqrt{\Delta_2}}\right) \phi(z_1 - \theta\sqrt{\mathcal{I}_1})\, dz_2\, dz_1.$$

Integrating over z_2, the double integral in the above equation becomes

$$\int_a^b \frac{1}{\mathcal{I}_2}(z_1\sqrt{\mathcal{I}_1} + \Delta_2\theta)\, \phi(z_1 - \theta\sqrt{\mathcal{I}_1})\, dz_1.$$

Using the fact that

$$\int_{-\infty}^{\infty} z_1\, \phi(z_1 - \theta\sqrt{\mathcal{I}_1})\, dz_1 = \theta\sqrt{\mathcal{I}_1},$$

we find that the bias in $\hat{\theta}$ is

$$E_\theta(\hat{\theta}) - \theta = \int_a^b \left(\frac{z_1\sqrt{\mathcal{I}_1}}{\mathcal{I}_2} - \frac{z_1}{\sqrt{\mathcal{I}_1}} + \frac{\Delta_2\theta}{\mathcal{I}_2}\right) \phi(z_1 - \theta\sqrt{\mathcal{I}_1})\, dz_1$$

$$= -\int_a^b \frac{\Delta_2}{\mathcal{I}_2\sqrt{\mathcal{I}_1}}(z_1 - \theta\sqrt{\mathcal{I}_1})\, \phi(z_1 - \theta\sqrt{\mathcal{I}_1})\, dz_1.$$

Hence, using $\phi'(u) = -u\,\phi(u)$, we obtain

$$E_\theta(\hat{\theta}) - \theta = \frac{\Delta_2}{\mathcal{I}_2\sqrt{\mathcal{I}_1}} \{\phi(b - \theta\sqrt{\mathcal{I}_1}) - \phi(a - \theta\sqrt{\mathcal{I}_1})\}.$$

For the case of equal group sizes, $\mathcal{I}_2 = 2\mathcal{I}_1$ and $\Delta_2 = \mathcal{I}_1$, so

$$E_\theta(\hat{\theta}) - \theta = \frac{\exp\{-(b - \theta\sqrt{\mathcal{I}_1})^2/2\} - \exp\{-(a - \theta\sqrt{\mathcal{I}_1})^2/2\}}{2\sqrt{(2\pi\mathcal{I}_1)}}.$$

It is straightforward to check that this expression is only equal to zero when θ happens to equal $(a + b)/(2\sqrt{\mathcal{I}_1})$.

9.3.2 *Bias-corrected Maximum Likelihood Estimator*

Whitehead (1986a) suggests adjusting the MLE, $\hat{\theta}$, by subtracting an estimate of its bias. Denote the bias function of the MLE by

$$b(\theta) = E_\theta(\hat{\theta}) - \theta.$$

Ideally one would subtract $b(\theta)$, the bias calculated at the true value of θ but, since θ is unknown, Whitehead proposes the estimator $\tilde{\theta}$ which is the solution to the equation

$$\tilde{\theta} = \hat{\theta} - b(\tilde{\theta}),$$

so the adjustment in this estimator is to subtract the bias of the MLE at the adjusted estimator itself. This estimator can be found by an iterative process, starting with $\hat{\theta}_{[1]} = \hat{\theta}$ and setting

$$\hat{\theta}_{[i]} = \hat{\theta} - b(\hat{\theta}_{[i-1]}), \quad i = 2, 3, \ldots,$$

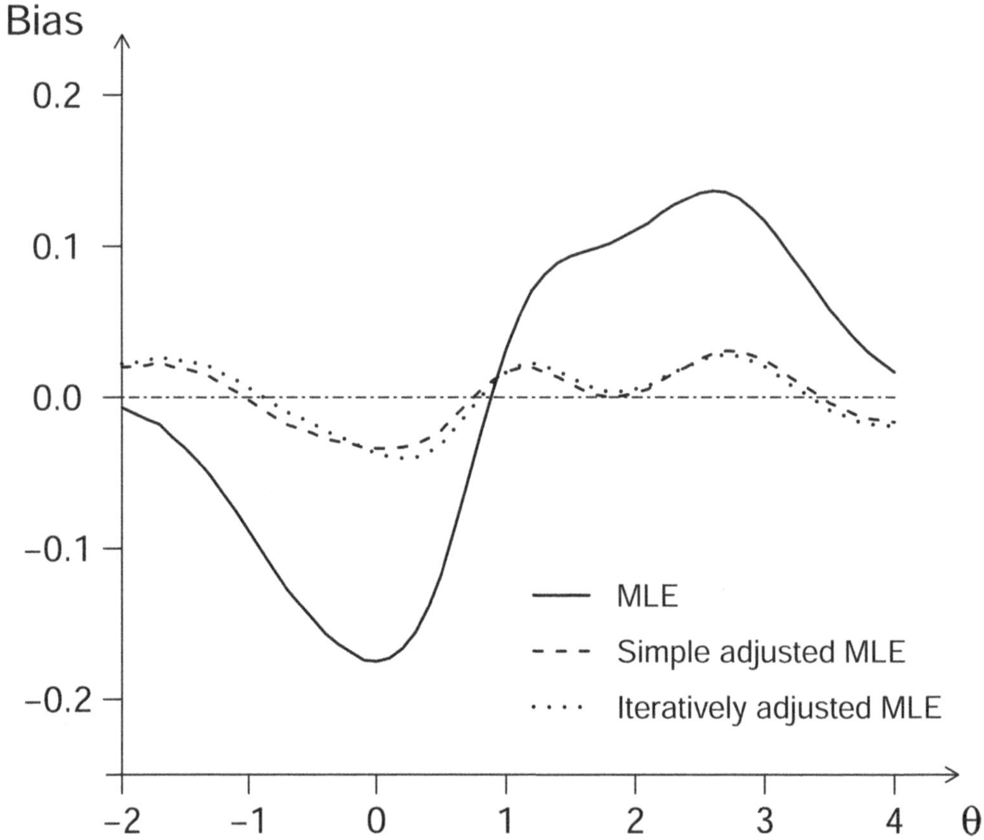

Figure 9.5 *Bias of the MLE, $\hat{\theta}$, simple adjusted MLE, $\bar{\theta}$, and iteratively adjusted, MLE $\tilde{\theta}$, for a Pampallona & Tsiatis one-sided test with $\Delta = 0$, four analyses, binding futility boundary, one-sided Type I error probability* 0.025, *and power* 0.8 *at* $\theta = 1$

until convergence is reached (in just a few steps). The formula (9.3) may be applied to re-use computations performed at a single value of θ. Note that, as the bias adjustment in $\tilde{\theta}$ is based on estimated value of θ, this estimator is still not completely unbiased.

A simpler approach is to use the estimator obtained after one step of the iterative process described above, namely,

$$\bar{\theta} = \hat{\theta} - b(\hat{\theta}).$$

We shall refer to $\tilde{\theta}$ as the "iteratively adjusted MLE" and to $\bar{\theta}$ as the "simple adjusted MLE".

We illustrate the results of applying these methods in the case of a Pampallona & Tsiatis one-sided test with $\Delta = 0$, four analyses and a binding futility boundary, that has one-sided Type I error probability 0.025 and power 0.8 at $\theta = 1$. Figure 9.5 shows the bias, as a function of θ for the MLE and the two versions of the adjusted MLE. The bias of the MLE ranges from -0.17 to 0.14, a significant amount given that the treatment effect at which power is set is equal to 1. The bias functions of the simple adjusted MLE and the iteratively adjusted MLE are almost identical and their bias, which ranges from -0.04 to 0.03, may now be regarded as negligible in the context of this testing problem.

In our calculations, we have computed the bias function using numerical integration. Wang & Leung (1997) note that such calculations can also be accomplished by simulation. So, for example, one may simulate the group sequential design under the MLE, $\hat{\theta}$, estimate the bias $b(\hat{\theta})$ from these simulations, and subtract this from $\hat{\theta}$ to obtain a value for the simple adjusted MLE.

9.3.3 *Minimum Variance Unbiased Estimators*

It is natural to enquire whether a uniformly minimum variance unbiased estimator (UMVUE) of the treatment effect θ can be found in the group sequential setting. Liu & Hall (1999) prove that the sufficient statistic (T, Z_T) is not complete and hence such an estimator does not exist. However, the same authors show there is a UMVUE in the class of estimators which do not require knowledge of the number of further analyses and their associated information levels that would have occurred had the study continued past the terminal analysis T. This is a natural restriction and we shall term this estimator the "UMVUE".

Emerson & Fleming (1990) first suggested use of the UMVUE, which they calculated by applying the familiar Rao-Blackwell technique to the unbiased estimate of θ formed from the first-stage data, $\hat{\theta}_1 = Z_1/\sqrt{\mathcal{I}_1}$. Emerson (1993) and Emerson & Kittelson (1997) describe the necessary calculations in detail. The UMVUE is $E\{\hat{\theta}_1 \,|\, (T, Z_T)\}$, and this expectation can be found from the conditional density of Z_1 given $(T, Z_T) = (k, z)$,

$$p(z_1 \mid (T, Z_T) = (k, z);\, \theta) \;=\; \frac{p(1, z_1; \theta)\, p(k, z \mid z_1;\, \theta)}{p(k, z; \theta)}. \tag{9.9}$$

Here, $p(k, z \,|\, z_1;\, \theta)$ denotes the conditional sub-density under θ of outcomes terminating at analysis k, given that $Z_1 = z_1$. Importantly, since we are conditioning on (T, Z_T), the sufficient statistic for θ, the conditional sub-density (9.9) does not depend on θ and $E\{\hat{\theta}_1 \,|\, (T, Z_T)\}$ may be computed assuming, say, $\theta = 0$. In Section 9.2 we gave a representation of the sequence $Z_1\sqrt{\mathcal{I}_1}, \ldots, Z_K\sqrt{\mathcal{I}_K}$ as partial sums of independent normal variates $Y_1, \ldots, Y_K$. It follows from this representation that $p(k, z \,|\, z_1;\, \theta)$ is the same as the sub-density at analysis $k-1$ arising in a "shifted" group sequential test with a maximum of $K-1$ analyses conducted at information levels $\mathcal{I}_2 - \mathcal{I}_1, \ldots, \mathcal{I}_K - \mathcal{I}_1$ and continuation regions

$$\frac{(\mathcal{C}_2\sqrt{\mathcal{I}_2} - z_1\sqrt{\mathcal{I}_1})}{\sqrt{(\mathcal{I}_2 - \mathcal{I}_1)}}, \;\ldots,\; \frac{(\mathcal{C}_K\sqrt{\mathcal{I}_K} - z_1\sqrt{\mathcal{I}_1})}{\sqrt{(\mathcal{I}_K - \mathcal{I}_1)}}$$

on the Z-statistic scale. Hence, $p(k, z \,|\, z_1; \theta)$ may be calculated using formulae (9.4) to (9.6).

Emerson & Fleming (1990) compare the UMVUE with several other estimators of θ calculated after four-stage Pocock and O'Brien & Fleming tests. In these examples, the lack of bias of the UMVUE is accompanied by a relatively high variance and Whitehead's bias-adjusted estimator, $\tilde{\theta}$, has a noticeably lower mean squared error. Thus, although it is impressive that a minimum variance unbiased estimator can be found for this problem, this estimator does not necessarily have great practical value.

9.3.4 *Bias Correction for MLE of a Secondary Endpoint*

Suppose a group sequential test monitors the primary endpoint of a trial but the treatment effect on a secondary endpoint is also estimated when the trial terminates, using the data available at this point. Then, if patient responses on the primary and secondary endpoints are correlated, bias may occur in the MLE of this secondary endpoint parameter. Whitehead (1986b) gives examples of this problem and addresses the analysis of secondary endpoints on termination of a group sequential study.

Suppose an individual's primary and secondary responses are bivariate normal with correlation ρ. For patient i on treatment A, the primary response is

$$X_{A1i} \sim N(\mu_{A1}, \sigma_1^2)$$

and the secondary response is

$$X_{A2i} \sim N(\mu_{A2}, \sigma_2^2)$$

with $\mathrm{Corr}(X_{A1i}, X_{A2i}) = \rho$. Similarly, patient i on treatment B has responses

$$X_{B1i} \sim N(\mu_{B1}, \sigma_1^2) \quad \text{and} \quad X_{B2i} \sim N(\mu_{B2}, \sigma_2^2)$$

and $\mathrm{Corr}(X_{B1i}, X_{B2i}) = \rho$. The treatment effects on the primary and secondary endpoints are $\theta_1 = \mu_{A1} - \mu_{B1}$ and $\theta_2 = \mu_{A2} - \mu_{B2}$, respectively.

Consider a group sequential design where the bias in the MLE $\hat{\theta}_1$ is

$$b_1(\boldsymbol{\theta}) = E_{\boldsymbol{\theta}}(\hat{\theta}_1) - \theta_1$$

when the true treatment effects are $\boldsymbol{\theta} = (\theta_1, \theta_2)$. Note that $E_{\boldsymbol{\theta}}(\hat{\theta}_1)$ depends on θ_1 and not on θ_2. Whitehead (1986b) shows that the MLE $\hat{\theta}_2$ has bias

$$b_2(\boldsymbol{\theta}) = E_{\boldsymbol{\theta}}(\hat{\theta}_2) - \theta_2 = \frac{\sigma_2}{\sigma_1} \rho \, b_1(\boldsymbol{\theta})$$

when the true treatment effects are $\boldsymbol{\theta} = (\theta_1, \theta_2)$. Interestingly, this bias depends on θ_1, not on θ_2. As for the primary endpoint, we can adjust the MLE, $\hat{\theta}_2$, by subtracting an estimate of its bias, giving the simple adjusted MLE

$$\bar{\theta}_2 = \hat{\theta}_2 - \frac{\sigma_2}{\sigma_1} \rho \, b_1(\hat{\boldsymbol{\theta}}).$$

Just as for estimation of θ_1, the bias in the MLE $\hat{\theta}_2$ is largely eliminated in the adjusted estimator. Indeed, after changing the vertical scale by the appropriate factor, plots of the bias functions of $\hat{\theta}_2$ and $\bar{\theta}_2$ have exactly the same form as the plots of the bias functions of $\hat{\theta}_1$ and $\bar{\theta}_1$ in Figure 9.5.

9.4 *P*-values

We return to the problem of testing the null hypothesis H_0: $\theta \leq 0$ against a one-sided alternative or H_0: $\theta = 0$ against a two-sided alternative in the group sequential tests described in previous chapters. Rather than simply announcing acceptance or rejection of H_0, we can enlarge on this by reporting the P-value of the observed data for testing H_0.

Let Ω be the sample space defined by the group sequential design, that is, the set of all pairs (k, z) where $z \notin \mathcal{C}_k$ so that the test can terminate at analysis k with $Z_k = z$. We denote the outcome of a study by (T, Z) where T is the analysis at which termination occurs and $Z = Z_T$ is the final Z-value. The P-value on termination of the trial is the minimum significance level under which a test defined on the sample space Ω can reject H_0 on seeing the observed outcome (t^*, z^*), say, with smaller P-values indicating stronger evidence against H_0. For a continuous response distribution, the P-value is uniformly distributed when $\theta = 0$, so $Pr_{\theta=0}\{P\text{-value} \leq p\} = p$ for all $0 \leq p \leq 1$.

The P-value for testing H_0 on observing $(T, Z) = (t^*, z^*)$ can also be stated as

$$Pr_{\theta=0}\{\text{Obtain } (T, Z) \text{ as extreme or more extreme than } (t^*, z^*)\},$$

where "extreme" refers to the ordering of Ω implicit in the construction of tests of H_0 at different significance levels. However, since the number of observations varies between different outcomes, the monotone likelihood ratio property does not hold on Ω. Consequently, there is no single natural way to order possible outcomes and one must choose between a number of orderings of Ω, all of which have intuitive appeal. We write $(t', z') \succ (t, z)$ to denote that (t', z') is above (t, z) in a given ordering. Four orderings which have received attention are described below.

A *Stage-wise ordering.* This ordering was first proposed by Armitage (1957) and later used by Siegmund (1978), Fairbanks & Madsen (1982) and Tsiatis, Rosner & Mehta (1984). It can be used if the continuation regions $\mathcal{C}_1, \ldots, \mathcal{C}_{K-1}$ are intervals, which is true for most of the designs in the previous chapters but not for the "inner wedge" designs of Chapter 6. We define the ordering in a counter-clockwise sense around the continuation region, considering in decreasing order of priority: the boundary crossed, the analysis at which stopping occurs, and the Z-value on termination. In the resulting ordering $(t', z') \succ (t, z)$ if any one of the following three conditions holds:

$$\begin{array}{lll} \text{(i)} & t' = t \text{ and } & z' \geq z, \\ \text{(ii)} & t' < t \text{ and } & z' \geq b_{t'}, \\ \text{(iii)} & t' > t \text{ and } & z \leq a_t. \end{array}$$

B *MLE ordering.* This was used by Armitage (1958a) in connection with a test for a binomial proportion, and investigated for normal data by Emerson & Fleming (1990). Outcomes are ordered according to the value of the MLE, $\hat{\theta} = Z_T/\sqrt{\mathcal{I}_T}$, with

$$(t', z') \succ (t, z) \text{ if } z'/\sqrt{\mathcal{I}_{t'}} > z/\sqrt{\mathcal{I}_t}.$$

This is sometimes termed the *sample mean ordering,* as it is orders outcomes by $\bar{X}_T$ in the one-sample problem and by $\bar{X}_{AT} - \bar{X}_{BT}$ in the two-sample problem.

C *Likelihood ratio (LR) ordering.* In this ordering,

$$(t', z') \succ (t, z) \text{ if } z' > z.$$

Chang (1989) shows this is the ordering induced by consideration of the signed likelihood ratio statistic for testing H_0: $\theta = 0$ against a general alternative.

D *Score test ordering.* This ordering declares

$$(t', z') \succ (t, z) \text{ if } z'\sqrt{\mathcal{I}_{t'}} > z\sqrt{\mathcal{I}_t}.$$

Rosner & Tsiatis (1988) show that this ordering ranks outcomes by values of the score statistic for testing H_0: $\theta = 0$.

For any given ordering, we can use the formulae given in Section 9.2 to compute a one-sided upper P-value on observing $(T, Z) = (t^*, z^*)$ as

$$Pr_{\theta=0}\{(T, Z) \succeq (t^*, z^*)\}$$

and a one-sided lower P-value is

$$Pr_{\theta=0}\{(T, Z) \preceq (t^*, z^*)\}.$$

The two-sided P value is twice the smaller of these two quantities. As an example, suppose we use the stage-wise ordering and the test terminates by exiting the upper boundary or by reaching the last analysis K. Then the one-sided upper P-value is

$$\int_{b_1}^{\infty} p(1, z; 0)dz + \ldots + \int_{b_{t^*-1}}^{\infty} p(t^* - 1, z; 0)dz + \int_{z^*}^{\infty} p(t^*, z; 0)dz,$$

where $p(k, z; \theta)$ is defined in (9.4). This ordering implies that when the test exits the upper boundary, termination at an earlier stage decreases the one-sided upper P-value and, similarly, when the lower boundary is crossed, earlier termination reduces the lower P-value. This is not necessarily the case for the other orderings.

Of the four orderings, we prefer (A), the stage-wise ordering. It automatically ensures that

(i) the P-value is less than the significance level α of the group sequential test if and only if H_0 is rejected,

an important consideration for internal consistency. Also, unlike the other orderings, the stage-wise ordering has the property that

(ii) the P-value does not depend on information levels or group sizes beyond the observed stopping stage T.

This is a key feature when increments in information are unpredictable since the P-values based on the other three orderings cannot be computed in such situations. The stage-wise ordering fits naturally with the error spending tests of Chapter 8, and we shall see in the next section that it has similar advantages in relation to confidence intervals computed on termination.

Although it is unfortunate that the definition of a P-value should depend on a choice of ordering on Ω, it should be stressed that different orderings yield very similar P-values for many outcomes. The choice of ordering has the greatest effect when a test stops with a value $Z_t = z_t$ a long way outside the continuation region $\mathcal{C}_t$. However, the probability of observing such large "overshoot" or "undershoot" is often small since the sample path is likely to have crossed an earlier boundary on its way to the value z_t at stage t.

9.5 Confidence Intervals

Equal tailed $(1-\alpha)$-level confidence intervals for θ can be derived by inverting a family of hypothesis tests with two-sided Type I error probability α. Consider an ordering of the sample space Ω, with the outcome variable (T, Z) as defined in the previous section, in which we write $(t', z') \succ (t, z)$ to denote that (t', z') is higher in the ordering than (t, z) and higher values of (T, Z) are typical of higher values of θ. For any given value θ_0 we can find pairs $(t_u(\theta_0), z_u(\theta_0))$ and $(t_\ell(\theta_0), z_\ell(\theta_0))$ such that

$$Pr_{\theta=\theta_0}\{(T, Z) \succeq (t_u(\theta_0), z_u(\theta_0))\} = \alpha/2 \tag{9.10}$$

and

$$Pr_{\theta=\theta_0}\{(T, Z) \preceq (t_\ell(\theta_0), z_\ell(\theta_0))\} = \alpha/2. \tag{9.11}$$

Then, the acceptance region

$$A(\theta_0) = \{(t, z) \colon (t_\ell(\theta_0), z_\ell(\theta_0)) \prec (t, z) \prec (t_u(\theta_0), z_u(\theta_0))\} \tag{9.12}$$

defines a two-sided hypothesis test of $H\colon \theta = \theta_0$ with Type I error probability α. Standard arguments imply that the set

$$\{\theta_0 \colon (T, Z) \in A(\theta_0)\} \tag{9.13}$$

obtained by inverting this family of tests is a $(1-\alpha)$-level, equal tailed confidence set for θ. This confidence set contains all values θ_0 for which the hypothesis $H\colon \theta = \theta_0$ is accepted.

If $Pr_\theta\{(T, Z) \succeq (t, z)\}$ is an increasing function of θ for each $(t, z) \in \Omega$, we say the distributions on Ω are *stochastically ordered* with respect to θ, and we shall refer to this as the *monotonicity condition*. In this case, $(t_\ell(\theta), z_\ell(\theta))$ and $(t_u(\theta), z_u(\theta))$ increase in the ordering on Ω as θ increases and the set (9.13) is an interval. When the observed value of (T, Z) is (t^*, z^*), this interval is (θ_L, θ_U), where

$$Pr_{\theta_L}\{(T, Z) \succeq (t^*, z^*)\} = Pr_{\theta_U}\{(T, Z) \preceq (t^*, z^*)\} = \alpha/2. \tag{9.14}$$

For a given ordering, the probabilities in (9.14) can be evaluated numerically using the recursive computation of sub-densities described in Section 9.2. Solutions θ_L and θ_U can be found by a search method such as successive bisection. If the monotonicity condition fails to hold, the set (9.13) may not be an interval and must be calculated directly.

Each ordering of Ω discussed in Section 9.4 in the context of P-values can be used to construct confidence intervals. The first two, (A) the stage-wise ordering and (B) the MLE ordering, are the simplest to apply since their definitions do not depend on the hypothesis being tested. The other two orderings are dependent on the hypothesized value of θ, and the order inequalities in (9.10), (9.11) and (9.12) should be treated as being with respect to the ordering for θ_0.

The likelihood ratio ordering, (C), for θ_0 states $(t', z') \succ (t, z)$ if

$$\mathrm{sign}(\hat{\theta}' - \theta_0)\, \frac{p(t', z';\, \hat{\theta}')}{p(t', z';\, \theta_0)} \;>\; \mathrm{sign}(\hat{\theta} - \theta_0)\, \frac{p(t, z;\, \hat{\theta})}{p(t, z;\, \theta_0)},$$

where $\hat{\theta}' = z'/\sqrt{\mathcal{I}_{t'}}$ and $\hat{\theta} = z/\sqrt{\mathcal{I}_t}$ are the respective MLEs of θ. Using (9.3), we see that this condition reduces to

$$z' - \theta_0\sqrt{\mathcal{I}_{t'}} \;>\; z - \theta_0\sqrt{\mathcal{I}_t}.$$

Finally, the score test ordering, (D), for θ_0 declares $(t', z') \succ (t, z)$ if

$$\left.\frac{\partial}{\partial\theta}\log p(t', z'; \theta)\right|_{\theta=\theta_0} \;>\; \left.\frac{\partial}{\partial\theta}\log p(t, z; \theta)\right|_{\theta=\theta_0},$$

which reduces to

$$z'\sqrt{\mathcal{I}_{t'}} - \theta_0\mathcal{I}_{t'} \;>\; z\sqrt{\mathcal{I}_t} - \theta_0\mathcal{I}_t$$

or, equivalently,

$$(\hat{\theta}' - \theta_0)\mathcal{I}_{t'} \;>\; (\hat{\theta} - \theta_0)\mathcal{I}_t.$$

Each of the orderings (A) to (D) defines a notion of distance of the outcome (T, Z) from the hypothesis H: $\theta = \theta_0$. Rosner & Tsiatis (1988) refer to (C) and (D) as the "standardized distance" and "distance" orderings, respectively.

Let us now consider some desiderata for confidence intervals following a group sequential test.

(i) *The confidence set given by* (9.13) *should be an interval.* This is guaranteed to occur if the monotonicity condition holds. Computation is simpler in this case and reduces to solving (9.14) for θ_L and θ_U. If the set (9.13) is not an interval, we can define θ_L and θ_U to be the infimum and supremum of this set but, even then, finding θ_L and θ_U is more difficult than in the monotone case. Also, the confidence interval (θ_L, θ_U) will be conservative, with its coverage probability exceeding $1 - \alpha$ for some values of θ.

(ii) *The confidence interval should agree with the original test.* If the original group sequential design is constructed as an α level, two-sided test of the null hypothesis H_0: $\theta = \theta_0$ for a certain θ_0, the $1-\alpha$ confidence interval should exclude θ_0 if and only if H_0 is rejected. Similarly, following an $\alpha/2$ level group sequential test of H_0: $\theta \le \theta_0$, a $1-\alpha$ confidence interval should lie above θ_0 if and only if the test rejects H_0: $\theta \le \theta_0$. This corresponds to the consistency property (i) of P-values described toward the end of Section 9.4.

(iii) *The confidence interval should contain the MLE,* $\hat{\theta} = Z/\sqrt{\mathcal{I}_T}$. Alternatively, we might ask that one of the other point estimates discussed in Section 9.3 should lie in the confidence interval.

(iv) *Narrower confidence intervals are to be preferred.* A related feature is that the probability of the confidence interval containing particular values of θ other than the true value should be small.

(v) *The confidence interval should be well-defined when information levels are unpredictable.* This ensures that confidence intervals can be obtained when the experimental design follows the error spending approach of Chapter 8.

Confidence intervals based on the four orderings (A) to (D) vary in their performance with respect to these desiderata.

With regard to desideratum (i), note first that it is tacitly assumed in the definition of the stage-wise ordering, (A), that the group sequential test has interval continuation regions: this is not the case for "inner wedge" designs and so ordering (A) cannot be used for such tests. For one-sided or two-sided tests with just an upper and a lower boundary, the stage-wise ordering is well-defined and satisfies the monotonicity condition, so (9.13) does yield an interval. We establish the monotonicity property by a coupling argument which relates outcomes under parameter values θ and θ'. Suppose $\theta' = \theta + \delta$ where $\delta > 0$. Under θ, the joint distribution of $\{Z_1, \ldots, Z_K\}$ is that of the sequence of variables $\{(Y_1 + \ldots + Y_k)/\sqrt{\mathcal{I}_k};\, k = 1, \ldots, K\}$ where the Y_k are independent and $Y_k \sim N(\Delta_k \theta, \Delta_k)$. If we define $Y'_k = Y_k + \Delta_k \delta$ and $Z'_k = (Y'_1 + \ldots + Y'_k)/\sqrt{\mathcal{I}_k}$, $k = 1, \ldots, K$, the sequence $\{Z'_1, \ldots, Z'_K\}$ has the joint distribution appropriate to θ'. Now, by construction, $Z'_k > Z_k$ for all k and hence $(T', Z') \succ (T, Z)$, where $T = \min\{k : Z_k \notin \mathcal{C}_k\}$, $Z = Z_T$, $T' = \min\{k : Z'_k \notin \mathcal{C}_k\}$ and $Z' = Z_{T'}$. Since (T, Z) and (T', Z') have the distributions of the sufficient statistic pair under θ and θ', respectively, we deduce that $Pr_\theta\{(T, Z) \succeq (t, z)\}$ increases with θ for any (t, z) and the monotonicity condition is satisfied. This proof is a special case of that used by Bather (1988) to establish monotonicity for the stage-wise ordering applied to response variables in the general exponential family, which includes the normal, binomial and many other distributions.

The same form of argument cannot be used to prove the monotonicity property for the MLE ordering since, in the above coupling, it is possible that $\theta' > \theta$ leads to $T' < T$ but $Z'_{T'}/\sqrt{\mathcal{I}_{T'}} < Z/\sqrt{\mathcal{I}_T}$ and so $(T', Z'_{T'}) \preceq (T, Z)$ in the MLE ordering. This happens when the Z_k sequence crosses a boundary with a particularly large overshoot and the coupled Z'_k sequence stops earlier with little overshoot and a smaller MLE of θ. Nevertheless, Emerson (1988, Sec. 4.2) has proved that $Pr_\theta\{\hat{\theta} > w\}$ is an increasing function of θ for any fixed w and the monotonicity condition is in fact satisfied for the MLE ordering. This property was shown for binomial responses by Armitage (1958a).

There is no result concerning monotonicity for orderings (C) and (D). Indeed, Emerson & Fleming (1990, p. 880) report numerical studies in which the set (9.13) occasionally fails to be an interval when ordering (C) is used. Rosner & Tsiatis (1988, p. 726) report a similar phenomenon arising with ordering (D). However such occurrences are rare and the occasional extension of a confidence set to a surrounding interval need not be regarded as a serious difficulty.

The consistency property (ii) holds for a one-sided test if every outcome in Ω for which H_0: $\theta \le \theta_0$ is rejected is higher in the ordering under θ_0 than every outcome where H_0 is accepted. For a two-sided test of H_0: $\theta = \theta_0$, outcomes where H_0 is accepted must be in the middle of the ordering under θ_0, higher than every outcome where H_0 is rejected in favor of $\theta < \theta_0$ and lower than every outcome for which H_0 is rejected in favor of $\theta > \theta_0$.

This consistency property clearly holds for intervals constructed using the stage-wise ordering. To see this in the case of a two-sided test of H_0: $\theta = \theta_0$, suppose a test with interval continuation regions $\mathcal{C}_k = (a_k, b_k)$ terminates at $(T, Z) = (t^*, z^*)$, rejecting H_0

after crossing the upper boundary. Then $z^* > b_{t^*}$ and

$$\begin{aligned} Pr_{\theta_0}\{(T,Z) \succeq (t^*,z^*)\} &= \sum_{k=1}^{t^*-1} \int_{b_k}^{\infty} p(k,u;\theta_0)du + \int_{z^*}^{\infty} p(t^*,u;\theta_0)du \\ &\leq \sum_{k=1}^{t^*} \int_{b_k}^{\infty} p(k,u;\theta_0)du \leq \sum_{k=1}^{K} \int_{b_k}^{\infty} p(k,u;\theta_0)du = \alpha/2. \end{aligned}$$

Similarly, $Pr_{\theta_0}\{(T,Z) \preceq (t^*,z^*)\} \leq \alpha/2$ if the lower boundary is crossed and H_0 rejected, and in neither case will θ_0 lie in the interval (θ_L, θ_U) defined by (9.14). Conversely, if the test stops to accept H_0 with $(T,Z) = (t^*,z^*)$, then $Pr_{\theta_0}\{(T,Z) \succeq (t^*,z^*)\} > \alpha/2$, $Pr_{\theta_0}\{(T,Z) \preceq (t^*,z^*)\} > \alpha/2$, and θ_0 does lie in (θ_L, θ_U).

In order for the MLE ordering to satisfy (ii) in a one-sided test of H_0: $\theta = \theta_0$ against $\theta > \theta_0$, values of $\hat{\theta}$ must be larger for all outcomes where H_0 is rejected than for those where it is accepted. For a two-sided test, we require values of $\hat{\theta}$ to be largest for those outcomes where H_0: $\theta = \theta_0$ is rejected in favor of $\theta > \theta_0$ and smallest for outcomes where H_0 is rejected in favor of $\theta < \theta_0$. Since $\hat{\theta} = Z/\sqrt{\mathcal{I}_T}$, this implies that a necessary and sufficient condition for a test with interval continuation regions $\mathcal{C}_k = (a_k, b_k)$ is

$$\frac{b_k}{\sqrt{\mathcal{I}_k}} \geq \frac{b_K}{\sqrt{\mathcal{I}_K}} \geq \frac{a_K}{\sqrt{\mathcal{I}_K}} \geq \frac{a_k}{\sqrt{\mathcal{I}_k}} \quad \text{for all } k = 1, \ldots, K-1. \tag{9.15}$$

This condition is satisfied for all the proposed group sequential tests of which we are aware. It is illustrated graphically in Figure 9.6, which shows the boundaries of two-sided tests with four analyses plotted on the $(S_k, \mathcal{I}_k)$ scale. In example (a), any ray representing constant $\hat{\theta} = S_k/\mathcal{I}_k = Z_k/\sqrt{\mathcal{I}_k}$ which passes through the upper or lower boundary cannot go through (a_K, b_K) where H_0 is accepted; thus (9.15) holds. In example (b), it *is* possible for a ray to pass through both a rejection and acceptance boundary and (9.15) is violated—but this boundary has a very unnatural shape for a group sequential test.

The consistency property (ii) does not necessarily hold true for the likelihood ratio (LR) ordering. Emerson & Fleming (1990, Table 1) give an example of a symmetric one-sided test of H_0: $\theta = -0.85$ versus H_1: $\theta = 0.85$ with four analyses and Type I error probability 0.05. If the final result is $(T, Z) = (4, \epsilon)$, where ϵ is small and positive, the test rejects H_0, but the equal tailed 90% confidence interval under the LR ordering is approximately $(-0.89, 0.89)$, which includes the rejected value of -0.85. Rosner & Tsiatis (1988, p. 726) report numerical studies in which the score test ordering also produces occasional inconsistencies between the confidence interval on termination and the outcome of the group sequential test.

With regard to property (iii), the stage-wise ordering does not always give a confidence interval containing the MLE of θ. Tsiatis, Rosner & Mehta (1984, Table 2) give examples where this can happen even for typical confidence levels such as 0.95. However the MLE is only excluded when the test stops at analysis 2 or later with a large overshoot of the stopping boundary and this usually has a very small probability of occurring. Other studies, such as those reported by Rosner & Tsiatis (1988) and Emerson & Fleming (1990), indicate that this phenomenon does not occur under orderings (B), (C) and (D) for standard confidence levels $1 - \alpha$.

The mean length of a confidence interval is obtained by averaging over all possible outcomes in Ω. This mean length depends on the group sequential design used, the confidence level $1 - \alpha$ and the true value of θ. Only limited studies based on numerical comparisons have been reported (e.g., Emerson & Fleming, 1990), and these suggest a slight advantage to confidence intervals based on the MLE and LR orderings but only of the order of 1% or 2%. Rosner & Tsiatis (1988, Tables 2 and 3) present a comparison between the stage-wise and score statistic orderings with respect to the probability that

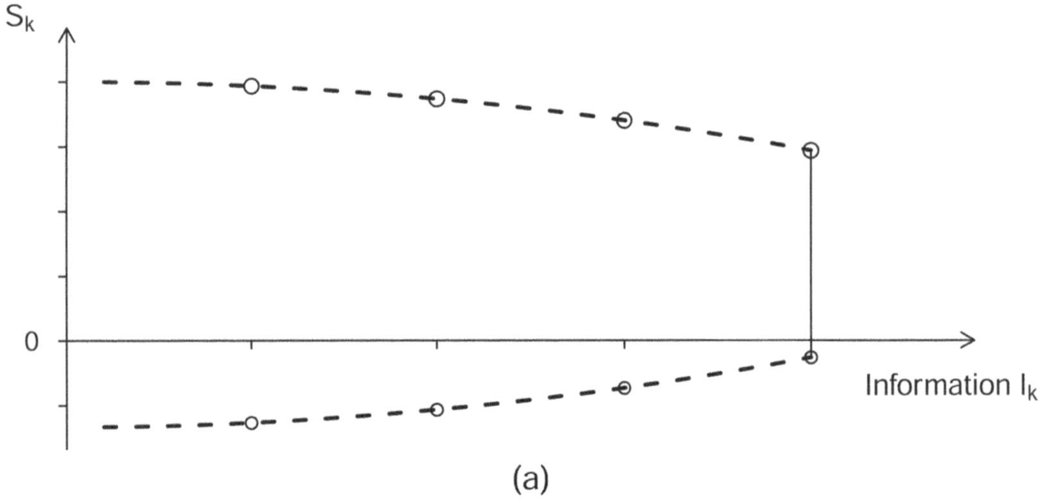

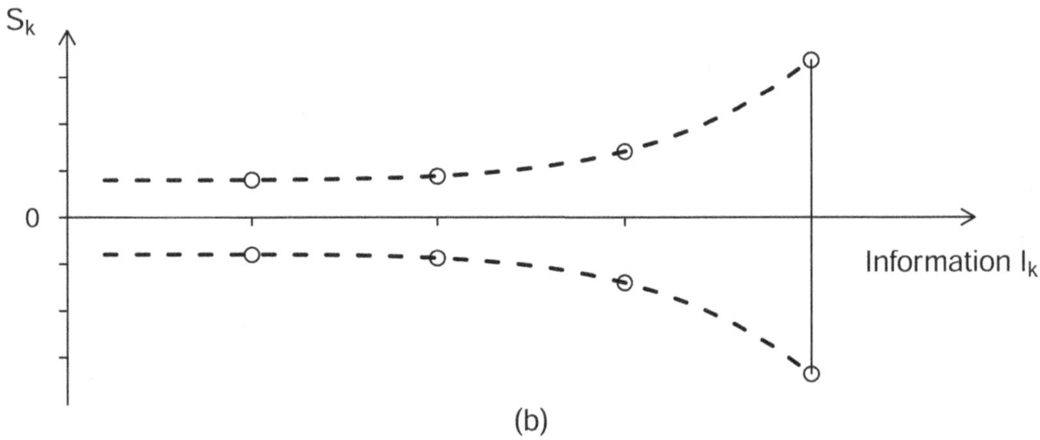

Figure 9.6 *Stopping boundaries where* (a) *the consistency condition does hold for the MLE ordering and* (b) *this condition does not hold*

confidence intervals contain certain incorrect values of θ. In the one situation investigated, the stage-wise ordering has smaller probabilities of covering an incorrect θ value when θ is close to zero and the score statistic ordering is better for large θ. However, in all cases the differences are small.

As remarked in the discussion of P-values in Section 9.4, the stage-wise ordering, (A), is the only one of the four orderings which can be used without knowledge of information levels beyond the observed stopping stage T. Hence this is really the only method available for use when information levels are unpredictable. Kim & DeMets (1987b) give details of the use of this ordering in conjunction with designs based on error spending functions. Emerson & Fleming (1990, Sec. 5) suggest a "patch" to the MLE ordering method in which confidence intervals are calculated assuming that there would have been one more analysis, i.e., $K = T + 1$, with a specified final information level. The authors claim this is a worst case scenario in some sense and present supporting numerical results, but it is not clear that similar results will always be found in other situations.

In summary, we believe confidence interval construction based on the stage-wise ordering to be the preferred method for designs with interval continuation regions. It produces true intervals, consistent with the decision of the group sequential test, and it is the only method available for use with unpredictable information sequences. It has the disadvantage of slightly longer interval lengths, on average, than some other methods. A notable property of this interval is that it reduces to the naive interval $(\hat{\theta}_1 - z_{\alpha/2}/\sqrt{\mathcal{I}_1},\, \hat{\theta}_1 + z_{\alpha/2}/\sqrt{\mathcal{I}_1})$ if the test stops at the first analysis. If $T \geq 2$, the interval is not symmetric about the MLE of θ but shifted toward zero, and in extreme cases it may not even contain the MLE. Although this shift may appear disturbing at first sight, Hughes & Pocock (1988) and Pocock & Hughes (1989) have argued that such shifts may have a desirable effect on bias or effectively incorporate prior information.

For designs with non-interval continuation regions, such as two-sided tests with inner wedges, the stage-wise method is not applicable and we recommend the MLE ordering method. Armitage (1957, p. 25) has suggested a hybrid ordering in which outcomes falling outside the lower and upper boundaries are ordered from each extreme by the stage-wise criterion, while, in between, the points in the inner wedge are ordered by the MLE criterion. We know of no investigations of the properties of this ordering.

Even though normality of the observations is assumed for the construction of the group sequential test, it may be desirable to avoid this assumption when analyzing the results. Chuang & Lai (1998) describe a nonparametric bootstrap procedure for constructing a robust confidence interval which involves resampling the stopped sequence of observations; they consider both the stage-wise and LR orderings of the sample space. Exact methods for inference following a sequential test may be developed for particular response distributions: examples of this will be seen in Sections 16.1.3, 16.2.2 and 16.3, where we describe the construction of confidence intervals following tests for binomial probabilities.

9.6 Median Unbiased Point Estimate

An alternative form of estimate for the treatment effect θ is a "median unbiased" estimate, whose distribution has median equal to the true value of θ. If (θ_L, θ_U) is an equal tailed $(1-\alpha)$-level confidence interval for θ, then

$$Pr\{\theta_L \geq \theta\} \;=\; Pr\{\theta_U \leq \theta\} \;=\; \alpha/2. \tag{9.16}$$

When $\alpha/2 = 0.5$, the confidence interval contracts to a single point, so $\theta_L = \theta_U$. If we then define the estimate $\hat{\theta}_{Med} = \theta_L = \theta_U$, it follows from (9.16) that

$$Pr\{\hat{\theta}_{Med} \leq \theta\} \;=\; Pr\{\hat{\theta}_{Med} \geq \theta\} \;=\; 0.5$$

and $\hat{\theta}_{Med}$ is a median unbiased estimate. A median unbiased estimate can be derived using any of the available methods for constructing a confidence interval for θ.

9.7 Discussion

The four papers by Robertson, Choodari-Oskooei et al. (2023a,b) and Robertson, Burnett et al. (2025a,b) review methods for obtaining point estimates and confidence intervals after sequential and adaptive trials. The authors provide a critical appraisal of methods that we have described in this chapter, report some additional approaches, and list examples of clinical trials that have applied these methods.

In this chapter we have considered inference on conclusion of a trial conducted using a fixed stopping rule. One may, however, ask what can be done if a rigid stopping rule is not being followed. Suppose a group sequential design has an efficacy boundary that will be rigorously applied and a non-binding futility boundary. Suppose also that the stage-wise ordering, (A), is to be used in calculating a P-value and confidence interval and the test has stopped after crossing the upper boundary, with rejection of H_0: $\theta \leq 0$. Then, it is natural to ignore the futility boundary in calculating a P-value and the lower endpoint of a confidence interval for θ on termination as this will ensure agreement with the outcome of the hypothesis test of H_0. In contrast, we recommend assuming the futility boundary is strictly applied when calculating the upper endpoint of a confidence interval for θ as this will ensure the confidence interval achieves the desired coverage level with some conservatism. We also recommend assuming the futility boundary is strictly applied when computing the bias function of the MLE of θ for use in calculating an adjusted MLE: this will lead to a slightly lower adjusted estimate than ignoring the futility boundary and avoid any suggestion that the estimate is artificially high.

There are further questions to address when a group sequential trial is stopped at an interim analysis but additional data accrue after the decision to terminate the trial has been made. We shall present methods for designing such a trial in Chapter 14 and we refer in Section 14.3 to methods for making inferences at the conclusion of the study.

The adaptive designs that we shall describe in later chapters present further challenges and we shall describe the methods available for inference on termination of such designs in Sections 23.8, 26.6, 27.6 and 28.6.

Another possibility is that investigators wish to report interval estimates at several times during the course of the trial. We present methods for calculating such intervals with overall protection of the Type I error rate in the next chapter.

Chapter 10

Repeated Confidence Intervals

10.1 Introduction

In Section 9.5 we discussed how a confidence interval for a parameter, θ, could be constructed upon termination of a group sequential procedure. In this chapter we are concerned with the construction of a sequence of interval estimates for θ that we can form at *any* interim look at the data.

The multiple-looks problem affects the construction of confidence intervals just as it affects error rates of hypothesis tests. Suppose, for example, independent normal observations $X_1, \ldots, X_{mK}$ with mean θ and known variance σ^2 are available sequentially. Suppose also that after each group k of m observations, $k = 1, \ldots, K$, we form the usual (or "naive") 95% confidence interval for θ, $(\bar{X}_{mk} - 1.96\,\sigma/\sqrt{\{mk\}},\ \bar{X}_{mk} + 1.96\,\sigma/\sqrt{\{mk\}})$, based on the observations accumulated so far. Then, the probability that all K intervals so formed contain the true value of θ is certainly less than 95%. These probabilities are listed in Table 10.1 for selected values of K and correspond to the probabilities in Table 2 of Armitage, McPherson & Rowe (1969).

Repeated confidence intervals (RCIs) for a parameter θ are defined as a sequence of intervals I_k, $k = 1, \ldots, K$, for which a simultaneous coverage probability is maintained at some level, $1 - \alpha$ say. The defining property of a $(1 - \alpha)$-level sequence of RCIs for θ is

$$Pr_\theta\{\theta \in I_k \text{ for all } k = 1, \ldots, K\} = 1 - \alpha \quad \text{for all } \theta. \tag{10.1}$$

Here, each I_k, $k = 1, \ldots, K$, is an interval computed from the information available at analysis k. We shall use the notation $I_k = (\underline{\theta}_k, \overline{\theta}_k)$ for these intervals.

The interval I_k provides a statistical summary of the information about the parameter θ at the kth analysis, automatically adjusted to compensate for repeated looks at the accumulating data. As such, it can be presented to a Data and Safety Monitoring Board (DSMB) to be considered with all other relevant information when discussing early termination of a study.

If τ is *any* random stopping time that takes values in $\{1, \ldots, K\}$, the guarantee of simultaneous coverage (10.1) implies that the probability I_τ contains θ must be at least $1 - \alpha$, i.e.,

$$Pr_\theta\{\theta \in I_\tau\} \geq 1 - \alpha \quad \text{for all } \theta. \tag{10.2}$$

This property shows that an RCI can be used to summarize information about θ on termination and the confidence level $1 - \alpha$ will be maintained, regardless of how the decision to stop the study was reached. In contrast, the methods of Chapter 9 for constructing confidence intervals on termination rely on a particular stopping rule being specified at the outset and strictly enforced.

When a study is monitored using RCIs, the intervals computed at interim analyses might also be reported at scientific meetings. The basic property (10.1) ensures that these interim results will not be "over-interpreted". Here, over-interpretation refers to the fact that, when the selection bias or optional sampling bias of reported results is ignored, data

DOI: 10.1201/9781584888482-10

Table 10.1 *Simultaneous coverage probabilities of naive 95% confidence intervals*

Number of analyses, K	*Probability that all K intervals contain θ*
1	0.95
2	0.92
3	0.89
4	0.87
5	0.86
10	0.81
20	0.75
50	0.68
∞	0

may seem more significant than warranted, and this can lead to adverse effects on accrual and drop-out rates, and pressure to unblind or terminate a study prematurely.

Experience shows that in many medical trials it is unrealistic to expect a designated stopping rule to be enforced under all circumstances, and sometimes there is no planned stopping rule at all. Both the efficacy and futility stopping boundaries of the group sequential designs discussed in earlier chapters are often used only as guidelines, not as strict rules (DeMets, 1984). Other factors, such as treatment side-effects, financial cost or reports of new scientific developments from outside the trial, will often influence the DSMB in their decision to stop or to continue a trial; see Pocock (1996) for a discussion of the role external evidence can play in monitoring a clinical trial. As an example, in the NSABP Breast Cancer Prevention Trial (Fisher et al., 1998), the O'Brien & Fleming (1979) statistical stopping boundary was crossed in April 1996 but, for very good reasons, the decision to stop the trial was not made until a later interim analysis in March 1998. The opposite happened in a CALGB 3×2 factorial trial reported by Goldhirsch et al. (1998), which examined the use of doxorubicin (Adriamycin) and Taxol as adjuvant therapy in breast cancer treatment. That trial was unblinded and results reported at the first interim analysis in April 1998 because of strong evidence of benefit of Taxol, even though the statistical stopping boundaries had not been crossed. Evidently, the decision to stop a clinical trial is complex and involves many subjective factors, indeed, Meier (1975, p. 526) describes the problem as political, rather than medical, legal or statistical. In such circumstances, the probability statements associated with conventional group sequential tests are of limited value since the stopping rule on which they rely may not be enforced. RCIs do not depend on a particular stopping rule and so remain valid in these situations.

The historical antecedent of repeated confidence intervals is the confidence sequence introduced by Robbins (1970) in his 1969 Wald Lectures. He defined an infinite sequence of intervals $\{I_n;\, n = 1, 2, \dots\}$, where I_n is calculated from the first n observations and, for a $(1 - \alpha)$-level sequence,

$$Pr_\theta\{\theta \in I_n \text{ for all } n = 1, 2, \dots\} = 1 - \alpha \quad \text{for all } \theta.$$

Initially the theory of confidence sequences had little practical impact, one reason being that the intervals turn out to be extremely wide and unacceptable to investigators accustomed to fixed sample intervals. However, sequences of RCIs are finite in number and much narrower. The development of RCIs is due to Jennison (1982) and Jennison & Turnbull (1984, 1989); similar ideas were introduced by Lai (1984).

Repeated confidence intervals are formed by inverting a family of two-sided group sequential tests with Type I error probability α, of the type described in Chapters 2, 4 and 8. We have seen in Chapter 3 that many common testing problems concerning a parameter θ lead to consideration of a sequence of test statistics $Z_1, \ldots, Z_K$ with the canonical joint distribution (3.1), given observed information levels $\mathcal{I}_1, \ldots, \mathcal{I}_K$. Here, θ might represent a difference, $\mu_A - \mu_B$, in mean response between two treatments as in Section 3.2.1, or the logarithm of a hazard ratio for comparing two survival distributions under the proportional hazards model of Section 4.5. Suppose we define

$$Z_k(\theta_0) = Z_k - \theta_0\sqrt{\mathcal{I}_k}, \quad k = 1, \ldots, K.$$

It follows from (3.1) that a two-sided group sequential test of the hypothesis H_0: $\theta = \theta_0$ with Type I error probability α has the form:

$$\text{Reject } H_0 \text{ at stage } k \text{ if } |Z_k(\theta_0)| \geq c_k(\alpha), \quad k = 1, \ldots, K, \tag{10.3}$$

where $c_1(\alpha), \ldots, c_K(\alpha)$ are critical values appropriate to the particular form of test and information sequence $\mathcal{I}_1, \ldots, \mathcal{I}_K$. The argument θ_0 of Z_k is displayed in this notation to signify the explicit dependence on the hypothesized value θ_0, whereas we have previously focused on a single θ_0, typically $\theta_0 = 0$, in any one problem.

The argument θ_0 in (10.3) indexes a family of hypothesis tests and the RCI sequence $\{I_k;\, k = 1, \ldots, K\}$ is defined from these by

$$I_k = \{\theta_0 \colon |Z_k(\theta_0)| < c_k(\alpha)\}, \quad k = 1, \ldots, K. \tag{10.4}$$

In inverting a family of group sequential tests to produce a sequence of RCIs, we are using the familiar duality between confidence sets and hypothesis tests. The construction of the test and the critical values $c_1(\alpha), \ldots, c_K(\alpha)$ ensure that

$$Pr_\theta\{|Z_k(\theta)| \geq c_k(\alpha) \text{ for some } k = 1, \ldots, K\} = \alpha.$$

Hence,

$$Pr_\theta\{\theta \in I_k \text{ for all } k = 1, \ldots, K\} = 1 - \alpha$$

and we see that $\{I_1, \ldots, I_K\}$ defined by (10.4) does indeed satisfy the RCI property (10.1). This property is satisfied approximately if $Z_1, \ldots, Z_K$ satisfy (3.1) only approximately.

When the information levels $\mathcal{I}_k$ do not depend on θ, as in the case of normally distributed data with a common variance, the RCI at stage k is simply

$$\left(\{Z_k - c_k(\alpha)\}/\sqrt{\mathcal{I}_k},\, \{Z_k + c_k(\alpha)\}/\sqrt{\mathcal{I}_k}\right), \quad k = 1, \ldots, K, \tag{10.5}$$

or writing $\hat{\theta}^{(k)}$ for $Z_k/\sqrt{\mathcal{I}_k}$, the maximum likelihood estimate of θ at analysis k,

$$\left(\hat{\theta}^{(k)} - c_k(\alpha)/\sqrt{\mathcal{I}_k},\, \hat{\theta}^{(k)} + c_k(\alpha)/\sqrt{\mathcal{I}_k}\right), \quad k = 1, \ldots, K. \tag{10.6}$$

If the information levels do depend on θ, approximate $(1 - \alpha)$-level RCIs may be obtained using $\mathcal{I}_k$ evaluated for the current estimate $\theta^{(k)}$ in (10.5) or (10.6). Examples of this are considered for binary and survival data in Chapters 16 and 17, respectively. Direct application of (10.4) when $\mathcal{I}_k$ depends on θ could possibly lead to a confidence set that is not an interval, although this should only happen in rare pathological cases; see Brookmeyer & Crowley (1982) for such an example. If this problem does arise, it can be resolved by defining the RCI, I_k, as the smallest interval containing the set defined in (10.4). In this case, we can always write I_k in the form $(\underline{\theta}_k, \overline{\theta}_k)$ and the RCI property (10.1) becomes

$$Pr_\theta\{\underline{\theta}_k < \theta < \overline{\theta}_k \text{ for all } k = 1, \ldots, K\} \geq 1 - \alpha. \tag{10.7}$$

The intervals we shall construct will be symmetric and satisfy

$$Pr_\theta\{\underline{\theta}_k < \theta \text{ for all } k = 1, \ldots, K\} = Pr_\theta\{\overline{\theta}_k > \theta \text{ for all } k = 1, \ldots, K\} \approx 1 - \alpha/2 \quad (10.8)$$

when the Z_k follow (3.1). The second equality here is only approximate because of the possibility that $\underline{\theta}_k > \theta$ for some k and $\overline{\theta}_{k'} < \theta$ at some other analysis k'. However, the probability of this event is negligible and the departure from equality can be ignored in practice; see Proschan (1999, Theorem 1). Later in this chapter, we shall use the notation $\underline{\theta}_k(\alpha/2)$ and $\overline{\theta}_k(\alpha/2)$ to denote $\underline{\theta}_k$ and $\overline{\theta}_k$ satisfying (10.8).

The coverage property (10.1) would still be satisfied if I_k were replaced by the intersection of $I_1, \ldots, I_k$, thereby giving narrower confidence intervals. However, the original sequence (10.4) is to be preferred, because the intervals are then functions of the sufficient statistics for θ at each stage. This also avoids the possibility of obtaining an empty confidence interval, although the probability of this occurring is very small; for a discussion of this point, see Freeman (1989).

The construction of a sequence of RCIs depends on the choice of the group sequential test used in (10.4) and the associated critical values $c_1(\alpha), \ldots, c_K(\alpha)$. One could use a Pocock test, an O'Brien & Fleming test or one of the other two-sided tests described in Chapters 2 and 4, or a two-sided error spending test from Chapter 8. The test chosen to construct RCIs is called the *parent test* and should be specified in the study protocol. It is not necessary for a stopping rule to be prescribed if RCIs are used to monitor a study. However, if a stopping rule is applied then, for consistency, it should relate to the parent test of the RCIs in a natural manner: we shall describe ways of doing this in Section 10.3. Note that if we adopt the error spending approach, the maximum number of analyses K need not be pre-specified.

10.2 Example: Difference of Normal Means

For illustration, consider the "prototype" problem of Chapter 2. We are interested in constructing RCIs for the difference in means, $\theta = \mu_A - \mu_B$, when comparing two treatments with normal response and common, known variance σ^2. Observations are taken in equally sized groups of m on each treatment. The information for θ at analysis k is $\mathcal{I}_k = mk/(2\sigma^2)$ and the standardized statistics for testing H_0: $\theta = \theta_0$ are

$$Z_k(\theta_0) = \frac{1}{\sqrt{(2mk\sigma^2)}}\left(\sum_{i=1}^{mk} X_{Ai} - \sum_{i=1}^{mk} X_{Bi} - mk\theta_0\right), \quad k = 1, \ldots, K.$$

Thus, at stage k, the interval I_k is obtained from (10.5) as

$$(\bar{X}_{Ak} - \bar{X}_{Bk} - c_k(\alpha)\sqrt{\{2\sigma^2/(mk)\}},\ \bar{X}_{Ak} - \bar{X}_{Bk} + c_k(\alpha)\sqrt{\{2\sigma^2/(mk)\}}), \quad (10.9)$$

where $\bar{X}_{Ak}$ and $\bar{X}_{Bk}$ denote the cumulative sample means of observations on treatments A and B, respectively. The form of I_k is the same as that of the usual or "unadjusted" interval

$$(\bar{X}_{Ak} - \bar{X}_{Bk} - z_{\alpha/2}\sqrt{\{2\sigma^2/(mk)\}},\ \bar{X}_{Ak} - \bar{X}_{Bk} + z_{\alpha/2}\sqrt{\{2\sigma^2/(mk)\}}),$$

except that the standard normal percentile $z_{\alpha/2} = \Phi^{-1}(1 - \alpha/2)$ is replaced by the critical value $c_k(\alpha)$ from the parent group sequential test.

The ratio of the width of interval I_k to that of the unadjusted interval is $c_k(\alpha)/z_{\alpha/2}$, and this varies with the parent test of the RCI sequence. Table 10.2 displays these ratios for RCIs constructed from Pocock (1977) and O'Brien & Fleming (1979) tests using $\alpha = 0.05$ and $K = 5$ and 10 groups of observations.

Table 10.2 *Ratios of widths of 95% RCIs, I_k, to unadjusted 95% confidence intervals*

	$K = 5$		$K = 10$	
Analysis	*Pocock*	*O'Brien & Fleming*	*Pocock*	*O'Brien & Fleming*
k	$c_k = 2.413$	$c_k = 2.040\sqrt{(5/k)}$	$c_k = 2.555$	$c_k = 2.087\sqrt{(10/k)}$
1	1.231	2.327	1.304	3.367
2	1.231	1.646	1.304	2.381
3	1.231	1.344	1.304	1.944
4	1.231	1.164	1.304	1.684
5	1.231	1.041	1.304	1.506
6			1.304	1.375
7			1.304	1.273
8			1.304	1.190
9			1.304	1.122
10			1.304	1.065

Similar results for the case $\alpha = 0.1$ appear in Table 2 of Jennison & Turnbull (1989). The same table could be drawn up just as easily for RCIs with a Wang & Tsiatis test or Haybittle-Peto test of Section 2.7 as the parent test, using the constants in Tables 2.9 and 2.13, respectively, to determine the values $c_k(\alpha)$. The widths of the RCIs with the Pocock parent test are a constant multiple of those of the unadjusted intervals whereas RCIs based on the O'Brien & Fleming test are very wide at the beginning but narrow rapidly and are quite close to the unadjusted interval at the last analysis. This feature of the O'Brien & Fleming-based RCIs is often quite desirable since the increased width early on may be appropriate in view of uncertainty about validity of model assumptions, and the fact that the final RCI, I_K, is almost the same as the unadjusted interval can be advantageous in explaining results. The Pocock-based RCIs could be recommended for situations where it is of equal importance to obtain precise intervals for θ at all K analyses.

Suppose now that we desire the final interval I_K to have a specified width δ. On examining (10.9), we see this will occur if we set the group size $m = 8\sigma^2 c_K^2(\alpha)/(K\delta^2)$. The ratio of the final sample size to that required for a fixed sample procedure yielding a confidence interval of the same width is $\{c_K(\alpha)/z_{\alpha/2}\}^2$, the square of the final entry in each column of Table 10.2. For instance, using an O'Brien & Fleming parent test with five analyses, this ratio is $1.041^2 = 1.084$, representing an 8.4% increase in maximum sample size over the fixed sample procedure. This increase in sample size can be viewed as the "cost" of obtaining the benefits of flexible group sequential monitoring and the ability to construct interval estimates for θ at interim analyses, whether or not a rigid stopping rule is applied.

10.3 Derived Tests: Use of RCIs to Aid Early Stopping Decisions

The most important use that a DSMB might make of RCIs presented at an interim analysis is in reaching a decision for early termination. As stated in Section 10.1, this is often a subjective decision based on many factors but an interval estimate for the primary endpoint, adjusted for optional sampling bias, will clearly play a large role. In this section we describe how RCIs can be employed in the formal construction of one-sided tests, two-sided tests and equivalence tests. We refer to these tests constructed from RCIs as "derived tests". The progression from parent test to derived test is displayed schematically in Figure 10.1.

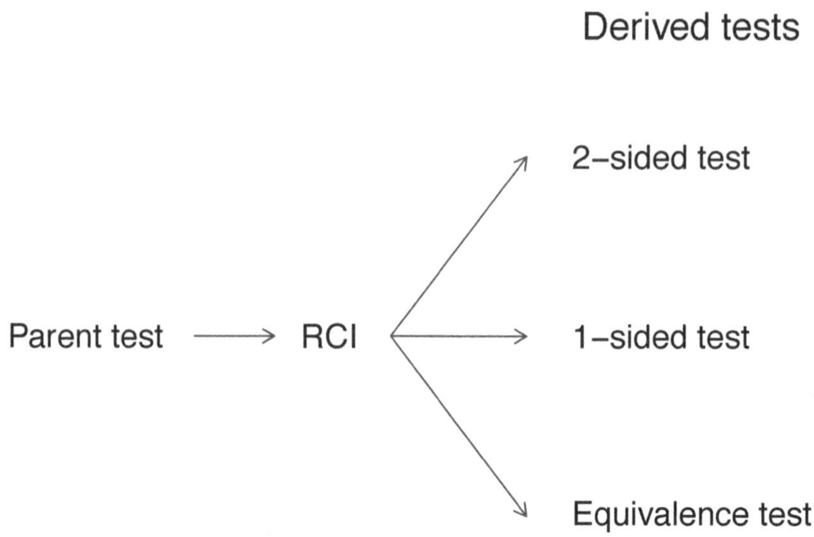

Figure 10.1 *RCIs, parent test and derived tests*

10.3.1 Two-sided Derived Tests

Suppose we wish to test a null hypothesis H_0: $\theta = \theta_0$ against the two-sided alternative $\theta \neq \theta_0$ and we are using a sequence of RCIs to monitor the study. Typically, θ_0 might be equal to zero, but we shall continue to discuss a general value θ_0. An obvious way to use the RCIs is to terminate the study early with rejection of H_0 if ever an RCI fails to include θ_0. By definition of the RCIs, this happens exactly when the parent test rejects H_0 and, thus, we recover the original parent test with two-sided Type I error rate α. In this case an RCI can be considered as an adjunct to the test of H_0: $\theta = \theta_0$, indicating which other values of θ are plausible, given the current data.

An RCI can be useful if opinion about an appropriate null hypothesis changes during the course of a study. Suppose we are comparing the efficacy of two treatments A and B, and as the study progresses, there is evidence of some serious adverse side-effects on treatment B. To compensate for this we might shift H_0 from $\theta_0 = 0$ to a non-zero value, thereby requiring more of an improvement in efficacy of treatment B to offset the extra discomfort or incapacitation caused by the side-effects. The new rule for termination is simply to stop if an RCI fails to include the new value of θ_0.

We would recommend caution in the practice of altering hypotheses during the course of a study, especially if this could cast doubt on the study's credibility. In order to provide protection from possible abuse, blinding should be maintained as much as possible and contingencies should be discussed at the planning stage and written into the protocol. However, it is clear that DSMBs *do* take notice of information from a variety of sources, often unexpected, and this can influence their attitudes to the originally stated goals. RCIs provide flexibility in the monitoring of a trial which can allow external information to be incorporated into the decision making process.

It is possible to derive two-sided tests from RCIs that are not just the parent test but also provide an "inner wedge" in the stopping boundary to allow early stopping in favor of H_0. These tests are similar to those described in Chapter 6. Suppose we desire a group sequential test of H_0: $\theta = \theta_0$ with Type I error probability α and power $1 - \beta$ at $\theta = \theta_0 \pm \delta$, where $0 < \beta < 0.5$. There are two ways to proceed.

In the first approach, we form a $(1-\alpha)$-level sequence of RCIs, $\{(\underline{\theta}_k, \overline{\theta}_k);\ k = 1, \ldots, K\}$ and define the formal stopping rule:

$$
\begin{array}{lll}
\text{At stage } k = 1, \ldots, K-1 & & \\
\quad \text{if } \underline{\theta}_k > \theta_0 + \delta \text{ or } \overline{\theta}_k < \theta_0 - \delta & & \text{stop, reject } H_0 \\
\quad \text{if } (\underline{\theta}_k, \overline{\theta}_k) \subset (\theta_0 - \delta, \theta_0 + \delta) & & \text{stop, accept } H_0 \\
\quad \text{otherwise} & & \text{continue to group } k+1, \\
\text{at stage } K & & \\
\quad \text{if } \underline{\theta}_K > \theta_0 + \delta \text{ or } \overline{\theta}_K < \theta_0 - \delta & & \text{stop, reject } H_0 \\
\quad \text{otherwise} & & \text{stop, accept } H_0.
\end{array}
$$

It follows from the defining relation (10.1) with $\theta = \theta_0$ that this test's Type I error probability is below α. (In fact, the Type I error probability is controlled conservatively since it would still be no greater than α if the criteria for rejecting H_0 at any $k \in \{1, \ldots, K\}$ were replaced by "$\underline{\theta}_k > \theta_0$ or $\overline{\theta}_k < \theta_0$".) Furthermore, by computing the power under different values of m, a group size can be found for which the power requirement is met.

The second procedure involves *two* sequences of RCIs for θ, one of level $(1-\alpha)$ and one of level $(1-2\beta)$. At each stage, H_0 is rejected if the current $(1-\alpha)$ RCI does not contain θ_0, and H_0 is accepted if the $(1-2\beta)$ RCI lies entirely within the interval $(\theta_0 - \delta, \theta_0 + \delta)$. The group size must be chosen to ensure termination at stage K. This is done by making sure the sum of the half-widths of the Kth $(1-\alpha)$-level and $(1-2\beta)$-level RCIs is equal to δ, since then precisely one of the conditions for rejecting H_0 or accepting H_0 is bound to be met. Again, it follows from (10.1) with $\theta = \theta_0$ that this test has Type I error probability at most α. To see that the power requirement is satisfied note that, writing $\underline{\theta}_k(\beta)$ and $\overline{\theta}_k(\beta)$ for the end points of the kth $(1-2\beta)$-level RCI, Equation (10.7) with $\theta = \theta_0 + \delta$ implies

$$
\begin{aligned}
& Pr_{\theta=\theta_0+\delta}\{\text{Accept } H_0\} \\
& \quad \leq \; Pr_{\theta=\theta_0+\delta}\{\underline{\theta}_k(\beta) > \theta_0 - \delta \text{ and } \overline{\theta}_k(\beta) < \theta_0 + \delta \text{ for some } k\} \\
& \quad \leq \; Pr_{\theta=\theta_0+\delta}\{\overline{\theta}_k(\beta) < \theta_0 + \delta \text{ for some } k\} \\
& \quad = \; 2\beta/2 \; = \; \beta.
\end{aligned}
$$

A similar argument establishes the power under $\theta = \theta_0 - \delta$.

The first of these procedures maintains the Type I error rate below α for any observed information sequence as long as the sequence of RCIs retains its overall coverage probability of $1-\alpha$, but differences from planned information levels will affect the power at $\theta_0 \pm \delta$. In the second procedure, a large value of $\mathcal{I}_K$ may produce shorter final RCIs than anticipated, allowing decisions both to accept H_0 and to reject H_0 according to the stated rule. The justification of this method holds true whichever decision is made in such cases. Intuitively, the data in these situations suggest that θ is between $\theta_0 - \delta$ and θ_0 or between θ_0 and $\theta_0 + \delta$, in which cases neither a Type I error nor a Type II error (at $\theta_0 \pm \delta$) can be made. In order to maximize power subject to a Type I error probability of at most α, one should reject H_0 whenever the formal rule permits, even when acceptance of H_0 is also allowed. If, on the other hand, $\mathcal{I}_K$ is lower than planned, a wide final RCI may contain θ_0 as well as $\theta_0 + \delta$ or $\theta_0 - \delta$. In this case, H_0 should be accepted in order to keep the Type I error probability at or below α.

10.3.2 *One-sided Derived Tests: Symmetric Case*

In Chapter 5, we considered one-sided hypothesis tests of H_0: $\theta \leq 0$ versus $\theta > 0$ designed to have Type I error probability α and power $1-\beta$ at $\theta = \delta$. We now show how RCIs can be used to monitor a study in a manner which meets these objectives. We first consider the case $\alpha = \beta$ where tests are symmetric about $\delta/2$. This may seem unusual as, in practice, it is common for β to be larger than α. However, as explained in Section 5.7.1, one may design a study with high power at a large treatment effect in order to achieve more modest power at a smaller treatment effect which represents the anticipated treatment effect or the minimally significant treatment difference. A consequence for the symmetric methods defined here is that tests are derived from RCIs with equal error probabilities in each tail.

Given a sequence of RCIs $(\underline{\theta}_k, \overline{\theta}_k)$, $k = 1, \ldots, K$, we allow the study to be terminated at the kth analysis with rejection of H_0 if $\underline{\theta}_k > 0$ or with acceptance of H_0 if $\overline{\theta}_k < \delta$. In order that this test has error probabilities equal to α at $\theta = 0$ and $\theta = \delta$, we need to use RCIs with confidence level $1 - 2\alpha$ rather than $1 - \alpha$. Then, by (10.8) with 2α in place of α, the probability of accepting H_0 if $\theta \geq \delta$ or of rejecting H_0 if $\theta \leq 0$ will be at most α.

To ensure a decision can be made at analysis K, the width of the final RCI must be no more than δ, so that the RCI for θ cannot contain both 0 and δ. Thus, from (10.5), we require $\mathcal{I}_K = 4c_K^2(2\alpha)/\delta^2$. In the comparison of two normal means considered in Section 10.2, we had $\mathcal{I}_K = mK/(2\sigma^2)$ and this condition reduces to requiring a group size of $m = 8\sigma^2 c_K^2(2\alpha)/(K\delta^2)$. If, in practice, a larger sample is accrued, the width of the final RCI may be less than δ and it is possible that both $\overline{\theta}_k < \delta$ and $\underline{\theta}_k > 0$: as explained for two-sided tests with an inner wedge, either decision is permissible in such a situation and one should choose to reject H_0 in order to maximize power while still satisfying the Type I error condition. If, however, the sample is small and a wide final RCI contains both 0 and δ, H_0 should be accepted to protect the Type I error probability.

We refer to this procedure as a "derived test". This test stands in its own right as a one-sided group sequential test with Type I error probability at most α and power at least $1 - \alpha$ at $\theta = \delta$. Hence, its maximum information level and expected information level under specific values of θ can be compared to those of symmetric tests as defined in Chapter 5. However, expressing the test in terms of RCIs means it is easy to use it flexibly when taking account of additional data or responding to changing study objectives. If a study continues past an interim analysis at which $\underline{\theta}_k > 0$ or $\overline{\theta}_k < \delta$, later RCIs are still valid and can be used subsequently in reaching decisions. Similarly, the terminal RCI is valid when a trial concludes before the derived test's formal stopping criterion is met.

Suppose a study is designed for $K = 5$ stages with error probabilities $\alpha = \beta = 0.025$. We could choose to base RCIs and, hence, derived tests on a Pocock parent test, for which $c_k = C_P(5, 0.05)$, or an O'Brien & Fleming parent test with $c_k = C_B(5, 0.05)\sqrt{(5/k)}$. The kth RCI is $(\underline{\theta}_k, \overline{\theta}_k) = (\hat{\theta}^{(k)} - c_k/\sqrt{\mathcal{I}_k}, \hat{\theta}^{(k)} + c_k/\sqrt{\mathcal{I}_k})$ and the stopping rule is

$$
\begin{array}{lll}
\text{At stage } k = 1, \ldots, K-1 & & \\
\quad \text{if } \underline{\theta}_k > 0 & \text{stop, reject } H_0 & \\
\quad \text{if } \overline{\theta}_k < \delta & \text{stop, accept } H_0 & \\
\quad \text{otherwise} & \text{continue to group } k+1, & \qquad (10.10) \\
\text{at stage } K & & \\
\quad \text{if } \underline{\theta}_K > 0 & \text{stop, reject } H_0 & \\
\quad \text{otherwise} & \text{stop, accept } H_0. &
\end{array}
$$

Note that, since $Z_k = \hat{\theta}^{(k)}\sqrt{\mathcal{I}_k}$, the condition $\underline{\theta}_k > 0$ is equivalent to $Z_k > c_k$.

Table 10.3 lists properties of these derived tests. The ratio $R = \{c_K(2\alpha)/z_\alpha\}^2$ is the maximum information level, $\mathcal{I}_K$, divided by the information level $\mathcal{I}_{f,1}$ required by the corresponding fixed sample test. The error rates are the attained Type I or Type II error probabilities (which are equal by symmetry, so a single number is shown). Also listed are expected information levels on termination, $E_\theta(\mathcal{I})$, for $\theta = 0$ or δ (which are identical by

Table 10.3 *Properties of one-sided derived tests for testing H_0: $\theta \le 0$ versus $\theta > 0$ with $K = 5$ analyses and nominal Type I and Type II error rates $\alpha = \beta = 0.025$. Parent tests are Pocock and O'Brien & Fleming tests. $R = \mathcal{I}_K/\mathcal{I}_{f,1}$ denotes the ratio of the maximum information level to that of the corresponding fixed sample test. Expected information levels on termination are expressed as percentages of $\mathcal{I}_{f,1}$. Optimal tests were found using the methods of Barber & Jennison (2002).*

Form of one-sided test	*Information ratio* R	*Expected information on termination** E_1	E_2	E_3	*Error rates*
Fixed sample size	1	100	100	100	0.0250
Pocock RCIs					
Derived test	1.52	57.5	85.3	66.0	0.0227
Optimal test with nominal α	1.52	55.2	80.8	63.7	0.0250
Optimal test with achieved error rates	1.52	56.9	84.1	65.9	0.0227
O'Brien & Fleming RCIs					
Derived test	1.08	67.1	86.3	72.5	0.0234
Optimal test with nominal α	1.08	58.7	81.4	65.3	0.0250
Optimal test with achieved error rates	1.08	61.8	85.2	68.6	0.0234

*$E_1 = 100\, E_0(\mathcal{I})/\mathcal{I}_{f,1} = 100\, E_\delta(\mathcal{I})/\mathcal{I}_{f,1}$, $E_2 = 100\, E_{\delta/2}(\mathcal{I})/\mathcal{I}_{f,1}$, $E_3 = 100 \int f(\theta) E_\theta(\mathcal{I})\, d\theta/\mathcal{I}_{f,1}$.

symmetry) and for $\theta = \delta/2$, as well as the integral

$$\int f(\theta) E_\theta(\mathcal{I})\, d\theta, \tag{10.11}$$

where $f(\theta)$ is the density of a $N(\delta/2, (\delta/2)^2)$ distribution for θ. These values of expected information are expressed as percentages of $\mathcal{I}_{f,1}$. We could compare the ratios R and information levels with the analogous results for the power family one-sided tests defined in Section 5.2. However, it is more interesting to compare expected information levels with those of optimal tests with the same maximum information level and 5 analyses, as described in Section 5.2.2. The methods of Barber & Jennison (2002), which we shall describe later in Chapter 12, were used to find the tests with symmetry about $\delta/2$ that minimize expected information under $\theta = 0$ or δ, under $\theta = \delta/2$, and integrated over θ as in (10.11). Table 10.3 shows the minimum possible values for these three criteria. Results are shown first for tests with Type I and Type II error probabilities equal to 0.025 and then for tests that have the lower error rates attained by the derived tests.

The derived tests are conservative, having error probabilities a little below the nominal $\alpha = \beta = 0.025$. Intuitively, since the RCI is designed to accommodate all possible stopping rules, the probability of reaching an incorrect decision is reduced when only the stopping rule of the derived test is applied. More formally, in the case of the Type I error rate, this conservatism arises because, if $\underline{\theta}_k > 0$ for some k, a Type I error will not be made if $\overline{\theta}_{k'} < \delta$ for some $k' < k$ so that the test terminates to accept H_0 before it can reject H_0 at analysis k. Similarly for the Type II error rate, if $\overline{\theta}_k < \delta$ for some k, a Type II error will not be made if $\underline{\theta}_{k'} > 0$ for some $k' < k$. The conservatism in the Type I error rate is analogous to that seen in one-sided tests with a non-binding futility boundary since, in both cases, sample paths that have crossed the futility boundary are considered when calculating the Type I error probability. Indeed, the efficacy boundary of a "derived test" satisfies exactly the condition needed for the efficacy boundary of a one-sided test with a non-binding futility boundary.

The conservatism of the derived tests is reflected in higher expected information than seen for optimal tests attaining the nominal Type I and Type II error probabilities of 0.025. If, however, the derived tests are compared with one-sided tests achieving the same error rates as they do, we see they have quite high efficiency, especially when one considers that the optimal results are obtained using a different test for each of the three expected information criteria. The derived test with a Pocock parent test does particularly well in terms of the integrated expected information (10.11).

10.3.3 One-sided Derived Tests: Asymmetric Case

We continue to consider one-sided tests of H_0: $\theta \le 0$ versus $\theta > 0$, designed to have Type I error probability α and power $1 - \beta$ at $\theta = \delta$, but we now suppose α and β are unequal. In creating a test, we define $\{\underline{\theta}_1(\alpha), \ldots, \underline{\theta}_K(\alpha)\}$ to be a $(1-\alpha)$-level lower confidence sequence for θ if

$$Pr_\theta\{\underline{\theta}_k(\alpha) < \theta \text{ for all } k = 1, \ldots, K\} \;=\; 1 - \alpha \quad \text{for all } \theta. \tag{10.12}$$

Similarly, $\{\overline{\theta}_1(\beta), \ldots, \overline{\theta}_K(\beta)\}$ is a $(1 - \beta)$-level upper confidence sequence if

$$Pr_\theta\{\overline{\theta}_k(\beta) > \theta \text{ for all } k = 1, \ldots, K\} \;=\; 1 - \beta \quad \text{for all } \theta. \tag{10.13}$$

These definitions imply that

$$Pr_\theta\{\underline{\theta}_k(\alpha) < \theta < \overline{\theta}_k(\beta) \text{ for all } k = 1, \ldots, K\} \ge 1 - \alpha - \beta,$$

with approximate equality if we ignore the very small probability that $\underline{\theta}_k(\alpha) > \theta$ for one value of k and $\overline{\theta}_{k'}(\beta) < \theta$ for some other k'. Thus, $\{(\underline{\theta}_k(\alpha), \overline{\theta}_k(\beta));\, k = 1, \ldots, K\}$ forms a $(1-\alpha-\beta)$-level RCI sequence. We can derive a one-sided test from these RCIs which meets the specified error conditions by continuing to sample as long as the interval $(\underline{\theta}_k(\alpha), \overline{\theta}_k(\beta))$ includes both $\theta = 0$ and $\theta = \delta$. The study stops at the first analysis k where $(0, \delta)$ is not wholly contained in $(\underline{\theta}_k(\alpha), \overline{\theta}_k(\beta))$ and we reject H_0 if $\underline{\theta}_k(\alpha) > 0$ or accept H_0 if $\overline{\theta}_k(\beta) < \delta$. Termination is ensured if we choose the group size m so that $\overline{\theta}_K(\beta) - \underline{\theta}_K(\alpha) = \delta$. By construction, we have

$$Pr_{\theta=0}\{\text{Reject } H_0\} \;<\; Pr_{\theta=0}\{\underline{\theta}_k(\alpha) > 0 \text{ for some } k\} \;=\; \alpha$$

and

$$Pr_{\theta=\delta}\{\text{Accept } H_0\} \;<\; Pr_{\theta=\delta}\{\overline{\theta}_k(\beta) < \delta \text{ for some } k\} \;=\; \beta.$$

Thus, error probabilities at $\theta = 0$ and $\theta = \delta$ are bounded by α and β, respectively.

Returning to the example of Section 10.2 and using the same definitions and notation, the $(1 - \alpha)$-level lower confidence sequence for $\theta = \mu_A - \mu_B$ is

$$\underline{\theta}_k(\alpha) \;=\; \bar{X}_{Ak} - \bar{X}_{Bk} - c_k(2\alpha)\sqrt{\{2\sigma^2/(mk)\}}, \quad k = 1, \ldots, K, \tag{10.14}$$

and the $(1 - \beta)$-level upper confidence sequence is

$$\overline{\theta}_k(\beta) \;=\; \bar{X}_{Ak} - \bar{X}_{Bk} + c_k(2\beta)\sqrt{\{2\sigma^2/(mk)\}}, \quad k = 1, \ldots, K. \tag{10.15}$$

Here $c_k(2\alpha)$, $k = 1, \ldots, K$, are critical values of a two-sided group sequential test, for example, a Pocock or O'Brien & Fleming test with Type I error probability equal to 2α. Similarly, $c_k(2\beta)$, $k = 1, \ldots, K$, are critical values of a two-sided group sequential test with Type I error probability 2β. By construction, the sequences (10.14) and (10.15) satisfy the conditions (10.12) and (10.13). If a Pocock test is used as the parent test, critical values are

$$c_k(2\alpha) = C_P(K, 2\alpha) \quad \text{and} \quad c_k(2\beta) = C_P(K, 2\beta),$$

where, for certain values of α and β, the constants $C_P(K, 2\alpha)$ and $C_P(K, 2\beta)$ can be found in Table 2.1. Using an O'Brien & Fleming parent test, constants $C_B(K, 2\alpha)$ and $C_B(K, 2\beta)$ can be taken from Table 2.3 to give

$$c_k(2\alpha) = C_B(K, 2\alpha)\sqrt{(K/k)} \quad \text{and} \quad c_k(2\beta) = C_B(K, 2\beta)\sqrt{(K/k)}.$$

When group sizes are liable to be unequal or unpredictable, the error spending approach described in Section 8.2 can be used to obtain the critical values for the parent tests, using separate error spending functions to derive $c_1(2\alpha), \ldots, c_K(2\alpha)$ and $c_1(2\beta), \ldots, c_K(2\beta)$.

With equal group sizes, a final sample size of

$$mK = 2\{c_K(2\alpha) + c_K(2\beta)\}^2 \sigma^2/\delta^2 \qquad (10.16)$$

on each treatment arm ensures that $\overline{\theta}_K - \underline{\theta}_K = \delta$ and thus the derived test will stop at or before stage K. By comparison, the required sample size for a fixed sample test is obtained by replacing $c_K(2\alpha)$ by z_α and $c_K(2\beta)$ by z_β in (10.16). Once again, a narrow final RCI may permit both acceptance and rejection of H_0, in which case a natural choice is to reject H_0 in order to maximize power while preserving a Type I error rate of at most α. If, instead, the final RCI is wider than planned and contains both 0 and δ, H_0 should be accepted. For more details, see Jennison & Turnbull (1993a).

In Table 10.4, we present properties of tests with $K = 5$ analyses, $\alpha = 0.025$ and $\beta = 0.1$. The parent tests are error spending tests from the Rho family, as defined in Section 8.2, with the same value of ρ in the spending functions for Type I and Type II error probability. The stopping rule is as in (10.10) but with $\underline{\theta}_k(\alpha)$ and $\overline{\theta}_k(\beta)$ in place of $\underline{\theta}_k$ and $\overline{\theta}_k$. Results are shown for derived tests when the parent tests have $\rho = 2$ and $\rho = 3$. The information ratio, R, and values E_1, E_2 and E_3 for expected information expressed as a percentage of $\mathcal{I}_{f,1}$ are as defined for Table 10.3, assuming once again that observed information levels are equally spaced and culminate at the value planned for $\mathcal{I}_5$.

As in the symmetric case, the derived tests are a little conservative. For the derived test with $\rho = 2$ in the parent test, the ratio R is considerably smaller than for the Pocock derived test in Table 10.3, while the value of R for the derived test with $\rho = 3$ in the parent test is similar to that for the O'Brien & Fleming derived test in Table 10.3. The derived tests with $\rho = 2$ achieves lower values of E_1, E_2 and E_3 than the derived test with $\rho = 3$ but both these tests are highly efficient as values of E_1, E_2 and E_3 are within 2% or 3% of the optimal possible for their maximum information level and attained error rates.

10.3.4 Equivalence Tests

In Section 7.4, we introduced the problem of a two-sided test of equivalence where θ is a treatment difference and it is desired to distinguish between equivalence, when $|\theta|$ is "small", and inequivalence, when $|\theta|$ is "large". In (7.2) and (7.3), the error probability requirements are expressed as

$$Pr_{\theta=\pm\delta}\{\text{Declare equivalence}\} \le \beta \qquad (10.17)$$

and

$$Pr_{\theta=0}\{\text{Do not declare equivalence}\} \le \alpha. \qquad (10.18)$$

Table 10.4 *Properties of one-sided derived tests for testing $H_0: \theta \le 0$ versus $\theta > 0$ with 5 analyses and nominal Type I error rate $\alpha = 0.025$ and Type II error rate $\beta = 0.1$ at $\theta = \delta$. Parent tests are error spending tests in the Rho family (8.6). $R = \mathcal{I}_{max}/\mathcal{I}_{f,1}$ denotes the ratio of the maximum information level to that of the corresponding fixed sample test. Expected information levels on termination are expressed as percentages of $\mathcal{I}_{f,1}$.*

Form of one-sided test	*Information ratio* R	*Expected information on termination** E_1	E_2	E_3	*Error rates* *Type I*	*Type II*
Fixed sample size	1	100	100	100	0.0250	0.1000
Error spending test, $\rho = 2$						
Derived test	1.18	66.5	83.4	70.7	0.0227	0.0917
Optimal test with nominal α and β	1.18	60.1	76.6	64.5	0.0250	0.1000
Optimal test with achieved error rates	1.18	64.3	82.1	69.0	0.0227	0.0917
Error spending test, $\rho = 3$						
Derived test	1.10	69.4	84.9	73.1	0.0235	0.0942
Optimal test with nominal α and β	1.10	62.4	78.3	66.5	0.0250	0.1000
Optimal test with achieved error rates	1.10	66.7	83.3	70.9	0.0235	0.0942

$^*E_1 = 100\,[\{E_0(\mathcal{I}) + E_\delta(\mathcal{I})\}/2]/\mathcal{I}_{f,1}$, $E_2 = 100\,E_{\delta/2}(\mathcal{I})/\mathcal{I}_{f,1}$, $E_3 = 100 \int f(\theta) E_\theta(\mathcal{I})\,d\theta/\mathcal{I}_{f,1}$.

The testing problem is similar to that of the usual two-sided test except that now declaring equivalence when $\theta = \pm\delta$ is treated as the key error and it is of paramount importance to ensure its probability does not exceed β. As noted in Section 7.4, some authors would interchange the symbols α and β but we use the above choice for consistency with our earlier notation for two-sided tests. In the case of bioequivalence testing, β is the probability of accepting a nonequivalent compound and the values of β and δ must be chosen to satisfy the appropriate regulatory agency. One would expect a manufacturer to choose the value of α to be small since the advantages of establishing equivalence of a new formulation to an existing one are so great.

In a fixed sample size trial, a convenient way to satisfy (10.17) is to conduct two one-sided tests with Type I error rates β (Schuirmann, 1987) and accept equivalence only if both H: $\theta = -\delta$ is rejected in favor of $\theta > -\delta$ and H: $\theta = \delta$ is rejected in favor of $\theta < \delta$. It can be seen that this rule accepts equivalence if and only if the $(1 - 2\beta)$-level confidence interval $(\underline{\theta}, \overline{\theta})$, for which $Pr_\theta\{\underline{\theta} < \theta\} = Pr_\theta\{\overline{\theta} > \theta\} = 1 - \beta$, is entirely contained within the range $(-\delta, \delta)$; see also Westlake (1976). The condition (10.18) can usually be satisfied by an appropriate choice of sample size or, more generally, information level.

In the group sequential setting, there are two approaches to deriving tests based on repeated confidence intervals. The first is a natural extension of the above idea. Suppose that $\{(\underline{\theta}_k, \overline{\theta}_k);\ k = 1, \ldots, K\}$ is a $(1-2\beta)$-level RCI sequence. Then, we can define the stopping rule:

At stage $k = 1, \ldots, K-1$

if $(\underline{\theta}_k, \overline{\theta}_k) \subset (-\delta, \delta)$	stop, accept equivalence
if $\underline{\theta}_k > \delta$ or $\overline{\theta}_k < -\delta$	stop, reject equivalence
otherwise	continue to group $k+1$,

at stage K

if $(\underline{\theta}_K, \overline{\theta}_K) \subset (-\delta, \delta)$	stop, accept equivalence
otherwise	stop, reject equivalence.

By construction, the probability of declaring equivalence when $\theta = \pm\delta$ is at most β. However, the sequence of sample sizes or information levels at analyses 1 to K should be planned to ensure that condition (10.18), concerning acceptance of H_0 when $\theta = 0$, is met.

The second approach involves *two* sequences of RCIs: one of level $(1-2\beta)$ and the other of level (1-α). At each stage, equivalence is accepted if the current $(1-2\beta)$-level RCI for θ lies entirely within the interval $(-\delta, \delta)$, and equivalence is rejected if the $(1-\alpha)$-level RCI does not contain 0; otherwise, the procedure continues to the next stage. The sample sizes or information levels must be chosen so that the widths of the final RCIs are small enough to ensure termination at stage K, specifically, the sum of the two half-widths should be no greater than δ. It is possible that the stated rule allows both acceptance and rejection of equivalence if the final RCI is narrower than planned. In this case, it is permissible to declare equivalence since the probability of wrongly declaring equivalence under $\theta = \pm\delta$ remains protected. On the other hand, if the final RCI is wide and contains 0 along with either δ or $-\delta$, equivalence should be rejected.

Jennison & Turnbull (1993b) describe these procedures more fully, illustrating their implementation and performance characteristics in a comparison of two normal means. In general, the second procedure has a higher maximum sample size and a somewhat higher expected sample size under $\theta = 0$, but a substantially lower expected sample size under $\theta = \pm\delta$ when compared to the first method. Jennison & Turnbull (1993b) also describe how to employ the error spending function approach when group sizes are unequal or unpredictable. They go on to propose an approximate procedure for handling a nuisance parameter by updating group sizes adaptively as new estimates of the nuisance parameter are obtained. In their example of a comparison between two binomial distributions with success probabilities p_1 and p_2, the parameter of interest is $\theta = p_2 - p_1$ and the average success probability $(p_1 + p_2)/2$ is a nuisance parameter.

10.4 Repeated P-values

A theory of "repeated P-values" can be developed analogously to that of repeated confidence intervals. At analysis k, a two-sided repeated P-value for the null hypothesis H_0: $\theta = \theta_0$ is defined as $P_k = \min\{\alpha : \theta_0 \notin I_k(\alpha)\}$, where $I_k(\alpha)$ is the current $(1-\alpha)$-level RCI for θ. In other words, P_k is that value of α for which the kth $(1-\alpha)$-level RCI has the null value, θ_0, as one of its endpoints. The construction ensures that, for any $p \in (0, 1)$, the overall probability under H_0 of ever seeing a repeated P-value less than or equal to p is no more than p and this probability is exactly p if all repeated P-values are always observed. Thus, the repeated P-value can be reported with the usual interpretation of a P-value, yet with protection against the multiple-looks effect. These P-values should not be confused with the significance levels of Section 9.4 which are only computed once, on the termination of a sequential test conducted according to a strictly enforced stopping rule.

10.5 Discussion

Repeated confidence intervals offer a frequentist data summary with the flexibility of application usually needed for interim monitoring of clinical trials. The presentation of this approach at a meeting of the Royal Statistical Society's Research Section led to a robust discussion and some initial skepticism about the methodology was expressed; see Jennison & Turnbull (1989). However, an examination of recent reports of clinical trials which cite this paper shows that RCIs have been adopted as a useful tool for monitoring clinical trials and summarizing their conclusions.

Koepcke (1989, p. 228S) and others have criticized RCIs for being too wide when compared with confidence intervals at termination of a group sequential test of the type described in Section 9.5. However, if the trial continues until the final stage, K, the O'Brien & Fleming based RCI is then only slightly wider than the fixed sample interval; also if stopping occurs earlier, the implication is that the RCI was precise enough to convey the information about θ needed to make that decision. In any case, in the absence of a rigidly followed stopping rule, the methods of Chapter 9 should not really be used to construct a confidence interval on termination.

In summary, RCIs provide interval estimates at each interim analysis "adjusted" for multiple looks. As such, they may be reported as interim estimates of treatment effect at scientific meetings, without the threat of over-interpretation. RCIs can be computed when responses are normally distributed, or follow binary, survival, multivariate or other distributions, with allowance for covariates if necessary. They can be based on error spending parent tests when group sizes are unequal or unpredictable. At interim analyses, RCIs can serve as an adjunct to a group sequential test enlarging on the simple "stop/continue" decision. On termination, RCIs are valid independently of the stopping rule applied, unlike the terminal confidence intervals of Chapter 9. A sequence of RCIs can be used as a basis for either an informal or formal stopping rule. In the latter case, their flexibility leads to a little conservatism but, with a suitable choice of parent test, the derived tests are efficient in terms of expected sample size. Moreover, RCIs can continue to be used even if a decision is indicated by a formal group sequential test and yet, for some reason, the study continues. The way in which the widths of a sequence of RCIs vary over successive analyses depends on the choice of parent test and this should be selected with a view to how the RCIs are likely to be used. Certainly, the chosen type of RCI should be clearly stated in a study protocol.

Jennison & Turnbull (1989) describe a number of applications of RCIs. For accumulating censored survival data, they construct RCIs for the median survival time using the Kaplan-Meier estimate of a survival distribution, and they present RCIs for a hazard ratio using log-rank and partial likelihood score statistics. Theory for the median survival time problem is given in Jennison & Turnbull (1985) and will be presented in Section 17.7; RCIs for a hazard ratio will be discussed in Section 17.5 and illustrated in an example in Section 17.6. For discrete data, Jennison & Turnbull (1989) show how to find RCIs for a log odds ratio in a set of 2×2 tables or stratified case-control data, based on the Mantel-Haenszel estimator, and they derive RCIs for the slope coefficient in an odds ratio regression model. A distributional result for group sequential Mantel-Haenszel estimates is proved in Jennison & Turnbull (1991c) and will be reported here in Section 16.4.

In Section 10.3.4 we derived two-sided tests of equivalence from a sequence of RCIs. One-sided equivalence tests, as described in Section 7.2, can also be created from RCIs; for details, see Durrleman & Simon (1990) and Jennison & Turnbull (1993a). Fleming (1992) describes several case studies illustrating the use of repeated confidence intervals in monitoring equivalence trials. For the two-sided equivalence testing problem, Hsu, Hwang, Liu & Ruberg (1994) propose alternate forms of RCI in connection with the first approach of Section 10.3.4. They replace the $(1-2\beta)$-level RCI as defined in (10.7) by the interval $(\min\{\underline{\theta}_k(\beta), c\}, \max(\overline{\theta}_k\{\beta), c\})$ for a chosen constant c. The usual choice is $c = 0$, or for a more general region of equivalence (δ_1, δ_2), one can take $c = (\delta_1 + \delta_2)/2$. This choice of RCI leads to generally narrower intervals when θ is close to c, the region of most interest, but wider intervals otherwise.

Lin & Wei (1991) discuss RCIs for the scale change parameter in a two-sample accelerated failure time model for survival data. Keaney & Wei (1994) show how to construct RCIs for the difference of two median response time distributions based on accumulating survival data. Lin, Wei & DeMets (1991) discuss exact methods for obtaining RCIs for comparing two treatments with dichotomous responses. Coe & Tamhane (1993) also treat exact RCIs for one-sample and two-sample binomial problems. We present methods for these binomial problems in Sections 16.1.4 and 16.2.3. Su & Lachin (1992, p. 1038) describe RCIs based on a multivariate Hodges-Lehmann estimator of group differences with repeated measures data.

RCIs can also be computed in group sequential trials with adaptive designs as described in later chapters of this book. Lehmacher & Wassmer (1999) show how to construct RCIs in an adaptive design using a sequential combination test (see Section 23.7); Brannath, König & Bauer (2006) use RCIs to construct "sequential flexible confidence intervals" for use in a flexible, adaptive design; Hartung & Knapp (2006) apply RCIs to create an adaptive trial design that allows switching between tests of noninferiority and superiority. Mehta et al. (2007) develop derived tests based on RCIs which accommodate adaptations by preserving the conditional probability of rejecting a null hypothesis (see Section 23.6).

Chapter 11

Stochastic Curtailment

11.1 Introduction

The use of "stochastic curtailment" in sequential monitoring evolved from the idea of simple curtailment whereby a trial can be terminated as soon as the ultimate result becomes inevitable. As an example, consider an industrial acceptance sampling scheme in which items are classified as defective or non-defective. A single-sampling scheme designates a lot as acceptable if fewer than c defectives are found in a random sample of n items from the lot. In the curtailed version of this procedure (see, for example, Alling, 1966), sampling proceeds sequentially and terminates as soon as the decision is inevitable, that is, when either the number of defectives reaches c or the number of non-defectives reaches $n - c + 1$. Clearly, this curtailed scheme always leads to the same conclusion as the fixed sample procedure. The two methods have identical operating characteristics, but the curtailed procedure has a lower expected sample size. Any fixed sample test with bounded contributions from each new observation, such as the Wilcoxon rank sum test or many other nonparametric tests, can be curtailed in a similar way; see Alling (1963) and Halperin & Ware (1974).

If outcome measures are not bounded, as in the case of normally distributed observations, or if greater savings in expected sample size are desired, a stochastic form of curtailment may be considered. Then, at any stage, the procedure is terminated if a particular decision, while not inevitable, is highly likely given the current data.

In this chapter, we describe three approaches to stochastic curtailment, the conditional power approach, the predictive power approach and a parameter free approach. We describe how stochastic curtailment can be used to aid the decision to abandon a trial for futility. Finally, we illustrate these methods in a case study involving survival data.

11.2 The Conditional Power Approach

Suppose the parameter of interest in a trial is θ and we wish to test H_0: $\theta = \theta_0$ against $\theta \neq \theta_0$ in a two-sided test or to test H_0: $\theta \leq \theta_0$ against $\theta > \theta_0$ in a one-sided test. We first designate a "reference test", which we shall call $\mathcal{T}$. This is typically a fixed sample test with Type I error probability α when $\theta = \theta_0$ and power $1 - \beta$ at values θ_1 and θ_2, either side of θ_0 for a two-sided test, or power $1 - \beta$ at an alternative $\theta = \theta_1 > \theta_0$ for a one-sided test. However, $\mathcal{T}$ could also be a group sequential test. At an interim stage, k, during the study, let $D(k)$ denote the data accumulated so far. The conditional power at stage k is defined as

$$p_k(\theta, D(k)) = Pr_\theta\{\mathcal{T} \text{ will reject } H_0 \,|\, D(k)\}, \qquad (11.1)$$

which we shall write as $p_k(\theta)$ when the conditioning event is clear. Initially, when $k = 0$, this is the usual power function. When the study has terminated under the original plan, this probability is either zero or one. At an interim stage k, the conditional power can be plotted as a function of θ and its values at θ_0 and θ_1 for a one-sided test, and θ_0, θ_1 and θ_2 for a two-sided test, as well as at $\hat{\theta}^{(k)}$, the current maximum likelihood estimate of θ, are of particular interest.

 DOI: 10.1201/9781584888482-11

The conditional power function is a useful device for communicating with clinical investigators. For instance, it can be used to illustrate the effects of low accrual or to aid the decision to abandon a study if the conditional power appears too low. The opposite situation, where a study may be terminated with a positive outcome when the conditional power is high, will be illustrated in Section 11.6.

Lan, Simon & Halperin's Stochastic Curtailment Procedure

A high value of $p_k(\theta_0)$ indicates that the reference test is likely to reject H_0 given the accrued data, even if H_0 is true. This led Lan, Simon & Halperin (1982) to suggest the formal rule: for a specified constant γ, terminate the trial at stage k and reject H_0 if $p_k(\theta_0) \geq \gamma$. The value of γ should be between 0.5 and 1 and values 0.8 or 0.9 are recommended. Similarly, early termination may be allowed to accept H_0 if $1 - p_k(\theta_1) \geq \gamma'$ in a one-sided test, or if both $1 - p_k(\theta_1) \geq \gamma'$ and $1 - p_k(\theta_2) \geq \gamma'$ in a two-sided test, where γ' is another constant between 0.5 and 1. The quantity $1 - p_k(\theta_1)$ has been termed the *futility index* by Ware, Muller & Braunwald (1985).

In using the above rules, the error probabilities of the test can be expected to rise since the final decision of the reference test $\mathcal{T}$ will sometimes be incorrectly anticipated. We refer to a formally defined test based on the above rules as a stochastically curtailed procedure (SCP). The error probabilities of an SCP with a given interim analysis schedule can be computed by numerical integration but Lan, Simon & Halperin (1982) use a simple argument to prove that the Type I error probability will not be more than α/γ. It can be shown that $\{p_k(\theta);\ k = 1, 2, \dots\}$ is a martingale with respect to the filtration defined by $\{D(1),\ D(2),\ \dots\}$. In fact, it is an example of a Doob's martingale; see Grimmett & Stirzaker (2020, p. 392 and 546). Let ν denote the stopping time of the reference test $\mathcal{T}$ (allowing the possibility that $\mathcal{T}$ is itself a sequential test). First, suppose $\gamma' = 1$ so there is no early stopping to accept H_0. In this case, the stage at which the SCP stops is

$$\tau \;=\; \min(\nu,\ \min\{k\colon\ p_k(\theta_0) \geq \gamma\}).$$

By the optional stopping theorem (e.g., Grimmett & Stirzaker, 2020, Sec. 12.5) we have

$$E_{\theta_0}\{p_\tau(\theta_0)\} \;=\; E_{\theta_0}\{p_0(\theta_0)\} = p_0(\theta_0) \;=\; \alpha.$$

But, from the definition of τ,

$$E_{\theta_0}\{p_\tau(\theta_0)\} \;\geq\; \gamma\, Pr_{\theta_0}\{p_k(\theta_0) \geq \gamma \text{ for some } k \leq \nu\} \;=\; \gamma\, Pr_{\theta_0}\{\text{SCP Rejects } H_0\}$$

and hence

$$Pr_{\theta_0}\{\text{SCP Rejects } H_0\} \;\leq\; \alpha/\gamma.$$

Now suppose $\gamma' < 1$. In this case, some trials that would have ended with rejection of H_0 will stop at an earlier stage so the Type I error probability is reduced, and thus it is still bounded above by α/γ.

A similar argument shows that the Type II error probability of the SCP is no more than β/γ', where β is the Type II error rate of the reference test. With these results in mind, we could ensure error probabilities of at most α and β for the SCP's Type I and II error probabilities by designing the reference test to have Type I error rate $\alpha\gamma$ and power $1 - \beta\gamma'$ at $\theta = \theta_1$ for a one-sided test, or at $\theta = \theta_1$ and $\theta = \theta_2$ for a two-sided test, a process that would involve increasing the final information level $\mathcal{I}_K$.

Calculating Conditional Power

We now illustrate the calculation of a conditional power function in the general formulation of Section 3.1. Suppose successive analyses yield a sequence of test statistics $Z_1, \dots, Z_K$ with the canonical joint distribution (3.1), given information levels $\mathcal{I}_1, \dots, \mathcal{I}_K$ for θ. Since Z_k is

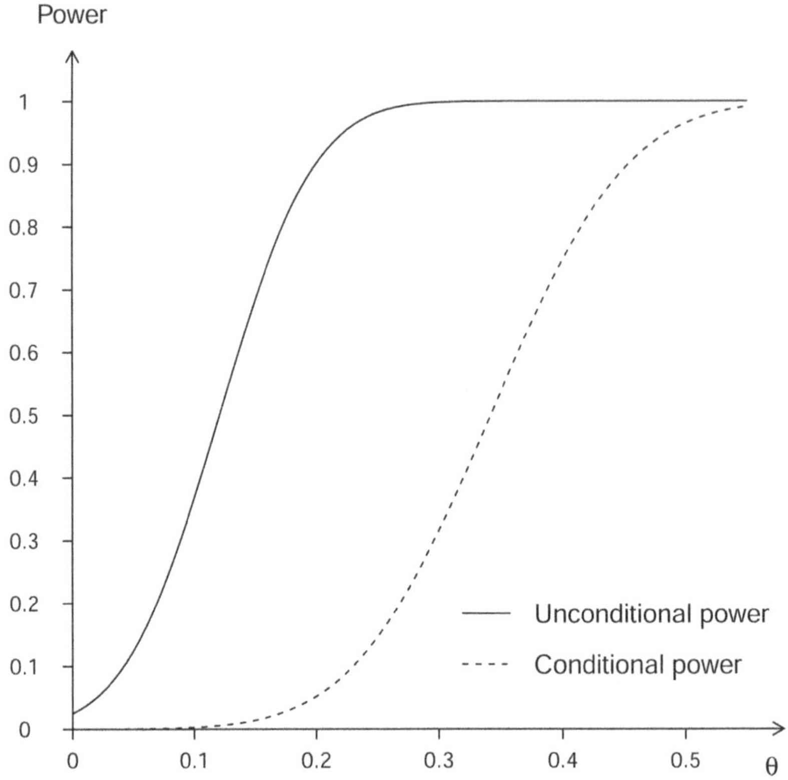

Figure 11.1 *Conditional and unconditional power curves for a one-sided test. The reference test* $\mathcal{T}$ *is a fixed sample test with Type I error rate* $\alpha = 0.025$ *and power* 0.9 *at* $\theta = \delta = 0.2$. *The conditional power curve is for the case that* $\hat{\theta}^{(k)} = -\delta/2 = -0.1$ *after half the observations have been taken.*

a sufficient statistic for θ at stage k, it can be used in place of $D(k)$ in (11.1). Suppose the reference test $\mathcal{T}$ is a one-sided, fixed sample test of H_0: $\theta \le 0$ with Type I error probability α and power $1-\beta$ at $\theta = \theta_1 = \delta$. The reference test $\mathcal{T}$ continues to the final stage K where it rejects H_0 if $Z_K \ge z_\alpha$ and accepts H_0 otherwise. Recall $z_\alpha = \Phi^{-1}(1-\alpha)$, where Φ is the standard normal cumulative distribution function. Power $1-\beta$ at $\theta = \delta$ is achieved by choosing the sample size to yield an information level $\mathcal{I}_K = \mathcal{I}_{f,1}$, as given by (5.1). For each $k = 1, \ldots, K-1$, the conditional distribution of Z_K given $Z_k = z_k$ is

$$Z_K \mid Z_k = z_k \sim N(z_k\sqrt{(\mathcal{I}_k/\mathcal{I}_K)} + \{\theta(\mathcal{I}_K - \mathcal{I}_k)/\sqrt{\mathcal{I}_K}\},\ 1 - \mathcal{I}_k/\mathcal{I}_K) \qquad (11.2)$$

and, hence, the conditional power at analysis k is

$$p_k(\theta) = \Phi\left\{\frac{z_k\sqrt{\mathcal{I}_k} - z_\alpha\sqrt{\mathcal{I}_K} + (\mathcal{I}_K - \mathcal{I}_k)\,\theta}{\sqrt{(\mathcal{I}_K - \mathcal{I}_k)}}\right\}. \qquad (11.3)$$

Suppose $\alpha = 0.025$ and $\mathcal{I}_K = 262.7$ so the fixed sample reference test $\mathcal{T}$ has power 0.9 at $\theta = \delta = 0.2$. The solid line in Figure 11.1, shows the power curve for the reference test $\mathcal{T}$, i.e., $p_k(\theta)$ for $k = 0$. The dashed line in Figure 11.1 is the conditional power function (11.3) calculated at an intermediate stage when half the observations have been taken, so $\mathcal{I}_k = \mathcal{I}_K/2 = 131.3$, and the current MLE of θ is $\hat{\theta}^{(k)} = -\delta/2 = -0.1$, so $z_k = -0.1\sqrt{131.3} = -1.146$. We see that under the alternative $\theta = 0.2$, the probability

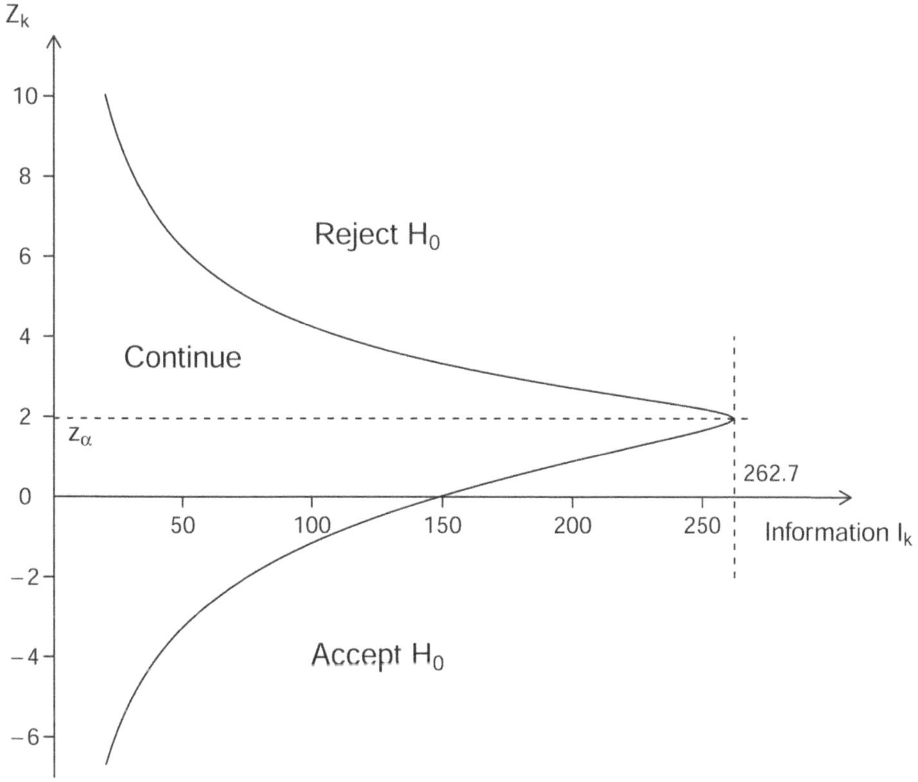

Figure 11.2 *Stopping boundary for a stochastically curtailed, one-sided test using Lan, Simon & Halperin's approach. The reference test is a fixed sample, one-sided test with Type I error probability* $\alpha = 0.025$ *and information level* $\mathcal{I}_{f,1} = 262.7$*, set to achieve power* 0.9 *when* $\theta = \delta = 0.2$*. The stochastic curtailment parameters are* $\gamma = \gamma' = 0.8$.

of rejecting H_0 if the trial goes to completion as planned has been reduced from 0.9 to approximately 0.05.

If we follow Lan, Simon & Halperin's approach and define a formal stopping rule, the condition for stopping to reject H_0 is $p_k(\theta_0) \geq \gamma$ for $\theta_0 = 0$, and this can be written as

$$Z_k \;\geq\; z_\alpha \sqrt{(\mathcal{I}_K/\mathcal{I}_k)} + z_{1-\gamma}\sqrt{\{(\mathcal{I}_K - \mathcal{I}_k)/\mathcal{I}_k\}}. \tag{11.4}$$

Similarly, the condition $p_k(\theta_1) \leq 1 - \gamma'$, with $\theta_1 = \delta$, for stopping to accept H_0 is

$$Z_k \;\leq\; z_\alpha \sqrt{(\mathcal{I}_K/\mathcal{I}_k)} - z_{1-\gamma'}\sqrt{\{(\mathcal{I}_K - \mathcal{I}_k)/\mathcal{I}_k\}} - \delta(\mathcal{I}_K - \mathcal{I}_k)/\sqrt{\mathcal{I}_k}. \tag{11.5}$$

This stopping boundary is shown in Figure 11.2 for $\alpha = 0.025$, $1 - \beta = 0.9$, $\delta = 0.2$, $\mathcal{I}_K = 262.7$, and $\gamma = \gamma' = 0.8$. Whatever the interim analysis schedule, the Type I error probability of this procedure will not exceed $0.025/0.8 = 0.03125$ and the power at $\theta = 0.2$ is guaranteed to be at least $1 - (0.1/0.8) = 0.875$. This boundary can be compared with the continuation regions shown in Figure 5.1 for two power family one-sided tests with $K = 4$ analyses and the same error probability specifications.

Suppose now that the reference test $\mathcal{T}$ is a two-sided fixed sample test of H_0: $\theta = 0$ with Type I error probability α and power $1 - \beta$ at $\theta = \pm\delta$. The test $\mathcal{T}$ only makes a decision at the final stage K, rejecting H_0 if $|Z_K| \geq z_{\alpha/2}$ and accepting H_0 otherwise. The sample size

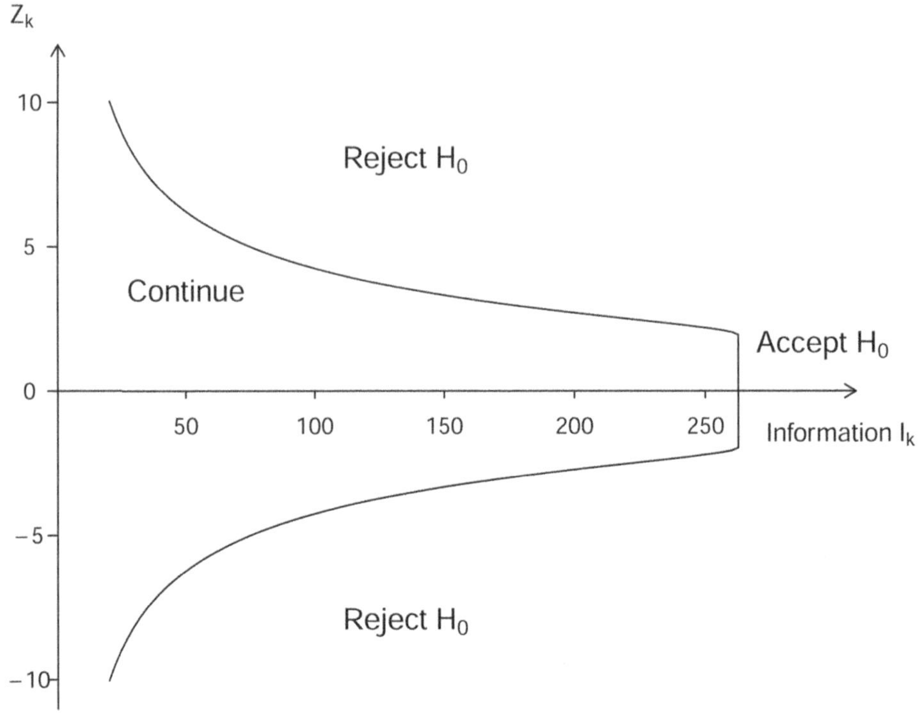

Figure 11.3 *Stopping boundary for a stochastically curtailed, two-sided test using Lan, Simon & Halperin's approach. The reference test is a fixed sample test with two-sided Type I error probability* $\alpha = 0.05$ *and information level* $\mathcal{I}_{f,2} = 262.7$*, set to achieve power* 0.9 *when* $\theta = \pm 0.2$*. The stochastic curtailment parameter is* $\gamma = 0.8$*.*

is chosen to yield an information level $\mathcal{I}_K = \mathcal{I}_{f,2}$, as given in (4.5). The conditional power function at analysis k when $Z_k = z_k$ is

$$p_k(\theta) = \Phi\left\{ \frac{z_k\sqrt{\mathcal{I}_k} - z_{\alpha/2}\sqrt{\mathcal{I}_K} + (\mathcal{I}_K - \mathcal{I}_k)\,\theta}{\sqrt{(\mathcal{I}_K - \mathcal{I}_k)}} \right\} + \Phi\left\{ \frac{-z_k\sqrt{\mathcal{I}_k} - z_{\alpha/2}\sqrt{\mathcal{I}_K} - (\mathcal{I}_K - \mathcal{I}_k)\,\theta}{\sqrt{(\mathcal{I}_K - \mathcal{I}_k)}} \right\}.$$

Suppose we use stochastic curtailment to define a formal stopping rule which rejects H_0 at analysis k if $p_k(0) \geq \gamma$. By the standard argument, the Type I error probability of this SCP is at most α/γ.

Figure 11.3 shows the stopping boundary when the reference test has two-sided Type I error probability $\alpha = 0.05$ and power 0.9 at $\theta = \pm\delta = \pm 0.2$, so $\mathcal{I}_K = 262.7$, and the SCP is defined using $\gamma = 0.8$. This group sequential procedure is directly comparable with the Pocock and O'Brien & Fleming tests with $K = 4$ groups of observations and the same error probability specifications, which were displayed in Figure 3.1. Note that we have not introduced early stopping to accept H_0 based on large values of $1 - p_k(\delta)$ and $1 - p_k(-\delta)$. If we were to do this, the SCP would have an inner wedge, and so would resemble the tests of Chapter 6.

In the case of a two-sided SCP defined without an inner wedge, if the reference test is formed with Type I error probability $\alpha\gamma$ instead of α, the SCP has Type I error probability at most α and can be compared directly with the two-sided tests of Chapter 2. Figure 11.3 shows that the boundaries are wide initially and then converge in a similar manner to the O'Brien & Fleming (1979) boundaries in Figures 2.2 and 3.1. In the special case $\gamma = 0.5$, the

SCP is the continuous-monitoring version of an O'Brien & Fleming procedure. This SCP also coincides with the test suggested by Samuel-Cahn (1980) and is related to proposals of Chatterjee & Sen (1973), Davis (1978) and Koziol & Petkau (1978); see Halperin et al. (1982, p. 322). If the number of analyses K is small, α/γ will be a very conservative bound on the Type I error probability of an SCP based on a reference test with Type I error probability α. What is gained from this conservatism is the ability to do unplanned interim analyses at arbitrary, even data-dependent, times. However, if the timing of analyses is fixed in advance, numerical methods may be used to obtain an exact value of the Type I error probability and less conservative boundaries can be derived.

11.3 The Predictive Power Approach

A criticism of the methods of the previous section is that they are based on conditional power calculated under values of θ which may not be supported by the current data. For instance, the criterion for stopping to reject H_0 involves $p_k(\theta)$ evaluated at $\theta = \theta_0$, a value of θ unlikely to be true if rejection is being considered. We now introduce a procedure which overcomes this problem by using a Bayesian calculation to average conditional power over values of θ which are consistent with the data, but still with a frequentist test in mind.

As before, suppose the reference test $\mathcal{T}$ is a one-sided test of H_0: $\theta \leq 0$ with Type I error probability α and power $1-\beta$ at $\theta = \theta_1$ or a two-sided test of H_0: $\theta = 0$ with power $1-\beta$ at $\theta = \theta_1$ and $\theta = \theta_2$. Let $\pi\{\theta \mid D(k)\}$ denote the posterior density of θ, given a prior distribution for θ and the data $D(k)$ accumulated by analysis k, so π reflects current beliefs about the value of θ. Instead of using the conditional power at particular θ values, we define the "predictive power" at stage k to be the integral of the conditional power $p_k(\theta)$, as defined in (11.1), weighted by the posterior density $\pi\{\theta \mid D(k)\}$,

$$P_k(D(k)) = \int p_k(\theta, D(k))\, \pi\{\theta \mid D(k)\}\, d\theta, \tag{11.6}$$

which we shall write as P_k when the conditioning event is clear. This approach has been advocated by Herson (1979), Choi, Smith & Becker (1985), and Spiegelhalter, Freedman & Blackburn (1986). The predictive power can be a useful quantity to communicate to applied researchers at an intermediate stage of a trial. As with conditional power, it may be used to define a formal stopping rule by stopping at stage k to reject H_0 if $P_k \geq \gamma$ or to accept H_0 if $P_k \leq 1-\gamma'$.

Consider again the first example of Section 11.2. Standardized statistics $Z_1, \ldots, Z_K$ are assumed to follow the joint distribution (3.1), given $\mathcal{I}_1, \ldots, \mathcal{I}_K$, and $\mathcal{T}$ is a fixed sample one-sided reference test with Type I error probability α and power $1-\beta$ at $\theta = \delta$, conducted with information level $\mathcal{I}_K = \mathcal{I}_{f,1}$. This test rejects H_0: $\theta = \theta_0$ if $Z_K \geq z_\alpha$. Suppose we use the improper prior, $\pi(\theta) = 1$ for all θ, to represent an initial lack of knowledge about θ. Then, at stage k, when the MLE of θ is $Z_k/\sqrt{\mathcal{I}_k}$, the posterior distribution for θ is

$$\theta \mid Z_k = z_k \sim N(z_k/\sqrt{\mathcal{I}_k},\, 1/\mathcal{I}_k) \tag{11.7}$$

and

$$Z_K \mid Z_k = z_k \sim N\left(\frac{z_k\sqrt{\mathcal{I}_K}}{\sqrt{\mathcal{I}_k}},\, \frac{\mathcal{I}_K - \mathcal{I}_k}{\mathcal{I}_k}\right). \tag{11.8}$$

Either by substituting the posterior density (11.7) of θ into (11.6) and carrying out the integration, or by working from the conditional distribution (11.8) of Z_K, we obtain

$$P_k = \Phi\left\{\frac{z_k\sqrt{\mathcal{I}_K} - z_\alpha\sqrt{\mathcal{I}_k}}{\sqrt{(\mathcal{I}_K - \mathcal{I}_k)}}\right\}. \tag{11.9}$$

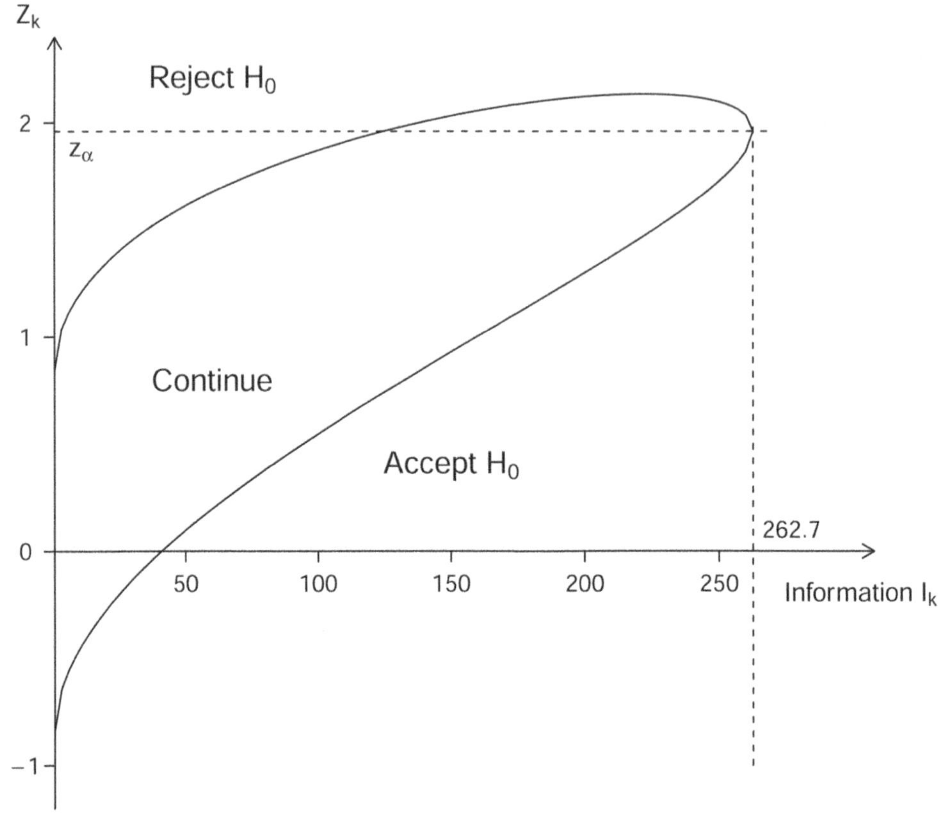

Figure 11.4 *Stopping boundary for a stochastically curtailed, one-sided test using the predictive power approach with a uniform prior. The reference test is a fixed sample one-sided test with Type I error probability* $\alpha = 0.025$ *and information level* $\mathcal{I}_{f,1} = 262.7$*, set to achieve power* 0.9 *when* $\theta = 0.2$*. The stochastic curtailment parameters are* $\gamma = \gamma' = 0.8$*.*

The argument of Φ in (11.9) is similar to that in (11.3) after θ has been replaced by its MLE, $\hat{\theta} = z_k/\sqrt{\mathcal{I}_k}$, but it is multiplied by a factor $\sqrt{(\mathcal{I}_k/\mathcal{I}_K)}$. Thus, if the argument of Φ is positive, the predictive power is less than the conditional power under $\theta = \hat{\theta}$, reflecting the uncertainty about the true θ accounted for in the predictive power calculation.

If the study is to be terminated to reject H_0 when $P_k \geq \gamma$ and to accept H_0 when $P_k \leq 1 - \gamma'$, the stopping criteria can be expressed as

$$\text{Reject } H_0 \text{ if } \quad Z_k \;\geq\; z_\alpha\sqrt{(\mathcal{I}_k/\mathcal{I}_K)} + z_{1-\gamma}\sqrt{\{(\mathcal{I}_K - \mathcal{I}_k)/\mathcal{I}_K\}} \tag{11.10}$$

and

$$\text{Accept } H_0 \text{ if } \quad Z_k \;\leq\; z_\alpha\sqrt{(\mathcal{I}_k/\mathcal{I}_K)} - z_{1-\gamma'}\sqrt{\{(\mathcal{I}_K - \mathcal{I}_k)/\mathcal{I}_K\}}. \tag{11.11}$$

Figure 11.4 displays these stopping boundaries when $\gamma = \gamma' = 0.8$ and the reference test $\mathcal{T}$ is the same one-sided test that led to the stopping rule based on conditional power shown in Figure 11.2.

Although the criteria (11.10) and (11.11) are similar in form to those for the conditional power stopping rule in (11.4) and (11.5), a comparison of Figures 11.4 and 11.2 shows that the predictive power approach gives much narrower boundaries for the same γ and γ'. Early stopping is permitted more readily since the conditional probabilities are based on an estimate of θ obtained from the data rather than the hypothesized values $\theta = \theta_0$ and $\theta = \delta$. The martingale argument made by Lan, Simon & Halperin (1982) for their procedure cannot be applied to bound the Type I error probability by α/γ and the Type II error probability by β/γ', but the effects of this early stopping on Type I and II error rates can be computed numerically using the methods described in Chapter 31 and these calculations show that, with four equally spaced analyses, the Type I error rate is 0.066 and the Type II error rate at $\theta = 0.2$ is 0.155. These error probabilities could be reduced by increasing γ and γ'. One might also design the reference test to have lower Type I and Type II error probabilities to start with, which would necessitate increasing the final information level $\mathcal{I}_K$, in order that the design with curtailment has acceptable Type I and II error rates.

In the two-sided testing problem of Section 11.2, the reference test has information level $\mathcal{I}_K = \mathcal{I}_{f,2}$ and rejects H_0: $\theta = 0$ if $|Z_K| \geq z_{\alpha/2}$. The criterion $P_k \geq \gamma$ now calls for a decision to stop and reject H_0 at stage k with $Z_k = z_k$ if

$$\Phi\left\{\frac{|z_k|\sqrt{\mathcal{I}_K} - z_{\alpha/2}\sqrt{\mathcal{I}_k}}{\sqrt{(\mathcal{I}_K - \mathcal{I}_k)}}\right\} + \Phi\left\{\frac{-|z_k|\sqrt{\mathcal{I}_K} - z_{\alpha/2}\sqrt{\mathcal{I}_k}}{\sqrt{(\mathcal{I}_K - \mathcal{I}_k)}}\right\} \geq \gamma. \tag{11.12}$$

The boundary using $\gamma = 0.8$ is shown in Figure 11.5. Here, the reference test has two-sided Type I error probability $\alpha = 0.05$ and $\mathcal{I}_K = 262.7$ in order to achieve power $1 - \beta = 0.9$ at $\theta = \pm 0.2$, as in the example used to illustrate the conditional power approach in Figure 11.3.

The continuation region for this two-sided test based on predictive power is similar to that for the one-sided case in that it is initially narrower than the continuation region obtained using the conditional power approach and there is greater opportunity for early stopping. Numerical computation shows that, with four equally spaced analyses, the Type I error probability is rather high at 0.147. The Type II error probability under $\theta = \pm 0.2$ is 0.076, so power is higher than that of the reference test, a consequence of the high Type I error rate. Again, the Type I error rate may be reduced by increasing γ and the reference test could be designed to have a lower Type I error probability to start with, increasing the final information level $\mathcal{I}_K$ accordingly, so that the design with curtailment has acceptable Type I and Type II error rates.

We have only used the criterion $P_k \geq \gamma$ for early stopping in this example. An early decision to accept H_0 if $P_k \leq 1 - \gamma'$ could also be permitted, in which case the stopping boundary would contain an inner wedge as in the tests of Chapter 6.

11.4 A Parameter-free Approach

The two previous approaches to stochastic curtailment require specification of θ, either a particular value at which to evaluate the conditional power, or a prior distribution for θ in the predictive power approach. A third method, proposed by Jennison (1992), dispenses with the need to specify θ at all. This approach has been further developed by Xiong (1995), Tan & Xiong (1996), and Tan, Xiong & Kutner (1998), who use the terms "sequential conditional probability ratio test" and "reverse stochastic curtailing" to describe their procedures.

Suppose statistics $Z_1, \ldots, Z_K$ follow the usual joint distribution (3.1), given $\mathcal{I}_1, \ldots, \mathcal{I}_K$, and we have a fixed sample, one-sided reference test with Type I error probability α and power $1 - \beta$ at $\theta = \delta$, conducted at information level $\mathcal{I}_K = \mathcal{I}_{f,1}$. This test will reject H_0: $\theta \leq \theta_0$ if $Z_K \geq z_\alpha$. Suppose the information level at stage k is $\mathcal{I}_k$ and we observe $Z_k = z$. A pertinent question is "If the reference test will eventually accept H_0, how likely is it that we should observe $Z_k \geq z$ at stage k?" This leads us to consider

$$\begin{aligned} \sup_{\tilde{z} \leq z_\alpha} Pr\{Z_k \geq z \mid Z_K = \tilde{z}\} &= Pr\{Z_k \geq z \mid Z_K = z_\alpha\} \\ &= 1 - \tilde{p}_k(z), \quad \text{say.} \end{aligned}$$

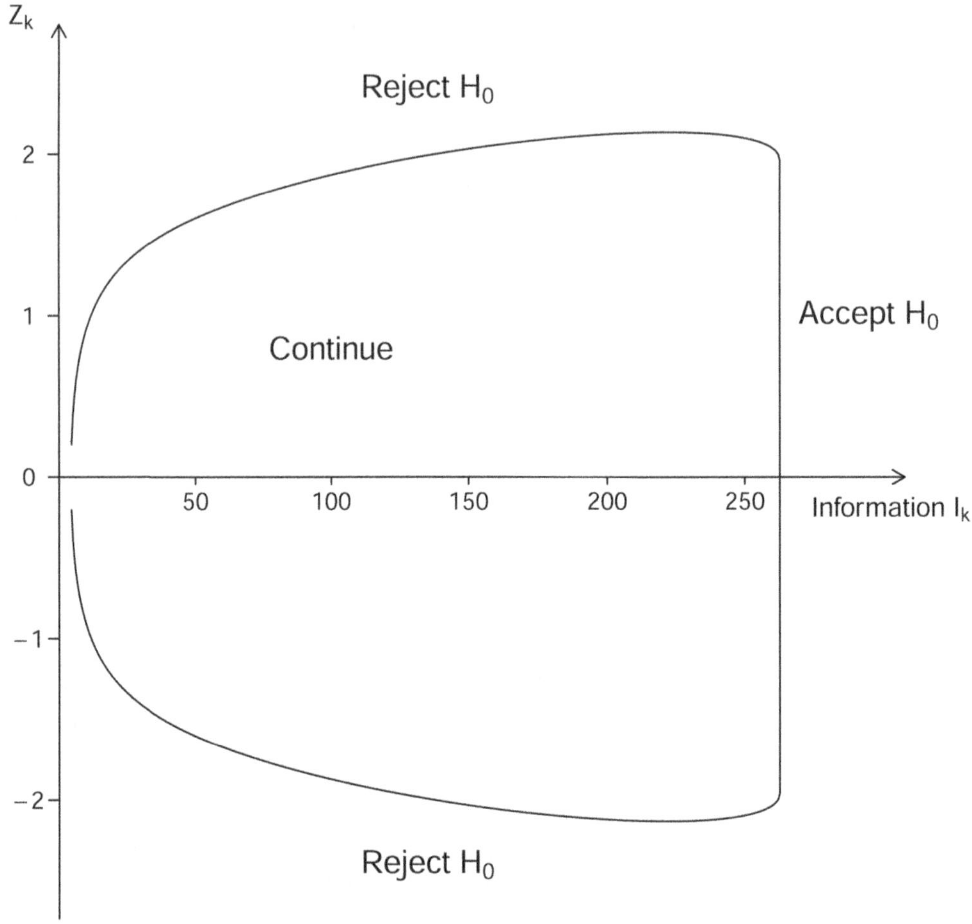

Figure 11.5 *Stopping boundary for a stochastically curtailed, two-sided test using the predictive power approach with a uniform prior. The reference test is a fixed sample test with two-sided Type I error probability* $\alpha = 0.05$ *and information level* $\mathcal{I}_{f,2} = 262.7$*, set to achieve power* 0.9 *when* $\theta = \pm 0.2$*. The stochastic curtailment parameter is* $\gamma = 0.8$*.*

Similarly, we can ask "If the reference test will eventually reject H_0, what is the probability of observing $Z_k \le z$ at stage k?" This conditional probability is at most

$$\sup_{\tilde{z} \ge z_\alpha} Pr\{Z_k \le z \mid Z_K = \tilde{z}\} \;=\; Pr\{Z_k \le z \mid Z_K = z_\alpha\} \;=\; \tilde{p}_k(z).$$

Now, if $S_k = Z_k\sqrt{\mathcal{I}_k}$, $k = 1, \ldots, K$, the conditional joint distribution of the process $\{S_k\}$ given $Z_K = z_\alpha$ is that of a Brownian bridge observed at discrete time points. In particular, the conditional distribution of Z_k given $Z_K = z_\alpha$ is

$$Z_k \mid Z_K = z_\alpha \;\sim\; N\Big(\frac{z_\alpha\sqrt{\mathcal{I}_k}}{\sqrt{\mathcal{I}_K}}, \frac{\mathcal{I}_K - \mathcal{I}_k}{\mathcal{I}_K}\Big)$$

and, hence,

$$\tilde{p}_k(z) \;=\; \Phi\left\{\frac{z\sqrt{\mathcal{I}_K} - z_\alpha\sqrt{\mathcal{I}_k}}{\sqrt{(\mathcal{I}_K - \mathcal{I}_k)}}\right\}, \quad k = 1, \ldots, K. \qquad (11.13)$$

As in the previous methods, a formal procedure can be created by stopping at stage k according to the following rules:

$$\text{Reject } H_0 \text{ if } \tilde{p}_k(Z_k) \geq \gamma \quad \text{and} \quad \text{Accept } H_0 \text{ if } \tilde{p}_k(Z_k) \leq 1 - \gamma'.$$

Here γ and γ' are specified constants, greater than 0.5. This procedure stops at stage k to reject H_0 if

$$Z_k \geq z_\alpha \sqrt{(\mathcal{I}_k/\mathcal{I}_K)} + z_{1-\gamma} \sqrt{\{(\mathcal{I}_K - \mathcal{I}_k)/\mathcal{I}_K\}} \tag{11.14}$$

and to accept H_0 if

$$Z_k \leq z_\alpha \sqrt{(\mathcal{I}_k/\mathcal{I}_K)} - z_{1-\gamma'} \sqrt{\{(\mathcal{I}_K - \mathcal{I}_k)/\mathcal{I}_K\}}. \tag{11.15}$$

Comparison with (11.10) and (11.11) shows these boundaries are exactly the same as those obtained in the predictive power approach using an improper flat prior! The quantity $\tilde{p}_k(z_k)$ is identical to the P_k in (11.6) and (11.9). Thus, for our previous example of a one-sided test, the stopping boundary in Figure 11.4 is also the boundary for the parameter-free method with $\gamma = \gamma' = 0.8$.

For the two-sided test, we consider the probability of observing a Z_k with an absolute value as large as that of the observed statistic, z say, given that the reference test eventually accepts H_0. If this conditional probability is low, we can stop to reject H_0. Thus we consider

$$\begin{aligned} \sup_{|\tilde{z}| \leq z_{\alpha/2}} Pr\{|Z_k| \geq |z| \mid Z_K = \tilde{z}\} &= Pr\{|Z_k| \geq |z| \mid Z_K = z_{\alpha/2}\} \\ &= Pr\{|Z_k| \geq |z| \mid Z_K = -z_{\alpha/2}\} \\ &= 1 - \tilde{p}_{k,2}(z), \quad \text{say.} \end{aligned} \tag{11.16}$$

A formal rule can be defined by stopping to reject H_0 if a value $Z_k = z_k$ is observed for which $1 - \tilde{p}_{k,2}(z_k)$ is less than $1 - \gamma$ for a chosen constant γ. A calculation analogous to that which produced (11.13) shows that H_0 is rejected if the observed value z_k satisfies

$$\Phi\left\{\frac{|z_k|\sqrt{\mathcal{I}_K} - z_{\alpha/2}\sqrt{\mathcal{I}_k}}{\sqrt{(\mathcal{I}_K - \mathcal{I}_k)}}\right\} - \Phi\left\{\frac{-|z_k|\sqrt{\mathcal{I}_K} - z_{\alpha/2}\sqrt{\mathcal{I}_k}}{\sqrt{(\mathcal{I}_K - \mathcal{I}_k)}}\right\} \geq \gamma. \tag{11.17}$$

This differs from condition (11.12) of the predictive power approach in the sign of the second term. Typically, this term is small unless $\mathcal{I}_k/\mathcal{I}_K$ is small, hence the parameter-free and predictive power boundaries grow more similar as $\mathcal{I}_k/\mathcal{I}_K$ increases.

Figure 11.6 shows the boundary for the two-sided testing problem considered earlier with $\alpha = 0.05$ and $\mathcal{I}_K = 262.7$, and using $\gamma = 0.8$. Comparing this with Figure 11.5, we see that the boundary of the parameter-free test is initially wider than the predictive power boundary but the two boundaries become indistinguishable as $\mathcal{I}_k$ increases. With four equally spaced analyses, the Type I error probability is 0.146 and the Type II error probability under $\theta = \pm 0.2$ is 0.077, almost exactly the same values as for the predictive power method. However, the error rates for the two methods do differ if earlier analyses, at which the stopping boundaries are quite different, are added.

Jennison (1992) reports the success of this parameter-free approach in reducing computation in a simulation study conducted to evaluate the error rates of a bootstrap hypothesis test. In each replication, it was required to determine whether the number of bootstrap samples with a certain property fell in a given range. A form of parameter-free stochastic curtailment was applied in the bootstrap sampling, for which the probability of reaching the opposite conclusion from the full bootstrap test was of the order of 10^{-5},

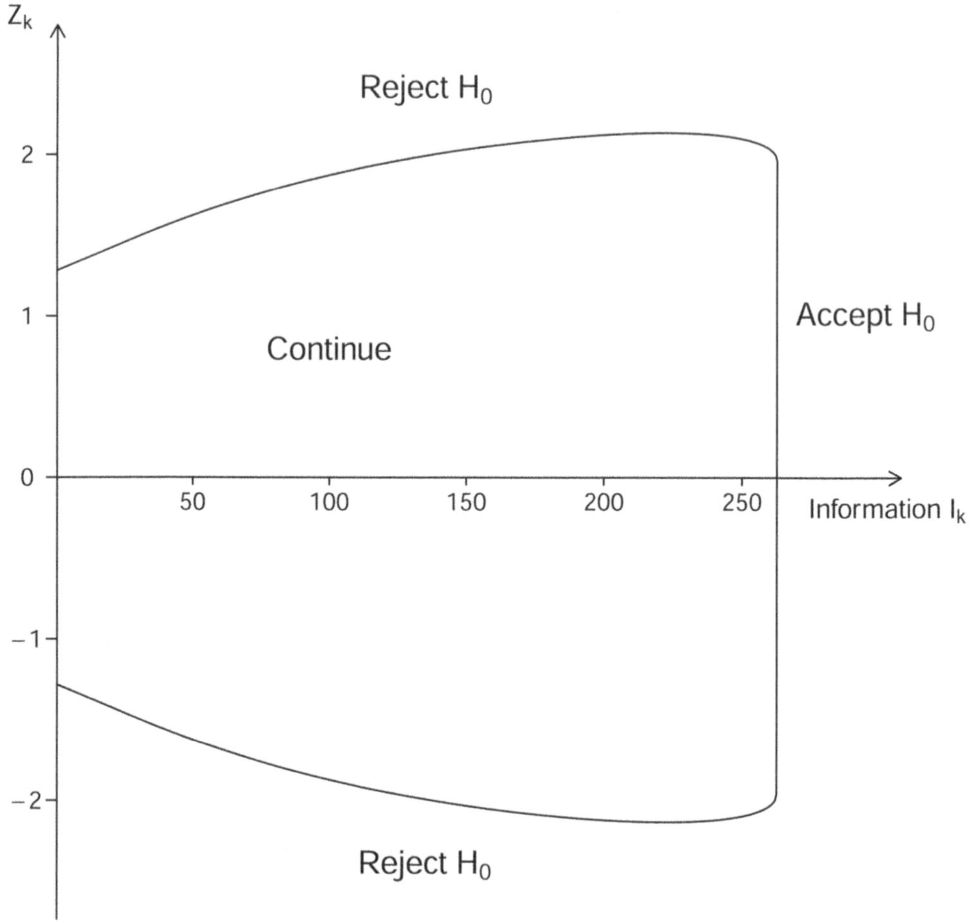

Figure 11.6 *Stopping boundary for a stochastically curtailed, two-sided test using the parameter-free approach. The reference test is a fixed sample test with two-sided Type I error probability* $\alpha = 0.05$ *and information level* $\mathcal{I}_{f,2} = 262.7$*, set to achieve power* 0.9 *at* $\theta = \pm 0.2$*. The stochastic curtailment parameter is* $\gamma = 0.8$*.*

a negligible amount in this context. This improved greatly on simple curtailment and computation was reduced dramatically, by a factor as high as 30 when the bootstrap sample size was 10,000.

The parameter-free method has been used by Stone et al. (2015) to create a method of audit sampling that can reduce the costs associated with the independent central review of diagnoses of disease progression in a trial where progression-free survival is the primary endpoint. In the audit sampling approach, a random sample of diagnoses is assessed and, where sufficient agreement is observed between local and central decisions, no further assessment is needed and the final analysis can be based on the local assessments of disease progression. Jänne et al. (2017) describe an application of this approach in the SELECT-1 clinical trial that compared chemotherapy regimens for patients with advanced lung cancer and they report that, given the consistency of results in the audit sample, the expense of an exhaustive review was avoided.

11.5 Using Conditional Power and Predictive Power to Define a Non-binding Futility Boundary

11.5.1 Calculating Conditional Power and Predictive Power in a Group Sequential Study

In previous chapters, we described one-sided group sequential tests in which the futility boundary is considered to be non-binding: in Section 5.3 the futility boundary was taken from a particular parametric family and in Section 8.4 the futility boundary was defined by a Type II error spending function. In practice, it is quite common for a trial to have a formal stopping rule for efficacy and an informal approach to stopping for futility. It is then up to the monitoring committee to decide whether to stop early for futility, taking into account conditional power and all other pertinent information; see, for example, Lachin (2005). We shall describe this informal approach to stopping for futility in a trial with an efficacy boundary defined through an error spending function.

Suppose a trial is designed to test H_0: $\theta \le 0$ against the one-sided alternative $\theta > 0$, with an error spending efficacy boundary and an informal approach to stopping for futility. Then, as information levels $\mathcal{I}_1, \mathcal{I}_2, \ldots$ are observed at interim analyses, upper boundary values $b_1, b_2, \ldots$ are calculated such that the overall Type I error probability will have the desired value α, in the absence of any early stopping for futility. Thus, if there may be up to K analyses,

$$Pr_{\theta=0}\{Z_k > b_k \text{ for some } 1 \le k \le K\} = \alpha.$$

At an interim analysis, conditional power can be computed by assuming plausible values for future information levels. The conditional power under treatment effect θ, given the current Z-statistic, $Z_k = z_k$, is defined as

$$p_k(\theta, z_k) = Pr_\theta\{Z_{k'} \ge b_{k'} \text{ for some } k < k' \le K \mid Z_k = z_k\}, \qquad (11.18)$$

where it is assumed there will be no early stopping for futility.

In calculating $p_k(\theta, z_k)$, we first find the efficacy boundary values $b_{k+1}, \ldots, b_K$, given the specified error spending function, the observed information levels $\mathcal{I}_1, \ldots, \mathcal{I}_k$, and the assumed future information levels $\mathcal{I}_{k+1}, \ldots, \mathcal{I}_K$. We then define the score statistics $S_k = Z_k\sqrt{\mathcal{I}_k}, \ldots, S_K = Z_K\sqrt{\mathcal{I}_K}$, and

$$\tilde{S}_{k'} = S_{k+k'} - S_k \quad \text{and} \quad \tilde{Z}_{k'} = \frac{\tilde{S}_{k'}}{\sqrt{\mathcal{I}_{k'} - \mathcal{I}_k}}, \quad \text{for } k' = 1, \ldots, K-k.$$

Now,

$$\begin{aligned}
p_k(\theta, z_k) &= Pr_\theta\{Z_{k+k'} \ge b_{k+k'} \text{ for some } 1 \le k' \le K-k \mid Z_k = z_k\} \\
&= Pr_\theta\{S_{k+k'} \ge b_{k+k'}\sqrt{\mathcal{I}_{k+k'}} \text{ for some } 1 \le k' \le K-k \mid S_k = z_k\sqrt{\mathcal{I}_k}\} \\
&= Pr_\theta\{\tilde{S}_{k'} \ge b_{k+k'}\sqrt{\mathcal{I}_{k+k'}} - z_k\sqrt{\mathcal{I}_k} \text{ for some } 1 \le k' \le K-k\} \\
&= Pr_\theta\{\tilde{Z}_{k'} \ge \frac{b_{k+k'}\sqrt{\mathcal{I}_{k+k'}} - z_k\sqrt{\mathcal{I}_k}}{\sqrt{\mathcal{I}_{k'} - \mathcal{I}_k}} \text{ for some } 1 \le k' \le K-k\}. \qquad (11.19)
\end{aligned}$$

It is straightforward to show that the sequence $\tilde{Z}_{k'}$, $k' = 1, \ldots, K-k$, has the canonical joint distribution (3.1) with information levels $\mathcal{I}_{k'} - \mathcal{I}_k$, $k' = 1, \ldots, K-k$. Thus $p_k(\theta, z_k)$ can be computed in the same way as the unconditional power (5.12), using the iterated integrals method of Section 3.5 but replacing K by $K-k$ and $\mathcal{I}_K$ by $\mathcal{I}_K - \mathcal{I}_k$ and using the new boundary values

$$\tilde{b}_{k'} = \frac{b_{k+k'}\sqrt{\mathcal{I}_{k+k'}} - z_k\sqrt{\mathcal{I}_k}}{\sqrt{\mathcal{I}_{k'} - \mathcal{I}_k}}, \quad k' = 1, \ldots, K-k.$$

Furthermore, predictive power can be calculated by integrating the conditional power $p_k(\theta, z_k)$ over the posterior distribution of θ.

One difficulty in implementing the above calculation is the requirement to predict the future information levels $\mathcal{I}_{k+1}, \ldots, \mathcal{I}_K$. Suppose that in an error spending design, the Type I error probability spending function is of the form $f(\mathcal{I}/\mathcal{I}_{max})$, where $\mathcal{I}_{max}$ is the target information level for the trial. Then, a simple approach is to assume there will be just one further analysis with information level $\mathcal{I}_{k+1} = \mathcal{I}_{max}$. In this case, the final boundary value b_{k+1} can be found by solving the equation

$$Pr_{\theta=0}\{Z_1 < b_1, \ldots, Z_k < b_k, Z_{k+1} \geq b_{k+1}\} = \alpha - f(\mathcal{I}_k/\mathcal{I}_{max}). \tag{11.20}$$

where $b_1, \ldots, b_k$ have already been calculated based on the observed information levels $\mathcal{I}_1, \ldots, \mathcal{I}_k$. Under these assumptions, the conditional power function is

$$p_k(\theta, z_k) = Pr_\theta\{Z_K \geq b_K \mid Z_k = z_k\} \tag{11.21}$$

where $K = k+1$, $\mathcal{I}_K = \mathcal{I}_{max}$ and, from (3.1), the conditional distribution of Z_K is

$$Z_K \mid Z_k = z_k \sim N(z_k\sqrt{(\mathcal{I}_k/\mathcal{I}_{max})} + \{\theta(\mathcal{I}_{max} - \mathcal{I}_k)/\sqrt{\mathcal{I}_{max}}\}, 1 - \mathcal{I}_k/\mathcal{I}_{max}).$$

This is precisely the conditional distribution (11.2) with $\mathcal{I}_K$ replaced by $\mathcal{I}_{max}$. However, now H_0 will be rejected if $Z_K > b_K$, so we replace z_α by b_K in Equation (11.3) to obtain the conditional power

$$p_k(\theta, z_k) = \Phi\left\{\frac{z_k\sqrt{\mathcal{I}_k} - b_K\sqrt{\mathcal{I}_{max}} + (\mathcal{I}_{max} - \mathcal{I}_k)\,\theta}{\sqrt{(\mathcal{I}_{max} - \mathcal{I}_k)}}\right\} \tag{11.22}$$

when $Z_k = z_k$ at analysis k. If, in fact, more interim analyses are to be conducted before the final analysis, the formula (11.22) will provide an over-estimate of conditional power. However, this is unlikely to have a significant impact when conditional power is considered as a guide to early stopping for futility.

11.5.2 Example: Stochastic Curtailment with an Error Spending Efficacy Boundary

A natural rule for terminating a trial for futility is to stop to accept H_0: $\theta \leq 0$ at any stage where the conditional power at a certain treatment effect θ or the predictive power, given a specified prior distribution for θ, is sufficiently small. We shall discuss how to apply such a rule as a guideline for early stopping in an example.

Consider a trial designed to test H_0: $\theta \leq 0$ against $\theta > 0$ with an error spending efficacy boundary. The Type I error probability is set at $\alpha = 0.025$ and this is to be spent according to the function

$$f(\mathcal{I}/\mathcal{I}_{max}) = \min\{\alpha(\mathcal{I}/\mathcal{I}_{max})^2, \alpha\},$$

where the target information level $\mathcal{I}_{max}$ is chosen to achieve power $1 - \beta = 0.9$ if the treatment effect is $\theta = \delta = 1$. Suppose up to 5 analyses are planned and, in designing the trial, equally spaced information levels are assumed. In this case, a maximum information level $\mathcal{I}_{max} = 11.1$ is required and the efficacy boundary that will arise under information levels $\mathcal{I}_k = (k/5)\mathcal{I}_{max}$, $k = 1, \ldots, 5$, is shown in Figure 11.7.

Suppose the observed information levels $\mathcal{I}_1$, $\mathcal{I}_2$ and $\mathcal{I}_3$ follow the assumed pattern and, at the third analysis, values $\hat{\theta}^{(3)} = 0.28$ and $Z_3 = 0.723$ are observed. The data point $(\mathcal{I}_3, Z_3) = (6.67, 0.723)$ is plotted in Figure 11.7. The conditional power at this analysis can be computed, as explained above, and the conditional power function is displayed in Figure 11.8. This conditional power function could be presented to the data monitoring committee to guide deliberations about whether to terminate the trial for futility.

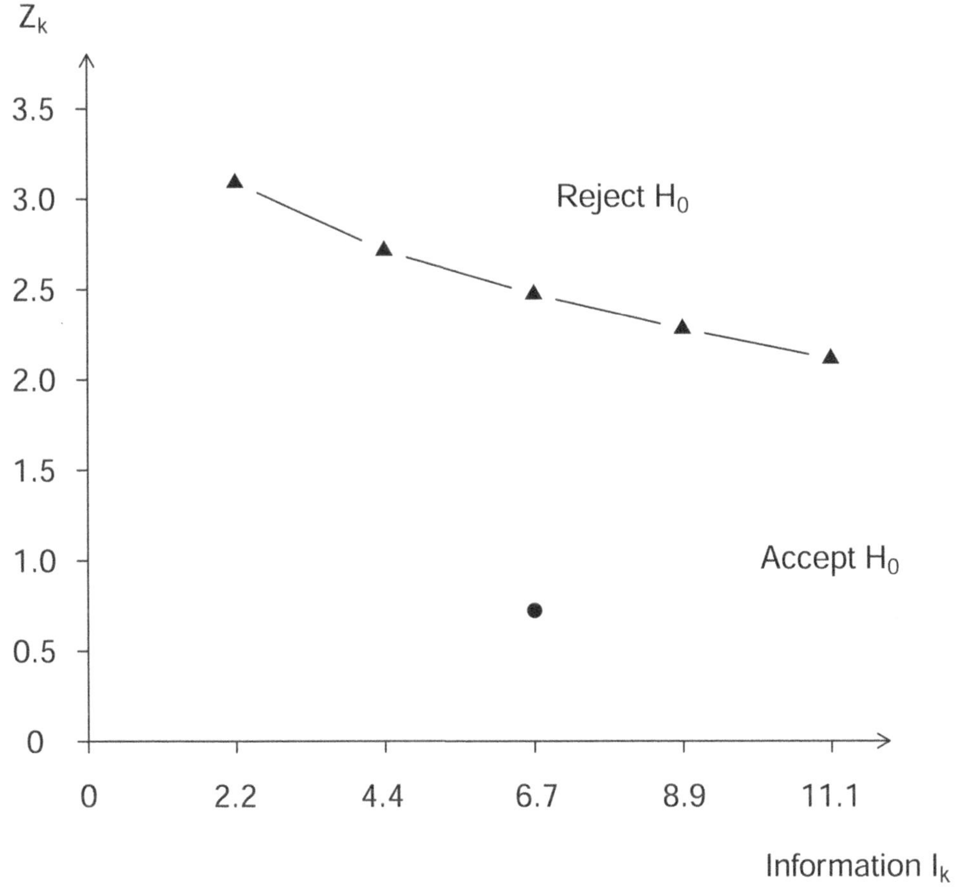

Figure 11.7 *Example: Efficacy boundary and data point. The one-sided group sequential test has five analyses and the efficacy boundary is defined through the error spending function* $f(\mathcal{I}/\mathcal{I}_{max}) = \min\{\alpha(\mathcal{I}/\mathcal{I}_{max})^2, \alpha\}$. *The maximum information level* $\mathcal{I}_{max} = 11.1$ *is chosen to give power* 0.9 *if* $\theta = \delta = 1$. *The data point observed at analysis* 3 *has* $(\mathcal{I}_3, Z_3) = (6.67, 0.723)$ *so* $\hat{\theta}^{(3)} = 0.28$.

A difficulty in making such a decision is that conditional power varies with θ and the true value of θ is unknown. Under the current estimate, $\hat{\theta}^{(3)} = 0.28$, conditional power is 0.082, while under $\theta = 1$, the effect size assumed in the original power calculation, it is much higher at 0.47. The 95% confidence interval for θ based on the current data (with no adjustment for sequential testing) is displayed in Figure 11.8 and this shows that values of θ between -0.5 and 1.0 are compatible with the observed data. Thus, assuming $\theta = \hat{\theta}^{(3)}$ could be unduly pessimistic, while focusing on the conditional power under $\theta = 1$ would ignore the fact that the true treatment effect may well be smaller than this. If one were to follow the approach of Lan, Simon & Halperin (1982), stopping to accept H_0 would require the conditional power under $\theta = \delta = 1$ to be less than $1 - \gamma'$ and, since $p_3(\delta, z_3) = 0.47$, this condition would not be met for a typical value such as $\gamma' = 0.8$ and the trial would continue.

Working with predictive power deals with the uncertainty about θ by integrating over the posterior distribution of θ given a certain prior distribution. Figure 11.9 shows the posterior distribution $\theta \sim N(0.28, 0.39^2)$ when an improper, flat prior is assumed, in which case the predictive power is 0.13. Information about θ at an interim analysis is limited: if

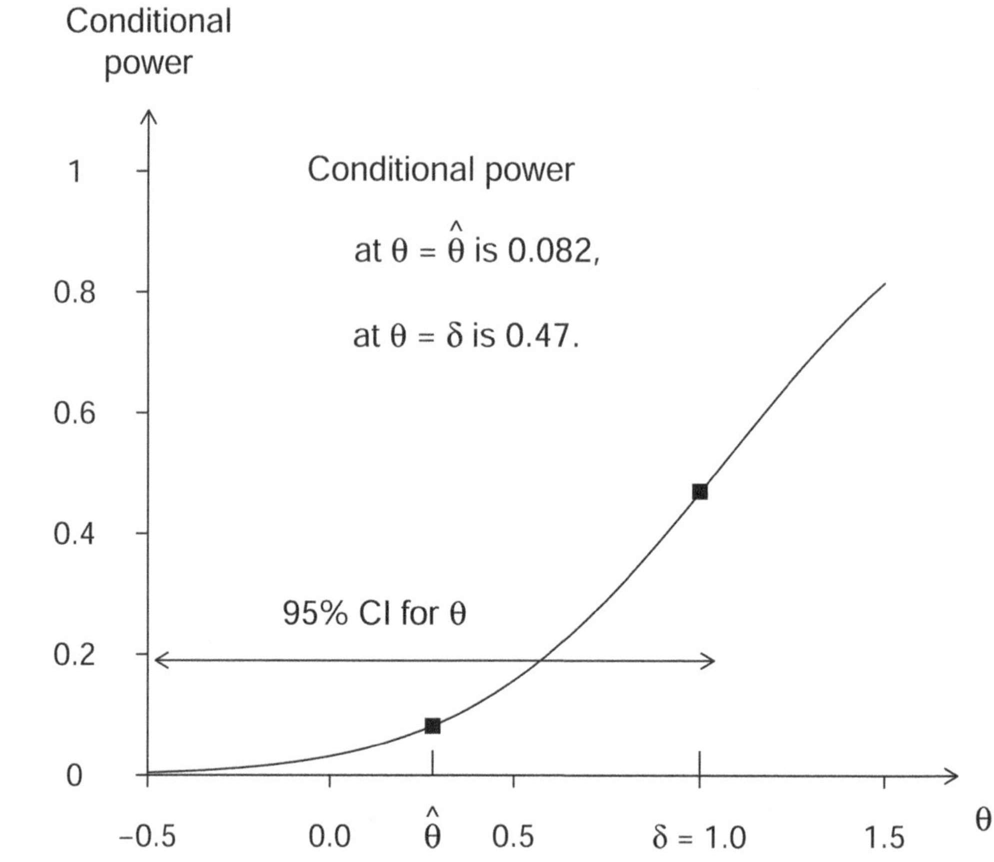

Figure 11.8 *Example: Conditional power as a function of* θ. *The one-sided group sequential test has five analyses and the efficacy boundary is defined through the error spending function* $f(\mathcal{I}/\mathcal{I}_{max}) = \min\{\alpha(\mathcal{I}/\mathcal{I}_{max})^2, \alpha\}$. *The estimated treatment effect at analysis 3 is* $\hat{\theta}^{(3)} = 0.28$.

the sample size is chosen to give power to distinguish between the cases $\theta = 0$ and $\theta = \delta$ at the end of the study, one should expect a 95% confidence interval based on 40 or 50% of these data to have width around $1.5\,\delta$, as was seen in Figure 11.8. Thus, the choice of prior has a significant impact on the predictive power. Suppose, for example, investigators' pre-trial opinions are captured by the prior distribution $\theta \sim N(0.6, 0.3^2)$, then Figure 11.9 shows the posterior distribution $\theta \sim N(0.48, 0.24^2)$ and the predictive power is 0.17.

As well as agreeing on an appropriate measure of the likelihood of success, a monitoring committee has to decide how high this likelihood has to be for the trial to continue. A key task here is to balance the saving in resources that accrue from early termination against the possibility that the trial would reach a positive result if it were allowed to continue. There can be both ethical and commercial issues to consider and it is not clear that a data monitoring committee is well placed to make the required judgement when presented with values for conditional power and predictive power. Lachin (2005) suggests taking the futility boundary of a formally defined group sequential test but expressing the futility stopping rule in terms of conditional power and providing this as guidance to the monitoring committee. We may, for example, define the futility boundary by a Type II error spending function as in Section 8.4. The target information level $\mathcal{I}_{max}$ can be set so that the test will have the desired power $1 - \beta$ at $\theta = \delta = 1$ when this futility stopping rule is followed but

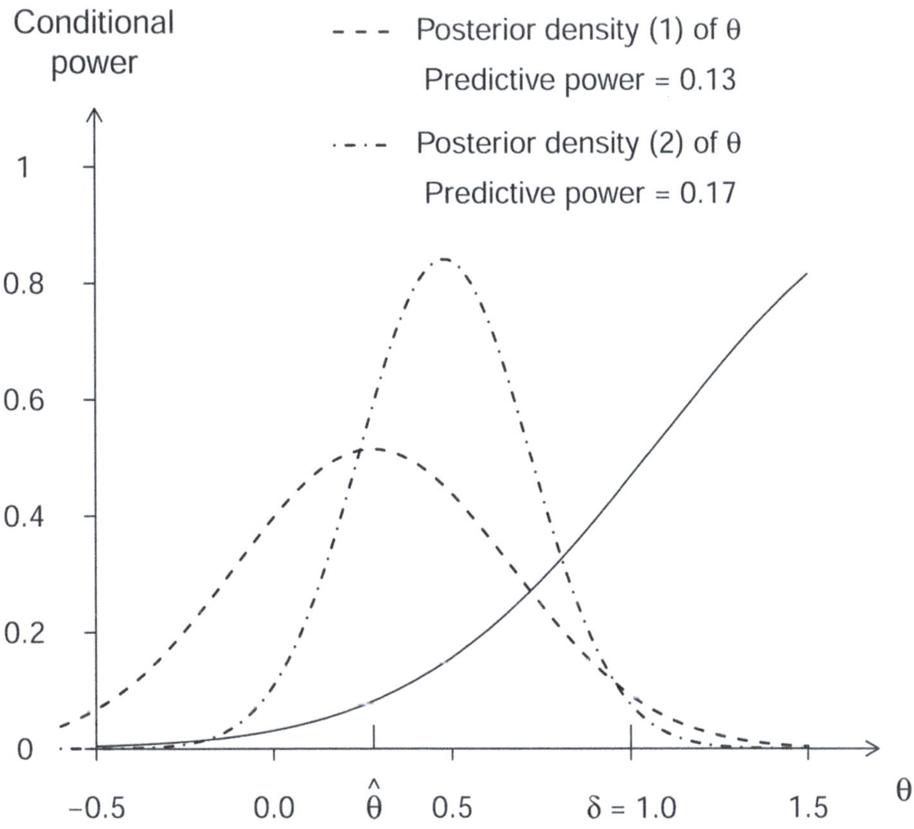

Figure 11.9 *Example: Predictive power. The one-sided group sequential test has five analyses and an efficacy boundary defined by the error spending function* $f(\mathcal{I}/\mathcal{I}_{max}) = \min\{\alpha(\mathcal{I}/\mathcal{I}_{max})^2, \alpha\}$. *Predictive power is calculated for* (1) *an improper, flat prior for* θ *and* (2) *the prior* $\theta \sim N(0.6, 0.3^2)$, *when* $\hat{\theta}^{(3)} = 0.28$ *at analysis* 3.

the monitoring committee has the flexibility to vary the stopping criteria in the light of information on, say, treatment safety or secondary endpoints.

In our example of a trial testing H_0: $\theta \leq 0$ against $\theta > 0$ with Type I error probability $\alpha = 0.025$ and power $1 - \beta = 0.9$ if $\theta = \delta = 1$, suppose a group sequential design has an efficacy boundary defined through a Type I error spending function

$$f(\mathcal{I}/\mathcal{I}_{max}) = \min\{\alpha(\mathcal{I}/\mathcal{I}_{max})^{\rho_1}, \alpha\}. \tag{11.23}$$

We denote by Design (i) the case where $\rho_1 = 2$ and there is no early stopping for futility, in which case the desired power is achieved by setting $\mathcal{I}_{max} = 11.1$. We shall also consider designs with a futility boundary defined by a Type II error spending function

$$g(\mathcal{I}/\mathcal{I}_{max}) = \min\{\beta(\mathcal{I}/\mathcal{I}_{max})^{\rho_2}, \beta\}. \tag{11.24}$$

Let Design (ii) be the case where $\rho_1 = 2$ and $\rho_2 = 2$, which requires $\mathcal{I}_{max} = 11.9$ and suppose Design (iii) has $\rho_1 = 2$ and $\rho_2 = 3$, which requires $\mathcal{I}_{max} = 11.5$. The resulting group sequential tests with non-binding futility boundaries are shown in Figure 11.10. We previously supposed a value $Z_3 = 0.723$ was observed when $\mathcal{I}_3 = 6.67$ at analysis 3 in a trial with no futility boundary, our Design (i). If the same value of Z_3 had been

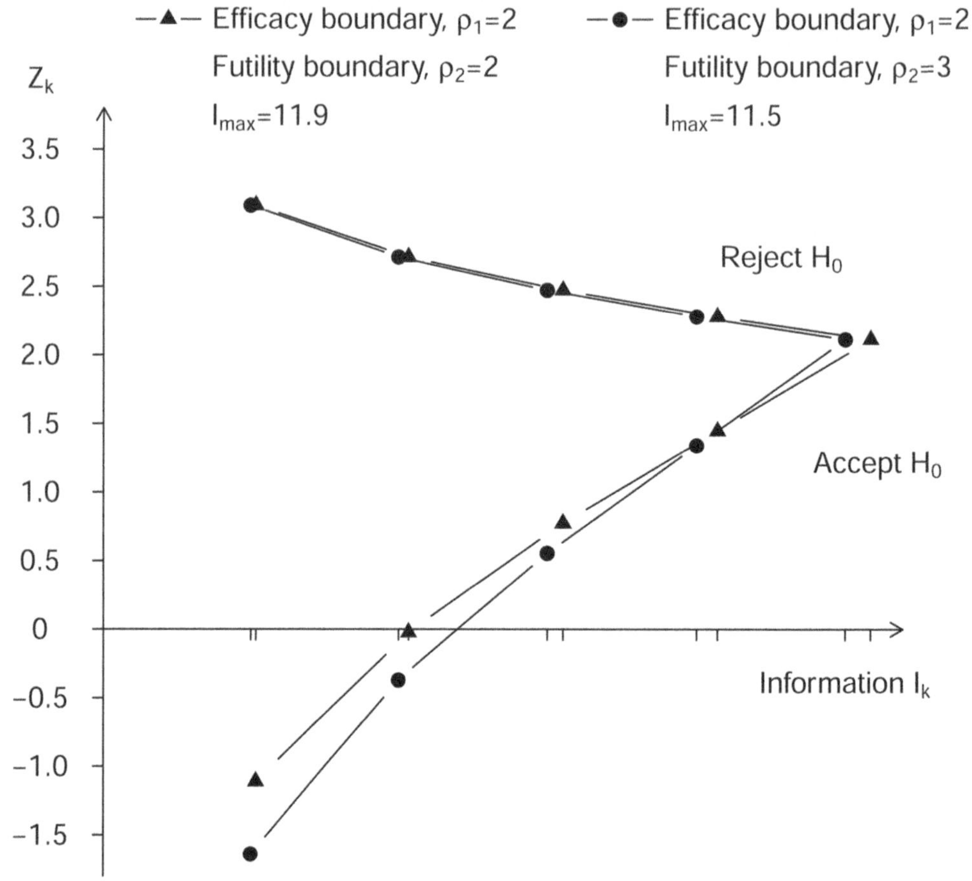

Figure 11.10 *Example: Two one-sided group sequential tests with a non-binding, error spending futility boundary. Each test has five analyses and an efficacy boundary defined by the error spending function* $f(\mathcal{I}/\mathcal{I}_{max}) = \min\{\alpha(\mathcal{I}/\mathcal{I}_{max})^2, \alpha\}$*. Futility boundaries are defined through the Type II error spending function* $g(\mathcal{I}/\mathcal{I}_{max}) = \min\{\beta(\mathcal{I}/\mathcal{I}_{max})^{\rho_2}, \beta\}$ *with* (ii) $\rho_2 = 2$ *and* (iii) $\rho_2 = 3$*. The maximum information levels* $\mathcal{I}_{max} = 11.9$ *and* $\mathcal{I}_{max} = 11.5$ *are chosen to give power* 0.9 *if* $\theta = \delta = 1$.

observed in Design (ii) when $\mathcal{I}_3 = 7.14$, it would have been below the futility boundary at $a_3 = 0.774$, indicating that the trial should terminate, while if $Z_3 = 0.723$ had been observed in Design (iii) when $\mathcal{I}_3 = 6.90$, it would have been above the futility boundary value, $a_3 = 0.556$.

Table 11.1 lists properties of the Designs (ii) and (iii) shown in Figure 11.10 when the formal futility stopping rules are obeyed, along with properties of Design (i) with no futility boundary. Since, in each design, $\mathcal{I}_{max}$ has been set to achieve the same power requirement, the main consideration in choosing between these designs is the trade-off between $\mathcal{I}_{max}$ and $E(\mathcal{I})$. Design (i), with no early stopping for futility, has high values of $E_\theta(\mathcal{I})$ for $\theta = 0$ and $\theta = 0.5$; Design (ii), with $\rho_1 = 2$ and $\rho_2 = 2$, spends the most Type II error probability

Table 11.1 *Properties of three one-sided group sequential tests with efficacy boundary defined by the Type I error spending function* (11.23) *with* $\rho_1 = 2$ *and* (i) *no futility boundary,* (ii) *a futility boundary defined by the Type II error spending function* (11.24) *with* $\rho_2 = 2$ *and* (iii) *a futility boundary defined by* (11.24) *with* $\rho_2 = 3$. *Tests have Type I error probability* $\alpha = 0.025$ *and five analyses with* $\mathcal{I}_k = (k/5)\,\mathcal{I}_{max}$, $k = 1, \ldots, 5$, *where in each case* $\mathcal{I}_{max}$ *is set to give power* 0.9 *at* $\theta = 1$. *Values of* $E_\theta(\mathcal{I})$ *are expressed as percentages of the information,* $\mathcal{I}_f$, *required in a fixed sample test.*

Design	$100\,\mathcal{I}_{\max}/\mathcal{I}_f$	$100\,E_\theta(\mathcal{I})/\mathcal{I}_f$			
		$\theta=0$	$\theta=0.5$	$\theta=1.0$	$\theta=1.5$
(i) Efficacy: $\rho_1 = 2$ only	105.8	105.2	96.7	70.5	46.8
(ii) Efficacy: $\rho_1 = 2$, Futility: $\rho_2 = 2$	113.3	59.2	80.9	70.6	48.2
(iii) Efficacy: $\rho_1 = 2$, Futility: $\rho_2 = 3$	109.4	64.1	83.0	70.1	47.5

at early analyses, reducing $E_{\theta=0}(\mathcal{I})$ considerably but with quite a high value of $\mathcal{I}_{max}$; Design (iii), with $\rho_1 = 2$ and $\rho_2 = 3$, spends Type II error more slowly but still achieves a substantial reduction in $E_{\theta=0}(\mathcal{I})$ with a maximum information level around 10% greater than that of a fixed sample test.

Suppose a futility stopping rule is specified as guidance for a monitoring committee but members of this committee wish to consider conditional power when deciding whether to stop for futility. Then, the values of conditional power for data points on the futility boundary may be calculated and presented to the committee. Table 11.2 shows these values for Designs (ii) and (iii). One may be surprised by the low values of $p_k(\hat{\theta}^{(k)}, a_k)$, the conditional power under the current estimate of θ at each analysis when $z_k = a_k$, but these reinforce the point that $\hat{\theta}^{(k)}$ is a very noisy estimate and predictions under $\theta = \hat{\theta}^{(k)}$ should be interpreted with caution. The values of $p_k(\delta, a_k)$ are more stable but they still vary between analyses. One might expect these thresholds for conditional power to increase as the trial continues and the potential savings in observed information decrease, however, the observed data also provide evidence about the true value of θ, with increasing accuracy as the trial continues, and this has an impact on the appropriate choice of a_k and, hence on $p_k(\delta, a_k)$: we shall return to this topic when we discuss optimal group sequential tests in the next chapter. Öhrn (2011) assessed the efficiency of futility stopping rules based on conditional and predictive power when the threshold used to determine whether the trial should stop for futility is constant across analyses and he concluded that stopping for futility when predictive power is less than a suitably specified constant can be an efficient option. Gallo, Mao & Shih (2014) conducted similar investigations and reached the same positive conclusion about designs that apply a constant threshold to predictive power, while noting that stopping rules based on a constant threshold for conditional power, calculated at either $\theta = \delta$ or $\theta = \hat{\theta}^{(k)}$, may be quite inefficient. Both Öhrn (2011) and Gallo, Mao & Shih (2014) considered predictive power arising from a flat, improper prior: in view of the results in Table 11.2 we would suggest using a more realistic, informative prior if a constant threshold is to be applied to the predictive power.

Our conclusions from this example are that it is desirable to provide a monitoring committee with a formal futility stopping rule that can be applied with some flexibility. This formal rule can be derived with due consideration of the pertinent ethical and commercial issues. Although the futility stopping rule is presented as a guideline, one would expect the monitoring committee to adhere to it in the absence of compelling data on safety or secondary endpoints and, in this case, a suitable choice of futility stopping rule will produce an efficient group sequential design. However, as long as the futility boundary is treated

Table 11.2 *Conditional power $p_k(\theta, z_k)$ and predictive power $P_k(z_k)$ at points $z_k = a_k$, $k = 1, \ldots, 4$, on the futility boundaries of two one-sided group sequential tests with an efficacy boundary defined by the Type I error spending function* (11.23) *with $\rho_1 = 2$, and futility boundary defined by the Type II error spending function* (11.24) *with* (ii) $\rho_2 = 2$ *and* (iii) $\rho_2 = 3$. *Tests have Type I error probability $\alpha = 0.025$ and five analyses with $\mathcal{I}_k = (k/5)\,\mathcal{I}_{max}$, $k = 1, \ldots, 5$, where in each case $\mathcal{I}_{max}$ is set to give power 0.9 at $\theta = 1$.*

	Analysis, k			
	1	2	3	4
Design (ii) Efficacy: $\rho_1 = 2$, Futility: $\rho_2 = 2$				
$p_k(\delta, a_k)$	0.55	0.41	0.51	0.39
$p_k(\hat{\theta}^{(k)}, a_k)$	0.00027	0.037	0.096	0.13
$P_k(a_k)$ for improper, flat prior	0.021	0.072	0.14	0.16
$P_k(a_k)$ for prior $\theta \sim (0.6, 0.3^2)$	0.16	0.14	0.19	0.17
Design (iii) Efficacy: $\rho_1 = 2$, Futility: $\rho_2 = 3$				
$p_k(\delta, a_k)$	0.48	0.33	0.39	0.30
$p_k(\hat{\theta}^{(k)}, a_k)$	0.000005	0.0087	0.043	0.084
$P_k(a_k)$ for improper, flat prior	0.0049	0.028	0.077	0.11
$P_k(a_k)$ for prior $\theta \sim (0.6, 0.3^2)$	0.10	0.080	0.12	0.12

as non-binding when defining the efficacy boundary, flexibility remains for the monitoring committee to use their discretion in following the guideline for futility stopping.

11.6 A Case Study with Survival Data

We now illustrate conditional power calculations using interim data from a randomized clinical trial designed to compare survival times of subjects on two treatment arms. These calculations are loosely adapted from those requested by the data and safety monitoring board (DSMB) at an interim analysis of the Nutritional Prevention of Cancer Trial (see Section 1.4.5 and Clark et al. (1996)) and they follow the methodology developed by Andersen (1987).

Consider a trial in which survival times are assumed to be exponentially distributed with means μ_A and μ_B for patients on treatments A and B, respectively. Let $\lambda_A = 1/\mu_A$ and $\lambda_B = 1/\mu_B$ denote the hazard rates on the two treatments and define $\psi = \lambda_B/\lambda_A$, the hazard ratio or "relative hazard". We suppose the null hypothesis H_0: $\psi = 1$ is to be tested against the two-sided alternative $\psi \neq 1$.

We first describe a fixed sample test of H_0. Let X_A denote the total number of patient-years observed on treatment A, that is the sum of the survival times until failure or censoring for each patient, and let N_A be the number of patients observed to fail on treatment A. Similarly, define X_B and N_B for patients on treatment B. We estimate the hazard rates by $\hat{\lambda}_A = N_A/X_A$ and $\hat{\lambda}_B = N_B/X_B$, and the hazard ratio by

$$\hat{\psi} = \hat{\lambda}_B/\hat{\lambda}_A.$$

A test of H_0: $\psi = 1$ can be based on the statistic $W = \log(\hat{\psi})$, which is approximately normally distributed with mean $\log(\psi)$ and variance $N_A^{-1} + N_B^{-1}$. Thus, a two-sided test with Type I error probability approximately equal to α rejects H_0 if

$$|W| \geq z_{\alpha/2}\sqrt{(N_A^{-1} + N_B^{-1})}.$$

In the notation of previous sections, with $\theta = \log(\psi)$, the observed information for θ is $\mathcal{I} = (N_A^{-1} + N_B^{-1})^{-1}$ and the standardized test statistic is $Z = W\sqrt{\mathcal{I}}$.

Given an initial estimate of λ_A, a specified Type I error probability α and power $1 - \beta$ to be achieved if $\psi = \psi^*$, it is possible to determine the number of person-years that should be observed in the study. The power of the above test at a given value of ψ is approximately

$$\Phi\left\{-z_{\alpha/2} + \frac{\log(\psi)}{\sqrt{(N_A^{-1} + N_B^{-1})}}\right\} + \Phi\left\{-z_{\alpha/2} - \frac{\log(\psi)}{\sqrt{(N_A^{-1} + N_B^{-1})}}\right\}. \tag{11.25}$$

If the study accrues a total of T person-years, divided equally between the two treatments, the expected values of N_A and N_B are $\lambda_A T/2$ and $\psi^* \lambda_A T/2$, respectively. Substituting these values for N_A and N_B in (11.25) yields an expression which can be evaluated to find the value of T that will give approximate power $1 - \beta$ when $\psi = \psi^*$. Knowledge of accrual and dropout rates can then be used to estimate the sample sizes and study duration needed to attain this value of T person-years. Various charts, nomograms and programs are available to aid this calculation; see, for example, George & Desu (1974), Schoenfeld & Richter (1982) and Makuch & Simon (1982).

In the Nutritional Prevention of Cancer Trial, treatment A was a control group and a failure rate of 3.14% per year was anticipated from previous experience so, with time measured in years, $\lambda_A \approx 0.0314$. A reduction in hazard rate of 25% was considered to be a meaningful improvement and hence the power requirement was set at $\psi^* = 0.75$. For a Type I error rate of $\alpha = 0.05$ and power $1 - \beta = 0.8$ at $\psi = \psi^* = 0.75$, it was determined that a total of $T = 14,116$ person-years was needed. Evaluating (11.25) as a function of ψ with $N_A = \lambda_A T/2$ and $N_B = \psi^* \lambda_A T/2$, where $\lambda_A = 0.0314$, $\psi^* = 0.75$ and $T = 14,116$, gives the power function for the original design shown as the solid line in Figure 11.11.

At an interim stage of the trial, we can assess conditional power by considering the distribution of $W = \log(\hat{\psi})$ at termination, given the current data. Let n_A be the number of patients on treatment A who have failed so far, x_A the total person-years observed on treatment A, and s_A the projected additional person-years for subjects on treatment A. Define n_B, x_B and s_B similarly for treatment B and let $s = s_A + s_B$. Define the random variables D_A and D_B to be the numbers of further deaths that will be observed on treatments A and B, respectively, in the remaining follow-up times of s_A and s_B person-years. Following Andersen (1987), at termination we shall have

$$W = \log(\hat{\psi}) = \log(\hat{\lambda}_B) - \log(\hat{\lambda}_A) = \log\left(\frac{n_B + D_B}{x_B + s_B}\right) - \log\left(\frac{n_A + D_A}{x_A + s_A}\right).$$

The conditional distribution of this statistic given the currently observed values n_A, x_A, n_B and x_B is approximately normal with mean

$$\mu_W = \log\left(\frac{n_B + \psi\,\lambda_A\,s_B}{x_B + s_B}\right) - \log\left(\frac{n_A + \lambda_A\,s_A}{x_A + s_A}\right)$$

and variance

$$\sigma_W^2 = \frac{\psi\,\lambda_A\,s_B}{(n_B + \psi\,\lambda_A\,s_B)^2} + \frac{\lambda_A\,s_A}{(n_A + \lambda_A\,s_A)^2}.$$

The test on termination will reject H_0 if

$$|W| \geq z_{\alpha/2}\sqrt{\{(n_A + D_A)^{-1} + (n_B + D_B)^{-1}\}} = c, \quad \text{say}. \tag{11.26}$$

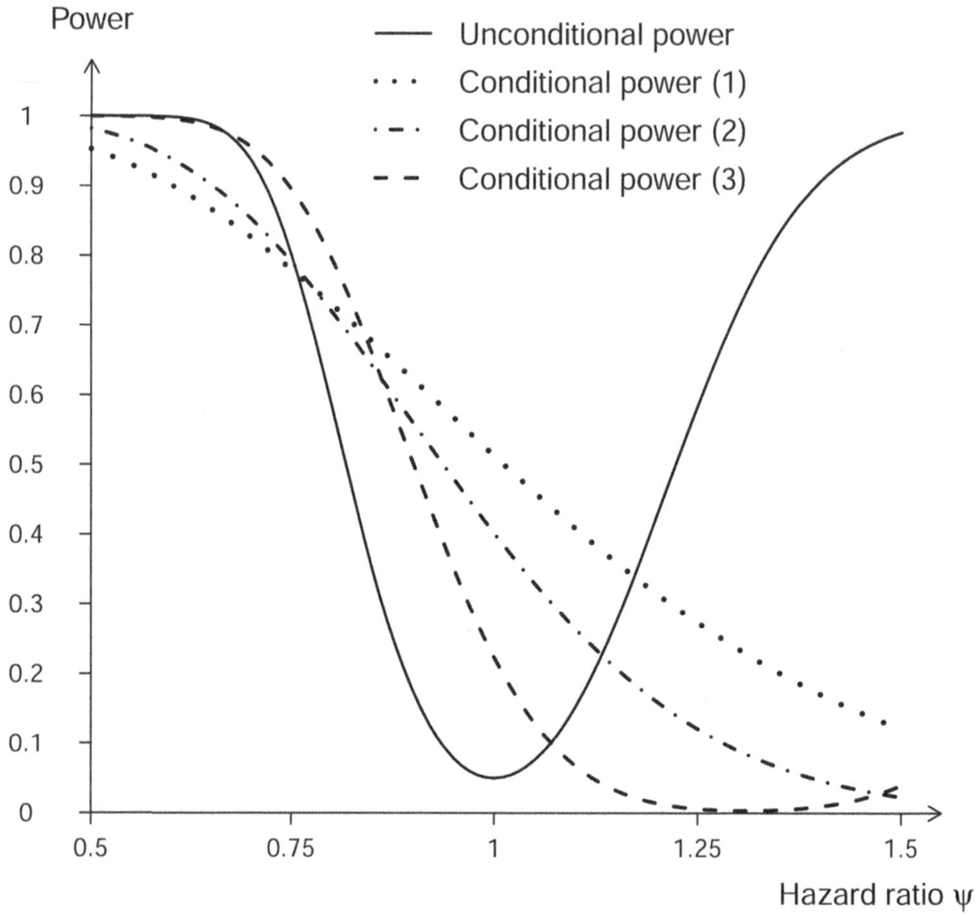

Figure 11.11 *Power and conditional power curves for the case study. The power function for the original design is shown by the solid line. The other curves represent conditional power functions under three different scenarios,* (1), (2) *and* (3), *for continuing after an interim analysis.*

The conditional power under a given value of ψ is, therefore,

$$\Phi\left\{\frac{-c-\mu_W}{\sigma_W}\right\}+\Phi\left\{\frac{-c+\mu_W}{\sigma_W}\right\}$$

and we can estimate this by substituting $\lambda_A = 0.0314$ in the formulae for μ_W and σ^2_W, and estimating D_A by $s_A\hat{\lambda}$ and D_B by $s_B\hat{\lambda}$ in (11.26), where $\hat{\lambda} = (n_A + n_B)/(x_A + x_B)$ is the current estimate of the common hazard rate under H_0: $\psi = 1$.

At an interim stage of the trial, the DSMB was faced with the mortality data displayed in Table 11.3. An analysis conducted at this interim point as if it were the final analysis yields an estimate of the hazard ratio

$$\hat{\psi} = \left(\frac{89}{3943}\right) / \left(\frac{118}{3896}\right) = 0.7452,$$

Table 11.3 *Interim mortality data for the case study*

Treatment arm	*Deaths*	*Person-years*
A: Control	$n_A = 118$	$x_A = 3896$
B: Active treatment	$n_B = 89$	$x_B = 3943$
Total	$n = 207$	$x = 7839$

Table 11.4 *Additional person-years under three possible scenarios for continuing the case study*

Scenario	*Additional person-years*			*Total person-years*		
	s_A	s_B	s	$x_A + s_A$	$x_B + s_B$	$x + s$
(1) *1 more year*	375	375	750	4271	4318	8589
(2) *2 more years*	815	815	1630	4711	4758	9469
(3) *Original plan*	3138	3139	6276	7034	7082	14116

indicating better observed survival on the active treatment. From this we find that $W = \log(0.7452) = -0.2940$, and the standardized test statistic is

$$Z = \frac{W}{\sqrt{\{1/118 + 1/89\}}} = -2.094,$$

for which the two-sided P-value is 0.036. Of course, this P-value is not adjusted for multiple looks and, for example, either a Pocock-type or O'Brien & Fleming-type 95% repeated confidence interval for ψ (see Chapter 10) would still contain the null value $\psi = 1$.

Even though the protocol indicated the trial should continue, the DSMB was concerned that, under the original plan, it would take many years for the study to reach a conclusion, and this could delay dissemination of an important positive result. They therefore asked that conditional power curves be calculated for a fixed sample test at termination under the following three scenarios:

(1) One more year of follow-up,

(2) Two more years of follow-up,

(3) Five more years of follow-up, as originally planned.

Table 11.4 shows the additional person-years and total person-years of follow-up estimated under each scenario, based on observed accrual and dropout rates and the numbers of subjects currently active.

Conditional power curves for each of the three scenarios were computed using the above formulae and these are plotted in Figure 11.11 along with the power function for the original design. All three conditional power curves are U-shaped but those for scenarios (1) and (2) do not start to rise on the right until θ reaches much higher values, around $\psi = 3$ for (2) and $\psi = 5$ for (1). Our interest is in the left hand part of the plot where we see that, under either $\psi^* = 0.75$ or the current estimate $\hat{\psi} = 0.7452$, there was a high conditional probability of a significant result: 0.78, 0.79 and 0.90 for scenarios (1), (2) and (3), respectively. Even under the null hypothesis, H_0: $\psi = 1$, rejection of H_0 was quite possible, with conditional probabilities 0.53, 0.41 and 0.22 under the three scenarios. After much discussion involving many issues, the DSMB decided to continue the trial. However, the trial was then terminated early at the next interim analysis.

We should note that in scenarios (1) and (2), conditional power was calculated assuming that a fixed sample test at level $\alpha = 0.05$ would be applied to the final set of data. Strictly speaking, such a test is inappropriate when the sample size is chosen in the light of earlier data. In Chapters 23 and 24, we shall present methodology for adaptive designs that can be used to obtain valid inferences after sample size modifications, even when the rule for determining the final sample size is not clearly defined in advance.

11.7 Bibliography and Notes

The idea of stochastic curtailment was first proposed by Lan, Simon & Halperin (1982). Earlier, Pasternack & Ogawa (1961) developed the related concept of *probability of reversal* of a test in the context of missing observations. This quantity relates to the question of whether the conclusion based on current data could be altered by the data yet to be observed. Pasternack (1984) discusses further the connection with curtailment procedures. Tan & Xiong (1996) describe the similar concept of a "discordant probability" in the context of the parameter-free stochastic curtailment procedure described in Section 11.4.

Details of the conditional power approach for comparing two treatment groups were given by Halperin et al. (1982) for dichotomous outcome variables; by Halperin et al. (1987) for slopes in longitudinal studies; by Andersen (1987) for exponentially distributed survival outcomes; by Pawitan & Hallstrom (1990) for survival data using the log-rank test; and by Lin, Yao & Ying (1999) for survival data using a proportional hazards regression model (see Section 17.4). The construction of repeated confidence intervals based on inverting an SCP is described in Davis & Hardy (1992).

Lan & Wittes (1988) present an application in which conditional power is evaluated under the current estimate $\hat{\theta}^{(k)}$ or at the limits of a confidence interval for θ based on the current data at analysis k. Pepe & Anderson (1992) use a similar idea in a one-sided testing procedure with two stages where conditional power is calculated at $\hat{\theta}^{(k)} + \text{s.e.}(\hat{\theta}^{(k)})$. Ware, Muller & Braunwald (1985) and Pepe & Anderson (1992) describe stochastic curtailment plans with special emphasis on stopping early if a negative result is indicated. Further discussion on the use of conditional power in decisions to stop early under H_0 is contained in Betensky (1997a,b). Applications of the predictive approach with binomial data have been described by Herson (1979), Choi, Smith & Becker (1985), Hilsenbeck (1988) and Johns & Andersen (1999). Snapinn et al. (2006) review various approaches for assessing futility, including stochastic curtailment, predictive power, predictive probability, and group sequential methods. They remark that, while abandoning a trial for futility can save time and money, there is usually no ethical imperative to do so, and stopping the trial early can leave secondary research questions unanswered. They therefore recommend

> "when planning a trial it is important for the study sponsor and data monitoring committee to carefully consider whether or not the trial should include a futility rule."

Dmitrienko & Wang (2006) discuss a Bayesian stochastic curtailment method for the predictive power approach of Section 11.3. They define a family of Bayesian efficacy and futility stopping rules for normal and binary responses and provide a comparison with stopping rules based on other conditional and predictive power approaches. The sensitivity of their stopping rules to the choice of prior distributions is examined and guidelines for choosing a prior distribution of the treatment effect are discussed. Xi, Gallo & Ohlssen (2017) consider the timing of futility analyses when there is the opportunity for one or two such analyses during a trial.

As noted in Section 11.5.2, Öhrn (2011) and Gallo, Mao & Shih (2014) found that stopping for futility when predictive power is less than a suitably specified constant can be an efficient option, and the properties of the efficient designs shown in Table 11.2 support the use of a realistic, informative prior in defining the predictive power. Whatever, approach is taken in using ideas of stochastic curtailment to define a futility stopping boundary, we agree with Gallo, Mao & Shih (2014), when they recommend

> "Futility rules should be determined on the basis of carefully considering the operating characteristics of a trial design and plan as a whole, rather than on intuitive interpretations of certain well-known conditional or predictive quantities."

Chapter 12

Optimal Group Sequential Tests

12.1 Optimizing within Classes of Group Sequential Designs

In previous chapters, we have described a variety of group sequential designs for one-sided and two-sided tests with early stopping to reject or accept the null hypothesis. Some tests have been defined through parametric descriptions of their boundaries, others through error spending functions. Since a key aim of interim monitoring is to terminate a study as soon as is reasonably possible under certain values of the treatment effect, it is of interest to find tests with optimal early stopping properties. These designs may be applied directly or used as benchmarks to assess the efficiency of designs that are attractive for other reasons. We have already presented properties of optimal one-sided tests in Table 5.4 in order to assess the efficiency of power family tests, and we noted in Chapter 8 that comparisons with these results show designs in the Rho family of error spending tests to be highly efficient. We made similar comparisons in Tables 10.3 and 10.4 to demonstrate the efficiency of one-sided tests derived from repeated confidence intervals. In this chapter we explain how to calculate stopping boundaries with specific optimality properties.

We shall work in the general group sequential setting described in Chapter 3, where standardized statistics Z_k have the canonical multivariate normal joint distribution (3.1). Optimality will be defined in terms of low values of $E_\theta(\mathcal{I}_T)$ under certain values of θ, where T denotes the analysis at which the study terminates and $\mathcal{I}_T$ is the observed information on termination. In a study where the time to each subject's response is small in comparison to the total study duration, observed information is roughly proportional to sample size and it is reasonable to seek designs that minimize $E_\theta(\mathcal{I}_T)$. In a time-to-event study where observed information is approximately proportional to the number of observed events, $\mathcal{I}_T$ can be a surrogate for the duration of the study and, again, minimizing $E_\theta(\mathcal{I}_T)$ may be an appropriate objective. If response is subject to a substantial delay, such as when the outcome is defined as the change in a clinical measurement from baseline to 12 months after treatment, the relationship between $\mathcal{I}_T$ and sample size is more complex: we shall define group sequential designs for a delayed response in Chapter 14 and there we shall discuss criteria to be applied when optimizing such designs.

The history of optimal sequential tests goes back to Wald & Wolfowitz's (1948) proof that the sequential probability ratio test (Wald, 1947) minimizes expected sample size at two parameter values subject to specified error probabilities being achieved at those points. This result concerned a sequential test with continuous monitoring and no upper limit on sample size. In contrast, current interest focuses on group sequential tests with a modest number of analyses and a restricted maximum sample size. Thus, we start by considering group sequential designs with a given inflation factor R, and a set number K of equally spaced information levels. In Section 12.2 we explain how to search in a class of designs with specified R, K, Type I error probability and power, to find the procedure which minimizes $E_\theta(\mathcal{I}_T)$ averaged over a set of θ values or integrated over a distribution for θ. Comparisons between optimal designs with different values of K and R will show the possible return from more frequent analyses or a higher maximum sample size.

 DOI: 10.1201/9781584888482-12

In practice, investigators may wish to see a number of different features in a good group sequential design rather than optimality in one specific sense. Nevertheless, the properties of optimal designs serve as useful benchmarks when judging the merits of a particular design: in Table 5.4 we listed optimal expected sample sizes to show how well one-sided tests in the power family perform under certain values of θ and, in Section12.2, we shall make further comparisons of these power family tests and two types of error spending test.

When the number of analyses is small, it may be advantageous to schedule these at unequally spaced intervals. For example, a very early analysis could help reduce expected sample size under an extreme effect size, either positive or negative. In Section 12.3 we extend the class of tests considered to allow unequally spaced information levels $\mathcal{I}_k$, $k = 1, \ldots, K$. Again, we explain how one can optimize within this class and we evaluate the benefits of designs with unequal group sizes.

In Section 12.4 we consider scheduling future analyses in a response-dependent manner so, for example, the next analysis may occur sooner when current data lie close to a stopping boundary. Schmitz (1993) refers to such designs as "sequentially planned decision procedures". We shall discuss how to find optimal designs in this wider class and we assess the benefits of this approach in terms of lower $E_\theta(\mathcal{I}_T)$ curves. Since these benefits turn out to be quite small, one might expect the complexity of these designs to prohibit their use in practice. However, this class of designs contains some popular proposals for adaptive trial designs that incorporate "sample size re-assessment", whereby the target sample size is modified in response to interim estimates of the effect size, θ. We shall see in Chapter 24 that understanding the possible benefits of such adaptive designs is helpful in assessing and refining proposed methods.

12.2 Optimizing Designs for a Fixed Sequence of Information Levels

The search for optimal group sequential tests requires a combination of analysis and numerical computation. Since we require tests to achieve specific values of Type I error and power, the search for optimal designs is a constrained optimization problem. Although Jennison (1987) showed this problem can be tackled directly, the optimization process becomes more complex as the number of analyses increases and it is increasingly difficult to guarantee that an optimum design has been found. A much more effective computational method, proposed for binomial and normal responses by Weiss (1962), involves creating a Bayes decision problem and choosing the prior distribution and costs so that the Bayes procedure is also optimal with respect to the desired frequentist criteria. Lai (1973) explained how to use this approach to minimize the maximum expected sample size for a normal response and Lorden (1976) applied this method to obtain numerical results, thereby solving the "Kiefer-Weiss" problem (Kiefer & Weiss, 1957) for this case. Eales (1991), Eales & Jennison (1992, 1995), Chang (1996) and Barber & Jennison (2002) have used this method to derive the one-sided and two-sided group sequential tests of a normal mean θ which minimize expected sample size at a particular value of θ, averaged over several values of θ, or integrated with respect to a normal density for θ. Öhrn & Jennison (2010) applied this approach to compute optimal group sequential designs for testing both superiority and non-inferiority. Hampson & Jennison (2013) extended this methodology to derive optimal group sequential designs when there is a delay in observing the response variable; see Chapter 14.

We introduce the general method in the following example. Consider a group sequential, one-sided test of H_0: $\theta \leq 0$ against H_A: $\theta > 0$ with Type I error probability α and power $1 - \beta$ at $\theta = \delta$. Assume analyses occur at equally spaced cumulative information levels $\mathcal{I}_k = (k/K)R\mathcal{I}_{f,1}$, $k = 1, \ldots, K$, where the fixed sample information level $\mathcal{I}_{f,1}$ is given by (5.1). A general group sequential test has the form (5.2). Suppose we wish to find the boundary values $(a_1, b_1), \ldots, (a_K, b_K)$ of the design that minimizes $E_\theta(\mathcal{I}_T)$ for $\theta = \delta/2$. As a device for solving this problem, we create a Bayes sequential decision problem with a prior distribution for θ on 0, $\delta/2$ and δ. We shall choose this prior, the loss function for incorrect decisions, and the costs of sampling under $\theta = 0$, $\delta/2$ and δ so that the procedure minimizing the overall Bayes risk will have the lowest possible value of $E_{\theta=\delta/2}(\mathcal{I}_T)$ among designs with the same error rates. To do this, we write code to solve the problem for a particular prior, loss function and sampling cost function, then we vary parameters in this problem until the optimal procedure has error rates equal to the specified α and β, and deduce that this must also be the optimal procedure for the original frequentist problem. This method of optimizing subject to error rate constraints is an application of the Lagrangian approach to constrained optimization. However, recognizing that the unconstrained problem is a Bayes sequential decision is important as this helps in implementing an efficient method of solution by the dynamic programming algorithm (Bellman, 1957).

Further details for this example are as follows: the parameter θ is given a three point prior distribution, $\pi(0) = \pi(\delta/2) = \pi(\delta) = 1/3$; on termination, the available decisions are D_0: '$\theta \leq 0$' and D_1: '$\theta > 0$', with loss function $L(D_1, 0) = d_1$, $L(D_0, \delta) = d_2$ and $L(D, \theta) = 0$ otherwise; the sampling cost is $c(\theta)$ per unit of information, giving a total sampling cost $c(\theta)\mathcal{I}_T$, and we set $c(0) = 0$, $c(\delta) = 0$ and $c(\delta/2) = 1$. The solution of this Bayes decision problem minimizes the expected total cost

$$\pi(\delta/2)E_{\theta=\delta/2}(\mathcal{I}_T) + \pi(0)Pr_{\theta=0}\{D_1\}\,d_1 + \pi(\delta)Pr_{\theta=\delta}\{D_0\}\,d_2$$

$$= \frac{1}{3}\{E_{\theta=\delta/2}(\mathcal{I}_T) + Pr_{\theta=0}\{D_1\}\,d_1 + Pr_{\theta=\delta}\{D_0\}\,d_2)\}. \tag{12.1}$$

For any pair of decision costs d_1 and d_2, the Bayes optimal problem can be found by dynamic programming. We explain the details of the computations involved in applying dynamic programming in Section 31.6.1; see also Jennison & Turnbull (2013). In order to solve our original problem, we search for values d_1^* and d_2^* for which the procedure minimizing (12.1) also satisfies

$$Pr_{\theta=0}\{D_1\} = \alpha \quad \text{and} \quad Pr_{\theta=\delta}\{D_0\} = \beta. \tag{12.2}$$

The optimal rule for d_1^* and d_2^* must minimize $E_{\theta=\delta/2}(\mathcal{I}_T)$ among all rules satisfying (12.2) since, otherwise, the superior design would have a lower value of (12.1) for the case $d_1 = d_1^*$ and $d_2 = d_2^*$. So, equating decisions D_0 and D_1 with accepting H_0 and rejecting H_0, respectively, we have the optimal group sequential test for our original problem. Table 5.4 showed properties of the tests minimizing $E_{\theta=\delta/2}(\mathcal{I}_T)$, and also of tests minimizing $\{E_{\theta=0}(\mathcal{I}_T) + E_{\theta=\delta}(\mathcal{I}_T)\}/2$, for the case $\alpha = 0.025$ and $\beta = 0.1$ with $K = 2$, 5, 10 and 20, and values of R between 1.01 and 1.3. In minimizing $\{E_{\theta=0}(\mathcal{I}_T) + E_{\theta=\delta}(\mathcal{I}_T)\}/2$, we give θ a two point prior distribution, $\pi(0) = \pi(\delta) = 1/2$, the loss function for the final decision has $L(D_1, 0) = d_1$ and $L(D_0, \delta) = d_2$, and the sampling cost per unit of information has $c(0) = c(\delta) = 1$.

Eales & Jennison (1992) and Barber & Jennison (2002) demonstrate how to choose the prior and cost of sampling function so that the above strategy gives group sequential tests minimizing a variety of optimality criteria involving either a weighted average of $E_\theta(\mathcal{I}_T)$ over a set of θ values or an integral of $E_\theta(\mathcal{I}_T)$ over a distribution for θ. Eales & Jennison (1995) follow a similar approach to find optimal two-sided tests, and optimal two-sided designs when early stopping is allowed to accept H_0 are derived by Eales (1991).

Minimizing an integral of $E_\theta(\mathcal{I}_T)$ provides a design with good efficiency across a range of θ values. As an example, suppose we wish to find the one-sided test minimizing $\int f(\theta)E_\theta(\mathcal{I}_T)d\theta$ where $f(\theta)$ is the density of a $N(\delta, \delta^2/4)$ distribution, placing weight primarily on values of θ between 0 and 2δ. Here, we specify a prior comprising point probability masses of 1/3 at $\theta = 0$ and δ plus a density $f(\theta)/3$. With D_0 and D_1 as previously defined, the loss function for decisions is $L(D_1, 0) = d_1$, $L(D_0, \delta) = d_2$ and $L(D, \theta) = 0$ otherwise. We take the sampling cost to be one per unit of observed information

Table 12.1 *Minimum possible values of $\int f(\theta)E_\theta(\mathcal{I}_T)d\theta$, where $f(\theta)$ is the density of a $N(\delta, \delta^2/4)$ distribution, for group sequential, one-sided tests with Type I error probability $\alpha = 0.025$, power 0.9 at $\theta = \delta$. Tests have K equally spaced information levels, and maximum information equal to R times $\mathcal{I}_{f,1}$, the information required for a fixed sample size test. Tabulated values are the minima of $\int f(\theta)E_\theta(\mathcal{I}_T)d\theta$ expressed as a percentage of $\mathcal{I}_{f,1}$.*

				R			
K	1.01	1.05	1.1	1.15	1.2	1.3	*Minimum over R*
2	79.3	74.7	73.8	74.1	74.8	77.1	73.8 at $R = 1.11$
3	74.8	69.0	67.0	66.3	66.1	66.6	66.1 at $R = 1.20$
4	72.5	66.5	64.2	63.2	62.7	62.5	62.5 at $R = 1.27$
5	71.1	65.1	62.7	61.5	60.9	60.5	60.5 at $R = 1.32$
10	68.2	62.1	59.5	58.2	57.5	56.7	56.4 at $R = 1.46$
20	66.8	60.6	58.0	56.6	55.8	54.8	54.2 at $R = 1.59$

under the continuous part of the prior distribution only; since this prior assigns probability zero to $\theta = 0$ and $\theta = \delta$, we can say the sampling cost is one per unit of information for all $\theta \notin \{0, \delta\}$ and 0 otherwise. This Bayes decision problem can be solved by dynamic programming and the solution will minimize the expected total cost

$$\frac{1}{3}\left(\int f(\theta)E_\theta(\mathcal{I}_T)d\theta + Pr_{\theta=0}\{D_1\}\, d_1 + Pr_{\theta=\delta}\{D_0\}\, d_2\right). \tag{12.3}$$

Hence, when we have found d_1^* and d_2^* such that this solution has the desired error rates α and β, we can conclude that the resulting group sequential design minimizes $\int f(\theta)E_\theta(\mathcal{I}_T)d\theta$ among all designs with these error rates. Properties of designs optimizing this criterion were reported in Table 5.5 of Jennison & Turnbull (2011) and we reproduce these results in Table 12.1.

The results in Table 12.1 show that the minimum expected sample size decreases initially as the inflation factor R increases, but starts to increase again as R gets larger. The value of R determines the maximum sample size, which must be planned for in case the trial continues to its final analysis. Thus, although we have noted the inflation factor R giving the minimum value of $\int f(\theta)E_\theta(\mathcal{I}_T)d\theta$ for each K, designs with lower values of R are likely to be preferred in practice. In fact, there are substantial benefits from sequential testing with R as low as 1.01 and most of the potential gains accrue as R is increased to 1.05 or 1.1. Efficiency also improves with the number of analyses K but, again, the first steps away from a fixed sample design provide most of the available benefits. Given the administrative effort needed to conduct each interim analysis, we would expect designs with up to 5 analyses to be appropriate in most studies.

A number of authors have optimized within families of group sequential tests containing designs with different inflation factors, R. Thus, in these cases, optimization is over R as well as other design parameters. Pocock (1982) followed such an approach when finding two-sided tests with minimum expected sample size at the effect size $\theta = \delta$ where a specific power $1-\beta$ is achieved. Wang & Tsiatis (1987) searched over their family of two-sided tests indexed by Δ, which we described in Section 2.7.1, to minimize expected sample size at $\theta = \delta$. Emerson & Fleming (1989) considered symmetric, one-sided tests with $\alpha = \beta$ and sought the parameter Δ for the "power family" tests defined in Section 5.2 which minimizes expected sample size at $\theta = 0$ and $\theta = \delta$, or at $\theta = \delta/2$. For both the Wang & Tsiatis (1987) tests and the power family considered by Emerson & Fleming (1989), R increases

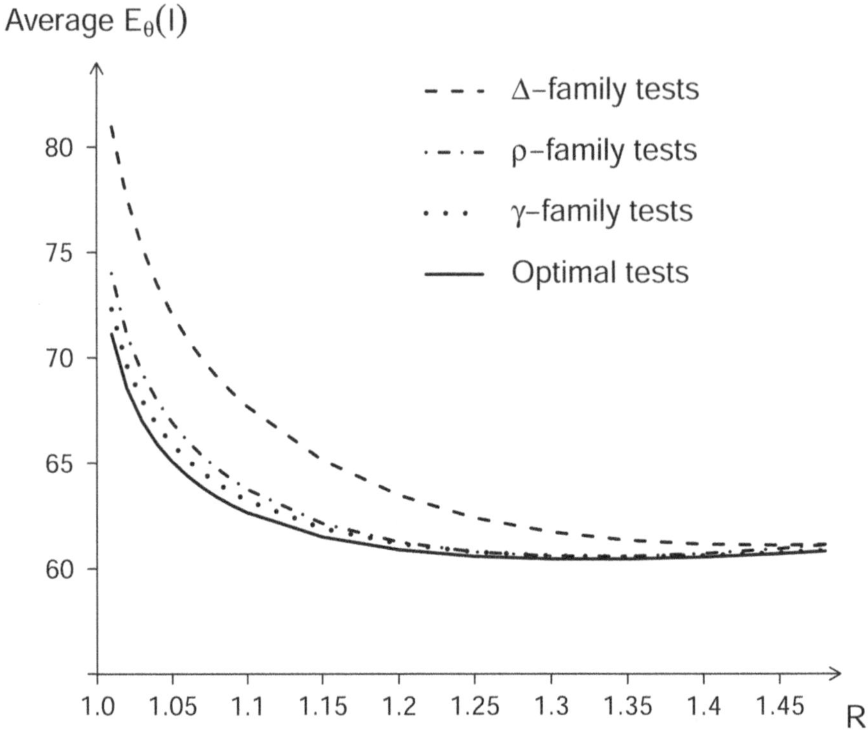

Figure 12.1 *Average expected information, $\int f(\theta) E_\theta(\mathcal{I}_T) d\theta$ as a percentage of $\mathcal{I}_{f,1}$ plotted against the inflation factor R for four families of tests with $K = 5$ equally spaced analyses, $\alpha = 0.025$ and $\beta = 0.1$. Here $f(\theta)$ is the density of a $N(\delta, \delta^2/4)$ distribution and $\mathcal{I}_{f,1}$ the information required for a fixed sample size test.*

monotonically with Δ so there is a single member of the family of tests with each value of R. The same is true for the extension of power family tests to the case where $\alpha \neq \beta$ proposed by Pampallona & Tsiatis (1994) if one sets $\Delta_1 = \Delta_2$ so there is a common shape parameter for the upper and lower boundaries. Anderson (2007) noted that high inflation factors are liable to arise when optimizing a function of $E_\theta(\mathcal{I}_T)$ within a family of error spending tests and chose to minimize $E_\theta(\mathcal{I}_T^2)$ instead, thereby obtaining optimal designs with lower inflation factors. We prefer to provide summaries of performance as maximum sample size is varied in order to help investigators find a satisfactory combination of expected sample size and maximum sample size.

We noted in Section 8.3.3 that certain types of error spending tests have better efficiency than one-sided tests in the power family proposed by Pampallona & Tsiatis (1994). Figure 12.1, reproduced from Figure 5.5 of Jennison & Turnbull (2011), illustrates this efficiency for two types of error spending tests: the Rho family, with Type I and II error spending functions

$$f(t) = \min\{\alpha, \alpha t^\rho\} \quad \text{and} \quad g(t) = \min\{\beta, \beta t^\rho\},$$

and the "Gamma family" with Type I and Type II error spending functions

$$f(t) = \min\{\alpha, \alpha(1 - e^{-\gamma t})/(1 - e^{-\gamma})\}$$

and

$$g(t) = \min\{\beta, \beta(1 - e^{-\gamma t})/(1 - e^{-\gamma})\}$$

based on Hwang, Shih & De Cani's (1990) proposal for a two-sided error spending function. Here $t = \mathcal{I}_k/\mathcal{I}_{max}$ is the information fraction at interim analysis k. In the figure, properties of these error spending tests are compared with tests in the Δ-family defined by Pampallona & Tsiatis (1994) and with optimized group sequential tests. All tests have Type I error rate $\alpha = 0.025$, power $1 - \beta = 0.9$ at $\theta = \delta$ and $K = 5$ equally spaced analyses. Values of $\int f(\theta) E_\theta(\mathcal{I}_T) d\theta$ are plotted against the inflation factor R, which is determined by the choice of ρ or γ for the error spending tests or by the parameter Δ for Pampallona & Tsiatis tests. The bottom curve is the minimum possible value of $\int f(\theta) E_\theta(\mathcal{I}_T) d\theta$ for each R for a group sequential test with $\alpha = 0.025$, $\beta = 0.1$ at $\theta = \delta$ and $K = 5$. These optimized tests provide a benchmark for assessing other types of test: while it is clear that both error spending families are highly efficient, we can see that the Pampallona & Tsiatis tests are somewhat sub-optimal.

In the objective functions considered above, $E_\theta(\mathcal{I}_T)$ serves as a surrogate for the expected sample size and, hence, the cost of the trial. When a pharmaceutical company brings a new drug to market, the economic value depends on the length of patent life remaining. Liu, Anderson & Pledger (2004) capture this aspect of the decision problem by maximizing a gain function comprising the benefit of a successful trial minus the cost of conducting the trial. The "benefit" term is positive when the trial concludes the new treatment to be superior to the control and this term decreases as the length of the trial increases, reducing the patent life remaining when the drug is brought to market. Although this objective is the result of economic considerations, it also captures the benefit to future patients of making an effective new treatment available as rapidly as possible.

It is noteworthy that, by construction, the optimal group sequential design with frequentist error probabilities α and β is the optimal solution of a Bayes sequential decision problem. We shall comment on this relationship further in our discussion of Bayesian sequential designs in Chapter 13. In the dynamic programming algorithm, calculation of the futility boundary at an interim analysis involves a comparison of the probability that H_0 will be rejected if the trial continues with the savings in sampling cost if the trial is stopped now with acceptance of H_0. These calculations are made under the posterior distribution of θ and so they involve the predictive power given the assumed prior. Thus, we have a solution to the problem posed in Chapter 11 of how to set thresholds for conditional power that operate consistently across analyses and produce a futility boundary with optimal properties.

12.3 Optimizing over Information Levels

Several authors have investigated how the group sizes, or increments in the information levels $\mathcal{I}_k$, should be chosen to produce the most efficient group sequential tests. Colton & McPherson (1976), Elashoff & Reedy (1984) and Patel (1996) addressed this question for 2-group designs, and Brittain & Bailey (1993) and Grayling & Mander (2021) considered 2-group and 3-group designs. Xi, Gallo & Ohlssen (2017) optimized the group sizes in trials with one or two futility analyses. Müller & Schäfer (1999) optimized the stopping boundary and the group sizes for two-sided equivalence tests with up to five stages.

A natural approach to this problem is to choose an objective function that combines values of $E_\theta(\mathcal{I}_T)$ over specified θ values, fix the number of groups K and inflation factor R, and write code that applies the methods of Section 12.2 to optimize the chosen objective function for a fixed sequence of information levels $\mathcal{I}_k$, $k = 1, \ldots, K$. Then, in an outer loop, one can search for the optimal sequence $\{\mathcal{I}_1, \ldots, \mathcal{I}_K\}$. Eales & Jennison (1992, Sec. 5) followed this approach for one-sided tests with $K = 2$, 3, 4 or 5 groups and, in our notation, Type I error rate 0.05 at $\theta = 0$ and power 0.95 at an alternative $\theta = \delta$. No restriction was placed on the final information level $\mathcal{I}_K$ and the algorithm of Nelder & Mead (1965) was used to search over sequences $\{\mathcal{I}_k\}$ to find the optimal design. Their results match those in earlier

Table 12.2 *Properties of tests with K analyses, Type I error rate $\alpha = 0.025$ and power $1-\beta = 0.8$ at $\theta = \delta$ which minimize criterion (12.4). The inflation factor R is capped at 1.2 if the minimum is not attained at a lower value.*

Optimizing $\{E_{\theta=0}(\mathcal{I}_T) + E_{\theta=\delta/2}(\mathcal{I}_T) + E_{\theta=\delta}(\mathcal{I}_T)\}/3$

K	Design	R	ρ	$\mathcal{I}_1$	$E_\theta(\mathcal{I}_T)$ at $\theta=0$	$\theta=\delta/2$	$\theta=\delta$	Average $E_\theta(\mathcal{I}_T)$
1	Fixed		1.00	100.0	100.0	100.0	100.0	100.0
2	ρ-family							
	A: Equal groups	1.14	1.00	57.0	67.8	83.9	82.7	78.1
	B: Optimal $\mathcal{I}_1$	1.18	0.80	52.5	66.0	84.0	83.4	77.8
	Optimal boundary							
	D: Equal groups	1.13	–	56.5	66.3	82.6	83.8	77.6
	E: Optimal $\mathcal{I}_1$	1.17	–	51.0	63.8	82.7	85.0	77.2
3	ρ-family							
	A: Equal groups	1.20	1.00	40.0	58.6	78.5	77.2	71.4
	B: Optimal $\mathcal{I}_1$	1.20	1.00	40.0	58.6	78.5	77.2	71.4
	C: Opt. $\mathcal{I}_1,\ldots,\mathcal{I}_K$	1.20	0.98	39.4	57.7	78.6	77.2	71.2
	Optimal boundary							
	D: Equal groups	1.20	–	40.0	55.8	76.7	78.3	70.3
	E: Optimal $\mathcal{I}_1$	1.20	–	36.6	54.9	76.8	78.8	70.2
	F: Opt. $\mathcal{I}_1,\ldots,\mathcal{I}_K$	1.20	–	36.4	54.6	76.8	78.7	70.1
4	ρ-family							
	A: Equal groups	1.20	1.13	30.0	55.1	76.2	74.7	68.6
	B: Optimal $\mathcal{I}_1$	1.20	1.14	33.7	55.4	75.9	74.3	68.5
	C: Opt. $\mathcal{I}_1,\ldots,\mathcal{I}_K$	1.20	1.11	32.9	54.4	76.0	74.4	68.3
	Optimal boundary							
	D: Equal groups	1.20	–	30.0	51.2	74.0	75.8	67.0
	E: Optimal $\mathcal{I}_1$	1.20	–	29.7	51.2	74.1	75.8	67.0
	F: Opt. $\mathcal{I}_1,\ldots,\mathcal{I}_K$	1.20	–	29.6	50.9	74.1	75.8	66.9

papers, in that only minor benefits accrue from using unequal group sizes to minimize one of the objective functions $E_{\theta=\delta/2}(\mathcal{I}_T)$, $\{E_{\theta=0}(\mathcal{I}_T) + E_{\theta=\delta}(\mathcal{I}_T)\}/2$ and $\int \pi(\theta)E_\theta(\mathcal{I}_T)d\theta$, where $\pi(\theta)$ represents a normal density with mean $\delta/2$ and standard deviation $\delta/2$. For designs with two or three analyses, Eales & Jennison (1992) found that having a particularly early first analysis could reduce the rather extreme objective function $\{E_{\theta=-\delta}(\mathcal{I}_T) + E_{\theta=2\delta}(\mathcal{I}_T)\}/2$ by up to 7% of the fixed sample size compared to designs with equally sized groups. However, these optimized designs had high values of $E_{\theta=0}(\mathcal{I}_T)$, $E_{\theta=\delta/2}(\mathcal{I}_T)$ and $E_{\theta=\delta}(\mathcal{I}_T)$, making them unsuitable for most applications.

In Table 12.2 we present results for different types of design that optimize the objective function

$$\{E_{\theta=0}(\mathcal{I}_T) + E_{\theta=\delta/2}(\mathcal{I}_T) + E_{\theta=\delta}(\mathcal{I}_T)\}/3. \qquad (12.4)$$

These designs have Type I error rate $\alpha = 0.025$, power 0.8 at $\theta = \delta$ and $K = 2$, 3 or 4 analyses. The designs considered are:

(A) Rho family error spending tests with equally spaced information levels $\mathcal{I}_1, \ldots, \mathcal{I}_K$,

(B) Rho family tests with an optimized choice of $\mathcal{I}_1$ followed by equally spaced $\mathcal{I}_2, \ldots, \mathcal{I}_K$,

(C) Rho family tests with an optimized choice of $\mathcal{I}_1, \ldots, \mathcal{I}_K$ (when $K \geq 3$),

(D) GSTs with optimized boundaries for equally spaced information levels,

(E) GSTs with an optimized choice of $\mathcal{I}_1$ followed by equally space $\mathcal{I}_2, \ldots, \mathcal{I}_K$,

(F) GSTs with optimized boundaries for optimized $\mathcal{I}_1, \ldots, \mathcal{I}_K$ (when $K \geq 3$).

The inflation factor R was capped at 1.2 but in several cases the optimum was achieved with a lower maximum information level. Results are expressed as percentages of the information $\mathcal{I}_{f,1}$ required by a fixed sample test. These results reinforce the conclusions of Eales & Jennison (1992) that, only small gains are to be made by optimizing group sizes. The results also demonstrate the efficiency of error spending designs using error spending functions in the Rho family.

Jennison & Turnbull (2006b) discussed the design of a trial when an extremely high treatment effect may be present. Tables 12.3 and 12.4 show results for designs that optimize the objective functions

$$\{E_{\theta=0}(\mathcal{I}_T) + E_{\theta=\delta}(\mathcal{I}_T) + E_{\theta=2\delta}(\mathcal{I}_T)\}/3 \tag{12.5}$$

and

$$\{E_{\theta=0}(\mathcal{I}_T) + E_{\theta=\delta}(\mathcal{I}_T) + E_{\theta=4\delta}(\mathcal{I}_T)\}/3 \tag{12.6}$$

The same six types of design are considered as in Table 12.2 but, unlike in Jennison & Turnbull (2006b), the inflation factor R is capped at 1.2.

The results in Table 12.3 show that, even when $E_{\theta=2\delta}$ is included in the objective function, there is little to be gained by optimizing group sizes. Table 12.4 shows that with $E_{\theta=4\delta}$ in the objective function, it may be worthwhile to optimize $\mathcal{I}_1$ and then continue with equally spaced information levels. However, these results confirm Eales & Jennison's (1992) observation that optimizing a design for high treatment effects is liable to impair its performance in less extreme situations.

If efficiency under extreme treatment effects is seen as important, adding an extra analysis can achieve this without harming efficiency at other effect sizes. In Table 12.3, the 3-group Rho family error spending design with equally spaced information levels is superior under all three θ values to the 2-group designs with optimized group sizes, and the 4-group Rho family error spending design with equally spaced information levels is uniformly superior to the 3-group designs with optimized group sizes. In Table 12.4, the 3-group Rho family error spending design with equally spaced information levels has a lower average $E_\theta(\mathcal{I}_T)$ than the 2-group designs with optimized group sizes and the 4-group Rho family error spending design with equally spaced information levels has an average $E_\theta(\mathcal{I}_T)$ only slightly higher than that of the optimized GST with optimized group sizes.

We conclude from these examples that group sequential designs with equally spaced information levels provide almost all of the benefits of early stopping. There may be other reasons why unequal group sizes are required, for example, a minimum sample size might be needed to demonstrate the safety of a new treatment. In that case, we would recommend conducting the first interim analysis when this minimum sample size has been reached and planning equally spaced information levels thereafter.

Table 12.3 *Properties of tests with K analyses, Type I error rate $\alpha = 0.025$ and power $1 - \beta = 0.8$ at $\theta = \delta$ which minimize criterion* (12.5)*. The inflation factor R is capped at 1.2 if the minimum is not attained at a lower value.*

	Optimizing $\{E_{\theta=0}(\mathcal{I}_T) + E_{\theta=\delta}(\mathcal{I}_T) + E_{\theta=2\delta}(\mathcal{I}_T)\}/3$							
K	Design	R	ρ	$\mathcal{I}_1$		$E_\theta(\mathcal{I}_T)$ at		Average
					$\theta=0$	$\theta=\delta$	$\theta=2\delta$	$E_\theta(\mathcal{I}_T)$
1	Fixed		1.00	100.0	100.0	100.0	100.0	100.0
2	ρ-family							
	A: Equal groups	1.09	1.36	54.5	68.0	83.3	56.4	69.3
	B: Optimal $\mathcal{I}_1$	1.20	0.69	43.1	64.6	86.2	48.9	66.6
	Optimal boundary							
	D: Equal groups	1.09	–	54.5	66.8	83.8	56.7	69.1
	E: Optimal $\mathcal{I}_1$	1.20	–	43.1	63.3	86.7	49.4	66.4
3	ρ-family							
	A: Equal groups	1.20	1.00	40.0	58.6	77.2	45.1	60.3
	B: Optimal $\mathcal{I}_1$	1.20	0.99	33.9	58.1	78.0	43.0	59.7
	C: Opt. $\mathcal{I}_1, \ldots, \mathcal{I}_K$	1.20	0.93	31.2	56.6	79.3	40.1	58.7
	Optimal boundary							
	D: Equal groups	1.19	–	39.7	56.7	77.7	45.1	59.8
	E: Optimal $\mathcal{I}_1$	1.20	–	32.7	55.7	78.8	42.6	59.0
	F: Opt. $\mathcal{I}_1, \ldots, \mathcal{I}_K$	1.20	–	31.0	55.1	79.5	40.9	58.5
4	ρ-family							
	A: Equal groups	1.20	1.13	30.0	55.1	74.7	40.0	56.6
	B: Optimal $\mathcal{I}_1$	1.20	1.13	28.7	55.1	74.8	39.8	56.5
	C: Opt. $\mathcal{I}_1, \ldots, \mathcal{I}_K$	1.20	1.05	25.8	53.4	76.1	36.7	55.4
	Optimal boundary							
	D: Equal groups	1.20	–	30.0	52.5	75.3	39.7	55.8
	E: Optimal $\mathcal{I}_1$	1.20	–	27.4	52.3	75.7	39.2	55.7
	F: Opt. $\mathcal{I}_1, \ldots, \mathcal{I}_K$	1.20	–	25.9	51.7	76.2	37.5	55.1

12.4 Procedures with Data-Dependent Group Sizes

The idea that the next group size or analysis time in a group sequential test should depend on current data has an intuitive appeal. It might seem reasonable to plan smaller group sizes when the current test statistic is close to a stopping boundary and larger group sizes when in the middle of the continuation region. Schmitz (1993) proposed such designs and called them "sequentially planned decision procedures". The group sequential test has the usual form (5.2) but now, at each analysis $k = 1, \ldots, K-1$, the next cumulative information level $\mathcal{I}_{k+1}$ and critical values (a_{k+1}, b_{k+1}) are chosen based on the current data.

In an optimal version of such a design, the rules for choosing each information level $\mathcal{I}_k$ and pair of critical values (a_k, b_k) are designed to minimize a criterion expressed in terms of $E_\theta(\mathcal{I}_T)$, subject to Type I and II error probability constraints, a maximum number of analyses K, and an upper limit $\mathcal{I}_K \leq R\mathcal{I}_{f,1}$ on the final information level. Jennison & Turnbull (2006a) extended the dynamic programming approach described in Section 12.2

Table 12.4 *Properties of tests with K analyses, Type I error rate $\alpha = 0.025$ and power $1 - \beta = 0.8$ at $\theta = \delta$ which minimize criterion* (12.6). *The inflation factor R is capped at* 1.2 *if the minimum is not attained at a lower value.*

	Optimizing $\{E_{\theta=0}(\mathcal{I}_T) + E_{\theta=\delta}(\mathcal{I}_T) + E_{\theta=4\delta}(\mathcal{I}_T)\}/3$							
K	Design	R	ρ	$\mathcal{I}_1$	$E_\theta(\mathcal{I}_T)$ at			Average
					$\theta=0$	$\theta=\delta$	$\theta=4\delta$	$E_\theta(\mathcal{I}_T)$
1	Fixed		1.00	100.0	100.0	100.0	100.0	100.0
2	ρ-family							
	A: Equal groups	1.08	1.46	54.0	68.3	83.5	54.0	68.6
	B: Optimal $\mathcal{I}_1$	1.20	0.64	32.7	67.0	91.1	32.7	63.6
	Optimal boundary							
	D: Equal groups	1.08	–	54.0	66.5	84.4	54.0	68.3
	E: Optimal $\mathcal{I}_1$	1.18	–	29.9	64.1	94.3	29.9	62.8
3	ρ-family							
	A: Equal groups	1.16	1.19	38.7	59.3	77.5	38.7	58.5
	B: Optimal $\mathcal{I}_1$	1.20	0.92	17.6	61.8	81.4	18.6	53.9
	C: Opt. $\mathcal{I}_1, \ldots, \mathcal{I}_K$	1.20	0.88	16.7	59.7	82.3	17.8	53.3
	Optimal boundary							
	D: Equal groups	1.14	–	38.0	56.4	78.8	38.0	57.7
	E: Optimal $\mathcal{I}_1$	1.20	–	18.2	56.3	82.1	19.4	52.6
	F: Opt. $\mathcal{I}_1, \ldots, \mathcal{I}_K$	1.20	–	17.9	55.7	82.3	19.2	52.4
4	ρ-family							
	A: Equal groups	1.20	1.13	30.0	55.1	74.7	30.0	53.3
	B: Optimal $\mathcal{I}_1$	1.20	1.09	15.3	56.7	76.5	17.2	50.1
	C: Opt. $\mathcal{I}_1, \ldots, \mathcal{I}_K$	1.20	1.03	14.4	54.9	77.4	16.3	49.5
	Optimal boundary							
	D: Equal groups	1.19	–	29.8	51.6	75.5	29.8	52.3
	E: Optimal $\mathcal{I}_1$	1.20	–	15.7	52.2	77.3	17.7	49.1
	F: Opt. $\mathcal{I}_1, \ldots, \mathcal{I}_K$	1.20	–	15.0	51.6	77.8	17.0	48.8

to solve this optimization problem. They specified a fine grid of possible information levels and defined $\mathcal{I}_{k+1}(z_k)$, $a_{k+1}(z_k)$ and $b_{k+1}(z_k)$ for each value of $\mathcal{I}_k$ and *every* z_k in the continuation region (a_k, b_k). Although the multiplicity of information levels dramatically increases the size of the state space for dynamic programming, computation is still feasible.

Comparisons between optimized adaptive designs and optimal group sequential tests with fixed group sizes allow us to evaluate the gain in efficiency that data-dependent information levels can provide. Jennison & Turnbull (2006a) compared three types of one-sided tests of H_0: $\theta \leq 0$ versus H_1: $\theta > 0$ with Type I error rate $\alpha = 0.025$, power $1 - \beta = 0.9$ at $\theta = \delta$, and maximum information $\mathcal{I}_K \leq R\,\mathcal{I}_{f,1}$, optimized for the efficiency criterion $\int f(\theta) E_\theta(\mathcal{I}_T)\, d\theta$, where $f(\theta)$ is the density of a $N(\delta, \delta^2/4)$ distribution, that we have considered previously. The first type of test has equally spaced information levels, $\mathcal{I}_k = (k/K)\, R\, \mathcal{I}_{f,1}$, $k = 1, \ldots, K$, as considered in Section 12.2; the second test has fixed but unequal group sizes, as in Section 12.3; and the third has a data-dependent choice of

Table 12.5 *Value of* $\int f(\theta)E_\theta(\mathcal{I}_T)\,d\theta$, *where* $f(\theta)$ *is the density of a* $N(\delta, \delta^2/4)$ *distribution, as a percentage of* $\mathcal{I}_{f,1}$ *for three types of optimized group sequential design. Tests have Type I error rate* $\alpha = 0.025$, *power* 0.9 *at* $\theta = \delta$, K *analyses and inflation factor* $R = 1.2$.

K	*Optimal design with K equal group sizes*	*Optimal design with K optimized group sizes*	*Optimal adaptive design of Schmitz*
1	100.0	100.0	100.0
2	74.8	73.2	72.5
3	66.1	65.6	64.8
4	62.7	62.4	61.2
5	60.9	60.5	59.2
6	59.8	59.4	58.0
8	58.3	58.0	56.6
10	57.5	57.2	55.9

group sizes. Table 12.5, which reports cases in Table 1 of Jennison and Turnbull (2006a) with $R = 1.2$ and $K = 1$ to 6, 8, and 10, shows the values of $\int f(\theta)E_\theta(\mathcal{I}_T)\,\mathrm{d}\theta$ achieved by the optimum version of each type of test, expressed as a percentage of the information $\mathcal{I}_{f,1}$ required for a fixed sample size test. The results show that designs in which group sizes are varied *adaptively* have a small advantage over optimal non-adaptive designs. On the other hand, such an adaptive procedure is more complex to administer and if such decreases in $E_\theta(\mathcal{I}_T)$ are deemed important, it may well be preferable to use a non-adaptive design with an additional analysis.

Similar conclusions follow when designs are optimized for other criteria. Jennison & Turnbull (2006b) found optimal adaptive designs for the criteria (12.5) and (12.6) considered in Section 12.2. They reported average $E_\theta(\mathcal{I}_T)$ values of 66.1, 57.8, 54.0 for $K = 2$, 3 and 4, respectively, for Schmitz designs optimizing (12.5) and 62.5, 51.8, 47.7 for Schmitz designs optimizing (12.6). Thus, there are improvements of between 0.3% and 1.1% of $\mathcal{I}_{f,1}$ over the "fully optimized" non-adaptive designs in Tables 12.3 and 12.4. However, in keeping with results in Table 12.5, they found that for $K = 2$ and 3, switching to a non-adaptive design with one additional analysis provided considerably higher efficiency gains than incorporating adaptive choice of group sizes. Lokhnygina & Tsiatis (2008) derived optimal two-stage adaptive designs for a normal response and reached the same conclusion that the benefit of choosing the second group size adaptively is minimal. Banerjee & Tsiatis (2006) computed optimal two-stage adaptive designs for a binary response and compared these with the two-stage designs of Simon (1989). They concluded that the benefits of adaptation are at best modest and noted that not knowing the second-stage sample size in advance may create difficulty in planning resources for a trial.

We note that, while the Schmitz designs are adaptive in the sense that future information levels are allowed to depend on current data, these designs are not "flexible". The rules for setting future information levels and stopping boundaries based on past and current information levels and Z-values are specified at the start of the study and these rules must be followed if error rates are to be properly controlled. In Chapter 24 we shall discuss flexible, adaptive procedures that offer greater flexibility in modifying sample size. Nevertheless, the results of this Section are informative about what such flexible designs can achieve when the decisions made about future sample sizes are based on information such as parameter estimates internal to the current trial.

Chapter 13

Bayesian Approaches

13.1 The Bayesian Paradigm

A very different framework for the design and analysis of group sequential trials arises when inference is made using Bayesian methods. In this paradigm, uncertainty about θ, the parameter or parameters of interest, is expressed by considering it as a random variable with a probability distribution. At the start of the trial, this is called the *prior* distribution, with density or probability mass function $\pi(\theta)$, say. As data accumulate, this prior is updated to form the *posterior* distribution, which summaries the current belief about the value of θ. The updating is determined by Bayes Law, which states that the posterior distribution, $p(\theta \,|\, data)$, is given by the normalized product of the prior distribution and the likelihood of the observed data, $L(data \,|\, \theta)$. Thus,

$$p(\theta \mid data) \;\propto\; L(data \mid \theta)\, \pi(\theta), \tag{13.1}$$

where the constant of proportionality is such that the posterior density integrates to unity if θ has a continuous distribution, or sums to unity if θ has a discrete distribution. At any stage of a trial, this posterior distribution can be used to draw inferences concerning θ. For instance, we may construct a *credible set* for θ with posterior probability equal to some pre-specified level $1 - 2\epsilon$. If the parameter θ is one-dimensional and $\epsilon = 0.025$, a 95% credible set, or Bayesian interval estimate, (θ_L, θ_U) satisfies (Lindley, 1965, p. 15)

$$Pr\{\theta_L < \theta < \theta_U \mid data\} \;=\; 0.95. \tag{13.2}$$

A one-sided or two-sided test for efficacy might be based on whether the interval excludes zero, or a test for equivalence could examine whether the interval lies completely within a specified range of equivalence.

To illustrate the calculations, consider the case of accumulating normal data giving rise to standardized statistics $Z_1, Z_2, \ldots$ with the canonical joint distribution (3.1). At any stage $k = 1, 2, \ldots$, the likelihood for θ is the normal density with mean θ and variance $\mathcal{I}_k^{-1}$ of the observed value $\hat{\theta}_k = Z_k/\sqrt{\mathcal{I}_k}$ (see Section 3.1). We shall discuss the choice of prior in detail later. For now, consider the convenient choice of a conjugate normal prior distribution $N(\mu_0, \sigma_0^2)$, where μ_0 and σ_0^2 are the specified prior mean and variance of θ. Using (13.1), the posterior distribution for θ at stage k is

$$N\left(\frac{\hat{\theta}_k \mathcal{I}_k + \mu_0\, \sigma_0^{-2}}{\mathcal{I}_k + \sigma_0^{-2}},\; \frac{1}{\mathcal{I}_k + \sigma_0^{-2}} \right) \tag{13.3}$$

and the 95% credible interval for θ is given by

$$\left(\frac{\hat{\theta}_k \mathcal{I}_k + \mu_0\, \sigma_0^{-2}}{\mathcal{I}_k + \sigma_0^{-2}} \;\pm\; 1.96\, \frac{1}{\sqrt{\{\mathcal{I}_k + \sigma_0^{-2}\}}} \right). \tag{13.4}$$

DOI: 10.1201/9781584888482-13

Note that this interval is "shrunk" toward the prior mean μ_0 whereas the usual fixed sample frequentist confidence interval would be centered on the estimate $\hat{\theta}_k$. If the prior mean is set as $\mu_0 = 0$, the credible interval for θ is shrunk toward zero, a property regarded as desirable by some authors; see Hughes & Pocock (1988) and Pocock & Hughes (1989).

The limiting case as $\sigma_0 \to \infty$ gives a *reference* or *non-informative* improper uniform prior and the posterior distribution for θ reduces to $N(\hat{\theta}_k, \mathcal{I}_k^{-1})$. In this case, the $(1-2\epsilon)$ credible interval (13.4) is numerically the same as a $(1-2\epsilon)$ confidence interval for θ unadjusted for multiple analyses, although its interpretation is, of course, different.

Unlike the frequentist confidence intervals of Chapters 9 and 10, which condition on parameter values rather than on the observed data, a Bayes credible interval at termination or at any interim analysis does not depend on the sampling scheme used to obtain the data. Thus, a final credible interval satisfies the likelihood principle; see, for example, Berger (1985, Ch. 4) or Robert (2007, Ch. 5). The same comments apply to Bayesian P-values (Lindley, 1965, p. 58) of the form $Pr\{\theta < 0 \,|\, data\}$ evaluated under the posterior distribution. Hence, the techniques for inference, both at interim analyses and upon termination, can be much more straightforward in the Bayesian paradigm.

13.2 Choice of Prior Distribution

A key feature of the Bayesian approach is that prior beliefs and external evidence can be summarized mathematically and expressed in the form of a prior distribution for the unknown parameters. The issues concerning the choice of prior distribution are well reviewed in Spiegelhalter, Freedman & Parmar (1994) and the accompanying published discussion. This choice is especially important in the context of group sequential monitoring where the outcome of a trial can be particularly sensitive to the assumed prior (Rosenbaum & Rubin, 1984). Ideally there would be a "single defendable prior" (Herson, 1994) but, almost by definition, a prior is subjective and there will be no single choice upon which all can agree.

One option is simply to report the likelihood function, enabling each consumer to combine this with their personal prior in order to make inferences and decisions. A more practical strategy is to base decisions on a collection of analyses. Kass & Greenhouse (1989, p. 312) assume that "the purpose of a trial is to collect data that bring to conclusive consensus at termination opinions that had been diverse and indecisive at the outset." Simultaneous consideration of a variety of priors can assist in this aim. Fayers, Ashby & Parmar (1997) state that the prior distributions "should reflect the level of skepticism that is expressed by those clinicians one seeks to influence."

Spiegelhalter, Freedman & Parmar (1994) list the following types of prior:

A clinical prior is intended to reflect expert opinion. Methods have been developed to elicit priors from clinicians knowledgeable in the field; see, for example, Freedman & Spiegelhalter (1983) and Chaloner et al. (1993). An example of this process appears in the description of the CHART trial by Parmar et al. (1994). A potential problem with the approach is that the chosen experts who are involved in the trial may be too optimistic about a new therapy; see, for example, Gilbert, McPeek & Mosteller (1977). Alternatively, clinical priors can be based on meta-analyses or overviews of historical information but, due to publication bias, these too may be overly optimistic.

A skeptical prior should represent one extreme in the range of opinion. As a working method of construction, Spiegelhalter, Freedman & Parmar (1994) suggest a symmetric prior centered at zero with only a small probability, 5% for example, that θ, the benefit of the new treatment, should exceed a specified value $\theta_1 > 0$.

An enthusiastic prior acts as a counterbalance to the skeptical prior. Operationally, an enthusiastic prior might be constructed by centering the distribution at a positive value θ_1, with only a small probability that $\theta < 0$. The presence of skeptical and enthusiastic priors in a collection can have the effect of lessening the pressure to stop a trial prematurely in the face of early positive or negative results, respectively.

Reference or *non-informative priors* (Bernardo, 1979) attempt to be objective in that they endeavor to represent a lack of any prior opinion. Thus, the Bayesian framework can be applied but it is the newly observed data that provide the evidence about a new treatment. In the normal data example we have been using, the non-informative prior is an improper uniform prior, the limit of a $N(\mu_0, \sigma_0^2)$ distribution as $\sigma_0 \to \infty$. This choice implies that the prior probability of θ lying in an interval around zero is the same as that for an equally wide interval centered at, say, 10^{10}, a property that might be regarded as very informative! However, the posterior density around $\theta = 10^{10}$ will become negligible as soon as any data are observed. For a fascinating discussion of the role of non-informative priors, see Irony & Singpurwalla (1997).

"Handicap" or *"pragmatic Bayes" priors* have been proposed by Grossman et al. (1994) and Spiegelhalter, Freedman & Parmar (1994) as a way to control frequentist properties of a Bayesian monitoring procedure. The prior is chosen so that, when a fixed maximum number of analyses is planned, the frequentist Type I error rate is controlled at a given level α. Grossman et al. (1994) term this a "unified" method in that it brings together Bayesian and frequentist concepts. In our normal data example, the unified method calls for a normal prior for the treatment effect θ centered at zero and with variance $\sigma_0^2 = (h\mathcal{I}_K)^{-1}$, where $\mathcal{I}_K$ is the information for θ that will be available at the final analysis. The constant h is termed the *handicap*.

This list of types of prior is not exhaustive. Other priors described in the literature include the prior of a "cautious reasonable skeptic" suggested by Kass & Greenhouse (1989) and the "open-minded" prior of George et al. (1994).

13.3 Stopping Rules

13.3.1 Stopping Rules Based on the Posterior Distribution of the Treatment Effect

We saw in Section 13.1 that the calculation of a Bayes credible set or a Bayesian P-value can be quite straightforward. Questions of experimental design and, in particular, the decision of when to stop a trial are more interesting. Suppose a trial is designed with a sequence of interim analyses. Then, given a prior for θ, one might define a stopping rule based on the posterior distribution of θ:

$$
\begin{array}{lll}
\text{At each analysis } k = 1, \ldots, K-1 & & \\
\quad \text{if } Pr\{\theta > 0 \mid data\} > 1 - \epsilon & \text{stop, declare } \theta > 0 & \\
\quad \text{if } Pr\{\theta < 0 \mid data\} > 1 - \epsilon & \text{stop, declare } \theta \leq 0 & \\
\quad \text{otherwise} & \text{continue to group } k+1, & \quad (13.5) \\
\text{after group } K & & \\
\quad \text{if } Pr\{\theta > 0 \mid data\} > 1 - \epsilon & \text{stop, declare } \theta > 0 & \\
\quad \text{otherwise} & \text{stop, do not declare } \theta > 0. &
\end{array}
$$

Here a typical value specified for ϵ might be 0.025. This rule calls for early stopping when a $1 - 2\epsilon$ credible region for θ excludes zero. Similar rules have been proposed by Mehta & Cain (1984), Berry (1985) and Freedman & Spiegelhalter (1989).

The sequential design does not affect the way in which the posterior distribution of θ is calculated and so the credible interval (13.4) is not affected by the fact that earlier analyses have been conducted. If the trial terminates at any stage with $Pr\{\theta > 0 \mid data\} > 1 - \epsilon$, the posterior probability that $\theta < 0$ at that point is less than ϵ, regardless of any stopping rule, formal or informal. Nevertheless, the frequentist properties of Bayesian designs are

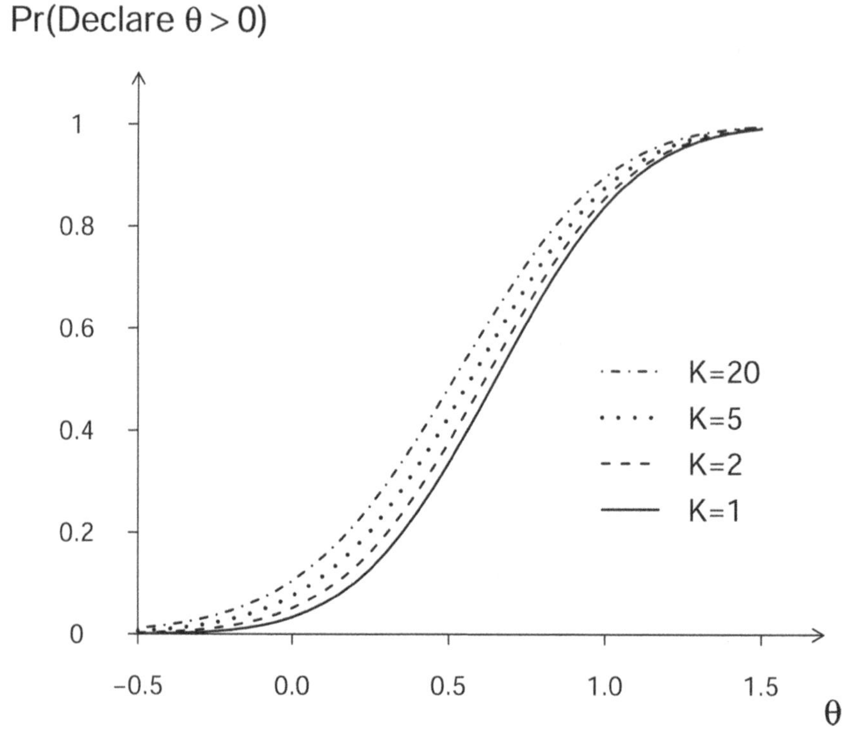

Figure 13.1 *Probability of declaring* $\theta > 0$ *for Bayesian procedures of the form (13.5) with prior* $\theta \sim N(0.5, 0.71^2)$, $\epsilon = 0.025$, *and 1, 2, 5 and 20 analyses.*

affected by the monitoring scheme applied. In the normal example with the non-informative prior, which arises in the limit as $\sigma_0 \to \infty$ in (13.3), the procedure (13.5) with $\epsilon = 0.025$ is equivalent to the repeated significance test with K analyses, which stops at the first analysis k where $|Z_k| > 1.96$. If $K = 1$ and no interim analyses are performed, the probability of concluding $\theta > 0$ when in fact $\theta = 0$ is 0.025. However, the effect of multiple looks at the data is exactly the same as that observed by Armitage, McPherson & Rowe (1969) for a repeated significance test so if, for example, $K = 5$ and up to five analyses are conducted, the probability of concluding $\theta > 0$ when $\theta = 0$ rises to 0.071.

As another example, suppose a trial has a total of K analyses, the maximum sample size is chosen to provide information $\mathcal{I}_K = 8$ for θ, and information levels at the interim analyses are $\mathcal{I}_1 = \mathcal{I}_K/K, \ldots, \mathcal{I}_{K-1} = (K-1)\mathcal{I}_K/K$. The prior $\theta \sim N(\mu_0, \sigma_0^2)$ is assumed with $\mu_0 = 0.5$ and $\sigma^2 = 0.5$, so the prior is equivalent to having an initial estimate of θ equal to 0.5 based on information $1/\sigma_0^2 = 2$ after starting with a non-informative prior. Figure 13.1 shows the conditional probability of declaring $\theta > 0$ given each value of θ in the range -0.5 to 1.5 when the number of analyses is $K = 1$, 2, 5 or 20 and the stopping rule (13.5) is applied with $\epsilon = 0.025$. In this example, the frequentist Type I error probability, $Pr\{\text{Declare } \theta > 0 \,|\, \theta = 0\}$, increases from 0.033 to 0.105 as the number of looks at the data increases from 1 to 20. Some proponents of the Bayesian approach would cite the likelihood principle and argue that such frequentist properties are largely irrelevant; see, for example, Berry (1987) and Berger & Berry (1988). Other "pragmatic" Bayesians, such as Spiegelhalter, Freedman & Parmar (1994), admit a concern for controlling the false positive rate and propose the use of priors that mollify the Type I error inflation which Bayesian procedures might otherwise produce. We have applied this pragmatic approach in

Table 13.1 *Type I error rates for designs with equally spaced analyses using the stopping rule* (13.5) *based on Bayesian posterior probabilities. The stopping rule is applied with* $\epsilon = 0.025$ *assuming (a) a non-informative prior and (b) a handicap prior with* $h = 0.249$ *appropriate to the design with 4 analyses. The final column shows the values of the "handicap"* h *needed to achieve a Type I error rate* $\alpha = 0.025$ *for each number of analyses* K.

Number of analyses K	*Type I error probability* (a) *Non-informative prior*	(b) *Handicap prior with* $h = 0.249$	*Handicap required for Type I error rate* $= 0.025$
1	0.025	0.014	0.000
2	0.042	0.019	0.163
3	0.054	0.023	0.217
4	0.063	0.025	0.249
5	0.071	0.027	0.271
10	0.097	0.032	0.334
20	0.124	0.037	0.382
50	0.160	0.043	0.429
∞	1.000	1.000	—

(a) Boundary: $|Z_k| > 1.96$, for $k = 1, \ldots, K$

(b) Boundary given by (13.6) with $h = 0.249$ and $\alpha = 0.05$.

our example with K equally spaced analyses and $\mathcal{I}_K = 8$. The "handicap prior" is specified as $\theta \sim N(0, \sigma_0^2)$ where $\sigma_0^2 = (h\,\mathcal{I}_K)^{-1}$ and the handicap h is computed so that the procedure, if rigidly followed, has Type I error rate equal to $\alpha = 0.025$. When applying the stopping rule (13.5) with $\epsilon = \alpha = 0.025$, the condition $Pr\{\theta > 0 \,|\, data\} > 1 - \epsilon$ at analysis k becomes

$$Z_k > \Phi^{-1}(1-\alpha)\sqrt{\{(1 + (K/k)h\}}, \quad \text{for each } k = 1, \ldots, K. \tag{13.6}$$

For $K = 1$, no handicap is needed and $h = 0$ gives the non-informative prior which arises as $\sigma_0 \to \infty$. In Table 13.1, we present properties of designs with K analyses for $K = 1$, 2, 3, 4, 5, 10 and 20, and for the limiting case as $K \to \infty$. The table shows the frequentist Type I error probabilities for these designs (a) assuming a non-informative prior, which is equivalent to a handicap $h = 0$, and (b) assuming a handicap prior with $h = 0.249$, the value needed for a design with 4 analyses to have Type I error probability 0.025. The Type I error rate can be interpreted in the Bayesian paradigm as the conditional probability, given $\theta = 0$, that the lower endpoint of the 95% credible interval on termination, based on the assumed prior, lies below zero. The results demonstrate that the frequentist Type I error probability of the Bayesian procedure is affected by the number of analyses performed. Using a handicap prior can ensure the Type I error probability has a particular value, 0.025 in this case, but this does require the design to be conducted with the number of analyses assumed when determining the handicap h (shown for each K in the final column of Table 13.1.)

Grossman et al. (1994) note that the stopping boundaries of their designs fall between those of Pocock (1977) and O'Brien & Fleming (1979) designs. We saw in Section 2.7.1 how the Wang & Tsiatis (1987) family provides a continuum of designs with Pocock and O'Brien & Fleming boundaries as extreme cases, and the handicap designs match members of this the family very closely. As an example, in the handicap design for $K = 4$, the critical values for Z_1 to Z_4 given by (13.6) are 2.77, 2.40, 2.26 and 2.19, while a Wang & Tsiatis boundary with $\Delta = 0.35$ has critical values 2.69, 2.42, 2.28 and 2.18. If one plots the power $P_\theta\{\text{Declare } \theta > 0\}$ as a function of θ for the two designs, the curves are indistinguishable and $E_\theta(\mathcal{I})$, the expected information on termination, is almost exactly equal for the two designs at all values of θ. The similarity between the end products of frequentist and Bayesian developments is noteworthy. We see this as a positive finding since it provides some reconciliation between the two philosophies. We shall return to this point in Section 13.3.5.

13.3.2 Sampling to a Foregone Conclusion

We noted in Table 13.1 that the false positive rate approaches unity as K tends to ∞. This phenomenon was termed "sampling to a foregone conclusion" by Cornfield (1966a,b), who reflected that "if one is concerned about the high probability of rejecting H_0, it must be because some probability of its truth is being entertained." Hence, for our normal example, he proposed using a mixture prior with a discrete probability mass p assigned to $\theta = 0$ and the remaining probability $1-p$ distributed normally with mean zero and variance σ_0^2. With this prior, on observing Z_k, the posterior odds in favor of the hypothesis H_0: $\theta = 0$ are defined as

$$\lambda = \frac{Pr\{\theta = 0 \mid data\}}{Pr\{\theta \neq 0 \mid data\}} = \frac{p}{1-p} RBO,$$

where RBO represents the "relative betting odds", the ratio of the posterior odds to the prior odds for H_0 (Cornfield, 1969). That is,

$$RBO = \sqrt{\{1 + \mathcal{I}_k \, \sigma_0^2\}} \exp\left\{ \frac{-Z_k^2}{2\{1 + (\mathcal{I}_k \, \sigma_0^2)^{-1}\}} \right\}.$$

This ratio is also known as the Bayes factor (Dickey, 1973). Suppose we apply the rule that stops the trial if $|Z_k| > 1.96$. Then, upon stopping with a large value of $\mathcal{I}_k$ and $Z_k \approx 1.96$, the posterior odds are approximately

$$\lambda = \{p/(1-p)\} \, \sigma_0 \sqrt{\mathcal{I}_k} \, \exp(-1.96^2/2),$$

which now favor H_0 and not H_1 if $\mathcal{I}_k$ is sufficiently large. Hence, if decisions are based on the value of the RBO, large numbers of analyses do not result in almost certain rejection of H_0 and such a procedure is not subject to the threat of "sampling to a foregone conclusion". Cornfield (1966b) went on to propose a stopping rule based on parallel stopping boundaries for the RBO or, equivalently, for the posterior odds λ. For such a procedure, stopping boundaries on the standardized Z_k scale diverge with sample size unlike the Pocock boundaries of Section (2.4), which are constant, or the O'Brien & Fleming boundaries of Section (2.5) that become narrower as the sample size increases.

Some Bayesians would disagree with the mixture prior approach of Cornfield, seeing the assignment of discrete mass to a single point as unrealistic; see remarks by Spiegelhalter & Freedman (1988, p. 461). Lachin (1981) addresses this and other issues by modifying the Cornfield model to contain a composite null hypothesis, replacing the discrete prior mass at $\theta = 0$ under H_0 by a continuous prior supported on a small interval around zero. Despite, these developments, it is fair to say that Cornfield's designs based on RBOs have not been adopted in practice and Bayesian recommendations for monitoring have been more along the lines of the rule (13.5)

13.3.3 Decisions Based on a Family or Priors

Carlin et al. (1993) describe the conduct of a toxoplasmic encephalitis prophylaxis trial treating patients with AIDS. Priors for the treatment effect on the primary endpoint were elicited from three AIDS clinicians, and two non-physicians involved in AIDS research. Figures in the paper show how the posterior distributions, starting from each prior, evolved over the course of the trial. Carlin & Sargent (1996) propose a "backward" Bayesian approach. They characterize the class of priors that would lead to a certain decision, such as stopping the trial and declaring the new treatment effective, conditional on the observed data. This information can be used in the deliberations of a data monitoring committee and is likely to be particularly helpful when the data and an assumed prior are in apparent conflict.

Greenhouse & Wasserman (1995) propose a "robust Bayesian methodology." Instead of taking a single prior, they propose using a class of priors indexed by a parameter γ, where $0 \leq \gamma \leq 1$. One such class is the family

$$\Gamma_\gamma = \{(1-\gamma)\,\pi(\theta) + \gamma\, Q;\ Q \in \mathcal{Q}\},$$

where $\pi(\theta)$ is, say, the clinical prior and $\mathcal{Q}$ is the set of all possible priors. For a given value of θ_T, bounds on the posterior probability $Pr\{\theta > \theta_T \mid data\}$ over the family Γ_γ can be plotted as γ varies from 0 to 1 and narrow bounds indicate a robustness to the choice of prior.

Fayers, Ashby & Parmar (1997) recommend presenting the posterior probabilities that the treatment effect θ exceeds a target value, θ_T say, in a two way table cross-classified by the choice of prior, for example, non-informative, skeptical or enthusiastic, and by a set of target values θ_T. These probabilities can then be used to guide the decisions of a data monitoring committee and communicated in scientific publications. Such a table could be useful to members of a data monitoring with diverse prior opinions committee as an aid to deciding when it is appropriate to stop the trial.

13.3.4 Bayes Decision-Theoretic Stopping Rules

In discussing the paper by Spiegelhalter, Freedman & Parmar (1994), Dennis Lindley (1994) commented:

> "The authors dismiss the use of expected utility as being unrealistic. In a sense this is right, for to have given a treatment of the decision aspect of clinical trials, in addition to the inference side, would have unreasonably lengthened the paper and demanded much research. Nevertheless, it must be recognized that clinical trials are not there for inference but to reach a decision, and the omission of their *raison d'être* is serious. ... It must be 25 years ago since I gave a seminar on sequential methods, including clinical trials, and was told, in the discussion, that the use of probability to express one's beliefs was unrealistic. The authors have demonstrated that, on the contrary, it is realistic. ... I hope that it will not take 25 years before the Royal Statistical Society has a paper, perhaps from a student of Spiegelhalter's, doing for utility what this does so well for probability."

More than 25 years have elapsed since then and progress has indeed been made. Anscombe (1963) had laid the foundations for a decision-theoretic approach in a "horizon problem" where the goal is to optimize the outcomes for a large number of future patients by running a sequential trial and giving the treatment selected at the end of this trial to all subsequent patients. Chernoff & Petkau (1981) developed this approach and calculated the Bayes optimal procedure for a continuous time version of the problem.

Berry & Ho (1988) formulated a Bayes sequential decision problem appropriate to a group sequential trial comparing a new treatment to a control, where early stopping is permitted for futility but not for efficacy. The possible decisions are

d_1 : Do not pursue drug approval
d_2 : Pursue drug approval

and, with θ denoting the benefit of the new treatment, the loss function for taking decision d_1 or d_2 when the true value of the treatment effect is θ is defined as

$$L(\theta, d_1) = 0 \qquad \text{for all } \theta,$$

$$L(\theta, d_2) = \begin{cases} -K\theta & \text{if } \theta > 0, \\ L & \text{if } \theta \leq 0. \end{cases}$$

In addition, a cost of 1 is incurred for each patient in the trial. Berry & Ho used backward induction (DeGroot, 1970, Ch. 12), to find the decision rule that minimizes the expected total cost, made up of the cost of sampling plus any loss incurred, averaged over their prior distribution for θ. The method of backward induction is also known as dynamic programming (Bellman, 1957) and, for consistency with discussion in other chapters, we shall refer to it by this name.

Carlin, Kadane & Gelfand (1998) describe a Bayesian approach to sequential decision making in clinical trials, specifying a prior distribution and loss function. They assert that solution by dynamic programming is not feasible due to the "explosion in bookkeeping complexity arising from the resulting alternating sequence of integrations and minimization" and they search for an optimal policy by an iterative search using forward simulation. In some important special cases, the problems tackled are similar to problems we referred to in Chapter 12 that were solved by Lorden (1976), Eales & Jennison (1992, 1995) and Chang (1996), using dynamic programming with computation growing linearly in the number of analyses. However, implementation of dynamic programming does become problematic when the sufficient statistic for model parameters at each stage is high dimensional. Computation can still be linear in the number of analyses but, in the discrete case, the set of possible states at each stage may be prohibitively large or, in the continuous case, the state space is high dimensional making numerical integration impractical. Brockwell & Kadane (2003) present a grid-based scheme for finding a near-optimal sequential design in which possible decisions are assessed by forward simulation. Müller et al. (2007) follow a similar approach to tackle a wider class of problems.

Lewis, Lipsky & Berry (2007) used dynamic programming to calculate Bayes optimal, decision theoretic, group sequential designs for a two-arm trial with a binary response. They compared the resulting Bayes procedures with the frequentist group sequential designs of Pocock (1977) and O'Brien & Fleming (1979). Since they wished to compare the Bayesian designs with frequentist designs, constants in the assumed loss function were chosen to yield specific error rates. The authors state that "the resulting trial designs could be viewed as a compromise between frequentist methods (which imply but do not make explicit an underlying loss function) and the Bayesian decision-theoretic approach."

Pertile, Forster & La Torre (2014) consider the problem of designing a trial to evaluate the cost-effectiveness of a new technology. Here the "technology" could range from a treatment for a disease to a campaign for improving public health, and the aim is to determine whether this innovation should be recommended for use in a healthcare system. The outcome measure is the net incremental monetary benefit (NIMB) of the new technology versus existing practice and an optimal design defines the action at each stage that maximizes the expected discounted NIMB minus research and investment costs. The authors follow a similar approach to Chernoff & Petkau (1981), solving a free boundary problem to find the optimal policy when monitoring is continuous. Chick, Forster & Pertile (2017) develop this approach further to handle the situation where there is a significant delay between the start of treatment and the time when a subject's response is observed. (We shall discuss group sequential tests for such a situation in Chapter 14.) Forster et al. (2021) apply their methods retrospectively to the design of the ProFHER trial that compared surgery against sling immobilization for the treatment of fracture of the humerus. It is assumed in this work that priors are an accurate reflection of initial beliefs and utilities represent real benefits and costs. Although some frequentist properties of procedures are reported in the examples, there is no suggestion that priors and utilities were chosen to produce particular results. Indeed, Type I error probabilities are not reported at all.

13.3.5 *Calibrated Bayes Procedures and the Relation between Efficient Bayesian and Frequentist Designs*

The relevance of the Type I error rate in Bayesian inference and whether account should be taken of multiple looks at the data has been a matter of debate. In commenting on the impact of interim analyses on inferences drawn from a sequential trial, Berry (1987, p. 119) states

> "there is no price to pay for looking at the data using a Bayesian approach."

On the other hand, Pocock & Hughes (1989) write

> "control of the Type I error is a vital aid to restricting the flood of false positives in the medical literature."

Ryan, Brock, Gates & Slade (2020) observe

> "there is some confusion as to whether control of type I error is required for Bayesian designs as this is a frequentist concept."

They quote Cox & Hinkley (1974, p. 45–46)

> "We should not follow procedures which for some possible parameter values would give, in hypothetical repetitions, misleading conclusions most of the time."

and they point out that, currently, regulators require demonstration of control of Type I error in late-phase trials.

Little (2006) argues for a "calibrated Bayes" approach that can capture the strengths of both Bayesian and frequentist approaches. The use of a handicap prior by Spiegelhalter, Freedman & Parmar (1994), which we presented in Section 13.3.1, is an example of this philosophy: inferences follow the Bayesian procedure given by (13.5) but with a prior chosen so that the Type I error probability, $Pr\{\text{Declare } \theta > 0 \,|\, \theta = 0\}$, has an acceptable value. Lewis, Lipsky & Berry (2007) choose constants in their loss function so that the resulting Bayes procedure has specific error rates. Shi & Yin (2019) modify a Bayes rule of the form (13.5) by letting the confidence level $1 - \epsilon$ depend on the stage k in such a way that the resulting procedure matches a certain error spending test, as described in Chapter 8.

Ventz & Trippa (2015) suggest a different way to obtain a calibrated procedure. They propose (1) formulating a Bayes decision problem with a prior and utility function and (2) specifying constraints on Type I and Type II error probabilities, then they search for the design with the highest possible expected utility, subject to meeting the error probability constraints. As an example of this approach, consider a trial with K analyses at information levels $\mathcal{I}_1, \ldots, \mathcal{I}_K$ where the final decision D can be

$$\begin{array}{ll} d_1: & \text{Do not pursue drug approval} \\ d_2: & \text{Pursue drug approval} \end{array}$$

Suppose utility is specified through a gain function made up of components

$$\begin{aligned} G(\theta, d_1) &= 0 \qquad \text{for all } \theta, \\ G(\theta, d_2) &= \begin{cases} K(\theta) & \text{if } \theta > 0, \\ -L(\theta) & \text{if } \theta \le 0, \end{cases} \end{aligned}$$

where θ is the treatment effect, plus a term $B(\theta, D, T)$, where T denotes the analysis at which the trial terminates, that reflects the additional benefit of reaching decision D at analysis T

(reflecting, for example, the commercial advantage of an early positive conclusion) minus the cost of treating patients up to that point in the trial. If the prior for the treatment effect is $\pi(\theta)$, the aim in the Bayes decision problem is to maximize the expected gain

$$-\int_{-\infty}^{0} L(\theta)\, Pr(D = d_2 \,|\, \theta)\, \pi(\theta)\, d\theta \;+\; \int_{0}^{\infty} K(\theta)\, Pr(D = d_2 \,|\, \theta)\, \pi(\theta)\, d\theta$$

$$+ \int_{-\infty}^{\infty} E\{B(\theta, D, T) \,|\, \theta\}\, \pi(\theta)\, d\theta. \qquad (13.7)$$

Suppose now that we impose the Type I and II error probability constraints

$$Pr\{D = d_2 \,|\, \theta = 0\} \;=\; \alpha \quad \text{and} \quad Pr\{D = d_1 \,|\, \theta = \delta\} = \beta \qquad (13.8)$$

for specified values of α, β and δ. We commented in Section 2.6 that, once the Type I and Type II error probabilities are specified, a test's power is largely determined over the range of possible parameter values. Our experience is that this statement is true to a high degree of accuracy for all reasonable group sequential tests. It follows that the constraints (13.8) are almost exactly equivalent to the conditions

$$\int_{-\infty}^{0} L(\theta)\, Pr(D = d_2 \,|\, \theta)\, \pi(\theta)\, d\theta \;=\; p_1 \quad \text{and} \quad \int_{0}^{\infty} K(\theta)\, Pr(D = d_2 \,|\, \theta)\, \pi(\theta)\, d\theta \;=\; p_2 \qquad (13.9)$$

for certain values of p_1 and p_2. These values of p_1 and p_2 can be found by, for example, evaluating the integrals in (13.9) with the function $Pr(D = d_2 \,|\, \theta)$ for a fixed sample test with Type I and II error rates given by (13.8). After making this approximation, the problem of maximizing the expected gain (13.7) subject to the constraint (13.8) becomes that of maximizing (13.7) subject to (13.9), which reduces to maximizing

$$\int_{-\infty}^{\infty} E\{B(\theta, D, T) \,|\, \theta\}\, \pi(\theta)\, d\theta \qquad (13.10)$$

subject to (13.9). This constrained optimization problem can be solved following the Lagrangian approach described in Section 12.2. At the conclusion of this process, the optimal design will have been found by maximizing

$$-\lambda_1 \int_{-\infty}^{0} L(\theta)\, Pr(D = d_2 \,|\, \theta)\, \pi(\theta)\, d\theta \;+\; \lambda_2 \int_{0}^{\infty} K(\theta)\, Pr(D = d_2 \,|\, \theta)\, \pi(\theta)\, d\theta$$

$$+ \int_{-\infty}^{\infty} E\{B(\theta, D, T) \,|\, \theta\}\, \pi(\theta)\, d\theta,$$

where the positive constants λ_1 and λ_2 are such that this design satisfies (13.9). Comparison with the original objective function (13.7) shows that the effect of introducing the constraints (13.9) has been to multiply the factors $L(\theta)$ and $K(\theta)$ in the initial problem by λ_1 and λ_2, respectively. Thus, the end product of applying Ventz & Trippa's approach has a simple interpretation as a calibrated Bayes optimal design.

Because of the almost exact equivalence of the constraints (13.8) and (13.9), this form of Bayes optimal design effectively maximizes (13.10) subject to the requirement that Type I error rate is at most α and power $1-\beta$ is achieved when $\theta = \delta$. In Chapter 12, we derived such optimal designs in a frequentist context for certain forms of $B(\theta, d_i, T)$, using a combination of numerical integration and dynamic programming. One implication of this is that the same computational methods can be used to find calibrated Bayes optimal designs. Another point is that the artificial looking priors that arose in some of the examples in Chapter 12 should not be seen as a cause for concern: placing point masses of probability at $\theta = 0$ and $\theta = \delta$

was a result of following the Lagrangian approach to apply the constraints (13.8) and it would be a simple exercise to replace these by constraints of the form (13.9) and produce essentially the same optimized design.

The convergence we have seen between Bayes optimal designs and optimized frequentist designs is no coincidence. The essence of the complete class theorems of Brown, Cohen & Strawderman (1980) is that, given the number of analyses in a trial and the information levels for θ at these analyses, the set of Bayes procedures is equal to the set of "admissible" frequentist procedures, where an admissible procedure is one that cannot be improved uniformly in terms of the probabilities $Pr\{\text{Declare } \theta > 0 \,|\, \theta\}$ or the expected value of a gain function such as our $B(\theta, D, T)$. These classes of procedure are large and more input is needed to choose a single design. In the frequentist setting, it is customary to specify the Type I error rate and power at a particular value of θ. In Bayes decision theory, the expected utility may penalize false positive and false negative decisions in a different way. However, if the Bayes procedure with a term (13.10) in its utility function is calibrated to control the Type I error rate and achieve a certain power, the resulting design should coincide with the frequentist procedure that optimizes the same objective (13.10).

13.4 Other Aspects of Bayesian Design

13.4.1 Target Sample Size

Spiegelhalter & Freedman (1986) argue against the "textbook" approach to setting the sample size of a clinical trial and present a Bayesian approach based on "pre-posterior" analyses using predictive distributions. Suppose θ represents the improvement offered by a new treatment and, upon termination of the trial, a decision will be based on the credible interval (θ_L, θ_U) satisfying (13.2). The decision will be that the new treatment is superior if $\theta_L > 0$, the new treatment is inferior if $\theta_U < 0$, and there is no firm conclusion if $\theta_L \leq 0 \leq \theta_U$. If $\pi(\theta)$ denotes the prior distribution for θ, then, for a given sample size, the conditional probability of concluding with a recommendation in favor of the new treatment given that it is indeed preferable, i.e., $\theta > 0$, can be computed as

$$\frac{\int_0^\infty Pr\{\theta_L > 0 \mid \theta\}\,\pi(\theta)\,d\theta}{\int_0^\infty \pi(\theta)\,d\theta}.$$

A similar expression can be computed for the conditional probability of finding $\theta_U < 0$ and rejecting the new treatment given that it is inferior, i.e., $\theta < 0$. A target sample size can then be determined by ensuring that both of these probabilities exceed some prescribed level, such as 90%. Alternatively, the target sample size can be based on the predictive probability of a conclusive trial,

$$1 - Pr\{\theta_L \leq 0 \leq \theta_U\} = 1 - \int_{-\infty}^{\infty} Pr\{\theta_L < 0 < \theta_U \mid \theta\}\,\pi(\theta)\,d\theta.$$

For further development of this approach, see Brown et al. (1987) and Moussa (1989).

Broglio, Connor & Berry (2014) present a Bayesian design for a confirmatory trial in which the final sample size is adapted in response to accumulating data. In this case, the inferences are ultimately based on a frequentist analysis of the final set of data. However, the decision to limit the sample size by stopping accrual of new patients to the trial is Bayesian, with accrual ending when the predictive probability of a positive outcome is either very high or very low. The final analysis is conducted after responses have been observed from all the "pipeline" subjects who had entered the trial when accrual ceased, but whose responses had not been observed at that time. Although, the title of this paper refers to "sample size selection", the procedures described are closely related to the group sequential tests for a delayed response that we shall present in Chapter 14. There, we shall obtain optimal frequentist designs as the solutions of Bayes decision problems. As we have seen in Section 13.3.5, for a specified trial structure and target utility, there should be convergence between optimized frequentist designs and suitably calibrated, efficient Bayes procedures.

13.4.2 Departures from the Anticipated Pattern of a Trial

In a Bayesian analysis, the posterior distribution of a parameter depends only on the prior and the observed data and there is no difficulty in calculating the posterior distribution of a treatment effect when a trial does not proceed as originally specified. The same is not true for a frequentist analysis where, for example, a P-value is defined relative to the sample space of possible outcomes: the error spending group sequential tests of Chapter 8 and the "combination tests" that we shall introduce in Chapter 22 are mechanisms for preserving the defining properties of frequentist analyses.

In the Bayesian paradigm, once the posterior distribution of θ has been calculated, it is straightforward to find a credible interval for θ and apply a rule such as (13.5) at an interim or final analysis. We would, however, qualify this simple picture. When a Bayesian procedure is calibrated to ensure a trial has a certain Type I error probability, assumptions are made about the pattern of analyses and this Type I error probability may not be maintained if that pattern changes. As an example, we saw in Table 13.1 that applying the rule (13.5) with a handicap prior with $h = 0.249$ gives a procedure with Type I error probability 0.025 if the trial has four equally spaced analyses, but this probability increases to 0.027 if there are five analyses and 0.032 if there are ten analyses.

Kunz et al. (2024) show that increases in Type I error may occur when a Bayesian analysis is conducted after the sample size of a trial is increased in a data-dependent manner after an unexpected disruption. However, the authors also comment that

> "One might argue that, since the Bayesian decision rules were acceptable for the original study design, it should also be acceptable to apply these in a modified design. For instance, if the loss function in a Bayesian decision theory rule provides the correct balance between the benefits of identifying an effective treatment and the risk of approving an ineffective treatment, then it should also do so when data come from a modified trial design."

13.4.3 Regulatory Views on Bayesian Trial Designs

Regulators have been open to Bayesian methods for some time. The International Conference on Harmonisation's Guideline E9 "Guidance on Statistical Principles for Clinical Trials" (FDA, 1998, Sec. 1.2) states

> "Because the predominant approaches to the design and analysis of clinical trials have been based on frequentist statistical methods, the guidance largely refers to the use of frequentist methods when discussing hypothesis testing and/or confidence intervals. This should not be taken to imply that other approaches are not appropriate: the use of Bayesian and other approaches may be considered when the reasons for their use are clear and when the resulting conclusions are sufficiently robust."

The FDA's "Guidance for the Use of Bayesian Statistics in Medical Device Clinical Trials" (FDA, 2010) recognizes there may often be significant prior information about a novel medical device and it is appropriate to use such information in a Bayesian analysis:

> "When good prior information on clinical use of a device exists, the Bayesian approach may enable this information to be incorporated into the statistical analysis of a trial. In some circumstances, the prior information for a device may be a justification for a smaller-sized or shorter-duration pivotal trial."

Nevertheless, Section 4.8 of the guidance does refer to frequentist error probabilities:

> "Because of the inherent flexibility in the design of a Bayesian clinical trial, a thorough evaluation of the operating characteristics should be part of the trial planning. This includes evaluation of:
>
> - probability of erroneously approving an ineffective or unsafe device (type I error rate),
> - probability of erroneously disapproving a safe and effective device (type II error rate)"

and

> "FDA considers type I error, along with other operating characteristics of the trial design, in evaluating submissions. We strive for reasonable control of the type I error rate."

The FDA (2019a) document "Adaptive Designs for Clinical Trials of Drugs and Biologics Guidance for Industry" is not particularly specific. In Section VI B on Bayesian Adaptive Designs it states

> "In general, the same principles apply to Bayesian adaptive designs as to adaptive designs without Bayesian features. Trial designs that use Bayesian adaptive features may rely on frequentist or Bayesian inferential procedures to support conclusions of drug effectiveness."

and, with reference to the process of obtaining feedback on a proposed design,

> "FDA's draft guidance for industry Interacting with the FDA on Complex Innovative Clinical Trial Designs for Drugs and Biological Products [final version published in 2020] provides recommendations on what information should be submitted to FDA to facilitate the review of trial design proposals that use Bayesian inference."

The European Medicines Agency published a Question and Answer document on complex clinical trials (EMA, 2022) which included Question 3: How to describe and explain Bayesian approaches in complex clinical trials? The answer states:

> "Results from Bayesian work have subsequently been assessed in regulatory submissions and enabled inference on pharmacokinetic and pharmacodynamic parameters as well as dose selection. Bayesian approaches may in some situations also be used for inference on efficacy and safety from clinical trials, including trials for regulatory submissions. However, there are few precedents with notable or successful uses of Bayesian approaches and thus critical points encountered are flagged in this answer. Furthermore, Bayesian methodologies have evolved in recent years, so that early engagement with regulators is highly recommended"

and it continues by citing the ICH guideline E9 (FDA, 1998) quoted above.

Our conclusion from these publications by regulators is that Bayesian methods can be acceptable but, given the longstanding practice of assessing evidence through P-values and confidence intervals, Bayesian procedures will be expected to have acceptable frequentist operating characteristics. This may fall short of precise calibration of a Type I error rate, but this issue and, in particular, the impact of multiple testing, should not be ignored.

13.4.4 More Complex Designs

Thus far, our discussion has focused on Phase III clinical trials comparing a new treatment to a control. When control of the frequentist Type I error probability is required in such a study, a Bayesian design has to be calibrated and at least some elements of the investigators' prior beliefs and utility function are replaced in order to produce an acceptable procedure. There are other points in the drug development process where Bayesian thinking can be applied without such interference. In early phase studies, Type I error control is not so crucial since the evidence to support a new drug application will come from later, confirmatory studies. When choosing a dose escalation rule in a Phase I first-in-human study or selecting doses for a sequence of subjects in a Phase II dose finding study, the objective is well-defined and decisions are made under uncertainty. Bayesian methods provide a natural and effective way to make such decisions; see, for example, Chapter 13 of Rosner, Laud & Johnson (2021) and references therein.

In later chapters, we shall consider the design of Phase III adaptive trials where multiple hypotheses may be tested and Type I error is to be controlled in an overall sense. Specifically a bound is set on the family-wise error rate (FWER), that is, the probability of rejecting at least one true null hypothesis. Examples of such studies include trials with multiple endpoints, seamless Phase II/III trials with treatment selection in Phase II, multi-arm multi-stage (MAMS) trials, and trials with identified sub-populations. Frequentist methods are available to control the FWER in trials with complex decision rules to add or drop treatments, or to focus attention on certain subgroups of patients. Then, freedom remains to try and define these decision rules in an optimal manner, allowing for uncertainty by integrating over the distributions of unknown parameters. The result is a hybrid procedure which controls the FWER by a standard, relatively simple procedure, and which optimizes remaining aspect of the design in a Bayesian framework. An example of this approach is the design of an adaptive enrichment trial that we shall describe in Section 28.3, where, at an interim analysis, the decision whether to restrict attention to a pre-specified sub-population is optimized by maximizing an expected utility under the posterior parameter distribution.

13.5 Bibliography and Notes

In June 1991, a conference on "Methodological and Ethical Issues in Clinical Trials" was organized jointly by the London School of Economics Centre for the Philosophy of the Natural and Social Sciences, and the Medical Section of the Royal Statistical Society. The presentations and discussion were published in a special issue of the journal *Statistics in Medicine* (1993, volume 12, issue 15/16). The conference started with presentations from Don Berry arguing for Bayesianism in clinical trials (Berry, 1993) and John Whitehead making the case for frequentism (Whitehead, 1993), and two days of lively debate followed. In his closing remarks, Peter Armitage said

> "We have heard much of Bayes and non-Bayes. I suspect that few converts have been made, but that is probably as well, since instant conversion is often less stable than a slower and more gradual change of view."

Since then, views have certainly changed. Bayesian methods have been adopted widely both by statisticians and by practitioners in a wide variety of application areas. To an extent, confirmatory clinical trials stand out as an area where strict control of Type I error rate remains an obstacle to full scale adoption of Bayesian inference. Nevertheless, there are numerous research articles on Bayesian methods. We conclude this chapter by noting a selected few of these.

First, there are substantial books on Bayesian clinical trial design, or more general applications in medicine. These include Berry, Carlin, Lee & Müller (2010), Yin (2012), Rosner, Laud & Johnson (2021) and Thall (2024). The review paper by Zhou & Ji (2024) provides an extensive bibliography with over 80 references. This paper offers three perspectives on sequential designs: the frequentist-oriented perspective, the subjective Bayesian perspective, and the calibrated Bayesian perspective. The designs presented in previous chapters fall in the first of these categories, the second category relates to methods described in Sections 13.3.1 and 13.3.4, and the third to methods described in the latter part of Section 13.3.1 and Section 13.3.5.

Emerson, Kittelson & Gillen (2007) explain how the stopping boundaries for a group sequential test designed from a frequentist perspective can be expressed in terms of posterior probabilities of hypotheses in a Bayesian model. Stallard et al. (2020) compare Bayesian and frequentist group sequential designs, noting that procedures from the two perspectives can have identical stopping rules and, given sufficient flexibility in the choice of design parameters, the classes of frequentist group sequential and Bayesian designs can be identical.

Bayes methods have applications beyond two-treatment comparison based on a single endpoint. Thall, Simon & Estey (1995) present methods for a single arm trial with multiple endpoints. Gajewski et al. (2022) describe a Bayesian group sequential design with two endpoints, in which the stopping rule focuses on the primary endpoint but ensures adequate data for the secondary endpoint. Christen et al. (2004), Orawo & Christen (2009) and Arjas & Gasbarra (2022) consider Bayesian designs for trials with multiple treatment arms in which treatments may be dropped or the whole trial terminated at an interim analysis.

Lee & Chu (2012) conducted an extensive literature review to assess how Bayesian methods are used in clinical trials. They found 121 articles in medical journals, 31 of which reported clinical trials with prospectively applied Bayesian design and analysis, 72 trials with a frequentist design and a Bayesian analysis, and 18 that used Bayesian methods to re-analyze data from a previously run clinical trial. They comment that

> "The relative merit of the Bayesian and frequentist approaches continues to be the subject of debate in statistics. However, more evidence can be found showing the convergence of the two camps, at least at the practical level."

In a similar exercise, Tidwell et al. (2019) reviewed 1020 oncology trials conducted at the University of Texas M.D. Anderson Cancer Center and found that 283 had Bayesian components. Gsponer et al. (2014) offer a tutorial in the design of Bayesian group sequential trials illustrated with examples from different phases of drug development, with various endpoints, and with informative priors.

Ryan et al. (2020) present a case study of Bayesian group sequential designs that could have been implemented in a phase III emergency medicine trial. The new designs incorporated information from previous studies in the prior distribution for the primary treatment effect and, in applying a new stopping rule to the data observed in PARAMEDIC2 trial, a decision in favor of the new treatment would have been reached with 1500 fewer patients out of a maximum sample size of 8000. Chiaruttini, Lorenzoni & Gregori (2025) present a Bayesian design for a medical device study in which information from historical data is incorporated in an informative prior. This informative prior is used as part of a mixture with a weakly informative prior to overcome a possible prior-data conflict. The authors conclude that their procedure

> "offers a robust, adaptive framework for medical device trials, balancing statistical rigor with clinical interpretability."

Chapter 14

Group Sequential Tests for Delayed Observations

14.1 The Pipeline Problem

In many clinical trials, the primary response is measured some time after the start of treatment. Thus, at each interim analysis there are subjects who have been treated but whose response is yet to be observed: we describe these patients as being "in the pipeline". In contrast, group sequential designs tend to be formulated assuming the response is immediate, or at least rapidly observed relative to the rate of recruitment of patients to the trial. Hampson & Jennison (2013) address this issue, surveying existing methods and proposing a new form of group sequential test to deal with delayed responses.

In their Example A, Hampson & Jennison describe a trial of a treatment for hypercholesterolemia in which the primary endpoint is the fall in serum cholesterol after four weeks. With 16 patients recruited per month, there will be 16 pipeline subjects at each interim analysis. If a group sequential test reaches its conclusion at an interim analysis, investigators are still expected to observe the responses of pipeline subjects. This raises the question as to how to analyze these data—and concerns that the extra data might undermine the evidence in favor of the new treatment that led to early stopping. On the other hand, having more information should, in general, be beneficial. The challenge is to use this information effectively, increasing the power to detect a positive treatment effect.

In studies with a time-to-event endpoint, right censoring is usually inevitable whenever the study terminates and, if early stopping occurs at an interim analysis, it is natural to cease follow-up of patients at that time. However, pipeline data can still be present: events may have occurred since data were locked prior to the interim analysis or there could be potential events awaiting adjudication by an external panel. A regulator may expect such data to be included in a final analysis. The methods we describe can be applied in such cases, although attention should be paid to the uncertainty about the number of events in the pipeline data and the final information level after these events are included.

In Section 14.2 we review earlier proposals for handling pipeline data and then describe Hampson & Jennison's (2013) proposal for a group sequential test that incorporates delayed responses, which they refer to as a delayed response group sequential test (DRGST). In Section 14.3, we describe error spending versions of DRGSTs which can handle unpredictable group sizes and provide an information monitoring framework to deal with nuisance parameters, such as an unknown response variance. We illustrate these methods in an error spending design for the hypercholesterolemia trial described above.

In Section 14.4 we discuss Hampson & Jennison's findings that a delay in observing the primary response tends to reduce the efficiency gains afforded by group sequential designs and we demonstrate the same pattern in error spending tests for problems with different waiting times for the final response. Hampson & Jennison noted that the loss of efficiency for larger delays can be ameliorated by incorporating data on a correlated short-term endpoint.

DOI: 10.1201/9781584888482-14

We describe their methods for fitting a joint model for the primary, long-term endpoint and a correlated short-term endpoint, and using the results to define a stopping rule for a DRGST which makes inferences on the primary endpoint.

14.2 A Delayed Response Group Sequential Test

Several methods have been suggested for incorporating data accrued after a group sequential test terminates. In Whitehead's (1992) "deletion" method, the final point of a group sequential boundary is computed ignoring the analysis at which the trial stopped and supposing the final analysis occurred with the pipeline data observed. Hall & Ding (2008) separate the final data into responses observed before and after the stopping decision and apply a combination test as proposed by Bauer & Köhne (1994). In a critical appraisal of these methods, Sooriyarachchi et al. (2003) found Whitehead's deletion method to be conservative with respect to Type I error and, with a modest amount of pipeline data, they observed that power was lower than that of a group sequential test that ignored the additional data; when the number of pipeline subjects is higher, the additional data did increase power but only by a small amount. With a moderate number of pipeline subjects, Hall & Ding's method leads to a greater loss of power than the deletion method.

In these earlier methods, the stopping rule is defined without regard to the possibility of additional data from "pipeline" subjects and, when they do occur, these data are treated as something of an afterthought. In contrast, Hampson & Jennison's (2013) delayed response group sequential tests anticipate the pipeline subjects and incorporate them into the design from the outset. In this case, the role of the stopping boundary is to decide when to terminate recruitment of patients to the trial, and an additional decision boundary point is introduced to determine whether to reject the null hypothesis after the pipeline subjects are observed.

We shall describe delayed response group sequential tests in the general setting introduced in Chapter 3 where θ represents the treatment effect and, in a one-sided test, the objective is to test H_0: $\theta \leq 0$ against $\theta > 0$ with Type I error rate α at $\theta = 0$ and power $1-\beta$ at $\theta = \delta$. In a K-stage trial, estimates $\hat{\theta}^{(k)}$ with associated information levels $\mathcal{I}_k$ are available at interim analyses $k = 1, \ldots, K-1$. If recruitment is terminated at interim analysis $k \leq K-1$, the estimate of θ after pipeline data have been observed is $\tilde{\theta}^{(k)}$ with associated information level $\tilde{\mathcal{I}}_k$. If the final analysis, K, is reached, we assume the responses of all patients recruited to the study are available at this point and we denote the final estimate of θ by $\tilde{\theta}^{(K)}$ and the associated information level by $\tilde{\mathcal{I}}_K$.

Define $Z_k = \hat{\theta}^{(k)}\sqrt{\mathcal{I}_k}$, $k = 1, \ldots, K-1$, and $\tilde{Z}_k = \tilde{\theta}^{(k)}\sqrt{\tilde{\mathcal{I}}_k}$, $k = 1, \ldots, K$. Then for $K \geq 3$, a K-stage, one-sided, delayed response group sequential test of H_0: $\theta \leq 0$ against $\theta > 0$ has the form:

At interim analysis $k = 1, \ldots, K-2$		
if $Z_k \leq a_k$ or $Z_k \geq b_k$	stop recruitment and proceed to decision analysis k,	
otherwise	continue recruitment and proceed to interim analysis $k+1$.	
At interim analysis $K-1$		
if $Z_{K-1} \leq a_{K-1}$ or $Z_{K-1} \geq b_{K-1}$	stop recruitment and proceed to decision analysis $K-1$,	(14.1)
otherwise	continue recruitment and proceed to decision analysis K.	
At decision analysis $k = 1, \ldots, K$		
if $\tilde{Z}_k \geq c_k$	reject H_0,	
if $\tilde{Z}_k < c_k$	accept H_0.	

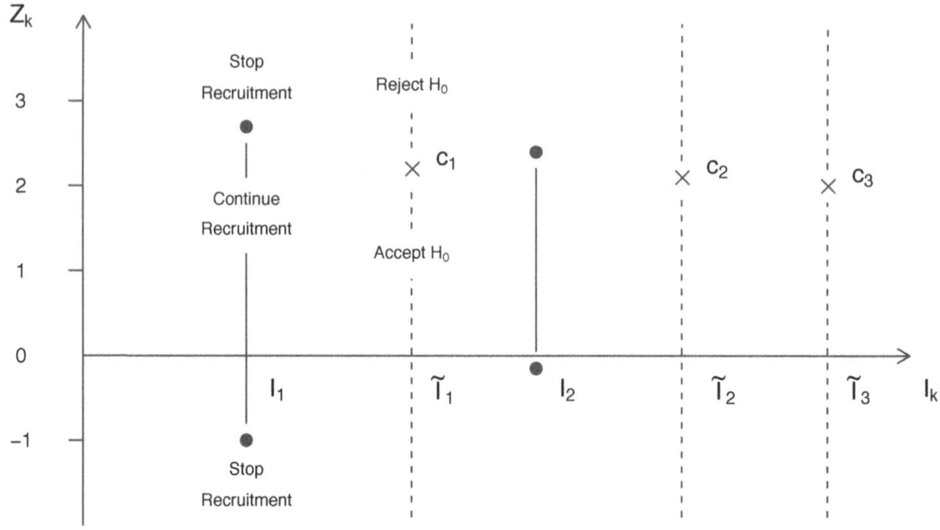

Figure 14.1 *A 3-stage delayed response group sequential test*

Figure 14.1 shows the boundary of a 3-stage delayed response group sequential test. The stopping boundaries a_k and b_k, $k = 1$ and 2, are represented by solid circles and the decision boundaries c_k, $k = 1$, 2 and 3, by crosses.

In the special case $K = 2$, the design simplifies to

At analysis 1

if $Z_1 \leq a_1$ or $Z_1 \geq b_1$	stop recruitment and proceed to decision analysis 1,
otherwise	continue recruitment and proceed to decision analysis 2. (14.2)

At decision analysis $k = 1$ or $k = 2$

if $\tilde{Z}_k \geq c_k$	reject H_0,
if $\tilde{Z}_k < c_k$	accept H_0.

We have assumed that once recruitment has been halted, it is not possible to restart recruitment and admit additional patients. We believe this is a realistic assumption for a large, multi-center trial where participating institutions are liable to disengage from a trial once it appears to have ended and renewing their interest in order to recruit more patients could be a challenging task.

As the study proceeds, at most one of the estimates $\tilde{\theta}^{(k)}$, $k = 1, \ldots, K$, will be observed and analyzed. The general theory implies that the canonical joint distribution holds for each potential sequence of estimates $\{\hat{\theta}^{(1)}, \ldots, \hat{\theta}^{(k)}, \tilde{\theta}^{(k)}\}$, $k = 1, \ldots, K-1$, and $\{\hat{\theta}^{(1)}, \ldots, \hat{\theta}^{(K-1)}, \tilde{\theta}^{(K)}\}$. Note that it is the unconditional distribution of each sequence that has the canonical form, not the distribution conditional on recruitment terminating at a particular analysis. The probability that recruitment terminates at interim analysis k and H_0 is rejected at the subsequent decision analysis can be found by integrating over sample paths under the joint distribution of $\{\hat{\theta}^{(1)}, \ldots, \hat{\theta}^{(k)}, \tilde{\theta}^{(k)}\}$. A design's Type I error probability and power can be found by combining such calculations.

In planning a K-stage study, the anticipated information levels $\mathcal{I}_k$, $k = 1, \ldots, K-1$, and $\tilde{\mathcal{I}}_k$, $k = 1, \ldots, K$, can be deduced from the expected recruitment rate and times of interim analyses. We can then calculate the Type I error probability and power of a study with these information levels and critical values (a_k, b_k), $k = 1, \ldots, K-1$, and c_k, $k = 1, \ldots, K$. Thus, in principle, it is possible to find combinations of analysis times and critical values that give a trial design with Type I error rate α at $\theta = 0$ and power $1 - \beta$ at $\theta = \delta$. For given information levels $\{\mathcal{I}_1, \tilde{\mathcal{I}}_1, \ldots, \mathcal{I}_{K-1}, \tilde{\mathcal{I}}_{K-1}, \tilde{\mathcal{I}}_K\}$, Hampson & Jennison (2013) show how to choose critical values $\{a_1, b_1, c_1, \ldots, a_{K-1}, b_{K-1}, c_{K-1}, c_K\}$ to optimize expected sample size or a combination of expected sample size and time to reach a decision, where expectation is under a certain value of θ or averaged over a distribution of θ values. It is an implicit property of such optimal designs that they make the best possible use of the data from pipeline subjects, maximizing the reduction in expected sample size or the increase in power that these data can provide. As a more practical option, Hampson & Jennison also propose error spending designs which achieve most of the efficiency of the optimized designs and it these designs that we describe next.

14.3 Error Spending Delayed Response Group Sequential Tests

We continue to consider the one-sided testing problem in which H_0: $\theta \le 0$ is tested against the alternative $\theta > 0$. The Type I error rate is set to be α under $\theta = 0$ and power $1 - \beta$ is desired when $\theta = \delta$. Error spending tests adapt to unpredictable information levels by spending Type I and Type II error probability as functions of the observed information levels. In the "maximum information" formulation, a target information level $\mathcal{I}_{max}$ is specified and it is assumed that, in the absence of early stopping, recruitment will continue until there are sufficient subjects to generate information level $\mathcal{I}_{max}$, so recruitment ceases at the first interim analysis k at which it is expected that $\tilde{\mathcal{I}}_k \ge \mathcal{I}_{\max}$. We shall return to the question of how to choose $\mathcal{I}_{max}$ to meet the power requirement after explaining how to implement an error spending design.

We start by defining error spending functions $f(t)$ and $g(t)$ for Type I and Type II errors, respectively, just as we did for the one-sided error spending tests with an immediate response discussed in Section 8.3. These functions must be non-decreasing and satisfy $f(0) = 0$, $g(0) = 0$, $f(t) = \alpha$ for $t \ge 1$ and $g(t) = \beta$ for $t \ge 1$. At interim analysis k, the values of $\mathcal{I}_1, \ldots, \mathcal{I}_k$ are known. Since the number of pipeline subjects who will contribute to $\tilde{\mathcal{I}}_k$ should also be known at this point, it is reasonable to proceed on the assumption that $\tilde{\mathcal{I}}_k$ is also known. Hampson & Jennison (2013) describe methods to create designs in which the cumulative probability under $\theta = 0$ of rejecting H_0 by analysis k is $f(\mathcal{I}_k/\mathcal{I}_{max})$ and the cumulative probability of accepting H_0 by analysis k under $\theta = \delta$ is $g(\mathcal{I}_k/\mathcal{I}_{max})$. Use of $\mathcal{I}_k$ rather than $\tilde{\mathcal{I}}_k$ here is motivated by the observation that pipeline subjects tend to make only a modest contribution to power relative to their numbers so $\mathcal{I}_k$ is the more meaningful summary of the information available at interim analysis k and the subsequent decision analysis.

In updating the design at an interim analysis k where $\tilde{\mathcal{I}}_k < \mathcal{I}_{max}$, we need to specify three constants a_k, b_k and c_k in the rules for terminating recruitment (if $Z_k < a_k$ or $Z_k > b_k$) and for deciding whether to reject H_0 at the subsequent decision analysis (if $\tilde{Z}_k \ge c_k$). With just two constraints on the cumulative Type I and Type II error probabilities to be satisfied, there is one degree of freedom left in choosing values for a_k, b_k and c_k. Hampson & Jennison propose two methods for choosing a_k, b_k and c_k. In their Method 1, they choose a_k and b_k to bring the cumulative Type I and II error probabilities up to $f(\mathcal{I}_k/\mathcal{I}_{max})$ and $g(\mathcal{I}_k/\mathcal{I}_{max})$ if the pipeline data are ignored, and then they choose c_k to maintain this cumulative Type I error probability when the pipeline data are considered, thereby reducing the Type II error probability to a value below $g(\mathcal{I}_k/\mathcal{I}_{max})$. In their Method 2, Hampson & Jennison choose b_k, as in Method 1, to bring the cumulative Type I error probability up to $f(\mathcal{I}_k/\mathcal{I}_{max})$ if the pipeline data are ignored, and then they select a_k and c_k so that the cumulative Type I and II error probabilities when the pipeline data are considered become $f(\mathcal{I}_k/\mathcal{I}_{max})$ and $g(\mathcal{I}_k/\mathcal{I}_{max})$, respectively.

When applying both Methods 1 and 2, it is quite common to find that $c_k < \Phi^{-1}(1-\alpha)$ and one might worry that a final Z-statistic $\tilde{Z} \in (c_k, \Phi^{-1}(1-\alpha))$ does not provide convincing evidence that $\theta > 0$. When discussing optimized delayed response group sequential tests, Hampson & Jennison suggest increasing c_k to $\Phi^{-1}(1-\alpha)$ to avoid this problem. This motivates a third approach for creating an error spending design, whereby c_k is set equal to $\Phi^{-1}(1-\alpha)$ and a_k and b_k are then chosen to satisfy the cumulative Type I and II error probability conditions.

Following this third approach, we shall consider error spending, delayed response group sequential tests where, at analyses $k = 1, \ldots, K-1$,

$$c_k = \Phi^{-1}(1-\alpha)$$

and a_k and b_k satisfy

$$Pr_{\theta=0}\{a_1 < Z_1 < b_1, \ldots, a_{k-1} < Z_{k-1} < b_{k-1}, (Z_k < a_k \text{ or } Z_k > b_k), \tilde{Z}_k \geq c_k\}$$

$$= f(\mathcal{I}_k/\mathcal{I}_{max}) - f(\mathcal{I}_{k-1}/\mathcal{I}_{max})$$

and

$$Pr_{\theta=\delta}\{a_1 < Z_1 < b_1, \ldots, a_{k-1} < Z_{k-1} < b_{k-1}, (Z_k < a_k \text{ or } Z_k > b_k), \tilde{Z} < c_k\}$$

$$= g(\mathcal{I}_k/\mathcal{I}_{max}) - g(\mathcal{I}_{k-1}/\mathcal{I}_{max}),$$

while at analysis K, c_K is chosen to satisfy

$$Pr_{\theta=0}\{a_1 < Z_1 < b_1, \ldots, a_{K-1} < Z_{K-1} < b_{K-1}, \tilde{Z}_K \geq c_K\} = \alpha - f(\mathcal{I}_{K-1}/\mathcal{I}_{max}).$$

Since only one critical value, c_K, is needed at the final analysis, this is chosen to fix the Type I error rate at α and the power achieved under $\theta = \delta$ will differ from the target value $1-\beta$, depending on the observed information levels. Interestingly, Lan, Lachin & Bautista (2003) had previously suggested following a group sequential error spending design with Type I error rate α then, if this test rejects H_0, applying the critical value $c_k = \Phi^{-1}(1-\alpha)$ in a final analysis with pipeline data included. The result of this approach is a conservative test of H_0 which does, at least, demonstrate that all available data have been considered in the final analysis.

Once a particular method of implementing an error spending test and error spending functions f and g have been chosen, the sample size required to achieve desired power can be found. For a given maximum sample size, n_{max}, the schedule of interim analyses must be specified. As an example, if there are expected to be m pipeline subjects at each of $K-1$ interim analyses, one may wish to conduct the interim analyses after observing $(n_{max}-m)/K, \ldots, (n_{max}-m)(K-1)/K$ responses. Then, if $\mathcal{I}_{max}$ denotes the information from n_{max} observations, the anticipated information levels are

$$\mathcal{I}_k = \frac{(n_{max}-m)}{n_{max}} \frac{k}{K} \mathcal{I}_{max}, \quad k = 1, \ldots, K-1, \tag{14.3}$$

$$\tilde{\mathcal{I}}_k = \left\{ \frac{(n_{max}-m)}{n_{max}} \frac{k}{K} + \frac{m}{n_{max}} \right\} \mathcal{I}_{max}, \quad k = 1, \ldots, K-1, \tag{14.4}$$

and

$$\tilde{\mathcal{I}}_K = \mathcal{I}_{max}. \tag{14.5}$$

Critical values $a_1, \ldots, a_{K-1}$, $b_1, \ldots, b_{K-1}$, and $c_1, \ldots, c_K$ can be found assuming these information levels and the achieved power under $\theta = \delta$ calculated. The task is then to carry out a numerical search for the value of n_{max} for which this achieved power is equal to $1 - \beta$. In practice, the information levels actually observed will vary from the assumed pattern but, as for the error spending designs of Chapter 8, power should be close to $1 - \beta$ if the target sample size of n_{max} is recruited and the associated information is close to the value of $\mathcal{I}_{max}$ assumed in the power calculation.

Hampson & Jennison (2013, Sec. 4.2) explain how one can calculate a P-value or confidence interval on termination of a delayed response group sequential test. A median unbiased estimate for the treatment effect can be obtained as the single point in a $(1 - \alpha)$-level confidence interval for θ when $\alpha = 0.5$, as described in Section 9.6. Hampson & Jennison's methods for constructing error spending designs do not generalize directly to the case of non-binding futility boundaries. Baayen, Blanche, Jennison & Ozenne (2025) address this case, presenting trial designs with a non-binding futility stopping rule and discussing the construction of a confidence interval and a median unbiased estimate for the treatment effect after such a trial.

Example: Hypercholesterolemia trial.

Consider the placebo controlled trial of a treatment for hypercholesterolemia referred to in Section 14.1. The primary endpoint, the fall in total serum cholesterol after four weeks of treatment, is assumed to be normally distributed with variance $\sigma^2 = 2$ and the treatment effect θ is the difference in mean response between the new treatment and the control. We wish to test H_0: $\theta \leq 0$ against $\theta > 0$ with Type I error rate $\alpha = 0.025$ and power $1 - \beta = 0.9$ at $\theta = 1$. A fixed sample test requires $n_f = 86$ subjects, allocated equally to the two treatments. We shall consider a group sequential design with a maximum sample size of 96, so 48 per treatment arm, to compensate for loss of power due to early stopping.

Suppose that responses are measured $\Delta T = 4$ weeks after randomization to treatment and there is no additional delay in preparing data for an analysis (if there were, we would simply increase the value of ΔT by this amount). We shall design a three-stage delayed response GST, planning ahead for responses that will be in the pipeline at each interim analysis. If subjects are recruited at a rate of 4 per week, the maximum sample size of $n_{max} = 96$ subjects will be recruited by $T_{max} = 24$ weeks and all observations will be available at 28 weeks. In this scenario, $\Delta T / T_{max} = 1/6$ and this fraction of the trial's maximum sample size, amounting to 16 observations, will be in the pipeline at each interim analysis.

In scheduling analyses, note that responses first become available after calendar time ΔT and the last subject is recruited by time T_{max}. For a K-stage design, Hampson & Jennison suggest the times of $K - 1$ potential interim analyses should be equally spaced within this interval at times

$$\Delta T + k\,(T_{max} - \Delta T)/K, \quad k = 1, \ldots, K - 1. \tag{14.6}$$

An interim analysis at T_{max} is unnecessary since the maximum number of patients will have been recruited at this point and a decision analysis at time $T_{max} + \Delta T$ is bound to follow. This pattern of analyses is equivalent to the scheme described above, leading to the information levels in (14.3), (14.4) and (14.5).

Consider a design with two interim analyses and a final analysis. The interim analyses are planned approximately 1/3 and 2/3 of the way from ΔT to T_{max} at times

$$T_1 = 11 \approx \Delta T + (T_{max} - \Delta T)/3 \quad \text{and} \quad T_2 = 17.5 \approx \Delta T + 2(T_{max} - \Delta T)/3$$

with possible decision analyses 4 weeks later at $\tilde{T}_1 = 15$ and $\tilde{T}_2 = 21.5$. We have adjusted these times to give even numbers of recruited patients so there can be equal numbers on each treatment. These timings lead to interim analyses and decision analyses with observed

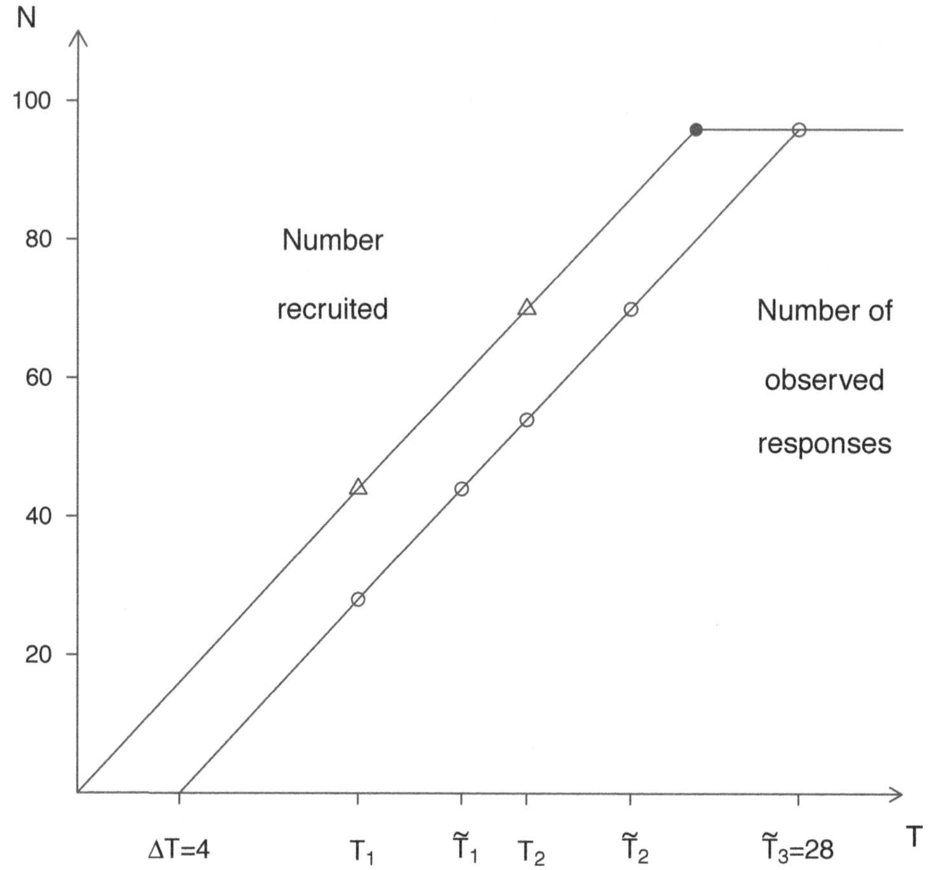

Figure 14.2 *Hypercholesterolemia trial: Numbers of patients recruited and observed responses*

numbers of responses $n_1 = 28$, $\tilde{n}_1 = 44$, $n_2 = 54$, $\tilde{n}_2 = 70$ and $\tilde{n}_3 = 96$. Figure 14.2 shows the number of subjects recruited and the number of observed responses over the course of the trial. The resulting information levels are $\mathcal{I}_1 = 3.5$, $\tilde{\mathcal{I}}_1 = 5.5$, $\mathcal{I}_2 = 6.75$, $\tilde{\mathcal{I}}_2 = 8.75$ and $\tilde{\mathcal{I}}_3 = 12.0$.

With Rho-family error spending functions of the form

$$f(t) = \min\{\alpha t^{\rho}, \alpha\} \quad \text{and} \quad g(t) = \min\{\beta t^{\rho}, \beta\},$$

as described in Section 8.3.2, we need $\rho = 1.219$ in order for Type II error probability under $\theta = \delta = 1$ to be 0.1. In this case, the critical values of the design are

$$(a_1, b_1) = (-0.380, 2.362),\ c_1 = 1.96\ \ (a_2, b_2) = (0.682, 2.222),\ c_2 = 1.96,\ \ c_3 = 2.049.$$

Figure 14.3 shows this boundary, which we see has the form of the generic boundary displayed in Figure 14.1.

It is instructive to know how the presence of pipeline subjects has affected the efficiency of a group sequential test. To do this, we can compare properties of the delayed response group sequential test with those of a fixed sample size design with 86 subjects that achieves the same Type I and Type II error rates and also a group sequential test that could have been applied if the response had been observed immediately after treatment. In the case of an

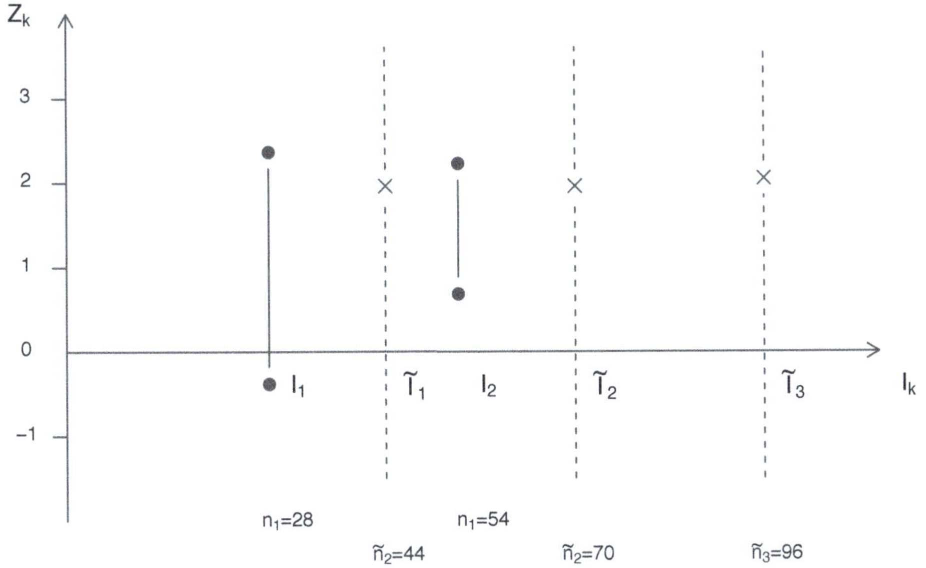

Figure 14.3 *Hypercholesterolemia trial: a 3-stage delayed response group sequential test*

immediate response, we have considered a group sequential test defined through Rho-family error spending functions: with a maximum sample size of 96 and equally spaced analyses, using Rho-family error spending functions with $\rho = 1.194$ produces a design that achieves Type I error probability 0.025 and power 0.8 when $\theta = 1$. The expected sample size functions $E_\theta(N)$ for these three designs are shown in Figure 14.4. We see there that the delayed response group sequential test is able to reduce the expected sample size below that of the fixed sample size design but the savings are between a third and two thirds of those of the group sequential test for an immediate response.

The efficiency of a delayed response group sequential test decreases as the number of pipeline subjects increases. As an overall measure of efficiency, we have calculated an average expected sample size

$$\text{Average } E_\theta(N) = \int \pi(\theta)\, E_\theta(N)\, d\theta,$$

where $\pi(\theta)$ is the density of a normal distribution for θ with mean 0.5 and standard deviation 0.5. Figure 14.5 shows values of this average expected sample size expressed as a percentage of the fixed sample size, $n_f = 86$, when the number of pipeline subject rises from 0 to 50. In the case we have considered, with $K = 3$ analyses and 16 pipeline subjects, the value of $100\,\text{Average } E_\theta(N)/n_f$ is 83.83 compared to the value of 70.1 for $K = 3$ analyses and an immediate response. As for the case of an immediate response, efficiency increases with the number of analyses. The results seen in Figure 14.5 are very similar to those shown in Figure 3 of Hampson & Jennison (2013) for optimized delayed response group sequential tests, demonstrating the efficiency of designs base on Rho-family error spending functions.

14.4 Utilizing Information on a Short-term Endpoint

Hampson & Jennison (2013, Sec. 5) propose using early information from pipeline subjects as a means to improve the efficiency of delayed response group sequential tests. They consider the case where a short-term response is recorded for each subject in addition to the primary

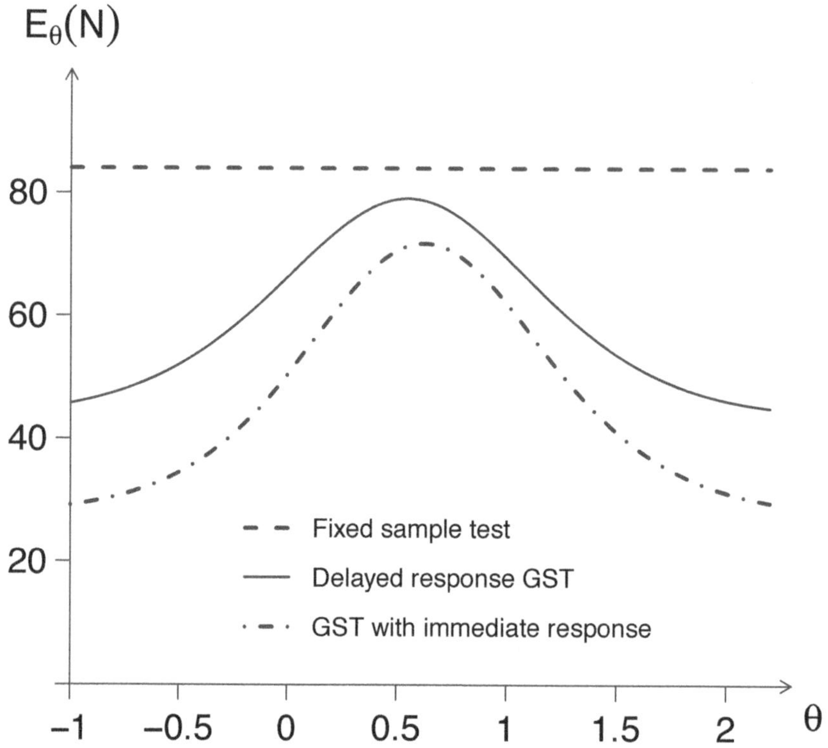

Figure 14.4 *Expected sample size functions for designs for the hypercholesterolemia trial*

endpoint so that, at interim analyses, the short-term response will be available for some of the pipeline subjects.

Let $(Y_{T,i}, X_{T,i})$ denote the pair of endpoints for subject i on treatment T for $T = A$ and B, and $i = 1, 2, \ldots$, where $Y_{T,i}$ is the short-term endpoint and $X_{T,i}$ the primary endpoint. Hampson & Jennison suppose $(Y_{T,i}, X_{T,i})$ has a bivariate normal distribution,

$$\begin{pmatrix} Y_{T,i} \\ X_{T,i} \end{pmatrix} \sim N\left\{ \begin{pmatrix} \mu_{T,1} \\ \mu_{T,2} \end{pmatrix}, \begin{pmatrix} \sigma_1^2 & \tau\sigma_1\sigma_2 \\ \tau\sigma_1\sigma_2 & \sigma_2^2 \end{pmatrix} \right\}.$$

If σ_1^2, σ_2^2 and τ are known, the parameter vector is $\boldsymbol{\beta} = (\mu_{A,1}, \mu_{B,1}, \mu_{A,2}, \mu_{B,2})$ and the treatment effect of interest is $\theta = \mu_{A,2} - \mu_{B,2}$. At interim analysis k, maximum likelihood estimation is used to compute the estimate $\widehat{\boldsymbol{\beta}}^{(k)}$ and its variance matrix $\mathrm{Var}(\widehat{\boldsymbol{\beta}}^{(k)})$, and hence $\hat{\theta}^{(k)}$ and $\mathcal{I}_k = \mathrm{Var}(\hat{\theta}^{(k)})$. If recruitment ceases at analysis k, the trial continues to decision analysis k, where the primary endpoint is available for all subjects and $\tilde{\theta}^{(k)}$ and $\tilde{\mathcal{I}}_k$ depend only on this endpoint. A delayed response group sequential test can be created in the usual way from the sequences $\{\hat{\theta}^{(1)}, \hat{\theta}^{(2)}, \ldots\}$ and $\{\mathcal{I}_1, \mathcal{I}_2, \ldots\}$ and the values of $\tilde{\theta}^{(k)}$ and $\tilde{\mathcal{I}}_k$ at the relevant decision analysis.

The incorporation of short-term data at analysis k increases the information $\mathcal{I}_k$ for the primary endpoint. This is simply due to the correlation τ between the two endpoints: it does not depend on any association between the means $\mu_{A,1}$ and $\mu_{A,2}$ on treatment A and $\mu_{B,1}$ and $\mu_{B,2}$ on treatment B. Hampson & Jennison show that this approach can be effective in terms of improving efficiency, particularly when $|\tau|$ is large and the short term endpoint is

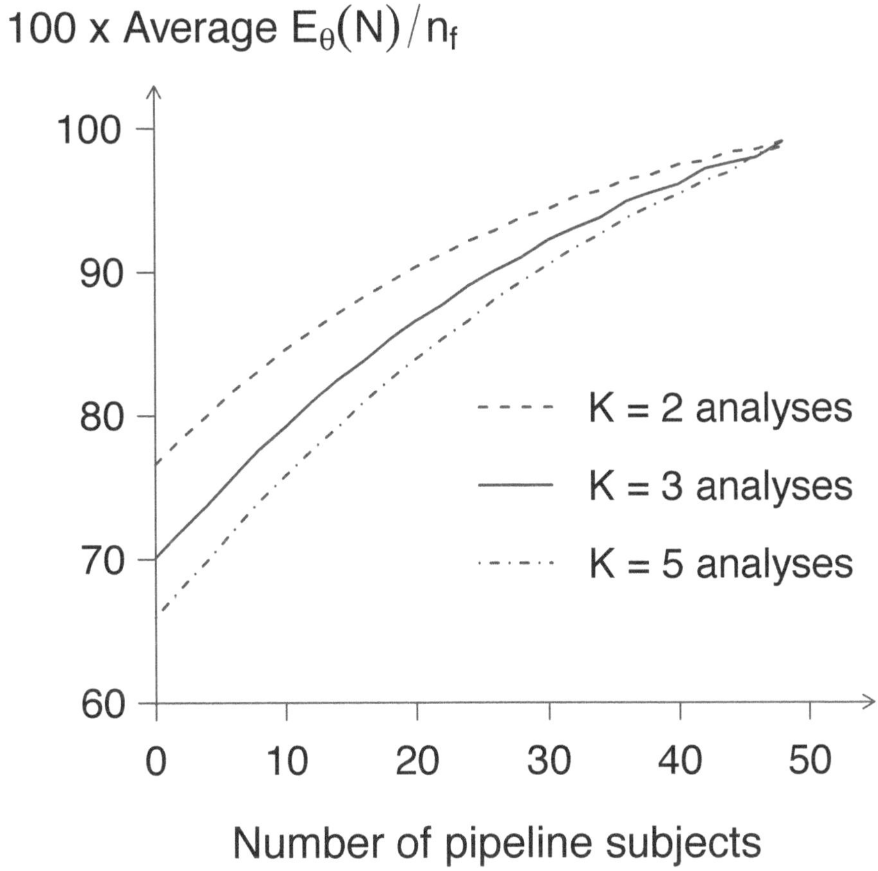

Figure 14.5 *Loss of efficiency of delayed response group sequential tests as pipeline size increases. Tests have a maximum sample size of* 96 *and Type I error probability* 0.025, *and are constructed using Rho-family error spending functions with the value of* ρ *chosen so that power is* 0.8 *when* $\theta = 1$.

measured early in the trial, so that the value of this endpoint is known for a large fraction of pipeline subjects. They also consider the case where σ_1^2, σ_2^2 and τ are unknown. They define information monitoring procedures which include sample size re-estimation as more accurate estimates of σ_2^2 are obtained and they conclude from a simulation study that this methodology is robust to uncertainty about the correlation parameter τ.

Kittelson, Sharples & Emerson (2005) have described methods for analyzing a clinical trial with longitudinal data that may include multiple repeated measurements of the key response variable. They propose several definitions for the overall treatment effect based on weighted combinations of the difference in mean response at each time point. Under an assumed multivariate normal model, maximum likelihood estimates of the overall treatment effect follow the canonical joint distribution of Section 3.1. Although Kittelson, Sharples & Emerson do not consider incorporating pipeline data obtained after stopping the trial into a final analysis, it would be a straightforward task to extend the methods we have described to the general case of multiple repeated measurements and an overall treatment effect defined as a linear combination of treatment differences.

14.5 Bibliography and Notes

Anderson (1964) noted the need to accommodate delayed responses in the final analysis of a sequential procedure. He described a method for doing this when a sequential probability ratio test was used to decide when to stop sampling but commented that a more general approach to the delayed response problem and the determination of an optimal stopping rule was desirable. Hampson & Jennison (2013) followed the general framework described by Anderson (1964) in deriving optimal stopping criteria and rules. They also proposed flexible error spending designs which achieve almost all the efficiency of the optimal designs.

Tsiatis & Davidian (2022) consider clinical trials where the outcome for each subject becomes known some time after treatment but this time-lag can differ between patients and, due to right censoring, only partial information is available for some subjects at each interim analysis. They use semiparametric theory to prove that estimates of the overall treatment effect follow the canonical joint distribution and, hence, construct group sequential tests. Given this joint distribution, it would not be difficult to create delayed response group sequential tests that would incorporate the complete data on all the study participants.

In discussing possible trial designs, we have supposed it would be impractical to restart recruitment once a trial has been stopped. If such a restart is permitted, more elaborate trial designs are possible and Schüürhuis, Konietschke & Kunz (2024) present methods that may be used in this case.

We conclude this chapter with a note on terminology. We have used the term "delayed response" to refer to an endpoint that is measured some time after treatment commences. In other cases, this term is used to describe a difference between treatments that occurs after some delay. As an example, in a study with overall mortality as the primary endpoint, survival curves for treatment and control groups may be very similar initially but start to separate several months after treatment; see, for example, Ghosh et al. (2022). The observations we describe as "pipeline data" are sometimes called "overrun data", but that name is also given to observations that take the observed information beyond the target $\mathcal{I}_{max}$ in a maximum information error spending design. We recommend use of the term "pipeline data" as this appears to be unambiguous. For a comprehensive discussion of terminology in clinical trials, see the paper by Asikanius et al. (2025).

Chapter 15

General Group Sequential Distribution Theory

15.1 Introduction

In Chapter 3 we showed that a wide variety of situations and data types lead to sequences of statistics with a common form of joint distribution. Consequently, the group sequential methods presented in Chapters 2 to 14 are widely applicable. In this chapter we prove general results which broaden further the applicability of these methods.

If observations follow a normal linear model, the distribution theory for a sequence of estimates is exact. We derive this theory in Section 15.3 and show that it extends to correlated observations, such as repeated measurements on subjects in a longitudinal study, as long as the correlation structure is known. If the covariance matrix in a normal linear model is completely known, Z-statistics can be created and these will have the canonical joint distribution (3.1). In most applications, a variance parameter σ^2 must be estimated from the observed data. Group sequential tests may then be based on a sequence of t-statistics and approximate procedures for doing this have been presented in Chapters 4 to 8. The results we shall present in Section 15.4 show that the joint distributions of sequences of t-statistics conform to a standard pattern and exact group sequential methods can be based on this joint distribution.

For non-normal data, the theory is asymptotic as sample size increases, but it usually provides a good approximation in small samples and this can serve as the basis of group sequential tests for parameters in, for example, a generalized linear model. We have already described such applications in Sections 3.4, 4.4 and 11.6. The theory for general parametric models is presented in Section 15.6. Statistics arising in the group sequential analysis of survival data share similar asymptotic properties but have a rather different derivation. In Chapter 17 we summarize theory for the log-rank test, which underlies the example in Section 4.5, and for semi-parametric Cox regression models for survival data subject to censoring.

15.2 A Standard Joint Distribution for Successive Estimates of a Parameter Vector

Consider a group sequential study with a maximum of K groups of observations. Let $\widehat{\boldsymbol{\beta}}^{(k)}$, $k = 1, \ldots, K$, denote the estimates of the parameter vector $\boldsymbol{\beta} = (\beta_1, \ldots, \beta_p)^T$ at successive analyses. It is not uncommon to find, at least approximately, that $(\widehat{\boldsymbol{\beta}}^{(1)}, \ldots, \widehat{\boldsymbol{\beta}}^{(K)})$ is multivariate normal with

$$\begin{cases} \widehat{\boldsymbol{\beta}}^{(k)} \sim N(\boldsymbol{\beta}, \mathcal{I}^{-1}(k, \boldsymbol{\beta})), & k = 1, \ldots, K, \\ \mathbf{Cov}(\widehat{\boldsymbol{\beta}}^{(k_1)}, \widehat{\boldsymbol{\beta}}^{(k_2)}) = \mathbf{Var}(\widehat{\boldsymbol{\beta}}^{(k_2)}) = \mathcal{I}^{-1}(k_2, \boldsymbol{\beta}), & k_1 \le k_2, \end{cases} \tag{15.1}$$

for certain $p \times p$ matrices $\mathcal{I}^{-1}(k, \boldsymbol{\beta})$, $k = 1, \ldots, K$. This is the multi-parameter generalization of the distribution of a sequence of scalar parameter estimates, $\hat{\theta}^{(k)}$, given in (3.2). It should

DOI: 10.1201/9781584888482-15

be noted that (15.1) is the joint distribution of $\{\widehat{\boldsymbol{\beta}}^{(1)}, \ldots, \widehat{\boldsymbol{\beta}}^{(K)}\}$ and $\mathcal{I}(k, \boldsymbol{\beta})$ is the inverse of the variance matrix of $\widehat{\boldsymbol{\beta}}^{(k)}$ when no stopping rule is applied. Properties of group sequential tests in which early stopping does occur follow from this distribution.

Suppose we wish to test the hypothesis H_0: $\boldsymbol{c}^T\boldsymbol{\beta} = \gamma$ for a given p-vector $\boldsymbol{c}$ and scalar constant γ. Define $\theta = \boldsymbol{c}^T\boldsymbol{\beta} - \gamma$ and

$$\mathcal{I}_k = \{\mathrm{Var}(\boldsymbol{c}^T\widehat{\boldsymbol{\beta}}^{(k)})\}^{-1} = (\boldsymbol{c}^T\mathcal{I}^{-1}(k, \boldsymbol{\beta})\boldsymbol{c})^{-1},$$

the Fisher information for $\boldsymbol{c}^T\boldsymbol{\beta}$ at analysis k. A group sequential test of H_0 can be based on the score statistics

$$S_k = \mathcal{I}_k(\boldsymbol{c}^T\widehat{\boldsymbol{\beta}}^{(k)} - \gamma), \quad k = 1, \ldots, K,$$

which follow a multivariate normal distribution with

$$\begin{cases} S_k \sim N(\theta\mathcal{I}_k, \mathcal{I}_k), & k = 1, \ldots, K, \\ \mathrm{Cov}(S_{k_1}, S_{k_2}) = \mathrm{Var}(S_{k_1}) = \mathcal{I}_{k_1}, & k_1 \le k_2, \end{cases} \tag{15.2}$$

as in (3.2). Alternatively, a test can be defined in terms of the Wald statistics,

$$Z_k = (\boldsymbol{c}^T\widehat{\boldsymbol{\beta}}^{(k)} - \gamma)\sqrt{\mathcal{I}_k}, \quad k = 1, \ldots, K,$$

which are jointly multivariate normal with

$$\begin{cases} Z_k \sim N(\theta\sqrt{\mathcal{I}_k}, 1), & k = 1, \ldots, K, \\ \mathrm{Cov}(Z_{k_1}, Z_{k_2}) = \sqrt{(\mathcal{I}_{k_1}/\mathcal{I}_{k_2})}, & k_1 \le k_2. \end{cases} \tag{15.3}$$

It is evident from (15.2) that the increments $S_1, S_2 - S_1, \ldots, S_K - S_{K-1}$ are independent. Hence, both sequences $S_1, \ldots, S_K$ and $Z_1, \ldots, Z_K$ are Markov. This property greatly simplifies the numerical computations for group sequential tests described in Chapter 31. Without the Markov structure, a general multivariate normal integration routine would be needed and calculation time would increase rapidly with the number of analyses K.

The distribution (15.3) of the standardized statistics Z_k is the canonical joint distribution (3.1). We showed this distribution to apply in the examples of parallel treatment studies and crossover trials in Sections 3.2.1 to 3.2.4. We now prove a theorem that shows these results hold for much more general linear mixed-effects normal models. This theorem helps to explain why the standard pattern occurs so frequently and also indicates when other behavior is liable to be found.

15.3 Normal Linear Models

We suppose univariate observations $X_1, X_2, \ldots$ are normally distributed with means depending linearly on an unknown parameter vector $\boldsymbol{\beta} = (\beta_1, \ldots, \beta_p)^T$. In a group sequential study with a maximum of K groups of observations, we denote the total number of observations in the first k groups by n_k, where $n_1 < \ldots < n_K$, and we denote the vector of observations available at the kth analysis by

$$\boldsymbol{X}^{(k)} = (X_1, \ldots, X_{n_k})^T, \quad k = 1, \ldots, K.$$

We assume the full vector of n_K observations, $\boldsymbol{X}^{(K)}$, has a multivariate normal distribution with design matrix $\boldsymbol{D}^{(K)}$ and non-singular covariance matrix $\boldsymbol{\Sigma}^{(K)}\sigma^2$, where $\boldsymbol{D}^{(K)}$ and $\boldsymbol{\Sigma}^{(K)}$ are known. At each analysis $k = 1, \ldots, K$, we observe

$$\boldsymbol{X}^{(k)} \sim N(\boldsymbol{D}^{(k)}\boldsymbol{\beta}, \boldsymbol{\Sigma}^{(k)}\sigma^2),$$

where $\boldsymbol{D}^{(k)}$ and $\boldsymbol{\Sigma}^{(k)}$ can be deduced from $\boldsymbol{D}^{(K)}$ and $\boldsymbol{\Sigma}^{(K)}$ by extracting the elements relating to the first n_k components of $\boldsymbol{X}^{(K)}$. We assume $\boldsymbol{\beta}$ is estimable at each stage $k = 1, \ldots, K$. In fact this may not always be true, especially at the earliest stages but we postpone discussion of this case to the end of the section. If $\boldsymbol{\beta}$ is estimable from $\boldsymbol{X}^{(k)}$, the maximum likelihood estimate (MLE) at stage k based on $\boldsymbol{X}^{(k)}$ is the generalized least squares estimate (LSE)

$$\widehat{\boldsymbol{\beta}}^{(k)} = (\boldsymbol{D}^{(k)T}\boldsymbol{\Sigma}^{(k)^{-1}}\boldsymbol{D}^{(k)})^{-1}\boldsymbol{D}^{(k)T}\boldsymbol{\Sigma}^{(k)^{-1}}\boldsymbol{X}^{(k)}, \quad k = 1, \ldots, K. \tag{15.4}$$

Theorem 15.1. Under the above model, the vectors $\widehat{\boldsymbol{\beta}}^{(1)}, \ldots, \widehat{\boldsymbol{\beta}}^{(K)}$ follow the multivariate normal joint distribution given by (15.1) with

$$\mathcal{I}^{-1}(k, \boldsymbol{\beta}) \; = \; \mathbf{Var}(\widehat{\boldsymbol{\beta}}^{(k)}) \; = \; \sigma^2(\boldsymbol{D}^{(k)T}\boldsymbol{\Sigma}^{(k)^{-1}}\boldsymbol{D}^{(k)})^{-1} \; = \; \sigma^2\boldsymbol{V}_k, \tag{15.5}$$

for $k = 1, \ldots, K$, where $\boldsymbol{V}_k = (\boldsymbol{D}^{(k)T}\boldsymbol{\Sigma}^{(k)^{-1}}\boldsymbol{D}^{(k)})^{-1}$ does not depend on $\boldsymbol{\beta}$. In particular,

$$\mathbf{Cov}(\widehat{\boldsymbol{\beta}}^{(k_1)}, \widehat{\boldsymbol{\beta}}^{(k_2)}) \; = \; \mathbf{Var}(\widehat{\boldsymbol{\beta}}^{(k_2)}) \quad \text{for } k_1 \le k_2.$$

Proof. As each $\widehat{\boldsymbol{\beta}}^{(k)}$ is a linear function of $\boldsymbol{X}^{(K)}$, elements of the vectors $\widehat{\boldsymbol{\beta}}^{(1)}, \ldots, \widehat{\boldsymbol{\beta}}^{(K)}$ have a multivariate normal joint distribution. It is well known that the marginal distribution of each $\widehat{\boldsymbol{\beta}}^{(k)}$ is that given in the theorem so it remains to determine the covariances of $\widehat{\boldsymbol{\beta}}^{(1)}, \ldots, \widehat{\boldsymbol{\beta}}^{(K)}$. Now, for $k_1 < k_2$, the data seen at analysis k_1 are a subset of the data at analysis k_2. We can, therefore, write

$$\widehat{\boldsymbol{\beta}}^{(k_1)} = \boldsymbol{A}^T\boldsymbol{X}^{(k_2)}$$

for some $n_{k_2} \times p$ matrix $\boldsymbol{A}$. Since $\widehat{\boldsymbol{\beta}}^{(k_1)}$ is an unbiased estimate of $\boldsymbol{\beta}$,

$$E(\boldsymbol{A}^T\boldsymbol{X}^{(k_2)}) \; = \; \boldsymbol{A}^T\boldsymbol{D}^{(k_2)}\boldsymbol{\beta} \; = \; \boldsymbol{\beta} \quad \text{for all } \boldsymbol{\beta}$$

and it follows that $\boldsymbol{A}^T\boldsymbol{D}^{(k_2)} = \boldsymbol{I}_p$, where $\boldsymbol{I}_p$ is the $p \times p$ identity matrix. Hence,

$$\begin{aligned}
\mathbf{Cov}(\widehat{\boldsymbol{\beta}}^{(k_1)}, \widehat{\boldsymbol{\beta}}^{(k_2)}) &= \mathbf{Cov}(\,\boldsymbol{A}^T\boldsymbol{X}^{(k_2)}, (\boldsymbol{D}^{(k_2)T}\boldsymbol{\Sigma}^{(k_2)^{-1}}\boldsymbol{D}^{(k_2)})^{-1}\boldsymbol{D}^{(k_2)T}\boldsymbol{\Sigma}^{(k_2)^{-1}}\boldsymbol{X}^{(k_2)}\,) \\
&= \boldsymbol{A}^T\,\mathbf{Var}(\boldsymbol{X}^{(k_2)})\,\boldsymbol{\Sigma}^{(k_2)^{-1}}\boldsymbol{D}^{(k_2)}(\boldsymbol{D}^{(k_2)T}\boldsymbol{\Sigma}^{(k_2)^{-1}}\boldsymbol{D}^{(k_2)})^{-1} \\
&= (\boldsymbol{D}^{(k_2)T}\boldsymbol{\Sigma}^{(k_2)^{-1}}\boldsymbol{D}^{(k_2)})^{-1}\sigma^2 \\
&= \mathbf{Var}(\widehat{\boldsymbol{\beta}}^{(k_2)}),
\end{aligned}$$

as required. □

It follows from Theorem 15.1 that, for $k \neq k'$,

$$\mathbf{Cov}(\boldsymbol{V}_k^{-1}\widehat{\boldsymbol{\beta}}^{(k)} - \boldsymbol{V}_{k-1}^{-1}\widehat{\boldsymbol{\beta}}^{(k-1)}, \boldsymbol{V}_{k'}^{-1}\widehat{\boldsymbol{\beta}}^{(k')} - \boldsymbol{V}_{k'-1}^{-1}\widehat{\boldsymbol{\beta}}^{(k'-1)}) = \mathbf{0}.$$

Hence, the process $\{\boldsymbol{V}_k^{-1}\widehat{\boldsymbol{\beta}}^{(k)};\; k = 1, \ldots, K\}$ has independent increments and we deduce that $\{\widehat{\boldsymbol{\beta}}^{(k)};\; k = 1, \ldots, K\}$ is a Markov sequence.

Consider the problem of testing the hypothesis H_0: $\boldsymbol{c}^T\boldsymbol{\beta} = \gamma$ for a given p-vector $\boldsymbol{c}$ and scalar constant γ. The standardized test statistic at analysis k is

$$Z_k \; = \; (\boldsymbol{c}^T\widehat{\boldsymbol{\beta}}^{(k)} - \gamma)\sqrt{\mathcal{I}_k}, \quad k = 1, \ldots, K, \tag{15.6}$$

where

$$\mathcal{I}_k = \{\mathrm{Var}(\boldsymbol{c}^T\widehat{\boldsymbol{\beta}}^{(k)})\}^{-1} = (\sigma^2\boldsymbol{c}^T\boldsymbol{V}_k\boldsymbol{c})^{-1} \tag{15.7}$$

is the Fisher information for $\boldsymbol{c}^T\boldsymbol{\beta}$ at analysis k. Theorem 15.1 shows that $(Z_1, \ldots, Z_K)$ has the canonical joint distribution (3.1) and so can be used with any of the group sequential procedures of Chapters 2 to 14. In particular, this justifies the two-sided tests for normal linear models presented in Section 3.3.

It remains to discuss the case where $\boldsymbol{c}^T\boldsymbol{\beta}$ is estimable but the whole of $\boldsymbol{\beta}$ cannot be estimated at the earliest analyses. This is likely to occur in longitudinal studies when initial analyses take place before accrual is complete and subject effects are not estimable for individuals yet to enter the study. An example was given in Section 3.3.4. In such a case, we can still use the same procedure based on the statistics $Z_1, \ldots, Z_K$ defined in (15.6) to test the hypothesis H_0: $\boldsymbol{c}^T\boldsymbol{\beta} = \gamma$, but now inverse matrices $\boldsymbol{V}_k = (\boldsymbol{D}^{(k)^T}\boldsymbol{\Sigma}^{(k)^{-1}}\boldsymbol{D}^{(k)})^{-1}$ in (15.4), (15.5) and (15.7) should be interpreted as generalized inverses. As long as $\boldsymbol{c}^T\boldsymbol{\beta}$ is estimable at each stage k, $\boldsymbol{c}^T\widehat{\boldsymbol{\beta}}^{(k)}$ and Z_k will be uniquely defined even if $\widehat{\boldsymbol{\beta}}^{(k)}$ is not.

Theorem 15.1 covers uncorrelated and correlated normal observations, including as special cases normal linear mixed-effects models with their particular covariance matrices. The covariance properties of estimates derived by Halperin et al. (1987), Reboussin et al. (1992) and Wu & Lan (1992) follow from this result. The restriction of the theorem to the sequence of MLEs, or equivalently generalized LSEs, is critical and explains why the estimates used by Armitage, Stratton & Worthington (1985, Sec. 4) and Geary (1988), which are ordinary LSEs rather than generalized LSEs, do not follow the standard pattern (15.1)

15.4 Normal Linear Models with Unknown Variance: Group Sequential *t*-Tests

We consider the same model as in the previous section, but now suppose the scale factor σ^2 in the covariance matrix is unknown. If $\boldsymbol{\beta}$ is estimable from $\boldsymbol{X}^{(k)}$, the MLE at stage k is still the generalized LSE given in (15.4). If $n_k > p$, the standard unbiased estimate of σ^2 based on the residuals at analysis k is

$$s_k^2 = (\boldsymbol{X}^{(k)} - \boldsymbol{D}^{(k)}\widehat{\boldsymbol{\beta}}^{(k)})^T\boldsymbol{\Sigma}^{(k)^{-1}}(\boldsymbol{X}^{(k)} - \boldsymbol{D}^{(k)}\widehat{\boldsymbol{\beta}}^{(k)})/(n_k - p), \tag{15.8}$$

which is marginally distributed as $\sigma^2\chi^2_{n_k-p}/(n_k - p)$ for each $k = 1, \ldots, K$. Suppose we are interested again in testing the hypothesis H_0: $\boldsymbol{c}^T\boldsymbol{\beta} = \gamma$ for a given p-vector $\boldsymbol{c}$ and scalar constant γ. At each analysis $k = 1, \ldots, K$, the estimate of $\boldsymbol{c}^T\boldsymbol{\beta}$ is

$$\boldsymbol{c}^T\widehat{\boldsymbol{\beta}}^{(k)} \sim N(\boldsymbol{c}^T\boldsymbol{\beta}, \boldsymbol{c}^T\boldsymbol{V}_k\boldsymbol{c}\,\sigma^2),$$

and the t-statistic for testing H_0 is

$$T_k = \frac{\boldsymbol{c}^T\widehat{\boldsymbol{\beta}}^{(k)} - \gamma}{\sqrt{\{\boldsymbol{c}^T\boldsymbol{V}_k\boldsymbol{c}\, s_k^2\}}}. \tag{15.9}$$

Marginally, T_k has a Student's t-distribution on $n_k - p$ degrees of freedom under H_0.

We have described how the significance level approach can be used to construct approximate two-sided tests (Section 4.6), one-sided tests (Section 5.6), tests with inner wedges (Section 6.2.4), and equivalence tests (Section 7.4.3) from a sequence of t-statistics T_k, $k = 1, \ldots, K$. However, to obtain exact properties for such tests we need to derive the exact joint distribution of $T_1, \ldots, T_K$. Unlike $Z_1, \ldots, Z_K$, this sequence of t-statistics is *not* Markov. Jennison & Turnbull (1997b, p. 305) present an argument showing why this is the case. However, the sequence of pairs $\{(\boldsymbol{c}^T\widehat{\boldsymbol{\beta}}^{(k)}, s_k^2);\ k = 1, \ldots, K\}$ *is* Markov and the following theorem establishes its joint distribution. This result can be used to compute exact properties of group sequential t-tests.

Theorem 15.2. Suppose

$$\boldsymbol{X}^{(K)} = (X_1, \ldots, X_{n_K})^T \sim N(\boldsymbol{D}^{(K)}\boldsymbol{\beta}, \boldsymbol{\Sigma}^{(K)}\sigma^2)$$

where the covariance matrix $\boldsymbol{\Sigma}^{(K)}\sigma^2$ is non-singular and at each analysis $k = 1, \ldots, K$ the first n_k elements of $X^{(K)}$ are available. Suppose also that $\boldsymbol{\beta}$ is estimable from $\boldsymbol{X}^{(1)} = (X_1, \ldots, X_{n_1})^T$ and let $\widehat{\boldsymbol{\beta}}^{(k)}$, $\boldsymbol{V}_k = \mathbf{Var}(\widehat{\boldsymbol{\beta}}^{(k)})/\sigma^2$ and s_k^2, $k = 1, \ldots, K$, be as defined by (15.4), (15.5) and (15.8), respectively.

Then, the sequence $\{(\boldsymbol{c}^T\widehat{\boldsymbol{\beta}}^{(k)}, s_k^2);\ k = 1, \ldots, K\}$ is Markov,

$$\boldsymbol{c}^T\widehat{\boldsymbol{\beta}}^{(1)} \sim N(\boldsymbol{c}^T\boldsymbol{\beta}, \boldsymbol{c}^T\boldsymbol{V}_1\boldsymbol{c}\sigma^2), \quad s_1^2 \sim \sigma^2\chi^2_{n_1-p}/(n_1 - p)$$

and, for $k = 1, \ldots, K-1$,

$$\boldsymbol{c}^T\widehat{\boldsymbol{\beta}}^{(k+1)} \mid \boldsymbol{c}^T\widehat{\boldsymbol{\beta}}^{(k)}, s_k^2 \sim N\left(\boldsymbol{c}^T\boldsymbol{\beta} + \frac{\boldsymbol{c}^T\boldsymbol{V}_{k+1}\boldsymbol{c}}{\boldsymbol{c}^T\boldsymbol{V}_k\boldsymbol{c}}\,\boldsymbol{c}^T(\widehat{\boldsymbol{\beta}}^{(k)} - \boldsymbol{\beta}),\ \boldsymbol{c}^T\boldsymbol{V}_{k+1}\boldsymbol{c}\sigma^2 - \frac{(\boldsymbol{c}^T\boldsymbol{V}_{k+1}\boldsymbol{c})^2}{\boldsymbol{c}^T\boldsymbol{V}_k\boldsymbol{c}}\,\sigma^2\right)$$

and

$$(n_{k+1} - p)s_{k+1}^2 \mid \boldsymbol{c}^T\widehat{\boldsymbol{\beta}}^{(k+1)}, \boldsymbol{c}^T\widehat{\boldsymbol{\beta}}^{(k)}, s_k^2 \sim (n_k - p)s_k^2 +$$

$$\{\boldsymbol{c}^T(\widehat{\boldsymbol{\beta}}^{(k+1)} - \widehat{\boldsymbol{\beta}}^{(k)})\}^2\{\boldsymbol{c}^T(\boldsymbol{V}_k - \boldsymbol{V}_{k+1})\boldsymbol{c}\}^{-1} + \sigma^2\chi^2_{n_{k+1}-n_k-1}$$

provided $\boldsymbol{c}^T(\widehat{\boldsymbol{\beta}}^{(k+1)} - \widehat{\boldsymbol{\beta}}^{(k)})$ is non-degenerate. If $\boldsymbol{c}^T\boldsymbol{\beta}$ is not estimable from the observations added between analyses k and $k+1$, then $\boldsymbol{c}^T\widehat{\boldsymbol{\beta}}^{(k+1)} = \boldsymbol{c}^T\widehat{\boldsymbol{\beta}}^{(k)}$ and

$$(n_{k+1} - p)s_{k+1}^2 \mid \boldsymbol{c}^T\widehat{\boldsymbol{\beta}}^{(k+1)}, \boldsymbol{c}^T\widehat{\boldsymbol{\beta}}^{(k)}, s_k^2 \sim (n_k - p)s_k^2 + \sigma^2\chi^2_{n_{k+1}-n_k}.$$

The proof of this theorem, which is rather lengthy, can be found in Jennison & Turnbull (1997b). They also treat the case where $\boldsymbol{c}^T\boldsymbol{\beta}$ is estimable but the whole of $\boldsymbol{\beta}$ is not estimable at early analyses. Note that σ^2 appears as a scale factor in the variance of each $\boldsymbol{c}^T\widehat{\boldsymbol{\beta}}^{(k)}$ and in the distribution of each s_k^2, $k = 1, \ldots, K$. Thus, the distribution of the sequence $\{T_k;\ k = 1, \ldots, K\}$ is independent of σ^2 under H_0: $\boldsymbol{c}^T\boldsymbol{\beta} = \gamma$, and any convenient value of σ^2, such as $\sigma^2 = 1$, can be used in calculating properties of a group sequential t-test under the null hypothesis.

15.5 Example: An Exact One-sample Group Sequential t-Test

An important special case of the general theory of Section 15.4 occurs when $X_1, X_2, \ldots$ are independent univariate $N(\mu, \sigma^2)$ random variables with scalar parameter μ. This was the situation considered by Jennison & Turnbull (1991b). In order to test the null hypothesis H_0: $\mu = \mu_0$, we use the above results with $\boldsymbol{c}^T\boldsymbol{\beta} = \mu$ and $\gamma = \mu_0$. Then

$$\boldsymbol{c}^T\widehat{\boldsymbol{\beta}} = \bar{X}^{(k)} = \frac{1}{n_k}\sum_{i=1}^{n_k} X_i, \quad s_k^2 = \frac{1}{n_k - 1}\sum_{i=1}^{n_k}(X_i - \bar{X}^{(k)})^2$$

and $\boldsymbol{c}^T\boldsymbol{V}_k\boldsymbol{c}$ is simply n_k^{-1} for $k = 1, \ldots, K$. The test statistic (15.9) reduces to the usual one-sample t-statistic,

$$T_k = \frac{\bar{X}^{(k)} - \mu_0}{\sqrt{(s_k^2/n_k)}}. \tag{15.10}$$

Suppose analyses are conducted after each new group of m observations, so $n_k = mk$, $k = 1, \ldots, K$. A general two-sided test of H_0: $\mu = \mu_0$ without early stopping in favor of H_0 stops at stage k to reject H_0 if

$$|T_k| \geq t_{mk-1, 1-\Phi(d_k)}, \tag{15.11}$$

for $k = 1, \ldots, K$, and H_0 is accepted if it has not been rejected by analysis K. Here, $t_{\nu,q}$ denotes the upper q tail-point of a t-distribution on ν degrees of freedom and Φ is the standard normal cdf. The sequence of constants $d_1, \ldots, d_K$ defines the testing boundary. These constants can be chosen so that the Type I error probability is exactly equal to a desired value α. If we adopt Pocock's approach of applying constant nominal significance levels at each stage, we set

$$d_k = Z_P(K, m, \alpha), \quad k = 1, \ldots, K,$$

whereas for an O'Brien & Fleming-type test, we set

$$d_k = Z_B(K, m, \alpha)\sqrt{(K/k)}, \quad k = 1, \ldots, K.$$

In either case, the constants $Z_P(K, m, \alpha)$ or $Z_B(K, m, \alpha)$ that lead to Type I error probability α can be computed using the distribution theory provided by Theorem 15.2 for this special case and nested numerical integration as described in Section 31.6.2; see also the appendix of Jennison & Turnbull (1991b). The tests described in Section 4.6 have the same general form but their Type I error rates are only approximately equal to α. Since those tests apply the same nominal significance level at each stage to a t-statistic as known variance tests apply to a Z-statistic, they correspond to using the constants $C_P(K, \alpha)$ or $C_B(K, \alpha)$ defined in Chapter 2 in place of $Z_P(K, m, \alpha)$ or $Z_B(K, m, \alpha)$, respectively, in the above definitions.

Table 15.1 gives values of $Z_P(K, m, \alpha)$ and $Z_B(K, m, \alpha)$ for $\alpha = 0.05$, $m = 1$, 2, 3, 5 and 10 and $K = 1$ to 10. The entries in this table are taken from Jennison & Turnbull (1991b, Table 1), where corresponding values for $\alpha = 0.01$ are also listed. If $m = 1$, it is assumed that no analysis is performed at the first stage as s_1^2 would be undefined: this explains the different pattern in the column for $m = 1$. The case $K = 1$ is a fixed sample test in which T_1 is compared with the usual critical value for a two-sided t-test, $t_{m-1,\alpha/2}$, and so $Z_P(1, m, \alpha)$ and $Z_B(1, m, \alpha)$ are both equal to $z_{\alpha/2}$. The final column shows limiting values of $Z_P(K, m, \alpha)$ and $Z_B(K, m, \alpha)$ as m increases to infinity and these are the values $C_P(K, \alpha)$ and $C_B(K, \alpha)$ for known σ^2 in Tables 2.1 and 2.3, respectively.

As an example, consider designing a test with two-sided Type I error probability $\alpha = 0.05$ and a maximum of $K = 5$ groups of 3 observations each. Suppose we choose a Pocock-type procedure so that $d_k = Z_P(5, 3, 0.05) = 2.457$ for $k = 1, \ldots, 5$, taking $Z_P(5, 3, 0.05)$ from Table 15.1. Since $\Phi(2.457) = 0.993$, the critical value for $|T_1|$ at stage 1 is $t_{2,0.007} = 8.36$. At stage 2, the critical value for $|T_2|$ is $t_{5,0.007} = 3.70$, and critical values at stages 3, 4 and 5 are 3.13, 2.92 and 2.81, respectively. Since the degrees of freedom increase with each stage, the critical values for $|T_k|$ decrease while the nominal significance levels remain constant. The "significance level" approach proposed by Pocock (1977) and described in Section 4.6 would use $d_k = C_P(5, 0.05) = 2.413$ for $k = 1, \ldots, 5$, where $C_P(5, 0.05)$ is taken from Table 2.1, and this yields critical values 7.86, 3.58, 3.05, 2.85 and 2.75 at stages 1 to 5 respectively. These values are smaller than those of the exact test, indicating that the actual Type I error rate of the approximate test is higher than the nominal 0.05. In fact, exact computation shows

Table 15.1 *Constants $Z_P(K, m, \alpha)$ and $Z_B(K, m, \alpha)$ for Pocock-type and O'Brien & Fleming-type two-sided repeated t-tests with exact Type I error probability* $\alpha = 0.05$. *Designs are for K groups of m observations and at each stage $k = 1, \ldots, K$, H_0 is rejected if Student's t-statistic* (15.10) *exceeds $t_{mk-1,1-\Phi(Z_P)}$ in the case of a Pocock-type test or $t_{mk-1,1-\Phi\{Z_B\sqrt{(K/k)}\}}$ for an O'Brien & Fleming-type test.*

K	Group size, m					
	$1^\dagger$	2	3	5	10	∞
			$Z_P(K, m, \alpha)$ for Pocock-type tests			
1^*	—	1.960	1.960	1.960	1.960	1.960
2	1.960	2.220	2.210	2.198	2.188	2.178
3	2.203	2.344	2.329	2.315	2.303	2.289
4	2.323	2.418	2.404	2.389	2.376	2.361
5	2.400	2.472	2.457	2.442	2.428	2.413
6	2.455	2.512	2.498	2.482	2.468	2.453
7	2.495	2.543	2.530	2.514	2.500	2.485
8	2.532	2.570	2.556	2.541	2.527	2.512
9	2.558	2.591	2.579	2.563	2.550	2.535
10	2.582	2.611	2.598	2.583	2.569	2.555
			$Z_B(K, m, \alpha)$ for O'Brien & Fleming-type tests			
1^*	—	1.960	1.960	1.960	1.960	1.960
2	1.960	1.995	1.993	1.989	1.984	1.977
3	2.047	2.034	2.027	2.019	2.012	2.004
4	2.077	2.058	2.050	2.041	2.033	2.024
5	2.097	2.075	2.065	2.056	2.048	2.040
6	2.112	2.084	2.077	2.068	2.061	2.053
7	2.117	2.094	2.088	2.078	2.071	2.063
8	2.128	2.100	2.096	2.086	2.079	2.072
9	2.133	2.105	2.102	2.093	2.087	2.080
10	2.138	2.109	2.108	2.099	2.093	2.087

* The case $K = 1$ corresponds to a fixed sample t-test on $m - 1$ degrees of freedom.
† If $m = 1$, it is assumed that no analysis is performed at the first stage.

the Type I error probability to be 0.056. Nevertheless, the approximation has performed quite well in this example, especially in view of the small group sizes.

We can also construct repeated confidence intervals for μ using the statistics given in (15.10). A $(1 - \alpha)$-level sequence of repeated confidence intervals is

$$(\bar{X}^{(k)} - c_k(\alpha)\sqrt{\{s_k^2/n_k\}},\ \bar{X}^{(k)} + c_k(\alpha)\sqrt{\{s_k^2/n_k\}}), \quad k = 1, \ldots, K,$$

where $c_k(\alpha) = t_{mk-1,1-\Phi(d_k)}$ is the critical value for $|T_k|$ stated in (15.11). The use of such intervals has been described in Chapter 10.

15.6 General Parametric Models: Generalized Linear Models

Suppose that observations X_i, $i = 1, \ldots, n_K$, are independent and we observe vectors $\boldsymbol{X}^{(k)} = (X_1, \ldots, X_{n_k})^T$ at analyses $k = 1, \ldots, K$. Denote the density or discrete distribution of X_i by $f_i(x_i;\boldsymbol{\beta})$, where $\boldsymbol{\beta}$ is a p-dimensional parameter vector. For example, $f_i(x_i;\boldsymbol{\beta})$ could specify a generalized linear model with linear predictor $d_{i1}\beta_1 + \ldots + d_{ip}\beta_p$. We define the column vector of efficient scores and the Fisher information matrix for each observation $i = 1, \ldots, n_K$ as

$$\boldsymbol{U}_i(X_i;\boldsymbol{\beta}) = \frac{\partial}{\partial\boldsymbol{\beta}} \log\{f_i(X_i;\boldsymbol{\beta})\} \quad \text{and} \quad \mathcal{I}_i(\boldsymbol{\beta}) = E\left\{\frac{-\partial}{\partial\boldsymbol{\beta}} \boldsymbol{U}_i^T(X_i;\boldsymbol{\beta})\right\},$$

respectively. The sum of scores and the information matrix at analysis k are then

$$\boldsymbol{U}(k,\boldsymbol{\beta}) = \sum_{i=1}^{n_k} \boldsymbol{U}_i(X_i;\boldsymbol{\beta}) \quad \text{and} \quad \mathcal{I}(k,\boldsymbol{\beta}) = \sum_{i=1}^{n_k} \mathcal{I}_i(\boldsymbol{\beta}).$$

We first review standard, fixed sample asymptotic theory applied at a single analysis, k. In the asymptotic setting we index sample sizes by n and suppose $n_k \to \infty$ and $\mathcal{I}(k,\boldsymbol{\beta}) \to \infty$ with $\mathcal{I}(k,\boldsymbol{\beta})/n$ converging to a fixed limit $\overline{\mathcal{I}}(k,\boldsymbol{\beta})$ as n increases. Under suitable regularity conditions on the $f_i(x_i;\boldsymbol{\beta})$, (see, for example, Cox & Hinkley, 1974, Ch. 9)

$$E\{\boldsymbol{U}_i(X_i;\boldsymbol{\beta})\} = \boldsymbol{0} \quad \text{and} \quad \mathcal{I}_i(\boldsymbol{\beta}) = \textbf{Var}\{\boldsymbol{U}_i(X_i;\boldsymbol{\beta})\} \tag{15.12}$$

for each $i = 1, \ldots, n_k$. With probability approaching 1 as $n \to \infty$, the MLE of $\boldsymbol{\beta}$ at analysis k satisfies

$$\boldsymbol{U}(k, \widehat{\boldsymbol{\beta}}^{(k)}) = \boldsymbol{0}. \tag{15.13}$$

Also, $\widehat{\boldsymbol{\beta}}^{(k)}$ is a consistent estimator of $\boldsymbol{\beta}$. Using (15.13) in a Taylor series expansion for each element of $\boldsymbol{U}(k,\boldsymbol{\beta})$ we obtain

$$n^{-1/2}\, U^j(k,\boldsymbol{\beta}) = n^{-1} \sum_{i=1}^{n_k} \left(\frac{-\partial}{\partial\boldsymbol{\beta}}\, U_i^j(X_i;\boldsymbol{\beta}_{jk}^*)\right)^T n^{1/2}\, (\widehat{\boldsymbol{\beta}}^{(k)} - \boldsymbol{\beta}) \tag{15.14}$$

for $j = 1, \ldots, p$. Here, $U^j(k,\boldsymbol{\beta})$ denotes the jth element of $\boldsymbol{U}(k,\boldsymbol{\beta})$, $U_i^j(X_i;\boldsymbol{\beta})$ is the jth element of $\boldsymbol{U}_i(X_i;\boldsymbol{\beta})$ and $\boldsymbol{\beta}_{jk}^*$ is a value of $\boldsymbol{\beta}$ on the line segment between $\boldsymbol{\beta}$ and $\widehat{\boldsymbol{\beta}}^{(k)}$. Again with suitable regularity conditions, the weak law of large numbers implies

$$n^{-1} \sum_{i=1}^{n_k} \frac{-\partial}{\partial\boldsymbol{\beta}} \boldsymbol{U}_i^T(X_i;\boldsymbol{\beta}) \longrightarrow \overline{\mathcal{I}}(k,\boldsymbol{\beta}) \tag{15.15}$$

in probability as $n \to \infty$, and a multivariate central limit theorem gives

$$n^{-1/2}\boldsymbol{U}(k,\boldsymbol{\beta}) \xrightarrow{\mathcal{D}} N(0, \overline{\mathcal{I}}(k,\boldsymbol{\beta})), \tag{15.16}$$

where $\xrightarrow{\mathcal{D}}$ denotes convergence in distribution. An important constraint on the experimental design for (15.15) and (15.16) to apply for a general response distribution is that the fraction of the total information contributed by any individual observation should decrease to zero as n increases. The further requirement that $\overline{\mathcal{I}}(k,\boldsymbol{\beta})$ be uniformly continuous in $\boldsymbol{\beta}$ in a neighborhood of the true value of $\boldsymbol{\beta}$ ensures that the difference between $\boldsymbol{\beta}_{jk}^*$ and $\boldsymbol{\beta}_j$ is asymptotically negligible and the standard result

$$n^{1/2}\, (\widehat{\boldsymbol{\beta}}^{(k)} - \boldsymbol{\beta}) \xrightarrow{\mathcal{D}} N(0, \overline{\mathcal{I}}^{-1}(k,\boldsymbol{\beta})) \quad \text{as } n \to \infty \tag{15.17}$$

follows from (15.14), (15.15) and (15.16).

The following theorem establishes the asymptotic *joint* distribution of $(\widehat{\boldsymbol{\beta}}^{(1)}, \ldots, \widehat{\boldsymbol{\beta}}^{(K)})$ under the regularity conditions used above in deriving the marginal distribution of each estimate $\widehat{\boldsymbol{\beta}}^{(k)}$. Thus, for any particular group sequential study, it suffices to check that standard, non-sequential asymptotic methods can be applied to deduce the asymptotic distribution of each individual $\widehat{\boldsymbol{\beta}}^{(k)}$. The sequential theory then follows automatically. In general, the regularity conditions that must be satisfied depend on the type of distribution the data follow and the form of dependence on the parameter vector $\boldsymbol{\beta}$. In the important case of generalized linear models, we refer the reader to McCullagh & Nelder (1989) for an account of asymptotic theory and for further references.

Theorem 15.3. Suppose observations X_i are independent with distributions $f_i(x_i; \boldsymbol{\beta})$, where $\boldsymbol{\beta}$ is p-dimensional, and observations $X_1, \ldots, X_{n_k}$ are available at analyses $k = 1, \ldots, K$. Let $\widehat{\boldsymbol{\beta}}^{(k)}$ denote the MLE of $\boldsymbol{\beta}$ based on $X_1, \ldots, X_{n_k}$ and $\mathcal{I}(k, \boldsymbol{\beta})$ the Fisher information for $\boldsymbol{\beta}$ summed over the first n_k observations. Suppose sample sizes are indexed by n and $n_k \to \infty$ and $\mathcal{I}(k, \boldsymbol{\beta}) \to \infty$ in such a way that

$$\mathcal{I}(k, \boldsymbol{\beta})/n \to \overline{\mathcal{I}}(k, \boldsymbol{\beta}) \quad \text{as } n \to \infty$$

for each $k = 1, \ldots, K$.

Suppose also that the distributions f_i are sufficiently regular that (15.12), (15.14), (15.15) and (15.16) hold, each $\overline{\mathcal{I}}(k, \boldsymbol{\beta})$ is uniformly continuous in $\boldsymbol{\beta}$ in a neighborhood of the true value of $\boldsymbol{\beta}$, and each $\widehat{\boldsymbol{\beta}}^{(k)}$ is consistent for $\boldsymbol{\beta}$ as $n \to \infty$.

Then, the joint distribution of $(\widehat{\boldsymbol{\beta}}^{(1)}, \ldots, \widehat{\boldsymbol{\beta}}^{(K)})$ is asymptotically multivariate normal,

$$n^{1/2}\,(\widehat{\boldsymbol{\beta}}^{(k)} - \boldsymbol{\beta}) \xrightarrow{\mathcal{D}} N(0, \overline{\mathcal{I}}^{-1}(k, \boldsymbol{\beta})) \quad \text{as } n \to \infty, \quad k = 1, \ldots, K,$$

and

$$\mathbf{Cov}_A(n^{1/2}\,\widehat{\boldsymbol{\beta}}^{(k_1)}, n^{1/2}\,\widehat{\boldsymbol{\beta}}^{(k_2)}) = \mathbf{Var}_A(n^{1/2}\,\widehat{\boldsymbol{\beta}}^{(k_2)}) \quad \text{for } k_1 \le k_2,$$

where $\mathbf{Var}_A$ and $\mathbf{Cov}_A$ denote asymptotic variance and covariance.

Proof. The increments in the score statistic, $n^{-1/2}\,\boldsymbol{U}(1, \boldsymbol{\beta})$ and

$$n^{-1/2}\,\{\boldsymbol{U}(k, \boldsymbol{\beta}) - \boldsymbol{U}(k-1, \boldsymbol{\beta})\}, \quad k = 2, \ldots, K,$$

depend on distinct sets of independent variables X_i and, hence, are independent. With (15.16), this implies that the increments have independent, asymptotically normal distributions with mean zero and variances equal to $\overline{\mathcal{I}}(1, \boldsymbol{\beta})$ and

$$\overline{\mathcal{I}}(k, \boldsymbol{\beta}) - \overline{\mathcal{I}}(k-1, \boldsymbol{\beta}), \quad k = 2, \ldots, K.$$

The sequence

$$\{n^{-1/2}\,\boldsymbol{U}(1, \boldsymbol{\beta}), \ldots, n^{-1/2}\,\boldsymbol{U}(K, \boldsymbol{\beta})\}$$

is, thus, asymptotically multivariate normal with mean zero,

$$\mathbf{Var}\,(n^{-1/2}\,\boldsymbol{U}(k, \boldsymbol{\beta})) = \overline{\mathcal{I}}(k, \boldsymbol{\beta}), \quad k = 1, \ldots, K,$$

and independent increments. This fact, together with (15.14) and (15.15), the consistency of $\widehat{\boldsymbol{\beta}}^{(k)}$ and the uniform continuity of each $\overline{\mathcal{I}}(k, \boldsymbol{\beta})$, implies that $\{n^{1/2}(\widehat{\boldsymbol{\beta}}^{(k)} - \boldsymbol{\beta});\, k = 1, \ldots, K\}$ has the same limiting joint distribution as

$$\{\overline{\mathcal{I}}^{-1}(k, \boldsymbol{\beta})\, n^{-1/2}\,\boldsymbol{U}(k, \boldsymbol{\beta});\, k = 1, \ldots, K\}$$

and the result follows. □

Because the asymptotic distribution of $(\widehat{\boldsymbol{\beta}}^{(1)}, \ldots, \widehat{\boldsymbol{\beta}}^{(K)})$ possesses the standard variance structure (15.1), group sequential tests can be constructed using boundaries computed for the normal case. For instance, a group sequential test of H_0: $\boldsymbol{c}^T\boldsymbol{\beta} = \gamma$ can be based on

$$S_k = \mathcal{I}_k\, n^{1/2}\, (\boldsymbol{c}^T\widehat{\boldsymbol{\beta}}^{(k)} - \gamma), \quad k = 1, \ldots, K,$$

where $\mathcal{I}_k = (\boldsymbol{c}^T\overline{\mathcal{I}}^{-1}(k, \boldsymbol{\beta})\boldsymbol{c})^{-1}$, and the vector $(S_1, \ldots, S_K)$ is asymptotically multivariate normal with independent increments. Furthermore, under a series of contiguous alternatives $\boldsymbol{\beta}^n$ for which $n^{1/2}\,(\boldsymbol{c}^T\boldsymbol{\beta}^n - \gamma) \to \theta$, S_k converges in distribution to $N(\theta\,\mathcal{I}_k, \mathcal{I}_k)$, $k = 1, \ldots, K$, and the standard distribution (15.2) applies. Alternatively, the standardized statistics

$$Z_k = \mathcal{I}_k^{1/2}\, n^{1/2}\, (\boldsymbol{c}^T\widehat{\boldsymbol{\beta}}^{(k)} - \gamma), \quad k = 1, \ldots, K,$$

are asymptotically multivariate normal with $Z_k \sim N(\theta\sqrt{\mathcal{I}_k}, 1)$ at each analysis k and $\mathrm{Cov}_A(Z_{k_1}, Z_{k_2}) = (\mathcal{I}_{k_1}/\mathcal{I}_{k_2})^{1/2}$ for $k_1 \le k_2$, under the same series of contiguous alternatives. This is the canonical joint distribution (3.1) for the vector $(Z_1, \ldots, Z_K)$ and hence tests designed for normal observations, as described in the preceding chapters, can be applied without modification.

The asymptotic theory for S_k and Z_k remains the same if $\overline{\mathcal{I}}(k, \boldsymbol{\beta})$ is replaced by a consistent estimate and, since $\overline{\mathcal{I}}^{-1}(k, \boldsymbol{\beta})$ is also the asymptotic variance of $n^{1/2}\widehat{\boldsymbol{\beta}}^{(k)}$, it suffices to have a consistent estimate of this variance. In finite samples we can use

$$(n\,\overline{\mathcal{I}}_k)^{-1} \simeq \mathrm{Var}(\boldsymbol{c}^T\widehat{\boldsymbol{\beta}}^{(k)})$$

to obtain standardized statistics

$$Z_k = \{\mathrm{Var}(\boldsymbol{c}^T\widehat{\boldsymbol{\beta}}^{(k)})\}^{-1/2}\, (\boldsymbol{c}^T\widehat{\boldsymbol{\beta}}^{(k)} - \gamma), \quad k = 1, \ldots, K,$$

which are approximately multivariate normal with

$$Z_k \sim N(\{\mathrm{Var}(\boldsymbol{c}^T\widehat{\boldsymbol{\beta}}^{(k)})\}^{-1/2}\, (\boldsymbol{c}^T\boldsymbol{\beta} - \gamma), 1)$$

and

$$\mathrm{Cov}(Z_{k_1}, Z_{k_2}) = \sqrt{\{\mathrm{Var}(\boldsymbol{c}^T\widehat{\boldsymbol{\beta}}^{(k_2)})/\mathrm{Var}(\boldsymbol{c}^T\widehat{\boldsymbol{\beta}}^{(k_1)})\}} \quad \text{for } k_1 \le k_2.$$

Alternatively, we can define score statistics corresponding to $n^{1/2}\, S_k$,

$$\widetilde{S}_k = \{\mathrm{Var}(\boldsymbol{c}^T\widehat{\boldsymbol{\beta}}^{(k)})\}^{-1}(\boldsymbol{c}^T\widehat{\boldsymbol{\beta}}^{(k)} - \gamma), \quad k = 1, \ldots, K,$$

which are approximately multivariate normal with

$$\widetilde{S}_k \sim N(\{\mathrm{Var}(\boldsymbol{c}^T\widehat{\boldsymbol{\beta}}^{(k)})\}^{-1}(\boldsymbol{c}^T\boldsymbol{\beta} - \gamma), \{\mathrm{Var}(\boldsymbol{c}^T\widehat{\boldsymbol{\beta}}^{(k)})\}^{-1})$$

and independent increments. Note that all we need here to calculate Z_k or $\widetilde{S}_k$ at analysis k is the MLE, $\widehat{\boldsymbol{\beta}}^{(k)}$, and an estimate of its variance at each analysis, both of which are usually provided by computer software programs for statistical model fitting.

Our general theory provides a basis for group sequential testing in a range of important applications, including the whole family of generalized linear models. One interesting application of the theory is to the sequential analysis of logistic regression models for the decay in germination rates of seeds, for which Whitehead (1989) proposed a sequential monitoring scheme based on the assumption that the sequence of score statistics is approximately multivariate normal with independent increments. Our results provide an asymptotic justification of this assumption to supplement the evidence of Whitehead's small sample simulations. In the next chapter we shall apply the theory to binary responses, including the analysis of logistic regression models in Section 16.6.

The general theory can also be extended to studies with correlated observations by replacing the distribution $f_i(x_i; \boldsymbol{\beta})$ by the conditional distribution of X_i given $X_1, \ldots, X_{i-1}$ and defining efficient scores $\boldsymbol{U}_i(X_i; \boldsymbol{\beta})$ in terms of this conditional distribution (see Cox & Hinkley, 1974, p. 299). The conclusion of Theorem 15.3 remains valid if it is permissible to apply a weak law of large numbers to sums of conditional information and a central limit theorem to sums of conditional scores to deduce the conditions required in the statement of Theorem 15.3.

We conclude this section by noting a remarkably simple proof that all asymptotically efficient estimators have the standard covariance structure (15.1). We consider the case $p = 1$ when β is a scalar, although the argument is easily generalized. Suppose the asymptotic covariance between $\hat{\beta}^{(k_2)}$ and $\hat{\beta}^{(k_2)} - \hat{\beta}^{(k_1)}$ is not zero; then a new estimate of the form $\hat{\beta}^{(k_2)} + \epsilon(\hat{\beta}^{(k_2)} - \hat{\beta}^{(k_1)})$, for ϵ close to zero and of the opposite sign to $\mathrm{Cov}(\hat{\beta}^{(k_2)}, \hat{\beta}^{(k_2)} - \hat{\beta}^{(k_1)})$, will have lower variance than $\hat{\beta}^{(k_2)}$, contradicting the asymptotic efficiency of $\hat{\beta}^{(k_2)}$. For example, Lai & Ying (1991) have proved that a modified version of the Buckley-James estimator (Buckley & James, 1979) is asymptotically efficient for the regression analysis of censored normal data and so the above argument establishes the asymptotic covariance structure of successive values of their estimate. In cases such as this, it remains to prove the asymptotic joint normality of the sequence of estimates. We note, conversely, that examples such as those described at the end of Section 15.3, in which the "standard distribution" does not apply, involve tests or estimates which are not founded on maximum likelihood methods.

15.7 Bibliography and Notes

The material in this chapter follows the development in Jennison & Turnbull (1997a,b). We have presented a general distribution theory for sequences of MLEs of a parameter vector. Examples include normal observations, both uncorrelated and correlated, and general parametric regression models. This theory forms the basis of a unified approach to group sequential analysis and supports the use of existing methods in a wide range of applications.

In fact, this general theory can be extended further. Tsiatis, Boucher & Kim (1995) obtain general results for parametric survival distributions when data are subject to right censoring. Scharfstein, Tsiatis & Robins (1997) show in great generality that score statistics in parametric and semi-parametric models have the independent increments structure (15.2). The asymptotic normality and efficiency of score statistics at individual analyses are required as conditions for the key Theorem 1 of Scharfstein et al. (1997) and, in general, these properties must be verified on a case-by-case basis to confirm that their theorem can be applied. The general theory for semi-parametric models applies to the proportional hazards regression model, where the baseline hazard function is an infinite-dimensional nuisance parameter. Applications to models for survival data will be discussed in Chapter 17.

The history of the use of the standard distribution (15.1) in the group sequential analysis of longitudinal data is interesting. Sequences of estimates have been shown to follow this standard distribution in some, but not all, cases. Armitage, Stratton & Worthington (1985) proved that the differences in mean response of two groups of subjects measured at three successive occasions satisfy (15.1) under a model assuming independent within-subject variation. However, these authors and Geary (1988) found that this property is lost when a within-subject autoregression term is added. Halperin et al. (1987) proved that estimates of the difference in the mean slopes of two groups in a random effects model follow the standard distribution even when subject entry is staggered and patterns of timing of observations for each subject are irregular. Lee & DeMets (1991) proposed a group sequential method for comparing rates of change in different forms of linear mixed-effects model and, subsequently, Reboussin, Lan & DeMets (1992) showed that their estimates follow (15.1). The standard distribution also arises in the asymptotic theory for generalized estimating equation models derived by Gange & DeMets (1996). Wu & Lan (1992) considered a group sequential procedure to compare areas under expected response change curves and, in important special cases with non-informative censoring, they showed that the sequence of score statistics follows (15.2) asymptotically see, for example, their equations (3.9) and (3.12).

There are however important applications where the distributions (15.1) and (15.2) fail to apply even asymptotically. Counter-examples are provided by the linear rank statistics of Lee & DeMets (1992), distribution-free multivariate Hodges-Lehmann estimators (Su & Lachin, 1992) and estimates obtained from "independence estimating equations" (Wei, Su & Lachin, 1990). Jennison & Turnbull (1991c) established conditions under which successive Mantel-Haenszel estimates of the odds ratio in stratified 2×2 tables are asymptotically multivariate normal and satisfy (15.1), but they note that other covariance structures occur if data do not accumulate in a specific manner (see Section 16.4).

In the group sequential analysis of survival data there are certain test statistics that also do not follow the standard distribution even asymptotically. Slud & Wei (1982) derive the asymptotic joint distribution of a sequence of modified-Wilcoxon scores, analogous to our $S_1, \ldots, S_K$, for comparing two sets of censored survival times and show that they can have correlated increments if patient entry is staggered. Lin (1991) notes that sequences of sums of linear rank statistics for analyzing multiple survival endpoints are asymptotically multivariate normal but have independent increments *only* if the limiting weight function is independent of survival time. Gu & Ying (1993, Sec. 4) consider the asymptotic joint distribution of successive Buckley-James score statistics (Buckley & James, 1979) for accumulating survival data with covariates, staggered entry and right censoring and they show this joint distribution differs from (15.2). The independent increment structure has been shown not to hold for certain robust score statistics used in semi-parametric multiplicative intensity regression models for recurrent event data in complex situations where there may be frailties or other dependencies present; see Cook & Lawless (1996, Sec. 5) and Jiang (1999). Burdon, Hampson & Jennison (2022) show that the asymptotic joint distribution of conditional score estimates of the treatment effect parameter in a joint model for survival time and a longitudinal variable does not have the standard correlation structure.

In Sections 15.3 and 15.4, we discussed group sequential procedures based on successive Z-statistics or t-statistics for testing a hypothesis of the form H_0: $\boldsymbol{c}^T\boldsymbol{\beta} = \gamma$ where $\boldsymbol{c}$ and $\boldsymbol{\beta}$ are vectors of dimension p. More generally we might consider hypotheses of the form H_0: $\boldsymbol{C}^T\boldsymbol{\beta} = \boldsymbol{0}$ where $\boldsymbol{C}$ is a $p \times q$ matrix of rank q. This leads to consideration of group sequential procedures based on successive χ^2 or F statistics. These have been considered by Jennison & Turnbull (1997b). Earlier, Jennison & Turnbull (1991b) treated the special case of testing the value of a multivariate normal mean with independent observations. In their Table 2, they provide tables of critical values for use with group sequential χ^2 procedures. This problem will be considered further in Chapters 19 and 21 as it is applicable to analysis of clinical trial data when there are multiple endpoints or more than two competing treatments.

Chapter 16

Binary Data

16.1 A Single Bernoulli Probability

16.1.1 Group Sequential Designs using the Normal Approximation

In this chapter we consider group sequential trials where the primary outcome is binary. We start by revisiting the one sample problem first considered in Section 4.4.1 and illustrated in an application to individual equivalence studies in Section 7.5. Here, individual responses $X_1, X_2, \ldots$ are independent Bernoulli random variables taking the value one, indicating "success", with probability p and zero, or "failure", with probability $1-p$. We denote this Bernoulli distribution with success probability p by $B(p)$. We shall assume responses are to be observed in up to K stages, with m_k observations in the kth group for $k = 1, \ldots, K$. We denote the cumulative sample size at the kth analysis by $n_k = m_1 + \ldots + m_k$ and define $S_k = X_1 + \ldots + X_{n_k}$ to be the total number of successes in the first k groups. The maximum likelihood estimate of p based on data available at analysis k is thus

$$\hat{p}^{(k)} = S_k/n_k.$$

Suppose we wish to test H_0: $p = p_0$ against a two-sided alternative. In Section 4.4.1 we explained how the standardized test statistics

$$Z_k = \frac{\hat{p}^{(k)} - p_0}{\sqrt{\{p_0(1-p_0)/n_k\}}}, \quad k = 1, \ldots, K, \tag{16.1}$$

may be used to construct such a two-sided test. There are, however, two problems with this approximation. First, the normal approximation is inaccurate if sample sizes are small or p is close to zero or one. Second, the variance of $\hat{p}^{(k)}$ is taken to be $p_0(1-p_0)/n_k$ in both Type I error and power calculations, whereas a variance of $p_A(1-p_A)/n_k$ should be used to calculate the power under the alternative $p = p_A$. In the example of Section 4.4.1, a test of H_0: $p = p_0 = 0.6$ with Type I error probability $\alpha = 0.05$ and power $1-\beta = 0.9$ at $p = p_0 \pm 0.2$ has up to four groups of 19 observations each, and H_0 is rejected at analysis k if $|Z_k| \geq C_P(4, 0.05) = 2.361$ for $k = 1, \ldots, 4$. When $p_A = 0.6 - 0.2 = 0.4$, $\mathrm{Var}(X_i) = p_A(1-p_A) = 0.24$, the same value as for $p = p_0 = 0.6$, hence, using the normal approximation with $\mathrm{Var}(X_i) = 0.24$, the power is found to be 0.906. However, if $p_A = 0.6 + 0.2 = 0.8$, we have $\mathrm{Var}(X_i) = 0.16$ and the normal approximation gives power equal to 0.939. The group size of 19 was calculated assuming $\mathrm{Var}(X_i) = 0.24$, obtained under $p = p_0$, holds for all values of p and, in this case, the requirement of power 0.9 or more at $p_A = 0.4$ and $p_A = 0.8$ is met — but this will not always be the case.

The same difficulties arise when using normal approximations to apply the one-sided tests of Chapter 5 to binary data. One common method of handling data whose variance is a function of the mean is to use a variance stabilizing transformation. For binary data, one would work with $\arcsin\{\sqrt{(\hat{p}^{(k)})}\}$ in place of $\hat{p}^{(k)}$, but even then this method is still approximate. Fortunately, it is possible to use exact methods for Bernoulli data, and we shall now describe these.

DOI: 10.1201/9781584888482-16

16.1.2 Exact Calculations for Binary Data

Since the cumulative numbers of successes, S_k, take integer values, it is easier to perform calculations in terms of these variables than the standardized statistics, Z_k, used previously. The two-sided and one-sided procedures of Chapters 4 and 5 are then determined by the maximum number of analyses K, the numbers of observations to be taken in each group, $m_1, \ldots, m_K$, and critical values for the $S_1, \ldots, S_K$. A general two-sided test has the form:

$$
\begin{array}{ll}
\text{After group } k = 1, \ldots, K-1 & \\
\quad \text{if } S_k \le a_k \text{ or } S_k \ge b_k & \text{stop, reject } H_0 \\
\quad \text{otherwise} & \text{continue to group } k+1, \\
\text{after group } K & \\
\quad \text{if } S_K \le a_K \text{ or } S_K \ge b_K & \text{stop, reject } H_0 \\
\quad \text{otherwise} & \text{stop, accept } H_0.
\end{array}
\tag{16.2}
$$

A one-sided test of H_0: $p \le p_0$ against $p > p_0$ has the general form:

$$
\begin{array}{ll}
\text{After group } k = 1, \ldots, K-1 & \\
\quad \text{if } S_k \le a_k & \text{stop, accept } H_0 \\
\quad \text{if } S_k \ge b_k & \text{stop, reject } H_0 \\
\quad \text{otherwise} & \text{continue to group } k+1, \\
\text{after group } K & \\
\quad \text{if } S_K \le a_K & \text{stop, accept } H_0 \\
\quad \text{if } S_K \ge b_K & \text{stop, reject } H_0.
\end{array}
\tag{16.3}
$$

In this case a_K must equal $b_K - 1$ to ensure a final decision. One-sided tests of H_0: $p \ge p_0$ against $p < p_0$ are similar but with the decisions to accept or reject H_0 interchanged. Note that these boundaries are defined in terms of sample sums and the critical values a_k and b_k are set on the S-scale, not on the Z-scale as previously.

Calculations for obtaining the operating characteristic and sample size distribution of a group sequential test for binary data are described by Schultz et al. (1973). Similar calculations appear earlier in the quality control literature on multi-stage acceptance sampling for attributes; see, for example, Wilson & Burgess (1971) and MIL-STD-105E (1989).

The number of successes in stage k has a binomial distribution with parameters m_k and p, and we denote the binomial probability by $B_{m_k}(y, p)$. Thus

$$
B_{m_k}(y, p) = \binom{m_k}{y} p^y (1-p)^{m_k - y}, \quad 0 \le y \le m_k, \tag{16.4}
$$

and $B_{m_k}(y, p) = 0$ otherwise. We define $C_k(y, p)$ to be the probability of reaching analysis k with $S_k = y$ when the success probability is p. At stage 1, $C_1(y, p) = B_{m_1}(y, p)$ for $0 \le y \le m_1$ and the following recursive formulae, analogous to (9.6) in the normal case, relate C_k to C_{k-1} for $k = 2, \ldots, K$. Let $E_k(y) = \max(a_{k-1} + 1, y - m_k)$ and $F_k(y) = \min(b_{k-1} - 1, y)$; then

$$
C_k(y, p) = \sum_{j=E_k(y)}^{F_k(y)} C_{k-1}(j, p) B_{m_k}(y - j, p), \quad y = 0, \ldots, n_k. \tag{16.5}
$$

The probabilities of crossing the lower and upper boundaries at stage k are, respectively,

$$
r_k^L(p) = \sum_{y=0}^{a_k} C_k(y, p) \quad \text{and} \quad r_k^U(p) = \sum_{y=b_k}^{n_k} C_k(y, p). \tag{16.6}
$$

Let $\pi(p)$ denote a test's power function, the probability of rejecting H_0 as a function of p. Then, for a one-sided test,

$$\pi(p) = \sum_{k=1}^{K} r_k^U(p), \tag{16.7}$$

while the power function of a two-sided test is

$$\pi(p) = \sum_{k=1}^{K} \{r_k^U(p) + r_k^L(p)\} = 1 - \sum_{y=a_K+1}^{b_K-1} C_K(y,p). \tag{16.8}$$

For both types of test, the Type I error probability is given by $\pi(p_0)$ and the expected sample size or average sample number (ASN) is

$$ASN(p) = \sum_{k=1}^{K-1} n_k\{r_k^U(p) + r_k^L(p)\} + n_K[1 - \sum_{k=1}^{K-1} \{r_k^U(p) + r_k^L(p)\}]. \tag{16.9}$$

To illustrate these calculations, consider the test with up to four groups of 19 observations described in Section 16.1.1. The procedure specified that sampling should stop with rejection of H_0: $p \le 0.6$ if $|Z_k| \ge 2.361$ at any stage $k = 1$, 2, 3 or 4. This criterion corresponds to critical values for the S_k of

$$(a_1, b_1) = (6, 17), \quad (a_2, b_2) = (15, 30), \quad (a_3, b_3) = (25, 43) \quad \text{and} \quad (a_4, b_4) = (35, 56).$$

Evaluating (16.8), we find the exact Type I error rate of the test to be $\pi(0.6) = 0.0517$, and at 0.6 ± 0.2 it has power $\pi(0.4) = 0.905$ and $\pi(0.8) = 0.946$. From (16.9), the ASNs at $p = 0.4$, 0.6 and 0.8 are 43.97, 74.24 and 42.36, respectively.

In choosing boundary values $\{(a_k, b_k); k = 1, \ldots, K\}$ for a one-sided or two-sided test, a useful strategy is to start with a procedure based on the normal approximation and then find the precise error rates for that design by exact calculation. If these error rates are acceptable, the test can be adopted, but if they are not, the design can be modified and its error rates checked again until a satisfactory test is obtained. Fleming (1982) has catalogued a collection of one-sided tests with two and three groups of observations for use in Phase II clinical trials. The procedures are for testing H_0: $p \le p_0$ versus $p > p_0$ with particular interest in the power at $p = p_A$ for pairs $(p_0, p_A) = (0.05, 0.2)$, $(0.1, 0.3)$, $(0.2, 0.4)$ and $(0.3, 0.5)$. All tests have one-sided Type I error probability of approximately 0.05, their maximum sample sizes range from 15 to 50, and their boundaries are similar in shape to those of the symmetric power family tests with parameter $\Delta = 0$ described in Section 5.2. Fleming (1982, Table 2) lists the exact Type I error probability, power and expected sample sizes under p_0 and p_A. Eight of these plans, all of which have approximate power 90% at $p = p_A$, are shown in Table 16.1.

16.1.3 Exact Methods for Inference upon Termination

A P-value or a confidence interval for p upon termination of a group sequential test can be constructed in a fashion analogous to that described in Sections 9.4 and 9.5 for normally distributed outcomes. Inferences are based on the pair of sufficient statistics (T, S_T), where $T \in \{1, \ldots, K\}$ is the stage at which the test stops and $S_T = X_1 + \ldots + X_{n_T}$ is the number of successes in the n_T observations taken up to this point. Suppose that we are using a one-sided group sequential test or a two-sided group sequential test with no inner wedge and we adopt the stage-wise ordering of outcomes (T, Z_T) specified in ordering A in Section 9.4. It is straightforward to define this ordering in terms of S_T rather than the standardized statistic Z_T. For a boundary of the form (16.2) or (16.3) defined through critical values

Table 16.1 *Selected two-stage and three-stage, one-sided binomial tests of* H_0*:* $p \leq p_0$ *versus* $p > p_0$ *taken from Fleming (1982). All tests are designed to have one-sided Type I error probability approximately equal to* $\alpha = 0.05$ *and power at* $p = p_A$ *approximately* $1 - \beta = 0.9$*. Exact Type I error probabilities and power are as tabulated.*

Two-stage procedures

Plan	1	2	3	4
(p_0, p_A)	(0.05, 0.2)	(0.1, 0.3)	(0.2, 0.4)	(0.3, 0.5)
n_K	40	35	45	50
Type I error rate	0.052	0.053	0.055	0.048
Power	0.922	0.920	0.909	0.894
(a_1, b_1, m_1)	(0, 4, 20)	(2, 6, 20)	(5, 10, 25)	(7, 14, 25)
(a_2, b_2, m_2)	(4, 5, 20)	(6, 7, 15)	(13, 14, 20)	(20, 21, 25)

Three-stage procedures

Plan	5	6	7	8
(p_0, p_A)	(0.05, 0.2)	(0.1, 0.3)	(0.2, 0.4)	(0.3, 0.5)
n_K	40	35	45	50
Type I error rate	0.046	0.063	0.057	0.049
Power	0.913	0.929	0.907)	0.887
(a_1, b_1, m_1)	(−, 4, 15)	(0, 5, 15)	(1, 8, 15)	(5, 12, 20)
(a_2, b_2, m_2)	(2, 5, 15)	(3, 6, 10)	(7, 11, 15)	(12, 17, 15)
(a_3, b_3, m_3)	(4, 5, 10)	(6, 7, 10)	(13, 14, 15)	(20, 21, 15)

$\{(a_k, b_k); k = 1, \ldots, K\}$ on the S-scale, we say $(k', s') \succ (k, s)$ if any one of the following three conditions holds:

(i) $k' = k$ and $s' \geq s$,

(ii) $k' < k$ and $s' \geq b_{k'}$,

(iii) $k' > k$ and $s \leq a_k$.

Let (k^*, s^*) be the observed value of (T, S_T). We can use the formulae developed in Section 16.1.2 to compute a one-sided upper P-value as

$$Pr_{p=p_0}\{(T, S_T) \succeq (k^*, s^*)\}$$

and a one-sided lower P-value as

$$Pr_{p=p_0}\{(T, S_T) \preceq (k^*, s^*)\}.$$

The two-sided P-value is twice the smaller of these two quantities. As an example, suppose the test terminates by exiting the upper boundary or by reaching the final analysis K. Then, the one-sided upper P-value is

$$Pr_{p=p_0}\{(T, S_T) \succeq (k^*, s^*)\} = \sum_{k=1}^{k^*-1} r_k^U(p_0) + \sum_{y=s^*}^{n_{k^*}} C_{k^*}(y, p_0),$$

where the $r_k^U(p_0)$, $k = 1, \ldots, k^* - 1$, and $C_{k^*}(y, p_0)$ are as defined in Section 16.1.2.

Derivation of a $(1-\alpha)$-level confidence interval for p is analogous to that for a normal mean in Section 9.5. The confidence interval is (p_L, p_U) where the lower and upper limits satisfy

$$Pr_{p=p_L}\{(T, S_T) \succeq (k^*, s^*)\} = \alpha/2 \tag{16.10}$$

and

$$Pr_{p=p_U}\{(T, S_T) \preceq (k^*, s^*)\} = \alpha/2. \tag{16.11}$$

Computation of p_L and p_U necessitates calculations similar to those for a one-sided P-value but repeated over a succession of values of p in order to find solutions to (16.10) and (16.11). Further details may be found in Jennison & Turnbull (1983) who list, in their Table 6, 80%, 90% and 95% confidence intervals for p following all possible outcomes of the three-stage designs described in Table 16.1. The same paper also tabulates 90%, 95% and 99% intervals for various multi-stage plans described in MIL-STD-105E (1989).

The 90% confidence intervals for p following three-stage plan number 6 of Table 16.1 are shown in Tables 16.2 and 16.3 in the column headed "Terminal CI"; these values were taken from Jennison & Turnbull (1983). For comparison, 90% confidence intervals calculated as if the same data had arisen in a fixed sample test are also listed in the column headed "Fixed sample CI". These fixed sample intervals, (p_{Lf}, p_{Uf}) say, can be obtained from the formulae

$$p_{Lf} = \frac{S_k}{S_k + (n_k - S_k + 1)\, F_{2(n_k - S_k + 1),\, 2S_k}(\alpha/2)}$$

and

$$p_{Uf} = \frac{(S_k + 1)\, F_{2(S_k+1),\, 2(n_k - S_k)}(\alpha/2)}{(n_k - S_k) + (S_k + 1)\, F_{2(S_k+1),\, 2(n_k - S_k)}(\alpha/2)}$$

with the adjustment that p_{Lf} is set equal to 0 if $S_k = 0$ and p_{Uf} to 1 if $S_k = n_k$. Here, $F_{\nu_1, \nu_2}(q)$ denotes the value exceeded with probability q by an F-statistic on ν_1 and ν_2 degrees of freedom. The above fixed sample confidence limits arise from the relation between the binomial, beta and F distributions (see, for example, Bowker & Lieberman, 1972, p. 467), which implies that the p_{Lf} and p_{Uf} satisfy

$$\sum_{i=S_k}^{n_k} \binom{n_k}{i} p_{Lf}^i (1 - p_{Lf})^{n_k - i} = \alpha/2$$

and

$$\sum_{i=0}^{S_k} \binom{n_k}{i} p_{Uf}^i (1 - p_{Uf})^{n_k - i} = \alpha/2.$$

Of course, for tests with $K \geq 2$ stages, (p_{Lf}, p_{Uf}) does not have the correct coverage probability, $1-\alpha$. However, under the sample space ordering A, which we are using here, the fixed sample interval does coincide with that constructed taking into account the group sequential design if the procedure stops at stage $k = 1$. Rows in Tables 16.2 and 16.3 containing the symbol ‡ represent outcomes that cannot be reached under the prescribed plan, either because the sampling rule specifies continuation to the next stage or because the outcome implies that stopping would have occurred at an earlier stage. The last column, headed "RCI", concerns repeated confidence intervals for p which we shall discuss in Section 16.1.4.

In the above construction of confidence intervals for p on termination of a group sequential test, equal error probabilities, $\alpha/2$, are allocated to each tail of the interval. In a fixed sample analysis, this approach is referred to as the Clopper-Pearson method (Clopper & Pearson, 1934), and this was the method used to obtain the fixed sample intervals in Tables 16.2 and 16.3. For the fixed sample problem, Sterne (1954) proposed

Table 16.2 90% *confidence intervals following a three-stage sampling plan*

Stage k	n_k	S_k	*Action*	*Fixed sample CI*	*Terminal CI*	*RCI*
1	15	0	Accept H_0	(0.0, 0.181)	(0.0, 0.181)	(0.0, 0.239)
1	15	1	Continue	(0.003, 0.279)	‡	(0.001, 0.342)
1	15	2	Continue	(0.024, 0.363)	‡	(0.013, 0.427)
1	15	3	Continue	(0.057, 0.440)	‡	(0.037, 0.503)
1	15	4	Continue	(0.097, 0.511)	‡	(0.069, 0.572)
1	15	5	Reject H_0	(0.142, 0.577)	(0.142, 0.577)	(0.107, 0.636)
1	15	6	Reject H_0	(0.191, 0.640)	(0.191, 0.640)	(0.150, 0.696)
1	15	7	Reject H_0	(0.244, 0.700)	(0.244, 0.700)	(0.197, 0.752)
1	15	8	Reject H_0	(0.300, 0.756)	(0.300, 0.756)	(0.248, 0.803)
1	15	9	Reject H_0	(0.360, 0.809)	(0.360, 0.809)	(0.304, 0.850)
1	15	10	Reject H_0	(0.023, 0.858)	(0.423, 0.858)	(0.364, 0.893)
1	15	11	Reject H_0	(0.489, 0.903)	(0.489, 0.903)	(0.428, 0.931)
1	15	12	Reject H_0	(0.560, 0.943)	(0.560, 0.943)	(0.497, 0.963)
1	15	13	Reject H_0	(0.637, 0.976)	(0.637, 0.976)	(0.573, 0.987)
1	15	14	Reject H_0	(0.721, 0.997)	(0.721, 0.997)	(0.658, 0.999)
1	15	15	Reject H_0	(0.819, 1.0)	(0.819, 1.0)	(0.761, 1.0)
2	25	0	‡	(0.0, 0.113)	‡	(0.0, 0.151)
2	25	1	Accept H_0	(0.002, 0.176)	(0.003, 0.199)	(0.001, 0.219)
2	25	2	Accept H_0	(0.014, 0.231)	(0.016, 0.238)	(0.010, 0.267)
2	25	3	Accept H_0	(0.034, 0.282)	(0.034, 0.284)	(0.024, 0.325)
2	25	4	Continue	(0.057, 0.330)	‡	(0.042, 0.367)
2	25	5	Continue	(0.082, 0.375)	‡	(0.069, 0.419)
2	25	6	Reject H_0	(0.110, 0.420)	(0.108, 0.412)	(0.092, 0.458)
2	25	7	Reject H_0	(0.139, 0.462)	(0.127, 0.445)	(0.116, 0.503)
2	25	8	Reject H_0	(0.170, 0.504)	(0.137, 0.470)	(0.150, 0.543)
2	25	9	Reject H_0	(0.202, 0.544)	(0.141, 0.489)	(0.176, 0.576)
2	25	10	Reject H_0	(0.236, 0.583)	(0.142, 0.501)	(0.204, 0.622)
2	25	11	Reject H_0	(0.270, 0.621)	(0.142, 0.508)	(0.243, 0.654)
2	25	12	Reject H_0	(0.305, 0.659)	(0.142, 0.510)	(0.272, 0.696)
2	25	13	Reject H_0	(0.341, 0.695)	(0.142, 0.511)	(0.304, 0.728)
2	25	14	Reject H_0	(0.379, 0.730)	(0.142, 0.511)	(0.346, 0.757)
2	25	15	‡	(0.417, 0.764)	‡	(0.378, 0.796)
2	25	16	‡	(0.456, 0.798)	‡	(0.424, 0.824)
2	25	17	‡	(0.496, 0.830)	‡	(0.457, 0.850)
2	25	18	‡	(0.538, 0.861)	‡	(0.497, 0.884)
2	25	19	‡	(0.580, 0.890)	‡	(0.542, 0.908)
2	25	20	‡	(0.625, 0.918)	‡	(0.581, 0.931)
2	25	21	‡	(0.670, 0.943)	‡	(0.633, 0.958)
2	25	22	‡	(0.718, 0.966)	‡	(0.675, 0.976)
2	25	23	‡	(0.769, 0.986)	‡	(0.733, 0.990)
2	25	24	‡	(0.824, 0.998)	‡	(0.781, 0.999)
2	25	25	‡	(0.887, 1.0)	‡	(0.849, 1.0)

‡ Stopping at this point (k, S_k) is not possible under the prescribed plan.

Table 16.3 90% *confidence intervals following a three-stage sampling plan (continued)*

Stage k	n_k	S_k	*Action*	*Fixed sample CI*	*Terminal CI*	*RCI*
3	35	0	‡	(0.0, 0.082)	‡	(0.0, 0.110)
3	35	1	‡	(0.001, 0.129)	‡	(0.001, 0.151)
3	35	2	‡	(0.010, 0.169)	‡	(0.008, 0.200)
3	35	3	‡	(0.024, 0.207)	‡	(0.019, 0.232)
3	35	4	Accept H_0	(0.040, 0.243)	(0.057, 0.286)	(0.033, 0.267)
3	35	5	Accept H_0	(0.058, 0.277)	(0.066, 0.297)	(0.047, 0.307)
3	35	6	Accept H_0	(0.077, 0.311)	(0.081, 0.315)	(0.069, 0.336)
3	35	7	Reject H_0	(0.098, 0.343)	(0.095, 0.335)	(0.088, 0.367)
3	35	8	Reject H_0	(0.119, 0.375)	(0.104, 0.351)	(0.107, 0.404)
3	35	9	Reject H_0	(0.141, 0.406)	(0.107, 0.363)	(0.125, 0.429)
3	35	10	Reject H_0	(0.164, 0.436)	(0.108, 0.369)	(0.150, 0.458)
3	35	11	Reject H_0	(0.187, 0.466)	(0.108, 0.372)	(0.173, 0.496)
3	35	12	Reject H_0	(0.211, 0.496)	(0.108, 0.373)	(0.195, 0.520)
3	35	13	Reject H_0	(0.236, 0.524)	(0.108, 0.374)	(0.214, 0.545)
3	35	14	Reject H_0	(0.260, 0.553)	(0.108, 0.374)	(0.243, 0.576)
3	35	15	Reject H_0	(0.286, 0.581)	(0.108, 0.374)	(0.268, 0.606)
3	35	16	‡	(0.312, 0.608)	‡	(0.291, 0.631)
3	35	17	‡	(0.338, 0.635)	‡	(0.312, 0.654)
3	35	18	‡	(0.365, 0.662)	‡	(0.346, 0.688)
3	35	19	‡	(0.392, 0.688)	‡	(0.369, 0.709)
3	35	20	‡	(0.419, 0.714)	‡	(0.394, 0.732)
3	35	21	‡	(0.447, 0.740)	‡	(0.424, 0.757)
3	35	22	‡	(0.476, 0.764)	‡	(0.455, 0.786)
3	35	23	‡	(0.504, 0.789)	‡	(0.480, 0.805)
3	35	24	‡	(0.534, 0.813)	‡	(0.504, 0.827)
3	35	25	‡	(0.564, 0.836)	‡	(0.542, 0.850)
3	35	26	‡	(0.594, 0.859)	‡	(0.571, 0.875)
3	35	27	‡	(0.625, 0.881)	‡	(0.596, 0.893)
3	35	28	‡	(0.657, 0.902)	‡	(0.633, 0.912)
3	35	29	‡	(0.689, 0.923)	‡	(0.664, 0.931)
3	35	30	‡	(0.723, 0.942)	‡	(0.693, 0.953)
3	35	31	‡	(0.757, 0.960)	‡	(0.733, 0.967)
3	35	32	‡	(0.793, 0.976)	‡	(0.768, 0.981)
3	35	33	‡	(0.831, 0.990)	‡	(0.800, 0.992)
3	35	34	‡	(0.871, 0.999)	‡	(0.849, 0.999)
3	35	35	‡	(0.918, 1.0)	‡	(0.890, 1.0)

‡ Stopping at this point (k, S_k) is not possible under the prescribed plan.

an alternative method based on inverting a family of tests of $p = p_0$ versus $p \neq p_0$ whose acceptance regions $\mathcal{A}(p_0)$ have probability content under p_0 of at least $1 - \alpha$ and consist of the most probable outcomes under $p = p_0$. Duffy & Santner (1987) extended Sterne's method to multi-stage designs and their Tables 2a to 2d enumerate intervals for the three-stage designs of Table 16.1. Duffy & Santner (1987, p. 89) note that their generalization suffers from several awkward features: interval endpoints are not necessarily non-decreasing with S_T, intervals may consist of just a single point, and computation of the intervals is not completely automated. Crow (1956) and Blyth & Still (1983) proposed a modification to Sterne's procedure which ensures interval confidence sets for p, and Duffy & Santner (1987, Sec. 4) extended this method to multi-stage designs, producing confidence intervals without the awkward features noted above. Duffy & Santner do not provide tables but Coe & Tamhane (1993, Table 2) list 90% intervals for the three-stage plan number 6 in our Table 16.1. In comparing different types of confidence interval, Coe & Tamhane note that those of Jennison & Turnbull (1983) tend to be wider and more conservative than the intervals obtained by Duffy & Santner's extension of the Crow-Blyth-Still method, although Jennison & Turnbull's intervals are perhaps easier to compute. It should also be noted that it may be inappropriate from a practical standpoint to exchange error probability freely between tails of a two-sided test or confidence interval when the errors associated with the two tails have quite different consequences. Although different in nature to confidence intervals upon termination, Coe & Tamhane also draw comparisons with repeated confidence intervals, the topic of our next section.

16.1.4 Repeated Confidence Intervals for a Bernoulli Probability p

Recall from Chapter 10 that a $(1 - \alpha)$-level sequence of RCIs for the parameter p is a sequence of K intervals, $I_1, \ldots, I_K$, with the property that, for all p,

$$Pr_p\{p \in I_k \text{ for all } k = 1, \ldots, K)\} = 1 - \alpha. \qquad (16.12)$$

Although RCIs can be used to define a group sequential stopping rule, they attain their properties without relying on a specific stopping rule being implemented. It is straightforward to use the normal approximation to construct such a sequence; see Jennison & Turnbull (1989, Sec. 5.1). At stage k the standardized statistic for testing a hypothesized success probability p is

$$Z_k(p) = \frac{S_k - n_k p}{\sqrt{\{n_k p(1-p)\}}}$$

and RCIs are given by

$$I_k = (p_{_{Lk}}, p_{_{Uk}}) = \{p \colon |Z_k(p)| \leq c_k\}, \quad k = 1, \ldots, K, \qquad (16.13)$$

where $c_1, \ldots, c_K$ are the critical values of a group sequential test with two-sided error probability α, termed the *parent* test in Chapter 10. With equal group sizes, we could use the Pocock or O'Brien & Fleming tests defined by constants in Tables 2.1 and 2.3, or we could use any of the other tests described in Chapter 2. If group sizes are unequal, the same critical values $c_1, \ldots, c_K$ can be used to give an approximate method, as described in Section 4.2, or the more accurate treatment based on the error spending approach of Chapter 8 may be preferred. In general, the choice of parent test should reflect the use anticipated for the RCIs in deciding when a group sequential study should terminate. Note that even though the information $\mathcal{I}_k = n_k/\{p(1-p)\}$ depends on p (see Section 4.4.1), the correlations $Corr(Z_{k_1}(p), Z_{k_2}(p)) = \sqrt{(n_{k_1}/n_{k_2})}$, $1 \leq k_1 \leq k_2 \leq K$, do not and hence the same sequence of constants $\{c_1, \ldots, c_K\}$ is applicable for all values of p.

Use of the normal approximation in constructing RCIs for p gives rise to the same difficulties noted in Section 16.1.1. Computation of exact RCIs for p is therefore desirable, and we shall adopt an error spending approach (see Section 8.2.1) in doing this. We start by specifying positive constants $\pi_1, \ldots, \pi_K$ summing to α, which represent the probability that a parent test should reject a hypothesized value of p at analyses $1, \ldots, K$. Then, for each value of p_0 we create a two-sided test of H_0: $p = p_0$ of the form (16.2) with critical values $a_k = a_k(p_0)$ and $b_k = b_k(p_0)$, $k = 1, \ldots, K$, defined for this particular p_0. Specifically, we choose the $a_k(p_0)$ to be the largest integers such that $r_k^L(p_0) \leq \pi_k/2$ and the $b_k(p_0)$ the smallest integers such that $r_k^U(p_0) \leq \pi_k/2$, where $r_k^L(p)$ and $r_k^U(p)$ are defined by (16.6). This family of tests, indexed by p_0, is then inverted to obtain the sequence of RCIs, as in Chapter 10. Since the π_k sum to α, each test in this family has two-sided Type I error probability of at most α so, for each p_0,

$$Pr_{p=p_0}\{a_k(p_0) < S_k < b_k(p_0) \text{ for } k = 1, \ldots, K\} \;\geq\; 1 - (\pi_1 + \ldots + \pi_K) \;=\; 1 - \alpha.$$

As the event $\{a_k(p_0) < S_k < b_k(p_0)\}$ corresponds exactly to the event that p_0 is contained in the kth RCI, it follows that the RCIs attain the confidence level $1 - \alpha$ conservatively.

Coe & Tamhane (1993) note that if interval endpoints are required to be accurate to three decimal places, it suffices to determine whether each of the values $p_0 = 0.0005$, 0.0015, $0.0025, \ldots, 0.9995$ is in each RCI and symmetry can be used to limit computations further to one half of these values with, say, $p_0 \leq 0.5$. The recursive method of calculation described in Section 16.1.2 may be used here to compute the values $a_k(p_0)$ and $b_k(p_0)$, $k = 1, \ldots, K$, for each p_0. We have followed this approach in calculating RCIs for outcomes arising in the three-stage plan number 6 of Table 16.1. The resulting 90% RCIs with error probability divided between the three stages as $\pi_1 = 0.0333$, $\pi_2 = 0.0334$ and $\pi_3 = 0.0333$ are listed in Tables 16.2 and 16.3 in the column headed "RCI". Since RCIs are valid and can be computed whether or not any particular stopping rule is adhered to, entries for RCIs appear in rows marked with a ‡ corresponding to outcomes that cannot occur under the original stopping rule. In contrast, the fixed sample intervals are not actually valid in the frequentist sense under any sequential stopping rule.

The above calculation produces conservative RCIs due to the discrete nature of the binomial distribution. This conservatism may be reduced a little by carrying forward the "unspent" parts of each π_k to later analyses. Thus, in the sequential test of H_0: $p = p_0$, once a_1 and b_1 are defined, error probabilities $\pi_1/2 - r_1^L(p)$ and $\pi_1/2 - r_1^U(p)$ can be carried forward to analysis 2, and there we set $a_2(p_0)$ to be the largest integer such that $r_2^L(p_0) \leq \pi_2/2 + \{\pi_1/2 - r_1^L(p)\}$ and we set $b_2(p_0)$ to be the smallest integer for which $r_2^U(p_0) \leq \pi_2/2 + \{\pi_1/2 - r_1^U(p)\}$. The process of carrying forward unspent error probability then continues in successive analyses.

Construction of RCIs for a Bernoulli probability p using exact binomial calculations was first proposed by Coe & Tamhane (1993). The procedure described above produces equal-tailed intervals, analogous to the Clopper-Pearson method for a fixed sample study, whereas Coe & Tamhane used a group sequential adaptation of the Blyth & Still modification to Sterne's fixed sample construction, resulting in intervals with unequal tail probabilities but of shorter width. Comparison of the RCIs in Tables 16.2 and 16.3 with those in Table 1 of Coe & Tamhane (1993) shows that the two types of RCI do differ, but not to an extent that is likely to be of great practical significance.

16.2 Two Bernoulli Probabilities

16.2.1 Exact Calculations of Operating Characteristics

We turn now to the problem of comparing two treatment arms, A and B, with respect to a binary outcome. We assume responses X_{A1}, X_{A2}, ... from subjects receiving treatment A

are independent and distributed as Bernoulli $B(p_A)$ and, similarly, responses X_{B1}, X_{B2}, ... on arm B are independent, Bernoulli $B(p_B)$. We shall consider two-sided tests of the null hypothesis H_0: $p_A = p_B$ against H_A: $p_A \neq p_B$ and one-sided tests of H_0: $p_A \leq p_B$ against H_A: $p_A > p_B$ or of H_0: $p_A \geq p_B$ against H_A: $p_A < p_B$.

We suppose the data are collected group sequentially with a maximum of K stages. For treatment arm A, the kth group size is m_{Ak}, at analysis k the cumulative sample size is $n_{Ak} = m_{A1} + \ldots + m_{Ak}$, and we denote the cumulative number of successes by $S_{Ak} = X_{A1} + \ldots + X_{Ak}$. The quantities m_{Bk}, n_{Bk}, and S_{Bk}, $k = 1, \ldots, K$, are defined analogously for treatment B.

For a given sequence of sample sizes n_{Ak} and n_{Bk}, we can define a group sequential test based on successive pairs (S_{Ak}, S_{Bk}) in terms of decision regions and a continuation region at each stage. For each $k = 1, \ldots, K$, we let $\Omega_k = \{0, 1, \ldots, n_{Ak}\} \times \{0, 1, \ldots, n_{Bk}\}$ denote the set of possible values for (S_{Ak}, S_{Bk}) and we partition Ω_k into three mutually exclusive sets, $\mathcal{C}_k$, $\mathcal{A}_{0k}$ and $\mathcal{A}_{Ak}$. The group sequential test of H_0 is then:

$$\begin{array}{ll} \text{After group } k = 1, \ldots, K & \\ \quad \text{if } (S_{Ak}, S_{Bk}) \in \mathcal{A}_{0k} & \text{stop and decide } H_0, \\ \quad \text{if } (S_{Ak}, S_{Bk}) \in \mathcal{A}_{Ak} & \text{stop and decide } H_A, \\ \quad \text{if } (S_{Ak}, S_{Bk}) \in \mathcal{C}_k & \text{continue to stage } k+1, \end{array}$$

where $\mathcal{C}_K = \emptyset$, the empty set, to ensure stopping by stage K. This formulation includes one-sided tests and two-sided tests with and without inner wedges to permit early stopping to accept H_0. In a two-sided test with no inner wedge, $\mathcal{A}_{0k} = \emptyset$ for $k = 1, \ldots, K-1$.

A procedure's operating characteristics can be calculated exactly in a fashion analogous to that for the one-sample problem of Section 16.1.2. Let $C_k(y_A, y_B; p_A, p_B)$ be the probability of reaching stage k with $S_{Ak} = y_A$ and $S_{Bk} = y_B$ under specified values of p_A and p_B. Then $C_1(y_A, y_B; p_A, p_B) = B_{m_{A1}}(y_A, p_A)\, B_{m_{B1}}(y_B, p_B)$, for $y_A = 0, \ldots, n_{A1}$ and $y_B = 0, \ldots, n_{B1}$, where $B_{m_k}(y, p)$ is the binomial probability mass function defined in (16.4). For $k = 2, \ldots, K$, the following recursive formulae are analogous to (9.6) in the normal case:

$$C_k(y_A, y_B; p_A, p_B) = \sum_{(i,j) \in \mathcal{C}_{k-1}} \sum C_{k-1}(i, j; p_A, p_B)\, B_{m_{Ak}}(y_A - i, p_A)\, B_{m_{Bk}}(y_B - j, p_B)$$

$$\text{for } y_A = 0, \ldots, n_{Ak}, \; y_B = 0, \ldots, n_{Bk}. \tag{16.14}$$

Recall that $B_m(y, p) = 0$ unless $y \in \{0, \ldots, m\}$, so there is no problem in having $y_A - i$ or $y_B - i$ negative or greater than m_{Ak} or m_{Bk}, respectively, in the above formula.

The expected total sample size is given by

$$\sum_{k=1}^{K} \left\{ (n_{Ak} + n_{Bk}) \sum_{(i,j) \notin \mathcal{C}_k} \sum C_k(i, j; p_A, p_B) \right\} \tag{16.15}$$

and the power function, the probability of rejecting H_0 as a function of p_A and p_B, is

$$\pi(p_A, p_B) = \sum_{k=1}^{K} \sum_{(i,j) \in \mathcal{A}_{Ak}} \sum C_k(i, j; p_A, p_B). \tag{16.16}$$

We define the size of the test to be its maximum Type I error probability over values of p_A and p_B satisfying H_0. For both one-sided and two-sided tests this should be attained when $p_A = p_B$, so the size of the test is

$$\sup_{p_A = p_B} \pi(p_A, p_B). \tag{16.17}$$

The quantities (16.15), (16.16) and (16.17) can be evaluated for any group sequential design. As an example, consider the test discussed by O'Brien & Fleming (1979, Sec. 5) which has three stages and group sizes $m_{A1} = m_{A2} = m_{A3} = 7$ on treatment A and $m_{B1} = m_{B2} = m_{B3} = 14$ on treatment B. As explained in Section 4.4.2, the standardized statistics for testing H_0: $p_A = p_B$ at analyses $k = 1$, 2 and 3 are

$$Z_k = (\hat{p}_A^{(k)} - \hat{p}_B^{(k)})\sqrt{\widehat{\mathcal{I}}_k} = \left(\frac{S_{Ak}}{n_{Ak}} - \frac{S_{Bk}}{n_{Bk}}\right)\{(n_{Ak}^{-1} + n_{Bk}^{-1})\,\tilde{p}_k\,(1-\tilde{p}_k)\}^{-1/2}, \quad (16.18)$$

where $\tilde{p}_k = (S_{Ak} + S_{Bk})/(n_{Ak} + n_{Bk})$. A two-sided O'Brien & Fleming test (with no early stopping to accept H_0) rejects H_0 at analysis k if

$$|Z_k| \geq C_B(3, 0.05)\sqrt{(3/k)} = 2.004\sqrt{(3/k)}.$$

Thus the test has continuation regions

$$\mathcal{C}_k = \{(S_{Ak}, S_{Bk})\colon |Z_k| < 2.004\sqrt{(3/k)}\}$$

at analyses $k = 1$ and 2, and acceptance region for H_0

$$\mathcal{A}_{03} = \{(S_{A3}, S_{B3})\colon |Z_3| < 2.004\}$$

at the third and final analysis.

The sets $\mathcal{C}_k$, $\mathcal{A}_{0k}$ and $\mathcal{A}_{Ak}$ can be described in terms of the pairs (S_{Ak}, S_{Bk}) that each contains. As an example, at stage 1, $\mathcal{C}_1$ contains outcomes satisfying $|Z_1| < 3.471$, and if $S_{A1} = 0$ this requires $S_{B1} \leq 11$, if $S_{A1} = 1$ this requires $S_{B1} \leq 12$, etc. Table 16.4 shows the continuation regions, $\mathcal{C}_1$ and $\mathcal{C}_2$, of this test and the final acceptance region $\mathcal{A}_{03}$.

Using the recursion (16.14) to evaluate (16.16) with $p_A = p_B = p$ over a fine grid of values for p we find the exact size of this test, as defined by (16.17), to be 0.0536 rather than the value 0.05, suggested by the normal approximation of Section 4.4.2. The same recursion can be used to calculate properties of the test at other values of p_A and p_B; for example, if $p_A = 0.4$ and $p_B = 0.1$, the test has power 0.797 and expected total sample size 53.30, and if $p_A = 0.1$ and $p_B = 0.4$, power is 0.743 and expected total sample size is 56.86, showing some reduction in both cases below the maximum sample size of 63.

If it is desired to ensure that the test's maximum Type I error probability does not exceed 0.05, one could enlarge the acceptance region $\mathcal{A}_{03}$ at the final stage. Increasing the critical value for Z_3 from 2.004 to 2.036 accomplishes this, yielding an exact size of 0.0499 but reducing power to 0.792 at $(p_A, p_B) = (0.4, 0.1)$ and to 0.734 at $(p_A, p_B) = (0.1, 0.4)$. The new acceptance region at stage 3, which differs in just six entries from the previous one, is shown in Table 16.5.

As for the one-sample problem, we recommend a general strategy for designing a group sequential test whereby an initial design is chosen based on the normal approximation to the distribution of the standardized statistics Z_k and employing one of the testing boundaries described in Chapters 2 to 7. This test's exact operating characteristics can be computed from (16.15), (16.16) and (16.17) using the recursion formulae (16.14). These properties are likely to be satisfactory for moderate or large group sizes, but if they are not, the initial design can be modified by trial and error until a satisfactory procedure is found. Lin, Wei & DeMets (1991) describe an alternative approach in which statistics Z_k, as defined in (16.18), are compared against critical values of the form used in Pocock or O'Brien & Fleming tests but with new values of the constants C_P and C_B, chosen to ensure the tests have exact size no greater than α.

Table 16.4 *Continuation regions at analyses 1 and 2 and acceptance region at analysis 3 of a three-stage test of H_0: $p_A = p_B$*

Analysis 1	S_{A1}	S_{B1}	S_{A1}	S_{B1}
Continuation region	0	0–11	4	0–14
$\mathcal{C}_1 = \{(S_{A1}, S_{B1}): \lvert Z_1\rvert < 3.471\}$	1	0–12	5	1–14
	2	0–13	6	2–14
	3	0–14	7	3–14
Analysis 2	S_{A2}	S_{B2}	S_{A2}	S_{B2}
Continuation region	0	0–9	8	6–25
$\mathcal{C}_2 = \{(S_{A2}, S_{B2}): \lvert Z_2\rvert < 2.454\}$	1	0–12	9	8–26
	2	0–15	10	9–27
	3	1–17	11	11–27
	4	1–19	12	13–28
	5	2–20	13	16–28
	6	3–22	14	19–28
	7	5–23		
Analysis 3	S_{A3}	S_{B3}	S_{A3}	S_{B3}
Acceptance region	0	0–7	11	12–32
$\mathcal{A}_{03} = \{(S_{A3}, S_{B3}): \lvert Z_3\rvert < 2.004\}$	1	0–10	12	13–33
	2	1–13	13	15–35
	3	1–16	14	17–36
	4	2–18	15	19–38
	5	3–21	16	21–39
	6	4–23	17	24–40
	7	6–25	18	26–41
	8	7–27	19	29–41
	9	9–29	20	32–42
	10	10–30	21	35–42

Table 16.5 *Acceptance region at analysis 3 of a modified three-stage test of H_0: $p_A = p_B$*

Analysis 3	S_{A3}	S_{B3}	S_{A3}	S_{B3}
Acceptance region	0	0–7	11	12–32
$\mathcal{A}_{03} = \{(S_{A3}, S_{B3}): \lvert Z_3\rvert < 2.036\}$	1	0–10	12	13–34
	2	0–13	13	15–35
	3	1–16	14	17–36
	4	2–19	15	19–38
	5	3–21	16	21–39
	6	4–23	17	23–40
	7	6–25	18	26–41
	8	7–27	19	29–42
	9	8–29	20	32–42
	10	10–30	21	35–42

16.2.2 Exact Methods for Inference upon Termination

The methods of Section 16.1.3 are readily extended to the two-treatment problem. Suppose we are using a one-sided test or two-sided test with no inner wedge and we adopt the stage-wise ordering of the sample space presented as ordering A in Section 9.4. Since the sufficient statistic (S_{Ak}, S_{Bk}) is two-dimensional, we still need to specify a criterion for ordering the sample space within each stage, and we shall order outcomes by the value of the standardized statistic Z_k given by (16.18).

We denote the upper part of the stopping region at stage k by $\mathcal{A}_k^+$ and the lower part by $\mathcal{A}_k^-$, for $k = 1, \ldots, K-1$. For a two-sided test of H_0: $p_A = p_B$ these regions are $\mathcal{A}_k^+ = \mathcal{A}_{Ak} \cap \{(S_{Ak}, S_{Bk})\colon Z_k > 0\}$ and $\mathcal{A}_k^- = \mathcal{A}_{Ak} \cap \{(S_{Ak}, S_{Bk})\colon Z_k < 0\}$, in the case of a one-sided test of H_0: $p_A \le p_B$ against $p_A > p_B$ they are $\mathcal{A}_k^+ = \mathcal{A}_{Ak}$ and $\mathcal{A}_k^- = \mathcal{A}_{0k}$, and in a test of H_0: $p_A \ge p_B$ against $p_A < p_B$ we have $\mathcal{A}_k^+ = \mathcal{A}_{A0}$ and $\mathcal{A}_k^- = \mathcal{A}_{Ak}$.

Let (k^*, z^*) denote the observed value of (T, Z_T). If (k^*, z^*) falls in the upper region, $\mathcal{A}_k^+$, or if $k^* = K$, we define

$$\begin{aligned} p_U(p_A, p_B) &= Pr_{(p_A,p_B)}\{(T, Z_T) \succeq (k^*, z^*)\} \\ &= \sum_{k=1}^{k^*-1} \sum_{(i,j)\,\in\,\mathcal{A}_k^+} C_k(i,j;\, p_A, p_B) + \sum\sum_{\{(i,j)\,\in\,\Omega_{k^*}\colon\, Z_{k^*}\,\ge\, z^*\}} C_{k^*}(i,j;\, p_A, p_B) \end{aligned}$$

and we set $p_L(p_A, p_B) = 1 - p_U(p_A, p_B) + Pr_{(p_A,p_B)}\{(T, Z_T) = (k^*, z^*)\}$. If (k^*, z^*) falls in the lower region, $\mathcal{A}_k^-$, $p_L(p_A, p_B)$ is defined analogously to $p_U(p_A, p_B)$ but the double sums over(i, j) in the above expression are now over $\{(i,j) \in \mathcal{A}_k^-\}$ and $\{(i,j) \in \Omega_{k^*}\colon Z_{k^*} \le z^*\}$ and $p_U(p_A, p_B) = 1 - p_L(p_A, p_B) + Pr_{(p_A,p_B)}\{(T, Z_T) = (k^*, z^*)\}$.

The above definitions imply that if the true success probabilities are $p_A = \tilde{p}_A$ and $p_B = \tilde{p}_B$ then, for any $0 < \epsilon < 1$,

$$Pr\{p_L(\tilde{p}_A, \tilde{p}_B) < \epsilon\} \le \epsilon \quad \text{and} \quad Pr\{p_U(\tilde{p}_A, \tilde{p}_B) < \epsilon\} \le \epsilon,$$

so $p_L(\tilde{p}_A, \tilde{p}_B)$ and $p_U(\tilde{p}_A, \tilde{p}_B)$ have the appropriate distribution to be one-sided P-values for testing H_0: $(p_A, p_B) = (\tilde{p}_A, \tilde{p}_B)$, with some conservatism due to the discreteness of the distribution of (T, Z_T). We shall use these two quantities as the basis for defining general P-values and confidence intervals.

Since the common success probability under H_0: $p_A = p_B$ is unknown, we allow for all possibilities in defining the one-sided upper and lower P-values for testing H_0 as

$$P_U = \sup_{p_A = p_B} p_U(p_A, p_B) \quad \text{and} \quad P_L = \sup_{p_A = p_B} p_L(p_A, p_B).$$

The two-sided P-value is twice the smaller of these two quantities.

Similar arguments show we can define a $(1-\alpha)$-level confidence interval for the difference $\delta = p_A - p_B$ as (δ_L, δ_U), where

$$\sup_{p_A} p_U(p_A, p_A - \delta_L) = \alpha/2 \quad \text{and} \quad \sup_{p_A} p_L(p_A, p_A - \delta_U) = \alpha/2.$$

The suprema in these two equations are taken over $p_A \subset [0,1]$ such that $p_A - \delta_L \in [0,1]$ and $p_A - \delta_U \in [0,1]$, respectively. The roots δ_L and δ_U can be found by evaluating p_U and p_L over a grid of p_A and δ values.

The same approach can be taken to obtain a confidence interval for the ratio of two success probabilities or for their odds ratio. A $(1-\alpha)$-level confidence interval for the ratio $\rho = p_A/p_B$ is (ρ_L, ρ_U) where

$$\sup_{p_A} p_U(p_A, p_A/\rho_L) = \alpha/2 \quad \text{and} \quad \sup_{p_A} p_L(p_A, p_A/\rho_U) = \alpha/2,$$

the suprema being taken over p_A values in the intervals $[0, \min(1, \rho_L)]$ and $[0, \min(1, \rho_U)]$, respectively. The end points of a $(1-\alpha)$-level confidence interval for the odds ratio $\psi = p_A(1-p_B)/\{p_B(1-p_A)\}$ are the solutions of

$$\sup_{p_A \in [0,1]} p_U\left(p_A, p_A/\{\psi_L(1-p_A)+p_A\}\right) = \alpha/2$$

and

$$\sup_{p_A \in [0,1]} p_L\left(p_A, p_A/\{\psi_U(1-p_A)+p_A\}\right) = \alpha/2.$$

16.2.3 Repeated Confidence Intervals

A $(1-\alpha)$-level sequence of RCIs, $I_1, \ldots, I_K$, for the difference in success probabilities $\delta = p_A - p_B$ is required to have the property

$$Pr_{(p_A, p_B)}\{\delta \in I_k \text{ for all } k = 1, \ldots, K\} \geq 1 - \alpha \tag{16.19}$$

for all p_A and p_B such that $p_A - p_B = \delta$. An approximate sequence obtained using a normal approximation is

$$I_k = (\hat{p}_A^{(k)} - \hat{p}_B^{(k)} \pm c_k \sqrt{\{\hat{p}_A^{(k)}(1-\hat{p}_A^{(k)})/n_{Ak} + \hat{p}_B^{(k)}(1-\hat{p}_B^{(k)})/n_{Bk}\}}), \quad k = 1, \ldots, K,$$

where $\hat{p}_A^{(k)} = S_{Ak}/n_{Ak}$, $\hat{p}_B^{(k)} = S_{Bk}/n_{Bk}$ and $c_1, \ldots, c_K$ are the critical values of a group sequential test for normally distributed statistics with two-sided Type I error probability α. This "parent" test could be one of those described in Chapter 2, for example, that of Pocock or O'Brien & Fleming, but if increments in information are unequal and unpredictable, it is more accurate to define a test by the error spending approach of Chapter 8 with information levels estimated by

$$\mathcal{I}_k = \left\{ \frac{\hat{p}_A^{(k)}(1-\hat{p}_A^{(k)})}{n_{Ak}} + \frac{\hat{p}_B^{(k)}(1-\hat{p}_B^{(k)})}{n_{Bk}} \right\}^{-1}, \quad k = 1, \ldots, K.$$

We now turn to the construction of exact RCIs for $\delta = p_A - p_B$. We first specify an increasing sequence of constants $0 \leq \phi_1 \leq \ldots \leq \phi_K = \alpha$ representing the cumulative Type I error probabilities at each analysis to be attained by the parent two-sided test. For each pair of probabilities (p_A^*, p_B^*), we form a group sequential test of H^*: $(p_A, p_B) = (p_A^*, p_B^*)$ which stops to reject H^* at stage if $Z_k \notin (a_k(p_A^*, p_B^*), b_k(p_A^*, p_B^*))$, for $k = 1, \ldots, K$, and which accepts H^* if stage K is reached without this hypothesis being rejected. Here the Z_k are the standardized statistics defined in (16.18), although these could be replaced by another choice of statistic such as $\hat{p}_A^{(k)} - \hat{p}_B^{(k)}$. For a fixed sequence of group sizes, the probabilities $C_k(y_A, y_B; p_A^*, p_B^*)$ defined in Section 16.2.1 can be used to construct the critical values $a_k(p_A^*, p_B^*)$ and $b_k(p_A^*, p_B^*)$. In doing this, we follow the approach of Stallard & Todd (2000) to handle the discrete nature of the response variable and aim to achieve the specified cumulative error probabilities, $\phi_1, \ldots, \phi_K$, at each analysis rather than the individual error probabilities for each analysis, $\pi_1, \ldots, \pi_K$, that were used in Chapter 8. Defining the events

$$\mathcal{G}_j = \{a_i(p_A^*, p_B^*) < Z_i < b_i(p_A^*, p_B^*) \text{ for } i = 1, \ldots, j-1 \text{ and } Z_j \leq a_j(p_A^*, p_B^*)\}$$

and

$$\mathcal{H}_j = \{a_i(p_A^*, p_B^*) < Z_i < b_i(p_A^*, p_B^*) \text{ for } i = 1, \ldots, j-1 \text{ and } Z_j \geq b_j(p_A^*, p_B^*)\},$$

for $j = 1, \ldots, K$, we find $a_k(p_A^*, p_B^*)$ and $b_k(p_A^*, p_B^*)$ for $k = 1, \ldots, K$ in turn, taking $a_k(p_A^*, p_B^*)$ to be the largest value of a_k such that

$$Pr_{(p_A^*, p_B^*)}\{\cup_{j=1,\ldots,k} \mathcal{G}_j\} \leq \phi_k/2$$

and taking $b_k(p_A^*, p_B^*)$ to be the smallest value of b_k satisfying

$$Pr_{(p_A^*, p_B^*)}\{\cup_{j=1,\dots,k} \mathcal{H}_j\} \leq \phi_k/2.$$

By construction, the Type I error probability of this test of H^* does not exceed α. We define

$$a_k(\delta) = \inf_{p_A - p_B = \delta} a_k(p_A, p_B) \quad \text{and} \quad b_k(\delta) = \sup_{p_A - p_B = \delta} b_k(p_A, p_B) \quad \text{for } k = 1, \dots, K.$$

Then, for any fixed δ between 0 and 1,

$$Pr_{(p_A, p_B)}\{a_k(\delta) < Z_k < b_k(\delta) \text{ for all } k = 1, \dots, K\} \geq 1 - \alpha$$

whenever $p_A - p_B = \delta$. Thus, we have an acceptance region for testing a hypothesized value of δ at each stage $k = 1, \dots, K$ with overall probability of wrong rejection of at most α. Inverting this family of group sequential tests of all possible values of δ, we obtain the RCI sequence

$$I_k = \{\delta\colon a_k(\delta) < Z_k < b_k(\delta)\}, \quad k = 1, \dots, K.$$

Lin, Wei & DeMets (1991) present a similar construction, inverting a family of tests of $p_A - p_B = \delta$ of the same general form as their tests of $p_A - p_B = 0$.

In calculating these RCIs, consideration of any one δ requires the construction of group sequential boundaries at each of a grid of values for p_A. Nevertheless, the computation involved is quite manageable, particularly when a study has a small number of groups of moderate size. Just as for confidence intervals upon termination, the above method may be adapted to give RCIs for the ratio or odds ratio between success probabilities p_A and p_B.

The methods we have described divide the error probability α equally between the two tails of each test of H^*: $(p_A, p_B) = (p_A^*, p_B^*)$ and, hence, between the two tails of the RCIs. Coe & Tamhane (1993) propose a construction based on a generalization of Blyth & Still's (1983) method which produces intervals that are not equal tailed but may be narrower.

16.2.4 Conservatism and Other Issues Caused by Nuisance Parameters

The procedures we have described for comparing two Bernoulli distributions have an element of conservatism due to the need to guarantee appropriate properties under all values of a nuisance parameter. As an example, the size of an exact test of H_0: $p_A = p_B$, as defined in (16.17), may occur at a common value of p_A and p_B that is unlikely to be true.

One way to produce less conservative tests is to restrict the range of values of the nuisance parameter over which the various suprema are taken. An alternative approach is to operate adaptively, treating an estimate of the nuisance parameter at each stage as if it were exactly correct. This approach was followed in Section 8.6.3 to create information monitoring designs to compare treatments with a binary response. These designs had the additional feature that the target sample size was updated in response to interim data in order to meet a specified power requirement. In Chapter 18 we shall explain how to create adaptive designs with a variety of response types where the sample size needed to achieve a given power depends on unknown nuisance parameters and in Section 18.4 we re-visit the comparison of two Bernoulli probabilities to illustrate sample size re-estimation in error spending tests. It may sometimes be feasible to use exact calculation to check whether adaptive procedures attain their nominal properties accurately but it is also possible, and often simpler, to use simulation to assess complex procedures

16.3 Exact Tests Based on the Odds Ratio in Multi-stage Trials

An alternative approach to the problem of Section 16.2 is to formulate a test of a null hypothesis $p_A = p_B$ or $p_A \leq p_B$ in terms of the odds ratio ψ between the response probabilities p_A and p_B, defined by

$$\psi = \frac{p_A(1-p_B)}{p_B(1-p_A)}.$$

Note that $\psi = 1$ precisely when $p_A = p_B$. Jung & Sargent (2014) have followed this approach to construct two-stage designs for Phase II trials. We shall show how the error spending approach of Chapter 8 can be employed for a general K-stage group sequential procedure. We shall focus on one-sided tests of H_0: $\psi \leq 1$ against the alternative $\psi > 1$. This corresponds to testing H_0: $p_A \leq p_B$ against $p_A > p_B$ only now we set the power requirement at a specific value of ψ.

Consider first a single stage trial. Let S_A denote the number of successes among m_A observations on treatment A and S_B the number of successes among m_B observations on treatment B. Let $h(s_A \,|\, m_A, m_B, s, \psi)$ be the conditional probability mass function of S_A given the total number of successes, $S = S_A + S_B$. Then

$$\begin{aligned} h(s_A \,|\, m_A, m_B, s, \psi) &= Pr(S_A = s_A \,|\, S = s) \\ &= \frac{\binom{m_A}{s_A}\binom{m_B}{s-s_A}\psi^{s_A}}{\sum_{i=\ell}^{u}\binom{m_A}{i}\binom{m_B}{s-i}\psi^{i}} \quad \text{for } \ell \leq s_A \leq u, \end{aligned} \qquad (16.20)$$

where $\ell = \max\{0, s - m_B\}$ and $u = \min\{s, m_A\}$, and $h(s_A \,|\, m_A, m_B, s, \psi) = 0$ otherwise. The denominator in (16.20) can instead be written as a sum over $i = 0$ to m_A with the convention that

$$\binom{a}{b} = 0 \quad \text{if } b < 0 \text{ or } b > a.$$

If $\psi = 1$ this is a hypergeometric distribution and for general ψ it is known as Fisher's noncentral hypergeometric distribution. Fisher (1935, p. 59) proposed using this conditional distribution to create a two-sided test of the null hypothesis $\psi = 1$ or, equivalently, $p_A = p_B$.

For a one-sided test of H_0: $\psi \leq 1$ against $\psi > 1$, we reject H_0 if the observed value of S_A is greater than or equal to the critical value c_α, defined as the smallest integer satisfying

$$Pr(S_A \geq c_\alpha \,|\, S = s, \psi = 1) = \sum_{s_A = c_\alpha}^{s} h(s_A \,|\, m_A, m_B, s, 1) \leq \alpha. \qquad (16.21)$$

An advantage of this approach is that only a single parameter ψ is being tested and there is no nuisance parameter to deal with when considering the Type I error rate. The total number of successes S is an ancillary statistic and the conditional Type I error probability given S is controlled at level α. The conditional power $Pr(S_A \geq c_\alpha \,|\, S = s, \psi)$ depends on the observed value of S as well as the sample sizes m_A and m_B. Overall power can be calculated under pairs of probabilities p_A and p_B with a specified odds ratio ψ by summing over the distribution of S.

We can now develop an exact, one-sided group sequential test of H_0: $\psi \leq 1$ versus $\psi > 1$, or equivalently of H_0: $p_A \leq p_B$ versus $p_A > p_B$, as an alternative to that of Section 16.2. Because the test depends on the total numbers of successes at each analysis and these numbers are unpredictable, it is appropriate to use an error spending approach, as described in Chapter 8. A K-stage group sequential has the general form

$$\begin{array}{lll} \text{After group } k = 1, \ldots, K-1 & & \\ \quad \text{if } S_{Ak} \le a_k(S_1, \ldots, S_k) & \text{stop, accept } H_0 & \\ \quad \text{if } S_{Ak} \ge b_k(S_1, \ldots, S_k) & \text{stop, reject } H_0 & \\ \quad \text{otherwise} & \text{continue to group } k+1, & (16.22) \\ \text{after group } K & & \\ \quad \text{if } S_{AK} < b_K(S_1, \ldots, S_K) & \text{stop, accept } H_0 & \\ \quad \text{if } S_{AK} \ge b_K(S_1, \ldots, S_K) & \text{stop, reject } H_0, & \end{array}$$

where S_{Ak} and S_{Bk} are the cumulative numbers of successes on treatments A and B at analysis k and $S_k = S_{Ak} + S_{Bk}$ for $k = 1 \ldots, K$. The notation in (16.22) indicates that the critical values a_k and b_k depend on the complete sequence of observed numbers of successes up to analysis k, as we shall see from the construction of the error spending boundaries.

Let m_{Ak} and m_{Bk} be the numbers of subjects in group k assigned to treatments A and B, respectively, and denote the total number of successes in group k by S_k with observed value s_k. Calculation of the stopping boundaries is conditional on the sequence $\{m_{A1}, m_{B1}, s_1, \ldots, m_{AK}, m_{BK}, s_K\}$ and the number of successes on treatment A in stage k follows the noncentral hypergeometric distribution (16.20) with $m_A = m_{Ak}$, $m_B = m_{Bk}$ and $s = s_k$.

For $k = 1, \ldots, K-1$, we denote the continuation region for S_{Ak}, the cumulative number of successes on treatment A at stage k, by

$$\mathcal{CR}_k = \{a_k(s_1, \ldots, s_k) + 1, \ \ldots \ , b_k(s_1, \ldots, s_k) - 1\}.$$

Let

$$C_k(s_{Ak}, \psi) = Pr\{\text{Reach stage } k \text{ with } S_{Ak} = s_{Ak}\}.$$

Then

$$C_1(s_{A1}, \psi) = h(s_{A1} \,|\, m_{A1}, m_{B1}, s_1, \psi)$$

and

$$C_k(s_{Ak}, \psi) = \sum_{j \in \mathcal{CR}_{k-1}} C_{k-1}(j, \psi)\, h(s_{Ak} - j \,|\, m_{Ak}, m_{Bk}, s_k, \psi) \quad \text{for } k = 2, \ldots, K. \quad (16.23)$$

Consider an error spending design with no futility boundary or a non-binding futility boundary, analogous to the situation described in Section 8.4. In the absence of a futility boundary, we set $a_1 = \ldots = a_{K-1} = -1$, so the properties of the procedure (16.22) depend only on the choice of $\{b_1, \ldots, b_K\}$. We shall follow the one-sided version of the error spending method of Slud & Wei (1982), described in Section 8.2.1, in which the number of analyses K and the way in which Type I error probability is partitioned between analyses are fixed in advance. Let the cumulative Type I error probability up to analysis k be ϕ_k for $k = 1, \ldots, K$, with $\phi_K = \alpha$. The efficacy boundary $\{b_1, \ldots, b_K\}$ is chosen to achieve these cumulative error rates as closely as possible, given the discrete distribution of the data.

In spending Type I error probability ϕ_1 at the first analysis, we define b_1 to be the smallest integer such that $Pr_{\psi=1}\{S_{A1} \ge b_1\} \le \phi_1$, that is the smallest b_1 for which

$$\sum_{s_{A1}=b_1}^{m_{A1}} C_1(s_{A1}, 1) \le \phi_1. \quad (16.24)$$

At each analysis $k = 2, \ldots, K$, we choose the boundary point b_k so that the cumulative Type I error probability is as close to ϕ_k as possible without exceeding this value. Thus b_k is the smallest integer such that

$$\sum_{s_{A1}=b_1}^{m_{A1}} C_1(s_{A1}, 1) + \ldots + \sum_{s_{A,k-1}=b_{k-1}}^{m_{A,k-1}} C_{k-1}(s_{A,k-1}, 1) + \sum_{s_{Ak}=b_k}^{m_{Ak}} C_k(s_{Ak}, 1) \leq \phi_k. \quad (16.25)$$

In working with the cumulative error probabilities $\phi_1, \ldots, \phi_K$, we have followed the approach of Stallard & Todd (2000) to handle the discrete nature of the response variable, as we did to construct RCIs in Section 16.2.3. The test will still be slightly conservative but less so than if we had assigned error probabilities $\pi_1, \ldots, \pi_K$ to each analysis and not carried forward "unspent" error probability to the next analysis.

Additional work is needed to design a study to achieve a certain power or to construct a futility boundary, since the information about ψ depends on the total numbers of successes S_k in each group k and this has a random distribution that depends on the unknown probabilities p_A and p_B. We can obtain information levels for ψ at each analysis using the theory presented in Section 15.6 and use these to construct efficacy and futility boundaries spending Type I and Type II error probability as a function of observed information. We define the likelihood function for ψ at analysis k as

$$\prod_{j=1}^{k} h(s_{Aj} \,|\, m_{Aj}, m_{Bj}, s_j, \psi). \quad (16.26)$$

Then, Theorem 15.3 implies the sequence of maximum likelihood estimates $(\hat{\psi}_1, \ldots, \hat{\psi}_K)$ is approximately multivariate normal with variances and covariances following the canonical joint distribution (3.1), and $\mathrm{Var}(\hat{\psi}_k)$ is the inverse of the Fisher information $\mathcal{I}_k$ for ψ at analysis k, derived from the likelihood function (16.26).

One may simply use the normal approximation to the distribution of $(\hat{\psi}_1, \ldots, \hat{\psi}_K)$ to define an error spending test with approximate Type I error probability α under $\psi = 1$ and power $1 - \beta$ at a specified odds ratio $\psi = \psi^* > 1$. Applying the appropriate inflation factor to a fixed sample size design gives the maximum information level $\mathcal{I}_{max}$ and Type I and Type II error probabilities are spent as functions of the observed information. Information monitoring, as described in Section 8.6, can be used to select group sizes in order to achieve the target information level at the final analysis. Under the normal approximation, efficacy and futility boundaries would be defined for the standardized test statistics

$$Z_k = \hat{\psi}_k \sqrt{\mathcal{I}_k}, \quad k = 1, \ldots, K.$$

However, we can combine aspects of this design with exact calculations to protect the Type I error probability more accurately. Suppose Type I and Type II error spending functions are specified and applying these to information levels obtained from the normal approximation gives cumulative Type I error probabilities $\phi_{1,1}, \ldots, \phi_{1,K}$ and cumulative Type II error probabilities $\phi_{2,1}, \ldots, \phi_{2,K}$. If the futility boundary is to be non-binding, efficacy boundary values $b_1, \ldots, b_K$ can be computed as before, solving (16.24) and (16.25) with cumulative Type I error probabilities $\phi_{1,1}, \ldots, \phi_{1,K}$. Then, the non-binding futility boundary values $a_1, \ldots, a_{K-1}$ are calculated as the smallest integers satisfying the inequalities

$$\sum_{s_{A1}=0}^{a_1} C_1(s_{A1}, \psi^*) + \ldots + \sum_{s_{A,k-1}=0}^{a_{k-1}} C_{k-1}(s_{A,k-1}, \psi^*) + \sum_{s_{Ak}=0}^{a_k} C_k(s_{Ak}, \psi^*) \leq \phi_{2,k}$$

for $k = 1, \ldots, K-1$, and we set $a_K = b_K - 1$ at the final analysis. Note that the upper boundary values appear in the definitions of the continuation regions $\mathcal{CR}_{k-1}$ so the values of $b_1, \ldots, b_{k-1}$ are involved in the calculation of a_k.

It is quite straightforward to apply the methods we have described to derive a one-sided group sequential test with a binding futility boundary or a two-sided group sequential test. Also, the probabilities $C_k(s_{Ak}, \psi)$ can be used to compute a P-value or confidence interval on termination, or a sequence of RCIs for ψ. In each case, the defining properties will be satisfied conditionally on the sequence of observed group sizes m_{Ak} and m_{Bk} and the total numbers of successes S_k.

16.4 The Odds Ratio and Multiple 2 × 2 Tables

The odds ratio is used in the analysis of stratified two-by-two tables when it can be assumed that a common odds ratio, ψ, occurs, within each stratum. Indexing strata by $j = 1, \ldots, J$, suppose all observations are independent, with those on treatment A in stratum j distributed as $B(p_{Aj})$ and those on treatment B as $B(p_{Bj})$, where $B(p)$ denotes a Bernoulli distribution with parameter p. In the constant odds ratio model it is assumed that

$$\frac{p_{Aj}(1 - p_{Bj})}{p_{Bj}(1 - p_{Aj})} = \psi, \quad j = 1, \ldots, J.$$

Let $\theta = \log(\psi)$ and suppose we wish to test H_0: $\theta = 0$ against $\theta \neq 0$ or H_0: $\theta \leq 0$ against $\theta > 0$. Note that when $\theta = 0$, we have $\psi = 1$ and $p_{Aj} = p_{Bj}$ for each $j = 1, \ldots, J$. In the special case when there is only one stratum, this reduces to the problem of comparing two Bernoulli probabilities addressed in Sections 16.2 and 16.3.

In a group sequential design, suppose that at the kth analysis totals of n_{Ajk} and n_{Bjk} observations have been observed in stratum j on treatments A and B, respectively, and set $N_{jk} = n_{Ajk} + n_{Bjk}$. For each analysis $k = 1, \ldots, K$ and stratum $j = 1, \ldots, J$, we define S_{Ajk} to be the sum of the n_{Ajk} responses on treatment A and S_{Bjk} the sum of the n_{Bjk} responses on treatment B. We shall use the well-known estimator of Mantel & Haenszel (1959). At each analysis $k = 1, \ldots, K$, the Mantel-Haenszel estimate of ψ is

$$\hat{\psi}^{(k)} = \sum_{j=1}^{J} \frac{S_{Ajk}(n_{Bjk} - S_{Bjk})}{N_{jk}} \Bigg/ \sum_{j=1}^{J} \frac{S_{Bjk}(n_{Ajk} - S_{Ajk})}{N_{jk}}$$

and we estimate θ by $\hat{\theta}^{(k)} = \log \hat{\psi}^{(k)}$. If stratum k has not yet been observed and $N_{jk} = 0$, the term $0/0$ in the above formula and in any subsequent formulae should be interpreted as zero. Robins, Breslow & Greenland (1986) propose an estimator of the asymptotic variance of $\hat{\psi}^{(k)}$, denoted V_{US} in their paper, which is consistent as cell sizes increase in a fixed number of strata or when cell sizes are bounded and the number of strata increases. Let

$$P_{j,k} = \frac{S_{Ajk} + n_{Bjk} - S_{Bjk}}{N_{jk}}, \quad Q_{j,k} = \frac{S_{Bjk} + n_{Ajk} - S_{Ajk}}{N_{jk}},$$

$$R_{j,k} = \frac{S_{Ajk}(n_{Bjk} - S_{Bjk})}{N_{jk}}, \quad U_{j,k} = \frac{S_{Bjk}(n_{Ajk} - S_{Ajk})}{N_{jk}},$$

$$R_{+,k} = \sum_{j=1}^{J} R_{j,k} \quad \text{and} \quad U_{+,k} = \sum_{j=1}^{J} U_{j,k}.$$

Then, the estimator of $\text{Var}(\hat{\theta}^{(k)})$ is

$$V_k = \frac{\sum_j P_{j,k} R_{j,k}}{2R_{+,k}^2} + \frac{\sum_j (P_{j,k} U_{j,k} + Q_{j,k} R_{j,k})}{2R_{+,k} U_{+,k}} + \frac{\sum_j Q_{j,k} U_{j,k}}{2U_{+,k}^2}$$

and we define

$$\mathcal{I}_k = V_k^{-1}, \quad k = 1, \ldots, K.$$

Here, $\mathcal{I}_k$ is not precisely the Fisher information for θ contained in the data at analysis k since the Mantel-Haenszel estimate is not 100% efficient for estimating θ. However, the Mantel-Haenszel estimate is highly efficient in many situations and has other attractive features, and $\mathcal{I}_k$ represents the effective information when θ is estimated in this way. The standardized statistics for testing H_0: $\theta = 0$ are

$$Z_k = \hat{\theta}^{(k)} \sqrt{\mathcal{I}_k}, \quad k = 1, \ldots, K.$$

Jennison & Turnbull (1991c) have shown that $\{Z_1, \ldots, Z_K\}$ follows, approximately, the canonical joint distribution (3.1) with information levels $\{\mathcal{I}_1, \ldots, \mathcal{I}_K\}$ for $\theta = \log(\psi)$ if (a) sample sizes increase within just one stratum, (b) there are multiple strata and sample sizes per treatment within each stratum grow at the same rate or (c) new strata are observed at each analysis but no new subjects are added to previous strata and the different strata form a random sample from a "super-population" of possible strata. Jennison & Turnbull (1989, Table 6) have used simulation to study the adequacy of the multivariate normal approximation to the distribution of $\{Z_1, \ldots, Z_K\}$ in the context of coverage probabilities of repeated confidence intervals. Their results illustrate its accuracy both for a fixed number of strata of increasing size and in matched pair comparisons which produce increasing numbers of strata of fixed size.

Given this standard distribution for $\{Z_1, \ldots, Z_K\}$, the group sequential tests of Chapters 2 to 11 can be used in the usual way. Good initial estimates of the p_{Aj} and p_{Bj} are required at the design stage to determine the sample sizes needed to satisfy a given power condition. If reliable estimates of the p_{Aj} and p_{Bj} are not available, information monitoring, as described in Chapter 8, or the adaptive methods we shall present in Chapter 18 may be used to achieve the desired information levels accurately.

Jennison & Turnbull (1989, Table 7) presented the data in Table 16.6 as an illustrative example with a fixed number of strata of increasing size. These are cumulative frequencies representing case-control data with six strata, collected in three stages. The frequencies at the third analysis are the Ille-et-Vilaine data set of Breslow & Day (1980, p. 137). The cumulative frequencies at the first and second analyses are hypothetical values which might have been observed had the study been conducted group sequentially. Table 16.7 shows the resulting sequences of 90% RCIs for the odds ratio ψ based on both a Pocock and an O'Brien & Fleming parent test with three analyses. The RCIs for θ are $\hat{\theta}_k \pm c_k \sqrt{V_k}$, $k = 1, \ldots, 3$, with constants $c_k = C_P(3, 0.1) = 1.992$ (from Table 2.1) and $c_k = C_B(3, 0.1)\sqrt{(3/k)} = 1.710\sqrt{(3/k)}$ (from Table 2.3), respectively, for the two types of RCI. The increments in information are approximately equal and so it is appropriate to follow the significance level approach of Section 4.2 in defining the parent test. The RCIs for ψ are obtained by exponentiating the endpoints of the corresponding intervals for θ. For both types of RCI, the first interval fails to include unity, indicating that an exposure effect might have been concluded at this point.

Jennison & Turnbull (1989, Table 9) also presented an example with an increasing number of strata of fixed sizes. These data are reproduced in Table 16.8. The frequencies at the third analysis form the Leisure World data set of Breslow & Day (1980, p. 174), while data at analyses 1 and 2 are hypothetical counts that might have been observed under a group sequential design. Each stratum is comprised of a case and four matched controls, and the exposure or non-exposure of each individual is recorded. Thus, for example, an exposed case for which 2 matched controls were exposed contributes 1 to the entry of "16" in row 5 and column 3 of the table, and it may also contribute to the entries of "5" and "10" in rows 1 and 3, depending on when this case was observed. Table 16.9 shows sequences of 90% RCIs for the odds ratio ψ based on both a Pocock and an O'Brien & Fleming parent

Table 16.6 *Hypothetical interim data for the Ille-et-Vilaine study (Breslow & Day, 1980, p. 137). Entries are cumulative counts of exposed and unexposed cases and controls in each stratum at three interim analyses.*

Analysis	*Stratum*	*Cumulative frequencies*			
		Cases		*Controls*	
		Exposed	*Unexposed*	*Exposed*	*Unexposed*
k	j	S_{Ajk}	$n_{Ajk}-S_{Ajk}$	S_{Bjk}	$n_{Bjk}-S_{Bjk}$
1	1	0	0	4	31
	2	2	2	8	60
	3	6	5	9	40
	4	20	19	5	30
	5	8	14	7	31
	6	2	3	0	9
2	1	1	0	6	66
	2	2	4	18	111
	3	18	14	19	89
	4	30	27	20	100
	5	11	22	12	57
	6	3	7	0	21
3	1	1	0	9	106
	2	4	5	26	164
	3	25	21	29	138
	4	42	34	27	139
	5	19	36	18	88
	6	5	8	0	31

Table 16.7 *Interim results for the data of Table 16.6 assuming the odds ratio ψ to be constant across the six strata*

Analysis	$\hat{\theta}_k = \log \hat{\psi}_k$	$\sqrt{V_k}$	$\hat{\psi}_k$	*Pocock test-based*	*O'Brien & Fleming*
k				*90% RCIs for ψ*	*90% RCIs for ψ*
1	1.60	0.336	4.93	(2.5, 9.6)	(1.8, 13.4)
2	1.58	0.227	4.85	(3.1, 7.6)	(3.0, 7.8)
3	1.64	0.189	5.16	(3.5, 7.5)	(3.7, 7.1)

test with $K = 3$ analyses. By the second analysis, both types of RCI fail to include unity, and so an early decision in favor of an exposure effect might have been made.

16.5 Case-control and Matched Pair Analyses

An important special case of the stratified designs discussed in Section 16.4 is the matched pair case-control or cohort study. The use of sequential methods in retrospective case-control studies has been described by O'Neill & Anello (1978) and by Pasternack & Shore (1982).

Table 16.8 *Hypothetical interim data for the Leisure World study (Breslow & Day, 1980, p. 174). Entries are cumulative counts of each type of matched set, as specified by the exposure or non-exposure of the case and the number of exposed controls, at three interim analyses.*

Analysis		*Cumulative frequencies*				
	Case	*Number of controls exposed*				
k		0	1	2	3	4
1	*Exposed*	0	8	5	5	1
	Unexposed	0	1	0	1	0
2	*Exposed*	2	14	10	9	2
	Unexposed	0	3	0	1	1
3	*Exposed*	3	17	16	15	5
	Unexposed	0	4	1	1	1

Table 16.9 *Interim results for the data of Table* 16.8 *assuming the odds ratio* ψ *to be constant across the six strata*

Analysis k	$\hat{\theta}_k = \log \hat{\psi}_k$	$\sqrt{V_k}$	$\hat{\psi}_k$	*Pocock test-based* 90% *RCIs for* ψ	*O'Brien & Fleming* 90% *RCIs for* ψ
1	2.28	0.831	9.74	(1.9, 51.1)	(0.8, 114.3)
2	2.07	0.538	7.90	(2.7, 23.1)	(2.6, 24.4)
3	2.14	0.463	8.46	(3.4, 21.3)	(3.8, 18.7)

If the jth pair has been observed by analysis k, stratum j consists of two samples of size 1, so $n_{Ajk} = n_{Bjk} = 1$, and otherwise stratum j is empty at analysis k.

Again assuming a common odds ratio ψ, the Mantel-Haenszel estimate $\hat{\psi}_k$ reduces to the McNemar (1947) estimate $\hat{\psi}_k = d_1(k)/d_2(k)$, where $d_1(k)$ is the number of discordant pairs of the first type, with $S_{Ajk} = 1$ and $S_{Bj} = 0$, and $d_2(k)$ the number of discordant pairs of the second type, with $S_{Ajk} = 0$ and $S_{Bj} = 1$, observed by the time of the kth analysis. Although the methods of Section 16.4 could be applied, it is easier to proceed directly. Let $d(k) = d_1(k) + d_2(k)$ be the total number of discordant pairs observed by the kth analysis. Then, conditional on $d(k)$, the variable $d_1(k)$ is binomial with success probability $p = \psi/(1 + \psi)$. Hence, group sequential procedures can be constructed using the methods of Section 16.1 with $S_k = d_1(k)$ and $n_k = d(k)$. In particular, RCIs for ψ can be obtained by transforming intervals for p obtained using the methods of Section 16.1.4.

16.6 Logistic Regression: Adjusting for Covariates

Consider the situation where binary observations $X_1, X_2, \ldots$ follow independent Bernoulli distributions with success probabilities given by a logit linear function of a parameter vector

$\boldsymbol{\beta} = (\beta_1, \ldots, \beta_r)$. Thus, for $i = 1, 2, \ldots$, the observation X_i is distributed as $B(p_i)$, where

$$\log\{p_i/(1-p_i)\} = \sum_{j=1}^{r} d_{ij}\beta_j$$

and $(d_{i1}, \ldots, d_{ir})$ is a vector of r covariates associated with subject i. Typically, $d_{i1} = 1$ for all i and β_1 represents an intercept term. We suppose that the data are collected group sequentially and for $k = 1, \ldots, K$, we denote the maximum likelihood estimate of $\boldsymbol{\beta}$ based on data accumulated by interim analysis k by $\widehat{\boldsymbol{\beta}}^{(k)}$. This is an example of a generalized linear model and, by the results of Section 15.6, estimates and related test statistics have, approximately, the standard group sequential distributions. In particular, proceeding as in Section 4.3.2, let $\boldsymbol{c}$ be a given vector of length r and $\mathcal{I}_k^{-1}$ a consistent estimate of the variance of $\boldsymbol{c}^T\widehat{\boldsymbol{\beta}}^{(k)}$. Then, if γ is a specified scalar, the standardized statistics

$$Z_k = (\boldsymbol{c}^T\widehat{\boldsymbol{\beta}}^{(k)} - \gamma)\sqrt{\mathcal{I}_k}, \quad k = 1, \ldots, K,$$

have approximately the canonical joint distribution (3.1) with $\theta = (\boldsymbol{c}^T\boldsymbol{\beta} - \gamma)$. Thus, we may construct group sequential tests of the hypothesis H_0: $\boldsymbol{c}^T\boldsymbol{\beta} = \gamma$ or repeated confidence intervals for $\boldsymbol{c}^T\boldsymbol{\beta}$.

An early example application of logistic regression modeling in a group sequential experiment described by Whitehead (1989) concerns a series of germination tests performed at periodic intervals to assess viability of seed samples. Other binary response regression models can also be employed in the group sequential framework. Jennison & Turnbull (1989, Sec. 5.5) propose a group sequential analysis of an odds ratio regression model in which the log odds ratio between a treatment and control group for an event of interest increases linearly in time, a reasonable model when the effect of treatment on a population is postulated to be gradual rather than immediate.

Returning to the Ille-et-Vilaine data of Table 16.6, we can analyze this data set with a logistic regression model in which the response variable is the indicator of exposure status, taking the value 1 if a subject is exposed and 0 otherwise. The first covariate distinguishes cases from controls with $d_{i1} = 1$ if subject i is a case and $d_{i1} = 0$ if subject i is a control. The next five variables denote the stratum for each subject, with $d_{ij} = 1$ if subject i belongs to stratum j and $d_{ij} = 0$ otherwise for $j = 2, \ldots, 6$. The absence of any interaction terms in the model implies a constant odds ratio across strata, and so we have the same model as was used in the analysis of this example in Section 16.4. However, we now use maximum likelihood estimation rather than the Mantel-Haenszel estimator. We are interested in the regression coefficient for the case/control indicator variable which is equal to the logarithm of the odds ratio. We therefore specify $\boldsymbol{c} = (1, 0, 0, 0, 0, 0)$. The logistic regressions can be readily performed using standard statistical software such as the `glm` function in R. Successive estimates for the log odds ratio based on cumulative data at analyses 1, 2 and 3 are 1.617, 1.597 and 1.670 with standard errors 0.335, 0.228 and 0.190, respectively. These estimates and standard errors are almost identical to those in Table 16.7 obtained using the Mantel-Haenszel estimates and, hence, so will be related RCIs for the odds ratio.

16.7 Poisson Data and Vaccine Trials

An important special case of the problem of comparing two Bernoulli probabilities described in Sections 16.2 and 16.3 is when the probabilities p_A and p_B are expected to be small and the consequent sample sizes are very large. We shall describe three examples of this type.

A Vaccine Efficacy Trial

In a vaccine efficacy trial, subjects are randomly allocated to receive the vaccine or a placebo inoculation. The incidence among subjects of the infection under study is tracked and rates of infection in the treated and control groups are compared. Denote the probabilities of becoming infected within a certain time from entry into the trial by p_V and p_C for subjects in the treated and control groups, respectively. We expect both these probabilities to be small. As an example, in the early results reported for the Phase III trial of the Pfizer COVID-19 vaccine, with over approximately 20,000 subjects per group, there were 162 cases among the controls and 8 cases in the vaccinated group; see Polack et al. (2020). The efficacy of a vaccine is measured by the quantity $V_E = 1 - (p_V/p_C)$, which can be viewed as the proportion of cases that would be avoided by vaccination. Assuming equal group sizes in the Pfizer example, the estimated of V_E is

$$\widehat{V_E} = 1 - \frac{8}{162} = 0.95,$$

which implies that 95% of cases would be prevented by this vaccine. In general, we are interested in testing the hypotheses

$$H_0\colon V_E \leq V_0 \quad \text{versus} \quad H_A\colon V_E > V_0,$$

where $0 < V_0 < 1$ and the vaccine cannot be considered clinically or economically worthwhile if $V_E \leq V_0$.

A vaccine safety trial

In a vaccine safety trial, the incidence of adverse reactions is monitored and rates are compared been subjects who were vaccinated and those in the control group who received a placebo inoculation. Denote the probabilities of having an adverse reaction within a certain time from entry into the trial by p_V and p_C for subjects in the treated and control groups, respectively. Again both probabilities should be very small. The relative risk of an adverse reaction is defined as $R = p_V/p_C$, and we wish to test

$$H_0\colon R \geq R_0 \quad \text{versus} \quad H_A\colon R < R_0,$$

where $R_0 > 1$ and the vaccine will be considered unsafe, regardless of any positive benefits, if $R \geq R_0$.

An environmental health monitoring study

Consider an environmental health monitoring study in which the incidence of babies born with birth defects is to be assessed in a community with a potential environmental hazard, such as a compromised water supply. Comparisons will be made with a matched control community of a similar size. The situation is analogous to that in a vaccine safety trial. Let p_A and p_B denote the probabilities of a birth defect in the exposed community and the control community, respectively. With the relative risk defined as $R = p_A/p_B$, suppose investigators wish to test

$$H_0\colon R \geq R_0 \quad \text{versus} \quad H_A\colon R < R_0,$$

where $R_0 > 1$ and action should be taken if $H_0\colon R \geq R_0$ cannot be rejected.

In all three of these examples, the small values of two Bernoulli probabilities preclude the use of the normal approximation. Instead, we can make use of the Poisson approximation to the binomial; see, for example, Grimmett & Stirzaker (2020, p. 94 and 145–149). Dragalin & Fedorov (2006) follow this approach and propose fixed sample size and group sequential tests. Suppose there are n_A and n_B subjects, respectively, in two treatment groups. We may approximate the distributions of S_A and S_B, the number of events in each group, by Poisson distributions with means $\lambda_A = n_A p_A$ and $\lambda_B = n_B p_B$. The conditional distribution of S_A given the total number of events $S_A + S_B = m$, say, is binomially distributed with a probability mass function $B_m(s_A, \pi)$ given by (16.4) with $\pi = \lambda_A/(\lambda_A + \lambda_B)$, for $s_A = 0, 1, \ldots, m$.

For ease of exposition, we consider the balanced case $n_A = n_B$, so $\pi = p_A/(p_A + _B)$. The hypothesis tests in the three examples can all be expressed in the form

$$H_0\colon \pi \geq \pi_0 \quad \text{versus} \quad H_A\colon \pi < \pi_0.$$

In the vaccine efficacy trial we set $p_A = p_V$, $p_B = p_C$, then

$$\pi = p_A/(p_A + p_B) = (1 - V_E)/(2 - V_E)$$

and $\pi_0 = (1 - V_0)/(2 - V_0)$. In the vaccine safety trial, we again set $p_A = p_V$, $p_B = p_C$, then

$$\pi = p_A/(p_A + p_B) = R/(R + 1)$$

and $\pi_0 = R_0/(R_0 + 1)$. In the environmental health monitoring study, we also have $\pi = R/(R + 1)$ and $\pi_0 = R_0/(R_0 + 1)$.

A test of the hypothesis H_0 can be constructed using the conditional distribution of S_A given $S = S_A + S_B$. Thus we have reduced the two-sample comparison to that of the one-sample problem described in Section 16.1. If results are monitored group sequentially, stopping boundaries can be constructed using the normal approximation as described in Section 16.1.1 or the exact methods of Sections 16.1.2 and 16.1.3. One could, for example, use one of the two-stage and three-stage plans displayed in Table 16.1 where now the cumulative sample sizes m_1, m_2 and m_3 in these designs correspond to the total numbers of observed events at each analysis time. For the three-stage plan number 6 in Table 16.1, the Tables 16.2 and 16.3 give details of confidence intervals upon termination and of RCIs for the parameter π, which may be used to give confidence intervals or RCIs for the relative risk R or vaccine efficiency V_E, as appropriate. Dragalin & Fedorov (2006) give details of the construction of fixed sample, curtailed and fully sequential procedures and present further examples of three-stage designs created specifically for vaccine trials.

In implementing a design in which interim analyses are to be conducted after specified numbers of events, we say the trial is "event-based". Operationally, this feature has implications for planning meetings of the monitoring committee. This is not unlike the situation of a therapeutic trial with a time-to-event endpoint, which we described in Section 4.5 and shall consider further in Chapter 17, where interim analyses are planned when certain numbers of events have been observed. If it is anticipated that the numbers of events at each analysis are likely to differ from the target values, error spending designs can be used.

For further discussion of methodology and a comprehensive review of the statistical and other issues concerning vaccine trials with special focus on the COVID-19 vaccine development process, we refer the reader to the papers by Chen & Ting (2020) and Silva, Maro & Kulldorff (2021).

16.8 Connection with Survival Analysis

A common situation where it is desired to compare two probabilities, as in Section 16.2, occurs in reliability tests or follow-up studies where two treatments are to be compared according to the proportions of subjects that have failed or responded within a fixed time horizon. If the actual times to failure or response are available, it is usually more appropriate to consider this as a problem with censored survival data. Group sequential procedures may then be based on the Kaplan-Meier (1958) estimates of the failure or response probabilities. This issue will be considered in Section 17.7 of the next chapter.

Chapter 17

Survival Data

17.1 Introduction

In many clinical trials, the primary response variable is a subject's survival time or the time to occurrence of a particular event such as relapse or remission of symptoms. These are elapsed times measured from an origin pertinent to each subject, often the date of entry into the study. Although time is typically measured on a chronological scale of days, weeks or months, this need not always be so. In some epidemiological studies cumulative exposure plays the role of survival time: in a veterinary study of dairy cattle the equivalent measure might be cumulative milk production, or in a reliability strength test it could be the cumulative stress or load applied. We shall phrase the methods of this chapter in terms of "survival" or "failure" time measured from entry to a study. However, the same methods can be applied quite generally for any time-to-event response.

An individual's survival time may be censored if failure has not occurred when that individual is last seen or when the data are analyzed. In Section 4.5, we introduced the basic concepts of survival models, including the survival function $S(t)$ and hazard rate $h(t)$ of a response time distribution. We also described the two-sided, group sequential log-rank test for testing equality of two survival distributions. Suppose two sets of subjects, A and B, have respective hazard rates $h_A(t)$ and $h_B(t)$, then the two-sided log-rank test is used to test the null hypothesis H_0: $h_A(t) = h_B(t)$ for all $t > 0$, or equivalently $S_A(t) = S_B(t)$ for $t > 0$, and this test is particularly powerful against "proportional hazards" alternatives where $h_B(t) = \lambda h_A(t)$ for some constant λ. In Section 4.5 we constructed group sequential tests following the significance level approach, which requires equal increments in information between successive analyses if the Type I error rate α is to be attained accurately. In a survival study, information is roughly proportional to the total number of observed failures and increments are usually unequal and unpredictable. The error spending approach of Chapter 8 is thus an attractive alternative, as we demonstrated in the example of Section 8.6.4.

In Section 17.2, we shall review the group sequential log-rank test described in Section 4.5, adding a modification to accommodate tied survival times. This test is further generalized in Section 17.3 to permit the inclusion of strata. In Section 17.4, we describe how to adjust the comparison of survival distributions for the effects of covariates through use of the proportional hazards regression model of Cox (1972) and we explain how to construct RCIs for a hazard ratio, λ, in Section 17.5. These methods are illustrated by a case study in Section 17.6. In Section 17.7, we show how successive Kaplan-Meier estimates of the survival function (Kaplan & Meier, 1958) can be used as a basis for group sequential tests and repeated confidence intervals concerning survival probabilities and quantiles.

17.2 The Log-rank Test

Consider the problem of testing the equality of survival distributions $S_A(t)$ and $S_B(t)$ for two treatment arms, A and B, based on accumulating survival data. At each analysis

DOI: 10.1201/9781584888482-17

we observe a failure or censoring time for each subject, measured from that subject's date of entry or randomization as defined in the study protocol. Let d_k, $k = 1, \ldots, K$, denote the total number of uncensored failures observed across both treatment arms when analysis k is conducted. Some of these times may be tied and we suppose that d'_k of the d_k failure times are distinct, where $1 \leq d'_k \leq d_k$. We denote these distinct failure times by $\tau_{1,k} < \tau_{2,k} < \ldots < \tau_{d'_k,k}$ and let $r_{iA,k}$ and $r_{iB,k}$ denote the numbers at risk on treatment arms A and B, respectively, just before time $\tau_{i,k}$. Finally, let $\delta_{iA,k}$ and $\delta_{iB,k}$ denote the numbers on treatment arms A and B, respectively, that fail at time $\tau_{i,k}$ and define $\delta_{i,k} = \delta_{iA,k} + \delta_{iB,k}$ for $i = 1, \ldots, d'_k$. If there are no ties, as assumed in Section 4.5, then $\delta_{i,k} = 1$ and either $\delta_{iA,k} = 1$ and $\delta_{iB,k} = 0$ or $\delta_{iA,k} = 0$ and $\delta_{iB,k} = 1$ for each pair i and k.

If the survival distributions $S_A(t)$ and $S_B(t)$ are equal, the conditional distribution of $\delta_{iB,k}$ given $r_{iA,k}$, $r_{iB,k}$ and $\delta_{i,k}$ is hypergeometric with expectation

$$e_{i,k} = \frac{r_{iB,k}\, \delta_{i,k}}{r_{iA,k} + r_{iB,k}}$$

and variance

$$v_{i,k} = \frac{r_{iA,k}\, r_{iB,k}\, \delta_{i,k}\, (r_{iA,k} + r_{iB,k} - \delta_{i,k})}{(r_{iA,k} + r_{iB,k} - 1)\, (r_{iA,k} + r_{iB,k})^2}.$$

The standardized log-rank statistic at analysis k is given by

$$Z_k = \frac{\sum_{i=1}^{d'_k} (\delta_{iB,k} - e_{i,k})}{\left(\sum_{i=1}^{d'_k} v_{i,k}\right)^{1/2}} \tag{17.1}$$

and the information $\mathcal{I}_k$ for the log hazard ratio is

$$\mathcal{I}_k = \sum_{i=1}^{d'_k} v_{i,k}. \tag{17.2}$$

In the absence of ties, the numerator of (17.1) reduces to the log-rank score statistic (4.8) and the information given by (17.2) is the same as (4.9).

The log-rank test has optimum power to detect alternatives when hazard rates in the two treatment arms are proportional, as described in Section 17.1. The sequence of log-rank statistics defined by (17.1) then has, approximately, the canonical joint distribution (3.1), given $\mathcal{I}_1, \ldots, \mathcal{I}_K$, with $\theta = \log(\lambda)$, the log hazard ratio; references to the derivation of this result were given in Section 4.5 and we provide further references in Section 17.9. Although an estimate of θ does not appear explicitly in the expressions for the log-rank test statistic, the estimate

$$\hat{\theta}^{(k)} = Z_k / \sqrt{\mathcal{I}_k}$$

at analysis k can be derived from the formulae for the canonical joint distribution.

Since the canonical joint distribution holds, the methods described in Chapter 8 can be used to construct group sequential error spending tests from the sequence of statistics Z_k and information levels $\mathcal{I}_k$. In designing a maximum information trial to meet a given power requirement, it is necessary to predict the information levels that will arise, especially that at the final possible analysis. Here, it is helpful to note that each $v_{i,k}$ is approximately $\delta_{i,k}/4$ if $r_{iA,k} \approx r_{iB,k}$ and either $\delta_{i,k} = 1$ or $\delta_{i,k}$ is small relative to $r_{iA,k} + r_{iB,k}$. Hence, $\mathcal{I}_k$ will be approximately equal to $d_k/4$ and the final information level will be close to one quarter of the total number of observed failures. In Section 8.2.3, we applied a two-sided group sequential log-rank test to the BHAT data in the setting of a maximum duration trial. We shall describe a further application, illustrating a one-sided test in a maximum information trial, in Section 17.6.

17.3 The Stratified Log-rank Test

Suppose now that each subject entering the trial can be classified as belonging to precisely one of L distinct categories or strata and it is desired to test the null hypothesis that survival distributions under treatments A and B are equal *within each stratum*. Specifically, if $h_{A\ell}(t)$ and $h_{B\ell}(t)$ denote the hazard rates for subjects in strata $\ell = 1, \ldots, L$ under treatments A and B respectively, then in a two-sided test, the null hypothesis is

$$H_0\colon\ h_{A\ell}(t) = h_{B\ell}(t) \ \text{ for all } t > 0 \text{ and } \ell = 1, \ldots, L.$$

The stratified log-rank test can be used in this situation and is particularly powerful against alternatives in which the hazard rates in the two treatment arms are proportional within each stratum with a common hazard ratio λ in each case, i.e., $h_{B\ell}(t) = \lambda h_{A\ell}(t)$ for all $t > 0$ and $\ell = 1, \ldots, L$. The test is described in the non-sequential setting in a number of textbooks on survival analysis, for example, Klein & Moeschberger (1997, Sec. 7.5).

Generalizing the notation of Sections 4.5 and 17.2, we let $d_{\ell k}$, $k = 1, \ldots, K$ and $\ell = 1, \ldots, L$, denote the number of uncensored failures in stratum ℓ observed when analysis k is conducted. Considering only those subjects in stratum ℓ, we suppose $d'_{\ell k}$ of the $d_{\ell k}$ failure times observed at analysis k are distinct and denote these times by $\tau_{1,\ell k} < \tau_{2,\ell k} < \ldots < \tau_{d'_{\ell k},\ell k}$. We also define $r_{iA,\ell k}$ and $r_{iB,\ell k}$, for $i = 1, \ldots, d'_{\ell k}$, to be the numbers at risk on each treatment arm, A and B, in stratum ℓ just before time $\tau_{i,\ell k}$. Finally, we let $\delta_{iA,\ell k}$ and $\delta_{iB,\ell k}$ denote the numbers that fail on treatments A and B at time $\tau_{i,\ell k}$ and define $\delta_{i,\ell k} = \delta_{iA,\ell k} + \delta_{iB,\ell k}$. In the absence of ties, $\delta_{i,\ell k} = 1$ and we must have either $\delta_{iA,\ell k} = 1$ and $\delta_{iB,\ell k} = 0$ or $\delta_{iA,\ell k} = 0$ and $\delta_{iB,\ell k} = 1$ for each triple i, ℓ and k.

Under H_0, the conditional distribution of $\delta_{iB,\ell k}$ given $r_{iA,\ell k}$, $r_{iB,\ell k}$ and $\delta_{i,\ell k}$ is hypergeometric with expectation

$$e_{i,\ell k} = \frac{r_{iB,\ell k}\, \delta_{i,\ell k}}{r_{iA,\ell k} + r_{iB,\ell k}}$$

and variance

$$v_{i,\ell k} = \frac{r_{iA,\ell k}\, r_{iB,\ell k}\, \delta_{i,\ell k}\, (r_{iA,\ell k} + r_{iB,\ell k} - \delta_{i,\ell k})}{(r_{iA,\ell k} + r_{iB,\ell k} - 1)\, (r_{iA,\ell k} + r_{iB,\ell k})^2}.$$

The stratified log-rank statistic at analysis k is defined as

$$Z_k = \frac{\sum_{\ell=1}^{L} \sum_{i=1}^{d'_{\ell k}} (\delta_{iB,\ell k} - e_{i,\ell k})}{\left(\sum_{\ell=1}^{L} \sum_{i=1}^{d'_{\ell k}} v_{i,\ell k}\right)^{1/2}} \tag{17.3}$$

and the information for the log hazard ratio is

$$\mathcal{I}_k = \sum_{\ell=1}^{L} \sum_{i=1}^{d'_{\ell k}} v_{i,\ell k}. \tag{17.4}$$

When $L = 1$ and there is only a single stratum, (17.3) reduces to (17.1) and (17.4) to (17.2). If, in addition, there are no ties, the numerator in (17.3) is equal to the log-rank score statistic given by (4.8) and the information given by (17.4) reduces to (4.9). Large sample results for the non-stratified case can be extended to show that, if the numbers of failures in each stratum are large, the sequence of stratified log-rank statistics $Z_1, \ldots, Z_K$ also has, approximately, the canonical joint distribution (3.1), given $\mathcal{I}_1, \ldots, \mathcal{I}_K$, with $\theta = \log(\lambda)$. Group sequential tests can, therefore, be based on the statistics Z_k and information levels $\mathcal{I}_k$ in the usual way. As for the log-rank statistic, the information level $\mathcal{I}_k$ can be expected to be close to a quarter of the total number of deaths observed by analysis k, although slightly lower values will occur if the numbers at risk on the two treatment arms become uneven in any one stratum. An application of the group sequential stratified log-rank test will be presented in Section 17.6.

17.4 Group Sequential Methods for Survival Data with Covariates

17.4.1 The Cox Regression Model

Suppose independent subjects $i = 1, 2, \ldots$ are observed, each having a column vector of covariates $\boldsymbol{x}_i = (x_{i1}, \ldots, x_{ip})^T$. In the proportional hazards regression model (Cox, 1972), the survival distribution for subject i is continuous with hazard rate

$$h_i(t) = h_0(t)\, e^{\boldsymbol{\beta}^T \boldsymbol{x}_i}, \quad t > 0, \tag{17.5}$$

where $h_0(t)$ represents an unknown baseline hazard function and $\boldsymbol{\beta} = (\beta_1, \ldots, \beta_p)^T$ is a column vector of parameters to be estimated. The log-linear link function between covariate values x_{ij} and the hazard rate $h_i(t)$ is not as restrictive as might seem at first, since interactions and transformed covariates may be included in the vector $\boldsymbol{x}$. Entry of individuals to a study will usually be staggered and survival, measured as elapsed time from entry to the study, can be subject to competing risk censoring, which we shall assume to be non-informative (see Klein & Moeschberger, 1997, p. 91) as well as "end-of-study" censoring at interim and final analyses.

We use the same notation as in previous sections. For $k = 1, \ldots, K$, we let d_k denote the total number of failures observed by the time the kth analysis is conducted. Supposing for now that there are no ties, we denote the survival times of the subjects observed to have failed at analysis k by $\tau_{1,k} < \tau_{2,k} < \ldots < \tau_{d_k,k}$. Let $\boldsymbol{x}^k_{(i)}$, $i = 1, \ldots, d_k$, denote the covariate vector associated with the individual who fails at time $\tau_{i,k}$ and define $\mathcal{R}_{i,k}$ to be the set of indices of individuals known at analysis k to have survived at least up to just before time $\tau_{i,k}$.

Following Cox (1972), we estimate the vector $\boldsymbol{\beta}$ at the kth analysis by maximizing the log partial likelihood

$$\mathcal{L}_k(\boldsymbol{\beta}) = \sum_{i=1}^{d_k} \left\{ \boldsymbol{\beta}^T \boldsymbol{x}^k_{(i)} - \log\left(\sum_{j \in \mathcal{R}_{i,k}} e^{\boldsymbol{\beta}^T \boldsymbol{x}_j} \right) \right\}. \tag{17.6}$$

This estimate, denoted $\widehat{\boldsymbol{\beta}}^{(k)}$, satisfies the score equations

$$S(k, \boldsymbol{\beta}) = \frac{\partial \mathcal{L}_k(\boldsymbol{\beta})}{\partial \boldsymbol{\beta}} = \sum_{i=1}^{d_k} \left(\boldsymbol{x}^k_{(i)} - \frac{\sum_{j \in \mathcal{R}_{i,k}} \boldsymbol{x}_j\, e^{\boldsymbol{\beta}^T \boldsymbol{x}_j}}{\sum_{j \in \mathcal{R}_{i,k}} e^{\boldsymbol{\beta}^T \boldsymbol{x}_j}} \right) = 0. \tag{17.7}$$

The expressions (17.6) and (17.7) can be adapted to accommodate ties; see, for example, Klein & Moeschberger (1997, Sec. 8.3).

The model (17.5) can be modified to include strata with different baseline hazard rates, as in the stratified proportional hazards model of Section 17.3. In the stratified Cox regression model, the hazard rate for subject i in stratum ℓ, with covariate vector $\boldsymbol{x}_i$, is

$$h_{i\ell}(t) = h_{0\ell}(t)\, e^{\boldsymbol{\beta}^T \boldsymbol{x}_i}, \quad t > 0, \tag{17.8}$$

where the baseline hazard rate, $h_{0\ell}$, depends on the stratum ℓ but the vector of regression coefficients, $\boldsymbol{\beta}$, remains constant across strata. If there are L strata, the log partial likelihood is a sum of L terms of the form (17.6) computed separately within each stratum. Differentiating this with respect to $\boldsymbol{\beta}$ produces a sum of L terms in the score equations corresponding to (17.7); for further details, see Klein & Moeschberger (1997, Sec. 9.3).

The score equations (17.7) must be solved numerically and many statistical computer software packages now include procedures to do this. The information matrix associated with $\widehat{\boldsymbol{\beta}}^{(k)}$ is $\mathcal{I}_k(\widehat{\boldsymbol{\beta}}^{(k)})$, where

$$\mathcal{I}_k(\boldsymbol{\beta}) = \frac{-\partial^2}{\partial \boldsymbol{\beta}^2} \mathcal{L}_k(\boldsymbol{\beta}).$$

This matrix and its inverse are routinely provided along with $\widehat{\boldsymbol{\beta}}^{(k)}$ in the output of statistical software packages.

It can be shown that the sequence of estimates $\widehat{\boldsymbol{\beta}}^{(1)}, \ldots, \widehat{\boldsymbol{\beta}}^{(K)}$ has, asymptotically, the standard multivariate normal joint distribution given by (15.1); see the references in Section 17.9. Hence, the sequence of standardized vector statistics

$$\mathcal{I}_1(\widehat{\boldsymbol{\beta}}^{(1)})^{1/2}\widehat{\boldsymbol{\beta}}^{(1)}, \ldots, \mathcal{I}_K(\widehat{\boldsymbol{\beta}}^{(K)})^{1/2}\widehat{\boldsymbol{\beta}}^{(K)}$$

has an asymptotic joint distribution that is the multivariate analogue to the canonical joint distribution (3.1). The remainder of this section describes applications of this result.

17.4.2 A Test of Equality of Two Survival Distributions with Adjustment for Covariates

We can now derive a generalization of the group sequential log-rank test incorporating adjustment for covariates. Suppose survival distributions are modeled by the Cox regression model (17.5) or its stratified version (17.8) and x_{i1} is a binary treatment indicator, so that the scalar coefficient β_1 represents the treatment effect. Our goal is to construct a two-sided group sequential test of H_0: $\beta_1 = 0$ in the presence of confounding variables $x_{i2}, \ldots, x_{ip}$ for each subject i. The maximum partial likelihood estimate of β_1 at stage k is $\hat{\beta}_1^{(k)}$ and the variance of this estimate is approximated by

$$v_k = [\{\mathcal{I}_k(\widehat{\boldsymbol{\beta}}^{(k)})\}^{-1}]_{11}, \tag{17.9}$$

the $(1,1)$ element of the inverse of the information matrix $\mathcal{I}_k(\widehat{\boldsymbol{\beta}}^{(k)})$; see Jennison & Turnbull (1997a, Sec. 5.2). An alternative variance estimator, which is also consistent under H_0 and can be more accurate in small samples is obtained by replacing $\widehat{\boldsymbol{\beta}}^{(k)}$ in (17.9) with the value of $\boldsymbol{\beta}$ that maximizes (17.6) when β_1 is constrained to be zero. Defining the standardized statistics

$$Z_k = \hat{\beta}_1^{(k)}/\sqrt{v_k}, \quad k = 1, \ldots, K, \tag{17.10}$$

the results stated at the end of Section 17.4.1 imply that $Z_1, \ldots, Z_K$ have, approximately, the canonical joint distribution (3.1) when we equate θ with β_1 and $\mathcal{I}_k$ with v_k^{-1}.

Note that if $p = 1$, so the only regression term is for the treatment effect, the assumed model is identical to the proportional hazards model of Section 17.2, or the stratified proportional hazards model of Section 17.3 in the case of a stratified Cox regression model. The statistics (17.10) are then asymptotically equivalent to the standardized log-rank statistics and either statistic may be used to define a group sequential test.

The effects of covariates on the observed information add to the difficulty of organizing interim analyses to occur at pre-specified information levels, and the error spending method of Chapter 8 is the obvious choice for constructing testing boundaries. If randomization and similar patterns of survival on the two treatment arms lead to balance in covariate values between treatments throughout a study, information for β_1 will be roughly one quarter of the number of observed failures, as in the log-rank test. Thus, in designing a study to reach a target information level $\mathcal{I}_{max}$, it might be decided to enroll a total of $4\mathcal{I}_{max}/\gamma$ subjects, where γ is the estimated probability that any one subject will be observed to fail by the time of the last scheduled analysis. Alternatively, a somewhat higher sample size could be chosen to protect against errors in estimating the baseline failure rate, the level of competing risk censoring, or the relation between number of failures and information.

17.4.3 Factorial Designs

Another application of the results of Section 17.4.1 is in making inferences about a linear combination of parameters of the form $\boldsymbol{d}^T\boldsymbol{\beta} = d_1\beta_1 + \ldots + d_p\beta_p$. This could be desirable if the x-variables represent factors in a factorial design with multiple treatments, a trend in clinical trials noted by Natarajan et al. (1996). The vector $\boldsymbol{d} = (d_1, \ldots, d_p)^T$ might then be chosen so that $\boldsymbol{d}^T\boldsymbol{\beta}$ represents a main effect, a linear trend, or possibly an interaction. We can test the hypothesis H_0: $\boldsymbol{d}^T\boldsymbol{\beta} = 0$ by defining the sequence of test statistics

$$Z_k = \boldsymbol{d}^T\widehat{\boldsymbol{\beta}}^{(k)}\sqrt{\mathcal{I}_k}, \quad k = 1, \ldots, K,$$

where $\mathcal{I}_k = \{\boldsymbol{d}^T\mathcal{I}_k^{-1}(\widehat{\boldsymbol{\beta}}^{(k)})\boldsymbol{d}\}^{-1}$. This sequence has, approximately, the canonical joint distribution (3.1), given $\mathcal{I}_1, \ldots, \mathcal{I}_K$, and group sequential tests can be constructed in the usual way. Note, however, that in a factorial trial involving multiple treatments, more than one main effect may be designated a primary endpoint to be monitored. We shall discuss group sequential tests for multivariate endpoints in Chapters 19 and 20.

17.5 Repeated Confidence Intervals for a Hazard Ratio

Consider again the problem of comparing the survival distributions for two treatment groups with adjustment for covariates, as described in Section 17.4.2. We can use the arguments of Section 10.1 to construct RCIs for β_1, the log hazard ratio between treatments. These RCIs are obtained by inverting a family of group sequential tests of hypotheses H_0: $\beta_1 = \widetilde{\beta}_1$, where $\widetilde{\beta}_1$ ranges over the real line. The RCI for β_1 is the set of values of $\widetilde{\beta}_1$ for which H_0 is currently accepted.

A group sequential test of H_0: $\beta_1 = \widetilde{\beta}_1$ can be based on the standardized statistics

$$Z_k(\widetilde{\beta}_1) = (\hat{\beta}_1^{(k)} - \widetilde{\beta}_1)/\sqrt{v_k}, \quad k = 1, \ldots, K,$$

where $\hat{\beta}_1^{(k)}$ and v_k are as defined in Section 17.4.2. A sequence of RCIs for β_1 with overall confidence level $1 - \alpha$ is then given by

$$(\hat{\beta}_1^{(k)} \pm c_k(\alpha)\sqrt{v_k}), \quad k = 1, \ldots, K, \tag{17.11}$$

where $\pm c_k(\alpha)$, $k = 1, \ldots, K$, are critical values for a group sequential, two sided test with Type I error probability α based on standardized statistics $Z_1, \ldots, Z_K$ with the canonical joint distribution (3.1) for information levels $\mathcal{I}_k = v_k^{-1}$. If an error spending approach is used, each $c_k(\alpha)$ is obtained as the solution of (8.1) with π_k equal to the Type I error probability allocated to analysis k. RCIs for the hazard ratio e^{β_1} are found by exponentiating the limits in (17.11). RCIs can be constructed in a similar fashion for linear combinations of regression parameters, as described in Section 17.4.3.

Small sample accuracy may be improved by tailoring the estimate of the variance of $\hat{\beta}_1^{(k)}$ to each hypothesized value $\widetilde{\beta}_1$. This is done by replacing $\widehat{\boldsymbol{\beta}}^{(k)}$ in (17.9) with the maximum partial likelihood estimate of $\boldsymbol{\beta}$ subject to the constraint $\beta_1 = \widetilde{\beta}_1$, thereby defining a new variance estimate, $v_k(\widetilde{\beta}_1)$ say. The kth RCI is then

$$\{\widetilde{\beta}_1\colon \hat{\beta}_1^{(k)} \in [\widetilde{\beta}_1 \pm c_k(\alpha)\sqrt{v_k(\widetilde{\beta}_1)}\,]\}.$$

Computation of this RCI is less straightforward since a separate v_k must be found for each value of $\widetilde{\beta}_1$ considered, and the constrained estimation of variance is not so readily available in standard software packages. In principle, the critical values $c_1(\alpha), \ldots, c_K(\alpha)$ also depend on $\widetilde{\beta}_1$, since they are defined for the particular information sequence $\mathcal{I}_k = \{v_k(\widetilde{\beta}_1)\}^{-1}$, $k = 1, \ldots, K$; however, it should usually suffice to calculate the $c_k(\alpha)$ using the single sequence of information levels $\mathcal{I}_k = v_k^{-1}$ with v_k taken from (17.9) for $k = 1, \ldots, K$.

It is instructive to examine more closely the case $p = 1$ when the only regression term, x_1, is a binary variable taking the value 0 for treatment A and 1 for treatment B (or equivalently 1 for A and 2 for B). We then have a two-sample comparison under the proportional hazards model of Section 17.2 or under a stratified proportional hazards model, as in Section 17.3. For simplicity, we consider the non-stratified proportional hazards model and suppose there are no ties; the stratified case is similar but with sums over strata in the formulae. We write $\theta = \beta_1$ so $\theta = \log(\lambda)$ if, as in Section 17.2, λ denotes the hazard ratio of treatment B to treatment A. Using the notation of Section 17.2, the partial likelihood for the data observed at analysis k is

$$\prod_{i=1}^{d_k} \left(\frac{r_{iA,k}}{r_{iA,k} + e^{\theta} r_{iB,k}} \right)^{\delta_{iA,k}} \left(\frac{e^{\theta} r_{iB,k}}{r_{iA,k} + e^{\theta} r_{iB,k}} \right)^{\delta_{iB,k}} . \tag{17.12}$$

The value $\hat{\theta}^{(k)}$ maximizing this expression is the maximum partial likelihood estimate, $\hat{\beta}_1^{(k)}$ in our previous notation for the Cox model. We can calculate $\hat{\theta}^{(k)}$ as the root of the score equation $S_k(\theta_0) = 0$, where

$$S_k(\theta_0) = \sum_{i=1}^{d_k} \left\{ \delta_{iB,k} - \frac{e^{\theta_0} r_{iB,k}}{r_{iA,k} + e^{\theta_0} r_{iB,k}} \right\} \tag{17.13}$$

is the score statistic for testing H_0: $\theta = \theta_0$ at analysis k, obtained as the derivative with respect to θ of the logarithm of (17.12) at $\theta = \theta_0$. The information for θ when $\theta = \theta_0$ is the second derivative of the log partial likelihood,

$$\mathcal{I}_k(\theta_0) = \sum_{i=1}^{d_k} \frac{e^{\theta_0} r_{iA,k} r_{iB,k}}{(r_{iA,k} + e^{\theta_0} r_{iB,k})^2} .$$

By definition, $\mathcal{I}_k(\hat{\theta}^{(k)})^{-1} = v_k$ and, hence, the RCI sequence (17.11) can also be written as

$$(\hat{\theta}^{(k)} \pm c_k(\alpha)\{\mathcal{I}_k(\hat{\theta}^{(k)})\}^{-1/2}), \quad k = 1, \ldots, K. \tag{17.14}$$

Tailoring the estimated variance of $\hat{\theta}^{(k)}$ to each hypothesized value of θ gives the alternative sequence

$$\{\theta_0 \colon \hat{\theta}^{(k)} \in [\,\theta_0 \pm c_k(\alpha)\{\mathcal{I}_k(\theta_0)\}^{-1/2}\,]\}, \quad k = 1, \ldots, K.$$

However, the sequence (17.14) is simpler to compute and Harrington (1989) notes that it is sufficient, in large samples, to base the RCI at analysis k on the single information estimate $\mathcal{I}_k(\hat{\theta}^{(k)})$.

If one wishes to avoid solving the score equations to find the maximum partial likelihood estimates $\hat{\theta}^{(k)}$, $k = 1, \ldots, K$, a sequence of RCIs can be constructed from the standardized log-rank statistics Z_k and associated information levels $\mathcal{I}_k$ given by (17.1) and (17.2) for the unstratified proportional hazards model or by (17.3) and (17.4) for the stratified model. These $\mathcal{I}_k$ are the same as $\mathcal{I}_k(0)$, $k = 1, \ldots, K$, in our current notation. Also, $Z_k = S_k(0)/\sqrt{\mathcal{I}_k}(0)$ where $S_k(0)$ is the log-rank score statistic for testing H_0: $\theta = 0$, referred to as S_k in Section 4.5. Under the assumption that the Z_k follow the canonical joint distribution (3.1) given information levels $\mathcal{I}_k$, the RCIs for θ are

$$(Z_k/\sqrt{\mathcal{I}_k} \pm c_k(\alpha)/\sqrt{\mathcal{I}_k}), \quad k = 1, \ldots, K, \tag{17.15}$$

where the constants $c_k(\alpha)$ are calculated using the observed values of $\mathcal{I}_1, \ldots, \mathcal{I}_K$. As noted in Section 17.2, $Z_k/\sqrt{\mathcal{I}_k}$ is the estimate of θ derived from the log-rank statistic at analysis k. These RCIs are reliable if θ is close to zero but, when the true value of θ is away from zero, one should expect the intervals (17.14) to attain their nominal coverage probabilities more accurately.

Substituting $Z_k = S_k(0)/\sqrt{\mathcal{I}_k(0)}$ into (17.15) with the approximation $\mathcal{I}_k(0) \approx d_k/4$, which is appropriate if the numbers at risk on the two treatment arms remain roughly equal and θ is close to zero, we obtain the RCIs for θ:

$$\left(\frac{4\,S_k(0)}{d_k} \pm \frac{2\,c_k(\alpha)}{\sqrt{d_k}} \right), \quad k = 1, \ldots, K. \tag{17.16}$$

This formula, first given in Jennison & Turnbull (1984, Eqs 1 and 2; 1989, Eq. 4.4), is attractive for its computational simplicity. However, simulation studies such as those reported in Jennison & Turnbull (1984, 1989) suggest that the intervals given by (17.16) may fail to maintain a coverage probability adequately close to $1-\alpha$ when θ is not close to zero or censoring patterns differ in the two treatment groups. The approximation $\mathcal{I}_k \approx d_k/4$ is useful in its own right and can help when designing a group sequential study to reach a target information level.

17.6 Example: A Clinical Trial for Carcinoma of the Oropharynx

17.6.1 Description of the Study

We shall illustrate the methods just described by applying them to a clinical trial conducted by the Radiation Therapy Oncology Group in the U.S.A. to investigate treatments of carcinoma of the oropharynx. We use the data from six of the larger institutions participating in this trial as recorded by Kalbfleisch & Prentice (2002, Appendix A.11). Subjects were recruited to the study between 1968 and 1972 and randomized to either a standard radiotherapy treatment or an experimental treatment in which the radiotherapy was supplemented by chemotherapy. The major endpoint was patient survival and patients were followed until around the end of 1973. Several baseline covariates, thought to have strong prognostic value, were also recorded.

The study was not monitored with a group sequential plan but, for purposes of illustration, we have reconstructed patients' survival times and their status, dead or censored, at times 720, 1080, 1440, 1800 and 2160 days from the beginning of 1968. As the central survival records would not have been updated continuously, our constructed data sets most likely resemble the information that would have been available at interim analyses conducted a month or two after these times, and so they are an approximation to the data that could have been studied by a monitoring committee meeting at dates a little after 2, 3, 4, 5 and 6 years from the start of the study. The longer waiting period to the first interim analysis is intended to compensate for the slow initial accrual of survival information while only a few patients had been entered into the trial.

Since the experimental treatment involved chemotherapy as well as radiotherapy, the researchers would have been looking for a substantive improvement in survival on this treatment in return for the additional discomfort and short term health risks. A one-sided testing formulation is, therefore, appropriate and we shall conduct our retrospective interim analyses as a group sequential test of the null hypothesis of no treatment difference against the one-sided alternative that the new combination therapy is superior to the standard treatment of radiotherapy alone. For the sake of illustration, we suppose the study was designed to achieve a Type I error probability of $\alpha = 0.025$ and power $1-\beta = 0.8$ when the log hazard ratio for the standard treatment to the experimental is equal to 0.5. By (5.1), our stated Type I error and power condition lead to a required information level

$$\mathcal{I}_{f,1} = \{\Phi^{-1}(0.975) + \Phi^{-1}(0.8)\}^2/0.5^2 = 31.40$$

for a single fixed sample test and we shall multiply this by the appropriate factor R to find the target maximum information level in a group sequential test.

17.6.2 A Group Sequential Stratified Log-rank Test

We first consider group sequential monitoring using the log-rank test. The study protocol allowed patients to receive additional medical care "as deemed prudent" and with no restrictions, except a prohibition of the study treatment, after the initial 90 day treatment period. Kalbfleisch & Prentice (2002, p. 120) note that decisions on such medical care were liable to be made on an institutional basis and so it is reasonable to allow for a possible effect on baseline survival patterns by stratifying the log-rank test with respect to institution.

For $k = 1, \ldots, 5$, let Z_k denote the standardized, stratified log-rank statistic (17.3) computed at analysis k and denote the associated information (17.4) by $\mathcal{I}_k$. We have taken treatment A here to be the experimental treatment and treatment B the standard, so the Z_k have positive expectation under the alternative that the experimental treatment improves survival. The stratified log-rank test is particularly appropriate if survival times follow a stratified proportional hazards model in which the hazard rate at time t is $h_{0\ell}(t)$ or $\lambda h_{0\ell}(t)$ for a subject in stratum ℓ on the experimental or standard treatment, respectively. For this model, we denote the log hazard ratio between treatments by $\theta = \log(\lambda)$. The following analysis is based on the assumption that $Z_1, \ldots, Z_5$ follow the standard joint distribution (3.1), given information levels $\mathcal{I}_1, \ldots, \mathcal{I}_5$.

The observed information level at each analysis depends primarily on the total number of deaths at that time and is therefore subject to considerable random variability. It is unlikely that observed information levels at interim analyses conducted on pre-specified dates will be equally spaced and we therefore adopt the error spending approach to control Type I and Type II error rates, following the description for one-sided tests in Section 8.3. We shall work in the maximum information paradigm of Section 8.2.2, using Rho family error spending functions with $\rho = 2$. Thus Type I error probability is spent according to the function $f(\tilde{t}) = \min\{\alpha\tilde{t}^2, \alpha\}$ and Type I error probability according to the function $g(\tilde{t}) = \min\{\beta\tilde{t}^2, \beta\}$, where $\tilde{t}$ represents the ratio of the current information level to $\mathcal{I}_{max}$, the target maximum information level (we denote this information fraction by $\tilde{t}$ to avoid confusion with t, an individual's survival time). This form of error spending function was considered in Section 8.3 and shown to provide good opportunity for early stopping with about a 10% increase in the maximum information level over that needed for a fixed sample size test.

With $\alpha = 0.025$, $\beta = 0.2$, $K = 5$ analyses planned, and a non-binding futility boundary, we multiply $\mathcal{I}_{f,1} = 31.40$, the information for θ required in a fixed sample test, by $R_{NBF}(5, 0.025, 0.2, 2) = 1.133$, taken from Table 8.12, to obtain the maximum information level

$$\mathcal{I}_{max} = 1.133 \times 31.40 = 35.58.$$

If numbers at risk remain roughly equal on the two treatment arms, each observed death contributes about 0.25 to the information for the log hazard ratio, so an information level of 35.58 would need a total of around 142 deaths. We shall assume this maximum information level appeared plausible to the organizers of our hypothetical group sequential trial. In principle, the target information level, $\mathcal{I}_{max}$, must be reached eventually but this is not easy to guarantee in a survival study as loss to follow-up or extended survival may staunch the flow of deaths after patient accrual is complete. As explained in Section 8.2.2, this problem of "under-running" can be handled by spending all the remaining Type I error probability at a pre-planned final analysis, whether or not $\mathcal{I}_{max}$ has been achieved.

We conduct our group sequential test according to a rule of the form (5.2), stopping at analysis k to conclude $\theta > 0$ with a recommendation for the experimental combination treatment if $Z_k \geq b_k$, and stopping to accept H_0: $\theta \leq 0$, concluding that the experimental treatment is no better than the control, if $Z_k \leq a_k$. The critical values a_k and b_k are found as the solutions to (8.17) and (8.19) and only become known as analysis k is reached.

Table 17.1 *Summary data and critical values for a group sequential, stratified log-rank test in the clinical trial for carcinoma of the oropharynx. Critical values a_k and b_k are for a test of $H_0: \theta \le 0$ against $\theta > 0$, where θ is the log hazard ratio and $\theta > 0$ implies the experimental treatment is superior to the control.*
The test uses stratified log-rank statistics (17.3) *and the associated information levels, $\mathcal{I}_k$, are given by* (17.4). *The test is constructed using Rho family error spending functions with $\rho = 2$ and a non-binding futility boundary. It is designed to have Type I error probability $\alpha = 0.025$ and power $1 - \beta = 0.8$ at $\theta = 0.5$ if $\mathcal{I}_1, \ldots, \mathcal{I}_5$ are equally spaced, concluding with $\mathcal{I}_5 = 35.58$.*

Analysis k	*Date*	*Number of patients entered*	*Number of deaths*	*Stratified log-rank score*	$\mathcal{I}_k$	a_k	b_k	Z_k
1	12/69	83	27	−2.41	5.43	−1.44	3.25	−1.04
2	12/70	126	58	−3.55	12.58	−0.23	2.78	−1.00
3	12/71	174	91	−5.56	21.11	0.75	2.46	−1.21
4	12/72	195	129	−4.05	30.55	1.64	2.20	−0.73
5	12/73	195	142	−5.03	33.28	2.09	2.09	−0.87

Observed information levels $\mathcal{I}_k$ and the resulting boundary values, a_k and b_k, are displayed in Table 17.1. We have included information levels and boundary values for all 5 analyses for completeness although, with early stopping, the later values might not in fact have become known. The observed values $\mathcal{I}_1, \ldots, \mathcal{I}_5$ are unequally spaced and $\mathcal{I}_5$ is less than the target of 35.58. Conditional on these information levels, numerical calculations under the approximation (3.1) to the distribution of $Z_1, \ldots, Z_5$ show the test attains a power of 0.789 when θ is 0.5, still very close to the design value of 0.8. By construction, the Type I error probability is exactly 0.025 under this normal approximation if the futility boundary is never applied, but the Type I error is a little lower at 0.0237 if the futility boundary is always applied.

Inspection of the statistics Z_k shows the lower boundary is crossed at the second analysis and the test would have stopped to accept H_0: $\theta \le 0$ at that point. In fact, the estimate of the log hazard ratio at analysis 2 is $\hat{\theta}_2 = -0.28$ with a standard deviation of 0.28, suggesting that survival may be worse on the experimental treatment than on the standard treatment. The one-sided group sequential test stops without resolving the question of whether $\theta = 0$ or $\theta < 0$. Assuming that an experimental treatment would not be used after a negative outcome to a trial, it is not necessary, or indeed appropriate, to continue a trial to answer such a question.

Instead of applying a formal stopping rule, the study could have been monitored by repeated confidence intervals $(\underline{\theta}_k, \overline{\theta}_k)$, $k = 1, \ldots, 5$. Suppose that observing $\underline{\theta}_k > 0$ will prompt a decision to reject H_0: $\theta < 0$ in favor of $\theta > 0$. Then, in order to control the Type I error probability at 0.025, we need $\{\underline{\theta}_k\}$ to be a 97.5% level lower confidence sequence for θ, that is

$$Pr_\theta(\underline{\theta}_k < \theta \text{ for all } k = 1, \ldots, 5) = 0.975 \quad \text{for all } \theta. \tag{17.17}$$

Similarly, suppose that observing $\overline{\theta}_k < 0.5$ will prompt a decision to accept H_0. Then, in order to maintain power 0.8 under $\theta = 0.5$, we require $\{\overline{\theta}_k\}$ to be an 80% level upper confidence sequence for θ, as then

$$Pr_{\theta=0.5}(\overline{\theta}_k > 0.5 \text{ for all } k = 1, \ldots, 5) = 0.8.$$

Table 17.2 *Repeated confidence intervals for θ, the log hazard ratio, using the stratified log-rank statistics* (17.3) *and associated information levels* (17.4) *in formula* (17.15) *for the oropharynx trial data.*
The lower endpoints $\underline{\theta}_k$ form a 97.5% lower confidence sequence for θ and the upper endpoints $\overline{\theta}_k$ form an 80% upper confidence sequence.

Analysis, k	$\mathcal{I}_k$	$\hat{\theta}^{(k)} = Z_k/\sqrt{\mathcal{I}_k}$	77.5% *RCI*, $(\underline{\theta}_k, \overline{\theta}_k)$
1	5.43	−0.45	(−1.84, 0.67)
2	12.58	−0.28	(−1.07, 0.28)
3	21.11	−0.26	(−0.80, 0.07)
4	30.55	−0.13	(−0.53, 0.07)
5	33.28	−0.15	(−0.51, 0.01)

Thus, to achieve the desired Type I and II error probabilities, it is appropriate to base decisions on a sequence of 77.75% RCIs for θ in which the endpoints of the intervals form a 97.5% lower confidence sequence and an 80% upper confidence sequence for θ. Table 17.2 shows this sequence of RCIs for the log hazard ratio, θ, in the stratified proportional hazards model. Under the assumption that $Z_1, \ldots, Z_5$ have the canonical joint distribution (3.1), these RCIs are computed from a generalization of formula (17.15) to allow for unequal error probabilities in the upper and lower tails. The RCIs are

$$(\hat{\theta}^{(k)} - c_{1k}(\alpha)/\sqrt{\mathcal{I}_k},\ \hat{\theta}^{(k)} + c_{2k}(\alpha)/\sqrt{\mathcal{I}_k}), \quad k = 1, \ldots, K, \tag{17.18}$$

where $\hat{\theta}^{(k)} = Z_k/\sqrt{\mathcal{I}_k}$ and the critical values $c_{1k}(\alpha)$ are taken from the upper boundary of a one-sided test with error spending function $f(\tilde{t}) = \min(0.025\,\tilde{t}^2, 0.025)$ and the values $c_{2k}(\beta)$ are taken from the upper boundary of a one-sided test with error spending function $f(\tilde{t}) = \min(0.2\,\tilde{t}^2, 0.2)$. Here $\tilde{t}$ is the information fraction $\mathcal{I}_k/\mathcal{I}_{max}$ with $\mathcal{I}_{max} = 35.58$, the target information level previously calculated for a one-sided group sequential test with a non-binding futility boundary.

With $\theta = 0.5$ specified as the treatment effect to be detected with power 0.8, the study could have been terminated at the second analysis since the RCI at that point, $(-1.07, 0.28)$, is well below 0.5. Formalizing such decisions into a rule in which the study is stopped the first time an RCI fails to contain either 0 or 0.5 gives a "derived test", as described in Section 10.3, and as long as the final RCI for θ has width at most 0.5 this derived test will have Type I and II error rates of less than 0.025 and 0.2, the overall error probabilities in the lower and upper tails, respectively, of the RCI sequence. In computing the RCIs, we have assumed that the full error probabilities $\alpha = 0.025$ and $\beta = 0.2$ are spent at analysis 5 in the error spending tests that produced the critical values $c_{1k}(\alpha)$ and $c_{2k}(\beta)$. Thus, under the normal approximation to the distribution of $\{Z_1, \ldots, Z_5\}$, equation (17.17) holds exactly and if the derived test were applied with a non-binding futility boundary, the Type I error rate would be controlled at 0.025. If, instead, the derived test were applied with a binding futility boundary, the Type I and Type II error rates would be 0.0237 and 0.211, respectively. Here, the Type I error requirement is met conservatively due to the presence of the futility boundary but the Type II error is higher than the desired value of 0.2 because the final information level, $\mathcal{I}_5 = 33.28$, is lower than the target, $\mathcal{I}_{max} = 35.58$.

The RCI approach has the advantage of flexibility as RCIs remain valid whether or not a formal stopping rule is obeyed. Thus, if researchers had decided to continue past the second analysis, despite the low upper limit of the RCI, it would still have been permissible to quote the RCIs at later analyses with the original overall confidence levels. In this case, it is likely that the study would have been terminated at the third analysis in favor of the null hypothesis. The discussion in the previous paragraph demonstrates the relationship between a derived test and a one-sided group sequential test with a non-binding futility boundary. In both cases it is possible to reject H_0: $\theta \leq 0$ when the study has continued past an analysis where a decision could have been made to accept H_0. The derived test, as it is described in Section 10.3 differs in that the Type II error rate is still protected if one accepts H_0 after continuing past an analysis where a decision could have been made to reject H_0 and, as a consequence, it needs a slightly higher maximum information level.

We have also calculated RCIs from formula (17.14), finding the maximum partial likelihood estimates $\hat{\theta}^{(k)}$ by fitting a Cox regression model, stratified by institution and with a single x-variable for treatment. Each of these estimates was within 0.01 of the corresponding estimate $Z_k/\sqrt{\mathcal{I}_k}$ in Table 17.2 and the information levels $\mathcal{I}_k(\hat{\theta}^{(k)}) = v_k^{-1}$ were one or two percent higher than the values of $\mathcal{I}_k$ in that table. Endpoints of the resulting intervals differed by at most 0.02 from the values shown in Table 17.2. On the other hand, there are substantial differences between the values $\mathcal{I}_k$ in Table 17.2 and the simple estimates of information, $d_k/4$, where d_k is the number of observed deaths at analysis k listed in Table 17.1, and so the simplest expression for RCIs (17.16) would not be reliable in this instance.

17.6.3 A Group Sequential Test using the Cox Regression Model to Include Covariates

Kalbfleisch & Prentice (2002, p. 4) remark on the importance of adjusting for possible imbalance with regard to baseline covariates when assessing the effect of treatment on survival. The stratified log-rank test does not make such adjustments, relying instead on randomization to balance subjects with respect to important prognostic variables. Use of this test can lead to an awkward situation if randomization happens to produce imbalance with respect to a variable that is known or suspected to have a substantial effect on survival.

As explained in Section 17.4.1, the proportional hazards regression model (Cox, 1972) allows covariates to be modeled either as variables in the regression term for the log hazard ratio or as stratification variables defining separate baseline hazard rates for each stratum. We have conducted a retrospective group sequential analysis assuming a model of the form

$$h_{i\ell}(t) = h_{0\ell}(t)\, e^{\boldsymbol{\beta}^T \boldsymbol{x}_{i\ell}}, \quad t > 0,$$

for the hazard rate of subject i in stratum ℓ, where $\boldsymbol{x}_{i\ell}$ is the vector of covariates for this subject. We have taken six strata, corresponding to the six participating institutions, and the following x variables which were identified as important prognostic indicators prior to the study's commencement:

$x_{i\ell 1}$	Treatment	1 = experimental, 2 = standard
$x_{i\ell 2}$	Gender	1 = male, 2 = female
$x_{i\ell 3}$	Condition	1 = no disability, 2 = restricted work 3 = assistance with self-care, 4 = bed confined
$x_{i\ell 4}$	T-staging	1, 2, 3, 4 with increasing size of primary tumor
$x_{i\ell 5}$	N-staging	0, 1, 2, 3 with increasing incidence of metastases
$x_{i\ell 6}$	Site 1	1 for faucial arch, 0 for other two sites
$x_{i\ell 7}$	Site 2	1 for tonsillar fossa, 0 for other two sites.

Table 17.3 *Group sequential estimates of Cox model parameters and their standard errors for the oropharynx trial data. The table shows estimates $\hat{\beta}_1^{(k)}, \ldots, \hat{\beta}_7^{(k)}$ at analyses $k = 1, \ldots, 5$, with the standard error of each estimate in parentheses.*

Analysis, k	*Treatment,* β_1	*Gender,* β_2	*Condition,* β_3	*T-staging,* β_4
1	−0.79 (0.49)	−0.95 (0.82)	1.24 (0.39)	0.31 (0.38)
2	−0.14 (0.30)	−0.76 (0.42)	1.30 (0.26)	0.32 (0.22)
3	−0.08 (0.23)	−0.50 (0.30)	1.19 (0.21)	0.32 (0.17)
4	0.04 (0.19)	−0.59 (0.24)	1.08 (0.17)	0.32 (0.14)
5	0.01 (0.18)	−0.40 (0.22)	1.04 (0.17)	0.25 (0.13)

Analysis, k	*N-staging,* β_5	*Site* 1, β_6	*Site* 2, β_7
1	0.73 (0.28)	−0.18 (0.30)	0.10 (0.20)
2	0.35 (0.15)	−0.32 (0.18)	−0.08 (0.11)
3	0.17 (0.10)	−0.10 (0.15)	−0.03 (0.08)
4	0.16 (0.09)	−0.11 (0.12)	−0.04 (0.07)
5	0.17 (0.08)	−0.13 (0.11)	−0.03 (0.06)

The first variable, $x_{i\ell 1}$ indicates the treatment received by patient i in stratum ℓ and its coefficient, β_1, is the log hazard ratio between the standard and experimental treatment, a positive value indicating an improvement in survival for the experimental treatment. Tumors were classified as occurring at one of three different sites, and the coefficients β_6 and β_7 represent differences in log hazard rates at each of the two named sites relative to the third, the pharyngeal tongue. Table 17.3 shows the estimates of each element of β and their associated standard errors at the five analyses, based on survival data up to times 720, 1080, 1440, 1800 and 2160 days from the beginning of 1968.

A group sequential test can be based on the sequence of estimates of the treatment effect, $\hat{\beta}_1^{(k)}$, $k = 1, \ldots, 5$, and information levels $\mathcal{I}_k$, which are simply the reciprocals of the squares of the standard errors of the $\hat{\beta}_1^{(k)}$. These estimates and information levels are shown in Table 17.4, as are the standardized test statistics $Z_k = \hat{\beta}_1^{(k)} \sqrt{\mathcal{I}_k}$, $k = 1, \ldots, 5$. In deriving a group sequential test, we assume the sequence $Z_1, \ldots, Z_5$ has the canonical joint distribution (3.1) with $\theta = \beta_1$, given the observed information levels $\mathcal{I}_1, \ldots, \mathcal{I}_5$. Table 17.4 shows the critical values a_k and b_k for a test of H_0: $\beta_1 \leq 0$ against $\beta_1 > 0$. As for the earlier stratified log-rank test, this test is designed to achieve a Type I error probability of $\alpha = 0.025$ and power $1 - \beta = 0.8$ if $\beta_1 = 0.5$ using error spending functions $f(\tilde{t}) = \min\{\alpha \tilde{t}^2, \alpha\}$ and $g(\tilde{t}) = \min\{\beta \tilde{t}^2, \beta\}$ for Type I and II errors, respectively, and treating the futility boundary as non-binding. Again, $\tilde{t}$ denotes the information fraction $\mathcal{I}_k/\mathcal{I}_{max}$ with $\mathcal{I}_{max} = 35.58$, the target information level previously calculated for a group sequential log-rank test. The table also shows a 77.5%-level RCI sequence, computed using the same formula (17.18) as for the log-rank analysis but using information levels $\mathcal{I}_1, \ldots, \mathcal{I}_5$ for the Cox model analysis in calculating the critical values $c_{1k}(\alpha)$ and $c_{2k}(\beta)$.

It is evident that the covariate adjustment has affected the standardized statistics, Z_k, the observed information levels, $\mathcal{I}_k$, and the RCI sequence. This time the test would have stopped at the third analysis, again in favor of H_0. Similarly, a "derived test" based on a sequence of the RCI sequence for β_1 would have stopped at the third analysis since the RCI is completely below 0.5 for the first time on this occasion.

Table 17.4 *Test statistics and critical values for a group sequential test based on a Cox regression model for the oropharynx trial data.*
Critical values a_k *and* b_k *are for a test of* H_0: $\beta_1 = 0$ *against the one-sided alternative* $\beta_1 > 0$ *with Type I and Type II error probabilities* $\alpha = 0.025$ *and* $\beta = 0.2$ *and a non-binding futility boundary. The test is constructed using error spending functions* $f(\tilde{t}) = \min\{\alpha\tilde{t}^2, \alpha\}$ *and* $g(\tilde{t}) = \min\{\beta\tilde{t}^2, \beta\}$, *with* $\mathcal{I}_{max} = 35.58$.
The sequence of RCIs is based on the same error spending functions and has overall coverage probability 0.775 *under the usual normal approximation. The lower endpoints* $\underline{\beta}_{1k}$ *form a* 97.5% *lower confidence sequence for* β *and the upper endpoints* $\overline{\beta}_{1k}$ *an* 80% *upper confidence sequence.*

Analysis, k	$\mathcal{I}_k$	a_k	b_k	$\hat{\beta}_1^{(k)}$	Z_k	77.5% *RCI*, $(\underline{\beta}_{1k}, \overline{\beta}_{1k})$
1	4.11	−1.77	3.40	−0.79	−1.60	(−2.47, 0.58)
2	10.89	−0.47	2.87	−0.14	−0.45	(−1.00, 0.51)
3	19.23	0.55	2.52	−0.08	−0.33	(−0.65, 0.30)
4	28.10	1.41	2.27	0.04	0.20	(−0.39, 0.27)
5	30.96	2.05	2.05	0.01	0.04	(−0.36, 0.17)

17.7 Survival Probabilities and Quantiles

A time-to-event variable is more informative than a binary outcome indicating whether or not failure has occurred within a fixed time period of, say, T years. However, it may sometimes be useful to consider a "two-year recurrence rate" or a "five-year survival rate" as a primary outcome. A simple proportion of those surviving the period of τ years will be a biased estimate of the survival rate if not all subjects are followed for the full period, while omitting those subjects with potential censoring times less than T years is inefficient. These difficulties are overcome by use of the Kaplan-Meier estimate (Kaplan & Meier, 1958) of the survival function $S(t)$.

As an example, consider a study that was conducted to investigate the effect of a short course of zidovudine on the mother-to-infant HIV (human immunodeficiency virus) transmission rate among HIV infected mothers in a developing country. The primary response here was the time from birth to the first positive PCR (polymerase chain reaction) test in the infant, indicating the presence of the HIV virus. For a non-breast feeding population, the proportion of infants with a positive PCR within 3 months of birth is a reasonable outcome measure since any perinatal HIV infection will generally lead to a positive PCR test within 3 months. For a breast feeding population, a horizon of 24 months would be a more suitable choice as infants have continued high exposure to infection. It is likely, because of ethical considerations, that such a trial would be uncontrolled, in which case we have a one-sample problem and must compare the infection rate by the specified time with the historical rate.

Consider a group sequential trial conducted to study the survival rate at a specific time τ. For $k = 1, \ldots, K$, let d_k denote the total number of failures observed by the time of the kth analysis, let d'_k be the number of distinct failure times, and denote these failure times by $t_{1,k} < t_{2,k} < \ldots < t_{d'_k,k}$. Here $d'_k \le d_k$ with $d'_k = d_k$ if there are no ties. Let $r_{i,k}$ denote the number at risk just before time $\tau_{i,k}$ and $\Delta_{i,k}$ the number of failures at $\tau_{i,k}$, for $i = 1, \ldots, d'_k$. If there are no ties, then each $\Delta_{i,k}$ is equal to one.

The Kaplan-Meier estimate of the survival probability $S_k(t)$ at time t based on data available at analysis k is

$$\widehat{S}_k(t) = \prod_{i:\, t_{i,k} \le t} \left(1 - \frac{\Delta_{i,k}}{r_{i,k}}\right) \quad \text{for } 0 < t \le t_{d'_k,k}$$

If the time τ is such that $0 < S(\tau) < 1$ and there is a positive probability of having an uncensored observation greater than τ, Jennison & Turnbull (1985) show that the sequence

$$Z_k = \{\widehat{S}_k(\tau) - S(\tau)\}/\sqrt{\text{Var}\{\widehat{S}_k(\tau)\}}, \quad k = 1, \ldots, K, \tag{17.19}$$

has, asymptotically, the canonical joint distribution (3.1) with $\theta = S(\tau)$ and information levels $\mathcal{I}_k = [\text{Var}\{\widehat{S}_k(\tau)\}]^{-1}$.

Greenwood's formula (Greenwood, 1926) provides a consistent estimate of the variance of $\widehat{S}_k(\tau)$,

$$\widehat{V}_k(\tau) = \widehat{S}_k^2(\tau) \sum_{i:\, t_{i,k} \le \tau} \frac{\Delta_{i,k}}{r_{i,k}(r_{i,k} - \Delta_{i,k})}. \tag{17.20}$$

Hence, a group sequential test of the hypothesis H_0: $S(\tau) = p_0$, where τ and p_0 are specified, can be based on the standardized statistics

$$Z_k = \{\widehat{S}_k(\tau) - p_0\}/\sqrt{\{\widehat{V}_k(\tau)\}}, \quad k = 1, \ldots, K, \tag{17.21}$$

and associated information levels $\mathcal{I}_k = \{\widehat{V}_k(\tau)\}^{-1}$. Since information depends on the number and times of observed failures, the error spending approach of Chapter 8 is to be recommended for constructing such tests.

Repeated confidence intervals for a survival probability can be obtained by inverting a family of two-sided tests, as described in Chapter 10. A sequence of $(1 - \alpha)$-level RCIs for $S(\tau)$ has the form

$$(\widehat{S}_k(\tau) - c_k(\alpha)\sqrt{\{\widehat{V}_k(\tau)\}},\, \widehat{S}_k(\tau) + c_k(\alpha)\sqrt{\{\widehat{V}_k(\tau)\}}), \quad k = 1, \ldots, K, \tag{17.22}$$

where $c_k(\alpha)$ and $-c_k(\alpha)$, $k = 1, \ldots, K$, are the critical values of a group sequential two-sided test with Type I error probability α based on standardized statistics with the joint distribution (3.1) when information levels are $\mathcal{I}_k = \{\widehat{V}_k(\tau)\}^{-1}$.

Thomas & Grunkemeier (1975, Sec. 4) propose an alternative estimate of the variance of $\widehat{S}_k(\tau)$ under H_0: $S(\tau) = p_0$, derived in a similar manner to Greenwood's formula but with estimation of $S(t)$, $0 < t \le \tau$, constrained to agree with the hypothesis H_0. At analysis k, the generalized nonparametric estimate of $S(t)$, $0 < t \le \tau$, satisfying $S(\tau) = p_0$ is

$$\widetilde{S}_k(t) = \prod_{i:\, \tau_{i,k} \le t} \left(1 - \frac{\Delta_{i,k}}{r_{i,k} + \eta_k}\right),$$

where η_k is the value for which this expression gives $\widetilde{S}_k(\tau) = p_0$. The "constrained variance" estimate of Thomas & Grunkemeier (1975, Eq. 4.5) is

$$\widetilde{V}_k(\tau) = \hat{p}_0^2 \sum_{i:\, \tau_{i,k} \le \tau} \frac{\Delta_{i,k}}{r_{i,k}\,(r_{i,k} + \eta_k - \Delta_{i,k})} \frac{\widehat{S}_k(\tau_{i,k}-)}{\widetilde{S}_k(\tau_{i,k}-)}, \tag{17.23}$$

where $\widehat{S}_k(t-)$ and $\widetilde{S}_k(t-)$ denote values of the functions $\widehat{S}_k$ and $\widetilde{S}_k$ just before time t. In the absence of censoring, $\widetilde{V}_k(\tau)$ simplifies to the correct binomial variance under H_0, namely $p_0(1 - p_0)/r_{1,k}$, whereas the Greenwood estimate $\widehat{V}_k(\tau)$ is equal to $\widehat{S}_k(\tau)\{1 - \widehat{S}_k(\tau)\}/r_{1,k}$. Simulations for fixed sample tests reported by Thomas & Grunkemeier (1975) and Barber & Jennison (1999) suggest that use of $\widetilde{V}_k(\tau)$ rather than $\widehat{V}_k(\tau)$ in test statistics (17.21) or RCIs (17.22) should lead to more accurate attainment of error rates and coverage probabilities, respectively. The constrained variance estimate, $\widetilde{V}_k(\tau)$, is not typically provided by statistical software packages, but the calculation is straightforward to program.

The Greenwood estimate, $\widehat{V}_k(\tau)$, is a little easier to calculate and is usually available in the output of standard statistical computer software for estimating survival curves.

Sometimes, interest is in a certain *quantile* of the survival distribution. As an example, in a trial of a vaccine for Stage IV melanoma patients it was desired to estimate the median survival time and to compare it with the figure of six months suggested by historical data. For $0 < p < 1$, we define the pth quantile of the survival distribution $S(t)$ to be $t_p = \inf\{t\colon S(t) \geq p\}$. Assuming $S(t)$ to be strictly decreasing in t, a group sequential test of $H_0\colon t_p = t^*$ for specified t^* and p is equivalent to a test of $H_0\colon S(t^*) = p$ and the test statistics (17.21) can be used with $\tau = t^*$ and $p_0 = p$.

We can also invert a family of group sequential tests with fixed p and varying t to obtain a sequence of RCIs for a quantile t_p. A $(1-\alpha)$-level sequence of RCIs for t_p has the form

$$\{t\colon |\widehat{S}_k(t) - p| \leq c_k(\alpha)\sqrt{V_k(t)}\}, \quad k = 1, \ldots, K, \tag{17.24}$$

where $c_k(\alpha)$ and $-c_k(\alpha)$, $k = 1, \ldots, K$, are the critical values for standardized statistics in a group sequential, two-sided test with Type I error probability α. For $V_k(t)$, we can use either the Greenwood estimate $\widehat{V}_k(t)$ given by (17.20) with $\tau = t$, which produces the interval proposed by Brookmeyer & Crowley (1982) in the fixed sample case, or the constrained variance estimate $\widetilde{V}_k(t)$ with $p_0 = p$ and the appropriate $\widetilde{S}_k$ in (17.23). The values $c_k(\alpha)$ should be derived for the observed information sequence $\mathcal{I}_k = \{V_k(t)\}^{-1}$, $k = 1, \ldots, K$. They will therefore vary with t, whichever variance estimate is used, but in practice it may well suffice to calculate their values using a single, representative value of t at each analysis. Jennison & Turnbull (1985, Tables 1, 2) present simulation results for the coverage probabilities of a single confidence interval or sequences of 5 or 10 RCIs for the median survival time. They find the intervals based on the constrained variance estimate to have error rates in each tail closer to the nominal value $\alpha/2$ than those based on the Greenwood estimate, which tend to be anti-conservative since estimated variances are often too small.

Lin, Shen, Ying & Breslow (1996) have further developed group sequential procedures that use Kaplan-Meier estimates to monitor the survival probability up to a fixed time. They give a detailed example of designing a trial to investigate a surgical treatment of Wilms tumor in which decisions are to be based on the two-year relapse free survival rate.

Analogous methods can also be used in a two-treatment comparison. If $S_A(t)$ and $S_B(t)$ denote survival functions on treatments A and B in a randomized trial, the null hypothesis $H_0\colon S_A(\tau) = S_B(\tau)$, for a given choice of τ, can be tested using the successive statistics

$$Z_k = \frac{\widehat{S}_{Ak}(\tau) - \widehat{S}_{Bk}(\tau)}{\sqrt{\{\widehat{V}_{Ak}(\tau) + \widehat{V}_{Bk}(\tau)\}}}, \quad k = 1, \ldots, K,$$

where $\widehat{S}_{Ak}(\tau)$ and $\widehat{S}_{Bk}(\tau)$ are Kaplan-Meier estimates of $S_A(\tau)$ and $S_B(\tau)$, respectively, at analysis k and $\widehat{V}_{Ak}(\tau)$ and $\widehat{V}_{Bk}(\tau)$ their estimated variances. The problem of comparing the pth quantiles of two survival distributions has been addressed by Keaney & Wei (1994).

17.8 Delays in Treatment Administration or Treatment Effect

17.8.1 Survival Trials with Treatment Delays in the Control Arm

In a certain class of trials patients are randomized to two arms, with the experimental treatment administered immediately in one arm but given after a delay in the other arm. An example is the CONCORDE trial (White et al., 1999) which compared immediate versus deferred administration of zidovudine in HIV patients. Branson & Whitehead (2002) describe a trial to compare immediate versus delayed thoracic radiotherapy in patients with inoperable non-small cell lung cancer. In the COURAGE trial (Boden et al., 2007), patients with stable coronary artery disease were randomized either to a treatment of immediate PCI (percutaneous coronary intervention) surgery or to an initial regimen of medical therapy involving intensive pharmacologic therapy and lifestyle intervention. When deemed appropriate, patients in the latter arm were allowed to switch treatments and undergo PCI later on. The primary endpoint was time to myocardial infarction or death from any cause. This trial was considered a *pragmatic* trial where the aim was to compare treatment management strategies, namely immediate PCI versus optimal medical treatment with the possibility of PCI later. The statistical methods described in this chapter to monitor and analyze the results are appropriate using the intention-to-treat principle, classifying all patients by the treatment arm to which they were randomized, regardless of subsequent treatment changes.

In a trial where treatment switching occurs, it may also be of interest to evaluate *biologic efficacy* (Sommer & Zeger, 1991), defined as the treatment effect that would have been seen if no switching were allowed. Any such analysis is complicated by the fact that switchers typically have a poorer prognosis than non-switchers, a form of informative censoring. Several model-based methods have been proposed to analyze such trials. These are compared by White (2006) who recommends the rank preserving structural failure time method (RPSFT) of Robins & Tsiatis (1991). An example of the application of this method is in the analysis of the trial of Sunitinib for patients with gastrointestinal cancer, reported by Demetri et al. (2012). At any interim stage of the trial, the RPSFT method can be used to produce an estimate of the biologic efficacy, along with standard errors. Because these estimators are not necessarily asymptotically efficient, successive estimates and test statistics may not follow the canonical joint distribution of Section 3.1 and this must be allowed for when constructing group sequential testing boundaries.

17.8.2 Survival Trials with a Delay in Treatment Effect

Although the log-rank test and Cox regression model have proved immensely powerful and are very widely applied, the performance of these methods relies on the proportional hazards assumption. Lin et al. (2020) and Roychoudhury et al. (2023) note that patterns of delayed treatment effects have been observed in immuno-oncology trials, with Kaplan-Meier survival curves separating some time after the start of treatment. Alternatively, an initially high treatment effect may diminish over time. Lin et al. and Roychoudhury et al. survey methods for analyzing such data including the Fleming-Harrington family of weighted log-rank test statistics (Fleming & Harrington, 1981; Harrington & Fleming, 1982), weighted Kaplan-Meier tests (Pepe & Fleming 1989), and maximum combination or "MaxCombo" tests that consider the maximum of several weighted log-rank test statistics which are, of course, correlated.

Li (1999) describes a group sequential version of the weighted Kaplan-Meier test. In order to construct a group sequential MaxCombo test, Ghosh et al. (2022) derive the asymptotic joint distribution of the sequence of values for a pair of weighted log-rank test statistics, taking weights from the Fleming-Harrington family or the "modestly weighted log-rank tests" proposed by Magirr & Burman (2019). A group sequential test with specified Type I error probability is then found by numerical computation.

17.9 Bibliography and Notes

The asymptotic joint distributions of successively computed log-rank statistics and Cox partial likelihood estimates have been proved under differing degrees of generality by Tsiatis (1981), Sellke & Siegmund (1983), Slud (1984), Tsiatis, Rosner & Tritchler (1985), Gu & Ying (1995), Jennison & Turnbull (1997a), Scharfstein, Tsiatis & Robins (1997), and

Bilias, Gu & Ying (1997). Jennison (1992) proposed a method using bootstrap sampling to improve the approximation to the small sample distribution of a log-rank statistic.

The weighted log-rank statistic generalizes (17.1) to

$$Z_k = \sum_{i=1}^{d'_k} w_{i,k}\,(\delta_{iB,k} - e_{i,k})\,/\sqrt{(\sum_{i=1}^{d'_k} w_{i,k}^2\, v_{i,k})},$$

incorporating weights $w_{1,k}, \ldots, w_{d'_k,k}$. Setting unit weights, $w_{i,k} = 1$ for all i and k, recovers the log-rank test. Weights $w_{i,k} = r_{iA,k} + r_{iB,k}$ yield Gehan's (1965) modification of the Wilcoxon test for censored data. This choice places greater emphasis on differences between observed and expected numbers of failures at earlier follow-up times, yielding a test that is sensitive to early differences between the two survival curves. However, Slud & Wei (1982) show that use of such weights leads to test statistics $Z_1, Z_2, \ldots$ which do *not* necessarily follow the canonical joint distribution (3.1) when there is staggered entry; see also the discussion in Section 15.7. Bilias, Gu & Ying (1997, Corollary 5.1) give the asymptotic distribution of successive weighted log-rank statistics and conditions on the weights for (3.1) to hold. The asymptotic joint distribution of a sequence of pairs of weighted log-rank statistics derived by Ghosh et al. (2022) follows from the asymptotic theory proved by Tsiatis (1982b) and is a bivariate generalization of the canonical joint distribution.

An example of the use of a group sequential Cox regression analysis is provided by Gu & Ying (1995) in an application to the prostate cancer survival data of Byar (1985). Lin, Yao & Ying (1999) present a thorough investigation into the use of methods for censored survival data in the context of the conditional power approach to stochastic curtailment and include an illustration with data from a colon cancer trial.

A joint model for a survival time and longitudinal measurements of a biomarker that affects survival is described by Tsiatis & Davidian (2001), who propose a conditional score estimator for the treatment effect in the survival time model. Burdon, Hampson & Jennison (2022) derive the asymptotic joint distribution of a sequence of estimates of the treatment effect in this joint model and, although the associated Z-statistics do not follow the canonical joint distribution (3.1), they show how these statistics can be used to construct a group sequential test with a specified Type I error rate.

The accelerated life model appears to be an attractive alternative to the proportional hazards model for censored survival data with covariates. However, it has proved cumbersome to deal with, both analytically and computationally. Lin (1992) has considered sequential procedures using this model. Tsiatis, Boucher & Kim (1995) prove results for a parametric survival model under quite general conditions, obtaining analogues to the nonparametric and semi-parametric results of this chapter. Lee & Sather (1995) consider group sequential methods for testing equality of cure rates in survival models where a fraction of each population is not susceptible to failure. They present both parametric and nonparametric procedures and illustrate these with data from a childhood leukemia trial.

Jennison & Turnbull (1984) provide a numerical example of the computation of RCIs for a hazard ratio based on a group sequential log-rank test. Jennison & Turnbull (1985) use the same set of data analyzed in Section 17.6 to illustrate construction of RCIs for a median survival time. Keaney & Wei (1994) illustrate their method of finding a sequence of RCIs for the difference in medians of two survival distributions with an application to AIDS (acquired immunodeficiency syndrome) data.

Chapter 18

Nuisance Parameters: Internal Pilot Studies, Sample Size Re-estimation and Information Monitoring

18.1 The Role of an Internal Pilot Phase

In setting the sample size for a study, one may find the power calculation requires knowledge of an unknown parameter, ϕ say. Examples include the variance of a normal response, the control response rate for binary outcomes, or the baseline event rate or censoring rate for survival data. Such parameters can only be estimated from the data to be collected in the study. One solution to this problem is to carry out a preliminary pilot study before commencing the main trial. There may be other uses for such a pilot study: checking whether a treatment process is feasible, refining the definition of a treatment, learning how to identify suitable participants, or ensuring that randomization and data collection run smoothly. However, if organizers are confident that their treatments and experimental procedures are already well defined, they may wish to launch into the main study without delay. Wittes & Brittain (1990) quote W.G. Cochran's warning that over-interpretation of results from a small pilot study, either positive or negative, may undermine support for the major investigation. Their solution is to forego an "external" pilot study but treat the first part of the main study as an "internal pilot" and re-calculate the overall sample size using estimates of the relevant parameters obtained from the internal pilot data. In Chapter 23, we shall present another approach using a so-called combination test and we address the case of an unknown response variance in Section 23.3. In that approach, data observed in the first stage of the trial are used to modify the target sample size of the second stage and summary statistics calculated separately from the two stages are combined in the final analysis.

Wittes & Brittain emphasize the need to estimate parameters governing sample size from the actual study population, and they cite examples of clinical trials where initial estimates of response variance, recruitment rate or mortality rate were wildly inaccurate. They note the efficiency gained by using data obtained during an internal pilot phase in the end of study analysis, whereas data from an external pilot study would normally be discarded from the main analysis.

Re-estimation of sample size from internal pilot data is simplest when the study will have a single analysis of the major outcome measure once sampling is complete. Thus, the experimental design is sequential only in that interim data are used to modify a fixed sample design. In this case, the mathematical details are quite straightforward. Suppose hypotheses and Type I and II error rates have been specified and the sample size needed for a fixed sample study can be expressed as a function $n(\phi)$ of the unknown parameter ϕ. Based on a pre-study estimate $\hat{\phi}_0$, the study commences with target sample size $n(\hat{\phi}_0)$. After a certain fraction of this sample has been obtained, a new estimate, $\hat{\phi}_1$, is found from

 DOI: 10.1201/9781584888482-18

the data available. The study then continues with $n(\hat{\phi}_1)$ as the target for its overall sample size, and all the data are analyzed at the end as if they had been collected in a fixed sample study. Variations on this scheme are possible. Wittes & Brittain (1990) treat $n(\hat{\phi}_0)$ as a minimum sample size and continue to take at least $n(\hat{\phi}_0)$ observations even if $n(\hat{\phi}_1) < n(\hat{\phi}_0)$. Gould (1992) proposes retaining the initial choice of sample size unless $n(\hat{\phi}_1) > \gamma\, n(\hat{\phi}_0)$, where $\gamma > 1$, and restricting sample size to a maximum of $\omega\, n(\hat{\phi}_0)$, where $\omega > 1$, suggesting possible values $\gamma = 1.33$ and $\omega = 2$ for the two constants. If $n(\hat{\phi}_1)$ is particularly high, organizers may decide to abandon a study on the grounds that it is not feasible to recruit sufficient subjects to achieve acceptable power: this is exactly what happened in the clinical trial of treatments for rheumatoid arthritis reported by Pedersen & Starbuck (1992) when the required sample size was re-estimated three years after the start of patient enrollment.

There is one important proviso when data from a study's internal pilot phase are used to re-calculate the overall sample size. If sample size is allowed to depend on the observed treatment effect estimate, it is possible to construct sampling rules which inflate the Type I error rate significantly: adapting the first type of example in Section 7 of Proschan, Follmann & Waclawiw (1992) to this setting gives a rule which can more than double the Type I error probability. Thus, it is crucial that the new sample size be based *only* on estimates of parameters controlling the variance of the major response variable and *not* on any comparison of observed responses between treatment groups. Gould (1992, 1995b), Gould & Shih (1992, 1998) and Shih (1992) advocate maintaining the blinding of treatment codes when analyzing the internal pilot data in order to guard against any possibility of inappropriate practices. Sample size must then be re-estimated from the set of response values pooled across treatment groups. Other authors, for example Herson & Wittes (1993), propose schemes in which a data analyst is allowed access to response variables and treatment codes, but only those data summaries relevant to the sample size calculation are revealed to the data monitoring committee, and even this information is not passed on to investigators administering the treatments to patients. Since these analyses of internal pilot data do not allow early termination on the grounds of an observed treatment difference, they may be regarded as "administrative analyses" in the nomenclature of Enas et al. (1989); indeed, study monitors might take the opportunity to make the checks on patient compliance, safety variables, etc., usually associated with an administrative analysis, when meeting to decide on a revised sample size.

Wittes & Brittain (1990) note that adjusting sample size on the basis of an estimated variance implies that the usual final test statistic does not have its standard distribution. This in turn will typically lead to a slight increase in the Type I error probability, above the specified value α. In the case of normal responses with an unknown variance σ^2, the intuitive reasoning is as follows. If σ^2 is underestimated at the interim stage, $n(\hat{\sigma}_1^2)$ will be low and the final estimate of σ^2 will have less chance to "recover" to be closer to the true value. On the other hand, if σ^2 is overestimated at the interim stage, $n(\hat{\sigma}_1^2)$ will be large and the additional data will give a more accurate estimate of σ^2. Thus, the final estimate of σ^2 is biased downwards and, since this estimate appears in the denominator of the t-statistic, results will tend to appear more significant than they should. A similar inflation of the Type I error rate occurs in studies involving binary endpoints. We shall examine this issue in more detail for applications of sample size re-estimation to binary data in Section 18.2.1 and to normal data in Section 18.2.2. Numerical investigations have shown this inflation to be modest in most cases. Kieser & Friede (2000) quantify the maximum excess of the Type I error rate over the nominal level α and apply this knowledge to calibrate the final test when using a sample size re-estimation rule, so as to guarantee that the Type I error probability does not exceed α.

The approach outlined above has been used successfully to produce tests with a specified power, often achieving this with an expected sample size close to that of the appropriate fixed sample test. Not surprisingly, the expected sample size increases and additional power is achieved as the number of observation in the pilot phase approaches or exceeds this fixed sample size. The general method can be used to update the target sample size or information level for many forms of response distribution. There are however practical limitations; for example, in a survival study the decision on when to terminate patient accrual may have to be taken before information on long-term survival is available. If there are doubts about the practicality of sample size re-estimation or its effects on Type I error or power in a particular application, it is advisable to investigate achieved error probabilities by simulations of the type we report later in this chapter.

Sample size re-estimation can also be combined with group sequential testing. Indeed, in the maximum information paradigm for error spending tests, sample size is automatically adjusted in order to reach the target set for the final information level. Problems may arise in implementing a group sequential test when observed information levels must be estimated and estimates at early analyses have significant variance. There is then a danger that error rates will stray from their intended values and care must be taken to avoid this. We shall address these issues in Section 18.4, where we discuss sample size re-estimation in group sequential tests.

18.2 Sample Size Re-estimation for a Fixed Sample Test

18.2.1 Binary Data

In this section we consider the comparison of binary responses from subjects randomized between two treatment arms. Suppose observations $X_{A1}, X_{A2}, \ldots$ and $X_{B1}, X_{B2}, \ldots$ are independent and take values 0 or 1 with $Pr\{X_{Ai} = 1\} = p_A$ and $Pr\{X_{Bi} = 1\} = p_B$ for $i = 1, 2, \ldots$. If n observations from each distribution are used to test H_0: $p_A = p_B$ against the two-sided alternative $p_A \neq p_B$, we have, approximately,

$$\bar{X}_A - \bar{X}_B \sim N(0, 2p(1-p)/n)$$

under H_0, where $\bar{X}_A = (X_{A1} + \ldots + X_{An})/n$, $\bar{X}_B = (X_{B1} + \ldots + X_{Bn})/n$ and p denotes the common value of p_A and p_B. Estimating p by

$$\hat{p} = \frac{1}{2n}\left(\sum_{i=1}^{n} X_{Ai} + \sum_{i=1}^{n} X_{Bi}\right), \tag{18.1}$$

we obtain the two-sided test with Type I error probability α which rejects H_0 if

$$|\bar{X}_A - \bar{X}_B| \geq z_{\alpha/2} \sqrt{\{2\hat{p}(1-\hat{p})/n\}}. \tag{18.2}$$

Here $z_{\alpha/2} = \Phi^{-1}(1-\alpha/2)$ denotes the upper $\alpha/2$ tail point of a standard normal distribution.

For general p_A and p_B, the distribution of $\bar{X}_A - \bar{X}_B$ is approximately

$$\bar{X}_A - \bar{X}_B \sim N(p_A - p_B, \{p_A(1-p_A) + p_B(1-p_B)\}/n).$$

Suppose $|p_A - p_B|$ is sufficiently large that we can neglect the probability of rejecting H_0 with the sign of $\bar{X}_A - \bar{X}_B$ opposite to that of $p_A - p_B$. The expectation of $\hat{p}$ defined by (18.1) is $(p_A + p_B)/2$, which we denote by $\bar{p}$. Substituting $\bar{p}$ for $\hat{p}$ in (18.2), we find the power of the above test under parameter values p_A and p_B is approximately

$$\Phi\left\{\frac{|p_A - p_B|\sqrt{n} - z_{\alpha/2}\sqrt{\{2\,\bar{p}(1-\bar{p})\}}}{\sqrt{\{p_A(1-p_A) + p_B(1-p_B)\}}}\right\}.$$

Using the further approximation $p_A(1-p_A)+p_B(1-p_B) \approx 2\bar{p}(1-\bar{p})$, we see that power $1-\beta$ is achieved at a specific pair of values (p_A, p_B) by a sample size

$$n = \frac{(z_{\alpha/2}+z_\beta)^2\{p_A(1-p_A)+p_B(1-p_B)\}}{(p_A-p_B)^2} \tag{18.3}$$

from each treatment arm. Replacing $p_A(1-p_A)+p_B(1-p_B)\}$ in (18.3) by $2\bar{p}(1-\bar{p})$ gives the alternative approximation

$$n = \frac{(z_{\alpha/2}+z_\beta)^2\, 2\bar{p}(1-\bar{p})}{(p_A-p_B)^2}, \tag{18.4}$$

which agrees with that given by Gould (1992) for a test of H_0: $p_A = p_B$ based on the χ^2 statistic for the 2×2 table of responses on the two treatments.

Suppose one wishes to achieve power $1-\beta$ if $p_B = p_A + \Delta$, where Δ is specified but p_A is not. We can substitute Δ for $p_A - p_B$ in the denominator of (18.3) or (18.4), but estimates of p_A and p_B are needed to evaluate the numerator of either function. In the internal pilot approach, we use initial estimates of response rates p_A and p_B that satisfy the alternative hypothesis $p_B = p_A + \Delta$ to calculate a first approximation to the required sample size. We then sample a fraction of these observations, n_0 say, from each treatment arm in the internal pilot phase. Estimates of p_A and p_B from pilot phase data are used to re-calculate the total sample size. If the new value for the sample size is n_1 per treatment, we sample a further $n_1 - n_0$ observations from each arm. Exceptions to this rule arise if n_0 already exceeds n_1, in which case no further observations are necessary, or if other constraints are set on the second stage sample size. When sampling is completed, the test of H_0 is applied as defined by (18.1) and (18.2) with n equal to the total sample size actually observed from each arm, or with the appropriate correction if sample sizes on the two arms are not exactly equal.

In analyzing the pilot phase data, Gould (1992) recommends the use of (18.4) with $\bar{p}$ estimated by

$$\frac{1}{2n_0}\left(\sum_{i=1}^{n_0} X_{Ai} + \sum_{i=1}^{n_0} X_{Bi}\right).$$

An advantage of this method is that $\bar{p}$ depends only on the pooled response data, and it is not necessary to unblind treatment allocations to find $\bar{p}$.

Herson & Wittes (1993) adopt a different approach in a problem where the X_{Ai}s and X_{Bi}s are responses on standard and experimental treatments, respectively. They use the pilot phase data just to obtain an estimate, $\hat{p}_A$, of the response rate on the standard treatment and calculate the required sample size by pairing this with the value of p_B that satisfies the alternative hypothesis. Thus, a test with approximate power $1-\beta$ at $p_B = p_A + \Delta$ is obtained by substituting $p_A = \hat{p}_A$ and $p_B = \hat{p}_A + \Delta$ in (18.3). This approach requires unblinding of treatment allocations, at least for the data analyst.

The power requirement can be specified in other ways, for example, in terms of the ratio p_A/p_B. If power $1-\beta$ is required when $p_B = p_A/\rho$, formula (18.4) becomes

$$n = \frac{(z_{\alpha/2}+z_\beta)^2\,(1-\bar{p})\,(\rho+1)^2}{2\bar{p}\,(\rho-1)^2} \tag{18.5}$$

and Gould's method of substituting an estimate of $\bar{p}$ from the pooled response data, without unblinding, can still be employed. Alternatively, one can follow Herson & Wittes' approach, writing (18.3) under $p_B = p_A/\rho$ as

$$n = \frac{(z_{\alpha/2}+z_\beta)^2\{p_A(1-p_A)+(p_A/\rho)(1-p_A/\rho)\}}{(p_A-p_A/\rho)^2} \tag{18.6}$$

and substituting the estimate $\hat{p}_A$ obtained from pilot phase observations on the first treatment arm. Gould (1992) also gives a formula for the sample size necessary to achieve a specific power under a given odds ratio $p_A(1-p_B)/\{p_B(1-p_A)\}$.

Tables 18.1 and 18.2 show the results of applying the two approaches described above to construct tests with two-sided Type I error rate $\alpha = 0.05$ and power $1 - \beta = 0.9$ when $p_B = p_A + \Delta$. Tests reported in Table 18.1 use Herson & Wittes' method, taking an estimate $\hat{p}_A$ from pilot phase data on the first treatment and calculating the overall sample size per treatment arm from (18.3) with $p_A = \hat{p}_A$ and $p_B = \hat{p}_A + \Delta$. Since p_B can be at most 1, $\hat{p}_A$ is truncated to $1 - \Delta$ if it exceeds this value. Note that the sample size formula involves $\hat{p}_A$ and Δ but does not depend on the responses from treatment arm B. This explains why the same means and standard deviations of the sample size distribution are seen in the top and bottom halves of Table 18.1. Table 18.2 concerns tests using Gould's method in which an estimate of $\bar{p}$ from the pooled pilot phase data is substituted into (18.4). As this calculation implicitly assumes the values $p_A = \bar{p} - \Delta/2$ and $p_B = \bar{p} + \Delta/2$, values of $\bar{p}$ are restricted to the interval $(\Delta/2, 1 - \Delta/2)$ and if response rates outside this range are observed, the estimate of $\bar{p}$ is shifted upward to $\Delta/2$ or downward to $1 - \Delta/2$ as necessary.

It is evident from the tables that Type I error rates are perturbed a little from their nominal value of 0.05, rising to around 0.06 in a few examples with low n_0 and low total sample size. Power is close to its nominal value of 0.9, lying between 0.86 and 0.93 for the Herson & Wittes approach and in the range 0.89 to 0.91 for Gould's method. These results are in general agreement with those reported by other authors and support Gould's (1992) claim that re-estimation of sample size "does not materially affect the Type I error rates".

Sample sizes under H_0: $p_A = p_B$ differ between the two methods since different alternatives satisfying $p_B = p_A + \Delta$ are considered in each case. In Herson & Wittes' approach p_A is set first and Δ added to give the value for p_B; we have followed this procedure, using the true value of p_A, in calculating the fixed sample sizes in Table 18.1. Gould's approach starts with a value for the mean response rate $\bar{p}$ and then sets the individual probabilities at $p_A = \bar{p} - \Delta/2$ and $p_B = \bar{p} + \Delta/2$; the fixed sample sizes in Table 18.2 are calculated by this method using the true value of $\bar{p}$. Note that although sample sizes under H_0 are consistently smaller in Table 18.2 than in Table 18.1, the position can be reversed in other situations not presented here. Differences between the fixed sample sizes in the two tables when $p_B = p_A + \Delta$ arise because of the different variance estimates in formulae (18.3) and (18.4), and evaluating the numerators of these formulae using estimates of p_A and $\bar{p}$, respectively, leads to further variation between the two methods.

Table 18.3 shows properties of tests constructed to satisfy a power requirement when $p_B = p_A/\rho$ for a specified constant ρ; the cases shown here were selected as representative examples from a more extensive study. In Herson & Wittes' tests, the pilot phase data are used to estimate p_A and the overall sample size per arm is found by substituting $\hat{p}_A$ for p_A in (18.6). If $\rho < 1$ and $\hat{p}_A > \rho$, $\hat{p}_A$ is truncated to ρ so that $\hat{p}_A/\rho$ does not exceed 1. If all observations on arm 1 in the pilot phase are zero, we set $\hat{p}_A = 0.5/n_0$ to ensure that (18.6) may still be used. Herson & Wittes' sample size formula involves $\hat{p}_A$ and ρ but not the responses from treatment arm B. Hence, the pattern of means and standard deviations for the sample size distribution is repeated in the second group of four Herson & Wittes tests when p_A and ρ take the same values as in the first four tests but p_B is different. Gould's tests are carried out using sample size formula (18.5) with $\bar{p}$ estimated from the pooled pilot phase data. For $\rho > 1$, any estimate of $\bar{p}$ greater than $(1+\rho)/(2\rho)$ is reduced to this value so that the value of p_A implicit in (18.5) does not exceed 1; similarly, for $\rho < 1$, high estimates of $\bar{p}$ are reduced to $(1+\rho)/2$ so that $p_B = 2\bar{p}/(1+\rho)$ is not greater than 1. If all pilot phase observations on both treatment arms are zero, we set $\bar{p} = 0.5/(2n_0)$. In all of these simulations the second stage sample size was restricted to 1000 observations per treatment.

Table 18.1 *Properties of "internal pilot" tests of two binary responses with power at* $p_B = p_A + \Delta$ *constructed using the approach of Herson & Wittes (1993).*
An initial sample of n_0 *is observed on each arm yielding estimates* $\hat{p}_A$ *and* $\hat{p}_B$*, and the overall sample size is calculated from* (18.3) *with* $\alpha = 0.05$, $\beta = 0.1$, $p_A = \hat{p}_A$ *and* $p_B = \hat{p}_A + \Delta$. *The tabulated fixed sample size gives a test with two-sided Type I error probability* $\alpha = 0.05$ *when* $p_A = p_B$ *and power* $1 - \beta = 0.9$ *under the given value of* p_A *and* $p_B = p_A + \Delta$.
Results are based on simulations with one million replicates: standard errors are 0.0002 *for Type I errors,* 0.0003 *for power, and well below* 0.1 *for the mean and standard deviation of the total sample size.*

Δ	p_A	p_B	*Pilot size* n_0	*Fixed sample size*	*Total sample size* *Mean*	*Standard deviation*	*Type I error probability*
0.3	0.1	0.1	10	39	36.8	9.6	0.059
	0.1	0.1	20	39	38.0	7.2	0.061
	0.2	0.2	20	48	46.3	6.1	0.056
	0.2	0.2	30	48	47.1	5.0	0.055
	0.3	0.3	30	53	51.4	2.9	0.053
	0.3	0.3	40	53	51.6	2.3	0.053
	0.4	0.4	30	53	51.2	3.5	0.054
0.2	0.1	0.1	30	79	77.6	16.6	0.056
	0.1	0.1	50	79	78.3	12.8	0.053
	0.2	0.2	50	106	103.9	11.8	0.053
	0.2	0.2	70	106	104.4	9.9	0.052
	0.3	0.3	60	121	119.4	6.4	0.052
	0.3	0.3	90	121	120.1	5.2	0.051
	0.4	0.4	60	127	124.3	3.0	0.050
							Power
0.3	0.1	0.4	10	39	36.8	9.6	0.897
	0.1	0.4	20	39	38.0	7.2	0.919
	0.2	0.5	20	48	46.4	6.1	0.898
	0.2	0.5	30	48	47.1	5.0	0.903
	0.3	0.6	30	53	51.4	2.9	0.879
	0.3	0.6	40	53	51.6	2.3	0.880
	0.4	0.7	30	53	51.2	3.5	0.859
0.2	0.1	0.3	30	79	77.6	16.6	0.921
	0.1	0.3	50	79	78.4	12.8	0.931
	0.2	0.4	50	106	103.9	11.8	0.910
	0.2	0.4	70	106	104.4	9.9	0.914
	0.3	0.5	60	121	119.4	6.4	0.898
	0.3	0.5	90	121	120.0	5.2	0.900
	0.4	0.6	60	127	124.3	3.0	0.886

Table 18.2 *Properties of "internal pilot" tests of two binary responses with power at $p_B = p_A + \Delta$ constructed using the approach of Gould (1992).*
An initial sample of n_0 is observed on each arm and the overall sample size is calculated from (18.4) with $\alpha = 0.05$, $\beta = 0.1$ and $\bar{p}$ estimated from the pooled pilot phase data. The tabulated fixed sample size gives a test with two-sided Type I error probability $\alpha = 0.05$ when $p_A = p_B$ and power $1 - \beta$ under $p_A = \bar{p} - \Delta/2$ and $p_B = \bar{p} + \Delta/2$, where $\bar{p}$ is the mean of the true p_A and p_B.
Results are based on simulations with one million replicates: standard errors are 0.0002 for Type I errors, 0.0003 for power, and well below 0.1 for the mean and standard deviation of the total sample size.

Δ	p_A	p_B	*Pilot size* n_0	*Fixed sample size*	*Total sample size* Mean	*Standard deviation*	*Type I error probability*
0.3	0.1	0.1	10	22	31.4	3.9	0.057
	0.1	0.1	20	22	30.6	2.2	0.056
	0.2	0.2	20	38	38.0	7.0	0.051
	0.2	0.2	30	38	37.8	6.0	0.049
	0.3	0.3	30	50	48.8	5.6	0.051
	0.3	0.3	40	50	49.2	4.6	0.051
	0.4	0.4	30	57	55.7	3.2	0.050
0.2	0.1	0.1	30	48	53.9	9.1	0.057
	0.1	0.1	50	48	53.8	6.4	0.049
	0.2	0.2	50	85	83.8	12.5	0.051
	0.2	0.2	70	85	84.6	9.5	0.050
	0.3	0.3	60	111	109.9	8.9	0.050
	0.3	0.3	90	111	110.2	7.1	0.051
	0.4	0.4	60	127	125.7	4.9	0.050
							Power
0.3	0.1	0.4	10	44	43.3	9.0	0.893
	0.1	0.4	20	44	43.5	7.1	0.901
	0.2	0.5	20	54	52.6	5.3	0.903
	0.2	0.5	30	54	52.9	4.3	0.907
	0.3	0.6	30	58	57.5	1.9	0.908
	0.3	0.6	40	58	57.8	1.6	0.910
	0.4	0.7	30	58	57.5	1.9	0.908
0.2	0.1	0.3	30	85	83.3	15.4	0.889
	0.1	0.3	50	85	83.9	12.1	0.897
	0.2	0.4	50	111	109.9	9.4	0.901
	0.2	0.4	70	111	110.1	8.0	0.903
	0.3	0.5	60	127	125.7	4.8	0.903
	0.3	0.5	90	127	126.0	3.9	0.904
	0.4	0.6	60	132	131.0	1.5	0.911

Table 18.3 *Properties of "internal pilot" tests of two binary responses with power at* $p_B = p_A/\rho$. *An initial sample of* n_0 *is observed on each arm. For Herson & Wittes' (1993) tests, the overall sample size is found from* (18.6) *with* $\alpha = 0.05$, $\beta = 0.1$, *and* p_A *estimated from the pilot phase data. In Gould's (1992) approach, formula* (18.5) *is used with an estimate of* $\bar{p}$ *from the pooled pilot phase data.*
The tabulated fixed sample size gives a test with two-sided Type I error probability $\alpha = 0.05$ *when* $p_A = p_B$ *and power* $1-\beta$ *under the given value of* p_A *and* $p_B = p_A/\rho$ *for Herson & Wittes' tests, and under values* p_A *and* p_B *with mean equal to the true* $\bar{p}$ *and satisfying* $p_B = p_A/\rho$ *for Gould's tests.*
Results are based on simulations with one million replicates: standard errors are 0.0002 *for Type I errors,* 0.0003 *for power and at most* 0.3, *but usually much less, for the mean and standard deviation of the total sample size.*

Herson & Wittes' tests

ρ	p_A	p_B	*Pilot size* n_0	*Fixed sample size*	*Total sample size* *Mean*	*Standard deviation*	*Type I error probability*
1.5	0.75	0.75	20	74	77.8	30.3	0.054
1.25	0.75	0.75	80	200	202.8	41.6	0.052
0.5	0.2	0.2	80	106	116.9	40.2	0.048
0.25	0.1	0.1	30	39	71.4	68.8	0.051
							Power
1.5	0.75	0.5	20	74	77.8	30.3	0.917
1.25	0.75	0.6	80	200	202.8	41.6	0.927
0.5	0.2	0.4	80	106	116.9	40.4	0.834
0.25	0.1	0.4	30	39	71.3	68.7	0.832

Gould's tests

ρ	p_A	p_B	*Pilot size* n_0	*Fixed sample size*	*Total sample size* *Mean*	*Standard deviation*	*Type I error probability*
1.5	0.75	0.75	20	44	46.3	16.1	0.052
1.25	0.75	0.75	80	142	143.6	26.1	0.050
0.5	0.2	0.2	80	190	195.7	40.8	0.050
0.25	0.1	0.1	30	132	165.6	115.5	0.051
							Power
1.5	0.75	0.5	20	79	82.4	26.7	0.882
1.25	0.75	0.6	80	205	207.1	34.6	0.894
0.5	0.2	0.4	80	111	113.2	19.3	0.909
0.25	0.1	0.4	30	44	47.3	14.4	0.936

Once again, Type I error rates are in close agreement with their nominal value, $\alpha = 0.05$. However, for this problem the power shows greater variation above and below its target value of 0.9. It is notable that the standard deviation of the total sample size is very high both under H_0 and when $p_B = p_A/\rho$. This variability in sample size can be attributed to the sensitivity of the formulae (18.6) and (18.5) to changes in p_A and $\bar{p}$, respectively, and the fact that both these quantities may be estimated quite poorly from the pilot phase data. The fact that Type I error rates are not affected in this situation lends further support to claims of robustness of the Type I error probability to sample size re-estimation.

The testing problem considered in this section has the special feature that variances of the estimates of response probability differ on the two treatment arms when $p_A \neq p_B$. Consequently, the variance of the final estimate of $p_A - p_B$ could be reduced by allocating observations unequally between the two treatment arms. This possibility has not been exploited in the methods reported here but it could prove a useful addition to two-stage experimental designs.

18.2.2 Normal Data with Unknown Variance

One of the most common difficulties in carrying out sample size calculations arises from the lack of an accurate estimate of the variance of normally distributed responses. Suppose X_{Ai} and X_{Bi}, $i = 1, 2, \ldots$, are responses of subjects allocated to treatments A and B, respectively, and these observations are independently distributed as $X_{Ai} \sim N(\mu_A, \sigma^2)$ and $X_{Bi} \sim N(\mu_B, \sigma^2)$, $i = 1, 2, \ldots$. We shall consider the problem of testing H_0: $\mu_A = \mu_B$ against the two-sided alternative $\mu_A \neq \mu_B$ with Type I error probability α and power $1 - \beta$ at $\mu_A - \mu_B = \pm\delta$.

If n responses have been observed on each treatment, the t-test with two-sided Type I error probability α rejects H_0 if

$$\frac{|\bar{X}_A - \bar{X}_B|}{\sqrt{(2s^2/n)}} \geq t_{2n-2,\alpha/2} \, ,$$

where $t_{\nu,q}$ denotes the upper q tail-point of a t-distribution on ν degrees of freedom. Here, $\bar{X}_A$ is the mean of $X_{A1}, \ldots, X_{An}$, $\bar{X}_B$ the mean of $X_{B1}, \ldots, X_{Bn}$ and s^2 the standard unbiased estimate of σ^2. When $\mu_A - \mu_B = \delta$, the statistic $(\bar{X}_A - \bar{X}_B)/\sqrt{(2s^2/n)}$ has a noncentral t-distribution on $2n - 2$ degrees of freedom with noncentrality parameter $\delta\sqrt{\{n/(2\sigma^2)\}}$, which we shall write as $T(2n-2; \delta\sqrt{\{n/(2\sigma^2)\}})$. Thus, in order to meet the power condition, n must be large enough that

$$Pr\{T(2n-2; \delta\sqrt{\{n/(2\sigma^2)\}}) \geq t_{2n-2,\alpha/2}\} \geq 1 - \beta. \tag{18.7}$$

Taking α, β and δ as fixed, we denote the smallest value of n satisfying (18.7) as $N_f(\sigma^2)$, and this represents the required number of observations per treatment arm to provide the stated error probabilities for a given true value of σ^2.

Wittes & Brittain (1990) propose the following general strategy for using an internal pilot study to re-estimate sample size. One first calculates the sample size, $N_f(\sigma_0^2)$, required if σ^2 were equal to an initial estimate of its value, σ_0^2. In the internal pilot phase, a fraction of this sample size is observed and s_0^2, the usual unbiased estimate of σ^2, is calculated from these data. The overall sample size is then re-estimated as $N_f(s_0^2)$ per treatment. At this point, various options are available. Wittes & Brittain recommend taking the maximum of $N_f(s_0^2)$ and the original figure $N_f(\sigma_0^2)$ as the new total sample size per treatment, partly out of concern that reducing the sample size below $N_f(\sigma_0^2)$ may lead to an increase in the Type I error probability. Gould & Shih (1992) suggest retaining the initial target of $N_f(\sigma_0^2)$ unless $N_f(s_0^2)$ is substantially higher, say by more than 25%, and they also propose truncating excessive values of $N_f(s_0^2)$ to a practical level such as $2N_f(\sigma_0^2)$. In assessing this

Table 18.4 *Properties of Wittes & Brittain (1990) and Birkett & Day (1994) two-sided tests with an "internal pilot" stage for two normal distributions with unknown variance.*
An initial sample of n_0 *is observed on each arm and the overall sample size,* n_1*, is set as the maximum of* n_0 *and* $N_f(s_0^2)$*, as defined by* (18.7) *with* $\alpha = 0.05$, $1-\beta = 0.9$ *and* $\delta = 1$.
Results are based on simulations with one million replicates: standard errors are at most 0.0001 *for Type I errors,* 0.0002 *for power and* 0.1 *for the mean and standard deviation of the total sample size,* n_1.

σ^2	*Fixed sample size per arm* $N_f(\sigma^2)$	*Pilot sample size per arm* n_0	*Total sample size per arm,* n_1 *Mean*	*St. dev.*	*Type I error rate*	*Power at* $\mu_A - \mu_B = \pm 1$
0.3	8	2	8.0	6.2	0.090	0.829
	8	3	8.0	4.4	0.076	0.893
	8	5	8.1	2.9	0.062	0.931
	8	10	10.2	0.7	0.050	0.977
0.5	12	2	12.2	10.4	0.085	0.781
	12	3	12.1	7.4	0.074	0.851
	12	5	12.1	5.2	0.065	0.897
	12	10	12.6	2.9	0.055	0.931
1.0	23	3	22.6	14.8	0.065	0.810
	23	5	22.6	10.5	0.060	0.862
	23	10	22.5	7.0	0.057	0.898
	23	20	23.4	3.8	0.053	0.921
2.0	44	3	43.5	29.7	0.058	0.786
	44	5	43.5	21.0	0.055	0.842
	44	10	43.5	14.0	0.054	0.880
	44	20	43.5	9.6	0.053	0.898
4.0	86	5	85.5	42.1	0.052	0.832
	86	10	85.5	28.0	0.052	0.871
	86	20	85.5	19.3	0.052	0.890
	86	40	85.6	13.5	0.051	0.899

method of sample size re-estimation, we have followed Birkett & Day's (1994) example and simply set the new total sample size per treatment to be the maximum of n_0 and $N_f(s_0^2)$, where n_0 is the sample size already observed in the pilot phase. In any event, if a total of n_1 observations per treatment is deemed appropriate after examining the pilot phase data, one continues to observe a further $n_1 - n_0$ responses on each arm and then conducts a test of H_0 using the full set of data. In this test, H_0 is rejected if

$$\frac{|\bar{X}_A - \bar{X}_B|}{\sqrt{(2s_1^2/n_1)}} \geq t_{2n_1-2,\alpha/2} \,,$$

where now $\bar{X}_A$ and $\bar{X}_B$ are the means of $X_{A,1}, \ldots, X_{A,n_1}$ and $X_{B,1}, \ldots, X_{B,n_1}$, respectively, and s_1^2 is the usual estimate of σ^2 calculated from all $2n_1$ observations.

Table 18.4 shows properties of tests of the above form when n_1 is set equal to the maximum of n_0 and $N_f(s_0^2)$. Values of s_0^2 and, hence, n_1 and s_1^2 were generated by simulation, and the conditional probability of rejecting H_0 was then found exactly as a tail point of the standard normal distribution. In the examples considered here, $\alpha = 0.05$ and tests are intended to achieve power $1-\beta = 0.9$ at $\mu_A - \mu_B = \pm 1$. The true values of σ^2 are such that the necessary fixed sample size ranges from 8 to 86 per treatment arm. The problem depends on σ^2 and δ only through σ^2/δ^2, so the reported results also apply to other situations with the same value of this ratio.

Examples with low values for n_0 and $N_f(\sigma^2)$ are included to demonstrate just where the method starts to break down. In Section 18.1 we noted Wittes & Brittain's concern that the negative bias in s_1^2, the final variance estimate, might inflate the Type I error rate. The tabulated results show that this problem does indeed occur, but only at extremely low sample sizes. For practical purposes, the attained Type I error rates may be regarded as sufficiently close to the target of 0.05 as long as the initial sample contains at least 20 observations on each treatment. Errors in estimating σ^2 from a small pilot phase have a noticeable effect on the attained power, and an initial sample of at least 10 observations per treatment appears necessary for adequate power to be reliably achieved. Of course, an initial sample greater than or equal to $N_f(\sigma^2)$ leads to attained power in excess of the specified $1 - \beta$.

We have obtained similar results in further simulation studies and have found the general pattern of results to be only slightly affected by variations to the sampling rule. We found that Wittes & Brittain's proposal of setting a lower limit on n_1 does indeed reduce inflation in the Type I error probability when n_0 and $N_f(\sigma^2)$ are very low. If sample sizes are high, the method can be simplified by replacing $N_f(\sigma^2)$ with the sample size formula for a known variance problem,

$$N = (z_{\alpha/2} + z_\beta)^2 \, \frac{2\sigma^2}{\delta^2}.$$

However, if either n_0 or $N_f(\sigma^2)$ is small, this can lead to slightly higher Type I error rates and, since the resulting sample size is uniformly lower, some loss of power.

Gould & Shih (1992) recommend a straightforward approach to study design, allowing reconsideration of the overall sample size at just one interim inspection of the data and only modifying the initial choice of sample size if this is clearly inadequate, for example, less than 80% of the new estimate. They suggest conducting this sample size analysis as soon as sufficient data are available to give a reliable estimate of σ^2 since, from an administrative perspective, changes in design are easier to implement early on in a study. In contrast, Sandvik et al. (1996) conclude that re-analysis of sample size should occur rather later if the major objective is to estimate σ^2 as accurately as possible, subject to a limit on the probability that the internal pilot sample exceeds the sample size needed for the entire study. Most of the standard deviations of n_1 shown in Table 18.4 are quite substantial, reflecting the high variability of estimates of σ^2 based on pilot phase data. Hall (1981) has suggested the addition of a small third study phase to bring the total sample size up to a value calculated using information on σ^2 from the pilot phase and the main body of observations. While this modification can assist in stabilizing the final choice of sample size, there may sometimes be practical difficulties in adjusting targets for patient accrual so late in the course of a study.

The methods discussed so far make no allowance for the fact that an estimate of variance is used in setting the overall sample size. It should not, therefore, be surprising that values for power at $\mu_A - \mu_B = \pm 1$ reported in Table 18.4 can be well below the target of 0.9 when σ^2 is estimated from a particularly small pilot sample. One may try to compensate for the possibility that an under-estimate of σ^2 has been observed; see, for example, Browne (1995) and Betensky & Tierney (1997). However, an elegant and precise way of dealing with the sampling distribution of s_0^2 is provided in Stein's (1945) two-stage test, which has Type I error rate exactly equal to its nominal level α and guarantees the specified power conservatively. In Stein's test, an estimate s_0^2 on $2n_0 - 2$ degrees of freedom is obtained from pilot phase data and the total sample size per treatment, n_1, is taken to be the maximum of n_0 and

$$(t_{2n_0-2,\alpha/2} + t_{2n_0-2,\beta})^2 \, \frac{2s_0^2}{\delta^2}. \tag{18.8}$$

Table 18.5 *Properties of Denne & Jennison's (1999a) two-sided tests with an "internal pilot" stage for two normal distributions with unknown variance.*
An initial sample of n_0 is observed on each arm and the overall sample size, n_1, is set as the maximum of n_0 and the expression (18.8). *On termination the test is conducted using rule* (18.9). *Results are based on simulations with one million replicates: standard errors are at most* 0.0001 *for Type I errors,* 0.0002 *for power and* 0.1 *for the mean and standard deviation of the total sample size, n_1.*

σ^2	*Fixed sample size per arm* $N_f(\sigma^2)$	*Pilot sample size per arm* n_0	*Total sample size per arm, n_1* *Mean*	*St. dev.*	*Type I error rate*	*Power at* $\mu_A - \mu_B = \pm 1$
0.3	8	2	23.5	22.9	0.055	0.861
	8	3	11.7	7.8	0.053	0.900
	8	5	8.9	3.9	0.056	0.933
	8	10	10.2	0.8	0.050	0.977
0.5	12	2	38.8	38.2	0.053	0.855
	12	3	19.1	13.1	0.047	0.875
	12	5	14.2	6.8	0.052	0.905
	12	10	12.9	3.3	0.053	0.933
1.0	23	3	37.6	26.2	0.041	0.862
	23	5	27.9	13.7	0.044	0.882
	23	10	24.1	7.8	0.051	0.903
	23	20	23.5	4.1	0.052	0.922
2.0	44	3	74.8	52.5	0.041	0.863
	44	5	55.3	27.4	0.042	0.875
	44	10	47.6	15.7	0.046	0.891
	44	20	44.8	10.2	0.050	0.901
4.0	86	5	110.1	54.9	0.044	0.877
	86	10	94.7	31.4	0.046	0.888
	86	20	89.1	20.3	0.048	0.895
	86	40	86.8	13.8	0.050	0.901

On termination, H_0: $\mu_A - \mu_B = 0$ is rejected if

$$\frac{|\bar{X}_A - \bar{X}_B|}{\sqrt{(2s_0^2/n_1)}} \geq t_{2n_0-2,\alpha/2}.$$

Use of the initial variance estimate, s_0^2, in this rule ensures that the test has the stated error probabilities but detracts from its appeal as a practical method. If the estimate s_1^2 calculated from the final data set differs substantially from s_0^2, it is difficult to justify a decision which relies on use of s_0^2, rather than s_1^2, in the final t-test.

Denne & Jennison (1999a) present a two-stage test that uses Stein's sample size formula (18.8), but with the natural choice of variance estimate, s_1^2, in the final test statistic. This test rejects H_0 if

$$\frac{|\bar{X}_A - \bar{X}_B|}{\sqrt{(2s_1^2/n_1)}} \geq t_{2n_0-2+2\epsilon(n_1-n_0),\alpha/2} \tag{18.9}$$

where the constant ϵ can be varied between 0 and 1 to fine-tune the test and a value $\epsilon = 1/8$ is recommended for accurate attainment of both Type I error and Type II error probabilities. Properties of this test are shown in Table 18.5. It is evident that Stein's sampling rule increases the expected sample size over the corresponding values in Table 18.4, especially under low values of n_0, and this helps increase power toward the target of 0.9.

It is surprising that some tests in Table 18.5 should have Type I error rate greater than $\alpha = 0.05$ and power less than $1 - \beta = 0.9$ when $\mu_A - \mu_B = \pm 1$, since Stein's original procedure uses the same samples to produce a test with Type I error rate exactly equal to α and power at least $1 - \beta$. Denne & Jennison (1999b) attribute this efficiency to the way in which Stein's test implicitly allocates Type I error conditionally on the value of s_0^2 and, hence, n_1. This distribution of conditional Type I error over values of n_1 happens to produce high overall power in this instance. In contrast, by using s_1^2 in the final t-test, the modified test evens out the conditional Type I error over values of n_1 and this reduces overall power. Denne & Jennison (1999b) provide further discussion of the conditional properties of Stein's test. Although a simple comparison of Type I error and power may suggest advantages to Stein's original procedure, there is little practical value in having a significance test reject the null hypothesis using one estimate of variance when further information, not used in the test, indicates that this estimate is too low. We would therefore recommend using the standard variance estimate from all available data on conclusion of a study incorporating sample size re-estimation.

18.3 Blinded Sample Size Re-estimation

Almost all the methods we have described thus far require data to be unblinded, so the treatment assignment of each subject is known, when sample size is re-calculated. As an example, the methods of Section 18.2.2 for normal data use the variance estimate s_0^2 which is formed from the within group variances in the internal pilot data. According to the FDA Guidance E9 (FDA, 1998, Sec. 2.3.1), blinding is a crucial design technique for avoiding bias in the conduct of clinical trials. In many large trials, an independent statistician may provide the information needed for sample size re-estimation, such as the variance estimate s_0^2 in the case of normal data, without any other party seeing unblinded data. However, in general, it is desirable to have methods available that do not require unblinding at all. Indeed, the European Medicines Agency (CHMP, 2007, Sec. 4.2.2) states:

> "Whenever possible, methods for blinded sample size reassessment that properly control the type I error should be used, especially if the sole aim of the interim analysis is the re-calculation of sample size. In cases where sample size needs to be reassessed based on unblinded data, sufficient justification should be made."

For the case of normal data, Gould & Shih (1992) proposed treating the pilot response data as coming from a single population having a distribution that is a mixture of two normal distributions with unknown means and a common unknown variance. If equal numbers of subjects are assigned to each treatment, the two elements of the mixture distribution each have weight 0.5. This three parameter model can be fitted by maximum likelihood.

Gould & Shih (1992) followed the approach of Dempster, Laird & Rubin (1977, Sec. 4.3), who describe how the EM algorithm may be used to find the maximum likelihood estimates of parameters for a finite mixture distribution, treating group identifiers as missing data. The variance estimate is then used to calculate the overall sample size, as in the unblinded case, and after the necessary additional data have been collected, treatment allocations are unblinded and a standard t-test performed.

Gould & Shih's methods have been criticized for being sensitive to the initial values used in the EM algorithm and for producing biased estimates. Friede & Kieser (2006, Sec. 5.5) observe there are situations where Gould & Shih's approach underestimates the variance substantially. Waksman (2007) explains that a step in Gould & Shih's implementation of the EM algorithm departs slightly from the original definition of Dempster, Laird & Rubin and this can cause convergence to parameter values that differ from the true maximum likelihood estimate. After correcting the method so that the true maximum likelihood estimate is found, Waksman notes that "the within-group standard deviation so obtained is negatively biased with a large standard deviation".

Xing & Ganju (2005) proposed a method for estimating the variance σ^2 from blinded data when treatment is allocated by the method of randomized blocks. Suppose subjects are considered in blocks of size b, where b is even, and within a block $b/2$ subjects are randomized to each treatment. Then the sample sum from each block has variance $b\sigma^2$ and an unbiased estimate of σ^2 can be obtained from the set of sample sums. Ganju & Xing (2009) discuss extensions of this method. A drawback is that, even with the smallest block size $b = 2$, the variance of the estimate of σ^2 is twice that of the estimate that would have been obtained from unblinded data.

Kieser & Friede (2003) discuss a simple estimate of variance based on pooled data. In our notation, suppose $X_{A,1}, \ldots, X_{A,n_1}$ and $X_{B,1}, \ldots, X_{B,n_1}$ have been observed and the pooled sample mean is

$$\bar{X} = \frac{1}{2n_1}\left\{\sum_{i=1}^{n_1} X_{A,i} + \sum_{i=1}^{n_1} X_{B,i}\right\}.$$

Then a pooled estimate of σ^2 is

$$s_p^2 = \frac{1}{2n_1 - 1}\left\{\sum_{i=1}^{n_1}(X_{A,i} - \bar{X})^2 + \sum_{i=1}^{n_1}(X_{B,i} - \bar{X})^2\right\}.$$

Denoting the difference in treatment means by $\theta = \mu_A - \mu_B$, this estimate has expectation

$$E(s_p^2) = \sigma^2 + \frac{n_1}{2(2n_1 - 1)}\,\theta^2,$$

so it is unbiased if $\theta = 0$ but, in general, has a small positive bias. In estimating the sample size needed to achieve power when $\theta = \delta$, Kieser & Friede (2003) consider the adjusted estimate

$$s_{adj}^2 = s_p^2 - \frac{n_1}{2(2n_1 - 1)}\,\delta^2,$$

which is unbiased if $\theta = \delta$. Kieser & Friede note that this adjusted variance estimator was proposed by Zucker et al. (1999) and an approximate version had previously been suggested by Gould & Shih (1992). Waksman (2007) reports extensive simulations of the maximum likelihood estimate and the pooled estimate s_p^2 and concludes that the s_p^2 "is a simpler and better alternative for most clinical trial situations".

Kieser & Friede (2003) consider the impact on the Type I error rate when estimates s_p^2 and s_{adj}^2 are used for sample size re-estimation. Applying numerical integration, they show that in a variety of situations, some with n_1 as small as 10, the maximum deviation of the Type I error rate from a target of $\alpha = 0.025$ is 0.0001. These results are quite remarkable, given the higher levels of Type I error rate inflation for procedures using an unblinded estimate of σ^2 that we saw in Tables 18.4 and 18.5 and the fact that blinded procedures have to manage with less information about σ^2. Friede & Miller (2012) describe "continuous monitoring" procedures that update sample size repeatedly based on variance estimates from unblinded data and they report similar results from a simulation study, with the target Type I error rate being achieved to a very high degree of accuracy.

A partial explanation of these findings is the following. With

$$\bar{X}_A = \frac{1}{n_1}\sum_{i=1}^{n_1} X_{A,i} \quad \text{and} \quad \bar{X}_B = \frac{1}{n_1}\sum_{i=1}^{n_1} X_{B_i},$$

define the "sums of squares" statistics

$$SS_1 = \frac{n_1}{2}(\bar{X}_A - \bar{X}_B)^2, \quad SS_2 = 2(n_1-1)s_0^2 = \sum_{i=1}^{n_1}(X_{A,i} - \bar{X}_A)^2 + \sum_{i=1}^{n_1}(X_{B,i} - \bar{X}_B)^2$$

and

$$SS_3 = (2n_1 - 1)s_p^2 = \sum_{i=1}^{n_1}(X_{A,i} - \bar{X})^2 + \sum_{i=1}^{n_1}(X_{B,i} - \bar{X})^2.$$

By standard analysis of variance arguments

$$SS_3 = SS_1 + SS_2.$$

If $\theta = \mu_A - \mu_B = 0$, then

$$SS_1 \sim \sigma^2\chi_1^2, \quad SS_2 \sim \sigma^2\chi_{2n_1-1}^2$$

and SS_1 and SS_2 are independent. It follows from the result of Exercise 11.10 of Stuart & Ord (1994, p. 407) in the case $r = 2$ that

$$\frac{SS_1}{SS_2} \text{ is independent of } SS_3 = SS_1 + SS_2. \tag{18.10}$$

Now, the t-statistic for testing H_0: $\theta = 0$ based on the pilot data is

$$T_1 = \frac{\bar{X}_A - \bar{X}_B}{\sqrt{2s_0^2/n_1}}$$

and thus

$$T_1^2 = \frac{SS_1}{s_0^2} = 2(n_1 - 1)\frac{SS_1}{SS_2}.$$

If $\theta = 0$ then, from (18.10), T_1^2 is independent of SS_3 and, therefore, of s_p^2 and, since the distribution of T_1 is symmetric about zero when $\theta = 0$, T_1 is independent of s_p^2. The adjusted estimate s_{adj}^2 differs from s_p^2 by a constant and so T_1 is also independent of s_{adj}^2.

In the case of sample size re-estimation based on the unblinded variance estimate s_0^2, a multiple of SS_2, there is a dependence between T_1, and the modified sample size and the inflation of the Type I error rate can be explained as a consequence of this dependence. The absence of such dependence when sample size re-estimation is based on a blinded variance estimate s_p^2 or s_{adj}^2 does not rule out the possibility of Type I error rate inflation completely, but it does eliminate the most obvious mechanism for this.

The same motivation for blinded sample size re-estimation arises in comparative studies with binary responses. However, in Section 18.2.1, we have already seen that the methods of Gould (1992) use pooled response data from the pilot stage so these designs can be implemented without unblinding. Friede & Kieser (2004) present an extensive computational study of the effect on Type I error power and sample size of an internal pilot design for binary data. They consider procedures using Pearson's chi-square test and conclude that, for the sample size rules they have investigated:

> "... if one is willing to accept the extent of liberality that the chi-square test shows in the fixed sample size design, one is not disturbed by the excess in Type I error rate observed under blinded sample size adaptation."

18.4 Sample Size Re-estimation in Group Sequential Tests

The maximum information formulation of the error spending method, described in Section 8.2.2, accommodates sample size re-estimation very naturally. We may distinguish between two different types of application. In case A, the information $\mathcal{I}_k$ at analysis k is affected by unknown parameters, but this information level can be accurately assessed at the time the analysis is conducted. In case B, information levels are functions of an unknown parameter, ϕ say, which is estimated more accurately as the trial continues and in this case some modifications to the basic approach become necessary.

As an example of case A, where information levels are readily available, consider a comparison of two survival distributions using a group sequential log-rank test. The observed information $\mathcal{I}_k$ at analysis k can be calculated from equation (17.2) and is roughly proportional to the total number of observed deaths. Thus, the level of information is affected by the accrual rate, the two survival functions, and the amount of competing risk censoring. The error spending methods of Chapter 8 can be used to conduct a group sequential test conditionally on the observed information levels $\{\mathcal{I}_1, \mathcal{I}_2, \dots\}$. A requirement in the maximum information paradigm is that information should eventually reach a pre-specified target, $\mathcal{I}_{max}$. In this instance, the maximum information that can become available is determined by the number of patients accrued and the total length of follow-up and changing these features of the experimental design forms the analogue of sample size re-estimation in other contexts. The papers by Scharfstein, Tsiatis & Robins (1997) and Scharfstein & Tsiatis (1998) illustrate how such design changes can be planned so that a target information level will ultimately be reached.

In case B, the quantity $\mathcal{I}_k = \mathcal{I}_k(\phi)$ depends on an unknown nuisance parameter ϕ. Mehta & Tsiatis (2001) and Tsiatis (2006) suggest using the procedure described in Section 8.2.2 but with $\mathcal{I}_k(\phi)$ replaced by $\mathcal{I}_k(\hat{\phi})$ at each stage, where $\hat{\phi}_k$ is the current estimate of ϕ. The overall target sample size is updated based on the current estimate $\hat{\phi}_k$ so that the maximum information will be achieved. When the sample sizes are large, Type I and II error rates are controlled approximately. However this control may not be achieved so accurately with small or moderate sample sizes. There are typically several types of approximation in play here. In the case of normal response with unknown variance, t-statistics are being monitored but the error spending boundary is computed using the joint distribution of Z statistics. Furthermore, only estimates of the observed information are available. The correlation between Z-statistics at analyses k_1 and k_2, where $k_1 < k_2$, is $\mathrm{Corr}(Z_{k_1}, Z_{k_2}) = \sqrt{(\mathcal{I}_{k_1}/\mathcal{I}_{k_2})}$. Suppose that in a balanced design, n_k denotes the cumulative sample size on each treatment arm at analysis k. While it is natural to approximate $\mathrm{Corr}(Z_{k_1}, Z_{k_2})$ by $\sqrt{(\hat{\mathcal{I}}_{k_1}/\hat{\mathcal{I}}_{k_2})}$, it is preferable to write the correlation as $\sqrt{(n_{k_1}/n_{k_2})}$ since this removes the ratio of two different estimates of σ^2. Surprisingly perhaps, simulation studies show that even with moderate to large sample sizes the estimated information may actually decrease from one stage to the next, due to a high increase in the estimate of σ^2. A pragmatic solution is simply to omit any analysis at which such a decrease occurs and to make a special adjustment to spend all remaining Type I error probability if this happens at the final analysis.

Continuing the discussion of a trial with normal data of unknown variance, we note that re-estimating the target sample size based on $\hat{\sigma}_k^2$ produces a downwards bias in later estimates of σ^2. This can lead to inflation of the Type I error rate, just as we saw for the Wittes & Brittain (1990) internal pilot procedure in Section 18.2.2. In fact, repeated re-estimation of sample size exacerbates the problem since it enhances the effect of stopping when the current estimate of σ^2 is unusually low. Suppose σ^2 is under-estimated at a particular analysis. Then, the observed information is over-estimated and the error spending rule will allocate an increased cumulative error probability; at the same time, the under-estimate of variance enhances the probability of observing a large standardized test statistic, increasing the chance of a Type I error. The net effect is that error spending methods allocate proportionately more Type I error probability to analyses where the variance estimate is low, and this can be expected to increase the overall probability of a Type I error. However, the impact on the Type I error rate will decrease as the overall sample size increases.

In our simulations we have seen the Type I error probability for a one-sided test increase from $\alpha = 0.025$ to 0.03 or more for a study with 5 analyses and a target degrees of freedom equal to 50. For more accurate approximations, Denne & Jennison (2000, Sec. 3) suggest a K-stage generalization of Stein's (1945) two-stage procedure that leads to group sequential t-tests constructed using error spending functions and with sample size updating.

Another example of case B is the comparison of the success probabilities p_A and p_B for two treatments with a binary outcome. Suppose a group sequential test is to be conducted to achieve a certain power when $p_A - p_B = \Delta$ and this implies a target maximum information level $\mathcal{I}_{max}$ given the form of group sequential test and specified error probabilities. If n_k observations are available on each treatment arm at analysis k, the variance of the usual estimate $\hat{p}_A - \hat{p}_B$ is approximately $2\bar{p}(1-\bar{p})/n_k$, where $\bar{p}$ is the mean of the true probabilities p_A and p_B. So, the information for $\theta = p_A - p_B$, which is approximately

$$\mathcal{I}_k = n_k/\{2\bar{p}(1-\bar{p})\},$$

depends on the unknown $\bar{p}$. Denoting the estimate of $\bar{p}$ based on the data at analysis k by $\tilde{p}_k$, we can estimate $\mathcal{I}_k$ by

$$\hat{\mathcal{I}}_k = \frac{n_k}{2\tilde{p}_k(1-\tilde{p}_k)}, \quad k = 1, 2, \ldots,$$

and construct an error spending test using these estimated information levels. Thus, at analysis k, future group sizes should be chosen to meet the target of a final sample n_{max} per treatment arm where

$$\frac{n_{max}}{2\tilde{p}_k(1-\tilde{p}_k)} = \mathcal{I}_{max}. \tag{18.11}$$

One complication in calculating the testing boundary at analysis k is that the current estimate of $\bar{p}$ is more reliable than earlier estimates used in constructing previous boundary points, and it may now be evident that error probabilities spent at previous analyses were not in fact equal to their desired values. It is not possible to go back and change the stopping rule at analyses $1, \ldots, k-1$. However, corrections to earlier inaccuracies can be made by using the current estimate, $\tilde{p}_k$, in approximating the joint distribution of the first k test statistics when the current boundary point is being chosen to give a specified cumulative error probability. Details of this calculation for tests on binary data are given by Jennison & Turnbull (1993a, Sec. 3.3) and Jennison & Turnbull (1993b, Sec. 3.3).

Zhou et al. (2014) present an excellent case study of a longitudinal trial in which repeated measurements are made on the primary endpoint. Inference may concern the treatment effect on the final measurement of this endpoint or a linear contrast of the treatment means across time. Assuming a multivariate normal model for the longitudinal data, the authors present methods for information-based sample size re-estimation that may be based on blinded or unblinded data. The testing procedure accommodates pipeline data on subjects with partial follow-up at an interim analysis. Simulation results demonstrate the ability of their methods to achieve a specific power when variances and correlations of the primary response variable and patient retention rates are initially unknown, with efficiency gains from the opportunity for early stopping.

18.5 Discussion

The review articles by Chuang-Stein et al. (2006), Friede & Kieser (2006) and Proschan (2009) contain further references on the topic of sample size re-estimation. These papers consider methods that modify sample size in response to interim estimates of both nuisance parameters and the treatment effect $\hat{\theta}$. Protecting the Type I error rate in the latter case requires the use of special methods, such as "combination tests" or tests that preserve the conditional Type I error probability after a sample size adaptation, which we shall introduce in Chapters 22 and 23. We shall see in Section 23.3 that these methods offer an alternative approach to testing when sample size is re-estimated for a nuisance parameter and this approach avoids the problems of Type I error rate inflation we have seen in this chapter.

Chapter 19

Multivariate Endpoints

19.1 Introduction

Interim analyses in clinical trials often include the examination of a number of patient outcome measures. As an example, Pocock, Geller & Tsiatis (1987) describe a crossover trial of chronic respiratory disease in which active drug and placebo were compared with respect to three lung function measurements: peak expiratory flow rate, forced expiratory volume and forced vital capacity. This is an example where the outcome variables might be regarded as *unordered* since one does not take priority over another. Another example occurred in the NINDS Stroke Trial (Tilley et al., 1996) in which treatment effect in patients with acute ischemic stroke was measured on four different ratings scales. In an extreme example, O'Brien (1984) described a study of diabetes patients in which improvements in nerve function were measured on 34 electromyographic variables.

In other trials there will be an ordering of importance of the outcome variables. Jennison & Turnbull (1993c) describe a clinical trial of an analgesic drug used to relieve arthritic pain where the primary endpoint was a measure of the amount of pain relief experienced by the patient but a secondary endpoint concerned a possible effect on the arthritic condition of the joint. It was thought that the two outcome measures might be related, if for no other reason than that the drug's success in relieving pain may lead the patient to be less careful in protecting the joint. When the primary outcome measure is a rare event, consideration of additional intermediate endpoints can lead to lower sample sizes or higher power. For instance, in evaluating therapies for acute myocardial infarction (MI), mortality will be the most important endpoint but other clinically meaningful endpoints that are correlated with an adverse long-term outcome, such as recurrent MI or ventricular dysfunction, may also be considered (Cannon, 1997).

Sometimes it is reasonable to adopt a strategy of assigning a composite score or summary scalar measure to each subject based on all of that individual's responses. The problem is then reduced to one of univariate response and the methods described in preceding chapters for monitoring a Z-statistic or a Student's t-statistic can be applied to the summary measure. This approach is suitable when the outcome variables all attempt to measure a common underlying response. In quality of life studies, several responses may be combined into a single index, based perhaps on a multivariate trait analysis. Another example is a study of efficacy of chemoprophylaxis for tuberculosis (TB) in HIV positive patients with two binary endpoints: suspected TB (a positive culture or smear) and confirmed TB. We could turn this into a single ordinal response with a score of 0 for no TB, 1 for suspected TB or 2 for confirmed TB. In a more complicated example, the Hormone Replacement Trial of the Women's Health Initiative considered a "weighted combined index" involving five binary outcomes: incidence of coronary heart disease, hip fracture, breast cancer, endometrial cancer and death from other causes. These outcomes were assigned weights using data external to the trial of 0.5, 0.18, 0.35, 0.15 and 1.00, respectively; see Freedman et al. (1996, Table 1).

DOI: 10.1201/9781584888482-19

Table 19.1 *Pocock tests: constants $C_P(K, 0.05/p)$ for two-sided tests with K groups of observations and Type I error probability $\alpha = 0.05/p$, for use with Bonferroni procedures with p endpoints*

	$C_P(K,0.05/p)$						
K	$p=1$	$p=2$	$p=3$	$p=4$	$p=5$	$p=6$	$p=10$
1	1.960	2.241	2.394	2.498	2.576	2.638	2.807
2	2.178	2.449	2.596	2.696	2.772	2.832	2.995
3	2.289	2.556	2.700	2.799	2.873	2.932	3.093
4	2.361	2.625	2.768	2.865	2.939	2.997	3.156
5	2.413	2.675	2.817	2.913	2.986	3.045	3.203
10	2.555	2.811	2.950	3.045	3.117	3.174	3.329

When the multiple outcomes are survival times, the time to the first occurring event can be used. Examples of such composite endpoints might include "total" or "all-cause" mortality, "total morbidity", "cancer mortality" or "cancer incidence". Cannon (1997) describes similar composite survival endpoints that have been used in large trials of acute MI. Gent (1997) discusses some of the issues that arise in the construction of composite event outcomes.

The strategy of using a summary score or composite endpoint obviates the problem of analyzing multiple endpoints. However, using a group sequential rule based on several outcome measures rather than a rule based on a single primary outcome measure is sometimes desirable for two reasons. First, specific stopping guidelines can be defined that recognize separately the presence of all important variables and, second, consideration of the multivariate structure of the observations can lead to more sensitive and powerful group sequential tests.

Setting aside essentially univariate methods based on a summary response for each subject, most multiple-endpoint procedures can be classified into two general categories. In the first, univariate procedures are applied to test for a treatment effect on each outcome variable separately with allowance for the multiplicity effect which would otherwise increase the family-wise error rate (FWER), defined as the probability of rejecting at least one true null hypothesis. In the second category, stopping decisions are based on a single "global" univariate statistic, constructed from the accumulated results on one, some, or all of the variables. Then, a step-down method can be employed to make inferences about individual response variables but care must still be taken; see Tang & Geller (1999) and also Section 20.5.3. In the next sections, we describe a variety of methods from both these categories. In Chapter 20 we give further consideration to multiple testing procedures for trials with a hierarchy of endpoints: primary, secondary and even tertiary. We return to the topic of multiple hypothesis testing in Chapters 25 to 29, where we present adaptive designs for trials with multiple treatment arms or multiple endpoints.

19.2 The Bonferroni Procedure

A simple strategy when monitoring multiple outcomes is to run separate univariate group sequential tests on each variable, terminating the study the first time any one of the individual tests stops to reject its null hypothesis, but using a Bonferroni (1936) adjustment to account for the multiplicity of tests. For instance, with p response variables, each group sequential test would be conducted at level α/p to restrict the overall probability of a Type I error to at most α. We need to expand the tables of Chapter 2 to obtain critical values for these stopping rules. Tables 19.1 and 19.2 give the constants needed to construct two-sided

Table 19.2 *O'Brien & Fleming tests: constants $C_B(K, 0.05/p)$ for two-sided tests with K groups of observations and Type I error probability $\alpha = 0.05/p$, for use with Bonferroni procedures with p endpoints*

	$C_B(K, 0.05/p)$						
K	$p = 1$	$p = 2$	$p = 3$	$p = 4$	$p = 5$	$p = 6$	$p = 10$
1	1.960	2.241	2.394	2.498	2.576	2.638	2.807
2	1.977	2.250	2.400	2.502	2.580	2.641	2.809
3	2.004	2.272	2.419	2.519	2.595	2.656	2.820
4	2.024	2.289	2.435	2.534	2.609	2.669	2.833
5	2.040	2.303	2.448	2.547	2.621	2.681	2.843
10	2.087	2.346	2.489	2.586	2.660	2.719	2.879

Pocock or O'Brien & Fleming tests with Type I error probability α/p for $\alpha = 0.05$ and $p = 1, \ldots, 6$ and $p = 10$. These two tables can be considered as extensions of Tables 2.1 and 2.3, respectively.

Example

Suppose a trial has $K = 4$ planned analyses and $p = 3$ outcome variables and we choose to run an O'Brien & Fleming test on each endpoint with a family-wise, two-sided Type I error probability of $\alpha = 0.05$. The critical values in each test are then

$$c_k = C_B(4, 0.05/3)\sqrt{(4/k)}, \quad k = 1, \ldots, 4,$$

where we find $C_B(4, 0.05/3) = 2.435$ from Table 19.2. The resulting values, $c_1 = 4.870$, $c_2 = 3.444$, $c_3 = 2.812$ and $c_4 = 2.435$, correspond to two-sided nominal significance levels of $\alpha'_1 = 0.000001$, $\alpha'_2 = 0.0006$, $\alpha'_3 = 0.0049$ and $\alpha'_4 = 0.0149$. At each analysis $k = 1, \ldots, 4$, the marginal two-sided significance level for each of the p variables is obtained and compared with α'_k, and stopping occurs if any one value falls below α'_k. If no Bonferroni adjustment for the multiple endpoints were made and we used $C_B(4, 0.05) = 2.024$ in place of $C_B(4, 0.05/3)$, the nominal significance levels for each endpoint would be 0.00005, 0.0042, 0.0194 and 0.0430 at the four analyses and each test on an individual outcome would have a Type I error rate of 0.05, giving a family-wise Type I error probability considerably larger than this.

In the above example, we have assumed the sequence of standardized test statistics for each endpoint follows the canonical joint distribution (3.1). This assumption is not crucial and methods for other types of distribution may be used, as long as the Type I error probability for each test is controlled at the specified level. It is not necessary to allocate the total Type I error probability α equally among the p endpoints. When one endpoint is designated as the primary outcome and the remaining $p - 1$ are secondary outcomes, Davis (1997) and Prentice (1997) discuss the merits of allocating $\alpha/2$ to the test of the primary endpoint and $\alpha/(2p - 2)$ to tests for each secondary endpoint. In a similar spirit, the Data Safety Monitoring Board of the NPC trial described in Section 1.4.5 (Clark et al., 1996) proposed using Haybittle-Peto boundaries of Section 2.7.2 with interim boundary values for standardized statistics of 3.0 for the primary endpoint (incidence of squamous cell carcinoma of the skin), 3.5 for secondary endpoints (incidence of other site-specific cancers, etc.) and 2.0 for all-cause mortality. This type of approach had been suggested by Cox (1989).

The Bonferroni procedure is very simple to carry out. The endpoints can be of differing types, continuous, binary, time-to-event, etc., and information levels for each variable do not need to be the same at a given analysis. Indeed, different boundaries (Pocock, O'Brien & Fleming, error spending, etc.) could be used for different endpoints. This procedure does not take into account how many outcome variables appear discrepant from the null hypothesis as judged by their standardized test statistics or nominal P-values: stopping occurs if just one of the p tests on different variables yields a decision, or if many of the tests reach significance. The Bonferroni approach can be inefficient because it does not take into account correlations between the response variables. In particular, it will be very conservative if outcomes are highly positively correlated, as might be expected in the example of a trial evaluating therapies for acute MI cited in Section 19.1. Such conservatism does not arise in the alternative methods considered in the following sections, which monitor a single summary statistic or monitor several outcomes in an integrated manner.

19.3 A Group Sequential Hotelling Test

Suppose a group sequential study with a maximum of K analyses yields a sequence of summary statistics $\boldsymbol{Y}_k$, $k = 1, \ldots, K$, where each $\boldsymbol{Y}_k = (Y_{1k}, \ldots, Y_{pk})^T$ is a column vector of length p. We shall consider tests of the null hypothesis H_0: $E(Y_{1k}) = \ldots = E(Y_{pk}) = 0$, for $k = 1, \ldots, K$, against a general alternative. We first assume that the covariance matrices of the $\boldsymbol{Y}_k$, denoted $\mathbf{Var}(\boldsymbol{Y}_k)$, $k = 1, \ldots, K$, are completely known so we can divide each element of $\boldsymbol{Y}_k$ by its standard deviation to create standardized statistics $\boldsymbol{Z}_k = (Z_{1k}, \ldots, Z_{pk})^T$, where each Z_{jk} has unit variance and mean zero under H_0. The relation between the $\boldsymbol{Z}_k$ and the original observations will depend on the underlying data structure. As an example, in a two-treatment parallel comparison $\boldsymbol{Z}_k$ is formed, component by component, as the vector analogue of the univariate Z_k defined in Section 3.2.1. Similarly, vector analogues of expressions for Z_k in Sections 3.2.2, 3.2.3 and 3.2.4 can be defined for the single sample problem, the two-treatment paired comparison and the two-period crossover design, respectively.

For $k = 1, \ldots, K$, let $\boldsymbol{\Sigma}_k$ denote the correlation matrix of $\boldsymbol{Y}_k$, and therefore of $\boldsymbol{Z}_k$. If the underlying data are multivariate normal, the quantity

$$\boldsymbol{Z}_k^T \boldsymbol{\Sigma}_k^{-1} \boldsymbol{Z}_k \qquad (19.1)$$

has a χ^2_p distribution under H_0. Jennison & Turnbull (1991b, Sec. 2.2; 1997b, Sec. 5) show that the sequence $\{\boldsymbol{Z}_k^T \boldsymbol{\Sigma}_k^{-1} \boldsymbol{Z}_k;\ k = 1, \ldots, K\}$ is Markov and derive its joint distribution. The constants in Table 2 of Jennison & Turnbull (1991b) for group sequential chi-squared tests can be used to construct boundary values for the statistics (19.1) which give tests with a stated Type I error probability α. The table gives values for $\alpha = 0.01$, 0.05 and 0.10, $p = 1$ to 5, and $K = 1$ to 10, 20 and 50.

Now suppose the covariance matrices of the summary statistics $\boldsymbol{Y}_k$ are not known. One possibility is that variances and covariances are partly known and each $\mathbf{Var}(\boldsymbol{Y}_k)$ can be written as the product of a known matrix and an unknown scale factor. In this case, a group sequential F-test can be used; see Jennison & Turnbull (1991b, Sec. 2.3; 1997b, Sec. 5).

When multiple endpoints arise in a clinical trial, it is likely that the matrices $\mathbf{Var}(\boldsymbol{Y}_k)$ are completely unknown. In this case, at analysis k, we create a Studentized statistic t_{jk} for each $j = 1, \ldots, p$ using an estimate of $Var(Y_{jk})$ calculated from the original observations. The form of this variance estimate will depend on the original data structure but, since only one endpoint is involved at a time, the variance estimation problem is the same as that for a single endpoint. As before, we denote by $\boldsymbol{\Sigma}_k$ the correlation matrix of $\boldsymbol{Y}_k$ but now the observed data are used to obtain an estimate of this matrix, $\widehat{\boldsymbol{\Sigma}}_k$. The quantity

$$\boldsymbol{t}_k^T \widehat{\boldsymbol{\Sigma}}_k^{-1} \boldsymbol{t}_k \qquad (19.2)$$

is an example of Hotelling's t-squared statistic (Hotelling, 1931) and this has, marginally, a Hotelling's $\mathcal{T}_p^2(\nu_k)$ distribution, where ν_k denotes the degrees of freedom for $\widehat{\boldsymbol{\Sigma}}_k$; see, for example, Chatfield & Collins (1980, Ch. 7). In the one-sample problem, $\widehat{\boldsymbol{\Sigma}}_k$ is the usual sample correlation matrix based on all the data accumulated so far and $\nu_k = n_k - 1$; for the paired treatment comparison, it is the within-pair sample correlation matrix and $\nu_k = n_k - 1$; for the two-treatment parallel design with common covariance matrices, $\widehat{\boldsymbol{\Sigma}}_k$ is the pooled within-treatment sample correlation matrix and $\nu_k = n_{Ak} + n_{Bk} - 2$; in the two-period crossover design, assuming the covariance matrix is the same for subjects on either treatment ordering, $\widehat{\boldsymbol{\Sigma}}_k$ is the sample correlation matrix pooled across both treatment orderings and again $\nu_k = n_{Ak} + n_{Bk} - 2$.

Jennison & Turnbull (1991b, Sec. 2.4) point out that an exact theory for the joint distribution of successive Hotelling's $\mathcal{T}^2$ statistics would be very complicated, involving joint distributions of repeated Wishart variables. As an alternative, we shall adopt the "significance level approach" of Section 4.2 and compare successive values of (19.2) with percentiles of the $\mathcal{T}_p^2(\nu)$ distribution corresponding to the significance levels of the normal theory tests of Chapter 2. Percentiles of Hotelling's $\mathcal{T}_p^2(\nu)$ distribution can be obtained from tables of the F-distribution by using the well-known relation

$$\mathcal{T}_p^2(\nu) = \frac{p\nu}{\nu - p + 1} F_{p,\,\nu-p+1}, \tag{19.3}$$

where $F_{\nu_1,\,\nu_2}$ denotes an F-distribution on ν_1 and ν_2 degrees of freedom; see, for example, Chatfield & Collins (1980, Eq. 6.26).

Example

Consider a study with $p = 3$ outcome variables and a maximum of $K = 4$ analyses. Suppose the covariance matrices of summary statistics are unknown and we observe the following results. Here, $\boldsymbol{t}_k$ is the vector of Studentized statistics at analysis k and $\widehat{\boldsymbol{\Sigma}}_k$ the estimated correlation matrix on ν_k degrees of freedom.

$$\textit{Analysis } 1 \qquad \nu_1 = 12 \qquad \boldsymbol{t}_1 = \begin{pmatrix} 2.01 \\ 1.92 \\ 1.39 \end{pmatrix} \qquad \widehat{\boldsymbol{\Sigma}}_1 = \begin{pmatrix} 1 & 0.20 & 0.02 \\ 0.20 & 1 & 0.03 \\ 0.02 & 0.03 & 1 \end{pmatrix}$$

$$\textit{Analysis } 2 \qquad \nu_2 = 25 \qquad \boldsymbol{t}_2 = \begin{pmatrix} 1.86 \\ 2.00 \\ 1.55 \end{pmatrix} \qquad \widehat{\boldsymbol{\Sigma}}_2 = \begin{pmatrix} 1 & 0.21 & 0.03 \\ 0.21 & 1 & 0.05 \\ 0.03 & 0.05 & 1 \end{pmatrix}$$

$$\textit{Analysis } 3 \qquad \nu_3 = 35 \qquad \boldsymbol{t}_3 = \begin{pmatrix} 1.89 \\ 2.15 \\ 1.53 \end{pmatrix} \qquad \widehat{\boldsymbol{\Sigma}}_3 = \begin{pmatrix} 1 & 0.17 & 0.02 \\ 0.17 & 1 & 0.04 \\ 0.02 & 0.04 & 1 \end{pmatrix}$$

$$\textit{Analysis } 4 \qquad \nu_4 = 48 \qquad \boldsymbol{t}_4 = \begin{pmatrix} 1.92 \\ 1.93 \\ 1.74 \end{pmatrix} \qquad \widehat{\boldsymbol{\Sigma}}_4 = \begin{pmatrix} 1 & 0.12 & 0.02 \\ 0.12 & 1 & 0.03 \\ 0.02 & 0.03 & 1 \end{pmatrix}$$

Calculation shows the successive $\mathcal{T}^2$ statistics (19.2) are 8.155, 8.193, 9.045 and 9.356. Suppose it was decided to base stopping criteria on significance levels from an O'Brien & Fleming test with Type I error probability $\alpha = 0.05$. From Table 2.3, we have $C_B(4, 0.05) = 2.024$ and the corresponding nominal significance levels are

$$\alpha_1' = 2\{1 - \Phi(4.048)\} = 0.00005,$$

Table 19.3 *Worksheet for a group sequential test using Hotelling's $\mathcal{T}^2$ statistic*

k	ν_k	$\mathcal{T}^2$	F	$Pr\{F_{3,\nu_k-2} > F\}$	*Nominal significance level,* α'_k
1	12	8.155	2.265	0.1433	0.0001
2	25	8.193	2.513	0.0837	0.0042
3	35	9.045	2.843	0.0527	0.0194
4	48	9.356	2.989	0.0406	0.0430

where Φ is the standard normal cdf, and similarly $\alpha'_2 = 0.0042$, $\alpha'_3 = 0.0194$ and $\alpha'_4 = 0.0430$. At analysis 1, we compute the significance level

$$Pr\{\mathcal{T}_3^2(\nu_1) > 8.155\} = Pr\{F_{3,10} > 2.265\} = 0.1433$$

and since this exceeds α'_1, the procedure continues to analysis 2. We can draw up a worksheet as displayed in Table 19.3 to assist in applying the stopping rule. The value F in this worksheet is the $\mathcal{T}^2$ statistic multiplied by $(\nu_k - p + 1)/(p\,\nu_k)$, i.e., by $(\nu_k - 2)/(3\,\nu_k)$, to produce a statistic with null distribution $F_{p,\,\nu_k-p+1} = F_{3,\nu_k-2}$. The results show that the test continues to the last analysis, $K = 4$, at which point it rejects the null hypothesis H_0: $E(Y_{1k}) = \ldots = E(Y_{pk}) = 0$, for $k = 1, \ldots, K$.

We may compare this test with the Bonferroni procedure of the previous section. For that test, since each t_{jk}, $j = 1, \ldots, 3$, is compared to the same critical value, the outcome of analysis k is determined by the maximum of $|t_{1k}|$, $|t_{2k}|$ and $|t_{3k}|$, which takes values 2.01, 2.00, 2.15 and 1.93 at analyses $k = 1$, 2, 3 and 4, respectively. These values have nominal significance levels 0.067, 0.056, 0.039 and 0.060 relative to the appropriate t-distributions and these significance levels do not come close to falling below the critical values $\alpha'_1 = 0.000001$, $\alpha'_2 = 0.0006$, $\alpha'_3 = 0.0049$ and $\alpha'_4 = 0.0149$ derived in the example of Section 19.2 for the Bonferroni procedure with an O'Brien & Fleming test. Thus the Bonferroni procedure is unable to reject the null hypothesis at any analysis. In fact, even if we applied significance levels of an O'Brien & Fleming procedure test with a Type I error rate of 0.05 for each endpoint separately, without any adjustment for multiplicity, the null hypothesis would not be rejected at any analysis. The $\mathcal{T}^2$ statistic combines the moderate evidence against H_0 from each of the three endpoints to produce a significant result, demonstrating the advantage of combining endpoints when each mean $E(Y_{jk})$ is a small distance from the value zero specified in H_0.

The Hotelling's $\mathcal{T}^2$ test is appropriate if any departure from H_0, in whatever direction, is of interest. It has found application in multivariate control charts in industrial quality control, where it is important to detect any change in distribution of the recorded variables. The test is not well suited to clinical trials where such an "omnibus" test, with power against all alternatives to H_0, is usually not appropriate. Suppose the above example concerned a two-treatment comparison with each endpoint defined so that a positive value indicates the superiority of one treatment and a negative value implies superiority of the other. If the value of $\boldsymbol{t}_3$ had been equal to $(1.89, -2.15, 1.53)^T$, we would have found $\mathcal{T}_3^2 = 12.379$, giving $F = 3.891$ and significance level $Pr\{F_{3,33} > 3.891\} = 0.017$, and the procedure would have stopped to reject H_0 at the third analysis. (Because of the positive correlations, the negative component of $\boldsymbol{t}_3$ now provides stronger evidence against H_0, even though its magnitude is unchanged.) However, these data show that one treatment has a better response in two endpoints and a poorer response in the other, which is not helpful as a recommendation for either treatment. The major interest in such a comparison is in alternatives where the treatment difference is in the same direction for all variables. By giving up power against alternatives where the null hypothesis fails because some endpoints respond well to the first treatment and others to the second, we can design procedures which are more sensitive to the most relevant alternatives: we consider such procedures in the next section.

19.4 A Group Sequential Version of O'Brien's Test

O'Brien (1984) proposed a test of the null hypothesis that a multivariate distribution has mean zero against alternatives under which means of all elements have the same sign. With suitably defined response variables, this is the appropriate formulation for detecting an improvement of one treatment over another across the range of outcome measures. O'Brien considered parametric and nonparametric test statistics but we shall pursue the parametric case here, and the multivariate normal response distribution in particular. Pocock, Geller & Tsiatis (1987) illustrate O'Brien's test in a number of examples and we follow their notation in this section. Dallow, Leonov & Roger (2008) discuss the application of O'Brien's methods in clinical trials and extend the methods to model the effects of covariates. Adaptation of O'Brien's test to a group sequential setting has been described by Tang, Gnecco & Geller (1989a) and Tang, Geller & Pocock (1993).

We first consider the case where variances and correlations of the different endpoints are known so that at each analysis k we can create a vector of p standardized summary statistics $\boldsymbol{Z}_k = (Z_{1k}, \ldots, Z_{pk})^T$, where each Z_{jk} has unit variance and mean zero under the null hypothesis H_0. We suppose that interest centers on alternatives for which each $E(Z_{jk})$ is positive. O'Brien's generalized least squares (GLS) statistic is then

$$\mathcal{Z}_k = \frac{\boldsymbol{J}^T \boldsymbol{\Sigma}_k^{-1} \boldsymbol{Z}_k}{(\boldsymbol{J}^T \boldsymbol{\Sigma}_k^{-1} \boldsymbol{J})^{1/2}}, \tag{19.4}$$

where $\boldsymbol{\Sigma}_k$ is the correlation matrix of $\boldsymbol{Z}_k$. The components of the vector $\boldsymbol{J}$ are positive and, typically, represent the relative differences of interest in each of the p endpoints, measured on the standardized scale. The common choice $\boldsymbol{J} = (1, \ldots, 1)^T$ is optimal for testing alternatives in which the improvements of the treatment are the same, on the standardized scale, for each endpoint. In this case, $\mathcal{Z}_k$ is a linear combination of Z_{1k} to Z_{pk} in which those variables that are less highly correlated with others receive correspondingly larger weights.

Other choices for $\boldsymbol{J}$ can be made if sensitivity of the test to departures from the null hypothesis in other directions is preferred. However, Tang, Geller & Pocock (1993, p. 25) point out that for certain choices of $\boldsymbol{J}$, and even for $\boldsymbol{J} = (1, \ldots, 1)^T$, the vector $\boldsymbol{J}^T \boldsymbol{\Sigma}_k^{-1}$ can have some negative components, leading to negative weights in $\mathcal{Z}_k$ for certain endpoints. In this case, if the mean of $\boldsymbol{Z}_k$ lies along particular directions in the positive or negative orthants, the GLS statistic (19.4) has expectation zero. Hence, in some parts of the alternative hypothesis where high power might be expected, tests have power only equal to α. To avoid this possibility, Tang, Geller & Pocock (1993) recommend a choice of $\boldsymbol{J}$ which guarantees that $\boldsymbol{J}^T \boldsymbol{\Sigma}_k^{-1}$ has no negative components and is also optimal with regard to maximizing the minimum power over the positive or negative orthant. In a non-sequential test, their choice is

$$\boldsymbol{J}^T = (\phi_1, \ldots, \phi_p)\, \boldsymbol{\Sigma} \tag{19.5}$$

where $\boldsymbol{\Sigma}$ is the correlation matrix of the vector of standardized statistics and ϕ_j^2 is the jth diagonal element of $\boldsymbol{\Sigma}^{-1}$. In a group sequential test, this $\boldsymbol{J}$ can be used with $\boldsymbol{\Sigma} = \boldsymbol{\Sigma}_k$ when the correlation matrices $\boldsymbol{\Sigma}_k$ do not vary with k, a typical property which we shall see in the following examples; in other cases, a value of $\boldsymbol{\Sigma}$ can be chosen to capture the key features of all the $\boldsymbol{\Sigma}_k$s.

Bolland et al. (2009) describe the group sequential design of the ICTUS trial for the treatment of acute ischemic stroke, which was based on the sum of the score statistics for three outcome measures at each analysis. Denoting the score statistic on scale j at analysis k by S_{jk} and the associated Fisher information by $\mathcal{I}_{jk}$, for $j = 1, \ldots, 3$, the combined Z-statistic at analysis k is a standardized version of

$$\sum_{i=1}^{3} S_{jk} = \sum_{i=1}^{3} Z_{jk}\sqrt{\mathcal{I}_{jk}},$$

where $Z_{jk} = S_{jk}/\sqrt{\mathcal{I}_{jk}}$, and so can be written in the form (19.4) by setting

$$\boldsymbol{J}^T = (\sqrt{\mathcal{I}_{1k}}, \sqrt{\mathcal{I}_{2k}}, \sqrt{\mathcal{I}_{3k}})^T \boldsymbol{\Sigma}_k$$

so that

$$\boldsymbol{J}^T\boldsymbol{\Sigma}_k^{-1} = (\sqrt{\mathcal{I}_{1k}}, \sqrt{\mathcal{I}_{2k}}, \sqrt{\mathcal{I}_{3k}})^T.$$

We have assumed here that the correlation matrix $\boldsymbol{\Sigma}_k$ of (Z_{1k}, Z_{2k}, Z_{3k}) and the ratios between $\mathcal{I}_{1k}$, $\mathcal{I}_{2k}$ and $\mathcal{I}_{3k}$ remain constant over analyses k.

In applications where each Z_{jk} is a multiple of a sample mean or of the difference in the means of two samples, it is straightforward to check that the sequence $\{\mathcal{Z}_1, \ldots, \mathcal{Z}_K\}$ has the canonical joint distribution (3.1) for suitably defined θ and $\mathcal{I}_1, \ldots, \mathcal{I}_K$, either exactly when observations are multivariate normal or approximately for other response distributions. In other problems where each $\boldsymbol{\Sigma}_k^{-1}\boldsymbol{Z}_k$ is a fixed linear function of an estimated parameter vector at analysis k, the same result can be deduced using the general theory of Chapter 15. Hence, in order to apply the methods of earlier chapters to the GLS statistic (19.4), we need only identify the quantities θ and $\mathcal{I}_k$ and interpret these in the particular problem being considered.

The One-Sample Problem

In the one-sample problem analogous to the one-sample univariate problem of Section 3.2.2, we observe independent, multivariate normal p-vectors $\boldsymbol{X}_1, \boldsymbol{X}_2, \ldots$ with common mean vector $\boldsymbol{\mu}$ and covariance matrix $\widetilde{\boldsymbol{\Sigma}}$, say. Suppose we wish to test H_0: $\boldsymbol{\mu} = \boldsymbol{\mu}_0$. If n_k observations have accumulated by analysis k, let $\bar{\boldsymbol{X}}_k = (\bar{X}_{1k}, \ldots, \bar{X}_{pk})^T$ denote the mean of these n_k observation vectors. The jth component of $\boldsymbol{Z}_k$ is

$$Z_{jk} = \frac{\bar{X}_{jk} - \mu_{0j}}{\sqrt{(\sigma_j^2/n_k)}},$$

where σ_j^2 is the variance of the jth element of $\boldsymbol{X}$, i.e., the jth diagonal element of $\widetilde{\boldsymbol{\Sigma}}$. We can also write

$$\boldsymbol{Z}_k = \boldsymbol{M}(\bar{\boldsymbol{X}}_k - \boldsymbol{\mu}_0)\sqrt{n_k}$$

where $\boldsymbol{M}$ is a diagonal matrix with jth diagonal entry $1/\sigma_j$, $j = 1, \ldots, p$, and all other entries equal to zero. Since, marginally,

$$\bar{\boldsymbol{X}}_k \sim N_p(\boldsymbol{\mu}, \widetilde{\boldsymbol{\Sigma}}/n_k),$$

where N_p denotes a p dimensional multivariate normal distribution, it follows that

$$\boldsymbol{Z}_k \sim N_p(\boldsymbol{M}(\boldsymbol{\mu} - \boldsymbol{\mu}_0)\sqrt{n_k}, \boldsymbol{M}\widetilde{\boldsymbol{\Sigma}}\boldsymbol{M}). \tag{19.6}$$

Thus, $\boldsymbol{\Sigma}_k$ in (19.4) does not depend on k and is equal to $\boldsymbol{M}\widetilde{\boldsymbol{\Sigma}}\boldsymbol{M}$. Writing $\boldsymbol{J}^T = (J_1, \ldots, J_p)$, define

$$\boldsymbol{\delta}^T = \boldsymbol{J}^T\boldsymbol{M}^{-1} = (J_1\sigma_1, \ldots, J_p\sigma_p), \tag{19.7}$$

then it follows from (19.4) and (19.6) that, marginally, $\mathcal{Z}_k \sim N(\theta\sqrt{\mathcal{I}_k}, 1)$, where

$$\theta = \frac{\boldsymbol{\delta}^T \widetilde{\boldsymbol{\Sigma}}^{-1} (\boldsymbol{\mu} - \boldsymbol{\mu}_0)}{\boldsymbol{\delta}^T \widetilde{\boldsymbol{\Sigma}}^{-1} \boldsymbol{\delta}} \quad \text{and} \quad \mathcal{I}_k = n_k\,(\boldsymbol{\delta}^T \widetilde{\boldsymbol{\Sigma}}^{-1} \boldsymbol{\delta}).$$

If $\boldsymbol{\mu} - \boldsymbol{\mu}_0$ is assumed to be a multiple of $\boldsymbol{\delta}$, we have the model $\boldsymbol{\mu} - \boldsymbol{\mu}_0 = \lambda\boldsymbol{\delta}$ and the GLS statistic can be written as

$$\mathcal{Z}_k = \hat{\lambda}_k / \sqrt{\{Var(\hat{\lambda}_k)\}},$$

where

$$\hat{\lambda}_k = \frac{\boldsymbol{\delta}^T \widetilde{\boldsymbol{\Sigma}}^{-1} (\bar{\boldsymbol{X}}_k - \boldsymbol{\mu}_0)}{\boldsymbol{\delta}^T \widetilde{\boldsymbol{\Sigma}}^{-1} \boldsymbol{\delta}}$$

is the GLS estimate of λ based on the data at analysis k.

Two-Sample Problems

In a balanced two-treatment comparison, suppose there are n_k multivariate normal p-vectors from each treatment at analysis k, with observations $\boldsymbol{X}_{A1}, \ldots, \boldsymbol{X}_{An_k}$ from treatment A distributed as $N_p(\boldsymbol{\mu}_A, \widetilde{\boldsymbol{\Sigma}})$ and $\boldsymbol{X}_{B1}, \ldots, \boldsymbol{X}_{Bn_k}$ from treatment B distributed as $N_p(\boldsymbol{\mu}_B, \widetilde{\boldsymbol{\Sigma}})$. Suppose response variables have been defined so that high values are desirable and consider a test of H_0: $\boldsymbol{\mu}_A = \boldsymbol{\mu}_B$ against alternatives in which each element of $\boldsymbol{\mu}_A - \boldsymbol{\mu}_B$ has the same sign. Let σ_j^2 be the variance of the jth element of each $\boldsymbol{X}_{Ai}$ or $\boldsymbol{X}_{Bi}$, i.e., the jth diagonal element of $\widetilde{\boldsymbol{\Sigma}}$. Using obvious notation for the means on treatments A and B, the jth component of the standardized statistic $\boldsymbol{Z}_k$ at analysis k is

$$Z_{jk} = \frac{\bar{X}_{Ajk} - \bar{X}_{Bjk}}{\sqrt{(2\sigma_j^2/n_k)}}$$

and $\mathcal{Z}_k$ is defined from $\boldsymbol{Z}_k = (Z_{1k}, \ldots, Z_{pk})^T$ by (19.4) with $\boldsymbol{\Sigma}_k$ the correlation matrix of $\boldsymbol{Z}_k$.

Defining $\boldsymbol{M}$ to be the matrix with diagonal entries $1/\sigma_1, \ldots, 1/\sigma_p$, and all other entries equal to zero, we have

$$\boldsymbol{Z}_k = \boldsymbol{M}(\bar{\boldsymbol{X}}_{Ak} - \bar{\boldsymbol{X}}_{Bk})\sqrt{(n_k/2)}$$

and

$$\boldsymbol{Z}_k \sim N_p(\boldsymbol{M}(\boldsymbol{\mu}_A - \boldsymbol{\mu}_B)\sqrt{(n_k/2)},\, \boldsymbol{M}\widetilde{\boldsymbol{\Sigma}}\boldsymbol{M}),$$

so again $\boldsymbol{\Sigma}_k$ does not depend on k. As for the one-sample problem, if we define $\boldsymbol{\delta}$ by (19.7), then $\mathcal{Z}_k$ has an interpretation as the standardized value of a generalized least squares estimator, this time of λ in the model $\boldsymbol{\mu}_A - \boldsymbol{\mu}_B = \lambda\boldsymbol{\delta}$ for given $\boldsymbol{\delta}$.

Similar calculations to those for the one-sample problem show that, marginally,

$$\mathcal{Z}_k \sim N(\theta\sqrt{\mathcal{I}_k}, 1), \quad k = 1, \ldots, K,$$

where

$$\theta = \frac{\boldsymbol{\delta}^T \widetilde{\boldsymbol{\Sigma}}^{-1} (\boldsymbol{\mu}_A - \boldsymbol{\mu}_B)}{\boldsymbol{\delta}^T \widetilde{\boldsymbol{\Sigma}}^{-1} \boldsymbol{\delta}} \quad \text{and} \quad \mathcal{I}_k = n_k\,(\boldsymbol{\delta}^T \widetilde{\boldsymbol{\Sigma}}^{-1} \boldsymbol{\delta})/2,$$

and the sequence $\mathcal{Z}_1, \ldots, \mathcal{Z}_K$ has the canonical joint distribution (3.1).

If $\boldsymbol{\mu}_A - \boldsymbol{\mu}_B$ is in the direction $\boldsymbol{\delta}$ then θ is positive, and if $\boldsymbol{\mu}_A - \boldsymbol{\mu}_B$ is in the opposite direction θ is negative. More generally, since every component of $\boldsymbol{J}$ is positive, θ will usually be positive when treatment A is superior across all endpoints, and certainly so if (19.5) is used to define $\boldsymbol{J}$. Similarly, θ will usually be negative when treatment B is superior across all endpoints. Thus, one-sided or two-sided tests, as described in Chapters 2 to 8, can be

Table 19.4 *Worksheet for a group sequential test using O'Brien's GLS statistic*

k	ν_k	$\mathcal{Z}_k$	$Pr\{\lvert t_{\nu_k}\rvert > \lvert\mathcal{Z}_k\rvert\}$	*Nominal significance level,* α'_k
1	12	2.817	0.0155	0.0001
2	25	2.843	0.0088	0.0042
3	35	2.974	0.0053	0.0194
4	48	3.055	0.0037	0.0430

constructed using the sequence of $\mathcal{Z}_k$ statistics. One-sided tests are appropriate when it is desired to show that one specific treatment is superior to the other on the basis of its combined performance over all endpoints. Two-sided tests should be used to detect an overall superiority of either treatment. In each case, a test will have high power when the superior treatment is equal to or better than the other with respect to all p endpoints. GLS statistics following the canonical joint distribution (3.1) can be derived in a similar fashion for multivariate data from paired sample, crossover or other designs, and used to construct group sequential tests in the usual way.

In practice, the covariance matrix of multivariate observations will often be unknown. As in the development in Section 19.3 of the procedure based on Hotelling's statistic, components of the standardized statistics $\boldsymbol{Z}_k$ can be replaced by t-statistics and an estimate $\widehat{\boldsymbol{\Sigma}}_k$ substituted for the correlation matrix. The GLS statistic (19.4) at analysis k then becomes (Pocock, Geller & Tsiatis, 1987)

$$\mathcal{Z}_k = \frac{\boldsymbol{J}^T \widehat{\boldsymbol{\Sigma}}_k^{-1} \boldsymbol{t}_k}{(\boldsymbol{J}^T \widehat{\boldsymbol{\Sigma}}_k^{-1} \boldsymbol{J})^{1/2}}. \tag{19.8}$$

In the two-treatment problem we have described, the p components of $\boldsymbol{t}_k$ are the two-sample t-statistics corresponding to each variable by variable comparison, and $\widehat{\boldsymbol{\Sigma}}_k$ is the pooled within-treatment estimate of the correlation matrix.

The statistic (19.8) is a combination of t-statistics. It does not itself have a Student's t-distribution, but it is asymptotically normal. Simulations reported by Pocock, Geller & Tsiatis (1987) for the fixed sample problem suggest that the normal approximation for (19.8) is good for small dimensions of $p = 2$ or 3, even when the degrees of freedom ν_k associated with $\widehat{\boldsymbol{\Sigma}}_k$ are as low as 10. However, they also report that the normal approximation is not nearly so accurate as the dimension p increases. More work is needed here to find critical values to achieve a specified Type I error probability.

The elements of $\boldsymbol{t}_k$ have greater variance than the standard normal variates $\mathcal{Z}_k$. Compensation for this additional variability can be made by basing critical values for successive statistics (19.8) on percentiles of Student's t-distribution with ν_k degrees of freedom when using significance levels taken from tests for normal data of known variance, in keeping with the "significance level approach" of Section 4.2. We have applied this procedure to construct a group sequential two-sided test for the example discussed earlier in this chapter. We used $\boldsymbol{J} = (1, 1, 1)^T$ in defining the statistics $\mathcal{Z}_k$ and based boundaries on an O'Brien & Fleming test with two-sided Type I error probability 0.05, supposing there is particular interest in rejecting H_0 when the elements of $\boldsymbol{t}_k$ are all positive or all negative.

The results of our computations are shown in Table 19.4. Values in the column headed $Pr\{|t_{\nu_k}| > |\mathcal{Z}_k|\}$ are the two-sided significance levels of the observed $\mathcal{Z}_k$ relative to a t_{ν_k} distribution. The test stops to reject H_0 if this significance level is less than the value α'_k in the final column. Using this procedure, the trial would stop to reject H_0 at the third analysis, one analysis earlier than the procedure based on Hotelling's $\mathcal{T}^2$. This is to be expected as the O'Brien GLS test is most sensitive to alternatives in which all the endpoints depart from H_0 in the same direction. On the other hand, if the sign of the second component of $\boldsymbol{t}_k$ is switched to create a negative value at every stage, the GLS statistics become $\mathcal{Z}_1 = 0.889$, $\mathcal{Z}_2 = 0.872$, $\mathcal{Z}_3 = 0.795$ and $\mathcal{Z}_4 = 1.020$, and the procedure continues to the last analysis, where it fails to reject H_0. In contrast, the Hotelling's $\mathcal{T}^2$ procedure stopped at analysis $k = 3$ to reject H_0 for this modified set of data.

In designing a trial, one may need to choose between a group sequential test based on a single particular endpoint and one based on the GLS statistic using multiple endpoints. Tang, Gnecco & Geller (1989a) show there can be substantial savings in sample size if the GLS statistic is used, rather than a single selected endpoint, when there are multiple correlated endpoints. In planning a group sequential study based on the GLS statistic, the maximum sample size needed to assure Type I error and power requirements depends on the correlation matrix of the standardized variables $\boldsymbol{Z}_k$, and this is typically unknown at the design stage. However, Tang, Gnecco & Geller (1989a) prove that the sample size needed is no greater than the smallest required for any of the corresponding univariate tests. Thus, if a good prior estimate of the correlation matrix is not available, a conservative strategy is to use that smallest maximum sample size and significant efficiency gains will still result.

19.5 Tests Based on Other Global Statistics

As an alternative to tests based on (19.8), Tang, Geller & Pocock (1993) describe group sequential tests based on the approximate likelihood ratio (ALR) statistic of Tang, Gnecco & Geller (1989b). A computational problem with the latter approach is that the statistics have a chi-bar-squared distribution (Robertson, Wright & Dykstra, 1988) which is not easily handled by standard numerical methods. The statistics are not asymptotically normally distributed, so critical values tabulated in earlier chapters are not applicable. However, simulation can be used to generate appropriate boundaries. Tang, Geller & Pocock (1993) provide details and give tables of critical values for sequences of ALR statistics for use in group sequential tests with Type I error probability $\alpha = 0.05$ and two boundary shapes. The first is a repeated significance test with constant significance level (Pocock, 1977); the second is a Lan & DeMets (1983) error spending test that allocates a Type I error probability of 0.01 to each interim analysis and all remaining error probability to the final analysis. The tables are for tests with 1 to 5 analyses and 2, 3, 4 and 5 outcome variables.

Nonparametric tests are of value when data depart substantially from normality. O'Brien (1984) considers rank-sum statistics as well as the parametric models that lead to the GLS statistic (19.4). Similarly, Wei & Johnson (1985) define tests in terms of linear combinations of U-statistics to combine results on separate endpoints. Su & Lachin (1992) describe a group sequential test based on a summary rank statistic and Lin (1991) proposes use of a summary statistic based on ranks in a group sequential test for multivariate survival data.

19.6 Tests Based on Marginal Criteria: Co-primary Endpoints

A problem with combining data from different endpoints is that the resulting summary measure may have little clinical meaning. If the endpoints are quite different in nature, it may not be appropriate to combine them at all: this could be the case if one outcome is an efficacy variable and a second is a safety variable measuring treatment toxicity. As an example, Kittelson et al. (2016) describe a graphical method to represent the bivariate relation of benefit and risk for antithrombotic drugs which aim to prevent a thromboembolic event but may increase the risk of major bleeding. In this approach, a curve representing a null hypothesis defines a region of benefit in the two-dimensional plane.

When endpoints are qualitatively different, it may be important to monitor each variable separately but still control the overall error rate of making any incorrect inferences. An alternative strategy whereby a group sequential test is used only for the efficacy variable has the defect of not providing any guidelines for monitoring the safety endpoint. Thus, we are led to consider methods that retain the identity of the different endpoints but are more efficient than the simple Bonferroni procedure described previously.

The FDA Guidance on Multiple Endpoints in Clinical Trials (FDA, 2022) uses the term "co-primary endpoints" to describe the case where a positive outcome is required on each endpoint in order to demonstrate the acceptability of a new treatment. In Section III.C.1, under the heading *"When Demonstration of Treatment Effects on Two or More Distinct Endpoints Is Recommended to Establish Clinical Benefit (Co-Primary Endpoints)"*, the Guidance states:

> "For some disorders, there are two or more different features that are so critically important to the disease under study that a drug will not be considered effective without demonstration of a treatment effect on all of these disease features. The term used in this guidance to describe this circumstance of multiple primary endpoints is co-primary endpoints. Multiple primary endpoints become co-primary endpoints when demonstrating an effect on each of the endpoints is critical to concluding that a drug is effective."

The Guidance describes an example concerning treatment of migraine headaches and then discusses vaccine studies:

> "Trials of combination vaccines are a situation in which co-primary endpoints are applicable. These vaccine trials are typically designed and powered for demonstration of a successful outcome on effectiveness endpoints for each pathogen against which the vaccine is intended to provide protection."

Previously, in a report from the Multiple Endpoints Expert Team of the Pharmaceutical Research and Manufacturers of America, Offen et al. (2007) had observed:

> "There are quite a few disorders for which regulatory agencies have required a treatment to demonstrate a statistically significant effect on multiple endpoints, each at the one-sided 2.5% level, before accepting the treatment's efficacy for the disorders. Depending on the correlation among the endpoints, this requirement could lead to a substantial reduction in the study's power to conclude the efficacy of a treatment."

In their discussion of a draft version of the FDA Guidance, Chuang-Stein & Li (2017) note that this approach to co-primary endpoints

> ". . . leads to a substantial loss of power when the number of co-primary endpoints exceeds 2, a fact acknowledged in the draft guidance."

and they go on to discuss ways of tackling the loss of power. However the final version of the FDA Guidance makes no concessions when it states:

> "There have been suggestions that the statistical testing criteria for each co-primary endpoint could be increased (e.g., testing at an α of 0.06 or 0.07) when the targeted α is 0.05 to accommodate the loss in statistical power arising from the need to show an effect on both endpoints. Increasing α for each co-primary endpoint is not acceptable because doing so may undermine the ability to interpret a treatment effect on each disease aspect considered critical to show that the drug is effective in support of approval."

The FDA Guidance goes on to acknowledge there are cases where it is sufficient to demonstrate a treatment effect on at least one of several primary endpoints, but it warns that a multiple testing procedure will then be required to control the overall Type I error probability:

> "This use of multiple endpoints creates a multiplicity problem because there are several ways for the study to successfully demonstrate a treatment effect. Control of the Type I error rate for the primary endpoint family is critical."

The European Medicines Agency has expressed similar views (CHMP, 2016). Section 5.5.1 of this guideline concerns cases where "statistical significance is needed for all primary endpoints" and states:

> "Here, interpretation of the results is most clear-cut because, in order to provide sufficient evidence of the clinically relevant efficacy, each null hypothesis on every primary variable has to be rejected at the same significance level (e.g. 0.05). For example, according to the CHMP Guideline on clinical investigation of medicinal products in the treatment of chronic obstructive pulmonary disease (EMA/CHMP/483572/2012), lung function would be insufficient as a single primary endpoint and should be accompanied by an additional co-primary endpoint, which should either be a symptom-based endpoint or a patient-related endpoint."

Most work to date on constructing stopping boundaries for monitoring multivariate test statistics has concentrated on the bivariate case and we shall focus on that case here. However, with some computational effort, methods can be extended in an obvious way to three or more endpoints. We begin by stating the multivariate analogue of the unified formulation of Chapter 4 that enables us to handle a variety of problems, including the one-sample problem, paired and unpaired two-treatment comparisons, and the two-period crossover design. We suppose the covariance matrix of the response vector is known and, as before in this chapter, a standardized p-variate statistic $\boldsymbol{Z}_k = (Z_{1k}, \ldots, Z_{pk})$ is created at each analysis $k = 1, \ldots, K$. The distribution of $\boldsymbol{Z}_k$ is assumed to be multivariate normal with mean $(\theta_1\sqrt{\mathcal{I}_{1k}}, \ldots, \theta_p\sqrt{\mathcal{I}_{pk}})$ and correlation matrix $\boldsymbol{\Sigma}_k$. The information levels $\mathcal{I}_{jk}$ and the matrix $\boldsymbol{\Sigma}_k$ are taken to be known and we suppose the unknown parameter vector $\boldsymbol{\theta} = (\theta_1, \ldots, \theta_p)$ lies in a p-dimensional parameter space Ω.

Consider a two-decision problem where we must decide whether to accept a new treatment over a standard or control based on trial data. Acceptance might mean proceeding to submit a New Drug Application to a regulatory body. For each variable j, the range of possible values of the parameter θ_j is partitioned into three exhaustive and non-overlapping sets in which, with regard to this variable:

(i) the new treatment is clearly acceptable,

(ii) the new treatment is clearly unacceptable,

(iii) the new treatment does not differ from the standard or control by a clinically significant amount.

This leads to a partition of the p-dimensional parameter space Ω into 3^p regions, which we need to collapse into just two regions corresponding to acceptance and rejection of the new treatment. This mapping will depend on the nature of the variables and on clinical assessment of the trade-offs between them.

In an equivalence trial, the new treatment may be deemed acceptable if $\boldsymbol{\theta}$ is in a region of the form

$$\Omega_A = \{\boldsymbol{\theta} \colon \epsilon_j < \theta_j < \delta_j \text{ for all } j = 1, \ldots, p\},$$

for specified pairs (ϵ_j, δ_j) with $\epsilon_j < \delta_j$, $j = 1, \ldots, p$. If for certain endpoints, it is desired only to show that a test formulation is at least as good as a standard or not worse by more than a certain amount, a "one-sided" criterion (see Section 7.2), then the corresponding constants δ_j may be set at $+\infty$.

For efficacy trials, Jennison & Turnbull (1993c) argue that, after a suitable translation of the parameters θ_j, the set of values of $\boldsymbol{\theta}$ for which the new treatment is acceptable can often be represented in the form

$$\Omega_A = \{\boldsymbol{\theta}: \theta_j > 0 \text{ for all } j = 1, \ldots, p\}. \quad (19.9)$$

Let Ω_R denote the complement of Ω_A in Ω, i.e., $\Omega_R = \Omega \setminus \Omega_A$. Then, if $\boldsymbol{\theta}$ is in Ω_R the new treatment should be rejected.

For a given testing procedure, fixed or sequential, let $\pi(\boldsymbol{\theta})$ denote the probability of deciding in favor of the new treatment for a given value of $\boldsymbol{\theta}$. We regard acceptance of the new treatment when $\boldsymbol{\theta} \in \Omega_R$ as a Type I error and impose the requirement that the size of the test, i.e., the maximum probability of accepting the new treatment for $\boldsymbol{\theta} \in \Omega_R$, should be at most α, that is,

$$\sup_{\boldsymbol{\theta} \in \Omega_R} \{\pi(\boldsymbol{\theta})\} \leq \alpha. \quad (19.10)$$

For an efficacy trial, with Ω_A given by (19.9), Jennison & Turnbull (1993c) suggest a procedure which combines univariate, one-sided group sequential tests defined separately for each variable. Tests may be chosen from the families described in Chapter 5 with the null hypothesis $H_{0,j}$ taken to mean that the new treatment does not improve on the standard treatment for endpoint j. Thus, a decision to accept $H_{0,j}$ indicates that the new treatment cannot be recommended overall, while a decision to reject $H_{0,j}$ shows the effect of the new treatment on endpoint j is acceptable.

Suppose that for each $j = 1, \ldots, p$, the univariate test for endpoint j is defined as in (5.2) with critical values a_{jk} and b_{jk}, $k = 1, \ldots, K$, for statistics Z_{j1} to Z_{jK}. In combining the p tests on different endpoints, the trial is stopped to accept the new treatment if all p univariate tests stop at a particular analysis with rejection of their hypotheses $H_{0,j}$, $j = 1, \ldots, p$, and the trial stops to reject the new treatment if any one univariate test accepts its hypothesis $H_{0,j}$. Thus, the sequential decision rule has the form:

At analysis $k = 1, \ldots, K-1$

- if $\boldsymbol{Z}_k \in \mathcal{A}_k$ stop, accept the new treatment
- if $\boldsymbol{Z}_k \in \mathcal{R}_k$ stop, reject the new treatment
- if $\boldsymbol{Z}_k \in \mathcal{C}_k$ continue to group $k+1$,

at analysis K

- if $\boldsymbol{Z}_K \in \mathcal{A}_K$ stop, accept the new treatment
- if $\boldsymbol{Z}_K \in \mathcal{R}_K$ stop, reject the new treatment,

where the acceptance, rejection and continuation regions at analysis k are

$$\mathcal{A}_k = \{\boldsymbol{Z}_k : Z_{jk} \geq b_{jk} \text{ for all } j = 1, \ldots, p\},$$
$$\mathcal{R}_k = \{\boldsymbol{Z}_k : Z_{jk} \leq a_{jk} \text{ for at least one } j = 1, \ldots, p\}$$

and

$$\mathcal{C}_k = \{\boldsymbol{Z}_k : Z_{jk} > a_{jk} \text{ for all } j = 1, \ldots, p, \text{ and } Z_{jk} < b_{jk} \text{ for at least one } j\}.$$

Since the univariate tests have $a_{jK} = b_{jK}$ for each j, the test is sure to terminate with a decision at analysis K.

In the case $p = 2$, the continuation regions are "L-shaped"; see Jennison & Turnbull (1993c, Fig. 3). It turns out that if each univariate test is defined to achieve Type I error probability α, the constraint (19.10) will be met, whatever the correlation matrix $\boldsymbol{\Sigma}_k$. This is because the supremum in (19.10) occurs when one component of $\boldsymbol{\theta}$, θ_j say, is zero and all the others are infinite (Jennison & Turnbull 1993c, Sec. 4) so the overall conclusion is determined by the outcome of the jth univariate test. The power function $\pi(\boldsymbol{\theta})$ and expected sample size of the overall test do depend on $\boldsymbol{\Sigma}_k$ and multivariate calculations are required to evaluate such properties, even if endpoints are uncorrelated. Computation of $\pi(\boldsymbol{\theta})$ is needed to establish the group size that will satisfy a power requirement of the form

$$\pi(\boldsymbol{\theta}) \geq 1-\beta \quad \text{whenever } \theta_j \geq \theta_j^* \text{ for all } j=1,\ldots,p$$

for specified values θ_j^*. These calculations involve a multivariate analogue of the recursion formulae we shall describe in Chapter 31. Further details are given in the appendix of Jennison & Turnbull (1993c). Hu et al. (2020) propose designs with curved regions rather than L-shaped regions and present methods for calculating their operating characteristics.

Cook & Farewell (1994) present a similar procedure for a bivariate outcome with one efficacy and one toxicity response. Their procedure is asymmetric in that a two-sided, rather than one-sided, group sequential test is proposed for monitoring the efficacy response. The Type I error probability is controlled at $\theta_1=\theta_2=0$ and bivariate calculations involving the correlation matrices $\boldsymbol{\Sigma}_k$ are needed to construct the testing boundaries. An error spending approach is taken in designing the boundaries, with the error spent at each stage apportioned between the two endpoints. When correlations are unknown, an adaptive procedure is proposed whereby the correlation matrix is estimated from the data currently available at each analysis. Results of a small simulation study reported by Cook & Farewell (1994) with $K=2$ analyses and $p=2$ uncorrelated endpoints suggest the adaptive procedure gives close to nominal error rates.

Cook (1994) also considers the case of a bivariate outcome and defines a procedure similar to that of Cook & Farewell (1994), except that both univariate tests are two-sided. The resulting bivariate group sequential test can be inverted to obtain repeated confidence *regions* for (θ_1, θ_2) analogous to the repeated confidence intervals described in Chapter 10 for a one-dimensional parameter. Cook (1994) goes on to describe how a group sequential equivalence test for a bivariate response can be derived from the repeated confidence regions, just as "derived tests" for a univariate response were created from RCIs in Section 10.3. It should be noted that repeated confidence regions generated by bivariate group sequential tests of the form considered in this section lead to rectangular repeated confidence regions. Elliptical repeated confidence regions can be obtained by inverting the group sequential Hotelling's $\mathcal{T}^2$ test of Section 19.3. However, for the reasons expressed at the beginning of this section concerning disadvantages of the use of global statistics, rectangular regions are to be preferred; for more discussion, see Cook (1994, p. 190).

Van Eekelen, de Hoop & van der Tweel (2016) stress the importance of defining objectives carefully when two co-primary endpoints are monitored in a group sequential design, noting that stopping the trial as soon as there is a significant result for one endpoint will decrease the probability of obtaining a positive outcome for the other endpoint. Schüler, Kieser & Rauch (2017) and Asakura, Evans & Hamasaki (2020) propose methods for defining futility boundaries in trials with co-primary endpoints. Hamasaki, Evans & Asakura (2018) review the design and analysis of trials with co-primary endpoints, covering both fixed sample and group sequential methods.

19.7 Bibliography and Notes

All the methods in this chapter are designed to control the overall family-wise Type I error rate. Cook & Farewell (1996) have argued that there are situations when this criterion may not be relevant and tests on the different endpoints should be designed with reference to marginal rather than family-wise Type I error rates; for example, if the marginal test results have separate implications on treatment evaluation and prescription. Cook (1996) describes a "coupled error spending" function approach, defining procedures that control either the family-wise or marginal Type I error rate. Sankoh, D'Agostino & Huque (2003) discuss how decision-making can also differ according to the purpose of a clinical trial, including the case where drug effectiveness is claimed if the *majority* of the endpoints are significant at the nominal level α. Quan, Luo & Capizzi (2005) consider studies in which there are both multiple endpoints and multiple doses of the new treatment, and they apply some of the multiple testing procedures that we shall present in the next chapter.

We have concentrated here on continuous outcomes that can be considered to be multivariate normally distributed, at least approximately. As in the univariate case, the ideas are readily adapted to other types of endpoint. Cook (1994) and Cook & Farewell (1994) have considered application of their procedures to bivariate failure time responses. Williams (1996) discusses several different methods for constructing repeated joint confidence regions for monitoring two hazard ratios with bivariate survival data. Group sequential designs for multinomial outcomes are considered by Cook (1994) and by Conaway & Petroni (1995). Bayesian monitoring procedures for multinomial data have been proposed by Thall, Simon & Estey (1995). Etzioni & Pepe (1994) also propose a Bayesian monitoring scheme for multinomial data, combined with frequentist inference upon termination using ideas similar to those described in Chapter 9. Kavelaars, Mulder & Kaptein (2020) describe a Bayesian model for the analysis of correlated binary outcomes based on the multivariate Bernoulli distribution and apply this model, with decision criterion reflecting the relative importance of the outcomes, in fixed sample size, group sequential and adaptive designs.

In this section we have concentrated on multiple outcomes, all of which can be used to influence early stopping. Another situation occurs when a univariate group sequential procedure is used to monitor a single response variable with parameter θ_1 and test statistics $\{Z_{1k}\}$ but, upon termination, it is desired to make inferences about the parameter θ_2 associated with a secondary response variable for which the test statistics $\{Z_{2k}\}$ are correlated with $\{Z_{1k}\}$. We discussed the estimation of such a parameter in Section 9.3.4 and we shall present methods for testing a secondary endpoint after a group sequential test for the primary endpoint in Section 20.10. A confidence interval for θ_2 will depend on the joint distribution of $\{(Z_{1k}, Z_{2k});\, k = 1, \ldots, K\}$ and the boundary used to monitor Z_{1k}. Whitehead (1986b) has proposed a method for constructing such a confidence interval, and the same computational methods referred to in Section 19.6 allow exact numerical calculation of the coverage probability of a confidence interval for θ_2 under specific values of θ_1; see the comments in Jennison & Turnbull (1993c, p. 750).

Pocock (1997) gives a comprehensive account of some of the statistical and practical issues in handling multiple outcomes, with examples from several clinical trials. Multivariate responses can also arise in industrial acceptance sampling (see Jackson & Bradley, 1961; Baillie, 1987a,b; and Hamilton & Lesperance, 1991), and the methods of this chapter can be applied there also.

Chapter 20

Multiple Hypotheses: Primary and Secondary Endpoints

20.1 Introduction

Suppose there are p outcome variables or "endpoints" of interest in a clinical trial. In this chapter, we discuss how to make inferences about such endpoints in a group sequential setting where data accumulate on all the endpoints. This situation is very common and is the subject of the European Medicines Agency's "Guideline on Multiplicity Issues in Clinical Trials" (CHMP, 2016) and the FDA (2022) guidance "Multiple Endpoints in Clinical Trials Guidance for Industry".

Often the stopping rule for a trial is based on monitoring one "primary" endpoint but it is still desired to make inferences about one or more additional endpoints. This is unlike the situation of the previous chapter where the stopping decision could depend on multiple endpoints. In a Phase III clinical trial, demonstration of a significant benefit on the primary endpoint is needed to support a regulatory claim for the treatment. However, if the treatment has also been shown to provide benefit for secondary outcomes, this information can be included in the label and may lead to increased adoption in clinical practice. In an oncology trial where overall survival (OS), also referred to as all-cause mortality, is the primary outcome, there may also be interest in progression-free survival (PFS), defined as the time from randomization until a certain increase in tumor size or death from any cause. In some oncology trials, such as for adjuvant colon cancer treatment, this order may be reversed. Because of low mortality, a clinical trial of sufficient power may require a prohibitively large sample size or length of follow-up. In such a case, a surrogate endpoint such as PFS could become the primary response variable on which a registration claim is based; see the FDA (2018) guidance "Clinical Trial Endpoints for the Approval of Cancer Drugs and Biologics Guidance for Industry". If the data also showed a significant improvement in OS it would be very worthwhile to the sponsor to be able to make this claim. This should not be done as a *post hoc* observation, instead OS should be listed in the protocol as a secondary endpoint. A tertiary endpoint in either case might be tumor response or a quality of life measure.

In Section 9.3.4, we discussed point estimation of the treatment effect on a secondary endpoint when a trial has stopped based on the primary endpoint. This chapter will address significance testing for a secondary endpoint. Before we treat this topic for group sequential procedures, we introduce methods for testing multiple hypotheses in the case of a single-stage or fixed sample size trial.

20.2 The Family-wise Error Rate (FWER)

Suppose we wish to test a "family" of null hypotheses, $H_1, \ldots, H_p$, where each hypothesis may or may not be rejected. We shall consider the specific case where the jth hypothesis concerns a treatment effect parameter θ_j and it is desired to test the null

DOI: 10.1201/9781584888482-20

hypothesis H_j: $\theta_j \leq 0$ against the one-sided alternative $\theta_j > 0$. In a fixed sample trial, we suppose each hypothesis H_j, $j = 1, \ldots, p$, has an associated P-value, P_j, which we term the "raw" or "unadjusted" P-value.

Strong control of the Family-wise error rate

A multiple comparisons procedure (MCP) is a rule that uses the observed data to decide which of the hypotheses $H_1, \ldots, H_p$ are rejected and which are not. We say such a procedure strongly controls the family-wise error rate at level α if the probability of rejecting at least one true hypothesis is no more than α whatever the values of $\theta_1, \ldots, \theta_p$, that is

$$Pr_{\boldsymbol{\theta}}\{\text{Reject } H_j \text{ for some } j \text{ with } \theta_j \leq 0\} \leq \alpha \quad \text{for all } \boldsymbol{\theta} = (\theta_1, \ldots, \theta_p).$$

The term "weak control" is used to indicate that an MCP controls the FWER under the global null configuration $\theta_1 = \ldots = \theta_p = 0$. For a good discussion of weak versus strong control, see Proschan & Brittain (2020).

20.3 Closed Testing Procedures (CTPs)

Marcus, Peritz & Gabriel (1976) introduced a very effective way of creating a test that strongly controls the FWER. The resulting method is called a "closed testing procedure" and it is constructed as follows.

For each subset J of $\{1, \ldots, p\}$, we define the intersection hypothesis

$$H_J = \cap_{j \in J} H_j.$$

Note that the set of all such hypotheses, $\mathcal{H} = \{H_J \colon J \subseteq \{1, \ldots, p\}\}$ is a closed set since the intersection of any two members of $\mathcal{H}$ is also a member of $\mathcal{H}$. Next, we construct a level α test of each intersection hypothesis H_J, i.e., a test which rejects H_J with probability at most α whenever all the hypotheses specified in H_J are true. In the overall testing procedure, we reject H_j if and only if the intersection hypothesis H_J is rejected at level α for every set J that contains the index j.

Theorem 20.1. Suppose hypotheses $H_1, \ldots, H_p$ are to be tested, level α tests are defined for each intersection hypothesis

$$H_J = \cap_{j \in J} H_j,$$

where $J \subseteq \{1, \ldots, p\}$, and an overall testing procedure rejects H_j if and only if the intersection hypothesis H_J is rejected for every set J that contains the index j. Then this procedure provides strong control of the family-wise error rate at level α.

Proof. The proof is elegant and short. For a given combination of true parameter values, denote by $\tilde{J}$ be the set of indices j of all true hypotheses H_j. For a family-wise error to be committed, the overall testing procedure must reject H_j for some $j \in \tilde{J}$ and this requires the intersection hypothesis $H_{\tilde{J}}$ to be rejected. Since $H_{\tilde{J}}$ is true,

$$Pr\{\text{Reject } H_{\tilde{J}}\} \leq \alpha$$

and, hence, the probability of a family-wise error is no greater than α. □

We have intentionally presented this theorem in a general form without specific reference to hypotheses concerning treatment effects on different endpoints. We shall use the result again in later chapters where the multiple comparisons concern multiple treatments or patient subgroups as well as multiple endpoints.

20.4 Testing Intersection Hypotheses: Bonferroni and Simes Tests

In order to apply Theorem 20.1, we need to know how to construct an α level test of any given intersection hypothesis H_J. Two such tests, the Bonferroni test and Simes' test, can be defined in terms of P-values for the individual hypotheses H_j. Suppose H_J is the intersection of m hypotheses, where $1 \leq m \leq p$, and the ordered, unadjusted P-values for these hypotheses are $P_{(1)} \leq \ldots \leq P_{(m)}$.

The Bonferroni test

The Bonferroni (1936) test at level α rejects H_J if $mP_{(1)} \leq \alpha$.

The Simes test

The Simes (1986) test at level α rejects H_J if

$$\min_{j=1,\ldots,m} \{mP_{(j)}/j\} \leq \alpha. \tag{20.1}$$

If the Bonferroni condition is met, the first term in the Simes condition is less than or equal to α, so the Simes test rejects H_J whenever the Bonferroni test does. However, there are cases where the Simes test rejects H_J but the Bonferroni test does not, so the Simes test is more powerful. On the other hand, the Bonferroni test has error rate no greater than α whatever the joint distribution of $P_1, \ldots, P_p$, whereas the Simes test is only applicable under certain conditions. The error rate of the Simes test is at most α if the tests of $H_1, \ldots, H_p$ are independent or if the test statistics are multivariate normal with positive correlations. More generally a type of positive dependence in the test statistics known as total positivity of order 2 ensures the Simes test is valid; see Sarkar & Chang (1997), Sarkar (1998) and Huque (2016). This latter condition is likely to be met if different endpoints measure overall improvement such as PFS and tumor response in an oncology trial. Alternatively, different endpoints may have overlapping definitions: for example, in a heart disease study, two endpoints might be occurrence of a Core MACE event (CV-related death, nonfatal stroke or nonfatal myocardial infarction) and occurrence of an extended MACE event, which also includes hospitalization for unstable angina or coronary revascularization.

In Section 21.4 we shall define another important intersection test proposed by Dunnett (1955). The Dunnett test is designed for comparing several treatments with a common control and so is not particularly relevant to the topic of testing for a treatment effect on multiple endpoints. However, we shall see it can be a very useful option in other multiple testing problems.

Burnett & Jennison (2021) showed that every multiple testing procedure that controls FWER can be expressed as a closed testing procedure with a particular set of tests for all possible intersection hypotheses. In the following sections, we shall show how certain procedures can be described as CTPs and this is often a useful way to demonstrate their control of the FWER.

20.5 Single-stage Testing Procedures

20.5.1 Structures for Hypotheses: Multiple Primary Hypotheses

Different hypotheses may be treated quite differently in a multiple testing procedure. We shall illustrate the principal structures for testing multiple hypotheses in the case of just three hypotheses. Although these examples are quite simple, our treatment of them will capture the essential ideas and the same structures can be used as building blocks to construct more complex procedures.

The first case we examine is that of several primary endpoints of equal importance. As an example, the Long-Term Odanacatib Fracture Trial (LOFT) reported by Bone et al.(2015)

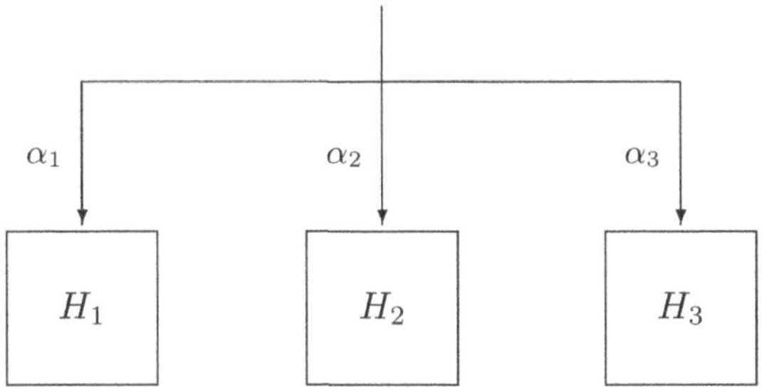

Figure 20.1 *Graphical depiction of a simple Bonferroni procedure with $p = 3$ hypotheses*

investigated the drug Odanacatib in the treatment of osteoporosis in postmenopausal women. There were three primary outcomes: radiologically determined vertebral fractures, clinical hip fractures, and clinical non-vertebral fractures. Note that in this example, the new treatment is to be regarded as beneficial if it leads to a decrease in *any* of the three types of fracture. Thus, it is appropriate to test a null hypothesis for each fracture type but with family-wise error control. This is qualitatively different from the case of "co-primary" endpoints discussed in Section 19.6 where a positive outcome is required on *all* endpoints in order to show that the new treatment is effective.

Suppose we wish to test p null hypotheses with the FWER less than or equal to a pre-specified value α. The Bonferroni test is a simple but useful method in which we test each hypothesis at significance level α/p. Then if all p null hypotheses are true,

$$\begin{aligned} Pr\{\text{Reject at least one of } H_1, \ldots, H_p\} &\leq Pr\{\text{Reject } H_1\} + \ldots + Pr\{\text{Reject } H_p\} \\ &= p(\alpha/p) = \alpha. \end{aligned}$$

If only some of the p null hypotheses are true then $Pr\{\text{Reject at least one true } H_j\} < \alpha$, so we see that the Bonferroni method provides strong control of the FWER. If the unadjusted P value associated with H_j is P_j, we can say the *adjusted* P-value is pP_j since H_j is rejected in a procedure with FWER equal to α if $pP_j \leq \alpha$.

If some hypotheses are of greater importance than others, one may prefer to use the *weighted* Bonferroni procedure. In this generalization of the Bonferroni test, significance levels α_j are associated with hypotheses H_j, $j = 1, \ldots, p$, where $\alpha_1 + \ldots + \alpha_p = \alpha$, and H_j is rejected if $P_j \leq \alpha_j$. The proof that this procedure protects the FWER is straightforward. Figure 20.1 provides a graphical representation of this Bonferroni procedure.

The Holm procedure

Since the Bonferroni procedure is known to be conservative, Holm (1979) suggested a Bonferroni procedure "with recycling". If equal importance is attached to the three hypotheses in our example, the Holm procedure is implemented as follows:

If $\min\{P_1, P_2, P_3\} > \alpha/3$, the procedure stops and no hypothesis is rejected.

If $\min\{P_1, P_2, P_3\} \leq \alpha/3$, the hypothesis with the smallest P-value is rejected and consideration moves on to the two remaining hypotheses, H_{j_1} and H_{j_2}, say.

If $\min\{P_{j_1}, P_{j_2}\} > \alpha/2$, the procedure stops and neither H_{j_1} nor H_{j_2} is rejected.

If $\min\{P_{j_1}, P_{j_2}\} \leq \alpha/2$, the hypothesis with the smaller P-value is rejected and the procedure continues to examine the remaining hypothesis, H_k, say.

The hypothesis H_k is rejected if and only if $P_k \leq \alpha$.

It can be shown directly that this procedure protects the FWER of α by considering the $2^3 = 8$ cases in which each hypothesis is true or false. If H_1, H_2 and H_3 are all true, a family-wise error occurs if any one hypothesis is rejected and this happens with probability at most $3 \times (\alpha/3) = \alpha$. If H_3 is false but H_1 and H_2 are true, then an error only occurs if H_1 or H_2 is rejected, which happens with probability at most $2 \times (\alpha/2) = \alpha$, and a similar argument applies if only H_2 is false or only H_3 is false. If H_1 and H_2 are both false but H_3 is true, then an error occurs if H_3 is rejected, which happens with probability at most α, and a similar argument applies if only H_1 is true or only H_2 is true. Finally, if H_1, H_2 and H_3 are all false, a family-wise error cannot occur and the FWER is zero.

An alternative way to show that the Holm procedure protects the FWER of α is to represent it as a closed testing procedure (CTP) as defined in Section 20.3. In this representation, the tests for each intersection hypothesis are:

Hypothesis	*Reject if*
H_1	$P_1 \leq \alpha$,
H_2	$P_2 \leq \alpha$,
H_3	$P_3 \leq \alpha$,
H_{12}	$P_1 \leq \alpha/2$ or $P_2 \leq \alpha/2$,
H_{13}	$P_1 \leq \alpha/2$ or $P_3 \leq \alpha/2$,
H_{23}	$P_2 \leq \alpha/2$ or $P_3 \leq \alpha/2$,
H_{123}	$P_1 \leq \alpha/3$ or $P_2 \leq \alpha/3$ or $P_3 \leq \alpha/3$.

Note that for simplicity we write H_{12} rather than $H_{\{1,2\}}$, etc., when there is no danger of any ambiguity. As an example, applying the closure principle with the above tests of intersection hypotheses, we reject H_1 overall if H_1, H_{12}, H_{13} and H_{123} are rejected in their level α tests and this happens in any of the following five situations:

1) $P_1 \leq \alpha/3$,
2) $P_1 \leq \alpha/2$ and $P_2 \leq \alpha/3$,
3) $P_1 \leq \alpha/2$ and $P_3 \leq \alpha/3$,
4) $P_1 \leq \alpha$ and $P_2 \leq \alpha/3$ and $P_3 \leq \alpha/2$,
5) $P_1 \leq \alpha$ and $P_2 \leq \alpha/2$ and $P_3 \leq \alpha/3$.

One can check that these conditions for rejecting H_1 correspond exactly to the Holm procedure. Similar conditions are required for the rejection of H_2 or H_3 and, again, these correspond to the Holm procedure.

20.5.2 The Holm Test as a Graphical Procedure

Figure 20.2(a) provides a visualization of the Holm procedure. In this graph, each box or "node" corresponds to a null hypothesis. The arrows between nodes represent the fact that when a hypothesis is rejected, that node is removed from the graph and the significance level currently associated with that node is passed to adjacent nodes that are still present in the graph. When there is more than one adjacent node, the label on the arrow indicates the fraction of the significance level passed in that direction.

In implementing the Holm procedure, hypotheses are rejected and fractions of α are recycled until no more hypotheses can be rejected. One can see that using this graph corresponds to the Holm procedure. As an example, if $P_3 \leq \alpha/3$, then H_3 is rejected, the node for H_3 is removed and the associated significance level $\alpha/3$ is divided into two

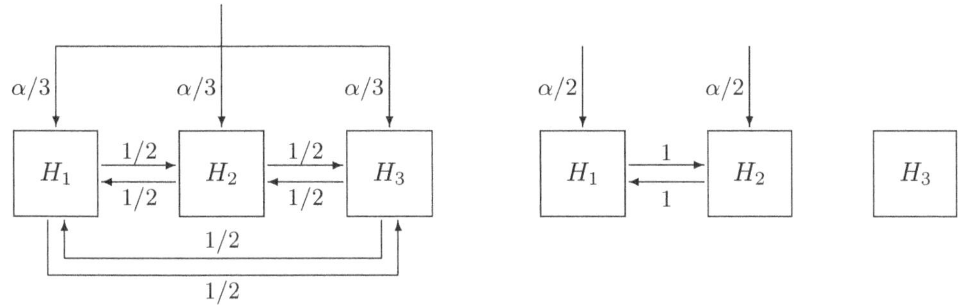

(a) Initial graph (b) Revised graph after H_3 is rejected

Figure 20.2 *Visualization of Holm procedure (Bonferroni with recycling) for* $p = 3$ *hypotheses*

halves which are passed on to the nodes for H_1 and H_2 so these hypotheses are now tested at level $\alpha/2$. The revised figure after H_3 has been rejected is displayed in Figure 20.2 (b). Now, if $P_1 \leq \alpha/2$, H_1 is rejected and its $\alpha/2$ is passed in full to the only remaining node, H_2, so in the final step H_2 is tested at level α. Similarly, if $P_2 \leq \alpha/2$, H_2 is rejected and its $\alpha/2$ is passed to H_1. Since the graphical procedure is equivalent to the Holm procedure, FWER is protected at level α.

Figure 20.2 can be modified in an obvious fashion to represent a weighted Holm procedure in which the error probability α is initially divided unequally among the hypotheses: in panel (a), one replaces the three entries $\alpha/3$ by α_1, α_2 and α_3 and the rest of the figure remains the same.

It is also a straightforward task to generalize the Holm procedure to test $p = 2$ or $p \geq 4$ hypotheses. For general p with hypotheses ordered so that $P_1 \leq \ldots \leq P_p$, H_1 is rejected if $P_1 \leq \alpha/p$. Then, for $j = 2, \ldots, p$, hypothesis H_j is rejected if $H_1, \ldots, H_{j-1}$ have all been rejected and $P_j \leq \alpha/(p + 1 - j)$.

The graphical displays of MCPs in this and subsequent sections are special cases of the very general methods described by Bretz, Maurer, Brannath & Posch (2009) and Burman, Sonesson & Guilbaud (2009), who prove that these methods protect the specified FWER.

20.5.3 Structures for Hypotheses: Hierarchical Endpoints

Suppose now that outcome variables can be arranged in order of importance. As an example, in a long-term trial for the treatment of chronic obstructive pulmonary disease (COPD) the primary efficacy outcome of interest may be "Time to death" (all-cause mortality) with "Time to first COPD exacerbation" of secondary importance, and a third endpoint "Number of COPD exacerbations". Dmitrienko & Tamhane (2007) review such hierarchical testing problems in clinical trials.

In general, we have p endpoints ordered by importance. Thus, endpoint 1 is primary, endpoint 2 is secondary, endpoint 3 is tertiary, and so on. We wish to test the p one-sided hypotheses

$$H_j\colon \theta_j \leq 0 \quad \text{vs} \quad \theta_j > 0, \qquad j = 1, \ldots, p.$$

A gatekeeping procedure

Let P_j be the unadjusted P-value associated with H_j, $1 \leq j \leq p$. In a step-down testing procedure we first consider H_1 and reject this hypothesis if $P_1 \leq \alpha$. If H_1 is rejected, we then test H_2, rejecting this hypothesis if $P_2 \leq \alpha$. Continuing this process, for general j we reject H_j if $P_1 \leq \alpha, \ldots, P_{j-1} \leq \alpha$, and then $P_j \leq \alpha$. Thus H_j is rejected if it and all

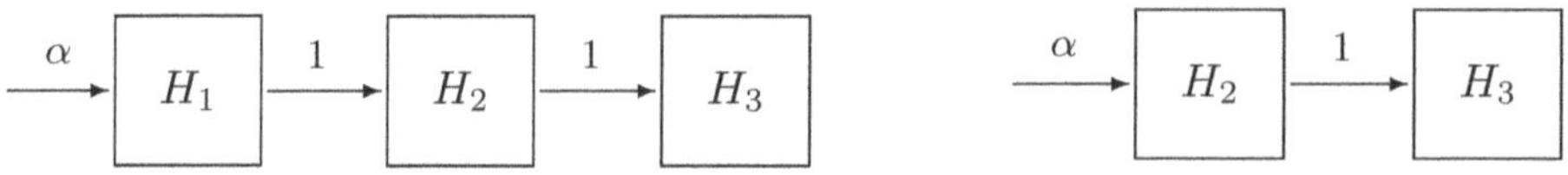

(a) Initial graph (b) Revised graph after H_1 is rejected

Figure 20.3 *Visualization of a gatekeeping procedure with* $p = 3$ *hypotheses*

preceding hypotheses in the hierarchy are rejected in their α level tests. This is termed a gatekeeping procedure.

It is a simple task to show the gatekeeping procedure protects the FWER. Let j^* be the smallest value of j for which H_j is true. Then a family-wise error occurs if and only if H_{j^*} is rejected in the overall procedure, so the probability of a family-wise error is

$$Pr\{H_{j^*} \text{ is rejected}\} \leq Pr\{P_{j^*} \leq \alpha\} \leq \alpha.$$

We can show this gatekeeping procedure is a CTP by defining appropriate tests of each intersection hypothesis $H_J = \cap_{j \in J} H_j$, where $J \subseteq \{1, \ldots, p\}$. To do this, let H_J be rejected if $P_{j^*} \leq \alpha$ where $j^* = \min\{j \in J\}$. This is a level α test since if H_J is true, then H_{j^*} is true and $Pr\{P_{j^*} \leq \alpha\} \leq \alpha$. In the CTP with these intersection tests, H_j is rejected overall if all hypotheses H_J with $j \in J$ are rejected and this occurs if $H_{1,j}, \ldots, H_{j-1,j}$ and H_j are rejected in their level α tests. This requires $P_i \leq \alpha$ for all $i = 1, \ldots, j$, which is the same as the condition for the gatekeeping procedure to reject H_j. Since we have shown the gatekeeping procedure is equivalent to a CTP with FWER at most α we have a second proof that the gatekeeping procedure controls the FWER at level α.

Figure 20.3 (a) gives a graphical representation of the gatekeeping procedure with $p = 3$ hypotheses. If H_1 is rejected, the node for H_1 is removed, the significance level α is passed to H_2 and the procedure continues with the revised graph shown in Figure 20.3 (b).

A fallback procedure

A modification of the simple gatekeeping procedure is to hold back a small fraction of the family-wise error probability α to permit the secondary hypothesis H_2 to be tested even if the primary hypothesis is not rejected. This method can be generalized to tertiary or lower level hypotheses. The resulting method is called a *fallback* procedure. The structure for the case of $p = 3$ hypotheses is displayed graphically in Figure 20.4. It is a fairly simple exercise to check that FWER is protected by considering the three cases: H_1 is true; H_1 is false and H_2 is true; H_1 and H_2 are false and H_3 is true.

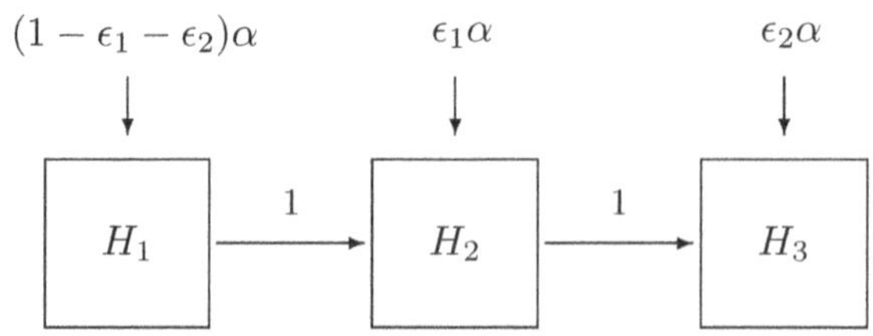

Figure 20.4 *Visualization of a fallback procedure with* $p = 3$ *hypotheses*

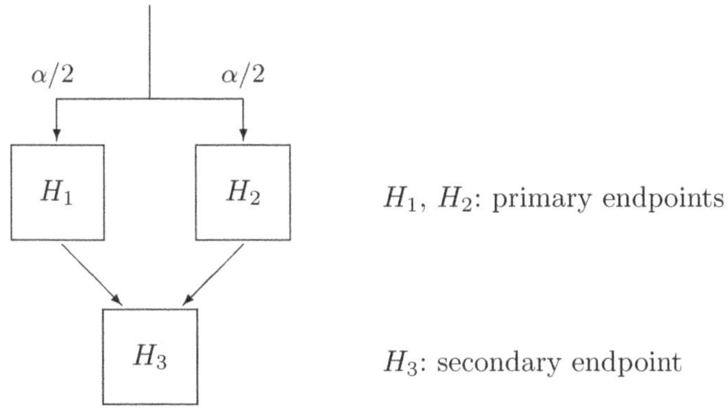

Figure 20.5 *Visualization of a procedure with two primary endpoints and a secondary endpoint*

20.5.4 Multiple Primary Endpoints and a Secondary Endpoint

The procedures we have seen for multiple primary endpoints and hierarchical testing can be combined to create more complex structures.

Suppose, for example, we have two primary endpoints and a secondary endpoint which we test if a positive result is obtained on either of the primary endpoints. Here we can apply a Bonferroni test to the primary endpoints and recycle family-wise error probability to the secondary endpoint when the test for a primary endpoint is positive. The depiction of this testing procedure in Figure 20.5 shows that the secondary endpoint is tested at significance level $\alpha/2$ if just one of H_1 and H_2 is rejected, and at level α if both H_1 and H_2 are rejected.

One can check that the FWER of this procedure is at most α by considering each of the eight possible combinations of true and false values for H_1, H_2 and H_3, as we did for the Holm procedure in Section 20.5.1. Alternatively, we can express the procedure as a CTP with suitably defined tests of H_1, H_2 and H_3, and their intersection hypotheses. These tests are:

Hypothesis	*Reject if*
H_1	$P_1 \leq \alpha/2$,
H_2	$P_2 \leq \alpha/2$,
H_3	$P_3 \leq \alpha$,
H_{12}	$\min(P_1, P_2) \leq \alpha/2$,
H_{13}	$\min(P_1, P_3) \leq \alpha/2$,
H_{23}	$\min(P_2, P_3) \leq \alpha/2$,
H_{123}	$\min(P_1, P_2) \leq \alpha/2$.

As an example, rejection of H_3 overall requires rejection of H_3, H_{13}, H_{23} and H_{123}, so we need

$$P_3 \leq \alpha, \quad \min(P_1, P_3) \leq \alpha/2, \quad \min(P_2, P_3) \leq \alpha/2 \quad \text{and} \quad \min(P_1, P_2) \leq \alpha/2$$

and this can be seen to agree with the procedure just described.

It is noticeable that in the graphical depiction of this procedure, the family-wise error is not always recycled. So, it is reasonable to ask whether it is possible to add more re-cycling,

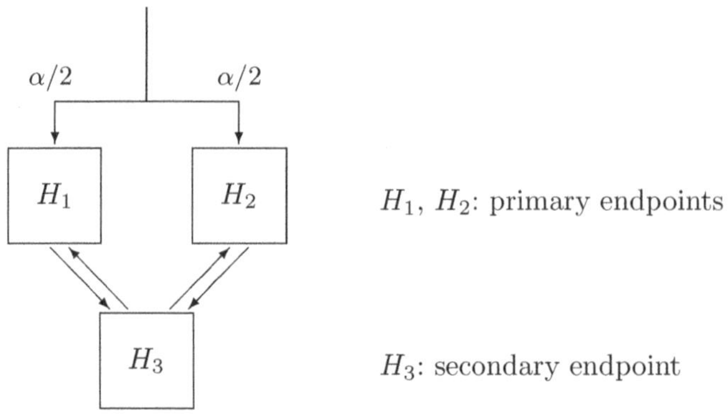

Figure 20.6 *Graph for a procedure with two primary endpoints, one secondary endpoint and additional recycling*

reduce conservatism, and increase power. In the above list of tests of intersection hypotheses we have

Hypothesis	*Reject if*
H_1	$P_1 \leq \alpha/2$,
H_2	$P_2 \leq \alpha/2$.

These are clearly conservative tests and if we replace them by

Hypothesis	*Reject if*
H_1	$P_1 \leq \alpha$,
H_2	$P_2 \leq \alpha$,

the closed testing procedure will still have FWER at most α. This modification corresponds to recycling error probability from the test of H_3 back to whichever of H_1 and H_2 has not been rejected at level $\alpha/2$. Figure 20.6. depicts the testing procedure with this extra recycling. The additional lines in the graph indicate that

If $P_1 \leq \alpha/2$, and $P_3 \leq \alpha/2$, then H_2 is tested at level α (rather than $\alpha/2$),

If $P_2 \leq \alpha/2$ and $P_3 \leq \alpha/2$, then H_1 is tested at level α (rather than $\alpha/2$).

The extra opportunities to reject H_1 and H_2 reduce conservatism and give greater power.

Another option is to increase power for tests of the primary endpoints before testing the secondary endpoint at all. This is achieved by recycling error probability from H_1 to H_2, and vice versa, before allocating any error probability to a test of H_3. Figure 20.7 presents a graphical representation of this testing procedure. Here, the dashed lines show one half of the family-wise error α being cycled through H_1, H_2 and on to H_3 while the solid lines show that the other half of α is cycled through H_2, H_1 and on to H_3.

20.5.5 General Networks of Hypotheses

More elaborate networks of hypotheses are possible. In the example in Figure 20.8, secondary endpoints H_4 and H_5 and the tertiary endpoint H_7 are of interest if either H_1 or H_2 is rejected, whereas secondary endpoint H_6 and tertiary endpoint H_8 are of interest if H_3

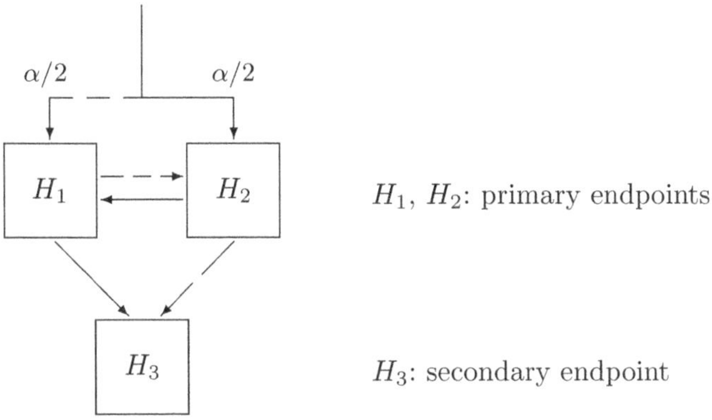

Figure 20.7 *Graph for a procedure that recycles first between primary endpoints*

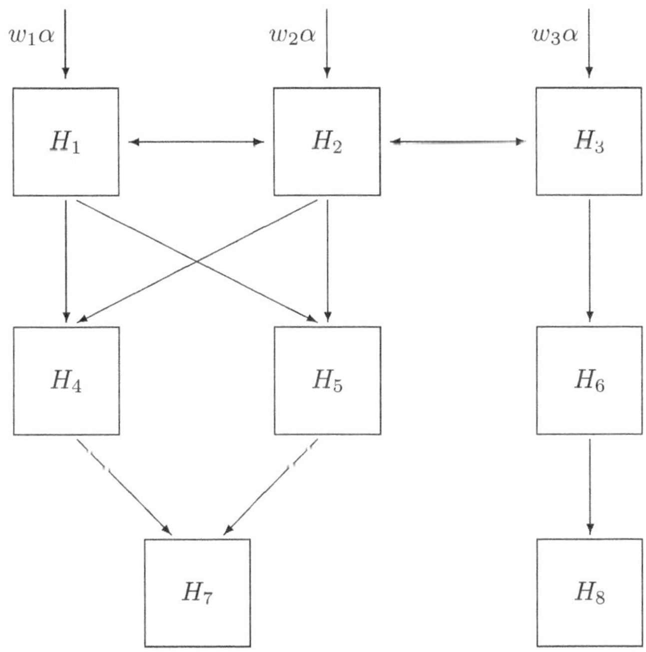

Figure 20.8 *An example of a general network of hypotheses: weights w_1, w_2 and w_3 sum to 1*

is rejected. As hypotheses are rejected, their nodes are removed and the significance level assigned to that node is propagated to add to the significance levels of neighboring nodes. The implementation of graphically defined procedures becomes more complicated as the number of nodes increases and we refer the reader to the paper by Bretz, Maurer, Brannath & Posch (2009) for the full details.

20.6 Using Simes' Test in a Closed Testing Procedure

We return to the situation of Section 20.5.1 with multiple primary endpoints and hypotheses H_1 to H_p to be tested. There, the Holm test was constructed using a closed testing procedure

Table 20.1 *Worksheet for numerical example of Simes' test in a Closed Testing Procedure with $p = 4$ and $\alpha = 0.05$*

Hypothesis	mP_j/j, $j = 1, \ldots, m$	*Simes P-value*
H_1	0.012	0.012
H_2	0.020	0.020
H_3	0.030	0.030
H_4	0.080	0.080
H_{14}	0.024, 0.080	0.024
H_{24}	0.040, 0.080	0.040
H_{34}	0.060, 0.080	0.060
H_{134}	0.036, 0.045, 0.080	0.036
H_{234}	0.060, 0.045, 0.080	0.045
H_{1234}	0.048, 0.040, 0.040, 0.080	0.040

The table contains results for the subset of intersection hypotheses needed to ascertain which hypotheses can be rejected overall in the CTP.

with a Bonferroni test for intersection hypotheses. Recall from Section 20.4 that the Simes test, which can be applied under conditions of positive dependence among the P-values, is more powerful than the Bonferroni test. It is straightforward to define a CTP using the Simes test for each intersection hypotheses. In order to apply this method, it would appear that all $2^p - 1$ intersection hypotheses have to be checked. However, Wright (1992) points out that this calculation can be simplified. As an example, if $p = 4$ and we wish to determine whether H_2 can be rejected in a test with FWER α, we need to check that hypotheses H_2, H_{12}, H_{23}, H_{24}, H_{123}, H_{124}, H_{234} and H_{1234} are significant at level α using the Simes criterion in each case. Without loss of generality, suppose $P_1 < P_3 < P_4$. Then the most challenging intersection of two hypotheses is H_{24} and the most challenging intersection of three hypotheses is H_{234}, so it suffices to check whether each of H_2, H_{24}, H_{234} and H_{1234} is rejected by the Simes test. We illustrate this method in a numerical example.

Example of applying Simes' test in a Closed Testing Procedure

Suppose we have $p = 4$ hypotheses and it happens that the P-values P_j increase with j, taking values $P_1 = 0.012$, $P_2 = 0.02$, $P_3 = 0.03$, $P_4 = 0.08$. We wish to find which hypotheses can be rejected in a test with FWER $\alpha = 0.05$. Since $P_1 < P_2 < P_3 < P_4$, we can ascertain whether each of H_1 to H_4 is rejected in the overall procedure by carrying out a Simes test on H_1, H_2, H_3, H_4, H_{14}, H_{24}, H_{34}, H_{134}, H_{234} and H_{1234}. The results of these tests are shown in Table 20.1. The entries mP_j/j in the middle section of the table are the terms appearing in the formula (20.1) for a Simes test with m hypotheses since $P_{(j)} = P_j$ for $j = 1, \ldots, 4$ in this example. The Simes P-value for each intersection hypothesis is the minimum of the terms mP_j/j. Since all intersection hypotheses involving H_1 and H_2 are rejected at level $\alpha = 0.05$, we reject H_1 and H_2 in the overall procedure. However, since H_{34} is not rejected at level $\alpha = 0.05$ neither H_3 nor H_4 can be rejected overall.

Note that in this example the Bonferroni procedure with $\alpha = 0.05$ would only reject H_1 as P_1 is the only P-value less than $\alpha/4 = 0.0125$. The Holm procedure also only rejects H_1 here because $P_1 < \alpha/4$ but P_2, P_3 and P_4 are all greater than $\alpha/3$.

In the computational approach we have described, the number of intersection hypotheses considered is of order p^2 as the number of hypotheses p increases. Hommel (1988) proposed a more streamlined computational method which is also of order p^2 and yields the same conclusions as the CTP with Simes tests for intersection hypotheses. Hochberg (1988) proposed a multiple testing procedure in which the computational requirement is only of order p. In Hochberg's method, we define

$$j' = \max_{1 \le j \le p} \left\{ j : P_{(j)} \le \frac{\alpha}{p - j + 1} \right\}. \tag{20.2}$$

If no j satisfies the inequality, $j' = 0$ and no hypotheses are declared significant. Otherwise, all the hypotheses $H_{(1)}, \ldots, H_{(j')}$ are rejected. However, Hommel (1989) showed that the Hochberg procedure is more conservative in that it rejects no more hypotheses and may reject fewer hypotheses than the CTP with Simes' tests for intersection hypotheses.

20.7 Adjusted P-values

So far, we have discussed the "raw" or "unadjusted" P-value P_j for hypothesis H_j, which is the P-value for that hypothesis test when there are no other hypotheses under consideration. The *adjusted* P-value (Wright, 1992) for hypothesis H_j, denoted $\tilde{P}_j$, takes into account the multiplicity of hypotheses and the particular MCP being applied. It is defined as the smallest value of the FWER at which that particular hypothesis would be rejected. As such, $\tilde{P}_j$ can be compared directly with any given significance level α in the knowledge that hypothesis H_j would be rejected in a MCP with FWER α if $\tilde{P}_j \le \alpha$.

Bonferroni adjusted P-values

For the simple Bonferroni MCP, the adjusted P-values are $\tilde{P}_j = p\,P_j$, $j = 1, \ldots, p$.

Adjusted P-values for the Holm procedure

For the Holm procedure, as before, we denote the ordered P-values $P_{(1)} \le \ldots \le P_{(p)}$ with corresponding hypotheses $H_{(1)}, \ldots, H_{(p)}$. Since the Holm test treats hypotheses symmetrically, the adjusted P-values must satisfy $\tilde{P}_{(1)} \le \ldots \le \tilde{P}_{(p)}$. Also, for $H_{(j)}$ to be rejected in a Holm procedure with FWER α, $H_{(1)}, \ldots, H_{(j-1)}$ must have been rejected, which requires $\alpha \ge \tilde{P}_{(j-1)}$, and at this point $H_{(j)}$ is tested at significance level $\alpha/(p-j+1)$ and so is rejected if $\alpha \ge (p-j+1)P_{(j)}$. Hence, $H_{(j)}$ is rejected in a procedure with FWER equal to α if $\alpha \ge \max\{\tilde{P}_{(j-1)}, (p-j+1)P_{(j)}\}$. It follows that the adjusted P-values can be expressed iteratively as

$$\begin{aligned} \tilde{P}_{(1)} &= p\,P_{(1)}, \\ \tilde{P}_{(j)} &= \max\{\tilde{P}_{(j-1)}, (p-j+1)P_{(j)}\}, \quad j = 2, \ldots, p. \end{aligned}$$

In the numerical example of Table 20.1 with $P_1 = 0.012$, $P_2 = 0.02$, $P_3 = 0.03$ and $P_4 = 0.08$, we see that

$$\begin{aligned} \tilde{P}_{(1)} &= 4 \times 0.012 = 0.048, \\ \tilde{P}_{(2)} &= \max(0.048, 3 \times 0.02) = 0.06, \\ \tilde{P}_{(3)} &= \max(0.06, 2 \times 0.03) = 0.06, \\ \tilde{P}_{(4)} &= \max(0.06, 0.08) = 0.08. \end{aligned}$$

Thus, if the specified FWER is $\alpha = 0.05$, only H_1 is significant, in agreement with our previous analysis.

Table 20.2 *Summary of adjusted P-values for the numerical example with* $p = 4$

		Adjusted P-values, $\tilde{P}_{(j)}$		
Hypothesis	$P_{(j)}$	*Bonferroni*	*Holm*	*CTP with Simes*
H_1	0.012	0.048	0.048	0.040
H_2	0.020	0.080	0.060	0.045
H_3	0.030	0.120	0.060	0.060
H_4	0.080	0.320	0.080	0.080

Adjusted P-values for the CTP using Simes intersection tests

In a closed testing procedure, hypothesis H_j is rejected overall if every intersection hypothesis H_J where $j \in J$ is rejected at level α. In applying Simes' test to the intersection hypotheses, we see from (20.1) that when J contains m indices, H_J is rejected at level α if

$$\min_{j \in J}\{mP_{(j)}/j\} \leq \alpha$$

and so we can ascribe a P-value to this test of H_J

$$\tilde{P}_J = \min_{j \in J}\{mP_{(j)}/j\}.$$

Now, H_j is rejected overall in a test with FWER α if

$$\max_{J:\, j \in J}\{\tilde{P}_J\} \leq \alpha.$$

Therefore, the adjusted P-value for H_j, which is the minimum value of α for which H_j is rejected overall, is

$$\tilde{P}_j = \max_{J:\, j \in J}\{\tilde{P}_J\}.$$

The information to calculate the adjusted P-values $\tilde{P}_1, \ldots, \tilde{P}_4$ in our example is available in Table 20.1 since the column headed "Simes P-value" contains the adjusted P-values $\tilde{P}_J$ for the key intersection hypotheses. As explained earlier, in the overall test of H_j we need only consider the most challenging intersection hypothesis of each size and these are the cases presented in Table 20.1. The adjusted P-values for individual hypotheses are

$$\begin{aligned}
\tilde{P}_{(1)} &= \max(0.012, 0.024, 0.036, 0.040) = 0.040, \\
\tilde{P}_{(2)} &= \max(0.020, 0.040, 0.045, 0.040) = 0.045, \\
\tilde{P}_{(3)} &= \max(0.030, 0.060, 0.045, 0.040) = 0.060, \\
\tilde{P}_{(4)} &= \max(0.080, 0.060, 0.045, 0.040) = 0.080.
\end{aligned}$$

In Table 20.2, we summarize results for the adjusted P-values when different methods are applied to our numerical example When each method is represented as a closed testing procedure, one can deduce the power ordering of the MCPs

$$\text{CTP with Simes} \succ \text{Holm} \succ \text{Bonferroni},$$

where "$\succ$" denotes "more powerful than". For instance, in the Holm method, the intersection hypothesis H_{12} is rejected if $P_1 \leq \alpha/2$ or $P_2 \leq \alpha/2$ whereas in the CTP with Simes method

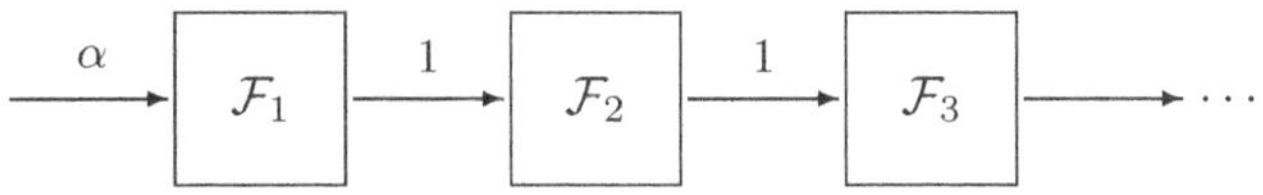

Figure 20.9 *A testing procedure with a hierarchy of parallel hypotheses*

H_{12} is also rejected if $\max(P_1, P_2) \leq \alpha$. This ordering is reflected in the adjusted P-values for the example which decrease as one moves though the Bonferroni, Holm and CTP with Simes procedures. It should, however, be remembered that the Bonferroni and Holm tests preserve the FWER α whatever the joint distribution of $P_1, \ldots, P_p$, whereas the Simes test requires the P-values P_j to be positively dependent, as explained in Section 20.4

20.8 Hierarchical Testing of Several Families of Hypotheses

We can generate more complex hypothesis networks by considering a hierarchy of families of hypotheses, where each family in the hierarchy consists of several hypotheses. Such a sequence of families of hypotheses is depicted in Figure 20.9. Here the node $\mathcal{F}_1$ represents a family of primary hypotheses, node $\mathcal{F}_2$ a family of secondary hypotheses, and node $\mathcal{F}_3$ a family of tertiary hypotheses. Testing for each family $\mathcal{F}_1$, $\mathcal{F}_2$, $\ldots$, proceeds sequentially. However we must distinguish between two cases.

(a) Suppose an analysis plan specifies that all hypotheses in $\mathcal{F}_\ell$ must be rejected in order for hypotheses $\mathcal{F}_{\ell+1}$ to be considered at all. In this case, tests within each family $\mathcal{F}_\ell$ should form a MCP with FWER α and any of the methods we have described, such as Bonferroni, Holm or a CTP with Simes tests for intersection hypotheses, may be used. If all the hypotheses tested in $\mathcal{F}_\ell$ are rejected, the whole error probability α is passed to the next family $\mathcal{F}_{\ell+1}$. However, if one or more hypothesis in $\mathcal{F}_\ell$ is not rejected, the testing process ends and no error probability is propagated to $\mathcal{F}_{\ell+1}$. In consequence, the adjusted P-value for any hypothesis in $\mathcal{F}_{\ell+1}$ must be at least as high as the largest adjusted P-value for a hypothesis in $\mathcal{F}_\ell$.

(b) In general, multiple testing methods defined through a graphical network are closed testing procedures that use weighted Bonferroni tests for the intersection hypotheses and it is not possible to use, say, a Simes test for intersection hypotheses in this framework; see Burmann, Sonesson & Guilbaud (2009) and Bretz, Maurer, Brannath & Posch (2009). However, the full generality of graphical procedures is available as long intersection hypotheses are tested by weighted Bonferroni tests. Then, if only some of the hypotheses in $\mathcal{F}_\ell$ are rejected, the significance levels associated with these hypotheses can be passed on to be used as the family-wise error rate for testing hypotheses in $\mathcal{F}_{\ell+1}$. Since the Holm procedure can be represented as a graphical method, it can be used in this context.

Figure 20.10 shows an example where $\mathcal{F}_1$ consists of a single primary hypothesis which must be rejected before a family $\mathcal{F}_2$ of secondary hypotheses can be considered. These secondary hypotheses can be tested using a Bonferroni or Holm test or a CTP with Simes tests for intersection hypotheses, as specified in the statistical analysis plan. More complex arrangements of hypotheses can be contemplated, for example, if a Bonferroni or Holm test is used for $\mathcal{F}_2$, a tertiary endpoint might require a particular pair of the secondary endpoints to be rejected but not others.

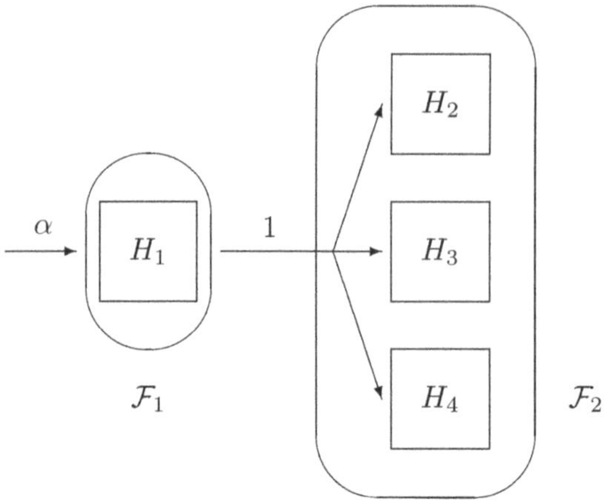

Figure 20.10 *A testing procedure with one primary hypothesis and three secondary hypotheses*

20.9 Controlling the False Discovery Rate (FDR): An Alternative Approach when there are Many Endpoints

In some studies, a very large number of primary endpoints may be specified. For example, in a Phase II rehabilitation trial for traumatic brain injury, there might be thirty or more endpoints arising from various assessment instruments involving clinical, psychological, physical and quality of life outcomes. In this case, the methods we have described which protect the FWER may require extremely small P-values to reject a null hypothesis, so it can be very difficult to obtain a statistically significant result even if there is a large treatment effect in one outcome variable. Benjamini & Hochberg (1995) (BH) proposed an alternative approach to multiple testing based on the false discovery rate (FDR). Instead of controlling the probability that at least one true null hypothesis should be rejected (a false positive result for that endpoint), the FDR approach controls the proportion of endpoints which are declared positive that are in fact negative. The FDR approach is appropriate for exploratory trials with many endpoints, especially when reducing Type II error is of high importance and results from such trials may be followed up in a confirmatory trial. Wason & Robertson (2021) argue that the FDR approach should be considered in multi-arm clinical trials testing distinct treatment arms. They comment that if a low proportion of treatments are effective, then just as for FWER methods, a high proportion of the positive results found will be true positives, while if a high proportion of treatments are effective, then FDR methods will identify a large fraction of these as positive.

Suppose we wish to follow the FDR approach and control the false discovery rate at level d where $0 < d < 1$. In the BH-FDR procedure, we denote the ordered P-values by $P_{(1)} \leq \ldots \leq P_{(p)}$ and label the corresponding hypotheses $H_{(1)}, \ldots, H_{(p)}$. We define

$$j' = \max_{1 \leq j \leq p} \left\{ j \colon P_{(j)} \leq \frac{d\,j}{p} \right\}. \tag{20.3}$$

If no j satisfies the inequality, $j' = 0$ and no hypotheses are declared significant. Otherwise, all hypotheses $H_{(1)}, \ldots, H_{(j')}$ are declared significant with a false discovery rate of d.

Benjamini & Hochberg (1995) proved that this procedure protects the FDR at level d under independence assumptions. Benjamini & Yekutieli (2001) extended this result to the case where there is a positive dependency among the test statistics, similar to the condition needed for the validity of the Simes procedure. This positive dependence assumption is

Table 20.3 *Numerical example with $p = 5$ comparing FWER and FDR procedures with $\alpha = 0.05$ and $d = 0.05$, respectively*

			Critical values	
j	$P_{(j)}$	*Bonferroni*	*Hochberg*	*BH-FDR*
1	0.011	0.010	0.010*	0.010*
2	0.012	0.010	0.0125*	0.020*
3	0.036	0.010	0.0167	0.030*
4	0.039	0.010	0.025	0.040*
5	0.410	0.010	0.050	0.050

Critical values with a superscript $*$ are those where $j < j'$ for the Hochberg or BH-FDR testing procedure, so $H_{(j)}$ is rejected. No hypotheses are rejected in the Bonferroni procedure.

often reasonable when testing multiple endpoints that reflect a general improvement in a patient's condition, or in comparing multiple treatments against a common control.

This method is similar to the Hochberg (1988) procedure for controlling FWER which rejects hypotheses $H_{(1)}, \ldots, H_{(j')}$ for j' defined by Equation (20.2). However, as Benjamini & Hochberg point out, if we set $d = \alpha$, the critical values for $P_{(j)}$ in (20.2) and (20.3) are the same when $j = 1$ and $j = p$ but those in (20.3) are higher for $2 \le j \le p - 1$. Thus the BH-FDR procedure will find at least as many significant hypotheses, and usually many more as p becomes large.

A simple illustrative example with just $p = 5$ endpoints using $\alpha = d = 0.05$ is displayed in Table 20.3. The ordered P-values are listed in the second column. None of these fall below the Bonferroni critical value of $\alpha/5 = 0.01$. Thus there can be no recycling, and no hypothesis is declared significant using either the Bonferroni or Holm procedures. On the other hand, the Hochberg procedure rejects both $H_{(1)}$ and $H_{(2)}$ while controlling the FWER at $\alpha = 0.05$. (The CTP with Simes' test for intersection hypotheses leads to the same result.) If, instead, we wish to guarantee an FDR of 5%, the BH-FDR procedure leads us to declare the four endpoints corresponding to $H_{(1)}$ to $H_{(4)}$ as significant because $P_{(4)} = 0.039$ is less than the critical value of 0.040. Note, it does not matter that $P_{(3)} = 0.036$ fails to exceed its critical value of 0.030.

In this example we have set $d = \alpha = 0.05$. However, values of d are usually chosen to be larger than typical false positive error rates α and a common choice might be $d = 0.2$. This would lead to some outcomes being labelled significant even when their P-values are greater than 0.05.

In the next sections, we shall return to discussing procedures that control FWER. However, it is good to be aware of usefulness of this alternative approach for exploratory trials with a large number of endpoints or multiple treatments.

20.10 Testing a Secondary Endpoint after a Group Sequential Test of a Primary Endpoint

20.10.1 A Cautionary Example

Consider a trial in which a group sequential test with Type I error probability α is used to monitor a primary endpoint. Suppose the efficacy boundary is crossed at an interim or final analysis so the null hypothesis $H_1\colon \theta_1 \le 0$ for the primary endpoint is rejected. Investigators now wish to test a secondary endpoint. It is tempting to argue that a hierarchical testing

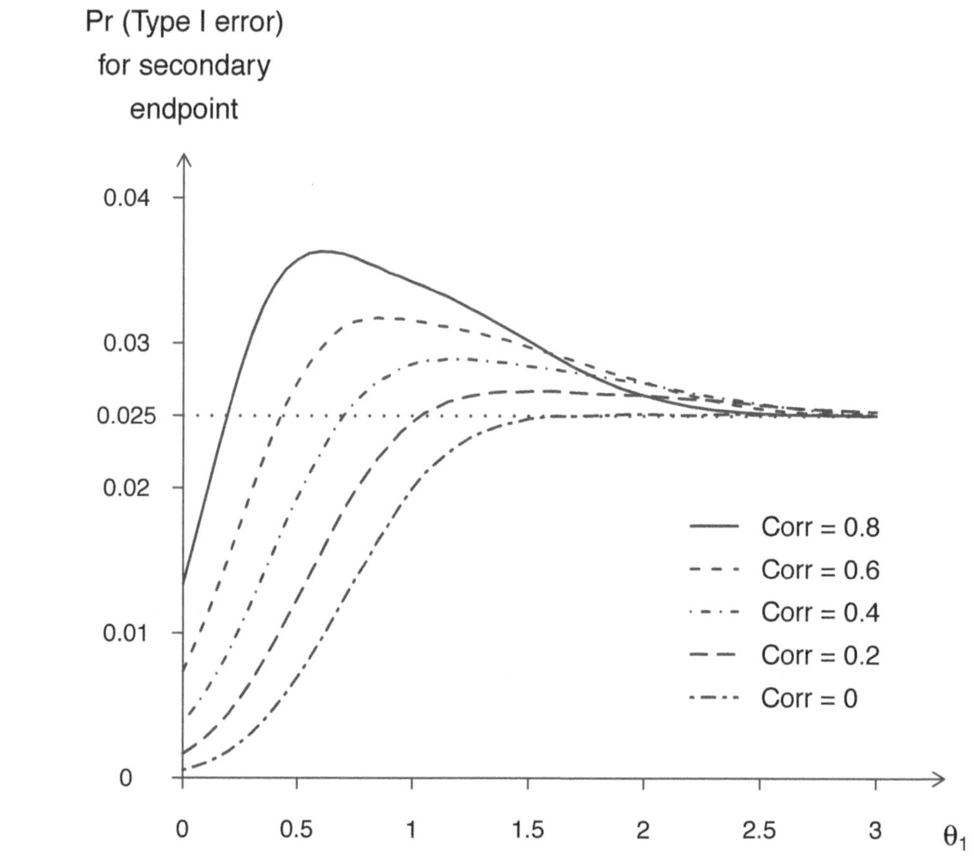

Figure 20.11 *Type I error probability when the secondary endpoint is tested naively in a fixed sample, level* $\alpha = 0.025$ *test after a positive outcome in a group sequential test for the primary endpoint. Results are based on a million simulations at each value of* θ_1.

procedure can be applied and, since the null hypothesis H_2: $\theta_2 \leq 0$ for the secondary endpoint is only tested once, a simple significance test at level α can now be applied making no "adjustment" for multiple looks and this will control the FWER at level α. Hung, Wang & O'Neill (2007) explain that this argument is specious since it overlooks the effects of any correlation between the primary and secondary endpoints and the data-dependent analysis time for the secondary endpoint. This is in keeping with the fact that a stopping rule for the primary endpoint induces a bias in the MLE of the treatment effect on a secondary endpoint, as we saw in Section 9.3.4.

Hung, Wang & O'Neill give an example of a 2-stage trial with $\alpha = 0.025$ where the probability of falsely rejecting H_2 can be as high as 0.04 as the correlation between primary and secondary endpoints approaches 1. As another example, consider a group sequential trial with four planned analyses, designed to have Type I error probability $\alpha = 0.025$ for the primary endpoint and power 0.8 if $\theta_1 = 1$. The futility boundary is non-binding and an error spending design in the Rho family with $\rho = 2$ is used, requiring an inflation factor $R = 1.117$ (from Table 8.12). If the test for the primary endpoint rejects H_1: $\theta_1 \leq 0$, then a naive approach is followed and a simple test of H_2: $\theta_2 \leq 0$ is conducted at significance level $\alpha = 0.025$ using the data at that point. Suppose the primary and secondary responses for each patient have a bivariate normal distribution. Then, the probability of rejecting H_2 when $\theta_2 = 0$ depends on the value of θ_1 and the correlation between the two endpoints. The Type I error probability for the test of H_2 is plotted in Figure 20.11 for the case

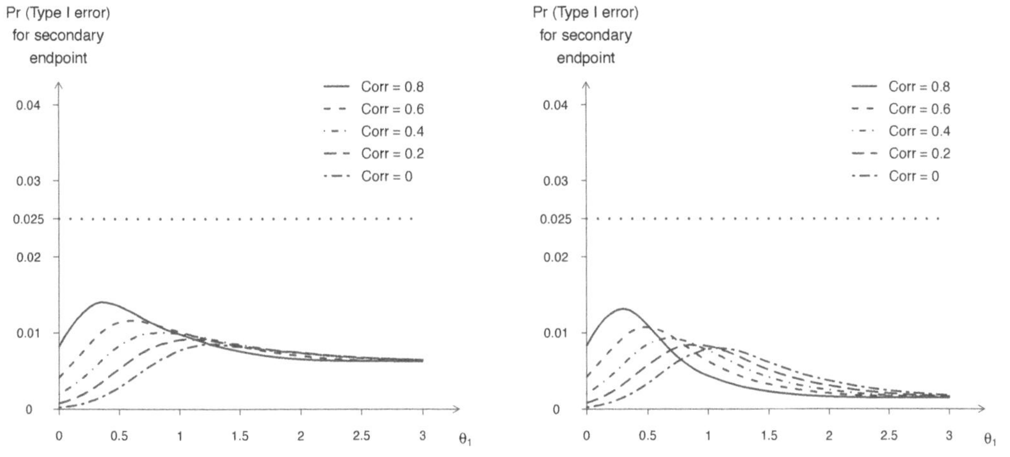

(a) Rho-family error spending test, $\rho = 1$ (b) Rho-family error spending test, $\rho = 2$

Figure 20.12 *Type I error probability for the test of a secondary endpoint after a group sequential test for a primary endpoint. An error spending GST from the Rho-family with $\alpha = 0.025$ and (a) $\rho = 1$ and (b) $\rho = 2$ is applied to the secondary endpoint at the analysis at which H_1 is rejected. Results are based on a million simulations at each value of θ_1.*

where information levels follow their planned values exactly. The plot shows clear evidence of Type I error inflation above the supposed value of 0.025, even for modest correlations between the two endpoints.

20.10.2 A Correct Test for a Secondary Endpoint after a GST for the Primary Endpoint

Tang & Geller (1999) and Maurer & Bretz (2013) explain how to carry out tests of multiple hypotheses in a group sequential trial with strong control of FWER. Tang & Geller define an MCP as a closed testing procedure and test all hypotheses and intersection hypotheses in level α group sequential tests. Maurer & Bretz consider an MCP defined by a graphical network and, whenever a test of a single hypothesis is performed at a certain significance level, a pre-specified group sequential test is applied with the appropriate significance level as its Type I error probability. Both these approaches can be applied to test a primary endpoint and a secondary endpoint hierarchically. The result is a gatekeeping procedure in which H_2 is tested in a level α group sequential test if and when H_1 has been rejected. Thus, if H_1 is rejected at analysis k, one compares the test statistic for H_2 with the critical value specified in its group sequential test at analysis k.

Following this approach in our example with four analyses, we need to specify a group sequential test of H_2: $\theta_2 \leq 0$ with Type I error probability $\alpha = 0.025$. Now, this testing boundary will only be employed at one analysis, and this could be early in the trial if there is a large treatment effect on the primary endpoint or later if that treatment effect is smaller. We therefore recommend an error spending test in the Rho family with $\rho = 1$ as this gives fairly constant critical values on the Z scale. Also, since only an efficacy boundary is applied in testing H_2, a test with no futility boundary should be used. To illustrate these points, we have applied error spending tests from the Rho family with $\rho = 1$ and $\rho = 2$, and no futility boundary, to test the secondary endpoint in our example. The plots in Figure 20.12 show the probability of rejecting H_2 when $\theta_2 = 0$ as a function of θ_1 for five different values of the correlation between the primary and secondary endpoints. It is clear that the Type I

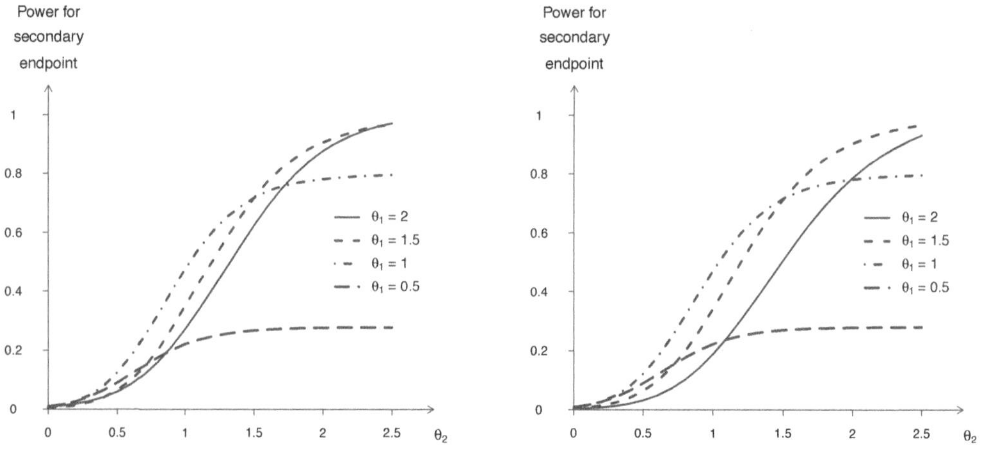

(a) Rho-family error spending test $\rho = 1$ (b) Rho-family error spending test $\rho = 2$

Figure 20.13 *Power for test of a secondary endpoint after a group sequential test for a primary endpoint. An error spending GST from the Rho-family with* $\alpha = 0.025$ *and (a)* $\rho = 1$ *and (b)* $\rho = 2$ *is applied to the secondary endpoint at the analysis at which* H_1 *is rejected. The correlation between primary and secondary endpoints is* 0.4. *Results are based on a million simulations at each combination of* θ_1 *and* θ_2.

error probability for the test of H_2 is protected with some conservatism. Under the highest values of θ_1, the test of the primary endpoint is likely to stop at an early analysis where the error spending test for H_2 with $\rho = 2$ has a high boundary and so the Type I error probability for testing H_2 is extremely small here.

When $\theta_2 > 0$, the probability of rejecting H_2 still depends on θ_1 and the correlation between primary and secondary endpoints. We have run a simulation study to find power curves for selected values of θ_1 when the variances of the primary and secondary endpoints are equal, so the information level for θ_2 is the same as that for θ_1 at each analysis, and the correlation between primary and secondary endpoints is 0.4. These power curves are shown in Figure 20.13. For both tests of H_2, we see that power is higher at $\theta_1 = 1$ and $\theta_1 = 1.5$ than at the extreme effect size of $\theta_1 = 2$. This is because $\theta_1 = 2$ leads to stopping at early analyses when less data has been gathered on the secondary endpoint. Again, the roughly constant critical values on the Z scale in the group sequential test for H_2 with $\rho = 1$ facilitates rejection of H_2 at whichever analysis the group sequential test for H_1 terminates. As a consequence, the power curves for the case $\rho = 1$ rise more steeply. For both $\rho = 1$ and $\rho = 2$, power curves for testing H_2 reach a maximum well below one if $\theta_1 = 1$ or $\theta_1 = 0.5$, due to the fact that the power of the test of H_1 is significantly less than 1 in these cases and H_2 is only tested after H_1 has been rejected.

20.10.3 Extending a Trial beyond a Stopping Point

Tang & Geller (1999) point out that the FWER is still protected if a trial continues after the stopping boundary for the primary endpoint has been crossed and the group sequential test for the secondary endpoint is applied. This may be desirable if there are signs that the null hypothesis H_2: $\theta_2 \leq 0$ may be rejected once further data are collected. Glimm, Maurer & Bretz (2010) define four testing strategies for primary and secondary endpoints in a two-stage trial which include cases where the study continues after rejecting H_1 but not H_2 at the interim analysis. Continuing a trial after establishing a positive treatment effect on the primary endpoint might be considered unethical because it does not protect patients on the inferior control arm. However, we are reminded by Liu & Anderson (2008) that the U.S. Code of Federal Regulations (21 CFR 312.21) states

> "Phase 3 studies are expanded controlled and uncontrolled trials. They are performed after preliminary evidence suggesting effectiveness of the drug has been obtained, and are intended to gather the additional information about effectiveness and safety that is needed to evaluate the overall benefit-risk relationship of the drug and to provide an adequate basis for physician labeling. ..."

This implies it may be allowable to extend a trial in order to obtain more information about other efficacy or safety variables which will enable a better informed judgement to be made about risks and benefits of the new treatment.

20.10.4 Addressing the Conservatism of Hierarchical Sequential Tests

Figure 20.12 shows that the procedures we have described have maximum Type I error probabilities below the nominal $\alpha = 0.025$. In the case of a sequential trial with just two analyses, Tamhane, Mehta & Liu (2010) propose lowering the testing boundary for the secondary endpoint while still protecting the FWER at level α. They show that a procedure's Type I error probability is highest when the correlation between primary and secondary endpoints is equal to 1 and they define the boundary for the secondary endpoint so that the maximum error probability for testing H_2 over all possible values of θ_1 is equal to α in this case.

Tamhane, Wu & Mehta (2012) extend the approach of Tamhane, Mehta & Liu (2010) for two-stage trials. They define the boundary for the secondary endpoint assuming the correlation between endpoints is equal to the lowest value in an upper confidence interval, doing this in a way that protects the FWER at level α under all possible parameter values of θ_1 and the correlation between endpoints. Tamhane, Gou, Jennison & Mehta (2018) extend the above methods to group sequential trials with more than two analyses.

20.11 Group Sequential Tests in General Multiple Testing Procedures

The methods of Tang & Geller (1999) and Maurer & Bretz (2013) allow group sequential stopping rules to be included in many types of multiple testing procedure. As noted previously, in Tang & Geller's approach, each hypothesis and intersection hypothesis in a closed testing procedure is tested in a level α group sequential test. Maurer & Bretz define procedures defined by a graphical network and whenever a hypothesis is to be tested, a group sequential test with the appropriate significance level is applied to the current data.

20.11.1 Hierarchical Testing of Endpoints after a Group Sequential Test of a Primary Endpoint

Suppose a trial's stopping rule is based on a test of H_1: $\theta_1 \leq 0$ against $\theta_1 > 0$ where θ_1 is the treatment effect on the primary endpoint. Let $H_2, \ldots, H_p$ be a sequence of hypotheses concerning treatment effects θ_2 for a secondary endpoint, θ_3 for a tertiary endpoint, etc. If H_1 is rejected, these hypotheses are to be tested hierarchically in a way that controls the FWER at level α. A general procedure with K analyses can be defined as follows.

First, we specify a group sequential testing boundary for the primary endpoint, $(a_{11}, b_{11}), \ldots, (a_{1K}, b_{1K})$ with $a_{1K} = b_{1K}$, that has Type I error probability α and the appropriate power. For simplicity, we suppose this test has a non-binding futility boundary. For each subsidiary endpoint $j = 2, \ldots, p$, we specify a group sequential efficacy boundary

$b_{j1}, \ldots, b_{jK}$ with Type I error probability α. Thus, for each $j = 1, \ldots, p$,

$$Pr_{\theta_j=0}\{Z_{jk} > b_{jk} \text{ for some } 1 \leq k \leq K\} = \alpha. \quad (20.4)$$

These could be parametric boundaries from the power family of Section 5.2, say. Alternatively they could be error spending boundaries as defined in Chapter 8. It is not necessary to use the same boundary for each endpoint j. However, we learnt from the example in Section 20.10.2 that boundaries with roughly constant values $b_{j1}, \ldots, b_{jK}$ are desirable for a secondary endpoint and the same is true for subsequent endpoints, so the upper half of a Pocock boundary or a Rho family error spending test with $\rho = 1$ are to be recommended.

The testing procedure is then:

After group $k = 1, \ldots, K-1$

if $Z_{1k} \geq b_{1k}$	stop, reject H_1 and continue hierarchical testing (see below)
if $Z_{1k} \leq a_{1k}$	stop, accept H_1
otherwise	continue to group $k+1$,

after group K

if $Z_K \geq b_K$	stop, reject H_1 and continue hierarchical testing (see below)
if $Z_K < a_K$	stop, accept H_1.

If H_1 has been rejected at analysis k, the hierarchical tests of $H_2 \ldots, H_p$ are as follows:

If $Z_{2k} \geq b_{2k}$	reject H_2 and continue to test H_3
If $Z_{2k} < b_{2k}$	stop and accept $H_2, \ldots, H_p$.

If H_2 has been rejected,

If $Z_{3k} \geq b_{3k}$	reject H_3 and continue to test H_4
If $Z_{3k} < b_{3k}$	stop and accept $H_3, \ldots, H_p$

If H_2 and H_3 have been rejected, continue until finally, if $H_2, \ldots, H_{p-1}$ have been rejected

If $Z_{pk} \geq b_{pk}$	reject H_p and stop
If $Z_{pk} < b_{pk}$	stop and accept H_p.

This procedure can be viewed as an example of the procedures proposed by Tang & Geller (1999) and Maurer & Bretz (2013), so control of FWER at level α follows from the general theory in those papers. For a more direct proof, suppose

$$j^* = \min\{j \in \{1, \ldots, p\}: H_j \text{ is true}\}.$$

Then, for a family-wise error to occur, all the hypotheses H_1 to H_{j^*} must be rejected. This requires $Z_{j^*k} \geq b_{j^*k}$ for some $k \in \{1, \ldots, K\}$ and by (20.4), this has probability no greater than α.

20.11.2 Example

In Table 20.4 we present a simple pedagogic example with $p = 4$ ordered endpoints observed in a group sequential procedure with a maximum of $K = 5$ analyses. The four endpoints are to be tested hierarchically with a FWER of $\alpha = 0.025$. For the primary endpoint, we choose a power family, one-sided test with a non-binding futility boundary, shape parameters

Table 20.4 *Worksheet for a group sequential test with four ordered endpoints*

Analysis k	*Boundary* (a_{1k}, b_{1k}) *for endpoint* 1	*Boundary* b_{jk} *for endpoints* $j = 2, 3, 4$	Z_{1k}	Z_{2k}	Z_{3k}	Z_{4k}
1	(−0.928, 4.562)	2.413	2.82	—	—	—
2	(0.314, 3.226)	2.413	3.05	—	—	—
3	(1.049, 2.634)	2.413	2.90	2.70	2.20	2.75
4	(1.595, 2.281)	2.413	—	—	—	—
5	(2.040, 2.040)	2.413	—	—	—	—

$\Delta_1 = \Delta_2 = 0$, and power 0.8 at a treatment effect $\theta_1 = 1$. From Table 5.8, we see this design has an inflation factor $\tilde{R}^* = 1.200$ so an information level $\mathcal{I}_{max}$=9.42 for θ_1 should be reached by analysis 5. Substituting $\tilde{C}_1' = 2.040$ and $\tilde{C}_2^* = 1.029$ in (5.11) and (5.13) gives the boundary values (a_{1k}, b_{1k}) listed in the second column of Table 20.4. The efficacy boundaries for the secondary, tertiary and fourth endpoints are those of a power family, one-sided test with no futility boundary and shape parameter $\Delta = 0.5$, so the critical values for Z_{jk} are constant over $k = 1, \ldots, K$. In Chapter 5 we recommended using lowers value of Δ in a one-sided test but this test has a different role here in the hierarchical procedure and a "Pocock-type" boundary is advantageous; see, for example, Tamhane, Mehta & Liu (2010) and Glimm, Maurer & Bretz (2010). This test has $C_1' = 2.413$ and hence $b_{jk} = 2.413$ for $j = 2$, 3 and 4, and $k = 1, \ldots, 5$.

In our hypothetical example, the standardized statistics for the primary endpoint at the first two analyses are $Z_{11} = 2.82$ and $Z_{12} = 3.05$. Since Z_{11} lies between −0.928 and 4.562, and Z_{12} lies between 0.314 and 3.226, the study proceeds to analysis 3 with no further testing of the other hypotheses. As the test statistics Z_{j1} and Z_{j2} for $j = 2$, 3 and 4 are not used in any tests, we have not shown these in Table 20.4. At stage 3, the observed value $Z_{13} = 2.90$ exceeds $b_{13} = 2.634$, so the trial stops and H_1 is rejected. We now examine the secondary endpoint and since $Z_{23} = 2.70 > b_{23} = 2.413$, we may also reject H_2. However, for the tertiary endpoint, $Z_{33} = 2.20 < b_{33} = 2.413$, and H_3 cannot be rejected. This means that any hypothesis lower in the hierarchy is not tested, so H_4 cannot be rejected despite the fact that the $Z_{43} = 2.75 > b_{43} = 2.413$.

As noted in Section 20.10.3, investigators would have the option of continuing the trial to see if H_3 or H_4 might be rejected at a later analysis. The main consideration here should be the likelihood of a positive outcome for the tertiary endpoint since, not only is this the more important endpoint, but H_3 has to be rejected in its sequential test before H_4 can be tested in the hierarchical procedure.

20.12 Group Sequential Tests for Multiple Primary Endpoints using the Holm Procedure

The LOFT trial cited earlier in Section 20.5.1 had three primary endpoints: incidences of radiologically determined vertebral fractures, clinical hip fractures, and clinical non-vertebral fractures. This trial was monitored using a group sequential procedure, in which all three endpoints were examined at each interim analysis along with a number of other variables. This is an example where a positive result on any of the three endpoints may lead to a treatment recommendation, whether or not the results for the other endpoints are positive.

Suppose a trial has p primary endpoints and these are to be tested in a group sequential version of the Holm procedure presented in Section 20.5. (Although we have seen that

Table 20.5 *Constants* $C_1'(K, \gamma, \Delta)$. *Using the constants* $C_1'(K, \gamma, 0.5)$ *in* (20.6) *gives Pocock-type boundaries while using* $C_1'(K, \gamma, 0)$ *in* (20.7) *gives O'Brien & Fleming-type boundaries, both satisfying* (20.5) *for* $K = 5$, *and selected values of* $\gamma = \alpha/j$ *with* $\alpha = 0.025$.

	$\gamma = \alpha$	$\gamma = \alpha/2$	$\gamma = \alpha/3$	$\gamma = \alpha/4$	$\gamma = \alpha/5$
$C_1'(5, \gamma, 0.5)$	2.413	2.675	2.817	2.913	2.986
$C_1'(5, \gamma, 0)$	2.040	2.303	2.448	2.547	2.621

a CTP with Simes tests for intersection hypotheses is more powerful, that method is only applicable when endpoints are independent or there is a positive association between endpoints.) For each endpoint $j = 1, \ldots, p$, denote the standardized statistics at analyses 1 to K by $Z_{j1}, \ldots, Z_{jK}$ and suppose these follow the canonical distribution (3.1) with $\theta = \theta_j$. Since hypotheses may be tested at different significance levels as the testing procedure is applied, it is helpful to specify for each endpoint a family of efficacy boundaries indexed by the Type I error rate γ. For each $\gamma \in (0, 1)$, the γ-level boundary $(b_{j1}(\gamma), \ldots, b_{jK}(\gamma))$ for endpoint j should satisfy

$$Pr_{\theta_j=0}\{Z_{jk} > b_{jk}(\gamma) \text{ for some } 1 \leq k \leq K\} = \gamma. \tag{20.5}$$

Stopping the trial for futility may be appropriate if results on all p endpoints are negative but we shall assume that any futility stopping rule is non-binding. Following Liu & Anderson (2008), we require that for each $j = 1, \ldots, p$, the family of efficacy boundaries is "well-ordered", that is

$$b_{jk}(\gamma') \leq b_{jk}(\gamma) \text{ for } \gamma' > \gamma, \text{ for each } k = 1, \ldots, K.$$

This ensures that, at any stage k, if a hypothesis H_j is tested at a larger significance level due to propagation of error from another rejected hypothesis, the critical value b_{jk} does not increase. It is easy to see that the well-ordered condition is satisfied by the power family boundaries with no futility boundary of Section 5.3. Maurer & Bretz (2013, Sec. 2.1) show this condition is also satisfied by the boundaries formed using error spending functions in the Rho family defined by (8.6).

It is convenient to use the same family of boundaries for each endpoint $j = 1, \ldots, p$. Two boundaries in the power family of particular interest are Pocock-type boundaries with

$$b_{jk}(\gamma) = C_1'(K, \gamma, 0.5), \quad k = 1, \ldots, K \tag{20.6}$$

and O'Brien & Fleming-type boundaries, for which

$$b_{jk}(\gamma) = C_1'(K, \gamma, 0)\sqrt{K/k}, \quad k = 1, \ldots, K. \tag{20.7}$$

Here, the constants $C_1'(K, \gamma, 0.5)$ and $C_1'(K, \gamma, 0)$ are chosen to satisfy the condition (20.5). Table 20.5 give the values of these constants for $K = 5$ and $\gamma = \alpha$, $\alpha/2$, $\alpha/3$, $\alpha/4$ and $\alpha/5$ when $\alpha = 0.025$.

To illustrate this form of procedure, consider a trial with three primary endpoints conducted group sequentially with a maximum of $K = 5$ analyses and suppose this trial is stopped at analysis $\tau \in \{1, \ldots, K\}$. This stopping time could be the stage at which

(a) a particular hypothesis H_j is rejected in the overall testing procedure,

(b) any hypothesis H_j is rejected in the overall testing procedure,

(c) all hypotheses, H_1, H_2 and H_3, are rejected in the overall testing procedure,

(d) any other reason based on accumulated data observed so far, including safety or risk/benefit profiles, as well as the three primary endpoint variables.

An MCP for testing the hypotheses H_1, H_2 and H_3 is best described by referring to Figure 20.2 in Section 20.5.2. In a sequential trial, the hypothesis tests to be carried out for each node are given by the family of efficacy boundaries at different significance levels for that endpoint. The graph evolves as before, except that now hypothesis H_j is rejected when the trial stops at analysis τ if

$$Z_{jk} \geq b_{jk}(\gamma) \quad \text{for some } 1 \leq k \leq \tau,$$

where γ is the significance level currently associated with that node. If a hypothesis H_j is rejected, the significance level associated with node H_j is propagated to neighboring nodes just as in the single stage procedures of Section 20.5.2.

As an example, suppose the trial has stopped at stage $\tau = 4$. Following the Holm procedure depicted in Figure 20.2, we look for any outcome variable $j \in \{1, 2, 3\}$ for which

$$Z_{jk} \geq b_{jk}(\alpha/3) \quad \text{for some } 1 \leq k \leq 4.$$

If no such variable exists, then no hypothesis is rejected. If such a j exists, the same question is asked of the other two endpoints but with critical values $b_{jk}(\alpha/2)$. If one of those two hypotheses is rejected, the remaining hypothesis is tested with critical values $b_{jk}(\alpha)$.

The general theory of Maurer & Bretz (2013) implies this procedure controls the FWER at level α. Alternatively, we can view the Holm procedure as a CTP, but with the tests of intersection hypotheses defined in Section 20.5.1 replaced by

Hypothesis	*Reject if*
H_1	$Z_{1k} > b_{1k}(\alpha)$ for some $1 \leq k \leq \tau$,
H_2	$Z_{2k} > b_{2k}(\alpha)$ for some $1 \leq k \leq \tau$,
H_3	$Z_{3k} > b_{3k}(\alpha)$ for some $1 \leq k \leq \tau$,
H_{12}	$Z_{1k} > b_{1k}(\alpha/2)$ or $Z_{2k} > b_{2k}(\alpha/2)$ for some $1 \leq k \leq \tau$,
H_{13}	$Z_{1k} > b_{1k}(\alpha/2)$ or $Z_{3k} > b_{3k}(\alpha/2)$ for some $1 \leq k \leq \tau$,
H_{23}	$Z_{2k} > b_{2k}(\alpha/2)$ or $Z_{3k} > b_{3k}(\alpha/2)$ for some $1 \leq k \leq \tau$,
H_{123}	$Z_{1k} > b_{1k}(\alpha/3)$, $Z_{2k} > b_{2k}(\alpha/3)$ or $Z_{3k} > b_{3k}(\alpha/3)$ for some $1 \leq k \leq \tau$.

As an example, for H_1 to be rejected in the CTP we need H_1, H_{12}, H_{13} and H_{123} all to be rejected and it is a simple task to check that this occurs under the same conditions as in the earlier description of the procedure.

Example

Table 20.6 shows testing boundaries and observed data in a trial with four primary endpoints and five analyses. A group sequential Holm test with FWER is $\alpha = 0.025$ was employed with Pocock-type boundaries. At each analysis k, the largest Z_{jk} was compared to $b_{jk}(\alpha/4) = 2.931$. The first time this critical value was exceeded was at analysis 2 where $Z_{42} = 2.94$. Although H_4 was rejected at this point, the trial continued and the error probability $\alpha/4$ for H_4 was propagated equally to the remaining three hypotheses. Since $Z_{23} > b_{23}(\alpha/3) = 2.817$, hypothesis H_2 was rejected at analysis 3. At the fourth analysis, $Z_{34} > b_{34}(\alpha/2) = 2.657$ and hypothesis H_3 was rejected. Given that positive results had

Table 20.6 *Worksheet for a Pocock-type group sequential Holm test with four primary endpoints. Entries Z_{jk} shown in bold are high enough that hypothesis H_j is rejected at analysis k in the group sequential Holm procedure.*

Analysis	*Efficacy boundaries* $j = 1, 2, 3, 4$				*Observed data*			
k	$b_{jk}(\alpha/4)$	$b_{jk}(\alpha/3)$	$b_{jk}(\alpha/2)$	$b_{jk}(\alpha)$	Z_{1k}	Z_{2k}	Z_{3k}	Z_{4k}
1	2.931	2.817	2.675	2.413	1.11	2.22	1.67	1.82
2	2.931	2.817	2.675	2.413	1.28	2.22	1.91	**2.94**
3	2.931	2.817	2.675	2.413	1.34	**2.92**	2.22	**3.37**
4	2.931	2.817	2.675	2.413	0.91	**2.86**	**2.77**	**3.24**
5	2.931	2.817	2.675	2.413	—	—	—	—

Table 20.7 *Worksheet for an O'Brien & Fleming-type group sequential Holm test with four primary endpoints. Entries Z_{jk} shown in bold are high enough that hypothesis H_j is rejected at analysis k in the group sequential Holm procedure.*

Analysis	*Efficacy boundaries* $j = 1, 2, 3, 4$				*Observed data*			
k	$b_{jk}(\alpha/4)$	$b_{jk}(\alpha/3)$	$b_{jk}(\alpha/2)$	$b_{jk}(\alpha)$	Z_{1k}	Z_{2k}	Z_{3k}	Z_{4k}
1	5.694	5.473	5.151	4.562	1.11	2.22	1.67	1.82
2	4.026	3.870	3.642	3.226	1.28	2.22	1.91	2.94
3	3.288	3.160	2.974	2.634	1.34	2.92	2.22	**3.37**
4	2.847	2.737	2.575	2.281	0.91	**2.86**	**2.77**	**3.24**
5	2.547	2.448	2.303	2.040	—	—	—	—

been obtained for three endpoints and the prospects of a positive result on the remaining endpoint were not good, the trial was stopped at analysis 4.

If O'Brien & Fleming-type efficacy boundaries had been specified for the four endpoints, the critical values $b_{jk}(\alpha/4)$, $b_{jk}(\alpha/3)$, $b_{jk}(\alpha/2)$ and $b_{jk}(\alpha)$ would have been those shown in Table 20.7. With the same observed values Z_{jk}, we see that H_2, H_3 and H_4 would still have been rejected, although some of these decisions would have occurred at later analyses.

A peculiar feature of these designs is that a trial may continue when a test statistic Z_{jk} is above the efficacy boundary. It is therefore possible that $Z_{jk'}$ falls below the boundary at a later analysis k' when more data on endpoint j have been obtained. As an example, suppose the observed value of Z_{24} in our example is changed to 2.35, with all other data remaining the same. Following the stated testing procedure with Pocock-type tests, we can still reject H_2 at analysis 4 and the FWER will be controlled at level α. Since H_2 has been rejected, it is also appropriate to test H_3 in a GST with significance level $\alpha/2$ at analysis 4, an important point for the design with Pocock-type boundaries since $z_{34} = 2.77 > b_{34}(\alpha/2)$ but $z_{34} < b_{34}(\alpha/3)$.

Regulatory guidelines require statistical analyses to include all randomized subjects, so additional data on an endpoint that has been rejected at an early analysis should be reported. In an extreme case, where a hypothesis H_j is not rejected in a fixed sample, level α test based on the final set of data, a claim that $\theta_j > 0$ lacks credibility. We therefore recommend adding the additional requirement that $Z_{j\tau} \geq \Phi^{-1}(1-\alpha)$ for H_j to be rejected when a study terminates at analysis τ. However, a higher value of Z_{jk} will have been needed for some $k < \tau$ in order to cross the efficacy boundary of a group sequential test with significance level $\alpha/2$ or less, so the probability of observing $Z_{j\tau} < \Phi^{-1}(1-\alpha)$ is small.

20.13 Group Sequential Designs with Multiple Primary Endpoints Using a Closed Testing Procedure with Simes Tests

In a trial with multiple primary endpoints, suppose it can be assumed that the observed values of each endpoint for a subject are positively associated. Then a CTP can be applied using Simes tests for the intersection hypotheses.

Following Tang & Geller (1999), we require a level α group sequential test for each intersection hypothesis H_J, where $J \subseteq \{1, \ldots, p\}$. We construct these tests from a sequence of P-values for each hypothesis H_j. Suppose a trial has K analyses. We start with a family of efficacy boundaries indexed by the Type I error rate γ, such that for each $\gamma \in (0,1)$, the γ-level boundary $(b_{j1}(\gamma), \ldots, b_{jK}(\gamma))$ for endpoint j satisfies

$$Pr_{\theta_j=0}\{Z_{jk} > b_{jk}(\gamma) \text{ for some } 1 \le k \le K\} = \gamma.$$

The "sequential P-value" for H_j at analysis k, as defined by Liu & Anderson (2008), is the smallest value of γ such that the GST with Type I error rate γ would have rejected H_j by analysis k, that is,

$$P_{jk} = \min\{\gamma;\ Z_{ji} > b_{ji}(\gamma) \text{ for some } 1 \le i \le k\}. \tag{20.8}$$

Now,

$$Pr_{\theta_j=0}\{P_{jK} \le \gamma\} = Pr_{\theta_j=0}\{Z_{ji} > b_{ji}(\gamma) \text{ for some } 1 \le i \le K\} = \gamma,$$

so the final P-value P_{jK} has a Uniform(0,1) distribution when $\theta_j = 0$. Furthermore, the sequence $P_{j1}, \ldots, P_{jK}$ is non-increasing and, in particular, $P_{jk} \ge P_{jK}$ for all $j < K$.

At each analysis k, the CTP is applied using the current P-values $P_{1k}, \ldots, P_{pk}$. The P-value for intersection hypothesis H_J is obtained by applying Simes formula (20.1) to the values P_{jk} for $j \in J$. Thus if J contains m indices,

$$P_{Jk} = \min_{j=1,\ldots,m} \{m P_{(jk)}/j\}, \tag{20.9}$$

where $P_{(1k)} \le \ldots \le P_{(mk)}$ are the ordered P-values for indices $j \in J$. The hypothesis H_j is rejected in the overall procedure if $P_{Jk} \le \alpha$ for all intersection hypotheses H_J where $j \in J$.

If this procedure were applied only at the final analysis K, then, as each P_{jK} has a Uniform(0,1) distribution when $\theta_j = 0$ and these P-values are positively associated, the FWER would be controlled at level α. Since the P-values P_{jk} can only decrease or stay constant as k increases, any H_j which is rejected in the overall procedure at an interim analysis will also be rejected at analysis K, and thus the overall procedure has FWER no greater than α. If the trial is terminated before analysis K, the CTP will be conducted using higher P values, fewer hypotheses may be rejected, and so the FWER will be decreased.

Example

Table 20.8 shows results from a trial with three primary endpoints and five planned analyses. The Z-statistics Z_{jk} are based on the cumulative data for endpoint j at analysis k. The sequential P-values are calculated using the definition (20.8) applied to the family of power family tests with $\Delta = 0$. These "O'Brien & Fleming-type" efficacy boundaries are of the form

$$b_{jk}(\gamma) = C_1'(\gamma)\sqrt{5/k}, \quad k = 1, \ldots, 5.$$

At each analysis, the P-value P_{Jk} for each intersection hypotheses H_J is calculated from P_{1k}, P_{2k} and P_{3k} using the Simes formula (20.9).

Table 20.8 *Worksheet for a group sequential CTP with three primary endpoints using O'Brien & Fleming type boundaries and Simes tests for intersection hypotheses. Entries shown in bold are high enough that hypothesis H_j is rejected at analysis k.*

Analysis		*Z-values*			*Sequential P-values*	
k	Z_{1k}	Z_{2k}	Z_{3k}	P_{1k}	P_{2k}	P_{3k}
1	1.01	0.65	1.82	0.485	0.583	0.295
2	1.55	0.44	3.45	0.226	0.583	0.017
3	2.83	1.37	**3.53**	0.017	0.198	0.004
4	**2.65**	1.02	**3.20**	0.010	0.198	0.002
5	—	—	—	—	—	—

Analysis		*Sequential P-values P_{Jk} for intersection hypotheses*					
k	H_1	H_2	H_3	H_{12}	H_{13}	H_{23}	H_{123}
1	0.485	0.583	0.295	0.583	0.485	0.583	0.583
2	0.226	0.583	0.017	0.452	0.034	0.034	0.051
3	0.017	0.198	**0.004**	0.034	**0.008**	**0.008**	**0.012**
4	**0.010**	0.198	**0.002**	**0.020**	**0.004**	**0.004**	**0.006**
5	—	—	—	—	—	—	—

We see from the table that no hypotheses can be rejected in the overall procedure at analyses 1 and 2. Even though $Z_{32} = 3.45$, the O'Brien & Fleming-type boundary is high at this point and sequential P-values for the intersection hypotheses H_{13}, H_{23} and H_{123} are all above $\alpha = 0.025$. At analysis 3, H_3 is rejected but the trial continues as results on endpoint 1 are promising. Then, H_1 is rejected at analysis 4 and the trial is terminated as there appears to be little chance of a positive outcome for endpoint 2.

Chapter 21

Multi-arm Multi-stage Trials

21.1 Introduction

A trial with three or more treatment arms and two or more analyses is referred to as a multi-arm multi-stage (MAMS) trial. There can be economies of scale when several competing treatments are compared in the same trial. Advantages over conducting multiple two-armed trials include savings in overhead expenses in setting up the trial, and logistical and administrative savings such as having a common protocol and sharing a DSMB. A multi-arm trial can have a single control group where separate trials would each need their own control arm. When there is a common pool of subjects, with similar eligibility requirements and time trends, there can be more direct comparisons between the experimental treatment arms with regard to efficacy, tolerability and adverse events. As an example, Follmann, Proschan & Geller (1994, p. 331) describe an example of a hypertension prevention trial in which subjects were randomized either to a control group or to one of three interventions, weight loss, sodium restriction or stress management, making a total of four treatment arms. The primary endpoint was change in diastolic blood pressure from a baseline measurement taken on entry to the study.

There has been much recent attention to large, long-term clinical trials with a "master protocol". In an "umbrella" trial, several experimental treatments are compared to a common control. A well-known example is the STAMPEDE trial for the treatment of prostate cancer in which five treatments were compared to standard-of-care in a multi-stage trial; see, for example, Sydes et al. (2009). By contrast, in a "basket" trial, a single therapy is evaluated for multiple diseases that share common molecular alterations or risk factors that may predict whether patients will respond to the given therapy. Two examples of basket trials are the study reported by Hyman et al. (2015) that assessed the drug Vemurafenib on multiple nonmelanoma cancers with BRAF V600 mutations, and the SUMMIT basket trial that studied the efficacy of Neratinib in patients with HER2-mutant, metastatic cervical cancer (Oaknin et al., 2020).

A key feature of these trial designs is the ability to drop poorly performing experimental treatments or patient subgroups at interim analyses and to modify the proportions of patients randomized to each treatment arm in response to observed data. We shall discuss adaptive MAMS designs in Chapter 27 after we have introduced general principles for adaptive trial design. In this chapter, we review earlier methods, which include some of the precursors to recent developments.

21.2 Global Tests

21.2.1 A J-treatment Comparison

We illustrate the methods for multi-armed trials in the simple case of a study with $J \geq 3$ treatment arms and equal sample sizes on each arm. Independent observations are available from each arm $j = 1, \dots, J$, and these are assumed to be normally distributed with unknown

DOI: 10.1201/9781584888482-21

mean μ_j and common variance σ^2. Observations are taken in groups of g per treatment arm. For $j = 1, \ldots, J$ and $k = 1, \ldots, K$, let $\bar{X}_{jk}$ denote the sample mean of the $n_k = gk$ responses from treatment arm j available at analysis k. Also, let s_k^2 denote the pooled within-arms estimate of σ^2 available at analysis k.

21.2.2 Group Sequential Chi-squared Tests

In some circumstances it may be appropriate to monitor a single global test statistic. In testing the hypothesis that the J normal means are equal, we can consider group sequential tests that monitor successive chi-squared and F-statistics. For known variance σ^2, the test of H_0: $\mu_1 = \ldots = \mu_J$ is based on the standardized between-arms sum of squares statistics

$$S_k = \frac{n_k}{\sigma^2} \sum_{j=1}^{J} (\bar{X}_{jk} - \bar{X}_{\cdot k})^2, \quad k = 1, \ldots, K. \tag{21.1}$$

Here $n_k = gk$, the cumulative sample size on each arm, and

$$\bar{X}_{\cdot k} = \frac{1}{J} \sum_{j=1}^{J} \bar{X}_{jk}$$

is the overall mean at analysis k.

We consider a group sequential chi-squared test which stops to reject H_0 the first time $S_k \geq c_k$ and accepts H_0 at the final analysis, K, if $S_K < c_K$. Critical values $c_1, \ldots, c_K$ are to be constructed so that the test has a specified Type I error probability α. The marginal distribution of S_k is chi-squared with $p = J - 1$ degrees of freedom and noncentrality parameter

$$n_k \sum_{j=1}^{J} (\mu_j - \bar{\mu})^2 / \sigma^2,$$

where $\bar{\mu} = (\mu_1 + \ldots + \mu_J)/J$. The joint distribution of $S_1, \ldots, S_K$ does not have the canonical distribution (3.1). However, by using the Helmert transformation (Stuart & Ord, 1994, Example 11.3), S_k can be written as a quadratic form in $J - 1$ independent normal variates. This allows us to apply the results of Jennison & Turnbull (1991b) who show that the sequence $S_1, S_2, \ldots$ is Markov and derive its joint distribution, in which S_k is a multiple of a noncentral chi-squared variate conditional on S_{k-1}. This joint distribution is quite tractable and can be used to calculate the critical values that give tests with specified Type I error probabilities.

For boundaries with constant nominal significance levels, analogous to the Pocock test of Section 2.4, we set $c_k = C_P(p, K, \alpha)$, say, for $k = 1, \ldots, K$, where $p = J - 1$. For boundaries analogous to those of the O'Brien &Fleming test of Section 2.5, we set $c_k = (K/k) C_B(p, K, \alpha)$, $k = 1, \ldots, K$. Values of the constants $C_P(p, K, \alpha)$ and $C_B(p, K, \alpha)$ are provided in Table 2 of Jennison & Turnbull (1991b) for $p = 1$ to 5, $K = 1$ to 10, 20 and 50, and $\alpha = 0.01$, 0.05 and 0.10. An abbreviated version of this table with $\alpha = 0.05$ only is shown in Table 21.1.

If the group sizes or the numbers of subjects allocated to each arm within groups are unequal, tests can be based on the statistics

$$S_k = \frac{1}{\sigma^2} \sum_{j=1}^{J} n_{jk} (\bar{X}_{jk} - \bar{X}_{\cdot k})^2, \quad k = 1, \ldots, K, \tag{21.2}$$

Table 21.1 *Constants $C_P(p, K, \alpha)$ and $C_B(p, K, \alpha)$ for, respectively, Pocock-type and O'Brien & Fleming-type repeated χ^2-tests of equality of J normal means. Tests have K analyses, the χ^2 statistic at each analysis has $p = J - 1$ degrees of freedom and the Type I error probability is $\alpha = 0.05$.*

Number of arms, J	*Degrees of freedom, p*	*Number of analyses, K*	$C_P(p, K, \alpha)^*$	$C_B(p, K, \alpha)^\dagger$
2	1	1	3.84	3.84
		2	4.74	3.91
		3	5.24	4.02
		4	5.58	4.10
		5	5.82	4.16
		6	6.02	4.21
		10	6.53	4.35
3	2	1	5.99	5.99
		2	7.08	6.02
		3	7.67	6.12
		4	8.06	6.20
		5	8.35	6.27
		6	8.58	6.33
		10	9.17	6.48
4	3	1	7.81	7.81
		2	9.04	7.83
		3	9.69	7.92
		4	10.13	7.99
		5	10.44	8.06
		6	10.69	8.11
		10	11.34	8.26
5	4	1	9.49	9.49
		2	10.82	9.50
		3	11.53	9.57
		4	12.00	9.64
		5	12.35	9.71
		6	12.62	9.77
		10	13.32	9.93
6	5	1	11.07	11.07
		2	12.49	11.08
		3	13.25	11.14
		4	13.75	11.21
		5	14.12	11.27
		6	14.41	11.33
		10	15.16	11.50

* Critical value c_k for the χ^2-statistic at analysis k is $C_P(p, K, \alpha)$, $k = 1, \ldots, K$

† Critical value c_k for the χ^2-statistic at analysis k is $(K/k)C_B(p, K, \alpha)$, $k = 1, \ldots, K$

where n_{jk} is the cumulative sample size on arm j at analysis k and

$$\bar{X}_{\cdot k} = \sum_{j=1}^{J} n_{jk}\bar{X}_{jk} / \sum_{j=1}^{J} n_{jk}$$

is the overall mean. The marginal distribution of S_k is still χ^2_{J-1} and, in keeping with the significance level approach of Section 4.2, critical values $c_1, \ldots, c_K$ obtained assuming equally sized groups can be applied to the data actually observed to give an approximate test. As long as the imbalances are not too severe, the Type I error probability should remain close to its nominal level.

Proschan, Follmann & Geller (1994) have shown how (21.1) can be generalized to accommodate unequal variances. Again, the exact tables are no longer applicable and the same significance level approach could be used to obtain an approximate test. Alternatively, Proschan, Follmann & Geller suggest using simulation to obtain critical values.

21.2.3 Group Sequential F-tests

In practice, the variance σ^2 will usually be unknown. In that case we replace σ^2 by its current estimate s_k^2 in (21.1) and monitor the F-statistics

$$F_k = \frac{n_k}{(J-1)s_k^2} \sum_{j=1}^{J} (\bar{X}_{jk} - \bar{X}_{\cdot k})^2, \quad k = 1, \ldots, K. \tag{21.3}$$

A group sequential F-test would then stop to reject H_0 the first time $F_k \geq c_k$, or accept H_0 on reaching the final analysis with $F_K < c_K$. Exact critical values $c_1, \ldots, c_K$ for which such a test has a specified Type I error probability α are not available in tables, but recursive formulae that could be used to compute them are given in Jennison & Turnbull (1991b, Sec. 2.3). Alternatively, approximate critical values can be obtained using the significance level approach of Chapter 4. In this prescription, standard tables of the F-distribution are used to find the values c_k satisfying

$$Pr\{F_{J-1,\, J(n_k-1)} \geq c_k\} = \alpha'_k, \quad k = 1, \ldots, K, \tag{21.4}$$

where F_{ν_1,ν_2} denotes an F-distribution on ν_1 and ν_2 degrees of freedom and α'_k is the nominal significance level at analysis k of a group sequential chi-squared test. The values α'_k, $k = 1, \ldots, K$, can be found using constants from, say, our Table 21.1 or Table 2 of Jennison & Turnbull (1991b). This approach could also be used when group sizes are unequal, with the obvious modifications to (21.3) and (21.4). Proschan, Follmann & Geller (1994) generalize the statistic (21.3) to accommodate unequal variances using separate (i.e., non-pooled) variance estimators within each treatment arm. They suggest using simulation to obtain critical values.

21.2.4 A Test for Trend

Another type of global statistic may be called for if treatment arms represent different levels of a single quantitative factor. For example, the treatments may be increasing levels of a particular drug, in which case it will be more appropriate to monitor a trend statistic of the form

$$Z_k = \frac{\sum_{j=1}^{J} (d_j - \bar{d})\bar{X}_{jk}}{\sqrt{\{\sum_{j=1}^{J} (d_j - \bar{d})^2 \sigma^2 / n_k\}}} \tag{21.5}$$

where $d_1 \leq \ldots \leq d_J$ are specified constants. These d_j could be the dose levels in the J treatment arms, or one could simply set $d_j = j$ for $j = 1, \ldots, J$. It is straightforward to construct group sequential tests based on (21.5) since $Z_1, \dots, Z_K$ have the canonical joint distribution (3.1) with

$$\theta = \sum_{j=1}^{J} (d_j - \bar{d})\mu_j \quad \text{and} \quad \mathcal{I}_k^{-1} = \sum_{j=1}^{J} (d_j - \bar{d})^2 \sigma^2 / n_k.$$

Hence, the methods described in Chapters 4 to 11 may be applied directly. In particular, when σ^2 is unknown, group sequential t-tests can be applied with σ^2 replaced by s_k^2, the pooled within-groups variance estimate at analysis k. Use of a test based on (21.5) will be more powerful against alternatives in which the means form a monotonic sequence than the "omnibus" global statistic (21.1).

The same methodology could be applied if it was desired to monitor some other single linear combination of treatment arm means. As an example, it might be desired to monitor the main effect of a particular factor when treatment arms have a factorial structure and all other factors are of secondary importance. A nonparametric form of group sequential trend test has been considered by Lin & Liu (1992).

21.3 Monitoring Pairwise Comparisons

When a study has multiple treatment arms, it is likely there will be interest in monitoring several statistics simultaneously, rather than following a single global statistic as in Section 21.2. Some examples are:

(i) *All pairwise comparisons* Here, all $p = J(J-1)/2$ pairwise differences between treatment means are monitored and the corresponding null hypotheses, H_{ij}: $\mu_i = \mu_j$, $i \neq j$, are tested simultaneously.

(ii) *Comparisons with a control* Here, each of J experimental treatments is compared with a control treatment. All J differences in means between the experimental treatment arms and the control arm are monitored simultaneously to test the $p = J$ null hypotheses H_j: $\mu_j = \mu_0$, $j = 1, \dots, J$, where μ_j is the mean response on treatment arm j and μ_0 is the mean response on the control arm.

(iii) *Factorial or fractional factorial designs* Here, interest is in the main effects and possibly some interactions between factors. Null hypotheses are defined stating that the main effect of each factor is zero. If interactions are also to be monitored, additions are made to the family of null hypotheses. We denote by p the total number of separate linear hypotheses to be tested.

An example of a factorial design is the Linxian study (Blot et al., 1993) which investigated the effectiveness of nutritional and lifestyle interventions in disease prevention. There, a fractional 2^4 design was employed to study the cancer prevention properties of four vitamin and mineral groups in a six year prospective intervention trial. Further examples of trials with a factorial design are given by Natarajan, Turnbull et al. (1996).

In each of the three cases described above, there are p hypotheses to be tested. As in Chapter 19, we define the family-wise error rate to be the probability of rejecting at least one true hypothesis. The family-wise error rate (FWER) is said to be controlled strongly if it is no more than a specified overall level α, whatever the configuration of means μ_j, $j = 1, \dots, J$. This requirement can be satisfied by a Bonferroni approach in which a group sequential test with Type I error probability α/p is run separately for each hypothesis and a hypothesis is rejected if it is rejected at any stage of the group sequential test. In examples (i) and (ii) the individual tests are two-sample tests, whereas in (iii) the appropriate linear contrast of treatment means is monitored in a similar way to the trend test statistic defined in Section 21.2.4. The Tables 19.1 and 19.2 are convenient sources of the constants needed to construct two-sided Pocock or O'Brien & Fleming tests at levels α/p for $\alpha = 0.05$. Additional constants for $\alpha = 0.01$ are provided in Tables 1 and 2 of Follmann, Proschan & Geller (1994).

Example 1

Suppose a trial with $J = 4$ treatment arms is monitored using a $K = 5$ stage O'Brien & Fleming boundary and the family-wise error rate is set at $\alpha = 0.05$. Suppose observations are normally distributed with known variance. If we are interested in all six pairwise comparisons, we should monitor $p = 6$ two-sample statistics and, from Table 19.2, the critical values for two-sided tests based on each of the six standardized statistics are $c_k = C_B(5, 0.05/6)\sqrt{(5/k)} = 2.681\sqrt{(5/k)}$ at analyses $k = 1, \ldots, 5$. On the other hand, if one of the treatments is a control, as in the hypertension trial example cited earlier, we might simply monitor the three pairwise differences with that control. In that case we use $C_B(5, 0.05/3) = 2.448$ from Table 19.2 and the critical values for two-sided tests based on each of the three standardized two-sample statistics are $c_k = 2.448\sqrt{(5/k)}$, $k = 1, \ldots, 5$.

If variances are unknown, the significance level approach of Section 4.2 can be employed. Then, at each analysis k, the critical value c_k is converted to the corresponding percentile of the appropriate t-distribution, i.e., with tail probability $1 - \Phi(c_k)$, where Φ is the cdf of a standard normal distribution.

The Bonferroni procedure is conservative. Assuming equal allocation of subjects across treatment arms, and equal known variances, Follmann, Proschan & Geller (1994, Tables 1 and 2) present critical values for procedures that have family-wise error rates equal to a specified value α. Based on 100,000 simulations, they tabulate constants needed to construct Pocock and O'Brien & Fleming boundaries to monitor p pairwise differences in situations (i) and (ii) above for $\alpha = 0.05$ and 0.01, $J = 2, \ldots, 5$, and $K = 1, \ldots, 6$. The discrepancies between exact and Bonferroni-based critical values are surprisingly modest, with constants C_P and C_B varying only in the second decimal place for all the situations considered. Given that the Bonferroni procedure is more flexible, allows unbalanced treatment allocation, and can accommodate different boundary shapes for different comparisons, it seems that little is lost in using the Bonferroni constants rather than their exact analogs.

If the test statistic for a null hypothesis corresponding to a particular pairwise difference exceeds its boundary at an interim analysis, leading to a declaration that one treatment arm is inferior, we may wish to drop that arm from the study for ethical or economic reasons. Whether or not this happens, the procedures of this section are unaffected and the overall Type I FWER is still protected. If we are in case (ii), and the difference between an experimental treatment and the control exceeds the boundary value in a positive direction, this could lead to a declaration that the control arm is inferior to that treatment. It might then be deemed unethical to randomize further subjects to the control arm and the trial would be terminated without a definitive evaluation of the other treatments. However, if the trial did continue with the control arm in place and further conclusions were reached using the specified boundaries, the Type I FWER would be protected at level α.

In the fixed sample case, Fisher's Least Significant Difference (LSD) approach provides a test for all pairwise differences. A level α chi-squared test or F-test is conducted to test the global null hypothesis that all means are equal. Then, if this is rejected, level α tests are applied to all pairwise differences of means. The Newman-Keuls procedure is similar but the global test is based on the range statistic; see Hochberg & Tamhane (1987) for further details. These methods have been generalized to the group sequential setting by Proschan, Follmann & Geller (1994). In the analog of the LSD procedure, the global statistic (21.1) or (21.3) is used initially to monitor the study at the specified error rate α. When, and only when, this statistic exceeds its boundary, pairwise differences are examined using boundary values for individual, level α group sequential tests. In the group sequential version of the

Newman-Keuls procedure, a sequential version of the range test is applied. At each analysis, if the difference between the largest and smallest sample means is found to be significant, the exercise is repeated comparing the largest mean to the second smallest and the second largest mean to the smallest, using critical values for the case of one fewer treatment arms. This process is repeated until no comparison is significant, the inferior treatment arms are dropped and the study continues to the next analysis.

Just as in the fixed sample case, the group sequential versions of the LSD and Newman-Keuls procedures only *weakly* protect the family-wise Type I error rate α when comparisons are made between four or more treatments. That is, the probability of at least one true hypothesis being rejected is no more than the specified α under the global null hypothesis that all treatment means are equal. For other configurations of means $\mu_j, \ldots, \mu_J$, this probability may exceed α. For example, this can happen under either procedure if there are four treatment arms with means $\mu_1 = \mu_2 << \mu_3 = \mu_4$. Then, one can expect the global null hypothesis to be rejected early on and there will be a probability of approximately 2α that at least one of the two true hypotheses $\mu_1 = \mu_2$ or $\mu_3 = \mu_4$ will be rejected. In the fixed sample case, Hochberg & Tamhane (1987, p. 69) show how the family-wise error can be strongly controlled at level α by reducing the nominal levels at which each component comparison is made; in principle, the same modification could be made to the group sequential versions. An alternative approach is to apply the closed testing procedure of Marcus, Peritz & Gabriel (1976), as described in Section 20.11, in a group sequential design. We shall describe such group sequential methods for the case of multiple comparisons with a control in the next section and return to this topic again in Chapters 26 and 27.

21.4 Multi-arm Multi-stage Procedures for Comparisons with a Control

21.4.1 A Single Stage Dunnett Test

We continue with a more detailed discussion of case (ii) of Section 21.3, where J experimental treatment arms are compared to a control in a multi-armed trial. Dunnett (1955) considered a single stage trial with normally distributed outcomes with means $\mu_1, \ldots, \mu_J$ for the experimental treatments and μ_0 for the control, and a common variance σ^2. Denoting the treatment mean differences by $\theta_j = \mu_j - \mu_0$, $j = 1, \ldots, J$, the global null hypothesis is

$$H_0\colon\ \theta_j \leq 0 \text{ for all } j = 1, \ldots, J. \tag{21.6}$$

Suppose there are n_0 observations on the control arm and a common sample size n on each of the J experimental arms. Standardized test statistics are

$$Z_j = \frac{(\bar{X}_j - \bar{X}_0)}{\sqrt{\{\sigma^2(1/n_0 + 1/n)\}}}, \quad j = 1, \ldots, J, \tag{21.7}$$

where $\bar{X}_0$ denotes the sample mean on the control arm and $\bar{X}_j$ is the sample mean on treatment arm j. The distribution of $(Z_1, \ldots, Z_J)$ is multivariate normal with unit variances and common correlation $\rho = n/(n_0 + n)$. The Dunnett test with Type I error probability α rejects the global null hypothesis H_0 if

$$\max\{Z_1, \ldots, Z_J\} \geq C_D(J, \rho, \alpha).$$

Calculation of critical values $C_D(J, \rho, \alpha)$ is facilitated by noting that, after conditioning on $\bar{X}_0$, the estimated treatment effects $\bar{X}_j - \bar{X}_0$, $j = 1, \ldots, J$, are independent and computation reduces to evaluating a one-dimensional integral numerically. Values of $C_D(J, \rho, \alpha)$ are available from tables such as those in Gupta, Nagel & Panchapakesan (1973) and computer programs, for example, the R package `DunnettTest`. In Table 21.2, we present

Table 21.2 *Critical values $C_D(J, \rho, \alpha)$ for Dunnett tests comparing J treatments with a control. Sample sizes on each treatment and the control are equal, so the correlation between Z-statistics is $\rho = 0.5$. Critical values are for one-sided tests with Type I error rate $\alpha = 0.025$.*

J	$C_D(J, 0.5, 0.025)$
1	1.960
2	2.212
3	2.349
4	2.442
5	2.511
6	2.567

values of $C_D(J, 0.5, 0.025)$ for $J = 1, \ldots, 6$, which are appropriate for the case $n_0 = n$, so that the $\mathrm{Corr}(Z_{j_1}, Z_{j_2}) = 1/2$ for all pairs (j_1, j_2).

Dunnnett's test can be used to create a closed testing procedure for the set of hypotheses

$$H_j\colon \theta_j \leq 0, \quad j = 1, \ldots, J,$$

with FWER α. Suppose samples sizes on all experimental treatment arms are equal and $\mathrm{Corr}(Z_{j_1}, Z_{j_2}) = \rho$ for all pairs (j_1, j_2). Then, for a subset $I \subseteq \{1, \ldots, J\}$ with J' elements, the intersection hypothesis is

$$H_I = \cap_{j \in I} H_j,$$

and, applying a Dunnett test, we reject H_I if

$$\max_{j \in \mathcal{I}} \{Z_j\} \geq C_D(J', \rho, \alpha).$$

In the global procedure, an individual hypothesis H_j is rejected overall if all intersection hypotheses H_I with $j \in I$ are rejected by their level α tests.

Because of the special nature of the Dunnett test statistic and the fact that $C_D(J', \rho, \alpha) < C_D(J, \rho, \alpha)$ for $J' < J$, the outcome of the closed testing procedure can be found by considering tests of at most J intersection hypotheses, rather than all $2^J - 1$ tests. This "step-down" version of the Dunnett procedure is described by Hochberg & Tamhane (1987, Example 4.2). First, the Z-statistics are ordered as $Z_{(1)} \leq \ldots \leq Z_{(J)}$ and we let $H_{(1)}, \ldots, H_{(J)}$ denote the corresponding individual hypotheses. We start by considering $H_{(J)}$: if $Z_{(J)} < C_D(J, \rho, \alpha)$, the largest intersection hypothesis $H_{\{1,\ldots,J\}}$ cannot be rejected and the procedure stops without rejecting any hypotheses overall; if $Z_{(J)} \geq C_D(J, \rho, \alpha)$ we can deduce that all intersection hypotheses involving $H_{(J)}$ are rejected at level α, and we reject $H_{(J)}$ overall and continue to investigate the remaining hypotheses. At each stage of the process, if there are J' individual hypotheses remaining, we repeat the above comparisons with J' in place of J so the new $H_{(J')}$ is rejected overall if $Z_{(J')} \geq C_D(J', \rho, \alpha)$ and otherwise the procedure stops with no more treatments declared significant.

Since $C_D(J', \rho, \alpha) < C_D(J, \rho, \alpha)$ for $J' < J$, a closed testing procedure in which tests of all intersection hypotheses use the common critical value $C_D(J, \rho, \alpha)$ in place of $C_D(J', \rho, \alpha)$ will be conservative. Thus, such a procedure will preserve the FWER α. This is equivalent to using the critical value $C_D(J, \rho, \alpha)$ in place of $C_D(J', \rho, \alpha)$ at each stage of the step-down Dunnett procedure. The rule for this conservative application of the Dunnett test is simply:

Reject all hypotheses H_j, $1 \leq j \leq J$, for which $Z_j \geq C_D(J, \rho, \alpha)$.

Some authors have used this conservative approach.

Table 21.3 *P-values for tests of intersection hypotheses H_I obtained by applying Dunnett and Simes tests to the data of Example 2*

H_I	*Dunnett**	*Simes†*
H_1	0.0035	0.0035
H_2	0.0094	0.0094
H_3	0.0188	0.0188
H_4	0.7088	0.7088
H_{12}	0.0066	0.0069
H_{13}	0.0066	0.0069
H_{14}	0.0066	0.0069
H_{23}	0.0176	0.0188
H_{24}	0.0176	0.0188
H_{34}	0.0344	0.0375
H_{123}	0.0096	0.0104
H_{124}	0.0096	0.0104
H_{134}	0.0096	0.0104
H_{234}	0.0249	0.0281
H_{1234}	0.0123	0.0139

* Entries for Dunnett procedure are $Pr\{\max(Z_i, i \in I) \geq \max(z_i, i \in I)\}$

† P-values for Simes tests are as defined in Section 20.4

Example 2

Suppose that $J = 4$ treatments are compared to a common control with equal sample sizes, $n_0 = n_1 = \ldots = n_4$, and the observed values of the Z-statistics are

$$z_1 = 2.70, \quad z_2 = 2.35, \quad z_3 = 2.08, \quad z_4 = -0.55.$$

We apply a closed testing procedure based on Dunnett tests with FWER $\alpha = 0.025$. Following the step-down method for implementing this procedure, we note that

$$\max\{z_1, \ldots, z_4\} \; = \; z_1 \; = \; 2.70 \; > \; C_D(4, 0.5, 0.025) \; = \; 2.442$$

and we conclude that the individual hypothesis H_1 can be rejected overall. We now turn to H_2 since z_2 is the second highest Z-value and we see that $z_2 = 2.35$ exceeds $C_D(3, 0.5, 0.025) = 2.349$, so H_2 can also be rejected overall. However, the next highest Z-value, $z_3 = 2.08$, does not exceed $C_D(2, 0.5, 0.025) = 2.212$ so H_3 is not rejected and this implies that H_4 is not rejected either. We note that z_1 is the only Z-value that exceeds $C_D(4, 0.5, 0.025) = 2.442$, so the conservative Dunnett procedure does not reject H_2, illustrating the loss of power in the conservative test.

For completeness, we have computed the P-values when Dunnett's test is applied to each of the $2^4 - 1$ intersection hypotheses and these are listed in the second column of Table 21.3. For simplicity we have written H_1 rather than $H_{\{1\}}$, H_{12} rather than $H_{\{1,2\}}$, etc., in this table. Defining

$$P_D(k, \rho, z) \; = \; Pr\{\max_{j=1,\ldots,k} Z_j \geq z\} \tag{21.8}$$

when $Z_1, \ldots, Z_k$ are multivariate normal with each $Z_j \sim N(0, 1)$ and $\text{Corr}(Z_{j_1}, Z_{j_2}) = \rho$ for $j_1 \neq j_2$, the Dunnett P-value for H_I in Table 21.3 is $P_D(k, 0.5, z)$, where k denotes the

Table 21.4 *Adjusted P-values for Example 1 using Dunnett and Simes tests for intersection hypotheses*

Hypothesis	*Unadjusted*	*Adjusted P-values*	
	P-values	*Dunnett*	*Simes*
H_j	P_j	$\tilde{P}_j$	$\tilde{P}_j$
H_1	0.0035	0.0123	0.0139
H_2	0.0094	0.0249	0.0281
H_3	0.0188	0.0344	0.0375
H_4	0.7088	0.7088	0.7088

number of hypotheses involved in H_I and z is the maximum observed Z-value for these hypotheses. Since the statistics Z_j are based on differences from a common control, they are positively correlated and Simes tests of intersection hypothesis are valid with some conservatism; see Simes (1986) and Sarkar & Chang (1997). The third column of Table 21.3 shows the P-values from these Simes tests.

Applying the closed testing procedure with FWER α, we reject an individual hypothesis H_j overall if the P-values for all intersection hypotheses involving H_j are less than or equal to α. When Dunnett tests are used, we see that all intersection hypotheses involving H_1 or H_2 are less than 0.025, so both H_1 and H_2 are rejected overall when the FWER is $\alpha = 0.025$, in agreement with the conclusion from the step-down approach. In the case of H_1, the largest value P-value is 0.0123, which comes from the test of $H_{\{1234\}}$, so according to the definition in Section 20.7, the adjusted P-value for H_1 is $\tilde{P}_1 = 0.0123$, a higher value than the simple $P_1 = 1 - \Phi(2.7) = 0.0035$ due to the adjustment for multiple comparisons. Similarly, the adjusted P-value for H_2 is $\tilde{P}_2 = 0.0249$ from the test of $H_{\{234\}}$. These adjusted P-values are listed in Table 21.4 along with those for hypotheses H_3 and H_4.

Since the Z_j, $j = 1, \ldots, J$, arise from comparisons with a common control, they are positively correlated and we may use Simes method to obtain a P-value for each intersection hypothesis test. Calculations using Simes tests are shown in the final columns of Tables 21.3 and 21.4. We see that in this example, the Simes method fails to reject H_2 at level $\alpha = 0.025$ and the adjusted P-values are higher. This situation can be reversed in other data sets and Simes method does have the advantage of requiring fewer assumptions in order to be valid.

Unequal sample sizes and unknown variance

In our example, we have assumed that the sample size n_0 on the control arm is the same as the common sample size n for each of the J experimental treatments, but this is not essential. The closed testing procedure using Dunnett's test remains the same for general values of $\rho = n/(n_0 + n)$. Indeed, Dunnett (1955) showed that, for a fixed total sample size, the greatest power is achieved by allocating subjects in the ratios $\sqrt{J} : 1 : \ldots : 1$ between the control and experimental arms. Dunnett's original paper also dealt with the case of normal responses with unknown variances so that the set of test statistics followed a multivariate t-distribution. The closed testing procedure we have presented generalizes to this case; see Hochberg & Tamhane (1987).

21.4.2 A Group Sequential Dunnett Test with Treatment Selection

Suppose now that the study described in the previous section is conducted group sequentially with K analyses. At analysis k, let n_{0k} be the cumulative sample size on the control arm and denote the cumulative sample sizes on the experimental treatment arms by n_{jk}, $j = 1, \ldots, J$. Then, the standardized test statistics at analysis k are

$$Z_{jk} = \frac{\bar{X}_{jk} - \bar{X}_{0k}}{\sqrt{\{\sigma^2(1/n_{0k} + 1/n_{jk})\}}}, \quad j = 1, \ldots, J,$$

and the associated information levels are

$$\mathcal{I}_{jk} = \left(\frac{\sigma^2}{n_{0k}} + \frac{\sigma^2}{n_{jk}}\right)^{-1}, \quad j = 1, \ldots, J.$$

In the multivariate version of the canonical joint distribution (3.1), we have

(i) $(Z_{11}, \ldots, Z_{1K}, \ldots, Z_{J1}, \ldots, Z_{JK})$ is multivariate normal,

(ii) $E(Z_{jk}) = \theta\sqrt{\mathcal{I}_{jk}}$, $j = 1, \ldots, J$, $k = 1, \ldots, K$,

(iii) $\mathrm{Var}(Z_{jk}) = 1$, $j = 1, \ldots, J$, $k = 1, \ldots, K$,

(iv) $\mathrm{Cov}(Z_{jk_1}, Z_{jk_2}) = \sqrt{(\mathcal{I}_{jk_1}/\mathcal{I}_{jk_2})}$, $j = 1, \ldots, J$, $1 \leq k_1 < k_2 \leq K$,

(v) $\mathrm{Cov}(Z_{j_1k_1}, Z_{j_2k_2}) = \sigma^2\sqrt{(\mathcal{I}_{j_1k_1}\mathcal{I}_{j_2k_2})}/n_{0k_2}$, $j_1 = 1, \ldots, J$, $j_2 = 1, \ldots, J$, $j_1 \neq j_2$, $1 \leq k_1 < k_2 \leq K$.

This distribution enables computation of the operating characteristics of any proposed procedure.

Stallard & Todd (2003) proposed a "keep the winner" group sequential procedure. Boundaries (a_k, b_k), $k = 1, \ldots, K$, are specified with $a_K = b_K$. At the first analysis, if the highest Z_{j1} falls below a_1 the trial is abandoned as futile and the global null hypothesis H_0: $\theta_1 \leq 0, \ldots, \theta_J \leq 0$ is accepted. Otherwise, the treatment with the highest Z_{j1}, $j = S$ say, is selected for further comparison with the control. The trial continues with new subjects randomized between treatment S and the control. At each analysis $k = 2, \ldots, K$, the trial stops with rejection of H_S: $\theta_S \leq 0$ if $Z_{Sk} \geq b_k$ and without rejection of any hypothesis if $Z_{Sk} \leq a_k$. The sequence of boundary values (a_k, b_k) is calculated using an error spending approach, with allowance for the selection bias in focusing on the maximum Z value, Z_{S1}, at the first analysis. Stallard & Friede (2008) show that this procedure controls the FWER at the specified level α and this property also holds if a different treatment is chosen at the first analysis, for reasons of safety perhaps. Wason, Stallard, Bowden & Jennison (2017) propose a design in which treatments are dropped during the trial with the numbers of treatments remaining after each analysis following a fixed pattern, and the one treatment remaining at the final analysis is declared superior to the control if the corresponding Z-statistic is sufficiently high. They prove that this procedure controls the FWER at the desired level.

Stallard & Friede (2008) extended the method of Stallard & Todd (2003) to allow more than one experimental treatment to continue past the first analysis, with the number of treatments remaining in the trial at any stage depending on the data observed thus far. At each analysis k, if Z_{jk} is low a treatment j may be dropped for futility or, if Z_{jk} is high, H_j may be rejected and sampling on treatment j discontinued since this treatment has already been shown to be effective. The authors prove their procedure strongly controls the FWER at level α if the number of treatments that proceed after each analysis is fixed in advance and they report a simulation study that suggests this property is preserved under a wider range of strategies for dropping or retaining treatments.

Magirr, Jaki & Whitehead (2012) develop this methodology further in proposing a multi-stage generalization of the Dunnett test to compare multiple treatments against a control. Each treatment arm j is continued as long as Z_{jk} falls within the upper and lower boundaries (a_k, b_k), $k = 1, \ldots, K$. The number of arms to be dropped or retained at each stage is data dependent and need not be specified in advance. The authors show their procedure guarantees strong control of the FWER α and they provide power and sample size calculations. Computations exploit the fact that, conditional on the control treatment means, sequences of statistics Z_{jk}, $k = 1 \ldots, K$, for different values of j are independent. For a trial with up to $K = 2$ or 3 stages, numerical integration is feasible, while Monte Carlo integration may be used for larger problems. Ghosh et al. (2017) tackle the same calculations using quasi-Monte Carlo simulation which allows them to obtain accurate results with only a moderate amount of computation. In this approach, the computation time is linear in both the number of stages and the number of treatment comparisons, and this is particularly useful as the number of stages increases to four or five, say. Wason, Magirr, Law & Jaki (2016) give details of the power calculations and an assessment of different design options in examples based on the TAILoR (TelmisArtan and InsuLin Resistance in HIV) trial, the design and outcome of which were ultimately reported by Pushpakom et al. (2020).

Wason & Jaki (2012) discuss the optimal design of multi-arm multi-stage trials. For a trial that compares J treatments against the control, with up to K analyses, they consider designs of the form proposed by Magirr, Jaki & Whitehead (2012) and use simulation and a stochastic search algorithm to find the designs that achieve a specified power with the minimum expected sample size under certain vectors of treatment effects. Their results demonstrate the significant gains in efficiency that multi-arm trials can provide compared to separate randomized trials of each treatment. They recommend designs that apply the triangular stopping boundary of Whitehead & Stratton (1983) and designs with the δ-minimax property, as defined by Wason, Mander & Thompson (2012), that minimize the maximum expected sample size over all possible vectors of treatment effects. We noted in Section 5.7 that the triangular test has a high maximum sample size, and the same is true of Wason & Jaki's δ-minimax designs. Thus, a design optimized with a particular maximum sample size might be preferred in order to balance the reduction in expected sample size against an increase in maximum sample size. Wason et al. (2016) note that power family boundaries of the form proposed by Wang & Tsiatis (1987) and Pampallona & Tsiatis (1994) for a two-armed trial, which we described in Chapters 2 and 5, offer close approximations to optimized multi-arm designs. Versions of these designs with lower values of the shape parameter Δ can give efficient procedures with a smaller maximum sample size. In view of the findings reported in Chapter 12, we recommend consideration of the ρ-family of error spending designs both for their efficiency and for the ability to adapt to unpredictable sample sizes on each treatment, noting that the error spending boundaries can be calculated using methods analogous to those of Magirr, Jaki & Whitehead (2012) or Ghosh et al. (2017).

Wason & Jaki (2012) also consider designs in which a higher proportion of patients is allocated to the control arm. They report that this can yield modest reductions in the total expected sample size but the optimal allocation ratios are less extreme than Dunnett's optimal ratio of $\sqrt{J} : 1 : \ldots : 1$. This is to be expected when the ratio is fixed throughout the trial since fewer treatments are compared with the control in the later stages.

A drawback of these designs is that details of the entire procedure, including sample sizes, stopping boundaries, and treatment selection rules, have to be specified in advance. Posch et al. (2005), Bretz et al. (2009), Wassmer (2011) and Magirr, Stallard & Jaki (2014) describe more flexible "adaptive" procedures in which such design features can be changed at interim analyses, based on accrued data or external information and the FWER α is still controlled strongly. Such adaptive MAMS procedures rely on the general theory of adaptive designs and we shall return to this topic in Chapter 27 when we can refer to the material presented in Chapters 22 to 25. We shall also discuss the topic of inference after an adaptive design in those chapters, for the case of seamless Phase II/III trials with treatment selection in Section 26.6 and for more general multi-arm multi-stage trials in Section 27.6.

21.5 Bibliography and Notes

Proschan, Follmann & Geller (1994) adapt methods for designing multi-armed trials to the case of comparing survival time responses. The more technical details are given in Appendix A.2 of Follmann, Proschan & Geller (1994).

Siegmund (1993) and Betensky (1996) propose a fully sequential design for comparing three treatments using a hybrid method, similar in spirit to the LSD procedure described in Section 21.3. Their procedure consists of two phases. In the first, a global statistic is monitored sequentially to test for equality of all three means. If this hypothesis is accepted, the trial terminates. Otherwise, the least promising treatment arm is dropped and, in the second phase, a sequential test is performed with the remaining two treatments. Betensky (1997c) adapts the same procedure to accommodate survival time endpoints.

Hughes (1993) proposes a group sequential procedure based on two-sample statistics for pairwise comparisons with the aim of dividing the treatment arms into two subsets, one labeled inferior, the other superior. At interim analyses, treatments determined to be in the inferior group are dropped. If, at termination, only one arm is contained in the superior set, it is identified as the best one; otherwise all treatments in that set are declared superior. The procedure is in the spirit of subset selection procedures, often used as screening methods to pick the most promising treatments for future study from a larger group of candidate treatments; see, for example, Bechhofer, Santner & Goldsman (1995, Ch. 3). In fact, in the ranking and selection literature, there is a considerable body of work on sequential and multi-stage procedures for identifying the best of several competing treatments. An indifference zone formulation is often adopted whereby, typically, designs must satisfy the requirement that the best treatment is selected with a pre-specified, high probability whenever its mean exceeds that of the next best by a stated amount. Bechhofer, Santner & Goldsman (1995, Ch. 2) survey the various procedures that have been proposed.

More recent research on multi-armed trials concerns adaptive designs and we shall survey that work in Chapters 26 and 27.

Chapter 22

From Group Sequential to Adaptive Designs: Flexible Sample Size Re-assessment

22.1 Adaptive Sample Size Modification Based on Accruing Outcome Data

The adjective "adaptive" has been used in several ways in connection with the statistical design of clinical trials. Here, we use the term specifically to refer to designs that permit changes in sample size during the course of a trial. These changes may be planned or unplanned, based on interim unblinded information on efficacy endpoints. The idea of changing sample size adaptively dates back to the paper "Multistage testing with adaptive designs" by Peter Bauer in the journal *Biometrie und Informatik in Medizin und Biologie* (Bauer, 1989). However, widespread interest in the ideas did not really begin until some ten years later. Since then, these methods have attracted considerable attention.

We begin this chapter with a motivating example described by Cui, Hung & Wang (1999), henceforth denoted "CHW". They present a case study of a Phase III group sequential clinical trial evaluating the effect of a new drug in preventing myocardial infarction (MI) in patients undergoing coronary artery bypass graft surgery. The originally planned maximum sample size of 600 subjects per arm was chosen to provide power 0.95 to show a 50% reduction in the incidence rate of MIs. However, at an interim analysis, the observed reduction was only about 25% and the conditional power (see Section 11.2) was deemed to be very low if the current trend continued. The sponsor believed a reduction of 25% would still be clinically meaningful and commercially viable. Thus a protocol modification was submitted to the FDA proposing an increase in sample size to increase the probability of a positive result if the reduction in incidence of MIs was only 25%. The FDA denied this request as it was feared it could substantially inflate the Type I error rate. The sponsor proceeded with the trial as planned and the eventual result was non-significant, as was anticipated from the interim results. CHW state

> "A natural solution to this dilemma is implementation of a valid inferential procedure that allows flexibility for adjusting sample size based on the updated estimate of treatment effect during the course of the trial"

and, in their paper, they go on to propose such a method.

Note that in this example, it was desired to change sample size and future information levels based on current or past treatment effect estimates or test statistics Z_k. This is different from the "information monitoring" methods described in Sections 8.6 and 18.4 where future group sizes are adjusted to take account of updated estimates of nuisance parameters that determine the information levels $\mathcal{I}_k$, but the target information level remains unchanged.

We have already seen in Section 8.5 that, if the number and timing of future analyses, and hence the information $\{\mathcal{I}_{k+1}, \mathcal{I}_{k+2}, \ldots\}$, are allowed to depend on observed outcomes $\{Z_1, \ldots, Z_k\}$, then the Type I error can be inflated above the nominal level α. Cui, Hung & Wang (1999, Table A1) illustrate this Type I error rate inflation in an error spending design

 DOI: 10.1201/9781584888482-22

with 5 analyses. In their example, data-dependent changes in the maximum information level $\mathcal{I}_5$ lead to Type I error rates as high as 0.038 compared to the nominal level, $\alpha = 0.025$. Proschan & Hunsberger (1995, p. 1316), have shown that in a two-stage design example, the Type I error rate can be more than doubled under certain rules for choosing the second stage sample size. Wang et al. (2013) show that inflation of the Type I error rate can also occur when sample size is modified based on a biomarker outcome occurring earlier than the primary clinical endpoint.

A key issue here is flexibility. We saw in Section 12.4 that a group sequential trial may be designed with data-dependent group sizes so that Type I and Type II error probabilities are attained when this design is adhered to strictly. Suppose, however, that an unexpected event occurs during a trial and investigators wish to make some ad hoc change that will affect future information increments, as in the example described by CHW. Whether these changes are motivated by a revised view of the minimal clinically significant treatment effect size, a desire to increase power, or new information from external sources, it is crucial that the Type I error probability is controlled.

We shall see that the methods that meet these requirements have applications beyond sample size re-assessment. In later chapters, we shall consider mid-trial adaptations that involve selecting a dose level, choosing between treatments, or focusing on a subgroup of the patient population. We shall make a distinction between rigid adaptive designs, where all possible design changes in response to accruing data are specified in advance, and flexible adaptive designs that can accommodate unplanned or ad hoc changes. The FDA (2019a) Guidance on Adaptive Designs for Clinical Trials of Drugs and Biologics indicates a strong preference that adaptively designed studies be prospectively planned, stating

> "Prospective planning should include prespecification of the anticipated number and timing of interim analyses, the type of adaptation, the statistical inferential methods to be used, and the specific algorithm governing the adaptation decision. Complete prespecification is important for a variety of reasons ..." (Sec. III.C, p. 8).

and

> "prospective planning should include prespecification of not only the statistical hypothesis testing method that will be used, but also the rule governing the sample size modification." (Sec. V.B, p. 14).

The Guidance does, however, acknowledge that "monitoring committee recommendations might occasionally deviate from the anticipated algorithm based on the totality of the data" but its goes on to say

> "If this type of flexibility is desired, the prespecified plan should acknowledge the possibility of deviations from the anticipated algorithm, outline factors that may lead to such deviations, and propose testing and estimation methods that do not rely on strict adherence to the algorithm." (Sec. III.C, p. 9).

22.2 Combination Test Statistics

If future information levels depend on previous outcome data, the conditional distribution of the Wald statistics $Z_1, \ldots, Z_K$ given information levels $\mathcal{I}_1, \ldots, \mathcal{I}_K$ is no longer the canonical joint distribution (3.1), even when $\theta = 0$, and the Type I error probability of a group sequential test based on the canonical distribution need not be the specified α. To illustrate this point, suppose $\mathcal{I}_1 = 1$ is fixed and we specify $\mathcal{I}_2 = 2$ if $Z_1 \geq 1$ and $\mathcal{I}_2 = 3$ if $Z_1 < 1$. Then, conditioning on the observed values of Z_1 and $\mathcal{I}_2$, we have that when $z_1 \geq 1$,

$$Z_2 \,|\, Z_1 = z_1, \mathcal{I}_2 = 2 \;\sim\; N(z_1/\sqrt{2},\, 1/2),$$

but when $z_1 < 1$,

$$Z_2 \,|\, Z_1 = z_1, \mathcal{I}_2 = 3 \;\sim\; N(z_1/\sqrt{3},\, 2/3).$$

Integrating over the distribution of Z_1 we find that when $\theta = 0$, Z_2 does not have a marginal $N(0,1)$ distribution, $E(Z_2) = 0.0314$ and $Pr\{Z_2 > \Phi^{-1}(0.975)\} = 0.0287$. So, the decision rule "Reject H_0: $\theta \leq 0$ if $Z_2 > 1.96$", that would give Type I error probability $\alpha = 0.025$ if $\{Z_1, Z_2\}$ followed the canonical joint distribution, has an inflated Type I error rate.

When future information levels can depend on previous outcome data in an unplanned, arbitrary way, it is not possible to compute frequentist properties, such as Type I or Type II error probabilities, of a procedure based on the usual statistics $\{Z_1, \ldots, Z_K\}$. We shall describe group sequential tests based on a new sequence of test statistics $\{\tilde{Z}_1, \ldots, \tilde{Z}_K\}$ which do have the canonical joint distribution (3.1) under $\theta = 0$ for appropriately defined information levels, even when group sizes and analysis times are modified in an unplanned manner in response to previous outcome data. It follows that we may use stopping boundaries $(a_1, b_1), \ldots, (a_K, b_K)$ from a standard group sequential design, and the Type I error will be protected at level α. We refer to the new Z-statistics $\{\tilde{Z}_1, \ldots, \tilde{Z}_K\}$ as *combination test statistics.*

Suppose a one-sided, group sequential trial is designed to test H_0: $\theta \leq 0$ against $\theta > 0$. At the planning stage it is assumed there will be K analyses with information levels $\mathcal{I}_1, \ldots, \mathcal{I}_K$ and boundaries $(a_1, b_1), \ldots, (a_K, b_K)$ in a group sequential test of the form (5.2) are computed for which the Type I error probability is α and power under $\theta = \delta$ is $1 - \beta$. We saw in Section 3.1 that if $Z_1, \ldots, Z_K$ follow the canonical joint distribution (3.1), then the score statistics $S_k = Z_k \sqrt{\mathcal{I}_k}$, $k = 1, \ldots, K$, are multivariate normal with independent increments

$$S_1 \sim N(\theta \mathcal{I}_1, \mathcal{I}_1) \quad \text{and} \quad S_k - S_{k-1} \sim N(\theta(\mathcal{I}_k - \mathcal{I}_{k-1}), \mathcal{I}_k - \mathcal{I}_{k-1}), \quad k = 2, \ldots, K.$$

Let

$$Z^{(1)} = \frac{S_1}{\sqrt{\mathcal{I}_1}} = Z_1 \quad \text{and} \quad Z^{(k)} = \frac{S_k - S_{k-1}}{\sqrt{(\mathcal{I}_k - \mathcal{I}_{k-1})}} = \frac{Z_k \sqrt{\mathcal{I}_k} - Z_{k-1}\sqrt{\mathcal{I}_{k-1}}}{\sqrt{(\mathcal{I}_k - \mathcal{I}_{k-1})}}, \quad k = 2, \ldots, K. \tag{22.1}$$

Then, under $\theta = 0$, the $Z^{(k)}$ are independent $N(0,1)$ random variables. Defining

$$w_1 = \sqrt{\frac{\mathcal{I}_1}{\mathcal{I}_K}} \quad \text{and} \quad w_k = \sqrt{\frac{\mathcal{I}_k - \mathcal{I}_{k-1}}{\mathcal{I}_K}}, \quad k = 2, \ldots, K, \tag{22.2}$$

the sequence of test statistics Z_k can be expressed in terms of the $Z^{(k)}$ as

$$Z_k = \frac{w_1 Z^{(1)} + \ldots + w_k Z^{(k)}}{\sqrt{(w_1^2 + \ldots + w_k^2)}}, \quad k = 1, \ldots, K. \tag{22.3}$$

The statistic $Z^{(k)}$ can be viewed as the Z-value arising from the data observed in stage k of the trial. This is seen very easily in the case of a trial with a single treatment arm and normally distributed responses, which we describe below. More generally, the new data in stage k lead to a change in the score statistic and this is translated into a Z-value according to the formulae in Equation (22.1).

Now suppose the trial is conducted and it is anticipated that a design change may occur during the trial. The observed information levels are unlikely to be exactly equal to the values used when planning the trial and they will be substantially different after a change in group sizes or analysis times. Let $\mathcal{I}_1^*, \ldots, \mathcal{I}_K^*$ denote the observed information levels, then we can calculate stage-wise Z-values $Z^{(k)}$ by substituting the observed Z-values and information levels $\mathcal{I}_k^*$ into (22.1). We may then use these $Z^{(k)}$ to define cumulative statistics

$$\tilde{Z}_k = \frac{w_1 Z^{(1)} + \ldots + w_k Z^{(k)}}{\sqrt{(w_1^2 + \ldots + w_k^2)}}, \quad k = 1, \ldots, K. \tag{22.4}$$

The reason for expressing the Z_k in the form (22.3) should now become clear. The variables $Z^{(k)}$ in (22.3) represent stage-wise Z-values when the information levels are those assumed in planning the trial while the variables $Z^{(k)}$ in (22.4) are the stage-wise Z-values based on observed data. In both cases, the $Z^{(k)}$ are independent $N(0,1)$ variables if $\theta = 0$. Hence, when $\theta = 0$, the joint distribution of $(\tilde{Z}_1, \ldots, \tilde{Z}_K)$ is the same as that of $(Z_1, \ldots, Z_K)$. Thus, if a group sequential test is created by applying the boundaries $(a_1, b_1), \ldots, (a_K, b_K)$ to the sequence $(\tilde{Z}_1, \ldots, \tilde{Z}_K)$, the Type I error probability will be α.

A complete description of how to design and run an adaptive group sequential using this methodology is as follows.

(a) Specify planned information levels $\mathcal{I}_1, \ldots, \mathcal{I}_K$ and compute group sequential stopping boundaries $(a_1, b_1), \ldots, (a_K, b_K)$ for which the Type I error probability is α and power under $\theta = \delta$ is $1 - \beta$ with these information levels. Define weights $w_1, \ldots, w_K$ according to (22.2), using the planned information levels.

(b) At each analysis k, either before or after a design modification, calculate the information level $\mathcal{I}_k^*$ and the Z-value Z_k based on the cumulative data, and substitute these in (22.1) to obtain $Z^{(k)}$.

(c) Calculate $\tilde{Z}_k$ from (22.4) and apply the group sequential test, comparing $\tilde{Z}_k$ against boundary values a_k and b_k.

In a highly influential paper, Bauer & Köhne (1994) made the proposal to combine P-values calculated from data observed in each stage of an adaptive trial with two or three stages. Lehmacher & Wassmer (1999) developed this idea using Z-values in place of P-values and presented the above prescription for designing a group sequential with the opportunity for sample size modification. This method can also be viewed as a version of that proposed by Cui, Hung & Wang (1999) as a "valid inferential procedure that allows flexibility for adjusting sample size ... during the course of the trial", but generalized to handle arbitrary information sequences $\mathcal{I}_1^*, \ldots, \mathcal{I}_K^*$.

Note that in the absence of any design modification, the combination test method handles differences between planned and observed information levels in a different way from an error spending test, but it still has the property that the Type I error probability is exactly α. The fact that the sequence $(\tilde{Z}_1, \ldots, \tilde{Z}_K)$ has the same canonical joint distribution as the original $(Z_1, \ldots, Z_K)$ when $\theta = 0$ is termed the *conditional invariance principle* by Brannath, Koenig & Bauer (2007, Sec. 2.1). It is simple to check that the above derivation also applies if design modifications are made on several occasions during the course of the trial, so the conditional invariance principle remains true. This property does *not* hold for $\theta \neq 0$. Indeed, the purpose of the design modification is usually to increase power by increasing the group sizes and, hence, the observed information levels, so it is important that the means of the $\tilde{Z}_k$ increase with the observed information levels when $\theta > 0$.

The following coupling argument shows that power increases with increasing information levels. Let $Z_A^{(1)}, \ldots, Z_A^{(K)}$, denote a sequence of stage-wise Z-statistics observed under the originally planned information sequence $\mathcal{I}_1, \ldots, \mathcal{I}_K$, so $Z_A^{(k)} \sim N(\Delta\mathcal{I}_k \sqrt{\theta}, 1)$ where $\Delta\mathcal{I}_k = \mathcal{I}_k - \mathcal{I}_{k-1}$. If $Z_B^{(1)}, \ldots, Z_B^{(K)}$, is a sequence of stage wise Z-statistics when the observed information sequence is $\mathcal{I}_1^*, \ldots, \mathcal{I}_K^*$, the conditional distribution of $Z_B^{(k)}$ given $\mathcal{I}_k^*$ and $\mathcal{I}_{k-1}^*$ is $N(\Delta\mathcal{I}_k^* \sqrt{\theta}, 1)$, where $\Delta\mathcal{I}_k^* = \mathcal{I}_k^* - \mathcal{I}_{k-1}^*$. In the coupling argument, we first generate the sequence $Z_A^{(1)}, \ldots, Z_A^{(K)}$ and then use these values to create

$$Z_B^{(k)} = Z_A^{(k)} + (\Delta\mathcal{I}_k^* - \Delta\mathcal{I}_k)\sqrt{\theta}, \quad k = 1, \ldots, K.$$

Regardless of how $\mathcal{I}_k^*$ is determined based on previous data or external information, each $Z_B^{(k)}$ has the correct conditional distribution given $\mathcal{I}_k^*$ and $\mathcal{I}_{k-1}^*$. Now, let $\{Z_{k,A}\}$ be the sequence of cumulative Z-statistics formed from the $Z_A^{(k)}$ and $\{\tilde{Z}_{k,B}\}$ the sequence formed from the $Z_B^{(k)}$, applying the formula (22.3) in both cases. If $\theta > 0$ and $\Delta\mathcal{I}_k^* \geq \Delta\mathcal{I}_k$ for each $k = 1\ldots,K$, then $Z_B^{(k)} \geq Z_A^{(k)}$ and hence $\tilde{Z}_{k,B} \geq Z_{k,A}$ for each k. Thus, when a group sequential test of form (5.2) is applied to both $\{Z_{k,A}\}$ and $\{\tilde{Z}_{k,B}\}$, rejection of H_0 for the sequence $\{Z_{k,A}\}$ automatically implies rejection of H_0 for the sequence $\{\tilde{Z}_{k,B}\}$. We therefore conclude that for any positive value of θ, the power of a trial is higher when information levels are increased beyond their originally planned values.

We have already proved that the proposed methods for a trial design with sample size modification control the Type I error probability at level α when $\theta = 0$. We can apply a similar coupling argument to show that the Type I error probability is less than α when $\theta < 0$. In this case, we define $Z_A^{(1)},\ldots,Z_A^{(K)}$ to be a sequence of stage-wise Z-statistics observed when $\theta = 0$, so these are independent and each $Z_A^{(k)} \sim N(0,1)$. Then, when considering the Type I error probability at $\theta = \theta' < 0$ of a trial with sample size modification, we set

$$Z_B^{(k)} = Z_A^{(k)} + \Delta\mathcal{I}_k^* \sqrt{\theta'}, \quad k = 1,\ldots,K.$$

As before, we apply the formula (22.3) to create cumulative Z-statistics $\{Z_{k,A}\}$ from the $Z_A^{(k)}$ and $\{\tilde{Z}_{k,B}\}$ from the $Z_B^{(k)}$. Then, for any sequence $\mathcal{I}_1^*,\ldots,\mathcal{I}_K^*$, we have $Z_B^{(k)} < Z_A^{(k)}$ and $\tilde{Z}_{k,B} < Z_{k,A}$ for each $k = 1,\ldots,K$. So, when a group sequential test is applied to both $\{Z_{k,A}\}$ and $\{\tilde{Z}_{k,B}\}$, rejection of H_0 for the sequence $\{\tilde{Z}_{k,B}\}$ implies rejection of H_0 for the sequence $\{Z_{k,A}\}$. Hence for a trial with or without sample size modification, the Type I error probability at any $\theta < 0$ is less than α.

Example: A study with a single treatment arm

To illustrate the connection between the sequences of statistics $\{Z_1,\ldots,Z_K\}$ and $\{\tilde{Z}_1,\ldots,\tilde{Z}_K\}$, consider a study with a single treatment arm yielding data $X_i \sim N(\theta,\sigma^2)$, $i = 1, 2,\ldots,$ where σ^2 is known. Suppose investigators wish to test $H_0\colon \theta \leq 0$ against $\theta > 0$ at two analyses with m_1 observations available at analysis 1 and $n_2 = m_1 + m_2$ observations at analysis 2. The information for θ is $\mathcal{I}_2 = m_1/\sigma^2$ at analysis 1 and $\mathcal{I}_2 = n_2/\sigma^2$ at analysis 2.

If the test proceeds to the second stage, we can write the final Z-statistic as

$$\begin{aligned} Z_2 &= \frac{\sum_{i=1}^{n_2} X_i}{\sqrt{n_2\sigma^2}} \\ &= \sqrt{\frac{m_1}{n_2}}\,\frac{\sum_{i=1}^{m_1} X_i}{\sqrt{m_1\sigma^2}} + \sqrt{\frac{m_2}{n_2}}\,\frac{\sum_{i=m_1+1}^{m_1+m_2} X_i}{\sqrt{m_2\sigma^2}} \\ &= w_1 Z^{(1)} + w_2 Z^{(2)}, \end{aligned} \tag{22.5}$$

where $w_1 = \sqrt{(m_1/n_2)}$, $w_2 = \sqrt{(m_2/n_2)}$, and $Z^{(1)}$ and $Z^{(2)}$ are the Z-values formed from data in the two stages separately. Thus, Z_2 is a weighted sum of the stage-wise Z-values $Z^{(1)}$ and $Z^{(2)}$ with weights w_1 and w_2 satisfying $w_1^2 + w_2^2 = 1$.

Suppose now that, at analysis 1, it is decided to change the second stage group size from m_2 to $\tilde{m}_2$. Define $\tilde{n}_2 = m_1 + \tilde{m}_2$ and denote the new Z-value based on the second stage alone as

$$\tilde{Z}^{(2)} = \frac{\sum_{i=m_1+1}^{m_1+\tilde{m}_2} X_i}{\sqrt{\tilde{m}_2\sigma^2}}.$$

The final Z-statistic based on the sufficient statistic for θ is

$$Z_2 = \frac{\sum_{i=1}^{\tilde{n}_2} X_i}{\sqrt{\tilde{n}_2\sigma^2}}.$$

However, if $\tilde{m}_2$, is a function of Z_1, then Z_2 has a complicated distribution, which is certainly not normal. So, (Z_1, Z_2) cannot have the canonical joint distribution (3.1). Instead, we define

$$\begin{aligned}\tilde{Z}_2 &= w_1 Z^{(1)} + w_2 \tilde{Z}^{(2)} \\ &= \sqrt{\frac{m_1}{n_2}} \frac{\sum_{i=1}^{m_1} X_i}{\sqrt{m_1 \sigma^2}} + \sqrt{\frac{m_2}{n_2}} \frac{\sum_{i=m_1+1}^{m_1+\tilde{m}_2} X_i}{\sqrt{\tilde{m}_2 \sigma^2}} \\ &= \frac{1}{\sqrt{n_2\,\sigma^2}} \left[\sum_{i=1}^{m_1} X_i + \frac{1}{\sqrt{\gamma}} \sum_{i=m_1+1}^{m_1+\tilde{m}_2} X_i \right], \qquad (22.6)\end{aligned}$$

where $\gamma = \tilde{m}_2/m_2$. If $\theta = 0$, both $Z^{(1)}$ and $\tilde{Z}^{(2)}$ have $N(0,1)$ distributions and, since the weights w_1 and w_2 are fixed constants with $w_1^2 + w_2^2 = 1$, we have $\tilde{Z}_2 \sim N(0,1)$ and the pair (Z_1, $\tilde{Z}_2$) does have the canonical joint distribution (3.1) if $\theta = 0$

We have described this example in the context of a two-stage, group sequential trial. Suppose instead that only a final analysis was intended after n_2 observations but an unplanned review after m_1 observations led to the decision to increase the final sample size. It is still possible to express the original test statistic Z_2 in the form (22.5) as a weighted combination of stage-wise Z-values. Hence, the same approach can be taken to modify the remaining sample size and conduct a final hypothesis test, controlling the Type I error probability at level α.

Typically, the second stage sample size is increased and $\gamma > 1$. This means that in the formula (22.5), observations in the second stage are downweighted in the computation of the test statistic $\tilde{Z}_2$. This may raise issues of credibility of the final analysis or of efficiency of the trial design. We shall discuss questions of credibility and efficiency further in Chapter 24. For now, we simply note that it is possible to modify a planned trial design while still controlling the Type I error rate.

22.3 Example of a Trial with Sample Size Re-assessment

22.3.1 An Example of a Group Sequential Trial

We illustrate the methods of Section 22.2 with an example. Consider a group sequential trial with five analyses comparing two treatments with treatment effect θ. Initially it is planned to test H_0: $\theta \le 0$ against $\theta > 0$ with Type I error probability $\alpha = 0.025$ and power $1 - \beta = 0.9$ at $\theta = \delta$. We suppose test statistics $Z_1, \ldots, Z_5$ are to be observed and, in the absence of any adaptation, these will have the canonical joint distribution (3.1) given information levels $\mathcal{I}_1, \ldots, \mathcal{I}_5$. A fixed sample size test for this problem requires information

$$\mathcal{I}_f = \{\Phi^{-1}(1-\alpha) + \Phi^{-1}(1-\beta)\}^2/\delta^2 = 10.51/\delta^2$$

as given by (5.1). The study is to be monitored using a one-sided, error spending test from the Rho family with index $\rho = 3$, as described in Section 8.3.1. The futility boundary is assumed to be binding, so values $(a_1, b_1), \ldots, (a_5, b_5)$ will be calculated to satisfy

$$Pr_{\theta=0}\{\text{Reject } H_0 \text{ with } Z_k \ge b_j \text{ for some } j \le k\} = (\mathcal{I}_k/\mathcal{I}_{max})^3\,\alpha$$

and

$$Pr_{\theta=\delta}\{\text{Accept } H_0 \text{ with } Z_k \le a_j \text{ for some } j \le k\} = (\mathcal{I}_k/\mathcal{I}_{max})^3\,\beta$$

for $k = 1, \ldots, 5$, where $\mathcal{I}_1, \ldots, \mathcal{I}_5$ are the observed information levels. At the design stage, equally-spaced information levels $\mathcal{I}_k = (k/5)\,\mathcal{I}_{max}$ are assumed and from Table 8.6, we see

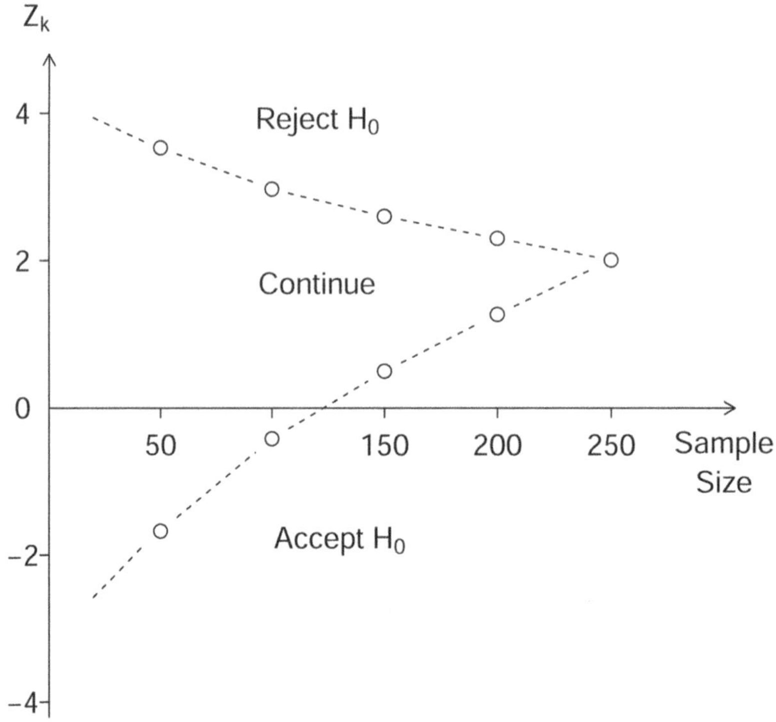

Figure 22.1 *A one-sided, group sequential test with 5 analyses, Type I error probability* 0.025 *and power* 0.9 *at* $\theta = \delta$, *using a Rho error-spending test with* $\rho = 3$

that a maximum information $\mathcal{I}_{max} = 1.049\,\mathcal{I}_f$ is needed for the boundaries to meet up at the final analysis with $a_5 = b_5$. Thus, the trial must be able to reach the information level

$$\mathcal{I}_{max} = 1.049\,\mathcal{I}_f = 11.02/\delta^2.$$

Suppose this value of $\mathcal{I}_{max}$ equates to 250 observations per treatment arm. Then the test has the stopping boundary given in Figure 22.1.

22.3.2 Adapting to External Information

Now suppose this trial has reached analysis 2 with observed information levels $\mathcal{I}_1 = \mathcal{I}_{max}/5$ and $\mathcal{I}_2 = 2\,\mathcal{I}_{max}/5$, and the stopping boundaries have not been crossed. At this time the external environment has changed and the trial's sponsor is interested in detecting the smaller effect size $\theta = 0.7\delta$ with high power. The sponsor, or the DSMB acting on the sponsor's behalf, examines the current data in order to investigate the conditional power function

$$p_2(\theta, z_2) = Pr\{\text{Reject } H_0 \text{ at analysis 3, 4 or 5} \,|\, Z_2 = z_2\},$$

computed under the assumption that future information levels will be $\mathcal{I}_3 = 3\,\mathcal{I}_{max}/5$, $\mathcal{I}_4 = 4\,\mathcal{I}_{max}/5$ and $\mathcal{I}_5 = \mathcal{I}_{max}$. This calculation can be carried out using the method for evaluating $p_k(\theta, z_k)$ in Equation (11.19) that we described in Section 11.5, but with the addition of a futility boundary. In Table 22.1, we present results of these computations under selected values of $\hat{\theta}_2$ and the associated values of z_2, which span the continuation region $(-0.41,\ 2.97)$ at analysis 2. Column 3 gives the conditional Type I error probability and column 4 the conditional power if $\theta = 0.7\delta$.

Table 22.1 *Conditional properties of a group sequential test when re-designing in response to external information. The original trial design has 5 analyses and uses a Rho family error spending design with $\rho = 3$, with Type I error rate 0.025 and power 0.9 if $\theta = \delta$. Sample size is increased adaptively to achieve conditional power 0.9 under $\theta = 0.7\delta$.*

$\hat{\theta}_2$	z_2	*Conditional type I error probability*	*Conditional power under $\theta = 0.7\delta$ before re-design*	γ	*Conditional power under $\theta = 0.7\delta$ after re-design*
$1.40\,\delta$	2.94	0.5707	0.9658	1.00	0.9658
$1.20\,\delta$	2.52	0.3856	0.9189	1.00	0.9189
$1.00\,\delta$	2.10	0.2329	0.8402	1.35	0.9000
$0.90\,\delta$	1.89	0.1743	0.7882	1.62	0.9000
$0.80\,\delta$	1.68	0.1272	0.7285	1.91	0.9000
$0.70\,\delta$	1.47	0.0907	0.6621	2.22	0.9000
$0.60\,\delta$	1.26	0.0630	0.5906	2.55	0.9000
$0.50\,\delta$	1.05	0.0426	0.5158	2.92	0.9000
$0.40\,\delta$	0.84	0.0279	0.4398	3.00	0.8699
$0.20\,\delta$	0.42	0.0109	0.2939	3.00	0.7630
$0.00\,\delta$	0.00	0.0036	0.1707	3.00	0.6095
$-0.19\,\delta$	−0.40	0.0011	0.0866	3.00	0.4337

We see that the conditional power under $\theta = 0.7\delta$ is quite high when $\hat{\theta}_2 > \delta$ but it falls off rapidly at lower values of $\hat{\theta}_2$. To counter this, an increase in sample size is proposed whereby the numbers of subjects in groups 3, 4 and 5 will be increased by a factor γ, chosen to increase the conditional power under $\theta = 0.7\delta$ to 0.9. Since Z_2 is known to investigators, data from the modified design must be analyzed in a way that is seen to protect the Type I error rate, and thus the method of Section 22.2 will be applied. In the modified design, if the conditional power under $\theta = 0.7\delta$ is already as high as 0.9, the sample size is unchanged, so $\gamma = 1$, while for lower values of z_2, the value of γ is capped at 3, corresponding to an increase in the maximum sample size from 250 to 550 subjects per treatment arm. Column 5 of Table 22.1 shows the resulting values of γ and column 6 the conditional power achieved under $\theta = 0.7\delta$. As an example, if the interim estimate at analysis 2 is $\hat{\theta} = 0.5\delta$, group sizes are increased by the factor $\gamma = 2.92$, giving 146 patients per treatment in each group, so the maximum sample size is increased to 538 per treatment arm and this leads to a rise in conditional power under $\theta = 0.7\delta$ from 0.516 to 0.9.

In panel (a) of Figure 22.2 we present plots of conditional power as a function of $\hat{\theta}_2$ under $\theta = \delta$ and $\theta = 0.7\delta$ for the original design, and under $\theta = 0.7\delta$ for the modified design with sample size re-assessment. It is clear that the increase in sample size has achieved its goal of increasing the conditional power, although the cap on the factor γ means that conditional power remains less than 0.9 for values of $\hat{\theta}_2$ below 0.48. The overall power curves for the same three designs are shown in panel (b) of Figure 22.2 and we see overall power for the design with sample size re-assessment is quite close to that of a design set up to have power 0.9 when $\theta = 0.7\delta$ from the start. The power at $\theta = 0.7\delta$ of the design with sample size re-assessment is, in fact, 0.842.

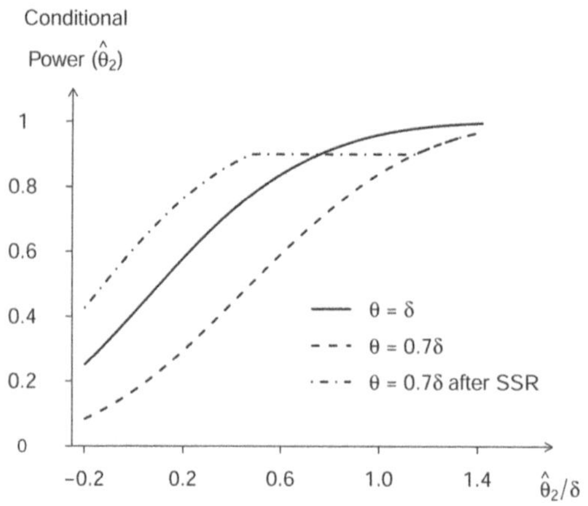

(a) Conditional power given $\hat{\theta}_2$

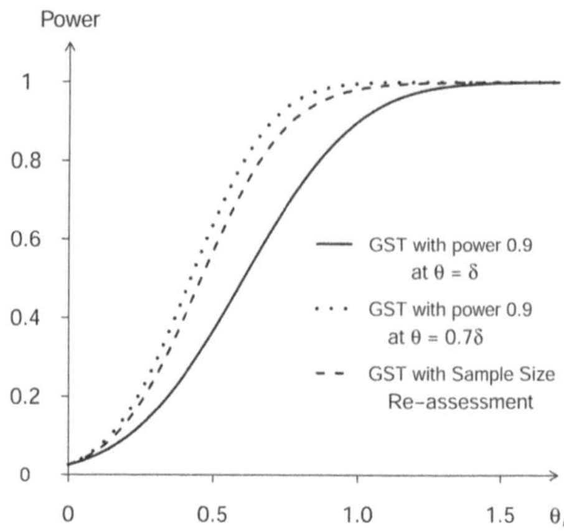

(b) Overall power

Figure 22.2 *Power functions of a group sequential trial with sample size re-assessment in response to external information. The original trial design has 5 analyses and uses a Rho family error spending design with* $\rho = 3$*, with Type I error rate* 0.025 *and power* 0.9 *if* $\theta = \delta$*. Sample size is increased adaptively to achieve conditional power* 0.9 *under* $\theta = 0.7\delta$*.*
Panel (a) shows conditional power at analysis 2 as a function of $\hat{\theta}_2$ *for: the original design when* $\theta = \delta$*; the original design when* $\theta = 0.7\delta$*; the design with sample size re-assessment and* $\theta = 0.7\delta$*. Panel (b) shows the overall power as a function of* θ *for: the original design; a non-adaptive design planned to have power* 0.9 *at* $\theta = 0.7\delta$*; the design with sample size re-assessment.*

22.3.3 Adapting to Internal Information

In the motivating example of Cui, Hung & Wang (1999) described at the beginning of this chapter, the reason for changing the effect size at which high power should be achieved was the low observed treatment effect at an interim analysis. Thus, in our worked example, one might wish to consider a procedure in which sample size is modified to achieve a specified conditional power under $\theta = \hat{\theta}_2$, the interim estimate, rather than under $\theta = 0.7\delta$. We have calculated properties of a procedure in which sample size is re-assessed on observing $Z_2 = z_2$ in the continuation region at analysis 2, and the remaining group sizes are increased by a factor γ that ensures

$$Pr_{\theta=\hat{\theta}_2}\{\text{Reject } H_0 \text{ at analysis 3, 4 or 5} \,|\, Z_2 = z_2\} \;=\; 0.9.$$

If the conditional power under $\theta = \hat{\theta}_2$ is already as high as 0.9, the sample size remains unchanged, so $\gamma = 1$, while for lower values of z_2, the value of γ is capped at 3.

Table 22.2 shows properties of this design where sample size is modified in response to internal data, specifically the treatment effect estimate $\hat{\theta}_2$ at analysis 2. Entries in the table are analogous to those in Table 22.1 but now the conditional powers in columns 4 and 6 are computed assuming $\theta = \hat{\theta}_2$. The role of the factor γ is as before, so, for example, if the interim estimate at analysis 2 is $\hat{\theta}_2 = 0.8\delta$, we have $\gamma = 1.46$, giving 73 patients per treatment in each group, so the maximum sample size is increased to 319 per treatment arm, and the conditional power under $\theta = 0.8$ rises from 0.805 to 0.9.

Panel (a) of Figure 22.3 shows plots of conditional power as a function of $\hat{\theta}_2$ under $\theta = \delta$ and $\theta = 0.7\delta$ for the original design, and under $\theta = 0.7\delta$ for the modified design with sample size re-assessment. Although the effect size $\theta = 0.7\delta$ no longer appears in the rule

Table 22.2 *Conditional properties of a group sequential test when re-designing in response to internal information. The original trial design has 5 analyses and uses a Rho family error spending design with* $\rho = 3$*, with Type I error rate* 0.025 *and power* 0.9 *if* $\theta = \delta$*. Sample size is increased adaptively to achieve conditional power* 0.9 *under* $\theta = \hat{\theta}_2$*.*

$\hat{\theta}_2$	z_2	*Conditional type I error probability*	*Conditional power under* $\theta = \hat{\theta}_2$ *before re-design*	γ	*Conditional power under* $\theta = \hat{\theta}_2$ *after re-design*
$1.40\,\delta$	2.94	0.5707	0.9998	1.00	0.9998
$1.20\,\delta$	2.52	0.3856	0.9959	1.00	0.9959
$1.00\,\delta$	2.10	0.2329	0.9597	1.00	0.9597
$0.90\,\delta$	1.89	0.1743	0.9036	1.00	0.9036
$0.80\,\delta$	1.68	0.1272	0.8051	1.46	0.9000
$0.70\,\delta$	1.47	0.0907	0.6621	2.22	0.9000
$0.60\,\delta$	1.26	0.0630	0.4908	3.00	0.8605
$0.50\,\delta$	1.05	0.0426	0.3213	3.00	0.6761
$0.40\,\delta$	0.84	0.0279	0.1825	3.00	0.4327
$0.20\,\delta$	0.42	0.0109	0.0365	3.00	0.0770
$0.00\,\delta$	0.00	0.0036	0.0036	3.00	0.0036
$-0.19\,\delta$	−0.40	0.0011	0.0002	3.00	0.0000

for sample size modification, it is a meaningful effect size under which we would hope the adaptive design has increased conditional power, and we see this is indeed the case. That conditional power under $\theta = 0.7\delta$ falls as $\hat{\theta}_2$ increases from 0.63 to 0.90 may seem surprising. This is a consequence of the fact that, for values of $\hat{\theta}_2$ in this interval, the conditional power under $\theta = \hat{\theta}_2$ after re-design is 0.9, and 0.7δ becomes smaller relative to $\hat{\theta}_2$ as $\hat{\theta}_2$ increases. Outside this interval, the sample size factor γ remains constant at 3 for $\hat{\theta}_2 < 0.63$ and at 1 for $\hat{\theta}_2 > 0.90$ and thus conditional power under $\theta = 0.7\delta$ increases with $\hat{\theta}_2$.

From the overall power curves for the three designs in panel (b) of Figure 22.3, we see that the design with sample size re-assessment has succeeded in increasing power and the overall power curve is very close to that of a design set up to have power 0.9 when $\theta = 0.7\delta$ from the start. The power at $\theta = 0.7\delta$ of the design with sample size re-assessment is, in fact, 0.880.

22.4 Discussion

In this chapter we have introduced key methodology for adaptive trial design in the context of sample size re-assessment. We shall return to this application of adaptive methods in Chapter 24, where we discuss issues of credibility and statistical efficiency in greater detail. Before that, in Chapter 23, we describe other approaches to constructing adaptive designs.

Proschan & Hunsberger (1995) proposed adaptive procedures that preserve the conditional Type I error probability after a design modification and, hence, control the overall Type I error rate at the specified level. The combination tests we have described in this chapter also have this property and the conditional Type I error probabilities shown in the third column of Tables 22.1 and 22.2 remain unchanged after a design modification. We shall explain in Section 23.5 that any testing procedure that controls the Type I error probability after a flexible adaptation must maintain the conditional Type I error

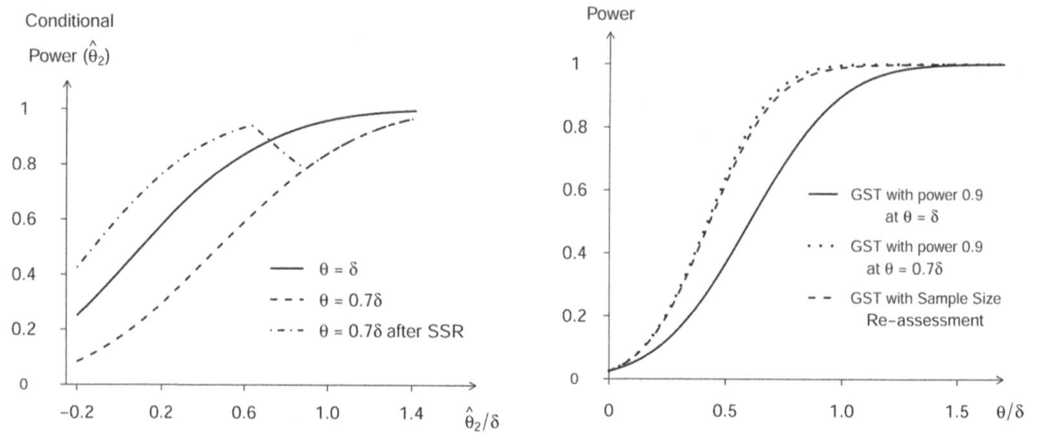

(a) Conditional power given $\hat{\theta}_2$ (b) Overall power

Figure 22.3 *Power functions of a group sequential trial with sample size re-assessment in response to internal information. The original trial design has* 5 *analyses and uses a Rho family error spending design with* $\rho = 3$, *with Type I error rate* 0.025 *and power* 0.9 *if* $\theta = \delta$. *Sample size is increased adaptively to achieve conditional power* 0.9 *under* $\theta = \hat{\theta}_2$.
Panel (a) shows conditional power at analysis 2 *as a function of* $\hat{\theta}_2$ *for: the original design when* $\theta = \delta$; *the original design when* $\theta = 0.7\delta$; *the design with sample size re-assessment and* $\theta = 0.7\delta$. *Panel (b) shows the overall power as a function of* θ *for: the original design; a non-adaptive design planned to have power* 0.9 *at* $\theta = 0.7\delta$; *the design with sample size re-assessment.*

probability, so the approach of Proschan & Hunsberger (1995) is closely aligned with combination test methods.

The opportunities for adaptive methods extend beyond sample size re-assessment in a two treatment comparison. In further chapters we shall describe applications that involve multiple hypothesis testing, where we shall use both combination tests and multiple testing procedures to control a family-wise error rate when flexible adaptations are envisaged. In Chapter 25 we explain how these two methodologies can be merged and in subsequent chapters, we discuss application of the resulting procedures in: seamless Phase II/III trials, in a multi-arm Phase III trial with treatment selection, in adaptive multi-arm multi-stage trials, and in enrichment trials that may adapt by re-focusing on a patient subgroup.

Chapter 23

Adaptive Combination Tests

23.1 Combining P-values from Independent Studies

The key ingredient in many adaptive procedures is the combination test. This is an idea borrowed from the methodology of meta-analysis, where results from different studies are combined to make an overall inference. However, here we shall be concerned with combining results from different stages in a group sequential trial.

We shall focus on one-sided tests of a hypothesis H_0: $\theta \leq 0$ against the alternative $\theta > 0$. In this case, two small P-values can be combined to provide stronger evidence against H_0. The same is not true when testing H_0: $\theta = 0$ against $\theta \neq 0$ as one small P-value may come from an estimate $\hat{\theta} > 0$ and the other from an estimate $\hat{\theta} < 0$. If a two-sided test is required, this should be formulated as two one sided tests, one of H_0: $\theta = 0$ versus $\theta > 0$ and the other of H_0: $\theta = 0$ versus $\theta < 0$. Then results for each hypothesis test can be combined separately before deciding whether or not to reject H_0: $\theta \neq 0$ in a two-sided test.

In an influential paper in the psychological literature, Rosenthal (1978) summarized the advantages and limitations of nine methods of combining P-values. Subsequently, Becker (1994) surveyed methodology for combining significance levels, listing eighteen different proposals in her Table 15.1. Goutis, Casella & Wells (1996) give a very general discussion on the use of P-values as a means to assess evidence from multiple studies. We shall consider just a few of the methods that are commonly used in the social sciences and which are natural to use in the clinical trials context.

Suppose data from K studies are to be combined and let $P^{(k)}$ denote the P-value from testing H_0: $\theta \leq 0$ versus $\theta > 0$ based on data from the kth study, where $1 \leq k \leq K$. We assume that $P^{(k)}$ has a standard uniform distribution, which we write as $P^{(k)} \sim U(0,1)$, when $\theta = 0$ and that the $P^{(k)}$ are independent.

The inverse chi-squared method

The earliest method for combining data from a number of studies, proposed by Fisher (1932), is based on the product of the P-values and is termed the "sum of logs" method or the "inverse chi-squared" method. Under $\theta = 0$, each $P^{(k)} \sim U(0,1)$ and, hence,

$$-\log P^{(k)} \;\sim\; \exp(1) \;\sim\; \frac{1}{2}\,\chi^2_2,$$

where $\exp(1)$ denotes the exponential distribution with mean 1. Since the $P^{(k)}$, $k = 1, \ldots, K$, are independent,

$$-\sum_{k=1}^{K} \log P^{(k)} \;\sim\; \frac{1}{2}\,\chi^2_{2K}.$$

Thus, an α-level test is obtained by rejecting H_0 if

$$-2\log\{P^{(1)} \ldots P^{(K)}\} \;\geq\; \chi^2_{2K}(\alpha) \tag{23.1}$$

DOI: 10.1201/9781584888482-23

or, equivalently, if

$$\prod_{k=1}^{K} P^{(k)} \leq \exp\left\{-\frac{1}{2}\chi^2_{2K}(\alpha)\right\},$$

where $\chi^2_{2K}(\alpha)$ denotes the upper α tail point of a χ^2 distribution with $2K$ degrees of freedom. An overall P-value is defined by finding the value of p for which

$$-2\log\{P^{(1)}\ldots P^{(K)}\} = \chi^2_{2K}(p).$$

As an example, suppose $K = 2$, $P^{((1)} = 0.060$ and $P^{(2)} = 0.055$. Then, if we conduct a test at significance level $\alpha = 0.025$, we have $P^{(1)}P^{(2)} = 0.0033$ which is less than $\exp\{-\chi^2_4(0.025)/2\} = 0.0038$. Thus, even though results from neither stage reach significance, they can "borrow strength" from each other and together yield a significant result at level $\alpha = 0.025$. The overall P-value is

$$1 - F_{\chi^2_4}(-2\,\log(0.0033)) = 0.022,$$

where $F_{\chi^2_4}$ denotes the cdf of a χ^2_4 distribution.

The inverse normal and weighted inverse normal methods

The most commonly used method in the social sciences is the weighted inverse normal or "weighted Z-score" method, dating back to the work of Stouffer et al. (1949). Suppose we have P-values $P^{(1)},\ldots,P^{(K)}$ from tests of H_0: $\theta \leq 0$ against $\theta > 0$ and let

$$Z^{(k)} = \Phi^{-1}(1 - P^{(k)}), \quad k = 1,\ldots,K.$$

If, under $\theta = 0$, the $P^{(k)}$ are independent and each $P^{(k)} \sim U(0,1)$, then the $Z^{(k)}$ are independent and each $Z^{(k)} \sim N(0,1)$. It follows that

$$\frac{1}{\sqrt{K}}\left(Z^{(1)} + \ldots + Z^{(K)}\right) \sim N(0,1)$$

when $\theta = 0$, and we can obtain an α-level test by rejecting H_0 if

$$\frac{1}{\sqrt{K}}\left(Z^{(1)} + \ldots + Z^{(K)}\right) \geq z_\alpha, \tag{23.2}$$

where $z_\alpha = \Phi^{-1}(1-\alpha)$ is the upper α tail point of a standard normal distribution.

If, as in the earlier example, we have $K = 2$, $P^{(1)} = 0.060$ and $P^{(2)} = 0.055$, then $Z^{(1)} = 1.555$ and $Z^{(2)} = 1.598$. The combined Z-value is $(1.555 + 1.598)/\sqrt{2} = 2.229$ and, since this exceeds $z_{0.025} = 1.960$, the null hypothesis H_0: $\theta \leq 0$ is rejected in a test at significance level $\alpha = 0.025$. Furthermore, the overall P-value for the combined data is $1 - \Phi(2.229) = 0.013$, the minimum value of α for which (23.2) is satisfied.

In the weighted inverse normal test described by Mosteller & Bush (1954), weights $w_1,\ldots,w_K$ satisfying $w_1^2 + \ldots + w_K^2 = 1$ are specified. These weights are fixed before observing the values of $Z^{(1)},\ldots,Z^{(K)}$. Then, under $\theta = 0$,

$$w_1 Z^{(1)} + \ldots + w_K Z^{(K)} \sim N(0,1)$$

and a level α test rejects H_0 if

$$w_1 Z^{(1)} + \ldots + w_K Z^{(K)} \geq z_\alpha. \tag{23.3}$$

Hedges, Cooper & Bushman (1992) argue that the Z-values and P-values "are already weighted" in some sense and so such adjustment may not be needed. However, if each

$Z^{(i)} \sim N(\theta\sqrt{\mathcal{I}_i}, 1)$, the most powerful test is that based on the statistic with weights $w_i \propto \sqrt{\mathcal{I}_i}$ and we shall use these weights in defining an adaptive group sequential design in Section 23.7.1.

The maximum P-value method

Goutis, Casella & Wells (1996, Table 1) include a method for combing data based on the maximum of K observed P-values. If $P^{(1)}, \ldots, P^{(K)}$ are independent with $U(0,1)$ distributions under $\theta = 0$, then

$$Pr\{\max(P^{(1)}, \ldots, P^{(K)}) \leq \alpha^{1/K}\} = \prod_{k=1}^{K} Pr\{P^{(k)} \leq \alpha^{1/K}\} = (\alpha^{1/K})^K = \alpha.$$

Thus, a level α test can be constructed by rejecting H_0 if

$$\max(P^{(1)}, \ldots, P^{(K)}) \leq \alpha^{1/K}. \tag{23.4}$$

We mention this approach since, in the context of clinical trials, the FDA's "two pivotal trial" rule (FDA, 1995) may be considered a special case of this method. This rule stipulates that in order to show that a new treatment is efficacious, statistically significant results must be obtained in two clinical trials. Li & Huque (2003, p. 623) note that if a significance level of 0.025 is required in each trial, this implies an overall significance level for the combination test of $0.025^2 = 0.000625$.

It should be understood that the requirement for two pivotal studies is not just to produce a small significance level but rather to ensure that findings can be replicated and are not subject to systematic bias or fraud. The FDA Statement notes that

> "While a second study may well be needed to replicate results demonstrated in a first study, in some instances, it is possible to replicate results within one large, well-designed, multi-center study. FDA emphasizes that this approach can be successful only when results are strong."

Commenting on this policy, Fleming & Richardson (2004, p. 668) observe

> "Given that the strength of evidence for declaring a single trial to be "positive" usually requires that a ... 1-sided P-value of 0.025 be obtained, the FDA often has viewed evidence of the effect of treatment on a major clinical end point to be "compelling," when provided by only a single trial, if the 1-sided P-value is bounded above by a value in the range of 0.0005–0.005 (i.e., achieving a strength of evidence in the range of 1.5–2 positive trials ...). Of course, the final decision regarding approval would depend on several factors, including effects on safety measures and secondary end points, issues related to quality of the conduct of the trial, and relevant evidence from related external trials."

23.2 Use of Combination Tests in Adaptive Designs

We now consider application to clinical trials of the methods for combining information described in Section 23.1. The proposal by Bauer & Köhne (1994) to apply such combination tests was a key step in the development of adaptive clinical trial design. Suppose a trial has a group sequential design with a maximum of K stages and an adaptive strategy is to be followed, whereby the conduct of each stage of the study is allowed to depend on the results of previous stages. We shall show how to construct a test of H_0: $\theta \leq 0$ against $\theta > 0$ by combining P-values or Z-values obtained from each stage. Although our objective is to combine data from separate stages of a single study rather than from a number of different studies, we shall see that the methods described in Section 23.1 can still be used, and this remains the case when adaptations based on previously observed data mean that stages of the single study are not independent.

When the primary endpoint for each patient is a measurement taken at a certain time after the start of treatment, it is straightforward to analyze the data that become available in a particular stage of the trial. We then define $Z^{(k)}$ to be the usual Wald Z-statistic and $\mathcal{I}^{(k)}$ the associated Fisher information based on the data observed in stage k alone. We also define $P^{(k)} = 1 - \Phi(Z^{(k)})$ as the corresponding P-value, where Φ denotes the standard normal cdf. In some cases, a P-value can be computed directly from the stage k data, for example, as the tail probability of a t-distribution when responses are normally distributed with unknown variance. Then the Z-value for stage k is calculated as $Z^{(k)} = \Phi^{-1}(1 - P^{(k)})$.

For more complex data types, such as a survival data, where information on a particular subject may accrue over several stages of the trial, we use the definition in (22.1) and set

$$Z^{(1)} = \frac{S_1}{\sqrt{\mathcal{I}_1}} = Z_1$$

and

$$Z^{(k)} = \frac{S_k - S_{k-1}}{\sqrt{(\mathcal{I}_k - \mathcal{I}_{k-1})}} = \frac{Z_k\sqrt{\mathcal{I}_k} - Z_{k-1}\sqrt{\mathcal{I}_{k-1}}}{\sqrt{(\mathcal{I}_k - \mathcal{I}_{k-1})}}, \quad k = 2, \ldots, K,$$

where S_k is the score statistic and Z_k the associated Z-value based on the cumulative data at analysis k, for $k = 1, \ldots, K$. In this general case, $\mathcal{I}^{(k)} = \mathcal{I}_k - \mathcal{I}_{k-1}$ is the increment in Fisher information between analyses $k-1$ and k. As before, we define $P^{(k)} = 1 - \Phi(Z^{(k)})$ as the corresponding P-value.

Since aspects of stage k of the trial, such as the sample size for that stage, are allowed to depend on outcomes observed in stages 1 to $k-1$, the joint distribution of $(Z^{(1)}, \ldots, Z^{(K)})$ or $(P^{(1)}, \ldots, P^{(K)})$ may be complex. However, we can still obtain key properties of these sequences for the cases $\theta = 0$, $\theta < 0$ and $\theta > 0$. In describing these properties, we denote by $D(k)$ all the information available at analysis k. Thus $D(k)$ includes $Z^{(j)}$, $P^{(j)}$ and $\mathcal{I}^{(j)}$ for $j = 1, \ldots, k-1$, as well as other information from within the trial or from external sources.

When $\theta = 0$

If $\theta = 0$ and the primary response is continuous, we assume that the conditional distribution of $P^{(k)}$ given the design of stage k of the trial is $U(0,1)$. Thus, we can write

$$P^{(k)} \,|\, D(k) \sim U(0,1).$$

Because this conditional distribution is the same for all values of $D(k)$ we can conclude that $P^{(k)} \sim U(0,1)$ unconditionally and $P^{(k)}$ is independent of $D(k)$. It follows that, when $\theta = 0$, the stage-wise P-values, $P^{(1)}, \ldots, P^{(K)}$, are independent and each follows a $U(0,1)$ distribution, a property referred to by Brannath, Koenig & Bauer (2007, Sec. 2.1) as the *conditional invariance principle.* This independence property implies that any of the combination tests of Section 23.1 can be applied to $(Z^{(1)}, \ldots, Z^{(K)})$ or $(P^{(1)}, \ldots, P^{(K)})$ and thus, when $\theta = 0$, the Type I error rate will be controlled at level α.

If the primary response is discrete, the conditional distribution $P^{(k)} \,|\, D(k)$, will be stochastically larger than $U(0,1)$, that is $Pr\{P^{(k)} \le p \,|\, D(k)\} \le p$ for $0 \le p \le 1$ with $Pr\{P^{(k)} \le p_j \,|\, D(k)\} = p_j$ at a finite, or possibly countable, set of values $0 = p_1 < p_2 < \ldots$. Brannath, Posch & Bauer (2002) refer to this property of being "conditionally larger than a uniform distribution" as "p-clud". In this case $P^{(1)}, \ldots, P^{(K)}$ may not be independent but we can still apply one of the combination tests of Section 23.1 to produce a test with Type I error rate no greater than α. The following argument, similar to the coupling argument presented in Section 22.2, justifies this claim. If $P^{(k)} \,|\, D(k)$ is p-clud, then it is possible to define a random variable $\tilde{P}^{(k)}$, based on the cumulative distribution function of $P^{(k)} \,|\, D(k)$, such that

$$\tilde{P}^{(k)} \le P^{(k)} \quad \text{and} \quad \tilde{P}^{(k)} \,|\, D(k) \sim U(0,1).$$

Specifically, if $P^{(k)} = p_j$ so p_{j-1} is the next lowest value that $P^{(k)}$ can take (with p_0 set as 0) and the conditional distribution $P^{(k)} \mid D(k)$ has probability $p_j - p_{j-1}$ at $P^{(k)} = p_j$, we set

$$\tilde{P}^{(k)} = p_{j-1} + (p_j - p_{j-1})\, Y,$$

where Y is a $U(0,1)$ random variable.

Since the conditional distribution of $\tilde{P}^{(k)}$ is the same for all values of $D(k)$, we can deduce that $\tilde{P}^{(1)}, \ldots, \tilde{P}^{(K)}$, are independent and each has a $U(0,1)$ distribution. Hence, if a level α combination test is applied to $\tilde{P}^{(1)}, \ldots, \tilde{P}^{(K)}$, the Type I error rate will be at most α. Now, if a combination test applied to $P^{(1)}, \ldots, P^{(K)}$ rejects H_0 then H_0 will also be rejected when that combination test is applied to $\tilde{P}^{(1)}, \ldots, \tilde{P}^{(K)}$, as $\tilde{P}^{(k)} \leq P^{(k)}$ for each k. Thus the test based on $P^{(1)}, \ldots, P^{(K)}$ rejects H_0 less often than the test based on $\tilde{P}^{(1)}, \ldots, \tilde{P}^{(K)}$, and this implies that the test based on $P^{(1)}, \ldots, P^{(K)}$ controls the Type I error rate at level α.

When $\theta < 0$

We assume that if $\theta < 0$, each conditional distribution $P^{(k)} \mid D(k)$, $k = 1, \ldots, K$, is stochastically larger than $U(0,1)$. For each $k = 1, \ldots, K$, let

$$F_k(p) = Pr\{P^{(k)} \leq p \mid D(k)\} \leq p.$$

If the distribution $P^{(k)} \mid D(k)$ is continuous, we set $\tilde{P}^{(k)} = F_k(P^{(k)})$ to create a $U(0,1)$ variable $\tilde{P}^{(k)}$. If the distribution $P^{(k)} \mid D(k)$ is discrete we have $Pr\{F_k(P^{(k)}) \leq p \mid D(k)\} \leq p$ for $0 \leq p \leq 1$ with $Pr\{F_k(P^{(k)}) \leq p_j \mid D(k)\} = p_j$ at a set of values $0 = p_1 < p_2 < \ldots$. We can then follow the same approach as for a discrete response variable when $\theta = 0$ to create a $U(0,1)$ variable $\tilde{P}^{(k)}$.

We now have a sequence of independent $U(0,1)$ variables $\tilde{P}^{(1)}, \ldots, \tilde{P}^{(K)}$ satisfying $\tilde{P}^{(k)} \leq P^{(k)}$ for $k = 1, \ldots, K$. Since a test based on $P^{(1)}, \ldots, P^{(K)}$ rejects H_0 less often than a test based on $\tilde{P}^{(1)}, \ldots, \tilde{P}^{(K)}$, we can conclude that the test based on $P^{(1)}, \ldots, P^{(K)}$ controls the Type I error rate at level α. This result is important as it shows that the Type I error rate is no greater than α over the whole null hypothesis, $\theta \leq 0$.

When $\theta > 0$

We assume that if $\theta > 0$, each conditional distribution $P^{(k)} \mid D(k)$, $k = 1, \ldots, K$, is stochastically smaller than $U(0,1)$. In the examples of Chapter 22, we used the equivalent property of the stage-wise Z-statistics, namely that the conditional distribution of $Z^{(k)}$ given $D(k)$ is stochastically larger than a $N(0,1)$ distribution when $\theta > 0$. Working on the P-value scale, we can apply the method used for the case $\theta < 0$ to create independent $U(0,1)$ variables $\tilde{P}^{(1)}, \ldots, \tilde{P}^{(K)}$ with $\tilde{P}^{(k)} \geq P^{(k)}$ for $k = 1, \ldots, K$. A test based on $\tilde{P}^{(1)}, \ldots, \tilde{P}^{(K)}$ has Type I error probability α and, since a test based on $P^{(1)}, \ldots, P^{(K)}$ rejects H_0 more often, we conclude that this test has power greater than α when $\theta > 0$. This is, of course, a minimal power requirement and a proposed design should be thoroughly investigated, often in a simulation study, to evaluate its power curve and other relevant properties.

In later chapters, we shall see how combination tests may be used to construct many different types of adaptive trial design. The preceding summary draws on a substantial literature, starting with the papers of Bauer (1989) and Bauer & Köhne (1994) which applied the inverse chi-squared test (23.1). Adaptive designs based on the weighted inverse normal test (23.3) were proposed by Fisher (1999). Although Bauer & Köhne (1994) incorporated some early stopping in their 2-stage and 3-stage procedures, it was the papers of Cui, Hung & Wang (1999) and Lehmacher & Wassmer (1999) that introduced adaptive modifications to more traditional group sequential tests using weighted normal statistics. As an example of yet another approach, Chang (2007) describes adaptive group sequential designs based on sums of the stage-wise P-values, $P^{(k)}$.

23.3 Modifying Sample Size in Response to a Variance Estimate

Consider a study comparing two treatments with normal responses $X_{Ai} \sim N(\mu_A, \sigma^2)$ and $X_{Bi} \sim N(\mu_B, \sigma^2)$ on treatments A and B, respectively, where the variance σ^2 is unknown. Denoting the treatment effect by $\theta = \mu_A - \mu_B$, suppose it is desired to test H_0: $\theta \leq 0$ against $\theta > 0$ with Type I error rate α and power $1 - \beta$ at $\theta = \delta$.

The investigators note that the sample size needed to meet their power requirement depends on the unknown variance σ^2. Anticipating that they may wish to adapt the sample size in response to an estimate of σ^2 obtained during the study, they decide to run a Bauer & Köhne (1994) two-stage design using the inverse chi-squared combination test to test H_0: $\theta \leq 0$ at the end of the second stage. They first calculate the sample size of n_0 per treatment arm that will give power $1 - \beta$ at $\theta = \delta$ if the variance is equal to an initial estimate σ_0^2 and stage 1 is conducted with a sample size of $n_1 = n_0/2$ per treatment.

Suppose the data from n_1 subjects per treatment in stage 1 yield estimates $\hat{\theta}_1$ and $\hat{\sigma}_1^2$, and the t-statistic for testing H_0 is

$$t_1 = \frac{\hat{\theta}_1}{\sqrt{\{2\,\hat{\sigma}_1^2/n_1\}}}.$$

This is converted to the P-value $P^{(1)} = Pr_{\theta=0}\{T_{2\,n_1-2} > t_1\}$ to be used in the combination test. The estimate $\hat{\sigma}_1^2$ is now used to plan the second stage of the trial by, for example, using it in place of σ_0^2 in the earlier sample size calculation. Alternatively, one might modify the second stage sample size to achieve a certain conditional power given t_1, assuming $\theta = \hat{\theta}_1$ and $\sigma^2 = \hat{\sigma}_1^2$.

However the stage 2 sample size of n_2 per treatment arm is chosen, the t-statistic

$$t_2 = \frac{\hat{\theta}_2}{\sqrt{\{2\,\hat{\sigma}_2^2/n_2\}}}$$

is computed, where $\hat{\theta}_2$ and $\hat{\sigma}_2^2$ are based on stage 2 data alone. This gives the stage 2 P-value $P^{(2)} = Pr_{\theta=0}\{T_{2\,n_2-2} > t_2\}$. By the usual argument for a combination test, $P^{(1)}$ and $P^{(2)}$ are independent Uniform(0,1) random variables when $\theta = 0$ and thus the inverse chi-squared combination test which rejects H_0 if

$$-2\ln(P^{(1)}P^{(2)}) > \chi^2_{4,\,1-\alpha}$$

has Type I error probability exactly equal to α.

We considered a similar problem in Section 18.2.2, where we described a two-sided test of H_0: $\theta = 0$ with the opportunity to update the sample size in response to an interim variance estimate. The attractive feature of the procedure based on a combination test is that it has Type I error rate exactly equal to the specified α, whereas the procedures of Section 18.2.2 achieve this only approximately.

It is straightforward to replace the inverse chi-squared test by a weighted inverse normal test in this example. In this case, the test after stage 2 rejects H_0 if

$$w_1 Z^{(1)} + w_2 Z^2 \geq \Phi^{-1}(1 - \alpha),$$

where $Z^{(1)} = \Phi^{-1}(1 - P^{(1)})$, $Z^{(2)} = \Phi^{-1}(1 - P^{(2)})$ and $w_1^2 + w_2^2 = 1$. One might choose equal weights $w_1 = w_2 = 1/\sqrt{2}$ in the expectation that the second stage sample size will remain at $n_1/2$ subjects per treatment arm, or one may set $w_2 > w_1$ in view of the possibility that the second stage sample size may be increased. The important point is that w_1 and w_2 should be specified before the first stage data are observed and they are not to be changed after seeing the interim estimate of σ^2.

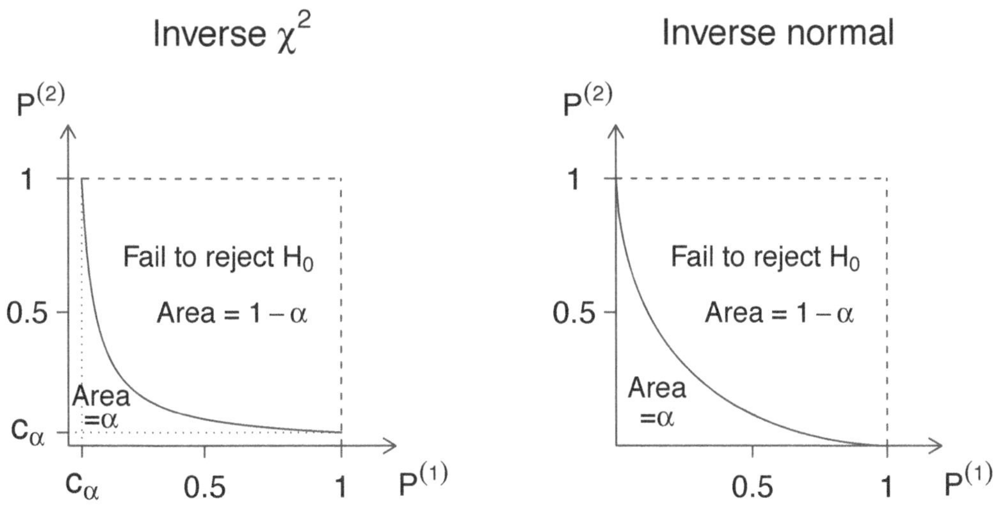

Figure 23.1 *Rejection regions on the P-value scale for two combination tests. The Type I error probability of both tests is* $\alpha = 0.2$.

Timmesfeld, Schäfer & Müller (2007) present an adaptive design that allows an increase in sample size, but which uses the standard t-statistic based on the whole data if adaptation does not happen. The procedure is somewhat complicated but use of the overall estimate of σ^2 does give a slight improvement in efficiency. It is also reassuring for a test to be based on sufficient statistics, which is not the case for the weighted inverse normal test, even if there is no change in the second stage sample size.

23.4 A Graphical Illustration of Two-stage Combination Tests

In this section we give a graphical representation of two-stage combination tests, similar to the graphics shown by Proschan (2003) and Jennison & Turnbull (2005). Recall that, under $\theta = 0$, the P-values $P^{(1)}$ and $P^{(2)}$ have independent $U(0,1)$ distributions. The range of $(P^{(1)}, P^{(2)})$ is the unit square $[0,1] \times [0,1]$ and an α level test rejects H_0 if $(P^{(1)}, P^{(2)}) \in \mathcal{C}$, where the critical region $\mathcal{C}$ is a subset of the unit square with area α. Different combination tests vary with respect to the choice of this subset.

From (23.1) with $K = 2$, Fisher's inverse χ^2 test, rejects H_0 if

$$P^{(1)} P^{(2)} \leq c_\alpha = \exp\left\{-\frac{1}{2}\chi_4^2(\alpha)\right\}. \tag{23.5}$$

From (23.3), the weighted inverse normal test rejects H_0 if

$$w_1 \Phi^{-1}(1 - P^{(1)}) + w_2 \Phi^{-1}(1 - P^{(2)}) \geq z_\alpha, \tag{23.6}$$

where the weights w_1 and w_2 satisfy $w_1^2 + w_2^2 = 1$. These rejection regions are illustrated in Figure 23.1, with $w_1 = w_2 = 1/\sqrt{2}$ in the case of the inverse normal test. We have used an unusually high error rate $\alpha = 0.2$ in this illustration so that c_α is large enough to be easily distinguishable from zero and there are discernible differences in the shapes of the two rejection regions.

We can also illustrate designs on the Z-scale by plotting the critical regions for $(Z^{(1)}, Z^{(2)})$ where $Z^{(1)} = \Phi^{-1}(1 - P^{(1)})$ and $Z^{(2)} = \Phi^{-1}(1 - P^{(2)})$. The critical regions for

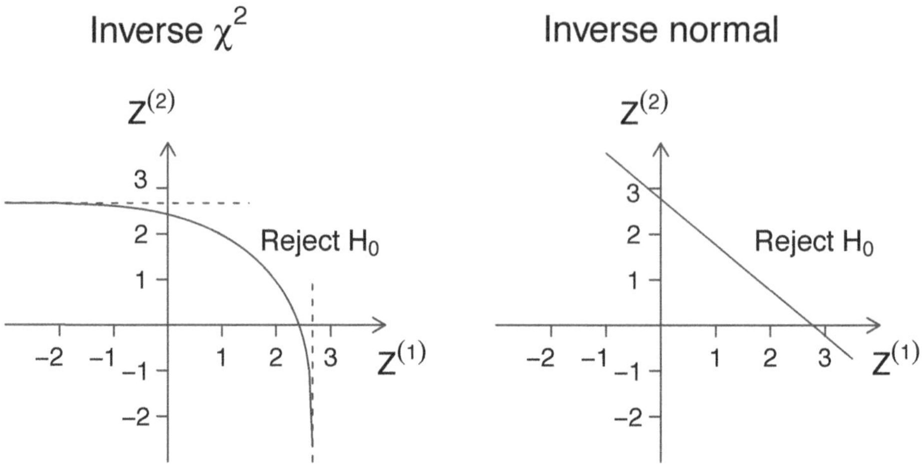

Figure 23.2 *Rejection regions on the Z-value scale for two combination tests. The Type I error probability of both tests is* $\alpha = 0.025$.

the two combination tests on this scale are shown in Figure 23.2, with a more conventional one-sided error rate, $\alpha = 0.025$, in this case.

Truncated two-stage designs

We note that the inverse χ^2 test will always reject H_0: $\theta \leq 0$ if $P^{(1)} \leq c_\alpha$, where c_α is defined by (23.5). Thus there is no need to conduct the second stage if $P^{(1)} \leq c_\alpha$ since rejection of H_0 is inevitable. Such automatic curtailment is not possible for the inverse normal test. However, both tests can be adapted to incorporate early stopping to declare efficacy or futility after the first stage.

For the inverse χ^2 test, we specify critical values c_1, d_1 and c_2, where $c_1 < d_1$ and $c_2 < c_1$, and define the two-stage procedure:

$$
\begin{array}{lll}
\text{At stage 1} & & \\
\quad \text{if } P^{(1)} \leq c_1 & \text{stop, reject } H_0 & \\
\quad \text{if } P^{(1)} \geq d_1 & \text{stop, accept } H_0 & \\
\quad \text{otherwise} & \text{continue to stage 2} & \quad (23.7) \\
\text{At stage 2} & & \\
\quad \text{if } P^{(1)}P^{(2)} \leq c_2 & \text{stop, reject } H_0 & \\
\quad \text{otherwise} & \text{stop, accept } H_0. &
\end{array}
$$

To ensure that the overall Type I error probability is equal to α, we need the area of the critical region in the unit square of $(P^{(1)}, P^{(2)})$ values to equal α. This condition will be fulfilled if

$$c_1 + \int_{c_1}^{d_1} \frac{c_2}{p_1} \, dp_1 \;=\; \alpha,$$

which can be achieved by choosing values c_1, d_1 and c_2 satisfying $c_\alpha \leq c_1 < \alpha$, $d_1 > \alpha$ and $c_2 = (\alpha - c_1)/\log(d_1/c_1)$.

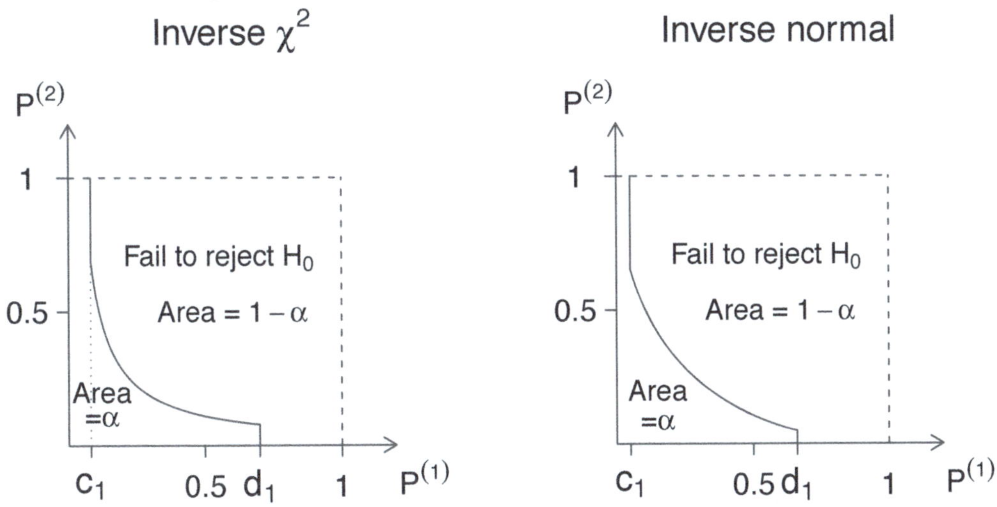

Figure 23.3 *Rejection regions on the P-value scale for two combination tests with truncation. The Type I error probability of both tests is* $\alpha = 0.2$.

We can incorporate early stopping in the weighted inverse normal method in a procedure expressed in terms of stage-wise Z-values as:

$$
\begin{array}{lll}
\text{At stage 1} & & \\
\quad \text{if } Z^{(1)} \geq b_1 & \text{stop, reject } H_0 & \\
\quad \text{if } Z^{(1)} \leq a_1 & \text{stop, accept } H_0 & \\
\quad \text{otherwise} & \text{continue to stage 2} & (23.8) \\
\text{At stage 2} & & \\
\quad \text{if } w_1 Z^{(1)} + w_2 Z^{(2)} \geq b_2 & \text{stop, reject } H_0 & \\
\quad \text{otherwise} & \text{stop, accept } H_0. &
\end{array}
$$

The overall Type I error probability is controlled at level α if a_1, b_1 and b_2 satisfy

$$1 - \Phi(b_1) + \int_{a_1}^{b_1} \left\{ 1 - \Phi\left(\frac{b_2 - w_1 z_1}{w_2} \right) \right\} \phi(z_1)\, dz_1 = \alpha.$$

Figure 23.3 shows the critical regions of combination tests drawn on the P-scale for a truncated inverse χ^2 design and for a truncated weighted inverse normal design with $w_1 = w_2 = 1/\sqrt{2}$. Once again, we have shown designs with the unusually high error probability $\alpha = 0.2$ to produce a more informative illustration. In this case, we have set $c_1 = 0.4\alpha$ and $d_1 = 0.7$ for the inverse χ^2 test and a_1 and b_1 are such that $c_1 = 0.25\alpha$ and $d_1 = 0.66$ for the inverse normal test. In Figure 23.4 we depict the same types of design on the Z-scale, but now the Type I error probability is $\alpha = 0.025$, and in this case $c_1 = 0.4\alpha$ and $d_1 = 0.45$ for the inverse χ^2 test while $a_1 = \Phi^{-1}(0.65)$, $b_1 = \Phi^{-1}(1 - 0.25\alpha)$ and $b_2 = \Phi^{-1}(1 - 0.0218)$ for the inverse normal test.

We may also view the truncated weighted inverse normal design as a type of group sequential test. If we write $Z_1 = Z^{(1)}$ and $Z_2 = w_1 Z^{(1)} + w_2 Z^{(2)}$, then the stopping rule (23.8) is equivalent to a rule of the form (5.2) defining a one-sided group sequential test with two analyses, and the stopping rule is of the form depicted in Figure 5.1 but with two analyses rather than four. However, it is important to note that the statistic

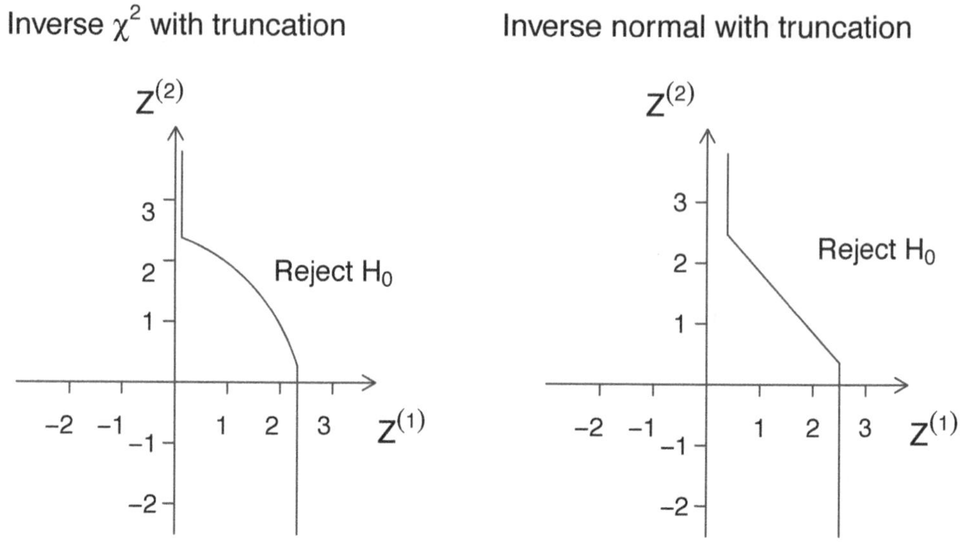

Figure 23.4 *Rejection regions on the Z-value scale for two combination tests with truncation. The Type I error probability of both tests is* $\alpha = 0.025$.

$Z_2 = w_1 Z^{(1)} + w_2 Z^{(2)}$ only equals the usual Z-statistic based on the cumulative data at analysis 2 when the ratio of the stage 1 and stage 2 sample sizes (or, more generally, the ratio of increments in information) is equal to w_1^2/w_2^2. It is a feature of adaptive designs that the ratio of sample sizes in two stages is variable while the weights w_1 and w_2 are fixed, so it is quite possible this condition will not be met.

23.5 The Conditional Type I Error Probability Function

Suppose the critical region of a two-stage combination test with possible truncation has the form

$$\mathcal{C} = \{(P^{(1)}, P^{(2)})\colon\, P^{(2)} \le \tilde{A}(P^{(1)})\}$$

for some non-increasing function $\tilde{A}(P^{(1)})$. This will be the case for a combination test in which $(P^{(1)}, P^{(2)}) \in \mathcal{C}$ implies that $(\tilde{P}^{(1)}, \tilde{P}^{(2)}) \in \mathcal{C}$ whenever $\tilde{P}^{(1)} \le P^{(1)}$ and $\tilde{P}^{(2)} \le P^{(2)}$, as in the tests illustrated in the left hand panels of Figures 23.1 and 23.3.

For the inverse χ^2 test with truncation, as defined in (23.7), we have

$$\tilde{A}(p_1) = \begin{cases} 1 & \text{for } p_1 \le c_1, \\ c_2/p_1 & \text{for } c_1 < p_1 < d_1, \\ 0 & \text{for } p_1 \ge d_1. \end{cases}$$

while, for the truncated weighted inverse normal combination test with weights w_1 and w_2, as defined in (23.8),

$$\tilde{A}(p_1) = \begin{cases} 1 & \text{for } p_1 \le 1 - \Phi(b_1), \\ 1 - \Phi\left[\{b_2 - w_1 \Phi^{-1}(1 - p_1)\}/w_2)\right] & \text{for } 1 - \Phi(b_1) < p_1 < 1 - \Phi(a_1), \\ 0 & \text{for } p_1 \ge 1 - \Phi(a_1). \end{cases} \tag{23.9}$$

Note that, for the critical region to have area α, the function $\tilde{A}$ must satisfy

$$\int_0^1 \tilde{A}(p_1)\, dp_1 = \alpha. \tag{23.10}$$

A non-increasing function $\tilde{A}(p)$ taking values on $[0, 1]$ and satisfying (23.10) is termed a "conditional Type I error function". Such a function defines the rejection region for a two-stage combination test with H_0 being rejected if and only if $P^{(2)}$, the second stage P-value, is less than $\tilde{A}(P^{(1)})$.

The overall testing procedure can be described by saying that $\tilde{A}(P^{(1)})$ is calculated from the stage 1 data and then, in the second stage, a hypothesis test is carried out at significance level $\tilde{A}(P^{(1)})$. The condition (23.10) ensures the unconditional Type I error probability is α, as required. The design of the second stage does not have to be decided in advance of the results of the first stage as long as the significance level $\tilde{A}(P^{(1)})$ is applied in stage 2. This connection between combination tests and the conditional error probability has been pointed out by Posch & Bauer (1999), Wassmer (2000, Sec. 3.1.3) and Proschan (2003) among others.

There are many possible choices for a conditional error function $\tilde{A}(P_1)$. Proschan & Hunsberger (1995) proposed a design in which it is possible to extend a study when the P-value after the sample size originally planned is small but not statistically significant. They proposed tests defined through a "circular" conditional error function

$$\tilde{A}(p_1) = \begin{cases} 1 & \text{for } p_1 \leq c, \\ 1 - \Phi[\sqrt{\{(\Phi^{-1}(1-c))^2 - (\Phi^{-1}(1-p_1))^2\}}] & \text{for } c < p_1 < d, \\ 0 & \text{for } p_1 \geq d, \end{cases}$$

where c and d are constants chosen to satisfy (23.10). Using this conditional error function, the second stage is not required if $P_1 < c$, in which case H_0 is rejected, or if $P_1 > d$, in which case H_0 is accepted.

23.6 Use of Conditional Type I Error Probability in Unplanned Adaptations

Müller & Schäfer (2001, 2004) present general methodology for flexible adaptive designs in which the overall Type I error rate is controlled by preserving the conditional Type I error probability before and after a design modification. They have termed this probability the "conditional rejection probability" (CRP).

As an example, consider a fixed sample study designed to test H_0: $\theta \leq 0$ against $\theta > 0$ at significance level $\alpha = 0.025$. The planned sample size per treatment arm is $N = 200$ and H_0 will be rejected if the standardized Wald statistic Z_N exceeds $\Phi^{-1}(1-\alpha) = 1.96$ or, equivalently, if the P-value P_N is less than 0.025. At an interim point in the trial after responses from $n_1 = 120$ subjects per treatment arm have been observed, an unplanned analysis is performed resulting in a Wald statistic $Z_1 = z_1$ and corresponding P-value $P_1 = p_1 = 1 - \Phi(z_1)$. Based on this information, the investigators wish to adjust the remaining sample size in order to increase the power of their study, as in the motivating example of Cui, Hung & Wang (1999) described in Section 22.1. It is decided that the remaining sample size will be increased from 80 to 120 subjects per treatment arm. Although this adaptation was unplanned, the CRP principle can be applied to control the Type I error at $\alpha = 0.025$. Under the original design, the final test statistic would have been

$$Z_N = \sqrt{\frac{n_1}{N}}\, Z_1 + \sqrt{\frac{N - n_1}{N}}\, Z^{(2)}$$

where $Z^{(2)}$ denotes the Z-statistic based on the observations from the remaining $N - n_1$ subjects per treatment arm. Under $\theta = 0$, we have $Z^{(2)} \sim N(0, 1)$ independently of Z_1, so the conditional Type I error probability given $Z_1 = z_1$ is

$$Pr_{\theta=0}\left\{\sqrt{\frac{n_1}{N}}\, Z_1 + \sqrt{\frac{N - n_1}{N}}\, Z^{(2)} > 1.96 \mid Z_1 = z_1\right\} = \Phi\left\{\frac{z_1\sqrt{n_1} - 1.96\sqrt{N}}{\sqrt{N - n_1}}\right\}.$$

Thus, the original design can be considered to have a conditional error function $\tilde{A}(p_1)$ at the time of the unplanned analysis obtained by substituting $N = 200$, $n_1 = 120$ and $z_1 = \Phi^{-1}(1-p_1)$ in the above expression. Applying the CRP, after the change in sample size, H_0 should be rejected if and only if $\tilde{P}^{(2)} \leq \tilde{A}(p_1)$ where $\tilde{P}^{(2)}$ denotes the one-sided P-value for testing H_0 based on the responses observed after the unplanned analysis. Another way to view this is to say that a test of H_0 at significance level $\tilde{A}(p_1)$ is conducted based on the second stage data alone. Since $\tilde{A}(p_1)$ is the conditional Type I error probability under $\theta = 0$ at the time of the adaptation, it follows from Equation (23.10) that the overall Type I error rate is controlled at level α.

This argument can be extended to other situations as long as the quantity $\tilde{A}(p_1)$ can be computed. As examples, Müller & Schäfer (2001, 2004) discuss adaptation of group sequential tests defined through an error spending function, and Irle & Schäfer (2012) consider studies with a time-to-event endpoint.

If an adaptation really is unplanned, the above method is essentially the only option to test a hypothesis with control of the Type I error rate. Jennison & Turnbull (2003, p. 977) prove that a testing procedure that controls the Type I error probability after a flexible, possibly unplanned, adaptation must maintain the conditional Type I error probability. Their reasoning is as follows. Consider a study in which adaptation occurs after an initial stage and let $D(1)$ denote the data observed up to this point. Let $\tilde{A}(D(1))$ be the conditional Type I error probability under $\theta = 0$ if the study were to continue without modification and let $\tilde{A}_X(D(1))$ be the conditional error function under a particular adaptive strategy for conducting the remainder of the study and analyzing the results. Let $\mathcal{D}$ be the set of outcomes $D(1)$ for which $\tilde{A}_X(D(1)) > \tilde{A}(D(1)$ and suppose $D(1)$ has probability greater than zero under $\theta = 0$. Now define another adaptive procedure that follows the stated adaptive re-design when $D(1) \in \mathcal{D}$ but continues with the original design when $D(1) \notin \mathcal{D}$. Given the flexibility available to investigators, this is a permissible design strategy. This new adaptive design has conditional error function $\tilde{A}_Y(D(1)) = \tilde{A}(D(1))$ for $D(1) \notin \mathcal{D}$ and $\tilde{A}_Y(D(1)) > \tilde{A}(D(1))$ for $D(1) \in \mathcal{D}$. Since the original design has Type I error probability α, we know that

$$E_{\theta=0}\{\tilde{A}(D(1))\} = \alpha$$

and hence we must have

$$E_{\theta=0}\{\tilde{A}_Y(D(1))\} > \alpha,$$

which means the Type I error rate for the second adaptive procedure is greater than α. We conclude that if an adaptive procedure may be applied flexibly, it must never increase the conditional Type I error probability if the overall Type I error rate is to be protected.

The CRP approach does have its limitations. In order to compute the function $\tilde{A}(p_1)$, we need to know how a study would proceed if no adaptation were made. Consider our initial example in which 120 responses on each treatment had been observed at an unplanned analysis. Suppose this analysis was conducted when investigators had expected to see 75% of the final set of data, i.e., 150 responses. It could be that the slower than anticipated recruitment rate would have led to a final sample size below that originally planned, giving a final sample size of $N = 160$ per treatment. Alternatively the study might have been extended to reach the original target, $N = 200$. Calculating $\tilde{A}(p_1)$ in these two different scenarios gives two different answers, so which of these should be used in applying the CRP approach? This problem is exacerbated in a survival study where future information levels depend on the numbers of observed failures, which are inherently unpredictable. Öhrn (2011, Ch. 3) considers this question and presents examples where the overall Type I error rate can be inflated by up to 10% if one applies the CRP approach, always calculating the conditional error rate from the future scenario that gives the highest value. This problem may be avoided by stating that a combination test with two or more stages will be used to test H_0 and specifying the weights for the data from each stage at the start of the trial.

23.7 Group Sequential Designs and Multi-stage Combination Tests

23.7.1 Lehmacher-Wassmer Adaptive Designs

We noted in Section 23.4 that multi-stage combination tests can be viewed as group sequential procedures whereby after each stage, a decision is made to either stop or to continue to the next stage. In Chapter 22 we described the method proposed by Lehmacher & Wassmer (1999) in which conventional group sequential boundaries are applied to the sequence of statistics arising from weighted inverse normal adaptive combination tests.

To recap, the Lehmacher-Wassmer method starts with a one-sided group sequential test of H_0: $\theta \le 0$ of the form

$$\begin{array}{ll} \text{After group } k = 1, \ldots, K-1 & \\ \quad \text{if } Z_k \ge b_k & \text{stop, reject } H_0 \\ \quad \text{if } Z_k \le a_k & \text{stop, accept } H_0 \\ \quad \text{otherwise} & \text{continue to group } k+1, \\ \text{after group } K & \\ \quad \text{if } Z_K \ge b_K & \text{stop, reject } H_0 \\ \quad \text{if } Z_K < a_K & \text{stop, accept } H_0, \end{array} \tag{23.11}$$

where $a_K = b_K$. Here the $\{a_k, b_k\}$ have been chosen so that the Type I error probability is equal to α when $\{Z_1, \ldots, Z_K\}$ have the canonical joint distribution (3.1) with $\theta = 0$ for a given sequence of cumulative information levels $\{\mathcal{I}_1, \ldots \mathcal{I}_K\}$. Stage-wise statistics $\{Z^{(1)}, \ldots, Z^{(K)}\}$ are defined which, under $\theta = 0$, are independent with $N(0,1)$ distributions. Setting weights $w_1 = \sqrt{\mathcal{I}_1}$ and $w_k = \sqrt{(\mathcal{I}_k - \mathcal{I}_{k-1})}$ for $k = 2, \ldots, K$, the Lehmacher-Wassmer procedure is based on the cumulative statistics

$$\tilde{Z}_k = (w_1 Z^{(1)} + \ldots + w_k Z^{(k)}) \,/\, (w_1^2 + \ldots + w_k^2)^{1/2}, \quad k = 1, \ldots, K, \tag{23.12}$$

and these have the same joint distribution under $\theta = 0$ as the original $Z_1, \ldots, Z_K$. Hence, boundary crossing probabilities under $\theta = 0$ are the same for both sequences and the Type I error probability α is maintained. When adaptation occurs, as in Section 23.2, the stage-wise statistics $Z^{(k)}$ retain their independent $N(0,1)$ distributions under $\theta = 0$. Hence the boundary crossing probabilities under $\theta = 0$ are not affected by the adaptation and the Type I error probability remains at level α.

One may view the Lehmacher-Wassmer procedure as beginning with a conventional group sequential design which differs from an error spending design in that the weights w_k are fixed and do not change in response to observed information levels. Nevertheless, the Type I error rate is protected exactly at level α. In normal circumstances, this initial design will be followed. However, it is permissible to make an adaptation at analysis k after previously observing $\tilde{Z}_1, \ldots, \tilde{Z}_k$ with $a_i < \tilde{Z}_i < b_i$ for each $i = 1, \ldots, k$. If an adaptation, such as an increase in group size, is made at this point, the statistics $Z^{(k+1)}, \ldots, Z^{(K)}$ will be based on revised sample sizes in future groups but $\tilde{Z}_{k+1}, \ldots, \tilde{Z}_K$ will retain their original distribution under $\theta = 0$ so, when the boundary values $\{a_{k+1}, b_{k+1}, \ldots a_K, b_K\}$ are applied, the Type I error probability will remain at α. Power will not stay the same, but should increase when group sizes are increased.

A significant advantage of this construction is that it allows investigators to make an initial choice of an efficient group sequential design knowing that under normal conditions this will be used throughout the trial, but with the reassurance that, if necessary, it will be possible to make modifications in a way that protects the Type I error rate.

23.7.2 Variance Spending Designs

Fisher (1998) proposed an adaptive design in which the final sample size is modified in response to observed data. He describes the method in an example with a single treatment arm that produces independent responses $X_i \sim N(\mu, \sigma^2)$, $i = 1, 2, \ldots$, where the null hypothesis H_0: $\mu \leq 0$ is to be tested against $\mu > 0$ with Type I error probability α. Monitoring is continuous and each observation X_i is assigned a weight $a_i(X_1, \ldots, X_{i-1})$ that depends on all the previously observed data. The process of assigning weights is such that with probability 1 there is a finite number m, determined by $X_1, \ldots, X_{m-1}$, for which

$$\sum_{i=1}^{m} a_i(X_1, \ldots, X_{i-1})^2 = 1.$$

The study is terminated when this sample size m is reached. Fisher defines a weighted test statistic, which has the standardized version

$$Z = \frac{1}{\sigma} \sum_{i=1}^{m} a_i(X_1, \ldots, X_{i-1})\, X_i. \tag{23.13}$$

He shows that, under $\mu = 0$,

$$Z \sim N(0, 1)$$

and uses this fact in the final hypothesis test, which rejects H_0 if $Z \geq \Phi^{-1}(1 - \alpha)$. Fisher refers to this as a *variance spending* design since the total variance of 1 is "spent" during the course of the trial through the variances $a_i(X_1, \ldots, X_{i-1})^2$ of terms in the sum (23.13). This approach should not be confused with the error spending methods of Chapter 8: in a trial with a single treatment arm, error spending methods apply equal weight to each observation but adjust stopping boundaries to the observed sample size at each analysis.

A special case of this variance spending design yields a two-stage combination test. To achieve this, we specify a first group size n_1, and values w_1 and w_2 satisfying $w_1^2 + w_2^2 = 1$. Observations in the first group are assigned weights

$$a_i = \frac{w_1}{\sqrt{n_1}}, \quad i = 1, \ldots, n_1,$$

and, after observing $X_1, \ldots, X_{n_1}$, we choose the second group size n_2 and assign these observations weights

$$a_i = \frac{w_2}{\sqrt{n_2}}, \quad i = n_1 + 1, \ldots, n_1 + n_2.$$

The statistic (23.13) is then the weighted inverse normal statistic $w_1 Z^{(1)} + w_2 Z^{(2)}$, where

$$Z^{(1)} = \frac{1}{\sigma} \sum_{i=1}^{n_1} \frac{w_1}{\sqrt{n_1}} X_i \quad \text{and} \quad Z^{(2)} = \frac{1}{\sigma} \sum_{i=n_1+1}^{n_1+n_2} \frac{w_2}{\sqrt{n_2}} X_i$$

are the Z-values computed from data in group 1 and group 2, respectively.

The same approach may be followed with more than two stages. Here, we specify that there will be $K > 2$ groups of observations and define weights $w_1, \ldots, w_K$ satisfying $w_1^2 + \ldots + w_K^2 = 1$. The size of the first group, n_1, is fixed, while for $k = 2, \ldots, K$, the sample size n_k for group k may depend on the responses observed in groups 1 to $k - 1$. Writing $m_k = n_1 + \ldots + n_k$, the weights in the variance spending formulation for responses in group k are

$$a_i = \frac{w_i}{\sqrt{n_k}}, \quad i = m_{k-1} + 1, \ldots, m_k.$$

Once again, the final statistic defined by (23.13) has a $N(0,1)$ distribution if $\mu = 0$ and rejecting H_0 if $Z \geq \Phi^{-1}(1-\alpha)$ gives a test with Type I error rate α.

This design with $K > 2$ stages can be viewed as a version of Lehmacher-Wassmer's K-stage design, but without the opportunity for early stopping. Shen & Fisher (1999) extend the variance spending approach to allow early stopping for futility, creating designs that achieve the Type I error rate requirement conservatively. The general form of the variance spending approach allows additional flexibility in that the number of remaining groups and their associated weights may be modified during the study. We shall explain how this feature may be incorporated in a Lehmacher-Wassmer design in the next Section.

23.7.3 Recursive Use of Adaptive Combination Tests

Brannath, Posch & Bauer (2002) define multi-stage combination tests by a recursive process, starting from a two-stage combination test. Suppose a study is designed initially with $K = 2$ stages and an inverse χ^2 test is to be used, so the null hypothesis H_0 will be rejected if

$$-2\log\{P^{(1)}P^{(2)}\} \geq \chi_4^2(\alpha),$$

where $P^{(1)}$ and $P^{(2)}$ are P-values based on first stage and second stage data, respectively. Now, after observing the first stage data and the P-value $P^{(1)}$, it is decided to run two further stages rather than one. Denote the P-values observed in these two stages by $Q^{(1)}$ and $Q^{(2)}$, and define $\tilde{P}^{(2)}$ as the solution to

$$-2\log(Q^{(1)}Q^{(2)}) = \chi_4^2(\tilde{P}^{(2)}).$$

If $Q^{(1)}$ and $Q^{(2)}$ are independent Uniform(0,1) random variables, independent of $P^{(1)}$, then $\tilde{P}^{(2)}$ is also Uniform(0,1) and independent of $P^{(1)}$, and we can create a level α test by rejecting H_0 if

$$-2\log\{P^{(1)}\tilde{P}^{(2)}\} \geq \chi_4^2(\alpha).$$

Note this is not the same as following (23.1) and rejecting H_0 if

$$-2\log\{P^{(1)}Q^{(1)}Q^{(2)}\} \geq \chi_6^2(\alpha),$$

and this approach would not give a valid α-level test. This illustrates the importance of pre-specifying the number of stages K and the combination rule to be used in combining data across stages.

Brannath, Posch & Bauer (2002) explain how to generalize this approach in a recursive strategy, leading to a trial with $K > 3$ stages. As an example, in a trial with a total of $K = 4$ stages, $Q^{(2)}$ would be replaced in the above description by a P value $\tilde{Q}^{(2)}$ based on P-values $R^{(1)}$ and $R^{(2)}$, say, from data observed in stages 3 and 4, respectively. Furthermore, early stopping at each analysis can be introduced by using truncated two-stage combination tests of the form (23.7) in place of the simple combination tests in the preceding description.

An alternative way to construct a group sequential design with a data-dependent number of groups is to start by defining a Lehmacher-Wassmer procedure with a specific number of groups K and a stopping boundary as defined in (23.11). Now suppose the study has continued to analysis k^* and the observed value of $\tilde{Z}_k^*$ is $\tilde{z}_k^*$ where $a_k^* < \tilde{z}_k^* < b_k^*$, and it is decided to change the total number of analyses from K to K'. Applying the conditional Type I error probability principle of Section 23.6, one can calculate

$$P^* = Pr_{\theta=0}\{H_0 \text{ will be rejected at an analysis } k \in \{k^*+1,\ldots,K\} \mid \tilde{Z}_k^* = \tilde{z}_k^*\}$$

under the original design. One then constructs a stopping boundary to apply to data at analyses k^*+1 to K' such that the probability under $\theta = 0$ of crossing the upper boundary and rejecting H_0 is equal to P^*.

23.7.4 The Benefits of Simplicity

Although the preceding methods for increasingly exotic adaptive designs are valid in that there are formal proofs that the Type I error probability is protected, a word of caution is appropriate. Regulators have expressed the view that more complex designs offer increased opportunity that erroneous conclusions may be reached. The European Medicines Agency (CHMP, 2007, Sec. 4.2.1) states:

> "In general, changes to the design of an ongoing phase III trial are not recommended even though, in an adaptive design, such changes could be introduced with full control of the type 1 error. If design changes are anticipated in a confirmatory clinical trial this would require pre-planning and a clear justification from an experimental point of view."

and

> "If an adaptive design is used, the number of design modifications should be limited."

The FDA Guidance for Industry on Adaptive Designs for Clinical Trials of Drugs and Biologics (FDA, 2019a, Sec. III C) is more positive but still states:

> "In general, as with any clinical trial, it is expected that the details of the adaptive design are completely specified prior to initiation of the trial and documented accordingly."

and it continues in Section VIII B:

> "To allow for a thorough FDA evaluation, the documented plan for a clinical trial with an adaptive design will necessarily be more complex than for a trial with a non-adaptive design. ... documentation submitted to the Agency prior to initiation of an adaptive design trial should include ... a detailed description of the adaptation plan, including the anticipated number and timing of interim analyses, the specific aspects of the design that may be modified, and the rule that will be used to make adaptation decisions."

Thus, we would recommend making use of methods that can achieve the desired objectives in as simple a manner as possible. If the only adaptation in a trial will be to stop at an interim analysis with a decision to reject or accept the null hypothesis, a group sequential test can be used and the error spending approach provides an excellent way of handling unpredictable group sizes or increments in information. If other adaptations are anticipated, a Lehmacher-Wassmer design combines the facility to stop and make an early decision with the possibility of modifying group sizes and information levels at future analyses in response to observed data, while controlling the Type I error rate in a precise and transparent manner. Also, as we shall see in the next section, methods for making more complete inferences on termination are available for Lehmacher-Wassmer designs.

23.8 Analysis Following an Adaptive Trial

23.8.1 Introduction

In Chapter 9 we presented methods to calculate P-values, confidence intervals and point estimates following a group sequential test. We now address the same questions following an adaptive trial. We shall focus on the case of a trial analyzed using a two-stage combination test or a multi-stage trial with an adaptive group sequential design of the form proposed by Lehmacher & Wassmer (1999). As in Chapter 9, the sample space is complex in that the observed sample size and estimated treatment effect both vary. An additional problem in a flexible adaptive design is that it may not be possible to say with certainty how the trial would have proceeded if different data had been observed.

Throughout this section, we shall suppose a study is conducted to investigate a treatment effect θ. In a study with K analyses, at the first analysis,

$$Z^{(1)} \sim N(\theta\sqrt{\mathcal{I}^{(1)}}, 1)$$

and, for $k = 2, \ldots, K$, the additional information arising between analyses $k-1$ and k provides a Z-value

$$Z^{(k)} \mid \mathcal{I}^{(k)} \sim N(\theta\sqrt{\mathcal{I}^{(k)}}, 1),$$

where $\mathcal{I}^{(k)} = \mathcal{I}_k - \mathcal{I}_{k-1}$ is the increment in information for θ between analyses $k-1$ and k, as described in Section 23.2. The adaptive nature of the design implies that $\mathcal{I}^{(k)}$ may depend on the previously observed $Z^{(1)}, \ldots, Z^{(k-1)}$ and $\mathcal{I}^{(1)}, \ldots, \mathcal{I}^{(k-1)}$, but $Z^{(k)}$ is conditionally independent of previous data given $\mathcal{I}^{(k)}$.

For the most part, we shall confine our discussion to trial designs based on weighted inverse normal combination statistics, including Lehmacher-Wassmer designs for studies with more than two analyses. Similar constructions are possible for other methods, such as those based on the inverse chi-squared test statistic; Wassmer & Brannath (2016, Ch. 8) describe these methods in considerable generality.

23.8.2 P-values and Confidence Intervals after an Adaptive Trial

23.8.2.1 P-value after a Two-stage Adaptive Trial

The one-sided P-value for testing H_0: $\theta \leq 0$ is defined as the probability under $\theta = 0$ of an outcome as extreme or more extreme than that observed. If $\theta = 0$, this P-value should have a Uniform(0,1) distribution or a conservative P-value may be stochastically larger than a Uniform(0,1) random variable.

As explained in Section 23.2, if $\theta = 0$, in a two-stage adaptive trial

$$Z^{(1)} \sim N(0,1), \quad Z^{(2)} \sim N(0,1),$$

and $Z^{(1)}$ and $Z^{(2)}$ independent. Suppose the trial has a truncated two-stage design with a weighted inverse normal combination rule as described in (23.8). Thus, the trial stops at analysis 1 to accept H_0 if $Z^{(1)} \leq a_1$ or to reject H_0 if $Z^{(1)} \geq b_1$. Then, if it continues to analysis 2, H_0 is rejected if

$$w_1 Z^{(1)} + w_2 Z^{(2)} \geq b_2.$$

We summarize the outcome from this design as

$$\begin{array}{ll} (1, Z^{(1)}) & \text{if the trial stops at analysis 1,} \\ (2, Z^{(1)}, Z^{(2)}) & \text{if the trial stops at analysis 2.} \end{array}$$

We shall write this outcome variable as (T, Z), where T denotes the analysis at which the trial terminates and

$$Z = (Z^{(1)}, \ldots, Z^{(T)}),$$

and we denote an observed outcome by (t, z). In determining whether one outcome is more "extreme" than another, we need to specify an ordering of the sample space. We define a stage-wise ordering similar to that of Section 9.4 where, for consistency with the weighted inverse normal hypothesis test, we say $(t', z') \succ (t, z)$ if any one of the following four conditions holds:

$$\begin{array}{lll} \text{(i)} & t' = t = 1 & \text{and} \;\; z'^{(1)} \geq z^{(1)}, \\ \text{(ii)} & t' = t = 2 & \text{and} \;\; w_1 z'^{(1)} + w_2 z'^{(2)} \geq w_1 z^{(1)} + w_2 z^{(2)}, \\ \text{(iii)} & t' = 1, \; t = 2 & \text{and} \;\; z'^{(1)} \geq b_1, \\ \text{(iv)} & t' = 2, \; t = 1 & \text{and} \;\; z^{(1)} \leq a_1. \end{array} \qquad (23.14)$$

On observing the outcome (t^*, z^*), the P-value for testing $H_0\colon \theta \le 0$ is

$$Pr_{\theta=0}\{(T, Z) \succ (t^*, z^*)\} \tag{23.15}$$

and this can be computed numerically using the fact that $Z^{(1)}$ and $Z^{(2)}$ are independent $N(0, 1)$ variates under $\theta = 0$.

The above approach can also be followed to define a P-value after a study conducted using a truncated two-stage design with an inverse χ^2 combination rule, as defined in(23.7). With $P^{(1)} = 1 - \Phi(Z^{(1)})$ and $P^{(2)} = 1 - \Phi(Z^{(2)})$, the four conditions under which $(t', z') \succ (t, z)$ become

$$\begin{array}{lll}
\text{(i)} & t' = t = 1 & \text{and } \; p'^{(1)} \le p^{(1)}, \\
\text{(ii)} & t' = t = 2 & \text{and } \; p'^{(1)}p'^{(2)} \le p^{(1)}p^{(2)}, \\
\text{(iii)} & t' = 1,\ t = 2 & \text{and } \; p'^{(1)} \le c_1, \\
\text{(iv)} & t' = 2,\ t = 1 & \text{and } \; p^{(1)} \ge d_1
\end{array} \tag{23.16}$$

and the P-value is defined by (23.15) using this ordering.

23.8.2.2 Confidence Interval after a Two-stage Adaptive Trial

We consider the construction of an equal tailed $(1 - \alpha)$-level confidence interval for the treatment effect θ after an adaptive trial. The approach parallels that described in Section 9.5 for calculating a confidence interval after a group sequential test. To do this, we create a family of tests of hypotheses $H\colon \theta = \theta_0$, each with two-sided Type I error probability α, and the confidence interval comprises the set of values θ_0 whose hypotheses are not rejected.

Suppose the trial has a truncated two-stage design with a weighted inverse normal combination rule as described in (23.8). Let (T, Z) denote the trial outcome with T the analysis at which the trial terminates and $Z = (Z^{(1)}, \ldots, Z^{(T)})$, and suppose outcomes are ordered as in (23.14). (A similar approach can be followed for a design based on the inverse chi-squared combination statistic). Then, for a given value of θ_0, we can find pairs $(t_u(\theta_0), z_u(\theta_0))$ and $(t_\ell(\theta_0), z_\ell(\theta_0))$ such that

$$Pr_{\theta=\theta_0}\{(T, Z) \succeq (t_u(\theta_0), z_u(\theta_0))\} = \alpha/2$$

and

$$Pr_{\theta=\theta_0}\{(T, Z) \preceq (t_\ell(\theta_0), z_\ell(\theta_0))\} = \alpha/2.$$

If $t_u(\theta_0) = 2$, there will be a set of values $z_u(\theta_0)$ with $w_1 z_u^{(1)}(\theta_0) + w_2 z_u^{(2)}(\theta_0)$ equal to some constant which are equally placed in the ordering (23.14) and any one of these may be selected as the pair $(t_u(\theta_0), z_u(\theta_0))$. Similarly, if $t_\ell(\theta_0) = 2$, a representative pair with $w_1 z_\ell^{(1)}(\theta_0) + w_2 z_\ell^{(2)}(\theta_0)$ equal to the appropriate value may be chosen.

The acceptance region

$$A(\theta_0) = \{(t, z)\colon (t_\ell(\theta_0), z_\ell(\theta_0)) \prec (t, z) \prec (t_u(\theta_0), z_u(\theta_0))\}$$

defines a two-sided hypothesis test of $H\colon \theta = \theta_0$ with Type I error probability α. Standard arguments imply that the set

$$\{\theta_0\colon (T, Z) \in A(\theta_0)\}$$

obtained by inverting this family of tests is a $(1 - \alpha)$-level, equal tailed confidence set for θ.

We note that the ordering (23.14) of the sample space is consistent with the group sequential two-sided test of $H_0\colon \theta = 0$ vs $\theta \ne 0$ or the one-sided test of $H_0\colon \theta \le 0$ vs $\theta > 0$.

In the first case, the acceptance region $A(0)$ arising in the construction of a $(1-\alpha)$-level confidence interval contains precisely those outcomes for which the group sequential test of H_0: $\theta = 0$ with two-sided Type I error probability α fails to reject H_0. Hence the $(1-\alpha)$-level confidence interval for θ on termination excludes $\theta = 0$ if and only if H_0 is rejected in the group sequential test. In the second case, the group sequential one-sided test of H_0: $\theta \leq 0$ with Type I error probability α rejects H_0 if and only if a $(1-2\alpha)$-level confidence interval lies completely above zero.

It is instructive to consider an alternative description of the above method. Consider the test of H: $\theta = \theta_0$. If $\theta = \theta_0$, we have

$$Z^{(1)} \sim N(\theta_0 \sqrt{\mathcal{I}^{(1)}}, 1)$$

and, when continuing to analysis 2,

$$Z^{(2)} \mid Z^{(1)}, \mathcal{I}^{(2)} \sim N(\theta_0 \sqrt{\mathcal{I}^{(2)}}, 1).$$

Let

$$Z_0^{(1)} = Z^{(1)} - \theta_0 \sqrt{\mathcal{I}^{(1)}} \quad \text{and} \quad Z_0^{(2)} = Z^{(2)} - \theta_0 \sqrt{\mathcal{I}^{(2)}}.$$

Then, under $\theta = \theta_0$, $Z_0^{(1)} \sim N(0,1)$, $Z_0^{(2)} \sim N(0,1)$ and $Z_0^{(1)}$ and $Z_0^{(2)}$ are independent. In the procedure defined by (23.8), the trial stops at analysis 1 if $Z_0^{(1)} \leq a_1 - \theta_0 \sqrt{\mathcal{I}^{(1)}}$ or $Z_0^{(1)} \geq b_1 - \theta_0 \sqrt{\mathcal{I}^{(1)}}$. We denote the outcome by (T, Z_0), where $T \in \{1, 2\}$ is the analysis at which the trial terminates and

$$Z_0 = (Z_0^{(1)}, \ldots, Z_0^{(T)}),$$

and we order the outcomes (t, z_0) by saying $(t', z_0') \succ (t, z_0)$ if any one of the following four conditions holds:

$$\begin{array}{llll}
\text{(i)} & t' = t = 1 & \text{and} & z_0'^{(1)} \geq z_0^{(1)}, \\
\text{(ii)} & t' = t = 2 & \text{and} & w_1 z_0'^{(1)} + w_2 z_0'^{(2)} \geq w_1 z_0^{(1)} + w_2 z_0^{(2)}, \\
\text{(iii)} & t' = 1,\ t = 2 & \text{and} & z_0'^{(1)} \geq b_1 - \theta_0 \sqrt{\mathcal{I}^{(1)}}, \\
\text{(iv)} & t' = 2,\ t = 1 & \text{and} & z_0^{(1)} \leq a_1 - \theta_0 \sqrt{\mathcal{I}^{(1)}}.
\end{array}$$

For each possible value of θ_0, we find the pairs $(t_u(\theta_0), z_u(\theta_0))$ and $(t_\ell(\theta_0), z_\ell(\theta_0))$ such that

$$Pr_{\theta=\theta_0}\{(T, Z_0) \succeq (t_u(\theta_0), z_u(\theta_0))\} = \alpha/2$$

and

$$Pr_{\theta=\theta_0}\{(T, Z_0) \prec (t_\ell(\theta_0), z_\ell(\theta_0))\} = \alpha/2,$$

then the acceptance region for a test of H: $\theta = \theta_0$ is

$$A(\theta_0) = \{(t, z_0) \colon (t_\ell(\theta_0), z_\ell(\theta_0)) \prec (t, z_0) \prec (t_u(\theta_0), z_u(\theta_0))\}$$

and a two-sided $1-\alpha$ level confidence interval for θ is

$$\{\theta_0 \colon (T, Z_0) \in A(\theta_0)\}.$$

Wassmer & Brannath (2016, p. 202) note that, since

$$w_1 z_0^{(1)} + w_2 z_0^{(2)} = w_1 z^{(1)} + w_2 z^{(2)} - \theta_0 (\sqrt{\mathcal{I}^{(1)}} + \sqrt{\mathcal{I}^{(2)}}),$$

the ordering of the sample space based on (T, Z_0) is identical to the previous ordering based on (T, Z). Thus, both approaches produce the same confidence interval for θ. In the special case where stopping at analysis 1 is not permitted, the second description shows that the test of H: $\theta = \theta_0$ is simply the weighted inverse normal combination test of this hypothesis.

23.8.2.3 P-value after a Multi-stage Adaptive Trial

We consider a Lehmacher-Wassmer design with K analyses designed to test H_0: $\theta \le 0$. This procedure allows adaptation at any analysis and is defined in terms of stage-wise statistics $\{Z^{(1)}, \ldots, Z^{(K)}\}$, which are independent with $N(0,1)$ distributions under $\theta = 0$. Weights $w_1, \ldots, w_K$ are pre-specified and the test applies the stopping rule (23.11), with each Z_k replaced by

$$\tilde{Z}_k = (w_1 Z^{(1)} + \ldots + w_k Z^{(k)}) \,/\, (w_1^2 + \ldots + w_k^2)^{1/2}, \quad k = 1, \ldots, K.$$

We define an outcome of this test to be $(T, \tilde{Z}_T)$ where T is the analysis at which the trial terminates. In a version of the stage-wise ordering of Section 9.4 appropriate to this design, we say that $(t', \tilde{z}') \succ (t, \tilde{z})$ if any one of the following three conditions holds:

$$\begin{array}{lll} \text{(i)} & t' = t \text{ and } & \tilde{z}' \ge \tilde{z}, \\ \text{(ii)} & t' < t \text{ and } & \tilde{z}' \ge b_{t'}, \\ \text{(iii)} & t' > t \text{ and } & \tilde{z} \le a_t. \end{array}$$

On observing the outcome $(t^*, \tilde{z}^*)$, the P-value for testing H_0: $\theta \le 0$ is

$$Pr_{\theta=0}\{(T, \tilde{Z}_T) \succ (t^*, \tilde{z}^*)\},$$

which can be computed numerically using the fact that $\tilde{Z}^{(1)}, \ldots, \tilde{Z}^{(K)}$ are independent $N(0,1)$ variates under $\theta = 0$. By construction, this P-value has exactly a Uniform(0,1) distribution when $\theta = 0$.

23.8.2.4 Confidence Interval after a Multi-stage Adaptive Trial

We now consider the construction of an equal tailed $(1-\alpha)$-level confidence interval for the treatment effect θ after an adaptive trial with more than two stages. This is a challenging problem and we shall focus on a specific trial design where a precise solution is possible. As in Section 23.8.2.3, we consider a trial with a Lehmacher-Wassmer design. Weights $w_1, \ldots, w_K$ are specified and the formula (23.12) is applied to define summary statistics $\tilde{Z}_k$. The stopping rule compares the $\tilde{Z}_k$ with boundary values (a_k, b_k), $k = 1, \ldots, K$. In addition, we suppose adaptation is only permitted at one pre-specified analysis, J, and a sequence of potential information levels, $\mathcal{I}_P^{(J+1)}, \ldots, \mathcal{I}_P^{(K)}$, associated with stages $J+1$ to K, is stipulated. The reason for this last condition will become apparent as we explain how a confidence interval can be calculated. In setting values for $\mathcal{I}_P^{(J+1)}, \ldots, \mathcal{I}_P^{(K)}$, we recommend using estimates of the information levels that would occur under the initial plans for the trial.

As for a two-stage trial, we shall create a family of tests of hypotheses H: $\theta = \theta_0$, each with two-sided Type I error probability α, and the confidence interval will comprise the set of values θ_0 whose hypotheses are not rejected. We denote the analysis at which the trial terminates by T and write $(T, \tilde{Z}_T)$ as the observed outcome. If the trial stops at an analysis $T \le J$, with $\tilde{Z}_T \le a_T$ or $\tilde{Z}_T \ge b_T$, only the stage-wise information levels $\mathcal{I}^{(1)}, \ldots, \mathcal{I}^{(T)}$ are observed. If $T > J$, then all of $\mathcal{I}^{(1)}, \ldots, \mathcal{I}^{(J)}$ are observed and, in this case, adaptation can take place at analysis J. Then, as the trial continues, new values $\mathcal{I}^{(J+1)}, \mathcal{I}^{(J+2)}, \ldots$ are observed until termination occurs.

In defining a stage-wise ordering for the test of H: $\theta = \theta_0$, we place outcomes with $T \le J$ and $\tilde{Z}_T \le a_T$ at the low end of the ordering and outcomes with $T \le J$ and $\tilde{Z}_T \ge b_T$ at the high end. Since no adaptation can occur before analysis J, for a given $\theta = \theta_0$, we know that $Z^{(1)}, \ldots, Z^{(J)}$ are independent with

$$Z^{(k)} \sim N(\theta_0 \sqrt{\mathcal{I}^{(k)}}, 1), \quad k = 1, \ldots, J.$$

Hence, for an outcome $(T, \tilde{Z}_T) = (t^*, \tilde{z}^*)$ with $t^* \leq J$, we can carry out computations to find the values θ_L and θ_U for which

$$Pr_{\theta_L}\{(T, \tilde{Z}_T) \succeq (t^*, \tilde{z}^*)\} = Pr_{\theta_U}\{(T, \tilde{Z}_T) \preceq (t^*, \tilde{z}^*)\} = \alpha/2,$$

and the $(1-\alpha)$-level confidence interval for θ is (θ_L, θ_U). Note that this calculation is possible even though we have not yet discussed how to order outcomes arising after adaptation at analysis J.

Now consider testing H: $\theta = \theta_0$ after an outcome $(T, \tilde{Z}_T) = (t^*, \tilde{z}^*)$ with $t^* > J$. Conditional on the sequence of values $\mathcal{I}^{(J+1)}, \ldots, \mathcal{I}^{(K)}$, the stage-wise Z-values $Z^{(J+1)}, \ldots, Z^{(K)}$ are independent with

$$Z^{(k)} \sim N(\theta_0\sqrt{\mathcal{I}^{(k)}}, 1), \quad k = J+1, \ldots, K,$$

and, given the value of $\tilde{Z}_J$, the distribution of $\tilde{Z}_{J+1}, \ldots, \tilde{Z}_K$ follows. Since the values $\mathcal{I}^{(J+1)}, \ldots, \mathcal{I}^{(K)}$ depend on the adaptation made at analysis J, the full distribution of outcomes $(t^*, \tilde{z}^*)$ with $t^* > J$ depends on the rule that determines this adaptation. We shall assume some flexibility is allowed in making the adaptation and so we need a methodology that protects the error rate for the test of H: $\theta = \theta_0$ whatever adaptation is made. In Section 23.6, we presented the argument of Jennison & Turnbull (2003) that the only way to protect the Type I error rate in a test of H_0: $\theta = 0$ after adaptation is to preserve the conditional probability under $\theta = 0$ of rejecting H_0. By the same reasoning, in a test of H: $\theta = \theta_0$, the probabilities under $\theta = \theta_0$ of rejecting this hypothesis in either direction must stay the same before and after adaptation.

The reason for pre-specifying the "potential information levels" $\mathcal{I}_P^{(J+1)}, \ldots, \mathcal{I}_P^{(K)}$ is that this allows an unambiguous definition of a testing procedure prior to any adaptation. Since the observed information levels may not have taken these values in the absence of any adaptation, we may regard the procedure with information levels $\mathcal{I}_P^{(J+1)}, \ldots, \mathcal{I}_P^{(K)}$ as a "reference test". Given an initial sequence of observed stage-wise information levels $\mathcal{I}^{(1)}, \ldots, \mathcal{I}^{(J)}$, and supposing the subsequent sequence of values $\mathcal{I}_P^{(J+1)}, \ldots, \mathcal{I}_P^{(K)}$, the reference test of H: $\theta = \theta_0$ is based on the stage-wise ordering of the sample space in which $(t', \tilde{z}') \succ (t, \tilde{z})$ if any one of the following three conditions holds:

$$\begin{array}{llll} \text{(i)} & t' = t & \text{and} & \tilde{z}' \geq \tilde{z}, \\ \text{(ii)} & t' < t & \text{and} & \tilde{z}' \geq b_{t'}, \\ \text{(iii)} & t' > t & \text{and} & \tilde{z} \leq a_t. \end{array} \tag{23.17}$$

We can carry out calculations to find the pairs $(t_u(\theta_0), \tilde{z}_u(\theta_0))$ and $(t_\ell(\theta_0), \tilde{z}_\ell(\theta_0))$ such that

$$Pr_{\theta=\theta_0}\{(T, \tilde{Z}_T) \succeq (t_u(\theta_0), \tilde{z}_u(\theta_0))\} = \alpha/2$$

and

$$Pr_{\theta=\theta_0}\{(T, \tilde{Z}_T) \preceq (t_\ell(\theta_0), \tilde{z}_\ell(\theta_0))\} = \alpha/2$$

when information levels are $\mathcal{I}_P^{(J+1)}, \ldots, \mathcal{I}_P^{(K)}$. Hence, we may compute the conditional error probabilities of the reference test give $\tilde{Z}_J = \tilde{z}_J$,

$$CP_1(\tilde{z}_J; \theta_0) = Pr_{\theta=\theta_0}(\text{Reject } H\text{: } \theta = \theta_0 \text{ in favor of } \theta < \theta_0 \text{ at an analysis } k \in \{J+1, \ldots, K\} \mid \tilde{Z}_J = \tilde{z}_j)$$

and

$$CP_2(\tilde{z}_J; \theta_0) = Pr_{\theta=\theta_0}(\text{Reject } H\text{: } \theta = \theta_0 \text{ in favor of } \theta > \theta_0 \text{ at an analysis } k \in \{J+1, \ldots, K\} \mid \tilde{Z}_J = \tilde{z}_j)$$

for values of $\tilde{z}_J \in (a_J, b_J)$.

We note first that if $t_\ell(\theta_0) \leq J$ and $\tilde{z}_\ell(\theta_0) \leq a_{t_\ell(\theta_0)}$, then rejection of H: $\theta = \theta_0$ in favor of $\theta < \theta_0$ at an analysis $k \in \{J+1, \ldots, K\}$ is not possible and $CP_1(\tilde{z}_J; \theta_0) = 0$. On the other hand, if $t_\ell(\theta_0) \leq J$ and $\tilde{z}_\ell(\theta_0) \geq b_{t_\ell(\theta_0)}$, rejection of H: $\theta = \theta_0$ in favor of $\theta < \theta_0$ after continuing to analysis J is certain and $CP_1(\tilde{z}_J; \theta_0) = 1$. Similarly, if $t_u(\theta_0) \leq J$ and $\tilde{z}_u(\theta_0) \leq a_{t_u(\theta_0)}$, then $CP_2(\tilde{z}_J; \theta_0) = 1$, while if $t_u(\theta_0) \leq J$ and $\tilde{z}_u(\theta_0) \geq b_{t_u(\theta_0)}$, then $CP_2(\tilde{z}_J; \theta_0) = 0$. In maintaining the conditional error probabilities of the reference test, we simply reject H: $\theta = \theta_0$ in favor of $\theta < \theta_0$ if $CP_1(\tilde{z}_J; \theta_0) = 1$ and do not reject H_0 in favor of $\theta < \theta_0$ if $CP_1(\tilde{z}_J; \theta_0) = 0$. Likewise we reject H: $\theta = \theta_0$ in favor of $\theta > \theta_0$ if $CP_2(\tilde{z}_J; \theta_0) = 1$ and do not reject H_0 in favor of $\theta > \theta_0$ if $CP_2(\tilde{z}_J; \theta_0) = 0$.

Suppose now that we observe $\tilde{Z}_J = \tilde{z}_J$ at analysis J, where $0 < CP_1(\tilde{z}_J; \theta_0) < 1$, and the trial continues until it terminates at analysis t^* with $\tilde{Z}_{t^*} = \tilde{z}_{t^*}$ and observed stage-wise information levels $\tilde{\mathcal{I}}^{(J+1)}, \ldots, \tilde{\mathcal{I}}^{(t^*)}$. We calculate the conditional probability of observing such a low outcome,

$$Pr_{\theta=\theta_0}\{(T, Z_T) \preceq (t^*, \tilde{z}_{t^*}) \mid \tilde{Z}_J = \tilde{z}_J, \tilde{\mathcal{I}}^{(J+1)}, \ldots, \tilde{\mathcal{I}}^{(t^*)}\},$$

where $\preceq$ refers to the ordering (23.17). If this probability is less than $CP_1(\tilde{z}_J; \theta_0)$, we reject H: $\theta = \theta_0$ in favor of $\theta < \theta_0$. Similarly, when $0 < CP_2(\tilde{z}_J; \theta_0) < 1$, we calculate

$$Pr_{\theta=\theta_0}\{(T, Z_T) \succeq (t^*, \tilde{z}_{t^*}) \mid \tilde{Z}_J = \tilde{z}_J, \tilde{\mathcal{I}}^{(J+1)}, \ldots, \tilde{\mathcal{I}}^{(t^*)}\}$$

and reject H: $\theta = \theta_0$ in favor of $\theta > \theta_0$ if this probability is less than $CP_2(\tilde{z}_J; \theta_0)$. Following this scheme, the conditional probability of rejecting H: $\theta = \theta_0$ in each direction is equal to that of the reference test and hence the test of H has the required error probabilities and, by inverting this family of tests, we obtain a $(1 - \alpha)$-level confidence interval for θ.

Somewhat strangely perhaps, it is implicit that the above procedure should be followed even when no features of the trial are changed at the "adaptation analysis" J. This is necessary since current knowledge may indicate the likely pattern of future information levels under the original design, conditional probabilities under these information levels will differ from those of the reference test, and conducting tests with these conditional error probabilities could undermine the procedure; see Öhrn (2011, Ch. 3) for further discussion of this point. We note, however, that this requirement has no impact for the test of H: $\theta = 0$ since the statistics $\tilde{Z}^{(J+1)}, \ldots, \tilde{Z}^{(K)}$ have independent $N(0,1)$ distributions under any sequence of information levels. A consequence of this distributional property is that, as we saw in a two-stage trial, the confidence interval for θ on termination is consistent with the outcome of the original group sequential test of H_0: $\theta = 0$ vs $\theta \neq 0$ or of H_0: $\theta \leq 0$ vs $\theta > 0$.

One might ask whether a simpler construction would be possible using Z-values

$$Z_0^{(k)} = Z^{(k)} - \theta_0 \sqrt{\mathcal{I}^{(k)}}, \quad k = 1, \ldots, K,$$

as we saw in the case of a two-stage trial. These statistics have the advantage that they are independent with $N(0,1)$ distributions under $\theta = \theta_0$ for any sequence of information levels. The problem is that the trial's stopping boundary is defined in terms of the combination statistics $\tilde{Z}_k$ formed from the unadjusted $Z^{(k)}$, and when this is converted to a boundary for combination statistics based on the $Z_0^{(k)}$, the stage-wise information levels appear in the formula.

Gao, Liu & Mehta (2013) describe a procedure for constructing a confidence interval for the treatment effect θ after an error spending group sequential test. During this test, adaptations are implemented by preserving the conditional Type I error probability under $\theta = 0$. The method for computing a confidence interval is described in terms of a "backwards image" of outcomes that occur after adaptation on the original sample space. Analysis of this method shows it is similar to the approach we have described for a Lehmacher-Wassmer design, in that tests of each hypothesis H: $\theta = \theta_0$ maintain the conditional error probability under $\theta = \theta_0$. (One may also apply the argument of Jennison & Turnbull (2003) to conclude this has to be the case if the confidence level $1-\alpha$ is to be achieved with a flexible adaptation rule.) A key difference in the method of Gao, Liu & Mehta (2013) is that predicted values of future information levels are used when computing the conditional error probability and, as we noted at the end of Section 23.6, this can be problematic when information levels are unpredictable. In our method for a Lehmacher-Wassmer design, we overcame this problem by introducing the "reference test" in which potential information levels $\mathcal{I}_P^{(J+1)}, \ldots, \mathcal{I}_P^{(K)}$ are assumed to arise after the adaptation analysis J.

23.8.3 Point Estimation after an Adaptive Trial

23.8.3.1 Estimation after a Two-stage Adaptive Trial with no Early Stopping

We saw in Section 9.3 that the maximum likelihood estimate (MLE) of the treatment effect θ after a group sequential trial is subject to bias. The same is true in an adaptive design, even when this does not permit early stopping. A point estimate of the treatment effect is often used by decision makers when they assess the cost and benefits of a new treatment and an estimate that may be positively biased is liable to treated skeptically. Thus, methods for eliminating, or at least reducing, bias in an estimated treatment effect are of practical importance.

Consider a trial comparing two treatments with responses $X_{Ai} \sim N(\mu_A, \sigma^2)$ and $X_{Bi} \sim N(\mu_B, \sigma^2)$ on treatments A and B, respectively. For simplicity, we shall assume the variance σ^2 is known. Denoting the treatment difference by $\theta = \mu_A - \mu_B$, suppose the trial is designed to test H_0: $\theta \leq 0$ against $\theta > 0$ with Type I error probability $\alpha = 0.025$. Assuming $\sigma^2 = 9$, a trial with a total sample size of 140 subjects on each treatment arm will have power 0.8 if $\theta = 1$.

Suppose the trial is conducted in two stages. In the first stage, responses from $n_1 = 70$ patients per treatment are observed and the estimate

$$\hat{\theta}^{(1)} = \bar{X}_A^{(1)} - \bar{X}_B^{(1)}$$

is obtained, where $\bar{X}_A^{(1)}$ and $\bar{X}_B^{(1)}$ are the mean responses on each treatment arm. The stage 2 sample size is modified adaptively, increasing by a factor of 3 if $\hat{\theta}^{(1)} < 0.5$. So, denoting the stage 2 sample size per arm by n_2, we have

$$n_2 = 70 \quad \text{if} \quad \hat{\theta}^{(1)} \geq 0.5 \quad \text{and} \quad n_2 = 210 \quad \text{if} \quad \hat{\theta}^{(1)} < 0.5.$$

A weighted inverse normal combination test with weights $w_1 = w_2 = 1/\sqrt{2}$ is used to test H_0: $\theta \leq 0$ and this protects the Type I error rate precisely at level α.

Suppose that a point estimate of θ is also required. Let

$$\hat{\theta}^{(2)} = \bar{X}_A^{(2)} - \bar{X}_B^{(2)},$$

where $\bar{X}_A^{(2)}$ and $\bar{X}_B^{(2)}$ are the means of the n_2 observations on each treatment arm in stage 2. Then, the MLE of θ is

$$\hat{\theta}_{ML} = \frac{n_1 \hat{\theta}^{(1)} + n_2 \hat{\theta}^{(2)}}{n_1 + n_2}$$

The bias function of this estimator,

$$b_{ML}(\theta) = E_\theta(\hat{\theta}_{ML}) - \theta, \qquad (23.18)$$

is displayed in Figure 23.5.

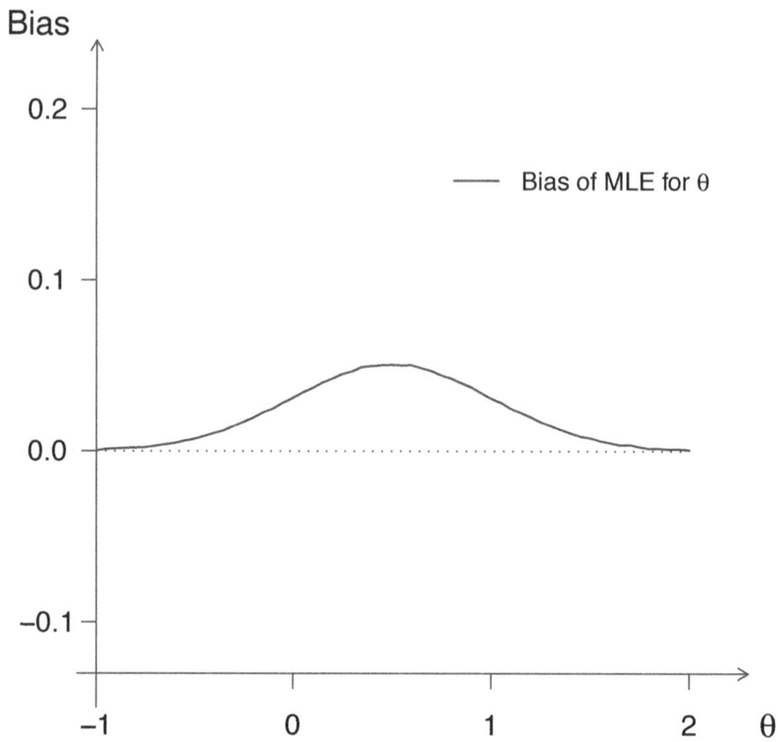

Figure 23.5 *Bias of the maximum likelihood estimator following a two-stage trial with adaptive sample size modification*

Brannath, König & Bauer (2006) have shown that, in a two-stage design of this type, the absolute value of this bias is at most 40% of the standard deviation of the first stage estimate. In our example this implies

$$|\,b_{ML}(\theta)\,| \;\leq\; 0.4\sqrt{(2\sigma^2/n_1)} \;=\; 0.4\sqrt{(2\times 9/70)} \;=\; 0.203.$$

In fact, the largest bias in our example is 0.051, which is considerably smaller than this maximum. We are able to calculate the bias here since we have assumed a particular rule for modifying the sample size. In a flexible adaptive trial, one would not have knowledge of this rule; also, other adaptive modifications of sample size could produce a larger bias.

It is not difficult to define an unbiased estimate of θ in a design with no early stopping, so that estimates $\hat{\theta}^{(1)}$ and $\hat{\theta}^{(2)}$ are always observed. Both $\hat{\theta}^{(1)}$ and $\hat{\theta}^{(2)}$ are unbiased estimates of θ. Hence, any estimate formed as a convex linear combination

$$\hat{\theta} \;=\; u\,\hat{\theta}^{(1)} + (1-u)\,\hat{\theta}^{(2)}, \tag{23.19}$$

where $u \in (0,1)$ is pre-specified, is an unbiased estimate of θ. Suppose, for example, that sample sizes n_1^* and n_1^* are anticipated in stages 1 and 2 and, accordingly, weights

$$w_1 \;=\; \sqrt{n_1^*/(n_1^*+n_1^*)} \quad \text{and} \quad w_2 \;=\; \sqrt{n_2^*/(n_1^*+n_1^*)}$$

are used to define the weighted inverse normal combination test. Then it would be natural to use $u = w_1^2$ and $1-u = w_2^2$ in (23.19) and this would give the MLE of θ, as defined in (23.18), if indeed $n_1 = n_1^*$ and $n_2 = n_2^*$.

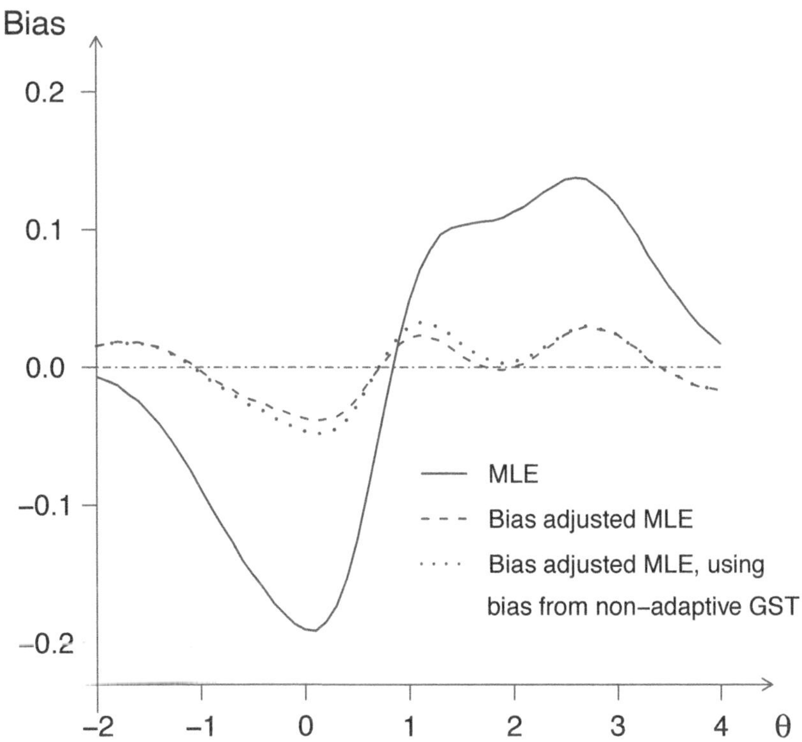

Figure 23.6 *Bias of the maximum likelihood estimator following a two-stage trial with adaptive sample size modification*

23.8.3.2 Estimation after a Multi-stage Adaptive Trial with Early Stopping

In Section 9.3.2 we described Whitehead's (1986a) method for adjusting the MLE on termination of a group sequential test. A similar approach can be followed to reduce the bias of the MLE following a group sequential test with adaptive sample size modification.

As in Section 9.3.2, we illustrate this methodology with the example of a Pampallona & Tsiatis one-sided test with $\Delta = 0$, four analyses, a binding futility boundary, one-sided Type I error probability 0.025, and power 0.8 at $\theta = 1$. In order to accommodate adaptive sample size modification, we apply this group sequential boundary in a Lehmacher-Wassmer design, as defined in Section 23.7.1. Suppose that adaptation can occur at analysis 2, where the boundary values are $a_2 = 0.664$ and $b_2 = 2.756$, and it is stipulated that future group sizes and associated increments in information are increased by a factor of 2 if $\tilde{Z}_2 \in (1.0, 2.0)$ but remain the same if the study continues with $\tilde{Z}_2 < 1.0$ or $\tilde{Z}_2 > 2.0$. Figure 23.6 shows the bias of the MLE on termination of this adaptive design.

In order to adjust the MLE, we ran 1 million simulations of the adaptive design under values of θ in the set $\{-4, -3.9, \ldots, 5.9, 6\}$ and estimated the bias function

$$b_{ML}(\theta) = E_\theta(\hat{\theta}_{ML}) - \theta$$

for each value of θ. The one-step bias adjusted estimator is defined as

$$\bar{\theta} = \hat{\theta}_{ML} - b(\hat{\theta}_{ML}). \tag{23.20}$$

In computing this estimator, we interpolated among our simulated values of $b_{ML}(\theta)$ to obtain a value for $b(\hat{\theta}_{ML})$. The bias of the adjusted estimator is shown as the dashed line in Figure 23.6. The results are remarkably similar to those seen in Figure 9.5 for estimates following a non-adaptive group sequential test. The bias of the unadjusted MLE can be as high as 14% or as low as −19% of the effect size $\theta = 1$ at which power was originally set, while the bias of the adjusted MLE is between −4% and 3% of this effect size.

Of course, in a flexible design one may not know how the sample size would have been modified if different data had been observed at the adaptation analysis. One can however, run simulations to compute the bias function $b_{ML}(\theta)$ under a plausible sample size rule and use values of this function in the formula (23.20) to obtain an adjusted estimate. When a study protocol specifies a guideline for sample size modification, this would be an obvious choice. Our investigations suggest that it suffices to use a rough approximation to the true sample size rule. As an extreme case, we calculated the bias of the MLE of θ following the group sequential design for our example with no adaptation. We found the bias function for the non-adaptive design to be very similar to that of the adaptive design and the bias of the adjusted estimate, shown as the dotted line in Figure 23.6, is very close to that of the estimate adjusted by the true bias function. We have observed similar results in other examples, demonstrating the robustness of this approach.

23.8.3.3 Median Unbiased Estimates

As noted in Section 9.6, an alternative form of estimate for the treatment effect θ is a "median unbiased" estimate, whose distribution has median equal to the true value of θ. If (θ_L, θ_U) is an equal tailed $(1-\alpha)$-level confidence interval for θ, then

$$Pr\{\theta_L \geq \theta\} = Pr\{\theta_U \leq \theta\} = \alpha/2. \tag{23.21}$$

If we define the estimate $\hat{\theta}_{Med} = \theta_L = \theta_U$ where (θ_L, θ_U), is a confidence interval with $\alpha/2 = 0.5$, it follows from (23.21) that

$$Pr\{\hat{\theta}_{Med} \leq \theta\} = Pr\{\hat{\theta}_{Med} \geq \theta\} = 0.5.$$

Thus $\hat{\theta}_{Med}$ is a median unbiased estimate.

A median unbiased estimate is available as long as we have a method for constructing a confidence interval for θ. We can, therefore, use the methods of Section 23.8.2.2 to calculate a median unbiased estimate after a two-stage adaptive trial with no early stopping, analyzed with a combination test. Similarly, we can apply the methods of Section 23.8.2.4 to construct a median unbiased estimate following a K-stage Lehmacher-Wassmer design with early stopping and an adaptation analysis J, provided that a sequence of potential information levels $\mathcal{I}_P^{(J+1)}, \ldots, \mathcal{I}_P^{(K)}$ is pre-specified.

Brannath, König & Bauer (2006) and Bretz, Koenig et al. (2009) discussed the bias and mean square error of estimators following a two-stage adaptive trial with no early stopping and found the median unbiased estimate to perform well. These comparisons included the unbiased estimate (23.19), but not a bias adjusted estimate as defined in Section 23.8.3.2.

23.9 Repeated Confidence Intervals in an Adaptive Trial

As discussed in Chapter 10, repeated confidence intervals (RCIs) can be used to monitor a trial and inform a decision to stop early without specifying a formal stopping rule. A $(1-\alpha)$-level sequence of RCIs is formed by inverting a family of two-sided group sequential tests with Type I error probability α. The requirement that the RCIs have the desired coverage probability if no early stopping occurs simplifies their construction when the trial is subject to adaptation.

We shall describe the method for calculating a sequence of RCIs when the two-sided group sequential tests of hypotheses $H\colon \theta = \theta_0$ are defined as two-sided Lehmacher-Wassmer tests of the form described in Section 23.7.1. At the design stage, it is necessary to specify the number of analyses K at which RCIs will be computed, weights $w_1, \ldots, w_K$ to be used in the Lehmacher-Wassmer procedure, and boundary values c_k, $k = 1 \ldots, K$. Given a sequence of statistics $Z^{(1)}, \ldots, Z^{(K)}$, which have independent $N(0,1)$ distributions under a null hypothesis H, the two-sided Lehmacher-Wassmer test of H is based on the combination statistics $\tilde{Z}_1, \ldots, \tilde{Z}_K$ defined by (23.12) and has the form

$$\begin{array}{ll} \text{After group } k = 1, \ldots, K-1 & \\ \quad \text{if } |\tilde{Z}_k| \geq c_k & \text{stop, reject } H \\ \quad \text{otherwise} & \text{continue to group } k+1, \\ \text{after group } K & \\ \quad \text{if } |\tilde{Z}_K| \geq c_K & \text{stop, reject } H \\ \quad \text{if } |\tilde{Z}_K| < c_K & \text{stop, accept } H. \end{array} \qquad (23.22)$$

Here, $c_1, \ldots, c_K$ are chosen so the probability of rejecting H when it is true is equal to α.

In calculating RCIs, we suppose the data at analysis 1 provide an estimate $\hat{\theta}^{(1)}$ with associated information level $\mathcal{I}^{(1)}$ and, for each $k = 2, \ldots, K$, the data accumulated between analyses $k-1$ and k produce a stage-wise estimate $\hat{\theta}^{(k)}$ with information level $\mathcal{I}^{(k)}$. Furthermore, we assume these estimates have the conditional distributions

$$\hat{\theta}^{(k)} \mid \hat{\theta}^{(1)}, \ldots, \hat{\theta}^{(k-1)}, \mathcal{I}^{(1)}, \ldots, \mathcal{I}^{(k)} \sim N(\theta, 1/\mathcal{I}^{(k)}),$$

or at least this is approximately true in large sample sizes. For the test of $H\colon \theta = \theta_0$, we define

$$Z^{(k)}(\theta_0) = (\hat{\theta}^{(k)} - \theta_0)\sqrt{\mathcal{I}^{(k)}}, \quad k = 1, \ldots, K.$$

From these, we form the combination statistics

$$\tilde{Z}^{(k)}(\theta_0) = (w_1 Z^{(1)}(\theta_0) + \ldots + w_k Z^{(k)}(\theta_0)) \,/\, (w_1^2 + \ldots + w_k^2)^{1/2}, \quad k = 1, \ldots, K,$$

which are used as the $\tilde{Z}_k$ in the sequential test of $H\colon \theta = \theta_0$, given by (23.22). Inverting this family of tests at analyses $k = 1, \ldots, K$ gives the intervals

$$I_k = \{\theta_0 \colon |\tilde{Z}^{(k)}(\theta_0)| < c_k\},$$

which form a $(1-\alpha)$-level sequence of RCIs.

Some algebra shows that I_k is centered at the estimate of θ,

$$\ddot{\theta}_k = \frac{\sum_{i=1}^{k} w_i \sqrt{\mathcal{I}^{(i)}}\, \theta^{(i)}}{\sum_{i=1}^{k} w_i \sqrt{\mathcal{I}^{(i)}}},$$

and this interval can be written as

$$I_k = \left(\ddot{\theta}_k - \frac{\sqrt{(\sum_{i=1}^{k} w_i^2)}}{\sum_{i=1}^{k} w_i \sqrt{\mathcal{I}^{(i)}}}\, c_k,\ \ddot{\theta}_k + \frac{\sqrt{(\sum_{i=1}^{k} w_i^2)}}{\sum_{i=1}^{k} w_i \sqrt{\mathcal{I}^{(i)}}}\, c_k \right).$$

A similar method for constructing RCIs is possible for a trial conducted using an error spending test with adaptation handled through the conditional rejection principle; see Mehta et al. (2007).

23.10 Bibliography and Notes

We have referred to the key papers of Bauer (1989), Bauer & Köhne (1994), Fisher (1998), Cui, Hung & Wang (1999) and Lehmacher & Wassmer (1999) that inspired the development of a wide variety of adaptive designs. Our discussion of the connections between combination tests and meta-analyses is based on the paper of Jennison & Turnbull (2005). The book by Wassmer & Brannath (2016) and the paper by Bauer et al. (2016), which looks back on 25 years of adaptive designs, provide comprehensive accounts of the history of this methodology.

Technical issues, such as measurability of rules for adaptation and principles of sequential inference are addressed by Liu, Proschan & Pledger (2002) and Liu & Anderson (2008). When adaptation is present, some of the most standard statistical properties can fail to hold, for example, Jennison & Turnbull (2003) give an example of an adaptive design whose power function is not monotone increasing. This explains the need for the careful arguments in Section 23.2 to show that the combination tests described there do protect the Type I error rate at all values of $\theta \leq 0$. Öhrn & Jennison (2010) make a more subtle argument to show that an adaptive design which tests for both superiority and non-inferiority protects its Type I error rates.

It took time for methods of inference following an adaptive trial to be developed. Our exposition draws on the description of such methods in Chapter 8 of Wassmer & Brannath (2016). While calculating a P-value on termination can be fairly straightforward, confidence intervals pose more significant problems.

Brannath, Mehta & Posch (2009) presented methods to compute a confidence interval after a two-stage adaptive design or a multi-stage design in which adaptation occurs at the penultimate analysis. Gao, Liu & Mehta (2013) present a general solution when adaptation occurs at an earlier analysis in a multi-stage design. These methods make use of the conditional rejection probability principle and so may be approximate when the future information levels that would have occurred with no adaptation have to be estimated. Our proposals in Section 23.8.2.4 also use the conditional rejection probability principle but avoid the need to predict future information levels by the introduction of a reference sequence of increments in information after the adaptive analysis.

Levin, Emerson & Emerson (2014) discuss the properties of different types of confidence interval following an adaptive trial. They assume adaptations follow fixed rules, which allows a greater variety of methods to be employed. They find that intervals with the best properties are those based on the likelihood ratio ordering of the sample space or the stage-wise ordering combined with the conditional rejection probability principle. However, even for a non-adaptive group sequential test, use of the likelihood ratio ordering is not feasible if information levels are unpredictable. Thus, these findings favor the stage-wise ordering combined with the conditional rejection probability principle. Robertson et al. (2025a,b) provide a survey of methods for constructing a confidence interval after an adaptive design and give examples of their use.

In an early paper on point estimation following an adaptive two-stage trial, Denne (2000) proposed bias adjustment using the iterated version of Whitehead's method. He comments that although the bias function $b_{ML}(\theta)$ depends on the exact strategy for choosing the second stage sample size, this bias function is very similar for different strategies so this is not a significant problem in practice. We have found a similar robustness for this approach in the examples reported in Section 23.8.3.2. A great many methods have been proposed for estimation following an adaptive trial. Nelson, Liu & Mehta (2022) compare several methods and illustrate these with case studies. Robertson et al. (2023a,b) survey these methods and give examples of their use in clinical trials.

Chapter 24

Sample Size Re-assessment Given an Interim Estimate of Treatment Effect

24.1 Introduction

The combination tests and related methods presented by Bauer & Köhne (1994), Proschan & Hunsberger (1995), Fisher (1998), Cui, Hung & Wang (1999) and Lehmacher & Wassmer (1999) paved the way for a new "adaptive" approach to clinical trial design. In the early 2000s, many types of design were proposed and pharmaceutical companies greeted these enthusiastically as tools that might increase success rates in the drug development process. The response from regulators was more cautious but they indicated support for new methods as long as the validity of a trial's conclusions, in particular the Type I error rate, could be protected.

In July 2006, the Wall Street Journal ran an article with the headline "FDA Signals It's Open to Drug Trials That Shift Midcourse" and reported that novel methods could:

- Route a larger share of patients to the treatment that seems to work the best
- Drop treatments that don't appear to be effective
- Add more of the type of patients who seem to be reacting best to a particular treatment
- Merge two different phases of drug development into one trial.

We shall discuss designs that incorporate these features in later chapters. However, Bauer et al. (2016, Sec. 3.3) note that in the early days of confirmatory adaptive designs, the main motivation for their use was the opportunity to perform unblinded sample size re-assessment, conditioning on the interim estimate of the treatment effect. In reviews of scientific advice letters from the European Medicines Agency (Elsässer et al., 2014) and of new drug applications in cardiovascular and CNS diseases submitted to the FDA (Wang, Peng & Hung, 2018), designs with unblinded sample size re-assessment were the most common type of adaptive designs proposed to the regulators.

Consider a trial comparing two treatments with normal responses of variance σ^2 where the treatment effect θ is defined as the difference in mean response between the two treatments. Suppose H_0: $\theta \leq 0$ is to be tested against $\theta > 0$ with one-sided Type I error probability α and power $1 - \beta$ when $\theta = \delta$. In a balanced fixed sample size study with n_f observations on each treatment, the information for θ is $n_f/(2\sigma^2)$. Thus, the required information, as given by Equation (5.1), is achieved by setting the sample size per treatment to be

$$n_f = \frac{2\,\{\Phi^{-1}(1-\alpha) + \Phi^{-1}(1-\beta)\}^2\,\sigma^2}{\delta^2}. \tag{24.1}$$

In Chapter 18 we discussed methods for adapting sample size as estimates of σ^2 become available during the course of a study. The topic we now consider is how to specify δ in the

DOI: 10.1201/9781584888482-24

formula (24.1) and how, in an adaptive design, sample size may be modified in response to an interim estimate of θ.

Some textbooks advocate choosing δ to represent the minimum "clinically relevant" or "commercially viable" effect size; see, for example, Pocock (1983, Ch. 9). The FDA Guidance E9 (FDA, 1998, Sec. 3.5) advises:

> "The treatment difference to be detected may be based on a judgement concerning the minimal effect which has clinical relevance in the management of patients or on a judgement concerning the anticipated effect of the new treatment, where this is larger."

Shun et al. (2001) take the notion of designing for the anticipated treatment effect further in stating:

> "To ensure an adequate power, the sample size is often based on clinically meaningful minimal delta. However, the use of this clinically meaningful minimal delta is limited in real life. Therefore, the focus is often on the true delta in today's clinical trial planning."

Adaptive designs offer the possibility of updating the final sample size by replacing δ in (24.1) with an interim estimate of θ or increasing the sample size to achieve a certain conditional power assuming the interim estimate is equal to the true value of θ. This approach was taken in the early papers of Proschan & Hunsberger (1995), Fisher (1998), and Cui, Hung & Wang (1999), then followed by others, for example, Denne (2001), Shun (2001), Li et al. (2002) and Hartung & Knapp (2003). A Bayesian version of this approach is also possible; see Moerbeek (2021). The term "sample size re-estimation" reflects the belief that a certain power can be achieved at the true value of θ if δ is replaced by an interim estimate of θ in the sample size formula. However, we saw in Chapter 11 that interim estimates of θ are highly variable, making such a conclusion unrealistic. We therefore prefer to use the term "sample size re-assessment" or "sample size modification" in describing these adaptive designs.

Proposals for adaptive clinical trial designs were not without controversy. The way in which data are combined across stages means that inference may not be a function of the sufficient statistic for the treatment effect θ and, hence, may not respect the likelihood principle. Also, while rules that have been suggested for modifying sample size in response to an interim estimate of θ may have intuitive appeal, they are not guaranteed to produce efficient designs in terms of power and expected sample size, particularly when compared with efficient group sequential designs with the same number of analyses and fixed group sizes. We shall discuss issues of credibility of adaptive designs with sample size re-assessment in Section 24.2 and we remark on the efficiency of such methods in Section 24.3. We comment on regulators' views on these issues in Section 24.4. A popular example of an adaptive design with sample size re-assessment is the "promising zone" design of Mehta & Pocock (2011). In Section 24.5 we recount Jennison & Turnbull's (2015) critical evaluation of this design, which includes a proposal for a more efficient sample size rule, and we note the modified version of the promising zone design presented by Hsaio, Liu & Mehta (2019) which incorporates features suggested by Jennison & Turnbull.

24.2 Credibility of Adaptive Designs with Sample Size Re-assessment

24.2.1 The Critique by Burman & Sonneson (2006)

In discussing designs with sample size re-assessment, we shall focus on the two-treatment comparison with normally distributed responses described in Section 24.1. However, in view of the general theory of Chapter 3, the conclusions from the analysis of this case extend to other testing problems. For simplicity, we suppose the variance σ^2 is known. We assume the treatment effect is $\theta = \mu_A - \mu_B$ and the null hypothesis H_0: $\theta \leq 0$ is to be tested against $\theta > 0$ with one-sided Type I error rate α. Suppose that in the initial design, the sample size is chosen to provide power $1 - \beta$ if θ is equal to a specified value δ, which requires $n_f = 2\{\Phi^{-1}(1-\alpha) + \Phi^{-1}(1-\beta)\}^2\sigma^2/\delta^2$ observations on each treatment. Anticipating possible adaptation, the statistical analysis plan states that an interim analysis will take place after n_1 observations on each treatment have been observed and the remaining sample size of n_2 observations per treatment may be modified from the original value $n_f - n_1$, depending on the interim estimate

$$\hat{\theta}^{(1)} = \frac{1}{n_1}\sum_{i=1}^{n_1} X_{Ai} - \frac{1}{n_1}\sum_{i=1}^{n_1} X_{Bi}.$$

Defining the estimate of θ from the second stage data

$$\hat{\theta}^{(2)} = \frac{1}{n_2}\sum_{i=n_1+1}^{n_1+n_2} X_{Ai} - \frac{1}{n_2}\sum_{i=n_1+1}^{n_1+n_2} X_{Bi},$$

the stage-wise Z-statistics are

$$Z^{(1)} = \hat{\theta}^{(1)}\sqrt{\mathcal{I}^{(1)}} \quad \text{and} \quad Z^{(2)} = \hat{\theta}^{(2)}\sqrt{\mathcal{I}^{(2)}}$$

where

$$\mathcal{I}^{(1)} = n_1/(2\sigma^2) \quad \text{and} \quad \mathcal{I}^{(2)} = n_2/(2\sigma^2).$$

Applying a weighted inverse normal combination test after both stages of the trial are completed, H_0: $\theta \leq 0$ is rejected if

$$\tilde{Z}_2 = w_1 Z^{(1)} + w_2 Z^{(2)} \geq \Phi^{-1}(1-\alpha), \qquad (24.2)$$

where the weights w_1 and w_2 satisfy $w_1^2 + w_2^2 = 1$. As explained in Section 23.2, $Z^{(1)}$ and $Z^{(2)}$ are independent under $\theta = 0$ by the conditional invariance principle and, as long as the weights w_1 and w_2 are fixed in advance, this test has Type I error probability α, whatever rule is used to determine n_2. In order for this procedure to give the standard hypothesis test when no adaptation occurs, we suppose these weights are specified as

$$w_1 = \sqrt{\frac{n_1}{n_f}} \quad \text{and} \quad w_2 = \sqrt{\frac{n_f - n_1}{n_f}}.$$

Burman & Sonesson (2006) note that when $n_2 \neq n_f - n_1$, the test statistic $\tilde{Z}_2$ assigns different weights to observations in the first and second stages and so it is not a function of the sufficient statistic $(n_1 + n_2, \hat{\theta})$ where

$$\hat{\theta} = \frac{1}{n_1+n_2}\sum_{i=1}^{n_1+n_2} X_{Ai} - \frac{1}{n_1+n_2}\sum_{i=1}^{n_1+n_2} X_{Bi}.$$

Thus, the test of H_0: $\theta \leq 0$ does not satisfy the principles of sufficiency, invariance or conditionality as defined by, for example, Cox & Hinkley (1974). This feature of the combination test may be expected to have an impact on efficiency, a topic we shall discuss in Section 24.3.

Burman & Sonneson give an intriguing numerical example, which we now re-express in the context of a two treatment comparison. Suppose, in our formulation, $\alpha = 0.05$, $\sigma^2 = 0.5$, the initial design has $n_f = 1000$ and an interim analysis is planned after $n_1 = 100$

observations per treatment. The final hypothesis test will be based on $\tilde{Z}_2$ of the form (24.2) with weights

$$w_1 = \sqrt{\frac{100}{1000}} \quad \text{and} \quad w_2 = \sqrt{\frac{900}{1000}}.$$

A fixed sample size trial with n_f per treatment has power 0.812 if the treatment effect θ is equal to 0.08. Suppose at the interim analysis the estimate

$$\hat{\theta}^{(1)} = -0.03$$

is observed, based on $n_1 = 100$ observations per treatment. The investigators are reluctant to expend further resources in this trial but they note that with a second stage sample size of just 1 observation per treatment there is a possibility of rejecting H_0, the conditional probability of this rejection being 0.033, 0.037 and 0.040 under $\theta = 0$, 0.05 and 0.08, respectively. They therefore decide to continue the trial with $n_2 = 1$. These two observations produce the estimate

$$\hat{\theta}^{(2)} = X_{A,101} - X_{B,101} = 2.5,$$

giving $Z^{(1)} = -0.03$, $Z^{(2)} = 2.5$ and

$$\tilde{Z}_2 = \sqrt{\frac{1}{10}}\, Z^{(1)} + \sqrt{\frac{9}{10}}\, Z^{(2)} = 2.28.$$

Hence, the combination test rejects H_0. This is despite the fact that the conventional estimate of θ based on the total of 101 observations per treatment,

$$\hat{\theta} = \frac{1}{101}\sum_{i=1}^{101} X_{Ai} - \frac{1}{101}\sum_{i=1}^{101} X_{Bi} = \frac{-3+2.5}{101} = -0.005,$$

is negative.

Several discussants of the paper by Burman & Sonesson acknowledged the concerns raised but defended the methods nevertheless. Proschan (2006) wrote "The greater the flexibility we allow, the more care we must exercise to ensure a sensible test" but went on to say "They are correct that such methods, if extremely abused, produce illogical conclusions, but that is no more a condemnation of adaptive methods than Jack the Ripper is a condemnation of cutlery". Proschan presented a proof that "there does not exist a sample size rule ... and test based on the sufficient statistic that maintains level α even if one does not follow the sample size rule" and concluded that strict adherence to the sufficiency principle "eliminates much of the flexibility afforded by adaptive methods". In a similar vein, Bauer (2006) stated that

> "Flexibility is needed in practice and the statistical price to be paid for the many options offered by the designs is known. The authors have summarized the ongoing discussion. Is wine sound? It is the way to drink it that matters. For adaptive designs it seems that it is the way they are used, presented, and interpreted that matters."

24.2.2 The Dual Test

Burman & Sonesson (2006) recommend use of a "dual test" as a way to maintain the credibility of the conclusion from an adaptive trial. This approach had been proposed previously by Denne (2001) and discussed by Posch, Bauer & Brannath (2003).

The dual test requires the null hypothesis to be rejected by both the combination test, pre-specified as a component of an adaptive design, and a level α Wald test which treats the final set of data as if it had been collected in a fixed sample size, non-adaptive design.

Thus, in the example of Section 24.2.1, H_0: $\theta \leq 0$ would not be rejected since the fixed sample size analysis has

$$Z = \frac{\hat{\theta}}{\sqrt{(2/101)}} = -0.035.$$

Since the combination test has Type I error probability α and the rejection of H_0 by the dual test requires rejection of H_0 by the combination test, the dual test protects the Type I error rate α conservatively.

24.2.3 Chen, DeMets & Lan's 50% rule

Chen, DeMets & Lan (2004) prove that in certain situations, there is no inflation of the Type I error rate when a standard, fixed sample size analysis is conducted after an increase in sample size based on unblinded interim data. Specifically, consider a two-stage adaptive trial planned to have cumulative information for θ equal to $\mathcal{I}_1$ and $\mathcal{I}_2$ at the interim and final analyses, respectively. Let $\hat{\theta}_1$ be the interim estimate of θ and $Z^{(1)}$ the associated Z-statistic. Suppose the second stage sample size is increased after observing an interim estimate of θ and the new final information level is $\mathcal{I}_2^* > \mathcal{I}_2$. Denoting the Z-statistic based on second stage data alone by $Z^{(2)}$, the combination test with weights $w_1 = \sqrt{(\mathcal{I}_1/\mathcal{I}_2)}$ and $w_2 = \sqrt{\{(\mathcal{I}_2 - \mathcal{I}_1)/\mathcal{I}_2\}}$ rejects H_0: $\theta \leq 0$ if

$$\sqrt{(\mathcal{I}_1/\mathcal{I}_2)}\, Z^{(1)} + \sqrt{\{(\mathcal{I}_2 - \mathcal{I}_1)/\mathcal{I}_2\}}\, Z^{(2)} \geq \Phi^{-1}(1-\alpha) \tag{24.3}$$

and has Type I error exactly equal to α. The Wald test, applied as if the data had been collected in a fixed sample size design, is based on the maximum likelihood estimate

$$\hat{\theta}_{ML} = \frac{Z^{(1)}\sqrt{\mathcal{I}_1} + Z^{(2)}\sqrt{(\mathcal{I}_2^* - \mathcal{I}_1)}}{\mathcal{I}_2^*}$$

and rejects H_0 if

$$\hat{\theta}_{ML}\sqrt{\mathcal{I}_2^*} \geq \Phi^{-1}(1-\alpha). \tag{24.4}$$

Suppose values $Z^{(1)} = z_1$ and $\hat{\theta}_1 = z_1/\sqrt{\mathcal{I}_1}$ are observed. From (11.3), with no adaptation, the conditional power under $\hat{\theta} = \theta_1$, given $Z^{(1)} = z_1$, is

$$\begin{aligned} p_1(\hat{\theta}_1, z_1) &= \Phi\left\{\frac{z_1\sqrt{\mathcal{I}_1} - \Phi^{-1}(1-\alpha)\sqrt{\mathcal{I}_2} + (\mathcal{I}_2 - \mathcal{I}_1)\,\hat{\theta}_1}{\sqrt{(\mathcal{I}_2 - \mathcal{I}_1)}}\right\} \\ &= \Phi\left\{\frac{z_1\sqrt{\mathcal{I}_1} - \Phi^{-1}(1-\alpha)\sqrt{\mathcal{I}_2} + (\mathcal{I}_2 - \mathcal{I}_1)\, z_1/\sqrt{\mathcal{I}_1}}{\sqrt{(\mathcal{I}_2 - \mathcal{I}_1)}}\right\} \\ &= \Phi\left\{\frac{z_1\mathcal{I}_2/\sqrt{(\mathcal{I}_1)} - \Phi^{-1}(1-\alpha)\sqrt{\mathcal{I}_2}}{\sqrt{(\mathcal{I}_2 - \mathcal{I}_1)}}\right\}. \end{aligned}$$

This conditional power under $\theta = \hat{\theta}_1$ is greater than or equal to 0.5 when

$$z_1 \geq \Phi^{-1}(1-\alpha)\sqrt{(\mathcal{I}_1/\mathcal{I}_2)}. \tag{24.5}$$

Chen, DeMets & Lan (2004) show that in this case, if (24.4) holds, then so does (24.3). It follows that if sample size is only increased when (24.5) is true, that is, when the conditional power under the current treatment effect estimate is at least 0.5, then the Wald test rejects H_0 on fewer occasions than the combination test and so the overall Type I error probability of the Wald test is no greater than α.

A procedure in which sample size is only increased when conditional power under $\hat{\theta}_1$ is at least 0.5 and a Wald test is applied after stage 2 can be thought of as applying the dual test since, in the situations where adaptation occurs, the Wald test is the more stringent part of the dual test.

24.2.4 Example

Again, we consider the two-treatment comparison with normally distributed responses that was described in Section 24.1. Suppose the treatment effect is θ and H_0: $\theta \leq 0$ is to be tested against $\theta > 0$ with one-sided Type I error probability $\alpha = 0.025$. We assume $\sigma^2 = 100$ and in the initial plan, power 0.9 is desired if the effect size is $\theta = 5$. This leads to an initial choice of sample size $n_f = 84$ subjects per treatment arm.

Now suppose an interim analysis is conducted after $n_1 = 42$ subjects per treatment and we observe $\hat{\theta}_1 = 3.2$, so $z_1 = 3.2/\sqrt{(2 \times 100/42)} = 1.466$. The conditional power under $\theta = \hat{\theta}_1 = 3.2$ is 0.564.

Scenario 1: No change in sample size

With second stage sample size $n_2 = 42$, suppose the estimated treatment effect based on stage 2 data alone is

$$\hat{\theta}^{(2)} = 2.7,$$

which gives an overall estimate $\hat{\theta}_2 = 2.95$ and overall Z-statistic

$$\tilde{Z}_2 = 2.95/\sqrt{(2 \times 100/84)} = 1.912.$$

In this case H_0 cannot be rejected in a test at level $\alpha = 0.025$. Since there is no adaptation, the combination test and the Wald test produce exactly the same Z-values.

Scenario 2: An increase in sample size

Now suppose that, in order to improve the probability of a positive result, the second stage sample size is increased by 50% to $n_2 = 63$. Since the conditional power under $\theta = \hat{\theta}_1 = 3.2$ is greater than 0.5, investigators follow the Chen, Lan & DeMets approach and analyze the combined data using a Wald test.

If the estimated treatment effect based on stage 2 data alone is

$$\hat{\theta}^{(2)} = 2.7,$$

the maximum likelihood estimate of θ is

$$\hat{\theta}_{ML} = (42 \times 3.2 + 63 \times 2.7)/105 = 2.9.$$

The resulting Z-statistic is

$$2.9/\sqrt{(2 \times 100/105)} = 2.101$$

and H_0 is rejected at level $\alpha = 0.025$.

If a combination test were applied, the Z-statistic based on stage 2 data would be

$$Z^{(2)} = 2.7/\sqrt{(2 \times 100/63)} = 1.515$$

and, since

$$Z^{(1)} + Z^{(2)} = (1/\sqrt{2})\, 1.466 + (1/\sqrt{2})\, 1.515 = 2.108,$$

this test also rejects H_0, in keeping with the result proved by Chen, Lan & DeMets.

There is, inevitably, a small range of values of $\hat{\theta}^{(2)}$ for which the combination test leads to rejection of H_0 but the Wald test does not. As an example, if $\hat{\theta}^{(2)} = 2.35$, the Z-statistics from these two tests are 1.97 and 1.95 respectively. If the combination test is regarded as acceptable, then one would expect the investigators to favor using that test in such an instance in order to increase the power when a positive treatment effect is present.

24.3 Efficiency of Adaptive Designs with Sample Size Re-assessment

24.3.1 Examples of Inefficiency

When sample size re-assessment based on the estimated treatment effect is incorporated in a trial design, the intention is usually to achieve a desired power with as small a sample size as possible. It is therefore important that the rule for modifying sample size should lead to an efficient procedure. In their review of group sequential and adaptive designs, Lai, Lavori & Shih (2012a) stress the importance of marginal or unconditional properties of a design, stating

> "the performance that really matters is the true overall power curve"

and emphasizing that expected sample size, or other measures of a design's efficiency, should be used to compare methods.

In their discussion of the paper by Burman & Sonesson (2006), Jennison & Turnbull (2006c) note that

> "Traditional group sequential tests also offer a methodology for reaching a data-dependent choice of final sample size and the efficiency of new adaptive methods should be assessed against this alternative approach."

They go on to compare a 10-stage adaptive design with sample size re-assessment of the form proposed by Shen & Fisher (1999) with group sequential tests with 5 and 10 analyses. The group sequential tests have similar power curves to the adaptive design, with slightly higher power at the largest treatment effects, but their expected sample size functions are lower by between 15 and 20%. Jennison & Turnbull attribute the inefficiency of the Shen & Fisher design to the rule for sample size modification, which follows the customary approach of aiming for a certain conditional power under the current estimate of the treatment effect. Similar comparisons are made in Jennison & Turnbull (2003) and Jennison & Turnbull (2006a) and, again, adaptive designs that aim for a certain conditional power under the current estimate of the treatment effect are out-performed by group sequential tests with essentially the same power curve.

24.3.2 Efficient Adaptive Designs with Sample Size Re-assessment

It is difficult to assess the properties of a flexible adaptive design when the way in which the study will be conducted is not completely specified. We therefore start by considering adaptive designs with fixed sample size re-assessment (SSR) rules. If stopping for efficacy or futility is permitted at each of K analyses and future sample size is re-assessed at each of the analyses 1 to $K-1$, such a design can be viewed as an example of a K-stage Schmitz (1993) design, as described in Section 12.4. If no changes in group size are allowed, we have a standard group sequential test (GST). Thus, the class of GSTs with K analyses is a subset of the class of adaptive designs with SSR and a maximum of K analyses. It follows that, if error probabilities and an optimality criterion are specified, the optimal K-stage design with SSR must be at least as good as the optimal K-stage GST.

This prompts one to ask how much can be gained in moving from a standard GST to an adaptive design with SSR. In fact, we have already answered this question in Section 12.4. Comparisons there were between designs with the same maximum number of analyses, K, that achieved type I error probability α under $\theta = 0$ and power $1 - \beta$ at $\theta = \delta$, and the optimality criterion was a function of the expected information on termination, $E_\theta(\mathcal{I})$. The figures in Table 12.5, taken from Jennison & Turnbull (2006a), and further results from Jennison & Turnbull (2006b) reported in Section 12.4 show that adding adaptive sample size re-assessment to a well-chosen group sequential design can reduce average sample size by about 1% of the fixed sample size needed to achieve the desired power. One might conclude from this that SSR is unnecessary since almost all the benefits of adaptive designs with

SSR can be achieved using simpler GSTs. On the other hand, we have demonstrated that efficient designs with SSR are possible and, if such a design is desired, we might study the form of the optimized adaptive designs found by Jennison & Turnbull (2006a,b) and mimic this in defining an efficient SSR rule.

In deriving an optimal Schmitz design to test H_0: $\theta \leq 0$ against $\theta > 0$, Jennison & Turnbull (2006a) solve a Bayes sequential decision problem. When continuing past interim analysis k, the choice of the next group size, or more generally the next increment in information, balances the cost of additional observations against the reduction in conditional error probabilities, taking expectations over the posterior distribution of θ when calculating these quantities. In these optimized procedures, future group sizes are highest when the current estimate $\hat{\theta}_k$ lies in the middle of the continuation region and lowest when $\hat{\theta}_k$ is near the upper or lower boundary. This behavior is qualitatively different from that of an SSR procedure that aims to achieve a fixed conditional power under $\theta = \hat{\theta}_k$ since, in this case, the sample size continues to increase as $\hat{\theta}_k$ decreases. Furthermore, integration over the posterior distribution of θ acknowledges the uncertainty in the estimate $\hat{\theta}_k$ and reduces the variation in possible group sizes compared to SSR procedures that treat the estimate $\hat{\theta}_k$ as the true value of θ.

The optimized adaptive SSR designs computed by Jennison & Turnbull (2006a) are complicated to describe. Furthermore, the Type I error probability is calculated assuming group sizes are exactly equal to the values specified by the procedure, which is unlikely to be true in a real application. If one also wishes to be able to respond to aspects of the data other than the estimated treatment effect, a more flexible approach is needed. We can, however, note that optimized designs were derived by minimizing a conditional expected loss, integrated over the posterior distribution of θ, and apply the same principle in defining an SSR rule. We shall follow such an approach in the example we present in Section 24.5 and show this can lead to an efficient adaptive SSR design.

Finally, we note that conclusions from the above discussion apply to flexible adaptive designs, where the SSR is not pre-specified. To see this, suppose we could enumerate every possible scenario, including internal and external data and the attitudes of the monitoring committee, that might occur during the course of a trial, and note the sample size modification that the monitoring committee would have applied in each scenario. This produces an SSR rule (which may involve randomized decisions given treatment effect estimates $\hat{\theta}_k$). While it is not feasible to carry out this process to identify an SSR rule, the point is that a certain SSR rule is actually being used. Thus, for a specified optimality criterion and any constraints one may wish to impose on the number of analyses or the maximum sample size, it is not possible for a flexible adaptive design to improve on the optimal Schmitz design. Hence the modest benefits that Schmitz designs can provide relative to standard GSTs are the most that may be achieved by adding sample size re-assessment to a well-chosen group sequential design.

24.4 Regulatory Views on Sample Size Re-assessment

Regulators have recognized the possibility that sample size may be modified in response to an interim estimate of the treatment effect.

The FDA Guidance on Adaptive Designs for Clinical Trials of Drugs and Biologics (FDA, 2019a, Sec. V B) states

> "Similar to a group sequential trial, a design with sample size adaptations based on comparative interim results can provide adequate power under a range of plausible effect sizes, and therefore, can help ensure that a trial maintains adequate power if the true magnitude of treatment effect is less than what was hypothesized, but still clinically meaningful."

The Guidance continues

> "If one carries out a hypothesis test at the end of the trial at the conventional 0.025 significance level, the Type I error probability can be more than doubled (Proschan and Hunsberger 1995). Therefore, one of a variety of available methods should be used to appropriately control the Type I error probability with this type of adaptive design. For example, hypothesis testing approaches have been developed based on combining test statistics or p-values from the different stages of the trial in a pre planned manner or through preservation of the conditional Type I error probability ..."

The FDA document also raises the issue of information leakage:

> "... there are additional challenges in maintaining trial integrity in the presence of sample size adaptations. For example, sample size modification rules are often based on maintaining the conditional probability of a statistically significant treatment effect at the end of the trial (often called the conditional power) at or near some desired level. In this scenario, knowledge of the adaptation rule and the adaptively chosen sample size allows a relatively straightforward back-calculation of the interim estimate of treatment effect. Therefore, additional steps should be taken to limit personnel with this detailed knowledge so that trial integrity can be maintained."

The EMA Reflection Paper on Adaptive Designs (CHMP, 2007, Sec. 4.2) is less encouraging with regard to adaptive designs:

> "In general, changes to the design of an ongoing phase III trial are not recommended even though, in an adaptive design, such changes could be introduced with full control of the type 1 error."

The reflection paper states

> "The option to re-assess sample size in an ongoing trial should not be seen as a substitute for careful planning. The relevance of a particular size of treatment effect should be discussed at the planning stage of the trial and not deferred to the point where interim results are already available. In general, consideration of what magnitude of treatment effect might be of clinical importance should not be influenced by trial results (interim or final)."

but the paper acknowledges that

> "In cases where good justification can be provided that uncertainty about the required sample size is not an indicator for insufficient research in earlier stages, sample size re-assessment based on results of an ongoing trial is an option."

with the qualification

> "Analysis methods that control the type I error must be pre-specified. ... However, if more than one sample size re-assessment becomes necessary, concerns would be raised that the experimental conditions are fluctuating and are not fully understood."

We noted earlier that Elsässer et al. (2014) and Wang, Peng & Hung (2018) reported designs with unblinded sample size re-assessment to be the most common type of adaptive designs proposed to European and U.S. regulators. Taking the positive comments in the above statements, we note that regulators have recognized that combination tests provide an acceptable mechanism to control the Type I error rate and they do not require the additional protection afforded by dual tests or the 50% rule of Chen, DeMets & Lan (2004).

24.5 The Promising Zone Design of Mehta & Pocock (2011)

24.5.1 Description of the Method

Mehta & Pocock (2011) applied several of the ideas we have described to define their "Promising Zone Design". This design has been applied in a number of trials; see, for example, the account of the Campion Phoenix trial reported by Bhatt & Mehta (2016). We shall describe this approach in the case of a two-treatment comparison with normally distributed responses, as described in Section 24.1. The treatment effect θ is the difference in mean responses for the two treatments and H_0: $\theta \le 0$ is to be tested against $\theta > 0$ with one-sided Type I error probability α.

Mehta & Pocock propose setting an initial target sample size of n_t, with $n_t/2$ randomized to each treatment, that will achieve power $1-\beta$ if $\theta = \delta$, where δ is an optimistic estimate of the treatment effect. However, it is recognized that power to detect a smaller treatment effect than δ is desirable. At an interim analysis with effect estimate $\hat{\theta}_1$ based on a total of m_1 observations, $m_1/2$ per treatment, the conditional power is calculated assuming $\theta = \hat{\theta}_1$. With a further $n_t - m_1$ responses to be observed, we denote this conditional power by

$$CP_{\hat{\theta}_1}(\hat{\theta}_1, n_t - m_1) \;=\; Pr_{\theta=\hat{\theta}_1}\{H_0 \text{ will be rejected} \mid \text{Interim estimate is } \hat{\theta}_1\}.$$

Mehta & Pocock divide the set of possible values for $\hat{\theta}_1$ into three zones and in the middle "promising zone", the sample size is increased to improve power. This middle zone is defined in such a way that the result of Chen, DeMets & Lan (2004) implies that increasing the sample size does not inflate the Type I error rate. The zones are

Unfavorable:	$CP_{\hat{\theta}_1}(\hat{\theta}_1, n_t - m_1) < 0.5$	Continue to n_t observations
Promising:	$0.5 \le CP_{\hat{\theta}_1}(\hat{\theta}_1, n_t - m_1) < 1-\beta$	Increase the final sample size so the conditional power becomes $1-\beta$
Favorable:	$CP_{\hat{\theta}_1}(\hat{\theta}_1, n_t - m_1) \ge 1-\beta$	Continue to n_t observations.

If $CP_{\hat{\theta}_1}(\hat{\theta}_1, n_t - m_1)$ is only a little higher than 0.5, the above prescription leads to a very high sample size and Mehta & Pocock impose an upper bound on the final sample size, such as $n_{\max} = 2n_t$. With this bound in place, the promising zone can be extended a little to include lower values of $\hat{\theta}_1$ while still protecting the Type I error rate at level α; see Gao, Ware & Mehta (2008).

24.5.2 An Illustrative Example

We illustrate the promising zone approach and discuss its properties by reviewing the example of a schizophrenia trial presented in Section 4.1 of Mehta & Pocock (2011). The primary efficacy endpoint in this study is the improvement from baseline to week 26 in the Negative Symptoms Assessment. Responses X_{Ai} on the new treatment and X_{Bi} on the comparator treatment are distributed as

$$X_{Ai} \sim N(\mu_A, \sigma^2) \quad \text{and} \quad X_{Bi} \sim N(\mu_B, \sigma^2), \quad i = 1, 2, \ldots,$$

where, from previous studies, the standard deviation σ is assumed to be known and equal to 7.5. The treatment effect is denoted by $\theta = \mu_A - \mu_B$.

The null hypothesis H_0: $\theta \le 0$ is to be tested against $\theta > 0$ with one-sided Type I error probability $\alpha = 0.025$. Mehta & Pocock note that an effect size $\theta = 1.6$ would be clinically meaningful and commercially viable but a large sample size would be needed to detect such an effect. Instead, they start with a fixed sample size design that achieves power $1-\beta = 0.8$ if $\theta = \delta = 2$. From Equation (2.3), the required total sample size is

$$n_t \;=\; 4\sigma^2\{\Phi^{-1}(1-\alpha) + \Phi^{-1}(1-\beta)\}^2/\delta^2 \;=\; 4 \times 7.5^2 \times (1.96 + 0.84)^2/4 \;=\; 442,$$

which implies 221 patients per treatment arm (the value of n_f in Equation 24.1). Under this design, H_0 will be rejected if

$$\frac{\bar{X}_A - \bar{X}_B}{\sqrt{\{4\sigma^2/n_t\}}} \ge \Phi^{-1}(1-\alpha),$$

where $\bar{X}_A$ and $\bar{X}_B$ are the sample means on treatments A and B, respectively.

At an interim analysis when $m_1 = 208$ observations are available, 104 per treatment arm, the interim estimate $\hat{\theta}_1$ is computed along with the z-value, $z_1 = \hat{\theta}_1 \surd\{m_1/(4\sigma^2)\}$. The sample size is re-assessed at this point and "favorable", "promising" and "unfavorable" zones are defined, based on conditional power. If a further m_2 subjects are observed the interim and final information levels will be $\mathcal{I}_1 = m_1/(4\sigma^2)$ and $\mathcal{I}_2 = (m_1 + m_2)/(4\sigma^2)$ and, from (11.3), conditional power under $\theta = \hat{\theta}_1$ of a final fixed sample size test will be

$$CP_{\hat{\theta}_1}(z_1, m_2) \;=\; \Phi\left\{\frac{z_1\surd m_1 - z_\alpha\surd(m_1 + m_2)}{\surd m_2} + \sqrt{\frac{m_2}{4\sigma^2}}\,\hat{\theta}_1\right\},$$

where $z_\alpha = \Phi^{-1}(1-\alpha)$.

Under the initial plan, the second stage sample size is $m_2 = n_t - m_1 = 234$, comprising 117 per treatment arm. We denote this value by $m_{2,0} = 234$. The favorable zone is defined as the set of outcomes satisfying

$$CP_{\hat{\theta}_1}(z_1, n_t - m_1) \geq 1 - \beta \tag{24.6}$$

so that the conditional power is at least $1-\beta$ under the original design, and in this case the study continues with $m_2(z_1) = m_{2,0}$ subjects. We denote by Z_U the minimum value of z_1 satisfying (24.6) so the favorable zone is the set of z_1 values (Z_U, ∞).

The promising zone contains values of z_1 below Z_U. In the upper part of the promising zone, the second stage sample size, $m_2(z_1)$, is the solution to

$$CP_{\hat{\theta}_1}(z_1, m_2) \;=\; 1 - \beta.$$

An upper limit $n_{\max} = 844$, double the original value of 422, is specified for the total sample size, and this region continues down to the value z_1 at which this limit is reached with $m_2(z_1) = 884 - 208 = 676$. We denote this value of m_2 by $m_2^* = 676$. In the lower part of the promising zone, $m_2(z_1)$ is set at m_2^* and the zone continues as long as this increase in sample size leads to a reduction in the conditional Type I error probability when the final hypothesis test is conducted as a fixed sample size test. At this point

$$CP_{\theta=0}(z_1, m_2^*) \;=\; CP_{\theta=0}(z_1, m_{2,0})$$

and this occurs when z_1 is equal to

$$Z_L \;=\; z_\alpha\,\frac{\surd\{m_2^*\, n_t\} - \surd\{m_{2,0}\, n_{\max}\}}{\surd\{m_2^*\, m_1\} - \surd\{m_{2,0}\, m_1\}}.$$

The unfavorable zone contains z_1 values in the interval $(-\infty, Z_L)$ and here the study continues with $m_2(z_1) = m_{2,0}$ further subjects, as originally planned.

If z_1 falls in the favorable zone or the unfavorable zone, the final test of H_0: $\theta \leq 0$ is unchanged from that of the initial design with a fixed total sample size of $n_t = 442$, while for values of z_1 in the promising zone the conditional Type I error probability is decreased. Hence this design protects the Type I error probability of $\alpha = 0.025$ conservatively.

Values of the promising zone limits Z_L and Z_U for this example are shown in Table 24.1. In the Chen, DeMets & Lan version of the design the promising zone contains z_1 values for which conditional power under $\theta = \hat{\theta}_1$ is at least 0.5. However, Mehta & Pocock propose following the approach of Gao, Ware & Mehta (2008), which takes account of the upper limit $n_{\max}$ and in this case they extend the promising zone to include cases where conditional power under $\theta = \hat{\theta}_1$ is as low as 0.365.

We shall denote the final sample size by the random variable $N = m_1 + m_2(Z_1)$. Figure 24.1 shows values of N that arise when using the extended promising zone based on the method of Gao, Ware & Mehta (2008). Also displayed is the density of $\hat{\theta}_1$ under $\theta = 1.6$

Table 24.1 *Promising zone limits for the illustrative example*

Scale	Promising zone limits		
	Lower limit		Upper limit
	Gao, Ware & Mehta	Chen, DeMets & Lan	
Conditional Power	0.365	0.5	0.8
Z-scale, z_1	1.172	1.345	1.765
Interim estimate, $\hat{\theta}_1$	1.219	1.398	1.835

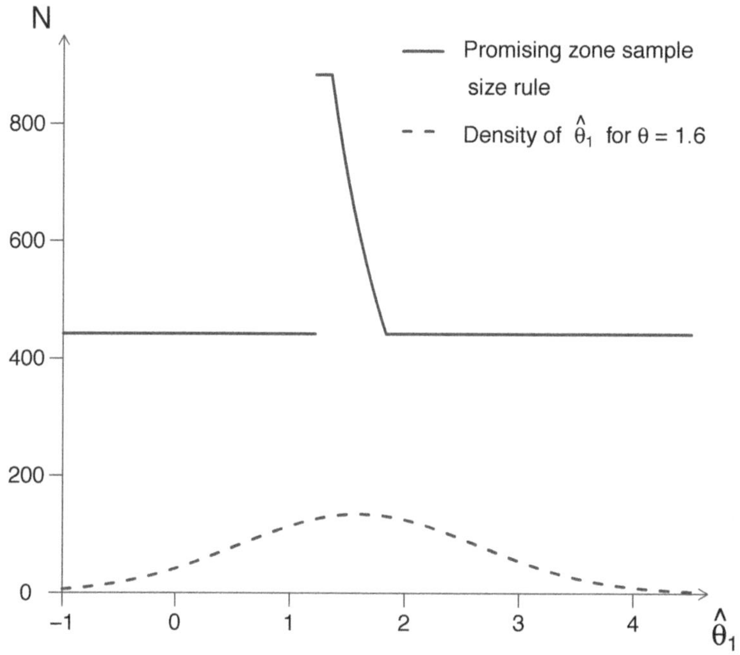

Figure 24.1 *Promising Zone Design: Sample size rule. The figure also shows a scaled version of the density of* $\hat{\theta}_1$ *when* $\theta = 1.6$.

(on a different vertical scale), which illustrates the variability in $\hat{\theta}_1$ at the first analysis. The distribution of $\hat{\theta}_1$ under other values of θ is shifted but has the same variance, so we can see that increases in the second stage sample size occur in a region of quite small probability regardless of the true value of θ.

Figure 24.2 presents operating characteristics of the promising zone design and the initial design with a fixed sample size of 442. Comparison of the power curves in the left panel of Figure 24.2 shows that there is only a small increase in power, for example, the power at $\theta = 1.6$ has been increased from 0.611 to 0.657 while power of 0.8 is now achieved at $\theta = 1.91$ rather than $\theta = 2.0$. The right panel of Figure 24.2 shows that this increase in power is accompanied by a rise in the expected sample size function.

In their discussion of this example, Jennison & Turnbull (2015) consider two other ways of achieving the same increase in overall power. A simple option is a fixed sample size design with 490 subjects. Jennison & Turnbull also suggest using a two-stage group sequential design defined by Type I and Type II error spending functions from the ρ-family of Section 8.3 with $\rho = 2$. This design has a maximum sample size of 514 and an interim

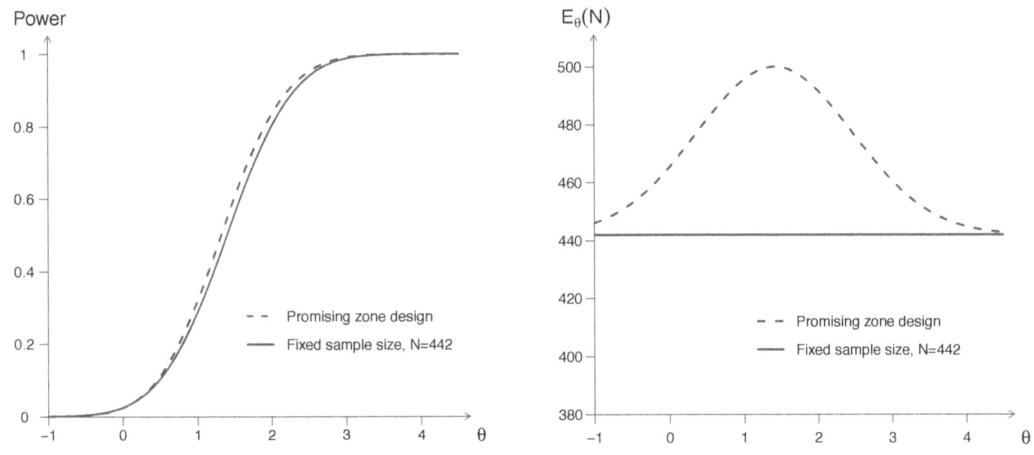

Figure 24.2 *Power and $E_\theta(N)$ curves for promising zone design and initial fixed sample size design with $N = 442$*

analysis after 208 response have been observed. The trial continues past the interim analysis if $0.12 < Z_1 < 2.54$, H_0 is rejected at the final analysis if $Z_2 \geq 2.00$, and the test of H_0 has Type I error rate $\alpha = 0.025$ under $\theta = 0$ and power 0.8 when $\theta = 1.9$. In Mehta & Pocock's example there are 208 "pipeline" subjects who have been treated at the time of the interim analysis but their response at 6 months is yet to be observed. These patients are counted in the total sample size N, even though their data is not used in the group sequential test when the trial stops at the first analysis.

The left hand panel of Figure 24.3 shows the sample size rules for the Mehta & Pocock design and the two designs suggested by Jennison & Turnbull. In the case of the two-stage group sequential design, the figure shows both the number of observed responses and the number of subjects randomized when the trial stops at the first analysis. By construction, the power curves of all three designs are essentially the same. The right hand panel of 24.3 shows the expected final sample size N as a function of θ. The fixed sample size design has lower values of $E_\theta(N)$ than the promising zone design for values of θ between 0.8 and 2.0, a range of particular interest to the investigators. The group sequential test is wasteful in that it does not use information from the pipeline subjects when the trial stops at the first analysis and its efficiency could be improved by using a "delayed response" design of the form described in Chapter 14. However, despite its wastefulness, this design has lower $E_\theta(N)$ than the promising zone design at all values of θ. The variability in sample size and the need to plan for the possibility of a sample size as high as 884 may also be seen as disadvantages of the promising zone design.

There are several reasons for the promising zone design's high expected sample size. The conservatism which follows from implementation of the Chen, DeMets & Lan approach has a small effect but it is the form of the sample size rule that is particularly influential. In the promising zone design, the second stage sample size is sensitive to the value of $\hat{\theta}_1$, and we have seen this is a highly variable estimate of θ. A small decrease in $\hat{\theta}_1$ in the upper part of the promising zone causes a dramatic increase in the sample size from $N = n_t$ to $N = n_{\max}$. In discussing Mehta & Pocock's paper, Glimm (2012) notes that, since $\hat{\theta}_1$ is a function of z_1, the conditional power at $\theta = \hat{\theta}_1$, $CP_{\hat{\theta}_1}(z_1, m_2)$ "uses the stage 1 data twice" and he comments that "the conditional power with the observed point estimate plugged in as if it were the true effect is an unstable estimate of the true conditional power. Hence, it can be rather dangerous to base a sample size re-estimation on it."

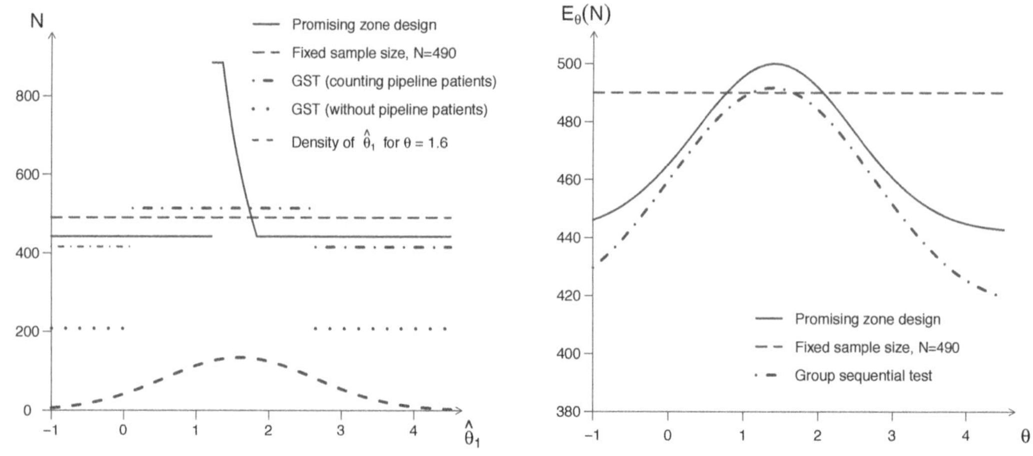

Figure 24.3 *(a) Sample size rules for the promising zone design, a fixed sample size design with* $N = 490$*, and a two-stage group sequential test (GST), along with a scaled version of the density of* $\hat{\theta}_1$ *when* $\theta = 1.6$*, (b)* $E_\theta(N)$ *functions for the three trial designs*

We noted in Section 24.3.2 that efficient designs with sample size re-assessment are possible in principle. However, for practical application, a simple and effective way to define a suitable sample size rule is needed. Jennison & Turnbull (2015) propose a method for doing this in the context of Mehta & Pocock's example. We shall describe the final version of this method discussed in Section 5 of their paper.

Jennison & Turnbull's procedure starts with a target of $n_t = 442$ for the final sample size and, as in Mehta & Pocock's design, an interim analysis is conducted when $m_1 = 208$ observations are available. The sample size rule, which we shall describe below, specifies the number m_2 of additional observations as a function of the first stage statistic z_1, with a minimum value of 208 to accommodate those subjects already in the "pipeline". After gathering the stage 2 data, a combination test is applied to the summary statistics z_1 and z_2 from the m_1 and m_2 subjects in stages 1 and 2, respectively, and H_0 is rejected if

$$w_1 z_1 + w_2 z_2 \geq \Phi^{-1}(0.975),$$

where $w_1 = \sqrt{(208/422)}$ and $w_2 = \sqrt{(234/422)}$, reflecting the initial plan for the trial. If $m_2 = 234$, so the final sample size is equal to n_t, this decision rule is that of a fixed sample size design with n_t observations. In other cases, the rule maintains the conditional Type I error probability of an initial trial design with a fixed sample size n_t. Using a combination test avoids the conservatism in Mehta & Pocock's design. Furthermore, sample size can be modified for any value of z_1, either increasing or decreasing m_2 relative to the original value of $n_t - m_1$.

Jennison & Turnbull (2015) address the competing goals of high conditional power and low sample size by specifying an effect size of interest θ^* and a parameter γ and choosing $m_2(z_1)$ to maximize the combined gain function

$$\text{Gain}(z_1, m_2) = CP_{\theta^*}(z_1, m_2) - \gamma\, m_2 \qquad (24.7)$$

subject to the constraint $208 \leq m_2 \leq m_2^*$, where the lower bound on m_2 is the number of "pipeline" subjects. They set $\theta^* = 1.6$ as this is stated to be a meaningful treatment effect at which higher power is desirable.

The parameter γ acts as a "rate of exchange" between sample size and conditional power. Although one could base the value of γ on the cost of treating each subject in the trial and

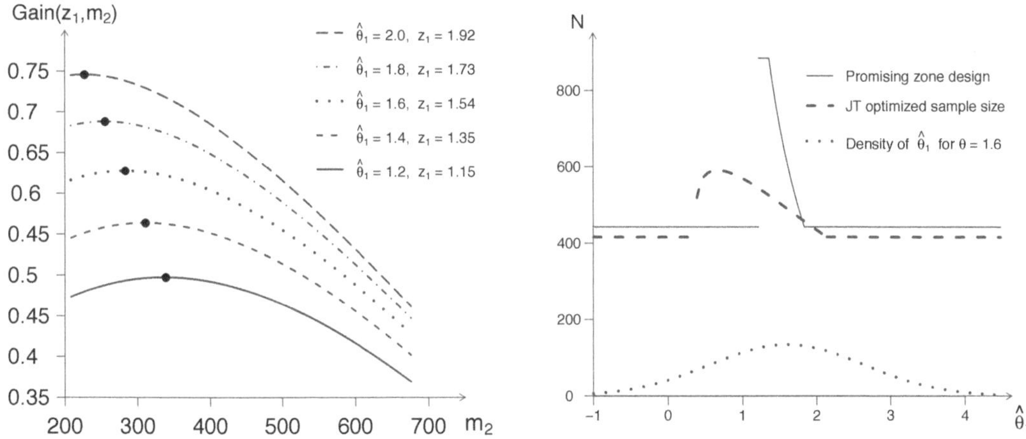

Figure 24.4 *The JT design: (a) finding the stage 2 sample size $m_2(z_1)$ by maximizing Gain(z_1, m_2), (b) sample size rules for the Mehta & Pocock and JT designs*

an estimate of the financial return from a positive outcome, this is certainly not required. Instead, γ can be regarded as a tuning parameter that controls the degree to which sample size may be increased when interim data are promising but not overwhelming. With such a criterion in place, interim data sets with different values of z_1 are treated consistently since the benefits of any increase in sample size are measured on a common scale. The value of γ can be chosen to achieve a specified overall power and Jennison & Turnbull do precisely this, setting $\gamma = 0.2449$ to obtain a design with power 0.658 at $\theta = 1.6$, matching the power of the Mehta & Pocock design seen in Figure 24.2. We shall refer to the resulting procedure as the JT design.

Maximizing the gain function (24.7) for each value of z_1 produces a design with an overall optimality property with respect to unconditional power. Let $\tilde{m}_2(z_1)$ denote a particular sample size rule. The unconditional power at $\theta = \theta^*$ of the procedure with this sample size rule is

$$Pr_{\theta^*}\{\text{Reject } H_0\} = \int CP_{\theta^*}(z_1, \tilde{m}_2(z_1))\, f_{\theta^*}(z_1)\, dz_1,$$

where $f_{\theta^*}(z_1)$ denotes the density of Z_1 under $\theta = \theta^*$. Furthermore,

$$Pr_{\theta^*}\{\text{Reject } H_0\} - \gamma E_{\theta^*}(N) = \int \{CP_{\theta^*}(z_1, \tilde{m}_2(z_1)) - \gamma \tilde{m}_2(z_1)\}\, f_{\theta^*}(z_1)\, dz_1. \quad (24.8)$$

Since the sample size rule obtained by maximizing (24.7) maximizes the integrand in the right hand side of (24.8) for every z_1, it must also maximize the left hand side of (24.8). It follows that this sample size rule yields the minimum possible value of $E_{\theta^*}(N)$ among all sample size rules that achieve the same power under $\theta = \theta^*$. Pilz, Kieser & Kunzmann (2020) point out that the converse is also true: finding the function $\tilde{m}_2(z_1)$ that maximizes the unconditional objective function (24.8) is a problem in the calculus of variations for which Pilz et al. (2019) have showed that solving the Euler-Lagrange equation reduces to the problem of maximizing the conditional objective (24.7) for each z_1.

It is straightforward to compute values of Gain(z_1, m_2) defined in (24.7) to find the value of m_2 that maximizes this function. The left hand panel of Figure 24.4 illustrates this process for five values of z_1. In each case, the gain function is unimodal and the optimal value of m_2 can be found by a simple numerical search. The resulting sample size rule for the JT design is depicted in the right hand panel of Figure 24.4, which shows the total sample size $N = 208 + m_2$. The discontinuity in the optimal m_2 at $\hat{\theta}_1 = 0.396$ arises because

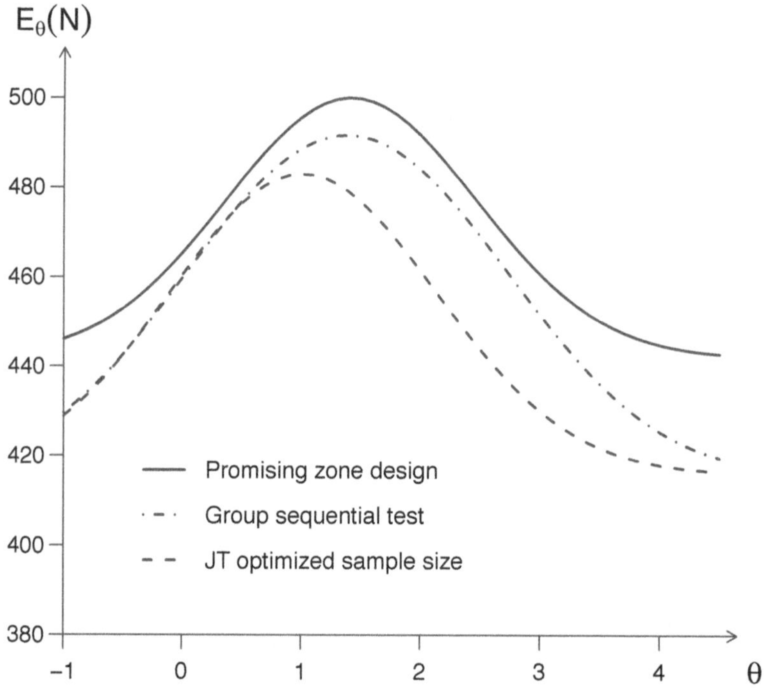

Figure 24.5 *Expected sample size for Mehta & Pocock and Jennison & Turnbull designs*

the gain function has derivative zero at $m_2 = 208$ for this value of $\hat{\theta}_1$ and the location of the maximum jumps from $m_2 = 311$ to the minimum permitted value, $m_2 = 208$, as $\hat{\theta}_1$ decreases past 0.396.

It is evident that the JT design's largest sample size is considerably smaller than the value $N = 884$ that can occur in Mehta & Pocock's design. The region in which the JT design increases the sample size above $n_t = 442$ is wider, extending down to $\hat{\theta}_1 = 0.396$, and the greatest increases occur at values of $\hat{\theta}_1$ for which the Mehta & Pocock design sticks to the original value of 442.

Figure 24.5 compares the expected sample size function for the JT design with that of the Mehta & Pocock design and the two-stage group sequential test described earlier. It is clear that the JT design has the lowest values of $E_\theta(N)$ at all values of θ and the improvement over the Mehta & Pocock design is particularly high when $\theta > 1$.

Jennison & Turnbull (2015) go on to construct a delayed response group sequential test (DRGST) of the form described in Chapter 14 and present this in their Figure 11. This test has two analyses, the first taking place after $n_1 = 208$ observed responses, Type I error probability 0.025, and power 0.658 when $\theta = 1.6$. If recruitment is halted at the first analysis, the study continues to obtain responses from the 208 pipeline patients, and the final sample size is 416. If recruitment continues, a further 102 patients are admitted to the trial, and the final sample size is 518. Jennison & Turnbull also derive an optimal adaptive DRGST in which the stage 2 sample size is a function of the interim estimate $\hat{\theta}_1$. This design can be regarded as an extension of the JT design in which the final decision rule is not restricted to a simple combination test. The power curves for the two new procedures match that of the JT design. The $E_\theta(N)$ curves of the adaptive DRGST and the JT design are nearly identical while that of the non-adaptive DRGST is a little higher but within 2% of the other two curves at all values of θ. The conclusion for this example is that Jennison & Turnbull's approach has produced a highly efficient procedure. If one prefers a simpler design with a choice between stopping recruitment at the first analysis or continuing with a fixed second stage sample size, then a non-adaptive DRGST can deliver almost all the potential savings in expected sample size.

Jennison & Turnbull's approach can be generalized to minimize an objective function involving a weighted sum of expected sample sizes $E_{\theta_i}(N)$ at selected effect sizes θ_i or an integral of $E_\theta(N)$ over θ. In the latter case, suppose $\pi(\theta)$ represents a prior distribution for θ. Then, one may choose to minimize an objective function of the form

$$Pr_{\theta^*}\{\text{Reject } H_0\} \;-\; \gamma \int \pi(\theta)\, w(\theta)\, E_\theta(N)\, d\theta,$$

where the weight function $w(\theta)$ indicates the importance attached to reducing $E_\theta(N)$ at each value of θ. If $\tilde{m}_2(z_1)$ is a particular sample size rule, this objective function can be written as

$$\int CP_{\theta^*}(z_1, \tilde{m}_2(z_1))\, f_{\theta^*}(z_1)\, dz_1 \;-\; \gamma \int \pi(\theta) w(\theta) \int \{m_1 + \tilde{m}_2(z_1)\}\, f_\theta(z_1)\, dz_1\, d\theta,$$

$$= \int \left[CP_{\theta^*}(z_1, \tilde{m}_2(z_1))\, f_{\theta^*}(z_1) - \gamma \int w(\theta)\, \{m_1 + \tilde{m}_2(z_1)\}\, \pi(\theta)\, f_\theta(z_1)\, d\theta \right] dz_1, \qquad (24.9)$$

where $f_\theta(z_1)$ denotes the density of Z_1 under treatment effect θ. Since m_1 is fixed, the optimal m_2 for a given z_1 is the value that maximizes

$$CP_{\theta^*}(z_1, m_2)\, f_{\theta^*}(z_1) - \gamma \int m_2\, w(\theta)\, \pi(\theta)\, f_\theta(z_1)\, d\theta,$$

and this may be found by a simple numerical search.

Note that in this formulation of the decision problem with prior distribution $\pi(\theta)$, the posterior distribution of θ has a density proportional to $\pi(\theta)\, f_\theta(z_1)$ and the integral in (24.9) is with respect to a multiple of this posterior density. This establishes the link to the optimized adaptive SSR designs referred to in Section 24.3.2, which Jennison & Turnbull (2006a) computed as the solutions to Bayes decision problems. The main difference here is that the Type I error rate is controlled at level α directly by use of a combination test, simplifying the optimization process. A further cosmetic difference is that the optimization problem is defined with the goal of maximizing an expected gain rather than minimizing an expected loss.

Hsiao, Liu & Mehta (2019) acknowledge the benefits of Jennison & Turnbull's approach, referring to the resulting design is as the "gold standard". They define a "constrained promising zone" design which adopts a similar approach to Jennison & Turnbull in defining the sample size rule but they impose a constraint that a specified minimum conditional power should be achieved if sample size is to be increased at the lower end of the promising zone. They comment that there may have been some confusion over terminology and state (p. 1184) that "any region for the interim statistic in which the sample size is increased should be considered a promising zone, and the corresponding design should be termed a promising zone design."

24.6 Discussion

In this chapter we have described adaptive designs with sample size re-assessment (SSR). Such designs are commonly used, indeed, one of the main uses of adaptive techniques such as combination tests or the conditional error function has been to facilitate unblinded sample size re-assessment. Edwards et al. (2020) present a review of trials using a "promising zone" design in which they found articles reporting 30 such trials.

We have noted that some commonly used forms of sample size rule lead to inefficient procedures and we have described a method to derive efficient sample size rules. The aim of SSR to achieve power with a small average sample size, is also the goal of group sequential tests (GSTs) and we recommend considering a GST with fixed group sizes as a simpler option.

An adaptive design with SSR can be regarded as a GST in which group sizes are adapted in response to previously observed data. Such adaptations can improve efficiency but the potential gains are small and come at the price of a more complex design. There is also a risk of information leakage if updates in sample size can be traced back to particular estimates of the treatment effect.

The example presented in Section 24.5.2 illustrates the strengths and weaknesses of different methods. In this example, the number of pipeline subjects at the interim analysis is unusually high. In consequence, a conventional GST, which does not use the pipeline data after stopping at the interim analysis, is wasteful. The SSR approach provides a conceptually simple way of dealing with these subjects. A similar strategy was previously proposed by Faldum & Hommel (2007). As we saw in the example, a good choice of sample size rule is crucial for the resulting design to have desirable properties. A delayed response GST with fixed group sizes can be regarded as a special case of an adaptive design with SSR, where just two values of the second stage sample size are permitted and we reported the findings of Jennison & Turnbull (2015) that this restriction has a minimal effect on efficiency.

Whatever design is favored, it is good practice to study its operating characteristics, in particular its power curve and expected sample size function, both to fine tune the design and to compare with other types of design. Menon et al. (2013) report a simulation study comparing properties of several two-stage SSR procedures including the methods of Cui, Hung & Wang (1999) and Chen, DeMets & Lan (2004). They also investigate robustness properties when an incorrect variance assumption is made at the interim analysis. Gaffney & Ware (2017) investigated the properties of various promising zone designs and concluded "clear superiority" in terms of efficiency and overall power of designs in which the interim analysis where sample size is re-assessed occurs relatively late in the study.

There is an argument that group sequential tests require a large up-front commitment to the pre-specified maximum sample size while adaptive designs do not. This initial commitment may be less of a concern for cooperative groups such as the National Institutes for Health or the Veterans Administration, or larger pharmaceutical companies. In this case, Emerson, Levin & Emerson (2011) argue that

> "minimizing the ASN [average sample number] is tantamount to maximizing the number of new therapies studied (and, hopefully, adopted) in a specified calendar time."

Mehta & Liu (2016) state that, in contrast,

> "Small pharmaceutical and biotechnology companies that are venture backed ... typically have limited resources for a single compound and rely on outside investment to fund their trials. Adaptive financing coupled with adaptive design might increase the chances that the trial will be adequately powered."

This fact is acknowledged in the FDA Guidance (FDA, 2019a, Sec. II C) which states

> "sponsors might be more willing to commit to a trial that allows planned design modifications based on accumulating information."

One may question whether there really is a difference between the "initial commitment" in a group sequential test with fixed group sizes and an adaptive design with SSR. In either case, one would expect resources to be secured before commencing a Phase III clinical trial, albeit with the understanding that the highest level of investment will only be called upon if interim results are sufficiently promising. In the case of a design with SSR, sample size will be increased when results are in the promising zone, while in the group sequential test, the trial will stop for futility if interim results are poor. Perhaps the difference is psychological, with a group sequential design viewed as starting with a large maximum sample but stopping early when appropriate, while an adaptive design with SSR has a small initial sample size that may be increased later. Lai, Lavori & Shih (2012a) contrast the "start small, then ask for more" philosophy of an SSR design with that of a group sequential design in which you "ask for what you might need, and hope you need less."

We have described SSR procedures for the case of normally distributed responses. Using the results of Chapters 3 and 15, the same methods can be applied in a very wide variety of situations where test statistics are approximately normally distributed. In particular, the promising zone approach has been extended to survival endpoints; see, for example Chen, Li & Lan (2015) and Liu & Lim (2017). However, Freidlin & Korn (2017) advise caution when SSR methods are employed with time-to-event endpoints and there may be unblinded access to short-term endpoints for subjects still at risk at the time of the interim analysis. Jenkins, Stone & Jennison (2011) note the danger of inflating the Type I error rate in such a situation but show how a combination test may be applied in a manner that avoids this problem: we shall follow their approach in the case study of Chapter 29. Irle & Schäfer (2012) address the same issue using an extension of the conditional rejection probability principle that we described in Section 23.6. Magirr, Jaki, Koenig & Posch (2016) discuss the analysis of survival data in an adaptive design with sample size reassessment. They comment on methods that ignore a subset of the observed events and they propose an alternative method which takes account of all event times.

Chapter 25

Adaptive Trials Testing Multiple Hypotheses: Closed Testing Procedures with Combination Tests

25.1 Adaptive Designs with Multiple Hypotheses

It is not uncommon for multiple hypotheses to be tested in a single clinical trial. In Chapter 20 we presented methods that control the family-wise Type I error rate (FWER) when such tests are conducted in either a fixed sample size or group sequential design. We are now ready to derive methods that control the FWER in adaptive trials that test multiple hypotheses. To do so, we shall bring together the closed testing procedures of Section 20.3 and the combination tests seen in Chapters 22 and 23.

In Chapter 21 we described multi-armed trials in which multiple treatments are compared with each other or with a common control and we shall discuss adaptive versions of such designs in Chapter 27. If p treatments are compared with a control and we denote the p treatment effects by $\theta_1, \ldots, \theta_p$, the null hypotheses H_j: $\theta \leq 0$, $j = 1, \ldots, p$, may be tested. Adaptation can occur at interim analyses when treatment arms are dropped from the study either because a treatment has been shown to be superior to the control or because interim results are unpromising.

A "seamless Phase II/III" trial, which we shall consider in Chapter 26, is a special case of a multi-arm multi-stage trial in which two phases of a drug development program are combined, merging the selection of the dose level for a new treatment and the testing of this dose against the control into a single study. If such a trial initially involves p dose levels with treatment effects $\theta_1, \ldots, \theta_p$, one of the null hypotheses H_j: $\theta \leq 0$, $j = 1, \ldots, p$, will be tested at the end of the Phase III stage. The choice of dose level to take forward to the Phase III stage and the associated null hypothesis will depend on data from the Phase II stage that are also included in the final test statistic.

In the "enrichment" trials that we shall describe in Chapter 28, interest is in a treatment effect in one or more subgroups of a heterogeneous patient population. Here $\theta_1, \ldots, \theta_p$ may denote treatment effects in each of p subgroups and all the null hypotheses H_j: $\theta \leq 0$, $j = 1, \ldots, p$, can be tested. Adaptation can occur when certain subgroups are dropped and the sample size for the remaining subgroups is increased, enriching these subgroups.

Yet another example is a superiority/non-inferiority trial where the primary objective is to show that a new treatment is superior to the control but, if this is not possible, it is still of value to demonstrate non-inferiority. In this case, there are two null hypotheses that may be tested, or more if there are several experimental treatments. Adaptation may take place if the focus shifts from superiority to non-inferiority and the sample size is then increased to give the test of non-inferiority adequate power; see, for example, Wang et al. (2001).

DOI: 10.1201/9781584888482-25

25.2 Creating a Closed Testing Procedure from Adaptive Combination Tests

Suppose hypotheses $H_1, \ldots, H_p$ are to be tested in a group sequential trial with K analyses where adaptive changes are possible at the first $K-1$ analyses. We wish to control the family-wise Type I error rate at level α while allowing flexible adaptations to be made during the trial. Following Bretz et al. (2006), we construct a closed testing procedure (CTP) in which a level α test is defined for each intersection hypothesis

$$H_J = \cap_{j \in J} H_j$$

where J is a subset of $\{1, \ldots, p\}$. In the overall procedure, the simple hypothesis H_j is rejected if and only if the intersection hypothesis H_J is rejected at level α for every set J that contains the index j.

Testing simple hypotheses

Suppose that for $j = 1, \ldots, p$ and $k = 1, \ldots, K$, the P-value for testing H_j based on the data in stage k alone is $P_j^{(k)}$. Let $C(P_j^{(1)}, \ldots, P_j^{(K)})$ denote the overall test statistic for H_j obtained by combining the stage-wise P-values. So if, for example, we use a weighted inverse normal combination test with weights $w_1, \ldots, w_K$,

$$C(P_j^{(1)}, \ldots, P_j^{(K)}) = 1 - \Phi\left\{\sum_{k=1}^{K} w_k\, \Phi^{-1}(1 - P_j^{(k)})\right\}.$$

Testing intersection hypotheses

For a given subset J of $\{1, \ldots, p\}$, we define $P_J^{(k)}$ to be a P-value for testing H_J based on the data in stage k alone using the elementary P-values $P_j^{(k)}$, $j \in J$. So if, for example, J is of size m and we use a Simes (1986) test, as described in in Section 20.4, then

$$P_J^{(k)} = \min_{j=1,\ldots,m} \{m P_{(j)}^{(k)}/j\}$$

where

$$P_{(1)}^{(k)} \le P_{(2)}^{(k)} \le \ldots \le P_{(m)}^{(k)}$$

are the ordered P-values with indices $j \in J$.

Combination tests

At the end of the study, we combine the stage-wise P-values $P_J^{(k)}$, $k = 1, \ldots, K$, to obtain the overall P-value for H_J. So, using a weighted inverse normal combination test with weights $w_1, \ldots, w_K$ gives

$$C(P_J^{(1)}, \ldots, P_J^{(K)}) = 1 - \Phi\left\{\sum_{k=1}^{K} w_k\, \Phi^{-1}(1 - P_J^{(k)})\right\}.$$

The closed testing procedure

In applying the closed testing procedure, it is a simple matter to check which intersection hypotheses H_J are rejected with

$$C(P_J^{(1)}, \ldots, P_J^{(K)}) \le \alpha.$$

We can then make the overall decisions, rejecting an elementary hypothesis H_j when H_J has been rejected for every set J containing the index j.

Table 25.1 *Worksheet for 4-stage adaptive combination tests with 3 hypotheses*

Hypothesis	*P-values*				*Combined P-value*
	Stage 1	*Stage 2*	*Stage 3*	*Stage 4*	
H_1	$P_1^{(1)}$	$P_1^{(2)}$	$P_1^{(3)}$	$P_1^{(4)}$	$C(P_1^{(1)}, P_1^{(2)}, P_1^{(3)}, P_1^{(4)})$
H_2	$P_2^{(1)}$	$P_2^{(2)}$	$P_2^{(3)}$	$P_2^{(4)}$	$C(P_2^{(1)}, P_2^{(2)}, P_2^{(3)}, P_2^{(4)})$
H_3	$P_3^{(1)}$	$P_3^{(2)}$	$P_3^{(3)}$	$P_3^{(4)}$	$C(P_3^{(1)}, P_3^{(2)}, P_3^{(3)}, P_3^{(4)})$
H_{12}	$P_{12}^{(1)}$	$P_{12}^{(2)}$	$P_{12}^{(3)}$	$P_{12}^{(4)}$	$C(P_{12}^{(1)}, P_{12}^{(2)}, P_{12}^{(3)}, P_{12}^{(4)})$
H_{13}	$P_{13}^{(1)}$	$P_{13}^{(2)}$	$P_{13}^{(3)}$	$P_{13}^{(4)}$	$C(P_{13}^{(1)}, P_{13}^{(2)}, P_{13}^{(3)}, P_{13}^{(4)})$
H_{23}	$P_{23}^{(1)}$	$P_{23}^{(2)}$	$P_{23}^{(3)}$	$P_{23}^{(4)}$	$C(P_{23}^{(1)}, P_{23}^{(2)}, P_{23}^{(3)}, P_{23}^{(4)})$
H_{123}	$P_{123}^{(1)}$	$P_{123}^{(2)}$	$P_{123}^{(3)}$	$P_{123}^{(4)}$	$C(P_{123}^{(1)}, P_{123}^{(2)}, P_{123}^{(3)}, P_{123}^{(4)})$

At the design stage, there is freedom to choose the ingredients of the above procedure. In testing intersection hypotheses, we can obtain the P-value $P_J^{(k)}$ for H_J at stage k from a Bonferroni or Simes test, or in cases where multiple treatments are compared to a control a Dunnett test. A weighted inverse normal or inverse χ^2 combination test can be applied to combine the P-values $P_J^{(1)}, \ldots, P_J^{(K)}$ across stages.

Table 25.1 shows the elements of this procedure for a trial with $K = 4$ stages and $p = 3$ hypotheses. For simplicity we have written P_1 rather than $P_{\{1\}}$, P_{12} rather than $P_{\{1,2\}}$, etc., in this table. We can define a final adjusted P-value for each elementary hypothesis H_j as

$$\tilde{P}_j = \max\{C(P_J^{(1)}, \ldots, P_J^{(K)}) \colon J \subseteq \{1, \ldots, p\} \text{ with } j \in J\}$$

so, for example, when $K = 4$ and $p = 3$,

$$\tilde{P}_1 = \max\{C(P_1^{(1)}, \ldots, P_1^{(4)}),\, C(P_{12}^{(1)}, \ldots, P_{12}^{(4)}),\, C(P_{13}^{(1)}, \ldots, P_{13}^{(4)}),\, C(P_{123}^{(1)}, \ldots, P_{123}^{(4)})\}.$$

Dropping a hypothesis

It may happen that at some stage it becomes no longer of interest to test one or more of the hypotheses H_j. In a trial comparing multiple treatments, one or more of the arms may be dropped for futility or, in an enrichment trial, a subgroup may be dropped and recruitment restricted to the remaining subgroups. In such cases, Bretz et al. (2006, p. 628) explain that the test of each H_J should be replaced in future stages by a test of $H_{\tilde{J}}$ where $\tilde{J}$ contains indices in J of hypotheses which are still under consideration.

Suppose, in the example of Table 25.1 that hypothesis H_1 is abandoned after stage 2, then we would set

$$P_{12}^{(3)} = P_2^{(3)}, \;\; P_{13}^{(3)} = P_3^{(3)} \;\text{ and }\; P_{123}^{(3)} = P_{23}^{(3)}$$

in stage 3. Because $P_{12}^{(3)}$, $P_{13}^{(3)}$ and $P_{123}^{(3)}$ are defined before any stage 3 data are observed, this is a legitimate change, and since $P_2^{(2)}$, $P_3^{(3)}$ and $P_{23}^{(3)}$ have Uniform(0,1) distributions or are stochastically larger than a Uniform(0,1) random variable under H_2, H_3 and H_{23}, respectively, the family-wise Type I error rate will be controlled at the specified level α.

Similarly, if hypotheses H_2 and H_3 are still being tested in stage 4, we set

$$P_{12}^{(4)} = P_2^{(4)}, \quad P_{13}^{(4)} = P_3^{(4)} \quad \text{and} \quad P_{123}^{(4)} = P_{23}^{(4)}$$

and, again, the family-wise Type I error rate remains controlled.

Early stopping to reject a hypothesis

It is also possible to allow early stopping to reject one or more of the elementary hypotheses. Consider a two-stage trial comparing two treatments against a control with FWER $\alpha = 0.025$. Suppose the null hypotheses are H_1: $\theta_1 \leq 0$ and H_2: $\theta_1 \leq 0$, a Dunnett test is used to test the intersection hypothesis H_{12} in each stage, and an inverse χ^2 test is used to combine P-values across stages. Then, for example, H_1 will be rejected overall if both H_1 and H_{12} are rejected at level $\alpha = 0.025$. As noted in Section 23.1, the critical value for this inverse χ^2 test is

$$\exp\{-\chi_4^2(0.025)/2\} = 0.0038,$$

so H_1 will be rejected if

$$P_1^{(1)} P_1^{(2)} \leq 0.0038$$

and

$$P_{12}^{(1)} P_{12}^{(2)} \leq 0.0038.$$

If values $P_1^{(1)} < 0.0038$ and $P_{12}^{(1)} < 0.0038$ are observed, rejection of H_1 is inevitable whatever the values of $P_1^{(2)}$ and $P_{12}^{(2)}$, so H_1 could be rejected overall and the first treatment dropped at this point. In this case the trial might be continued to test the other experimental treatment. Since $P_{12}^{(1)} < 0.0038$, H_{12} will be rejected however $P_{12}^{(2)}$ is defined so, unlike the case where a treatment arm is dropped for futility, we do not need to consider how to re-define $P_{12}^{(2)}$.

If a weighted inverse normal combination test is used, it is not possible to predict the outcome of a final hypothesis test with certainty. However, early stopping for a positive outcome can be achieved by using a Lehmacher-Wassmer combination test, as defined in Section 23.7.1, with the lower boundary set at $-\infty$. In testing a generic null hypothesis H: $\theta \leq 0$ in a trial with K stages, weights $w_1, \ldots, w_K$ are pre-specified and the test is based on the cumulative statistics

$$\tilde{Z}_k = (w_1 Z^{(1)} + \ldots + w_k Z^{(k)}) \,/\, (w_1^2 + \ldots + w_k^2)^{1/2}, \quad k = 1, \ldots, K,$$

where each $Z^{(k)}$ is calculated from stage k data and has a $N(0,1)$ distribution if $\theta = 0$. The null hypotheses H is rejected at analysis k if $\tilde{Z}_k \geq b_k$, where the critical values b_k are chosen so that the Type I error probability of this group sequential test is α when $\theta = 0$. In our multiple testing problem, we create a test of each intersection hypothesis H_J by defining

$$Z_J^{(k)} = \Phi^{-1}(1 - P_J^{(k)})$$

and

$$\tilde{Z}_{k,J} = (w_1 Z_J^{(1)} + \ldots + w_k Z_J^{(k)}) \,/\, (w_1^2 + \ldots + w_k^2)^{1/2}, \quad k = 1, \ldots, K.$$

Then, H_J is rejected at stage k if $\tilde{Z}_{k,J} \geq b_k$.

If H_J has been rejected for every J that contains the index j^*, the elementary hypothesis H_{j^*} can be rejected in the closed testing procedure and no further data need be collected in relation to hypothesis H_{j^*}. If this occurs at analysis k and the trial continues, it is not necessary to consider how to re-define stage-wise test statistics for intersection hypotheses H_J where $j^* \in J$ since these intersection hypotheses have already been rejected.

The above prescription for creating a closed testing procedure that uses adaptive combination tests to test the intersection hypotheses is both simple and powerful. We shall demonstrate its versatility in a variety of applications in the next four chapters.

Chapter 26

Treatment Selection and Testing: Seamless Phase II/III Trials

26.1 Introduction

There are several forms of clinical trial in which multiple treatments are compared to a control. In Section 21.4.2 we discussed group sequential trials comparing two or more treatments against a control. In Chapter 29 we shall present a case study of a Phase III trial in which two versions of a new treatment are compared with a control and one version of the new treatment is dropped at an interim analysis. In Chapter 27 we shall consider a general form of adaptive, multi-arm trial in which there is flexibility in deciding when treatment arms may be dropped.

In this chapter, we discuss the specific case of a "seamless Phase II/III" trial with treatment selection. As we explained in Section 1.1, in Phase II, or in some cases Phase IIb, several variants of a new treatments are compared to a control. In a trial for a specific drug, these variants could be differing dose levels, frequencies of dosing or methods of administration, or they could be different combination therapies involving the new drug. At the end of Phase II, a single treatment is selected and the Phase III trial is run as a confirmatory study to demonstrate superiority of the selected treatment against the control. A seamless Phase II/III trial combines the process of selecting and testing a treatment into a single study.

In the traditional approach, investigators plan their Phase III trial in the knowledge of Phase II results. This information may be used in deciding on the precise treatment definition, the target population, and the primary endpoint. The treatment effect observed in Phase II may also guide the choice of sample size for the Phase III trial. The release of positive results from Phase II may help recruitment to the Phase III trial. However, planning and gaining approval for the Phase III trial after the completion of Phase II can be time-consuming. It is possible to plan both phases jointly and gain approval for these before the Phase II trial even though the results of these studies will be analyzed separately. In particular, the test of whether the selected treatment is superior to the control will be based only on Phase III data. This is known as an "operationally seamless" Phase II/III design. Taking the idea of combining phases a step further, gains in efficiency can be made by using the combined data from Phase II and Phase III in the final hypothesis test. Maca et al. (2006) describe this as an "inferentially seamless" Phase II/III design. Such a design is also known as an "adaptive seamless design" (ASD). Cuffe et al. (2014) present examples of trials with operationally and inferentially seamless Phase II/III designs with discussion of the benefits gained in these designs as well as the logistical issues addressed in their implementation.

Both operationally seamless and inferentially seamless designs have the advantage that a single, overarching protocol document can be approved by the appropriate regulatory and ethical authorities, and a single monitoring committee can be appointed for the whole trial.

DOI: 10.1201/9781584888482-26

In addition, the time gap, or "white space", between the two phases can be significantly reduced, and momentum and enthusiasm for subject recruitment at the trial centers is less likely to be lost.

Since an inferentially seamless ASD uses both Phase II data and Phase III data in the final hypothesis test, it is standard practice for such a study to be treated as a single clinical trial. Thus, once trial data have been observed there is a "firewall" between the independent data monitoring committee and the study sponsor. Guidelines must, therefore, be provided for the data monitoring committee to follow when making the important decisions:

Whether or not to proceed to Phase III,

Selecting the treatment to take forward to the Phase III stage, considering efficacy and safety data,

Using the estimated treatment effect, response rates and response variance to set the Phase III sample size.

Since the monitoring committee will not be able to consult the study sponsor for clarification of the guidelines or to discuss how to respond to unexpected outcomes, the guidance must be detailed and the monitoring committee should have the opportunity to talk through possible scenarios at their kick-off meeting before they have seen any trial data.

Given the level of advance planning required, running an ASD trial can be challenging. Since the rules for adapting the Phase III stage of the trial design based on results from the Phase II stage are fixed in advance, there is less flexibility to respond to unanticipated outcomes in the Phase II stage. Members of the monitoring committee may be reluctant to take on the responsibility of making decisions about the Phase III stage of the trial: they are not stakeholders and these decisions will have serious consequences for the sponsor. Nevertheless, ASD trials have been conducted; for some examples, see Geiger et al. (2012) who report a trial in type 2 diabetes mellitus and Palermos et al. (2022) who review seamless Phase II/III trials, including a number of ASD trials, in lung cancer.

In Section 26.2, we shall describe inferentially seamless Phase II/III designs following the methods of Bretz et al. (2006), using combination tests in multiple testing procedures, as described in Chapter 25. Then, in Section 26.3.1 we show how the earlier proposal by Thall, Simon & Ellenberg (1988) can be viewed as a procedure of that form. In Section 26.4 we discuss the design choices for an ASD trial and make recommendations for the type of combination test to use and the method for testing intersection hypotheses. In Section 26.5 we discuss the use of short-term endpoints in the treatment selection decision after Phase II. In Section 26.6 we describe methods for producing point estimates and confidence intervals following a seamless Phase II/III trial with treatment selection and, at the end of this section, we present a method to create an upper confidence interval for the treatment effect of the selected treatment that will be consistent with the hypothesis test comparing this treatment with the control. We discuss ASDs with a single experimental treatment in Section 26.7 before concluding the chapter with comments on some additional references in Section 26.8.

26.2 An Adaptive Seamless Design Using a Closed Testing Procedure with Combination Tests

26.2.1 Hypothesis Testing in a Seamless Phase II/III Design

Consider a Phase II/III trial in which p treatments are compared with a control in stage 1. For $i = 1, \ldots, p$, let θ_i denote the effect size of treatment i versus the control, where a positive value of θ_i indicates that treatment i is superior and define the hypotheses H_i: $\theta_i \leq 0$ for $i = 1, \ldots, p$. Suppose that at the conclusion of stage 1, the Phase II part of the trial, the observed treatment effect estimates are $\hat{\theta}_{1,i}$, $i = 1, \ldots, p$, and treatment i^* is selected to advance to stage 2, the Phase III part of the trial. Although $\hat{\theta}_{1,i^*}$ will typically be the largest among $\hat{\theta}_{1,1}, \ldots, \hat{\theta}_{1,p}$, the procedure is flexible in that other choices of i^* are allowed, informed by safety data or other considerations. As a point of nomenclature, note that in the remainder of this chapter, we shall sometimes refer to the Phase II part of a trial as "stage 1" and the Phase III part as "stage 2".

After stage 2, we wish to test hypothesis H_{i^*}: $\theta_{i^*} \leq 0$ at significance level α, but with allowance for the selection bias caused by choosing H_{i^*} from p null hypotheses. We require a testing procedure that controls the family-wise error rate at level α so that, for all possible combinations of treatment effects $\boldsymbol{\theta} = (\theta_1, \ldots, \theta_p)$,

$$P_{\boldsymbol{\theta}}\{\text{Reject a true } H_i \text{ in the two-stage procedure}\} \leq \alpha.$$

This can be viewed as a special case of the framework of Chapter 25 with p hypotheses and $K = 2$ stages. Let $P_1^{(1)}, \ldots, P_p^{(1)}$, denote the P-values for hypotheses $H_1, \ldots, H_p$ from stage 1 data and denote the P-value for H_{i^*} from stage 2 data by $P_{i^*}^{(2)}$. After stage 1, we are only interested in testing H_{i^*}: $\theta_{i^*} \leq 0$ and the other hypotheses are dropped from consideration. In the terminology of Chapter 25, we require tests of intersection hypotheses H_J for subsets $J \subseteq \{1, \ldots, p\}$ that contain the index i^*.

We specify a method for testing intersection hypotheses and use this method to form P-values $P_J^{(1)}$ from $P_i^{(1)}$, $i = 1, \ldots, p$. Then, because there is no information from other treatments in stage 2, we simply set

$$P_J^{(2)} = P_{i^*}^{(2)}$$

for all sets J containing i^*. Finally, we apply a combination tests to each pair $P_J^{(1)}$ and $P_J^{(2)}$ where $i^* \in J$, and the overall closed testing procedure rejects H_{i^*} if H_J is rejected at level α for every subset J of $\{1, \ldots, p\}$ that contains i^*.

The usual options are available for testing intersection hypotheses in stage 1. Suppose H_J is the intersection of m simple hypotheses and denote the ordered P-values for these m hypotheses by

$$P_{(1)}^{(1)} \leq P_{(2)}^{(1)} \leq \ldots \leq P_{(m)}^{(1)}.$$

A simple Bonferroni adjustment gives

$$P_J^{(1)} = m\, P_{(1)}^{(1)}.$$

Because the individual tests involve comparisons with a common control, there is a positive dependence between the P-values $P_1^{(1)}, \ldots, P_p^{(1)}$, so the more powerful Simes (1986) method that we presented in Section 20.4 may be used, giving

$$P_J^{(1)} = \min_{i=1,\ldots,m} (m\, P_{(i)}^{(1)}/i).$$

Another option is Dunnett's (1955) method, which sets

$$P_J^{(1)} = Pr\{\, P_{(1)}^{(1)} \leq p_{(1)}^{(1)} \}$$

where $p_{(1)}^{(1)}$ denotes the observed value of $P_{(1)}^{(1)}$ and the probability is calculated assuming $\theta_i = 0$ for all $i \in J$. Details of this calculation for the case of normal responses with n subjects on each treatment arm and n_0 on the control arm are given in Section 21.4.1.

Since each $P_J^{(1)}$, where $i^* \in J$, is combined with the same stage 2 P-value, $P_J^{(2)} = P_{i^*}^{(2)}$, the largest combined P-value will be that involving the maximum value of $P_J^{(1)}$ over sets J that contain i^*. Thus this maximum value of $P_J^{(1)}$ is the key data summary to carry forward to the final hypothesis test and we refer to this as the adjusted stage 1 P-value.

Table 26.1 *Observed P-values and Z-values in the pedagogic example. On the strength of its Phase II performance, treatment $i^* = 4$ was selected to go forward to Phase III.*

	Phase II results				
	Control	*Treatment 1*	*Treatment 2*	*Treatment 3*	*Treatment 4*
n	100	100	100	100	100
$P_j^{(1)}$		0.20	0.04	0.05	0.03
$Z_j^{(1)}$		0.84	1.75	1.64	1.88

	Phase III results	
	Control	Treatment 4
n	500	500
$P_j^{(2)}$		0.04
$Z_j^{(2)}$		1.75

26.2.2 An Illustrative Example

We illustrate the above methods with a small pedagogic example taken from Jennison & Turnbull (2007). Summary data are shown in Table 26.1 with $P_J^{(k)}$ as defined above and

$$Z_J^{(k)} = \Phi^{-1}(1 - P_J^{(k)}).$$

Here, Phase II comprises a control and four experimental treatment groups, each with 100 subjects. On the basis of these results, treatment $i^* = 4$ is selected to be to be compared against the control in Phase III with 500 new subjects on each arm. In this case, it is the treatment with the smallest P-value that is selected, but the method remains valid if another of the treatments had been selected, perhaps because it had a superior safety profile.

We shall examine whether treatment 4 should be recommended as superior to the control using the one-sided significance level $\alpha = 0.025$.

Testing intersection hypotheses in stage 1

Suppose first that we use Simes' method to test each intersection hypothesis. The sets J containing $i^* = 4$ are $\{4\}$, $\{1, 4\}$, $\{2, 4\}$, $\{3, 4\}$, $\{1, 2, 4\}$, $\{1, 3, 4\}$, $\{2, 3, 4\}$ and $\{1, 2, 3, 4\}$. In finding the maximum $P_J^{(1)}$, we consider the subsets of size 1, 2, 3 and 4 with the highest P-values, namely $\{4\}$, $\{1, 4\}$, $\{1, 3, 4\}$ and $\{1, 2, 3, 4\}$. Calculations show that the maximum $P_J^{(1)}$ comes from $J = \{1, 3, 4\}$ and is equal to

$$3\, P_3^{(1)}/2 = 0.075.$$

We carry this value forward to combine with the P-value from stage 2 data in the final hypothesis test.

If responses are normally distributed with known variances, we may use Dunnett's method to test each intersection hypothesis. In this case

$$P_J^{(1)} = Pr\{\max_{i \in J} Z_i^{(1)} \geq \max_{i \in J} z_i^{(1)}\}$$

where $z_i^{(1)}$ denotes the observed value of $Z_i^{(1)}$ for $i = 1, \ldots, p$ and the probability is calculated assuming $\theta_i = 0$ for all $i \in J$. Since $Z_4^{(1)} = 1.88$ is the largest Z-value, it follows that the maximum $P_J^{(1)}$ for a set J containing i^* is

$$Pr\{\max(Z_1, Z_2, Z_3, Z_4) \geq 1.88\}$$

where $(Z_1, \ldots, Z_4)$ is a multivariate normal random variable with each $Z_i \sim N(0,1)$ and $\text{Corr}(Z_i, Z_{i'}) = 0.5$ for $i \neq i'$. Calculation using, for example, the R package `DunnettTest` shows this probability to be 0.092 and we carry this value forward to combine with the P-value from stage 2 data.

Combining stage 1 and stage 2 P-values

(a) Using the inverse χ^2 *method*

If Simes' method is used, the adjusted stage 1 P-value is 0.075 and the inverse χ^2 combination statistic is

$$-2\log(0.075 \times 0.04) = 11.62.$$

Comparing this with a χ_4^2 distribution gives the significance level

$$Pr\{\chi_4^2 > 11.6\} = 0.0204.$$

Thus treatment 4 is found to be superior to the control in a test with family-wise error rate $\alpha = 0.025$.

If Dunnett's method is used, the adjusted stage 1 P-value is 0.092 and the inverse χ^2 combination statistic is

$$-2\log(0.092 \times 0.04) = 11.21$$

which has significance level

$$Pr\{\chi_4^2 > 11.2\} = 0.0243,$$

so again treatment 4 is found to be superior to the control in the closed testing procedure.

(b) Using the weighted inverse normal method

We use pre-specified weights proportional to the square roots of the sample sizes in the two stages, that is

$$w_1 = \sqrt{(100/600)} \quad \text{and} \quad w_2 = \sqrt{(500/600)}.$$

With the adjusted stage 1 P-value 0.075 from Simes' method, the combination statistic is

$$\sqrt{(1/6)}\,\Phi^{-1}(1 - 0.075) + \sqrt{(5/6)}\,\Phi^{-1}(1 - 0.04) = 2.19$$

and this has significance level $1 - \Phi(2.19) = 0.014$.

Using the adjusted stage 1 P-value 0.092 from Dunnett's method, we have

$$\sqrt{(1/6)}\,\Phi^{-1}(1 - 0.092) + \sqrt{(5/6)}\,\Phi^{-1}(1 - 0.04) = 2.14$$

and this has significance level $1 - \Phi(2.14) = 0.016$.

Here, all tests are significant at one-sided level $\alpha = 0.025$. If only the stage 2 data had been used, to test H_4, the results would not have been significant since $P_4^{(2)} > 0.025$. Thus, including the stage 1 results in the final analysis has helped demonstrate the effectiveness of the selected treatment.

In this example, different choices for the form of intersection test and combination test have led to the same conclusion but in general they may not. This raises the question as to which choices give the most efficient procedure. We may also ask if it is possible to quantify the value of the stage 1 data, given that they are down-weighted by the adjustment for multiple testing in stage 1. The answer to this question is important in deciding whether the gains from using Phase II data in the final hypothesis test compensate for the logistical complications of running a seamless design. Another interesting question is how to divide resources optimally between stages 1 and 2. We shall return to these questions in Section 26.4 after noting some other approaches to constructing seamless Phase II/III trial designs.

26.3 Other Seamless Phase II/III Trial Designs

26.3.1 Thall, Simon and Ellenberg's procedure

The trial design proposed by Thall, Simon & Ellenberg (1988) (hereafter TSE) preceded the approach we have just described based on combination tests and a closed testing procedure. Jennison & Turnbull (2007) apply TSE's approach to a trial in which normal responses are observed. They assume responses are normally distributed with variance σ^2, mean μ_0 for the control treatment and means μ_i, $i = 1, \ldots, p$, for p experimental treatments. In stage 1, n_1 observations are taken on the control arm and each of the experimental arms leading to estimated treatment effects $\hat{\theta}_{1,i} = \hat{\mu}_{1,i} - \hat{\mu}_{1,0}$, $i = 1, \ldots, p$. If the estimated treatment effect of the best performing treatment is high enough, the trial continues on to stage 2 and n_2 observations are taken on the selected treatment and the control.

Constants C_1 and C_2 are specified and, if $\hat{\theta}_{1,i^*}$ is the highest treatment effect estimate in stage 1, then

If $\hat{\theta}_{1,i^*} < C_1$, the trial stops and the global null hypothesis H_0: $\theta_1 \leq 0, \ldots, \theta_p \leq 0$ is accepted,

If $\hat{\theta}_{1,i^*} \geq C_1$, treatment i^* is selected and the trial proceeds to stage 2.

After observing n_2 responses on treatment i^* and the control in stage 2, we denote the estimated treatment effect from these data by $\hat{\theta}_{2,i^*}$ and form the overall statistic

$$T_{i^*} = \frac{n_1 \hat{\theta}_{1,i^*} + n_2 \hat{\theta}_{2,i^*}}{n_1 + n_2}, \quad (26.1)$$

the simple estimate of θ_{i^*} obtained by pooling stage 1 and stage 2 data. Then

If $T_{i^*} < C_2$, we accept H_0: $\theta_{i^*} \leq 0$,

If $T_{i^*} \geq C_2$, we reject H_0 and conclude that $\theta_{i^*} > 0$.

Type I error probability and power of the TSE procedure

In defining the Type I error probability, TSE say treatment i^* is "chosen" if it is selected at the end of stage 1 and H_0: $\theta_{i^*} \leq 0$ is rejected in the final analysis. TSE define the Type I error rate as the maximum value of

$$Pr_{\boldsymbol{\theta}}\{\text{Any experimental treatment is "chosen"}\}$$

over vectors $\boldsymbol{\theta}$ with $\theta_1 \leq 0, \ldots, \theta_p \leq 0$ and they note this occurs when $\theta_1 = \ldots = \theta_p = 0$. The Type I error rate requirement is that this maximum probability should be no greater than a specified value α.

In discussing power, TSE set effect sizes $0 < \delta_1 < \delta_2$ where δ_1 represents a marginal treatment effect and δ_2 a clinically significant effect. They say treatment i is "acceptable" if $\theta_i \geq \delta_2$. The power for a given treatment effect vector $\boldsymbol{\theta}$ is defined as

$$Pr_{\boldsymbol{\theta}}\{\text{An acceptable treatment is "chosen"}\}$$

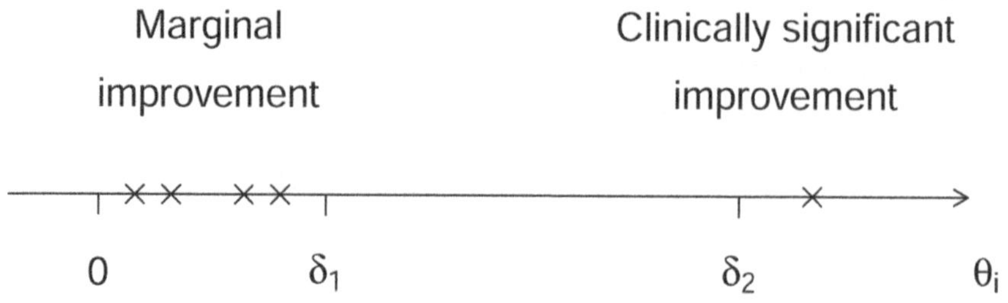

Figure 26.1 *A configuration of treatment effects with one "acceptable" treatment*

and the power requirement is that this probability should be at least $1 - \beta$ for all vectors $\boldsymbol{\theta}$ in which at least one treatment is acceptable and no θ_i lies in the interval (δ_1, δ_2). An example of a set of treatment effects with one "acceptable" treatment is shown in Figure 26.1 TSE show that the minimum power in such cases occurs when one θ_i is equal to δ_2 and the others are all equal to δ_1. Without loss of generality, we take this to be the case

$$\boldsymbol{\theta}^* = (\delta_1, \dots, \delta_1, \delta_2).$$

TSE require power to be at least $1 - \beta$ in this "least favorable configuration".

The sample sizes n_1 and n_2 and the critical values C_1 and C_2 determine the operating characteristics of the procedure. Numerical integration for the cases $\boldsymbol{\theta} = (0, \dots, 0)$ and $\boldsymbol{\theta} = \boldsymbol{\theta}^*$ enables parameters n_1, n_2, C_1 and C_2 to be found that satisfy the specified Type I error and power requirements. TSE go on to determine combinations of these parameters that minimize expected sample size averaged over the cases $\boldsymbol{\theta} = (0, \dots, 0)$ and $\boldsymbol{\theta} = \boldsymbol{\theta}^*$.

The TSE method is a closed testing procedure with family-wise error rate α

Although TSE did not discuss the family-wise error rate of their design, Jennison & Turnbull (2006d) show that the TSE procedure controls FWER at level α, that is

$$Pr_{\boldsymbol{\theta}}\{\text{A treatment } i^* \text{ with } \theta_{i^*} \le 0 \text{ is "chosen"}\} \le \alpha$$

for all treatment effect vectors $\boldsymbol{\theta}$, including cases where $\theta_i > 0$ for some indices i. The argument starts from the case $\boldsymbol{\theta} = (0, \dots, 0)$, where the Type I error rate is α, and proceeds by showing that reducing some effect sizes to values $\theta_i < 0$ decreases the family-wise error probability, then increasing other values to $\theta_i > 0$ decreases the error rate since there are fewer treatments with $\theta_i \le 0$.

Jennison & Turnbull (2006d) also show that the TSE procedure can be formulated as a closed testing procedure. In doing this, they define level α tests of intersection hypotheses H_J for sets J containing i^*, the index of the treatment taken forward to stage 2. If the selected treatment is always that with the highest $\hat{\theta}_{1,i}$, the result is precisely the TSE procedure and it follows that this procedure has FWER α.

If a treatment i^* is selected for stage 2 when $\hat{\theta}_{i^*} > C_1$ but $\hat{\theta}_{i^*}$ is not the maximum $\hat{\theta}_{1,i}$, the TSE procedure can be applied, rejecting H_0: $\theta_{i^*} \le 0$ if T_{i^*} defined by (26.1) is greater than C_2. It is not difficult to show that the FWER is still controlled at level α. However, the closed testing formulation defined by Jennison & Turnbull may be preferred as it can lead to a lower critical value for T_{i^*}, increasing the chance to reject H_0: $\theta_{i^*} \le 0$ while still protecting the FWER.

Applying TSE's method in the illustrative example

We illustrate TSE's methods with the data displayed in Table 26.1. Suppose that in this case the responses from the four treatments and control are normally distributed with common variance $\sigma^2 = 12.5$, so the variance of each $\hat{\theta}_{1,i}$ is $25/n_1$ and the variance of $\hat{\theta}_{2,4}$ is $25/n_2$. Let the critical value at stage 1 be $C_1 = 0$, so the trial would have been abandoned as futile if no experimental treatment mean had exceeded that of the control in stage 1. The stage 2 critical value C_2 is chosen to satisfy the requirement

$$Pr_{\boldsymbol{\theta}}\{\hat{\theta}_{1,i^*} > C_1 = 0 \text{ and } T_{i^*} > C_2\} = 0.025 \qquad (26.2)$$

when $\boldsymbol{\theta} = (0,0,0,0)$. In calculating this probability, the integral over the multivariate normal distribution of $(\hat{\theta}_{1,1}, \ldots, \hat{\theta}_{1,4})$ is most easily evaluated by first conditioning on $\hat{\mu}_0$, the mean response for the control treatment, after which the estimates $\hat{\theta}_{1,i} = \hat{\mu}_{1,i} - \hat{\mu}_{1,0}$, $i = 1, \ldots, 4$, can be treated as independent. Solving (26.2) yields the value $C_2 = 0.449$.

From Table 26.1, in the stage 1 data we have $i^* = 4$ and $Z_4^{(1)} = 1.88$ so

$$\hat{\theta}_{1,i^*} = \hat{\theta}_{1,4} = Z_4^{(1)} \sqrt{2\sigma^2/n_1} = 0.940.$$

In the stage 2 data we have $Z_4^{(2)} = 1.75$ and

$$\hat{\theta}_{2,4} = Z_4^{(2)} \sqrt{2\sigma^2/n_2} = 0.391.$$

Combining these gives the statistic

$$T_{i^*} = \frac{100\,\hat{\theta}_{1,i^*} + 500\,\hat{\theta}_{2,i^*}}{600} = 0.483$$

and, since $T_{i^*} > C_2 = 0.449$, the hypothesis H_0: $\theta_4 \leq 0$ is rejected and treatment 4 is declared superior to the control.

Note that the standard Z-statistic based on pooled data from treatment 4 is

$$\frac{T_{i^*}}{\sqrt{2\sigma^2/600}} = 2.365.$$

However, in allowing for selection bias in choosing the treatment to study in stage 2, it was necessary for this statistic to exceed

$$\frac{C_2}{\sqrt{2\sigma^2/600}} = 2.200$$

rather than $\Phi^{-1}(1-\alpha) = 1.960$ in order to reject H_0: $\theta_4 \leq 0$.

26.3.2 Extensions of the Thall, Simon and Ellenberg method

As noted above, the TSE method is flexible in that a treatment i^* with $\hat{\theta}_{1,i^*}$ less than the highest $\hat{\theta}_{1,i}$ may be selected for the Phase III stage. Other aspects of the design, as originally presented, are less flexible, in particular, the sample sizes in stage 2 are fixed and cannot be modified after seeing the first stage data. The closed testing procedure based on a weighted inverse normal combination tests and Dunnett tests for intersection hypotheses is very close to the TSE method but also allows adaptations to the stage 2 design after observing stage 1 data.

The papers of Thall, Simon & Ellenberg (1988, 1989) actually concern binary responses, although a normal approximation is applied to give procedures similar to those we have described. Schaid, Wieand & Therneau (1990) describe a two-stage design with a survival endpoint in which all treatments with sufficiently high estimated treatment effects proceed to the second stage and data are pooled across stages in the final hypothesis tests. With the option of more than one treatment being tested in the final analysis, a different approach is needed and these authors use a Bonferroni correction to control the family-wise Type I error rate.

The multi-stage procedure proposed by Stallard & Todd (2003) includes the TSE procedure as a special case when there are just two stages. In the first stage, p experimental treatments are compared with a control. If the highest treatment effect estimate exceeds a certain threshold, the relevant experimental treatment continues to be compared with the control in the second stage. This second stage has a group sequential stopping rule with boundaries similar to those of Chapter 5 and Section 8.3, but with an upwards adjustment to take account of the selection bias in stage 1. Stallard & Todd give details of the computations required to create a design with specified family-wise Type I error rate and power. As for the TSE method, the procedure is conservative if a lower performing treatment is selected to enter the second stage.

26.3.3 Further Methods

We noted in Section 21.4.2 that Stallard & Friede (2008) proposed an extension of the procedure of Stallard & Todd (2003) in which experimental treatments are dropped from the trial over several stages. This method controls the FWER if the number of treatments included at each stage is specified in advance, and simulation studies suggest FWER is controlled conservatively when the treatments are dropped based on the observed data in a practically relevant way.

Koenig et al. (2008) suggest a two-stage procedure using the "conditional rejection probability" approach that we described in Section 23.6 in conjunction with Dunnett's test of intersection hypotheses. The procedure is adaptive in that more than one of the experimental treatments may be carried forward into the second stage and the rule for selecting treatments and choosing second stage sample sizes need not be specified in advance.

Friede & Stallard (2008) compare properties of adaptive seamless designs, including the methods of Stallard & Friede (2008) and Koenig et al. (2008).

26.4 Efficiency of Adaptive Seamless Designs

Jennison & Turnbull (2007) report a simulation study investigating the power of different Phase II/III designs in which a single experimental treatment may proceed to the Phase III stage for comparison with the control. They consider studies with four experimental treatments and a control where the stage 1 and 2 sample sizes, n_1 and n_2, are fixed and all designs have FWER α. They suppose the treatment effect vector is of the form $\boldsymbol{\theta} = (0, 0, 0, \delta)$ and consider power as δ varies for the following designs:

(i) The TSE test,

(ii) A closed testing procedure (CTP) using a weighted inverse normal combination test and a Simes test for intersection hypotheses,

(iii) A CTP using an inverse χ^2 combination test and a Simes test for intersection hypotheses,

(iv) A "conventional" test that uses Phase III data only in the final hypothesis test.

Jennison & Turnbull's simulation results show that: test (i), the TSE test, has the highest power; test (ii), the CTP using a weighted inverse normal combination test and Simes' test, has almost exactly the same power as test (iv), the conventional test; finally, test (iii), the CTP using an inverse χ^2 combination test and Simes' test, has the lowest power. The differences in performance of the designs are significant. In some comparisons an increase in overall sample size of 5% would be needed to raise the inferior design's power to match that of a superior design.

Since the conventional test does not use Phase II data in the final hypothesis test, it can be performed without the complexities of a seamless design. It is surprising that a seamless design should be *less* efficient than this simple option. This illustrates the point that it is important to fine tune an innovative design and check that it has the desired properties before applying it in practice.

Hampson & Jennison (2015) continued the investigation of this example. They added two more methods for conducting the final hypothesis test: CTPs using either the weighted inverse normal test or the inverse χ^2 combination test with Dunnett's test for intersection hypotheses. In calculations with a wide variety of forms of treatment effect vector $\boldsymbol{\theta}$, they found the CTP with an inverse normal test and Dunnett's test to have very similar power to the TSE test. For a given type of combination test, using Dunnett's test for intersection hypotheses yielded higher power than Simes' test, the difference in power varying with the form of $\boldsymbol{\theta}$. The clear conclusion is that the highest power is achieved by applying either the TSE test or a CTP with a weighted inverse normal test and Dunnett's test for intersection hypotheses.

Given the final set of data from a seamless design, Hampson & Jennison derive an optimal test of the hypothesis $\boldsymbol{\theta} = (0,0,0,0)$ against alternatives of the form $\boldsymbol{\theta} = (\gamma,\gamma,\gamma,1)\delta$ for a fixed $\gamma < 1$. They show that when $\gamma < 0.5$, the optimal decision rule involves a linear combination of $\hat{\theta}_{1,1}$, $\hat{\theta}_{1,2}$, $\hat{\theta}_{1,3}$ and $\hat{\theta}_{1,4}$ with *negative* weights on the three lower values. In the case $\gamma = 0.5$, the weights on the three lower values are zero and it follows that the TSE test is optimal here. The negative weights when $\gamma < 0.5$ are a consequence of the correlations between $\hat{\theta}_{1,i}$ and $\hat{\theta}_{1,i^*}$ for $i \neq i^*$ caused by the common control arm: positive values of the $\hat{\theta}_{1,i}$, $i \neq i^*$, may be due to lower than average responses on the control arm and this possibility detracts from the evidence for the case $\theta_{i^*} > 0$ provided by a positive value of $\hat{\theta}_{1,i^*}$. Calculations are more complex for a general alternative, $\boldsymbol{\theta} = (\gamma_1, \gamma_2, \gamma_3, 1)\delta$, but negative weights can still arise for small values of γ_1, γ_2 and γ_3. This helps explain the poor performance of procedures that use Simes' test since this method treats high values of $\hat{\theta}_{1,i}$ for $i \neq i^*$ as evidence in favor of treatment i^* when an optimized rule does exactly the opposite.

Hampson & Jennison (2015) assess the value of Phase II data in the final hypothesis test by finding the number of additional Phase III observations that would be needed for a test based on Phase III data alone to achieve the same power. They conclude that when a weighted inverse normal test with Dunnett's test for intersection hypotheses is used,

> "... stage 1 data on treatment i^* should typically be viewed as offering around 50% to 70% of their face value as stage 2 observations in many situations."

Hampson & Jennison also discuss how a fixed number of subjects should be divided between the two stages of a trial. The results presented in Section 7 of their paper show that an optimal choice depends on the unknown vector of treatment effects $\boldsymbol{\theta}$. They conclude that

> "... no single choice is close to ideal in all scenarios. Rather, investigators should consider the most likely scenarios ... for their trial and choose group sizes that will give the best average power across these cases."

26.5 Seamless Phase II/III Trials with a Short-term Endpoint in Phase II

Many Phase II trials investigate a short-term endpoint rather than the longer-term primary endpoint that will be studied in Phase III. For example, the short-term endpoint might be the improvement on a performance scale over eight weeks while the primary endpoint is the two-year improvement. In such cases, it is still possible to conduct an inferentially seamless Phase II/III trial if treatment continues for the Phase II patients receiving the control and the treatment selected for Phase III and the long-term endpoint is measured. Stallard (2011) reviews seamless Phase II/III trial designs, including methods that use short-term endpoints in Phase II.

Todd & Stallard (2005) extend the method of Stallard & Todd (2003) to create a seamless Phase II/III design in which treatment selection after Phase II is based on measurements of a short-term endpoint. In the subsequent Phase III part of the trial, patients are randomized to the selected treatment and the control. The long-term endpoint is observed for all patients, including those on treatments dropped after Phase II. A group sequential stopping rule is applied during Phase III and a joint model for the short-term and long-term endpoints is used to ensure FWER is protected at level α, allowing for the effect of selecting the best performing treatment in Phase II.

Stallard (2010) proposes a design similar to that of Todd & Stallard (2005) but now treatment selection after Phase II is based on measurements of the short-term endpoint from all Phase II subjects and measurements of the long-term endpoint from some of these. Again, a joint model for the short-term and long-term endpoints is used to ensure FWER is protected at level α.

Friede et al. (2011) consider a trial design in which it is not feasible to continue treating Phase II patients other than those on the selected treatment and the control after a treatment has been selected for Phase III. They propose a closed testing procedure that applies a combination test to data on the long-term endpoint from the cohorts admitted in Phase II and in Phase III. Since the long-term endpoint is only available for those Phase II subjects on the selected treatment and the control, a Dunnett test is used for intersection hypotheses tests based on Phase II data. In this case, a joint model for Phase II and Phase III data is not needed to calculate a group sequential boundary for the Phase III stage that protects the FWER, but simulations from an assumed model are used in making design choices for the study. A design of the type proposed by Friede et al. (2011) was used in the trial of treatments for multiple sclerosis described by Chataway et al. (2011).

Kunz et al. (2015) compare the designs proposed by Stallard (2010) and Friede et al. (2011). They conclude that the best choice of method depends on the treatment effects on short-term and long-term endpoints. If the treatment effect for the short-term endpoint is less than that for the long-term endpoint, it is particularly helpful to have some data on the long-term endpoint when selecting a treatment. There will, of course, be uncertainty about the likely size of treatment effects on both endpoints. Thus, the quote from Hampson & Jennison (2015) at the end of Section 26.4 is just as relevant here: investigators should consider the most likely scenarios for their trial and choose a design that will give the best average performance across these cases.

26.6 Inference Following a Seamless Phase II/III Trial

In Chapter 9 we discussed how to define P-values, confidence intervals and point estimates on the conclusion of a group sequential test and in Section 23.8 we addressed similar questions for an adaptive trial in which a single hypothesis is tested. Making such inferences after an adaptive trial with multiple parameters and hypotheses is more challenging. Here, we survey methods proposed for adaptive designs with treatment selection and we present a method for calculating a confidence interval for the treatment effect of the selected treatment at the end of a two-stage design. The review papers by Robertson, Choodari-Oskooei et al. (2023a,b) and Robertson, Burnett et al. (2025a,b) include descriptions of some additional methods for inference following a seamless Phase II/III trial. Some of the ideas discussed here will also be relevant to inference after a multi-arm multi-stage trial, which we discuss in Chapter 27, and to the adaptive enrichment designs of Chapter 28.

26.6.1 Point Estimation

In a Phase II/III trial with treatment selection, it is the estimate of the treatment effect of the selected treatment, θ_{i^*} in our notation, that will be reported. Here, θ_{i^*} is not simply a parameter since the index i^* is a function of the observed data. The bias of an estimator $\hat{\theta}_{i^*}$ of the selected treatment effect θ_{i^*} is defined as

$$E_{\boldsymbol{\theta}}(\hat{\theta}_{i^*} - \theta_{i^*}).$$

Shen (2001) considered the problem of defining such an estimator when each θ_i is the difference in success rates between treatment i and the control. Denoting the success rate on treatment i by p_i, for $i = 1, \ldots, p$, and the success rate on the control by p_0, we have $\theta_i = p_i - p_0$, $i = 1, \ldots, p$. When the total number of observations on the control is fixed, estimating p_0 is straightforward and the problem of estimating θ_{i^*} reduces to finding a good estimate of p_{i^*}. Shen demonstrated that the bias of the maximum likelihood estimator of p_{i^*} can be significant. In order to guard against over-estimation of the effect of the selected treatment, Shen suggests calculating an estimate of the maximum possible bias for plausible parameter values and subtracting this from the maximum likelihood estimate. This approach to bias adjustment is similar to that introduced by Whitehead (1986a) to estimate a treatment effect after a group sequential test, which we also employed in an adaptive setting in Section 23.8.3.2. Stallard & Todd (2005) adapt this method to produce an approximately unbiased estimator of the selected treatment effect in a multi-stage trial in which one treatment is selected at the first analysis and the trial continues to compare that treatment with the control. Posch et al. (2005) present results that quantify the bias of the maximum likelihood estimator of the effect of the selected treatment but they do not discuss alternative estimators.

Cohen & Sackrowitz (1989) considered the problem of estimating the mean of a treatment selected in the first stage of a two-stage design. Suppose that responses on treatments 1 to p are normally distributed with means $\mu_1, \ldots, \mu_p$. We denote the ordered first stage estimates by $\hat{\mu}_{(1)} > \ldots > \hat{\mu}(p)$ and let the vector $Q = (q_1, \ldots, q_p)$ contain the indices of the treatments that produced these estimates, so $\hat{\mu}_{(1)} = \hat{\mu}_{q_1}, \ldots, \hat{\mu}_{(p)} = \hat{\mu}_{q_p}$. Then an estimator of μ_{q_1} is said to be conditionally unbiased given Q if

$$E\{\hat{\mu}_{q_1} - \mu_{q_1} \mid Q = (q_1, \ldots, q_p)\} = 0 \quad \text{for every } (q_1, \ldots, q_p).$$

Noting that the index q_1 is itself a random variable, the estimate $\hat{\mu}_{q_1}$ based on the second stage data for treatment q_1 is an unbiased estimate of μ_{q_1}. Cohen & Sackrowitz applied Rao-Blackwellization to this estimate to derive the Uniformly Minimum Variance Conditionally Unbiased Estimate (UMVCUE) of μ_{q_1} given Q. Bowden & Glimm (2008) extended the result of Cohen & Sackrowitz to the case of unequal sample sizes per arm in stage 1. Bretz, Koenig et al. (2009) used similar methods to obtain an UMVCUE when the sample sizes per treatment in stages 1 and 2 are unequal. Robertson et al. (2016a) obtain an UMVCUE when the sample sizes per treatment within a stages are unequal.

Luo, Wu & Xiong (2010) discuss an adaptive Phase II/III trial with a binary response and use the method of conditional moments to derive an estimator for the success rate of the selected treatment. Their simulation results show the new estimator to have only a small bias and a lower mean square error (MSE) than a Rao-Blackwellized estimator.

Kimani, Todd & Stallard (2013) argue that, when a trial design includes the possibility of stopping for futility at the interim analysis, it is of interest to consider the properties of an estimate conditional on the fact that the trial continues to stage 2. They use a similar strategy to that of Cohen & Sackrowitz to derive a UMVCUE for the treatment effect of the selected treatment conditional on proceeding to stage 2 with that treatment. They also apply Whitehead's (1986a) approach to construct a separate bias adjusted estimator, subtracting an estimate of the conditional bias given that the trial proceeds to stage 2 from the maximum likelihood estimate of the selected treatment effect.

Carreras & Brannath (2013) consider shrinkage estimation of the mean of the selected treatment, extending the empirical Bayes approach of Lindley (1962) and Hwang (1993). They report simulations comparing several estimators which show that, while the UMVCUE has zero bias, it has a high MSE compared to other estimators. In the examples presented, they show that the bias-adjusted estimator of Stallard & Todd (2005) over-corrects the bias and their proposed shrinkage estimator has a smaller MSE. Bowden, Brannath & Glimm (2014) develop the idea of a shrinkage estimator, further and describe new methods that improve on the performance of the estimator proposed by Carreras & Brannath (2013). Pickard & Chang (2014) propose a bootstrap method for bias adjustment that can be applied to a variety of response distributions and they show the resulting estimator has properties comparable with Carreras & Brannath's shrinkage estimator.

Brückner, Titman & Jaki (2017) and Khan, Kimani, Glimm & Stallard (2023) consider Phase II/III trials with a time to event endpoint, assuming a proportional hazards model. Defining the treatment effect of an experimental treatment as the log hazard ratio between that treatment and the control, Brückner, Titman & Jaki note that all the methods they consider tend to overcorrect the bias, and only the shrinkage methods reduce the mean squared error. Khan et al. derive an approximate UMVCUE for the treatment effect of the selected treatment. Although other estimators are less variable, Khan et al. argue that unbiased estimators are desired in confirmatory Phase II/III clinical trials and they recommend the UMVCUE or a certain bias-adjusted estimator which they show has similar properties.

26.6.2 Confidence Intervals

Stallard & Todd (2005) present a method for calculating simultaneous confidence intervals for the set of treatment effects $\theta_1, \ldots, \theta_p$ after a group sequential Phase II/III trial in which a single experimental treatment is selected at analysis 1. They specify an ordering of the sample space based on the analysis at which termination occurs and the final value of the test statistic for the selected treatment. For each possible vector of treatment effects $\boldsymbol{\theta}_0$, they define a P-value for testing $\boldsymbol{\theta} = \boldsymbol{\theta}_0$ and show this has a Uniform(0,1) distribution. Hence, the set of vectors $\boldsymbol{\theta}_0$ with P-values in the range $(\alpha/2, 1-\alpha/2)$ is an exact $(1-\alpha)$ confidence region for $\boldsymbol{\theta}$. Stallard & Todd note that, given the manner in which this region is defined, it must always be unbounded.

Stallard & Todd (2005) also suggest methods for calculating a confidence interval for the effect of the selected treatment, θ_{i^*} in our notation. In one approach, the treatment effects of non-selected treatments, θ_i, $i \neq i^*$, are assumed to equal the values estimated at the first analysis and simulation results show the resulting confidence intervals for θ_{i^*} to be very conservative. In a second approach in which the θ_i, $i \neq i^*$, are assumed to be zero, the coverage rate of the resulting confidence intervals is below the desired level in some cases. Furthermore, the error probabilities in the two tails of the confidence interval are unequal and the probability that the lower endpoint is above the true treatment effect can be considerably higher than $\alpha/2$.

Posch et al. (2005) present a method for calculating a confidence region for the vector of treatment effects $\boldsymbol{\theta}$ after a Phase II/III trial with treatment selection. They consider a two-stage design with a closed testing procedure in which combination tests are applied to data from the two stages. In our notation, hypotheses $H_i(\tilde{\theta}_i)$: $\theta_i \leq \tilde{\theta}_i$ are defined for general values of $\tilde{\theta}_i$, and tests are then specified for intersection hypotheses of the form

$$H(\tilde{\boldsymbol{\theta}}) = \cap_{i=1}^{p} H_i(\tilde{\theta}_i),$$

which state that $\theta_i \leq \tilde{\theta}_i$ for $i = 1, \ldots, p$. An upper confidence region for $\boldsymbol{\theta}$ is the set of all vectors $\tilde{\boldsymbol{\theta}}$ for which $H(\tilde{\boldsymbol{\theta}})$ is accepted. This confidence region will, by construction, agree with the original multiple testing procedure in that it will exclude parameter values rejected in the original procedure. Posch et al. note that a rectangular confidence region may be formed by taking the smallest rectangle that contains the calculated confidence region, but they observe that the result may then be inconsistent with some of the original hypothesis testing decisions.

Bebu, Dragalin & Luta (2013) have proposed methods to construct a confidence interval based on a conditional likelihood calculation. Magirr, Jaki, Posch & Klinglmueller (2013) describe a general method for finding a confidence region for the parameter vector after applying a closed testing procedure in a two-stage adaptive trial. Unlike the simultaneous confidence intervals proposed by Posch et al. (2005), the confidence intervals derived by Magirr et al. (2013) are guaranteed to be compatible with the outcomes of the closed testing procedure: we shall discuss this method further in the context of enrichment designs in Section 28.6.2. Kimani, Todd & Stallard (2014) compare properties of confidence intervals after a Phase II/III clinical trial with treatment selection constructed using the methods we have described, as well as proposals by Sampson & Sill (2005), Wu et al. (2010) and Neal et al. (2011). Robertson et al. (2025a) review confidence interval methodology for adaptive trial designs in general, including trials with treatment selection.

26.6.3 *An Informative Confidence Interval for the Treatment Effect of the Selected Treatment*

We conclude this section with a detailed description of a method for constructing a confidence interval for the effect of the selected treatment after a two-stage adaptive design with treatment selection. Suppose p experimental treatments are compared with the control in the Phase II stage and the trial continues to compare a single selected treatment, i^*, against the control in Phase III. We wish to obtain a level $1 - \alpha$ upper confidence interval for θ_{i^*}, which we shall denote (θ_{L,i^*}, ∞). The defining property of this interval is

$$Pr_{\boldsymbol{\theta}}\{\theta_{i^*} \in (\theta_{L,i^*}, \infty)\} \geq 1 - \alpha \quad \text{for all } \boldsymbol{\theta}, \tag{26.3}$$

where $\boldsymbol{\theta}$ is fixed and both i^* and θ_{L,i^*} are random variables. In order to exploit the equivalence between a confidence interval and a family of hypothesis tests, we create a family of tests for the hypotheses

$$H_i(\tilde{\theta}_i)\colon \theta_i \leq \tilde{\theta}_i, \quad i = 1, \ldots, p, \;\; \tilde{\theta}_i \in R, \tag{26.4}$$

with a family-wise error rate of α. Then, if i^* is the treatment selected in Phase II, we define (θ_{L,i^*}, ∞) to contain all values of $\tilde{\theta}_{i^*}$ for which $H_{i^*}(\tilde{\theta}_{i^*})$ is accepted in the closed testing procedure. If $\theta_{i^*} \notin (\theta_{L,i^*}, \infty)$, then $H_{i^*}(\theta_{i^*})$ must have been rejected in the closed testing procedure and, since this represents a family-wise error, the total probability of such an outcome is at most α. Hence, requirement (26.3) is satisfied.

In defining level α tests of the elementary hypotheses $H_i(\tilde{\theta}_i)$ and the intersection hypotheses that can be formed from them, we use a combination test to combine P-values from the Phase II and Phase III data. We shall describe the method using Dunnett tests for intersection hypotheses and inverse normal combination tests. For a given i, the hypotheses $H_i(\tilde{\theta}_i)$ are nested with

$$\cap_{j=1}^{k} H_i(\tilde{\theta}_j) = H_i(\min_{j=1,\ldots,k} \tilde{\theta}_j).$$

Hence, we need only consider general intersection hypotheses of the form

$$H_J(\tilde{\boldsymbol{\theta}}_J) = \cap_{j \in J} H_j(\tilde{\theta}_j)$$

where J is a subset of $\{1, \ldots, p\}$ and $\tilde{\boldsymbol{\theta}}_J$ contains a set of values $\tilde{\theta}_j$ for $j \in J$. For simplicity, we suppose the Phase II sample sizes are n_1 on each treatment and the control and the treatment effect estimates based on the Phase II data are distributed as

$$\hat{\theta}_{1,j} \sim N(\theta_j, 1/\mathcal{I}^{(1)}),$$

where $\mathcal{I}^{(1)} = n_1/(2\sigma^2)$. Then, if treatment i^* is selected at the end of Phase II and the Phase III sample sizes are n_2 on treatment i^* and the control, the Phase III data provide the estimate

$$\hat{\theta}_{2,i^*} \sim N(\theta_{i^*}, 1/\mathcal{I}^{(2)}),$$

where $\mathcal{I}^{(2)} = n_2/(2\sigma^2)$. Let

$$Z_j^{(1)}(\tilde{\theta}_j) = (\hat{\theta}_{1,j} - \tilde{\theta}_j)\sqrt{\mathcal{I}^{(1)}} \quad \text{for each } j = 1, \ldots, p \text{ and } \tilde{\theta}_j \in R.$$

To calculate the Phase II P-value for $H_J(\tilde{\boldsymbol{\theta}}_J)$: $\cap_{j \in J} H_j(\tilde{\theta}_j)$, note that if $\theta_j = \tilde{\theta}_j$ for $j \in J$, the $Z_j^{(1)}(\tilde{\theta}_j)$ are multivariate normal with

$$Z_j^{(1)}(\tilde{\theta}_j) \sim N(0, 1) \text{ for each } j \in J$$

and

$$\text{Corr}(Z_j^{(1)}(\tilde{\theta}_{j_1}), Z_j^{(1)}(\tilde{\theta}_{j_2})) = 0.5 \text{ for } j_1 \neq j_2.$$

Thus, if J contains k indices and values $Z_j^{(1)}(\tilde{\theta}_j) = z_j^{(1)}(\tilde{\theta}_j)$ are observed for $j \in J$, we can express the Dunnett P-value for $H_J(\tilde{\boldsymbol{\theta}}_J)$ in the notation of Equation (21.8) as

$$\Pr\{\max(Z_j^{(1)}(\tilde{\theta}_j), j \in J) \geq \max(z_j^{(1)}(\tilde{\theta}_j), j \in J)\} = P_D(k, 0.5, \max(z_j^{(1)}(\tilde{\theta}_j), j \in J)),$$

and this can be calculated in the usual way.

Suppose treatment i^* is selected in Phase II and, at the end of the trial, we wish to determine whether $H_i(\tilde{\theta}_i)$: $\theta_{i^*} \leq \tilde{\theta}_{i^*}$ is rejected in the closed testing procedure. This will be the case if each intersection hypothesis $H_J(\tilde{\boldsymbol{\theta}}_J)$, where $i^* \in J$ and $\tilde{\boldsymbol{\theta}}_J$ includes the value $\tilde{\theta}_{i^*}$ for θ_{i^*}, is rejected by its level α test. If the observed value of $Z_{i^*}^{(2)}(\tilde{\theta}_{i^*})$ is

$$z_{i^*}^{(2)}(\tilde{\theta}_{i^*}) = (\hat{\theta}_{2,i^*} - \tilde{\theta}_{i^*})\sqrt{\mathcal{I}^{(2)}},$$

the Phase III P-value is the same for all these intersection hypotheses $H_J(\tilde{\boldsymbol{\theta}}_J)$ and equal to

$$P_{i^*}^{(2)}(\tilde{\theta}_{i^*}) = 1 - \Phi\{z_{i^*}^{(2)}(\tilde{\theta}_{i^*})\}. \tag{26.5}$$

Thus, whether $H_i(\tilde{\theta}_i)$: $\theta_{i^*} \leq \tilde{\theta}_{i^*}$ is rejected in the closed testing procedure depends on the Phase II data through the largest P-value for an intersection hypothesis $H_J(\tilde{\boldsymbol{\theta}}_J)$ where $i^* \in J$ and $\tilde{\boldsymbol{\theta}}_J$ specifies that $\theta_{i^*} = \tilde{\theta}_{i^*}$. This P-value occurs when $J = \{1, \ldots, p\}$ and, for each $j \neq i^*$, $\tilde{\theta}_j$ is large so that $z_j^{(1)}(\tilde{\theta}_j)$ is small and

$$\max(z_j^{(1)}(\tilde{\theta}_j), j \in J) = z_{i^*}^{(1)}(\tilde{\theta}_{i^*}),$$

in which case the P-value for $H_J(\tilde{\boldsymbol{\theta}}_J)$ is

$$P_D(p, 0.5, z_{i^*}^{(1)}(\tilde{\theta}_{i^*}). \tag{26.6}$$

Thus, $H_i(\tilde{\theta}_i)$: $\theta_{i^*} \leq \tilde{\theta}_{i^*}$ is rejected in the closed testing procedure when the combination test applied to the P-values (26.5) and (26.6) is significant at level α. Suppose a weighted inverse normal combination test is applied with pre-specified weights w_1 and w_2. Then, $H_{i^*}(\tilde{\theta}_{i^*})$: $\theta_{i^*} \leq \tilde{\theta}_{i^*}$ is rejected if

$$w_1\, \Phi^{-1}\{1 - P_D(p,\, 0.5,\, z_{i^*}^{(1)}(\tilde{\theta}_{i^*}))\} + w_2\, z_{i^*}^{(2)}(\tilde{\theta}_{i^*}) \;\geq\; \Phi^{-1}(1-\alpha) \qquad (26.7)$$

and (θ_{L,i^*}, ∞) consists of all values $\tilde{\theta}_{i^*}$ for which this inequality holds.

In the case where the treatment selected after Phase II has the largest value of $\hat{\theta}_{1,i}$, the interval (θ_{L,i^*}, ∞) is certain to be consistent with the outcome of the closed testing procedure with FWER α that only considers the hypotheses H_i: $\theta_i \leq 0$, $i = 1, \ldots, p$, and does not attempt to provide a confidence interval for θ_{i^*}. In such a procedure, H_{i^*}: $\theta_{i^*} \leq 0$ is rejected if (26.7) holds for $\tilde{\theta}_{i^*} = 0$, which is precisely the condition for the confidence interval (θ_{L,i^*}, ∞) to lie above zero.

The above method is closely related to that proposed by Posch et al. (2005) to obtain a joint confidence region for $\boldsymbol{\theta}$ and, hence, a confidence interval for each θ_i. That methods inverts a family of tests of hypotheses of the form $\cap_{i=1,\ldots,p}\, \theta_i \leq \tilde{\theta}_i$, and the same hypotheses arise as intersection hypotheses in our closed testing procedure. However, restricting attention to a confidence interval for θ_{i^*} and not finding a confidence region for $\boldsymbol{\theta}$ as an intermediate step simplifies computation. By choosing a Dunnett test for intersection hypotheses, we ensure that the test of $\cap_{i=1,\ldots,p}\, \theta_i \leq 0$ and the confidence interval for θ_{i^*} depend on Phase II data only through $\hat{\theta}_{1,i^*}$ and this is the source of the consistency between tests and confidence intervals. The same property holds if the Dunnett test is replaced by a Bonferroni test or a Šidák test (Šidák, 1967), as proposed by Posch et al. (2005), but not if a Simes test is used. In view of the efficiency of designs using a Dunnett test and an inverse normal combination test reported by Hampson & Jennison (2015), we recommend this approach and note the additional benefit that a method is available to produce a consistent confidence interval for θ_{i^*}.

Example

We return to the example described in Table 26.1, assuming responses on treatment i are distributed as $N(\mu_i, \sigma^2)$, responses in the control group are $N(\mu_0, \sigma^2)$, $\sigma^2 = 12.5$, and $\theta_i = \mu_i - \mu_0$, $i = 1, \ldots, 4$. It follows that $\mathcal{I}^{(1)} = 4.0$ and $\mathcal{I}^{(2)} = 20.0$, the estimated treatment effects based on Phase II data are

$$\hat{\theta}_{1,1} = 0.421, \;\; \hat{\theta}_{1,2} = 0.875, \;\; \hat{\theta}_{1,3} = 0.822, \;\; \hat{\theta}_{1,4} = 0.940$$

each with standard deviation 0.5, and the Phase III data provide the estimate

$$\hat{\theta}_{2,4} = 0.391$$

with standard deviation 0.224. We seek an upper $(1-\alpha)$ level confidence interval for the effect of the selected treatment, $\theta_{i^*} = \theta_4$, with $\alpha = 0.025$. With $i^* = 4$, substituting

$$z_4^{(1)}(\tilde{\theta}_4) \;=\; (0.940 - \tilde{\theta}_4)/0.5 \quad \text{and} \quad z_4^{(2)}(\tilde{\theta}_4) \;=\; (0.391 - \tilde{\theta}_4)/0.224$$

in (26.7), we find the largest value of $\tilde{\theta}_4$ for which the inequality holds is 0.036. Thus the upper confidence interval for θ_4 is

$$(\theta_{L,4}, \infty) \;=\; (0.036, \infty).$$

The maximum likelihood estimate of θ_4 based on the pooled data from Phase II and Phase III is

$$\frac{1}{6}\,\hat{\theta}_1^{(1)} + \frac{5}{6}\,\hat{\theta}_1^{(2)} \;=\; 0.483$$

with variance

$$\left(\frac{1}{6}\right)^2 0.5^2 + \left(\frac{5}{6}\right)^2 0.224^2 = 0.0417,$$

so the naive upper $(1-\alpha)$ level confidence interval for θ_4, ignoring the treatment selection process, has lower end point

$$0.482 - \Phi^{-1}(1-\alpha)\sqrt{0.0417} = 0.083.$$

Taking proper account of the treatment selection based on Phase II data reduces this value to 0.036.

Computing the lower endpoint of an upper $1-\alpha$ level confidence interval with $\alpha = 0.5$ yields a median unbiased estimate of θ_{i^*}, θ_{M,i^*} say, with the property

$$Pr_{\boldsymbol{\theta}}\{\theta_{M,i^*} \leq \theta_{i^*}\} \geq 0.5 \quad \text{for all } \boldsymbol{\theta},$$

where $\boldsymbol{\theta}$ is fixed and i^* and θ_{M,i^*} are random variables. In the example, we find

$$\theta_{M,i^*} = 0.427,$$

a little lower than the maximum likelihood estimate $\hat{\theta}_{i^*} = 0.483$.

Extensions of the method

The method we have described has a simple extension to the case where more than one experimental treatment is taken forward to the Phase III stage. Suppose, for example, treatments i_1 and i_2 continue to Phase III. Then, we need tests based on Phase III data of

$$H_{i_1}(\tilde{\theta}_{i_1})\colon \theta_{i_1} \leq \tilde{\theta}_{i_1} \quad \text{and} \quad H_{i_2}(\tilde{\theta}_{i_2})\colon \theta_{i_2} \leq \tilde{\theta}_{i_2}, \quad \text{for } \tilde{\theta}_{i_1} \text{ and } \tilde{\theta}_{i_2} \in R,$$

and the associated intersection hypotheses. Let

$$z_{i_1}^{(2)}(\tilde{\theta}_{i_1}) = (\hat{\theta}_{2,i_1} - \tilde{\theta}_{i_1})\sqrt{\mathcal{I}^{(2)}} \quad \text{and} \quad z_{i_2}^{(2)}(\tilde{\theta}_{i_2}) = (\hat{\theta}_{2,i_2} - \tilde{\theta}_{i_2})\sqrt{\mathcal{I}^{(2)}}.$$

Applying a Dunnett test for each intersection hypothesis, we see that the largest Phase III P-value for an intersection hypothesis involving $H_{i_1}(\tilde{\theta}_{i_1})$ occurs in tests of $H_{i_1}(\tilde{\theta}_{i_1}) \cap H_{i_2}(\tilde{\theta}_{i_2})$ with high values of $\tilde{\theta}_{i_2}$ and is equal to

$$P_D(2,\, 0.5,\, z_{i_1}^{(2)}(\tilde{\theta}_{i_1})).$$

Thus, $H_{i_1}(\tilde{\theta}_{i_1})\colon \theta_{i_1} \leq \tilde{\theta}_{i_1}$ is rejected in the closed testing procedure if

$$w_1\, \Phi^{-1}\{1 - P_D(p,\, 0.5,\, z_{i_1}^{(1)}(\tilde{\theta}_{i_1}))\} + w_2\, \Phi^{-1}\{1 - P_D(2,\, 0.5,\, z_{i_1}^{(2)}(\tilde{\theta}_{i_1}))\} \geq \Phi^{-1}(1-\alpha) \quad (26.8)$$

and (θ_{L,i_1}, ∞), the $(1-\alpha)$ level upper confidence interval for θ_{i_1}, consists of all values $\tilde{\theta}_{i_1}$ for which this inequality holds. Similarly, (θ_{L,i_2}, ∞) consists of all values $\tilde{\theta}_{i_2}$ for which

$$w_1\, \Phi^{-1}\{1 - P_D(p,\, 0.5,\, z_{i_2}^{(1)}(\tilde{\theta}_{i_2}))\} + w_2\, \Phi^{-1}\{1 - P_D(2,\, 0.5,\, z_{i_2}^{(2)}(\tilde{\theta}_{i_2}))\} \geq \Phi^{-1}(1-\alpha). \quad (26.9)$$

The method can also be applied without modification when the selected treatment, i^*, does not have the highest value of $\hat{\theta}_j^{(1)}$ in Phase II. The largest Phase II P-value for an intersection hypothesis $H_J(\tilde{\boldsymbol{\theta}}_J)$ where $i^* \in J$ and $\tilde{\boldsymbol{\theta}}_J$ specifies $\theta_{i^*} = \tilde{\theta}_{i^*}$ is still given by (26.6) and the closed testing procedure rejects $H_{i^*}(\tilde{\theta}_{i^*})\colon \theta_{i^*} \leq \tilde{\theta}_{i^*}$ if the inequality (26.7) holds.

In all the cases we have described the confidence interval for θ_{i^*} will, by construction, be consistent with the test of the null hypothesis $H_{i^*}(0)\colon \theta_{i^*} \leq 0$, that is, the lower limit

θ_{L,i^*} will be above zero precisely when $H_{i^*}(0)$ is rejected. If more than one experimental treatment is taken forward, the set of confidence intervals will be consistent with the set of tests of null hypotheses $H_i(0)$: $\theta_i \leq 0$.

It is of interest to consider whether the outcome of a test of $\theta_{i^*} \leq 0$ may differ from an analysis which did not attempt to provide a confidence interval for θ_{i^*}. In that case, the closed testing procedure would only operate on the hypotheses $H_i(0)$: $\theta_i \leq 0$, $i = 1, \ldots, p$. We commented previously on the case when a single treatment i^*, with the highest value of $\hat{\theta}_i^{(1)}$, is selected. In that case, the largest Phase II P-value for an intersection hypothesis that includes $H_{i^*}(0)$ is

$$P_D(p\,, 0.5,\, \hat{\theta}_{i^*}^{(1)} \sqrt{\mathcal{I}^{(1)}})$$

and the closed testing procedure rejects $H_{i^*}(0)$ if

$$w_1\, \Phi^{-1}\{1 - P_D(p,\, 0.5,\, \hat{\theta}_{1,i^*} \sqrt{\mathcal{I}^{(1)}}\} \,+\, w_2\, \hat{\theta}_{2,i^*} \sqrt{\mathcal{I}^{(2)}} \;\geq\; \Phi^{-1}(1 - \alpha),$$

and this is also the condition that determines whether θ_{L,i^*} will be above zero. Thus, as we noted earlier, including the elements needed to construct a confidence interval for θ_{i^*} has not affected the outcome of the test of $\theta_{i^*} \leq 0$.

Now suppose treatment j^* has the highest estimated treatment effect in Phase II but this treatment is rejected for reasons of safety and treatment i^*, with the second highest value of $\hat{\theta}_i^{(1)}$, is selected for Phase III. When only the hypotheses $H_i(0)$, $i = 1, \ldots, p$, are considered in the closed testing procedure the Phase II P-value for the test of $H_J = \cap_{j \in J} H_j(0)$, where $J = \{1 \ldots, p\}$, is

$$P_D(p,\, 0.5,\, \hat{\theta}_{1,j^*} \sqrt{\mathcal{I}^{(1)}}),$$

and the largest Phase II P-value for an intersection hypothesis that includes $H_{i^*}(0)$ is

$$\max\{P_D(p,\, 0.5,\, \hat{\theta}_{1,j^*} \sqrt{\mathcal{I}^{(1)}}),\; P_D(p - 1\,, 0.5,\, \hat{\theta}_{1,i^*} \sqrt{\mathcal{I}^{(1)}})\}.$$

This is strictly less than the P-value

$$P_D(p\,, 0.5,\, \hat{\theta}_{1,i^*} \sqrt{\mathcal{I}^{(1)}})$$

that arises in the analysis where a confidence interval for θ_{i^*} will be provided. Thus, when the selected treatment does not have the highest estimated treatment effect in Phase II, it is possible that the closed testing procedure rejects $H_{i^*}(0)$ in the analysis that does not provide a confidence interval for θ_{i^*} but the analysis that also provides a confidence interval does not reject $H_{i^*}(0)$.

Another case of interest is when the two treatments, i_1 and i_2 say, with the highest estimated treatment effects in Phase II are taken forward to Phase III. In the analysis that produces confidence intervals for θ_{i_1} and θ_{i_2}, the conditions for rejecting $H_{i_1}(0)$ and $H_{i_2}(0)$ are given by (26.8) with $\tilde{\theta}_{i_1} = 0$ and (26.9) with $\tilde{\theta}_{i_2} = 0$, respectively. However, if confidence intervals are not to be calculated, the closed testing procedure only needs to operate on the set of hypotheses $H_i(0)$: $\theta_i \leq 0$, $i = 1, \ldots, p$. Then in (26.8) with $\tilde{\theta}_{i_1} = 0$, we can replace the Phase II adjusted P-value $P_D(p,\, 0.5,\, z_{i_1}^{(1)}(0))$ by

$$\max\{P_D(p,\, 0.5,\, \max[z_{i_1}^{(1)}(0),\, z_{i_2}^{(1)}(0)]),\; P_D(p - 1\,, 0.5,\, z_{i_1}^{(1)}(0))\}$$

and the Phase III adjusted P-value $P_D(2,\, 0.5,\, z_{i_1}^{(2)}(0))$ by

$$\max\{P_D(2,\, 0.5,\, \max[z_{i_1}^{(2)}(0),\, z_{i_2}^{(2)}(0)]),\; 1 - \Phi(z_{i_1}^{(2)}(0))\},$$

and we can make similar replacements for $P_D(p, 0.5, z_{i_2}^{(1)}(0))$ and $P_D(2, 0.5, z_{i_2}^{(2)}(0))$ in (26.9). These modifications will increase the left hand sides of (26.8) and (26.9), increasing the probability of rejecting $H_{i_1}(0)$ or $H_{i_2}(0)$. So, again a null hypothesis of "no treatment effect" may be rejected in the analysis that does not provide confidence intervals while the analysis that also provides confidence intervals does not reject that hypothesis.

Our discussion has focused on level $(1-\alpha)$ upper confidence interval for θ_{i^*} as this aligns with the one-sided testing problem when new treatments are compared with a control. It is straightforward to extend the above method to give an upper bound on θ_{i^*} and, hence, a $(1-2\alpha)$ two-sided confidence interval for θ_{i^*}. For this, we define the family of hypotheses

$$H_i(\tilde{\theta}_i)\colon \theta_i \geq \tilde{\theta}_i, \quad i = 1, \ldots, p, \quad \tilde{\theta}_i \in R,$$

and use the same form of closed testing procedure with, say, Dunnett tests of intersection hypotheses and weighted inverse normal combination tests.

26.7 Adaptive Seamless Designs with One Experimental Treatment: Go/NoGo Trials

The two-stage methods we have described in this chapter have involved treatment selection at the end of the first stage. However, seamless designs may also be employed when only one experimental treatment is compared to placebo. Lai, Lavori & Shih (2012b) observe that in oncology, few Phase II studies are randomized with internal controls. They comment that

> "It is well known that the success rate of phase III cancer clinical trials is low. This indicates that preliminary data at the end of phase II studies are inadequate for determining whether to launch phase III trials and how to design them."

They argue that large, controlled Phase II trials are needed but, with multiple research protocols competing for a limited patient population, they are difficult to achieve. Instead, they suggest merging Phases II and III:

> "Being able to include the phase II study as an internal pilot for the confirmatory phase III trial may be the only feasible way for a randomized phase II cancer trial of such sample size and scope to be conducted."

Lai, Lavori & Shih (2012b) propose a seamless Phase II/III design for cancer trials in which subjects in Phase II provide a short-term endpoint, such as tumor response, to inform the Go/NoGo decision. If the decision is to proceed, these patients continue into Phase III and additional patients are recruited. The final conclusion is based on a longer-term endpoint, such as overall survival or progression-free survival, from all subjects combined. Lai, Lavori & Shih propose a semiparametric joint model for the binary short-term endpoint and the survival endpoint. They suggest monitoring the trial with a group sequential generalized likelihood ratio test, as defined by Lai & Shih (2004), in which the first interim analysis uses the binary endpoint and, if the decision is to continue, remaining analyses switch to the survival endpoint.

A Bayesian approach to the same problem has been described by Inoue, Thall & Berry (2002) assuming exponentially distributed survival times. There, the decision to stop early, continue or shift to the Phase III stage is assessed repeatedly during Phase II.

Teng et al. (2018) consider the case when the same time-to-event endpoint is analyzed in both Phase II and Phase III, assuming a proportional hazards model with hazard rate $h_A(t)$ for the control group and $h_B(t) = \lambda h_A(t)$ for the treatment group, as described in Section 17.1. Hazard ratios HR_{eff} and HR_{ineff} are specified such that the new treatment is deemed effective if $\lambda = HR_{eff}$ and ineffective if $\lambda = HR_{ineff}$. Teng et al. evaluate seamless trial designs in terms of three probability of success criteria,

$$\begin{aligned}
\text{PoS}_{Go} &= Pr\{\text{Go decision after Phase II} \mid \lambda = HR_{eff}\} \\
\text{PoS}_{Suc} &= Pr\{\text{Go decision after Phase II and succeed in Phase III} \mid \lambda = HR_{eff}\} \\
\text{PoS}_{NoGo} &= Pr\{\text{NoGo decision after Phase II} \mid \lambda = HR_{ineff}\}
\end{aligned}$$

where success in Phase III occurs if the new treatment is found to be superior to control at the conclusion of the Phase III stage. The authors describe a procedure to optimize Phase II and Phase III sample sizes while meeting stated probability of success requirements. They also consider the addition of an interim analysis in Phase II. In a subsequent paper, Teng et al. (2020) propose an adaptive version of this procedure with the added feature that sample size re-assessment methods, as described in Chapter 24, are used to determine the Phase III sample size.

The seamless designs described in this section should not be confused with the 2-in-1 designs proposed by Chen et al. (2018) and extended by Chen, Li & Deng (2020). These designs allow the option of seamlessly expanding a Phase II trial into a Phase III trial. At an interim analysis during the Phase II trial, data on an intermediate endpoint are evaluated. At this point:

The study can be abandoned as futile;

If results are equivocal, the Phase II trial is continued to its planned conclusion and, if the final results justify it, a new Phase III trial can be planned;

If the interim Phase II results are promising, the trial is expanded seamlessly into a Phase III trial, increasing the sample size without changing the eligibility, randomization or treatment protocols, and data from all subjects are combined in the final analysis.

The rules for the expansion option are pre-specified and hypothesis tests are defined in a way that controls the probability of an overall Type I error in either the Phase II or Phase III version of the trial.

26.8 Bibliography and Notes

The origins of seamless Phase II/III designs can be traced to the papers of Kieser, Bauer & Lehmacher (1999), who combined multiple testing with adaptive interim analyses, and of Bauer & Kieser (1999), who addressed the specific case of comparing multiple treatments with a control in a two-stage trial where the set of treatments may be reduced after the first stage.

We have focused on procedures in which a single treatment proceeds to the Phase III stage and the family-wise error rate is protected through use of a closed testing procedure and combination tests. Other strategies are possible: see, for example, Bischoff & Miller (2005), Kelly, Stallard & Todd (2005), Liu & Pledger (2005), Sampson & Sill (2005), Wang (2006), Schmidli, Bretz & Racine-Poon (2007), Kimani, Stallard & Hutton (2009) and Posch, Maurer & Bretz (2011).

A number of reviews of the methodology and assessment of the potential benefits are available. Maca et al. (2006) discuss the operational aspects of ASDs and provide several examples of trials that used these methods. Thall (2008) surveys different approaches to defining ASDs, concluding that

> "in many circumstances, a properly designed phase 2–3 trial utilizes resources much more efficiently and provides much more reliable inferences than conventional methods."

Stallard & Todd (2011) describe different approaches in a unified framework, noting their similarities and differences. Kimani, et al. (2012) offer practical advice for the implementation of a seamless Phase II/III design, proposing a Bayesian approach in which a utility function is optimized when choosing the hypotheses to test in the later stages of the trial.

Cuffe et al. (2014) present four case studies of ASDs and comment on the varying success that was achieved. They conclude that

> "Committing to phase III plans and study designs before phase II begins introduces extra risk to drug development. However, seamless pivotal studies may be an attractive option when the clinical setting and development programme allow, for example, when revisiting dose selection."

There is a clear connection between the methods we have discussed in this chapter and the multi-arm multi-stage designs we shall present in Chapter 27, particularly when more than one experimental treatment may continue into the Phase III stage. There are also similarities between designs incorporating treatment selection and the enrichment designs we shall present in Chapter 28, where the decision is made to continue studying the new treatment in the full population or to focus on a selected a sub-population in the second part of a trial.

We conclude this chapter with some recommendations from Bretz, Koenig et al. (2009) that address practical aspects of adaptive trial designs and which have particular relevance to seamless Phase II/III trials:

> "In order to retain the validity of a confirmatory trial, a strict type I error rate control is mandatory for an acceptable adaptive design. Proper statistical methodology should thus be applied ... This requirement also implies that the experimental questions and hypotheses to be investigated are well specified upfront in the study protocol. The number of potential adaptations should generally be kept to a minimum; explorative, hypothesis generating adaptive designs in a confirmatory drug development phase are not acceptable to regulatory agencies.
>
> Because of their complexity, both in methodology and logistics, adaptive designs require a careful preplanning. This includes the amount of pre-specification required in the study protocol as well as the conduct of extensive simulations for a good understanding of the operational characteristics. To strictly control the overall type I error rate, it is mandatory to pre-specify the design of the first stage and how the information across the stages is combined. In addition, it is essential to concisely describe the study objectives, the response variables and the type of adaptive designs (e.g. treatment selection, selection of a pre-specified sub-population, etc.) ... The decision rules allowing early stopping for success have to be stated explicitly in the protocol. For other adaptations, like treatment selection or dropping a dose, this is not the case, since the type I error rate is not affected by the nature of these rules if appropriate statistical methods are used.
>
> An issue which is still under some debate is the degree of sponsor involvement in the interim decision process. ... Current regulatory guidance in more traditional monitoring settings, such as in group sequential designs, holds that sponsors should not have access to interim data while trials are ongoing. One concern in the context of adaptive designs is that unanticipated complexities might not fit a pre-specified algorithm that can be implemented without sponsor participation. One model on how to implement flexible interim decisions in practice is to include a proposal of potential interim decision rules in the DMC charter with the understanding that the DMC has the discretion to deviate from them as necessary and involve the sponsor based on the following principles: a clear rationale for a sponsor involvement; sponsor representatives properly distanced from trial operations; clear understanding

by all parties involved of the issues and potential risks; and documentation of the processes followed with restrictive firewalls put in place. The general aim should be a minimal sponsor exposure sufficient to make decisions, meaning that the smallest possible number of sponsor representatives should only get involved at the adaptation point with the minimally relevant information. Such an approach would minimize the sponsor's involvement and associated information leakage but still guarantee the sponsor's interest in case of unexpected emerging results."

Chapter 27

Adaptive Multi-arm Multi-stage Phase III Trials

27.1 Introduction

We have already discussed multi-arm multi-stage (MAMS) trials in Chapter 21. As outlined in Section 21.1, such designs offer considerable savings in time and cost compared to running several traditional Phase III two-armed trials. The seamless Phase II/III trial designs described in Chapter 26 may be considered as a special case in which all but one of the experimental treatments are dropped after stage 1. Here, we shall consider a Phase III clinical trial in which several experimental treatments are compared to a common control. The goal of such a trial may be to identify all treatments that are superior to the control, or investigators may be content to show a single new treatment is efficacious. The methods we present here can be used for either of these objectives. Indeed, their flexibility allows the objective to be refined in light of information on treatment safety or a secondary endpoint.

A number of strategies have been proposed for dropping unpromising treatments during a multi-arm trial and we presented some of these in Section 21.4. In this chapter, we follow the approach taken by Posch et al. (2005) and Bretz et al. (2009) for two-stage trials and by Wassmer (2011) and Magirr, Stallard & Jaki (2014) for multi-stage trials. These authors applied a closed testing procedure and combination tests to produce flexible designs that control the family-wise error rate (FWER). The construction of the resulting designs follows the general adaptive approach described in Chapter 25.

Consider a trial comparing J new treatments with a control. For each $j = 1, \ldots, J$, suppose the parameter θ_j is a measure of the difference in efficacy between treatment j and the control and we wish to test the null hypothesis, H_j: $\theta_j \leq 0$, which implies treatment j is no better than the control, against the alternative $\theta_j > 0$. Our objective is to design an adaptive trial in which the J null hypotheses are tested with FWER at most α.

Following the scheme depicted in Table 25.1, we define the $2^J - 1$ intersection hypotheses

$$H_I = \cap_{j \in I} H_j,$$

where $I \subseteq \{1, \ldots, J\}$. For each H_I, we compute P-values $P_I^{(k)}$ based on data observed in stage k for $k = 1, \ldots, K$. We denote the corresponding stage-wise Z-values by

$$Z_I^{(k)} = \Phi^{-1}(1 - P_I^{(k)}).$$

For a given intersection hypothesis H_I, we use the weighted inverse normal combination rule to find *cumulative* Z-values and P-values at stage k,

$$Z_{I,k} = \frac{\sum_{i=1}^{k} w_i Z_I^{(i)}}{\sqrt{(\sum_{i=1}^{k} w_i^2)}} \quad \text{and} \quad P_{I,k} = 1 - \Phi(Z_{I,k}),$$

where the weights w_i are pre-specified and satisfy $w_1^2 + \ldots + w_K^2 = 1$.

DOI: 10.1201/9781584888482-27

We use the construction of Lehmacher & Wassmer (1999), as described in Section 23.7, to define group sequential boundaries against which to compare these cumulative Z-values in testing each intersection hypothesis H_I. We treat the lower futility boundary $(a_1, \ldots, a_K)$ as non-binding so the probability that the upper boundary $(b_1, \ldots, b_K)$ is crossed when H_I is true is at most α, whether or not the futility boundary is applied; see Sections 5.3 and 8.4.

In a sequential implementation of the closed testing procedure, the elementary hypothesis H_j may be rejected at stage k if all intersection hypotheses H_I where $I \ni j$ are rejected. In order for all the relevant group sequential tests to reject their null hypotheses at stage k, we require $Z_{I,k} > b_k$ for all H_I with $I \ni j$. It is, therefore, useful to define the *adjusted* cumulative Z-value at stage k for each hypothesis H_j, $j = 1, \ldots, J$, as

$$\tilde{Z}_{j,k} = \min_{I \ni j} Z_{I,k}$$

so the criterion to reject H_j at stage k in the closed testing procedure is simply

$$\tilde{Z}_{j,k} > b_k.$$

We similarly define the adjusted cumulative P-values,

$$\tilde{P}_{j,k} = \max_{I \ni j} P_{I,k} = 1 - \Phi(\tilde{Z}_{j,k}).$$

Strictly speaking, a procedure that rejects H_j at stage k if all intersection hypotheses H_I where $I \ni j$ have been rejected at *or before* stage k will still protect the FWER at level α. However, we shall proceed here to work with decision rules in terms of the $\tilde{Z}_{j,k}$ and $\tilde{P}_{j,k}$.

The motivation of a MAMS trial is to use resources efficiently, so it is to be expected that recruitment to a treatment arm will cease after a positive outcome for that treatment. Treatments may also be dropped for futility, either for lack of efficacy or because of a high rate of adverse events. Regarding efficacy, it is natural to consider dropping treatment j at stage k if $Z_{I,k} < a_k$ for some $I \ni j$ as it is unlikely this H_I will eventually be rejected, which is a requirement for H_j to be rejected overall in the closed testing procedure. This gives the futility stopping criterion in terms of the adjusted cumulative Z-value,

$$\tilde{Z}_{j,k} = \min_{I \ni j} Z_{I,k} < a_k.$$

After some treatment arms have been discontinued, it may still be necessary to test intersection hypotheses involving these treatments. Suppose the set I contains the indices of some discontinued treatments and let $\tilde{I}$ be the subset of I comprising labels of treatments for which data are still being collected. We follow Hommel (2001, Sec. 4.1) and Bretz et al. (2006, p. 628), and define stage-wise Z-values or P-values for H_I at future stages to be those calculated for $H_{\tilde{I}}$. As an example, if treatment 1 is dropped after stage 1, but treatments 2 and 3 remain, we set $Z^{(2)}_{\{1,2,3\}}$ equal to $Z^{(2)}_{\{2,3\}}$ and take $P^{(2)}_{\{1,2,3\}}$ to be $P^{(2)}_{\{2,3\}}$.

27.2 An Example

The following example illustrates the calculations involved in a MAMS design, as described above. Suppose a multi-stage trial is to be conducted to compare $J = 5$ treatments with a control. The five treatments may be increasing doses of a particular drug or five distinct therapies. We decide to run the trial in $K = 4$ stages.

Table 27.1 shows hypothetical Z-values, $Z_j^{(k)}$, for testing treatment j against control based on data observed in each stage k of the trial. Entries are missing where a treatment has been dropped from the trial. Table 27.2 shows the equivalent P-values, $P_j^{(k)} = 1 - \Phi(Z_j^{(k)})$.

Table 27.1 *MAMS trial, Example 1: Stage-wise Z-values,* $Z_j^{(k)}$

	Stage k			
Treatment j	1	2	3	4
1	−0.060	0.270	—	—
2	1.129	0.866	1.461	−0.265
3	0.813	1.054	2.724	1.860
4	2.401	2.651	—	—
5	2.998	—	—	—

Table 27.2 *MAMS trial, Example 1: Stage-wise P-values,* $P_j^{(k)}$

	Stage k			
Treatment j	1	2	3	4
1	0.5239	0.3936	—	—
2	0.1294	0.1932	0.0720	0.6045
3	0.2081	0.1459	0.0032	0.0314
4	0.0082	0.0040	—	—
5	0.0014	—	—	—

We suppose the one-sided test with non-binding futility boundary to be used in the Lehmacher-Wassmer method is the member of the power family described in Section 5.3 with $\alpha = 0.025$, $1 - \beta = 0.9$ and parameter $\Delta = 0$. This test has upper boundary

$$b_k = 2.024(k/4)^{-1/2}, \quad k = 1, \ldots, 4,$$

and lower boundary

$$a_k = 3.410(k/4)^{1/2} - 1.386(k/4)^{-1/2}, \quad k = 1, \ldots, 4.$$

The same group sequential design was considered in in Example (c) of Section 5.5.2 and the boundary values are those in the final column of Table 5.11, namely

$$(b_1, b_2, b_3, b_4) = (4.049,\ 2.863,\ 2.337,\ 2.024)$$

and

$$(a_1, a_2, a_3, a_4) = (-1.068,\ 0.451,\ 1.353,\ 2.024).$$

We now explain how calculations are made in this example using either Dunnett's test or the Simes test to obtain P-values for intersection hypotheses.

Using Dunnett's test

We assume the primary endpoint follows a normal distribution with the same variance for all treatments and sample sizes in the control group and each of the five new treatment arms are equal within each stage. Results of the relevant calculations are shown in Table 27.3.

Stage 1

Consider first the data at the end of stage 1. The Z-values for the Dunnett test for each intersection hypothesis H_I, where $I \subseteq \{1, 2, 3, 4, 5\}$, are shown in the second column of

Table 27.3 *MAMS trial, Example 1: Stage-wise and cumulative Z-values for intersection hypotheses. For each H_I, a Dunnett test is applied to obtain the Z-value $Z_I^{(k)}$ based only on data from stage k. The cumulative Z-value $Z_{I,k}$ is the result of applying an inverse normal combination test with equal weights to the values $Z_I^{(1)}, \dots, Z_I^{(k)}$ from stages 1 to k.*

Index set	Stage 1	Stage 2		Stage 3		Stage 4	
I	$Z_I^{(1)} = Z_{I,1}$	$Z_I^{(2)}$	$Z_{I,2}$	$Z_I^{(3)}$	$Z_{I,3}$	$Z_I^{(4)}$	$Z_{I,4}$
{1}	−0.060	0.270	0.148	—	—	—	—
{2}	1.129	0.866	1.411	1.461	1.995	−0.265	1.595
{3}	0.813	1.054	1.320	2.724	2.651	1.860	3.225
{4}	2.401	2.651	3.572	—	—	—	—
{5}	2.998	—	—	—	—	—	—
{1,2}	0.799	0.514	0.929	1.461	1.602	−0.265	1.255
{1,3}	0.457	0.718	0.831	2.724	2.251	1.860	2.879
{1,4}	2.160	2.425	3.242	—	—	—	—
{1,5}	2.790	0.270	2.164	—	—	—	—
{2,3}	0.799	0.718	1.073	2.502	2.320	1.585	2.802
{2,4}	2.160	2.425	3.242	1.461	3.491	−0.265	2.891
{2,5}	2.790	0.866	2.585	1.461	2.954	−0.265	2.426
{3,4}	2.160	2.425	3.242	2.724	4.220	1.860	4.585
{3,5}	2.790	1.054	2.718	2.724	3.792	1.860	4.214
{4,5}	2.790	2.651	3.847	—	—	—	—
{1,2,3}	0.607	0.522	0.798	2.502	2.096	1.585	2.608
{1,2,4}	2.017	2.290	3.045	1.461	3.330	−0.265	2.751
{1,2,5}	2.665	0.514	2.248	1.461	2.679	−0.265	2.188
{1,3,4}	2.017	2.290	3.045	2.724	4.059	1.860	4.445
{1,3,5}	2.665	0.718	2.392	2.724	3.526	1.860	3.984
{1,4,5}	2.665	2.425	3.599	—	—	—	—
{2,3,4}	2.017	2.290	3.045	2.502	3.931	1.585	4.196
{2,3,5}	2.665	0.718	2.392	2.502	3.398	1.585	3.735
{2,4,5}	2.665	2.425	3.599	1.461	3.782	−0.265	3.143
{3,4,5}	2.665	2.425	3.599	2.724	4.512	1.860	4.837
{1,2,3,4}	1.914	2.193	2.004	2.502	3.816	1.585	4.097
{1,2,3,5}	2.576	0.522	2.191	2.502	3.233	1.585	3.592
{1,2,4,5}	2.576	2.290	3.441	1.461	3.653	−0.265	3.031
{1,3,4,5}	2.576	2.290	3.441	2.724	4.382	1.860	4.725
{2,3,4,5}	2.576	2.290	3.441	2.502	4.254	1.585	4.476
{1,2,3,4,5}	2.506	2.193	3.323	2.502	4.157	1.585	4.393

Table 27.4 *MAMS trial, Example 1: Adjusted cumulative Z-values, $\tilde{Z}_{j,k}$ and P-values, $\tilde{P}_{j,k}$. The adjusted cumulative Z-value associated with treatment j is the minimum cumulative Z-value, based on Dunnett tests of intersection hypotheses H_I with $j \in I$. Similarly, the adjusted cumulative P-value associated with treatment j is the maximum cumulative P-value.*

Treatment	Stage 1	Stage 2	Stage 3	Stage 4
j	$\tilde{Z}_{j,1}$	$\tilde{Z}_{j,2}$	$\tilde{Z}_{j,3}$	$\tilde{Z}_{j,4}$
1	−0.060	0.148	—	—
2	0.607	0.798	1.602	1.255
3	0.457	0.798	2.096	2.608
4	1.914	2.904	—	—
5	2.506	—	—	—
	$\tilde{P}_{j,1}$	$\tilde{P}_{j,2}$	$\tilde{P}_{j,3}$	$\tilde{P}_{j,4}$
1	0.524	0.441	—	—
2	0.272	0.212	0.055	0.105
3	0.324	0.212	0.018	0.005
4	0.028	0.002	—	—
5	0.006	—	—	—

Table 27.3. As an example, in testing $H_{\{1,2,3\}}$ after stage 1, the Dunnett statistic is $\max(-0.060, 1.129, 0.813) = 1.129$ and calculation shows $Pr\{\max(Z_1, Z_2, Z_3) > 1.129\} = 0.272$ when Z_1, Z_2 and Z_3 are standard normal random variables with $\text{Corr}(Z_1, Z_2) = \text{Corr}(Z_1, Z_3) = \text{Corr}(Z_2, Z_3) = 1/2$. Thus $Z^{(1)}_{\{1,2,3\}} = \Phi^{-1}(1 - 0.272) = 0.607$. Since this is only the first stage, there is no combination across stages to perform and $Z_{I,1} = Z_I^{(1)}$ for each I.

The adjusted cumulative Z-values and P-values at stage 1 for each of the elementary hypotheses H_j can now be obtained as

$$\tilde{Z}_{j,1} = \min_{I \ni j} Z_{I,j}, \quad j = 1, \ldots, 5,$$

and these are displayed in the second column of Table 27.4. As an example, the adjusted cumulative Z-value, $\tilde{Z}_{1,1}$ is

$$\min\{Z_{\{1\},1}, Z_{\{1,2\},1}, Z_{\{1,3\},1}, \ldots, Z_{\{1,2,3,4,5\},1}\} = Z_{\{1\},1} = -0.060.$$

For the other adjusted cumulative Z-values, $\tilde{Z}_{2,1} = 0.607$ comes from the test of the intersection hypothesis $H_{\{1,2,3\}}$, $\tilde{Z}_{3,1} = 0.457$ from the test of $H_{\{1,3\}}$, $\tilde{Z}_{4,1} = 1.914$ from the test of $H_{\{1,2,3,4\}}$, and $\tilde{Z}_{5,1} = 2.506$ from the test of $H_{\{1,2,3,4,5\}}$.

We now compare these adjusted cumulative Z-values with the boundary $(a_1, b_1) = (-1.068, 4.049)$. Treatment 1 is the worst performing, but its adjusted cumulative Z-value of -0.060 does not fall below a_1 and we decide to continue that arm. Treatment 5 is the best performing but its adjusted cumulative Z-value of 2.506 does not exceed $b_1 = 4.049$. Suppose, however, that treatment 5 exhibits unacceptable toxicity, perhaps because its dose level is too high, and it is decided to drop this arm.

We continue testing the treatments 1 to 4. In this flexible design we may decide to change the sample sizes for the next stage, remembering that a simple application of the Dunnett test requires a common sample size for each treatment and control within each stage.

Stage 2

The observed Z-values, $Z_j^{(2)}$, based only on the stage 2 data are shown in the third column of Table 27.1 and the associated P-values, $P_j^{(2)}$ are shown in Table 27.2. As before, we form the stage 2 intersection test statistics $Z_I^{(2)}$ for all subsets I of $\{1,2,3,4,5\}$. These are displayed in the third column of Table 27.3. Note that, even though treatment 5 has been dropped, we still compute Z-values for intersection hypotheses involving that treatment. As noted previously, if $5 \in I$, we take $Z_I^{(2)}$ to be the same as $Z_{I/5}^{(2)}$, thus $Z_{\{1,5\}}^{(2)} = Z_{\{1\}}^{(2)}$, $Z_{\{1,2,5\}}^{(2)} = Z_{\{1,2\}}^{(2)}$, etc.

Next, for each intersection hypothesis H_I, the Z-values $Z_I^{(1)}$ and $Z_I^{(2)}$ from stages 1 and 2, are combined to form the cumulative Z-value, $Z_{I,2}$. In this example, we use an inverse normal combination test statistic with equal weights. Note that the originally specified weights must be used, even if the stage 2 sample sizes are changed in response to observed data. Thus, $Z_{I,2} = (Z_I^{(1)} + Z_I^{(2)})/\sqrt{2}$ for each I, so

$$Z_{\{1\},2} = (-0.060 + 0.270)/\sqrt{2} = 0.148, \quad Z_{\{1,2\},2} = (0.799 + 0.514)/\sqrt{2} = 0.929, \quad \text{etc.}$$

The adjusted cumulative Z-value for treatment j after stage 2, $\tilde{Z}_{j,2}$, is now formed in the same way as after stage 1, with

$$\tilde{Z}_{j,2} = \min_{I \ni j} Z_{I,2} \quad \text{for } j = 1, \ldots, 4.$$

These adjusted cumulative Z-values and the associated P-values, $\tilde{P}_{j,2}$, are shown in the third column of Table 27.4. So, for example,

$$\tilde{Z}_{4,2} = \min\{Z_{\{4\},2}, Z_{\{1,4\},2}, Z_{\{2,4\},2}, \ldots, Z_{\{1,2,3,4,5\},2}\} = Z_{\{1,2,3,4\},2} = 2.904.$$

The other adjusted cumulative Z-values are $\tilde{Z}_{1,2} = 0.148$, coming from the test of $H_{\{1\}}$, $\tilde{Z}_{2,2} = 0.798$ from the test of $H_{\{1,2,3\}}$, and $\tilde{Z}_{3,2} = 0.798$, also from the test of $H_{\{1,2,3\}}$.

Examining these results, we may decide to drop treatment 1 for lack of efficacy since its adjusted cumulative Z-value, $\tilde{Z}_{1,2} = 0.148$, is below the futility boundary value, $a_2 = 0.451$. Conversely, for treatment 4, we have

$$\tilde{Z}_{4,2} = 2.904 > 2.863 = b_2$$

and we can cease recruitment to this treatment, rejecting the elementary hypothesis H_4: $\theta_4 \leq 0$ in the overall testing procedure. It is still of interest to continue the comparison of treatments 2 and 3 with the control to see if one of these is efficacious, perhaps with fewer adverse side effects than treatment 4. Again, we may choose to adapt the common sample size for treatments 2 and 3 and the control in the next stage.

Stages 3 and 4

Suppose stage 3 is conducted with treatments 2 and 3 and the control. Upon examining the data after this stage, the adjusted cumulative Z-values $\tilde{Z}_{2,3} = 1.602$ and $\tilde{Z}_{3,3} = 2.096$ are below the upper boundary value, $b_3 = 2.337$ and the trial continues with the same treatments to the fourth and final stage.

After stage 4, the trial concludes with the adjusted cumulative Z-values and P-values shown in the final column of Table 27.4 (obtained from the cumulative Z-values in the last column of Table 27.3). At this point, $\tilde{Z}_{3,4} = 2.608$ exceeds the final boundary value, $b_4 = 2.024$ and we may declare a significant positive effect for treatment 3. However $\tilde{Z}_{2,4} = 1.255 < 2.024$, so the hypothesis H_2: $\theta_2 \leq 0$ cannot be rejected.

In summary, in this example of an adaptive MAMS procedure:

Treatment 1 was dropped for futility after stage 2;

Treatment 2 was declared non-significant after completion of all four stages;

Treatment 3 was declared efficacious after completion of all four stages;

Treatment 4 was stopped early after stage 2 and declared efficacious;

Treatment 5 was dropped for unacceptable toxicity after stage 1.

These results might be typical of a trial in which the treatments are increasing doses of a drug. In that case, the positive results for treatments 3 and 4 could lead to a new drug application for both dose levels in the expectation that the lower dose will be prescribed first, with the higher dose available for patients who do not respond to the lower dose. In other cases, investigators may wish to recommend just one of the treatments 3 and 4 and this decision may depend on other features such as ease of administration, the levels of adverse side-effects, or cost of the drugs.

Using Simes' test

Instead of using the Dunnett test for the intersection hypotheses, we could use Simes' method as described in Section 20.4. Simes' method requires fewer assumptions about the response distribution and can be applied as long as valid P-values are available and there is a positive association between P-values from comparisons of different treatments with the control (see Sarkar & Chang, 1997).

We shall work through the preceding example applying Simes tests to the stage-wise Z-values and P-values shown in Tables 27.1 and 27.2. We shall find that in this analysis, the conclusions differ in that treatment 4 is not declared efficacious after stage 2. This leads to additional values $Z_4^{(3)} = 1.444$ and $P_4^{(3)} = 0.0744$ when observations on treatment 4 continue into stage 3, which should be added to the data in Tables 27.1 and 27.2.

The Z-values, $Z_I^{(k)}$, for tests of intersection hypotheses using Simes' method are shown in Table 27.5, along with the resulting cumulative Z-values, $Z_{I,k}$. The adjusted cumulative Z-values $\tilde{Z}_{I,j}$ and P-values $\tilde{P}_{I,j}$ are presented in Table 27.6.

Stage 1

To illustrate the calculation of a Simes' P-value, consider the test of $H_{\{1,2\}}$, based on the values $P_1^{(1)} = 0.5239$ and $P_2^{(1)} = 0.1294$ from Table 27.2. In this case,

$$P_{\{1,2\}}^{(1)} = \min(2 \times 0.1294, 0.5239) = 0.2588$$

and the corresponding Z-value is $Z_{\{1,2\}}^{(1)} = \Phi^{-1}(1 - 0.2588) = 0.647$. As a further example where H_I involves three hypotheses, consider the test of $H_{\{1,2,3\}}$, based on $P_1^{(1)} = 0.5239$, $P_2^{(1)} = 0.1294$ and $P_3^{(1)} = 0.2081$. Here, we have

$$P_{\{1,2,3\}}^{(1)} = \min(3 \times 0.1294, (3/2) \times 0.2081, 0.5239) = 0.3121$$

and $Z_{\{1,2,3\}}^{(1)} = \Phi^{-1}(1 - 0.3121) = 0.490$.

The resulting adjusted cumulative Z-values and P-values are displayed in Table 27.6. As before, all the stage 1 adjusted cumulative Z-values are below the upper boundary value $b_1 = 4.409$ and above the lower boundary $a_1 = -1.068$. Again, we assume treatment 5 is stopped for safety reasons and the study continues with treatments 1 to 4 in stage 2.

Table 27.5 *MAMS trial, Example 1: Stage-wise and cumulative Z-values for intersection hypotheses. For each H_I, a Simes test is applied to obtain the Z-value $Z_I^{(k)}$ based only on data from stage k. The cumulative Z-value $Z_{I,k}$ is the result of applying an inverse normal combination test with equal weights to the values $Z_I^{(1)}, \ldots, Z_I^{(k)}$ from stages 1 to k.*

Index set	Stage 1	Stage 2		Stage 3		Stage 4	
I	$Z_I^{(1)} = Z_{I,1}$	$Z_I^{(2)}$	$Z_{I,2}$	$Z_I^{(3)}$	$Z_{I,3}$	$Z_I^{(4)}$	$Z_{I,4}$
{1}	−0.060	0.270	0.148	—	—	—	—
{2}	1.129	0.866	1.411	1.461	1.995	-0.265	1.595
{3}	0.813	1.054	1.320	2.724	2.651	1.860	3.225
{4}	2.401	2.651	3.572	1.444	3.750	—	—
{5}	2.998	—	—	—	—	—	—
{1,2}	0.647	0.288	0.661	1.461	1.383	-0.265	1.066
{1,3}	0.212	0.548	0.537	2.724	2.011	1.860	2.672
{1,4}	2.136	2.408	3.213	1.444	3.457	—	—
{1,5}	2.780	0.270	2.157	—	—	—	—
{2,3}	0.813	0.866	1.187	2.487	2.405	1.531	2.848
{2,4}	2.136	2.408	3.213	1.444	3.457	-0.265	2.861
{2,5}	2.780	0.866	2.578	1.461	2.949	-0.265	2.421
{3,4}	2.136	2.408	3.213	2.487	4.059	1.860	4.445
{3,5}	2.780	1.054	2.711	2.724	3.786	1.860	4.209
{4,5}	2.780	2.651	3.840	1.444	3.969	—	—
{1,2,3}	0.490	0.554	0.738	2.487	2.038	1.531	2.531
{1,2,4}	1.968	2.256	2.987	1.444	3.272	-0.265	2.702
{1,2,5}	2.646	0.288	2.075	1.461	2.538	-0.265	2.065
{1,3,4}	1.968	2.256	2.987	2.487	3.874	1.860	4.285
{1,3,5}	2.646	0.548	2.258	2.724	3.417	1.860	3.889
{1,4,5}	2.646	2.408	3.573	1.444	3.751	—	—
{2,3,4}	1.968	2.256	2.987	2.339	3.789	1.531	4.047
{2,3,5}	2.646	0.866	2.483	2.487	3.463	1.531	3.765
{2,4,5}	2.646	2.408	3.573	1.444	3.751	-0.265	3.116
{3,4,5}	2.646	2.408	3.573	2.487	4.353	1.860	4.700
{1,2,3,4}	1.843	2.143	2.818	2.339	3.651	1.531	3.928
{1,2,3,5}	2.547	0.554	2.192	2.487	3.226	1.531	3.559
{1,2,4,5}	2.547	2.256	3.396	1.444	3.607	-0.265	2.991
{1,3,4,5}	2.547	2.256	3.396	2.487	4.208	1.860	4.575
{2,3,4,5}	2.547	2.256	3.396	2.339	4.123	1.531	4.336
{1,2,3,4,5	2.468	2.143	3.261	2.339	4.013	1.531	4.240

Table 27.6 *MAMS trial, Example 1: Adjusted cumulative Z-values,* $\tilde{Z}_{j,k}$ *and P-values,* $\tilde{P}_{j,k}$. *The adjusted cumulative Z-value associated with treatment j is the minimum cumulative Z-value, based on Simes tests of intersection hypotheses* H_I *with* $j \in I$. *Similarly, the adjusted cumulative P-value associated with treatment j is the maximum cumulative P-value.*

Treatment	Stage 1	Stage 2	Stage 3	Stage 4
j	$\tilde{Z}_{j,1}$	$\tilde{Z}_{j,2}$	$\tilde{Z}_{j,3}$	$\tilde{Z}_{j,4}$
1	−0.060	0.148	—	—
2	0.490	0.661	1.383	1.066
3	0.212	0.537	2.011	2.531
4	1.843	2.818	3.272	—
5	2.468	—	—	—
	$\tilde{P}_{j,1}$	$\tilde{P}_{j,2}$	$\tilde{P}_{j,3}$	$\tilde{P}_{j,4}$
1	0.524	0.441	—	—
2	0.312	0.254	0.083	0.143
3	0.416	0.296	0.022	0.006
4	0.033	0.002	0.001	—
5	0.007	—	—	—

Stage 2

The stage 2, adjusted cumulative Z-values and P-values in Table 27.6 are a little different from those in Table 27.4 that used the Dunnett intersection tests. In particular, the inverse normal combination test statistic for testing $H_{\{1,2,3,4\}}$ at stage 2 is

$$\tilde{Z}_{\{1,2,3,4\},2} = (1.843 + 2.143)/\sqrt{2} = 2.818$$

and this is the value taken by $\tilde{Z}_{\{4,2\}}$, the adjusted cumulative Z-value for treatment 4. Since $\tilde{Z}_{\{4,2\}} < b_2 = 2.863$, we cannot reject H_4: $\theta_4 \leq 0$ and declare treatment 4 superior to the control at this stage. Again, we assume treatment 1 is stopped for futility and, hence, the trial continues to stage 3 with treatments 2, 3 and 4 and the control.

Stages 3 and 4

With the additional value $Z_4^{(3)} = 1.444$ in the stage 3 data, the cumulative Z-statistics are shown in Table 27.5 and the adjusted cumulative Z-statistics in Table 27.6. Now we see

$$\tilde{Z}_{4,3} = 3.272 > 2.337 = b_3$$

and the hypothesis H_4: $\theta_4 \leq 0$ can be rejected after stage 3. We suppose the trial continues to the fourth and final stage with treatments 2 and 3 and the control.

After stage 4, the trial concludes with the adjusted cumulative Z-values and P-values shown in the final column of Table 27.6. At this point, $\tilde{Z}_{3,4} = 2.531$ exceeds the final boundary value, $b_4 = 2.024$, and we may declare a significant positive effect for treatment 3. However $\tilde{Z}_{2,4} = 1.066 < 2.024$, so the hypothesis H_2: $\theta_2 \leq 0$ cannot be rejected.

Table 27.7 *MAMS trial, Example 2: Stage-wise Z-values* $Z_j^{(k)}$ *in a MAMS trial where treatment 3 is added after stage 1*

	Stage k		
Treatment j	1	2	3
1	−0.098	−0.685	—
2	0.862	0.864	2.048
3	—	1.615	2.125

Comparing Dunnett and Simes tests

The two versions of this trial have led to the same conclusions with treatments 3 and 4 declared efficacious. With the data presented here, use of Dunnett tests led to an earlier decision to declare treatment 4 efficacious. This is not unexpected as the Dunnett test is particularly well suited to comparing several treatments with a common control and is liable to produce a procedure with higher power in a scenario such as this where the best treatment is superior both to the control and to the other new treatments. Furthermore, a seamless Phase II/III trial with treatment selection at the first analysis can be viewed as a special case of a MAMS design and the results of Hampson & Jennison (2015) that we reported in Section 26.4 showed the Simes test is not well suited to a trial comparing multiple treatments with a control.

27.3 Adding a Treatment

So far, we have considered the option of dropping treatments at interim analyses for efficacy, futility or toxicity. However, we might also consider *adding* a treatment during the trial while still testing null hypotheses for all treatments in a way that will still protect the FWER. Hommel (2001, Sec. 4.4) and Posch et al. (2005) describe a method to achieve this.

Suppose a trial starts with J experimental arms and a control arm and a MAMS design has been specified, as described in the preceding sections. Then, during the course of the trial, a new experimental treatment becomes available and it is decided to introduce this as an additional treatment arm. In testing H_{J+1}: $\theta_{J+1} \leq 0$, the null hypothesis for this additional treatment, the FWER must still be protected at level α, even when the decision to add the new treatment is informed by results already observed in the trial. The way to achieve this is to analyze the trial data as if the inclusion of treatment $J+1$ had been planned all along, retaining the stage-wise Z-values and P-values that were calculated before the addition of the new treatment, and defining Z-values and P-values for intersection hypotheses involving treatment $J+1$ for stages that have already been completed.

We illustrate this methodology with a hypothetical example of a trial initially designed to compare two experimental arms, treatments 1 and 2, with a control arm. The Dunnett method is used to test the intersection hypothesis $H_{\{1,2\}}$ and results are combined across stages using an inverse normal combination rule with equal weights. Group sequential testing follows the Lehmacher-Wassmer approach implemented using the one-sided test with three analyses and a non-binding futility boundary from the power family described in Section 5.3 with $\alpha = 0.025$, $1-\beta = 0.9$ and parameter $\Delta = 0$. This test has upper boundary

$$(b_1, b_2, b_3) = (3.471,\ 2.454,\ 2.004)$$

and lower boundary

$$(a_1, a_2, a_3) = (-0.421,\ 1.078,\ 2.004).$$

The data observed in this example are shown in Table 27.7.

Table 27.8 *MAMS trial, Example 2: Stage-wise and cumulative Z-values for intersection hypotheses. For each H_I, a Dunnett test is applied to obtain the Z-value $Z_I^{(k)}$ based only on data from stage k. The cumulative Z-value $Z_{I,k}$ is the result of applying an inverse normal combination test with equal weights to the values $Z_I^{(1)}, \ldots, Z_I^{(k)}$ from stages 1 to k or, in the case $I = \{3\}$, to the values $Z_I^{(2)}, \ldots, Z_I^{(k)}$ from stages 2 to k.*

Index set	Stage 1	Stage 2		Stage 3	
I	$Z_I^{(1)} = Z_{I,1}$	$Z_I^{(2)}$	$Z_{I,2}$	$Z_I^{(3)}$	$Z_{I,3}$
{1}	−0.098	−0.685	−0.554	—	—
{2}	0.862	0.864	1.220	2.048	2.179
{3}	—	1.615	1.615	2.125	2.645
{1,2}	0.510	0.512	0.723	2.048	1.772
{1,3}	−0.098	1.322	0.866	2.125	1.934
{2,3}	0.862	1.322	1.545	1.867	2.339
{1,2,3}	0.510	1.150	1.174	1.867	2.036

Stage 1

During stage 1, the original trial design with treatments 1 and 2 and the control arm is followed. At the end of this stage, we have $Z_{\{1\}}^{(1)} = -0.098$ and $Z_{\{2\}}^{(1)} = 0.862$. Applying Dunnett's method, the Z-value for the intersection hypothesis $H_{\{1,2\}}$ is $Z_{\{1,2\}}^{(1)} = 0.510$. When treatment 3 is added at the end of stage 1, Z-values for the intersection hypotheses $H_{\{1,3\}}$, $H_{\{2,3\}}$, and $H_{\{1,2,3\}}$ are required, even though there is no data from treatment 3 to use at this point. If $H_{\{1,3\}}$ is true, then so is $H_{\{1\}}$ and it is legitimate to use $Z_{\{1\}}^{(1)}$ as the stage 1 Z-value for $H_{\{1,3\}}$. Similarly, we use $Z_{\{2\}}^{(1)}$ and $Z_{\{1,2\}}^{(1)}$ as the stage 1 Z-values for $H_{\{2,3\}}$ and $H_{\{1,2,3\}}$, respectively. As there is no stage 1 data for treatment 3, $H_{\{3\}}$ cannot be tested at this point, so there is no value for $Z_{\{3\}}^{(1)}$. The resulting values for each $Z_I^{(1)}$ are shown in column 2 of Table 27.8.

Stage 2

We suppose all three treatment arms are observed in stage 2 of the trial. Once treatment 3 has been added, the stage-wise Z-values, $Z_I^{(2)}$, for each intersection hypothesis H_I are calculated as if the trial had been designed to compare three treatments against the control from the outset. The results are shown in column 3 of Table 27.8.

Combination tests with equal weights for each stage are used to produce the cumulative Z-values $Z_{I,2}$ shown in column 4 of Table 27.8 for all H_I apart from $H_{\{3\}}$. Since there is no statistic $Z_{\{3\}}^{(1)}$, we take $Z_{\{3\}}^{(2)} = 1.615$ as the cumulative Z-statistic for $H_{\{3\}}$ after stage 2.

Table 27.9 shows the adjusted cumulative Z-values and P-values obtained, in the usual way, from the cumulative Z-values in Table 27.8. Since $\tilde{Z}_{1,1}$ and $\tilde{Z}_{2,1}$ are between a_1 and b_1, both treatments 1 and 2 continued to stage 2. However, at stage 2, $\tilde{Z}_{1,2} < a_2$ and we assume this treatment is dropped at that point, so only treatments 2 and 3 and the control are observed in stage 3.

Stage 3

The stage-wise Z-values for $Z_I^{(3)}$ are shown in column 5 of Table 27.8 but with no entry for $Z_{\{1\}}^{(3)}$ as treatment 1 has now been dropped. The adjusted cumulative Z-values and P-values

Table 27.9 *MAMS trial, Example 2: Adjusted cumulative Z-values $\tilde{Z}_{j,k}$ and P-values, $\tilde{P}_{j,k}$. The adjusted cumulative Z-value associated with treatment j is the minimum cumulative Z-value, based on Dunnett tests of intersection hypotheses H_I with $j \in I$. Similarly, the adjusted cumulative P-value associated with treatment j is the maximum cumulative P-value.*

Treatment	Stage 1	Stage 2	Stage 3
j	$\tilde{Z}_{j,1}$	$\tilde{Z}_{j,2}$	$\tilde{Z}_{j,3}$
1	−0.098	−0.554	—
2	0.510	0.723	1.772
3	—	0.866	1.934
	$\tilde{P}_{j,1}$	$\tilde{P}_{j,2}$	$\tilde{P}_{j,3}$
1	0.539	0.710	—
2	0.305	0.235	0.038
3	—	0.193	0.027

in this table are obtained in the usual way but, since there is no $Z^{(1)}_{\{3\}}$, we obtain $Z_{\{3\},3}$ from $Z^{(2)}_{\{3\}}$ and $Z^{(3)}_{\{3\}}$ using a combination test with weights $w_2 = w_3 = 1/\sqrt{2}$. The final pair of adjusted cumulative Z-values is shown in column 4 of Table 27.9. Since both $\tilde{Z}_{2,3}$ and $\tilde{Z}_{3,3}$ are less than $b_3 = 2.004$, neither treatment can be declared superior to control according to the closed testing procedure.

One might wonder if a separate two-armed trial comparing treatment 3 and the control should have been contemplated. The unadjusted combination test statistic $Z_{\{3\},3} = 2.645$ would be highly significant in a separate trial but, in the multi-arm trial, the adjusted statistic $\tilde{Z}_{3,3}$ is lower due to the low value of $Z_{\{1,3\},3}$ for testing the intersection hypothesis $H_{\{1,3\}}$. The multiplicity adjustment and, in this case, the loss of a potentially significant result for treatment 3, is the price to be paid for the convenience of adding treatment 3 to an existing trial. This loss of power is noted by Posch et al. (2005, p. 3701) who state that there can be "a serious penalty for the rejection of the new null hypothesis."

There are advantages to adding an arm to a MAMS trial that is already underway. As well as the efficiency gain from sharing the control arm of the existing trial there are logistical and administrative benefits from joining an existing trial: the protocol is established, participating clinical centers are enrolled, the data collection process and reporting system for adverse events are in place. Cohen et al. (2015) conducted an in depth review of trials that had added a treatment arm and concluded

> "Only a very small number of trials were identified to have added arms in practice, indicating that although this type of amendment may be advantageous, it is very rarely implemented. Of the trials that had added an arm, some failed to adequately address the statistical issues and suffered from lack of power and difficulties of interpretability. However, it is clear that this type of amendment is desirable and advantageous, with the statistical and logistical issues seeming by no means insurmountable."

Umbrella, basket, and platform trials are becoming established as methods for comparing multiple treatments or the effect of a treatment on multiple indications, and dropping and adding treatments is a key feature of such designs; see, for example Simon (2017) for a review of umbrella, basket, and platform designs for oncology clinical trials. For further reading on this topic, see Elm et al. (2012), Ventz et al. (2018), Lee et al. (2019), Lee et al. (2021), Burnett, König & Jaki (2024) and Greenstreet et al. (2024).

27.4 Power, Sample Size and Efficient Designs

27.4.1 Definitions of Power

There are various ways in which one may define success when testing the set of hypotheses $H_j\colon \theta_j \leq 0$, $j = 1, \ldots, J$, about treatment effects $\theta_1, \ldots, \theta_J$. Senn & Bretz (2007) and Bretz, Koenig et al. (2009, Sec. 4.1) consider several definitions of power. Some possibilities are:

A *Global power*

The probability of rejecting H_j for at least one treatment j.

B *Disjunctive power*

The probability of rejecting H_j for at least one treatment j with $\theta_j > 0$.

C *Conjunctive power*

The probability of rejecting H_j for all treatments j with $\theta_j > 0$.

D *Power for the best treatment*

The probability of rejecting H_{j^*} where $\theta_{j^*} = \max\{\theta_1, \ldots, \theta_J\} > 0$.

E *Power with selection of the best treatment*

When only one treatment is selected at an interim analysis, the probability of selecting treatment j^* where $\theta_{j^*} = \max\{\theta_1, \ldots, \theta_J\} > 0$, and then rejecting H_{j^*}.

If the aim is to find the most effective treatment, then aims [D] and [E] may be the most appropriate. However, Bretz, Koenig et al. (2009, p. 1192) note that in reality, the overall quality of a treatment can depend on safety, economic and other considerations, in which case aim [C] may be more appropriate. Definition [A] of "global power" is not to be recommended since it treats false positive results as successes.

27.4.2 Achieving a Specified Power

A power requirement can be specified by stating the definition of power to be used and the level of power to be achieved under a certain configuration of treatment effects $\theta_1, \ldots, \theta_J$. For a given form of adaptive design, simulation may be used to determine the sample size per treatment in each stage of the trial that is needed to produce the desired power.

Posch et al. (2005, Sec. 6) illustrate this process in designing a trial to compare three treatments with a control, using a closed testing procedure with Simes tests for intersection hypotheses within each stage and a weighted inverse normal combination test for each hypothesis and intersection hypothesis. The response is binary, the trial has one interim analysis at which either one treatment is selected or the trial stops for futility. Simulation results show the disjunctive power and the probability of rejecting the null hypothesis for each individual treatment.

Wassmer (2011) suggests simplifying the design process by assuming a simple Bonferroni correction will be applied to adjust for multiple comparisons. Since a more sophisticated procedure, using a Dunnett test for example, will have higher power, one may go on to search for a smaller sample size that achieves the required power, taking the sample size for the Bonferroni procedure as a starting point.

27.4.3 Efficient Trial Designs

Wason et al. (2016) note that MAMS trials are implicitly efficient compared to multiple two-arm trials because

> "A shared control group can be used, instead of a separate control group for each treatment;
>
> a direct head-to-head comparison of treatments is conducted, minimising biases that can be introduced from making comparisons between treatments tested in separate trials;
>
> the use of interim analyses allows ineffective treatments to be dropped early, or early stopping of the trial if one treatment is clearly superior."

Options remain within the class of MAMS designs. In Section 21.4.2 we discussed efficiency of MAMS designs with clearly defined criteria for dropping or retaining treatments. It is more difficult to define an efficient *flexible* design since the way in which decisions will be made during the trial is not completely predictable. One can, however, aim to be efficient in what are deemed to be the most likely scenarios. Bretz, Koenig et al. (2009, Sec 4.3) suggest giving probabilities to certain events, for example, the treatment with the best efficacy outcomes at an interim analysis may be assigned a probability 0.3 of being dropped due to poor safety results.

Comparisons of different types of design have been reported by, for example, Bretz, Koenig et al. (2009, Sec. 5), Wassmer (2011, Sec. 3), Magirr, Stallard & Jaki (2014, Sec. 3) and Ghosh et al. (2017, Sec. 6). Results in these papers and our own investigations support certain conclusions: multi-arm trials are more efficient than separate two-arm trials; in testing intersection hypotheses, the Bonferroni adjustment leads to less power than a Simes test, which in turn is less powerful than a Dunnett test; the choice of weights used in a combination test has only a modest impact on efficiency. Increasing the proportion of subjects allocated to the control arm can lead to small gains in power but such an allocation ratio may be undesirable for other reasons, in particular, when the goal of a trial is to gather evidence to support the efficacy and safety of a selected treatment, a large sample size on the new treatment is desirable to detect serious but rare safety events. In an inferentially seamless Phase II/III trial, the contribution from Phase II data to the final analysis increases with the size of the Phase II component: Bretz, Koenig et al. (2009) recommend that the size of the Phase II stage relative to the Phase III stage should be larger than is customary in current practice.

It is informative to compare the flexible designs we have described in the chapter with the more constrained procedures of Chapter 21. Wassmer (2011) contrasts the flexible approach with that of Stallard & Todd (2003), in which the number of treatment arms selected at an interim analysis is pre-specified and the selection rule is to select the treatment arms with the highest responses. An advantage of Stallard & Todd's methods is that sufficient test statistics are used, rather than combination tests of Dunnett P-values for intersection hypotheses. However, Friede & Stallard (2008) find this advantage does not result in a substantial gain in power Wassmer concludes "in our opinion this property is not really convincing and a more flexible approach should be preferred."

Ghosh et al. (2017, Sec. 6) reach a different conclusion. They consider trial designs with three equally spaced analyses in which 3, 4, and 5 treatments are tested against a control. The number of treatments retained after each analysis is data dependent but follows a fixed rule. A testing procedure of the form proposed by Magirr, Jaki & Whitehead (2012) is compared with methods that apply a closed testing procedure, using combination tests to combine P-values across stages. The results of Ghosh et al. show that the former method dominates those based on combining P-values: in the most extreme case, a design with five treatment arms and a control, the difference in disjunctive power is 0.835 versus 0.750. These findings add weight to the advice given by Bretz, Koenig, et al. (2009):

> "When it comes to the decision, whether or not to apply an adaptive design, it is important to ensure that the finally chosen design adequately addresses the primary study objectives. The conduct of extensive simulations, similar to the ones reported here, are then of critical importance to understand and compare the operating characteristics of different design options and interim decision strategies. In the end, the final decision depends on many considerations and not all of them are of statistical nature .... But well-planned and extensive clinical trial scenario evaluations are likely to support the interdisciplinary discussions. Substantial lead time for their conduct should be devoted at the planning stage."

27.5 Cumulative MAMS Designs

Ghosh, Liu & Mehta (2020) describe a method for constructing flexible multi-arm multi-stage designs which they refer to as the "cumulative MAMS approach". This is a closed testing procedure in which the test of each intersection hypothesis H_I is based on the maximum score statistic for treatments $j \in I$, calculated from the cumulative data at each analysis. A group sequential test of each H_I is defined under an initial plan in which all treatments remain in the trial. Then, if a treatment is dropped or sample sizes for future stages are modified, the group sequential boundary for testing H_I is updated so as to preserve the conditional Type I error probability when $\theta_j = 0$ for all $j \in I$. A key difference from the "stage-wise MAMS" designs we have described thus far is that decisions at each stage are based on the sufficient statistics for each treatment effect. As a consequence of this, the computations required are substantial, but the quasi-Monte Carlo integration methods of Ghosh et al. (2017) make the cumulative MAMS approach feasible.

Ghosh, Liu & Mehta (2020) report significant gains in power compared with stage-wise MAMS designs that apply a combination test to each intersection hypothesis H_I. They note that the increases in power can translate into large sample size savings. As an example, in one case of a three-stage design comparing three treatments with a control, the maximum sample size is 388, the average sample size is 343, the cumulative MAMS design has disjunctive power 0.647, but the stage-wise MAMS design only has power 0.592. In order for the stage-wise MAMS design to achieve power 0.647, the maximum sample size would have to be increased to 448 and the average sample size would rise to 424.

Ghosh, Liu & Mehta suggest several reasons for the high efficiency of their designs. They conjecture that the way in which the correlation between treatment effect estimates is handled plays a significant role. Thus, the cumulative MAMS approach may also improve the efficiency of two-arm trials investigating the treatment effect in several sub-populations where estimates for overlapping sub-populations are correlated. In view of the significant gains in efficiency already demonstrated, further development of this approach in designs of trials comparing multiple treatments or with multiple sub-populations may be particularly rewarding.

27.6 Inference on Termination of a MAMS Trial

A seamless Phase II/III trial with treatment selection can be considered as a special case of a MAMS design with two stages. Thus, the methods that we discussed in Section 26.6 for estimating the treatment effect of a treatment selected in such a trial, have the potential to be extended to more general MAMS designs. However, in many cases, derivations rely on special features of a two-stage design and the generalization to trials with three or more stages is either difficult or impossible.

Bowden & Glimm (2014) extend methods for calculating conditionally unbiased estimates to MAMS designs with more than 2 stages. Here, the conditioning is on the sets of treatments that remain in the trial after each analysis and an ordering within each set of treatments dropped at the same analysis. Stallard & Kimani (2018) derive uniform minimum variance conditionally unbiased estimates of the treatment effects for those treatments remaining at the end of the trial, where conditioning is on the sets of treatments remaining after each analysis. Stallard & Kimani require the rule for dropping treatments to be specified in advance and they comment that their method may not yield conditionally unbiased estimators when a more flexible approach is taken, observing that "In a fully flexible approach, it is not clear that it is even possible to define the bias, as this would be an expectation over an unspecified sample space."

The construction of confidence intervals for treatment effects after a flexible MAMS trial, with control of the family-wise error rate, is challenging. One possible solution is based on the method described in Section 26.6.3, which is a particular implementation of the scheme proposed by Posch et al. (2005). In a MAMS trial comparing J treatment arms against a control, with treatment effects denoted by θ_j, $j = 1, \ldots, J$, we set up hypotheses

$$H_j(\tilde{\theta}_j)\colon \theta_j \leq \tilde{\theta}_j, \quad j = 1, \ldots, J, \;\; \tilde{\theta}_j \in R,$$

and define a group sequential closed testing procedure for these hypotheses with FWER α. For each $j = 1, \ldots, J$, we find the set of values $\tilde{\theta}_j$ for which every intersection hypothesis that includes $H_j(\tilde{\theta}_j)$ is rejected by the closed testing procedure, and the upper confidence interval $(\theta_{L,j}, \infty)$ for θ_j is defined as the complement of this set. It follows that

$$Pr_{\boldsymbol{\theta}}\{\theta_j \in (\theta_{L,j}, \infty) \text{ for all } 1 \leq j \leq J\} \geq 1 - \alpha \quad \text{for all } \boldsymbol{\theta} = (\theta_1, \ldots, \theta_J).$$

As we saw in Section 26.6.3, due to the nesting of hypotheses concerning a particular θ_j, it suffices to consider general intersection hypotheses of the form

$$H_J(\tilde{\boldsymbol{\theta}}_J) = \cap_{j \in J} H_j(\tilde{\theta}_j),$$

where J is a subset of $\{1, \ldots, p\}$ and $\tilde{\boldsymbol{\theta}}_J$ contains a set of values $\tilde{\theta}_j$ for $j \in J$.

For each intersection hypothesis $H_J(\boldsymbol{\theta}_J)$, we create a group sequential test based on a sequence of weighted inverse normal combination test statistics with pre-specified weights $w_1, \ldots, w_K$, comparing these with the upper boundary, $(b_1, \ldots, b_K)$, of a Lehmacher-Wassmer group sequential test with Type I error rate α. In each stage k for which treatment j remains in the trial, let

$$z_j^{(k)}(\tilde{\theta}_j) = (\hat{\theta}_{k,j} - \tilde{\theta}_j)\sqrt{\mathcal{I}^{(k)}},$$

where $\hat{\theta}_{k,j}$ is the estimate of θ_j based on stage k data alone and the information $\mathcal{I}^{(k)}$ is the reciprocal of the variance of this estimate. We assume sample sizes on each treatment and the control are equal, so the same information $\mathcal{I}^{(k)}$ applies to each treatment j and the correlation between $z_j^{(k)}$ values for two different treatments is 0.5.

Suppose m_k treatments with indices in the set J_k remain in the trial at stage k. We use the Dunnett test to calculate a P-value for each intersection hypothesis. If $j \in J_k$, the largest stage k P-value for testing an interaction hypothesis $H_J(\tilde{\boldsymbol{\theta}}_J)$ that includes $H_j(\tilde{\theta}_j)$ arises when all the $\tilde{\theta}_i$, $i \in J_k$ and $i \neq j$, are large. This Dunnett P-value is

$$P_{j,k}(\tilde{\theta}_j) = P_D(m_k, 0.5, z_j^{(k)}(\tilde{\theta}_j)) \tag{27.1}$$

as defined in Equation (21.8). The corresponding Z-value is

$$Z_{j,k}(\tilde{\theta}_j) = \Phi^{-1}(1 - \tilde{P}_{j,k}(\tilde{\theta}_j))$$

and the adjusted cumulative Z-value is the weighted combination of Z-values up to this point

$$\tilde{Z}_{j,k}(\tilde{\theta}_j) = \frac{\sum_{i=1}^{k} w_i Z_{j,i}(\tilde{\theta}_j)}{\left\{\sum_{i=1}^{k} w_i^2\right\}^{1/2}}.$$

To find the upper confidence interval $(\theta_{L,j}, \infty)$ for θ_j, we consider values of $\tilde{Z}_{j,k}(\tilde{\theta}_j)$ after the last stage k in which treatment j was in the trial. The lower end point of the interval is the value of $\tilde{\theta}_j$ for which

$$\tilde{Z}_{j,k}(\tilde{\theta}_j) = b_k,$$

where b_k is the Lehmacher-Wassmer boundary point.

The confidence intervals produced by this method are valid but they may have some drawbacks. When a MAMS design is used to test the family of hypotheses $H_j(0)$: $\theta_j \leq 0$ with FWER α, it is possible that $H_j(0)$ is rejected but the above method gives an interval $(\theta_{L,j}, \infty)$ for θ_j with $\theta_{L,j} < 0$. This is a natural consequence of expanding the family of hypotheses to include tests of $H_j(\tilde{\theta}_j)$: $\theta_j \leq \tilde{\theta}_j$ for all $\tilde{\theta}_j \in R$. The inconsistency between tests and confidence intervals can be resolved by starting with the larger family of hypotheses and, hence, only rejecting $H_j(0)$ when the test based on stage-wise Z-values defined by (27.1) rejects $H_j(0)$.

Another possible drawback is that if $H_j(0)$: $\theta_j \leq 0$ is rejected at an early analysis k with $\tilde{Z}_{j,k}(\tilde{\theta}_j)$ only a little larger than b_k, the lower confidence limit $\theta_{L,j}$ will be only a little larger than zero. Indeed, if $\tilde{Z}_{j,k}(\tilde{\theta}_j) = b_k$, $\theta_{L,j}$ will be exactly zero. This is in contrast to the confidence intervals of Section 9.5 for the case where a single treatment is compared with a control: there the lower endpoint of the interval is highest when the trial terminates to reject the null hypothesis at the first analysis. Given the flexibility desired in a MAMS design, it may not be possible to base a confidence interval on an ordering on the space of possible outcomes, as in Section 9.5. However, further progress in this direction would be desirable.

27.7 Bibliography and Notes

An early presentation of a MAMS trial design is given by Royston, Parmar & Qian (2003). These authors consider the design of a multi-arm clinical trial with survival outcomes and a particular application in a study investigating treatments for ovarian cancer. The MRC STAMPEDE trial reported by Sydes, Parmar, James et al. (2009) compared treatments for prostate cancer and had 5 stages and 4 treatment arms. This was one of the first examples of a trial that used multiple arms and stages synchronously.

Parmar, Sydes, Cafferty et al. (2017) give an illuminating overview of platform, umbrella and basket trials. They define an umbrella protocol as "one which aims to include patients with a single disease divided into subsets, often with different therapies being explored in each of these subsets." In contrast, a basket protocol is "one which includes patients with many different diseases, but the patients are linked because there is a treatment which targets a pathway which is common to all these patients." Lawler, Hochman & Zarychanski (2022) write of experiences with adaptive platform trials assessing treatments for patients hospitalized for COVID-19. They write

> "The term platform generally implies the use of a master protocol as a foundation to study multiple treatments for 1 disease; for example, specifying the study population, key enrollment features, outcome measures, etc, on top of which multiple interventions can be studied."

Korn & Freidlin (2017) discuss advantages and disadvantages of different types of umbrella trial, basket trial and master protocol. They comment in particular on: the BATTLE-2

Phase II trial that compared four treatments for advanced non-small cell lung cancer; the I-SPY 2 Phase II trial that assessed treatments for breast cancer in eight patient subgroups; and the FOCUS4 Phase II/III trial for patients with advanced or metastatic colorectal cancer that used an umbrella design in which each patient was assigned to one of five sub-studies.

In contrast to the MAMS designs we have described, a more complex master protocol may see new treatments added and new subsets of patients defined as time progresses. How the Type I error rate should be controlled in such a framework has been a point of debate. Bratton, Parmar, Phillips et al. (2016), argue for the control of pairwise error (PWER). Wason et al. (2016) write

> "Although we recommend that the FWER of the design should be specified and controlled in confirmatory trials, there are contrary opinions. Freidlin et al. (2008) advocate not adjusting multi-arm trials for multiple testing at all when the different arms correspond to different treatments. ... Although this argument has merit, we feel that the situation of conducting a MAMS trial is conceptually quite different to running a series of separate trials."

and they go on to note that

> "The MRC STAMPEDE trial (Sydes, Parmar & James, 2009) does not explicitly control or specify the FWER, but instead controls the pairwise type-I error rate, i.e. the type-I error rate of a test of one experimental treatment against the control treatment. Since this pairwise type-I error rate is low (0.013) and early stopping for efficacy is not allowed, it is likely that the overall FWER is low. For exploratory MAMS trials (for example in phase II), controlling the FWER would not be required by regulatory bodies. However, we believe that the FWER is a more relevant quantity than the pairwise type-I error rate associated with each experimental treatment. The FWER provides the maximum probability of recommending an ineffective treatment, which is important if a phase III trial is to be carried out subsequently. An additional reason to consider designing such trials with FWER control is due to the increased use of phase II studies as the second pivotal study when making a confirmatory claim."

It is clear that there is substantial current interest in platform, umbrella and basket trials and research continues on novel aspects of such designs. We anticipate significant developments in this area as challenges are identified and problems solved.

Chapter 28

Enrichment Designs: Targeting Therapies to Selected Sub-populations

28.1 Introduction

We consider trials comparing a new treatment with a control, where the target patient population is heterogeneous and, in view of the mode of the action of the new treatment or results from previous trials, it is suspected that only certain subgroups may benefit from the new treatment. Such subgroups may be determined by demographic, clinical or genomic characteristics.

Byar & Corle (1977) posed the question "which treatment is best for which kinds of patients?" and proposed fitting a model with treatment by covariate interactions to answer this question. Simon & Wang (2006) discuss the use of genetic biomarkers in oncology that can help identify those patients for whom a particular treatment will be effective. Fournier et al. (2010) report a meta-analysis of trials involving antidepressant medication for depression: they conclude that the efficacy of antidepressant treatment varies as a function of symptom severity, with little or no benefit for patients with mild, moderate and severe baseline symptoms, but a large benefit for patients with very severe symptoms.

Mehta et al. (2014, Table I) list a number of oncology products approved in the USA for selected populations. They note

> "The dilemma for the investigator planning a Phase III confirmatory trial for a targeted therapy is whether to open enrollment to all patients regardless of biomarker status or to restrict enrollment to a targeted subgroup based on a biological understanding of the mechanism of action. Restricting enrollment to the targeted subgroup without sufficient empirical evidence of lack of efficacy in the non-targeted subgroup may deny a large segment of the population access to a potentially beneficial treatment. On the other hand, by running a large trial in a heterogeneous population, the treatment effect may be diluted, resulting in an underpowered study. One way to resolve this dilemma is to start out by enrolling all patients, regardless of biomarker status. At a suitable time point, an interim analysis is performed, and a decision is taken to either continue enrollment to both subgroups, continue enrollment to the targeted subgroup only (population enrichment), or terminate the trial for futility."

In this chapter we shall describe methods for conducting an "enrichment" design, whose purpose is to address the issues outlined above. We shall focus on the case where patient subgroups are defined in advance. Although subgroup analyses are regularly reported, positive findings in subgroups defined after analyzing response data are often regarded with suspicion since they may be the result of "data dredging" or "cherry picking"; see, for example, the comments of Yusuf et al. (1991), Pocock et al. (2002) and Wang, O'Neill & Hung (2007). Peto (1990) illustrates this point neatly in an analysis of the ISIS-2 study that showed the effect of experimental treatments for acute myocardial infarction to be

DOI: 10.1201/9781584888482-28

significant for most patients but not for those born under the astrological signs Libra and Gemini.

In Section 28.2, we follow the general approach of Chapter 25 in explaining how a closed testing procedure and combination tests can be used to create a two-stage enrichment design in which certain subgroups are dropped after the first stage and sample sizes in remaining groups increased. With pre-defined rules for dropping sub-populations and increasing the sample size and no break between the stages, we might label the stages as Phase II and Phase III and term this a "seamless Phase II/III trial", similar to the seamless designs of Chapter 26 but with subgroup selection rather than treatment selection. In Section 28.3 we consider the potential gains from using an adaptive enrichment design. In the process, we discuss the possible goals of such designs and methods for optimizing designs to achieve specific goals. In Section 28.4 we describe enrichment designs where the enrichment decision is based on a short-term endpoint and in Section 28.5 we consider trials where there may be multiple subgroups, also considering group sequential designs with multiple analyses. In Section 28.6 we discuss methods for producing point estimates and confidence intervals following an adaptive enrichment trial. We report on the views expressed by regulators on enrichment designs in Section 28.7 and conclude with some additional references and discussion in Section 28.8.

28.2 Enrichment Designs Using a Closed Testing Procedure with Combination Tests

28.2.1 An Enrichment Design and Testing Procedure

Consider a clinical trial conducted to compare an experimental treatment against a control with patients drawn from a defined population Ω_1. Suppose this population is heterogeneous and subgroups $\Omega_2, \ldots, \Omega_p$ are specified, based on a biomarker or other patient characteristics. These $p-1$ sub-populations may be disjoint or nested, or they could overlap with each other in various ways. We assume that within each sub-population, patients are assigned randomly to the two treatments and, for $i = 1, \ldots, p$, we denote the treatment effect, a measure of the difference in response distributions between the experimental and control treatments in Ω_i, by θ_i. The parameter θ_1, sometimes referred to as the treatment effect for "all comers", can be viewed as a weighted average of treatment effects over the whole population Ω_1.

Tests of the hypotheses H_i: $\theta_i \leq 0$, $i = 1, \ldots, p$, are to be conducted where the alternative $\theta_i > 0$ indicates that the experimental treatment is superior to the control in sub-population i. The testing procedure should protect the family-wise error rate (FWER) at a specified level α. In a two-stage enrichment design, patients are recruited from the full population, Ω_1, in the first stage. Based on the data observed in stage 1, a set of sub-populations is chosen to go forward to stage 2. We denote the set of indices of these sub-populations by I^* and the union of selected sub-populations by

$$\Omega^* = \bigcup_{i \in I^*} \Omega_i.$$

This set Ω^* may or may not be one of the original sub-populations, $\Omega_1, \ldots, \Omega_p$. Schmidli et al. (2006) call the two stages of such a trial Phase II and Phase III and refer to this type of study as another example of a seamless Phase II/III trial.

In order to protect the FWER, we follow the approach of Chapter 25 and apply a closed testing procedure with combination tests across the two stages. Schmidli et al. (2006) describe this approach in the case of a full population Ω_1 and one sub-population Ω_2, but the method extends easily to more sub-populations.

Testing intersection hypotheses

Let $P_1^{(1)}, \ldots, P_p^{(1)}$, denote the P-values for testing the hypotheses $H_1, \ldots, H_p$ from stage 1 data. If it has been decided to test hypotheses H_i with $i \in I^*$ in stage 2, we denote the P-values for these hypotheses based on stage 2 data by $P_i^{(2)}$, $i \in I^*$.

We could construct P-values for intersection hypotheses in each stage using the simple Bonferroni method. However, since treatment effect estimates for disjoint sub-populations are independent and estimates for overlapping sub-populations are positively correlated, the more powerful method of Simes (1986), as described in Section 20.4, can be used. After stage 1, for each $J \subseteq \{1, \ldots, p\}$ we define a P-value $P_J^{(1)}$ for the intersection hypothesis $H_J = \cap_{j \in J} H_j$ based on $\{P_j^{(1)} \colon j \in J\}$. If H_J is the intersection of m simple hypotheses and the ordered P-values for these m hypotheses are

$$P_{(1)}^{(1)} \leq P_{(2)}^{(1)} \leq \ldots \leq P_{(m)}^{(1)},$$

Simes' method gives

$$P_J^{(1)} = \min_{i=1,\ldots,m} (m\, P_{(i)}^{(1)}/i).$$

After stage 2 we need to define a P-value for each intersection hypothesis H_J where J contains an index $i \in I^*$. For those cases where $J \subseteq I^*$, we have a P-value $P_i^{(2)}$ for each $i \in J$ and we can apply Simes' method to these values, just as in stage 1. If J also contains indices that are not in I^*, we follow the procedure described in Section 25.2 and set

$$P_J^{(2)} = P_{J \cap I^*}^{(2)}.$$

The combination test

At the end of the study, we create an overall P-value for each intersection hypothesis H_J where J contains an index $i \in I^*$. Using a weighted inverse normal combination test with pre-specified positive weights w_1 and w_2 satisfying $w_1^2 + w_2^2 = 1$ gives the overall P-value

$$C(P_J^{(1)}, P_J^{(2)}) = 1 - \Phi\{w_1\, \Phi^{-1}(1 - P_J^{(1)}) + w_2\, \Phi^{-1}(1 - P_J^{(2)})\}.$$

In applying the closed testing procedure, it is a simple matter to check which intersection hypotheses H_J are rejected with

$$C(P_J^{(1)}, P_J^{(2)}) \leq \alpha$$

and we can then make overall decisions, rejecting an elementary hypothesis H_j when H_J has been rejected for every set J containing the index j.

Within this process, it is necessary to decide which sub-population hypotheses should be tested in stage 2. This choice will reflect a trade-off between increasing power to demonstrate a treatment effect in a smaller population and retaining the prospect of a positive result in a larger population. The quote from Mehta et al. (2014) in Section 28.1 summarizes the dilemma for investigators. In practice, the decision should be based on the magnitudes of the observed treatment effects in different subgroups and on the relevant medical and economic considerations. The closed testing procedure is flexible in that it controls the FWER however this decision is made.

28.2.2 An Example with $K = 2$ Stages and $p = 4$ Populations.

The following illustrative example is taken from Jennison & Turnbull (2007, Sec. 6). In this example, the population Ω_1 and sub-populations Ω_2, Ω_3 and Ω_4 are

Table 28.1 *Observed P-values in the example of sub-population selection. In view of stage 1 data, stage 2 recruitment is restricted to "Men only".*

	Stage 1 results	*Stage 2 results*
Full population	$P_1^{(1)} = 0.20$	
All men	$P_2^{(1)} = 0.10$	$P_2^{(2)} = 0.11$
Men over 50	$P_3^{(1)} = 0.03$	$P_3^{(2)} = 0.08$
Men who smoke	$P_4^{(1)} = 0.03$	$P_4^{(2)} = 0.03$

Ω_1: The entire population,

Ω_2: Men only,

Ω_3: Men over 50,

Ω_4: Men who are smokers.

The elementary null hypotheses are H_i: $\theta_i \leq 0$, $i = 1, \ldots, 4$, where θ_i is the treatment effect in population Ω_i. Table 28.1 shows the observed P-values for tests of H_1 to H_4 based on stage 1 and stage 2 data. For simplicity we have written H_1 rather than $H_{\{1\}}$, H_{12} rather than $H_{\{1,2\}}$, etc., in this table and in Table 28.2. In their example, Jennison & Turnbull (2007) explain that, in view of the high value of $P_1^{(1)}$, recruitment in stage 2 was restricted to "Men only" and only the hypotheses H_2, H_3 and H_4 were to be tested at the end of the trial.

Table 28.2 *Example: Tests of intersection hypotheses*

Intersection	*Stage-wise P-values*		*Combined*	
hypothesis, H_J	$P_J^{(1)}$	$P_J^{(2)}$	Z_J	P_J
H_1	0.20	—	—	—
H_2	0.10	0.11	1.77	0.038
H_3	0.03	0.08	2.32	0.010
H_4	0.03	0.03	2.66	0.004
H_{12}	0.20	0.11*	1.46	0.072
H_{13}	0.06	0.08*	2.09	0.018
H_{14}	0.06	0.03*	2.43	0.008
H_{23}	0.06	0.11	1.97	0.025
H_{24}	0.06	0.06	2.20	0.014
H_{34}	0.03	0.06	2.43	0.008
H_{123}	0.09	0.11*	1.82	0.035
H_{124}	0.09	0.06*	2.05	0.020
H_{134}	0.045	0.06*	2.30	0.011
H_{234}	0.045	0.09	2.15	0.016
H_{1234}	0.06	0.09*	2.05	0.020

$Z_J = 1 - \Phi\{\sqrt{0.5}\,\Phi^{-1}(1 - P_J^{(1)}) + \sqrt{0.5}\,\Phi^{-1}(1 - P_J^{(2)})\}$,

$P_J = 1 - \Phi(Z_J)$

*Value for $P_{\{1\}\cup I}^{(2)}$ in stage 2 equals $P_I^{(2)}$ for $I \subseteq \{2, 3, 4\}$.

Table 28.2 shows the results of analyzing these data following the method described above, using Simes tests for intersection hypotheses and a weighted inverse normal test with weights $w_1 = w_2 = \sqrt{0.5}$ that were specified before observing any data. The choice $w_1 = w_2$ is appropriate if the two stages are to have equal sample sizes. However, the analysis remains valid if the stage 2 sample size is modified in light of stage 1 data. Since there is no value for $P_1^{(2)}$, the stage 2 P-value for a hypothesis H_J where J contains the index 1 is set equal to the P-value for the hypothesis $H_{J'}$ where $J' = J/\{1\}$.

In applying the closed testing procedure, a hypothesis H_i is rejected at the global level $\alpha = 0.025$ if all intersection hypothesis H_J with $i \in J$ are rejected at this level. Since all hypotheses H_J with $4 \in J$ have a combined P-value $P_J < 0.025$, the elementary hypothesis H_4 is rejected in favor of $\theta_4 > 0$ in the closed testing procedure with FWER $\alpha = 0.025$. However, since $H_{\{1,2,3\}} = 0.035$, neither H_2 nor H_3 is rejected.

Adjusted combined P-values.

Applying the definition given in Section 20.7, the adjusted P-value, $\tilde{P}_i$, for testing H_i with protection of the family-wise error rate is the maximum combined P-value for an intersection hypothesis H_J with $i \in J$. Thus for testing H_2, which concerns the sub-population Ω_2 of "All men", we have the adjusted P-value

$$\tilde{P}_2 = \max\{P_2, P_{12}, P_{23}, P_{24}, P_{123}, P_{124}, P_{234}, P_{1234}\} = 0.072.$$

In testing H_3 for sub-population Ω_3 comprising "Men over 50", we obtain

$$\tilde{P}_3 = \max\{P_3, P_{13}, P_{23}, P_{34}, P_{123}, P_{134}, P_{234}, P_{1234}\} = 0.035$$

and for the test of H_4 for sub-population Ω_3, "Men who smoke", we have

$$\tilde{P}_4 = \max\{P_4, P_{14}, P_{24}, P_{34}, P_{124}, P_{134}, P_{234}, P_{1234}\} = 0.020.$$

It follows from the definition that for each of H_2, H_3 and H_4, the adjusted P-value is less than 0.025 if and only if the elementary hypothesis is rejected in the overall testing procedure. So, these adjusted P-values are consistent with the fact that only H_4 is rejected by the closed testing procedure.

28.3 Assessing the Benefits of Enrichment Designs

Burnett & Jennison (2021) report a study of the benefits that can be achieved by introducing adaptive enrichment into a trial design when a new treatment is expected to be particularly effective in a certain sub-population. They consider a Phase III trial comparing a new therapy, treatment A, against a control, treatment B. The patient population contains a sub-population S_1 of biomarker positive patients in which a high treatment effect is anticipated, and the complement of this sub-population is S_2. For $j = 1$ and 2, responses for patients in S_j on treatment A follow a $N(\mu_{Aj}, \sigma^2)$ distribution, while responses on treatment B are distributed as $N(\mu_{Bj}, \sigma^2)$. The treatment effects in sub-populations S_1 and S_2 are $\theta_1 = \mu_{A1} - \mu_{B1}$ and $\theta_2 = \mu_{A2} - \mu_{B2}$, respectively.

When S_1 represents a fraction λ of the full population, the overall treatment effect in the full population is $\theta_3 = \lambda\theta_1 + (1-\lambda)\theta_2$. Burnett & Jennison write $\boldsymbol{\theta} = (\theta_1, \theta_2)$, noting that $\boldsymbol{\theta}$ determines the value of θ_3. They assume the investigators are interested in testing H_1: $\theta_1 \le 0$ against $\theta_1 > 0$ and H_3: $\theta_3 \le 0$ against $\theta_3 > 0$ but the hypothesis H_2: $\theta_2 \le 0$ is not to be tested.

The trial is conducted in two stages. On completion of stage 1, estimates $\hat{\theta}_1^{(1)}$, $\hat{\theta}_2^{(1)}$ and $\hat{\theta}_3^{(1)} = \lambda\hat{\theta}_1^{(1)} + (1-\lambda)\hat{\theta}_2^{(1)}$ are obtained and a decision is made about which patients should be recruited in stage 2. If it is decided to continue with sub-population S_1 only, the total sample size is kept fixed and the number of patients from sub-population S_1 in stage 2 is increased accordingly, enriching that subgroup.

Tests of H_1 and H_3 are defined using a closed testing procedure to control the FWER, so that

$$Pr_{\boldsymbol{\theta}}\{\text{Reject at least one true null hypothesis}\} \leq \alpha \quad \text{for all } \boldsymbol{\theta}.$$

We denote by $P_1^{(1)}$ and $P_3^{(1)}$ the P-values for testing H_1 and H_3 based on stage 1 data. Since $P_1^{(1)}$ and $P_3^{(1)}$ come from nested groups of patients, these P-values are positively associated and Simes' test gives a valid, but conservative, P-value for testing the intersection hypothesis H_{13}, namely

$$P_{13}^{(1)} = \min\{2\min(P_1^{(1)}, P_3^{(1)}),\ \max(P_1^{(1)}, P_3^{(1)})\}.$$

If enrichment occurs, $P_1^{(2)}$ is obtained from the stage 2 data and we set $P_{13}^{(2)} = P_1^{(2)}$. If the trial continues with the full population we obtain $P_1^{(2)}$ and $P_3^{(2)}$ from the stage 2 data and use Simes' test to obtain

$$P_{13}^{(2)} = \min\{2\min(P_1^{(2)}, P_3^{(2)}),\ \max(P_1^{(2)}, P_3^{(2)})\}.$$

In applying weighted inverse normal combination tests with the pre-specified weights w_1 and w_2, we first define

$$\begin{aligned} Z_1^{(1)} &= \Phi^{-1}(1 - P_1^{(1)}), & Z_1^{(2)} &= \Phi^{-1}(1 - P_1^{(2)}), \\ Z_3^{(1)} &= \Phi^{-1}(1 - P_3^{(1)}), & Z_3^{(2)} &= \Phi^{-1}(1 - P_3^{(2)}), \\ Z_{13}^{(1)} &= \Phi^{-1}(1 - P_{13}^{(1)}), & Z_{13}^{(2)} &= \Phi^{-1}(1 - P_{13}^{(2)}), \end{aligned}$$

and then set the combined Z-values to be

$$Z_1^{(c)} = w_1 Z_1^{(1)} + w_2 Z_1^{(2)}, \quad Z_3^{(c)} = w_1 Z_3^{(1)} + w_2 Z_3^{(2)} \quad \text{and} \quad Z_{13}^{(c)} = w_1 Z_{13}^{(1)} + w_2 Z_{13}^{(2)}.$$

Finally, the closed testing procedure rejects H_1 if

$$Z_1^{(c)} \geq \Phi^{-1}(1-\alpha) \quad \text{and} \quad Z_{13}^{(c)} \geq \Phi^{-1}(1-\alpha)$$

and H_3 is rejected if

$$Z_3^{(c)} \geq \Phi^{-1}(1-\alpha) \quad \text{and} \quad Z_{13}^{(c)} \geq \Phi^{-1}(1-\alpha).$$

The stage 1 data are summarized as $X_1 = (\hat{\theta}_1^{(1)}, \hat{\theta}_1^{(2)})$ and the enrichment decision is determined by the rule d, where

$$\begin{aligned} d(X_1) = 1 &\quad \Rightarrow \quad \text{Enrich in stage 2,} \\ d(X_1) = 2 &\quad \Rightarrow \quad \text{Do not enrich in stage 2.} \end{aligned}$$

If $d(X_1) = 1$, enrichment occurs and $X_2 = \hat{\theta}_1^{(2)}$, while if $d(X_1) = 2$ enrichment does not occur and $X_2 = (\hat{\theta}_1^{(2)}, \hat{\theta}_2^{(2)})$. In either case we write $X = (X_1, d(X_1), X_2)$ to summarize the full set of data at the end of the study and the decision taken at the interim analysis. We define $\mathcal{R}_1$ as the indicator variable of the event that H_1 is rejected but H_3 is not rejected, and $\mathcal{R}_3$ as the indicator that H_3 is rejected, where both $\mathcal{R}_1$ and $\mathcal{R}_3$ are functions of X.

Burnett & Jennison (2021) define a gain function

$$G(\boldsymbol{\theta}, X) = \lambda\theta_1\mathcal{R}_1 + \theta_3\mathcal{R}_3 \tag{28.1}$$

that rewards decisions in favor of the new treatment, weighted by the size of the population for which a positive treatment effect is found and the size of the treatment effect in that population. Letting $\pi(\boldsymbol{\theta})$ denote the density function of a prior distribution for $\boldsymbol{\theta}$, they use simulation to derive the decision rule that maximizes the expected gain

$$E\{G(\boldsymbol{\theta}, X)\} = \int \pi(\boldsymbol{\theta}) E_{\boldsymbol{\theta}}\{G(\boldsymbol{\theta}, X)\}\, d\boldsymbol{\theta}.$$

Burnett & Jennison then compare the expected gain for the optimized adaptive design with that of two designs which do not have adaptive enrichment. They consider two such designs with the same total sample size as the adaptive enrichment design. The first is a Fixed Full population (FF) design which has a single stage and a fraction λ of patients from sub-population S_1, and which uses a closed testing procedure to test H_1 and H_3. The second is a Fixed Sub-population (FS) design in which all patients are recruited from the sub-population S_1 and only H_1 is tested.

Burnett & Jennison first consider examples in which $\lambda = 0.5$, the adaptive analysis takes place after half the total sample size has been observed, weights $w_1 = w_2 = \sqrt{0.5}$ are used in the combination test, and the prior π places probability mass 1 at a single point $\boldsymbol{\theta}_0 = (\theta_{0,1}, \theta_{0,2})$. They find the FF design to be the best of the three designs when $\theta_{0,3} = 0.5(\theta_{0,1} + \theta_{0,2})$ is large or when $\theta_{0,1}$ is only a little larger than $\theta_{0,2}$. The FS design is the best when $\theta_{0,2}$ is small and $\theta_{0,1}$ is substantially larger than $\theta_{0,2}$. In the region of $\boldsymbol{\theta}_0$ values where the adaptive enrichment (AE) design is optimal, the increases in expected gain over the better of the FF and FS designs is modest. Since the prior already determines the values of $\boldsymbol{\theta}$ in these cases, the benefits of adaptation in this case arise from adapting to the observed data and how they will affect the final hypothesis tests, rather than to knowledge about $\boldsymbol{\theta}$.

One might expect adaptation to be more beneficial when the prior for $\boldsymbol{\theta}$ is dispersed and the interim data provide useful information about the true value of $\boldsymbol{\theta}$. However, after studying a variety of examples with a bivariate normal prior for $\boldsymbol{\theta}$, Burnett & Jennison state they have "failed to find an example where the AE design is vastly superior to both the FS and FF designs." They attribute this to the fact that the interim data provide only limited information about $\boldsymbol{\theta}$, so in cases where the prior variance is high, considerable uncertainty about θ_1 and θ_2 remains at the interim analysis, while if the prior variance is low, the interim data have little impact on the posterior distribution of θ_1 and θ_2 and results are similar to those for one-point priors. On a more positive note, they remark that "the AE design shows greater selectivity and is less likely to conclude the new treatment is beneficial to the full population when the treatment effect in S_2 is small or absent altogether." Burnett & Jennison also note that investigators may be unwilling to restrict attention to the sub-population S_1 from the outset, even if a large treatment effect in S_2 is deemed unlikely, and then an AE design offers the opportunity to adjust when unfavorable results in S_2 are observed at an interim analysis.

Burnett & Jennison also consider trials with a range of sub-population sizes and they explore the effect of making the enrichment decision at different times during the trial. These results still do not provide an example where the optimized AE design is vastly superior to both the FF and FS designs.

Ondra et al. (2019) had considered trial designs for the same testing problem in which recruitment rates in the two sub-populations could differ from the proportions in the underlying population and, in a two-stage design, recruitment rates in stage 2 were adapted in the light of stage 1 data. They used a simple Bonferroni correction to control the FWER, testing both H_1: $\theta_1 \leq 0$ and H_3: $\theta_3 \leq 0$ at significance level $\alpha/2$. With this approach to multiple testing, a design can be optimized using numerical integration and dynamic programming. Ondra et al. proposed two utility functions: the first, representing a societal perspective, is similar to (28.1) but includes a cost of sampling proportional to the number of patients in the trial; the second, from a sponsor perspective, replaces $\boldsymbol{\theta}$ by an estimate $\hat{\boldsymbol{\theta}}$ since the price of the new drug will be based on the estimated effect size. The inclusion of a cost per patient prompts early stopping for futility at the interim analysis when both $\hat{\theta}_3^{(1)}$ and $\hat{\theta}_1^{(1)}$ are low. The results presented by Ondra et al. show that optimized two-stage designs with adaptive enrichment can have significantly higher expected gain than optimized one-stage designs. We note, however, that this improvement is due, in part, to the cost savings from early stopping and it is difficult to assess the specific contribution made by adaptive enrichment.

Ballarini et al. (2021) consider an example where the new treatment may be particularly beneficial to either sub-population. Thus, with the above notation, investigators wish to test H_1: $\theta_1 \leq 0$ against $\theta_1 > 0$ and H_2: $\theta_2 \leq 0$ against $\theta_2 > 0$. A closed testing procedure is applied with a weighted Bonferroni test for the intersection hypothesis and the conditional error rate approach, as described in Section 23.6, is used to combine results across stages. The total sample size for the trial is fixed but recruitment rates in the two sub-populations may differ from the proportions in the underlying population and recruitment rates in stage 2 are adapted in response to stage 1 data. With $\mathcal{R}_1$ and $\mathcal{R}_2$ the indicator variables for the events that H_1 is rejected and H_2 is rejected, respectively, the gain function is defined as

$$\lambda \mathcal{R}_1 + (1 - \lambda)\mathcal{R}_2, \tag{28.2}$$

where λ is the prevalence of sub population 1 in the underlying population. Assuming a bivariate normal prior for $\boldsymbol{\theta}$, Ballarini et al. compute adaptive enrichment designs that protect the FWER at level α and maximize the expected gain, optimizing over the partition of the total sample size between stages 1 and 2, the allocation ratio in stage 1, the allocation ratio in stage 2 conditional on stage 1 data, and the weights in the weighted Bonferroni test for the intersection hypothesis. The results for single stage designs in their Figure 2 and for two-stage designs in their Figure 5 show that optimal design parameters are sensitive to the mean and variance of the prior distribution for $\boldsymbol{\theta}$. However, in the examples presented, the maximum expected gain for the optimized single-stage design is very close to that of the optimized two-stage design. Thus, there is a no clear evidence of an adaptive enrichment design improving on the performance of a well-chosen non-adaptive design.

Rosenblum, Fang & Liu (2020) consider the problem addressed by Ballarini et al. (2021) but with an additional test of H_3: $\theta_3 \leq 0$ against $\theta_3 > 0$, where $\theta_3 = \lambda\theta_1 + (1-\lambda)\theta_2$. We shall describe this procedure in terms of the notation we have used previously. In stage 1, $n_1^{(1)}$ and $n_2^{(1)}$ observations are taken from sub-populations S_1 and S_2, respectively, and treatment effects $\hat{\theta}_1^{(1)}$ and $\hat{\theta}_2^{(1)}$ are observed. Stage 2 sample sizes $n_1^{(2)}$ and $n_2^{(2)}$ are then selected from a list of L possible choices. If $n_1^{(2)} = 0$, sub-population S_1 is dropped; if $n_2^{(2)} = 0$, sub-population S_2 is dropped; if both $n_1^{(2)}$ and $n_2^{(2)}$ are zero the trial stops either for futility or for efficacy. After observing the treatment effect estimates $\hat{\theta}_1^{(2)}$ and $\hat{\theta}_2^{(2)}$ in stage 2, the overall estimates

$$\begin{aligned}
\hat{\theta}_1^{(c)} &= \frac{n_1^{(1)}}{n_1^{(1)} + n_1^{(2)}}\,\hat{\theta}_1^{(1)} + \frac{n_1^{(2)}}{n_1^{(1)} + n_1^{(2)}}\,\hat{\theta}_1^{(2)}, \\
\hat{\theta}_2^{(c)} &= \frac{n_2^{(1)}}{n_2^{(1)} + n_2^{(2)}}\,\hat{\theta}_2^{(1)} + \frac{n_2^{(2)}}{n_2^{(1)} + n_2^{(2)}}\,\hat{\theta}_2^{(2)}, \\
\hat{\theta}_3^{(c)} &= \lambda\,\hat{\theta}_1^{(c)} + (1-\lambda)\,\hat{\theta}_2^{(c)}
\end{aligned}$$

are obtained. Note that these are functions of the sufficient statistics for $\boldsymbol{\theta} = (\theta_1, \theta_2)$. The standardized test statistics $Z_1^{(c)}$, $Z_2^{(c)}$ and $Z_3^{(c)}$ are formed from these overall estimates and the decisions whether or not to reject H_1, H_2 or H_3 are based on these standardized statistics.

Rosenblum, Fang & Liu impose requirements that the FWER should be at most α and stipulate a minimum power that should be achieved in certain scenarios. They then seek the design that minimizes the expected total sample size averaged over a prior distribution for $\boldsymbol{\theta}$, subject to these constraints. Here, optimization is over the stage 1 sample sizes $n_1^{(1)}$ and $n_2^{(1)}$, the choice of stage 2 sample sizes from the L possible choices for every pair $(\hat{\theta}_1^{(1)}, \hat{\theta}_2^{(1)})$, and the rule for deciding which of the hypotheses H_1, H_2 and H_3 should be rejected at the end of the trial. The optimal design is found by converting the problem into a linear programming problem. This conversion involves discretizing the probability distributions in stages 1 and 2 on very fine grids, representing the FWER and power conditions as a set of linear constraints, and expressing the expected total sample size as a linear function. Despite the very high dimensionality of the problem, Rosenblum, Fang & Liu are able to find a solution using sparse linear programming techniques.

In the illustrative examples presented by Rosenblum, Fang & Liu, the optimized designs appear to offer significant improvements over certain more conventional procedures. It is difficult to discern whether the good performance is due to a well-chosen stopping rule or a more efficient analysis of the final data. It would be interesting to see whether designs produced by this method can make noticeable increases in expected gain for the problems tackled by Ondra et al. (2019), Burnett & Jennison (2021) and Ballarini et al. (2021). Designs that follow the general framework of Chapter 25, using closed testing procedures and combination tests, are attractive since they guarantee protection of the FWER while allowing optimization of remaining design parameters and flexibility in implementation. If efficiency gains are possible, it would be of interest to identify their source and investigate whether similar improvements might be achieved by modifying the form of the closed testing procedure or the combination tests in a flexible design.

It is interesting to note the role of optimization in this discussion of the potential benefits of adaptive enrichment designs. When comparing different types of design, it is important to fine tune these so that the best version of each type is considered. This principle applies to standard types of design as well as to complex new proposals. Since designs can be judged by multiple criteria, it is important to clarify the objectives of a study, as far as is possible, before trying to assess different trial designs. Defining a gain function such as (28.1) or (28.2) is a way to achieve this.

28.4 Adaptations Based on an Intermediate Endpoint

It may be that the interim decision to drop or enrich a sub-population is based on a short-term endpoint that is informative about the primary endpoint. As an example, in a psychotherapy trial comparing treatments for subjects with post-traumatic stress disorder (PTSD), the primary endpoint may be the score on the clinical administered PTSD scale (CAPS-5 score) twelve months after the initiation of therapy. At an early interim analysis when only a few patients have completed 12 months follow-up, the CAPS-5 score at 3 months may be available for many of the patients recruited thus far. Evidence of a treatment effect on the score at 3 months would suggest there is likely to be a treatment effect at 12 months. Also, the scores for the same subject at months 3 and 12 are liable to be correlated. In another example, Jenkins, Stone & Jennison (2011) consider a trial where overall survival is the endpoint of interest but, with only a few deaths occurring by the time of an interim analysis, an enrichment decision is made based on progression-free survival. Again, an increase in overall survival times on a new treatment is likely to be preceded by an improvement in progression-free survival.

A partitioning scheme that can cause bias

	$\boldsymbol{P^{(1)}}$	$\boldsymbol{P^{(2)}}$
Stage 1 cohort	Overall survival (during stage 1)	**Overall survival (during stage 2)**
Stage 2 cohort		Overall survival (during stage 2)

A partitioning scheme that avoids bias

Stage 1 cohort	Overall survival (during stage 1)	Overall survival (during stage 2)	$\boldsymbol{P^{(1)}}$
Stage 2 cohort		Overall survival (during stage 2)	$\boldsymbol{P^{(2)}}$

Figure 28.1 *Partitioning data in a combination test for survival data*

Applying the usual methods to analyze a trial where adaptations are made based on a short-term endpoint can be problematic. The standard way of partitioning data for use in a combination test is shown in the first part of Figure 28.1 with follow up on the initial cohort of patients after the adaptation analysis contributing to the second stage P-values. In a trial comparing two treatments in multiple sub-populations, let ϕ_i and θ_i denote the treatment effects on short-term and long-term endpoints, respectively, in sub-population i. Suppose that at an interim analysis where enrichment may occur, we have estimates $\hat{\phi}_i^{(1)}$ and $\hat{\theta}_i^{(1)}$ and the information for θ_i at this point is $\mathcal{I}_i^{(1)}$. At the second and final analysis, we have the estimate $\hat{\theta}_i^{(2)}$ with information level $\mathcal{I}_i^{(2)}$ based on the cumulative data from the whole trial.

The score statistics for θ_i at these two analyses are $S_1^{(i)} = \hat{\theta}_i^{(1)}\mathcal{I}_i^{(1)}$ and $S_2^{(i)} = \hat{\theta}_i^{(2)}\mathcal{I}_i^{(2)}$. It is tempting to form a combination test from the stage-wise Z-values

$$Z_i^{(1)} = S_1^{(i)}/\sqrt{\mathcal{I}_i^{(1)}} \quad \text{and} \quad Z_i^{(2)} = (S_i^{(2)} - S_i^{(1)})/\sqrt{(\mathcal{I}_i^{(2)} - \mathcal{I}_i^{(1)})}.$$

In order for such a test to be valid, we require $Z_i^{(1)}$ and $Z_i^{(2)}$ to be independent $N(0,1)$ random variables when $\theta_i = 0$. In simpler designs, this follows from the "independent increments" property of the score statistics,

$$S_i^{(2)} - S_i^{(1)} \mid \mathcal{I}_i^{(2)} - \mathcal{I}_i^{(1)} \sim N(\theta_i(\mathcal{I}_i^{(2)} - \mathcal{I}_i^{(1)}), (\mathcal{I}_i^{(2)} - \mathcal{I}_i^{(1)})). \tag{28.3}$$

However, this may not be the case in an adaptive enrichment design. Consider those subjects whose short-term endpoint is observed at analysis 1 and whose long-term endpoint becomes available at analysis 2. These subjects contribute to both $\hat{\phi}_i^{(1)}$ and to the increment $S_i^{(2)} - S_i^{(1)}$, and we expect these contributions to be correlated. Since the estimate $\hat{\phi}_i^{(1)}$ is used in making the enrichment decision at analysis 1, short-term responses from these subjects also have an impact on the value of $\mathcal{I}_i^{(2)}$. As a consequence, the property (28.3) may fail to hold and the guarantee of error rate control is lost. Bauer & Posch (2004) identified this problem in a two-stage adaptive design with short-term and long-term time-to-event endpoints, and described an extreme scenario where it is clear that Type I error inflation will occur. Jenkins, Stone & Jennison (2011, Appendix B) present simulation results that show a significant bias in stage 2 P-values can occur in their example with progression-free survival and overall survival endpoints.

Jenkins, Stone & Jennison (2011) propose a method to define P-values for use in a combination test that do have the desired distributions under the null hypothesis. Their proposal is illustrated in the second part of Figure 28.1. The key feature is that for each $i = 1$ and 2, the P-values $P_i^{(1)}$ and $P_i^{(2)}$ are associated with the two cohorts of patients admitted before and after the interim analysis. Thus when subjects whose data contribute to the enrichment decision at the interim analysis are followed up after that analysis, their subsequent survival data are used to calculate the "stage 1" P-values, $P_1^{(1)}$ and $P_2^{(1)}$.

Under this approach, $\boldsymbol{P^{(1)}} = (P_1^{(1)}, P_2^{(1)})$ and $\boldsymbol{P^{(2)}} = (P_1^{(2)}, P_2^{(2)})$ are based on data from separate groups of patients, simplifying considerably the conditional distribution of $\boldsymbol{P^{(2)}}$ given the stage 2 design. In order for $\boldsymbol{P^{(1)}}$ to have the correct null distribution, the amount of survival data collected on stage 1 subjects should not depend on the short-term responses from these subjects, since these may be predictive of overall survival. Thus, stage 1 subjects should be followed for a pre-specified time, irrespective of the short-term results and the enrichment decision that is made. As long as this requirement is satisfied, stage 1 and stage 2 P-values can be combined using the methods previously described in Section 28.1.

28.5 Enrichment Designs with Multiple Sub-populations

28.5.1 Group Sequential Enrichment Designs

Magnusson & Turnbull (2013) propose group sequential designs in which enrichment may occur at the first analysis. In describing their methods, we modify the notation to match that used in this chapter. Let $S_1, \ldots, S_p$ be disjoint sub-populations and, in a trial comparing a new treatment with a control, let θ_i be the treatment effect in sub-population i. In order for their procedures to protect the FWER, Magnusson & Turnbull assume each $\theta_i \geq 0$. Thus, for sub-population i, the null hypothesis H_i: $\theta_i = 0$ is to be tested against $\theta_i > 0$.

The trial starts with recruitment from the full population but, at the first analysis, some sub-populations are dropped. If all sub-populations are dropped, the trial stops for futility. The trial continues with recruitment restricted to S^*, the union of the remaining sub-populations. The motivation for this type of design is that the inclusion of sub-populations that do not respond well to the new treatment dilutes the treatment effect and removing such sub-populations at an early analysis will allow the trial to continue with a population that has a higher treatment effect.

Let $\mathcal{P}^*$ denote the set of indices of the sub-populations in S^*. Then, a new target sample size is specified and a test is conducted of the single hypothesis

$$H_{\mathcal{P}^*} = \cap_{i \in \mathcal{P}^*} H_i$$

or, equivalently, $H_{\mathcal{P}^*}$: $\theta_{\mathcal{P}^*} = 0$, where $\theta_{\mathcal{P}^*}$ is the average treatment effect in S^*. The trial continues in a group sequential manner with a maximum of K analyses. At each analysis k, a score statistic is formed by summing over $i \in \mathcal{P}^*$ the efficient score statistics for testing H_i based on the cumulative data in sub-population i, including that observed at analysis 1.

The sub-population selection rule and the group sequential stopping rule are based on a sequence of upper and lower boundary values, $(\ell_1, u_1), \ldots, (\ell_K, u_K)$ derived using an error spending approach. Calculations are similar to those described in Section 8.3 but with some important modifications. Following the approach of Stallard & Facey (1996), spending functions $\alpha_U^*(t)$ and $\alpha_L^*(t)$ are defined where α_U^* increases from 0 to α and α_L^* increases from 0 to $1-\alpha$ as t increases from zero to 1. Constants $0 < t_1 < \ldots < t_K = 1$ are specified and boundaries computed so that, when $\theta_i = 0$ for all $i \in \{1, \dots, p\}$, the probability of stopping to reject any H_{S^*} at or before analysis k is $\alpha_U^*(t_k)$ and the probability of stopping for futility at or before analysis k is $\alpha_L^*(t_k)$. Here, probabilities are calculated for each possible case $S^* = s^*$ and the results are then summed to give the overall probability. Since the family-wise Type I error rate is calculated assuming specific rules for dropping treatments and specifying sample sizes after adaptation at analysis 1, it is essential that these rules are clearly stated and followed precisely during a trial.

Magnusson & Turnbull (2013) consider first stage selection rules for two situations.

No a priori ordering of sub-populations

In this case, there is no strong initial opinion about which of the sub-populations $S_1, \dots, S_p$ are most likely to respond to the new treatment. A Z-statistic $Z_i^{(1)}$ for testing H_i: $\theta_i = 0$ against $\theta_i > 0$ is calculated for each sub-population S_i, $i = 1, \dots, p$, all the sub-populations with $Z_i^{(1)} < \ell_1$ are eliminated, and those with $Z_i^{(1)} \geq \ell_1$ are retained and their union is S^*.

Ordered sub-populations

In this case, sub-populations are ordered with respect to their anticipated response and it is assumed that $\theta_1 \geq \theta_2 \geq \dots \geq \theta_p$. Such an assumption may be appropriate when the sub-populations are defined in terms of a biomarker measured at baseline, and treatment is expected to be more effective at higher values of the biomarker. Then, if B denotes the biomarker value, constants $\lambda_1 > \lambda_2 > \dots > \lambda_{p-1}$ are defined, sub-population S_1 comprises subjects with $B \geq \lambda_1$, S_i contains subjects with $\lambda_{i-1} > B \geq \lambda_i$ for $i = 2, \dots, p-1$ and S_p contains subjects with $B < \lambda_{p-1}$. Magnusson & Turnbull define $r = \max\{i\colon Z_i^{(1)} > \ell_1\}$ and set S^* equal to the union of $S_1, \dots, S_r$. Thus, sub-population S_i is retained if $Z_i^{(1)} > \ell_1$ or if $Z_{i'}^{(1)} > \ell_1$ for some $i' < i$, and otherwise S_i is eliminated.

A proof that procedures for both these cases control the FWER at level α is given in the appendix of Magnusson & Turnbull (2013). As noted earlier, this proof relies on the assumption that each $\theta_i \geq 0$.

An example

Magnusson & Turnbull (2013, Sec. 6.1) describe an example in which there are three disjoint sub-populations with no obvious ordering with respect to the anticipated treatment effect. The trial has two stages and the planned sample size is divided equally between stages 1 and 2. If the trial continues past the first stage but one or more populations are eliminated, all the planned observations will be allocated to the sub-populations remaining in the second stage, enriching these sub-populations. The trial has 15 possible outcomes, rejecting or not rejecting H_{S^*} for each non-empty subset S^* of the full population, or stopping for futility at the interim analysis. The error spending functions are defined as

$$\alpha_U^*(0.5) = 0.0125, \quad \alpha_U^*(1) = 0.025, \quad \alpha_L^*(0.5) = 0.4875 \quad \text{and} \quad \alpha_L^*(1) = 0.975$$

and the resulting boundary values are

$$(\ell_1, u_1) = (0.7962, 2.7625) \quad \text{and} \quad (\ell_2, u_2) = (2.5204, 2.5204).$$

Sample size calculations show that responses from 748 patients are required to achieve the desired power.

Boundary values for $Z_i^{(1)}$ and $Z_2^{(1)}$ are shown in Figure 28.2. For comparison, the figure also shows the boundary of a simple group sequential design that tests for a treatment effect in a single population. The higher boundary for the adaptive enrichment design reflects the multiplicity adjustment needed to address the selection bias in choosing the set of sub-populations S^* based on first stage data.

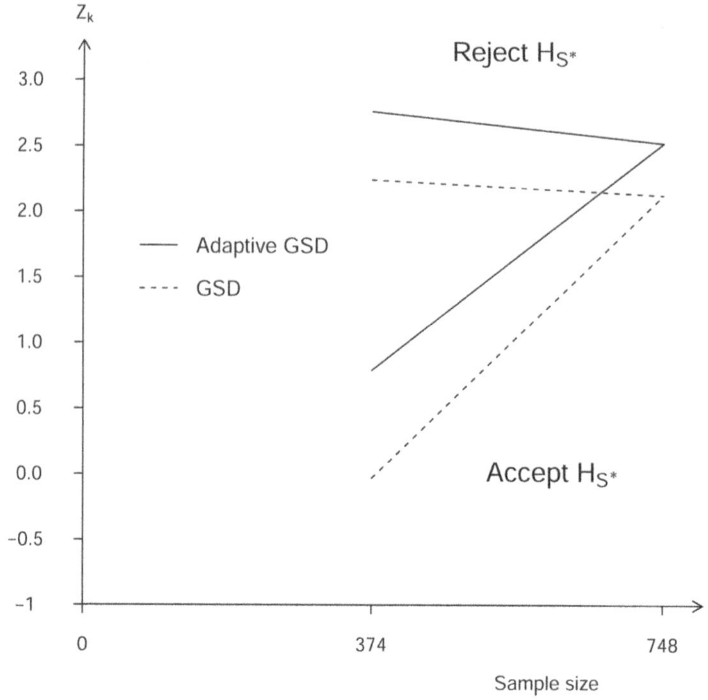

Figure 28.2 *Group sequential boundaries showing the effect of sub-population selection*

Magnusson & Turnbull use a hypothetical set of data to compare the outcome from their procedure with that of a design constructed using the general approach of Section 28.2. In the general approach, a closed testing procedure is applied with Simes' test for intersection hypotheses and an inverse normal combination tests to combine results across the two stages. In the hypothetical example, sub-population S_1 is dropped at the interim analysis for both procedures. In the general approach, a positive result is obtained for sub-population S_3 only. In contrast, Magnusson & Turnbull's procedure yields a positive result for the union of sub-populations S_2 and S_3. This apparently stronger result may be attributed to the use of sufficient statistics in the hypothesis tests.

28.5.2 Further Designs for Nested Sub-populations

Stallard (2023) proposes adaptive enrichment designs for the case where sub-populations are defined by values of a continuous biomarker. We shall describe these procedures with notation adapted to match that used so far in this chapter. Thresholds $\lambda_1 > \lambda_2 > \ldots > \lambda_{p-1}$ are defined, then for $i = 1, \ldots, p-1$, sub-population S_i comprises all subjects with $B \geq \lambda_i$, and S_p is the full population. Note that this definition of $S_1, \ldots, S_p$ differs from that used in Magnusson & Turnbull's designs, where S_1 to S_p were disjoint sets of subjects. Letting θ_i denote the average treatment effect in S_i, i.e., for all subjects with $B \geq \lambda_i$, Stallard assumes the monotonicity property

$$\theta_1 \geq \theta_2 \geq \ldots \geq \theta_p. \tag{28.4}$$

The hypotheses H_i: $\theta_i \leq 0$, $i = 1, \ldots, p$, are to be tested against the alternatives $\theta_i > 0$. The subsets S_1 to S_p are nested with $S_1 \subset S_2 \subset \ldots \subset S_p$ and, since the ordering (28.4) is assumed, we have

$$H_1 \Rightarrow H_2 \Rightarrow \ldots \Rightarrow H_p.$$

The trial is conducted in two stages with sub-population selection after stage 1 and enriched sampling from the selected sub-population in stage 2. For $i = 1, \ldots, p$, denote the estimate of θ_i after stage 1 by $\hat{\theta}_i^{(1)}$ and the associated information level by $\mathcal{I}_i^{(1)}$, and set $Z_i^{(1)} = \hat{\theta}_i^{(1)} \sqrt{\mathcal{I}_i^{(1)}}$. Stallard presents six options for selecting the sub-population S_M, where $M \in \{1, \ldots, p\}$, to be sampled in stage 2. In each rule, a statistic T_i is calculated for each sub-population $i = 1, \ldots, p$, for example,

$$T_i = \begin{cases} \hat{\theta}_i^{(1)} \sqrt{\mathcal{I}_i^{(1)}}, \text{ i.e., } Z_i^{(1)} & \text{in rule 1,} \\ \hat{\theta}_i^{(1)} & \text{in rule 2,} \\ \hat{\theta}_i^{(1)} \mathcal{I}_i^{(1)} & \text{in rule 3,} \end{cases}$$

and M is defined as the value of i that maximizes T_i, i.e.,

$$M = \arg\max_{i \in \{1, \ldots, p\}} T_i.$$

In order to define a closed testing procedure, we need level α tests of each elementary hypothesis H_i and of each intersection hypothesis H_J where $J \subseteq \{1, \ldots, p\}$. Writing

$$M(i) = \arg\max_{j \in \{i, \ldots, p\}} T_j,$$

we define the stage 1 P-value $P_i^{(1)}$ for H_i as

$$P_i^{(1)} = Pr\{Z_{M(i)}^{(1)} \geq z_{m(i)}^{(1)}\} \text{ when } \theta_i = \ldots = \theta_p = 0, \tag{28.5}$$

where $z_{m(i)}^{(1)}$ is the observed value of $Z_{M(i)}^{(1)}$. Following Jennison (2023a) in his discussion of Stallard's paper, for each $J \subseteq \{1, \ldots, p\}$, we define

$$P_J^{(1)} = P_{\min(J)}^{(1)},$$

where $\min(J) = \min\{i : i \in J\}$.

Note that when the sub-populations with the highest biomarker values are small in size, their treatment effect estimates have a high variance so the highest $\hat{\theta}_i^{(1)}$ is not necessarily associated with sub-population S_1. For similar reasons, when trying to show that $\theta_i > 0$, it can be advantageous to use information from subjects with biomarker values below λ_i in the test of H_i: $\theta_i \leq 0$, and this motivates the definition of $P_i^{(1)}$.

When sub-population S_M is selected for stage 2, let $P_M^{(2)}$ be the P-value for testing H_M: $\theta_M \leq 0$ based on data from the subjects in sub-population M recruited in stage 2. We set

$$P_J^{(2)} = P_M^{(2)}$$

for all subsets J of $\{1, \ldots, p\}$ that contain M. In the closed testing procedure we require a level α test of the intersection hypothesis H_J for each subset J of $\{1, \ldots, p\}$ that contains the index M. A specific form of combination test is applied to each pair of P-values, $P_J^{(1)}$ and $P_J^{(2)}$, and H_M is rejected in the overall closed testing procedure if all these tests are significant at level α. Since all these intersection hypotheses have the same P-value in stage 2, the outcome for the overall test of H_M depends on the largest stage 1 P-value,

$$\max_{J:\, M \in J} P_J^{(1)} = \max_{i \leq M} P_i^{(1)},$$

and this is the single summary statistic that needs to be carried forward.

The calculation of (28.5) is an interesting task. In the case of Stallard's rules 1 to 3, letting $t_{m(i)}$ denote the observed value of $T_{M(i)}$, we need to calculate

$$\sum_{j=1}^{p} Pr\{M(i) = j \text{ and } Z_j^{(1)} \geq z_{m(i)}^{(1)}\}$$

$$= \sum_{j=1}^{p} Pr\{Z_j^{(1)} \geq z_{m(i)}^{(1)} \text{ and } T_k < t_{m(i)} \text{ for } k = i, \ldots, j-1, j+1, \ldots, p\} \qquad (28.6)$$

under $\theta_i = \ldots = \theta_p = 0$. The estimates $\hat{\theta}_1^{(1)}, \ldots, \hat{\theta}_p^{(1)}$ are based on increasing sets of data and, thus, have the same canonical joint distribution as estimates observed at successive analyses in a group sequential test, which is given by (3.2). Since $Z_j^{(1)} \sim N(0,1)$, the jth term in (28.6) can be written as

$$\int_{z_{m(i)}^{(1)}}^{\infty} \phi(z_j)\, Pr\{T_k < t_{m(i)} \text{ for } k = i, \ldots, j-1, j+1, \ldots, p \mid Z_j^{(1)} = z_j\}\, dz_j,$$

where ϕ is the probability density function of a standard normal distribution. The conditional distribution of $\{T_1, \ldots, T_{j-1}, T_{j+1}, \ldots, T_p\}$ given $Z_j^{(1)} = z_j$ is multivariate normal and Markov, and thus can be computed by numerical integration in a similar way to the calculations for group sequential tests described in Chapter 31.

Suppose it is assumed that each $\theta_i \geq 0$ for $i = 1, \ldots, p$. Then, the assumption (28.4) implies that whenever H_i is true, H_{i+1} to H_p are also true and thus, $\theta_i = \ldots = \theta_p = 0$. It follows that $P_i^{(1)}$, as defined in (28.5), is a valid P-value for testing H_i. For selection rules 1 and 2, a coupling argument can be used to show that $P_i^{(1)}$ is a valid P-value for testing H_i without making the assumption that each $\theta_i \geq 0$. To do this, we create a set of treatment effect estimates under $0 \geq \theta_i' \geq \ldots \geq \theta_p'$ by first generating estimates $\hat{\theta}_i^{(1)}, \ldots, \hat{\theta}_p^{(1)}$ under $\theta_i = \ldots = \theta_p = 0$, then transform these by adding θ_j' to $\hat{\theta}_j^{(1)}$ for each $j = 1, \ldots, p$, and we observe that this transformation reduces the value of $Z_{M(i)}^{(1)}$. The same coupling argument fails for selection rule 3 as $Z_{M(i)}^{(1)}$ may increase when the value of $M(i)$ changes after the transformation. Jennison (2023a) noted that Stallard's rules 4, 5 and 6 do not protect the FWER in general but they do if one is willing to make the assumption that each $\theta_i \geq 0$.

28.5.3 Bayesian Hierarchical Modeling

Bayesian methods have been used to design and analyze clinical trials with multiple sub-populations, especially in Phase II oncology trials; see, for example, Thall et al. (2003), Chugh et al. (2009) and Berry et al. (2013). A hierarchical Bayesian model is natural and appealing to use because, while it is recognized that responses may be of differing magnitudes in the various sub-populations, there is a prior belief that there should be some common effect. With a hierarchical model, the "borrowing of strength" in treatment comparisons for different sub-populations is advantageous when there is a moderate beneficial effect of treatment over all sub-populations, but this approach may mask a strong effect that occurs in only one sub-population.

In a group sequential enrichment design the posterior distributions of the individual sub-population effect sizes can be computed at each interim analysis. Any sub-population for which a certain feature of this posterior distribution falls below a threshold value can be dropped and resources diverted to the sub-populations that remain. If no sub-populations remain, the trial can be abandoned as futile. Jones et al. (2019) describe a trial with this type of design that investigated the effects of a combination of gemcitabine and docetaxel with ontuxizumab (MORAb-004) in patients with metastatic soft-tissue sarcoma.

The outcome of a Phase II trial will affect whether a new treatment proceeds to a Phase III study and this is primarily an internal decision for a company or a research organization. Thus, while it is still important to guard against false positive conclusions in Phase II, it may not be crucial to control the FWER precisely. The opposite is true in a Phase III trial when results are to be presented to an external audience. Demonstrating that a complex Bayes procedure protects frequentist error rates can be challenging: it is often difficult to identify the set of parameter values at which the maximum Type I error rate occurs and if so, extensive simulations may be required to show that FWER is controlled in a large number of scenarios.

28.6 Inference Following an Adaptive Enrichment Trial

28.6.1 Point Estimation

In the designs we have considered, a sub-population is selected at an interim analysis and recruitment continues from this sub-population only. Suppose that, of the initial hypotheses, a subset H_i: $\theta_i \leq 0$ for $i \in I^*$, is specified, then we may wish to provide point estimates and confidence intervals for the treatment effects θ_i, $i \in I^*$. Kimani, Todd & Stallard (2015) consider two-stage designs in which treatment effects in the full population and one specific sub-population are investigated, and an enrichment decision may restrict attention to the sub-population in stage 2. Let θ_S denote the treatment effect in the sub-population and θ_{S^C} the treatment effect in the complement of this sub-population. Kimani, Todd & Stallard use Rao-Blackwellization to derive the uniform minimum variance unbiased estimate (UMVUE) of θ_S conditional on enrichment and to derive UMVUEs of θ_S and θ_{S^C} conditional on the trial continuing without enrichment. They compare the performance of these estimators with naive estimators based on two-stage sample means, and conclude that the unbiased estimators have higher mean square error (MSE) than the naive estimators. Overall, they recommend using the unbiased estimator of θ_S when enriching to the sub-population but they recommend the naive estimators of θ_S and θ_{S^C} when the study continues with the full population.

Li, Chen, Li & Beckman (2017) consider a trial investigating the treatment effect on the primary endpoint in a number of sub-populations when enrichment decisions are based on an intermediate endpoint. They estimate the treatment effect in each sub-population that continues to stage 2 from a simple analysis of data pooled across the two stages. They find that, in the worst cases, the bias in the treatment effect estimate for a selected sub-population can be as high as 6% or 7% of the true treatment effect and they conclude that, for the types of trial design considered "The bias is generally small or negligible except for the worst case scenario." Chiu, Koenig, Posch & Jaki (2018) state a different conclusion after their study of one-stage and two-stage trial designs with 2 or 3 sub-populations:

> "... we have assessed the bias of the MLE under various scenarios. We find that, almost always, bias is positive leading to an overenthusiastic estimate of the true treatment effect. While for some settings, the size of the bias can be viewed as negligible, it can become large under other situations. The challenge clearly being that one will usually not know if one is in one of these extreme situations."

Kunzmann, Benner & Kieser (2017) consider two-stage designs in which stage 1 data are used to decide whether the full population or a pre-specified subgroup is studied in stage 2. They investigate the performance of six different estimators of the treatment effect in the selected population: the MLE, two forms of empirical Bayes estimator, a parametric bootstrap estimator, and two hybrid estimators which use the minimum variance conditionally unbiased estimator (UMCVUE) similar to that defined in Section 26.6.1 when this is possible, and otherwise the MLE or a conditional moment estimator (CME). They observe that the most consistent reduction of bias over the range of scenarios considered is achieved by the hybrid method combining the UMCVUE and CME. However, this improved bias control comes at the price of a larger MSE than that of other estimators in many scenarios.

Kimani et al. (2018) consider two-stage adaptive designs in which the sub-population to be studied in stage 2 is defined in terms of a biomarker, with the threshold for inclusion determined by stage 1 data. They compare estimators of the treatment effect in the selected population: the naive maximum likelihood estimator, two forms of unbiased estimator, two bias adjusted estimators, and two shrinkage estimators. They report that, in various scenarios, the bias of the naive estimator of the treatment effect in the selected sub-population is substantial and can be negative or positive. They recommend the UMVCUE in the form proposed by Robertson, Prevost & Bowden (2016b) but note that the elimination of bias is accompanied by an increase in MSE. Di Stefano et al. (2022) compare estimators in an enrichment design with a time-to-event endpoint: the naive estimator, a conditionally unbiased estimator, two bias adjusted estimators, and two shrinkage estimators. Their results show the bias adjusted estimators to have the desirable properties of a small bias and a moderate mean square error, comparable to that of the naive estimator.

28.6.2 Confidence Intervals

Rosenblum (2013) considers trial designs in which a sub-population is selected at an interim analysis, the trial continues with recruitment in that sub-population only, then on termination, a $(1-\alpha)$-level confidence interval is required for the treatment effect in the selected sub-population. Rosenblum proposes using the standard $(1-\alpha)$-level confidence interval based on the final set of data for the selected sub-population but with the width increased by a factor c that guarantees the desired coverage probability. He presents an algorithm to find the combination of treatment effects which requires the largest expansion factor c for a given decision rule and this factor is then applied for all trial outcomes.

Kimani et al. (2020) follow the approach of Magirr, Jaki, Posch & Klinglmueller (2013) in constructing confidence intervals for the treatment effects of interest after an adaptive enrichment trial with a time to event endpoint. They note these confidence intervals may be "non-informative". As an example, if sub-populations i and j are studied in the second stage of a two-stage design and H_j: $\theta_j \leq 0$ is rejected but H_i: $\theta_i \leq 0$ is not rejected in a multiple testing procedure with FWER α, then the one-sided $(1-\alpha)$-level upper confidence interval for θ_j may be $(0, \infty)$.

The origin of such non-informative confidence intervals can be attributed to the process of first defining a procedure for testing hypotheses H_i: $\theta_i \leq 0$ with FWER α, then adding tests of the hypotheses $H_j(\tilde{\theta}_j)$: $\theta_j \leq \tilde{\theta}_j$, for $\tilde{\theta}_j > 0$ while continuing to reject the hypotheses H_i: $\theta_i \leq 0$, $i = 1, \ldots, p$, whenever this would have occurred in the original procedure. Suppose, for example, the tests of H_i: $\theta_i \leq 0$, $i = 1, \ldots, p$, are represented graphically in the form shown in Section 20.5 and the testing process ends with all the error probability α passed to a set of hypotheses, none of which is rejected. Then, there is no error probability to pass on to the tests of hypotheses $H_j(\tilde{\theta}_j)$: $\theta_j \leq \tilde{\theta}_j$, for $\tilde{\theta}_j > 0$, and thus none of these positive values can be excluded from a $(1-\alpha)$-level upper confidence interval. The methods of Kimani et al. (2020) and Magirr et al. (2013) have a parallel feature. Suppose an intersection hypothesis $H_{I,J}$ states that $\theta_i \leq 0$ for $i \in I$ and $\theta_j \leq \tilde{\theta}_j$, where $\tilde{\theta}_j > 0$, for $j \in J$, and the hypotheses in H_I are still being tested in stage k. Then, the P-value for $H_{I,J}$ based on stage k data is greater than or equal to the P-value for H_I based on these data. If this P-value for H_I is high, the P-value for $H_{I,J}$ cannot be reduced, even by an extremely high value of $\hat{\theta}_j$.

Hayter & Hsu (1994) discussed the issue of non-informative confidence intervals and proposed methods to avoid their occurrence, noting that this is achieved at the cost of some power in testing the set of hypotheses H_i: $\theta_i \le 0$, $i = 1, \ldots, p$, with a given FWER. Brannath & Schmidt (2014) and Schmidt & Brannath (2014) present further methods that enable multiple testing and the construction of informative upper confidence intervals for those parameters θ_i for which the null hypothesis $\theta_i \le 0$ has been rejected. A general solution to the problem of non-informative confidence intervals is to include hypotheses $H_j(\tilde{\theta}_j)$: $\theta_j \le \tilde{\theta}_j$, for $\tilde{\theta}_j \in R$, in the set of hypotheses to be tested in the initial closed testing procedure. While this approach may produce fewer rejections of hypotheses H_i: $\theta_i \le 0$, it will provide an informative confidence interval whenever such a rejection occurs. The method we now describe for an adaptive enrichment design parallels the procedure introduced in Section 26.6.3 for adaptive designs with treatment selection and is closely related to the methods presented by Posch et al. (2005) for obtaining a joint confidence region for a parameter vector $\boldsymbol{\theta}$ and, hence, a confidence interval for each parameter θ_i.

28.6.3 Informative Confidence Intervals for the Treatment Effects of Sub-populations

For simplicity, we consider a two-stage, adaptive enrichment trial where interest is in the treatment effects θ_S in a pre-specified sub-population and θ_F in the full population. We present a method that provides upper confidence intervals for θ_S and θ_F when recruitment from the full population continues throughout the trial and for θ_S if enrichment occurs. We denote the upper confidence intervals for θ_S and θ_F, respectively, by $(\theta_{L,S}, \infty)$ and $(\theta_{L,F}, \infty)$, setting $\theta_{L,F} = -\infty$ when enrichment occurs. Writing $\boldsymbol{\theta} = (\theta_S, \theta_F)$, the defining property of these intervals is

$$Pr_{\boldsymbol{\theta}}\{\theta_S \in (\theta_{L,S}, \infty) \text{ and } \theta_F \in (\theta_{L,F}, \infty)\} \ge 1 - \alpha \quad \text{for all } \boldsymbol{\theta}. \tag{28.7}$$

In exploiting the equivalence between a confidence interval and a family of hypothesis tests, we create a family of tests for the hypotheses

$$H_S(\tilde{\theta}_S)\colon \theta_S \le \tilde{\theta}_S,\ \tilde{\theta}_S \in R, \quad \text{and} \quad H_F(\tilde{\theta}_F)\colon \theta_F \le \tilde{\theta}_F,\ \tilde{\theta}_F \in R, \tag{28.8}$$

with a family-wise error rate of α.

In defining level α tests of the elementary hypotheses $H_S(\tilde{\theta}_S)$ and $H_F(\tilde{\theta}_F)$, and the intersection hypotheses that can be formed from them, we shall use an inverse normal combination test with weights w_1 and w_2 to combine P-values from the stage 1 and stage 2 data. Since the hypotheses $H_S(\tilde{\theta}_S)$ are nested with

$$\cap_{j=1}^{k} H_S(\tilde{\theta}_S) = H_S(\min_{j=1,\ldots,k} \tilde{\theta}_S)$$

and the $H_F(\tilde{\theta}_F)$ are nested with

$$\cap_{j=1}^{k} H_F(\theta_F) = H_F(\min_{j=1,\ldots,k} \tilde{\theta}_F),$$

we need only consider intersection hypotheses of the form

$$H_S(\tilde{\theta}_S),\ H_F(\tilde{\theta}_F) \text{ and } H_{SF}(\tilde{\boldsymbol{\theta}}) = H_S(\tilde{\theta}_S) \cap H_F(\tilde{\theta}_F).$$

Suppose the sub-population S represents a fraction λ of the full population and, in stage 1, $(\hat{\theta}_S^{(1)}, \hat{\theta}_F^{(1)})$ is bivariate normal with

$$\hat{\theta}_S^{(1)} \sim N(\theta_S, 1/(\lambda \mathcal{I}^{(1)})),\ \hat{\theta}_F^{(1)} \sim N(\theta_F, 1/\mathcal{I}^{(1)}) \text{ and } \mathrm{Cov}(\hat{\theta}_S^{(1)}, \hat{\theta}_F^{(1)}) = 1/\mathcal{I}^{(1)}.$$

We define

$$Z_S^{(1)}(\tilde{\theta}_S) = (\hat{\theta}_S^{(1)} - \tilde{\theta}_S)\surd\{\lambda \mathcal{I}^{(1)}\} \quad \text{and} \quad P_S^{(1)}(\tilde{\theta}_S) = 1 - \Phi\{Z_S^{(1)}(\tilde{\theta}_S)\} \quad \text{for } \tilde{\theta}_S \in R,$$

$$Z_F^{(1)}(\tilde{\theta}_F) = (\hat{\theta}_F^{(1)} - \tilde{\theta}_F)\surd \mathcal{I}^{(1)} \quad \text{and} \quad P_F^{(1)}(\tilde{\theta}_F) = 1 - \Phi\{Z_F^{(1)}(\tilde{\theta}_F)\} \quad \text{for } \tilde{\theta}_F \in R.$$

If the trial continues with the full population, we suppose $(\hat{\theta}_S^{(2)}, \hat{\theta}_F^{(2)})$ is bivariate normal with

$$\hat{\theta}_S^{(2)} \sim N(\theta_S, 1/(\lambda \mathcal{I}^{(2)})), \quad \hat{\theta}_F^{(2)} \sim N(\theta_F, 1/\mathcal{I}^{(2)}) \quad \text{and} \quad \text{Cov}(\hat{\theta}_S^{(2)}, \hat{\theta}_F^{(2)}) = 1/\mathcal{I}^{(2)},$$

and define

$$Z_S^{(2)}(\tilde{\theta}_S) = (\hat{\theta}_S^{(2)} - \tilde{\theta}_S)\surd\{\lambda \mathcal{I}^{(2)}\} \quad \text{and} \quad P_S^{(2)}(\tilde{\theta}_S) = 1 - \Phi\{Z_S^{(2)}(\tilde{\theta}_S)\} \quad \text{for } \tilde{\theta}_S \in R,$$

$$Z_F^{(2)}(\tilde{\theta}_F) = (\hat{\theta}_F^{(2)} - \tilde{\theta}_F)\surd \mathcal{I}^{(2)} \quad \text{and} \quad P_F^{(2)}(\tilde{\theta}_F) = 1 - \Phi\{Z_F^{(2)}(\tilde{\theta}_F)\} \quad \text{for } \tilde{\theta}_F \in R.$$

However, if enrichment occurs, we observe $\hat{\theta}_S^{(2)} \sim N(\theta_S, 1/\mathcal{I}^{(2)})$ and set

$$Z_S^{(2)}(\tilde{\theta}_S) = (\hat{\theta}_S^{(2)} - \tilde{\theta}_S)\surd \mathcal{I}^{(2)} \quad \text{and} \quad P_S^{(2)}(\tilde{\theta}_S) = 1 - \Phi\{Z_S^{(2)}(\tilde{\theta}_S)\} \quad \text{for } \tilde{\theta}_S \in R.$$

Suppose that within each stage, we test intersection hypotheses $H_{SF}(\tilde{\boldsymbol{\theta}})$ using Simes' test. Then, with $\tilde{\boldsymbol{\theta}} = (\tilde{\theta}_S, \tilde{\theta}_F)$, the stage 1 P-value for $H_{SF}(\tilde{\boldsymbol{\theta}})$ is

$$P_{SF}^{(1)}(\boldsymbol{\theta}) = \min\left[2\min\{P_S^{(1)}(\tilde{\theta}_S),\, P_F^{(1)}(\tilde{\theta}_F)\},\, \max\{P_S^{(1)}(\tilde{\theta}_S),\, P_F^{(1)}(\tilde{\theta}_F\}\right].$$

In order for the lower confidence limit for θ_S to be above $\tilde{\theta}_S$, we need to reject $H_{SF}(\tilde{\boldsymbol{\theta}})$ for the specified $\tilde{\theta}_S$ and all values of $\tilde{\theta}_F$. However, for large values of $\tilde{\theta}_F$, $P_F^{(1)}(\tilde{\theta}_F)$ is large and

$$P_{SF}^{(1)}(\boldsymbol{\theta}) = 2P_S^{(1)}(\tilde{\theta}_S).$$

Similarly, if the trial continues in the full population, we find

$$P_{SF}^{(2)}(\boldsymbol{\theta}) = 2P_S^{(2)}(\tilde{\theta}_S)$$

for large values of $\tilde{\theta}_F$ and, in the weighted inverse normal combination test, we require

$$w_1\Phi^{-1}\{2(1 - \Phi(Z_S^{(1)}(\tilde{\theta}_S)))\} + w_2\Phi^{-1}\{2(1 - \Phi(Z_S^{(2)}(\tilde{\theta}_S)))\} \geq \Phi^{-1}(1 - \alpha)$$

in order to reject $H_S(\tilde{\theta}_S)$ in the closed testing procedure and exclude $\tilde{\theta}_S$ from the upper confidence interval. It is evident that we would have obtained the same interval if we had used a simple Bonferroni adjustment and, due to the correlation between $Z_S^{(1)}(\tilde{\theta}_S)$ and $Z_F^{(1)}(\tilde{\theta}_F)$ and between $Z_S^{(2)}(\tilde{\theta}_S)$ and $Z_F^{(2)}(\tilde{\theta}_F)$, this is a conservative approach.

A more efficient approach is to test the intersection hypothesis using a Dunnett-like test in stage $k = 1$, and in stage $k = 2$ when the trial continues with the full population. The test of $H_{SF}(\tilde{\boldsymbol{\theta}})$ is based on the statistic

$$\max\{Z_S^{(k)}(\tilde{\theta}_S),\, Z_F^{(k)}(\tilde{\theta}_F)\}$$

and we define the P-value $P_{SF}^{(k)}(\tilde{\boldsymbol{\theta}})$ when $\tilde{\boldsymbol{\theta}} = (\tilde{\theta}_S, \tilde{\theta}_F)$ as

$$P_{SF}^{(k)}(\tilde{\theta}_S, \tilde{\theta}_F) = Pr_{(\tilde{\theta}_S, \tilde{\theta}_F)}\{\max\{Z_S^{(k)}(\tilde{\theta}_S), Z_F^{(k)}(\tilde{\theta}_F)\} \geq \max\{z_S^{(k)}(\tilde{\theta}_S), z_F^{(k)}(\tilde{\theta}_F)\}\}, \quad (28.9)$$

where $z_S^{(k)}(\tilde{\theta}_S)$ and $z_F^{(k)}(\tilde{\theta}_F)$ are the observed values of $Z_S^{(k)}(\tilde{\theta}_S)$ and $Z_F^{(k)}(\tilde{\theta}_F)$. This probability can be calculated numerically using the fact that $(Z_S^{(k)}(\tilde{\theta}_S), Z_F^{(k)}(\tilde{\theta}_F))$ is bivariate normal with

$$Z_S^{(k)}(\tilde{\theta}_S) \sim N(0,1), \quad Z_F^{(k)}(\tilde{\theta}_F) \sim N(0,1) \quad \text{and} \quad \text{Corr}(Z_S^{(k)}(\tilde{\theta}_S), Z_F^{(k)}(\tilde{\theta}_F)) = \sqrt{\lambda}$$

when $\boldsymbol{\theta} = (\tilde{\theta}_S, \tilde{\theta}_F)$. We define the "Dunnett-like" P-value $P_D(z, \rho)$ as

$$P_D(z, \rho) \;=\; Pr\{\max(Z_1, Z_2) \geq z\}$$

when (Z_1, Z_2) is bivariate normal with $Z_1 \sim N(0,1)$, $Z_2 \sim N(0,1)$ and $\text{Corr}(Z_1, Z_2) = \rho$. Note that $z_F^{(k)}(\tilde{\theta}_F) \to -\infty$ as $\tilde{\theta}_F \to \infty$ and thus

$$\begin{aligned}\max_{\tilde{\theta}_F} P_{SF}^{(k)}(\tilde{\theta}_S, \tilde{\theta}_F) &= \max_{\tilde{\theta}_F} Pr_{(\tilde{\theta}_S, \tilde{\theta}_F)}\{\max\{Z_S^{(k)}(\tilde{\theta}_S), Z_F^{(k)}(\tilde{\theta}_F)\} \geq z_S^{(k)}(\tilde{\theta}_S)\} \\ &= P_D(z_S^{(k)}(\tilde{\theta}_S), \sqrt{\lambda}).\end{aligned}$$

Similarly, $z_S^{(k)}(\tilde{\theta}_S) \to -\infty$ as $\tilde{\theta}_S \to \infty$ and

$$\begin{aligned}\max_{\tilde{\theta}_S} P_{SF}^{(k)}(\tilde{\theta}_S, \tilde{\theta}_F) &= \max_{\tilde{\theta}_S} Pr_{(\tilde{\theta}_S, \tilde{\theta}_F)}\{\max\{Z_S^{(k)}(\tilde{\theta}_S), Z_F^{(k)}(\tilde{\theta}_F)\} \geq z_F^{(k)}(\tilde{\theta}_F)\} \\ &= P_D(z_F^{(k)}(\tilde{\theta}_F), \sqrt{\lambda}).\end{aligned}$$

When enrichment occurs after stage 1, we set

$$P_{SF}^{(2)}(\tilde{\theta}_S, \tilde{\theta}_F) \;=\; P_S^{(2)}(\tilde{\theta}_S) \;=\; 1 - \Phi\{Z_S^{(2)}(\tilde{\theta}_S)\} \quad \text{for all } \tilde{\theta}_F$$

and, thus, we reject $H_S(\tilde{\theta}_S)$ if

$$w_1 \Phi^{-1}\{1 - P_D(Z_S^{(1)}(\tilde{\theta}_S), \sqrt{\lambda})\} + w_2 Z_S^{(2)}(\tilde{\theta}_S) \;>\; \Phi^{-1}(1-\alpha),$$

so the upper $(1-\alpha)$-level confidence interval for θ_S is $(\theta_{L,S}, \infty)$ where $\theta_{L,S}$ is the solution to

$$w_1 \Phi^{-1}\{1 - P_D(Z_S^{(1)}(\tilde{\theta}_{L,S}), \sqrt{\lambda})\} + w_2 Z_S^{(2)}(\tilde{\theta}_{L,S}) \;=\; \Phi^{-1}(1-\alpha).$$

Following a similar construction when enrichment does not occur, we obtain the confidence intervals $(\theta_{L,S}, \infty)$ and $(\theta_{L,F}, \infty)$ where $\theta_{L,S}$ is the solution to

$$w_1 \Phi^{-1}\{1 - P_D(Z_S^{(1)}(\tilde{\theta}_{L,S}), \sqrt{\lambda})\} + w_2 \Phi^{-1}\{1 - P_D(Z_S^{(2)}(\tilde{\theta}_{L,S}), \sqrt{\lambda})\} \;=\; \Phi^{-1}(1-\alpha)$$

and $\theta_{L,F}$ is the solution to

$$w_1 \Phi^{-1}\{1 - P_D(Z_F^{(1)}(\tilde{\theta}_{L,F}), \sqrt{\lambda})\} + w_2 \Phi^{-1}\{1 - P_D(Z_F^{(2)}(\tilde{\theta}_{L,F}), \sqrt{\lambda})\} \;=\; \Phi^{-1}(1-\alpha).$$

We illustrate these methods in a set of examples with the observed data shown in Table 28.3.

In Example 1, enrichment occurred. The unadjusted MLE of θ_S is $\hat{\theta}_S = 0.90$ and, applying the method we have just described, the 97.5% upper confidence interval for θ_S is $(0.13, \infty)$. That the lower limit is above zero reflects the fact that the elementary hypothesis $H_S(0)$: $\theta_S \leq 0$ is rejected in the closed testing procedure for the set of hypotheses $H_S(\tilde{\theta}_S)$, $\tilde{\theta}_S \in R$, and $H_F(\tilde{\theta}_F)$, $\tilde{\theta}_F \in R$. Using a Simes test or a Bonferroni test for each intersection hypothesis, gives the 97.5% upper confidence interval $(0.10, \infty)$, which illustrates the higher power of the Dunnett-like test.

Table 28.3 *Examples of summary data from two-stage adaptive enrichment trials. Estimated treatment effects and associated information are shown for the specified sub-population S, the complement of this sub-population S^c, and for the full population F. Sub-population S comprises half of the full population, so $\lambda = 0.5$.*

	Stage 1 results			Stage 2 results		
Example 1						
	$\hat{\theta}_S^{(1)} = 1.1$	$\hat{\theta}_{S^c}^{(1)} = 0.5$	$\hat{\theta}_F^{(1)} = 0.8$	$\hat{\theta}_S^{(2)} = 0.8$		
	$\mathcal{I}_S^{(1)} = 2.5$	$\mathcal{I}_{S^c}^{(1)} = 2.5$	$\mathcal{I}_F^{(1)} = 5.0$	$\mathcal{I}_S^{(2)} = 5.0$		
Example 2						
	$\hat{\theta}_S^{(1)} = 1.3$	$\hat{\theta}_{S^c}^{(1)} = 0.4$	$\hat{\theta}_F^{(1)} = 0.85$	$\hat{\theta}_S^{(2)} = 1.0$	$\hat{\theta}_{S^c}^{(2)} = 0.3$	$\hat{\theta}_F^{(2)} = 0.65$
	$\mathcal{I}_S^{(1)} = 2.5$	$\mathcal{I}_{S^c}^{(1)} = 2.5$	$\mathcal{I}_F^{(1)} = 5.0$	$\mathcal{I}_S^{(2)} = 2.5$	$\mathcal{I}_{S^c}^{(2)} = 2.5$	$\mathcal{I}_F^{(2)} = 5.0$
Example 3						
	$\hat{\theta}_S^{(1)} = 1.3$	$\hat{\theta}_{S^c}^{(1)} = 0.4$	$\hat{\theta}_F^{(1)} = 0.85$	$\hat{\theta}_S^{(2)} = 1.0$	$\hat{\theta}_{S^c}^{(2)} = 0.1$	$\hat{\theta}_F^{(2)} = 0.55$
	$\mathcal{I}_S^{(1)} = 2.5$	$\mathcal{I}_{S^c}^{(1)} = 2.5$	$\mathcal{I}_F^{(1)} = 5.0$	$\mathcal{I}_S^{(2)} = 2.5$	$\mathcal{I}_{S^c}^{(2)} = 2.5$	$\mathcal{I}_F^{(2)} = 5.0$
Example 4						
	$\hat{\theta}_S^{(1)} = 1.3$	$\hat{\theta}_{S^c}^{(1)} = 0.4$	$\hat{\theta}_F^{(1)} = 0.85$	$\hat{\theta}_S^{(2)} = 1.0$	$\hat{\theta}_{S^c}^{(2)} = -0.3$	$\hat{\theta}_F^{(2)} = 0.35$
	$\mathcal{I}_S^{(1)} = 2.5$	$\mathcal{I}_{S^c}^{(1)} = 2.5$	$\mathcal{I}_F^{(1)} = 5.0$	$\mathcal{I}_S^{(2)} = 2.5$	$\mathcal{I}_{S^c}^{(2)} = 2.5$	$\mathcal{I}_F^{(2)} = 5.0$

In Example 2, the trial continues in the full population and the unadjusted MLEs are $\hat{\theta}_S = 1.15$ and $\hat{\theta}_F = 0.75$. Applying our method, the 97.5% upper confidence interval for θ_S is $(0.12, \infty)$ and that for θ_F is $(0.02, \infty)$.

Example 3 has the same first stage data as Example 2 but a lower value for $\hat{\theta}_{S^c}^{(2)}$ and hence $\hat{\theta}_F^{(2)}$. Unadjusted MLEs are $\hat{\theta}_S = 1.15$ and $\hat{\theta}_F = 0.70$. The 97.5% upper confidence interval for θ_S is $(0.12, \infty)$ and that for θ_F is $(-0.03, \infty)$. So, in this case, the closed testing procedure does not reject $H_F(0)$: $\theta_F \leq 0$.

In Example 4, the value of $\hat{\theta}_{S^c}^{(2)}$ is reduced further. The unadjusted MLEs are $\hat{\theta}_S = 1.15$ and $\hat{\theta}_F = 0.6$, the 97.5% upper confidence interval for θ_S is $(0.12, \infty)$ and that for θ_F is $(-0.13, \infty)$. Again, the closed testing procedure does not reject $H_F(0)$: $\theta_F \leq 0$.

It is instructive to consider the results that would have been obtained in Examples 2, 3 and 4 if we had started by defining a closed testing procedure for the two elementary hypotheses $H_S(0)$: $\theta_S \leq 0$ and $H_F(0)$: $\theta_F \leq 0$ only, and then followed the approach of Magirr et al. (2013) or Kimani et al. (2020) to add upper confidence intervals for θ_S and θ_F. In applying this method, we have used the Dunnett-like test for the intersection hypothesis $H_{SF}(0,0)$. In Example 2, both methods reject $H_S(0)$ and $H_F(0)$ and yield identical intervals for θ_S and θ_F. In Example 3, both approaches lead to rejection of $H_S(0)$ and a confidence interval for θ_S of $(0.12, \infty)$. The second method, which starts by testing only $H_S(0)$ and $H_F(0)$, also rejects $H_F(0)$ but the ensuing upper confidence interval of $(0, \infty)$ for θ_F is non-informative. In Example 4, both methods reject $H_S(0)$ but not $H_F(0)$. While the first method gave the upper confidence interval $(0.12, \infty)$ for θ_S, the second method gives the non-informative interval $(0, \infty)$. This is in keeping with the statement of Magirr et al. (2013) that

> "these lower bounds will only be informative if all hypotheses considered in the second stage of testing are rejected, and even this may be insufficient"

and these authors go on to comment

> "Furthermore, in situations where informative lower confidence bounds are deemed to be more important than the possibility of rejecting as many individual null hypotheses as possible, it would be sensible to use the intervals [of Posch et al. (2005)] instead of applying the closed test procedure."

For further discussion of methods designed to produce an informative confidence interval when a null hypothesis is rejected, we refer the reader to the papers by Brannath & Schmidt (2014) and Schmidt & Brannath (2014). The first paper presents weighted Bonferroni tests in which the weights depend continuously on the parameter values. The second paper extends multiple testing procedures using the graphical representation of Bretz, Maurer et al. (2009): after each rejection of a null hypothesis, a proportion of the significance level is kept to test non-null values of the parameter and construct an informative confidence interval. In both cases, the authors argue that in a clinical trial with multiple treatments, rejection of a null hypothesis is only of value if accompanied by an informative confidence interval for the treatment effect. Hence, a small loss of power, particularly for secondary hypotheses, is an acceptable price to pay for stronger statements about treatments that have been shown to have a positive effect.

Methods for producing estimates and confidence intervals have been extended to trials with a time-to-event outcome; see the papers by Kimani, Todd, Renfro et al. (2020) and Di Stefano, Pannaux, Correges, et al. (2022).

28.7 Regulatory Views on Enrichment Trials

The FDA Guidance on Enrichment Strategies for Clinical Trials (FDA, 2019b) considers strategies "intended to increase the efficiency of drug development and support precision medicine, i.e., tailoring treatments to those patients who will benefit based on clinical laboratory, genomic, and proteomic factors." In one form of enrichment trial, only biomarker-positive patients are enrolled but, if approved, a new drug could be available to the wider patient population. This prompts the comment in Section II of the Guidance that

> "The principal concerns with the use of enrichment strategies relate to the generalizability and applicability of the study results. When considering use of an enrichment design, sponsors should consider ... whether the drug might be useful in a broader population."

The Guidance goes on to state in Section VI B that

> "FDA encourages inclusion of some predictive marker-negative patients in most trials intended to provide primary effectiveness support, unless earlier studies have established that the marker-negative patients do not respond or a strong mechanistic rationale makes it clear that they will not respond"

and in Section VII A

> "In general, then, FDA's regulatory standards do not bar the Agency from approving drugs whose effectiveness has been demonstrated primarily or even solely in enriched populations, and FDA will seek to ensure truthful labeling that does not overstate the likelihood of a response, the magnitude of the response, or the predictiveness of the enrichment factor."

Section VI D of the Guidance concerns designs in which adaptive enrichment may occur and, after noting that

> "Such changes should be prospectively planned and would often need appropriate type I error rate control to account for interim, unblinded analyses of the accumulating data as well as analyses of multiple subgroups"

the Guidance states

> "In a study that includes both marker-positive and -negative patients, and with a primary endpoint that is the effect in the marker-positive group, a planned interim look could reveal ... that the marker-negative population has a much lower response than the marker-positive group. Additional enrollment of marker-negative patients could be reduced or stopped entirely, provided an adaptive enrichment strategy was planned and was appropriately accounted for in statistical analyses."

While noting that the decision to use an enrichment design is largely left to the sponsor of the investigation, the Guidance recommends in Section VI B that

> "Given the potentially complex interpretation of studies using enrichment designs, plans to use them should be discussed with FDA early in development."

The FDA's Guidance on Adaptive Designs for Clinical Trials of Drugs and Biologics (FDA, 2019a, Sec. V C) also addresses adaptive enrichment designs, stating

> "a trial might enroll subjects from the overall trial population up through an interim analysis, at which time a decision will be made based on prespecified criteria whether to continue enrollment in the overall population or to restrict future enrollment to the targeted subpopulation."

The general from of trial design and analysis that we described in Section 28.2 is recommended

> "Therefore, statistical hypothesis testing methods should account for both sources of multiplicity. For example, one approach is to combine test statistics or p-values from the different stages of the trial in a preplanned manner, while also using an appropriate multiple testing procedure (Wassmer and Brannath 2016). Such an approach could potentially also accommodate adaptations to the sample size or to the proportion of patients enrolled from a particular subpopulation"

The designs with two sub-populations discussed in Section 28.3 have the property that one may reject H_3: $\theta_3 \leq 0$, and conclude there is a positive treatment effect in the full population, based primarily on the outcomes observed in sub-population S_1. Indeed, it is possible to reject H_3 when the overall estimate of θ_2 is negative. In their investigations of adaptive enrichment designs, Ondra et al. (2019) address this issue by adding a requirement to their testing procedures that a sufficiently positive Z-test statistic be observed in each sub-population in order to conclude there is a treatment effect in the full population. Implications for the labeling of a new drug still remain since, even if regulatory approval is limited to a targeted sub-population, there is potential for off-label use in the complementary sub-population.

In Section 28.5, we considered situations where a sub-population was characterized by having a biomarker value above some threshold and, when the treatment effect increases with the biomarker value, there is the possibility that an overall effect may be shown in a selected sub-population when the treatment effect is small, or even absent, for patients with the lowest biomarker values. Section V C of the FDA's 2019 Guidance states

> "if the baseline characteristic that is thought to modify the treatment effect is not binary in nature, any threshold or thresholds used to define subpopulations should be appropriately justified."

The possible impact of such considerations was evident when the FDA's Oncologic Drugs Advisory Committee met in September 2024. After assessing evidence for the treatment of gastric cancer by PD-L1 (programmed death-ligand 1) inhibitors such as pembrolizumab, the committee recommended that such treatment be restricted to those subjects with combined positive scores (CPS) for PD-L1 expression greater than 1.

28.8 Bibliography and Notes

The concept of a trial design with "enrichment" has a long history. Temple (1994) described trial designs in which the treatment effect in the study population is enhanced by taking steps to increase the proportion of potential responders or eliminating patients who are seen to respond to placebo in a preliminary stage of the study. Follmann (1997) proposed a form of adaptive enrichment design motivated by experiences in a study of interventions to prevent and treat hypertension.

Brannath et al. (2009) followed the approach of Schmidli et al. (2006), using a closed testing procedure and combination tests to construct a Phase II/III seamless design with the opportunity for adaptive enrichment. In this case, the study had a survival endpoint analyzed by a stratified log-rank test and the enrichment decision was determined by a Bayesian decision rule based on predictive probabilities. Mehta et al. (2009, p. 603) used the same tools to create a group sequential design with sample size re-estimation and adaptive enrichment.

Lai, Lavori & Liao (2014) present plans for a trial to investigate endovascular therapy for stroke patients with six nested sub-populations and the potential to restrict recruitment to a subgroup of patients at two interim analyses. Lai, Lavori & Tsang (2019) describe the application of such methods in the Diffusion and Perfusion Imaging Evaluation for Understanding Stroke Evolution (DEFUSE) 3 trial.

Variations on the approach we described in Section 28.2 are possible and elements of these designs may be tuned to improve performance. Götte, Donica & Mordenti (2015) optimized within specific classes of rules for making an enrichment decision. Sugitani, Posch, Bretz & Koenig (2018) proposed a generalization of the method proposed by Spiessens & Debois (2010) for testing nested hypotheses and combined this with several forms of combination test. The articles by Stallard et al. (2014) and Ondra et al. (2016) review a variety of methods and contain further references.

Turnbull (2017) provides a variety examples illustrating how complexities of an adaptive design can pose logistical problems. A particular feature of enrichment designs is the need to determine the sub-population to which each patient belongs and there may be delays in this process. Turnbull reports an enrichment trial in oncology where enrollment of patients with any of five histologic subtypes could be stopped early for efficacy or futility. However, due to faster accrual and slower communications than anticipated, accrual was essentially completed before the first meeting of the data monitoring committee.

In this chapter, we have focused on prospective trials where the biomarker used in enrichment decisions is pre-defined. This is typically the case in confirmatory trials but a more flexible approach may sometimes be desirable. Freidlin & Simon (2005) note that in oncology, tumor types are heterogeneous and only a subset of the patients with a given cancer may benefit from a targeted agent. To address this, they propose "adaptive signature designs" for Phase III trials. The objective in the first stage is to identify a subset of patients most likely to benefit from the new agent, then the treatment effect in this subset is tested using only patients randomized in the remainder of the trial. Simon & Simon (2013) propose an adaptive enrichment design for a study with a binary outcome in which a classifier is developed to indicate patients with a higher probability of a positive response on the new treatment than on the control. Recruitment is then restricted to such patients. At the end of the study, an exact test of the null hypothesis of no treatment difference is conducted based on the sum of the number of responses on the new treatment and the number of non-responses on the control. Simon & Simon (2018) develop this design further, using

Bayesian methods to model the probability of a positive response for enrollment decisions but retaining the original frequentist test for a treatment difference.

Chapter 29

Adaptive Seamless Design: A Case Study with Intermediate and Long-term Survival Endpoints

29.1 Motivating Example: The GATSBY Trial

The example we present in this chapter is motivated by the GATSBY trial, as reported by Thuss-Patience et al. (2017). The planning of a study of this form is described by Carreras, Gutjahr & Brannath (2015) and we shall cover similar design issues in our discussion. The GATSBY trial investigated the efficacy and safety of trastuzumab emtansine for the treatment of patients with previously treated, HER2-positive, advanced gastric cancer. The primary endpoint was overall survival (OS), the elapsed time from the date of randomization to date of death from all causes. If no death was reported, OS for that subject was considered as censored at the last date they were known to be alive. The adaptive seamless Phase II/III design (ASD) that was employed is depicted in the figure below.

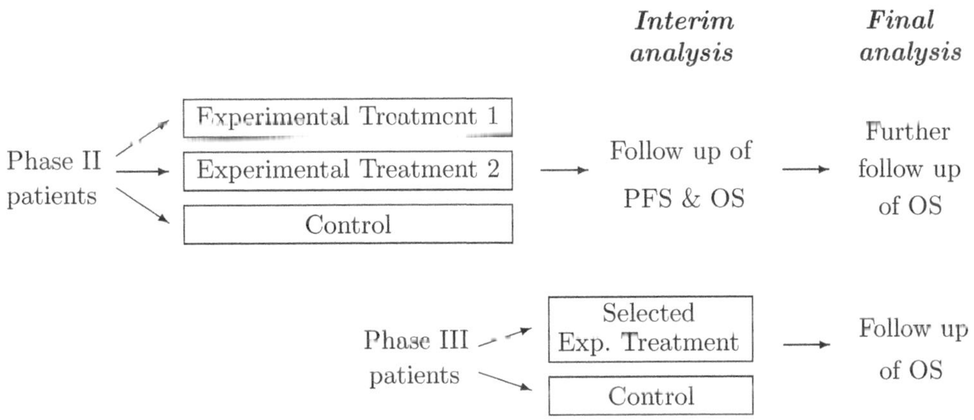

Figure 29.1 *Plan of an adaptive seamless Phase II/III trial*

In the Phase II stage, groups receiving the high and low doses of trastuzumab emtansine are compared with a control group in which patients receive the standard of care, either taxane or paclitaxel at the discretion of their physician. At an interim analysis, one of the doses of trastuzumab emtansine is selected for the Phase III stage. Additional subjects are then recruited and these are randomized between the selected experimental arm and the control arm.

Because the OS data are immature at the interim decision point, the treatment selection is based primarily on pharmacokinetic data and the intermediate endpoint of

DOI: 10.1201/9781584888482-29

progression-free survival (PFS). PFS is defined as the elapsed time from randomization to disease progression or death from any cause, whichever occurs first. PFS is also subject to censoring when neither progression nor death is known to have occurred.

The final analysis at the conclusion of the Phase III stage tests the null hypothesis that OS on the selected treatment is no better than OS on the control. Just as in Chapter 26, this analysis must take into account the selection bias from choosing the experimental treatment that performed better in Phase II. There is a further issue that subjects who are still alive at the end of Phase II will continue to be followed for OS in Phase III. This can lead to a similar problem as that seen for enrichment designs in Section 28.4. Suppose experimental treatment i^* is selected based on good PFS outcomes. By definition, PFS is positively associated with OS and a consequence of treatment selection is that the further contribution of Phase II patients on experimental treatment i^* to the log-rank statistic used to test OS is liable to be positively biased.

29.2 A Closed Testing Procedure and Combination Test for Survival Data

29.2.1 A Proportional Hazards Model

We assume a proportional hazards model for OS, as defined in Sections 4.5 and 17.2. For $i = 1$ and 2, let λ_i denote the hazard ratio for experimental treatment i versus the control, where a value $\lambda_i > 1$ implies the experimental treatment has superior survival. Defining $\theta_1 = \log(\lambda_1)$ and $\theta_2 = \log(\lambda_2)$, we are interested in testing the hypotheses H_1: $\theta_1 \leq 0$ against $\theta_1 > 0$ and H_2: $\theta_2 \leq 0$ against $\theta_2 > 0$.

Since we have two hypotheses, we shall define a testing procedure that controls the family-wise error rate (FWER) at a specified level α. We can do this by applying a closed testing procedure formed from level α tests of H_1, H_2 and the intersection hypothesis H_{12} that states $\theta_1 \leq 0$ and $\theta_2 \leq 0$. We shall use a weighted inverse normal combination test to combine information from before and after the interim analysis at which one of the experimental treatments is selected.

29.2.2 Defining Stage-wise Z-values

The ingredients of the desired closed testing procedure can be summarized as the following set of Z-values:

	Phase II data	*Phase III data*
H_1	$Z_1^{(1)}$	$Z_1^{(2)}$
H_2	$Z_2^{(1)}$	$Z_2^{(2)}$
H_{12}	$Z_{12}^{(1)}$	$Z_{12}^{(2)}$

A natural choice would be to base $Z_1^{(1)}$, $Z_2^{(1)}$ and $Z_{12}^{(1)}$ on data observed up to the interim analysis at which an experimental treatment is selected, and to base $Z_1^{(2)}$, $Z_2^{(2)}$ and $Z_{12}^{(2)}$ on data observed after that point. The statistic $Z_{12}^{(1)}$ could be formed from $Z_1^{(1)}$ and $Z_2^{(1)}$ using, for example, Simes' test. Since only one of $Z_1^{(2)}$ and $Z_2^{(2)}$ is observed, $Z_{12}^{(2)}$ would be set equal to that value.

However, there is a problem with this scheme. For the combination test of H_{12} to be valid, the conditional distribution of $Z_{12}^{(2)}$ under H_{12}, given the Phase II statistics $Z_1^{(1)}$, $Z_2^{(1)}$ and $Z_{12}^{(1)}$, should be $N(0, 1)$ or stochastically smaller than this. As noted above, since the experimental treatment i^* that goes forward to Phase III is selected based on PFS data, the contribution of continuing patients on treatment i^* to the log-rank statistic $Z_{12}^{(2)} = Z_{i^*}^{(2)}$

Phase II cohort	Overall survival (during stage 1)	Overall survival (during stage 2)	$Z_1^{(1)}, Z_2^{(1)}, Z_{12}^{(1)}$
Phase III cohort		Overall survival (during stage 2)	$Z_{i^*}^{(2)}, Z_{12}^{(2)}$

Figure 29.2 *Partitioning data to avoid bias in a combination test for survival data*

is liable to produce a positive bias. A solution to this problem is to follow the proposal of Jenkins, Stone & Jennison (2011), as previously described in Section 28.4. In this approach, we define cohorts 1 and 2 to consist of patients who enter the study during the Phase II and Phase III stages, respectively. We define $Z_1^{(1)}$ and $Z_2^{(1)}$ to be the log-rank statistics comparing experimental treatments 1 and 2 against the control based on the total OS data for the first cohort, from before *and after* the interim analysis at which treatment selection occurs. The intersection test statistic $Z_{12}^{(1)}$ is formed from $Z_1^{(1)}$ and $Z_2^{(1)}$ using, for example, Simes' test. Then, if experimental treatment i^* is selected for the Phase III stage, we define $Z_{i^*}^{(2)}$ to be the log-rank statistic comparing this treatment against the control in the second cohort, and we set $Z_{12}^{(2)} = Z_{i^*}^{(2)}$. This scheme for partitioning the data is depicted in Figure 29.2.

Since $Z_{i^*}^{(2)}$ comes from a completely new set of subjects, under $\theta_{i^*} = 0$ the conditional distribution of $Z_{i^*}^{(2)}$ given data from the Phase II cohort is $N(0,1)$, up to the usual, large sample approximation for a log-rank statistic. It follows that $Z_{i^*}^{(2)} \sim N(0,1)$ unconditionally when $\theta_{i^*} = 0$ and, by the argument of Section 23.2, $Z_{i^*}^{(2)}$ is stochastically smaller than a $N(0,1)$ random variable when $\theta_{i^*} < 0$. Since $Z_{12}^{(2)} = Z_{i^*}^{(2)}$, this statistic is $N(0,1)$, or stochastically smaller, when H_{12} holds.

Because $Z_1^{(1)}$, $Z_2^{(1)}$ and $Z_{12}^{(1)}$ involve data observed after the interim analysis at which treatment selection occurs, conditions must be imposed to ensure these have the desired distributions under H_1, H_2 and H_{12}. The conditions stipulated by Jenkins, Stone & Jennison (2011) are that the body of data on which $Z_1^{(1)}$ and $Z_2^{(1)}$ are based should not depend in any way on the OS or PFS data seen at the interim analysis. This can be achieved by requiring all cohort 1 subjects, on both experimental treatments and the control, to be followed up to a fixed calendar time or until a fixed number of events have occurred. Options for setting the time of the final analysis so as to satisfy such a constraint are discussed further in Carreras, Gutjahr & Brannath (2015, Sec. 3.2).

29.2.3 Testing the Intersection Hypothesis

In dealing with data from the Phase II cohort, we need to define the statistic $Z_{12}^{(1)}$.

Simes' test

We have already mentioned Simes' test as a possible choice. This is appropriate since the comparison with a common control means that $Z_1^{(1)}$ and $Z_2^{(1)}$ are positively associated. Using Simes' test, we define the P-values

$$P_1^{(1)} = 1 - \Phi(Z_1^{(1)}) \quad \text{and} \quad P_2^{(1)} = 1 - \Phi(Z_2^{(1)}),$$

set

$$P_{12}^{(1)} = \min\{2\min(P_1^{(1)}, P_2^{(1)}), \max(P_1^{(1)}, P_2^{(1)})\},$$

and define

$$Z_{12}^{(1)} = \Phi^{-1}(1 - P_{12}^{(1)}).$$

Dunnett's test

If $\theta_1 = 0$ and $\theta_2 = 0$, the survival distributions for the three treatment arms are identical and, in the large sample approximation, $(Z_1^{(1)}, Z_2^{(1)})$ is bivariate normal with mean $(0,0)$ and variance matrix

$$\begin{pmatrix} 1 & 0.5 \\ 0.5 & 1 \end{pmatrix}.$$

We can, therefore, compute a Dunnett P-value

$$P_{12}^{(1)} = Pr\{\max(Z_1^{(1)}, Z_2^{(1)}) > \max(z_1^{(1)}, z_2^{(1)})\},$$

where $z_1^{(1)}$ and $z_2^{(1)}$ are the observed values of $Z_1^{(1)}$ and $Z_2^{(1)}$, as explained in Section 21.4. We then set $Z_{12}^{(1)} = \Phi^{-1}(1 - P_{12}^{(1)})$. For a more precise method of calculating the correlation between $Z_1^{(1)}$ and $Z_2^{(1)}$ given the observed numbers of events in each treatment group, see Section 3 of Di Scala & Glimm (2011).

A pooled data comparison

Since the survival distributions of the three treatment arms are identical when $\theta_1 = 0$ and $\theta_2 = 0$, we could test H_{12} by applying a log-rank test to compare the pooled data from the two experimental treatment arms with the control arm.

We shall report a simulation study assessing these three options for defining $Z_{12}^{(1)}$ in Section 29.3. Since only one experimental treatment arm, treatment i^* say, is retained in the Phase III stage, there is no issue in how to define $Z_{12}^{(1)}$ and we simply set $Z_{12}^{(2)} = Z_{i^*}^{(2)}$.

29.2.4 Combination Tests

Having defined the set of Z-statistics, using these in combination tests of H_1, H_2 and H_{12} is relatively straightforward. In each case, we apply a weighted inverse normal combination test with pre-defined weights w_1 and w_2 satisfying $w_1^2 + w_2^2 = 1$. When treatment i^* is selected for the Phase III stage, we require positive results in the level α tests of H_{i^*} and H_{12} in order to reject H_{i^*} overall in the closed testing procedure, and this is the case if

$$w_1 Z_{i^*}^{(1)} + w_2 Z_{i^*}^{(2)} \geq \Phi^{-1}(1 - \alpha) \tag{29.1}$$

and

$$w_1 Z_{12}^{(1)} + w_2 Z_{12}^{(2)} \geq \Phi^{-1}(1 - \alpha). \tag{29.2}$$

A note on early stopping

In the procedure we have described, there is no allowance for early stopping at the interim analysis. In the main, this is because $Z_1^{(1)}$ and $Z_2^{(1)}$ are not yet known at this time. In principle, one could divide the data from Phase II patients into two parts, giving sequences of three Z-values for each hypothesis, and define a stopping rule based on the first set of Z-values. However, since only a small number of deaths are likely to have occurred before the interim analysis, the information about OS would be small and an early positive decision highly unlikely. Of course, if PFS and OS were sufficiently poor on the experimental treatment arms, a decision to stop early for futility could always be made without inflating the Type I error.

29.3 Design Considerations

We have conducted a simulation study to investigate the proposed design. This is a standard part of the process for planning a complex trial design and, with time-to-event endpoints, there are many scenarios to consider and simulations for each scenario can be time consuming. We simplified this task by simulating summary statistics directly, rather than generating PFS and OS data for individual patients. We assumed that overall survival followed a proportional hazards model with log hazard ratios θ_1 and θ_2 between experimental treatments 1 and 2 and the control. For progression-free survival, we simulated results as if the time to PFS followed a proportional hazards model with log hazard ratios ψ_1 and ψ_2 between experimental treatments 1 and 2 and the control. In fact, given the relation between time to progression and overall survival time, it is difficult, if not impossible, to define a joint model for progression and survival times in which both these proportional hazards assumptions hold. However, even if the proportional hazards assumption does not hold it is still possible to summarize the PFS data by estimates $\hat{\psi}_1$ and $\hat{\psi}_2$ of assumed log hazard ratios, and this is how our simulations should be interpreted.

We applied the large sample approximation that a standardized log-rank statistic based on d observed events is, approximately,

$$N(\beta\surd\{d/4\}, 1)$$

when the log hazard ratio is β, and we used this result with β taking the value of θ_1, θ_2, ψ_1 or ψ_2. We denote the correlation between log-rank statistics for OS and PFS when comparing a particular experimental treatment with the control by ρ.

For $i = 1$ and 2, denote the number of OS events for patients in the Phase II cohort on experimental treatment i and the control combined by n_i. Let Y_i denote the standardized log-rank statistic when comparing experimental treatment i with the control based on the PFS data available at the interim analysis, and denote the number of PFS events in experimental treatment i and the control combined by m_i. Then, as a consequence of our assumptions, we have, approximately,

$$\begin{pmatrix} Z_i^{(1)} \\ Y_i \end{pmatrix} \sim N\left(\begin{pmatrix} \theta_i\surd\{n_i/4\} \\ \psi_i\surd\{m_i/4\} \end{pmatrix}, \begin{pmatrix} 1 & \rho \\ \rho & 1 \end{pmatrix} \right).$$

Assuming patients are randomized equally to each experimental treatment and the control, we also have

$$\text{Corr}(Z_1^{(1)}, Z_2^{(1)}) = 0.5 \quad \text{and} \quad \text{Corr}(Y_1, Y_2) = 0.5.$$

Choice of test statistic for the intersection hypothesis

Our first use of simulations is to compare the options for testing the intersection hypothesis H_{12} in Phase II. Here, we have assumed that, in the Phase II cohort, a total of 300 OS events were observed with 100 on the control arm, so

$$n_1 = n_2 = 200,$$

and the number of PFS events observed at the interim analysis was also 300, with 100 on the control arm, so

$$m_1 = m_2 = 200.$$

We assume the correlation between log-rank statistics for OS and PFS for a particular treatment comparison to be $\rho = 0.6$. We set $\psi_1 = \theta_1$ and $\psi_2 = \theta_2$ for several pairs of values for θ_1 and θ_2. Tests in the closed testing procedure are carried out at the significance level $\alpha = 0.025$, so the FWER is also controlled at this level. The trial is conducted following the procedure described in Section 29.2 with weights $w_1 = w_2 = \surd 2$ in the combination tests.

Table 29.1 *Comparisons of three methods for testing the intersection hypothesis H_{12} in the Phase II stage of an adaptive seamless Phase II/III design with OS and PFS endpoints. The table shows the probabilities of a successful outcome, $P(1) = Pr$(Select treatment 1 and reject H_1) and $P(2) = Pr$(Select treatment 2 and reject H_2), using three different tests of H_{12}, as well as the expected gain, defined as $\theta_1 \times P(1) + \theta_2 \times P(2)$. All results are based on 1,000,000 simulations.*

			$P(1)$			E(Gain)	
$\theta_1 = \psi_1$	$\theta_2 = \psi_2$	Pooled	Simes	Dunnett	Pooled	Simes	Dunnett
0.3	0.0	0.78	0.85	0.87	0.234	0.256	0.261
0.3	0.1	0.79	0.81	0.82	0.239	0.246	0.249
0.3	0.2	0.69	0.69	0.69	0.238	0.238	0.239
0.3	0.25	0.59	0.58	0.58	0.251	0.250	0.250
0.3	0.295	0.48	0.48	0.48	0.276	0.274	0.274

For each scenario simulated, we note

$$P(1) = Pr(\text{Select treatment 1 and reject } H_1 \text{ overall}),$$

$$P(2) = Pr(\text{Select treatment 2 and reject } H_2 \text{ overall})$$

and

$$E(\text{Gain}) = \theta_1 \times P(1) + \theta_2 \times P(2).$$

Here "Gain" represents a possible utility, in which the value of a positive outcome is taken to be proportional to the effect size of the recommended treatment. The simulation results are shown in Table 29.1. Note that a log hazard ratio of $\theta = 0.3$ corresponds to a hazard ratio of $\lambda = e^{0.3} = 1.35$. in which case the hazard rate for the new treatment is $e^{-0.3} = 0.74$ times that for the control.

The first point to note is the poor performance of the procedure using the pooled data to test H_{12}. This may seem surprising as the pooled analysis gives the most efficient test of H_{12} on its own. The drawback is a lack of consonance with the tests of H_1 and H_2. Since a high value of $Z_{12}^{(1)}$ is not necessarily accompanied by a very high value of either $Z_1^{(1)}$ or $Z_2^{(1)}$ it is possible that the overall test given by (29.2) may reject H_{12} but the test (29.1) does not reject H_{i^*}, so H_{i^*} is not rejected overall in the closed testing procedure.

Our next observation is that somewhat better results are obtained using a Dunnett test rather than a Simes procedure to test H_{12} in the Phase II stage. This is in keeping with the findings of Hampson & Jennison (2015) cited in Section 26.4 and results of investigations into MAMS designs reported in Section 27.4.3.

Efficiency comparison with a simpler trial design

Our next use of simulations is to ascertain whether the adaptive design provides a worthwhile improvement on a simpler, non-adaptive design. Given previous findings, we focus on the adaptive design using a Dunnett test for the intersection hypothesis in Phase II.

A possible non-adaptive design is shown in Figure 29.3. Here, patients are randomized between all three treatment arms throughout the trial and a closed testing procedure is applied on conclusion of the trial. In comparing different tests of the intersection hypothesis H_{12}, we again found the Dunnett test to give the best power, so we also use this method for the non-adaptive design.

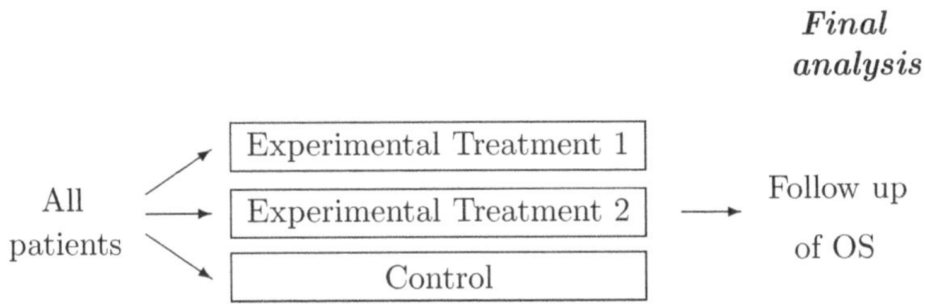

Figure 29.3 *Plan of a non-adaptive trial*

Now, it is possible that both H_1 and H_2 are rejected and, in this case, we suppose the treatment with the higher observed effect size is to be chosen for registration and marketing. Accordingly, with $\hat{\theta}_1$ and $\hat{\theta}_2$ denoting estimates of θ_1 and θ_2 from the final data, we define

$$P(1) = Pr(\hat{\theta}_1 > \hat{\theta}_2 \text{ and } H_1 \text{ is rejected overall}),$$

$$P(2) = Pr(\hat{\theta}_2 > \hat{\theta}_1 \text{ and } H_2 \text{ is rejected overall}),$$

and, as before, $E(\text{Gain}) = \theta_1 \times P(1) + \theta_2 \times P(2)$.

In our simulations we have supposed that, in the non-adaptive design, a total of 600 OS events are observed between the three treatment arms, maintaining the same total number of events as in the adaptive design. As before, we take $\rho = 0.6$ and $\alpha = 0.025$. Results of these simulations are shown in Table 29.2. We see from these results that the adaptive design performs particularly well when θ_1 is substantially greater than θ_2, achieving a higher value of $P(1)$ and higher $E(\text{Gain})$. When θ_1 and θ_2 are closer, the adaptive design may make an incorrect choice of experimental treatment but, overall, a higher value of $P(2)$ and a lower value of $P(1)$ combine to give the adaptive design a higher value of $E(\text{Gain})$ than the non-adaptive design.

A conclusion is emerging as to when the adaptive design is effective. First, the treatment effects on PFS, ψ_1 and ψ_2, must be reliable indicators of θ_1 and θ_2, the treatment effects on OS. Then, adequate information on PFS must be available at the time of the interim analysis to reliably select the treatment with the better PFS. For the case $\theta_1 = 0.3$ and $\theta_2 = 0.1$, we considered additional scenarios with

$$\psi_1 = \gamma\theta_1 \quad \text{and} \quad \psi_2 = \gamma\theta_2,$$

Table 29.2 *Comparisons of a non-adaptive design and an adaptive seamless Phase II/III design. Both designs test for a treatment effect on OS. The adaptive seamless design makes a treatment selection based on PFS at an interim analysis. The table shows probabilities $P(1)$ and $P(2)$ of a successful outcome, suitably defined for each design, and expected gain, defined as $\theta_1 \times P(1) + \theta_2 \times P(2)$. All results are based on 1,000,000 simulations.*

		Non-adaptive			Adaptive		
$\theta_1 = \psi_1$	$\theta_2 = \psi_2$	$P(1)$	$P(2)$	$E(\text{Gain})$	$P(1)$	$P(2)$	$E(\text{Gain})$
0.3	0.0	0.78	0.00	0.235	0.87	0.00	0.261
0.3	0.1	0.78	0.01	0.234	0.82	0.02	0.249
0.3	0.2	0.70	0.12	0.234	0.69	0.16	0.239
0.3	0.25	0.60	0.25	0.244	0.58	0.30	0.250
0.3	0.295	0.47	0.43	0.267	0.48	0.45	0.274

Table 29.3 *Further comparisons of a non-adaptive design and an adaptive seamless Phase II/III design. Both designs test for a treatment effect on OS. The adaptive seamless design makes a treatment selection based on PFS at an interim analysis. Parameters ξ and γ govern the amount of information on PFS available at the interim analysis where one of the experimental treatments is selected. The table shows probabilities $P(1)$ and $P(2)$ of a successful outcome, suitably defined for each design, and expected gain, defined as $\theta_1 \times P(1) + \theta_2 \times P(2)$. All results are based on 1,000,000 simulations.*

			Non-adaptive			Adaptive	
γ	ξ	$P(1)$	$P(2)$	$E(\text{Gain})$	$P(1)$	$P(2)$	$E(\text{Gain})$
1.5	1.2				0.88	0.00	0.264
1.2	1.1				0.86	0.01	0.258
1.0	1.0	0.78	0.01	0.234	0.82	0.02	0.249
0.9	0.9		for all γ and ξ		0.79	0.03	0.241
0.8	0.8		(PFS is not used)		0.76	0.04	0.231
0.7	0.7				0.72	0.05	0.220
0.6	0.6				0.67	0.06	0.208

and, in the adaptive design, with the number of PFS events at the interim analysis equal to $\xi \times 300$. Other aspects of the trial remained as before. Results with selected values of γ and ξ are shown in Table 29.3. It is evident from these results that the adaptive design works well if the hazard ratio for PFS is large enough and there are enough PFS events to ensure reliable treatment selection. When this is the case, the adaptive design offers the chance to select the better treatment and focus on this treatment in the Phase III stage of the trial. However, when these conditions are not met, the adaptive design may perform less well than a simple, non-adaptive design.

29.4 The Planning Process

General comments on simulations

The main difficulty when planning a trial of this type is that the parameters which determine whether an adaptive design will have superior operating characteristics to a non-adaptive design are unknown. Investigators must, therefore, consider which scenarios are most likely and decide if one or other design is superior in the majority of these situations. The criteria by which designs are to be judged need to be set. We have shown results for a simple gain function but other definitions of "gain" or "utility" are, of course, possible.

A third option for the trial would have been to choose one of the experimental treatments at the outset and conduct a two-arm trial comparing this treatment against the control. In a Bayesian decision framework, this may turn out to be the preferred option under a prior distribution for (θ_1, θ_2) that captures investigators' beliefs and expectations. However, in such a decision framework, it would be advisable to consider the relative risks in the two experimental treatments of adverse events that could lead to termination of the trial for safety reasons. A design that includes both experimental treatments has more protection against such eventualities.

We have presented a simple approach to running simulations in which summary statistics are generated directly. A more thorough approach would be to generate times to progression and death for each patient and conduct a proper survival analysis. An advantage of our approach is that the simulations run very quickly. Furthermore, as long as the important aspects of the data generating process have been modeled, a simple set of simulation results can make it easier to detect the key factors that determine when different design choices are to be preferred.

The role of the data monitoring committee

Where a decision such as selecting a treatment arm is to made during the trial, it is the independent data monitoring committee (IDMC) that will have this task. It is therefore important to give clear instructions of how such a decision should be made: once the IDMC has seen trial data, a firewall will be in place and it will be too late to ask for clarification as to whether, for example, pharmacokinetic data or PFS should have the primary role in differentiating treatments. The kick-off meeting between the trial sponsor and the IDMC, which is held before any data have been seen, offers an ideal opportunity to address such issues.

The GATSBY trial, which we outlined in Section 29.1, used the ASD design with the method of Jenkins, Stone & Jennison (2011) to underpin combination tests. Thuss-Patience et al. (2017) report that in the first stage, between 3 September 2012 and 14 October 2013, 70 patents were assigned to high dose trastuzumab emtansine, 75 to low dose, and 37 to the control group. Note that patients were randomized to the treatment arms in the ratios 2:2:1. At the pre-planned interim analysis on 14 Oct 2013, the IDMC selected the low dose to proceed to the second stage based on all information available, including OS, PFS and safety data. In the second stage, a further 153 patients were assigned to low dose trastuzumab emtansine and 80 to the control. At the final analysis in June 2015, the results were negative showing no significant benefit in OS for the low dose treatment over the control. This finding was consistent with that for PFS, the secondary efficacy endpoint. This negative outcome raises the question as to whether the trial could have been terminated early. It is customary for an IDMC to meet regularly to review safety data, so it should not be difficult for that group to consider a futility stopping rule at several occasions during the course of a trial.

29.5 Bibliography and Notes

Koenig et al. (2008) describe adaptive Dunnett tests for trials with treatment selection. They combine data from stages using the conditional error approach that we described in Section 23.6. Irle & Schäfer (2012) proposed methods for survival data designed to overcome the problem of a potential bias discussed in Section 29.2.2. Their procedure is also based on the conditional error approach. However, determining the conditional probability of a future event based on survival data can be problematic, since the final number of events and, hence, the information level for a log hazard ratio, is not known at the time this probability has to be calculated. The approach of Jenkins, Stone & Jennison (2011), which uses a combination test with pre-specified weights, avoids this problem. Magirr, Jaki, Koenig & Posch (2016) note that when sample size is reassessed at an adaptive analysis, the above methods may fail to include a significant subset of the observed events and they propose an alternative test that incorporates all event times.

Friede et al. (2011) also consider a seamless Phase II/III trial design with treatment selection based on both short-term and long-term responses. They give an example of a trial comparing treatments for multiple sclerosis. The primary, long-term outcome is the Extended Disability Scale score at three years post randomization but a short term response, based on a magnetic resonance imaging metric, is available at one year. Although these are not survival or time-to-event endpoints, they are measured after lengthy time intervals. When the treatment selection decision is made, only a short-term response is available for some subjects but these patients will go on to provide a long-term response later. In view of the correlation between short-term and long-term outcomes and the fact that the treatment selection decision is based on promising short-term response data, patients on the selected treatment are likely to produce results that are biased toward a positive treatment effect. Friede et al. (2011) follow a similar approach to Jenkins, Stone & Jennison (2011) and apply a combination test to the long-term response data from the cohorts of patients admitted before and after the interim decision point.

Chapter 30

Response Adaptive Treatment Assignment

30.1 Response Adaptive Randomization

The concept of adaptive randomization dates back to the work of Thompson (1933), who proposed a method to allocate a higher fraction of individuals in a study to the better of two treatments. Subsequently, Robbins (1952) introduced the "two-armed bandit" problem in which it is desired to maximize the expected return when presented with two Bernoulli payoffs with unknown and unequal success probabilities: this model can be used to describe a clinical trial where the outcome for each patient is "success" or "failure", the treatment allocated to each new patient is based on results observed so far, and the aim is to maximize the number of patients who receive the superior treatment. Cornfield, Halperin & Greenhouse (1969) addressed clinical trial design directly and proposed an adaptive randomization procedure "in which an increasing proportion of patients is assigned to the better of two treatments as evidence for it accumulates". Many more proposals for designs with adaptive randomization have followed. Robertson, Lee et al. (2023) provide a review and commentary of response adaptive randomization (RAR) methods for clinical trials. We shall give a brief summary of the wide variety of RAR designs in Section 30.4.

In Section 30.2 we present a simple approach to incorporating RAR in a clinical trial comparing two treatments, where the intent is to reduce the number of patients receiving the inferior treatment, referred to as the "inferior treatment number" (ITN). We shall focus on one-sided tests, first assuming no time trend in patient response, then with a stratified analysis that accounts for an arbitrary time trend. In Section 30.3 we assess the potential benefits of such methods in an illustrative example. There, we observe that a group sequential stopping rule is effective in reducing both the average total sample size and the ITN. Furthermore, a design that incorporates both a group sequential stopping and adaptive randomization is better still. However, in a comparison of a new treatment against a control where a positive treatment effect is anticipated, similar results can be achieved using a fixed randomization rule that allocates a higher fraction of subjects to the new treatment throughout the trial.

30.2 A Multi-stage Response Adaptive Design

30.2.1 Response Adaptive Randomization in a Group Sequential One-sided Test

Consider the unpaired two treatment comparison of Section 3.2.1. It is not unusual for unequal randomization or drop-out of some subjects to produce unbalanced numbers of observations on the two treatments. Here, we consider intentional imbalance between treatments achieved by allowing treatment allocation in each stage to depend on previously observed outcomes. Such designs are said to use stage-wise adaptive sampling rules.

Suppose responses $X_{Ai} \sim N(\mu_A, \sigma_A^2)$, $i = 1, 2, \ldots$, are observed for subjects allocated to a new treatment A and responses $X_{Bi} \sim N(\mu_B, \sigma_B^2)$, $i = 1, 2, \ldots$, for subjects on the control treatment B, where the variances σ_A^2 and σ_B^2 are known but not necessarily equal. In each

DOI: 10.1201/9781584888482-30

stage $k = 1, \ldots, K$ we observe m_{Ak} patients on treatment arm A and m_{Bk} on treatment arm B. The cumulative numbers of subjects observed by analysis k on arms A and B are

$$n_{Ak} = \sum_{j=1}^{k} m_{Aj} \quad \text{and} \quad n_{Bk} = \sum_{j=1}^{k} m_{Bj}$$

and we denote the cumulative sample means by $\bar{X}_A^{(k)}$ and $\bar{X}_B^{(k)}$, respectively.

Let $\theta = \mu_A - \mu_B$ and suppose we wish to test H_0: $\theta \leq 0$ against $\theta > 0$ in a group sequential one-sided test. The standardized statistic at analysis k for testing H_0 is

$$Z_k = (\bar{X}_A^{(k)} - \bar{X}_B^{(k)})\sqrt{\mathcal{I}_k},$$

where

$$\mathcal{I}_k = (\sigma_A^2/n_{Ak} + \sigma_B^2/n_{Bk})^{-1}. \qquad (30.1)$$

In Chapter 5 we saw how to specify an information sequence $\mathcal{I}_1, \ldots, \mathcal{I}_K$ and a boundary for $Z_1, \ldots, Z_K$ to satisfy Type I and II error probability requirements. If we opt for equal increments of information, so $\mathcal{I}_k = (k/K)\mathcal{I}_K$, $k = 1, \ldots, K$, the final information level $\mathcal{I}_K$ is determined using constants tabulated in Chapter 5.

In defining adaptive sampling rules, we shall assume $\mathcal{I}_1, \ldots, \mathcal{I}_K$ are pre-specified and design the sampling rules to produce these information levels. Thus, when planning stage k, group sizes m_{Ak} and m_{Bk} must be chosen to satisfy (30.1) with the pre-specified value of $\mathcal{I}_k$. There is still flexibility in the choice of m_{Ak} and m_{Bk} and we allow these to depend on the accumulated data through the current estimate of the treatment effect,

$$\hat{\theta}^{(k-1)} = \bar{X}_A^{(k-1)} - \bar{X}_B^{(k-1)}.$$

We show in Section 30.2.3 that, under such a sampling scheme, $\{Z_1, \ldots, Z_K\}$ have the canonical joint distribution (3.1) conditional on $\{\mathcal{I}_1, \ldots, \mathcal{I}_K\}$, and hence the required error probabilities are attained. However, the distribution (3.1) can fail to hold and error rates may vary from their intended values if m_{Ak} and m_{Bk} are allowed to depend on $\bar{X}_A^{(k-1)}$ and $\bar{X}_B^{(k-1)}$ in other ways.

The aim of an adaptive allocation rule is to reduce the number of subjects assigned to the inferior treatment arm. Whether this is arm A or B is unknown initially, but at any stage during the trial the current estimate of θ provides an indication of the relative merits of the two treatments which can be used to determine adaptive treatment assignments. Reducing the expected number of subjects allocated to the inferior treatment, the ITN, usually leads to an increase in the total expected sample size or "average sample number" (ASN). A large ASN is also undesirable since it implies a longer elapsed time before the trial can be concluded. As a device to balance the competing goals of reducing the ITN and ASN, one can aim to minimize the expected value of a loss function of the form

$$L(\theta) = u(\theta)\, n_{A\tau} + v(\theta)\, n_{B\tau}, \qquad (30.2)$$

where τ denotes the stage at which the study terminates and $n_{A\tau}$ and $n_{B\tau}$ are the cumulative sample sizes on arms A and B at this point. Supposing a high response is desirable, the inferior treatment is A if $\theta < 0$ and B if $\theta > 0$. Thus, the weights $u(\theta)$ and $v(\theta)$ in (30.2) should be strictly positive with $v(\theta)$ increasing in θ for $\theta > 0$ and $u(\theta)$ increasing as θ decreases for $\theta < 0$. Following Jennison & Turnbull (2001b), we shall consider the particular case where

$$u(\theta) = \begin{cases} 1 & \text{for } \theta \geq 0, \\ a^{-\theta/\delta} & \text{for } \theta \leq 0, \end{cases} \quad \text{and} \quad v(\theta) = \begin{cases} a^{\theta/\delta} & \text{for } \theta \geq 0, \\ 1 & \text{for } \theta \leq 0, \end{cases}$$

where δ is the value of θ appearing in the test's power condition and a is a positive constant chosen to reflect the emphasis placed on reducing the inferior treatment number rather than the total sample size. Thus

$$L(\theta) = \begin{cases} n_{A\mathcal{T}} + a^{\theta/\delta}\, n_{B\mathcal{T}} & \text{for } \theta \geq 0, \\ a^{-\theta/\delta}\, n_{A\mathcal{T}} + n_{B\mathcal{T}} & \text{for } \theta \leq 0. \end{cases} \tag{30.3}$$

Since successive information levels are subject to the constraint (30.1), the allocation ratio minimizing (30.2) is

$$\frac{n_{A\mathcal{T}}}{n_{B\mathcal{T}}} = \frac{\sigma_A}{\sigma_B}\, w(\theta), \tag{30.4}$$

where

$$w(\theta) = \sqrt{\frac{v(\theta)}{u(\theta)}} = a^{\theta/(2\delta)}.$$

Although this optimal ratio depends on the unknown θ, we can attempt to achieve it approximately using successive estimates of θ in the following adaptive rule.

30.2.2 An Adaptive Sampling Rule

In order to satisfy (30.1) and (30.4) at the first stage, set

$$\begin{aligned} m_{A1} &= \sigma_A\{\sigma_A + \sigma_B w(\tilde{\theta})\}\,\mathcal{I}_1, \\ m_{B1} &= \sigma_B\{\sigma_A + \sigma_B w(\tilde{\theta})\}\,\mathcal{I}_1/w(\tilde{\theta}), \end{aligned} \tag{30.5}$$

where $\tilde{\theta}$ is a preliminary estimate of θ. If no such estimate of θ is available, one can use $w(\tilde{\theta}) = 1$ in (30.5). Of course, these allocations must be rounded to integer values.

Suppose the test has not terminated at stage $k-1$ and we are planning stage k, where $2 \leq k \leq K$. Substituting the current estimate of θ in (30.4), the aim is to choose m_{Ak} and m_{Bk} so that the cumulative sample sizes at analysis k will be in the ratio

$$\frac{n_{Ak}}{n_{Bk}} = \frac{\sigma_A}{\sigma_B}\, w(\hat{\theta}^{(k-1)})$$

and achieve the specified information level by satisfying (30.1). This leads to group sizes

$$\begin{aligned} m_{Ak} &= \sigma_A\{\sigma_A + \sigma_B w(\hat{\theta}^{(k-1)})\}\,\mathcal{I}_k - n_{A\,k-1}, \\ m_{Bk} &= \sigma_B\{\sigma_A + \sigma_B w(\hat{\theta}^{(k-1)})\}\,\mathcal{I}_k/w(\hat{\theta}^{(k-1)}) - n_{B\,k-1}, \end{aligned} \tag{30.6}$$

which are again rounded to integers. If either of these values for m_{Ak} and m_{Bk} is negative, that allocation is set to zero and sufficient observations are taken on the other arm to satisfy (30.1).

One may prefer to set a minimum sample size, m^* say, for both arms at each stage. In this case, if either m_{Ak} or m_{Bk} as determined by (30.6) falls below m^*, that allocation is set to m^* and sufficient observations are taken on the other arm to satisfy (30.1). One reason for imposing a minimum group size on each treatment arm is to preserve at least some randomization in the treatment allocation and thereby maintain an element of blinding. Another reason is concern about possible bias due to time trends in the patient population or treatment administration, although in this case a better strategy is to include such time trends in a statistical model and use a suitable sequential procedure of the form we shall discuss in Section 30.2.4.

30.2.3 Validity of Adaptive Multi-stage Procedures

We now outline the theoretical results that justify the above multi-stage adaptive designs. We present results in terms of the score statistics $S_k = Z_k\sqrt{\mathcal{I}_k}$, $k = 1, \dots, K$, which were introduced in Section 3.1, and the joint distribution of $\{Z_1, \dots, Z_K\}$ follows directly from that of $\{S_1, \dots, S_K\}$. The procedures we have described are valid if the conditional joint distribution of $\{Z_1, \dots, Z_K\}$ given $\{\mathcal{I}_1, \dots, \mathcal{I}_K\}$ satisfies (3.1). We shall first produce an intermediate result concerning the joint distribution of $S_1, \dots, S_K$ when $\mathcal{I}_k$ may depend on $S_1, \dots, S_{k-1}$. We then obtain (3.1) by the simple device of restricting sampling rules so that they produce a pre-specified sequence of information levels, $\mathcal{I}_1, \dots, \mathcal{I}_K$.

The joint distribution of $S_1, \dots, S_K$

Suppose responses $X_{Ai} \sim N(\mu_A, \sigma^2)$ and $X_{Bi} \sim N(\mu_B, \sigma^2)$, $i = 1, 2, \dots,$ are observed on treatments A and B, respectively. For simplicity, we have assumed a common variance for the two response types but the results we present here extend to the case of unequal variances, σ_A^2 and σ_B^2, with only minor modifications. For $k = 1, \dots, K$, we have

$$\begin{aligned}
\hat{\theta}^{(k)} &= \bar{X}_A^{(k)} - \bar{X}_B^{(k)}, \\
\mathcal{I}_k &= (\sigma^2/n_{Ak} + \sigma^2/n_{Bk})^{-1}, \\
Z_k &= (\bar{X}_A^{(k)} - \bar{X}_B^{(k)})\sqrt{\mathcal{I}_k}
\end{aligned}$$

and $S_k = Z_k\sqrt{\mathcal{I}_k}$.

Under a typical adaptive sampling rule, the choice of n_{Ak} and n_{Bk} will be based on the estimated treatment effect, $\hat{\theta}^{(k-1)} = S_{k-1}/\mathcal{I}_{k-1}$, and the current cumulative sample sizes, $n_{A\,k-1}$ and $n_{B\,k-1}$. In general, we can allow more than this: n_{Ak} and n_{Bk} may depend on $n_{A1}, \dots, n_{A\,k-1}, n_{B1}, \dots, n_{B\,k-1}$ and $S_1, \dots, S_{k-1}$ but *not* on the scaled sample sum

$$W_{k-1} = \frac{1}{\sigma^2}\left(\sum_{i=1}^{n_{A,k-1}} X_{Ai} + \sum_{i=1}^{n_{B,k-1}} X_{Bi}\right).$$

Under these conditions, the joint distribution of $S_1, S_2, \dots$ is given by

$$S_1 \mid \mathcal{I}_1 \sim N(\mathcal{I}_1\theta, \mathcal{I}_1) \tag{30.7}$$

and

$$S_k - S_{k-1} \mid \mathcal{I}_1, \dots, \mathcal{I}_k, S_1, \dots, S_{k-1} \sim N(\{\mathcal{I}_k - \mathcal{I}_{k-1}\}\theta, \mathcal{I}_k - \mathcal{I}_{k-1}) \quad \text{for } k \geq 2. \tag{30.8}$$

The key step in proving this result is to decompose $S_k - S_{k-1}$ as

$$\begin{aligned}
S_k - S_{k-1} &= \frac{n_{Bk}}{n_{Ak} + n_{Bk}}\frac{1}{\sigma^2}\sum_{n_{A,k-1}+1}^{n_{Ak}} X_{Ai} - \frac{n_{Ak}}{n_{Ak} + n_{Bk}}\frac{1}{\sigma^2}\sum_{n_{B,k-1}+1}^{n_{Bk}} X_{Bi} \\
&\quad + \frac{n_{A,k-1}\, n_{Bk} - n_{B,k-1}\, n_{Ak}}{(n_{A,k-1} + n_{B,k-1})\,(n_{Ak} + n_{Bk})} W_{k-1}.
\end{aligned} \tag{30.9}$$

The first two terms in this decomposition involve only new data in the kth group of observations while the third term is a function of previously observed data. Note that if the ratio n_{Ak}/n_{Bk} remains constant across analyses, the third term is identically zero. Since the first two terms and the third term are based on separate sets of data, they are conditionally independent of each other given $n_{A,k-1}$, n_{Ak}, $n_{B,k-1}$ and n_{Bk}. It can also be shown that W_{k-1} is conditionally independent of $S_1, \ldots, S_{k-1}$ given $n_{A1}, \ldots, n_{A,k-1}$ and $n_{B1}, \ldots, n_{B,k-1}$ with a

$$N(\{n_{A,k-1}\,\mu_A + n_{B,k-1}\,\mu_B\}/\sigma^2,\ \{n_{A,k-1} + n_{B,k-1}\}/\sigma^2)$$

conditional distribution. The result (30.8) follows from this expression for $S_k - S_{k-1}$ as the sum of three independent normal variables and some straightforward algebra. For full details we refer the reader to the complete proof provided by Jennison & Turnbull (2001b); this paper also extends the above theory to general normal linear models, showing that the sequence of score statistics for a linear combination of parameters, $\boldsymbol{c}^T\boldsymbol{\beta}$, satisfies (30.7) and (30.8) with $\mathcal{I}_k$ the information for $\boldsymbol{c}^T\boldsymbol{\beta}$ at analysis k, as long as the adaptive sampling rule depends on each $\widehat{\boldsymbol{\beta}}^{(k)}$ only through $\boldsymbol{c}^T\widehat{\boldsymbol{\beta}}^{(k)}$.

Inspection of the decomposition (30.9) shows why the choice of n_{Ak} and n_{Bk} should not be allowed to depend on W_{k-1}. Suppose, for example, $n_{A,k-1} = n_{B,k-1}$ and let a and b be a pair of integers with $b > a > n_{A,k-1}$. Then $(n_{Ak} = a,\ n_{Bk} = b)$ and $(n_{Ak} = b,\ n_{Bk} = a)$ both lead to the same $\mathcal{I}_k$ but the factor multiplying W_{k-1} in (30.9) is positive in the first case and negative in the second. If $(n_{Ak} = a,\ n_{Bk} = b)$ is chosen whenever W_{k-1} is greater than a specified value and $(n_{Ak} = b,\ n_{Bk} = a)$ otherwise, the distribution of the increment $S_k - S_{k-1}$ will be stochastically greater than its distribution when a choice between these two cases is made independently of W_{k-1}.

Achieving the Canonical Joint Distribution (3.1)

In order to ensure that a group sequential test maintains its error probabilities when used in conjunction with an adaptive sampling rule, we need the conditional distribution of $Z_k = S_k/\sqrt{\mathcal{I}_k}$, $k = 1, \ldots, K$, given $\mathcal{I}_1, \ldots, \mathcal{I}_K$ to have the canonical form (3.1). It is important to realize that the joint distribution of $S_1, \ldots, S_K$ given by (30.7) and (30.8) is not sufficient to ensure this conditional distribution. Suppose, for example, a sampling rule produces one value of $\mathcal{I}_2$ if S_1 is positive and another if S_1 is negative, then it is clear that the conditional distribution of S_1 given either value of $\mathcal{I}_2$ is not normal.

However, (30.7) and (30.8) do imply (3.1) if $\mathcal{I}_1, \ldots, \mathcal{I}_K$ are fixed independently of the values of $S_1, \ldots, S_K$. Thus, if a sampling rule is designed to ensure a pre-specified sequence of information levels, (3.1) is obtained automatically. The procedures described in Sections 30.2.1 and 30.2.2 are defined in this way, hence (3.1) holds and the tests have the same error probabilities as non-adaptive tests with the same information levels and boundaries for $Z_1, \ldots, Z_K$.

The restriction to a fixed sequence $\mathcal{I}_1, \ldots, \mathcal{I}_K$ can be relaxed a little to allow for perturbations in observed information due to randomness in actual group sizes. The key requirement that must be met is that $\mathcal{I}_k$ should be statistically independent of $S_1, \ldots, S_{k-1}$, for $k = 2, \ldots, K$. Then, if information levels are variable, the error spending approach of Chapter 8 can be applied.

30.2.4 A Multi-stage Adaptive Design with Time Trends

The response adaptive designs described thus far can lead to considerable imbalance in the sample sizes m_{Ak} and m_{Bk} within a group and this can cause problems if there is a possibility of time trends in the responses; see, for example, Peto (1985, p. 32). The ECMO study (Ware, 1989, p. 302) was criticized because a lack of concurrent control subjects in the second stage "raised the possibility of non-comparability of treatment groups". One solution is to use the extension of the preceding methods to general normal linear regression models (see Jennison & Turnbull, 2001b) with linear or quadratic time trends incorporated as covariates, along with other concomitant variables, as in the two-treatment comparison adjusted for covariates described in the second example of Section 3.3.2. An alternative approach, which allows for arbitrary time trends in responses, is to stratify the data analysis by stage.

Suppose responses are independent and normally distributed as before, except now the observations at stage k have means $\mu_A + \gamma_k$ for treatment A and $\mu_B + \gamma_k$ for treatment B. We wish to test H_0: $\theta \leq 0$ against $\theta > 0$, where $\theta = \mu_A - \mu_B$, with $\gamma_1, \ldots, \gamma_K$ treated as unknown nuisance parameters. For $k = 1, \ldots, K$, let $\bar{X}_A^{(k)}$ denote the mean of the m_{Ak} responses on treatment arm A in the kth group, and similarly define $\bar{X}_B^{(k)}$ for treatment arm B. In each group $k = 1, \ldots, K$, the difference in mean responses from arms A and B, $\bar{X}_A^{(k)} - \bar{X}_B^{(k)}$, provides an unbiased estimate of θ. The maximum likelihood estimate of θ based on data accumulated up to stage k is the weighted average

$$\hat{\theta}^{(k)} = \frac{1}{w_1 + \ldots + w_k} \sum_{i=1}^{k} w_i(\bar{X}_A^{(i)} - \bar{X}_B^{(i)}) \tag{30.10}$$

where the weights

$$w_k = \left(\frac{\sigma_A^2}{m_{Ak}} + \frac{\sigma_B^2}{m_{Bk}} \right)^{-1}, \quad k = 1, \ldots, K, \tag{30.11}$$

are the reciprocals of the variances of the estimates $\bar{X}_A^{(k)} - \bar{X}_B^{(k)}$. Note that

$$Var(\hat{\theta}^{(k)}) = (w_1 + \ldots + w_k)^{-1}, \quad k = 1, \ldots, K.$$

Defining

$$\mathcal{I}_k = \sum_{i=1}^{k} w_i, \quad k = 1, \ldots, K,$$

the standardized statistic for testing H_0 at stage k is $Z_k = \hat{\theta}^{(k)} \sqrt{\mathcal{I}_k}$. Assuming the sampling rule is designed to achieve pre-specified information levels $\{\mathcal{I}_1, \ldots, \mathcal{I}_K\}$, the joint distribution of $\{Z_1, \ldots, Z_K\}$ does not depend on the parameters $\gamma_1, \ldots, \gamma_K$ and follows the canonical joint distribution (3.1). Hence, as for the unstratified analysis, an information sequence $\{\mathcal{I}_1, \ldots, \mathcal{I}_K\}$ and group sequential boundary for $\{Z_1, \ldots, Z_K\}$ can be chosen to give a test meeting stated Type I error probability and power requirements.

As before, we obtain a sampling rule by attempting to minimize a loss function of the form (30.2) subject to attaining a fixed sequence of information levels $\mathcal{I}_1, \ldots, \mathcal{I}_K$. This constraint implies that m_{Ak} and m_{Bk} must satisfy (30.11) with $w_k = \mathcal{I}_k - \mathcal{I}_{k-1}$, or simply $w_k = \mathcal{I}_K / K$ in the case of equal information increments. If θ were known, the group sizes m_{Ak} and m_{Bk} would be set in the optimal ratio

$$\frac{m_{Ak}}{m_{Bk}} = \frac{\sigma_A}{\sigma_B} w(\theta)$$

where $w(\theta) = \sqrt{\{v(\theta)/u(\theta)\}}$. This leads to allocations for the kth group of

$$\begin{aligned} m_{Ak} &= \sigma_A\{\sigma_A + \sigma_B w(\theta)\} (\mathcal{I}_k - \mathcal{I}_{k-1}), \\ m_{Bk} &= \sigma_B\{\sigma_A + \sigma_B w(\theta)\} (\mathcal{I}_k - \mathcal{I}_{k-1})/w(\theta), \end{aligned} \tag{30.12}$$

for $k = 1, \ldots, K$, where $\mathcal{I}_0$ is taken to be zero. Since θ is unknown, this sampling rule must be applied using successive estimates of θ. In the first stage, allocations are as in (30.5) and in subsequent stages, $k = 2, \ldots, K$, we use (30.12) with θ replaced by the estimate $\hat{\theta}^{(k-1)}$ defined by (30.10).

One should expect this method to produce smaller reductions in ITN than the previous procedure which made no allowance for time trends, since there is no longer a mechanism for redressing sub-optimal allocation of observations in earlier groups when improved estimates of θ become available.

30.2.5 Methods for an Unknown Variance

If the variances σ_A^2 and σ_B^2 are unknown, analogous procedures can be based on an approximate, unpooled, two-sample t-statistic with group sequential boundaries defined through significance levels of the t-distribution as in Sections 4.6, 5.6 and 6.2.4. Allocations of observations at stages $k = 2, \ldots, K$ can be based on (30.6) with σ_A and σ_B replaced by estimates from data obtained in the previous $k-1$ stages. At the initial stage, preliminary estimates of σ_A and σ_B, as well as θ, must be used in (30.5). Note that if σ_A^2 and σ_B^2 are unequal, the optimal ratio (30.4) differs from unity even when $u(\theta) \equiv v(\theta)$. Consequently, adaptive allocation of observations to treatments can reduce both ASN and ITN, so adaptive sampling procedures may still be of benefit when only a low ASN is deemed important.

30.3 An Illustrative Example

We illustrate RAR methods using the example presented by Jennison (2023b) in his discussion of the paper by Robertson et al. (2023). The context is the one-sided test described at the start of Section 30.2. We assume equal variances, $\sigma_A^2 = \sigma_B^2 = \sigma^2$, and require Type I error probability $\alpha = 0.025$ when $\theta = 0$ and power $1 - \beta = 0.9$ at $\theta = \delta$. A fixed sample test requires information $\mathcal{I}_{f,1}$ as defined by (5.1). In presenting results, we shall assume σ^2 and δ are such that a fixed sample test based on responses from 100 subjects per treatment arm achieves the specified error rates. Then, the values we present for expected sample sizes will also represent a percentage of the fixed sample size in the case of general σ^2 and δ.

In this example, the study has four interim analyses and a final analysis. To begin with, we suppose the interim analyses are used to make decisions on randomization ratios but there is no early stopping. Thus, the information level at the final analysis is $\mathcal{I}_5 = \mathcal{I}_{f,1}$ and $\mathcal{I}_k = (k/5)\,\mathcal{I}_5$ for $k = 1, \ldots, 4$. Power 0.9 is specified at the treatment effect $\theta = \delta$. The adaptive randomization scheme is based on the loss function (30.3) with $a = 4$, which implies a target of

$$n_{A\tau}/n_{B\tau} = a^{\theta/(2\delta)}$$

and, in particular,

$$\begin{aligned} n_{A\tau} &= n_{B\tau} && \text{when } \theta = 0, \\ n_{A\tau} &= \sqrt{2}\, n_{B\tau} && \text{when } \theta = \delta/2, \\ n_{A\tau} &= 2\, n_{B\tau} && \text{when } \theta = \delta. \end{aligned}$$

We assume the response for each subject is observed a certain time after randomization with the effect that responses from subjects randomized in a given stage are not available until the analysis one stage later. Thus, no responses are observed at the first interim analysis, responses from subjects randomized in stage 1 are observed at the time of the second interim analysis, and so forth.

We apply the methods of Section 30.2.4 to allow for time trends, so the data analysis is stratified by stage. Since no estimate of θ is available until analysis 2, equal randomization is applied in stages 1 and 2. Then, with the estimate

$$\hat{\theta}^{(2)} = (\bar{X}_A^{(1)} - \bar{X}_B^{(1)}),$$

we choose m_{A3} and m_{B3} to satisfy

$$\frac{m_{A3}}{m_{B3}} = w(\hat{\theta}^{(2)}) = a^{\hat{\theta}^{(2)}/(2\delta)}.$$

and

$$\left(\frac{\sigma^2}{m_{A3}} + \frac{\sigma^2}{m_{B3}}\right)^{-1} = \frac{1}{5}\mathcal{I}_{f,1}.$$

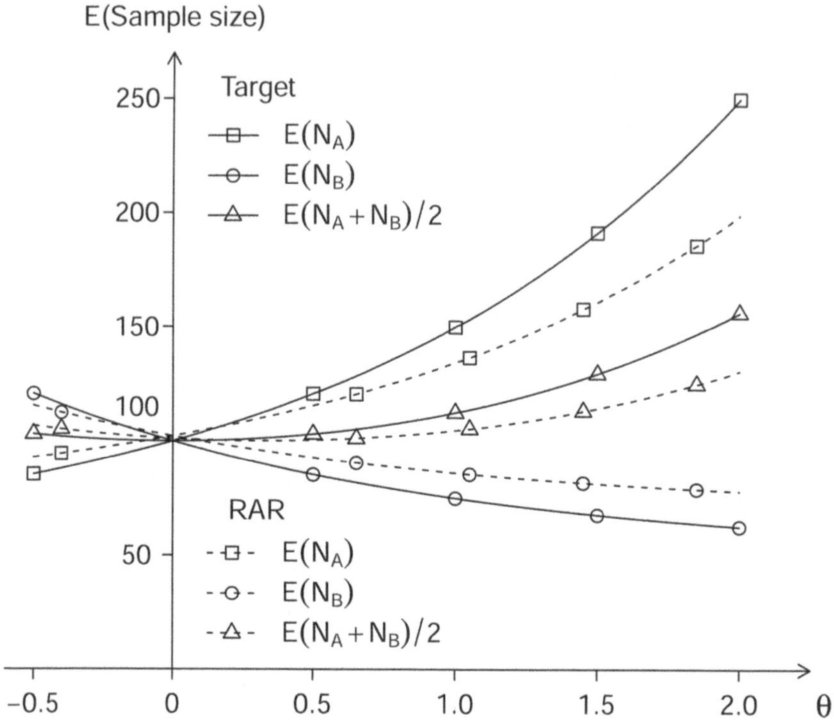

Figure 30.1 *Example: Target values of $E(N_A)$ and $E(N_B)$ for an RAR design, and values obtained in an RAR design when the data analysis is stratified by group and there is a one-group delay in observing responses*

Similarly, m_{A4} and m_{B4} are based on

$$\hat{\theta}^{(3)} = \frac{1}{2}\{(\bar{X}_A^{(1)} - \bar{X}_B^{(1)}) + (\bar{X}_A^{(2)} - \bar{X}_B^{(2)})\}$$

and m_{A5} and m_{B5} based on

$$\hat{\theta}^{(4)} = \frac{1}{3}\{(\bar{X}_A^{(1)} - \bar{X}_B^{(1)}) + (\bar{X}_A^{(2)} - \bar{X}_B^{(2)}) + (\bar{X}_A^{(3)} - \bar{X}_B^{(3)})\}.$$

The performance of this RAR design is shown in Figure 30.1. Here, we have taken $\delta = 1$ and $E(N_A)$ and $E(N_B)$ denote the expected numbers of patients allocated to treatments A and B, respectively. The target values for $E(N_A)$ and $E(N_B)$ are found from

$$\frac{m_{Ak}}{m_{Bk}} = \frac{\sigma_A}{\sigma_B} w(\theta) = 4^{(\theta/2\delta)} = 4^{\theta/2}, \quad k = 1, \ldots, 5.$$

It is evident that the RAR design has reduced the ITN and there are concomitant increases in the sample size on the superior treatment and in the ASN. Because the analysis is stratified to allow for time trends and allocation ratios are not modified based on the observed treatment effect until group 3, the ITN is not reduced as far as the target values. In the same plot for an RAR design with an immediate response, where allocation ratios are modified from group 2 onwards, the discrepancy between $E(N_A)$ and $E(N_B)$ values and the target curves is approximately halved. Results for an RAR design with an unstratified analysis would show values of $E(N_A)$ and $E(N_B)$ almost identical to the target, even when there is a one-group delay in observing responses.

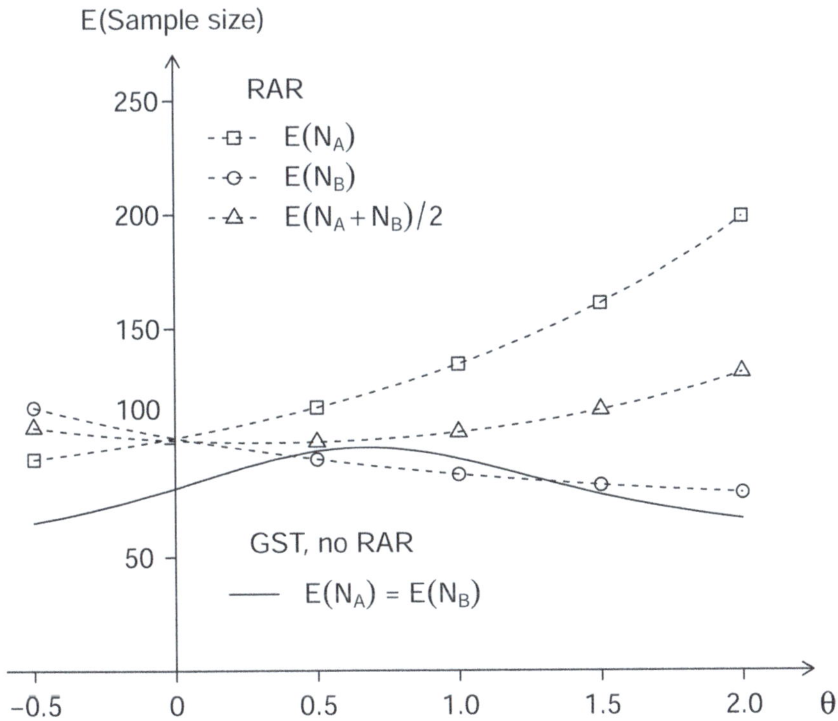

Figure 30.2 *Example: Values of $E(N_A)$ and $E(N_B)$ attained (i) by an RAR design with no early stopping and (ii) by a group sequential design with equal randomization*

Suppose now that the trial has a group sequential design with early stopping for efficacy or futility but no adaptive randomization. We consider a one-sided group sequential test from the power family of Section 5.2 with parameter $\Delta = 0$ and five equally spaced analyses. The boundary for $Z_1, \ldots, Z_5$ is given by (5.3) and information levels are $\mathcal{I}_k = (k/K)\mathcal{I}_5$, $k = 1, \ldots, 5$, where $\mathcal{I}_5$ is given by (5.4). The constants $\tilde{C}_1(5, 0.025, 0.1, 0) = 1.987$ and $\tilde{C}_2(5, 0.025, 0.1, 0) = 1.400$ are obtained from Table 5.2 and the inflation factor is 1.091. A binding futility boundary is assumed and patients are randomized equally between the two treatment arms. To maintain comparability with the previous case, we suppose there is a one-group delay in observing responses, so there are "pipeline" patients at each interim analysis who have been treated but whose responses not yet observed, as discussed in Chapter 14. For simplicity, we have not incorporated data from the pipeline subjects in the final analysis but these subjects are counted in the observed sample sizes N_A and N_B.

Figure 30.2 compares the performance of this non-adaptive, group sequential design with the previous RAR design. The plots show that for some values of θ, the group sequential design has the lower ITN and this is achieved without increasing the sample size on the superior treatment. In the corresponding plot for the case of an immediate response, we would see that the group sequential design has a lower ITN than the RAR design at all values of θ. This should not be surprising: when the interim estimate of θ is sufficiently large or small to affect the randomization ratio in the RAR design, it is liable to lead to early stopping in the group sequential design. We conclude that if it is deemed important to reduce the ITN, then a design with early stopping should be employed. The next step is to consider a design with both early stopping and RAR.

It is a straightforward task to follow the methods described in Section 30.2.4 to combine a group sequential stopping rule and an adaptive randomization scheme. The solid curves in Figure 30.3 show the values of $E(N_A)$ and $E(N_B)$ for the design that combines the

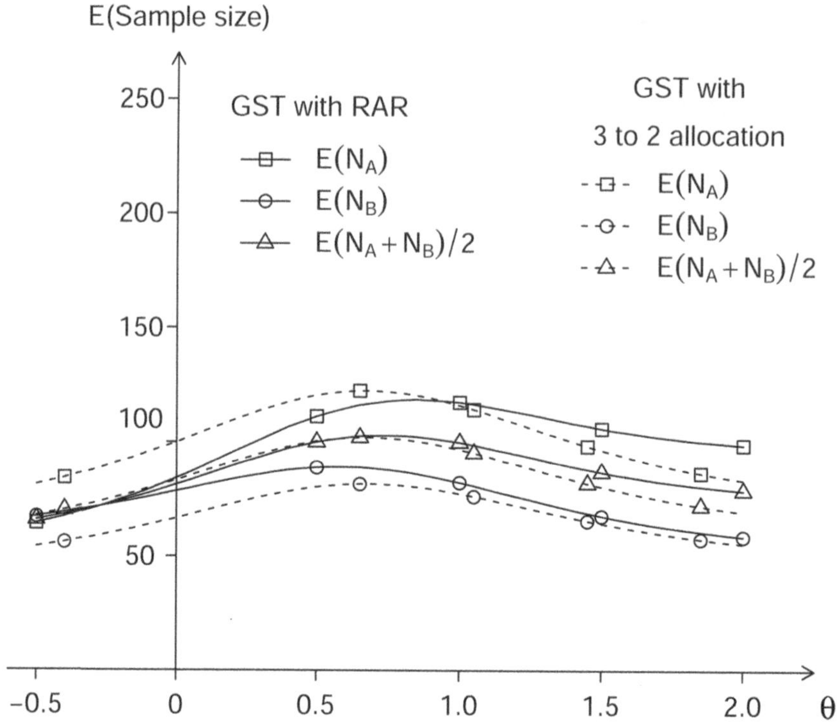

Figure 30.3 *Example: Values of $E(N_A)$ and $E(N_B)$ attained (i) by a group sequential design incorporating RAR and (ii) by a group sequential design with a fixed randomization ratio of 3:2 to treatments A and B, respectively.*

stopping boundary and RAR rule used in previous cases. Once again, we have assumed there is a one-group delay in observing responses, so allocation ratios are modified from group 3 onwards and when the trial stops at an interim analysis, the pipeline subjects are counted in the observed sample sizes. The ITN for the new procedure is uniformly lower than that for both the designs shown in Figure 30.2.

We note that the adaptive randomization has little impact for negative values of θ since if $\hat{\theta}^{(k)} < 0$ at an analysis $k \geq 2$, the futility boundary will be crossed and the trial will terminate. Also, the observed allocation ratio is fairly constant for effect sizes $\theta > 0.5\,\delta$. This suggests that one might simplify the design by using a fixed allocation ratio, for example,

$$\frac{m_{Ak}}{m_{Bk}} = 1.5,$$

randomizing 3 subjects to the new treatment for every 2 on the control. The properties of such a design are shown as the dashed curves in Figure 30.3. The one drawback of this design is that when θ is negative, so the new treatment is the inferior treatment, the ITN is increased. However, the futility stopping boundary ensures the trial is likely to stop early when $\theta < 0$ and, in any case, the evidence leading to a Phase III trial should imply that a treatment effect substantially less than zero is unlikely. The results from this sequence of designs raise questions about the value of response adaptive randomization. The final design with a group sequential stopping rule and a 3:2 allocation ratio is simple to plan and to implement and, in the context of a Phase III trial comparing a new treatment to the current standard, meets the objective of reducing ITN with a modest increase in ASN.

Interestingly, the conclusion here is in keeping with the recommendation of Korn & Freidlin (2011), who wrote "we find no benefits to outcome-adaptive randomization over 1:1 randomization, and we recommend the latter. If it is thought that the patient accrual rates will be substantially higher because of the possibility of a higher proportion of patients being randomly assigned to the experimental treatment (because the trial will be more attractive to patients and clinicians), we recommend using a fixed 2:1 randomization instead of an outcome-adaptive randomization."

30.4 Bibliography and Notes

There is considerable literature on statistical methodology for response-adaptive designs in sequential clinical trials. Early work has been reviewed by Hoel, Sobel & Weiss (1975), Simon (1977), Iglewicz (1983), Hardwick (1989) and Rosenberger & Lachin (1993). A variety of research topics are presented in twenty papers contained in the Proceedings of a 1992 conference on adaptive designs (Flournoy & Rosenberger, 1995). The books by Hu & Rosenberger (2006) and Atkinson & Biswas (2014) concern response-adaptive designs and work in this area is reviewed by Robertson, Lee et al. (2023).

The earliest ideas are due to Thompson (1933), Robbins (1952) and, specifically in the context of clinical trials, Anscombe (1963) and Colton (1963). Cornfield, Halperin & Greenhouse (1969) generalized these ideas to multi-stage designs. Zelen (1969) introduced the "play-the-winner" (PW) design for comparing two dichotomous response distributions and this was extended to "randomized play-the-winner" (RPW) rules by Wei & Durham (1978). Rosenberger (1993, 1996) summarizes further developments and the use of biased coin randomization schemes for normal and survival outcomes. Yao & Wei (1996) discuss RPW rules in a group sequential setting; using simulation, they retrospectively examine what would have happened had these rules been applied to the binomial data of the ACTG076 maternal HIV transmission trial and the survival end-point of a VACURG prostatic cancer trial.

Robbins & Siegmund (1974) discuss adaptive allocation rules similar to those of Section 30.2 for the two-sample normal problem. They use a sequential probability ratio test (SPRT) with continuous monitoring and are able to establish error rates by standard likelihood ratio arguments rather than the more detailed theory described in Section 30.2.3 for an arbitrary group sequential test. They then apply martingale arguments to obtain results on expected sample sizes. The SPRT with adaptive allocation was developed further by Louis (1975), who explicitly considered a loss function expressed as a weighted linear combination of ASN and ITN, and by Hayre (1979) who first used the loss function (30.2) with weights depending on the unknown difference θ. Louis (1977) and Hayre & Turnbull (1981a) apply similar adaptive SPRT-based designs to exponential survival data subject to censoring; in this case, the variance depends on the mean and the goals of reducing ASN and ITN are not necessarily in conflict, so adaptive sequential designs can be of particular benefit. Hayre & Turnbull (1981a) also consider other distributions, especially the binomial. It should be remarked that this class of designs is not only valuable in the hypothesis testing framework, but it is also helpful for point and interval estimation; see Hayre & Turnbull (1981b,c).

If there are three or more treatment arms and the experimenter's goal is to select the best treatment, precise analysis of adaptive procedures becomes more difficult. References in this area include Turnbull, Kaspi & Smith (1978), Jennison, Johnstone & Turnbull (1982) and Coad (1995). The problem can be alternately modeled as a multi-armed bandit; see Gittins, Glazebrook & Weber (2011).

Although many papers have been published on the methodology of adaptive sampling, there are few examples of its implementation in practice. Both logistical and ethical issues have been cited for this lack of usage; the arguments can be found in Bailar (1976), Simon (1977), Armitage (1985), Peto (1985), Royall (1991), Rosenberger & Lachin (1993) and particularly in Ware (1989) and the accompanying discussion. However, there have been a few trials where a response-adaptive design has been used, notably the ECMO trials reported by Cornell, Landenberger & Bartlett (1986) and Ware (1989) and two trials of anti-depression drugs sponsored by Eli Lilly and Co. described by Tamura et al. (1994) and Andersen (1996), the latter being an interesting example of a four-armed adaptive trial.

There has been an interest in applying RAR methods in the multi-arm multi-stage (MAMS) designs that we described in Chapter 27. Dunnett (1955) showed that when comparing K treatments with a common control in a fixed sample size study, the total sample size is minimized by allocating $\sqrt{K}$ subjects to the control for every one on each of the K new treatments. Wassmer (2011) and Wason & Jaki (2012) investigated the impact of allocating a higher proportion of patients to the control in a MAMS design and found the optimal ratio to closer to 1 than $\sqrt{K}$. This finding can be attributed to the fact that the number of experimental treatments remaining in the trial decreases in the later stages after treatments have been dropped for futility or because they have been shown to be superior to the control. Azher, Grayling & Wason (2025) considered a generalization of the type of MAMS design considered by Wason & Jaki (2012) in which the proportion of patients allocated to the control varies from stage to stage according to a pre-specified pattern. The motivation here is quite different from that behind the RAR schemes of Sections 30.2 and 30.3 which aim to reduce the number of subjects on an inferior treatment rather than the total sample size.

Wason & Trippa (2014) compare the impact of adaptive randomization (AR) in a trial when all treatments are retained until the final analysis against the benefits of dropping treatments during a multi-stage trial with equal randomization to those treatments remaining at each stage. The methods are assessed according to the total sample size and the numbers allocated to inferior treatments. The authors report that

> "AR is more efficient than MAMS designs when there is an effective experimental treatment, whereas if none of the experimental treatments is effective, then MAMS designs slightly outperform AR."

Lin & Bunn (2017) consider trial designs with a different form of AR rule and again compare these with MAMS designs. Their conclusions match those of Wason & Trippa as they find the MAMS designs to be more efficient when there are no effective treatments but the performance of designs with AR becomes superior as the probability of a treatment arm being dropped increases. These results suggest a hybrid approach combining AR and early decision making for some experimental treatments could be successful. In principle, it should be possible to adapt allocation ratios to the number of treatments remaining in the trial and their estimated treatment effects using the form of MAMS design presented in Chapter 27, that is, using a closed testing procedure to control FWER with Lehmacher-Wassmer group sequential combination tests for each intersection hypothesis. However, the methods must be applied carefully to avoid inflation of the family-wise error rate; see Graf et al. (2014).

Response adaptive randomization was a feature of some of the platform and umbrella trials mentioned in Section 27.7. Korn & Freidlin (2017) give a critical appraisal of the use of RAR in the BATTLE-2 and I-SPY 2 trials. For BATTLE-2 they note that RAR has led to a loss of power and a design with 120 patients allocated equally to the four treatments could have achieved similar power to the RAR design with 200 patients. Moreover, they estimate that the treatment outcomes for patients in the smaller trial with equal allocation would be no worse than those in the RAR design. For I-SPY 2, they express concern that treatment imbalance in some of the 10 signature subgroups that were analyzed could have led to a lack of power.

Chapter 31

Numerical Computations for Group Sequential and Adaptive Tests

31.1 Introduction

Numerical calculations for sequential tests with a discrete response date back to the two-stage acceptance sampling plans of Dodge & Romig (1929) and the multi-stage plans introduced by the Columbia University Research Group (Freeman et al., 1948) which developed into the United States military standard, MIL-STD-105E (1989). Computations for binary data were discussed in detail in Chapter 16. The computational task is greater in the analogous calculations for a continuous response where multiple integrals replace multiple sums. Armitage & Schneiderman (1958), Schneiderman (1961), Dunnett (1961) and Roseberry & Gehan (1964) computed properties of two-stage and three-stage procedures for normal observations with known variance and Armitage, McPherson & Rowe (1969) tackled sequential tests in earnest, obtaining accurate results for repeated significance tests with up to 200 analyses. Group sequential designs were proposed in the context of drug-screening trials with a dichotomous outcome by Schultz et al. (1973) and for clinical trials with continuous response variables by McPherson (1974) and Pocock (1977). Although the practical difference between sequential and group sequential analysis is substantial, the computations carried out by McPherson and Pocock are essentially the same as those for Armitage's (1975) repeated significance test with continuous monitoring and a small maximum sample size.

Subsequent research has produced a variety of group sequential testing boundaries, including error spending procedures to handle unpredictable group sizes, and methods for making frequentist inferences in the form of P-values and confidence intervals on termination of a group sequential test. In some instances, analytic approximations to the error probabilities and expected sample sizes of group sequential tests are available but, in general, direct numerical computation is needed to obtain accurate properties. This accuracy is particularly important when results are used in optimization procedures. Numerical computation can also be used in creating and assessing certain types of adaptive design. However, for the more complex adaptive designs described in Chapters 26 to 28, it is common to control the Type I error rate conservatively by using closed testing procedures and combination tests, and to assess the performance of a given design by simulation.

Some details of how numerical methods can be used to compute properties of group sequential tests are provided in the early papers of Armitage, McPherson & Rowe (1969) and McPherson & Armitage (1971). Here, we elaborate on these methods in the case of normally distributed observations with known variance. We shall work with the standardized statistics $Z_1, \ldots, Z_K$, assuming these have the canonical joint distribution (3.1) given an information sequence $\mathcal{I}_1, \ldots, \mathcal{I}_K$. Jennison (1994) presented a description of group sequential computations in terms of the sample sums, or the score statistics $Z_k\sqrt{\mathcal{I}_k}$ in our terminology;

DOI: 10.1201/9781584888482-31

we use similar notation in this chapter but for densities and conditional densities of different variables, so care is required if these two approaches are to be studied together.

In Section 31.2 we describe how to compute the probability that a group sequential test crosses a given boundary at a certain analysis k. We apply this basic calculation in Section 31.3 to find properties of a specified test, and explain how to construct testing boundaries and choose sample sizes to attain particular Type I and II error probabilities. These methods underpin the derivation of group sequential tests with various optimality properties presented in Chapter 12. The implementation of error spending tests is dealt with in Section 31.4, where we also discuss how to find the target maximum information level needed to attain specified Type I error and power. In Section 31.5 we describe computations of the frequentist data summaries on termination of a group sequential test that were presented in Chapter 9. We conclude with a discussion of extensions to the case of normal data with unknown variance and other response distributions, and comment on computer software currently available for designing and implementing group sequential trials.

As noted in Section 1.6, we have made some of our R programs available at the website `https://people.bath.ac.uk/mascj/` and there are a number of software packages, both commercial and non-commercial, to assist the user in applying group sequential and adaptive methods.

31.2 The Basic Calculation

Suppose a group sequential test with a maximum of K analyses is defined in terms of standardized statistics $Z_1, \ldots, Z_K$ following the canonical joint distribution (3.1) for a given sequence of information levels $\mathcal{I}_1, \ldots, \mathcal{I}_K$. Let $\Delta_k = \mathcal{I}_k - \mathcal{I}_{k-1}$ for $k = 2, \ldots, K$. Then, $Z_1 \sim N(\theta\sqrt{\mathcal{I}_1}, 1)$ and for each $k = 2, \ldots, K$,

$$Z_k\sqrt{\mathcal{I}_k} - Z_{k-1}\sqrt{\mathcal{I}_{k-1}} \sim N(\theta\Delta_k, \Delta_k) \qquad (31.1)$$

independently of $Z_1, \ldots, Z_{k-1}$.

We shall consider tests with continuation regions of the form (a_k, b_k), $k = 1, \ldots, K$. Most commonly used one-sided and two-sided hypothesis tests are of this form. The two-sided tests which permit early stopping to accept the null hypothesis discussed in Chapter 6 have continuation regions composed of two disjoint intervals and modified versions of the methods we shall describe can be applied to those tests.

A fundamental quantity to compute for a group sequential test is the probability of exiting by a specific boundary at a particular analysis. For each $k = 1, \ldots, K$, define

$$\psi_k(a_1, b_1, \ldots, a_k, b_k; \theta) = Pr_\theta\{a_1 < Z_1 < b_1, \ \ldots, a_{k-1} < Z_{k-1} < b_{k-1}, Z_k > b_k\} \qquad (31.2)$$

and

$$\xi_k(a_1, b_1, \ldots, a_k, b_k; \theta) = Pr_\theta\{a_1 < Z_1 < b_1, \ \ldots, a_{k-1} < Z_{k-1} < b_{k-1}, Z_k < a_k\}. \qquad (31.3)$$

These quantities will appear later in formulae for a test's error probabilities and expected sample size (more precisely, the expected information level on termination). They will also be used in constructing error spending boundaries and in computing P-values and confidence intervals on termination.

To evaluate (31.2) and (31.3), note that Z_1 has density

$$f_1(z_1; \theta) = \phi(z_1 - \theta\sqrt{\mathcal{I}_1}),$$

where $\phi(x) = \exp(-x^2/2)/\sqrt{(2\pi)}$ is the density at x of a standard normal variate. It follows from (31.1) that, for $k = 2, \ldots, K$, the conditional distribution of Z_k given $Z_1 = z_1, \ldots, Z_{k-1} = z_{k-1}$ depends only on z_{k-1} and has density

$$f_k(z_{k-1}, z_k; \theta) = \frac{\sqrt{\mathcal{I}_k}}{\sqrt{\Delta_k}} \phi\left(\frac{z_k\sqrt{\mathcal{I}_k} - z_{k-1}\sqrt{\mathcal{I}_{k-1}} - \theta\Delta_k}{\sqrt{\Delta_k}}\right). \tag{31.4}$$

Hence, for each $k = 2, \ldots, K$,

$$\begin{aligned}
\psi_k(a_1, b_1, \ldots, a_k, b_k; \theta) \\
&= \int_{a_1}^{b_1} \cdots \int_{a_{k-1}}^{b_{k-1}} \int_{b_k}^{\infty} f_1(z_1; \theta) f_2(z_1, z_2; \theta) \ldots f_k(z_{k-1}, z_k; \theta)\, dz_k \ldots dz_1 \\
&= \int_{a_1}^{b_1} \cdots \int_{a_{k-1}}^{b_{k-1}} f_1(z_1; \theta) f_2(z_1, z_2; \theta) \ldots f_{k-1}(z_{k-2}, z_{k-1}; \theta) \\
&\qquad \times e_{k-1}(z_{k-1}, b_k; \theta)\, dz_{k-1} \ldots dz_1,
\end{aligned} \tag{31.5}$$

where

$$e_{k-1}(z_{k-1}, b_k; \theta) = \Phi\left(\frac{z_{k-1}\sqrt{\mathcal{I}_{k-1}} + \theta\Delta_k - b_k\sqrt{\mathcal{I}_k}}{\sqrt{\Delta_k}}\right). \tag{31.6}$$

Similarly, $\xi_k(a_1, b_1, \ldots, a_k, b_k; \theta)$ can be written as a $(k-1)$-fold multiple integral of the form (31.5) but with $e_{k-1}(z_{k-1}, b_k; \theta)$ replaced by

$$e'_{k-1}(z_{k-1}, a_k; \theta) = 1 - \Phi\left(\frac{z_{k-1}\sqrt{\mathcal{I}_{k-1}} + \theta\Delta_k - a_k\sqrt{\mathcal{I}_k}}{\sqrt{\Delta_k}}\right).$$

In Chapter 3 we defined functions $g_k(z_k; \theta)$, $k = 1, \ldots, K$, representing the sub-densities of $Z_1, \ldots, Z_K$, their integrals being less than unity for $k > 1$ due to early stopping at stages $1, \ldots, k-1$. Formally,

$$g_1(z_1; \theta) = f_1(z_1; \theta)$$

and, for a test with interval continuation regions,

$$g_k(z_k; \theta) = \int_{a_{k-1}}^{b_{k-1}} g_{k-1}(z_{k-1}; \theta)\, f_k(z_{k-1}, z_k; \theta)\, dz_{k-1}, \quad k = 2, \ldots, K.$$

Then, for example, we can write

$$\begin{aligned}
\psi_k(a_1, b_1, \ldots, a_k, b_k; \theta) &= \int_{b_k}^{\infty} g_k(z_k; \theta)\, dz_k \\
&= \int_{a_{k-1}}^{b_{k-1}} \int_{b_k}^{\infty} g_{k-1}(z_{k-1}; \theta)\, f_k(z_{k-1}, z_k; \theta)\, dz_k\, dz_{k-1} \\
&= \int_{a_{k-1}}^{b_{k-1}} g_{k-1}(z_{k-1}; \theta)\, e_{k-1}(z_{k-1}, b_k; \theta)\, dz_{k-1}.
\end{aligned}$$

In evaluating integrals numerically, we use a quadrature rule to replace each integral by a weighted sum. Thus, we approximate the integral of a function $q(z)$ from u to l as

$$\int_u^l q(z) dz \approx \sum_{i=1}^{m} w(i)\, q(z(i)),$$

where the number of points at which q is evaluated and their locations $z(i)$ and weights $w(i)$, $i = 1, \ldots, m$, are chosen to ensure a sufficiently accurate approximation. We shall discuss the choice of grid points and weights later. First, we explain how the multiple summations in formulae such as (31.5) can be carried out efficiently. The integral on the right hand side of (31.5) is approximated by a $(k-1)$-fold sum to give

$$\psi_k(a_1, b_1, \ldots, a_k, b_k; \theta)$$

$$\approx \sum_{i_1=1}^{m_1} \cdots \sum_{i_{k-1}=1}^{m_{k-1}} w_1(i_1)\, f_1(z_1(i_1); \theta)\, w_2(i_2)\, f_2(z_1(i_1), z_2(i_2); \theta)$$

$$\ldots w_{k-1}(i_{k-1})\, f_{k-1}(z_{k-2}(i_{k-2}), z_{k-1}(i_{k-1}); \theta)\, e_{k-1}(z_{k-1}(i_{k-1}), b_k; \theta). \quad (31.7)$$

For given $\{z_1(1), \ldots, z_1(m_1)\}, \ldots, \{z_K(1), \ldots, z_K(m_K)\}$, we define functions $h_1, \ldots, h_K$ by

$$h_1(i_1; \theta) = w_1(i_1)\, f_1(z_1(i_1); \theta), \quad i_1 = 1, \ldots, m_1,$$

and

$$h_k(i_k; \theta) = \sum_{i_{k-1}=1}^{m_{k-1}} h_{k-1}(i_{k-1}; \theta)\, w_k(i_k)\, f_k(z_{k-1}(i_{k-1}), z_k(i_k); \theta),$$

for each $i_k = 1, \ldots, m_k$ and $k = 2, \ldots, K$. Summing over i_1 for each i_2, the right hand side of (31.7) reduces to

$$\sum_{i_2=1}^{m_2} \cdots \sum_{i_{k-1}=1}^{m_{k-1}} h_2(i_2; \theta)\, w_3(i_3) f_3(z_2(i_2), z_3(i_3); \theta) \ldots$$

$$w_{k-1}(i_{k-1})\, f_{k-1}(z_{k-2}(i_{k-2}), z_{k-1}(i_{k-1}); \theta)\, e_{k-1}(z_{k-1}(i_{k-1}), b_k; \theta)$$

and further summations over $i_2, \ldots, i_{k-2}$ in turn yield

$$\psi_k(a_1, b_1, \ldots, a_k, b_k; \theta) \approx \sum_{i_{k-1}=1}^{m_{k-1}} h_{k-1}(i_{k-1}; \theta)\, e_{k-1}(z_{k-1}(i_{k-1}), b_k; \theta). \quad (31.8)$$

The functions $h_1, \ldots, h_{K-1}$ can be evaluated in sequence and then used in formulae such as (31.8) to compute quantities of interest.

The progression from (31.7) to (31.8) parallels the iterative definition of the sub-densities $g_k(z_k; \theta)$ and it follows that the values $h_k(i_k; \theta)$, $i_k = 1, \ldots, m_k$, can be used to integrate the product of $g_k(z_k; \theta)$ and a function of z_k. For $1 \leq k \leq K$ and a general function $q(z_k)$,

$$\int_{a_1}^{b_1} \cdots \int_{a_k}^{b_k} f_1(z_1; \theta)\, f_2(z_1, z_2; \theta)\, \ldots\, f_k(z_{k-1}, z_k; \theta)\, q(z_k)\, dz_k \ldots dz_1$$

$$= \int_{a_k}^{b_k} g_k(z_k; \theta)\, q(z_k)\, dz_k \approx \sum_{i_k=1}^{m_k} h_k(i_k; \theta)\, q(z_k(i_k)). \quad (31.9)$$

The number of arithmetic operations needed to compute the functions $h_2, \ldots, h_K$ is of order $m_1 m_2 + \ldots + m_{K-1} m_K$; a lower order of operations is required to evaluate h_1 and any one-dimensional sum such as (31.8). By exploiting the special structure of the integrands in formulae such as (31.5), we have kept the total computational effort *linear* in K, whereas it would rise exponentially with K for general K-fold integrands.

We now consider the choice of grid points and their associated weights for each of the nested numerical integrations. In an expression such as (31.5), for each $j = 1, \ldots, k-1$

the integral with respect to z_j has a positive integrand bounded above by the marginal density of Z_j in the absence of early stopping, which is normal with mean $\theta\sqrt{\mathcal{I}_j}$ and unit variance. This can be seen by noting that the integrand is increased if all other ranges of integration are extended to the whole real line and, in this case, since variables z_1 to z_{j-1} are integrated over their full range, the result is the marginal density of Z_j. It is therefore reasonable to apply a numerical integration rule over z_j which is efficient for integrating the $N(\theta\sqrt{\mathcal{I}_j}, 1)$ density. The implementation of Simpson's rule described below provides $O(m^{-4})$ convergence as m, the number of grid points used in integrating this density over the real line or a subset of it, increases (see Chandler & Graham, 1988). This convergence rate also holds if the normal density is multiplied by a smooth bounded function, as is the case in expressions such as (31.5). We have observed this rate of convergence empirically in calculations using increasing numbers of grid points.

We start by defining grid points and weights to integrate a function of z_j over the whole real line. Since the tails of the normal distribution decrease rapidly, it is appropriate to concentrate grid points in the center and allocate a smaller proportion, more widely spaced, to the tails. We have found it most effective to place two thirds of the grid points within three standard deviations of the normal mean and the remaining third in the tails. We create a set of $6r-1$ values, $x_1, \ldots, x_{6r-1}$, defining elements on which to apply Simpson's rule, as follows:

$$x_i = \begin{cases} \theta\sqrt{\mathcal{I}_j} + \{-3 - 4\log(r/i)\}, & i = 1, \ldots, r-1, \\ \theta\sqrt{\mathcal{I}_j} + \{-3 + 3(i-r)/(2r)\}, & i = r, \ldots, 5r, \\ \theta\sqrt{\mathcal{I}_j} + \{3 + 4\log(r/(6r-i))\}, & i = 5r+1, \ldots, 6r-1. \end{cases}$$

Here, the logarithmic spacing between the outer values and the expansion of the interval (x_1, x_{6r-1}) with r is required for $O(m^{-4})$ accuracy as m, the total number of grid points, increases. Other values may be used for the width of the middle interval, the factor multiplying $\log(r/i)$ and $\log(r/(6r-i))$, and the distribution of points between the middle interval and the tails. However, we have found the above scheme to work as well as any.

In integrating the function $q(z_j)$ from $-\infty$ to ∞, we use $m_j = 12r - 3$ grid points. Odd numbered grid points are the x_i defined above, so $z_j(i_j) = x_{(i_j+1)/2}$ for $i_j = 1, 3, \ldots, m_j$, and we define the even numbered grid points as the midpoints of their neighbors,

$$z_j(i_j) = (z_j(i_j - 1) + z_j(i_j + 1))/2, \quad i_j = 2, 4, \ldots, m_j - 1.$$

We approximate the integral of $q(z_j)$ between the two odd numbered grid points $z_j(2i-1)$ and $z_j(2i+1)$ according to Simpson's rule as

$$\frac{d}{6}\, q(z_j(2i-1)) + \frac{4d}{6}\, q(z_j(2i)) + \frac{d}{6}\, q(z_j(2i+1)),$$

where d is the interval length, $z_j(2i+1) - z_j(2i-1)$. Hence, the weights required for integrating over the whole real line are

$$w_j(i_j) = \begin{cases} \frac{1}{6}\,(z_j(3) - z_j(1)), & i_j = 1, \\ \frac{1}{6}\,(z_j(i_j+2) - z_j(i_j-2)), & i_j = 3, 5, \ldots, m_j - 2, \\ \frac{4}{6}\,(z_j(i_j+1) - z_j(i_j-1)), & i_j = 2, 4, \ldots, m_j - 1, \\ \frac{1}{6}\,(z_j(m_j) - z_j(m_j-2)), & i_j = m_j, \end{cases} \tag{31.10}$$

and using these gives the approximation

$$\int_{-\infty}^{\infty} q(z_j)\, dz_j \;\approx\; \sum_{i_j=1}^{m_j} w_j(i_j)\, q(z_j(i_j)).$$

The rapid decay of the tails of the normal density ensures that the absence of any contribution from values of z_j outside the range $(z_j(1), z_j(m_j))$ does not cause a significant lack of accuracy.

In order to integrate over a finite range (a_j, b_j) we take the values $x_1, \ldots, x_{6r-1}$ defined above, trim off those less than a_j or greater than b_j, and introduce additional points at a_j and b_j where this is appropriate. If $a_j < x_1$, we make no adjustment at the left hand end; otherwise, we replace the largest x_i less than a_j by a_j and discard points x_i below this. Similarly, if $b_j > x_{6r-1}$, we do nothing at the right hand end, but otherwise we replace the smallest x_i greater than b_j by b_j and discard points x_i above this. The terms $4\log(r/i)$ and $4\log(r/(6r-i))$ in the definition of the x_i values mean that the tails of the sequence $x_1, \ldots, x_{6r-1}$ are widely spaced, increasingly so as r increases, and this ensures the $O(m^{-4})$ accuracy of the numerical integral when a_j is large and negative or b_j is large and positive.

The x_i values resulting from the above process comprise our odd numbered grid points, $\{z_j(i_j);\, i_j = 1, 3, \ldots, m_j\}$, say, where $m_j \leq 12r - 3$. We calculate even numbered grid points as the midpoints of their neighbors, as before, and weights for all points are again given by (31.10). The approximation to the integral is then

$$\int_{a_j}^{b_j} q(z_j)\, dz_j \approx \sum_{i_j=1}^{m_j} w_j(i_j)\, q(z_j(i_j)).$$

Using $r = 16$ in the above quadrature rule, giving a maximum of just less than 200 grid points per integral, we have found computed probabilities to be within 10^{-6} of their true values and about one decimal place of accuracy is lost in halving the value of r. Accuracy may be reduced if some increments in information are very small and we recommend using a higher value of r if $(\mathcal{I}_j - \mathcal{I}_{j-1})/\mathcal{I}_j$ is less than 0.01 for any j. Nevertheless, we have found the above method to provide accurate results over a wide range of group sequential tests with as many as $K = 200$ analyses.

31.3 Error Probabilities and Sample Size Distributions

31.3.1 Properties of a Given Test

The values of $\psi_k(a_1, b_1, \ldots, a_k, b_k; \theta)$ and $\xi_k(a_1, b_1, \ldots, a_k, b_k; \theta)$ for $k = 1, \ldots, K$ determine the distribution of the stopping time and associated decision for a group sequential test. Hence, we can obtain from them the test's error probabilities and expected information on termination for any value of θ.

Suppose, for example, a two-sided test of H_0: $\theta = 0$ has K analyses at information levels $\mathcal{I}_1, \ldots, \mathcal{I}_K$ with critical values $(a_1, b_1), \ldots, (a_K, b_K)$ for $Z_1, \ldots, Z_K$. Then, the test's Type I error probability is

$$Pr_{\theta=0}\{\text{Reject } H_0\} = \sum_{k=1}^{K} \{\psi_k(a_1, b_1, \ldots, a_k, b_k; 0) + \xi_k(a_1, b_1, \ldots, a_k, b_k; 0)\}. \qquad (31.11)$$

If $\delta > 0$ is large enough that we can ignore the probability of crossing the lower boundary, the test's power when $\theta = \delta$ is

$$\sum_{k=1}^{K} \psi_k(a_1, b_1, \ldots, a_k, b_k; \delta) \qquad (31.12)$$

and power when $\theta = -\delta$ has the same form but with ξ_k rather than ψ_k and $-\delta$ in place of δ. The expected information on termination under a given value of θ, a quantity closely related to the test's expected sample size, is

$$\sum_{k=1}^{K-1} \{\psi_k(a_1, b_1, \ldots, a_k, b_k; \theta) + \xi_k(a_1, b_1, \ldots, a_k, b_k; \theta)\}\, \mathcal{I}_k$$

$$+ [1 - \sum_{k=1}^{K-1} \{\psi_k(a_1, b_1, \ldots, a_k, b_k; \theta) + \xi_k(a_1, b_1, \ldots, a_k, b_k; \theta)\}]\, \mathcal{I}_K.$$

For a one-sided test of H_0: $\theta = 0$ against $\theta > 0$ with information levels $\mathcal{I}_1, \ldots, \mathcal{I}_K$ and boundary $(a_1, b_1), \ldots, (a_K, b_K)$, the Type I error probability and power at $\theta = \delta$ are

$$\sum_{k=1}^{K} \psi_k(a_1, b_1, \ldots, a_k, b_k; 0) \quad \text{and} \quad \sum_{k=1}^{K} \psi_k(a_1, b_1, \ldots, a_k, b_k; \delta)$$

respectively. The expected information on termination is given by the same expression as for the two-sided test.

If a test is to be evaluated under a number of values of θ, one may take advantage of the property

$$g_k(z_k; \theta_2) \;=\; g_k(z_k; \theta_1)\, l(z_k, \mathcal{I}_k; \theta_2, \theta_1) \tag{31.13}$$

where

$$l(z_k, \mathcal{I}_k; \theta_2, \theta_1) \;=\; \exp\{(\theta_2 - \theta_1) z_k \sqrt{\mathcal{I}_k} - (\theta_2^2 - \theta_1^2)\mathcal{I}_k/2\}$$

is the likelihood ratio for the observation $Z_k = z_k$ under parameter values θ_2 and θ_1. This result is closely related to the likelihood ratio identity (9.3) and it can be derived in a similar way from the recursive formulae for the functions $g_1, \ldots, g_K$. Using this with (31.9) we obtain

$$\int_{a_k}^{b_k} g_k(z_k; \theta_2)\, q(z_k)\, dz_k \approx \sum_{i_k=1}^{m_k} h_k(i_k; \theta_1)\, l(z_k(i_k), \mathcal{I}_k; \theta_2, \theta_1)\, q(z_k(i_k))$$

for a general function $q(z_k)$.

As an example, we can write probabilities of crossing the upper boundary under $\theta = \theta_2$ for each $k = 2, \ldots, K$ as

$$\psi_k(a_1, b_1, \ldots, a_k, b_k; \theta_2)$$

$$\approx \sum_{i_{k-1}=1}^{m_{k-1}} h_{k-1}(i_{k-1}; \theta_1)\, l(z_{k-1}(i_{k-1}), \mathcal{I}_{k-1}; \theta_2, \theta_1)\, e_{k-1}(z_{k-1}(i_{k-1}), b_k; \theta_2)$$

as an alternative to (31.8). This allows us to calculate the functions $h_1, \ldots, h_{k-1}$ under one particular value $\theta = \theta_1$ and then use these to find probabilities $\psi_k(a_1, b_1, \ldots, a_k, b_k; \theta)$ under other values of θ at little additional cost. In doing this, for each $j = 1, \ldots, k-1$, a common set of grid points $z_j(i_j)$ is used for calculations under different values of θ. The results obtained are precisely those that would have been produced by direct application of Simpson's rule on these grid points to the integral representation of $\psi_k(a_1, b_1, \ldots, a_k, b_k; \theta)$ given in (31.5). Our remarks on numerical accuracy in Section 31.2 were based on the assumption that grid points are chosen for the particular value of θ being investigated. Since some loss of numerical accuracy is to be expected using similar numbers of grid points placed elsewhere over the ranges of $z_1, \ldots, z_{k-1}$, care should be taken to ensure there are adequate numbers of grid points in the appropriate regions for all θ values addressed in a common calculation.

31.3.2 Designing a Test with Specified Error Probabilities

The calculations described in Section 31.3.1 can be used to derive group sequential tests with specific properties. As an example, consider the Wang & Tsiatis family of two-sided tests of H_0: $\theta = 0$ introduced in Section 2.7.1, which contains Pocock and O'Brien & Fleming tests as special cases. The test with parameter Δ has boundaries for Z_k of the form

$$a_k = -c\,(k/K)^{\Delta-1/2} \quad \text{and} \quad b_k = c\,(k/K)^{\Delta-1/2}, \quad k = 1, \dots, K,$$

and the constants $C_{WT}(K, \alpha, \Delta)$ tabulated in Section 2.7.1 are the values of c for which the test with parameter Δ and K equally spaced analyses has two-sided Type I error probability α. Consideration of the outcome for a particular sample path $Z_1, \dots, Z_K$ shows that the Type I error probability decreases as c increases. Hence a simple numerical search, using formula (31.11) to evaluate the Type I error rate for any given value of c, may be used to find $C_{WT}(K, \alpha, \Delta)$ for specified K, α and Δ.

In Chapter 2, we also tabulated constants R_P, R_B and R_{WT} for use in designing a test with the correct sample size to achieve power $1 - \beta$ at $\theta = \pm\delta$. Once again, these constants can be found by a numerical search, this time over the maximum information level. Suppose, in the Wang & Tsiatis family, K, α and Δ are specified and $C_{WT}(K, \alpha, \Delta)$ has been found. We need to find the value $\mathcal{I}_{max}$ such that (31.12) is equal to $1 - \beta$ when $\mathcal{I}_k = (k/K)\mathcal{I}_{max}$, $k = 1, \dots, K$. The value of $\mathcal{I}_{max}$ affects the distribution of $\{Z_1, \dots, Z_K\}$, and hence the power, through the means of $Z_1, \dots, Z_K$ and it can be shown that (31.12) increases with $\mathcal{I}_{max}$. Thus, it is straightforward to find the $\mathcal{I}_{max}$ required for given K, α, β and Δ. Taking the ratio of this value of $\mathcal{I}_{max}$ to the information level (4.5) required in a fixed sample test gives $R_{WT}(K, \alpha, \beta, \Delta)$.

The above derivation of two-sided tests is simplified by the separate treatment of Type I error and power. However, the task of finding constants $\tilde{C}_1(K, \alpha, \beta, \Delta)$, $\tilde{C}_2(K, \alpha, \beta, \Delta)$ and $\tilde{R}(K, \alpha, \beta, \Delta)$ for one-sided tests in the power family of Chapter 5 does not reduce to two separate problems. Instead, a pair of constants $\tilde{C}_1$ and $\tilde{C}_2$ must be found by a two-dimensional search. As explained in Section 31.3.1, there is no difficulty in computing the Type I error rate and power of the test defined by a particular pair of constants. The required values of $\tilde{C}_1$ and $\tilde{C}_2$ may therefore be obtained using a standard numerical routine to minimize the sum of squared differences between a test's Type I error rate and the target value α and between its power at $\theta = \delta$ and the target $(1 - \beta)$, the solution being found when this sum of squares is sufficiently close to zero. The constant $\tilde{R}$ is determined by $\tilde{C}_1$ and $\tilde{C}_2$ according to the formula stated in Section 5.2.1.

31.4 Tests Defined by Error Spending Functions

31.4.1 Implementation

The error spending tests of Chapter 8 require computation both in their design and in their implementation. In applying the error spending approach, new boundary points are calculated in response to the observed information level at each new analysis, as explained in Section 8.7. There, the expressions to be calculated were given in the form of integrals: the methods of Section 31.2 enable us to evaluate these integrals numerically.

For a two-sided test of H_0: $\theta = 0$, the boundary values for Z_1, Z_2, ... are $\pm c_1$, $\pm c_2, \dots$ and these are found in sequence as the study continues. At analysis k, the values $c_1, \dots, c_{k-1}$ will have already been computed and we find c_k as the solution to

$$\int_{c_k}^{\infty} g_k(z; 0)\, dz \;=\; \pi_k/2$$

or, equivalently,

$$\psi_k(-c_1, c_1, \dots, -c_k, c_k; 0) \;=\; \pi_k/2,$$

where π_k is the two-sided Type I error probability assigned to analysis k. Applying (31.8), the problem becomes that of finding the value c_k for which

$$\sum_{i_{k-1}=1}^{m_{k-1}} h_{k-1}(i_{k-1};0)\, e_{k-1}(z_{k-1}(i_{k-1}), c_k; 0) \;=\; \pi_k/2. \tag{31.14}$$

The $h_{k-1}(i_{k-1};0)$ can be computed in the usual way and, since the left hand side of (31.14) decreases as c_k increases, it is easy to find c_k by a numerical search. In fact, the derivative of the left hand side of (31.14) is readily computed since

$$\frac{d}{dc_k}\, e_{k-1}(z_{k-1}(i_{k-1}), c_k; 0) \;=\; -f_k(z_{k-1}(i_{k-1}), c_k; 0)$$

and using this derivative information in, say, a Newton-Raphson iteration can improve the speed of this search considerably.

The one-sided tests of Sections 8.3 and 8.4 are defined by two error spending functions, one under the null hypothesis $\theta = 0$ and the other under the alternative $\theta = \delta$. For a design with a binding futility boundary, as described in Section 8.3, values a_k and b_k are found as the solutions to

$$\int_{b_k}^{\infty} g_k(z;0)\, dz \;=\; \pi_{1,k} \quad \text{and} \quad \int_{-\infty}^{a_k} g_k(z;\delta)\, dz \;=\; \pi_{2,k},$$

where $\pi_{1,k}$ and $\pi_{2,k}$ are the Type I and II error probabilities, respectively, assigned to analysis k. To obtain a_k and b_k numerically, we solve

$$\sum_{i_{k-1}=1}^{m_{k-1}} h_{k-1}(i_{k-1};0)\, e_{k-1}(z_{k-1}(i_{k-1}), b_k; 0) \;=\; \pi_{1,k}$$

and

$$\sum_{i_{k-1}=1}^{m_{k-1}} h_{k-1}(i_{k-1};\delta)\, \{1 - e_{k-1}(z_{k-1}(i_{k-1}), a_k; \delta)\} \;=\; \pi_{2,k},$$

a similar task to finding the value c_k satisfying (31.14) for a two-sided test.

Implementing an error spending design with a non-binding futility boundary, as described in Section 8.4, is a little more complicated since Type I and II error probabilities are calculated under different versions of the lower boundary $\{a_1, a_2, \ldots\}$. A simple solution is to perform the calculations at each analysis in two steps. At analysis k, boundary values $(a_1, b_1), \ldots, (a_{k-1}, b_{k-1})$ have already been calculated. The efficacy boundary point b_k is computed assuming there is no futility boundary, so $a_1, \ldots, a_{k-1}$ are replaced by $-\infty$ for this calculation. Then, the futility boundary point a_k is computed with the previous futility boundary points in place.

31.4.2 Design

We explained in Section 8.2.2 how a two-sided error spending test can be designed in the "maximum information" paradigm so as to achieve a specified power if information levels are equally spaced and reach the chosen target information level $\mathcal{I}_{max}$ at analysis K. As we vary $\mathcal{I}_{max}$, setting $\mathcal{I}_k = (k/K)\mathcal{I}_{max}$, $k = 1, \ldots, K$, the error probabilities allocated to each analysis and the sequence of critical values for $Z_1, \ldots, Z_K$ remain the same. The situation is therefore identical to that arising for parametric two-sided tests in Section 31.3.2 and the same type of search can be used to find the value of $\mathcal{I}_{max}$ for which power $1 - \beta$ is attained at $\theta = \pm\delta$.

A different approach is needed to design a one-sided error spending test. Again, for design purposes, it is convenient to assume equally spaced information levels $\mathcal{I}_k = (k/K)\mathcal{I}_{max}$, $k = 1, \ldots, K$, will be observed. Initially, we must specify the Type I and Type II error spending functions $f(t)$ and $g(t)$, as defined in Section 8.3.1. Then, for a given target information level $\mathcal{I}_{max}$ and observed information sequence $\mathcal{I}_k = (k/K)\mathcal{I}_{max}$, we find values (a_k, b_k), $k = 1, \ldots, K$, to give cumulative Type I error probabilities $f(k/K)$ under H_0: $\theta = 0$ and Type II error probabilities $g(k/K)$ under $\theta = \delta$ at each analysis k. If the value of $\mathcal{I}_{max}$ is too low, this calculation will end with $a_K < b_K$, since the desired overall Type I and Type II error probabilities cannot be achieved simultaneously. If $\mathcal{I}_{max}$ is too high, the boundaries will cross, ending with $a_K > b_K$ or with $a_k > b_k$ at an earlier analysis. Learning which of these outcomes occurs for each proposed $\mathcal{I}_{max}$ is sufficient to implement a bisection search for the value of $\mathcal{I}_{max}$ under which the test terminates neatly at analysis K with $a_K = b_K$.

31.5 Analysis Following a Group Sequential Test

31.5.1 Point Estimation

The bias adjusted estimators discussed in Section 9.3 require evaluation under certain θ values of the bias of the MLE $\hat{\theta} = Z_T/\sqrt{\mathcal{I}_T}$, where T denotes the stage at which the study stops and Z_T is the value of the standardized statistic at this point. This bias is $E_\theta(\hat{\theta}) - \theta$ where $E_\theta(\hat{\theta})$, given by (9.8), can be written as

$$E_\theta(\hat{\theta}) = \sum_{k=1}^{K} \left\{ \int_{-\infty}^{a_k} g_k(z_k;\theta)\, \frac{z_k}{\sqrt{\mathcal{I}_k}}\, dz_k + \int_{b_k}^{\infty} g_k(z_k;\theta)\, \frac{z_k}{\sqrt{\mathcal{I}_k}}\, dz_k \right\}.$$

A typical lower integral in this sum can be written as

$$\int_{a_{k-1}}^{b_{k-1}} \int_{-\infty}^{a_k} g_{k-1}(z_{k-1};\theta)\, f_k(z_{k-1}, z_k;\theta)\, \frac{z_k}{\sqrt{\mathcal{I}_k}}\, dz_k\, dz_{k-1}$$

and it follows from the definition of f_k in equation (31.4) and properties of the normal distribution (see Todd, Whitehead & Facey, 1996, p. 455) that this is equal to

$$\int_{a_{k-1}}^{b_{k-1}} g_{k-1}(z_{k-1};\theta)\, r_{k-1}(z_{k-1}, a_k;\theta)\, dz_{k-1}, \tag{31.15}$$

where

$$\begin{aligned} r_{k-1}(z_{k-1}, a_k;\theta) &= \frac{-\sqrt{\Delta_k}}{\mathcal{I}_k}\, \phi\left(\frac{a_k\sqrt{\mathcal{I}_k} - z_{k-1}\sqrt{\mathcal{I}_{k-1}} - \theta\Delta_k}{\sqrt{\Delta_k}} \right) \\ &\quad + \frac{(z_{k-1}\sqrt{\mathcal{I}_{k-1}} + \theta\Delta_k)}{\mathcal{I}_k}\, \Phi\left(\frac{a_k\sqrt{\mathcal{I}_k} - z_{k-1}\sqrt{\mathcal{I}_{k-1}} - \theta\Delta_k}{\sqrt{\Delta_k}} \right). \end{aligned}$$

Using (31.9), we can evaluate the integral (31.15) numerically as

$$\sum_{i_{k-1}=1}^{m_{k-1}} h_{k-1}(i_{k-1};\theta)\, r_{k-1}(z_{k-1}(i_{k-1}), a_k;\theta).$$

The upper integrals from b_k to ∞ can be computed in a similar manner and these results may then be combined to give $E_\theta(\hat{\theta})$.

31.5.2 P-values

We defined frequentist P-values on termination of a group sequential test in Section 9.4. Suppose, for example, a test with boundary values $(a_1, b_1), \ldots, (a_K, b_K)$ for $Z_1, \ldots, Z_K$ stops after crossing the upper boundary at analysis k^* with $Z_{k^*} = z^*$. The one-sided upper P-value for testing H_0: $\theta = 0$ based on the stage-wise ordering is then

$$\sum_{j=1}^{k^*-1} \psi_j(a_1, b_1, \ldots, a_j, b_j; 0) + \psi_{k^*}(a_1, b_1, \ldots, a_{k^*-1}, b_{k^*-1}, a_{k^*}, z^*; 0),$$

which can be calculated numerically by the methods of Section 31.2. The one-sided lower P-value is found in the same manner and the two-sided P-value is twice the smaller of these two quantities.

Slightly different computations are required to find P-values for the other three orderings of the sample space discussed in Section 9.4. However, in each case the set of outcomes more extreme than that observed is a union of at most K semi-infinite intervals and upper and lower P-values are sums of terms ψ_j or ξ_j, as for the stage-wise ordering.

31.5.3 Confidence Intervals

In Section 9.5 we defined confidence intervals for a parameter θ on termination of a group sequential test, based on a specified ordering of the sample space. In computing such a confidence interval, the main task is to find values θ_L and θ_U satisfying (9.14) for the specified ordering. In the search for the roots θ_L and θ_U, the calculations for each θ investigated are the same as those needed to find one-sided upper and lower P-values, respectively, for testing a hypothesized value for θ.

Since the searches for θ_L and θ_U involve repeated calculations under a sequence of quite similar θ values, this is an ideal situation in which to use the likelihood ratio relation (31.13) to save computational effort.

An alternative way of reducing computation is to calculate derivatives of probabilities with respect to θ along with the probabilities themselves and exploit this derivative information in more efficient root-finding algorithms. Starting from the first integral expression for $\psi_k(a_1, b_1, \ldots, a_k, b_k; \theta)$ in (31.5), we have

$$\frac{d\psi_k}{d\theta} = \int_{a_1}^{b_1} \cdots \int_{a_{k-1}}^{b_{k-1}} \int_{b_k}^{\infty} \{f_1'(f_2 \ldots f_k) + f_1(f_2 \ldots f_k)'\}\, dz_k \ldots dz_1,$$

where, for conciseness, we have suppressed the arguments of ψ_k and $f_1, \ldots, f_K$. In this and subsequent formulae in this section, all primes denote derivatives with respect to θ.

Let $u_1(z_1; \theta) = f_1'(z_1; \theta)$ and $v_1(z_1; \theta) = f_1(z_1; \theta)$, and define functions u_k and v_k for $k = 2, \ldots, K$ by

$$u_k(z_k; \theta) = \int_{a_{k-1}}^{b_{k-1}} \{u_{k-1}(z_{k-1}; \theta)\, f_k(z_{k-1}, z_k; \theta) + v_{k-1}(z_{k-1}; \theta)\, f_k'(z_{k-1}, z_k; \theta)\}\, dz_{k-1}$$

and

$$v_k(z_k; \theta) = \int_{a_{k-1}}^{b_{k-1}} v_{k-1}(z_{k-1}; \theta)\, f_k(z_{k-1}, z_k; \theta)\, dz_{k-1}.$$

Then, again suppressing function arguments,

$$\begin{aligned}\frac{d\psi_k}{d\theta} &= \int_{a_1}^{b_1} \cdots \int_{a_{k-1}}^{b_{k-1}} \int_{b_k}^{\infty} \{u_1(f_2 \ldots f_k) + v_1(f_2 \ldots f_k)'\}\, dz_k \ldots dz_1 \\ &= \ldots \\ &= \int_{a_j}^{b_j} \cdots \int_{a_{k-1}}^{b_{k-1}} \int_{b_k}^{\infty} \{u_j(f_{j+1} \cdots f_k) + v_j(f_{j+1} \cdots f_k)'\}\, dz_k \ldots dz_j \\ &= \ldots \\ &= \int_{a_{k-1}}^{b_{k-1}} \int_{b_k}^{\infty} \{u_{k-1} f_k + v_{k-1} f_k'\}\, dz_k\, dz_{k-1}.\end{aligned}$$

The numerical evaluation of $d\psi_k/d\theta$ parallels that of ψ_k itself, each integral over a continuation region (a_j, b_j) being replaced by a weighted sum over grid points $z_j(i_j)$, $i_j = 1, \ldots, m_j$. Values of the functions $u_1, \ldots, u_K$ and $v_1, \ldots, v_K$ at the selected grid points are computed recursively. In fact, v_j is precisely the function g_j defined in Section 31.2 and the approximation to $v_j(i_j)$ multiplied by weight $w_j(i_j)$ is our previous $h_j(i_j)$. The approximation to $u_j(i_j)$ multiplied by $w_j(i_j)$ is a new function, readily obtained by additional calculations alongside those giving h_j. For further details of these computations see Jennison (1994).

31.6 Further Applications of Numerical Computation

31.6.1 Calculating Optimal Tests

In Chapter 12 we explained how to find an optimal group sequential design by using dynamic programming to solve a related Bayes decision problem. We now describe the numerical computations employed in the dynamic programming algorithm. As in Chapter 12, we consider a one-sided test of H_0: $\theta \leq 0$ vs H_A: $\theta > 0$ with Type I error probability α under $\theta = 0$ and power $1 - \beta$ when $\theta = \delta$. We fix the number of analyses K and cumulative information levels $\mathcal{I}_1, \ldots, \mathcal{I}_K$, and assume standardized statistics $Z_1, \ldots, Z_K$ follow the canonical joint distribution (3.1). In this illustrative example, our objective is to find the group sequential test that minimizes $E_{\theta=\delta/2}(\mathcal{I}_T)$, the expected information on termination when $\theta = \delta/2$. We shall describe the approach proposed by Eales & Jennison (1992) for tests with $\alpha = \beta$ and extended to designs with $\alpha \neq \beta$ by Barber & Jennison (2002). For a more discursive account of these methods, see Jennison & Turnbull (2013).

As described in Chapter 12, we start by formulating a Bayes sequential decision problem in which θ has the three point prior distribution, $\pi(0) = \pi(\delta/2) = \pi(\delta) = 1/3$, and a decision D_0: '$\theta \leq 0$' or D_1: '$\theta > 0$' is to be made with loss function $L(D_1, 0) = d_1$, $L(D_0, \delta) = d_2$ and $L(D, \theta) = 0$ otherwise. The sampling cost is $c(\theta)$ per unit of information, giving a total sampling cost $c(\theta)\mathcal{I}_T$, where $c(0) = 0$, $c(\delta) = 0$ and $c(\delta/2) = 1$. The solution of this Bayes decision problem minimizes the expected total cost

$$\pi(0) Pr_{\theta=0}\{D_1\}\, d_1 + \pi(\delta) Pr_{\theta=\delta}\{D_0\}\, d_2 + \pi(\delta/2) E_{\theta=\delta/2}(\mathcal{I}_T). \tag{31.16}$$

Once we have composed a routine to compute the Bayes optimal procedure that minimizes (31.16), we find the values of d_1^* and d_2^* for which the resulting test also satisfies

$$Pr_{\theta=0}\{D_1\} = \alpha \quad \text{and} \quad Pr_{\theta=\delta}\{D_0\} = \beta, \tag{31.17}$$

and we conclude this is the optimal group sequential test for our original problem.

In solving the Bayes decision problem of minimizing (31.16) for given d_1 and d_2, we need to determine the constants $a_1, \ldots, a_K$ and $b_1, \ldots, b_K$ such that stopping occurs at analysis k with decision D_0 if $Z_k \le a_k$ and decision D_1 if $Z_k \ge b_k$. For $k = 1, \ldots, K$, let $\pi^{(k)}(\theta \mid z_k)$ denote the posterior distribution of θ given $Z_k = z_k$. The minimum expected loss from stopping and making a decision at stage k when $Z_k = z_k$ is

$$\gamma^{(k)}(z_k) = \min\{\pi^{(k)}(0 \mid z_k)\, d_1,\, \pi^{(k)}(\delta \mid z_k)\, d_2\},$$

achieved by making decision D_1 when $\pi^{(k)}(0 \mid z_k)\, d_1 \le \pi^{(k)}(\delta \mid z_k)\, d_2$ and decision D_0 when $\pi^{(k)}(\delta \mid z_k)\, d_2 < \pi^{(k)}(0 \mid z_k)\, d_1$.

Let $\beta^{(k)}(z_k)$ denote the expected additional loss from continuing to sample at stage k when $Z_k = z_k$ and then proceeding optimally (not including the cost of information $\mathcal{I}_k$ which has already been incurred). Then, for $k = 1, \ldots, K-2$,

$$\beta^{(k)}(z_k) = (\mathcal{I}_{k+1} - \mathcal{I}_k)\pi^{(k)}(\delta/2 \mid z_k) + \int \min\{\beta^{(k+1)}(z_{k+1}), \gamma^{(k+1)}(z_{k+1})\}\, dF^{(k+1)}(z_{k+1} \mid z_k)$$

and

$$\beta^{(K-1)}(z_{K-1}) = (\mathcal{I}_K - \mathcal{I}_{K-1})\pi^{(K-1)}(\delta/2 \mid z_{K-1}) + \int \gamma^{(K)}(z_K)\, dF^{(K)}(z_K \mid z_{K-1}),$$

where $F^{(k+1)}(z_{k+1} \mid z_k)$, $k = 1, \ldots, K-1$, is the conditional cumulative distribution function of Z_{k+1} given $Z_k = z_k$.

Comparing the possible options when $Z_k = z_k$, we find it is: optimal to stop with decision D_1 if $z_k \ge b_k$ where

$$\beta^{(k)}(b_k) = \gamma^{(k)}(b_k) = \pi^{(k)}(0 \mid z_k)\, d_1;$$

optimal to stop with decision D_0 if $z_k \le a_k$ where

$$\beta^{(k)}(a_k) = \gamma^{(k)}(a_k) = \pi^{(k)}(\delta \mid z_k)\, d_2;$$

and optimal to continue to analysis $k+1$ if $a_k < z_k < b_k$.

The functions $\beta^{(k)}(z_k)$ can be found recursively, starting at $k = K-1$. Values of $\beta^{(k)}(z_k)$ should be computed for a grid of values of z_k including a_k and b_k, and these will be used in evaluating $\beta^{(k-1)}(z_{k-1})$ by numerical integration, first to find a_{k-1} and b_{k-1} by solving

$$\beta^{(k-1)}(b_{k-1}) = \gamma^{(k-1)}(b_{k-1}) = \pi^{(k-1)}(0 \mid z_{k-1})\, d_1$$

and

$$\beta^{(k-1)}(a_{k-1}) = \gamma^{(k-1)}(a_{k-1}) = \pi^{(k-1)}(\delta \mid z_{k-1})\, d_2,$$

and then to calculate $\beta^{(k-1)}(z_{k-1})$ at a grid of values of z_{k-1}. The same grids of points and Simpson's rule weights can be used here as in Section 31.2.

The above strategy can be adapted to a variety of problems. In Section 12.2 we defined the prior distribution for θ and cost of sampling function needed to create designs minimizing

$$\int f(\theta)\, E_\theta(\mathcal{I}_T)\, d\theta$$

where $f(\theta)$ is the density of a $N(\delta, \delta^2/4)$ distribution; in Section 12.3 we considered designs minimizing a variety of objective functions of the form

$$\sum_i w_i\, E_{\theta_i}(\mathcal{I}_T).$$

Jennison & Turnbull (2006a) used similar methods to obtain optimal "Schmitz" designs with data-dependent group sizes; Hampson & Jennison (2013) optimized group sequential designs for a delayed response where additional "pipeline" data are observed after stopping recruitment at an interim analysis; Öhrn (2011, Ch. 4) derived optimal group sequential tests with a non-binding futility boundary. In all these cases, the combination of numerical integration and dynamic programming leads to accurate and efficient calculations.

31.6.2 Other Response Distributions

The general approach we have described for normal data with known variance can be used to compute properties of group sequential tests for many other continuous response distributions. The key requirement is that the test should be defined in terms of a sequence of statistics, $T_1, \ldots, T_K$ say, which form a Markov process with tractable conditional distributions $T_{k+1} \,|\, T_k$. Denoting the density of T_1 by $f_1(t_1)$ and the conditional density of T_{k+1} given $T_k = t_k$ by $f_{k+1}(t_k, t_{k+1})$, $k = 1, \ldots, K-1$, the probability of exiting the upper boundary at analysis k is

$$\int_{a_1}^{b_1} \cdots \int_{a_{k-1}}^{b_{k-1}} \int_{b_k}^{\infty} f_1(t_1)\, f_2(t_1, t_2) \,\ldots\, f_k(t_{k-1}, t_k)\, dt_k \ldots dt_1,$$

an expression with the same form as that for $\psi_k(a_1, b_1, \ldots, a_k, b_k; \theta)$ in (31.5). This multiple integral can be evaluated numerically in a similar manner by a succession of summations. An example where this method is used to compute properties of group sequential χ^2 tests is described by Jennison & Turnbull (1991b).

In a group sequential t-test the appropriate statistic T_k is bivariate. Suppose, in the analysis of a normal linear model described in Section 15.4, that it is desired to test whether the scalar parameter $\theta = \boldsymbol{c}^T \boldsymbol{\beta}$ is equal to zero. The estimators at analysis k of θ and the variance scale factor σ^2 have marginal distributions

$$\hat{\theta}^{(k)} \sim N(\theta, \Gamma_k \sigma^2) \quad \text{and} \quad s_k^2 \sim \sigma^2 \chi^2_{\nu_k} / \nu_k,$$

respectively, for appropriately specified Γ_k and ν_k. Let $T_{1k} = \hat{\theta}^{(k)}/\sigma$ and $T_{2k} = \nu_k s_k^2/\sigma^2$, then the results of Section 15.4 imply that the sequence $\{(T_{1k}, T_{2k});\ k = 1, \ldots, K\}$ is Markov and, under H_0: $\theta = 0$,

$$T_{1,k+1} \mid T_{1k}, T_{2k} \sim N\left(\frac{\Gamma_{k+1}}{\Gamma_k} T_{1k},\ \Gamma_{k+1} - \frac{\Gamma_{k+1}^2}{\Gamma_k} \right)$$

and

$$T_{2,k+1} \mid T_{1,k+1}, T_{1k}, T_{2k} \sim T_{2k} + \frac{(T_{1,k+1} - T_{1k})^2}{\Gamma_k - \Gamma_{k+1}} + \chi^2_{\nu_{k+1} - \nu_k - 1}.$$

Probabilities concerning the sequence of t-statistics, $T_{1k}/\sqrt{(\Gamma_k T_{2k}/\nu_k)}$, $k = 1, \ldots, K$, can be calculated numerically from these conditional distributions in much the same way as for normal data with known variance. Care must be taken in integrating over the conditional distribution for $T_{2,k+1}$ if the degrees of freedom $\nu_{k+1} - \nu_k - 1$ are low, especially if this number of degrees of freedom is odd since then the χ^2 density has an infinite $\{(\nu_{k+1} - \nu_k - 2)/2\}th$ derivative at zero. Calculations of this type reported by Jennison & Turnbull (1991b) produced the constants displayed in Table 15.1.

Bivariate statistics arise naturally in studies with two distinct endpoints. Jennison & Turnbull (1993c) describe group sequential procedures for monitoring efficacy and safety endpoints in a clinical trial. In their prototype problem, each pair of observations (Y_{1i}, Y_{2i}) has a bivariate normal distribution with $E(Y_{1i}) = \theta_1$, $E(Y_{2i}) = \theta_2$ and $\text{Corr}(Y_{1i}, Y_{2i}) = \rho$. A problem formulation with an element of equivalence testing (see Chapter 7) for the safety endpoint leads to a test between the hypotheses H_0: '$\theta_1 \leq 0$ or $\theta_2 \leq 0$' and the alternative H_A: '$\theta_1 > 0$ and $\theta_2 > 0$'; with suitably defined response variables and parameters θ_1 and θ_2, H_A represents situations in which the new treatment performs satisfactorily with regard to both efficacy and safety considerations. Again, a Markov sequence of statistic pairs can be constructed and probabilities of crossing boundaries calculated by iterated numerical integration.

The same computational approach may be used with more than two response variables but at a greatly increased computational cost. If each 1-dimensional integral is replaced by a sum over m terms, the number of arithmetic operations required to assess a testing boundary for a d-variate response is of order m^{2d} but the numerical error, using Simpson's rule for each integral, remains of order m^{-4}. Thus, for high values of d, Monte Carlo simulation provides a better rate of convergence per numerical operation

31.6.3 Calculations for Multi-arm Multi-stage Trials

In a trial comparing J treatments with a control, let $\hat{\theta}_j^{(k)}$ denote the estimate of treatment effect j at analysis k for $j = 1 \ldots, J$ and $k = 1, \ldots, K$. Probabilities of particular outcomes of the trial can be calculated by following the vector $(\hat{\theta}_1^{(k)}, \ldots, \hat{\theta}_J^{(k)})$, which follows a Markov process, through analyses $k = 1, \ldots, K$. If each 1-dimensional integral is replaced by a sum over m terms, the number of arithmetic operations required to compute the Type I error probability or power under a certain set of treatment effects is of order m^{2J}, and this is already a daunting task for the case $J = 3$. In the MAMS procedures described by Magirr, Jaki & Whitehead (2012), each sequence $(\hat{\theta}_j^{(1)}, \ldots, \hat{\theta}_j^{(K)})$ is compared to a group sequential boundary, treatments may be dropped for futility, and the trial stops completely if one treatment crosses the efficacy boundary. For a design with a small number of analyses, numerical integrals are computed by summing over sequences of estimates of the mean of the control arm and utilizing the fact that, conditional on the control arm data, the sequences $(\hat{\theta}_j^{(1)}, \ldots, \hat{\theta}_j^{(K)})$ for different treatments j are independent. Following this approach, the computation required is of order m^{K+2} and Magirr, Jaki & Whitehead are able to present results for designs with $K = 2$ or 3 analyses comparing $J = 4$ treatments with a control.

We noted in Section 21.4.2 that Ghosh et al. (2017) carry out the same calculations using quasi-Monte Carlo simulation and this reduces the computation time dramatically. In this approach, the required computation is linear in the number of stages and the number of treatment comparisons. Ghosh et al. present results that show computations for a trial comparing 5 treatment arms against a control over 5 analyses to be perfectly feasible.

References

Sections in which the references are cited are shown in square brackets.

Alling, D.W. (1963). Early decision in the Wilcoxon two-sample test. *J. Amer. Statist. Assoc.*, **58**, 713–720. [11.1]

Alling, D.W. (1966). Closed sequential tests for binomial probabilities. *Biometrika*, **53**, 73–84. [11.1]

Altman, D.G. (1991). *Practical Statistics for Medical Research*, London: Chapman and Hall. [1.1]

Andersen, J.S. (1996). Clinical trial designs—made to order. *J. Biopharm. Statist.*, **6**, 515–522. [30.4]

Andersen, P.K. (1987). Conditional power calculations as an aid in the decision whether to continue a clinical trial. *Contr. Clin. Trials*, **8**, 67–74. [11.6, 11.7]

Anderson, K. (2025). gsDesign: Group Sequential Design. R package version 3.6.7. https://CRAN.R-project.org/package=gsDesign. [1.6]

Anderson, K.M. (2007). Optimal spending functions for asymmetric group sequential designs. *Biometrical J.*, **49**, 337–345. [12.2]

Anderson, S. and Hauck, W.W. (1990). Consideration of individual bioequivalence. *J. Pharmacokin. Biopharm.*, **18**, 259–273. [7.5]

Anderson, T.W. (1960). A modification of the sequential probability ratio test to reduce the sample size. *Ann. Math. Statist.*, **31**, 165–197. [1.3]

Anderson, T.W. (1964). Sequential analysis with delayed observations. *J. Amer. Statist. Assoc.*, **59**, 1006–1015. [14.5]

Anscombe, F.J. (1963). Sequential medical trials. *J. Amer. Statist. Assoc.*, **58**, 365–383. [13.3, 30.4]

Anthonisen, N.R. (1989). Lung Health Study. *Amer. Rev. Resp. Dis.*, **140**, 871–872. [1.4]

Arjas, E. and Gasbarra, D. (2022). Adaptive treatment allocation and selection in multi-arm clinical trials: a Bayesian perspective. *BMC Med. Res. Meth.*, **22**:50, 1–18. [13.5]

Armitage, P. (1954). Sequential tests in prophylactic and therapeutic trials, *Quart. J. Med.*, **23**, 255–274. [1.3]

Armitage, P. (1957). Restricted sequential procedures. *Biometrika*, **44**, 9–26. [1.3, 9.4, 9.5]

Armitage, P. (1958a). Numerical studies in the sequential estimation of a binomial parameter. *Biometrika*, **45**, 1–15. [9.4, 9.5]

Armitage, P. (1958b). Sequential methods in clinical trials. *Amer. J. Public Health*, **48**, 1395–1402. [1.3]

Armitage, P. (1975). *Sequential Medical Trials*, 2nd ed., Oxford: Blackwell. [1.3, 1.4, 1.6, 31.1]

Armitage, P. (1985). The search for optimality in clinical trials. *Int. Statist. Rev.*, **53**, 15–24. [30.4]

Armitage, P. (1993). Interim analyses in clinical trials. In *Multiple Comparisons, Selection, and Applications in Biometry*, (Ed., F.M. Hoppe), New York: Marcel Dekker, 391–402. [1.3]

Armitage, P., Berry, G. and Matthews, J.N.S. (2001). *Statistical Methods in Medical Research*, 4th ed., Oxford: Blackwell. [1.1]

Armitage, P., McPherson, C.K. and Rowe, B.C. (1969). Repeated significance tests on accumulating data. *J. Roy. Statist. Soc. A*, **132**, 235–244. [1.3, 2.4, 3.5, 8.7, 9.2, 10.1, 13.3, 31.1]

Armitage, P. and Schneiderman, M. (1958). Statistical problems in a mass screening program. *Ann. New York Acad. Sci.*, **76**, 896–908. [1.3, 31.1]

Armitage, P., Stratton, I.M. and Worthington, H.V. (1985). Repeated significance tests for clinical trials with a fixed number of patients and variable follow-up. *Biometrics*, **41**, 353–359. [1.4, 15.3, 15.7]

Aroian, L.A. (1968). Sequential analysis, direct method. *Technometrics*, **10**, 125–132. [1.3]

Aroian, L.A. and Robison, D.E. (1969). Direct methods for exact truncated sequential tests of the mean for a normal distribution. *Technometrics*, **11**, 661–675. [1.3]

Asakura, K., Evans, S.R. and Hamasaki, T. (2020). Interim monitoring for futility in clinical trials with two co-primary endpoints using prediction. *Statist. Biopharm. Res.*, **12**, 164–175. [19.6]

Ascher. H. and Feingold, H. (1984). *Repairable Systems Reliability: Modeling, Inference, Misconceptions and their Causes*, New York: Marcel Dekker. [1.4]

Asikanius, E., Hofner, B., Hampson, L.V., Wassmer, G., Jennison, C. et al. (2025) Clinical trials with interim analyses: standardizing terminology to increase clarity. *Trials*, **26**:247, 1–14. [14.5]

ASQ/ANSI(2018a). *Z1.4-2003 (R2018), Sampling Procedures and Tables for Inspection by Attributes*, Milwaukee, Wisconsin: ASQ Quality Press. [1.2]

ASQ/ANSI (2018b). *Z1.9-2003 (R2018), Sampling Procedures and Tables for Inspection by Variables for Percent Nonconforming*, Milwaukee, Wisconsin: ASQ Quality Press. [1.2]

Atkinson, A.C. and Biswas, A. (2014). *Randomised Response-Adaptive Designs in Clinical Trials.* Boca Raton: Chapman and Hall/CRC. [1.5, 30.4]

Azher, R.A., Grayling, M.J. and Wason, J.M. (2025). Efficient stage-wise allocation ratios in multi-arm multi-stage designs. *Contemp. Clin. Trials*, **153**:107908, 1–7. [30.4]

Baayen, C., Blanche, P., Jennison, C. and Ozenne, B. (2025) Design and analysis of group sequential trials for repeated measurements when pipeline data occurs: a tutorial. *Statist. Med.*, **44**:e70130, 1–24. [14.3]

Bailar, J.C. (1976). Patient assignment algorithms: an overview. In *Proceedings of the 9th International Biometric Conference, Vol. 1*, Raleigh: Biometric Society, 189–206. [30.4]

Baillie, D.H. (1987a). Multivariate acceptance sampling—some applications to defence procurement. *The Statistician*, **36**, 465–478. (Correction **37**, 97 and **38**, 315) [19.7]

Baillie, D.H. (1987b). Multivariate acceptance sampling. In *Frontiers in Statistical Quality Control, Vol. 3*, (Eds., H.J. Lenz, G.B. Wetherill and P.T. Wilrich), Wuerzberg-Vienna: Physica-Verlag. [19.7]

Ballarini, N.M., Burnett, T., Jaki, T., Jennison, C. et al. (2021). Optimizing subgroup selection in two-stage adaptive enrichment and umbrella designs. *Statist. Med.*, **40**, 2939–2956. [28.3]

Banerjee, A. and Tsiatis, A.A. (2006). Adaptive two-stage designs in phase II clinical trials. *Statist. Med.*, **25**, 3382–3395. [12.4]

Barber, S. and Jennison, C. (1999). Symmetric tests and confidence intervals for survival probabilities and quantiles of censored survival data. *Biometrics*, **55**, 430–436. [17.7]

Barber, S. and Jennison, C. (2002). Optimal asymmetric one-sided group sequential tests. *Biometrika*, **89**, 49–60. [5.1, 5.2, 8.8, 10.3, 12.2, 31.6]

Barnard, G.A. (1946). Sequential tests in industrial statistics. *J. Roy. Statist. Soc. Suppl.*, **8**, 1–26. [1.3]

Bartky, W. (1943). Multiple sampling with constant probability. *Ann. Math. Statist.*, **14**, 363–377. [1.3]

Bather, J.A. (1985). On the allocation of treatments in sequential medical trials. *Int. Statist. Rev.*, **53**, 1–13. [1.3]

Bather, J.A. (1988). Stopping rules and ordered families of distributions. *Seq. Analysis*, **7**, 111–126. [9.5]

Bauer, P. (1989). Multistage testing with adaptive designs. *Biometrie und Informatik in Medizin und Biologie*, **20**, 130–148. [1.3, 22.1, 23.2, 23.10]

Bauer, P. (2006). Discussion of 'Are flexible designs sound?' (by C-F. Burman and C. Sonesson). *Biometrics*, **62**, 676–678. [24.2]

Bauer, P. and Kieser, M. (1999). Combining different phases in the development of medical treatments within a single trial. *Statist. Med.*, **18**, 1833–1848. [26.8]

Bauer, P. and Köhne, K. (1994). Evaluation of experiments with adaptive interim analyses. *Biometrics*, **50**, 1029–1041. (Correction *Biometrics* (1996), {**52**, 380.) [1.3, 1.5, 14.2, 22.2, 23.2, 23.3, 23.10, 24.1]

Bauer P. and Posch M. (2004). Modification of the sample size and the schedule of interim analyses in survival trials based on data inspections (Letter to the Editor). *Statist. Med.*, **23**, 1333–1335. [28.4]

Bauer, P., Bretz, F., Dragalin, V., König, F. and Wassmer, G. (2016). Twenty-five years of confirmatory adaptive designs: opportunities and pitfalls (with discussion). *Statist. Med.*, **35**, 325–367. [1.3, 1.6, 23.10, 24.1]

Baum, C.W. and Veeravalli, V.V. (1994). A sequential procedure for multihypothesis testing. *IEEE Trans. Inform. Theory*, **40**, 1994–2007. [1.3]

Bebu, I., Dragalin, V. and Luta, G. (2013). Confidence intervals for confirmatory adaptive two-stage designs with treatment selection. *Biometrical J.*, **55**, 294–309. [26.6]

Bechhofer, R.E., Kiefer, J. and Sobel, M. (1968). *Sequential Identification and Ranking Procedures: with Special Reference to Koopman-Darmois Populations*, Chicago: University of Chicago Press. [1.3]

Bechhofer, R.E., Santner, T.J. and Goldsman, D.M. (1995). *Design and Analysis of Experiments for Statistical Selection, Screening, and Multiple Comparisons*, New York: Wiley. [21.5]

Beck, G.J., Berg R.L., Coggins C.H. et al. (1991). Design and statistical issues of the Modification of Diet in Renal Disease Trial. The Modification of Diet in Renal Disease Study Group *Contr. Clin. Trials*, **12**, 566–586. [1.1]

Becker, B.J. (1994). Combining significance levels. In *The Handbook of Research Synthesis*, (Eds., H. Cooper and L.V. Hedges), New York: Russell Sage Foundation, 215–230. [23.1]

Bellman, R. (1957) *Dynamic Programming*, Princeton: Princeton University Press. [12.2, 13.3]

Benjamini, Y. and Hochberg, Y. (1995). Controlling the false discovery rate: a practical and powerful approach to multiple testing. *J. Roy. Statist. Soc. B*, **57**, 289–300. [20.9]

Benjamini, Y. and Yekutieli, D. (2001). The control of the false discovery rate in multiple testing under dependency. *Ann. Statist.*, **29**, 1165–1188. [20.9]

Berger, J.O. (1985). *Statistical Decision Theory and Bayesian Analysis*, 2nd ed., New York: Springer-Verlag. [13.1]

Berger, J.O. and Berry, D.A. (1988). Statistical analysis and the illusion of objectivity. *Amer. Scientist*, **76**, 159–165. [13.3]

Bernardo, J.M. (1979). Reference posterior distributions for Bayesian inference (with discussion). *J. Roy. Statist. Soc. B*, **41**, 113–147. [13.2]

Berry, D.A. (1985). Interim analysis in clinical trials: classical vs. Bayesian approaches. *Statist. Med.*, **4**, 521–526. [13.3]

Berry, D.A. (1987). Interim analyses in clinical trials: the role of the likelihood principle. *The Amer. Statistician*, **41**, 117–122. [13.3]

Berry, D.A. (1993). A case for Bayesianism in clinical trials. *Statist. Med.*, **12**, 1377–1393. [13.5]

Berry, D.A. and Fristedt, B. (1985). *Bandit Problems: Sequential Allocation of Experiments*, London: Chapman and Hall. [1.3]

Berry, D.A. and Ho, C.H. (1988). One-sided sequential stopping boundaries for clinical trials: a decision theoretic approach. *Biometrics*, **44**, 219–227. [1.3, 13.3]

Berry, S.M., Broglio, K.R., Groshen, S. and Berry, D.A. (2013). Bayesian hierarchical modeling of patient subpopulations: efficient designs of phase II oncology clinical trials. *Clin. Trials*, **10**, 720–734. [28.5]

Berry, S.M., Carlin, B.P., Lee, J.J. and Müller, P. (2010). *Bayesian Adaptive Methods for Clinical Trials.* Boca Raton: Chapman and Hall/CRC. [13.5]

Betensky, R.A. (1996). An O'Brien-Fleming sequential trial for comparing three treatments. *Ann. Statist.*, **24**, 1765–1791. [21.5]

Betensky, R.A. (1997a). Conditional power calculations for early acceptance of H_0 embedded in sequential tests. *Statist. Med.*, **16**, 465–477. [11.7]

Betensky, R.A. (1997b). Early stopping to accept H_0 based on conditional power: approximations and comparisons. *Biometrics*, **53**, 794–806. [11.7]

Betensky, R.A. (1997c). Sequential analysis of censored survival data with three treatment groups. *Biometrics*, **53**, 807–822. [21.5]

Betensky, R.A. (1998). Construction of a continuous stopping boundary from an alpha spending function. *Biometrics*, **54**, 1061–1071. [8.5]

Betensky, R.A. and Tierney, C. (1997). An examination of methods for sample size recalculation during an experiment. *Statist. Med.*, **16**, 2587–2598. [18.2]

Bhatt, D.L. and Mehta, C. (2016). Adaptive designs for clinical trials *New Engl. J. Med.*, **375**, 65–74. [24.5]

Bilias, Y., Gu, M. and Ying, Z. (1997). Towards a general asymptotic theory for Cox model with staggered entry. *Ann. Statist.*, **25**, 662–682. [17.9]

Birkett, M.A. and Day, S.J. (1994). Internal pilot studies for estimating sample size. *Statist. Med.*, **13**, 2455–2463. [18.2]

Bischoff, W. and Miller, F. (2005). Adaptive two-stage test procedures to find the best treatment in clinical trials *Biometrika*, **92**, 197–212. [26.8]

Blot, W.J. and Meeter, D.A. (1973). Sequential experimental design procedures. *J. Amer. Statist. Assoc.*, **68**, 586–593. [1.3]

Blot, W.J., Li, J., Taylor, P.R. et al. (1993). Nutrition intervention trials in Linxian, China: supplementations with specific vitamin/mineral combinations, cancer incidence, and disease specific mortality in the general population. *J. Nat. Cancer Inst.*, **85**, 1483–1491. [1.1, 21.3]

Blyth, C.R. and Still, H.A. (1983). Binomial confidence intervals. *J. Amer. Statist. Assoc.*, **78**, 108–116. [16.1, 16.2]

Boden, W.E., O'Rourke, R.A., Teo, K.K. et al. (2007). Optimal medical therapy with or without PCI for stable coronary disease. *New Engl. J. Med.*, **356**, 1503–1516. [17.8]

Bolland, K., Whitehead, J., Cobo, E. and Secades, J.J. (2009). Evaluation of a sequential global test of improved recovery following stroke as applied to the ICTUS trial of citicoline. *Pharmaceut. Statist.*, **8**, 136–149. [19.4]

Bone, H. G., Dempster, D. W., Eisman, et al. (2015). Odanacatib for the treatment of postmenopausal osteoporosis: development history and design and participant characteristics of LOFT, the Long-Term Odanacatib Fracture Trial. *Osteoporosis Int.*, **26**, 699–712. [20.5]

Bonferroni, C.E. (1936). Teoria statistica delle classi e calcolo delle probabilita, *Pubbl. del R Ist. Super. di Sci. Econom. e Commerciali di Firenze*, **8**, 3–62. [19.2, 20.4]

Bowden, J., Brannath, W. and Glimm, E. (2014). Empirical Bayes estimation of the selected treatment mean for two-stage drop-the-loser trials: a meta-analytic approach. *Statist. Med.*, **33**, 388–400. [26.6]

Bowden, J. and Glimm, E. (2008). Unbiased estimation of selected treatment means in two-stage trials. *Biometrical J.*, **50**, 515–527. [26.6]

Bowden, J. and Glimm, E. (2014). Conditionally unbiased and near unbiased estimation of the selected treatment mean for multistage drop-the-losers trials. *Biometrical J.*, **56**, 332–349. [27.6]

Bowker, A.H. and Lieberman, G.J. (1972). *Engineering Statistics*, 2nd ed., Englewood Cliffs: Prentice-Hall. [16.1]

Box, G.E.P., Hunter, W.G. and Hunter, J.S. (2005). *Statistics for Experimenters*, 2nd ed., New York: Wiley. [1.1]

Braitman, L.E. (1993). Statistical estimates and clinical trials. *J. Biopharm. Statist.*, **3**, 249–256. [9.1]

Brannath, W., Bauer, P, Maurer, W. and Posch, M. (2003). Sequential tests for noninferiority and superiority. *Biometrics*, **59**, 106–114. [7.3]

Brannath, W., König F, and Bauer P. (2006). Estimation in flexible two stage designs. *Statist. Med.*, **25**, 3366–3381. [10.5, 23.8]

Brannath, W., Koenig, F. and Bauer, P. (2007). Multiplicity and flexibility in clinical trials. *Pharmaceut. Statist.*, **6**, 205–216. [22.2, 23.2]

Brannath, W., Mehta, C.R. and Posch, M. (2009). Exact confidence bounds following adaptive group sequential tests. *Biometrics*, **65**, 539–546. [23.10]

Brannath, W., Posch, M. and Bauer, P. (2002). Recursive combination tests. *J. Amer. Statist. Assoc.*, **97**, 236–244. [23.2, 23.7]

Brannath, W. and Schmidt, S. (2014). A new class of powerful and informative simultaneous confidence intervals. *Statist. Med.*, **33**, 3365–3386. [28.6]

Brannath, W., Zuber, E., Branson, M. et al. (2009). Confirmatory adaptive designs with Bayesian decision tools for a targeted therapy in oncology. *Statist. Med.*, **28**, 1445–1463. [28.8]

Branson, M. and Whitehead, J. (2002) Estimating a treatment effect in survival studies in which patients switch treatment. *Statist. Med.*, **21**, 2449–2463. [17.8]

Bratton, D.J., Parmar, M.K., Phillips, P.P. and Choodari-Oskooei, B. (2016). Type I error rates of multi-arm multi-stage clinical trials: strong control and impact of intermediate outcomes. *Trials*, **17**:309, 1–8. [27.7]

Breslow, N.E. and Day, N.E. (1980). *Statistical Methods in Cancer Research, Vol. I: The Analysis of Case-Control Studies* Lyon: Intern. Agency for Research on Cancer. [1.1, 16.4]

Breslow, N.E. and Day, N.E. (1987). *Statistical Methods in Cancer Research, Vol. II: The Design and Analysis of Cohort Studies*, Lyon: Intern. Agency for Research on Cancer. [1.1]

Bretz, F., Hothorn, T. and Westfall, P. (2011). *Multiple Comparisons Using R*, Boca Raton: Chapman and Hall/CRC. [1.3]

Bretz, F., Koenig, F., Brannath, W., Glimm, E. and Posch, M. (2009). Adaptive designs for confirmatory clinical trials. *Statist. Med.*, **28**, 1181–1217. [21.4, 23.8, 26.6, 26.8, 27.1, 27.4]

Bretz F., Maurer W., Brannath W. and Posch, M. (2009). A graphical approach to sequentially rejective multiple test procedures. *Statist. Med.*, **28**, 586–604. [20.5, 20.8, 28.6]

Bretz, F., Schmidli, H., König, F., Racine, A. and Maurer, W. (2006). Confirmatory seamless phase II/III clinical trials with hypotheses selection at interim: general concepts. *Biometrical J.*, **48**, 623–634. [1.3, 25.2, 26.1, 27.1]

Brittain, E.H. and Bailey, K.R. (1993). Optimization of multistage testing times and critical values in clinical trials. *Biometrics*, **49**, 763–772. [12.3]

Brockwell, A.E. and Kadane, J.B. (2003). A gridding method for Bayesian sequential decision problems. *J. Comp. Graph. Statist.*, **12**, 566–584. [13.3]

Broglio, K.R., Connor, J.T. and Berry, S.M. (2014). Not too big, not too small: a Goldilocks approach to sample size selection. *J. Biopharm. Statist.*, **24**, 685–705. [13.4]

Brookmeyer, R. and Crowley, J. (1982). A confidence interval for the median survival time. *Biometrics*, **38**, 29–41. [10.1, 17.7]

Bross, I. (1952). Sequential medical plans. *Biometrics*, **8**, 188–205. [1.3]

Bross, I. (1958). Sequential clinical trials. *J. Chron. Dis.*, **8**, 349–365. [1.3]

Brown, B.W., Herson, J., Atkinson, E.N. and Rozell, M.E. (1987). Projection from previous studies: a Bayesian and frequentist compromise. *Contr. Clin. Trials*, **8**, 29–44. [13.4]

Brown, L.D., Cohen, A. and Strawderman, W.E. (1980). Complete classes for sequential tests of hypotheses. *Ann. Statist.*, **8**, 377–398. (Correction **17**, 1414–1416). [13.3]

Browne, R.H. (1995). On the use of a pilot sample for sample size determination. *Statist. Med.*, **14**, 1933–1940. [18.2]

Brückner, M., Titman, A. and Jaki, T. (2017). Estimation in multi-arm two-stage trials with treatment selection and time-to-event endpoint. *Statist. Med.*, **36**, 3137–3153. [26.6]

Buckley, J. and James, I. (1979). Linear regression with censored data. *Biometrika*, **66**, 429–436. [15.6, 15.7]

Burdette, W.J. and Gehan, E.A. (1970). *Planning and Analysis of Clinical Studies*, Springfield, Illinois: Charles C. Thomas. [1.1]

Burdon, A.J., Hampson, L.V. and Jennison, C. (2022). Joint modelling of longitudinal and time-to-event data applied to group sequential clinical trials. *arXiv preprint arXiv:2211.16138, Submitted for publication.* [15.7, 17.9]

Burman, C-F. and Sonesson, C. (2006). Are flexible designs sound? (with discussion) *Biometrics*, **62**, 664–683. [1.3, 24.2, 24.3]

Burman, C-F., Sonesson, C. and Guilbaud, O. (2009). A recycling framework for the construction of Bonferroni-based multiple tests. *Statist. Med.*, **28**, 739–761. [20.5, 20.8]

Burnett, T. and Jennison, C. (2021). Adaptive enrichment trials: what are the benefits? *Statist. Med.*, **40**, 690–711. [1.3, 20.4, 28.3]

Burnett, T., König, F. and Jaki, T. (2024). Adding experimental treatment arms to multi-arm multi-stage platform trials in progress. *Statist. Med.*, **43**, 3447–3462. [27.3]

Buyse, M.E., Staquet, M.J. and Sylvester, R.J. (1984). *Cancer Clinical Trials, Methods and Practice*, Oxford: Oxford University Press. [1.1]

Byar, D. P. (1985). Prognostic variables for survival in a randomized comparison of treatments for prostatic cancer. In *Data*, (Eds., by D. F. Andrews and A. M. Herzberg), New York: Springer-Verlag, 261–274. [17.9]

Byar, D.P. and Corle, D.K., 1977. Selecting optimal treatment in clinical trials using covariate information. *J. Chron. Dis.*, **30**, 445–459. [28.1]

Canner, P.L. (1977). Monitoring treatment differences in long-term clinical trials. *Biometrics*, **33**, 603–615. [1.3, 2.3]

Cannon, C.P. (1997). Clinical perspectives on the use of composite endpoints. *Contr. Clin. Trials*, **18**, 517–529. [19.1]

Carlin, B.P., Chaloner, K., Church, T. et al. (1993). Bayesian approaches for monitoring clinical trials with an application to toxoplasmic encephalitis prophylaxis. *The Statistician*, **42**, 355–367. [13.3]

Carlin, B.P., Kadane, J.B. and Gelfand, A.E. (1998). Approaches for optimal sequential decision analysis in clinical trials. *Biometrics*, **54**, 964–975. [13.3]

Carlin, B.P. and Sargent, D.J. (1996). Robust Bayesian approaches for clinical trial monitoring. *Statist. Med.*, **15**, 1093–1106. [13.3]

Carreras, M. and Brannath, W. (2013). Shrinkage estimation in two-stage adaptive designs with midtrial treatment selection. *Statist. Med.*, **32**, 1677–1690. [26.6]

Carreras, M., Gutjahr, G. and Brannath, W. (2015). Adaptive seamless designs with interim treatment selection: a case study in oncology, *Statist. Med.*, **34**, 1317–1333. [29.1, 29.2]

Chaloner, K., Church, T., Louis, T.A. and Matts, J.P. (1993). Graphical elicitation of a prior distribution for a clinical trial. *The Statistician*, **42**, 341–353. [13.2]

Chandler, G.A. and Graham, I.G. (1988). The convergence of Nyström methods for Weiner-Hopf equations. *Numerische Mathematik*, **52**, 345–364. [31.2]

Chang, M. (2007). Adaptive design method based on sum of *p*-values. *Statist. Med.*, *26*, 2772–2784. [23.2]

Chang, M. (2016). *Adaptive Design Theory and Implementation using SAS and R*, 2nd ed., Boca Raton: Chapman and Hall/CRC. [1.6]

Chang, M.N. (1989). Confidence intervals for a normal mean following a group sequential test. *Biometrics*, **45**, 247–254. [9.2, 9.4]

Chang, M.N. (1996). Optimal designs for group sequential clinical trials. *Commun. Statist.–Theory and Methods*, **25**, 361–379. [12.2, 13.3]

Chang, M.N., Hwang, I.K. and Shin, W.J. (1998). Group sequential designs using both type I and type II error probability spending functions. *Commun. Statist.–Theory and Methods*, **27**, 1323–1339. [8.3, 8.8]

Chataway, J., Nicholas, R., Todd, S. et al. (2011). A novel adaptive design strategy increases the efficiency of clinical trials in secondary progressive multiple sclerosis. *Multiple Sclerosis J.*, **17**, 81–88. [26.5]

Chatfield, C. and Collins, A.J. (1980). *Introduction to Multivariate Analysis*, London: Chapman and Hall. [19.3]

Chatterjee, S.K. and Sen, P.K. (1973). Nonparametric testing under progressive censoring. *Calcutta Statist. Assoc. Bull.*, **22**, 13–50. [11.2]

Checkoway, H., Pearce, N.E. and Kriebel, D. (2004). *Research Methods in Occupational Epidemiology*, Oxford: Oxford University Press. [1.1]

Chen, C., Anderson, K., Mehrotra, D.V., Rubin, E.H. and Tse, A. (2018). A 2-in-1 adaptive phase 2/3 design for expedited oncology drug development. *Contemp. Clin. Trials*, **64**, 238–42. [26.7]

Chen, C., Li, W. and Deng, Q. (2020). Extensions of the 2-in-1 adaptive design. *Contemp. Clin. Trials*, **95**:106053, 1–7. [26.7]

Chen, J. and Ting, N. (2020). Design considerations for vaccine trials with a special focus on COVID-19 vaccine development. *J. Data Sci*, **18**, 550–580. [16.7]

Chen Y.H.J., DeMets, D.L and Lan K.K.G. (2004). Increasing the sample size when the unblinded interim result is promising. *Statist. Med.*, **23**, 1023–1038. [24.2, 24.4, 24.5, 24.6]

Chen, Y.H.J., Li, C. and Lan K.K.G. (2015). Sample size adjustment based on promising interim results and its application in confirmatory clinical trials. *Clin. Trials*, **12**, 584–595. [24.6]

Chernoff, H. (1972). *Sequential Analysis and Optimal Design*, Philadelphia: SIAM. [1.3]

Chernoff, H. and Petkau, A.J. (1981). Sequential medical trials involving paired data. *Biometrika*, **68**, 119–132. [13.3]

Chiaruttini, M.V., Lorenzoni, G. and Gregori, D. (2025). Bayesian dynamic borrowing in group-sequential design for medical device studies. *BMC Med. Res. Meth.*, **25**:78, 1–10. [13.5]

Chick, S., Forster, M. and Pertile, P. (2017). A Bayesian decision theoretic model of sequential experimentation with delayed response. *J. Roy. Statist. Soc. B*, **79**, 1439–1462. [13.3]

Chinchilli, V.M. (1996). The assessment of individual and population bioequivalence. *J. Biopharm. Statist.*, **6**, 1–14. [7.1]

Chiu, Y.D., Koenig, F., Posch, M. and Jaki, T. (2018). Design and estimation in clinical trials with subpopulation selection. *Statist. Med.*, **37**, 4335–4352. [28.6]

CHMP: Committee for Human Medicinal Products. (2016). *Guideline on Multiplicity Issues in Clinical Trials* (EMA/CHMP/44762/2017), London, UK: European Medicines Agency. [19.6, 20.1]

CHMP: Committee for Medicinal Products for Human Use. (2007). *Reflection Paper on Methodological Issues in Confirmatory Clinical Trials Planned with an Adaptive Design* (EMA/CHMP/EWP/2459/02), London, UK: European Medicines Agency. [18.3, 23.7, 24.4]

Choi, S.C., Smith, P.J. and Becker, D.P. (1985). Early decision in clinical trials when treatment differences are small: experience of a controlled trial in head trauma. *Contr. Clin. Trials*, **6**, 280–288. [1.3, 11.3, 11.7]

Chow, S-C. and Chang, M. (2012). *Adaptive Design Methods in Clinical Trials*, 2nd ed., Boca Raton: Chapman and Hall/CRC. [1.3]

Chow, S-C. and Liu, J-P. (2008). *Design and Analysis of Bioavailability and Bioequivalence Studies*, 3rd ed., Boca Raton: Chapman and Hall/CRC. [1.1, 7.4]

Chow, S-C. and Liu, J-P. (2014). *Design and Analysis of Clinical Trials: Concepts and Methodologies*, 3rd ed., New York: Wiley. [1.1]

Chow, Y.S. and Robbins, H. (1965). On the asymptotic theory of fixed-width sequential confidence intervals for the mean. *Ann. Math. Statist.*, **36**, 457–462. [8.6]

Christen, J.A., Müller, P., Wathen, K. and Wolf, J. (2004). Bayesian randomized clinical trials: A decision-theoretic sequential design. *Canadian J. Statist..*, **32**, 387–402. [13.5]

Chuang-Stein, C., Anderson, K., Gallo, P. and Collins, S. (2006). Sample size reestimation, a review and recommendations. *Drug Inform. J.*, **40**, 475–484. [18.5]

Chuang, C-S. and Lai, T.L. (1998). Resampling methods for confidence intervals in group sequential trials. *Biometrika*, **85**, 317–332. [9.5]

Chuang-Stein, C. and Li, J. (2017). Changes are still needed on multiple co-primary endpoints. *Statist. Med.*, **36**, 4427–4436. [19.6]

Chugh, R., Wathen, J.K., Maki, R.G. et al. (2009). Phase II multicenter trial of Imatinib in 10 histologic subtypes of sarcoma using a Bayesian hierarchical statistical model. *J. Clin. Oncol.*, **27**, 3148–3153. [28.5]

Clark, L.C., Combs, G.F., Turnbull, B.W. et al. (1996). Effects of selenium supplementation for cancer prevention in patients with carcinoma of the skin: a randomized clinical trial. *J. Amer. Med. Assoc.*, **276**, 1957–1963. [1.4, 11.6, 19.2]

Clopper, C.J. and Pearson, E.S. (1934). The use of confidence or fiducial limits illustrated in the case of the binomial. *Biometrika*, **26**, 404–413. [16.1]

Coad, D.S. (1995). Sequential allocation involving several treatments. In *Adaptive Designs* (Eds., N. Flournoy and W.F. Rosenberger), Hayward, California: Institute of Mathematical Statistics, 95–109. [30.4]

Cochran, W.G. and Cox, G.M. (1957). *Experimental Design*, 2nd ed., New York: Wiley. [1.1]

Coe, P.R. and Tamhane, A.C. (1993). Exact repeated confidence intervals for Bernoulli parameters in a group sequential clinical trial. *Contr. Clin. Trials*, **14**, 19–29. [10.5, 16.1, 16.2]

Cohen, A. and Sackrowitz, H.B. (1989). Two stage conditionally unbiased estimators of the selected mean. *Statist. Prob. Letters*, **8**, 273–278. [26.6]

Cohen, D.R., Todd, S., Gregory, W.M. and Brown, J.M. (2015). Adding a treatment arm to an ongoing clinical trial: a review of methodology and practice. *Trials*, **16**:179, 1–9. [27.3]

Colton, T. (1963). A model for selecting one of two medical treatments. *J. Amer. Statist. Assoc.*, **58**, 388–400. [30.4]

Colton, T. and McPherson, K. (1976). Two-stage plans compared with fixed-sample-size and Wald SPRT plans. *J. Amer. Statist. Assoc.*, **71**, 80–86. [12.3]

Conaway, M.R. and Petroni, G.R, (1995). Bivariate sequential designs for Phase II trials. *Biometrics*, **51**, 656–664. [19.7]

Cook, R.J. (1994). Interim monitoring of bivariate responses using repeated confidence intervals. *Contr. Clin. Trials*, **15**, 187–200. [19.6, 19.7]

Cook, R.J. (1996). Coupled error spending functions for parallel bivariate sequential tests. *Biometrics*, **52**, 442–450. [19.7]

Cook, R.J. and Farewell, V.T. (1994). Guidelines for monitoring efficacy and toxicity responses in clinical trials. *Biometrics*, **50**, 1146–1152. [19.6, 19.7]

Cook, R.J. and Farewell, V.T. (1996). Multiplicity considerations in the design and analysis of clinical trials. *J. Roy. Statist. Soc. A*, **159**, 93–110. [19.7]

Cook, R.J. and Lawless, J.F. (1996). Interim monitoring of longitudinal comparative studies with recurrent event responses. *Biometrics*, **52**, 1311–1323. [15.7]

Cornell, R.G., Landenberger, B.D. and Bartlett, R.H. (1986). Randomized play-the-winner clinical trials. *Commun. Statist.–Theory and Methods*, **15**, 159–178. [30.4]

Cornfield, J. (1966a). Sequential trials, sequential analysis and the likelihood principle. *The Amer. Statistician*, **20**, 18–23. [13.3]

Cornfield, J. (1966b). A Bayesian test of some classical hypotheses—with applications to sequential clinical trials. *J. Amer. Statist. Assoc.*, **61**, 577–594. [13.3]

Cornfield, J. (1969). The Bayesian outlook and its application (with discussion). *Biometrics*, **25**, 617–657. [13.3]

Cornfield, J., Halperin, M. and Greenhouse, S.W. (1969). An adaptive procedure for sequential clinical trials. *J. Amer. Statist. Assoc.*, **64**, 759–770. [30.1, 30.4]

Cox, D.R. (1972). Regression models and life-tables (with discussion). *J. Roy. Statist. Soc. B*, **34**, 187–220. [1.5, 3.4, 17.1, 17.4, 17.6]

Cox, D.R. (1975). Partial likelihood. *Biometrika*, **62**, 269–276. [3.4]

Cox, D.R. (1989). Discussion of 'Interim analyses: the repeated confidence interval approach' (by C. Jennison and B.W. Turnbull). *J. Roy. Statist. Soc. B*, **51**, 338. [19.2]

Cox, D.R. and Hinkley, D.V. (1974). *Theoretical Statistics*, London: Chapman and Hall. [9.3, 13.3, 15.6, 24.2]

Cox, D.R. and Oakes, D. (1984). *Analysis of Survival Data*, London: Chapman and Hall. [1.3, 1.4]

CPMP: Committee for Proprietary Medicinal Products. (2000). *Points to Consider on Switching Between Superiority and Non-inferiority* (CPMP/EWP/482/99), London, UK: European Agency for the Evaluation of Medicinal Products. [7.3]

Crow, E.L. (1956). Confidence intervals for a proportion. *Biometrika*, **43**, 423–435. [16.1]

Cuffe, R.L., Lawrence, D., Stone, A. and Vandemeulebroecke, M. (2014). When is a seamless study desirable? Case studies from different pharmaceutical sponsors. *Pharmaceut. Statist.*, **13**, 229–237. [26.1, 26.8]

Cui, L., Hung, H.M.J. and Wang, S-J. (1999). Modification of sample size in group sequential clinical trials. *Biometrics*, **55**, 853–857. [1.3, 1.5, 22.1, 22.2, 22.3, 23.2, 23.6, 23.10, 24.1, 24.6]

Cutler, S.J. Greenhouse, S.W. Cornfield, J. and Schneiderman, M.A. (1966). The role of hypothesis testing in clinical trials. *J. Chron. Dis.*, **19**, 857–882. [1.3, 9.1]

Dallow, N.S., Leonov, S.L. and Roger, J.H. (2008). Practical usage of O'Brien's OLS and GLS statistics in clinical trials. *Pharmaceut. Statist.*, **7**, 53–68. [19.4]

Davis, B.R. and Hardy, R.J. (1992). Repeated confidence intervals and prediction intervals using stochastic curtailment. *Commun. Statist.–Theory and Methods*, **21**, 351–368. [11.7]

Davis, C.E. (1978). A two sample Wilcoxon test for progressively censored data. *Commun. Statist.–Theory and Methods*, **7**, 389–398. [11.2]

Davis, C.E. (1997). Secondary endpoints can be validly analyzed, even if the primary endpoint does not provide clear statistical significance. *Contr. Clin. Trials*, **18**, 557–560. [19.2]

DeGroot, M.H. (1970). *Optimal Statistical Decisions*. New York: McGraw-Hill. [13.3]

Demetri, G.D., Garrett, C.R., Schoffski. P. et al. (2012). Complete longitudinal analyses of the randomised, placebo-controlled Phase III trial of sunitinib in patients with gastrointestinal stromal tumor following imatinib failure. *Clin. Cancer Res.*, **18**, 3170–3179. [17.8]

DeMets, D.L. (1984). Stopping guidelines vs stopping rules: a practitioner's point of view. *Commun. Statist.–Theory and Methods*, **13**, 2395–2417. [10.1]

DeMets, D.L. and Gail, M.H. (1985). Use of logrank tests and group sequential methods at fixed calendar times. *Biometrics*, **41**, 1039–1044. [4.5]

DeMets, D.L. and Ware, J.H. (1980). Group sequential methods for clinical trials with one-sided hypothesis. *Biometrika*, **67**, 651–660. [5.1]

DeMets, D.L. and Ware, J.H. (1982). Asymmetric group sequential boundaries for monitoring clinical trials. *Biometrika*, **69**, 661–663. [5.1]

DeMets, D.L., Hardy, R., Friedman, L.M. and Lan, K.K.G. (1984). Statistical aspects of early termination in the Beta-Blocker Heart Attack Trial. *Contr. Clin. Trials*, **5**, 362–372. [8.2]

DeMets, D.L. and Ware, J.H. (1982). Asymmetric group sequential boundaries for monitoring clinical trials. *Biometrika*, **69**, 661–663. [5.1]

Dempster, A.P., Laird, N.M. and Rubin, D.B. (1977). Maximum likelihood from incomplete data via the EM algorithm (with discussion). *J. Roy. Statist. Soc. B*, **39**, 1–38. [18.3]

Denne, J.S. (2000). Estimation following extension of a study on the basis of conditional power. *J. Biopharm. Statist.*, **10**, 131–144. [23.10]

Denne, J.S. (2001). Sample size recalculation using conditional power. *Statist. Med.*, **20**, 2645–2660. [24.1, 24.2]

Denne, J.S. and Jennison, C. (1999a). Estimating the sample size for a t-test using an internal pilot. *Statist. Med.*, **18**, 1575–1585. [18.2]

Denne, J.S. and Jennison, C. (1999b). Improving the post-experimental properties of Stein's two-stage procedure. *Seq. Analysis*, **18**, 43–56. [18.2]

Denne, J.S. and Jennison, C. (2000). A group sequential t-test with updating of sample size. *Biometrika*, **87**, 125–134. [18.4]

Di Scala, L. and Glimm, E. (2011). Time-to-event analysis with treatment arm selection at interim. *Statist. Med.*, **30**, 3067–3081. [29.2]

Di Stefano, F., Pannaux, M., Correges, A., et al. (2022). A comparison of estimation methods adjusting for selection bias in adaptive enrichment designs with time-to-event endpoints. *Statist. Med.*, **41**, 1767–1779. [28.6]

Dickey, J. (1973). Scientific reporting and personal probabilities: Student's hypothesis. *J. Roy. Statist. Soc. B*, **35**, 285–305. [13.3]

Dimairo, M., Pallmann, P., Wason, J. et al. (2020). The Adaptive designs CONSORT Extension (ACE) statement: a checklist with explanation and elaboration guideline for reporting randomised trials that use an adaptive design. *BMJ*, **369**:m115. [1.3]

Dmitrienko, A. and Tamhane, A.C. (2007). Gatekeeping procedures with clinical trial applications. *Pharmaceut. Statist.*, **6**, 171–180. [20.5]

Dmitrienko, A. and Wang, M-D. (2006). Bayesian predictive approach to interim monitoring in clinical trials. *Statist. Med.*, **25**, 2178–2195. [11.7]

Dobbins, T.W. and Thiyagarajan, B. (1992). A retrospective assessment of the 75/75 rule in bioequivalence. *Statist. Med.*, **11**, 1333–1342. [7.5]

Dodge, H.F. and Romig, H.G. (1929). A method for sampling inspection. *Bell Syst. Tech. J.*, **8**, 613–631. [1.3, 31.1]

Dragalin, V. and Fedorov, V. (2006). Multistage designs for vaccine safety studies. *J. Biopharm. Statist.*, **16**, 539–553. [16.7]

Duffy, D.E. and Santner, T.J. (1987). Confidence intervals for a binomial parameter based on multistage tests. *Biometrics*, **43**, 81–93. [1.3, 16.1]

Dunnett, C.W. (1955). A multiple comparison procedure for comparing several treatments with a control. *J. Amer. Statist. Assoc.*, **50**, 1096–1121. [20.4, 21.4, 26.2, 30.4]

Dunnett, C.W. (1961). The statistical theory of drug screening. In *Quantitative Methods in Pharmacology*, (Ed., H. DeJonge), Amsterdam: North-Holland, 212–231. [1.3, 31.1]

Dunnett, C.W. and Gent, M. (1977). Significance testing to establish equivalence between treatments, with special reference to data in the form of 2 × 2 tables. *Biometrics*, **33**, 593–602. [1.1, 7.1, 7.4]

Dunnett, C.W. and Gent, M. (1996). An alternative to the use of two-sided tests in clinical trials. *Statist. Med.*, **15**, 1729–1738. [7.3]

Durrleman, S and Simon, R. (1990). Planning and monitoring of equivalence studies. *Biometrics*, **46**, 329–336. [1.1, 7.6, 10.5]

Eales, J.D. (1991). *Optimal Group Sequential Tests*, Ph.D. Thesis, University of Bath, U.K. [12.2]

Eales, J.D. and Jennison, C. (1992). An improved method for deriving optimal one-sided group sequential tests. *Biometrika*, **79**, 13–24. [1.3, 5.1, 8.3, 8.8, 12.2, 12.3, 13.3, 31.6]

Eales, J.D. and Jennison, C. (1995). Optimal two-sided group sequential tests. *Seq. Analysis*, **14**, 273–286. [1.3, 2.8, 12.2, 13.3]

Edwards, J.M., Walters, S.J., Kunz, C. and Julious, S.A. (2020). A systematic review of the "promising zone" design. *Trials*, **21**:1000, 1–10. [24.6]

Elashoff, J.D. and Reedy, T.J. (1984). Two-stage clinical trial stopping rules. *Biometrics*, **40**,(3), 791–795. [12.3]

Elashoff, R.M. and Beal, S. (1976). Two-stage screening designs applied to chemical-screening programs with binary data. *Ann. Rev. of Biophys. and Bioeng.*, **5**, 561–587. [1.3]

Elfring, G.L. and Schultz, J.R. (1973). Group sequential designs for clinical trials. *Biometrics*, **29**, 471–477. [1.3, 2.3]

Elm, J.J., Palesch, Y.Y., Koch, G.G. et al. (2012). Flexible analytical methods for adding a treatment arm mid-study to an ongoing clinical trial. *J. Biopharm. Statist.*, **22**, 758–772. [27.3]

Elsäßer, A., Regnstrom, J., Vetter, T., et al. (2014). Adaptive clinical trial designs for European marketing authorization: a survey of scientific advice letters from the European Medicines Agency. *Trials*, **15**:383, 1–9. [24.1, 24.4]

EMA: European Medicines Agency. (2022). *Complex clinical trials—Questions and answers*, Amsterdam. [13.4]

Emerson, S.S (1988). *Parameter Estimation following Group Sequential Hypothesis Testing*, Ph.D. Thesis, University of Washington, Seattle. [9.3, 9.5]

Emerson, S.S. (1993). Computation of the uniform minimum variance unbiased estimator of a normal mean following a group sequential trial. *Comp. and Biomed. Res.*, **26**, 68–73. [9.3]

Emerson, S.S. and Fleming, T.R. (1989). Symmetric group sequential designs. *Biometrics*, **45**, 905–923. [1.3, 5.1, 5.2, 5.5, 6.1, 12.2]

Emerson, S.S. and Fleming, T.R. (1990). Parameter estimation following group sequential hypothesis testing. *Biometrika*, **77**, 875–892. [3.5, 9.2, 9.3, 9.4, 9.5]

Emerson, S.S. and Kittelson, J.M. (1997). A computationally simpler algorithm for the UMVUE of a normal mean following a group sequential test. *Biometrics*, **53**, 365–369. [9.3]

Emerson, S.S., Kittelson, J.M. and Gillen, D.L. (2007). Bayesian evaluation of group sequential clinical trial designs. *Statist. Med.*, **26**, 1431–1449. [13.5]

Emerson, S.S., Levin, G.P. and Emerson, S.C. (2011). Comments on 'Adaptive increase in sample size when interim results are promising: a practical guide with examples' (by C.R. Mehta and S.J. Pocock). *Statist. Med.*, **30**, 3285–3301. [24.6]

Enas, G.G., Dornseif, B.E., Sampson, C.B. et al. (1989). Monitoring versus interim analysis of clinical trials: a perspective from the pharmaceutical industry. *Contr. Clin. Trials*, **10**, 57–70. [1.2, 2.7, 18.1]

Ernisse, B., Rogers, S.K., DeSimio, M.P. and Raines, R.A. (1997). Complete automatic target cuer/recognition system for tactical forward-looking infrared images. *Optical Engineering*, **36**, 2593–2603. [1.3]

Etzioni, R. and Pepe, M.S. (1994). Monitoring of a pilot toxicity study with two adverse outcomes. *Statist. Med.*, **13**, 2311–2321. [19.7]

Facey, K.M. (1992). A sequential procedure for a phase II efficacy trial in hypercholesterolemia. *Contr. Clin. Trials*, **13**, 122–133. [1.4]

Fairbanks, K. and Madsen, R. (1982). P values for tests using a repeated significance test design. *Biometrika*, **69**, 69–74. [1.3, 9.4]

Faldum, A. and Hommel, G. (2007) Strategies for including patients recruited during interim analysis of clinical trials. *J. Biopharm. Statist.*, **17**, 1211–1225. [24.6]

Fayers, P.M., Ashby, D. and Parmar, M.K.B. (1997). Bayesian data monitoring in clinical trials. *Statist. Med.*, **16**, 1413–1430. [13.2, 13.3]

FDA: Food and Drug Administration. (1985). New drug and antibiotic regulations. *Federal Register*, **50**, 7452–7519. [1.2]

FDA: Food and Drug Administration. (1988). *Guideline for the Format and Content of the Clinical and Statistical Sections of an Application*, Center for Drug Evaluation and Research, Food and Drug Administration, Rockville, Maryland. [1.2]

FDA: Food and Drug Administration. (1992). *Guideline for Postmarketing Reporting of Adverse Drug Experiences*, Center for Drug Evaluation and Research, Food and Drug Administration, Rockville, Maryland. [1.2]

FDA: Food and Drug Administration. (1995). Statement regarding the demonstration of effectiveness of human drug products and devices. *Federal Register*, **60**, 39180–39181. [23.1]

FDA: Food and Drug Administration. (1998). International Conference on Harmonisation; Guidance on statistical principles for clinical trials. *Federal Register*, **63**, 49583–49598. [1.2, 13.4, 18.3, 24.1]

FDA: Food and Drug Administration. (2010). *Guidance for the Use of Bayesian Statistics in Medical Device Clinical Trials*, Center for Biologics Evaluation and Research, Food and Drug Administration, Rockville, Maryland. [13.4]

FDA: Food and Drug Administration. (2018). *Clinical Trial Endpoints for the Approval of Cancer Drugs and Biologics Guidance for Industry*, Center for Drug Evaluation and Research, Food and Drug Administration, Silver Spring, Maryland. [20.1]

FDA: Food and Drug Administration. (2019a). *Adaptive Designs for Clinical Trials of Drugs and Biologics Guidance for Industry*, Center for Drug Evaluation and Research, Food and Drug Administration, Silver Spring, Maryland. [Preface, 1.3, 5.3, 13.4, 22.1, 23.7, 24.4, 24.6, 28.7]

FDA: Food and Drug Administration. (2019b). *Enrichment Strategies for Clinical Trials to Support Determination of Effectiveness of Human Drugs and Biological Products Guidance for Industry*, Center for Drug Evaluation and Research, Food and Drug Administration, Silver Spring, Maryland. [28.7]

FDA: Food and Drug Administration. (2020). *Interacting with the FDA on Complex Innovative Clinical Trial Designs for Drugs and Biological Products Guidance for Industry*, Center for Drug Evaluation and Research, Food and Drug Administration, Silver Spring, Maryland. [Preface, 13.4]

FDA: Food and Drug Administration. (2022). *Multiple Endpoints in Clinical Trials Guidance for Industry*, Center for Drug Evaluation and Research, Food and Drug Administration, Silver Spring, Maryland. [19.6, 20.1]

Fisher, B., Costantino, J.P., Wickerham, D.I. et al. (1998). Tamoxifen for prevention of breast cancer: report of the National Surgical Adjuvant Breast and Bowel Project P-1 Study. *J. Nat. Cancer Inst.*, **90**, 1371–1388. [10.1]

Fisher, L.D. (1998). Self-designing clinical trials. *Statist. Med.*, **17**, 1551–1562. [1.3, 23.2, 23.7, 23.10, 24.1]

Fisher, R.A. (1932). *Statistical Methods for Research Workers*, 4th ed., Edinburgh: Oliver and Boyd. [23.1]

Fisher, R.A. (1935). The logic of inductive inference (with discussion). J. Roy. Statist. Soc., **98**, 39–82. [16.3]

Fleiss, J.L. (1986). *The Design and Analysis of Clinical Experiments*, New York: Wiley. [1.1]

Fleming, T.R. (1982). One-sample multiple testing procedures for phase II clinical trials. *Biometrics*, **38**, 143–151. [1.3, 16.1]

Fleming, T.R. (1992). Evaluating therapeutic interventions: some issues and experiences (with discussion). *Statist. Sci.*, **7**, 428–456. [1.6, 10.5]

Fleming, T.R. (2008). Current issues in non-inferiority trials. *Statist. Med.*, **27**, 317–332. [7.1, 7.2]

Fleming, T.R. and DeMets, D.L. (1993). Monitoring of clinical trials: issues and recommendations. *Contr. Clin. Trials*, **14**, 183–197. [2.7, 8.5]

Fleming, T.R. and Harrington, D.P. (1981). A class of hypothesis tests for one and two sample censored survival data. *Commun. Statist.–Theory and Methods*, **10**, 763–794. [17.8]

Fleming, T.R. and Richardson, B.A. (2004). Some design issues in trials of microbicides for the prevention of HIV infection. *J. Infectious Dis.*, **190**, 666–674. [23.1]

Fleming, T.R., Harrington, D.P. and O'Brien, P.C. (1984). Designs for group sequential tests. *Contr. Clin. Trials*, **5**, 348–361. [1.3, 2.8, 8.5]

Flournoy, N. and Rosenberger, W.F., Eds. (1995). *Adaptive Designs*, IMS Lecture Notes Monograph Series number 25, Hayward, California: Institute of Mathematical Statistics. [30.4]

Follmann, D. (1997). Adaptively changing subgroup proportions in clinical trials. *Statistica Sinica*, **7**, 1085–1102. [28.8]

Follmann, D.A., Proschan, M.A. and Geller, N.L. (1994). Monitoring pairwise comparisons in multi-armed clinical trials. *Biometrics*, **50**, 325–336. [21.1, 21.3, 21.5]

Forster, M., Brealey, S., Chick, S. et al. (2021). Cost-effective clinical trial design: application of a Bayesian sequential model to the ProFHER pragmatic trial. *Clin. Trials*, **18**, 647–656. [13.3]

Fournier, J.C., DeRubeis, R.J., Hollon, S.D. et al. (2010). Antidepressant drug effects and depression severity: a patient-level meta-analysis. *J. Amer. Med. Assoc.*, **303**, 47–53. [28.1]

Freedman, L., Anderson, G., Kipnis, V. et al. (1996). Approaches to monitoring the results of long-term disease prevention trials: examples from the Women's Health Initiative. *Contr. Clin. Trials*, **17**, 509–525. [19.1]

Freedman, L.S. and Spiegelhalter, D.J. (1983). The assessment of subjective opinion and its use in relation to stopping rules for clinical trials. *The Statistician*, **32**, 153–160. [13.2]

Freedman, L.S. and Spiegelhalter, D.J. (1989). Comparison of Bayesian with group sequential for monitoring clinical trials. *Contr. Clin. Trials*, **10**, 357–367. [13.3]

Freedman, L.S., Spiegelhalter, D.J. and Parmar, M.K.B. (1994). The what, why and how of Bayesian clinical trials monitoring. *Statist. Med.*, **13**, 1371–1383. [13.3]

Freeman, H.A., Friedman, M., Mosteller, F. and Wallis, W.A. (1948). *Sampling Inspection.* New York: McGraw-Hill. [1.3, 31.1]

Freeman, P.R. (1989). Discussion of 'Interim analyses: the repeated confidence interval approach' (by C. Jennison and B.W. Turnbull). *J. Roy. Statist. Soc. B*, **51**, 339. [10.1]

Freidlin, B. and Korn, E.L. (2017). Sample size adjustment designs with time-to-event outcomes: a caution. *Clin. Trials*, **14**, 597–604. [24.6]

Freidlin, B., Korn, E.L., Gray, R. and Martin, A. (2008). Multi-arm clinical trials of new agents: some design considerations. *Clin. Cancer Res.*, **14**, 4368–4371. [27.7]

Freidlin, B. and Simon, R. (2005). Adaptive signature design: an adaptive clinical trial design for generating and prospectively testing a gene expression signature for sensitive patients. *Clin. Cancer Res*, **11**, 7872–7878. [28.8]

Freiman, J.A., Chalmers, T.C., Smith, H. and Kuebler, R.R. (1992). The importance of beta, the type II error and sample size in the design and interpretation of the randomized control trial: survey of two sets of "negative" trials. In *Medical Uses of Statistics*, 2nd ed., (Eds., J.C. Baillar and F. Mosteller), Chapter 19. Boca Raton: CRC Press. [1.1]

Friede, T. and Kieser, M. (2004). Sample size recalculation for binary data in internal pilot study designs. *Pharmaceut. Statist.*, **3**, 269–279. [18.3]

Friede, T. and Kieser, M. (2006). Sample size recalculation in internal pilot study designs: a review. *Biometrical J.*, **48**, 537–555. [18.5]

Friede, T. and Miller, F. (2012). Blinded continuous monitoring of nuisance parameters in clinical trials. *J. Roy. Statist. Soc. C: App. Statist.*, **61**, 601–618. [18.3]

Friede, T., Parsons, N., Stallard, N. et al. (2011). Designing a seamless phase II/III clinical trial using early outcomes for treatment selection: an application in multiple sclerosis. *Statist. Med.*, **30**, 1528–1540. [26.5, 29.5]

Friede, T. and Stallard, N. (2008). A comparison of methods for adaptive treatment selection. *Biometrical J.* **50**, 767–781. [26.3, 27.4]

Friedman, L.M., Furberg, C.D., DeMets, D.L. et al. (2015). *Fundamentals of Clinical Trials*, 5th ed., New York: Springer. [1.1]

Fu, K.S. (1968). *Sequential Methods in Pattern Recognition and Learning*, New York: Academic Press. [1.3]

Gaffney, M. and Ware, J.H. (2017). An evaluation of increasing sample size based on conditional power. *J. Biopharm. Statist.*, **27**, 797–808. [24.6]

Gail, M.H., DeMets, D.L. and Slud, E.V. (1982) Simulation studies on increments of the two-sample logrank score test for survival time data, with application to group sequential boundaries. In *Survival Analysis*, (Eds., J. Crowley and R.A. Johnson), Hayward, California: Institute of Mathematical Statistics, 287–301. [4.5]

Gajewski, B.J., Kimler, B.F., Koestler, D.C. et al. (2022). A novel Bayesian adaptive design incorporating both primary and secondary endpoints for randomized IIB chemoprevention study of women at increased risk for breast cancer. *Trials*, **23**:981, 1–11. [13.5]

Gallo, P., Mao, L. and Shih, V.H. (2014). Alternative views on setting clinical trial futility criteria. *J. Biopharm. Statist.*, **24**, 976–993. [11.5, 11.7]

Gange, S.J. and DeMets, D.L. (1996). Sequential monitoring of clinical trials with correlated responses. *Biometrika*, **83**, 157–167. [15.7]

Ganju, J. and Xing, B. (2009). Re-estimating the sample size of an on-going blinded trial based on the method of randomization block sums. *Statist. Med.*, **28**, 24–38. [18.3]

Gao, P., Liu, L. and Mehta, C. (2013). Exact inference for adaptive group sequential designs. *Statist. Med.* **32**, 3991–4005. [23.8, 23.10]

Gao, P., Ware, J.H. and Mehta, C. (2008). Sample size re-estimation for adaptive sequential design in clinical trials. *J. Biopharm. Statist.*, **18**, 1184–1196. [24.5]

Gart, J.J., Krewski, D., Lee, P.N. et al. (1986). *Statistical Methods in Cancer Research, Vol. III–The Design and Analysis of Long-term Animal Experiments.* Intern. Agency for Research on Cancer, Lyon. [1.1]

Geary, D.N. (1988). Sequential testing in clinical trials with repeated measurements. *Biometrika*, **75**, 311–318. [15.3, 15.7]

Gehan, E.A. (1965). A generalized Wilcoxon test for comparing arbitrarily singly-censored samples. *Biometrika*, **52**, 203–224. [17.9]

Geiger, M.J., Skrivanek, Z., Gaydos, B. et al. (2012). An adaptive, dose-finding, seamless phase 2/3 study of a long-acting glucagon-like peptide-1 analog (dulaglutide): trial design and baseline characteristics. *J. Diabetes Sci. Technol.*, **6**, 1319–1327. [26.1]

Gent, M. (1997). Discussion: some issues in the construction and use of clusters of outcome events. *Contr. Clin. Trials*, **18**, 546–549 [19.1]

George, S.L. and Desu, M.M. (1974). Planning the size and duration of a clinical trial studying the time to some critical event. *J. Chron. Dis.*, **27**, 15–24. [11.6]

George, S.L., Li, C., Berry, D.A. and Green, M.R. (1994). Stopping a clinical trial early: frequentist and Bayesian approaches applied to a CALGB trial in non-small-cell lung cancer. *Statist. Med.*, **13**, 1313–1327. [13.2]

Ghosh, B.K. (1991). A brief history of sequential analysis. In *Handbook of Sequential Analysis*, (Eds., B.K. Ghosh and P.K. Sen), New York: Marcel Dekker, 1–19. [1.3]

Ghosh, B.K. and Sen, P.K., Eds. (1991). *Handbook of Sequential Analysis*, New York: Marcel Dekker. [1.3]

Ghosh, P., Liu, L. and Mehta, C. (2020). Adaptive multiarm multistage clinical trials. *Statist. Med.*, **39**, 1084– 1102. [27.5]

Ghosh, P., Liu, L., Senchaudhuri, P. et al. (2017). Design and monitoring of multi-arm multi-stage clinical trials. *Biometrics*, **73**, 1289–1299. [21.4, 27.4, 27.5, 31.6]

Ghosh, P., Ristl, R., König, F. et al. (2022). Robust group sequential designs for trials with survival endpoints and delayed response. *Biometrical J.*, **64**, 343–360. [14.5, 17.8, 17.9]

Gilbert, J.P., McPeek, B. and Mosteller, F. (1977). Statistics and ethics in surgery and anaesthesia. *Science*, **198**, 684–689. [13.2]

Gittins, J.C., Glazebrook, K. and Weber, R. (2011). *Multi-Armed Bandit Allocation Indices*, 2nd ed. Chichester: Wiley. [1.3, 30.4]

Glimm, E. (2012). Comments on 'Adaptive increase in sample size when interim results are promising: A practical guide with examples' (by C.R. Mehta and S.J. Pocock). *Statist. Med.*, **31**, 98–99. [24.5]

Glimm, E., Maurer, W. and Bretz, F. (2010). Hierarchical testing of multiple endpoints in group sequential trials. *Statist. Med.*, **29**, 219–228. [20.10, 20.11]

Goldhirsch, A., Coates, A.S., Castiglione-Gertsch, M. and Gelber, R.D. (1998). New treatments for breast cancer: breakthroughs for patient care or just steps in the right direction? *Ann. Oncology*, **9**, 973–976. [10.1]

Götte, H., Donica, M. and Mordenti, G. (2015). Improving probabilities of correct interim decision in population enrichment designs. *J. Bioopharm. Statist.*, **25**, 1020–1038. [28.8]

Gould, A.L. (1983). Abandoning lost causes (early termination of unproductive clinical trials). *Proc. Biopharm. Sec.*, Amer. Statist. Assoc., Washington, D.C., 31–34. [1.2, 1.3, 5.1, 6.1, 8.2]

Gould, A.L. (1992). Interim analyses for monitoring clinical trials that do not materially affect the Type I error rate. *Statist. Med.*, **11**, 55–66. [18.1, 18.2, 18.3]

Gould, A.L. (1995a). Group sequential extensions of a standard bioequivalence testing procedure. *J. Pharmacokin. Biopharm.*, **23**, 57–86. [7.6]

Gould, A.L. (1995b). Planning and revising the sample size for a trial. *Statist. Med.*, **14**, 1039 1051. [18.1]

Gould, A.L. and Pecore, V.J. (1982). Group sequential methods for clinical trials allowing early acceptance of H_0 and incorporating costs. *Biometrika*, **69**, 75–80. [1.3, 6.1]

Gould, A.L. and Shih, W.J. (1992). Sample size re-estimation without unblinding for normally distributed outcomes with unknown variance. *Commun. Statist.-Theory and Methods*, **21**, 2833–2853. [18.1, 18.2, 18.3]

Gould, A.L. and Shih, W.J. (1998). Modifying the design of ongoing trials without unblinding. *Statist. Med.*, **17**, 89–100. [18.1]

Goutis, C., Casella, G. and Wells, M.T. (1996). Assessing evidence in multiple hypotheses. *J. Amer. Statist. Assoc.*, **91**, 1268–1277. [23.1]

Graf, A. C. and Bauer, P. (2011). Maximum inflation of the type 1 error rate when sample size and allocation rate are adapted in a pre-planned interim look. *Statist. Med.*, **30**, 1637–1647. [8.5]

Graf, A.C., Bauer, P., Glimm, E. & Koenig, F. (2014). Maximum type I error rate inflation in multiarmed clinical trials with adaptive interim sample size modifications. *Biometrical J.*, **56**, 614–630. [30.4]

Grayling, M.J. and Mander, A.P. (2021). Accounting for variation in the required sample size in the design of group-sequential trials. *Contemp. Clin. Trials*, **107**:106459, 1–5. [12.3]

Greenhouse, J.B. and Wasserman, L. (1995). Robust Bayesian methods for monitoring clinical trials. *Statist. Med.*, **14**, 1379–1391. [13.3]

Greenstreet, P., Jaki, T., Bedding, A. et al. (2024). A multi-arm multi-stage platform design that allows preplanned addition of arms while still controlling the family-wise error. *Statist. Med.*, **43**, 3613–3632. [27.3]

Greenwood M. (1926). *The natural duration of cancer*, Reports on Public Health and Medical Subjects. **33**, 1-26. London: Her Majesty's Stationery Office. [17.7]

Grimmett, G.R. and Stirzaker, D.R. (2020). *Probability and Random Processes*, 4th ed., Oxford: Oxford University Press. [11.2, 16.7]

Grossman J., Parmar M.K.B., Spiegelhalter, D.J. and Freedman, L.S. (1994). A unified method for monitoring and analyzing controlled trials. *Statist. Med.*, **13**, 1815–1826. [13.2, 13.3]

Gsponer, T., Gerber, F., Bornkamp, B., et al. (2014). A practical guide to Bayesian group sequential designs. *Pharmaceut. Statist.*, **13**, 71–80. [13.5]

Gu, M. and Ying, Z. (1993). Sequential analysis for censored regression data. *J. Amer. Statist. Assoc.*, **88**, 890–898. [15.7]

Gu, M. and Ying, Z. (1995). Group sequential methods for survival data using partial likelihood score processes with covariate adjustment. *Statistica Sinica*, **5**, 793–804. [17.9]

Gupta, S.S., Nagel, K. and Panchapakesan, S. (1973). On the order statistics from equally correlated normal random variables. *Biometrika*, **6**, 403–413. [21.4]

Hall, P. (1981). Asymptotic theory of triple sampling for sequential estimation of a mean. *Ann. Statist.*, **9**, 1229–1238. [18.2]

Hall, W.J. and Ding, K. (2008). Sequential tests and estimates after overrunning based on p-value combination. In *Pushing the Limits of Contemporary Statistics: Contributions in Honor of Jayanta K. Ghosh*, (eds B. Clarke and S. Ghosal), Beachwood: Institute of Mathematical Statistics, **3**, 3–45. [14.2]

Halperin, M., Lan, K.K.G., Ware, J.H., Johnson, N.J. and DeMets, D.L. (1982). An aid to data monitoring in long-term clinical trials. *Contr. Clin. Trials*, **3**, 311–23. [11.2, 11.7]

Halperin, M., Lan, K.K.G., Wright, E.C. and Foulkes, M.A. (1987). Stochastic curtailing for comparison of slopes in longitudinal studies. *Contr. Clin. Trials*, **8**, 315–326. [11.7, 15.3, 15.7]

Halperin, M. and Ware, J.H. (1974). Early decision in a censored Wilcoxon two-sample test for accumulating survival data. *J. Amer. Statist. Assoc.*, **69**, 414–422. [11.1]

Hamasaki, T., Evans, S.R. and Asakura, K. (2018). Design, data monitoring, and analysis of clinical trials with co-primary endpoints: a review. *J. Biopharm. Statist.*, **28**, 28–51. [19.6]

Hamilton, D.C. and Lesperance, M.L. (1991). A consulting problem involving bivariate acceptance sampling by variables. *Canadian J. Statist.*, **19**, 109–117. [19.7]

Hampson L.V. and Jennison C. (2013). Group sequential tests for delayed responses (with discussion). *J. Roy. Statist. Soc. B*, **75**,3–54. [1.5, 12.2, 14.1, 14.2, 14.3, 14.4, 14.5, 31.6]

Hampson, L.V. and Jennison, C. (2015). Optimizing the data combination rule for seamless phase II/III clinical trials. *Statist. Med.*, **34**, 39–58. [26.4, 26.5, 26.6, 27.2, 29.3]

Hardwick, J. (1989). Recent progress in clinical trial designs that adapt for ethical purposes. Comment on ‘Investigating therapies of potentially great benefit: ECMO’ (by J.H. Ware). *Statist. Sci.*, **4**, 327–336. [30.4]

Harrington, D. (1989). Discussion of ‘Interim analyses: the repeated confidence interval approach’ (by C. Jennison and B.W. Turnbull). *J. Roy. Statist. Soc. B*, **51**, 346–347. [17.5]

Harrington, D.P. and Fleming, T.R. (1982). A class of rank test procedures for censored survival data. *Biometrika*, **69**, 553–566. [17.8]

Harrington, D.P., Fleming, T.R. and Green, S.J. (1982). Procedures for serial testing in censored survival data. In *Survival Analysis*, (Eds., J. Crowley and R.A. Johnson), Hayward, California: Institute of Mathematical Statistics, 269–286. [4.5]

Hartung, J. and Knapp, G. (2003). A new class of completely self-designing clinical trials. *Biometrical J.*, **45**, 3–19. [24.1]

Hartung, J. and Knapp, G. (2006). Repeated confidence intervals in self-designing clinical trials and switching between noninferiority and superiority. *Biometrical J.*, **48**, 697–709. [10.5]

Hauck, W.W., Preston, P.E. and Bois, F.Y. (1997). A group sequential approach to crossover trials for average bioequivalence. *J. Biopharm. Statist.*, **7**, 87–96. [7.4, 7.6]

Haybittle, J.L. (1971). Repeated assessment of results in clinical trials of cancer treatment. *Brit. J. Radiology*, **44**, 793–797. [2.7]

Hayre, L.S. (1979). Two-population sequential tests with three hypotheses. *Biometrika*, **66**, 465–74. [30.4]

Hayre, L.S. and Turnbull, B.W. (1981a). A class of simple approximate sequential tests for adaptive comparison of two treatments. *Commun. Statist.–Theory and Methods*, **10**, 2339–2360. [30.4]

Hayre, L.S. and Turnbull, B.W. (1981b). Sequential estimation in two-armed exponential clinical trials. *Biometrika*, **68**, 411–416. [30.4]

Hayre, L.S. and Turnbull, B.W. (1981c) Estimation of the odds ratio in the two-armed bandit problem. *Biometrika*, **68**, 661–668. [30.4]

Hayter, A.J. and Hsu, J.C. (1994). On the relationship between stepwise decision procedures and confidence sets. *J. Amer. Statist. Assoc.*, **89**, 128–136. [28.6]

Heart Special Project Committee. (1988). Organization, review and administration of cooperative studies (Greenberg Report): a report from the Heart Special Project Committee to the National Advisory Council, May 1967. *Contr. Clin. Trials*, **9**, 137–148. [1.3]

Hedges, L.V., Cooper, H.M. and Bushman, B.J. (1992). Testing the null hypothesis in meta-analysis: a comparison of combined probability and confidence interval procedures. *Psychol. Bull.*, **111**, 188–194. [23.1]

Herson, J. (1979). Predictive probability early termination plans for phase II clinical trials. *Biometrics*, **35**, 775–783. [1.3, 11.3, 11.7]

Herson, J. (1994). Discussion of ‘Bayesian approaches to randomized trials’ (by D.J. Spiegelhalter, L.S. Freedman and M.K.B. Parmar). *J. Roy. Statist. Soc. A*, **157**, 396–397. [13.2]

Herson, J. and Wittes, J. (1993). The use of interim analysis for sample size adjustment. *Drug Inf. J.*, **27**, 753–760. [18.1, 18.2]

Hewett, J.E. and Spurrier, J.D. (1983). A survey of two stage tests of hypotheses: theory and application. *Commun. Statist.–Theory and Methods*, **12**, 2307–2425. [1.3]

Hilsenbeck, S.G. (1988). Early termination of a phase II clinical trial. *Contr. Clin. Trials*, **9**, 177–188. [11.7]

Hochberg, Y. (1988). A sharper Bonferroni procedure for multiple tests of significance. *Biometrika*, **75**, 800–802. [20.6, 20.9]

Hochberg, Y. and Tamhane, A.C. (1987). *Multiple Comparison Procedures*, New York: Wiley. [1.3, 21.3, 21.4]

Hoel, D.G., Sobel, M. and Weiss, G.H. (1975). A survey of adaptive sampling for clinical trials. In *Perspectives in Biometrics*, (Ed., R. Elashoff), New York: Academic Press, 29–61. [30.4]

Holm, S. (1979). A simple sequentially rejective multiple test procedure. *Scand. J. Statist.*, **6**, 65–70. [20.5]

Hommel, G. (1988). A stagewise rejective multiple test procedure based on a modified Bonferroni test. *Biometrika*, **75**, 383–386. [20.6]

Hommel, G. (1989). A comparison of two modified Bonferroni procedures. *Biometrika*, **76**, 624–625. [20.6]

Hommel, G. (2001). Adaptive modifications of hypotheses after an interim analysis. *Biometrical. J.*, **43**, 581–589. [27.1, 27.3]

Hotelling, H. (1931). The generalization of Student's ratio. *Ann. Math. Statist.*, **2**, 360–378. [19.3]

Hsaio, S.T., Liu, L. and Mehta, C.Y. (2019). Optimal promising zone designs. *Biometrical J.*, **61**, 1175–1186. [24.1, 24.5]

Hsu, J.C. (1996). *Multiple Comparisons: Theory and Methods*, Boca Raton: Chapman and Hall/CRC. [1.3]

Hsu J.C., Hwang J.T.G., Liu H.K. and Ruberg, S.J. (1994). Confidence intervals associated with tests for bioequivalence. *Biometrika*, **81**, 103–114. [10.5]

Hu, F. and Rosenberger, W.F. (2006). *The Theory of Response-Adaptive Randomization in Clinical Trials*, Hoboken, New Jersey: Wiley. [1.5, 30.4]

Hu, J., Blatchford, P.J., Goldenberg, N.A, and Kittelson, J.M. (2020). Group sequential designs for clinical trials with bivariate endpoints. *Statist. Med.*, **39**, 3823–3839. [19.6]

Hughes, M.D. (1993). Stopping guidelines for clinical trials with multiple treatments. *Statist. Med.*, **12**, 901–915. [21.5]

Hughes, M.D. and Pocock, S.J. (1988). Stopping rules and estimation problems in clinical trials. *Statist. Med.*, **7**, 1231–1242. [9.5, 13.1]

Hung, H. M. J., Wang, S. J. and O'Neill, R. (2007). Statistical considerations for testing multiple endpoints in group sequential or adaptive clinical trials. *J. Biopharm. Statist*, **17**, 1201–1210. [20.10]

Huque, M.F. (2016). Validity of the Hochberg procedure revisited for clinical trial applications. *Statist. Med.*, **35**, 5–20. [20.4]

Hwang, I.K., Shih, W.J. and De Cani, J.S. (1990). Group sequential designs using a family of type I error probability spending functions. *Statist. Med.*, **9**, 1439–1445. [8.2, 8.3, 8.8, 12.2]

Hwang, J.T. (1993). Empirical Bayes estimation for the means of the selected populations. *Sankhyā*, **55**, 285–304. [26.6]

Hyman, D.M., Puzanov, I., Subbiah, V. et al. (2015). Vemurafenib in multiple nonmelanoma cancers with BRAF V600 mutations. *New Engl. J. Med.*, **373**, 726–736. [21.1]

ICH Harmonised Guideline. (2025). *Adaptive Designs for Clinical Trials E20*, Draft version, 25 June 2025. [Preface 2, 1.3]

Iglewicz, B. (1983). Alternative designs: sequential, multi-stage, decision theory and adaptive designs. In *Cancer Clinical Trials: Methods and Practice*, (Eds., M.E. Buyse, M.J. Staquet and R.J. Sylvester), Oxford: Oxford University Press, 312–334. [30.4]

Inoue, L.Y.T., Thall, P.F. and Berry, D.A. (2002). Seamlessly expanding a randomized phase II trial to phase III. *Biometrics*, **58**, 823–831. [26.7]

Irle, S. and Schäfer, H. (2012). Interim design modifications in time-to-event studies. *J. Amer. Statist. Assoc.*, **107**, 341–348. [23.6, 24.6, 29.5]

Irony, T.Z. and Singpurwalla, N.D. (1997). Non-informative priors do not exist A dialogue with José M. Bernardo (with discussion). J. Statist. Planning and Inf., **65**, 159–1789. [13.2]

Jackson, J.E. and Bradley, R.A. (1961). Sequential χ^2- and T^2-tests and their application to an acceptance sampling problem. *Technometrics*, **3**, 519–534. [19.7]

Jänne, P.A., van den Heuvel, M.M., Barlesi, F. et al. (2017). Selumetinib plus docetaxel compared with docetaxel alone and progression-free survival in patients with KRAS-mutant advanced non–small cell lung cancer: the SELECT-1 randomized clinical trial. *J. Amer. Med. Assoc.* **317**, 1844–1853. [11.4]

Jenkins, M., Stone, A. and Jennison, C. (2011). An adaptive seamless phase II/III design for oncology trials with subpopulation selection using correlated survival endpoints. *Pharmaceut. Statist.*, **10**, 347–356. [24.6, 28.4, 29.2, 29.4, 29.5]

Jennison, C. (1982). *Sequential Methods for Medical Experiments*, Ph.D. Thesis, Cornell University, Ithaca. [10.1]

Jennison, C. (1987). Efficient group sequential tests with unpredictable group sizes. *Biometrika*, **74**, 155–165. [5.1, 8.3, 8.8, 12.2]

Jennison, C. (1992). Bootstrap tests and confidence intervals for a hazard ratio when the number of observed failures is small, with applications to group sequential survival studies. In *Computing Science and Statistics, Vol. 22*, (Eds., C. Page and R. LePage), New York: Springer-Verlag, 89–97. [4.5, 11.4, 17.9]

Jennison, C. (1994). Numerical computations for group sequential tests. In *Computing Science and Statistics, Vol. 25*, (Eds., M. Tarter and M.D. Lock), Interface Foundation of America, 263–272. [31.1, 31.5]

Jennison, C. (2023a). Discussion on 'Adaptive enrichment designs with a continuous biomarker' (by N. Stallard). *Biometrics*, **79**, 26–30. [28.5]

Jennison, C. (2023b). Discussion of 'Response-adaptive randomization in clinical trials: from myths to practical considerations' (by D.S. Robertson, K.M. Lee, B.C. López-Kolkovska and S.S. Villar). *Statistical Science*, **38**, 219–223. [30.3]

Jennison, C., Johnstone, I.M. and Turnbull, B.W. (1982). Asymptotically optimal procedures for sequential adaptive selection of the best of several normal means. In *Statistical Decision Theory and Related Topics III, Vol. 2*, (Eds., S.S. Gupta and J.O. Berger), New York: Academic Press, 55–86. [30.4]

Jennison, C. and Turnbull, B.W. (1983). Confidence intervals for a binomial parameter following a multistage test with application to MIL-STD 105D and medical trials. *Technometrics*, **25**, 49–58. [1.3, 16.1]

Jennison, C. and Turnbull, B.W. (1984). Repeated confidence intervals for group sequential clinical trials. *Contr. Clin. Trials*, **5**, 33–45. [1.3, 4.5, 10.1, 17.5, 17.9]

Jennison, C. and Turnbull, B.W. (1985). Repeated confidence intervals for the median survival time. *Biometrika*, **72**, 619–625. [10.5, 17.7, 17.9]

Jennison, C. and Turnbull, B.W. (1989). Interim analyses: the repeated confidence interval approach (with discussion). *J. Roy. Statist. Soc. B*, **51**, 305–361. [1.3, 8.2, 10.1, 10.2, 10.5, 16.1, 16.4, 16.6, 17.5]

Jennison, C. and Turnbull, B.W. (1990). Statistical approaches to interim monitoring of medical trials: a review and commentary. *Statist. Sci.*, **5**, 299–317. [8.2]

Jennison, C. and Turnbull, B.W. (1991a). Group sequential tests and repeated confidence intervals. In *Handbook of Sequential Analysis*, (Eds., B.K. Ghosh and P.K. Sen), New York: Marcel Dekker, 283–311. [8.5]

Jennison, C. and Turnbull, B.W. (1991b). Exact calculations for sequential t, χ^2 and F tests. *Biometrika*, **78**, 133–141. [4.6, 15.5, 15.7, 19.3, 21.2, 31.6]

Jennison, C. and Turnbull, B.W. (1991c). A note on the asymptotic joint distribution of successive Mantel-Haenszel estimates of the odds ratio based on accumulating data. *Seq. Analysis*, **10**, 201–209. [10.5, 15.7, 16.4]

Jennison, C. and Turnbull, B.W. (1993a). One-sided sequential tests to establish equivalence between treatments with special reference to normal and binary responses. In *Multiple Comparisons, Selection, and Applications in Biometry*, (Ed., F.M. Hoppe), New York: Marcel Dekker, 315–330. [7.6, 10.3, 10.5, 18.4]

Jennison, C. and Turnbull, B.W. (1993b). Sequential equivalence testing and repeated confidence intervals, with applications to normal and binary response. *Biometrics*, **49**, 31–43. [7.6, 10.3, 18.4]

Jennison, C. and Turnbull, B.W. (1993c). Group sequential tests for bivariate response: interim analyses of clinical trials with both efficacy and safety endpoints. *Biometrics*, **49**, 741–752. [1.4, 19.1, 19.6, 19.7, 31.6]

Jennison, C. and Turnbull, B.W. (1997a). Group-sequential analysis incorporating covariate information. *J. Amer. Statist. Assoc.*, **92**, 1330–1341. [3.4, 4.3, 15.7, 17.4, 17.9]

Jennison, C. and Turnbull, B.W. (1997b). Distribution theory of group sequential t, χ^2 and F-tests for general linear models. *Seq. Analysis*, **16**, 295–317. [15.4, 15.7, 19.3]

Jennison, C. and Turnbull, B.W. (2001a) On group sequential tests for data in unequally sized groups and with unknown variance. *J. Statist. Planning and Inf.*, **96**, 263–288. [5.5, 6.1, 7.4]

Jennison, C. and Turnbull, B.W. (2001b). Group sequential tests with outcome-dependent treatment assignment. *Seq. Analysis*, **20**, 209–234. [30.2]

Jennison, C. and Turnbull, B.W. (2003) Mid-course sample size modification in clinical trials based on the observed treatment effect. *Statist. Med.*, **22**, 971–993. [1.3, 23.6, 23.8, 23.10, 24.3]

Jennison, C. and Turnbull, B.W. (2005). Meta-analyses and adaptive group sequential designs in the clinical development process. *J. Biopharm. Statist.*, **15**, 537–558. [23.4, 23.10]

Jennison, C. and Turnbull, B.W. (2006a). Adaptive and nonadaptive group sequential tests. *Biometrika*, **93**, 1–21. [12.4, 24.3, 24.5, 31.6]

Jennison, C, and Turnbull B.W. (2006b). Efficient group sequential designs when there are several effect sizes under consideration. *Statist. Med.* **25**, 917–932. [12.3, 12.4, 24.3]

Jennison C, and Turnbull B.W. (2006c). Discussion of 'Are flexible designs sound?' (by C-F. Burman and C. Sonesson). *Biometrics*, **62**, 670–673. [24.3]

Jennison, C. and Turnbull, B.W. (2006d). Confirmatory seamless phase II/III clinical trials with hypotheses selection at interim: opportunities and limitations. *Biometrical J.*, **48**, 650–655. [26.3]

Jennison, C. and Turnbull, B.W. (2007). Adaptive seamless designs: selection and prospective testing of hypotheses. *J. Biopharm. Statist.*, **17**, 1135–1161. [26.2, 26.3, 26.4, 28.2]

Jennison, C. and Turnbull, B. W. (2011). From group sequential to adaptive designs. In *Handbook of Adaptive Designs in Pharmaceutical and Clinical Development*, (Eds. S-C. Chow and A. Pong), Chapter 5. Boca Raton: Chapman and Hall/CRC. [12.2]

Jennison, C. and Turnbull, B. W. (2013). Interim monitoring of clinical trials: decision theory, dynamic programming and optimal stopping. *Kuwait J. Science*, **40**, 43–59. [12.2, 31.6]

Jennison, C. and Turnbull, B. W. (2015). Adaptive sample size modification in clinical trials: start small then ask for more? *Statist. Med.*, **34**, 3793–3810. [24.1, 24.5, 24.6]

Jiang, W. (1999). Group sequential procedures for repeated events data with frailty. *J. Biopharm. Statist.*, **9**, 379–399. [15.7]

Jiang, W., Turnbull, B.W. and Clark, L.C. (1999). Semiparametric regression models for repeated events with random effects and measurement error. *J. Amer. Statist. Assoc.*, **94**, 111–124. [1.4]

Johns, D. and Andersen, J.S. (1999). Use of predictive probabilities in phase II and phase III clinical trials. *J. Biopharm. Statist.*, **9**, 67–79. [11.7]

Jones, B. and Kenward, M.G. (2014). *Design and Analysis of Cross-Over Trials*, 3rd ed., London: Chapman and Hall. [1.4, 7.4]

Jones, R.L., Chawla, S.P., Attia, S. et al. (2019). A phase 1 and randomized controlled phase 2 trial of the safety and efficacy of the combination of gemcitabine and docetaxel with ontuxizumab (MORAb-004) in metastatic soft-tissue sarcomas. *Cancer*, **125**, 2445–2454. [28.5]

Ju, H.L. (1997). On TIER method for assessment of individual bioequivalence. *J. Biopharm. Statist.*, **7**, 63–85. [7.5]

Jung, S.H. and Sargent, D.J. (2014). Randomized phase II clinical trials. *J. Biopharm. Satist.*, **24**, 802–816. [16.3]

Kalbfleisch, J.D. and Prentice, R.L. (2002). *The Statistical Analysis of Failure Time Data*, 2nd ed., New York: Wiley. [1.3, 17.6]

Kaplan, E.L. and Meier, P. (1958). Nonparametric estimation from incomplete observations. *J. Amer. Statist. Assoc.*, **53**, 457–481. [1.5, 16.8, 17.1, 17.7]

Kass, R.E. and Greenhouse, J.B. (1989). Comment on 'Investigating therapies of potentially great benefit: ECMO' (by J.H. Ware). *Statist. Sci.*, **4**, 310–317. [13.2]

Kavelaars, X., Mulder, J. and Kaptein, M. (2020). Decision-making with multiple correlated binary outcomes in clinical trials. *Statist. Meth. Med. Res.*, **29**, 3265–3277. [19.7]

Keaney, K.M. and Wei, L.J. (1994). Interim analyses based on median survival times. *Biometrika*, **81**, 279–286. [10.5, 17.7, 17.9]

Kelly, P.J., Stallard, N. and Todd, S. (2005). An adaptive group sequential design for phase II/III clinical trials that select a single treatment from several. *J. Biopharm. Statist.*, **15**, 641–658. [26.8]

Khan, J.N., Kimani, P.K., Glimm, E. and Stallard, N. (2023). Adjusting for treatment selection in phase II/III clinical trials with time to event data. *Statist. Med.*, **42**, 146–163. [26.6]

Kiefer, J. and Weiss, L. (1957). Some properties of generalized sequential probability ratio tests. *Ann. Math. Statist.*, **28**, 57–74. [1.3, 12.2]

Kieser, M., Bauer, P. and Lehmacher, W. (1999). Inference on multiple endpoints in clinical trials with adaptive interim analyses *Biometrical J.*, **41**, 261–277. [26.8]

Kieser, M., and Friede, T. (2000). Re-calculating the sample size in internal pilot study designs with control of the type I error rate. *Statist. Med.*, **19**, 901–911. [18.1]

Kieser, M. and Friede, T. (2003). Simple procedures for blinded sample size adjustment that do not affect the type I error rate. *Statist. Med.*, **22**, 3571–3581. [18.3]

Kilkenny, C., Browne, W.J., Cuthill, I.C. et al. (2010). Improving bioscience research reporting: the ARRIVE guidelines for reporting animal research. *J. Pharmacology and Pharamacotherapeutics*, **1**, 94–99. [1.1]

Kim, K., Boucher, H. and Tsiatis, A.A. (1995). Design and analysis of group sequential logrank tests in maximum duration versus information trials. *Biometrics*, **51**, 988–1000. [8.2]

Kim, K., Bretz F., Cheung, Y.K.K. and Hampson, L.V., Eds. (2021). *Handbook of Statistical Methods for Randomized Controlled Trials*, Boca Raton: Chapman and Hall/CRC. [1.1]

Kim, K. and DeMets, D.L. (1987a). Design and analysis of group sequential tests based on the type I error spending rate function. *Biometrika*, **74**, 149–154. [1.3, 8.2]

Kim, K. and DeMets, D.L. (1987b). Confidence intervals following group sequential tests in clinical trials. *Biometrics*, **43**, 857–864. [1.3, 9.5]

Kim, K. and DeMets, D.L. (1992). Sample size determination for group sequential clinical trials with immediate response. *Statist. Med.*, **11**, 1391–1399. [1.4]

Kimani, P.K., Glimm, E., Maurer, W. et al. (2012). Practical guidelines for adaptive seamless phase II/III clinical trials that use Bayesian methods. *Statist. Med.*, **31**, 2068–2085. [26.8]

Kimani, P.K., Stallard, N. and Hutton, J.L. (2009). Dose selection in seamless phase II/III clinical trials based on efficacy and safety. *Statist. Med.*, **28**, 917–936 [26.8]

Kimani, P.K., Todd, S., Renfro, L.A., Glimm, E. et al. (2020). Point and interval estimation in two-stage adaptive designs with time to event data and biomarker-driven subpopulation selection. *Statist. Med.*, **39**, 2568–2586. [28.6]

Kimani, P.K., Todd, S., Renfro, L.A. and Stallard, N. (2018). Point estimation following two-stage adaptive threshold enrichment clinical trials. *Statist. Med.*, **37**, 3179–3196. [28.6]

Kimani, P.K., Todd, S. and Stallard, N. (2013). Conditionally unbiased estimation in phase II/III clinical trials with early stopping for futility. *Statist. Med.*, **32**, 2893–2910. [26.6]

Kimani, P.K., Todd, S. and Stallard, N. (2014). A comparison of methods for constructing confidence intervals after phase II/III clinical trials. *Biometrical J.*, **56**, 107–128. [26.6]

Kimani, P.K., Todd, S. and Stallard, N. (2015). Estimation after subpopulation selection in adaptive seamless trials. *Statist. Med.* **34**, 2581–2601. [28.6]

Kirkwood, T.B.L. (1981). Bioequivalence testing—a need to rethink. (Reader Reaction) *Biometrics*, **37**, 589–591. [7.6]

Kittelson, J.M., Sharples, K. and Emerson S.S. (2005). Group sequential clinical trials for longitudinal data with analyses using summary statistics. *Statist. Med.*, **24**, 2457–2475. [14.4]

Kittelson J.M., Steg, P.G., Halperin, J.L. et al. (2016). Bivariate evaluation of thromboembolism and bleeding in clinical trials of anticoagulants in patients with atrial fibrillation. *Thrombosis and Haemostasis*, **116**, 544–553. [19.6]

Klein, J.P. and Moeschberger, M.L. (1997). *Survival Analysis: Techniques for Censored and Truncated Data*, New York: Springer. [17.3, 17.4]

Kleinbaum, D.G., Kupper, L.L. and Morgenstern, H. (1982). *Epidemiologic Research: Principles and Quantitative Methods*, New York: Wiley. [1.1]

Koenig, F., Brannath, W., Bretz, F. and Posch, M. (2008). Adaptive Dunnett tests for treatment selection. *Statist. Med.*, **27**, 1612–1625. [26.3, 29.5]

Koepcke, W. (1989). Analyses of group sequential clinical trials. *Contr. Clin. Trials*, **10**, 222S–230S. [10.5]

Kolata, G.B. (1979). Controversy over study of diabetes drugs continues for nearly a decade: a bitter altercation raises major issues facing clinical scientists and regulatory agencies. *Science*, **203**, 986–990. [1.3]

Korn E.L. and Freidlin, B. (2011). Outcome-adaptive randomization: is it useful? *J. Clin. Oncol.*, **29**, 771–776. [30.3]

Korn, E.L. and Freidlin, B. (2017). Adaptive clinical trials: advantages and disadvantages of various adaptive design elements. *J. Natl. Cancer. Inst.*, **109**, 1–6. [27.7, 30.4]

Koyama, T., Sampson, A.R. and Gleser, L.J. (2005). A framework for two-stage adaptive procedures to simultaneously test non-inferiority and superiority. *Statist. Med.*, **24**, 2439–2456. [7.3]

Koziol, J.A. and Petkau, A.J. (1978). Sequential testing of the equality of two survival distributions using the modified Savage statistic. *Biometrika*, **65**, 615–623. [11.2]

Kunz, C.U., Friede, T., Parsons, N. et al. (2015) A comparison of methods for treatment selection in seamless phase II/III clinical trials incorporating information on short-term endpoints. *J. Biopharm. Statist.*, **25**, 170–189. [26.5]

Kunz, C.U., Tarima, S., Rosner, G.L., Emsley, R. et al. (2024). Bayesian and frequentist approaches to rescuing disrupted trials: a report from the NISS Ingram Olkin forum series on unplanned clinical trial disruptions. *Statist. in Biopharm. Res.*, **16**, 383–402. [13.4]

Kunzmann, K., Benner, L. and Kieser, M. (2017). Point estimation in adaptive enrichment designs. *Statist. Med.*, **36**, 3935–3947. [28.6]

Lachin, J.M. (1981). Sequential clinical trials for normal variates using interval composite hypotheses. *Biometrics*, **37**, 87–101. [13.3]

Lachin, J.M. (2005). A review of methods for futility stopping based on conditional power. *Statist. Med.*, **24**, 2747–2764. [11.5]

Lai, T.L. (1973). Optimal stopping and sequential tests which minimize the maximum sample size. *Ann. Statist.*, **1**, 659–673. [1.3, 12.2]

Lai, T.L. (1984). Incorporating scientific, ethical and economic considerations into the design of clinical trials in the pharmaceutical industry: a sequential approach. *Commun. Statist.-Theory and Methods*, **13**, 2355–2368. [10.1]

Lai, T.L. (2002). Asymptotic optimality of generalized sequential likelihood ratio tests in some classical sequential testing problems. *Seq. Analysis*, **21**, 219–247. [1.3]

Lai, T.L., Lavori, P.W. and Liao, O.Y.W. (2014). Adaptive choice of patient subgroup for comparing two treatments. *Contemp. Clin. Trials*, **39**, 191–200. [28.8]

Lai, T.L., Lavori, P.W. and Shih, M-C. (2012a). Adaptive trial designs. *Ann. Rev. Pharm. Toxicol.*, **52**, 101–110. [24.3, 24.6]

Lai, T.L., Lavori, P.W. and Shih, M-C. (2012b). Sequential design of phase II-III cancer trials. *Statist. Med.*, **31**, 1944–1960. [26.7]

Lai, T.L., Lavori, P.W. and Tsang, K.W, (2019). Adaptive enrichment designs for confirmatory trials. *Statist. Med.*, **38**, 613–624. [28.8]

Lai, T.L. and Shih, M-C. (2004). Power, sample size and adaptation considerations in the design of group sequential clinical trials. *Biometrika*, **91**, 507–528. [26.7]

Lai, T.L., Shih, M-C. and Zhu, G. (2006). Modified Haybittle-Peto group sequential designs for testing superiority and non-inferiority hypotheses in clinical trials. *Statist. Med.*, **25**, 1149–1167. [7.3]

Lai, T.L. and Ying, Z. (1991). Large sample theory of a modified Buckley-James estimator for regression analysis with censored data. *Ann. Statist.*, **19**, 1370–1402. [15.6]

Lan, K.K.G. and DeMets, D.L. (1983). Discrete sequential boundaries for clinical trials. *Biometrika*, **70**, 659–663. [1.3, 8.2, 8.8, 19.5]

Lan, K.K.G. and DeMets, D.L. (1989a). Group sequential procedures: calendar versus information time. *Statist. Med.*, **8**, 1191–1198. [8.2]

Lan, K.K.G. and DeMets, D.L. (1989b). Changing frequency of interim analysis in sequential monitoring. *Biometrics*, **45**, 1017–1020. [8.5]

Lan, K.K.G. and Lachin, J.M. (1990). Implementation of group sequential logrank tests in a maximum duration trial. *Biometrics*, **46**, 759–770. [8.2]

Lan, K.K.G., Lachin, J.M. and Bautista, O. (2003). Over-ruling a group sequential boundary—a stopping rule versus a guideline. *Statist. Med.*, **22**, 3347–3355. [14.3]

Lan, K.K.G., Reboussin, D.M. and DeMets, D.L. (1994). Information and information fractions for design and sequential monitoring of clinical trials. *Commun. Statist.-Theory and Methods*, **23**, 403–420. [3.3, 8.2]

Lan, K.K.G., Rosenberger, W.F. and Lachin, J.M. (1993). Use of spending functions for occasional or continuous monitoring of data in clinical trials. *Statist. Med.*, **12**, 2219–2231. [8.2]

Lan, K.K.G., Simon, R. and Halperin, M. (1982). Stochastically curtailed tests in long-term clinical trials. *Seq. Analysis*, **1**, 207–219. [1.3, 11.2, 11.3, 11.5, 11.7]

Lan, K.K.G. and Wittes, J. (1988). The B-value: a tool for monitoring data. *Biometrics*, **44**, 579–585. [11.7]

Lan, K.K.G. and Zucker, D.M. (1993). Sequential monitoring of clinical trials: the role of information and Brownian motion. *Statist. Med.*, **12**, 753–766. [3.1]

Lee, J.W. and DeMets, D.L. (1991). Sequential comparison of changes with repeated measurements data. *J. Amer. Statist. Assoc.*, **86**, 757–762. [15.7]

Lawler, P.R., Hochman, J.S. and Zarychanski, R. (2022). What are adaptive platform clinical trials and what role may they have in cardiovascular medicine? *Circulation*, **145**, 629–632. [27.7]

Lee, J.J. and Chu, C.T. (2012). Bayesian clinical trials in action. *Statist. Med.*, **31**, 2955–2972. [13.5]

Lee, J.W. and DeMets, D.L. (1992). Sequential rank tests with repeated measurements in clinical trials. *J. Amer. Statist. Assoc.*, **87**, 136–142. [15.7]

Lee, J.W. and Sather, H.N. (1995). Group sequential methods for comparison of cure rates in clinical trials. *Biometrics*, **51**, 756–763. [17.9]

Lee, K.M., Brown, L.C., Jaki, T. et al. (2021). Statistical consideration when adding new arms to ongoing clinical trials: the potentials and the caveats. *Trials*, **22**:203, 1–10. [27.3]

Lee, K.M., Wason, J. and Stallard, N. (2019). To add or not to add a new treatment arm to a multiarm study: a decision-theoretic framework. *Statist. Med.*, **38**, 3305–3321. [27.3]

Lee, Y.J., Staquet, M., Simon, R. et al. (1979). Two-stage plans for patient accrual in phase II cancer clinical trials. *Cancer Treatment Reports*, **63**, 1721–1726. [1.3]

Lehmacher, W. and Wassmer, G. (1999). Adaptive sample size calculations in group sequential trials. *Biometrics*, **55**, 1286–1290. [1.3, 1.5, 10.5, 22.2, 23.2, 23.7, 23.8, 23.10, 24.1, 27.1]

Levin, G.P., Emerson, S.C. and Emerson, S.S. (2014). An evaluation of inferential procedures for adaptive clinical trial designs with pre-specified rules for modifying the sample size. *Biometrics*, **70**, 556–567. [23.10]

Lewis, R.J., Lipsky, A.M. and Berry, D.A. (2007). Bayesian decision-theoretic group sequential clinical trial design based on a quadratic loss function: a frequentist evaluation. *Clin. Trials*, **4**, 5–14. [13.3]

Li, W., Chen, C., Li, X. and Beckman, R.A. (2017). Estimation of treatment effect in two-stage confirmatory oncology trials of personalized medicines. *Statist. Med.*, **36**, 1843–1861. [28.6]

Li, G., Shih, W.J., Xie, T. and Lu, J. (2002). A sample size adjustment procedure for clinical trials based on conditional power. *Biostatistics*, **3**, 277–287. [24.1]

Li, Q.H. and Huque, M.F. (2003). A decision rule for evaluating several independent clinical trials collectively. *J. Biopharma. Statist.*, **13**, 621–628. [23.1]

Li, Z. (1999). A group sequential test for survival trials: an alternative to rank-based procedures. *Biometrics*, **55**, 277–283. [17.8]

Lin, D.Y. (1991). Nonparametric sequential testing in clinical trials with incomplete multivariate observations. *Biometrika*, **78**, 123–131. [15.7, 19.5]

Lin, D.Y. (1992). Sequential log rank tests adjusting for covariates with the accelerated life model. *Biometrika*, **79**, 523–529. [17.9]

Lin, D.Y. and Liu, P.Y. (1992). Nonparametric sequential tests against ordered alternatives in multiple-armed clinical trials. *Biometrika*, **79**, 420–425. [21.2]

Lin, D.Y., Shen, L., Ying, Z. and Breslow, N.E. (1996). Group sequential designs for monitoring survival probabilities. *Biometrics*, **52**, 1033–1041. [17.7]

Lin, D.Y. and Wei, L.J. (1991). Repeated confidence intervals for a scale change in a sequential survival study. *Biometrics*, **47**, 289–294. [10.5]

Lin, D.Y., Wei, L.J. and DeMets, D.L. (1991). Exact statistical inference for group sequential trials. *Biometrics*, **47**, 1399–1408. [10.5, 16.2]

Lin, D.Y., Yao, Q. and Ying, Z. (1999). A general theory on stochastic curtailment for censored survival data. *J. Amer. Statist. Assoc.*, **94**, 510–521. [11.7, 17.9]

Lin, J. and Bunn, V. (2017). Comparison of multi-arm multi-stage design and adaptive randomization in platform clinical trials. *Contemp. Clin. Trials*, **54**, 48–59. [30.4]

Lin, R.S., Lin, J., Roychoudhury, S. et al. (2020). Alternative analysis methods for time to event endpoints under nonproportional hazards: a comparative analysis. *Statist. Biopharm. Res.*, **12**, 187–198. [17.8]

Lindley, D.V. (1962). Discussion on 'Confidence sets for the mean of a multivariate normal distribution' (by C.M. Stein). *J. Roy. Statist. Soc. B*, **24**, 285–287. [26.6]

Lindley, D.V. (1965). *Introduction to Probability and Statistics from a Bayesian Viewpoint; Part 2, Inference*, Cambridge: Cambridge University Press. [13.1]

Lindley, D.V. (1994). Discussion of 'Bayesian approaches to randomized trials' (by D.J. Spiegelhalter, L.S. Freedman and M.K.B. Parmar). *J. Roy. Statist. Soc. A*, **157**, 393. [13.3]

Liu, A. and Hall, W.J. (1999). Unbiased estimation following a group sequential test. *Biometrika*, **86**, 71–78. [9.3]

Liu, Q. and Anderson, M. (2008). On adaptive extensions of group sequential trials for clinical investigations. *J. Amer. Statist. Assoc.*, **103**, 1621–1630. [20.10, 20.12 20.13, 23.10]

Liu, Q., Anderson, K.M. and Pledger, G.W. (2004). Benefit-risk evaluation of multi-stage adaptive designs. *Seq. Analysis*, **23**, 317–331. [12.2]

Liu, Q. and Pledger, G.W. (2005). Phase 2 and 3 combination designs to accelerate drug development. *J. Amer. Statist. Assoc.*, **100**, 493–502. [26.8]

Liu, Q., Proschan, M.A. and Pledger, G.W. (2002). A unified theory of two-stage adaptive designs. *J. Amer. Statist. Assoc.*, **97**, 1034–1041. [23.10]

Liu, Y. and Lim, P. (2017). Sample size increase during a survival trial when interim results are promising. *Commun. Statist.–Theory and Methods*, **46**, 6846–6863. [24.6]

Little, R.J. (2006). Calibrated Bayes: a Bayes/frequentist roadmap. *The Amer. Statistician*, **60**, 213–223. [13.3]

Lokhnygina, Y. and Tsiatis, A.A. (2008). Optimal two-stage group-sequential designs. *J. Statist. Planning and Inf.*, **138**, 489–499. [12.4]

Lorden, G. (1976). 2-SPRT's and the modified Kiefer-Weiss problem of minimizing an expected sample size. *Ann. Statist.*, **4**, 281–291. [1.3, 5.7, 12.2, 13.3]

Louis, T.A. (1975). Optimal allocation in sequential tests comparing the means of two Gaussian populations. *Biometrika*, **62**, 359–369. [30.4]

Louis, T.A. (1977). Sequential allocation in clinical trials comparing two exponential survival curves. *Biometrics*, **33**, 627–634. [30.4]

Luo, X., Wu, S.S. and Xiong, J. (2010). Parameter estimation following an adaptive treatment selection trial design. *Biometrical J.*, **52**, 823–835. [26.6]

Maca, J., Bhattacharya, S., Dragalin, V. et al. (2006). Adaptive seamless phase II/III designs—background, operational aspects, and examples. *Drug Inf. J.*, **40**, 463–473. [26.1, 26.8]

Madsen, R.W. and Fairbanks, K.B. (1983). *P* values for multistage and sequential tests. *Technometrics*, **25**, 285–293. [1.3]

Magirr, D. and Burman, C.F. (2019). Modestly weighted logrank tests. *Statist. Med.*, **38**, 3782–3790. [17.8]

Magirr, D., Jaki, T., Koenig, F. and Posch, M. (2016). Sample size reassessment and hypothesis testing in adaptive survival trials. *PloS one*, **11**:e0146465, 1–14. [24.6, 29.5]

Magirr, D., Jaki, T., Posch, M. and Klinglmueller, F. (2013). Simultaneous confidence intervals that are compatible with closed testing in adaptive designs. *Biometrika*, **100**, 985–996. [26.6, 28.6]

Magirr, D., Jaki, T. and Whitehead, J. (2012). A generalized Dunnett test for multi-arm, multi-stage clinical studies with treatment selection. *Biometrika*, **99**, 494–501. [21.4, 27.4, 31.6]

Magirr, D., Stallard, N. and Jaki, T. (2014). Flexible sequential designs for multi-arm clinical trials. *Statist. Med.*, **33**, 3269–3279. [21.4, 27.1, 27.4]

Magnusson, B.P. and Turnbull, B.W. (2013). Group sequential enrichment design incorporating subgroup selection. *Statist. Med.*, **32**, 2695–2714. [28.5]

Makuch, R.W. and Simon, R.M. (1982) Sample size requirements for comparing time-to-failure among k treatment groups. *J. Chron. Dis.*, **35**, 861–867. [11.6]

Mantel, N. (1966). Evaluation of survival data and two new rank order statistics arising in its consideration. *Cancer Chemotherapy Reports*, **50**, 163–170. [4.5]

Mantel, N. and Haenszel, W. (1959). Statistical aspects of the analysis of data from retrospective studies of disease. *J. Nat. Cancer Inst.*, **22**, 719–748. [1.5, 16.4]

Marcus R., Peritz, E. and Gabriel, K.R. (1976). On closed testing procedures with special reference to ordered analysis of variance. *Biometrika*, **63**, 655–660. [20.3, 21.3]

Marcus, M. and Swerling, P. (1962). Sequential detection in radar with multiple resolution elements. *IRE Trans. Inform. Th.*, **IT-8**, 237–245. [1.3]

Martinez, H., Habicht, J-P., Garza, C. and Mota, F. (1996). Clinical trial of a rice-powder oral rehydration beverage. *Food and Nutrition Bulletin*, **17**, 1–9. [1.4]

Maurer, W. (2016). Comments on 'Twenty-five years of confirmatory adaptive designs: opportunities and pitfalls' (by P. Bauer, F. Bretz, V. Dragalin et al.) *Statist. Med.* **35**, 362-3. [1.3]

Maurer, W. and Bretz, F. (2013). Multiple testing in group sequential trials using graphical approaches. *Statist. Biopharm. Res.*, **5**, 311–320. [20.10, 20.11, 20.12]

McCullagh, P. and Nelder, J.A. (1989). *Generalized Linear Models*, 2nd ed., London: Chapman and Hall. [15.6]

McNemar Q. (1947). Note on the sampling error of the difference between correlated proportions or percentages. *Psychometrika*, **12**, 153–157. [1.5, 16.5]

McPherson, K. (1974). Statistics: the problem of examining accumulating data more than once. *New Engl. J. Med.*, **290**, 501–502. [1.3, 2.3, 31.1]

McPherson, C.K. and Armitage, P. (1971). Repeated significance tests on accumulating data when the null hypothesis is not true. *J. Roy. Statist. Soc. A*, **134**, 15–25. [31.1]

Mehta, C.R., Bauer, P., Posch, M. and Brannath, W. (2007). Repeated confidence intervals for adaptive group sequential trials. *Statist. Med.*, **26**, 5422–5433. [10.5, 23.9]

Mehta, C.R. and Cain, K.C. (1984). Charts for the early stopping of pilot studies. *J. Clin. Oncology*, **2**, 676–682. [13.3]

Mehta, C.R., Gao, P., Bhatt, D.L. et al. (2009). Optimizing trial design: sequential, adaptive and enrichment strategies. *Circulation* **119**, 597-605. [28.8]

Mehta, C.R. and Liu, L. (2016) An objective re-evaluation of adaptive sample size re-estimation: commentary on 'twenty-five years of confirmatory adaptive designs'. *Statist. Med.*, **35**, 350–358. [24.6]

Mehta, C.R. and Pocock, S.J. (2011). Adaptive increase in sample size when interim results are promising: a practical guide with examples. *Statist. Med.*, **30**, 3267–3284. [1.5, 24.1, 24.5]

Mehta, C., Schäfer, H., Daniel, H. and Irle, S. (2014). Biomarker driven population enrichment for adaptive oncology trials with time to event endpoints. *Statist. Med.*, **33**, 4515–4531. [28.1, 28.2]

Mehta C.R. and Tsiatis A. A. (2001) Flexible sample size considerations using information-based interim monitoring. *Drug Inf. J.*, **35**, 1095–1112. [8.1, 8.6, 18.4]

Meier, P. (1975). Statistics and medical experimentation. *Biometrics*, **31**, 511–529. [1.3, 10.1]

Menon, S., Massaro, J., Pencina, M.J. et al. (2013). Comparison of operating characteristics of commonly used sample size re-estimation procedures in a two-stage design. *Commun. Statist.–Simulation and Computation*, **42**, 1140–1152. [24.6]

Metzler, C.M. (1974). Bioavailability—a problem in equivalence. *Biometrics*, **30**, 309–317. [7.6]

Miké, V. and Stanley, K.E., Eds. (1982). *Statistics in Medical Research: Methods and Issues with Applications in Cancer Research*, New York: Wiley. [1.1]

MIL-STD-105E (1989). *Military Standard Sampling Procedures and Tables for Inspection by Attributes.* U.S. Dept. of Defense, Washington D.C. [1.2, 1.3, 16.1, 31.1]

MIL-STD-781C (1977). *Reliability Design Qualification and Production Acceptance Tests: Exponential Distribution.* U.S. Dept. of Defense, Washington D.C. [1.4]

Miller, R.G. (1998). *Survival Analysis*, 2nd ed., New York: Wiley. [1.3]

Moerbeek, M. (2021). Bayesian updating: increasing sample size during the course of a study. *BMC Med. Res. Meth.*, **21**:137, 1–11. [24.1]

Moher, D., Hopewell, S., Schulz, K.F. et al. (2012). CONSORT 2010 explanation and elaboration: updated guidelines for reporting parallel group randomised trials. *Int. J. Surgery*, **10**, 28–55. [1.3]

Morikawa, T. and Yoshida, M. (1995). A useful testing strategy in phase III trials: combined test of superiority and test of equivalence. *J. Biopharm. Statist.*, **5**, 297–306. [7.3]

Mosteller, F. and Bush, R.R. (1954). Selected quantitative techniques. In *Handbook of Social Psychology, Vol. 1*, (Ed. G. Lindzey), Cambridge, MA: Addison-Wesley, 289–334. [23.1]

Moussa, M.A.A. (1989). Exact, conditional, and predictive power in planning clinical trials. *Contr. Clin. Trials*, **10**, 378–385. [13.4]

Müller, P., Berry, D.A., Grieve, A.P., et al. (2007). Simulation-based sequential Bayesian design. *J. Statist. Planning and Inf.*. **137**, 3140–3150. [13.3]

Müller, H-H. and Schäfer, H. (1999). Optimization of testing times and critical values in sequential equivalence testing. *Statist. Med.* **18**, 1769–1788. [12.3]

Müller, H-H. and Schäfer, H. (2001). Adaptive group sequential designs for clinical trials: combining the advantages of adaptive and of classical group sequential approaches. *Biometrics*, **57**, 886–891. [1.5, 23.6]

Müller, H-H. and Schäfer, H. (2004). A general statistical principle for changing a design any time during the course of a trial. *Statist. Med.*, **23**, 2497–2508. [1.5, 23.6]

Natarajan, R., Turnbull, B.W., Slate, E.H. and Clark, L.C. (1996). A computer program for sample size and power calculations in the design of multi-arm and factorial clinical trials with survival time endpoints. *Comp. Meth. and Prog. in Biomed.*, **49**, 137–147. [17.4, 21.3]

Neal, D., Casella, G., Yang, M.C.K. and Wu, S.S. (2011). Interval estimation in two-stage, drop-the-losers clinical trials with flexible treatment selection. *Statist. Med.*, **30**, 2804–2814. [26.6]

Nelder, J.A. and Mead, R. (1965). A simplex method for function minimization. *The Computer J.* **7**, 308–313. [12.3]

Nelson, B.S., Liu, L. and Mehta, C. (2022). A simulation-based comparison of estimation methods for adaptive and classical group sequential clinical trials. *Pharmaceut. Statist.*, **21**, 599–611. [23.10]

Ng, T-H. (2003). Issues of simultaneous tests for noninferiority and superiority. *J. Biopharm. Statist.*, **13**, 629–639. [7.3]

Oaknin, A., Friedman, C.F., Roman, L.D. et al. (2020). Neratinib in patients with HER2-mutant, metastatic cervical cancer: findings from the phase 2 SUMMIT basket trial. *Gynecologic Oncology*, **159**, 150–156. [21.1]

O'Brien, P.C. (1984). Procedures for comparing samples with multiple endpoints. *Biometrics*, **40**, 1079–1087. [1.4, 19.1, 19.4, 19.5]

O'Brien, P.C. and Fleming, T.R. (1979). A multiple testing procedure for clinical trials. *Biometrics*, **35**, 549–556. [1.3, 2.3, 2.5, 2.6, 4.1, 5.1, 8.2, 8.8, 10.1, 10.2, 11.2, 13.3, 16.2]

Offen, W., Chuang-Stein, C., Dmitrienko, A. et al. (2007). Multiple co-primary endpoints: medical and statistical solutions: a report from ehe multiple endpoints expert team of the Pharmaceutical Research and Manufacturers of America. *Drug Inf. J.*, **41**, 31–46. [19.6]

Ohman Strickland, P.A. and Casella, G. (2003). Conditional inference following group sequential testing. *Biometrical J.*, **45**, 515–526. [9.3]

Öhrn, F. (2011). *Group Sequential and Adaptive Methods—Topics with Applications to Clinical Trials*, Ph.D. Thesis, University of Bath, U.K. [11.5, 11.7, 23.6, 23.8, 31.6]

Öhrn, F. and Jennison, C. (2010). Optimal group-sequential designs for simultaneous testing of superiority and non-inferiority. *Statist. Med.*, **29**, 743–759. [7.3, 12.2, 23.10]

Ondra, T., Dmitrienko, A., Friede. T. et al. (2016). Methods for identification and confirmation of targeted subgroups in clinical trials: a systematic review. *J. Biopharm. Statist.*, **26**, 99–119. [28.8]

Ondra, T., Jobjörnsson, S., Beckman, R.A. et al. (2019). Optimized adaptive enrichment designs. *Statist. Meth. Med. Res.*, **28**, 2096–2111. [28.3, 28.7]

O'Neill, R.T. and Anello, C. (1978). Case-control studies: a sequential approach. *Amer. J. Epidem.*, **108**, 415–424. [16.5]

Orawo, L.A.O. and Christen, J.A. (2009). Bayesian sequential analysis for multiple-arm clinical trials. *Statist. and Comp.*, **19**, 99–109. [13.5]

Palermos, D., Sergentanis, T.N., Gavriatopoulou, M. et al. (2022). Lung cancer clinical trials with a seamless phase II/III design: systematic review. *J. Clin. Med.*, **11:7176**, 1–14. [26.1]

Pampallona, S. and Tsiatis, A.A. (1994). Group sequential designs for one-sided and two-sided hypothesis testing with provision for early stopping in favor of the null hypothesis. *J. Statist. Planning and Inf.*, **42**, 19–35. [5.1, 5.2, 6.1, 6.2, 8.3, 12.2, 21.4]

Pampallona, S., Tsiatis, A.A. and Kim, K., (2001). Interim monitoring of group sequential trials using spending functions for the type I and type II error probabilities. *Drug Inf. J.*, **35**, 1113–1121. [8.3, 8.8]

Parmar, M.K.B., Spiegelhalter, D.J., Freedman, L.S. and the CHART Steering Committee (1994). The CHART trials: Bayesian design and monitoring in practice. *Statist. Med.*, **13**, 1297–1312 [13.2]

Parmar, M.K., Sydes, M.R., Cafferty, F.H. et al. (2017). Testing many treatments within a single protocol over 10 years at MRC Clinical Trials Unit at UCL: multi-arm, multi-stage platform, umbrella and basket protocols. *Clin. trials*, **14**, 451–461. [27.7]

Pasternack, B.S. (1984). A note on data monitoring, incomplete data and curtailed testing. *Contr. Clin. Trials*, **5**, 217–222. [11.7]

Pasternack, B.S. and Ogawa, J. (1961). The probability of reversal associated with a test procedure when data are incomplete. *J. Amer. Statist. Assoc.*, **56**, 125–134. [11.7]

Pasternack, B.S. and Shore, R.E. (1982). Sample sizes for individually matched case-control studies: a group sequential approach. *Amer. J. Epidem.*, **115**, 778–784. [16.5]

Patel, H.I. (1996). Some optimal and non-optimal two-stage designs using an alpha spending function. *Statist. Med.*, **15**, 1739–1745. [12.3]

Pawitan, Y. and Hallstrom, A. (1990). Statistical interim monitoring of the Cardiac Arrhythmia Suppression Trial. *Statist. Med.*, **9**, 1081–1090. [11.7]

Peace, K.E., Ed. (1992). *Biopharmaceutical Sequential Statistical Applications*, New York: CRC Press. [1.6]

Pedersen, R. and Starbuck, R.R. (1992). Interim analysis in the development of an anti-inflammatory agent: sample size reestimation and conditional power analysis. In *Biopharmaceutical Sequential Statistical Applications*, (Ed., K.E. Peace), New York: CRC Press, 303–314. [18.1]

Pepe, M.S. and Anderson, G.L. (1992). Two-stage experimental designs: early stopping with a negative result. *J. Roy. Statist. Soc. C: App. Statist.*, **41**, 181–190. [11.7]

Pepe, M.S. and Fleming, T.R. (1989). Weighted Kaplan-Meier statistics: a class of distance tests for censored data. *Biometrics*, **45**, 497–507. [17.8]

Pertile, P., Forster, M. and La Torre, D. (2014). Optimal Bayesian sequential sampling rules for the economic evaluation of health technologies. *J. Roy. Statist. Soc. A*, **177**, 419–438. [13.3]

Peto, R. (1985). Discussion of papers by J.A. Bather and P. Armitage. *Int. Statist. Rev.*, **53**, 31–34, [30.2, 30.4]

Peto, R. (1990). Misleading subgroup analyses in GISSI (letter). *Amer. J Cardiol.*, **66**, 771–771. [28.1]

Peto, R. and Peto, J. (1972). Asymptotically efficient rank invariant procedures (with discussion). *J. Roy. Statist. Soc. A*, **135**, 185–206. [4.5]

Peto, R., Pike, M.C., Armitage, P. et al. (1976). Design and analysis of randomized clinical trials requiring prolonged observation of each patient. I. Introduction and design. *Brit. J. of Cancer*, **34**, 585–612. [2.7]

Piantadosi, S. (2017). *Clinical Trials: A Methodologic Perspective*, 3rd ed., New York: Wiley. [1.1]

Pickard, M.D. and Chang, M. (2014). A flexible method using a parametric bootstrap for reducing bias in adaptive designs with treatment selection. *Statist. Biopharm. Res.*, **6**, 163–174. [26.6]

Pilz, M., Kieser, M. and Kunzmann, K. (2020). Comments on 'Adaptive sample size modification in clinical trials: start small then ask for more?' (by C. Jennison and B,W, Turnbull). *Statist. Med.*, **39**, 97–98. [24.5]

Pilz, M., Kunzmann, K., Herrmann, C. et al. (2019). A variational approach to optimal two-stage designs. *Statist Med.*, **38**, 4159–4171. [24.5]

Pinheiro, J.C. and DeMets, D.L. (1997). Estimating and reducing bias in group sequential designs with Gaussian independent increment structure. *Biometrika*, **84**, 831–845. [9.3]

PMA Biostatistics and Medical Ad Hoc Committee on Interim Analysis (1993). Interim analysis in the pharmaceutical industry: the PMA position. *Contr. Clin. Trials*, **14**, 160–173. [1.2]

Pocock, S.J. (1977). Group sequential methods in the design and analysis of clinical trials. *Biometrika*, **64**, 191–199. [1.3, 2.3, 2.4, 2.6, 4.1, 4.6, 5.1, 8.2, 8.8, 10.2, 13.3, 15.5, 19.5, 31.1]

Pocock, S.J. (1982). Interim analyses for randomized clinical trials: the group sequential approach. *Biometrics*, **38**, 153–162. [2.8, 12.2]

Pocock, S.J. (1983). *Clinical Trials: A Practical Approach*, New York: Wiley. [1.1, 3.3, 24.1]

Pocock, S.J. (1996). The role of external evidence in data monitoring of a clinical trial (with discussion). *Statist. Med.*, **15**, 1285–1297. [10.1]

Pocock, S.J. (1997). Clinical trials with multiple outcomes: a statistical perspective on their design, analysis, and interpretation. *Contr. Clin. Trials*, **18**, 530–545. [19.7]

Pocock, S.J., Assmann, S.E., Enos, L.E. and Kasten, L.E. (2002). Subgroup analysis, covariate adjustment and baseline comparisons in clinical trial reporting: current practice and problems. *Statist. Med.*, **21**, 2917–2930. [28.1]

Pocock, S.J., Geller, N.L. and Tsiatis, A.A. (1987). The analysis of multiple endpoints in clinical trials. *Biometrics*, **43**, 487–498. [19.1, 19.4]

Pocock, S.J. and Hughes, M.D. (1989). Practical problems in interim analyses, with particular regard to estimation. *Contr. Clin. Trials*, **10**, 209S–221S. [9.5, 13.1, 13.3]

Polack, F.P., Thomas, S.J., Kitchin, N. et al. (2020). Safety and efficacy of the BNT162b2 mRNA Covid-19 vaccine. *New Eng. J. Med.*, **383**, 2603–2615. [16.7]

Posch, M. and Bauer, P. (1999). Adaptive two stage designs and the conditional error function. *Biometrical J.*, **41**, 689–696. [23.5]

Posch, M., Bauer, P. and Brannath, W. (2003). Issues in designing flexible trials. *Statist. Med.*, **22**, 953–969. [24.2]

Posch, M., Koenig, F., Branson, M. et al. (2005). Testing and estimation in flexible group sequential designs with adaptive treatment selection. *Statist. Med.*, **24**, 3697–3714. [21.4, 26.6, 27.1, 27.3, 27.4, 27.6, 28.6]

Posch, M., Maurer, W. and Bretz, F. (2011). Type I error rate control in adaptive designs for confirmatory clinical trials with treatment selection at interim. *Pharmaceut. Statist.*, **10**, 96–104. [26.8]

Prentice, R.L. (1997). On the role and analysis of secondary endpoints in clinical trials. *Contr. Clin. Trials*, **18**, 561–567. [19.2]

Prentice, R.L., Williams, B.J. and Peterson, A.V. (1981). On the regression analysis of multivariate failure time data. *Biometrika*, **68**, 373–379. [1.4]

Proschan, M.A. (1999). Properties of spending function boundaries. *Biometrika*, **86**, 466–473. [8.2, 10.1]

Proschan, M.A. (2003). The geometry of two-stage tests. *Statistica Sinica*, **13**, 163–177. [23.4, 23.5]

Proschan, M.A. (2006). Discussion of 'Are flexible designs sound?' (by C-F. Burman and C. Sonesson). *Biometrics*, **62**, 674–676. [24.2]

Proschan, M.A. (2009) Sample size re-estimation in clinical trials. *Biometr. J.*, **51**, 348–357. [18.5]

Proschan, M. A., and Brittain, E. H. (2020). A primer on strong vs weak control of familywise error rate. *Statist. Med.*, **39**, 1407–1413. [20.2]

Proschan, M.A., Follmann, D.A and Geller, N.L. (1994). Monitoring multi-armed trials. *Statist. Med.*, **13**, 1441–1452. [21.2, 21.3, 21.5]

Proschan, M.A., Follmann, D.A. and Waclawiw, M.A. (1992). Effect of assumption violations on type I error rate in group sequential monitoring. *Biometrics*, **48**, 1131–1143. [4.2, 8.5, 18.1]

Proschan, M. A. and Hunsberger, S. A. (1995). Designed extension of studies based on conditional power. *Biometrics*, **51**, 1315–1324. [1.3, 1.5, 8.5, 22.1, 22.4, 23.5, 24.1, 24.4]

Proschan, M.A., Lan, K.K.G. and Wittes, J.T. (2006). *Statistical Monitoring of Clinical Trials: A Unified Approach.* New York: Springer. [1.6, 3.1]

Purich, E. (1980). Bioavailability/bioequivalency regulations: an FDA perspective. In *Drug Absorption and Disposition: Statistical Considerations*, (Ed., K.S. Albert), Amer. Pharmaceut. Assoc., Washington D.C., 115–137. [7.5]

Pushpakom, S., Kolamunnage-Dona, R., Taylor, C. et al. (2020). TAILoR (TelmisArtan and InsuLin Resistance in Human Immunodeficiency Virus [HIV]): an adaptive-design, dose-ranging phase IIb randomized trial of telmisartan for the reduction of insulin resistance in HIV-positive individuals on combination antiretroviral therapy. *Clinical Infectious Diseases*, **70**, 2062–2072. [21.4]

Quan, H., Luo, X. and Capizzi, T. (2005). Multiplicity adjustment for multiple endpoints in clinical trials with multiple doses of an active treatment. *Statist. Med.*, **24**, 2151–2170. [19.7]

Reboussin, D.M., Lan, K.K.G. and DeMets, D.L. (1992). Group sequential testing of longitudinal data. *Tech. Report No. 72*, Dept. of Biostat., University of Wisconsin. [15.3, 15.7]

Robbins, H. (1952). Some aspects of the sequential design of experiments. *Bull. Amer. Math. Soc.*, **58**, 527–535. [1.5, 30.1, 30.4]

Robbins, H. (1970). Statistical methods related to the law of the iterated logarithm. *Ann. Math. Statist.*, **41**, 1397–1409. [10.1]

Robbins, H. and Siegmund, D.O. (1974). Sequential tests involving two populations. *J. Amer. Statist. Assoc.*, **69**, 132–139. [30.4]

Robert, C.P. (2007). *The Bayesian Choice: From Decision-Theoretic Foundations to Computational Implementation* New York: Springer. [13.1]

Robertson, D.S., Burnett, T., Choodari-Oskooei, B. et al. (2025a). Confidence intervals for adaptive trial designs I: a methodological review. *Statist. Med.*, **44**:e70174, 1–18. [9.7, 23.10, 26.6]

Robertson, D.S., Burnett, T., Choodari-Oskooei, B., et al. (2025b). Confidence intervals for adaptive trial designs II: case study and practical guidance. *Statist. Med.*, **44**:e70202, 1–21. [9.7, 23.10, 26.6]

Robertson, D.S., Choodari-Oskooei, B., Dimairo, M. et al. (2023a). Point estimation for adaptive trial designs I: a methodological review. *Statist. Med.*, **42**, 122–145. [9.7, 23.10, 26.6]

Robertson, D.S., Choodari-Oskooei, B., Dimairo, M. et al. (2023b). Point estimation for adaptive trial designs II: practical considerations and guidance. *Statist. Med.*, **42**, 2496–2520. [9.7, 23.10, 26.6]

Robertson, D. S., Lee, K. M., López-Kolkovska, B. C., and Villar, S. S. (2023). Response-adaptive randomization in clinical trials: from myths to practical considerations (with discussion). *Statistical Science*, **38**, 185–239. [30.1, 30.3, 30.4]

Robertson, D.S., Prevost, A.T. and Bowden, J. (2016a). Unbiased estimation in seamless phase II/III trials with unequal treatment effect variances and hypothesis-driven selection rules. *Statist. Med.*, **35**, 3907–3922. [26.6]

Robertson, D.S., Prevost, A.T. & Bowden, J. (2016b). Accounting for selection and correlation in the analysis of two-stage genome-wide association studies. *Biostatistics*, **17**, 634–649. [28.6]

Robertson, T., Wright, F.T. and Dykstra, R.L. (1988). *Order Restricted Inference*, New York: Wiley. [19.5]

Robins, J.M. and Tsiatis, A.A. (1991). Correcting for non-compliance in randomized trials using rank preserving structural failure time models. *Commun. Statist.–Theory and Methods*, **20**, 2609–2631. [17.8]

Robins, J., Breslow, N. and Greenland, S. (1986). Estimators of the Mantel-Haenszel variance consistent in both sparse-data and large-strata limiting models. *Biometrics*, **42**, 311–323. [16.4]

Rocke, D.M. (1984). On testing for bioequivalence. *Biometrics*, **40**, 225–230. [7.6]

Roseberry, T.D. and Gehan, E.A. (1964). Operating characteristic curves and accept-reject rules for two and three stage screening procedures. *Biometrics*, **20**, 73–84. [1.3, 31.1]

Rosenbaum, P.R. and Rubin, D.B. (1984). Sensitivity of Bayes inference with data-dependent sampling rules. *The Amer. Statistician*, **38**, 106–109. [13.2]

Rosenberger, W.F. (1993). Asymptotic inference with response-adaptive treatment allocation designs. *Ann. Statist.*, **21**, 2098–2107. [30.4]

Rosenberger, W.F. (1996). New directions in adaptive designs. *Statist. Sci.*, **11**, 137–149. [30.4]

Rosenberger, W.F. and Lachin, J.M. (1993). The use of response-adaptive designs in clinical trials. *Contr. Clin. Trials*, **14**, 471–484. [30.4]

Rosenblum, M. (2013). Confidence intervals for the selected population in randomized trials that adapt the population enrolled. *Biometrical J.*, **55**, 322–340. [28.6]

Rosenblum, M. (2015). Adaptive randomized trial designs that cannot be dominated by any standard design at the same total sample size. *Biometrika*, **102**, 191–202. [1.3]

Rosenblum, M., Fang, E.X. and Liu, H. (2020). Optimal, two-stage, adaptive enrichment designs for randomized trials, using sparse linear programming. *J. Roy. Statist. Soc. B*, **82**, 749–772. [28.3]

Rosenthal, R. (1978). Combining results of independent studies. *Psychol. Bull.*, **85**, 185–193. [23.1]

Rosner, G.L., Laud, P.W. and Johnson, W.O. (2021). *Bayesian Thinking in Biostatistics*. Boca Raton: Chapman and Hall/CRC. [13.4, 13.5]

Rosner, G.L. and Tsiatis, A.A. (1988). Exact confidence intervals following a group sequential trial: a comparison of methods. *Biometrika*, **75**, 723–729. [9.4, 9.5]

Rothman, K.J., Greenland, S. and Lash, T.L., Eds. (2008). *Modern Epidemiology*, 3rd ed., Philadelphia: Lippincott Williams & Wilkins [1.1]

Royall, R.M. (1991). Ethics and statistics in randomized clinical trials (with discussion). *Statist. Sci.*, **6**, 52–88. [30.4]

Roychoudhury, S., Anderson, K.M., Ye, J. and Mukhopadhyay, P. (2023). Robust design and analysis of clinical trials with nonproportional hazards: a straw man guidance from a cross-pharma working group. *Statist. Biopharm. Res.*, **15**, 280–294. [17.8]

Royston, P., Parmar, M.K. and Qian, W. (2003). Novel designs for multi-arm clinical trials with survival outcomes with an application in ovarian cancer. *Statist. Med.* **22**, 2239–2256. [27.7]

Ryan, E.G., Brock, K., Gates, S. and Slade, D. (2020). Do we need to adjust for interim analyses in a Bayesian adaptive trial design? *BMC Med. Res. Meth.*, **20**:150, 1–9. [13.3]

Ryan, E.G., Stallard, N., Lall, R., et al. (2020). Bayesian group sequential designs for phase III emergency medicine trials: a case study using the PARAMEDIC2 trial. *Trials*, **21**:84, 1–13. [13.5]

Sampson, A.R. and Sill, M.W. (2005). Drop-the-losers design: normal case (with discussion). *Biometrical J.*, **47**, 257–281. [26.6, 26.8]

Samuel-Cahn, E. (1980). Comparisons of sequential two-sided tests for normal hypothesis. *Commun. Statist.-Theory and Methods*, **9**, 277–290. [11.2]

Sandvik, L., Erikssen, J., Mowinckel, P. and Rødland, E.A. (1996). A method for determining the size of internal pilot studies. *Statist. Med.*, **15**, 1587–1590. [18.2]

Sankoh, A.J., D'Agostino, R.B. and Huque, M.F. (2003). Efficacy endpoint selection and multiplicity adjustment methods in clinical trials with inherent multiple endpoint issues. *Statist. Med.*, **22**, 3133–3150. [19.7]

Sarkar, S. (1998). Some probability inequalities for ordered MTP_2 random variables: a proof of Simes conjecture. *Ann. Statist.*, **26**, 494–504. [20.4]

Sarkar, S. and Chang, C-K. (1997). The Simes method for multiple hypotheses testing with positively dependent test statistics. *J. Amer. Statist. Assoc.*, **92**, 1601–1608. [20.4, 21.4, 27.2]

Schaid, D.J., Wieand, S. and Therneau, T.M. (1990). Optimal two-stage screening designs for survival comparisons. *Biometrika*, **77**, 507–513. [26.3]

Schall, R. (1995). Assessment of individual and population bioequivalence using the probability that bioavailabilities are similar. *Biometrics*, **51**, 615–626. [7.5]

Scharfstein, D.O. and Tsiatis, A.A. (1998). The use of simulation and bootstrap in information-based group sequential studies. *Statist. Med.*, **17**, 75–87. [8.2, 18.4]

Scharfstein, D.O., Tsiatis, A.A. and Robins, J.M. (1997). Semiparametric efficiency and its implication on the design and analysis of group sequential studies. *J. Amer. Statist. Assoc.*, **92**, 1342–1350. [8.2, 15.7, 17.9, 18.4]

Schmidli, H., Bretz, F., Racine, A. and Maurer, W. (2006). Confirmatory seamless phase II/III clinical trials with hypotheses selection at interim: applications and practical considerations. *Biometrical J.*, **48**, 635–643. [1.3, 28.2, 28.8]

Schmidli, H., Bretz, F. and Racine-Poon, A. (2007). Bayesian predictive power for interim adaptation in seamless phase II/III trials where the endpoint is survival up to some specified timepoint. *Statist. Med.*, **26**, 4925–4938. [26.8]

Schmidt, S. and Brannath, W. (2014). Informative simultaneous confidence intervals in hierarchical testing. *Methods Inf. in Med.*, **53**, 278–283. [28.6]

Schmitz, N. (1993). *Optimal Sequentially Planned Decision Procedures*, Lecture Notes in Statistics, 79. New York: Springer-Verlag. [12.1, 12.4, 24.3]

Schneiderman, M. (1961). Statistical problems in the screening search for anticancer drugs by the National Cancer Institute of the United States. In *Quantitative Methods in Pharmacology*, (Ed., H. De Jonge), Amsterdam: North-Holland, 232–246. [1.3, 31.1]

Schnurr, P.P., Chard, K.M. Ruzek, J.I. et al. (2015). Design of VA Cooperative Study #591: CERV-PTSD, Comparative Effectiveness Research in Veterans with PTSD. *Contemp. Clin. Trials*, **41**, 75–84. [2.1]

Schoenfeld, D.A. and Richter, J.R. (1982). Nomograms for calculating the number of patients needed for a clinical trial with survival as an endpoint. *Biometrics*, **38**, 163–170. [11.6]

Schoenfeld, D.A. and Zheng, H. (2024). seqmon: Group Sequential Design Class for Clinical Trials. R package version 2.5. https://CRAN.R-project.org/package=seqmon. [1.6]

Schuirmann, D.J. (1987). A comparison of the two one-sided tests procedure and the power approach for assessing the equivalence of average bioavailability. *J. Pharmacokin. Biopharm.*, **15**, 657–680. [10.3]

Schüler, S., Kieser, M. and Rauch, G. (2017). Choice of futility boundaries for group sequential designs with two endpoints. *BMC Med. Res. Meth.*, **17**:119, 1–10. [19.6]

Schultz, J.R., Nichol, F.R., Elfring, G.L. and Weed, S.D. (1973). Multiple-stage procedures for drug screening. *Biometrics*, **29**, 293–300. [1.3, 16.1, 31.1]

Schüürhuis, S., Konietschke, F. and Kunz, C.U. (2024). A two-stage group-sequential design for delayed treatment responses with the possibility of trial restart. *Statist. Med.*, **43**, 2368–2388. [14.5]

Sellke, T. and Siegmund, D. (1983). Sequential analysis of the proportional hazards model. *Biometrika*, **70**, 315–326. [17.9]

Selwyn, M.R., Dempster, A.P. and Hall, N.R. (1981). A Bayesian approach to bioequivalence for the 2 × 2 changeover design. *Biometrics*, **37**, 11–21. [7.6]

Senn S. and Bretz F. (2007). Power and sample size when multiple endpoints are considered. *Pharmaceut. Statist.*, **6**, 161–170. [27.4]

Shapiro, S.H. and Louis, T.A., Eds. (1983). *Clinical Trials: Issues and Approaches*, New York: Marcel Dekker. [1.1]

Shen, L. (2001). An improved method of evaluating drug effect in a multiple dose clinical trial. *Statist. Med.*, **20**, 1913–1929. [26.6]

Shen, Y. and Fisher, L. (1999). Statistical inference for self-designing clinical trials with a one-sided hypothesis. *Biometrics*, **55**, 190–197. [23.7, 24.3]

Shewhart, W.A. (1931). *Economic Control of Quality of Manufactured Product.* New York: Van Nostrand. [1.3]

Shi, H. and Yin, G. (2019). Control of type I error rates in Bayesian sequential designs. *Bayesian Analysis*, **14**, 399-425. [13.3]

Shih, W.J. (1992). Sample size reestimation in clinical trials. In *Biopharmaceutical Sequential Statistical Applications*, (Ed., K.E. Peace), New York: CRC Press, 285–301. [18.1]

Shih, W.J. and Aisner, J. (2015). *Statistical Design and Analysis of Clinical Trials: Principles and Methods*, Boca Raton: Chapman and Hall/CRC. [1.1]

Shih, W.J., Quan, H. and Li, G. (2004). Two-stage adaptive strategy for superiority and non-inferiority hypotheses in active controlled clinical trials. *Statist. Med.*, **23**, 2781–2798. [7.3]

Shun, Z. (2001). Sample size reestimation in clinical trials. *Drug Inf. J.*, **35**, 1409–1422 [24.1]

Shun, Z., Yuan, W., Brady, W.E. and Hsu, H. (2001). Type I error in sample size reestimations based on observed treatment difference. *Statist. Med.*, **20**, 497–513. [24.1]

Šidák, Z. (1967). Rectangular confidence regions for the means of multivariate normal distributions. *J. Amer. Statist. Assoc.*, **62**, 626–633. [26.6]

Siegmund, D. (1978). Estimation following sequential tests. *Biometrika*, **65**, 341–349. [1.3, 9.4]

Siegmund, D. (1979). Corrected diffusion approximations in certain random walk problems. *Adv. App. Prob.*, **11**, 701–719. [5.7]

Siegmund, D. (1985). *Sequential Analysis*, Springer-Verlag, New York. [1.3, 3.5, 9.2]

Siegmund, D. (1993). A sequential clinical trial for comparing three treatments. *Ann. Statist.*, **21**, 464–483. [21.5]

Silva, I.R., Maro, J. and Kulldorff, M. (2021). Exact sequential test for clinical trials and post-market drug and vaccine safety surveillance with Poisson and binary data. *Statist. Med.*, **40**, 4890–4913. [16.7]

Simes, J.R. (1986). An improved Bonferroni procedure for multiple tests of significance. *Biometrika*, **73**, 751–754. [20.4, 21.4, 25.2, 26.2, 28.2]

Simon, R. (1977). Adaptive treatment assignment methods and clinical trials. *Biometrics*, **33**, 743–749. [30.4]

Simon, R. (1986). Confidence intervals for reporting results of clinical trials. *Ann. Internal Medicine*, **105**, 429–435. [1.1]

Simon, R. (1989). Optimal two-stage designs for phase II clinical trials. *Contr. Clin. Trials*, **10**, 1–10. [12.4]

Simon, R. (1993). Why confidence intervals are useful tools in clinical therapeutics. *J. Biopharm. Statist.*, **3**, 243–248. [9.1]

Simon, R. (2017). Critical review of umbrella, basket, and platform designs for oncology clinical trials. *Clin. Pharm. and Therapeutics*, **102**, 934–941. [27.3]

Simon, N. and Simon, R. (2013). Adaptive enrichment designs for clinical trials. *Biostatistics*, **14**, 613–625. [28.8]

Simon, N. and Simon, R. (2018). Using Bayesian modeling in frequentist adaptive enrichment designs. *Biostatistics*, **19**, 27–41. [28.8]

Simon, R. and Wang, S.J. (2006). Use of genomic signatures in therapeutics development in oncology and other diseases. *Pharmacogenomics J.*, **6**, 166–173. [28.1]

Slud, E.V. (1984). Sequential linear rank tests for two-sample censored survival data. *Ann. Statist.*, **12**, 551–571. [17.9]

Slud, E.V. and Wei, L.J. (1982). Two-sample repeated significance tests based on the modified Wilcoxon statistic. *J. Amer. Statist. Assoc.*, **77**, 862–868. [1.3, 8.2, 8.5, 8.7, 8.8, 15.7, 16.3, 17.9]

Snapinn, S., Chen, M-G., Jiang, Q. and Koutsoukos, T. (2006). Assessment of futility in clinical trials. *Pharmaceut. Statist.*, **5**, 273–281. [11.7]

Sobel, M. and Wald, A. (1949). A sequential decision procedure for choosing one of three hypotheses concerning the unknown mean of a normal distribution. *Ann. Math. Statist.*, **20**, 502–522. [6.3]

Sommer, A. and Zeger, S.L. (1991). On estimating efficacy from clinical trials. *Statist. Med.*, **10**, 45–52. [17.8]

Sooriyarachchi, M. R., Whitehead, J., Matsushita,T. et al. (2003). Incorporating data received after a sequential trial has stopped into the final analysis: implementation and comparison of methods. *Biometrics*, **59**, 701–709. [14.2]

Spiegelhalter, D.J. and Freedman, L.S. (1986). A predictive approach to selecting the size of a clinical trial, based on subjective clinical opinion. *Statist. Med.*, **5**, 1–13. [13.4]

Spiegelhalter, D.J. and Freedman, L.S. (1988). Bayesian approaches to clinical trials. In *Bayesian Statistics, Vol. 3*, (Eds., J.M. Bernardo, M.H. DeGroot, D.V. Lindley and A.F.M. Smith), Oxford: Oxford University Press, 453–477. [13.3]

Spiegelhalter, D.J., Freedman, L.S. and Blackburn, P.R. (1986). Monitoring clinical trials: conditional or predictive power? *Contr. Clin. Trials*, **7**, 8–17. [1.3, 11.3]

Spiegelhalter, D.J., Freedman, L.S. and Parmar, M.K.B. (1994). Bayesian approaches to randomized trials (with discussion). *J. Roy. Statist. Soc. A*, **157**, 357–416. [1.3, 13.2, 13.3]

Spiessens, B. and Debois, M. (2010). Adjusted significance levels for subgroup analyses in clinical trials. *Contemp. Clin. Trials*, **31**, 647–656. [28.8]

Spitzer, W.O., Sackett, D.L., Sibley, J.C. et al. (1974). The Burlington randomized trial of the nurse practitioner. *New Engl. J. Med.*, **290**, 251–256. [1.1]

Stallard, N. (2010). A confirmatory seamless phase II/III clinical trial design incorporating short-term endpoint information. *Statist. Med.*, **29**, 959–971. [26.5]

Stallard, N. (2011). Group-sequential methods for adaptive seamless phase II/III clinical trials. *J. Biopharm. Statist.*, **21**, 787–801. [26.5]

Stallard, N. (2023). Adaptive enrichment designs with a continuous biomarker. *Biometrics*, **79**, 9–19. [28.5]

Stallard, N. and Facey, K.M. (1996). Comparison of the spending function method and the Christmas tree correction for group sequential trials. *J. Biopharm. Statist.*, **6**, 361–373. [8.8, 28.5]

Stallard, N. and Friede, T. (2008). A group-sequential design for clinical trials with treatment selection. *Statist. Med.*, **27**, 6209–6227. [21.4, 26.3]

Stallard, N., Hamborg, T., Parsons, N. and Friede, T. (2014). Adaptive designs for confirmatory clinical trials with subgroup selection. *J. Biopharm. Statist.*, **24**, 168–187. [28.8]

Stallard, N. and Kimani, P.K. (2018). Uniformly minimum variance conditionally unbiased estimation in multi-arm multi-stage clinical trials. *Biometrika*, **105**, 495–501. [27.6]

Stallard, N. and Todd, S. (2000). Exact sequential tests for single samples of discrete responses using spending functions. *Statist. Med.*, **19**, 3051–3064. [16.2, 16.3]

Stallard, N. and Todd, S. (2003). Sequential designs for phase III clinical trials incorporating treatment selection. *Statist. Med.*, **22**, 689–703. [21.4, 26.3, 26.5, 27.4]

Stallard, N. and Todd, S. (2005). Point estimates and confidence regions for sequential trials involving selection. *J. Statist. Planning and Inf.*, **135**, 402–419. [26.6]

Stallard, N. and Todd, S. (2011). Seamless phase II/III designs. *Statist. Meth. Med. Res.*, **20**, 623–634. [26.8]

Stallard, N., Todd, S., Ryan, E.G. and Gates, S. (2020). Comparison of Bayesian and frequentist group-sequential clinical trial designs. *BMC Med. Res. Meth.*, **20**:4, 1–14. [13.5]

Stein, C. (1945). A two-sample test for a linear hypothesis whose power is independent of the variance. *Ann. Math. Statist.*, **16**, 243–258. [18.2, 18.4]

Sterne, T.E. (1954). Some remarks on confidence or fiducial limits. *Biometrika*, **41**, 275–278. [16.1]

Stone, A., Macpherson, E., Smith, A. and Jennison C. (2015). Model free audit methodology for bias evaluation of tumour progression in oncology. *Pharmaceut. Statist.*, **14**, 455–463. [11.4]

Stouffer, S.A., Suchman, E.A., DeVinney, L.C. et al. (1949). *The American Soldier: Adjustment during Army Life*, Princeton University Press. [23.1]

Stuart, A. and Ord, J.K. (1994). *Kendall's Advanced Theory of Statistics, Vol. 1: Distribution Theory*, 6th ed., London: Griffin. [18.3, 21.2]

Su, J.Q. and Lachin, J.M. (1992). Group sequential distribution-free methods for the analysis of multivariate observations. *Biometrics*, **48**, 1033–1042. [10.5, 15.7, 19.5]

Sugitani, T., Posch, M., Bretz, F. and Koenig, F. (2018). Flexible alpha allocation strategies for confirmatory adaptive enrichment clinical trials with a prespecified subgroup. *Statist. Med.*, **37**, 3387–3402 [28.8]

Sydes, M.R., Parmar, M.K.B., James, N.D. et al. (2009) Issues in applying multi-arm multi-stage methodology to a clinical trial in prostate cancer: the MRC STAMPEDE trial. *Trials*, **10**:39, 1–16. [21.1, 27.7]

Tamhane, A.C., Gou, J., Jennison, C. and Mehta, C.R. (2018) A gatekeeping procedure to test a primary and a secondary endpoint in a group sequential design with multiple interim looks. *Biometrics*, **74**, 40–48. [20.10]

Tamhane, A. C., Mehta, C. R. and Liu, L. (2010). Testing a primary endpoint and a secondary endpoint in a group sequential design. *Biometrics*, **66**, 1174–1184. [20.10, 20.11]

Tamhane, A.C., Wu, Y. and Mehta, C.R. (2012). Adaptive extensions of a two-stage group sequential procedure for testing primary and secondary endpoints (I): unknown correlation between the endpoints. *Statis. Med.* **31**, 2027–2040. [20.10]

Tamura, R.N., Faries, D.E., Andersen, J.S. and Heiligenstein, J.H. (1994). A case study of an adaptive clinical trial in the treatment of out-patients with depressive disorder. *J. Amer. Statist. Assoc.*, **89**, 768–776. [30.4]

Tan, M. and Xiong, X. (1996). Continuous and group sequential conditional probability ratio tests for phase II clinical trials. *Statist. Med.*, **15**, 2037–2051. [11.4, 11.7]

Tan, M., Xiong, X. and Kutner, M.H. (1998). Clinical trial designs based on sequential conditional probability ratio tests and reverse stochastic curtailing. *Biometrics*, **54**, 682–695. [11.4]

Tang, D-I. and Geller, N.L. (1999). Closed testing procedures for group sequential clinical trials with multiple endpoints. *Biometrics*, **55**, 1188–1192. [19.1, 20.10, 20.11, 20.13]

Tang, D-I., Geller, N.L. and Pocock, S.J. (1993). On the design and analysis of randomized clinical trials with multiple endpoints. *Biometrics*, **49**, 23–30. [19.4, 19.5]

Tang, D-I., Gnecco, C. and Geller, N.L. (1989a). Design of group sequential clinical trials with multiple endpoints. *J. Amer. Statist. Assoc.*, **84**, 776–779. [19.4]

Tang, D-I., Gnecco, C. and Geller, N.L. (1989b). An approximate likelihood ratio test for a normal mean vector with nonnegative components with application to clinical trials. *Biometrika*, **76**, 577–583. [19.5]

Temple, R.J. (1994). Special study designs: early escape, enrichment, studies in non-responders. *Commun. Statist.-Theory and Methods*, **23**, 499–531. [28.8]

Teng, Z., Liang, L, Liu, G, Liu, Y. (2018). Optimal seamless phase 2/3 oncology trial designs based on Probability of Success (PoS). *Statist. Med.*, **37**, 4097-4113. [26.7]

Teng, Z, Tian, Y, Liu, Y, Liu, G. (2020). Seamless phase 2/3 oncology trial design with flexible sample size determination. *Statist. Med.*, **39**, 2373–2386. [26.7]

Thall, P.F. (2008). A review of phase 2-3 clinical trial designs. *Lifetime Data Analysis*, **14**, 37–53. [26.8]

Thall, P.F. (2024) *Bayesian Precision Medicine.* Boca Raton: Chapman and Hall/CRC. [13.5]

Thall, P.F., Simon, R. and Ellenberg, S.S. (1988). Two-stage selection and testing designs for comparative clinical trials. *Biometrika*, **75**, 303–310. [26.1, 26.3]

Thall, P.F., Simon, R. and Ellenberg, S.S. (1989). A two-stage design for choosing among several experimental treatments and a control in clinical trials. *Biometrics*, **45**, 537–547. [26.3]

Thall, P.F., Simon, R.M. and Estey, E.H. (1995). Bayesian sequential monitoring designs for single-arm clinical trials with multiple outcomes. *Statist. Med.*, **14**, 357–379. [13.5, 19.7]

Thall, P.F., Wathen, J.K., Bekele, B.N. et al. (2003). Hierarchical Bayesian approaches to phase II trials in diseases with multiple subtypes. *Statist. Med.*, **22**, 763–780. [28.5]

The TIMI Study Group (1985). The thrombolysis in myocardial infarction (TIMI) trial: phase I findings. *New Engl. J. Med.*, **312**, 932–936. [1.4]

Thomas, D.R. and Grunkemeier, G.L. (1975). Confidence interval estimation of survival probabilities for censored data. *J. Amer. Statist. Assoc.*, **70**, 865–871. [17.7]

Thompson, W.R. (1933). On the likelihood that one unknown probability exceeds another in view of the evidence of two samples. *Biometrika*, **25**, 285–294. [30.1, 30.4]

Thuss-Patience, P.C., Shah, M.A., Ohtsu, A. et al. (2017). Trastuzumab emtansine versus taxane use for previously treated HER2-positive locally advanced or metastatic gastric or gastro-oesophageal junction adenocarcinoma (GATSBY): an international randomised, open-label, adaptive, phase 2/3 study. *Lancet Onco.*, **18**, 640–653. [1.5, 29.1, 29.4]

Tidwell, R.S.S., Peng, S.A., Chen, M. et al. (2019). Bayesian clinical trials at The University of Texas MD Anderson Cancer Center: an update. *Clin. Trials*, **16**, 645–656. [13.5]

Tilley, B.C., Marler, J., Geller, N.L. et al. (1996). Use of a global test for multiple outcomes instroke trials with application to the National Institute of Neurological Disorders and Stroke t-PA stroke trial. *Stroke*, **27**, 2136–2142. [19.1]

Timmesfeld, N., Schäfer, H. and Müller, H-H. (2007). Increasing the sample size during clinical trials with t-distributed test statistics without inflating the type I error rate. *Statist. Med.*, **26**, 2449–2464. [23.3]

Todd, S., and Stallard, N. (2005). A new clinical trial design combining phases II and III: sequential designs with treatment selection and a change of endpoint. *Drug Inf. J.*, **39**, 109–118. [26.5]

Todd, S., Whitehead, J. and Facey, K.M. (1996). Point and interval estimation following a sequential clinical trial. *Biometrika*, **83**, 453–461. [31.5]

Treadwell, J.R., Uhl, S., Tipton, K. et al. (2012). Assessing equivalence and noninferiority. *J. Clin. Epidem.*, **65**, 1144–1149. [1.1]

Tsiatis, A.A. (1981). The asymptotic joint distribution of the efficient scores test for the proportional hazards model calculated over time. *Biometrika*, **68**, 311–315. [17.9]

Tsiatis, A.A. (1982a). Group sequential methods for survival analysis with staggered entry. In *Survival Analysis*, (Eds., J. Crowley and R.A. Johnson), Hayward, California: Institute of Mathematical Statistics, 257–268. [4.5]

Tsiatis, A.A., (1982b). Repeated significance testing for a general class of statistics used in censored survival analysis. *J. Amer. Statist. Assoc.*, **77**, 855–861. [17.9]

Tsiatis, A.A, (2006). Information-based monitoring of clinical trials. *Statist. Med.*, **25**, 3236–3244. [8.1, 18.4]

Tsiatis, A.A., Boucher, H. and Kim, K. (1995). Sequential methods for parametric survival models. *Biometrika*, **82**, 165–173. [15.7, 17.9]

Tsiatis, A.A. and Davidian, M. (2001). A semiparametric estimator for the proportional hazards model with longitudinal covariates measured with error. *Biometrika*, **88**, 447–458. [17.9]

Tsiatis, A.A. and Davidian, M. (2022). Group sequential methods for interim monitoring of randomized clinical trials with time-lagged outcome. *Statist. Med.*, **41**, 5517–5536. [14.5]

Tsiatis, A.A., Rosner, G.L. and Mehta, C.R. (1984). Exact confidence intervals following a group sequential test. *Biometrics*, **40**, 797–803. [1.3, 9.1, 9.4, 9.5]

Tsiatis, A.A., Rosner, G.L. and Tritchler, D.L. (1985). Group sequential tests with censored survival data adjusting for covariates. *Biometrika*, **72**, 365–373. [17.9]

Turnbull, B.W. (2017). Adaptive designs from a Data Safety Monitoring Board perspective: some controversies and some case studies. *Clin. Trials*, **14**, 462–469. [28.8]

Turnbull, B.W., Kaspi, H. and Smith R.L. (1978). Sequential adaptive sampling rules for selecting the best of several normal populations. *J. Statist. Comput. Simul.*, **7**, 133–150. [30.4]

Tygstrup, N., Lachin, J.M. and Juhl, E., Eds. (1982). *The Randomized Clinical Trial and Therapeutic Decisions*, New York: Marcel Dekker. [1.1] van Eekelen, R., de Hoop, E. and van der Tweel, I., (2016). Simultaneous sequential monitoring of efficacy and safety led to masking of effects. *J. Clin. Epidem.*, **76**, 155–165. [19.6]

Veeravalli, V.V. and Baum, C.W. (1996). Hybrid acquisition of direct sequence CDMA signals. *Int. J. Wireless Inform. Networks*, **3**, 55–65. [1.3]

Ventz, S., Cellamare, M., Parmigiani, G. and Trippa, L. (2018). Adding experimental arms to platform clinical trials: randomization procedures and interim analyses. *Biostatistics*, **19**, 199–215. [27.3]

Ventz, S. and Trippa, L. (2015). Bayesian designs and the control of frequentist characteristics: a practical solution. *Biometrics*, **71**, 218–226. [13.3]

Waksman J.A. (2007). Assessment of the Gould–Shih procedure for sample size re-estimation. *Pharmaceut. Statist.*, **6**, 53–65. [18.3]

Wald, A. (1947). *Sequential Analysis*, New York: Wiley. [1.3, 5.1, 12.1]

Wald, A. and Wolfowitz, J. (1948). Optimum character of the sequential probability ratio test. *Ann. Math. Statist.*, **19**, 326–339. [1.3, 12.1]

Wallis, W.A. (1980). The Statistical Research Group, 1942–1945. *J. Amer. Statist. Assoc.*, **75**, 320–330. [1.3]

Wang, J. (2006). An adaptive two-stage design with treatment selection using the conditional error function approach. *Biometrical J.*, **48**, 679–689. [26.8]

Wang, S-J., Brannath, W. Brückner, M. et al. (2013). Unblinded adaptive statistical information design based on clinical endpoint or biomarker. *Statist. Biopharm. Res.*, **5**, 293–310. [8.5, 22.1]

Wang, S-J., Hung, H.M.J., Tsong, Y. and Cui, L. (2001). Group sequential test strategies for superiority and non-inferiority hypotheses in active controlled clinical trials. *Statist. Med.*, **20**, 1903–1912. [7.3, 25.1]

Wang S-J., O'Neill R.T. and Hung H.M.J. (2007). Approaches to evaluation of treatment effect in randomized clinical trials with genomic subset. *Pharmaceut. Statist.*, **6**, 227–244. [28.1]

Wang, S-J., Peng, H. and Hung, H-M.J. (2018). Evaluation of the extent of adaptation to sample size in clinical trials for cardiovascular and CNS diseases. *Contemp. Clin. Trials*, **67**, 31–36. [24.1, 24.4]

Wang, S.K. and Tsiatis, A.A. (1987). Approximately optimal one-parameter boundaries for group sequential trials. *Biometrics*, **43**, 193–200. [1.3, 2.3, 2.7, 8.2, 9.2, 12.2, 13.3, 21.4]

Wang, Y.G. and Leung, D.H.Y. (1997). Bias reduction via resampling for estimation following sequential tests. *Seq. Analysis*, **16**, 249–267. [9.3]

Ware, J.H. (1989). Investigating therapies of potentially great benefit: ECMO (with discussion). *Statist. Sci.*, **4**, 298–340. [1.3, 30.2, 30.4]

Ware, J.H., Muller, J.E. and Braunwald, E. (1985). The futility index: an approach to the cost-effective termination of randomized clinical trials. *Amer. J. Med.*, **78**, 635–643. [11.2, 11.7]

Wason, J.M.S. and Jaki, T. (2012). Optimal design of multi-arm multi-stage trials. *Statist. Med.* **31**, 4269–4279. [21.4, 30.4]

Wason, J., Magirr, D., Law, M. and Jaki, T. (2016). Some recommendations for multi-arm multi-stage trials. *Statist. Meth. Med. Res.*, **25**, 716–727. [21.4, 27.4, 27.7]

Wason, J.M.S, Mander, A.P. and Thompson, S.G. (2012). Optimal multistage designs for randomised clinical trials with continuous outcomes. *Statist. Med.*, **31**, 301–312. [21.4]

Wason, J.M.S. and Robertson, D.S. (2021). Controlling type I error rates in multi-arm clinical trials: a case for the false discovery rate. *Pharmaceut. Statist.*, **20**, 109–116. [20.9]

Wason, J., Stallard, N., Bowden, J. and Jennison, C. (2017). A multi-stage drop-the-losers design for multi-arm clinical trials. *Statist. Meth. Med. Res.*, **26**, 508–524. [21.4]

Wason, J.M.S. and Trippa, L. (2014). A comparison of Bayesian adaptive randomization and multi-stage designs for multi-arm clinical trials. *Statist. Med.*, **33**, 2206–2221. [30.4]

Wassmer, G. (2000). Basic concepts of group sequential and adaptive group sequential test procedures. *Statist. Papers*, **41**, 253–279. [23.5]

Wassmer, G. (2011). On sample size determination in multi-armed confirmatory adaptive designs. *J. Biopharm. Statist.*, **21**, 802–817. [21.4, 27.1, 27.4, 30.4]

Wassmer, G. and Brannath, W. (2016). *Group Sequential and Confirmatory Adaptive Designs in Clinical Trials*, Heidelberg: Springer. [1.3, 1.6, 23.8, 23.10, 28.7]

Wassmer, G. and Palke, F. (2025). rpact: Confirmatory Adaptive Clinical Trial Design and Analysis. R package version 4.2.0. https://CRAN.R-project.org/package=rpact. [1.6]

Wei, L.J. and Durham, S. (1978). The randomized play-the-winner rule in medical trials. *J. Amer. Statist. Assoc.*, **73**, 840–843. [30.4]

Wei, L.J. and Johnson, W.E. (1985). Combining dependent tests with incomplete repeated measurements. *Biometrika*, **72**, 359–364. [19.5]

Wei, L.J., Su, J.Q. and Lachin, J.M. (1990). Interim analyses with repeated measurements in a sequential clinical trial. *Biometrika*, **77**, 359–364. [15.7]

Weiss, L. (1962). On sequential tests which minimize the maximum expected sample size. *J. Amer. Statist. Assoc.*, **57**, 551–566. [12.2]

Westlake, W.J. (1972). Use of confidence intervals in analysis of comparative bioavailability trials. *J. Pharmaceut. Sci.*, **61**, 1340–1341. [7.6]

Westlake, W.J. (1976). Symmetrical confidence intervals for bioequivalence trials. *Biometrics*, **32**, 741–744. [7.6, 10.3]

Westlake, W.J. (1979). Statistical aspects of comparative bioavailability trials. *Biometrics*, **35**, 273–280. [7.4]

Westlake, W.J. (1981). Bioequivalence testing—a need to rethink (Reader Reaction Response). *Biometrics*, **37**, 591–593. [7.6]

Wetherill, G.B. and Glazebrook, K.D. (1986). *Sequential Methods in Statistics*, 3rd ed., London: Chapman and Hall. [1.3]

White, I.R. (2006). Estimating treatment effects in randomised trials with treatment switching. *Statist. Med.*, **25**, 1619–22. [17.8]

White, I.R., Babiker, A.G., Walker, S. and Darbyshire, J.H. (1999). Randomization-based methods for correcting for treatment changes: examples from the Concorde trial. *Statist. Med.*, **18**, 2617–2634. [17.8]

Whitehead, J. (1986a). On the bias of maximum likelihood estimation following a sequential test. *Biometrika*, **73**, 573–581. [1.3, 9.3, 23.8, 26.6]

Whitehead, J. (1986b). Supplementary analysis at the conclusion of a sequential clinical trial. *Biometrics*, **42**, 461–471. [9.3, 19.7]

Whitehead, J. (1989). Sequential methods for monitoring declining quality, with applications to long-term storage of seeds. *Biometrics*, **45**, 13–22. [15.6, 16.6]

Whitehead, J. (1992). Overrunning and underrunning in sequential clinical trials. *Contr. Clin. Trials*, **13**, 106–121. [14.2]

Whitehead, J. (1993). The case for frequentism in clinical trials. *Statist. Med.*, **12**, 1405–1413. [13.5]

Whitehead, J. (1997). *The Design and Analysis of Sequential Clinical Trials*, Revised 2nd ed., Chichester: Wiley. [1.6, 3.1, 3.3, 5.1, 5.7, 6.1, 6.3, 8.8]

Whitehead, J. and Stratton, I. (1983). Group sequential clinical trials with triangular continuation regions. *Biometrics*, **39**, 227–236 [1.3, 1.5, 5.1, 5.7, 6.1, 6.3, 21.4]

Williams, P.L. (1996). Sequential monitoring of clinical trials with multiple survival endpoints. *Statist. Med.*, **15**, 2341–2357. [19.7]

Wilson, E.B. and Burgess, A.R. (1971). Multiple sampling plans viewed as finite Markov chains. *Technometrics*, **13**, 371–383. [16.1]

Wittes, J. and Brittain, E. (1990). The role of internal pilot studies in increasing efficiency of clinical trials. *Statist. Med.*, **9**, 65–72. [18.1, 18.2, 18.4]

Wright, S.P. (1992). Adjusted *P*-Values for simultaneous inference. *Biometrics*, **48**, 1005–1013. [20.6, 20.7]

Wu, M.C. and Lan, K.K.G. (1992). Sequential monotoring for comparison of changes in a response variable in clinical studies. *Biometrics*, **48**, 765–779. [1.4, 15.3, 15.7]

Wu, S.S., Wang, W. and Yang, M.C.K. (2010) Interval estimation for drop-the-losers designs. *Biometrika*, **97**, 405–418. [26.6]

Xi, D., Gallo, P. and Ohlssen, D. (2017). On the optimal timing of futility interim analyses. *Statist. Biopharm. Res.*, **9**, 293–301. [11.7, 12.3]

Xing, B. and Ganju, J. (2005). A method to estimate the variance of an endpoint from an on-going blinded trial. *Statist. Med.*, **24**, 1807–1814. [18.3]

Xiong, X. (1995). A class of sequential conditional probability ratio tests. *J. Amer. Statist. Assoc.*, **90**, 1463–1473. [11.4]

Yao, Q. and Wei, L.J. (1996). Play the winner for phase II/III trials. *Statist. Med.*, **15**, 2413–2423. [30.4]

Yin, G. (2012). *Clinical Trial Design: Bayesian and Frequentist Adaptive Methods.* New York: Wiley. [13.5]

Yusuf, S., Wittes, J., Probstfield, J. and Tyroler, H.A. (1991). Analysis and interpretation of treatment effects in subgroups of patients in randomized clinical trials. *J. Amer. Med. Assoc.*, **266**, 93–98. [28.1]

Zelen, M. (1969). Play the winner rule and the controlled clinical trial. *J. Amer. Statist. Assoc.*, **64**, 131–146. [1.3, 30.4]

Zhou, J., Adewale, A., Shentu, Y. et al. (2014). Information-based sample size re-estimation in group sequential design for longitudinal trials. *Statist. Med.*, **33**, 3801–3814. [18.4]

Zhou, T. and Ji, Y. (2024). On Bayesian sequential clinical trial designs. *New Engl. J. Statist. Data Science*, **1**, 136–151. [13.5]

Zucker, D.M., Wittes, J.T., Schabenberger, O. and Brittain, E. (1999). Internal pilot studies II: comparison of various procedures. *Statist. Med.*, **18**, 3493–3509. [18.3]

Index

For Product Safety Concerns and Information please contact our EU representative GPSR@taylorandfrancis.com Taylor & Francis Verlag GmbH, Kaufingerstraße 24, 80331 München, Germany